Nanobiomaterials HANDBOOK

Nanobiomaterials HANDBOOK

Edited by
BALAJI SITHARAMAN

CRC Press
Taylor & Francis Group
Boca Raton London New York

CRC Press is an imprint of the
Taylor & Francis Group, an **informa** business

CRC Press
Taylor & Francis Group
6000 Broken Sound Parkway NW, Suite 300
Boca Raton, FL 33487-2742

First issued in paperback 2017

ISBN-13: 978-1-4200-9466-4 (hbk)
ISBN-13: 978-1-138-07652-5 (pbk)

Library of Congress Cataloging-in-Publication Data

Nanobiomaterials handbook / editor, Balaji Sitharaman.
p. ; cm.
Includes bibliographical references and index.
ISBN 978-1-4200-9466-4 (hardcover : alk. paper)
1. Nanotechnology. 2. Nanostructured materials. 3. Biomedical engineering. I. Sitharaman, Balaji.
[DNLM: 1. Nanostructures. 2. Biocompatible Materials. 3. Nanotechnology. QT 36.5]

R857.N34N3257 2011
610.28--dc23 2011023348

Visit the Taylor & Francis Web site at
http://www.taylorandfrancis.com

and the CRC Press Web site at
http://www.crcpress.com

Contents

Preface

The *Nanobiomaterials Handbook* aims to provide a comprehensive overview of the field of nanobiomaterials with a broad introduction for those who are unfamiliar with the subject and as a useful reference for advanced professionals.

Nanobiomaterials are poised to play a central role in nanobiotechnology and make significant contributions to biomedical research and health care. This assessment is based on numerous demonstrations that nanobiomaterials can exhibit distinctive nanoscopic characteristics (e.g., mechanical, electrical, and optical properties) suitable for a variety of biological applications. Advances in nanobiomaterials require a multidisciplinary approach spanning fields in physical sciences (e.g., chemistry), biological sciences (e.g., molecular biology), engineering (e.g., chemical engineering), and medicine with considerable interaction and collaboration among ethicists, regulatory bodies, and industry.

The introduction defines the field of nanobiomaterials and discusses its scope, current status, and future prospects. This is followed by an in-depth survey of nanobiomaterials. It also provides a comprehensive overview of the various synthesis and processing techniques important for developing bionanomaterials and explores the unique nanoscopic physicochemical properties of nanobiomaterials.

Next, a detailed survey of potential applications of nanobiomaterials is presented. Here the emphasis is on the unique challenges in the design, fabrication, and evaluation of biomaterials for a particular application/field. For instance, key physical properties necessary for developing bionanomaterials for molecular imaging applications are completely different than those for gene therapy. Even within a specific field such as molecular imaging, the specification can vary depending on imaging modality. This information should also help identify key necessary parameters for the development of nanomaterials for a particular application.

Finally, this handbook also provides a detailed overview of the interactions between bionanomaterials/biological systems and the biocompatibility issues associated with bionanomaterials. The physical interface between biological systems and bionanomaterials shares a number of common (e.g., similar size scales) as well as complementary (e.g., inorganic/organic versus biological composition) attributes. Understanding the interactions between biological systems and bionanomaterials at this interface offers opportunities for significant breakthroughs in fundamental and applied biosciences and is necessary in assessing the biological response of bionanomaterials and, thus, its biocompatibility. Since bionanomaterials constitute a diverse heterogeneous group with applications that entail their direct or indirect contact with humans, the various aspects of biocompatibility associated with biomaterials and regulatory guidelines are also included.

It has been a pleasure serving as the editor of this exciting endeavor. I would like to thank my colleagues who have contributed the various chapters. I would also like to thank the editorial staff at CRC Press/Taylor & Francis Group, particularly Arun Kumar, project manager, Glenon Butler, project editor, Jennifer Ahringer, the project coordinator, and Michael Slaughter, the executive editor, for skillfully steering the process of getting this handbook published. Finally, I would like to thank my undergraduate student Sunny Patel at Stony Brook University for helping me in reviewing and formatting the manuscripts.

Editor

Balaji Sitharaman is an assistant professor of biomedical engineering at Stony Brook University, Stony Brook, New York. He received his BS (2000) from the Indian Institute of Technology, Kharagpur, and his MA and PhD (2005) from Rice University, Houston, Texas, where he also completed his postdoctoral research (2005–2007) as the J. Evan Attwell-Welch Postdoctoral Fellow at the Richard E. Smalley Institute for Nanoscale Science and Technology. Dr. Sitharaman's research program is at the interface of nanotechnology and regenerative and molecular medicine, and synergizes the advancements in each of these fields to tackle problems related to diagnosis/treatment of disease and tissue regeneration. He has authored over 50 publications and 10 patents. He has received several awards for his research, including NIH Director's New Innovator Award from the National Institute of Health, the Idea Award from the Department of Defense, the Carol M. Baldwin Breast Cancer Research Award from the Carol Baldwin Foundation, and the George Kozmetsky Award from the Nanotechnology Foundation of Texas.

Contributors

Jeyarama S. Ananta
Department of Chemistry
Rice University
Houston, Texas

Joel M. Anderson
Department of Biomedical Engineering
University of Alabama at Birmingham
Birmingham, Alabama

Dian Respati Arifin
Cellular Imaging Section
Division of MR Research
Russel H. Morgan Department of Radiology and Radiological Science
Institute for Cell Engineering
The Johns Hopkins University School of Medicine
Baltimore, Maryland

Debra T. Auguste
School of Engineering and Applied Sciences
Harvard University
Cambridge, Massachusetts

Pramod K. Avti
Department of Biomedical Engineering
Stony Brook University
Stony Brook, New York

Daniel A. Balazs
Department of Chemical and Biomolecular Engineering
Tulane University
New Orleans, Louisiana

Rinti Banerjee
School of Biosciences and Bioengineering
Centre for Research in Nanotechnology and Science
Indian Institute of Technology
Mumbai, India

Harinder K. Bawa
Department of Chemistry, Chemical Biology and Biomedical Engineering
Stevens Institute of Technology
Hoboken, New Jersey

J. Michael Berg
College of Veterinary Medicine
Texas A&M University
College Station, Texas

David M. Berube
Department of Communication
North Carolina State University
Raleigh, North Carolina

Karen Briley-Saebo
Department of Radiology
Mount Sinai School of Medicine
New York

Alexandra H. Brozena
Department of Chemistry and Biochemistry
University of Maryland
College Park, Maryland

Jeff Bulte
Cellular Imaging Section
Department of Radiology, Biomedical Engineering, and Chemical & Biomolecular Engineering
Institute for Cell Engineering
The Johns Hopkins University School of Medicine
Baltimore, Maryland

Nathaniel C. Cady
College of Nanoscale Science and Engineering
University at Albany
Albany, New York

Yukti Choudhury
Institute of Bioengineering and Nanotechnology
Singapore, Singapore

Hailin Cong
University of California at Davis
Davis, California

Henry Du
Stevens Institute of Technology
Hoboken, New Jersey

M. Eswaramoorthy
Chemistry and Physics of Materials Unit
Jawaharlal Nehru Centre for Advanced Scientific Research
Bangalore, India

Shankar J. Evani
Department of Biomedical Engineering
University of Texas at San Antonio
San Antonio, Texas

Nicholas M. Fahrenkopf
College of Nanoscale Science and Engineering
University at Albany
Albany, New York

John P. Fisher
Department of Bioengineering
University of Maryland
College Park, Maryland

Akhilesh K. Gaharwar
Department of Biomedical Engineering
Purdue University
West Lafayette, Indiana

Zhonggao Gao
Chinese Academy of Medical Sciences
Beijing, People's Republic of China

Theoni K. Georgiou
Department of Chemistry
University of Hull
Hull, United Kingdom

Hamidreza Ghandehari
Department of Pharmaceutics and Pharmaceutical Chemistry
and
Department of Bioengineering
and
Utah Center for Nanomedicine
Nano Institute of Utah
University of Utah
Salt Lake City, Utah

Brian M. Gillette
Department of Biomedical Engineering
Columbia University
New York, New York

W.T. Godbey
Department of Chemical and Biomolecular Engineering
Tulane University
New Orleans, Louisiana

Michael C. Hacker
Institute of Pharmacy
University of Leipzig
Leipzig, Germany

Lothar Helm
Institute of Chemical Sciences and Engineering
Ecole Polytechnique Fédérale de Lausanne
Lausanne, Switzerland

Heather Herd
Department of Bioengineering
and
Utah Center for Nanomedicine
Nano Institute of Utah
University of Utah
Salt Lake City, Utah

Peter-Georg Hoffmeister
Institute of Pharmacy
University of Leipzig
Leipzig, Germany

Rudi Hötzel
Institute of Pharmacy
University of Leipzig
Leipzig, Germany

Albena Ivanisevic
Department of Biomedical Engineering and Chemistry
Purdue University
West Lafayette, Indiana

Hamsa Jaganathan
Weldon School of Biomedical Engineering
Purdue University
West Lafayette, Indiana

John Jansen
Department of Peridontology and Biomaterials
Nijmegen Medical Center
Radboud University
Nijmegen, the Netherlands

Dinesh Jegadeesan
Department of Chemistry
University of Toronto
Toronto, Ontario, Canada

Mingji Jin
Chinese Academy of Medical Sciences
Beijing, People's Republic of China

Ho-Wook Jun
Department of Biomedical Engineering
University of Alabama at Birmingham
Birmingham, Alabama

Maung Kyaw Khaing Oo
Stevens Institute of Technology
Hoboken, New Jersey

Kyobum Kim
Department of Bioengineering
Rice University
Houstan, Texas

Meenakshi Kushwaha
Department of Biomedical Engineering
University of Alabama at Birmingham
Birmingham, Alabama

Edwin Lamers
Department of Peridontology and Biomaterials
Nijmegen Medical Center
Radboud University
Nijmegen, the Netherlands

Elvin Lee
Department of Chemistry, Chemical Biology and Biomedical Engineering
Stevens Institute of Technology
Hoboken, New Jersey

Aaron Lifland
Department of Biomedical Engineering
Georgia Tech and Emory University
Atlanta, Georgia

Dong Jin Lim
Department of Biomedical Engineering
University of Alabama at Birmingham
Birmingham, Alabama

Seong Loong Lo
Institute of Bioengineering and Nanotechnology
Singapore

Angelique Louie
Department of Biomedical Engineering
University of California at Davis
Davis, California

Michael L. Matson
Department of Chemistry
Rice University
Houston, Texas

Tingrui Pan
Department of Biomedical Engineering
University of California at Davis
Davis, California

Minal Patel
Centre for Hearing and Deafness
University at Buffalo
Buffalo, New York

Sunny C. Patel
Department of Biomedical Engineering
Stony Brook University
Stony Brook, New York

Niccola N. Perez
Department of Biomedical Engineering
Columbia University
New York, New York

Marjan Rafat
School of Engineering and Applied Sciences
Harvard University
Cambridge, Massachusetts

Anand K. Ramasubramanian
Department of Biomedical Engineering
University of Texas at San Antonio
San Antonio, Texas

Phillip Z. Rice
College of Nanoscale Science and Engineering
University at Albany
Albany, New York

Philip J. Santangelo
Department of Biomedical Engineering
Georgia Tech and Emory University
Atlanta, Georgia

F. Kyle Satterstrom
School of Engineering and Applied Sciences
Harvard University
Cambridge, Massachusetts

Christie Sayes
College of Veterinary Medicine
Texas A&M University
College Station, Texas

Patrick J. Schexnailder
Department of Biomedical Engineering
Purdue University
West Lafayette, Indiana

Gudrun Schmidt
Department of Biomedical Engineering
Purdue University
West Lafayette, Indiana

Michaela Schulz-Siegmund
Institute of Pharmacy
University of Leipzig
Leipzig, Germany

Samuel K. Sia
Department of Biomedical Engineering
Columbia University
New York, New York

Balaji Sitharaman
Department of Biomedical Engineering
Stony Brook University
Stony Brook, New York

Helmut Strey
Department of Biomedical Engineering
Stony Brook University
Stony Brook, New York

Eva Toth
Centre de Biophysique Moléculaire
Centre National de la Recherche Scientifique
Orleans, France

Chuqiao Tu
Department of Biomedical Engineering
University of California at Davis
Davis, California

Alicia Vandersluis
Department of Chemistry, Chemical Biology and Biomedical Engineering
Stevens Institute of Technology
Hoboken, New Jersey

Prasant Varghese
Department of Biomedical Engineering
Columbia University
New York, New York

Youssef Zaim Wadghiri
Department of Radiology
Center for Biomedical Imaging
New York University School of Medicine
New York, New York

Frank Walboomers
Department of Peridontology and Biomaterials
Nijmegen Medical Center
Radboud University
Nijmegen, the Netherlands

Bin Wang
Department of Chemistry
Marshall University
Huntington, West Virginia

Hongjun Wang
Department of Chemistry, Chemical Biology and Biomedical Engineering
Stevens Institute of Technology
Hoboken, New Jersey

Shu Wang
Institute of Bioengineering and Nanotechnology
and
Department of Biological Sciences
National University of Singapore
Singapore

YuHuang Wang
Department of Chemistry and Biochemistry
University of Maryland
College Park, Maryland

Lon J. Wilson
Department of Chemistry
Rice University
Houston, Texas

Jin-Oh You
School of Engineering and Applied Sciences
Harvard University
Cambridge, Massachusetts

Xiaojun Yu
Department of Chemistry, Chemical Biology and Biomedical Engineering
Stevens Institute of Technology
Hoboken, New Jersey

Xiujuan Zhang
Department of Chemical and Biomolecular Engineering
Tulane University
New Orleans, Louisiana

Chiara Zurla
Department of Biomedical Engineering
Georgia Tech and Emory University
Atlanta, Georgia

1

Nanobiomaterials: Current Status and Future Prospects

Pramod K. Avti
Stony Brook University

Sunny C. Patel
Stony Brook University

Balaji Sitharaman
Stony Brook University

1.1 Introduction

One of the exciting advancements in the fields of biomaterial science and engineering is its ability to engineer new materials at the nanoscale level for various biological applications. Recent technological advancements in the development of sophisticated experimental methods for electron and scanning probe microscopy as well as for x-ray, neutron, and optical spectroscopies have led to the exploration of materials in the nanoscale range. Nanoscale materials can exhibit distinctive mechanical, electrical, and optical properties compared to other microscopic or macroscopic structures. Nanotechnology-based approaches are being explored for a variety of biomedical applications such as for drug delivery, bioimaging, tissue engineering, and biosensors. A substantial number of these approaches employ nanoscale materials or nanobiomaterials for developing unique functionalities required by these biomedical systems. In this chapter, we provide a broad perspective of the field of nanobiomaterials. The chapter also provides the reader a bird's-eye view of the various aspects of nanobiomaterials discussed in the handbook. We introduce the various nanobiomaterials, discuss their prospective applications, and finally present their future prospects.

1.1.1 Important Definitions

Nano is derived from the Greek word "nano" meaning "dwarf." Nanotechnology is a relatively new field of science broadly defined as research and technology development at length scales between 1 and 100 nm intended to create materials, gain fundamental insights into their properties, and to use the nanoscale materials as components or building blocks to create novel structures or devices (Nat Nanotechnol 2006). At these length scales, materials show unique properties and functions. However, in certain cases, the length scales for these novel properties may be under 1 nm (down to 0.1 nm for atomic and molecular manipulation) or over 100 nm (up to 300 nm in case of nanopolymers and nanocomposites). Nanotechnology is also referred to as a convergent technology in which the boundaries separating discrete disciplines become blurred. Biochemists, materials scientists, electrical engineers, and molecular biologists may all be considered experts in the field if they are involved in the development of nanosized structures.

1.2 Types of Nanobiomaterials

1.2.1 Metallic Nanobiomaterials

1.2.1.1 Gold Nanoparticles

The commonly synthesized gold nanoparticles (AuNPs) are spherical particles, nanorods, nanoshells, and nanocages (Figure 1.1) (Huang and El-Sayed 2010). Gold nanoparticles due to their enhanced and tunable optical properties, facile synthesis techniques, and relatively good biocompatibility have been used for biomedical applications. The ability to tune their optical properties such as surface plasmon absorption and scattering, or near-infrared (IR) fluorescence can achieved by controlling the size, shape, composition, and structure of the nanoparticle (Lee and El-Sayed 2005; Jain et al. 2006). AuNPs can convert absorbed light into heat via a series of nonradiative processes (Link et al. 1999, 2000).

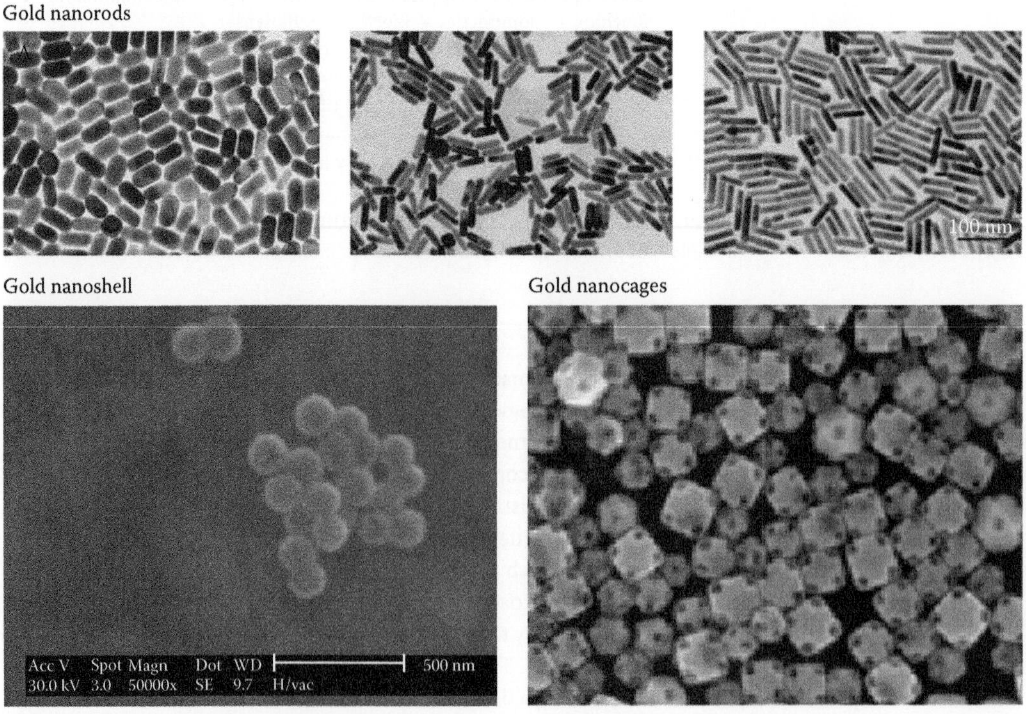

FIGURE 1.1 Representative HRTEM images of gold nanoparticles.

Thus, they have been used for drug delivery, photothermal therapy, cell tracking, and sensing applications (Giljohann et al. 2010). For instance, gold nanoshells have been shown to improve contrast in optical coherence tomography (OCT) in vivo and for tumor therapy by near-IR photothermal ablation combining diagnostic and treatment applications (Gobin et al. 2007). PEGylated nanogels containing gold nanoparticles have been employed as a fluorescence-based apoptosis sensor, where activated caspase-3 leads to release of FITC molecules (Oishi et al. 2009). Oligonucleotides tagged to gold nanoparticles have been reported to be more resistant to nuclease degradation with higher affinity toward complimentary strands against simple oligonucleotides. These complexes have been shown to regulate protein expression with no cytotoxic effects (Rosi et al. 2006). Gene regulation with siRNA tagged to gold nanoparticles has been demonstrated, which enter cells without the help of transfection agents (Giljohann et al. 2009).

1.2.1.2 Gadolinium-Based Nanoparticles

The synthesis of gadolinium nanoparticles from gadolinium chloride hydrate, gadolinium acetate hydrate, and gadolinium acetylacetonate hydrate form oxides with a diameter of 2–15 nm (Figure 1.2). Using thermal decomposition of metal precursors with oxygen-containing ligands, colloidal cubic gadolinium oxide nanorings and nanoplates (Paek et al. 2007), gadolinium phosphate ($GdPO_4$) nanorods with diameter of 20–30 nm (major axis) and 6–15 nm (minor axis) (Hifumi et al. 2006; Chang and Mao 2007) can be synthesized.

Lanthanide ion gadolinium (Gd^{3+}), due to its favorable magnetic properties (seven unpaired electrons, very large magnetic moment, symmetric electronic ground state, eight coordinated water molecules, and relatively long electron spin-relaxation times), is commonly used as magnetic resonance imaging (MRI) contrast agents. As naked Gd^{3+} ions are toxic, the most widely used approach to sequester its toxicity is to use multidentate ligands (chelates) that can coordinate with the Gd^{3+} ion. Gadolinium nanoparticles with different composition have been coordinated to form two main chelate structures (called as first generation structures): diethylenetriaminepentaacetic acid (Gd-DTPA) and 1,4,7,10-tetraazacyclododecane-1,4,7,10-tetraacetic acid (DOTA). Further efforts have led to the second-generation structures called affinity-targeted contrast agents that can target specific organ or disease and show lower toxicity, improved clearance properties, and high relaxivities (a measure of MRI contrast agent imaging efficacy) (Sipkins et al. 1998; Spinazzi et al. 1999; Zheng et al. 2005). Gd-DTPA complexed with albumin is widely used as blood pool enhancing agent because they show increased retention in the vascular compartment. However, it is not widely used for clinical purpose because of its inadequate stability, low elimination, and low water solubility (Schmiedl et al. 1986). Gd polymeric (e.g., dextran and polylysine derivatives)-based complexes have shown to increase circulation lifetime and relaxivity (Brasch et al. 1994, 1997). Gd_2O_3-polysiloxane-based nanoparticles tagged with organic

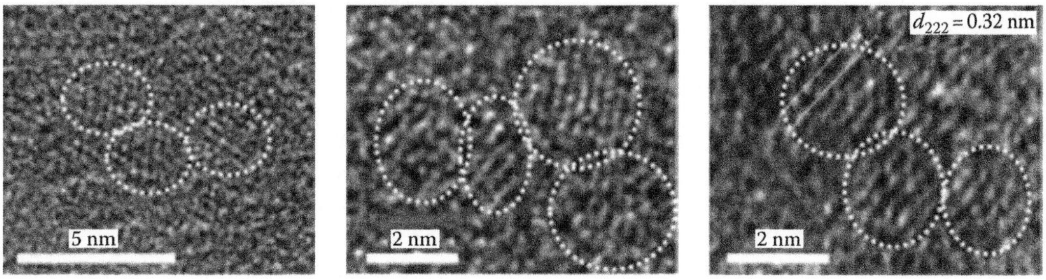

(A) (B) (C)

FIGURE 1.2 HRTEM micrographs of ultrasmall gadolinium oxide nanoparticles synthesized from (A) gadolinium chloride hydrate, (B) gadolinium acetate hydrate, and (C) gadolinium acetylacetonate hydrate as Gd(III) ion precursors. Particle diameters are nearly monodisperse and estimated to be ~1 nm for samples A and C and ~1.5 nm for sample B. Also provided is the lattice distance (d_{222}).

dye molecules have been shown to possess combined fluorescence and MRI capabilities in vivo (Bridot et al. 2007). These nanoparticles can also be used for neutron capture therapy to treat tumors. Viral capsid conjugated with gadolinium chelates to their inner surface has been demonstrated to possess high relaxivities and can be attached with target molecules (Datta et al. 2008). Recently, carbon nanomaterials such as carbon nanotubes and fullerenes have been used to encapsulate Gd^{3+} ions to reduce its toxicity and have shown MRI efficacies 40–90 times larger than current clinically used contrast agents (Sitharaman et al. 2005; Tóth et al. 2005; Bolskar 2010).

1.2.1.3 Iron Oxide Nanoparticles

Iron oxide nanoparticles are broadly divided into two types based on their size: (a) superparamagnetic iron oxides (SPIOs) with sizes greater than 50 nm (e.g., Endorem® and Resovist® are some of the clinically used SPIOs) and (b) ultrasmall superparamagnetic iron oxides (USPIOs) with sizes smaller than 50 nm (e.g., Ferumoxtran-10, ferucarbotran, very small superparamagnetic iron oxide particles [VSOP], feruglose [Clariscan®], ferumoxytol, and SH U 555C [Supravist]) (Wang et al. 2001; Neuwelt et al. 2009). These SPIOs are widely used as MRI contrast agents (Gossuin et al. 2009). SPIOs' clinical targets are liver tumors and metastasis whereas, USPIOs are widely used as blood pool agents for MR angiography. USPIOs generally have an iron oxide central core of size 10 nm, and a surrounding coating of organic–inorganic material. The size of the central (iron oxide) core controls the r_2/r_1 relaxivity ratio, and subsequently the MRI signal (Gossuin et al. 2009). The surrounding coating materials include dextran, starch, albumin, silicones, organic siloxanes, poly(lactic acid), poly(e-caprolactone) and polyalkylcyanoacrylate, poly(ethylene glycol), arabinogalactan, glycosaminoglycan, and sulfonated styrene-divinlybenzene (Zhang et al. 2002; Mornet et al. 2004; Arias et al. 2005; Flesch et al. 2005; Corot et al. 2006; Gomez-Lopera et al. 2006). The charge and nature of the coating material determine the stability, biodistribution, metabolism, pharmacokinetics, and pharmacodynamcis of USPIO agents (Corot et al. 2006). Recently, VSOPs, particle size 4–8 nm, have been developed as a new class of MRI contrast agents. As the name suggests, VSOP nanoparticles are smaller than SPIO and USPIO. Depending on the size and composition of iron oxide nanoparticle, they can be used for a variety of clinical applications such as detection of metastases, metastatic lymph nodes, inflammatory diseases, and degenerative diseases (Anzai et al. 1994a,b; Rogers et al. 1994; Bellin et al. 1998; Harisinghani et al. 2003). Other applications of these magnetic nanoparticles include cellular labeling and separation. Cells expressing a specific ligand in any diseased state could be identified by tagging the iron oxide nanoparticles with antibody. These iron oxide-labeled cells can then be separated by a process called magnetophoresis (Winoto-Morbach et al. 1994; Tchikov et al. 2001). Iron oxide magnetic nanoparticles have also been used to transfect the vector DNA and antisense oligonucleotides in vitro for effective gene therapy using a process called magnetofection. This process increases the efficiency of conventional transfection methods and decreases the toxicity (Chen et al. 2009; Namgung et al. 2010). Finally, iron oxide nanoparticles show potential in cancer treatment by magnetic hyperthermia (Liu et al. 2005; Balivada et al. 2010).

1.2.2 Ceramic Nanobiomaterials

Over the past few decades, significant advances in the field of ceramics have led to the development of nanobiomaterials for dental implants, hip replacements and tissue engineering scaffolds. Ceramic nanobiomaterials, which include alumina, zirconia, hydroxyapatite tricalcium phosphate, and silicon nitride, have many favorable characteristics such as high wear resistance, chemical stability, low density, and biocompatibility. Nanocrystalline hydroxyapatite has been reported for various applications like coatings to improve biocompatibility of titanium alloy (Bigi et al. 2007; Sato et al. 2008), as injectable pastes for bone substitution with good osteoconductive properties (Laschke et al. 2007), and as antibody delivery agents for bone infections (Rauschmann et al. 2005). Bioactive glass-based ceramic scaffolds have been synthesized with controlled rate of degradation (Chen et al. 2006) and shows promise for orthopedic applications in the future.

1.2.3 Semiconductor-Based Nanobiomaterials

The commonly studied semiconductor nanoparticles are cadmium (Cd)-based and are represented as CdE, where E = sulfide, selenide, and telluride. These nanoparticles are commonly called as quantum dots having core–shell binary structure. These types of materials are used for fluorescence bioimaging because of their unique electronic, size, and shape characteristics (Gerion et al. 2001; Lu et al. 2002; Klimov et al. 2004; Srivastava et al. 2010). Replacing Cd^{2+} with Co^{2+} or Mn^{2+} or preparing Fe_3O_4/CdSe and Fe_2O_3/CdSe imparts high magnetic properties to these nanoparticles (Schwartz et al. 2003; Gu et al. 2004; Norberg et al. 2004; Kwon et al. 2005). Surface modifications with biocompatible and biodegradable moieties such as polymers, chitosan, cellulose, and hydroxyapetite allow these nanoparticles to be transported into cells without affecting the core structure properties. These particles could be used as drug delivery systems (Huang and Lee 2006; Wang et al. 2007), and for bioimaging (Sharma et al. 2006; Arias et al. 2007; Weng et al. 2008), multiplexed gene and protein expression analysis (Shingyoji et al. 2005; Eastman et al. 2006; Zhang et al. 2006), and stem cell tracking (Lei et al. 2009).

1.2.4 Organic/Carbon-Based Nanobiomaterials

Nanosized structures created using organic material have attracted considerable interest in material and life sciences (Debuigne et al. 2000; Lu et al. 2002; Wang et al. 2003). Some of the well-known carbon-based nanomaterials used in biomedical applications include carbon nanotubes (CNTs) and fullerenes. These structures are hydrophobic, and thus, various surface modification methods have been used to water solubilize these carbon nanostructures for biomedical applications (Nakamura and Isobe 2003; Liu et al. 2008; Sitharaman et al. 2008; Prencipe et al. 2009). The unique photophysical and photochemical properties of these nanomaterials have been harnessed for various biomedical applications such as cell tracking, MRI contrast agents, microwave, photoacoustic, near-IR and Raman imaging, radiotracers, pressure sensors, gene and protein microarray, and tissue engineering (Guiseppi-Elie et al. 2002; Patolsky et al. 2004; Doorn et al. 2005; Sitharaman et al. 2005, 2007, 2008; Tóth et al. 2005; Bolskar 2008; Chen et al. 2008; Strano and Jin, 2008; Pramanik et al. 2009; Mashal et al. 2010).

Fullerene or buckyball is a molecule composed entirely of carbon, in form of a hollow geodesic structure. Fullerenes have been reported to show antioxidant and antiviral activity (Krusic et al. 1991; Brettreich and Hirsch 1998; Schuster et al. 2000). Photoirradiation of fullerenes in the presence of molecular oxygen generated highly toxic free radicals, such as singlet oxygen making them suitable as photosensitizers for photodynamic therapy (Iwamoto and Yamakoshi 2006).

CNTs are cylindrical graphene structures with unique physical and chemical properties such as high mechanical strength, electrical conductivity, and thermal conductivity (Sinha and Yeow 2005). Semiconducting single-walled CNTs display near-IR fluorescence. The IR spectrum between 900 and 1300 nm is an important optical window for biomedical applications because of its lower optical absorption (greater penetration depth of light) and small autofluorescent background. Additionally, CNTs display good photostability.

1.2.5 Organic–Inorganic Hybrid Nanobiomaterials

The organic–inorganic hybrid nanomaterials can be categorized into three main types depending on the type of materials used for forming either the core or the shell of the hybrid. These are as follows:

1. Inorganic core and organic shell nanoparticles. They have an inorganic core surrounded by an outer layer of covalently linked organic layers. The organic layer determines the chemical properties of the hybrids and their interaction with the surrounding environment. The physical properties of the hybrids depend upon the type, size, and shape of the inorganic core. Some of the examples of this kind of hybrid nanoparticles include SiO_2/PAPBA (poly(3-aminophenylboronic acid) (Zhang et al. 2006), Ag_2S/PVA (polyvinylalcohol), CuS/PVA (Kumar et al. 2002; Francoise et al. 2006)

Ag_2S/PANI (polyaniline) (Jing et al. 2007a,b), and TiO_2/cellulose. The organic coatings like PVA and PANI prevent the oxidation of the inorganic core while cellulose improves the pigment properties. Some of the above complexes are used in the preparation of dental brace materials and fillers. Based on the material's physical properties such as hardness, elasticity, and thermal expansion, these products can easily penetrate into the teeth cavity and harden under the influence of blue light (Wolter et al. 1992, 1994, 1996, 1998; Firla 1999).

2. The organic core and inorganic shell hybrids have the shell made of metals (silver, gold), silica, or silicone. The organic core is made of polymers, polyethylene, or polylactide (Lu and Lin 2003; Liu et al. 2005; Zhang et al. 2006). Due to their excellent strength and high resistance to corrosion and abrasion, they are widely used in joint replacements.
3. Dendrimer-based organic–inorganic hybrids have either a metallic (gold, silver, copper) or semiconductor quantum dots (cadmium sulfur (CdS) or cadmium selenium (CdSe)–based core (Balogh and Tomalia 1998; Esumi et al. 1998; Lemon and Crooks 2000; Wu et al. 2005). These hybrid nanobiomaterials allow controlled surface chemistry to obtain desired biocompatibility and nonimmunogenic properties and show potential as probes for fluorescence imaging, x-ray computed tomography (CT), and MRI (Shi et al. 2007, 2008; Shukla et al. 2008; Jang et al. 2009; Medina and El-Sayed 2009).

1.2.6 Silica-Based Nanobiomaterials

The different types of silica-based nanoparticles are monoliths, rodlike particles, fibers, hollow and solid nanospheres, silica nanotubes, and mesoporous silica nanoparticles (MSN) (Jafelicci et al. 1999; Miyaji et al. 1999; Fan et al. 2003; Miyazaki et al. 2004; Blasi et al. 2005; Deng et al. 2006; Han et al. 2006). Silica nanoparticles complexed with SPIO could be used as magnetic hyperthmia agents (Matín-Saavedra et al. 2010). Silica nanoparticles coated onto Fe_2O_3, CdSe quantum dots, and Au nanoparticles have been developed as novel MRI and optical imaging contrast agents for live cell imaging (Bottini et al. 2007; Gerion et al. 2007; Selvan et al. 2007). Lanthanide-doped silica nanoparticles have been recently used for multiplexed immunoassays (Murray et al. 2010). Mesoporous silica nanoparticles (MSN), due to honeycomb-like porous structures, can be used for loading large quantities of drugs and biosensing molecules (Slowing et al. 2007). Recent improvements in MSN's porosity, particle size control, and stability combined with low cytotoxicity make them efficient drug delivery platforms (Slowing et al. 2008).

1.2.7 Polymeric Nanobiomaterials and Nanocomposites

The polymers used in the development of polymeric nanobiomaterials are either of biological or synthetic polymers. Biological polymers include (a) polysaccharides (starch, alginate, chitin/chitosan, hyaluronic acid derivatives) and (b) proteins (collagen, fibrin gel, silk). Synthetic polymers include poly(lactic acid) (PLA), poly(glycolic acid) (PGA), poly(ε-caprolactone) (PCL), poly (hydroxyl butyrate) (PHB). The advantage of using biological polymer for the development of nanobiomaterials is their biocompatibility, which assists in cell adhesion and tissue regeneration. However, these polymers have poor mechanical properties. Synthetic polymers in general have better mechanical strength than biological polymers. Further, they can be synthetically manipulated to allow biological degradation. However, they show lower biocompatibility compared to biological polymers. Polymeric nanocomposites are composites in which nanomaterials are used as fillers to improve the polymer's bulk or surface properties. These nanomaterials, whether of biological or chemical origin, have gained much attention due to role in improving the physicochemical properties of the polymer matrix. The common nanostructures used as filler in polymer matrix are hydroxyapitite (HA), metal nanoparticles (alumoxane), and carbon-based nanoparticles (carbon nanotubes). Hydroxyapatite-based nanocomposites have been used for bone tissue engineering applications due to their osteoinductive properties. Hydroxyapatite/polylactic acid nanocomposites were shown to exhibit good osteoblast cell adherence and proliferation in vitro necessary bone repair and growth (Kim et al. 2006). Electrospun hydroxyapatite/chitosan-based novel nanofibers were recently reported to stimulate bone formation to a higher degree compared to chitosan (Zhang et al. 2008). Functionalized ultrashort single-walled carbon nanotubes (SWCNT) polypropylene fumarate nanocomposites has been shown to possess good cell adherence and osteoconductive properties (Shi et al. 2007; Sitharaman et al. 2008).

1.2.8 Biological Nanobiomaterials

1.2.8.1 Lipoprotein-Based Nanomaterials

Lipoproteins are the assembly of proteins and lipids into spherical nanostructures involved in the transport of water-insoluble lipids in the blood. Lipoproteins contain an apolar core of triglyceride and cholesteryl esters surrounded by phospholipid monolayer shell containing apolipoprotein and unesterified cholesterol. There are five different types of lipoproteins based on their density: (1) high-density lipoproteins (HDL, 5–15 nm), (2) low-density lipoproteins (LDL, 18–28 nm), (3) intermediate density lipoproteins (IDL, 25–50 nm), (4) very-low-density lipoprotein (VLDL, 30–80 nm), and (5) chylomicrons (100–1000 nm). Lipoprotein-based nanoparticles are completely biodegradable, biocompatible and stable in blood circulation. The hydrophobic core of the plasma-derived LDL could be used for incorporating lipophilic drugs for drug delivery to tumor target sites expressing LDL receptor (Rensen et al. 2001). A synthetic LDL-based nanoparticle has been developed for drug delivery to treat GBM tumors (Nikanjam et al. 2007). In vivo imaging of cancer in live animals by a near-IR dye–functionalized LDL nanoparticle has been reported, opening more avenues to alter these nanoparticles with respect to their size, degree of ligand conjugation, and functional groups (Chen et al. 2007).

1.2.8.2 Peptide Nanoparticles

Peptide-based nanoparticles are the small peptide sequences that have the flexibility in developing into desired biophysical characteristics by self-assembly or by binding with various other nanomaterials. The self-assembly of peptides may lead to various structure formations such as oligomeric coiled-α-coil helix (trimeric and pentameric) and icosahedron (Figure 1.3A) (Raman et al. 2006; Fraysse-Ailhas et al. 2007). Further, these peptide nanoparticles can be functionalized with other targeting moities (Figure 1.3B). The central core of peptide nanoparticles with a diameter of 6–10 nm is suitable for encapsulation of quantum dots as contrast agents for fluorescence imaging, gold nanoparticles as contrast agents for electron microscopy, and/or gadolinium and iron nanoparticles as probes for MRI. Trimeric coiled-coil structures allow binding of specific oligonucleotide sequences for effective gene delivery and therapy. Pentameric domains allow the incorporation of small hydrophobic molecules such as vitamins and lipophilic drugs. Recently, it has been shown that self-assembly of amphiphilic peptides forms a core–shell peptide nanoparticle having strong antimicrobial properties (Liu et al. 2009).

Since the discovery of the first cell-penetrating peptide Penetratin (Derossi et al. 1994), efforts have focused on exploiting such vectors for intracellular delivery of proteins and nucleotides incapable of crossing the cellular membrane due to their hydrophilic nature. Thus, peptide-based cationic nanoparticles have been developed for gene delivery (Wiradharma et al. 2008). Cell-penetrating peptides have been tagged to quantum dots for targeting and imaging tumor vasculature in vivo in mouse xenograft model (Cai et al. 2006). Tat peptide-tagged SPIO nanoparticles have better permeability into cells and can be exploited for in vivo MRI (Koch et al. 2005). Multifunctional Au nanoparticles tagged with antisense oligonucleotides and peptides have been reported for potential gene therapy application (Patel et al. 2008).

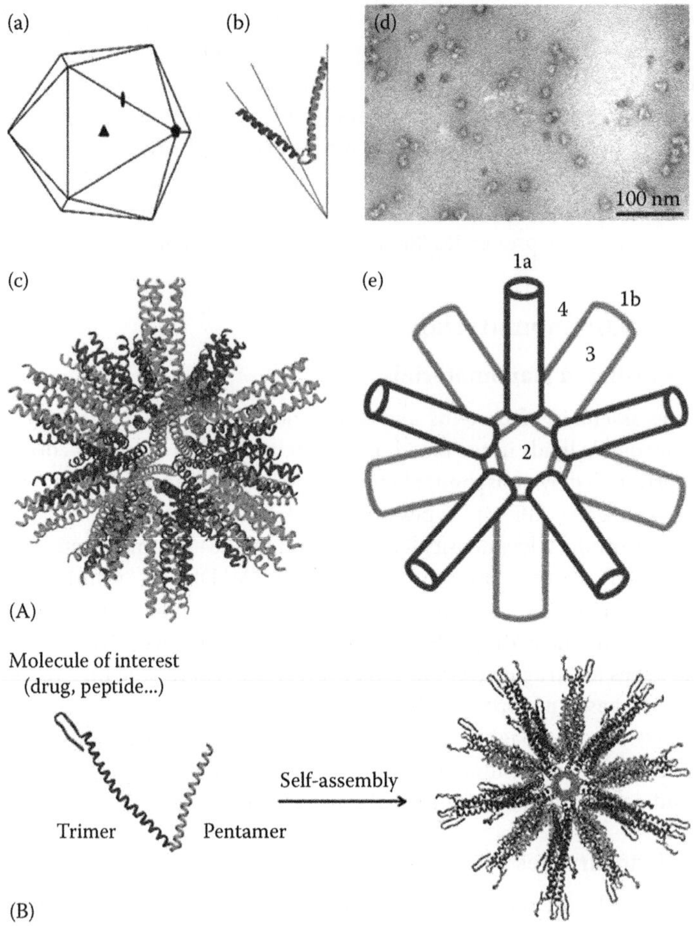

FIGURE 1.3 **(See color insert.)** (A) Design of peptide nanoparticles: (a) icosahedron with the symbols of the three different symmetry elements; (b) monomeric building block composed of a trimeric (blue) and pentameric coiled-coil-α-helix; (c) model of an assembled nanoparticle with icosahedral symmetry; (d) electron micrograph of the peptide nanoparticles functionalized with somatostatin; and (e) schematic diagram of the particle with possible modification sites. (B) Functionalized nanoparticles (right) formed by self-assembly of peptides (right). The trimeric domain can be modified by a ligand or by a drug.

1.2.9 Biologically Directed/Self-Assembled Nanobiomaterials

1.2.9.1 Virus Nanobiomaterials

Large numbers of viruses have been used in the field of nanobiotechnology because of their ability to target variety of molecules, proteins, and peptides with potential applications in the field of vaccines, molecular and electronic materials, biomaterials, and bioimaging (Singh et al. 2006). Some of the viruses that have been used for these applications are cowpea mosaic virus (CPMV), cowpea chlorotic mottle virus (CCMV), vault nanocapsules, hepatitis B cores, heat shock protein cages, MS2 bacteriophages, and M13 bacteriophages (Flenniken et al. 2006). CPMV has been used for vaccine applications. CCMV coat protein has been expressed in the yeast *Pichia pastoris*, and isolated, purified, and cage structures have been developed. These cages have the metal-binding domain wherein Tb (Manchester et al. 2006) and Gd^{3+} could be bound (Basu et al. 2003). The paramagnetic Gd^{3+}-bound nanoparticles show higher relaxivity than protein-bound Gd^{3+} chelates (Allen et al. 2005) and have been used for in vivo small animal MRI. Canine parvovirus (CPV), a natural pathogen of dogs, has a gene-delivery vehicle called adenoassociated virus (AAV) that has been used for targeting tumors (Tsao et al. 1991).

1.3 Unique Challenges in the Design, Fabrication, and Evaluation of Bionanomaterials from a Device Development and Applications Point of View

1.3.1 Biochips

Biochips provide a remarkable feature for multiplexing, enabling analysis of hundreds and thousands of different DNA or proteins simultaneously on a miniaturized device. This device could be used for analysis of single nucleotide polymorphism, mutational analysis, genetic diseases, genotyping, protein–DNA interactions (Warren et al. 2000; Bulyk 2006; Shai 2006; Wang et al. 2007). The DNA biochip efficiency depends on the surface type, the sequence of capture probes, size of the arrayed probes, immobilization reaction (tethering to the surface), method of hybridization, and detection (Beaucage 2001; Pirrung 2002; Seliger et al. 2003). Progress in the development of semiconductor nanomaterials has opened up the possibility of immobilizing a variety of oligonucleotide sequences (Halperin et al. 2004; Sethi et al. 2009). Most of the DNA biochips use a fluorescence-based detection strategy using organic fluorophore dyes (Cy3, Cy5, HEX, TAMRA, TET, etc.), which typically display a lot of background noise during detection and analysis. Developments in semiconductor nanocrystals has led to the replacement of the conventional dyes due to distinct advantages over classical dyes such as high extinction coefficient, high quantum yield, resistance to photobleaching, which can lead to an improvement in the sensitivity of detection (Resch-Genger et al. 2008).

1.3.2 Biomimetics

Biomimetic materials consist entirely of synthetic polymers, metal, or ceramics with surface or bulk modifications rendering the material biocompatible and suitable for tissue implants or tissue engineering. Numerous composites have been developed consisting of both biological and organic–inorganic nanocomposites (Shin et al. 2003; Webster and Ahn 2007; Khang et al. 2008; Ma 2008). These materials are developed to mimic the tissue and generate 3D scaffold that supports specific cell functions including cell growth, adhesion, differentiation, expression of tissue-specific genes while avoiding toxic reactions and immune response (Goodman et al. 2009; Von der Mark et al. 2010). For hard tissue metallic- and ceramic-based nanobiomaterials have been used for bone and teeth implants (Webster and Ahn 2007). These nanobiomaterials mimic the persistence and stable adhesiveness, biocompatibility, and support of natural structures found in vivo.

1.3.3 BioNEMs

Nanoelectromechanical systems (NEMS) are the nanoscopic devices less than 100 nm in length and have the ability to combine electrical and mechanical components. The NEMs fabricated with new nanobiomaterials act as biofunctionalized nanoelectromechanical systems (BioNEMs) for biological and clinical applications. BioNEMs sense analyte-induced changes that measurably alter *dynamical* device properties, and therefore are referred to as intelligent nanodevices having ability for sensing, processing, and/or actuating functions. The changes in the dynamical device properties include alterations of the nanomechanical device properties (especially force constant), changes to the device damping, or direct imposition of additional forces to the device. The biofunctionalized nanoelectromechanical systems (BioNEMs) are used as tweezers for handling single molecule manipulation (nanomolecules) such as DNA and proteins (Bustamente et al. 2003) to provide invaluable information about the molecule conformation (Strick et al. 1996), chromatin organization (Bancaud et al. 2006), or biomolecular interaction dynamics (Strick et al. 2000; Gore et al. 2006). Generally, the tweezer tips are fabricated such that they allow detection of the changes in the electric field applied in molecular solutions to understand their dynamics.

1.3.4 Biosensor

Biosensors are the devices that have biosensitive layers for the detection of analyte and have biological recognition elements (bioreceptor) along with the physiochemical detection system. Biological (e.g., antibody, peptides, enzymes, proteins, viruses) and synthetic (e.g., metallic, inorganic) nanoparticles can be used in designing complex biosensors with multivalent and 3D interactions (Whaley et al. 2000). The bioreceptors or biological sensing elements are typically a protein, enzyme, antibody, or nucleic acid that allow specific recognition high-affinity binding to the analyte of interest. This reduces the interference from other components in the complex sampling mixtures. A biosensor uses a specific biochemical mechanism for recognition, which can be transduced into measurable optical, mechanical (cantilever), electrochemical, or mass-sensitive signals for detection. Based on the type of the detection system, the nanobiosensors can be categorized into optical-, electrical-, surface-enhanced Raman scattering (SERS)-based system (Vo-Dinh et al. 2006; Jain 2007). Based on the type of the bioreceptor used, the nanosensors can be further categorized into antibody, viral, or DNA- or FRET-based biosensors (Jain 2003).

Cantilevers transduce a biochemical reaction into a mechanical motion on the nanometer scale (~10 nm), measured by deflecting the laser beam light from the cantilever surface. The advantage of using nanocantilever-based sensors is that they provide fast, label-free recognition of specific DNA sequences useful for the detection of single-nucleotide polymorphisms, oncogenes, and genotyping; act as a good alternative for PCR; and complement gene and protein microarray. These can also be used for the detection of viruses, bacteria, and pathogens (Gupta et al. 2006).

In case of viral-based nanosensors, supramolecular assembly of magnetic viral nanoparticles changes the optical or magnetic properties of the sensor system during viral analyte detection (Perez et al. 2003). Herpes simplex virus and adenovirus have been used to trigger the molecular assembly of nanosensors for the detection of viruses (as few as five virus particles) in the serum samples. Glucose oxidase (GOx) biosensor, an amperometric sensor, contains electrode-immobilized GOx enzyme sensitive to the redox reactions (Clark and Lyons 1962; Wang 2001). Within the GOx sensor, proteins can fulfill two different functional roles, namely the specific recognition of the analyte molecule and the transduction of the recognition event into an electrochemical signal. These kinds of sensors could be extended to various analytes, through the use of suitable enzymes with matching substrate specificity. The sensitivity and specificity could be improved for a variety of analytes by taking advantage of recent progress in protein engineering (Schulze et al. 2003).

1.3.5 Therapeutics—Drug and Gene Delivery

Drug and gene delivery system include organic, inorganic, polymeric, and lipid-based nanobiomaterials (Fattal and Barratt 2009). These nanobiomaterials could further be engineered to be stimuli-responsive. Further, the nanoparticles could be surface-functionalized to bind to the receptors to target cells and/or tissues.

Nanobiomaterials have been used as controlled release reservoirs for drug delivery. These drug-delivery systems can be synthesized with controlled composition, shape, size, and morphology. Their surface properties can be manipulated to increase solubility, immunocompatibility, and cellular uptake. The limitations of current drug delivery systems include suboptimal bioavailability, limited targeting capabilities, and potential cytotoxicity. Promising and versatile nanoscale drug delivery systems include nanoparticles, nanocapsules, nanotubes, nanogels, and dendrimers. They can be used to deliver both small-molecule drugs and various classes of biomacromolecules, such as peptides, proteins, plasmid DNA, and synthetic oligodeoxynucleotides.

Antisense oligonucleotide (AS-ODN) and small interfering RNA (siRNA) have also shown promise as gene delivery and therapeutic agents. However, their direct use is limited because of a number of contributing factors such as their sequence size, length, charge, half-life, or stability in solutions (Fattal and Barratt 2009). To overcome some of these limitations, their sequences have been modified by functional groups such as sulfur, boron, methyl or amino group by replacing the nonbridging oxygen of the phosphodiester backbone. These changes have resulted in resistance to RNase degradation (Agrawal 1999), increased circulation half-life (Zhang et al. 1995; Geary et al. 2001), and decreased the excretion without affecting their silencing efficacy (Braasch et al. 2003; Harborth et al. 2003). Other approaches used to over come some of the above limitations include use of lipid-based nanoparticles as carriers for increasing cellular permeability (Li and Szoka 2007), polymers such as poly(ethylene glycol) (PEG) and poly(ethyleneimine) (PEI) that provide hydrophilic shield and minimizes the interaction with negative plasma proteins by reducing aggregate formation and uptake by mononuclear phagocytic system (Opanasopit et al. 2002; Owens and Peppas 2006). Specific targeting groups such as Arg-Gly-Asp (RGD) peptides can also been coupled to above polymers to attain specific target delivery. The resistance of AS-ODN and SiRNA toward nuclease degradation can also be achieved by linking with dendrimers such as poly(amidoamine) (PAMAM) whose high density of positive charge are able to condense the nucleic acids (Zhou et al. 2006; Shen et al. 2007).

1.3.6 Bioimaging/Molecular Imaging

The use of nanoparticles has boosted the development of diagnostics agents for bioimaging. Recently, increased attention is being devoted to the development of nanoparticles as multimodal agents for diagnosis, imaging, and therapy. For the optical imaging systems, quantum dots (QDs) have been extensively used as probes due to their high brightness, longer photostability, and size-tunable narrow emission spectra. Targeting the QDs with various antibodies could allow the diagnosis different pathologies (Wang et al. 2007). Manganese (Mn (II)) dendritic nanoparticles developed as MRI contrast agent increased hydrophobicity and relaxivities (Bertin et al. 2009). Functionalized dendrimers (PAMAM) complexed with gadolinium and rhodamine B (particle size of 11 nm) for multimodal MRI and fluorescence imaging studies have been shown to cross the blood brain barrier (BBB) (Sarin et al. 2008). SPIO nanoparticles have been explored for various bioimaging applications. Skaat and Margel have designed fluorescent magnetic iron oxide nanoparticles bearing rhodamine or Congo red for amyloid-b (Ab) fibril detection (Skaat and Margel 2009). VSOPs have the capability of visualizing subtle macrophage infiltration into active neuroinflammatory plaques. Veiseh et al. (2009) have designed a nanoprobe (NPCPCTX-Cy5.5) comprised of an iron oxide nanoparticle coated with a PEGylated chitosan, to which a targeting ligand (chlorotoxin (CTX)) and a near-IR fluorophores (NIRF, Cy.5.5) were conjugated. CTX as a tumor-targeting ligand selectively binds to a variety of cancers including glioma, medulloblastoma, prostate cancer, sarcoma, and intestinal cancer (Veiseh et al. 2007). Carbon-based nanomaterials such as single-walled CNTs and fullerenes encapsulating gadolinium have been developed as MRI contrast agents (Sitharaman et al. 2005; Tóth et al. 2005).

1.3.7 Regenerative Medicine/Tissue Engineering

Tissue engineering can be considered as a subfield of regenerative medicine. This emerging field seeks to combine materials and engineering principles to improve the biological properties of a tissue. Scaffolds are porous biomaterials and play a pivotal role in the tissue engineering paradigm by providing temporary structural support, guiding cells to grow, assisting the transport of essential nutrients and waste products, and facilitating the formation of functional tissues and organs (Langer and Vacanti 1993). In general, polymers and ceramics are widely used in the development of tissue engineering scaffolds. Depending on the tissue of interest, appropriate nanobiomaterials need to be chosen in the development of these scaffolds. For instance, nanophase ceramics, especially nanohydroxyapatite (HA, a native component of bone), are widely used in bone tissue engineering scaffolds due to their documented ability to promote mineralization. The nanometer grain sizes and high surface fraction of grain boundaries in nanoceramics increase osteoblast functions (such as adhesion, proliferation, and differentiation) (Webster et al. 2000). Three-dimensional porous scaffolds made of nanohydroxyapatite and polymers also allow improvement in the mechanical properties (compressive moduli of 46–81 MPa) (Nukavarapu et al. 2008). Carbon nanomaterials such as single-walled CNTs also have been shown to reinforce biodegradable polymer scaffolds and improve the osteoinductive properties of the scaffolds (Sitharaman et al. 2008). However, the nanobiomaterials used for regeneration of bone tissue may not be suitable for the development of cardiovascular implants. For instance, a significant problem with vascular stents is the overgrowth of smooth muscle cells compared to endothelial cells. The improvement in the endothelial cell functions with greater synthesis of elastin and collagen can be achieved using metallic nanobiomaterials (Choudhary et al. 2007).

1.4 Characterizing the Interaction of Bionanomaterials with Biological Systems

Given that the sizes of functional elements in biology are in the nanometer scale range, it is not surprising that nanomaterials interact with biological systems at the molecular level (Bogunia-Kubik and Sugisaka 2002). In addition, nanomaterials have novel electronic, optical, magnetic, and structural properties that cannot be obtained from either individual molecules or bulk materials. These unique features can be tuned precisely to explore biological phenomena through numerous innovative techniques. One of the major goals of biology is to address the spatial-temporal interactions of biomolecules at the cellular and integrated systems level (Emerich and Thanos 2003). Nanomaterials can interact with biological systems at the single molecular level with high specificity, and it would be beneficial to understand these interactions. For instance, understanding the interaction between a particular nanobiomaterial and stem cells could lead to the development of mechanisms to control the intrinsic signals (e.g., growth factors and signaling molecules) underlying embryonic and adult stem cell behavior (Emerich and Thanos 2003; Green et al. 2009).

1.5 Assessment of Biocompatibility and Biological Response Toward Nanobiomaterials

An important issue that requires investigation for a nanobiomaterial under consideration for clinical development is its biocompatibility. There are numerous routes for nanomaterials to enter into the body. These include through the dermal layer, through the lungs via the respiratory system, and through the intestinal tract (Hussain et al. 2001). Once the nanomaterials are internalized, they are up taken by the cells through energy-dependent cellular uptake pathways such as endocytosis, and more specifically, phagocytosis and macropinocytosis. Phagocytosis is an endocytic process, known as "cell eating" where the nanomaterials are engulfed into the cell by the cellular membrane, and this cellular membrane invaginates the substance, and the nanomaterial remains in a vacuole, phagosome. Pinocytosis is also a nonspecific endocytic process, known as "cell drinking," and forms smaller vesicles. The pathway and route of entry are dependent on the particle size, even at the nanolevel. For instance, nanomaterial aggregation can be problematic and pulmonary fibrosis and cancer can be induced due to the shape of the nanomaterial (Greim et al. 2001).

The toxicity of nanomaterials is highly dependent on the material. For instance, toxicity levels of carbon nanomaterials are partially dependent on the aspect ratio (Magrez et al. 2006), but the actual toxic effects of carbon nanotubes is still a controversial topic. Studies shows the cytotoxicity of the carbon-based nanomaterials can be attributed to the size, dispersing agents used, aggregate formation, and metal impurities (Kang et al. 2008, 2009; Kolosnjaj-Tabi et al. 2010). Functionalization and surface modification can also affect the toxicity levels of nanomaterials. The agglomeration of nanomaterials can be one of the critical factors that affect their toxicity (Takagi et al. 2008). Some of toxic effects of nanobiomaterials in vivo could include oxidative stress, inflammation, granulomas, and fibrosis (Shvedova et al. 2005; Lam et al. 2006; Li et al. 2007; Muller et al. 2008).

1.6 Commercial Prospects and Future Challenges

Concomitant to the recent research advancements in the development of nanobiomaterials, there has been a substantial activity to translate some of the promising nanobiomaterials from bench to bedside. Analysis of issued patents related to nanobiomaterials clearly indicates this trend. The number of issued patents U.S. involving nanobiomaterials is gradually increasing since 1995 (Figure 1.4). In the United States, the number of issued patents increased fivefold between 2000 and 2009 (Figure 1.4A). Of the total 61 U.S. patents issued till date, related to nanobiomaterials, 31 of them have been patented for application (51%), 16 for the methods of preparation (26%), and 14 of them for the composition (23%) (Figure 1.4B). The nanobiomaterials used in top biomedical applications have been patented for tissue engineering (42%) followed by drug delivery (24%) (Figure 1.4C).

Several companies are involved in the commercialization of nanobiomaterial-based products for laboratory and clinical use as shown in Table 1.1. A recent report by Global Industry Analysts titled

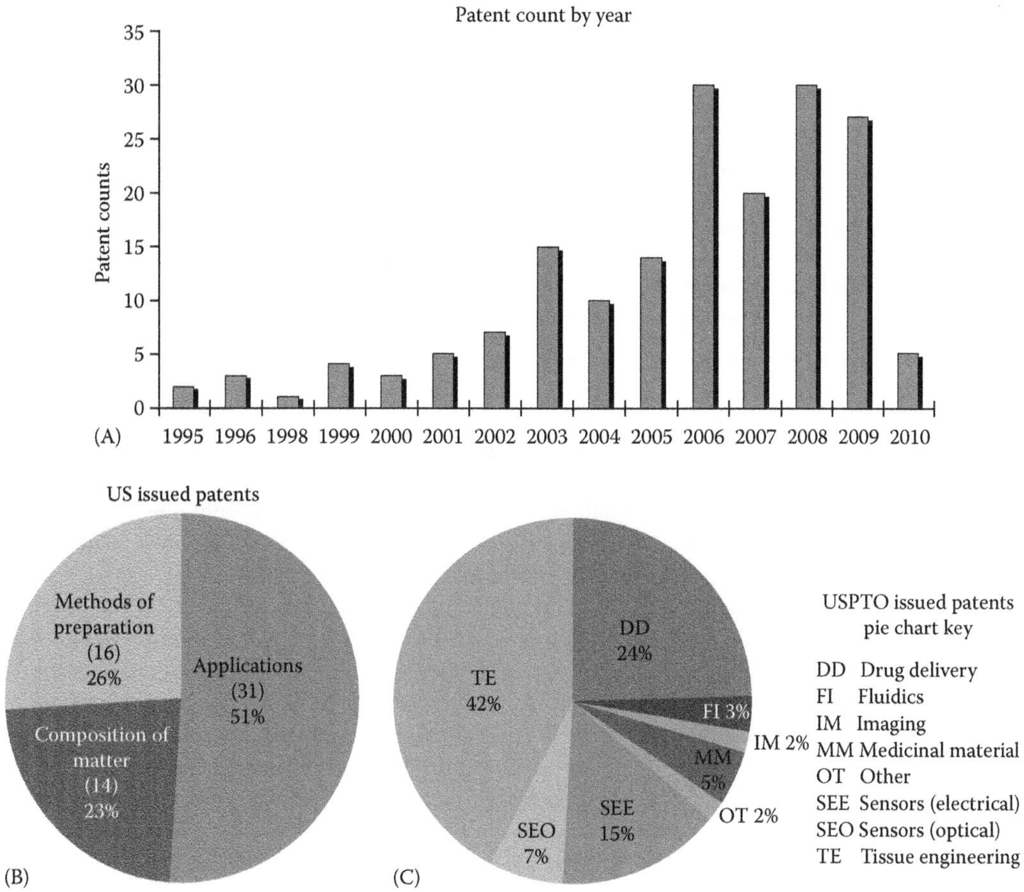

FIGURE 1.4 (A) Yearwise comparison of number of issued patents in the United States. (B) Percentage of patents based on the category. (C) Percentage of patents based on the applications.

TABLE 1.1 Commercialized Nanobiomaterials for Clinical Use

Company	Product Name	Nanomaterial	Indication
Sivida, Perth, Australia	Biosilicon™	Porous silicon	Drug delivery
Innovative Bioceramix, Vancouver, Canada	iRoot® and Bioaggregate®	White hydraulic bioceramic paste, White hydraulic cement mixture	Permanent dental root canal filling and sealing
Namos, Dresden, Germany	Namodots ZnS: Cu-1,2 and 5 ZnS:Mn-1,2 and 5	Zinc sulphide particles doped with copper and manganese	Laboratory use neither medical nor pharmaceutical products
3DM, Medical technology, Cambridge, MA	PuraMatrix™	Peptide hydrogel	Tissue repair, cell therapies, and drug deliveries
Genialab, Braunschwelg, Germany	GeniaBeads®	Hydrogel beads of chitosan	Wound healing
Organogenesis, Canton, OH	Apligraf®	Bilayered collagen gels	Dermal matrix for organogenesis
DePuy Orthapedics, Warsaw, IN	Healos®	Crosslinked collagen fibers coated with hydroxylapatite	Bone graft substitute in spinal fusions
Pfizer, New York, NY	Gelfilm®	Absorbable gelatin implant	Neurosurgery, thoracic and ocular surgery
Thermogenesis, Cordova, CA	CryoSeal®	Fibrin Sealant System	Autologous fibrin selant
Fidia, Abano Terme, Italy	Hyalgan® and Hyalubrix®	Hyaluronan	Viscoelastic gel for surgery and wound healng
Integra, Wheaton, IL	Integra®	Chondroitin sulfate	Scaffold for dermal regeneration
Biogums, Knowsley, United Kingdom	Gelrite®	Gellan gum	A novel ophthalmic vehicle
AMAG Pharmaceuticals, Lexington, MA	Ferridex	Iron oxide nanoparticles	MRI contrast agent for liver lesions
Nanosphere, Northbrook, IL	Verisens™ Prostate-specific antigen	Gold nanoparticles functionalized with DNA and antibodies	Detection and quantification of prostate-specific antigen (PSA)

"Nanomedicine: A Global Strategic Business Report" estimates that the nanobiomaterial product developments lead the nanomedicine market with a potential $160 billion market by the year 2015. A 2001 National Science Foundation (NSF) study predicted that nanotechnology applications will be used to improve quality of life and expects that half of pharmaceutical production spends about $180 billion on nanomedicine. As of mid-2006, 130 nanotech-based drugs and delivery systems and 125 devices or diagnostic tests are in preclinical, clinical, or commercial development, and 75% of these are developed in the United States (Kageyama 2005). The combined market for nano-enabled medicine (drug delivery, therapeutics and diagnostics) has jumped from just over $1 billion in 2005 to almost $10 billion in 2010. Another recent NSF study predicts that nanotechnology will produce half of the pharmaceutical industry product line by 2015. During 2005, Lux Research Inc. estimates that of all the nanotech funding, ~$1.6 billion was devoted to nano-enabled medical uses and industry contributed to 8% of this funding. It further pointed out that the majority of Fortune 500 companies are investing in nanotech R&D even though top pharmaceutical companies were still not heavily involved in nanobiomaterial-based product development due to the uncertainties in FDA approval (Boyd 2006). The market for the nanodrug delivery systems is expected to grow to $8.6 billion in 2010 from $980 million during 2005.

As the commercial applications of nanobiomaterials increase, the industry and the scientific community hold a greater responsibility toward ethical and regulatory issues regarding use of novel nanobiomaterials. Robust regulations and guidelines outlining the limitations on human exposure levels, environmental leakage issues, and waste disposal must be specifically established for each material rather than issuing general outlines. Nanobiomaterials present new challenges for assessments of biostability, biocompatibility, pharmacology, and biodistribution. Nevertheless, the development of nanobiomaterial-based biomedical technologies represents a challenging, but potentially rewarding opportunity to develop the next generation biomedical products.

References

Agrawal, S. 1999. Importance of nucleotide sequence and chemical modifications of antisense oligonucleotides. *Biochim. Biophys. Acta* 1489:53–68.

Allen, M., J. W. Bulte, L. Liepold, G. Basu, H. A. Zywicke, J. A. Frank, M. Young, and T. Douglas. 2005. Paramagnetic viral nanoparticles as potential high-relaxivity magnetic resonance contrast agents. *Magn. Reson. Med.* 54:807.

Ananta, J. S., Godin, B., Sethi, R., Moriggi, L., Liu, X., Serda, R. E., Krishnamurthy, R., Muthupillai, R., Bolskar, R. D., Helm, L., Ferrari, M., Wilson, L. J., and P. Decuzzi. 2010. Geometrical confinement of gadolinium-based contrast agents in nanoporous particles enhances π contrast. *Nat. Nanotechnol.* 5(11):815–821.

Anzai, Y., S. McLachlan, M. Morris, R. Saxton, and R. B. Lufkin. 1994a. Dextran-coated superparamagnetic iron oxide, an MR contrast agent for assessing lymph nodes in the head and neck. *AJNR Am. J. Neuroradiol.* 15(1):87–94.

Anzai, Y., K. E. Blackwell, S. L. Hirschowitz, J. W. Rogers, Y. Sato, W. T. Yuh, V. M. Runge, M. R. Morris, S. J. McLachlan, and R. B. Lufkin. 1994b. Initial clinical experience with dextran-coated superparamagnetic iron oxide for detection of lymph node metastases in patients with head and neck cancer. *Radiology* 192(3):709–715.

Arias, J. L., V. Gallardo, S. A. Gomez-Lopera, and A. V Delgado. 2005. Loading of 5-fluorouracil to poly(ethyl-2-cyanoacrylate) nanoparticles with a magnetic core. *J. Biomed. Nanotech.* 1:214–223.

Arias, J. L., M. López-Viota, M. A. Ruiz, J. López-Viota, and A. V. Delgado. 2007. Development of carbonyl iron/ethylcellulose core/shell nanoparticles for biomedical applications. *Int. J. Pharm.* 339(1–2):237.

Balivada, S., R. S. Rachakatla, H. Wang, T. N. Samarakoon, R. K. Dani, M. Pyle, F. O. Kroh et al. 2010. A/C magnetic hyperthermia of melanoma mediated by iron(0)/iron oxide core/shell magnetic nanoparticles: A mouse study. *BMC Cancer* 10:119.

Balogh, L. and D. A. Tomalia. 1998. Poly(amidoamine) dendrimer-templated nanocomposites. 1. Synthesis of zerovalent copper nanoclusters. *J. Am. Chem. Soc.* 120:7355–7356.

Bancaud, A., N. Silva, M. Barbi, G. Wagner, J. Allemand, J. Mozziconacci, C. Lavelle et al. 2006. Structural plasticity of single chromatin fibers revealed by torsional manipulation. *Nat. Struct. Mol. Biol.* 13:444–450.

Basu, G., M. Allen, D. Willits, M. Young, and T. Douglas. 2003. Melanoma and lymphocyte cell-specific targeting incorporated into a heat shock protein cage architecture. *J. Biol. Inorg. Chem.* 8:721.

Beaucage S. L. 2001. Strategies in the preparation of DNA oligonucleotide arrays for diagnostic applications. *Curr. Med. Chem.* 8:1213–1244.

Bellin, M. F., C. Roy, K. Kinkel, D. Thoumas, S. Zaim, D. Vanel, C. Tuchmann et al. 1998. Lymph node metastases: Safety and effectiveness of MR imaging with ultrasmall superparamagnetic iron oxide particles–initial clinical experience. *Radiology* 207:799–808.

Bertin, A., J. Steibel, A. I. Michou-Gallani, J. L. Gallani, and D. Felder-Flesch. 2009. Development of a dendritic manganese-enhanced magnetic resonance imaging (MEMRI) contrast agent: Synthesis, toxicity (in vitro) and relaxivity (in vitro, in vivo) studies. *Bioconjug. Chem.* 20(4):760–767.

Bigi, A., N. Nicoli-Aldini, B. Bracci, B. Zavan, E. Boanini, F. Sbaiz, S. Panzavolta et al. 2007. In vitro culture of mesenchymal cells onto nanocrystalline hydroxyapatite-coated Ti13Nb13Zr alloy. *J. Biomed. Mater. Res. Part A* 82A:213.

Blasi, L., L. Longo, G. Vasapollo, R. Cingolani, R. Rinaldi, T. Rizzello, R. Acierno, and M. Maffia. 2005. Characterization of glutamate dehydrogenase immobilization on silica surface by atomic force microscopy and kinetic analyses. *Enzyme Microb. Technol.* 36:818–823.

Bogunia-Kubik, K. and M. Sugisaka. 2002. From molecular biology to nanotechnology and nanomedicine. *Biosystems* 65(2–3):123–138.

Bolskar, R. D. 2008. Gadofullerene MRI contrast agents. *Nanomed. (Lond.)* 3(2):201–213.

Bottini, M., F. D'Annibale, A. Magrini, F. Cerignoli, Y. Arimura, M. I. Dawson, E. Bergamaschi, N. Rosato, A. Bergamaschi, and T. Mustelin. 2007. Silica nanoparticles as hepatotoxicants. *Int. J. Nanomed.* 2:227.

Boyd, R. S. 2006. "Scientists race to create bionic arm: Federal government wants better prostheses for wounded soldiers, Knight Ridder, May 29, May 2006. On the Internet: http://www.charlotte.com/mld/charlotte/news/14691684.htm

Braasch, D. A., S. Jensen, Y. Liu, K. Kaur, K. Arar, M. White, and D. R. Corey. 2003. RNA interference in mammalian cells by chemically-modified RNA. *Biochemistry* 42:7967–7975.

Brasch, R., C. Pham, D. Shames, T. Roberts, K. V. Djke, N. V. Bruggen, J. Mann, S. Ostrowitzki, and S. Melnyk. 1997. Assessing tumor angiogenesis using macromolecular MR imaging contrast media. *J. Magn. Reson. Imaging* 7(1):68–74.

Brasch, R. C., D. M. Shames, F. M. Cohen, R. Kuwatsuru, M. Neuder, J. S. Mann, V. Vexler, A. Muhler, and W. Rosenau. 1994. Quantification of capillary permeability to macromolecular magnetic resonance imaging contrast media in experimental mammary adenocarcinomas. *Invest. Radiol.* 29(Suppl 2): S8–S11.

Brettreich, M. and A. Hirsch. 1998. A highly water soluble dendro[60]fullerene. *Tetrahedron Lett.* 39(18):2731–2734.

Bridot, J. L., A. C. Faure, S. Laurent, C. Riviere, C. Billotey, B. Hiba, M. Janier et al. 2007. Hybrid gadolinium oxide nanoparticles: Multimodal contrast agents for in vivo imaging. *J. Am. Chem. Soc.* 129:5076.

Bulyk, M. L. 2006. DNA microarray technologies for measuring protein–DNA interactions. *Curr. Opin. Biotechnol.* 17:422–430.

Bustamente, C., Z. Bryant, and S. B. Smith. 2003. Ten years of tension: Single-molecule DNA mechanics. *Nature* 421:423–427.

Cai, W., D. W. Shin, K. Chen, O. Gheysens, Q. Cao, S. X. Wang, S. S. Gambhir, and X. Chen. 2006. Peptide-labeled near-infrared quantum dots for imaging tumor vasculature in living subjects. *Nano Lett.* 6:669.

Chang, C. and D. Mao. 2007. Thermal dehydration kinetics of a rare earth hydroxide, $Gd(OH)_3$. *Int. J. Chem. Kinet.* 39:75–81.

Chen, C. B., J. Y. Chen, and W. C. Lee. 2009. Fast transfection of mammalian cells using superparamagnetic nanoparticles under strong magnetic field. *J. Nanosci. Nanotechnol.* 9(4):2651–2659.

Chen, J., I. R. Corbin, H. Li, W. Cao, J. D. Glickson, and G. Zheng. 2007. Ligand conjugated low-density lipoprotein nanoparticles for enhanced optical cancer imaging in vivo. *J. Am. Chem. Soc.* 129:5798.

Chen, Z., S. M. Tabakman, A. P. Goodwin, M. G. Kattah, D. Daranciang, X. Wang, G. Zhang et al. 2008. Protein microarrays with carbon nanotubes as multicolor Raman labels. *Nat. Biotechnol.* 26(11):1285–1292.

Chen, Q. Z., I. D. Thompson, and A. R. Boccaccini. 2006. In vitro study of the antibacterial activity of bioactive glass-ceramic scaffolds. *Biomaterials* 27:2414.

Choudhary, S., K. M. Haberstroh, and T. J. Webster. 2007. Enhanced functions of vascular cells on nanostructured Ti for improved stent applications. *Tissue Eng. Part A* 13(7):1421.

Clark, L. C. and C. Lyons. 1962. Electrode systems for continuous monitoring in cardiovascular surgery. *Ann. N. Y. Acad. Sci.* 102:29–45.

Corot, C., P. Robert, J. M. Idée, and Port, M. 2006. Recent advances in iron oxide nanocrystal technology for medical imaging. *Adv. Drug Deliv. Rev.* 58:1471–504.

Datta, A., J. M. Hooker, M. Botta, M. B. Francis, S. Aime, and K. N. Raymond. 2008. Use of YbIII-centered near-infrared (nir) luminescence to determine the hydration state of a 3,2-HOPO-based MRI contrast agent. *J. Am. Chem. Soc.* 130:2546.

Debuigne, F., L. Jeunieau, M. Wiame, and J. B. Nagy. 2000. Synthesis of organic nanoparticles in different W/O microemulsions. *Langmuir* 16:7605.

Deng, A. D., M. Chen, S. X. Zhou, B. You, and L. M. Wu. 2006. A novel method for the fabrication of monodisperse hollow silica spheres. *Langmuir* 22:6403–6407.

Derossi, D., A. H. Joliot, G. Chassaing, and A. Prochiantz. 1994. Cell Internalization of the third helix of the Antennapedia homeodomain is receptor-independent. *J. Biol. Chem.* 269:10444.

Doorn, S. K., L. Zheng, M. J. O'Connell, Y. Zhu, S. Huang, and J. J. Liu. 2005. Raman spectroscopy and imaging of ultralong carbon nanotubes. *J. Phys. Chem. B* 109:3751–3758.

Eastman, P. S., W. Ruan, M. Doctolero, R. Nuttall, G. de Feo, J. S. Park, J. S. Chu, P. Cooke, J. W. Gray, S. Li, and F. F. Chen. 2006. Qdot nanobarcodes for multiplexed gene expression analysis. *Nano Lett.* 6(5):1059.

Emerich, D. F. and C. G. Thanos. 2003. Nanotechnology and medicine. *Expert Opin. Biol. Ther.* 3(4):655–663.

Esumi, K., A. Suzuki, N. Aihara, K. Usui, and K. Torigoe. 1998. Preparation of gold colloids with UV irradiation using dendrimers as stabilizer. *Langmuir* 14:3157–3159.

Fan, J., J. Lei, L. M. Wang, C. Z. Yu, B. Tu, and D. Y. Zhao. 2003. Rapid and highcapacity immobilization of enzymes based on mesoporous silicas with controlled morphologies. *Chem. Commun.* 17:2140–2141.

Fattal, E. and G. Barratt. 2009. Nanotechnologies and controlled release systems for delivery and antisense oligonucleotides and small interfering RNA. *Br. J. Pharmacol.* 157(2):179–194.

Firla, M. T. 1999. Dental Spiegel 8:48.

Flenniken, M. L., D. Willits, A. L. Harmsen, L. O. Liepold, A. G. Harmsen, M. J. Young, and T. Douglas. 2006. Melanoma and lymphocyte cell-specific targeting incorporated into a heat shock protein cage architecture. *Chem. Biol.* 13:161.

Flesch, C., E. Bourgeaut-Lami, S. Mornet, E. Duguet, C. Delaite, and P. Dumas. 2005. Synthesis of colloidal superparamagnetic nanocomposites by grafting poly(e-caprolactone) from the surface of organosilane-modified maghemite nanoparticles. *J. Polym. Sci. Part A: Polym. Chem.* 43:3221–3231.

Francoise, Q., C. Didier, D. R. Francesco, and G. Corine. 2006. Core-shell copper hydroxide-polysaccharide composites with hierarchical macroporosity. *Prog. Solid State Chem.* 34:161–169.

Fraysse-Ailhas, C., A. Graff-Meyer, P. Rigler, C. Mittelhozer, S. Raman, U. Aebi, and P. Burkhard. 2007. Peptide nanoparticles for drug delivery applications. *Eur. Cells Mater.* 14:115.

Geary R. S., T. A. Watanabe, L. Truong, S. Freier, E. A. Lesnik, and N. B. Sioufi. 2001. Pharmacokinetic properties of 2′-O-(2-methoxyethyl)-modified oligonucleotide analogs in rats. *Pharmacol. Exp. Ther.* 296:890–897.

Gerion, D., J. Herberg, R. Bok, E. Gjersing, E. Ramon, R. Maxwell, J. Kurhanewicz et al. 2007. Porous polymersomes with encapsulated Gd-labeled dendrimers as highly efficient MRI contrast agents. *J. Phys. Chem. C* 111:12542.

Gerion, D., F. Pinaud, S. C. Williams, W. J. Parak, D. Zanchet, S. Weiss, and A. P. Alivisatos. 2001. Synthesis and properties of biocompatible water-soluble silica-coated CdSe/ZnS semiconductor quantum dots. *J. Phys. Chem. B* 105:8861.

Giljohann, D. A., D. S. Seferos, W. L. Daniel, M. D. Massich, P. C. Patel, and C. A. Mirkin. 2010. Gold nanoparticles for biology and medicine. *Angew. Chem. Int. Ed. Engl.* 49(19):3280–3294.

Giljohann, D. A., D. S. Seferos, A. E. Prigodich, P. C. Patel, and C. A. Mirkin. 2009. Gene regulation with polyvalent siRNA–nanoparticle conjugates. *J. Am. Chem. Soc.* 131:2072.

Gobin, A. M., M. H. Lee, N. J. Halas, W. D. James, R. A. Drezek, and J. L. West. 2007. Near-infrared resonant nanoshells for combined optical imaging and photothermal cancer therapy. *Nano Lett.* 7:1929.

Gomez-Lopera, S. A., J. L. Arias, V. Gallardo, and A. V. Delgado. 2006. Colloidal stability of magnetite/poly(lactic acid) core/shell nanoparticles. *Langmuir* 22:2816–2821.

Goodman, S. B., B. E. Gomez, M. Takagi, and Y. T. Konttinen. 2009. Biocompatibility of total joint replacements: A review. *J. Biomed. Mater. Res. A* 90:603–618.

Gore, J., Z. Bryant, M. D. Stone, M. Nollman, N. R. Cozzarelli, and C. Bustamente. 2006. Mechanochemical analysis of DNA gyrase using rotor bead tracking. *Nature* 439:1010–1104.

Gossuin, Y., P. Gillis, A. Hocq, Q. L. Vuong, and A. Roch. 2009. Properties of superparamagnetic particles. *Wiley Interdiscip. Rev. Nanomed. Nanobiotechnol.* 1(3):299–310.

Green D. E., J. P. Longtin, and B. Sitharaman. 2009. The effect of nanoparticle-enhanced photoacoustic stimulation on multipotent marrow stromal cells. *ACS Nano*. 3(8):2065–2072.

Greim, H., P. Borm, R. Schins, K. Donaldson, K. Driscoll, A. Hartwig, E. Kuempel, G. Oberdörster, and G. Speit. 2001. Toxicity of fibers and particles. Report of the workshop held in Munich, Germany. *Inhal. Toxicol.* 13:737.

Gu, H., R. Zheng, X. Zhang, and B. Xu. 2004. Facile one-pot synthesis of bifunctional heterodimers of nanoparticles: A conjugate of quantum dot and magnetic nanoparticles. *J. Am. Chem. Soc.* 126:5664.

Guiseppi-Elie, A., C. H. Lei, and R. H. Baughman. 2002. Direct electron transfer of glucose oxidase on carbon nanotubes. Nanotechnology 13(5):559–564.

Gupta, A. K., P. R. Nair, D. Akin, M. R. Ladisch, S. Broyles, M. A. Alam, and Bashir, R. 2006. Anomalous resonance in a nanomechanical biosensor. *Proc. Natl. Acad. Sci. U.S.A.* 103:13362–13367.

Halperin, A., A. Buhot, and E. B. Zhulina. 2004. Sensitivity, specificity, and the hybridization isotherms of DNA chips. *Biophys. J.* 86:718–730.

Han, Y., S. S. Lee, and J. Y. Ying, 2006. Pressure-driven enzyme entrapment in siliceous mesocellular foam. *Chem. Mater.* 18:643–649.

Harborth, J., S. M. Elbashir, K. Vandenburgh, H. Manninga, S. A. Scaringe, K. Weber, and T. Tuschi. 2003. Sequence, chemical, and structural variation of small interfering RNAs and short hairpin RNAs and the effect on mammalian gene slicing. *Antisense Nucleic Acid Drug Dev.* 13:83–105.

Harisinghani, M. G., J. Barentsz, P. F. Hahn, W. M. Deserno, S. Tabatabaei, C. Hulsbergen van de Kaa, J. De la Rosette, and R. Weissleder. 2003. Noninvasive detection of clinically occult lymphnode metastases in prostate cancer. *N. Engl. J. Med.* 348:2491–2499.

Hifumi, H., S. Yamaoka, A. Tanimoto, D. Citterio, and K. Suzuki. 2006. Gadolinium-based hybrid nanoparticles as a positive MR contrast agents. *J. Am. Chem. Soc.* 128:15090–15091.

Huang, X. and M. A. El-Sayed. 2010. Gold nanoparticles: Optical properties and implementation in cancer diagnosis and photothermal therapy. *J. Adv. Res.* 1:13–28.

Huang, C. Y. and Y. D. Lee. 2006. Core-shell type of nanoparticles composed of poly[(n-butyl cyanoacrylate)-co-(2-octyl cyanoacrylate)] copolymers for drug delivery application: Synthesis, characterization and in vitro degradation. *Int. J. Pharm.* 325:132.

Hussain, N., V. Jaitley, and A. T. Florence. 2001. Recent advances in the understanding of uptake of microparticulates across the gastrointestinal lymphatics. *Adv. Drug Deliv. Rev.* 50:107.

Iwamoto Y. and Yamakoshi Y. 2006. A highly water-soluble C60-NVP copolymer: A potential material for photodynamic therapy. *Chem. Commun. (Camb.)* 46:4805–4807.

Jafelicci, Jr. M., M. R. Davolos, F. J. Santos, and S. J. de Andrade. 1999. Hollow silica particles from microemulsion. *J. Non-Cryst. Solids* 247:98–102.

Jain, K. K. 2003. Current status of molecular biosensors. *Med. Device Technol.* 14:10–15.

Jain, K. K. 2007. Applications of nanobiotechnology in clinical diagnostics. *Clin. Chem.* 53(11):2002–2009.

Jain, P. K., K. S. Lee, I. H. El-Sayed, and M. A. El-Sayed. 2006. Calculated absorption and scattering properties of gold nanoparticles of different size, shape, and composition: Applications in biological imaging and biomedicine. *J. Phys. Chem. B* 110(14):7238–7248.

Jang, W. D., K. M. Kamruzzaman Selim, C. H. Lee, and I. K. Kang. 2009. Bioinspired application of dendrimers: From bio-mimicry to biomedical applications. *Prog. Poly. Sci.* 34:1–23.

Jing, S., S. Xing, L. Yu, Y. Wu, and C. Zhao. 2007a. Synthesis and characterization of Ag/polyaniline core–shell nanocomposites based on silver nanoparticles colloid. *Mater. Lett.* 61:2794–2797.

Jing, S., S. Xing, L. Yu, and C. Zhao. 2007b. Synthesis and characterization of Ag/polypyrrole nanocomposites based on silver nanoparticles colloid. *Mater. Lett.* 61:4528–4530.

Kageyama, Y., Remote Control Device "Controls" Humans, Associated Press, October 26, 2005. On the Internet http://www.sfgate.com/cgi-bin/article.cgi?f=/n/a/2005/10/25/financial/f133702D73.DTL

Kang, S., Mauter, M. S., and Elimelech, M. 2008. Physicochemical determinants of multiwalled carbon nanotube bacterial cytotoxicity. *Environ. Sci. Technol.* 42(19):7528–7534.

Kang, S., Mauter, M. S., and Elimelech, M. 2009. Microbial cytotoxicity of carbon-based nanomaterials: Implications for river water and wastewater effluent. *Environ. Sci. Technol.* 43(7):2648–2653.

Khang, D., J. Carpenter, Y. W. Chun, R. Pareta, and T. J. Webster. 2008. Nanotechnology for regenerative medicine. *Biomed Microdevices* 19:19.

Kim, H. W., H. H. Lee, and J. C. Knowles. 2006. Electrospun poly(L-lactide-co-ε-caprolactone)/polyethylene oxide/hydroxyapaite nanofibrous membrane for guided bone regeneration. *J. Biomed. Mater. Res. Part A* 79A:643.

Klimov, V. I., L. P. Balet, M. Achermann, J. A. Hollingsworth, and H. Kim. 2004. Synthesis and characterization of Co/CdSe core/shell nanocomposites: Bifunctional magnetic-optical nanocrystals. *J. Am. Chem. Soc.* 127:544.

Koch, A. M., F. Reynolds, H. P. Merkle, R. Weissleder, and L. Josephson. 2005. Transport of surface-modified nanoparticles through cell monolayers. *Chem. Bio. Chem.* 6:337.

Kolosnjaj-Tabi, J., K. B. Hartman, S. Boudjemaa, J. S. Ananta, G. Morgant, H. Szwarc, L. J. Wilson, and F. Moussa. 2010. In vivo behavior of large doses of ultrashort and full-length single-walled carbon nanotubes after oral and intraperitoneal administration to Swiss mice. *ACS Nano* 4(3):1481–1492.

Krusic, P. J., E. Wasserman, P. N. Keizer, J. R. Morton, and K. F. Preston. 1991. Radical reactions of C60. *Science* 254(5035):1183–1185.

Kumar, R. V., O. Palchik, Y. Koltypin, Y. Diamant, and A. Gedanken. 2002. Sonochemical synthesis and characterization of Ag2S/PVA and CuS/PVA nanocomposite. *Ultrasonics Sonochem.* 9:65–70.

Kwon, K. W. and M. Shim. 2005. γ-Fe2O3/II–VI sulfide nanocrystal heterojunctions. *J. Am. Chem. Soc.* 127:10269.

Lam, C. W., J. T. James, R. McCluskey, S. Arepalli, and R. L. Hunter. 2006. A review of carbon nanotube toxicity and assessment of potential occupational and environmental health risks. *Crit. Rev. Toxicol.* 36(3):189–217.

Langer, R. and J. P. Vacanti. 1993. Tissue engineering. *Science* 260(5110):920–926.

Laschke, M. W., K. Witt, T. Pohlemann, and M. D. Menger. 2007. Injectable nanocrystalline hydroxyapatite paste for bone substitution: In vivo biocompatibility and vascularization. *J. Biomed. Mater. Res. Part B: Appl. Biomater.* 82B:494.

Lee, K. S. and M. A. El-Sayed. 2005. Dependence of the enhanced optical scattering efficiency relative to that of absorption for gold metal nanorods on aspect ratio, size, end-cap shape and medium refractive index. *J. Phys. Chem. B* 109(43):20331–20338.

Lei, Y., H. Tang, M. Feng, and B. Zou. 2009. Applications of fluorescent quantum dots to stem cell tracing in vivo. *J. Nanosci. Nanotechnol.* 9(10):5726.

Lemon, B. I. and R. M. Crooks. 2000. Preparation and characterization of dendrimer-encapsulated CdS semiconductor quantum dots. *J. Am. Chem. Soc.* 122:12886–12887.

Li, J. G., W. X. Li, J. Y. Xu, X. Q. Cai, R. L. Liu, Y. J. Li, Q. F. Zhao, and Q. N. Li. 2007. Comparative study of pathological lesions induced by multiwalled carbon nanotubes in lungs of mice by intratracheal instillation and inhalation. *Environ. Toxicol.* 22(4):415–421.

Li, W. and F. C. Szoka Jr. 2007. Lipid-based nanoparticles for nucleic acid delivery. *Pharm. Res.* 24:438–449.

Link, S., C. Burda, M. B. Mohamed, B. Nikoobakht, and M. A. El-Sayed. 1999. Laser photothermal melting and fragmentation of gold nanorods: Energy and laser pulse-width dependence. *J. Phys. Chem. A* 103(9):1165–1170.

Link, S., C. Burda, B. Nikoobakht, and M. A. El-Sayed. 2000. Laser-induced shape changes of colloidal gold nanorods using femtosecond and nanosecond laser pulses. *J. Phys. Chem. B* 104(26):6152–6163.

Liu, Z., C. Davis, W. Cai, L. He, X. Chen, and H. Dai. 2008. Circulation and long-term fate of functionalized, biocompatible single-walled carbon nanotubes in mice probed by Raman spectroscopy. *Proc. Natl. Acad. Sci. U.S.A.* 105:1410–1415.

Liu, L., K. Xu, H. Wang, P. K. Tan, W. Fan, S. S. Venkatraman, L. Li, and Y. Y. Yang. 2009. Self-assembled cationic peptide nanoparticles as an efficient antimicrobial agent. *Nat. Nanotechnol.* 4(7):457–463.

Liu, X., B. Xu, Q. S. Xia, T. D. Zhao, and J. T. Tang. 2005. A method of showing thermal effect of iron oxide nanoparticles in alternating magnetic field. *Ai Zheng* 24(9):1148–1150.

Lu, L., R. M. Jones, D. McBranch, and D. Whitten. 2002. Surface-enhanced surperquenching of Cyanine dyes as J-aggregates on laponite clay nanoparticles. *Langmuir* 18:7706–7713.

Lu, S. Y. and I. H. Lin. 2003. Rational design and fabrication of ZnO nanotubes from nanowire templates in a microwave plasma system. *J. Phys. Chem. B* 107:6974.

Ma, P. X. 2008. Biomimetic materials for tissue engineering. *Adv. Drug Deliv. Rev.* 60:184–198.

Magrez, A., S. Kasas, V. Salicio, N. Pasquier, J. W. Seo, M. Celio, S. Catsicas, B. Schwaller, and L. Forró. 2006. Cellular toxicity of carbon-based nanomaterials. *Nano Lett.* 6(6):1121–1125.

Manchester, M. and P. Singh. 2006. Virus-based nanoparticles (VNPs): Platform technologies for diagnostic imaging. *Adv. Drug Deliv. Rev.* 58:1505.

Martín-Saavedra, F. M., E. Ruíz-Hernández, A. Boré, D. Arcos, M. Vallet-Regí, and N. Vilaboa. 2010. Magnetic mesoporous silica spheres for hyperthermia therapy. *Acta Biomater.* 6(12):4522–4531.

Mashal, A., B. Sitharaman, X. Li, P. K. Avti, A. V. Sahakian, J. H. Booske, and S. C. Hagness. 2010. Toward carbon-nanotube-based theranostic agents for microwave detection and treatment of breast cancer: Enhanced dielectric and heating response of tissue-mimicking materials. *IEEE Trans. Biomed. Eng.* 57(8):1831–1834.

Medina, S. H. and M. E. H. El-Sayed. 2009. Dendrimers as carriers for delivery of chemotherapeutic agents. *Chem. Rev.* 109:3141–3157.

Miyaji, F., S. A. Davis, J. P. H. Charmant, and S. Mann. 1999. Organic crystal templating of hollow silica fibers. *Chem. Mater.* 11:3021–3024.

Miyazaki, M., J. Kaneno, R. Kohama, M. Uehara, K. Kanno, M. Fujii, H. Shimizu, and H. Maeda. 2004. Preparation of functionalized nanostructures on microchannel surface and their use for enzyme microreactors. *Chem. Eng. J.* 101:277–284.

Mornet, S., S. Vasseur, F. Grasset, and E. Duguet. 2004. Magnetic nanoparticle design for medical diagnosis and therapy. *J. Mater. Chem.* 14:2161–2175.

Muller, J., I. Decordier, P. H. Hoet, N. Lombaert, L. Thomassen, F. Huaux, D. Lison, and M. Kirsch-Volders. 2008. Clastogenic and aneugenic effects of multi-wall carbon nanotubes in epithelial cells. *Carcinogenesis* 29:427–433.

Murray, K., Y. C. Cao, S. Ali, and Q. Hanley. 2010. Lanthanide doped silica nanoparticles applied to multiplexed immunoassays. *Analyst* 135(8):2132–2138; Nanotechnology 2006. *Nature Nanotechnol.* 1(1):8–10.

Nakamura, E. and Isobe H. 2003. Energetics of water permeation through fullerene membrane. *Acc. Chem. Res.* 36:807–815.

Namgung, R., K. Singha, M. K. Yu, S. Jon, Y. S. Kim, Y. Ahn, I. K. Park, and W. J. Kim. 2010. Hybrid superparamagnetic iron oxide nanoparticle-branched polyethylenimine magnetoplexes for gene transfection of vascular endothelial cells. *Biomaterials* 31(14):4204–4213.

Neuwelt, E. A., B. E. Hamilton, C. G. Varallyay, W. R. Rooney, R. D. Edelman, P. M. Jacobs, and S. G. Watnick. 2009. Ultrasmall superparamagnetic iron oxides (USPIOs): A future alternative magnetic resonance (MRI) contrast agents for patients at risk for nephrogenic systemic fibrosis (NSF). *Kidney Int.* 75(5):465–474.

Nikanjam, M., E. A. Blakely, K. A. Bjornstad, X. Shu, T. F. Budinger, and T. M. Forte. 2007. Synthetic nano-low density lipoprotein as targeted drug delivery vehicle for glioblastoma multiforme. *Int. J. Pharm.* 328:86.

Norberg, N. S., K. R. Kittilstved, J. E. Amonette, R. K. Kukkadapu, D. A. Schwartz, and D. R. Gamelin, 2004. Synthesis of colloidal Mn2+:ZnO quantum dots and high-Tc ferromagnetic nanocrystalline thin films. *J. Am. Chem. Soc.* 126(30):9387.

Nukavarapu, S. P., S. G. Kumbar, J. L. Brown, N. R. Krogman, A. L. Weikel, M. D. Hindenlang, L. S. Nair, H. R. Allcock, and C. T. Laurencin. 2008. Polyphosphazene/nano-hydroxyapatite composite microsphere scaffolds for bone tissue engineering. *Biomacromolecules* 9:1818.

Oishi, M., A. Tamura, T. Nakamura, and Y. Nagasaki. 2009. A smart nanoprobe based on fluorescence-quenching PEGylated nanogels containing gold nanoparticles for monitoring the response to cancer therapy. *Adv. Funct. Mater.* 19:827.

Opanasopit, P., M. Nishikawa, and M. Hashida. 2002. Factors affecting drug and gene delivery: Effects of interaction with blood components. *Crit. Rev. Ther. Drug Carrier Syst.* 19:191–233.

Owens, D. E. and N. A. Peppas. 2006. Opsonization, biodistribution, and pharmacokinetics of polymeric nanoparticles. *Int. J. Pharm.* 307:93–102.

Paek, J., C. H. Lee, J. Choi, S. Y. Choi, A. Kim, J. W. Lee, and K. Lee. 2007. Gadolinium oxide nanorings and nanoplates: Anisotropic shape control. *Cryst. Growth Des.* 7:1378–1380.

Patel, P. C., D. A. Giljohann, D. S. Seferos, and C. A. Mirkin. 2008. Peptide antisense nanoparticles. *Proc. Natl. Acad. Sci. U.S.A.* 105:17222.

Patolsky, F., Weizmann Y., and Willner I. 2004. Long-range electrical contacting of redox enzymes by SWCNT connectors. *Angew. Chem. Int. Ed.* 43:2113–2117.

Perez, J. M., F. J. Simeone, Y. Saeki, L. Josephson, and R. Weissleder. 2003. Viral-induced self-assembly of magnetic nanoparticles allows the detection of viral particles in biological media. *J. Am. Chem. Soc.* 125:10192–10193.

Pirrung, M. C. 2002. How to make a DNA chip. *Angew. Chem. Int. Ed.* 41:1276–1289.

Pramanik, M., H. K. Song, M. Swierczewska, D. Green, B. Sitharaman, and L. V. Wang. 2009. In vivo carbon nanotube-enhanced non-invasive photoacoustic mapping of the sentinel lymph node. *Phys. Med. Biol.* 54(11):3291–3301.

Prencipe, G., S. M. Tabakman, K. Welsher, Z. Liu, A. P. Goodwin, L. Zhang, J. Henry, and H. Dai. 2009. PEG branched polymer for functionalization of nanomaterials with ultralong blood circulation. *J. Am. Chem. Soc.* 131:4783–4787.

Raman, S. K., G. Machaidze, A. Lustig, U. Aebi, and P. Burkhard. 2006. Structure-based design of peptides that self-assemble into regular polyhedral nanoparticles. *Nanomedicine* 2:95–102.

Rauschmann, M. A., T. A. Wichelhaus, V. Stirnal, E. Dingeldein, L. Zichner, and R. Schnettler. 2005. Nanocrystalline hydroxyapatite and calcium sulphate as biodegradable composite carrier material for local delivery of antibiotics in bone infections. *Biomaterials* 26:2677.

Rensen, P. C. N., R. L. A. D. de Vrueh, J. Kuiper, M. K. Bijsterbosch, E. A. L. Biessen, and T. J.C. V. Berkel. 2001. Recombinant lipoproteins: Lipoprotein-like lipid particles for drug targeting. *Adv. Drug Deliv. Rev.* 47:251–276.

Resch-Genger, U., M. Grabolle, S. Cavaliere-Jaricot, R. Nitschke, and T. Nann. 2008. Quantum dots versus organic dyes as fluorescent labels. *Nat. Methods* 5:763–775.

Rogers, J. M., J. Lewis, and L. Josephson. 1994. Visualization of superior mesenteric lymph nodes by the combined oral and intravenous administration of the ultrasmall superparamagnetic iron oxide AMI-227. *Magn. Reson. Imaging* 12:1161–1165.

Rosi, N. L., D. A. Giljohann, C. S. Thaxton, A. K. R. Lytton-Jean, M. S. Han, and C. A. Mirkin. 2006. Nanoparticle-based bio-bar codes for the ultrasensitive detection of proteins. *Science* 312:1027.

Sarin, H., A. S. Kanevsky, H. Wu, K. R. Brimacombe, S. H. Fung, A. A. Sousa, S. Auh et al. 2008. Effective transvascular delivery of nanoparticles across the blood–brain tumor barrier into malignant glioma cells. *J. Transl. Med.* 6:80.

Sato, M., A. Aslani, M. A. Sambito, N. M. Kalkhoran, E. B. Slamovich, and T. J. Webster. 2008. Nanocrystalline hydroxyapatite/titania coatings on titanium improves osteoblast adhesion. *J. Biomed. Mater. Res. Part A* 84A:265.

Schmiedl, U., M. D. Ogan, M. E. Moseley, and R. C. Brasch. 1986. Comparison of the contrast-enhancing properties of albumin-(Gd-DTPA) and Gd-DTPA at 2.0 T: An experimental study in rats. *Am. J. Roentgenol.* 147(6):1263–1270.

Schulze, H., S. Vorlova, F. Villatte, T. T. Bachmann, and R. D. Schmid. 2003. Design of acetylcholinesterases for biosensor applications. *Biosens. Bioelectron.* 18:201–209.

Schuster, D. I., S. R. Wilson, A. N. Kirschner, R. F. Schinazi, S. Schlueter-Wirtz, P. Tharnish, T. Barnett et al. 2000. Evaluation of the anti-HIV potency of a water-soluble dendrimeric fullerene. *Proc. Electrochem. Soc.* 9:267–270.

Schwartz, D. A., N. S. Norberg, Q. P. Nguyen, J. M. Parker, and D. R. Gamelin. 2003. Magnetic quantum dots: Synthesis, spectroscopy, and magnetism of Co^{2+}- and Ni^{2+}-doped ZnO nanocrystals. *J. Am. Chem. Soc.* 125:13205.

Seliger, H., M. Hinz, and E. Happ. 2003. Arrays of immobilized oligonucleotides—Contributions to nucleic acids technology. *Curr. Pharm. Biotechnol.* 4:379–395.

Selvan, S. T., P. K. Patra, C. Y. Ang, and J. Y. Ying, 2007. Synthesis of silica-coated semiconductor and magnetic quantum dots and their use in the imaging of live cells. *Angew. Chem. Int. Ed.* 46:2448.

Sethi, D., R. P. Gandhi, P. Kumar, and K. C. Gupta. 2009. Chemical strategies for immobilization of oligonucleotides. *Biotechnol. J.* 4:1513–1529.

Shai, R. M. 2006. Microarray tools for deciphering complex diseases. *Front. Biosci.* 11:1414–1424.

Sharma, P., S. Brown, G. Walter, S. Santra, and B. Moudgil. 2006. Advances in colloid and interface science nanoparticles for bioimaging. *Adv. Coll. Interf. Sci.* 123–126:471–485.

Shen, X. C., J. Zhou, X. Liu, J. Wu, F. Ou, and Z. L. Zhang. 2007. Importance of size-to-change ratio in construction of stable and uniform nanoscale RNA/dendrimer complexes. *Org. Biomol. Chem.* 5:3674–3681.

Shi X., S. Wang, S. Meshinchi, M. E. Van Antwerp, X. Bi, I. Lee, and J. R. Baker, Jr. 2007. Dendrimer-entrapped gold nanoparticles as a platform for cancer-cell targeting and imaging. *Small* 3:1245–1252.

Shi, X., S. H. Wang, S. D. Swanson, S. Ge, Z. Cao, M. E. Van Antwerp, K. J. Landmark, and J. R. Baker, Jr. 2008. Dendrimer-functionalized shell-crosslinked iron oxide nanoparticles for in-vivo magnetic resonance imaging of tumors. *Adv. Mater.* 20:1671–1678.

Shin, H., S. Jo, and A. G. Mikos. 2003. Biomimetic materials for tissue engineering. *Biomaterials* 24:4353–4364.

Shingyoji, M., D. Gerion, D. Pinkel, J. W. Gray, and F. Chen. 2005. Quantum dots-based reverse phase protein microarray. *Talanta* 67(3):472.

Shukla, R., E. Hill, X. Shi, J. Kim, M. C. Muniz, K. Sun, and J. R. Baker Jr. 2008. Tumor microvasculature targeting with dendrimer-entrapped gold nanoparticles. *Soft Matter* 4:2160–2163.

Shvedova, A., E. Kisin, R. Mercer, A. Murray, V. J. Johnson, A. Potapovich, Y. Tyurina et al. 2005. Unusual inflammatory and fibrogenic pulmonary responses to single walled carbon nanotubes in mice. *Am. J. Physiol. Lung Cell. Mol. Physiol.* 289:L698–L708.

Singh, P., M. J. Gonzalez, and M. Manchester. 2006. Viruses and their uses in nanotechnology. *Drug Dev. Res.* 67:23.

Sinha, N. and J. T. Yeow. 2005. Carbon nanotubes for biomedical applications. *IEEE Trans. Nanobiosci.* 4(2):180–195.

Sipkins, D. A., D. A. Cheresh, M. R. Kazemi, L. M. Nevin, M. D. Bednarski, and K. C. Li. 1998. Detection of tumor angiogenesis in vivo by alphaVbeta3-targeted magnetic resonance imaging. *Nat. Med.* 4:623–626.

Sitharaman, B., K. R. Kissell, K. B. Hartman, L. A. Tran, A. Baikalov, I. Rusakova, Y. Sun et al. 2005. Superparamagnetic gadonanotubes are high-performance MRI contrast agents. *Chem. Commun. (Camb.)* 31:3915–3917.

Sitharaman, B., Tran L. A., Pham Q. P., Bolskar R. D., Muthupillai R., Flamm S. D., Mikos A. G., and L. J. Wilson. 2007. Gadofullerenes as nanoscale magnetic labels for cellular MRI. *Contrast Media Mol. Imaging* 2(3):139–146.

Sitharaman, B., Zakharian, T. Y., Saraf, A., Misra, P., and Ashcroft, J. 2008. Water-soluble fullerene (C60) derivatives as nonviral gene-delivery vectors. *Mol. Pharm.* 5:567–578.

Skaat, H. and S. Margel. 2009. Synthesis of fluorescent-maghemite nanoparticles as multimodal imaging agents for amyloid-beta fibrils detection and removal by a magnetic field. *Biochem. Biophys. Res. Commun.* 386(4):645–649.

Slowing, I. I., B. G. Trewyn, S. Giri, and V. S. Y. Lin. 2007. Mesoporous silica nanoparticles for drug delivery and biosensing applications. *Adv. Funct. Mater.* 17:1225.

Slowing, I. I., J. L. Vivero-Escoto, C. W. Wu, and V. S. Y. Lin. 2008. Mesoporous silica nanoparticles as controlled release drug delivery and gene transfection carriers. *Adv. Drug Deliv. Rev.* 60:1278.

Spinazzi, A., V. Lorusso, G. Pirovano, and M. Kirchin. 1999. Safety, tolerance, biodistribution and MR imaging enhancement of the liver with Gd-BOPTA: Results of clinical pharmacologic and pilot imaging studies in non-patient and patient volunteers. *Acad. Radiol.* 6:282–291.

Srivastava S., A. Santos, K. Critchley, K. S. Kim, P. Podsiadlo, K. Sun, J. Lee, C. Xu, G. D. Lilly, S. C. Glotzer, and N. A. Kotov. 2010. Light-controlled self-assembly of semiconductor nanoparticles into twisted ribbons. *Science* 327(5971):1355–1359.

Strano, M. S. and H. Jin. 2008. Where is it heading? Single-particle tracking of single-walled carbon nanotubes. *ACS Nano.* 9:1749–1752.

Strick, T. R., J. F. Allemand, D. Bensimon, A. Bensimon, and V. Croquettel. 1996. The elasticity of a single supercoiled DNA molecule. *Science* 271:1835–1837.

Strick, T. R., V. Croquette, and D. Bensimon. 2000. Single-molecule analysis of DNA uncoiling by a type II topoisomerase. *Nature* 404:901–904.

Takagi, A., A. Hirose, T. Nishimura, N. Fukumori, A. Ogata, N. Ohashi, S. Kitajima, and J. Kanno. 2008. Induction of mesothelioma in p53+/− mouse by intraperitoneal application of multi-wall carbon nanotube. *J. Toxicol. Sci.* 33(1):105–116.

Tchikov, V., S. Winoto-Morbach, M. Krönke, M. Kabelitz, and S. Schutze. 2001. Adhesion of immunomagneticpartic les targeted to antigens and cytokine receptors on tumor cells determined by magnetophoresis. *J. Magn. Magn. Mater.* 225:285–293.

Tóth, E., R. D. Bolskar, A. Borel, G. González, L. Helm, A. E. Merbach, B. Sitharaman, and L. J. Wilson. 2005. Water-soluble gadofullerenes: Toward high-relaxivity, pH-responsive MRI contrast agents. *J. Am. Chem. Soc.* 127(2):799–805.

Tsao, J., M. S. Chapman, M. Agbandje, W. Keller, K. Smith, H. Wu, M. Luo, T. J. Smith, M. G. Rossmann, and R. W. Compans. 1991. The three-dimensional structure of canine parvovirus and its functional implications. *Science* 251:1456.

Veiseh, M., P. Gabikian, S. B. Bahrami, O. Veiseh, M. Zhang, R. C. Hackman, A. C. Ravanpay et al. 2007. Tumor paint: A chlorotoxin: Cy5.5 bioconjugate for intraoperative visualization of cancer foci. *Cancer Res.* 67:6882–6888.

Veiseh, O., C. Sun, C. Fang, N. Bhattarai, J. Gunn, F. Kievit, K. Du et al. 2009. Specific targeting of brain tumors with an optical/magnetic resonance imaging nanoprobe across the blood–brain barrier. *Cancer Res.* 69:6200–6207.

Vo-Dinh, T., P. Kasili, M. Wabuyele. 2006. Nanoprobes and nanobiosensors for monitoring and imaging individual living cells. *Nanomedicine* 2(1):22–30.

Von der Mark, K., J. Park, S. Bauer, and P. Schmuki. 2010. Nanoscale engineering of biomimetic surfaces: Cues from the extracellular matrix. *Cell Tissue Res.* 339(1):131–153.

Wang, J. 2001. Glucose biosensors: 40 years of advances and challenges. *Electroanalysis* 13:983–988.

Wang, X., Y. Dua, J. Luoa, B. Lina, and J. F. Kennedy. 2007. Chitosan/organic rectorite nanocomposite films: Structure, characteristic and drug delivery behaviour. *Carbohyd. Poly.* 69:41.

Wang, W., J. J. Han, L. Q. Qang, L. S. Li, W. Shaw, and A. D. Q. Li. 2003. Dynamic π–π stacked molecular assemblies emit from green to red colors. *Nano Lett.* 3:455.

Wang, Y. X., S. M. Hussain, and G. P. Krestin. 2001. Superparamagnetic iron oxide contrast agents: Physicochemical characteristics and applications in MR imaging. *Eur. Radiol.* 11(11):2319–2331.

Wang L., R. Luhm, and M. Lei. 2007. SNP and mutation analysis. *Adv. Exp. Med. Biol.* 593:105–116.

Warren, C. L., N. C. S. Kratochvil, K. E. Hauschild, S. Foister, M. L. Brezinski, P. B. Dervan, G. N. Phillips Jr., and A. Z. Ansari. 2000. Defining the sequence-recognition profile of DNA binding molecules. *Proc. Natl. Acad. Sci. U.S.A.* 103:867–872.

Webster, T. J. and E. S. Ahn. 2007. Nanostructured biomaterials for tissue engineering bone. *Adv. Biochem. Eng. Biotechnol.* 103:275–308.

Webster, T. J., C. Ergun, R. H. Doremus, R. W. Siegel, and R. Bizios. 2000. Enhanced functions of osteoblasts on nanophase ceramics. *Biomaterials* 21(17):1803–1810.

Weng, K. C., C. O. Noble, B. Papahadjopoulos-Sternberg, F. F. Chen, D. C. Drummond, D. B. Kirpotin, D. Wang, Y. K. Hom, B. Hann, and J. W. Park. 2008. Targeted tumor cell internalization and imaging of multifunctional quantum dot-conjugated immunoliposomes in vitro and in vivo. *Nano Lett.* 8(9):2851.

Whaley, S. R., D. S. English, E. L. Hu, P. F. Barbara, and A. M. Belcher. 2000. Selection of peptides with semiconductor binding specificity for directed nanocrystal assembly. *Nature* 405(6787):665–668.

Winoto-Morbach, S., V. Tchikov, and W. J. Müller-Ruchholtz. 1994. Magnetophoresis: I. Detection of magnetically labeled cells. *J. Clin. Lab. Anal.* 8:400.

Wiradharma, N., M. Khan, Y. W. Tong, S. Wang, and Y. Y. Yang. 2008. Design and evaluation of peptide amphiphiles with different hydrophobic blocks for simultaneous delivery of drugs and gene. *Adv. Funct. Mater.* 18:943.

Wolter, H., W. Glaubitt, and K. Rose. 1992. Sol–gel-processed SiO_2/TiO_2/poly(vinylpyrrolidone) composite materials for optical waveguides. *Mater. Res. Soc. Symp. Proc.* 271:719.

Wolter, H., W. Storch, and C. Gellermann. 1996. *Mater. Res. Soc. Symp. Proc.* 435:67.

Wolter, H., W. Storch, and H. Ott. 1994. New inorganic/organic copolymers (ormocer®s) for dental applications. *Mater. Res. Soc. Symp. Proc.* 346:143.

Wolter, H., W. Storch, S. Schmitzer, W. Geurtzen, G. Leuhausen, and R. Maletz. 1998. In *Werkstoffe für die Medizintechnik*, Vol. 4, eds. H. Plack and H. Stallforth, Wiley-VCH, Weinheim, Germany, p. 245.

Wu X. C., A. M. Bittner, and K. Kern. 2005. Synthesis, structure and optical properties of CdS/dendrimers nanocomposites. *J. Phys. Chem. B* 109:230–239.

Zhang, R., R. B. Diasio, Z. Lu, T. Liu, Z. Jiang, and W. M. Galbraith. 1995. Pharmacokinetics and tissue distribution in rats of an oligodcoxnucleotide phosphorothioate developed as a therapeutic agent for human immunodeficiency virus type-1. *Biomed. Pharmacol.* 49:929–939.

Zhang, Y., N. Kohler, and M. Zhang. 2002. Functionalisation of magnetic nanoparticles for applications in biomedicine. *Biomaterials* 23:1553–1561.

Zhang, T., J. L. Stilwell, D. Gerion, L. Ding, O. Elboudwarej, P. A. Cooke, J. W. Gray, A. P. Alivisatos, and F. F. Chen. 2006. Cellular effect of high doses of silica-coated quantum dot profiled with high throughput gene expression analysis and high content cellomics measurements. *Nano Lett.* 6(4):800.

Zhang, Y., J. R. Venugopal, A. El-Turki, S. Ramakrishna, B. Su, and C. T. Lim. 2008. Biocomposites containing natural polymers and hydroxyapatite for bone tissue engineering. *Biomaterials* 29:4314.

Zheng, Q., H. Dai, M. E. Merritt, C. Malloy, C. Y. Pan, and W. H. Li. 2005. A new class of macrocyclic lanthanide complexes for cell labeling and magnetic resonance imaging applications. *J. Am. Chem. Soc.* 127:16178–16188.

Zhou, J., J. Wu, N. Hafdi, J. P. Behr, P. Erbacher, and L. Peng. 2006. PAMAM dendrimers for efficient siRNA delivery and potent gene silencing. *Chem. Commun. (Camb.)* 22:2362–2364.

2

Multifunctional Gold Nanoparticles for Cancer Therapy

Maung Kyaw Khaing Oo
Stevens Institute of Technology

Henry Du
Stevens Institute of Technology

Hongjun Wang
Stevens Institute of Technology

2.1 Introduction

Noble gold nanoparticles (GNPs) have been broadly explored for targeted drug delivery, medical diagnosis and imaging, and monitoring of cancer treatment due to their inherent chemical stability, biocompatibility, and astonishing optical properties in the nanometer scale. They can be readily functionalized with a variety of drugs via surface modification at the molecular level. They can exhibit either positive or negative charge characteristics, depending on the synthesis route. As many drugs carry net negative charge, positively charged GNPs are particularly useful for allowing drug conjugation on the particle surface via electrostatic attraction without complex wet chemistry. A new approach to fabricating positively charged GNPs, developed in our group, will be described as an example. The nanometer dimension of GNPs makes it possible for their cellular uptake through cell membrane. The plasmonic resonance under light irradiation and resultant enhancement of the local electromagnetic field also renders GNPs as an excellent platform for label-free molecular finger printing, using surface-enhanced Raman scattering (SERS). This high local field around GNPs and their aggregates has shown to be responsible for the generation of elevated reactive oxygen species (ROS) from photosensitizers. ROS are a critical mediator in causing the apoptosis of cancerous cells. Consequently, GNPs can be used to formulate an efficient photodynamic therapy (PDT) for cancer treatment. Clearly, GNPs can serve as a magnificent multifunctional vehicle for a variety of biomedical applications, not to mention a vast range of their other utilities. This chapter focuses mainly on many aspects of the intriguing properties of GNPs, their synthesis, and currently projected applications, particularly for cancer diagnosis and therapy.

2.2 Beauty of Gold Nanoparticles

2.2.1 Biocompatibility

Biocompatibility is a critical design criterion in the selection of any type of materials, including nanoparticles, for biomedical applications. Many approaches have been developed to study the biocompatibility. For nanoparticles, the simplest and most effective way to determine the biocompatibility is to first evaluate their cytotoxicity in connection with their size, shape, chemical composition, and specific interactions with cells. Several questions need to be addressed for GNPs: whether GNPs can cause adverse effects inside a biological system, how GNPs are internalized by cells, and, subsequently, where they localize inside the cells. Although it has been shown that the effect of GNPs on various types of human cells is size and concentration dependent (Connor et al. 2005; Patra et al. 2007; Gannon et al. 2008; Qu and Lu 2009), generally speaking, GNPs exhibit minimal cytotoxicity over a large range of concentrations. More convincing results have been obtained from the culture of GNPs with macrophages (Shukla et al. 2005). Macrophages, as one of the principal immune effector cells, play essential roles as secretory, phagocytic, and antigen-presenting cells in the immune system. The cytotoxicity measurement of GNPs with RAW264.7 macrophage cells (Shukla et al. 2005) shows that GNPs are not cytotoxic. They do not elicit the secretion of proinflammatory cytokines TNF-α and IL1-β. These observations suggest their suitability for nanomedicine application. As a matter of fact, the noncytotoxic, nonimmunogenic, and biocompatible properties of GNPs indicate the excellent prospects for their applications in nanoimmunology, nanomedicine, and nanobiotechnology.

2.2.2 Chemical Stability

Gold is chemically stable and is resistant to surface oxidation, which is the reason why gold jewelries remain shining even after centuries. These properties together with the high surface-to-volume ratio determine that GNPs are important nanomaterials used as reaction platforms or as nanocatalysts in many chemical reactions (Tsunoyama et al. 2004; Yan et al. 2006). It should be noted that other materials such as silver and platinum may have similar catalysis properties, but silver is too reactive to be used and platinum is more expensive than gold (Safavi et al. 2008). Due to the high chemical stability, GNPs are even used in the CO oxidation under acidic condition (Chiang et al. 2005). The sustainable properties of GNPs at extreme conditions are critical for their applications in a biological system, where the coexistence of proteins, enzymes, salts, and other biomolecules can be very corrosive and invasive.

2.2.3 Cellular Uptake of Gold Nanoparticles and the Parameters Involved

Numerous studies have clearly demonstrated the free uptake of gold nanostructures by a variety of cells, including the cancerous cells (Yang et al. 2005a; Chithrani et al. 2006; Khan et al. 2007; Zhu et al. 2008; Mandal et al. 2009). However, the mechanism still remains elusive. Insightful understanding of the GNPs uptake as well as their intracellular fate is very important to assess the particle-induced cellular events, and to design multifunctional nanoparticles for target drug delivery and imaging. It is particularly important in this case to accumulate nanoparticles in the target cells in a controllable manner for improving the diagnostic sensitivity and therapeutic efficiency (Hong et al. 2006; Klostranec and Chan 2006). The uptake of GNPs is considered as part of the endocytotic process. Endocytosis is ubiquitous and important to eukaryotic cells, through which the cells can uptake the extracellular nutrients and regulate the

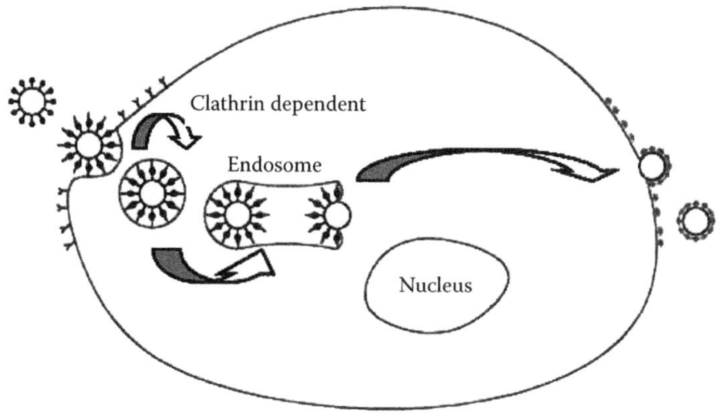

FIGURE 2.1 Schematic illustration of the receptor-mediated endocytosis of gold nanoparticles.

cell-surface receptors. However, the toxins, viruses, and microorganisms also gain their entry into the cells via this pathway. With many different processes identified in the endocytosis, the receptor-mediated endocytosis is probably the most important mechanism, in which the plasma membrane binds specific macromolecules on the smaller particles by means of specialized receptors, invaginates around those particles, and then pinches off to form small vesicles (see Figure 2.1). During the invagination, specific proteins (either clathrin or caveolin) may be required to polymerize into a spherical shell around the particle, meanwhile, the protein-independent invagination may also present as reported (Nichols and Lippincott-Schwartz 2001; Lakadamyali et al. 2004). A consensus has been reached that the cellular uptake of GNPs is mediated by cell-surface receptors, and the following procedures are most likely involved: (1) reaching the cell surface via diffusion or Brownian motion, (2) adsorption to cell membrane via receptor–ligand interaction, (3) invagination with the assistance of either clathrin or caveolin or through the lateral diffusion of mobile receptors (Chithrani and Chan 2007; Nativo et al. 2008), (4) intracellular trafficking via the vesicles, and (5) being excreted out of the cells via the receptor-mediated exocytosis (budding). With respect to the large variability in GNPs, it is certainly expected that the uptake process must be very complex and closely associated with the properties of GNPs. Indeed, many nanoparticle-relate parameters such as size (Bao and Bao 2005; Gao et al. 2005), shape (Bao and Bao 2005), and surface properties (Pugh and Heller 1960; Bao and Bao 2005; Gao et al. 2005; Sun et al. 2005, 2006) have been continuously highlighted to influence the GNP uptake, the uptake time, and the intracellular fate of GNPs.

Among all the variables, surface chemistry of nanoparticles is probably the most important and it determines which receptors are recruited to the particle surface as well as the further invagination for uptake. In this regard, it is possible to modify the particle surface with specific ligands for those receptors expressed specifically in the target cells and therefore to achieve the target delivery. For example, the immobilization of GNPs with Herceptin antibody (*abbv.* Her, a specific ligand to the ErbB2 receptor) can significantly enhance the uptake of Her–GNPs by ErbB2 overexpressing human breast cancer SK-BR-3 cells in comparison to unmodified nanoparticles (Jiang et al. 2008). In addition, other surface properties such as surface charge and hydrophobicity also control the uptake of GNPs, for example, positively charged nanoparticles are generally found to have higher uptake efficiencies than neutral and negatively charged particles, and hydrophobic surface of GNPs is not favorable for the uptake (Sun et al. 2005, 2006). Apart from surface properties, physical dimensions of the GNPs also play a determinant role in regulating the uptake kinetics and saturation concentrations. Several studies have investigated the size effect on the cellular uptake (Chithrani et al. 2006; Jiang et al. 2008). Interestingly, the uptake half-life of 50 nm GNPs is faster than either 14 or 74 nm GNPs (Chithrani et al. 2006). A possible explanation to this observation is illustrated in Figure 2.2 (Chithrani et al. 2006). During the uptake, a critical step is the appropriate formation of vesicles, that is, the membrane wraps and encloses a particle, which is a result of the competition between thermodynamic driving force for wrapping and the receptor diffusion kinetics. For 50 nm

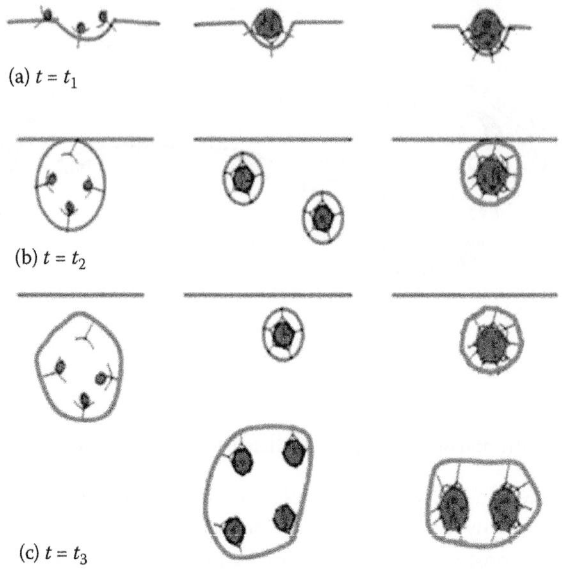

FIGURE 2.2 Schematic postulation of the size-dependent process of GNP uptake and fate. (Adapted from Chithrani, B.D. and Chan, W.C.W., *Nano Lett.*, 7(6), 1542, 2007. With permission.)

particles, the receptor–ligand interaction can produce enough free energy to drive the particles into the cell for wrapping. However, for those particles smaller than 50 nm, the receptor–ligand interaction is not enough to complete the wrapping. Instead, the small nanoparticles can aggregate to form large clusters prior to uptake. For nanoparticles larger than 50 nm, the wrapping is slower because more receptors are required and they are not available for binding. As a result, only a small number of particles are taken up. A further confirmation of the dimension effect on cellular uptake comes from a recent study on gold nanorods, in which the rate of uptake decreases with increasing aspect ratio, and spherical shape is most favorable to cellular uptake (Chithrani and Chan 2007). During the cellular uptake, many parameters such as ratio of adhesion, cell membrane stretching, and the cell membrane's bending energy may also be involved. As part of the uptake, the intracellular fate of GNPs is also important for both intracellular diagnosis and target delivery of specific drug. In general, the uptake of GNPs are entrapped in the early and later endosomes/lysosomes (Figure 2.3) (Shamsaie et al. 2008) and the number of particles accumulated in these compartments increases over the incubation time, from 2–3 particles after 120 min to 4–6 particles after 180 min and to larger lysosomal nanoaggregates after overnight incubation (Kneipp et al. 2006). Although most of GNPs are entrapped in endosomes during the trafficking, surface modification of GNPs with liposomes or polyethylene glycols can lead to their escape to cytosol and nucleus envelop (Arnida and Hamidreza 2010). The differential entrapment of GNPs in various organelles inside the cells can be used to deliver various drugs or plasmid for targeted control of the cellular behavior via regulating the expression of specific proteins, for example, the controlled differentiation of stem cells via gene transfection (Ferreira 2009). Despite the tremendous progress in understanding cellular uptake and intracellular fate of GNPs, extensive efforts are still required to further investigate the intracellular events at the presence of GNPs (Jiang et al. 2008), especially in the regenerative medicine.

2.2.4 Optical Properties

When matter is exposed to excitation light, a number of processes can occur. The light can be absorbed, reflected, and scattered. The scattered light can keep the same wavelength as the incident light (Mie or Rayleigh scattering), longer wavelength than the incident one (stoke Raman), or shorter wavelength

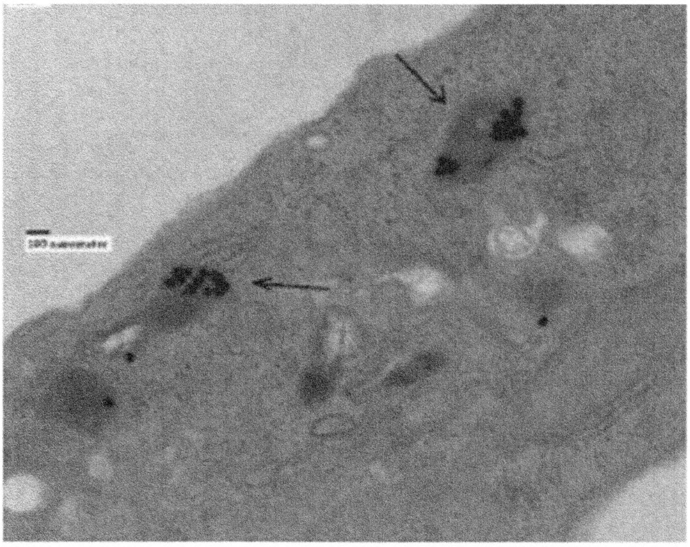

FIGURE 2.3 TEM image of the entrapped GNPs in the membrane bound lysosomal/endosomal compartments, scale = 100 nm. (Adapted from Shamsaie, A. et al., *Chem. Phys. Lett.*, 461(1–3), 131, 2008. With permission.)

than the incident one (antistoke Raman). The absorbed light can be reemitted (i.e., fluorescence). The local electromagnetic field of the incident light can be enhanced by nanoparticles. Thus, the enhancement of electromagnetic field would result in spectroscopic signals from the molecules on the material surface, for example, SERS.

In the case of GNPs, all these processes are enhanced strongly as a result of the unique interaction between the photons and the free electrons of the particles. When GNPs are exposed to light radiation, the energy of photons causes the collective oscillation of the conduction-band electrons at the surface of the particle, with respect to the ionic core of the nanoparticles. The coherent oscillation of the free electrons of metal in resonance with the electromagnetic field is called the surface plasmon resonance (SPR). Both theoretical and experimental discussion of the SPR can be found in earlier and recent literature reports (Hutter et al. 2001; Jain et al. 2006; Njoki et al. 2007; El-Brolossy et al. 2008; Guo et al. 2009). For gold nanospheres, this resonance occurs in the visible spectral region at approximately 520 nm, which is the origin of the brilliant red color of the nanoparticles in solution. For gold nanorods, the free electrons oscillate along both the long and short axis of nanorods, which results in a stronger resonance band in the near infrared (NIR) region and a weaker band in the visible region (similar to the nanospheres), respectively. The excitation of the SPR results in the enhancement of the photophysical properties of GNPs (Huang et al. 2007). Figure 2.4 summarizes the major optical processes that occur to the interaction between light and GNPs with closest surrounding medium.

2.2.4.1 Single Nanoparticles

It is expected that most of the GNPs would remain monodispersed in the solution due to the surface charge-induced repulsion. In this regard, fully understanding the light interaction with individual nanostructures is a fundamental issue in optoelectronics and nanophotonics. This becomes particularly important in many applications (such as biosensing, cancer therapy, and all-optical signal processing) relying on surface-bound optical excitation with single metallic nanoparticles. Recently, results have been reported to detect plasmons as resonance peaks in energy-loss spectra of subnanometer electron beams rastered on monodispersed silver nanotriangles (Nelayah et al. 2007). Clearly, the common resonance energy value at three different corners provides further evidence that the corner feature results in the electron energy loss (EEL) spectra, which is truly the signature of eigenmodes. The spatial extension

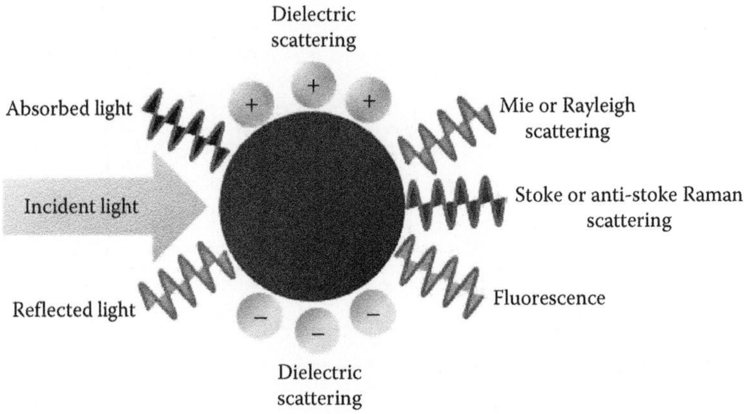

FIGURE 2.4 Schematic illustration of nanoparticles interacting with light.

involves the entire triangular silver prism instead of just the effect of a trivial local field enhancement at each triangle corner, as demonstrated in Nelayah et al. (2007). Taken together, this result represents a significant improvement in the spatial resolution with which plasmonic modes can be imaged, and provides a powerful tool in the development of nanometer-level optics.

2.2.4.2 Dimer

Surface plasmon of nanoparticles has a direct relationship to particle morphology, interparticles distance, and excitation polarization. Increasing evidence has clearly demonstrated that the formation of GNP dimers or aggregates is beneficial to the surface plasmon and, as a matter of fact, the interparticle distance plays a critical role in determining the SPR (Nelayah et al. 2007; Venkata et al. 2007; Yu et al. 2008). Based on the results obtained from the surface plasmon excitation in pairs of identical GNPs by optical transmission spectroscopy, it was found that with the decrease of interparticle distance, the SPR shifts to longer wavelengths in the case of polarization direction parallel to the long particle pair axis; otherwise a blue shift is observed for the orthogonal polarization. These experimental findings can be explained by a dipolar interaction mechanism (Rechberger et al. 2003). It is also necessary to mention that the dependence of surface plasmon on the interparticle distance allows us to manipulate the extinction coefficient and, therefore, to meet different needs.

2.2.5 Surface Enhanced Raman Scattering

Identification and structural characterization, including monitoring the structural changes of molecules, are the major function in biophysical/biochemical spectroscopy. Vibrational spectroscopic techniques such as Raman spectroscopy, which provides high structural information, are of particular interest. However, the extremely small cross section of the Raman process, which is approximately 12–14 orders of magnitude below fluorescence cross section, has significantly limited the application of Raman spectroscopy. In this regard, it is critical for the Raman signals to be enhanced in order to achieve low limits of detection. Experimental results have shown that enormously strong Raman signal can be produced when the molecules get attached to various rough or nanosized metal surfaces. This significant enhancement by rough metal surface or nanoparticles is called SERS. This enhancement probably is a result of collective contributions from electromagnetic field enhancement, chemical first layer effect, and geometrical enhancement as a consequence of increased surface area (Kneipp et al. 2002b). Compared to fluorescence, SERS can offer some new interesting aspects, which is widely used especially as a single-molecule spectroscopy tool in biophysics. One of the most spectacular applications of single-molecule SERS is rapid DNA sequencing in which specific DNA fragments down to single structural base can be detected using the Raman spectroscopic characterization without the use of fluorescent or radioactive labels (Kneipp et al. 1998). Another useful application of SERS in biophysics comes from its capability of providing the information on molecules residing on the surfaces or involving in the interface processes. For example, "SERS-active" silver or gold electrodes with a defined potential can be used as a model environment to study biologically relevant processes, such as charge transfer transitions in cytochrome C (Murgida and Hildebrandt 2001; Niki et al. 2002). Reviews on the studies of biological molecules by SERS studies have been summarized in the 1980s and early 1990s (Therese et al. 1991; Sokolov et al. 1993). SERS, especially through the use of biocompatible GNPs, opens up exciting opportunities in the field of biophysical and biomedical spectroscopy, which provides ultrasensitive detection and characterization of biophysically/biomedically relevant molecules and processes as well as a vibrational spectroscopy with extremely high spatial resolution.

2.3 Gold Nanoparticle Synthesis

2.3.1 One Step Approach

GNP colloidal solution can be directly synthesized by chemical reduction of metal salts, photolysis or radiolysis of metal salts, ultrasonic reduction of metal salts, and displacement of ligands from organometallic compounds (Roucoux et al. 2002). However, the instability of colloidal gold is a major obstacle to the practical applications simply because the gold colloids have a high tendency to aggregate in solution. One of the common strategies is to protect the colloids with stabilizer, which can be absorbed onto the particle surface and as a result prevent colloids from agglomeration (Rao et al. 2000; Bönnemann and Richards 2001). Commonly used protective agents include thiols, surfactants, and polymers. Different from small molecules, the stabilization of colloids with polyelectrolytes is a combination effect from both steric and electrostatic stabilization, namely electrosteric stabilization, as confirmed by Pugh et al. (1960). Apart from the stabilization, the addition of protective agents in colloid solution can also influence the particle size and introduce functionality to particle surfaces (Rao et al. 2000; Bönnemann and Richards 2001). Necessary to mention, depending on the reducing agent, the surface charge of colloidal gold could be either positive or negative (Sun et al. 2005, 2006) and the surface charge can partly stabilize the colloid solution via the intrinsic electrostatic repulsion.

2.3.2 Seed Mediation Approach

GNPs can be synthesized by not only one step method but also seed mediation method that involves two steps. First, small spherical particles (seed) with average diameters between 5 and 20 nm are prepared by varying the ratio of gold ion concentration to stabilizer/reductant. Second, desired nanoparticles size, such as 20–110 nm, can be formed by depositing the Au reduced from Au (III) ions onto the surface of seed particles. The kinetics of particle formation has been reported and this method can rapidly yield GNPs with improved monodispersity, sphericity, and excellent reproducibility (Sau et al. 2001). The final size of the particles has a great dependence on the size of the "seed" and the total amount of precursor ions to be reduced (Schmid 2002). Theoretically, the smaller the starting seed is, the lower the desired particle size limit is. This correlation allows preparing particles over a broad size range. Additionally, the particle size is closely related to the reducing agent used. Smaller particles are generally produced by stronger reducing agents, such as $NaBH_4$, phosphorus, tetrakis (hydroxymethyl) phosphonium chloride, or via radiolytic method (Slot and Geuze 1981; Siegfried et al. 1994; Sarathy et al. 1997; Henglein and Meisel 1998; Grunwaldt et al. 1999; Henglein 1999). The nature of particle stabilizer, solvent, as well as reaction conditions (e.g., pH, temperature, stirring speed) play a crucial role in determining the final particle size. The combination of seed-mediated growth of colloidal GNPs with radiolysis has proven to be more effective in controlling the size with improved monodispersity (Brown and Natan 1998; Henglein and Meisel 1998; Brown et al. 1999; Henglein 1999). In this attempt, an iterative growth strategy was followed, that is, particles grown in the former step were immediately used as seeds for the following growth step.

2.3.3 Shape-Controlled Synthesis

Technologically, the control of metal nanoparticle size, shape, and structure is tremendously important due to the close correlation between these parameters and the optical, electrical, and catalytic properties. On a nanometer scale, metals tend to nucleate and grow into twinned and multiply twinned particles (MTPs) with their surfaces bounded by the lowest-energy {111} facets, that is, face-centered cubic (FCC). To obtain uniform gold nanoboxes with a truncated cubic shape, silver cubes can be used as sacrificial templates and then replaced with pure gold from an aqueous $HAuCl_4$ solution following the reaction

$$3Ag(s) + HAuCl_4(aq) \rightarrow Au(s) + 3AgCl(aq) + HCl(aq)$$

The above example just showed the possibility of controlling the morphology of gold nanostructures. Actually, efforts have been extensively made to investigate the possible control of the size and shape of gold nanostructures by optimizing the synthesis conditions and exploring various synthesis methods. By far, various gold nanostructures such as nano-cube (Wiley et al. 2005), nanowire (Pei et al. 2004), nano-rod (Kim et al. 2002; Sau et al. 2001), branched nanoparticles (Hao et al. 2004), triangle nanoparticles (Ramaye et al. 2005), and plate nanometer structure (He and Shi 2005) have been successfully fabricated. Compared to other methods, wet chemical synthesis has the unique superiority especially in forming the nanoparticles of regular shape at a high yield (Sau and Murphy 2004). These regular nanoparticles can be further assembled into two- or three-dimensional nanometer structures (Kim et al. 2001; Sau and Murphy 2005) or used as building blocks to build future nanometer electronic device (Remacle et al. 1998; Norris et al. 2001; Nguyen et al. 2002; Yamanea et al. 2004; Li et al. 2005; Yang et al. 2005b). Despite the successful synthesis of nanoparticles with different sizes and shapes (Watzky and Finke 1997; Jana and Peng 2003), it remains a great challenge to synthesize small nanoparticles with uniform and well-controlled shape, which requires the insightful understanding of the reaction complexity with multiple components (Pileni et al. 1999; Ngo et al. 2004). One of the typical approach is to first fabricate the silver nanocubes by means of a modified polyol process that involved the reduction of silver nitrate with ethylene glycol in the presence of a capping regent such as poly vinly pyrolidone (PVP), and then these silver nanocubes react with $HAuCL_4$ to get gold nanocages (Sun and Xia 2002). Furthermore, well-controlled pores can be developed at the corners of these nanocages (Chen et al. 2006). Figure 2.5 shows both SEM and TEM (insets) images of sharp corners and truncated corners of the Ag/Au alloy nanocages through the galvanic replacement of Ag nanocubes. Figure 2.6 illustrates adjustability of extinction spectra by controlling the galvanic replacement reaction of these Ag nanocubes with various amounts of $HAuCl_4$. Another approach for shape-controlled synthesis is to grow appropriate amount of gold seeds by stirring the solutions containing desired amounts of cetyl trimethylammonium bromide (CTAB), $HAuCl_4$, and ascorbic acid (AA) and then age for certain period (normally in minutes) to obtain different shapes and sizes (Sun and Xia 2002; Fu et al. 2007).

2.3.4 Synthesis of Negatively Charged Gold Nanoparticles

Synthesizing GNPs with negative surface charge using wet chemistry approach involves chemical reducing agents such as sodium citrate (Turkevitch et al. 1951) and sodium borohydride (Male et al. 2007). Generally, the resultant nanoparticles are negatively charged due to the adsorption of negative ions onto the particle surface. Many properties of colloid suspensions are dependent on the particle sizes. A series of monodisperse suspensions with the same chemical composition but with different particle sizes can be used to study the particle-size-dependent phenomena, such as Brownian motion, light scattering, sedimentation, and electrophoresis of small particles. A standard procedure to obtain a monodispersed colloid suspension can be found in the literature (Frens 1973). Briefly, 10^{-2} wt% of $HAuCl_4$ and 1 wt% of sodium citrate were mixed under the boiling condition until the color of the solution became red. It has been noticed that changes of the relative ratios of reactants could lead to the relative rate changes of two independent processes in metal particle synthesis—nucleation and growth.

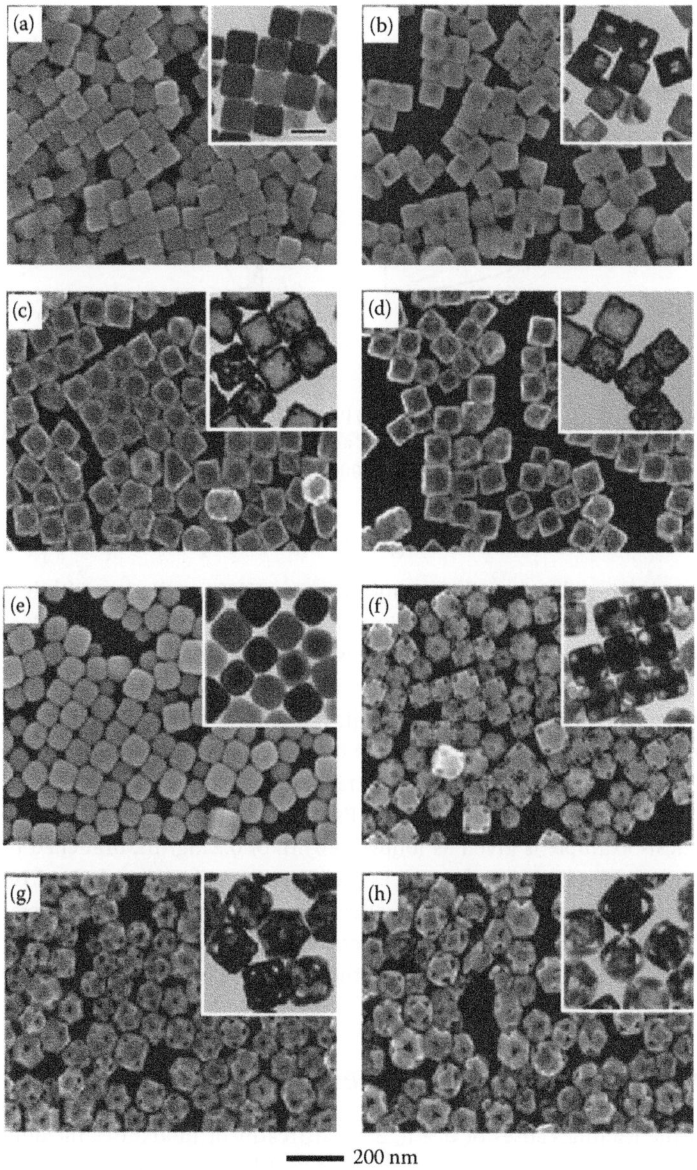

FIGURE 2.5 SEM images of four different stages of galvanic replacement reaction with Ag nanocubes as the sacrificial template. (a–d) Ag nanocubes with sharp corners titrated with various amount of 0.1 mM $HAuCl_4$, 0, 0.6, 1.6, and 3.0 mL, respectively. (e–h) Ag nanocubes with truncated corners reacted with the same volumes of 0.1 mM $HAuCl_4$ as for the sharp cubes. (Inset) TEM image of each respective sample. Scale bar = 100 nm for all TEM images. (Adapted from Chen, J. et al., *J. Am. Chem. Soc.*, 128(46), 14776, 2006. With permission.)

Nucleation and growth are independent processes; however, they are interlocked. As demonstrated in a previous study (Turkevich 1985), when chlorauric solution was treated at room temperature with hydroxylamine hydrochloride, no gold colloid was produced unless proper nuclei were present. Inoculation of such a growth solution with a controlled number of appropriate nuclei could result in the growth of particles into a desired size. Clearly, proper nucleation is an initial yet critical step for further growth of the particles. The size of nanoparticles is controlled by both nucleation and growth processes.

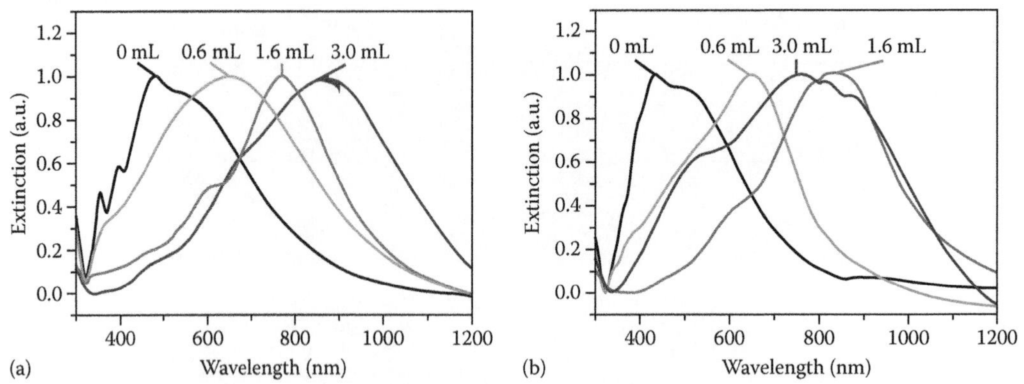

FIGURE 2.6 UV-vis-NIR spectra of two different types of Ag nanocubes titrated with different volumes of 0.1 mM $HAuCl_4$: (a) Ag nanocubes with sharp corners and (b) Ag nanocubes with truncated corners. (Adapted from Chen, J. et al., *J. Am. Chem. Soc.*, 128(46), 14776, 2006. With permission.)

2.3.5 Synthesis of Positively Charged Gold Nanoparticles

It has been reported that reduction of silver nitrate under UV irradiation with branched polyethyleneimine (BPEI) and 4-(2-hydroxyethyl)-1-piperazineethanesulfonic acid (HEPES) can lead to the formation of positively charged silver nanoparticles (Tan et al. 2007). The mechanism for reduction of Ag+ ions with the BPEI/HEPES mixtures involves the oxidative cleavage of BPEI chains, resulting in the formation of positively charged BPEI fragments enriched with amide groups, and the production of formaldehyde, which serves as a reducing agent for Ag+ ions. The resultant silver nanoparticles are positively charged due to protonation of surface amino groups. Following the same approach, it is possible to synthesis positively charged GNPs using BPEI/HEPES mixtures (Khaing Oo et al. 2008a,b). A synthesis example is given as follows: 40 mL of 0.2 mg/mL branched polyethyleneimine (BPEI) (molecular weight = 10,000, Polysciences Inc., Warrington, PA) and 40 mL of 0.01% of $HAuCl_4$ are mixed and stirred for 5 min in an ice bath. It is then placed under a 400 W metal halide UV lamp (Cure Zone 2) for 1 h until the color changes from yellow to dark red upon the completion of the reduction reaction. This synthesis route leads to the formation of GNPs with an average diameter of 25 nm and average zeta potential of +30 mV.

2.4 Nanoparticle-Assisted Cancer Treatment

Cancer is one of the leading causes of death. Several treatment modalities including chemotherapy, radiotherapy, gene therapy, hyperthermia therapy, and PDT have been developed and applied clinically. Among different therapies, one of the emerging challenges is to confine the treatment only to the tumor tissue without damaging the surrounding healthy ones. In this regard, nanobiotechnologies deliver many promising aspects to overcome the confronted challenges in cancer therapy, for example, target delivery of drug using nanoparticles to tumor tissues can improve the treatment specificity (Qian et al. 2008). Figure 2.7 exhibits GNPs with a diameter of 60 nm encoded with a Raman reporter and stabilized with a layer of thiol-PEG. Meanwhile, the embedding or encapsulation of cancer drug in nanoparticles can protect the drug from degradation by the host system and, therefore, prolong the treatment effect. The combination of nanoparticle-assisted drug delivery with other treatments such as radiotherapy allows the achievement of maximum treatment efficiency. GNPs, due to its good biocompatibility and readily functionalized surface, can be used as carriers to target delivery drugs or genes to cancer cells. In addition, the superior optical and plasmonic properties of GNPs can be used as a diagnostic tool, and the combination of diagnostics and therapeutics potentially enables the personalized management of cancer (Jain 2005).

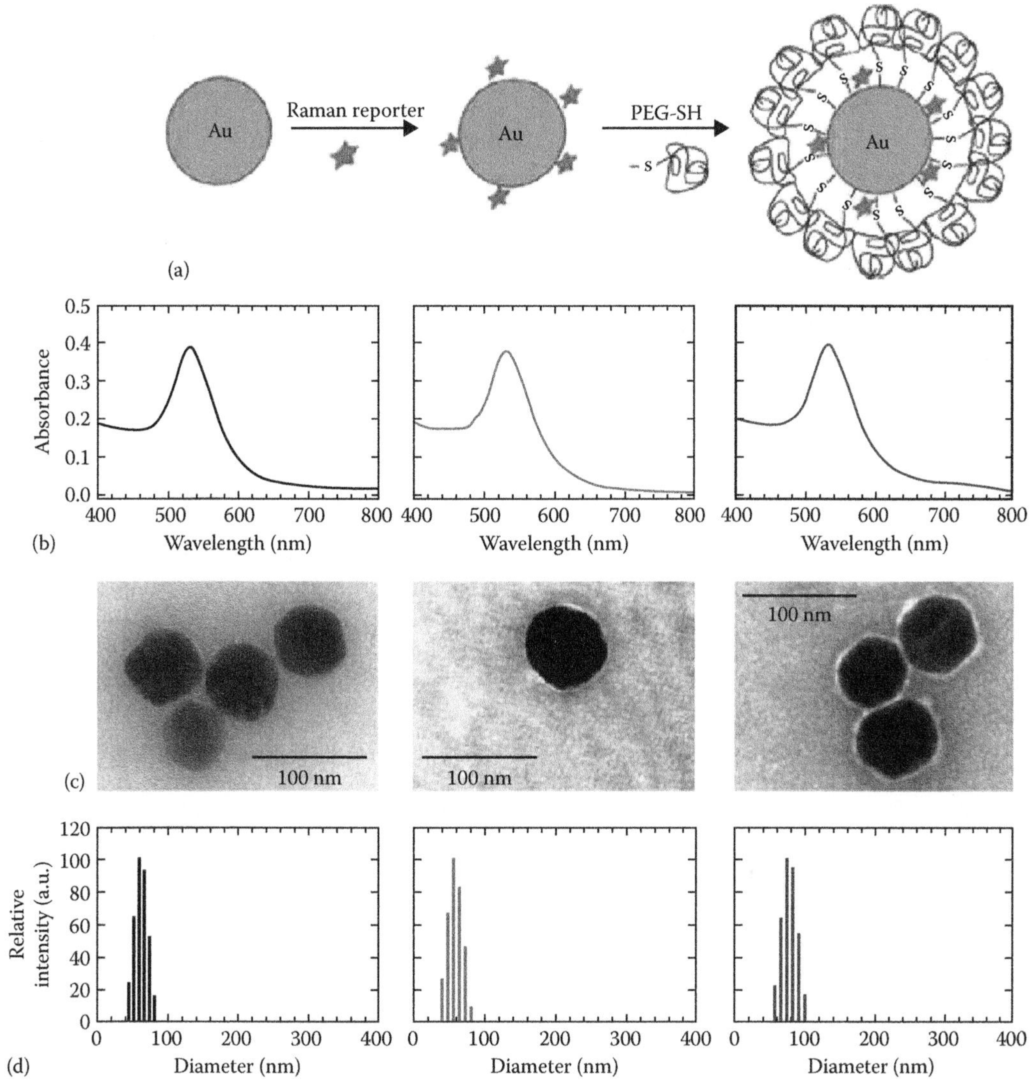

FIGURE 2.7 Design, preparation, and properties of pegylated GNPs for in vivo tumor targeting and spectroscopic detection. (a) Preparation and schematic structures of the original gold colloid, a particle encoded with a Raman reporter, and a particle stabilized with a layer of thiol-polyethyleneglycol (thiol-PEG). Approximately $1.4–1.5 \times 10^4$ reporter molecules (e.g., malachite green) are adsorbed on each 60 nm gold particle, which is further stabilized with 3.0×10^4 thiol-PEG molecules. (b) Optical absorbance, (c) transmission electron microscopy (TEM), and (d) dynamic light scattering size data obtained from the original, Raman-encoded, and PEG-stabilized GNPs as shown in (a). (Adapted from Qian, X. et al., *Nat. Biotechnol.*, 26(1), 83, 2008. With permission.)

2.4.1 Nanoparticle-Based Photodynamic Therapy

PDT, a minimal invasive cancer treatment modality, has proven to have low morbidity, minimum functional disturbance, good tolerance, and the ability to be repeatedly used at the same site (Hopper 2000). It includes the administration of a photosensitive drug (also called photosensitizer, PS) and subsequent irradiation with appropriate light to produce ROS for the destruction of the neoplastic tissue (Palumbo 2007).

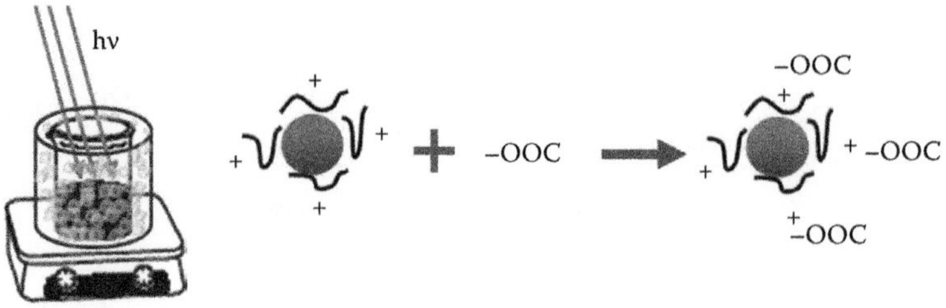

FIGURE 2.8 Schematic illustration of the synthesis of GNPs conjugated with 5-ALA. (Adapted from Pharma Focus Asia, 31–34, 2008).

Photofrin® (PF) and 5-aminolevulinic acid (5-ALA) are among the widely used PS in clinical applications. Compare to PF (Kessel and Thompson 1987; Spikes 1990; Baas et al. 1995; Orenstein et al. 1996), 5-ALA as a protoporphyrin IX (PpIX) precursor for PDT offers many advantages including low dark toxicity to cells, rapid clearance from the body (24–48 h), and rapid conversion into endogenous porphyrins, that is, PpIX, via the heme cycle (Castano et al. 2004). However, the zwitterionic nature (Merclin and Beronius 2004) and hydrophilicity of 5-ALA greatly limit its penetration through tissues such as intact skin, nodular skin lesions and through cell membranes (Peng et al. 1995, 1997), leading to an inconsistent accumulation of PpIX in tumor cells. Thus, further improvement in 5-ALA penetration through the cell membrane and targeted delivery to tumor cells has the potential to enhance the efficacy and specificity of PDT. The use of GNPs as the vehicle for 5-ALA delivery represents multifold advantages, coming not only from its biocompatibility, surface plasmon resonance, and autofluorescence, but particularly from the recent demonstration that immobilization of PS on the particle surface is better for ROS formation (Wieder et al. 2006). In addition, the high accumulation of GNPs in tumor tissue through the rich permeable vasculature around these tissues (Paciotti et al. 2004) represents another critical benefit in the use of GNPs. Our previous study showed that positive GNPs with 5-ALA absorbed onto their surface via electrostatic interaction (Figure 2.8) had a high accumulation in fibrosarcoma cells (cancer cells) and significantly promoted the formation of ROS, and, as a consequence, a high destruction rate of fibrosarcoma cells was achieved (Khaing Oo et al. 2008a). In order to assess the PDT effect, MTT assay (ASTM E2526-08 standard method for estimating of cytotoxicity) was performed on both normal human dermal fibroblasts (NHDF) and fibrosarcoma (WT) cells with different PDT treatment. The results showed that no cytotoxicity was observed in both NHDF and WT cells treated with GNPs (GNPs) (Figure 2.9). However, only about 60% cells survived in the 5-ALA PDT group for both NHDF and WT, as compared to the control. Moreover, the survival of WT cells was significantly decreased in the 5-ALA-GNPs group with about 30% survival. In the study on the selectivity of PDT with 5-ALA-GNPs, a coculture system was used to simulate the in vivo circumstances where cancer cells coexist with healthy cells. A significant decrease of green-labeled cells (WT) was observed in the 5-ALA-GNPs group (Figure 2.10). In contrast, NHDF (nonlabeled) became confluence after culture for 24 h, and cells retained their spindle shape. Clearly, with the assistance of GNPs, specific destruction of fibrosarcoma cells by 5-ALA PDT can be achieved with minimal damage to normal fibroblasts. Efforts have also been made to chemically immobilize PS molecules such as porphyrin onto the GNP surface and used for PDT treatment (Lü et al. 2008). More applications of GNPs in PDT are expected especially with better understanding of the ROS enhancement mechanism and further in vivo studies.

2.4.2 Nanoparticle-Based Hyperthermia Therapy

Upon the excitation of light (photons), GNPs can generate heat (Boyer et al. 2002; Cognet et al. 2003; Hu and Hartland 2003; Pitsillides et al. 2003; Maillard et al. 2004; Skirtach et al. 2005). The generation of heat is significantly enhanced when the incident photon energy is close to the plasmon frequency of GNPs. Richardson et al. (2006) have tried to quantify the heat produced by GNPs. In their study,

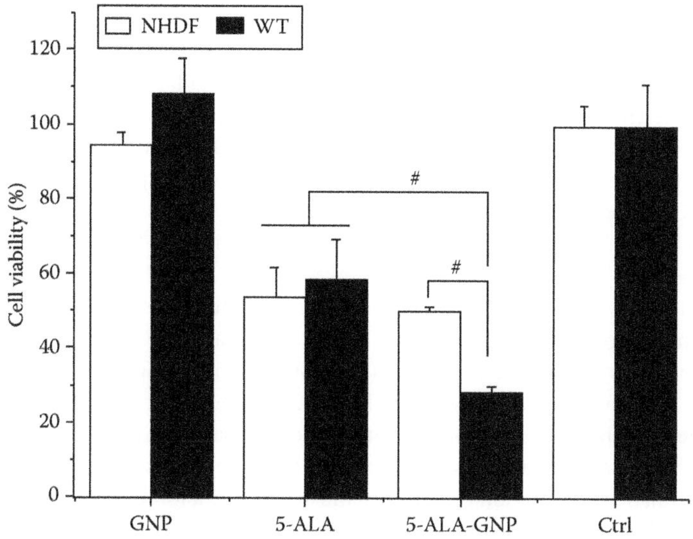

FIGURE 2.9 Viability of NHDF and WT cells determined by the MTT assay after various PDT treatments where cells were incubated with 5-ALA, GNPs and 5-ALA-GNPs for 4 h, irradiated for 1 min under a 150 W halogen light and then cultured for 24 h. Cells without treatment were used as control. The data are representative of three separate experiments. # $p < 0.001$. (Adapted from Khaing Oo, M.K. et al., *Nanomedicine*, 3(6), 777, 2008. With permission of Future Medicine Ltd.)

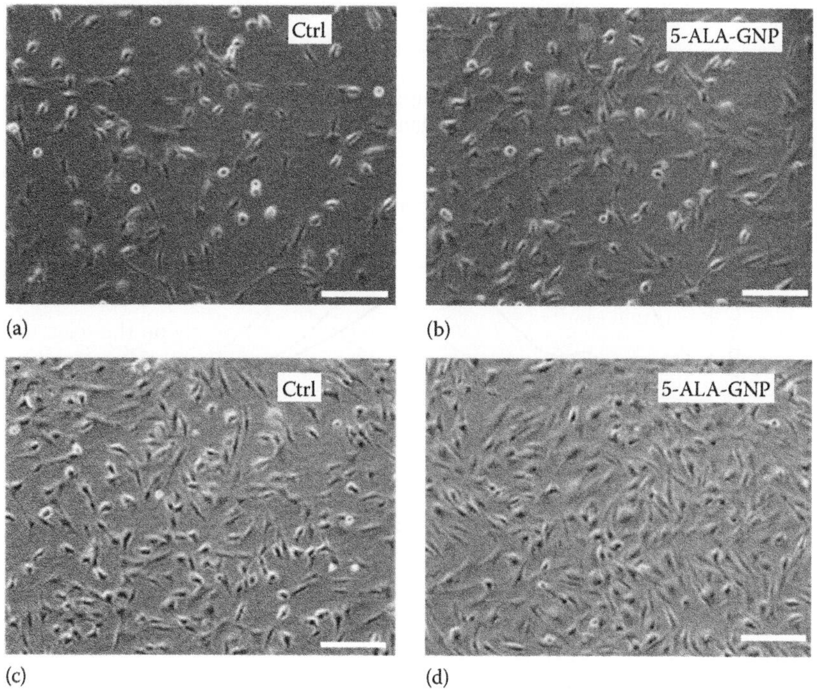

FIGURE 2.10 Merged bright field and fluorescent microscopy images of the cocultured NHDF (nonlabeled) and WT (green) before PDT treatment (a and b) and after 1 min irradiation and further cultured for 24 h (c and d). Scale: 200 μm. (Adapted from Khaing Oo, M.K. et al., *Nanomedicine*, 3(6), 777, 2008. With permission of Future Medicine Ltd.)

GNPs embedded in an ice matrix were optically excited and the heat generation and melting processes were monitored at the nanoscale. It was found that the ice melting intensity greatly depended on the temperature and position in the matrix, which is ascribed to the fact that nanoparticles form small complexes with different geometry and each complex has a unique thermal response. Theoretical calculations and experimental data are combined to make a quantitative measure of the amount of heat generated by optically excited GNPs and agglomerates. For example, at experimental set point of −21.2°C for single 50 nm GNP, heat flux intensity is 9.6 μW under 4 mW of 530 nm wavelength laser excitation (Richardson et al. 2006). The information obtained from this study can be used to design thermal ablative therapy for cancer.

Similarly, applications using gold nanoshells with tunable optical resonances for thermal therapy of tumors have been explored (Hirsch et al. 2003; Hauck and Chan 2007; Lal et al. 2008). The synthesis protocol was developed by Halas and colleagues (Jain 2005). Various stages in the growth of gold metallic shells on silica nanoparticles can be found in Figure 2.11. Their optical resonances can be predicted, depending on the core/shell size ratio or absolute size of the particles (Figure 2.12). In these applications, gold nanoshells that have a strong absorption of the tissue transmissible near infrared light are used. Due to the enhanced generation of heat by nanoshells, only a low dose of NIR light is sufficient to produce the required heat. A recent study (Hirsch et al. 2003) showed that the viability of human carcinoma cells incubated with nanoshells and exposed to NIR was significantly reduced. In contrast, cells without nanoshells exhibited no viability loss under the same treatment condition. Similar result was

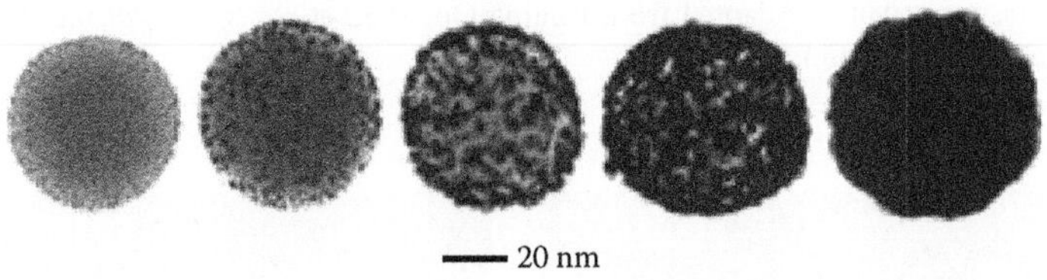

FIGURE 2.11 Transmission electron microscope images of gold/silica nanoshells during shell growth. (Adapted from Jain, K. K., *Technol. Cancer Res. Treat.*, 4(4), 407, 2005. With permission.)

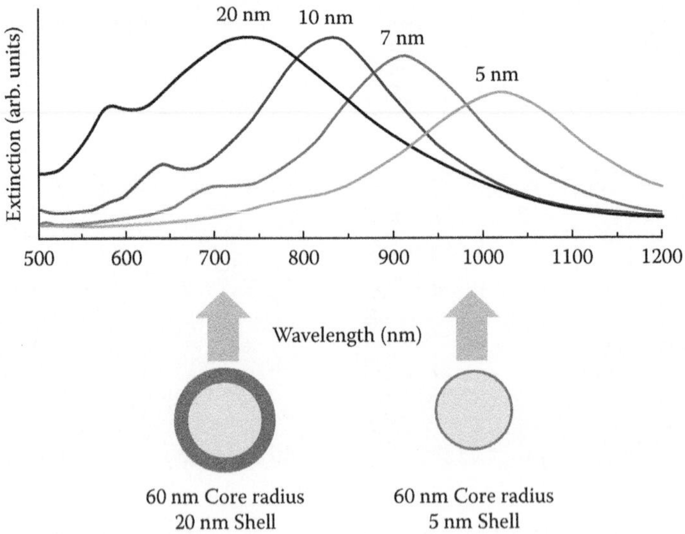

FIGURE 2.12 Optical resonance of gold shell-silica core nanoshells as a function of their core/shell ratio. Respective spectra correspond to the nanoparticles depicted beneath. (Adapted from Halas, N., *Opt. Phot. News*, 13(8), 26, 2002. With permission.)

observed for in vivo studies using nanoshells. In the other study on delivery of tumor necrosis factor-α (TNF-α) with or without GNPs, it was shown that hyperthermia significantly delayed tumor growth and reduced tumor cell survival (Visaria et al. 2006). Furthermore, gold nanorods are highly effective at transducing NIR light into heat and localized hyperthermia agents (Huff et al. 2007). All these findings have demonstrated a good correlation between the fundamental mechanisms and enhanced efficiency of GNP-based hyperthermia.

2.5 Diagnosis and Detection Techniques

The unique optical properties of GNPs may be used to develop biosensors for living cells. Studies have shown that GNPs even without the antibody functionalization have a high tendency to accumulate in tumorous cells compared to the healthy counterparts (Patra et al. 2007). Although the mechanism behind this is not completely understood, possible explanation could come from the differences between tumorous cells and healthy cells in the cell membrane surface charge, thiol concentration, membrane structure, as well as the local pH, any of which may determine the affinity of GNPs. The use of GNPs as a simple and inexpensive tool for cancer detection has been explored as summarized in the following sections.

2.5.1 Surface Plasmon Resonance Scattering

As discussed in Section 2.4, SPR of GNPs can be used to develop SPR image and spectra for cancer cell detection. Such method was explored in Richardson et al. (2006). Target delivery of GNPs specifically to cancer cells can be advantageous in increasing the specificity and enhancing the SPR signals. Functionalizing the GNPs with antibodies specific to antigens on cancer cell membranes is considered as a compelling approach. Following this strategy, a very recent study was done to correlate the specificity of antibody-conjugated GNPs with better SPR detection (El-Sayed et al. 2005), in which colloidal gold particles conjugated with or without monoclonal anti-epidermal growth factor receptor (anti-EGFR) antibodies were used. The results showed that GNPs with anti-EGFR antibody specifically and homogeneously bound to the surface of malignant oral epithelial cell lines (HOC 313 clone 8 and HSC 3) with 600% greater affinity than to the noncancerous epithelial cell line (HaCaT). This specific and homogeneous binding of GNPs to malignant cells yield a relatively sharper SPR absorption band with a great red shift compared to that observed with the noncancerous cells. In contrast, the bare GNPs without any modification dispersed and aggregated within the cell cytoplasm of both malignant and nonmalignant cells without clear preference and the SPR adsorption in both malignant and noncancerous cells was comparable. These results suggest that SPR scattering imaging or SPR absorption spectroscopy generated from antibody conjugated GNPs can be used as a diagnosis and detection tool for oral epithelial cancer cells. Based on this study, it could be very interesting to further determine whether this approach can be applied to other types of cancer cells.

2.5.2 Gold Nanoparticle–Based SERS and SERS Tag for Cellular Detection

For the study of cellular and subcellular systems, a wide range of analytical methods have been used with the fluorescence techniques as the most common tool. Fluorescence-based imaging is a highly attractive methodology for the study of organelle dynamics, identifying subcellular compartments and monitoring biological kinetics (Olson et al. 2005). Over the last two decades, Raman spectroscopy has become an increasingly important technology with the ability to study the biophysics and biochemical processes involved in cells (Puppels et al. 1990; Ramser et al. 2005; Van Manen et al. 2005; Jess et al. 2006; Krafft et al. 2006; Lee et al. 2007). Since Raman spectroscopy is based on vibrational transitions, where frequency shifts are associated with specific molecular vibrations within the sample of interest, it enables the identification of polarizable bio/chemical species, the elucidation of molecular structure, and the investigation of interface reactions in a nondestructive manner. In addition, unlike fluorescence-based techniques, Raman spectroscopy does not require labeling dyes and since water is almost Raman "transparent," the technique is ideally suitable for analyzing cell-based biological systems and has the potential to be used for the study of cellular dynamics of a large range of signaling processes. The integration of microfluidics, Raman scattering, and confocal microspectroscopy has been successfully used to characterize in situ single living cells (Zhang et al. 2008).

Raman spectroscopy provides fingerprint of molecular vibrations. It has been employed as a tool for cancer diagnosis. Lieber et al. (2008) recently reported the possibility of distinguishing normal, basal cell carcinoma (BCC), squamous cell carcinoma (SCC), and melanoma tissues from patients using Raman spectra, respectively. The correlation of Raman spectra with tissues of each pathophysiological state yielded 100% accuracy for diseased tissue detection with a high specificity. These results indicate the potential of using Raman microspectroscopy for skin cancer detection and provide a clear rationale for future clinical studies.

Raman spectroscopy, however, suffers greatly from its inherent poor sensitivity compared to, for example, fluorescence, which can be 12–14 orders of magnitude more sensitive. As a consequence, despite advances in detector technologies, the technique is far less than ideal for direct detection of intracellular components, which normally present at low concentrations and may need tens of seconds to minutes for spectral acquisition (Puppels et al. 1990). With the help of SERS (Fleischmann et al. 1974), Raman intensity can be dramatically increased by 10^5–10^{10} in the presence of metallic nanostructures either on a substrate surface or in a colloidal solution (Lee et al. 2007). For cellular and subcellular analysis with SERS, colloid nanoparticles (e.g., gold) are normally used in consideration of its convenience for incorporation within the living cells (Kneipp et al. 2006). Efficient loading of GNPs into cells is critical for SERS detection and it can be achieved through general incubation (fluid-phase uptake) or ultrasonication-assisted uptake (Kneipp et al. 2002a, 2006). For Raman spectroscopic analysis, the cells usually need to be fixed at a specific location. The most common method is to seed or grow nanoparticle-loaded cells on a stable substrate, for example, a microscope coverslip. Other approaches such as optical tweezers have also been reported (Jess et al. 2006). Due to the high tendency to aggregate in a cellular environment, GNP-based SERS is often used for qualification instead of quantification. However, the latest results from Shamsaie et al. (2008) showed the possibility to quantify the local concentrations of a dinitrophenol derivative (DAMP) accumulated in the lysosome of MCF10 epithelial cells by normalizing the DAMP signature peak with the vibron/plasmon coupling signal from aggregated gold. With DAMP as a model molecule, the modified SERS technique allows the precise quantification of exogenous chemicals in living human cells and can be extended to different environments utilizing different types of nanoparticles beyond the intracellular scheme.

In vivo early detection of tumor, especially noninvasively, is of a great benefit to cancer therapy. The high sensitivity and uniqueness of SERS enabled by GNPs can be used for in vivo tumor targeting and detection. In order to guarantee the maximum accumulation of GNPs in tumor, pegylated GNPs conjugated with tumor-specific ligands such as single-chain variable fragment (ScFv) antibodies are used (Qian et al. 2008). In addition, it has also been found that small-molecule Raman reporters such as organic dyes could be further stabilized during the pegylation. The successful use of these pegylated SERS nanoparticles as tumor tags for in vivo detection of human tumor in xenograft models has been recently reported (Qian et al. 2008), in which the signal was considerably brighter than semiconductor quantum dots with light emission in the NIR window.

Although SERS is a powerful technique for analyzing a variety of molecules and molecular structures, it remains a challenging task to acquire detailed molecular information from biological organisms due to their complexity. Furthermore, even within single cell, the SERS signal varies with great dependence on the laser focus and Raman collection location (Kneipp et al. 2002a). Meanwhile, the measurement also greatly depends on the experimental setup and conditions (Kahraman et al. 2007). In addition, the type of noble metal, sizes, and aggregation properties of nanoparticles, as well as the wavelength of laser has an enormous impact on SERS signals. In order to obtain comparable and reproducible results, the experimental conditions must be well defined with the development of a standard protocol for each situation. The use of GNPs of various sizes with or without conjugated SERS tag for cancer cell detection is summarized in Table 2.1.

TABLE 2.1 Gold Nanoparticle-Based SERS Application

No.	Size (nm)	Shape	Conjugation	Excitation (nm)	Cell Type	Reference
1	60	Sphere	Antibody	785	Tu686, H520 and In vivo	Qian et al. (2008)
2	50	Sphere	None	633	MCF10	Shamsaie et al. (2008)
3	60	Sphere	None	830	HT29	Kneipp et al. (2002a)
4	30–50	Sphere/cluster	None	786	IRPT, J774	Kneipp et al. (2006)
5	20–50	Sphere	None	1046 (two-photon)	J774	Kneipp et al. (2006)
6	1–300	Intracellularly grown	None	785	MCF10	Shamsaie et al. (2007)
7	N.A.	nanoaggregates	pMBA	830	NIH/3T3	Kneipp et al. (2007)
8	60	Sphere	ICG	680, 786, 830	R3327	Kneipp et al. (2005)

2.6 Prospective

The advancement of nanotechnology has introduced a new dimension in cancer therapy and diagnosis as highlighted in this chapter, which can significantly improve the existing paradigms and lead to the development of new strategies. However, many aspects such as the potential impact of nanoparticles on cell fate and their biological distribution remain unclear or under the investigation. Traditional bulk materials with good bioinertness or biocompatibility cannot be simply assumed to behave similarly when they are utilized at the nanoscale. Comprehensive investigation of the biosafety of these nanomaterials becomes critical prior to any clinical applications. For example, GNPs are compatible to most cells without causing adverse effects. This finding is, however, only applicable to cases with low gold concentrations. GNPs at high concentrations, somehow, significantly slow down the cell proliferation and exhibit certain cytotoxicity (Zhang et al. 2009). Obviously, the studies on the interaction between cell/tissue and GNPs as well as a full understanding of the cellular uptake mechanism of GNPs allow us to better design GNPs for applications in biological systems and to explore other yet-to-be-foreseen application modality.

2.7 Conclusions

In this chapter, key properties of GNPs have been discussed along with their synthesis and use in cancer therapy (PDT and hyperthermia therapy) and diagnosis (SPR scattering, Raman microscopy, and SERS tag). Clearly, GNPs possess many excellent attributes that include good tissue permeability, good biocompatibility, readily functionalized surface, and unique optical properties, which render them multiple functionalities for clinical applications. As an example, they can be used for target delivery of photosynthetizers, enhanced generation of ROS, and SERS-based monitoring of cancer cell destruction, all in a single PDT treatment scheme. The potential of GNPs for biomedical applications goes well beyond the examples presented in this chapter. Its future is limited only by our imagination.

References

Arnida, M. A. and G. Hamidreza. 2010. Cellular uptake and toxicity of gold nanoparticles in prostate cancer cells: A comparative study of rods and spheres. *J. Appl. Toxicol.* 30(3):212–217.

Baas, P., I. Van Mansom, H. Van Tinteren, F. A. Stewart, and N. Van Zandwijk. 1995. Effect of N-acetylcysteine on photofrin-induced skin photosensitivity in patients. *Lasers Surg. Med.* 16(4):359–367.

Bao, G. and X. R. Bao. 2005. Shedding light on the dynamics of endocytosis and viral budding. *Proc. Natl. Acad. Sci. U.S.A.* 102(29):9997–9998.

Boyer, D., P. Tamarat, A. Maali, B. Lounis, and M. Orrit. 2002. Photothermal imaging of nanometer-sized metal particles among scatterers. *Science* 297(5584):1160–1163.

Bönnemann, H. and M. R. Richards 2001. Nanoscopic metal particles—Synthetic methods and potential applications. *Eur. J. Inorg. Chem.* 2001(10):2455–2480.

Brown, K. R. and M. J. Natan. 1998. Hydroxylamine seeding of colloidal Au nanoparticles in solution and on surfaces. *Langmuir* 14(4):726–728.

Brown, K. R., D. G. Walter, and M. J. Natan. 1999. Seeding of colloidal Au nanoparticle solutions. 2. Improved control of particle size and shape. *Chem. Mater.* 12(2):306–313.

Castano, A. P., T. N. Demidova, and M. R. Hamblin. 2004. Mechanisms in photodynamic therapy: Part one-photosensitizers, photochemistry and cellular localization. *Photodiagn. Photodyn. Ther.* 1(4):279–293.

Chen, J., J. M. McLellan, A. Siekkinen, Y. Xiong, Z.-Y. Li, and Y. Xia. 2006. Facile synthesis of gold-silver nanocages with controllable pores on the surface. *J. Am. Chem. Soc.* 128(46):14776–14777.

Chiang, C.-W., A. Wang, B.-Z. Wan, and C.-Y. Mou. 2005. High catalytic activity for CO oxidation of gold nanoparticles confined in acidic support Al-SBA-15 at low temperatures. *J. Phys. Chem. B* 109(38):18042–18047.

Chithrani, B. D. and W. C. W. Chan. 2007. Elucidating the mechanism of cellular uptake and removal of protein-coated gold nanoparticles of different sizes and shapes. *Nano Lett.* 7(6):1542–1550.

Chithrani, B. D., A. A. Ghazani, and W. C. W. Chan. 2006. Determining the size and shape dependence of gold nanoparticle uptake into mammalian cells. *Nano Lett.* 6(4):662–668.

Cognet, L., C. Tardin, D. Boyer, D. Choquett, P. Tamarat, and B. Lounis. 2003. Single metallic nanoparticle imaging for protein detection in cells. *Proc. Natl. Acad. Sci. U.S.A.* 100(20):11350–11355.

Connor, E. E., J. Mwamuka, A. Gole, C. J. Murphy, and M. D. Wyatt. 2005. Gold nanoparticles are taken up by human cells but do not cause acute cytotoxicity. *Small* 1(3):325–327.

Cotton, T. M., J.-H. Kim, and G. D. Chumanov. 1991. Application of surface-enhanced Raman spectroscopy to biological systems. *J. Raman Spectrosc.* 22(12):729–742.

El-Brolossy, T. A., T. Abdallah, M. B. Mohamed, S. Abdallah, K. Easawi, S. Negm, and H. Talaat. 2008. Shape and size dependence of the surface plasmon resonance of gold nanoparticles studied by photoacoustic technique. *Eur. Phys. J. Spec. Top.* 153(1):361–364.

El-Sayed, I. H., X. Huang, and M. A. El-Sayed. 2005. Surface plasmon resonance scattering and absorption of anti-EGFR antibody conjugated gold nanoparticles in cancer diagnostics: Applications in oral cancer. *Nano Lett.* 5(5):829–834.

Ferreira, L. 2009. Nanoparticles as tools to study and control stem cells. *J. Cell. Biochem.* 108(4):746–752.

Fleischmann, M., P. J. Hendra, and A. J. McQuillan. 1974. Raman spectra of pyridine adsorbed at a silver electrode. *Chem. Phys. Lett.* 26(2):163–166.

Frens, G. 1973. Controlled nucleation for the regulation of the particle size in monodisperse gold suspensions. *Nat. Phys. Sci.* 241:20–22.

Fu, Y. Z., Y. K. Du, P. Yang, J. R. Li, and L. Jiang. 2007. Shape-controlled synthesis of highly monodisperse and small size gold nanoparticles. *Sci. China Ser. B: Chem.* 50(4):494–500.

Gannon, C., C. Patra, R. Bhattacharya, P. Mukherjee, and S. Curley. 2008. Intracellular gold nanoparticles enhance non-invasive radiofrequency thermal destruction of human gastrointestinal cancer cells. *J. Nanobiotechnol.* 6(1):2.

Gao, H., W. Shi, and L. B. Freund. 2005. Mechanics of receptor-mediated endocytosis. *Proc. Natl. Acad. Sci. U.S.A.* 102(27):9469–9474.

Grunwaldt, J.-D., C. Kiener, C. Wögerbauer, and A. Baiker. 1999. Preparation of supported gold catalysts for low-temperature CO oxidation via "size-controlled" gold colloids. *J. Catal.* 181(2):223–232.

Guo, H., F. Ruan, L. Lu, J. Hu, J. Pan, Z. Yang, and R. Bin. 2009. Correlating the shape, surface plasmon resonance, and surface-enhanced Raman scattering of gold nanorods. *J. Phys. Chem. C* 113(24):10459–10464.

Halas, N. 2002. The optical properties of nanoshells. *Opt. Phot. News* 13(8):26–30.

Hao, E., R. C. Bailey, G. C. Schatz, J. T. Hupp, and S. Li. 2004. Synthesis and optical properties of "branched" gold nanocrystals. *Nano Lett.* 4(2):327–330.

Hauck, T. S. and W. C. W. Chan. 2007. Gold nanoshells in cancer imaging and therapy: Towards clinical application. *Nanomedicine* 2(5):735–738.

He, Y. and G. Shi. 2005. Surface plasmon resonances of silver triangle nanoplates: Graphic assignments of resonance modes and linear fittings of resonance peaks. *J. Phys. Chem. B* 109(37):17503–17511.

Henglein, A. 1999. Radiolytic preparation of ultrafine colloidal gold particles in aqueous solution: Optical spectrum, controlled growth, and some chemical reactions. *Langmuir* 15(20):6738–6744.

Henglein, A. and D. Meisel. 1998. Radiolytic control of the size of colloidal gold nanoparticles. *Langmuir* 14(26):7392–7396.

Hirsch, L. R., R. J. Stafford, J. A. Bankson, S. R. Sershen, B. Rivera, R. E. Price, J. D. Hazle, N. J. Halas, and J. L. West. 2003. Nanoshell-mediated near-infrared thermal therapy of tumors under magnetic resonance guidance. *Proc. Natl. Acad. Sci. U.S.A.* 100(23):13549.

Hong, R., G. Han, J. M. Fernandez, B.-J. Kim, N. S. Forbes, and V. M. Rotello. 2006. Glutathione-mediated delivery and release using monolayer protected nanoparticle carriers. *J. Am. Chem. Soc.* 128(4):1078–1079.

Hopper, C. 2000. Photodynamic therapy: A clinical reality in the treatment of cancer. *Lancet Oncol.* 1(4):212–219.

Hu, M. and G. V. Hartland. 2003. Heat dissipation for Au particles in aqueous solution: Relaxation time versus size. *J. Phys. Chem. B* 107(5):1284–1284.

Huang, X., P. K. Jain, I. H. El-Sayed, and M. A. El-Sayed. 2007. Gold nanoparticles: Interesting optical properties and recent applications in cancer diagnostics and therapy. *Nanomedicine* 2(5):681–693.

Huff, T. B., L. Tong, Y. Zhao, M. N. Hansen, J.–X. Cheng, and A. Wei. 2007. Hyperthermic effects of gold nanorods on tumor cells. *Nanomedicine* 2(1):125–132.

Hutter, E., J. H. Fendler, and D. Roy. 2001. Surface plasmon resonance studies of gold and silver nanoparticles linked to gold and silver substrates by 2-aminoethanethiol and 1,6-hexanedithiol. *J. Phys. Chem. B* 105(45):11159–11168.

Jain, K. K. 2005. Nanotechnology-based drug delivery for cancer. *Technol. Cancer Res. Treat.* 4(4):407–416.

Jain, P. K., K. S. Lee, I. H. El-Sayed, and M. A. El-Sayed. 2006. Calculated absorption and scattering properties of gold nanoparticles of different size, shape, and composition: Applications in biological imaging and biomedicine. *J. Phys. Chem. B* 110(14):7238–7248.

Jana, N. R. and X. Peng. 2003. Single-phase and gram-scale routes toward nearly monodisperse Au and other noble metal nanocrystals. *J. Am. Chem. Soc.* 125(47):14280–14281.

Jess, P. R. T., V. Garcés-Chávez, D. Smith, M. Mazilu, L. Paterson, A. Riches, C. S. Herrington, W. Sibbett, and K. Dholakia. 2006. Dual beam fibre trap for Raman micro-spectroscopy of single cells. *Opt. Express* 14(12):5779–5791.

Jiang, W., Y. S. KimBetty, J. T. Rutka, and C. W. ChanWarren. 2008. Nanoparticle-mediated cellular response is size-dependent. *Nat. Nanotechnol.* 3(3):145–150.

Kahraman, M., M. M. Yazici, F. Sahin, and M. Culha. 2007. Experimental parameters influencing surface-enhanced Raman scattering of bacteria. *J. Biomed. Opt.* 12(5):054015–054016.

Kessel, D. and P. Thompson. 1987. Purification and analysis of hematoporphyrin and hematoporphyrin derivative by gel exclusion and reverse-phase chromatography. *Photochem. Photobiol.* 46(6):1023–1025.

Khaing Oo, M. K., X. Yang, H. Du, and H. Wang. 2008a. 5-Aminolevulinic acid-conjugated gold nanoparticles for photodynamic therapy of cancer. *Nanomedicine* 3(6):777–786.

Khaing Oo, M. K., X. Yang, H. Wang, and H. Du. 2008b. 5-Aminolevulinic acid conjugated gold nanoparticles for cancer treatment. In *Technical Proceedings of the 2008 NSTI Nanotechnology Conference,* Vol. 2, pp. 12–15.

Khan, J. A., B. Pillai, T. K. Das, Y. Singh, and S. Maiti. 2007. Molecular effects of uptake of gold nanoparticles in HeLa cells. *ChemBioChem* 8(11):1237–1240.

Kim, F., S. Kwan, J. Akana, and P. Yang. 2001. Langmuir–Blodgett nanorod assembly. *J. Am. Chem. Soc.* 123(18):4360–4361.

Kim, F., J. H. Song, and P. Yang. 2002. Photochemical synthesis of gold nanorods. *J. Am. Chem. Soc.* 124(48):14316–14317.

Klostranec, J. M. and W. C. W. Chan. 2006. Quantum dots in biological and biomedical research: Recent progress and present challenges. *Adv. Mater.* 18(15):1953–1964.

Kneipp, K., A. S. Haka, H. Kneipp, K. Badizadegan, N. Yoshizawa, C. Boone, K. E. Shafer-Peltier, J. T. Motz, R. R. Dasari, and M. S. Feld. 2002a. Surface-enhanced Raman spectroscopy in single living cells using gold nanoparticles. *Appl. Spectrosc.* 56:150–154.

Kneipp, K., H. Kneipp, I. Itzkan, R. R. Dasari, and M. S. Feld. 2002b. Surface-enhanced Raman scattering and biophysics. *J. Phys. Condens. Matter* 14(18):R597–R624.

Kneipp, K., H. Kneipp, V. Bhaskaran Kartha, R. Manoharan, G. Deinum, I. Itzkan, R. R. Dasari, and M. S. Feld. 1998. Detection and identification of a single DNA base molecule using surface-enhanced Raman scattering (SERS). *Phys. Rev. E* 57(6):R6281.

Kneipp, J., H. Kneipp, and K. Kneipp. 2006. Two-photon vibrational spectroscopy for biosciences based on surface-enhanced hyper-Raman scattering. *Proc. Natl. Acad. Sci. U.S.A.* 103(46):17149–17153.

Kneipp, J., H. Kneipp, M. McLaughlin, D. Brown, and K. Kneipp. 2006. In vivo molecular probing of cellular compartments with gold nanoparticles and nanoaggregates. *Nano Lett.* 6(10):2225–2231.

Kneipp, J., H. Kneipp, W. L. Rice, and K. Kneipp. 2005. Optical probes for biological applications based on surface-enhanced Raman scattering from indocyanine green on gold nanoparticles. *Anal. Chem.* 77(8):2381–2385.

Kneipp, J., H. Kneipp, B. Wittig, and K. Kneipp. 2007. One- and two-photon excited optical pH probing for cells using surface-enhanced Raman and hyper-Raman nanosensors. *Nano Lett.* 7(9):2819–2823.

Krafft, C., T. Knetschke, R. H. W. Funk, and R. Salzer. 2006. Studies on stress-induced changes at the subcellular level by Raman microspectroscopic mapping. *Anal. Chem.* 78(13):4424–4429.

Lakadamyali, M., M. J. Rust, and X. Zhuang. 2004. Endocytosis of influenza viruses. *Microbes Infect.* 6(10):929–936.

Lal, S., S. E. Clare, and N. J. Halas. 2008. Nanoshell-enabled photothermal cancer therapy: Impending clinical impact. *Acc. Chem. Res.* 41(12):1842–1851.

Lee, S., S. Kim, J. Choo, S. Y. Shin, Y. H. Lee, H. Y. Choi, S. Ha, K. Kang, and C. H. Oh. 2007. Biological imaging of HEK293 cells expressing PLCγ1 using surface-enhanced Raman microscopy. *Anal. Chem.* 79(3):916–922.

Li, Q., E. C. Walter, W. E. Van Der Veer, B. J. Murray, J. T. Newberg, E. W. Bohannan, J. A. Switzer, J. C. Hemminger, and R. M. Penner. 2005. Molybdenum disulfide nanowires and nanoribbons by electrochemical/chemical synthesis. *J. Phys. Chem. B* 109(8):3169–3182.

Lieber, C. A., S. K. Majumder, D. Billheimer, D. L. Ellis, and A. Mahadevan-Jansen. 2008. Raman microspectroscopy for skin cancer detection in vitro. *J. Biomed. Opt.* 13(2):024013–024019.

Lü, F., L. Tianjun, and W. Li. 2008. Synthesis, characterization and cell-uptake of porphyrin-capped gold nanoparticle. In Peng, Y. and X. Weng (eds.), *7th Asian-Pacific Conference on Medical and Biological Engineering*, 22–25 April, Beijing, China, Vol. 19, pp. 186–189.

Maillard, M., M.-P. Pileni, S. Link, and M. A. El-Sayed. 2004. Picosecond self-induced thermal lensing from colloidal silver nanodisks. *J. Phys. Chem. B* 108(17):5230–5234.

Male, K. B., J. Li, C. C. Bun, S.-C. Ng, and J. H. T. Luong. 2007. Synthesis and stability of fluorescent gold nanoparticles by sodium borohydride in the presence of mono-6-deoxy-6-pyridinium-β-cyclodextrin chloride. *J. Phys. Chem. C* 112(2):443–451.

Mandal, D., A. Maran, M. Yaszemski, M. Bolander, and G. Sarkar. 2009. Cellular uptake of gold nanoparticles directly cross-linked with carrier peptides by osteosarcoma cells. *J. Mater. Sci. Mater. Med.* 20(1):347–350.

Merclin, N. and P. Beronius. 2004. Transport properties and association behaviour of the zwitterionic drug 5-aminolevulinic acid in water: A precision conductometric study. *Eur. J. Pharm. Sci.* 21(2–3):347–350.

Murgida, D. H. and P. Hildebrandt. 2001. Proton-coupled electron transfer of cytochrome c. *J. Am. Chem. Soc.* 123(17):4062–4068.

Nativo, P., I. A. Prior, and M. Brust. 2008. Uptake and intracellular fate of surface-modified gold nanoparticles. *ACS Nano* 2(8):1639–1644.

Nelayah, J., M. Kociak, O. Stephan, F. J. G. de Abajo, M. Tence, L. Henrard, D. Taverna, I. Pastoriza-Santos, L. M. Liz-Marzan, and C. Colliex. 2007. Mapping surface plasmons on a single metallic nanoparticle. *Nat. Phys.* 3(5):348–353.

Ngo, Q., B. A. Cruden, A. M. Cassell, G. Sims, M. Meyyappan, J. Li, and C. Y. Yang. 2004. Thermal interface properties of Cu-filled vertically aligned carbon nanofiber arrays. *Nano Lett.* 4(12):2403–2407.

Nguyen, T. Q., M. L. Bushey, L. E. Brus, and C. Nuckolls. 2002. Tuning intermolecular attraction to create polar order and one-dimensional nanostructures on surfaces. *J. Am. Chem. Soc.* 124(50):15051–15054.

Nichols, B. J. and J. Lippincott-Schwartz. 2001. Endocytosis without clathrin coats. *Trends Cell Biol.* 11(10):406–412.

Niki, K., Y. Kawasaki, Y. Kimura, Y. Higuchi, and N. Yasuoka. 2002. Surface-enhanced Raman scattering of cytochromes c3 adsorbed on silver electrode and their redox behavior. *Langmuir* 3(6):982–986.

Njoki, P. N., I-Im S. Lim, D. Mott, H.-Y. Park, B. Khan, S. Mishra, R. Sujakumar, J. Luo, and C.-J. Zhong. 2007. Size correlation of optical and spectroscopic properties for gold nanoparticles. *J. Phys. Chem. C* 111(40):14664–14669.

Norris, D. J., N. Yao, F. T. Charnock, and T. A. Kennedy. 2001. High-quality manganese-doped ZnSe nanocrystals. *Nano Lett.* 1(1):3–7.

Olson, K. J., H. Ahmadzadeh, and E. A. Arriaga. 2005. Within the cell: Analytical techniques for subcellular analysis. *Anal. Bioanal. Chem.* 382(4):906–917.

Orenstein, A., G. Kostenich, L. Roitman, Y. Shechtman, Y. Kopolovic, B. Ehrenberg, and Z. Malik. 1996. A comparative study of tissue distribution and photodynamic therapy selectivity of chlorin e6, Photofrin II and ALA-induced protoporphyrin IX in a colon carcinoma model. *Br. J. Cancer* 73(8):937–944.

Paciotti, G. F., L. Myer, D. Weinreich, D. Goia, N. Pavel, R. E. McLaughlin, and L. Tamarkin. 2004. Colloidal gold: A novel nanoparticle vector for tumor directed drug delivery. *Drug Deliv.* 11(3):169–183.

Palumbo, G. 2007. Photodynamic therapy and cancer: A brief sightseeing tour. *Expert Opin. Drug Deliv.* 4(2):131–148.

Patra, H. K., S. Banerjee, U. Chaudhuri, P. Lahiri, and A. K. Dasgupta. 2007. Cell selective response to gold nanoparticles. *Nanomed. Nanotechnol. Biol. Med.* 3(2):111–119.

Pei, L., K. Mori, and M. Adachi. 2004. Formation process of two-dimensional networked gold nanowires by citrate reduction of $AuCl_4^-$ and the shape stabilization. *Langmuir* 20(18):7837–7843.

Peng Q., T. Warloe, K. Berg, J. Moan, M. Kongshaug, K. E. Giercksky, and T. M. Nesland. 1997. 5-Aminolevulinic acid-based photodynamic therapy: Clinical research and future challenges. *Cancer* 79(12):2282–2308.

Peng, Q., T. Warloe, J. Moan, H. Heyerdahl, H. Steen, K. Giercksky, and J. Nesland. 1995. ALA derivative-induced protoporphyrin IX build-up and distribution in human nodular basal cell carcinoma. *Photochem. Photobiol.* 61:82S.

Pileni, M. P., B. W. Ninham, T. Gulik-Krzywicki, J. Tanori, I. Lisiecki, and A. Filankembo. 1999. Direct relationship between shape and size of template and synthesis of copper metal particles. *Adv. Mater.* 11(16):1358–1362.

Pitsillides, C. M., E. K. Joe, X. Wei, R. R. Anderson, and C. P. Lin. 2003. Selective cell targeting with light-absorbing microparticles and nanoparticles. *Biophys. J.* 84(6):4023–4032.

Pugh, T. L and W. Heller. 1960. Coagulation and stabilization of colloidal solutions with polyelectrolytes. *J. Polym. Sci.* 47(149):219–227.

Puppels, G. J., F. F. M. de Mul, C. Otto, J. Greve, M. Robert-Nicoud, D. J. Arndt-Jovin, and T. M. Jovin. 1990. Studying single living cells and chromosomes by confocal Raman microspectroscopy. *Nature* 347(6290):301–303.

Qian, X., X.-H. Peng, D. O. Ansari, Q. Yin-Goen, G. Z. Chen, D. M. Shin, L. Yang, A. N. Young, M. D. Wang, and S. Nie. 2008. In vivo tumor targeting and spectroscopic detection with surface-enhanced Raman nanoparticle tags. *Nat. Biotechnol.* 26(1):83–90.

Qu, Y. and X. Lu. 2009. Aqueous synthesis of gold nanoparticles and their cytotoxicity in human dermal fibroblasts-fetal. *Biomed. Mater.* 4(2):025007.

Ramaye, Y., S. Neveu, and V. Cabuil. 2005. Ferrofluids from prism-like nanoparticles. *J. Magn. Magn. Mater.* 289:28–31.

Ramser, K., J. Enger, M. Goksör, D. Hanstorp, K. Logg, and M. Käll. 2005. A microfluidic system enabling Raman measurements of the oxygenation cycle in single optically trapped red blood cells. *Lab Chip: Miniaturisation Chem. Biol.* 5(4):431–436.

Rao, C. N. R., G. U. Kulkarni, P. J. Thomas, and P. P. Edwards. 2000. Metal nanoparticles and their assemblies. *Chem. Soc. Rev.* 29(1):27–35.

Rechberger, W., A. Hohenau, A. Leitner, J. R. Krenn, B. Lamprecht, and F. R. Aussenegg. 2003. Optical properties of two interacting gold nanoparticles. *Opt. Commun.* 220(1–3):137–141.

Remacle, F., C. P. Collier, G. Markovich, J. R. Heath, U. Banin, and R. D. Levine. 1998. Networks of quantum nanodots: The role of disorder in modifying electronic and optical properties. *J. Phys. Chem. B* 102(40):7727–7734.

Richardson, H. H., Z. N. Hickman, A. O. Govorov, A. C. Thomas, W. Zhang, and M. E. Kordesch. 2006. Thermooptical properties of gold nanoparticles embedded in Ice: Characterization of heat generation and melting. *Nano Lett.* 6(4):783–788.

Roucoux, A., J. Schulz, and H. Patin. 2002. Reduced transition metal colloids: A novel family of reusable catalysts? *Chem. Rev.* 102(10):3757–3778.

Safavi, A., G. Absalan, and F. Bamdad. 2008. Effect of gold nanoparticle as a novel nanocatalyst on luminol-hydrazine chemiluminescence system and its analytical application. *Anal. Chim. Acta* 610(2):243–248.

Sarathy, K. V., G. Raina, R. T. Yadav, G. U. Kulkarni, and C. N. R. Rao. 1997. Thiol-derivatized nanocrystalline arrays of gold, silver, and platinum. *J. Phys. Chem. B* 101(48):9876–9880.

Sau, T. K. and C. J. Murphy. 2004. Room temperature, high-yield synthesis of multiple shapes of gold nanoparticles in aqueous solution. *J. Am. Chem. Soc.* 126(28):8648–8649.

Sau, T. K. and C. J. Murphy. 2005. Self-assembly patterns formed upon solvent evaporation of aqueous cetyltrimethylammonium bromide-coated gold nanoparticles of various shapes. *Langmuir* 21(7):2923–2929.

Sau, T. K., A. Pal, N. R. Jana, Z. L. Wang, and T. Pal. 2001. Size controlled synthesis of gold nanoparticles using photochemically prepared seed particles. *J. Nanoparticle Res.* 3(4):257–261.

Schmid, G. 2002. Large clusters and colloids. Metals in the embryonic state. *Chem. Rev.* 92(8):1709–1727.

Shamsaie, A., J. Heim, A. A. Yanik, and J. Irudayaraj. 2008. Intracellular quantification by surface enhanced Raman spectroscopy. *Chem. Phys. Lett.* 461(1–3):131–135.

Shamsaie, A., M. Jonczyk, J. Sturgis, J. Paul Robinson, and J. Irudayaraj. 2007. Intracellularly grown gold nanoparticles as potential surface-enhanced Raman scattering probes. *J. Biomed. Opt.* 12(2):020502–3.

Shukla, R., V. Bansal, M. Chaudhary, A. Basu, R. R. Bhonde, and M. Sastry. 2005. Biocompatibility of gold nanoparticles and their endocytotic fate inside the cellular compartment: A microscopic overview. *Langmuir* 21(23):10644–10654.

Siegfried, S., P. Halbig, H. Grau, and U. Nickel. 1994. Reproducible preparation of silver sols with uniform particle size for application in surface-enhanced Raman spectroscopy. *Photochem. Photobiol.* 60(6):605–610.

Skirtach, A. G., C. Dejugnat, D. Braun, A. S. Susha, A. L. Rogach, W. J. Parak, H. Mohwald, and G. B. Sukhorukov. 2005. The role of metal nanoparticles in remote release of encapsulated materials. *Nano Lett.* 5(7):1371–1377.

Slot, J. W. and H. J. Geuze. 1981. Sizing of protein A-colloidal gold probes for immunoelectron microscopy. *J. Cell Biol.* 90(2):533–536.

Sokolov, K., P. Khodorchenko, A. Petukhov, I. Nabiev, G. Chumanov, and T. M. Cotton. 1993. Contributions of short-range and classical electromagnetic mechanisms to surface-enhanced Raman scattering from several types of biomolecules adsorbed on cold-deposited island films. *Appl. Spectrosc.* 47:515–522.

Spikes, J. D. 1990. Chlorins as photosensitizers in biology and medicine. *J. Photochem. Photobiol. B: Biol.* 6(3):259–274.

Sun, X., S. Dong, and E. Wang. 2005. One-step preparation of highly concentrated well-stable gold colloids by direct mix of polyelectrolyte and $HAuCl_4$ aqueous solutions at room temperature. *J. Colloid Interface Sci.* 288(1):301–303.

Sun, X., S. Dong, and E. Wang. 2006. One-step polyelectrolyte-based route to well-dispersed gold nanoparticles: Synthesis and insight. *Mater. Chem. Phys.* 96(1):29–33.

Sun, Y., and Y. Xia. 2002. Shape-controlled synthesis of gold and silver nanoparticles. *Science* 298(5601):2176–2179.

Tan, S., E. Melek, A. Attygalle, H. Du, and S. Sukhishvili. 2007. Synthesis of positively charged silver nanoparticles via photoreduction of $AgNO_3$ in branched polyethyleneimine/HEPES solutions. *Langmuir* 23(19):9836–9843.

Tsunoyama, H., H. Sakurai, N. Ichikuni, Y. Negishi, and T. Tsukuda. 2004. Colloidal gold nanoparticles as catalyst for carbon–carbon bond formation: Application to aerobic homocoupling of phenylboronic acid in water. *Langmuir* 20(26):11293–11296.

Turkevich, J. 1985. Colloidal gold. Part I. *Gold Bullet.* 18(3):86–91.

Turkevitch, J., P. C. Stevenson, and J. Hillier. 1951. A study of the nucleation and growth processes in the synthesis of colloidal gold. *Discuss. Faraday Soc.* 11:55–75.

Van Manen, H. J., Y. M. Kraan, D. Roos, and C. Otto. 2005. Single-cell Raman and fluorescence microscopy reveal the association of lipid bodies with phagosomes in leukocytes. *Proc. Natl. Acad. Sci. U.S.A.* 102(29):10159–10164.

Venkata, P. G., M. M. Aslan, M. P. Menguc, and G. Videen. 2007. Surface plasmon scattering by gold nanoparticles and two-dimensional agglomerates. *J. Heat Transfer* 129(1):60–70.

Visaria, R. K., R. J. Griffin, B. W. Williams, E. S. Ebbini, G. F. Paciotti, C. W. Song, and J. C. Bischof. 2006. Enhancement of tumor thermal therapy using gold nanoparticle–assisted tumor necrosis factor-delivery. *Mol. Cancer Ther.* 5(4):1014.

Watzky, M. A. and R. G. Finke. 1997. Transition metal nanocluster formation kinetic and mechanistic studies. A new mechanism when hydrogen is the reductant: Slow, continuous nucleation and fast autocatalytic surface growth. *J. Am. Chem. Soc.* 119(43):10382–10400.

Wieder, M. E., D. C. Hone, M. J. Cook, M. M. Handsley, J. Gavrilovic, and D. A. Russell. 2006. Intracellular photodynamic therapy with photosensitizer-nanoparticle conjugates: Cancer therapy using a "Trojan horse." *Photochem. Photobiol. Sci.* 5(8):727–734.

Wiley, B., Y. Sun, B. Mayers, and Y. Xia. 2005. Shape-controlled synthesis of metal nanostructures: The case of silver. *Chem. Eur. J.* 11(2):454–463.

Yamanea, K., K. Yakushijia, F. Ernulta, M. Matsuura, S. Mitani, K. Takanashi, and H. Fujimori. 2004. Inverse tunnel magnetoresistance associated with coulomb staircases in micro-fabricate granular systems. *J. Magn. Magn. Mater.* 6:272–276.

Yan, W., S. Brown, Z. Pan, S. M. Mahurin, S. H. Overbury, and S. Dai. 2006. Ultrastable gold nanocatalyst supported by nanosized non-oxide substrate. *Angew. Chem. Int. Ed.* 45(22):3614–3618.

Yang, P.-H., X. Sun, J.-F. Chiu, H. Sun, and Q.-Y. He. 2005a. Transferrin-mediated gold nanoparticle cellular uptake. *Bioconjug. Chem.* 16(3):494–496.

Yang, G., L. Tan, Y. Yang, S. Chen, and G. Y. Liu. 2005b. Single electron tunneling and manipulation of nanoparticles on surfaces at room temperature. *Surf. Sci.* 589(1–3):129–138.

Yu, K., K. L. Kelly, N. Sakai, and T. Tatsuma. 2008. Morphologies and surface plasmon resonance properties of monodisperse bumpy gold nanoparticles. *Langmuir* 24(11):5849–5854.

Zhang, X. D., M. L. Guo, H. Y. Wu, Y. M. Sun, Y. Q. Ding, X. Feng, and L. A. Zhang. 2009. Irradiation stability and cytotoxicity of gold nanoparticles for radiotherapy. *Int. J. Nanomedicine* 4:165–173.

Zhang, X., H. Yin, J. M. Cooper, and S. J. Haswell. 2008. Characterization of cellular chemical dynamics using combined microfluidic and Raman techniques. *Anal. Bioanal. Chem.* 390(3):833–840.

Zhu, Z.-J., P. S. Ghosh, O. R. Miranda, R. W. Vachet, and V. M. Rotello. 2008. Multiplexed screening of cellular uptake of gold nanoparticles using laser desorption/ionization mass spectrometry. *J. Am. Chem. Soc.* 130(43):14139–14143.

3

Synthesis, Processing, and Characterization of Ceramic Nanobiomaterials for Biomedical Applications

Michaela Schulz-Siegmund
University of Leipzig

Rudi Hötzel
University of Leipzig

Peter-Georg Hoffmeister
University of Leipzig

Michael C. Hacker
University of Leipzig

3.1 Introduction

Advances in chemistry, physics, engineering, and material sciences have enabled the preparation, synthesis, and manufacturing of materials on the nanometer scale, which offers tremendous opportunities to control material properties, to mimic hierarchical structures of biological composites by engineered materials, and to adjust interactions of a material with biological molecules or a biological system (Zhang and Webster 2009). The prefix "nano" in nanomaterials typically defines structures that are smaller than 100 nm in at least one dimension. Due to a very large surface-to-volume ratio, the unique properties of nanomaterials originate from cohesive and/or adhesive interactions at the surface of the material in combination with the materials' bulk properties. In contrast, the properties of materials at a larger size scale are dominated almost exclusively by their bulk characteristics. The surface of nanoceramics, for example, is more active in terms of dissolution and recrystallization processes and the interaction with organic molecules as compared to micrometer-sized crystallites. Also, ceramics, which in general suffer from low elasticity, may offer significant ductility before failure when synthesized at the nanoscale (Karch et al. 1987). Nanoscaled ceramics can be sintered at a lower temperature, which reduces processing problems associated with high temperature. Due to strong interactions with organic molecules, nano-sized ceramics can exhibit bioactivity and affect the adhesion, proliferation, and differentiation of cells in direct contact (Webster et al. 1999; LeGeros 2008).

The chapter focuses on providing a summary of ceramic nanostructures that are used as biomaterials. The term "biomaterials" defines natural and synthetic nonviable materials that are intended to interact with biological systems for any diagnostic or biomedical purpose, including the treatment, augmentation, or replacement of any tissue or function of the organism (Hench 1980; Veerapandian and Yun 2009). Ceramic materials by definition are typically obtained from nonmetallic inorganic solids through the application of heat and include crystalline and amorphous materials. Ceramic materials include inorganic oxides, non-oxides, and composites. Certain minerals, especially calcium phosphate minerals, will also be discussed in this chapter as they are the predominant inorganic component of hard tissues. By mass, calcium is the most abundant metal in the human body (Frieden 1972). Due to the physiological properties, calcium minerals are frequently used as biomaterials for orthopedic, maxillofacial, and dental applications.

3.2 Ceramic Nanobiomaterials

Several ceramics and minerals have been used as nanobiomaterials in various applications including implant coatings, nanocomposite materials for implants, and cell-carrying scaffolds as well as drug delivery systems and diagnostic devices (Vallet-Regi 2001; Dorozhkin 2010a). In order to be used as a biomaterial, the ceramic nanostructures should exhibit good biocompatibility. A biocompatible material typically can be further classified into three major groups according to the degree of interaction of the material with host tissue (Vallet-Regi 2001). When there is no significant interaction or remodeling, the material is classified as bioinert. This property is favorable when a diagnostic or drug delivery application is envisioned or when a device for permanent replacement of a certain organ or tissue function is designed. Biomaterials that are chemically degraded or metabolized in any way prior to renal or biliary elimination are classified as biodegradable or bioresorbable and preferably used in drug delivery and regenerative applications. Materials with the ability to directly or indirectly influence the development of cells or tissues in contact or close proximity are generally categorized as bioactive. Bioactive materials can be especially beneficial in regenerative applications or to promote hard tissue implant integration. As far as interactions with bony tissue are considered, a bioactive material can either be osteoconductive or osteoinductive (Habibovic and de Groot 2007; Kalita et al. 2007). The latter describes a material's ability to support osteogenesis, the formation of new bone, even in an ectopic environment *in vivo*. Osteoconduction describes the process of guided growth of bony tissue along the surface of a biomaterial implant away from the initial bone–biomaterial interface.

Compared to other nanobiomaterials, ceramics can combine excellent biocompatibility and bioactivity with mechanical properties appropriate to be used in load-bearing orthopedic applications (Kalita et al. 2007). Looking at the mechanical properties of ceramics more closely, this group of materials is characterized by high stiffness and high resistance to wear and corrosion but low toughness and resilience (Murugan and Ramakrishna 2006). Polymers on the other hand are less stiff but more flexible and resilient. The degradation properties of polymeric materials can be varied over a wide range and chemical surface modifications can be achieved relatively easy (Gunatillake and Adhikari 2003; Drotleff et al. 2004). This makes polymer–ceramic composites very attractive biomaterials as such materials combine the flexibility and resorption properties of a polymer with the mechanical strength and bioactivity of a ceramic biomaterial (Kalita et al. 2007; Dorozhkin 2010b). Metals and alloys are a group of biomaterials with excellent strength and toughness but also significantly higher densities than ceramics. The bioactivity of metals and alloys is typically low, which makes a surface coating with a ceramic nanophase very attractive.

For more than 50 years, specialty bioceramics such as alumina, zirconia, hydroxyapatite, di- and tricalcium phosphates, and bioactive glasses have evolved and found applications in many biomedical applications, particularly as bone substitutes and components of orthopedic, dental, and maxillofacial implants due to the above mentioned advantages (Hench 1998). An overview of important ceramics and minerals that have been used as nanobiomaterials is given in the following paragraphs. The most prominent class is the calcium phosphates, because such minerals constitute the vast majority of inorganic mass in the human body.

3.2.1 Calcium Phosphate Minerals

Calcium phosphates are chemically stable low density minerals composed of ions commonly found in physiological environment (Table 3.1) (Kalita et al. 2007; Dorozhkin 2010b). In general, they exhibit excellent biocompatibility, which is likely due to their compositional similarity with bone mineral and is the main reason for the copious *in vivo* applications these materials have been used in. Today, calcium phosphates are probably the most important biomaterials in dentistry and orthopedics. This versatile group of biomaterials exists in different forms and phases depending on temperature, partial pressure of water, and the presence of impurities. Hydroxyapatite/Hydroxylapatite (HA), β-tricalcium phosphate (β-TCP), α-tricalcium phosphate (α-TCP), biphasic calcium phosphate (BCP), monocalcium phosphate monohydrate (MCPM), and unsintered apatite are different calcium phosphate minerals with different chemical and mechanical properties (Kalita et al. 2007; Dorozhkin 2010a). HA, which resembles the mineral phase of bone closest, has strong bioactivity and resists hydrolytic degradation. Tricalcium phosphates, on the other hand, are resorbable minerals that, upon hydrolysis, are transformed into more stable calcium phosphates, such as HA *in vivo*.

In traditional applications, calcium phosphates were used because of their biocompatible chemistry (Dorozhkin 2010a). When it became clear that the unique mechanical properties of bone tissue originated from an evolved interplay between mineral nanocrystals and collagen microfibers (Fratzl et al. 2004), the dimensional component of the minerals also became a design criterion. The apatite crystals, which occur in the form of plates or needles, are about 40–60 nm long, 20 nm wide, and 1.5–5 nm thick (Kalita et al. 2007). The crystals, which constitute two-thirds of the bone mass, are oriented along the long axis of the collagen fibers forming a continuous phase.

Calcium phosphate minerals, besides being well biocompatible, are classified as bioactive materials (LeGeros 2008). In contact with biological tissues such as bone, porous constructs, or coated surfaces have been shown to be osteoconductive. It has also been observed that certain calcium phosphates have osteoinductive properties. In part, the bioactivity of these materials is attributed to the surface roughness

TABLE 3.1 Physical Properties of Selected Calcium Phosphate Phases

Compound	Chemical Formula	Ca/P Ratio	Crystal Structure	Density [g·cm^{-3}]	Solubility $-\log(K_s)$ (25°C)
Dicalcium phosphate dihydrate	$CaHPO_4 \cdot 2H_2O$	1	Monoclinic, *Ia*	2.32	6.59
α-Tricalcium phosphate (α-TCP)	$Ca_3(PO_4)_2$	1.5	Monoclinic, $P2_1/a$	2.86	28.9
β-Tricalcium phosphate (β-TCP)	$Ca_3(PO_4)_2$	1.5	Pure hexagonal, rhombohedral, *R*3cH	3.07	25.5
Hydroxyapatite (HA)	$Ca_{10}(PO_4)_6(OH)_2$	1.67	Hexagonal, $P6_3/m$ or monoclinic, $P2_1/b$	3.16	116.8
Tetracalcium phosphate	$Ca_4P_2O_9$	2	Monoclinic, $P2_1$	3.05	38–44

Source: Dorozhkin, S.V., *Acta Biomater.*, 6(3), 715, 2010; Kalita, S.J. et al., *Mater. Sci. Eng. C Biomim. Mater. Sens. Syst.*, 27(3), 441, 2007.

of nanostructured materials (Dorozhkin 2010b). The osteoinductive properties, however, cannot be directly attributed to the ceramic and are often described as intrinsic (Habibovic and de Groot 2007; LeGeros 2008). The current understanding is that a specific topography comprising interconnected porosities and concavities allows for the absorption, entrapment, and concentration of osteogenic factors such as bone morphogenetic proteins from surrounding body fluids (*in vivo*) or serum-containing media (*in vitro*).

3.2.1.1 Hydroxyapatite/Hydroxylapatite

Hard and mineralized tissues such as bone, dentin, and enamel all contain calcium phosphate minerals from the apatite group as the predominant inorganic component (Bigi et al. 1997; LeGeros 2002). These apatite minerals are formed from calcium, phosphorous, oxygen, and one or more channel-filling ion(s), such as chloride, fluoride, or hydroxyl ions. Depending on the exact chemical composition, the apatite's properties vary, making this group of minerals very flexible. Substitutions in the chemical composition affect the structure and key properties, including solubility, hardness, brittleness, thermal stability, and also optical properties. HA [$Ca_5(PO_4)_3(OH)$] is an hydroxyl-containing apatite with very specific structure and properties. HA is often described as the main apatite in bone and enamel and due to its natural occurrence a candidate biomaterial (Dorozhkin 2010b). Taking a closer look, bone apatite, enamel apatite, and dentin apatite all slightly differ from the chemical structure of HA; and these differences determine the different properties of these tissues (Wopenka and Pasteris 2005). Via ionic substitution, enamel apatite in contrast to bone apatite has become resistant to dissolution. In comparison to synthetic HA, bone and dentin apatite have also been identified to be less ordered in structure and to contain predominantly carbonate anions instead of hydroxyl ions, which affects the morphology of the apatite. The following approximated formula has been proposed to characterize bone apatite: $(Ca,X)_{10}(PO_4,HPO_4,CO_3)_6(OH,Y)_2$, with X representing cations (magnesium, sodium, strontium ions) that can substitute for the calcium ions and Y representing anions (chloride or fluoride ions) that can substitute for the hydroxyl group (LeGeros 2002). Pure carbonated HA (cHA), also known as dahllite, is represented by the formula $Ca_{10}(PO_4,CO_3)_6(OH)_2$.

HA is a bioactive ceramic, and because of its excellent stability above pH 4.3, the ideal calcium phosphate phase for application inside human body when nondegradability is required (Kalita et al. 2007). Obviously, traditional applications of HA powder and particulates are in bone repair and as coatings for metallic prosthesis to improve hard tissue integration, but it is also used for controlled drug release. HA possesses a hexagonal structure and has a stoichiometric Ca/P ratio of 1.67 (Table 3.1), which among all pure calcium phosphate phases is closest to mineralized human tissue. Like for most ceramics, the mechanical strength and fracture toughness of pure HA is significantly lower compared to bone (Santos et al. 1996). These properties can be improved by enhanced densification through the use of different sintering techniques. Such strategies include the addition of a low-melting secondary phase, for example glasses, as a binder (Georgiou and Knowles 2001) and the use of nanoscale powders for better densification due to their large surface area. Besides improved sinterability and enhanced densification, nano-HA (nHA) is also expected to have better bioactivity than coarser crystals (Webster et al. 2001a).

A number of techniques, including sol–gel synthesis, solid state reactions, precipitation, hydrothermal reaction, microemulsion synthesis, and mechanochemical routes have been used to fabricate nHA powders (Kalita et al. 2007; Dorozhkin 2010b). A selection of these techniques is described in the "Fabrication of ceramic nanobiomaterials" section. The sol–gel method has recently gained interest for synthesis of calcium phosphates nanoparticles due to its unique advantages. The method is capable of improving chemical homogeneity and offers almost molecular-level control by mixing of the calcium and phosphorus precursors and reducing synthesis temperature in comparison with conventional methods. Nanopowders with different Ca/P ratios have been produced by altering the quantity and the composition of precursors and processing variables. Another common preparation method for nHA powders is chemical precipitation through aqueous solutions of calcium chloride and ammonium hydrogen phosphate. It has been observed that the crystallinity and crystallite size of nHA increases with temperature and ripening time (Pang and Bao 2003). Particle morphology was found to correlate with crystallinity. Ceramic nanoparticles with regular shape, smooth surface, clear contour, and low water content were obtained by higher crystallinity of HA.

The synthesis of carbonate containing HA has become an important target in biomaterials with the objective to more closely mimic hard tissue apatite (Landi et al. 2003). Depending of whether the hydroxyl or the phosphate ions of HA are substituted, one commonly refers to an A-type or B-type carbonation, respectively. In human mineralized tissues, the B-type is the preferred form of cHA. Highly pure B-type cHA nanopowder has been produced with a wet-chemical synthesis from calcium nitrate, ammonium hydrogen phosphate and sodium carbonate. The compressive strength of sintered cHA porous bodies was about twice the strength of analogous HA matrices.

3.2.1.2 Other Calcium Phosphates

Calcium phosphates besides HA that have been processed into nano-sized structures include α-TCP, β-TCP, tetracalcium phosphate, dicalcium phosphate dehydrate, dicalcium phosphate anhydrous, and octacalcium phosphate (Table 3.1) (Kalita et al. 2007; Dorozhkin 2010b).

Out of both tricalcium phosphates, β-TCP, also known as β-whitlockite, degrades more slowly and is used as a bioresorbable calcium phosphate ceramic in biomedical applications, particularly in orthopedics. X-ray patterns reveal that β-TCP has a pure hexagonal crystal structure. Nano-sized β-TCP powders have been synthesized utilizing a variety of methods similar to those for the fabrication of HA. The conventional methods include solid-state processes and wet-chemical methods (Bow et al. 2004; Kalita et al. 2007). As for all wet synthesis strategies of calcium phosphates, a calcium source and a phosphate source are used as starting material. For nano-β-TCP, calcium acetate and phosphoric acid are typical sources. During the process, phase transition involving calcium hydrogen phosphate and intermediate amorphous calcium phosphate phases occurs until β-TCP is formed at increased aging times. It has been observed that the incorporation of carbonate favors the formation of the β-TCP phase (Bow et al. 2004).

3.2.2 Aluminum Oxides

The most prominent aluminum ceramic is aluminum(III) oxide (Al_2O_3, alumina) (Vallet-Regi 2001; Rahaman et al. 2007). Alumina exists in different modifications, cubic γ-Al_2O_3 and trigonal α-Al_2O_3 (Yang et al. 2009c). γ-Al_2O_3 is often used as a raw material for further processing. Both modifications have different dissolution behaviors. Whereas the γ-form is soluble in strong acid and base, the α-form is insoluble and bioinert. At temperatures above 800°C, the γ-form is transformed into the stable α-form. Traditional applications of alumina are in dentistry, in anthroplasty, and in the treatment of hand and elbow fractures. Similar to nHA, nanophase alumina showed higher bioactivity as compared to grains larger than 100 nm (Webster et al. 2001a). The adhesion of osteoblasts to nano-sized alumina substrates was significantly improved over conventional alumina (Webster et al. 1999). The effects were especially pronounced when serum was present, suggesting a contribution of plasma proteins that absorb to the nanostructured ceramic surfaces (Webster et al. 2001b). Alumina nanopowders can be prepared by plasma spraying of liquid precursors (Karthikeyan et al. 1997) or flame aerosol technology (Pratsinis 1998). With atomic layer deposition, it is possible to deposit uniform alumina nanofilms on zirconia nanoparticles without affecting size distribution and surface area of the particles (Hakim et al. 2005).

Alumoxanes or more precisely carboxylatoalumoxanes ($[Al(O)_x(OH)_y(O_2CR)_z]_n$ with $2x + y + z = 3$ and R = C1–C13) can be prepared as nano-sized particles by the reaction of pseudo-boehmite ($[Al(O)(OH)]_n$) with carboxylic acids (RCOOH) in an environmentally benign process (Landry et al. 1995; Callender et al. 1997). Depending on the alkyl substituents, the physical properties of the alumoxanes may range from insoluble crystalline powders to powders that readily form solutions or gels in hydrocarbon solvents. Upon thermolysis, the carboxylate-alumoxanes can be converted to alumina. The particle size of the carboxylate–alumoxane can comfortably be selected by the choice of carboxylic acid and by solution pH (Vogelson and Barron 2001). The alkyl residues of the alumoxane nanoparticles can be used to balance the hydrophilicity of the ceramic in a way that the dispersibility of the nanoparticles in hydrophobic matrices, for example, polymer bulks, is significantly improved. This is a key requisite for the successful fabrication of ceramic-polymer composites (Kim and O'Shaughnessy 2002; Horch et al. 2004). Furthermore, the residues can be chemically modified and reactive moieties introduced to yield hybrid nanoparticles that can be covalently integrated in composite polymer networks.

3.2.3 Titanium Oxides

Titanium(IV) oxide, titanium dioxide (TiO_2), or titania has eight modifications and is classified as a bioinert material (Vallet-Regi 2001; Rahaman et al. 2007). Titania is the dominant oxide on the surface of passivated titanium and formed upon contact of the pristine metal with air (Gotman 1997). In bioliquids, calcium and phosphate ions are also incorporated into the oxide layers on titanium, forming calcium titanium phosphate and other mineral deposits (Hanawa et al. 1998). The preparation of full-density nanostructured titanium dioxide ceramics is difficult, because of rapid grain growth during sintering (Lee et al. 2003). Titania nanoparticles can be obtained by flame aerosol technology (Pratsinis 1998). Single-crystalline titanium dioxide nanotubes with lengths up to a few hundred nanometers have been synthesized via the hydrolysis of TiF_4 at low pH and 60°C (Liu et al. 2002). Positive effects of individual titania nanoparticles on the differentiation of neural stem cells toward neuron have been described (Liu et al. 2010). The biological effects of titanium dioxide nanostructures on implant surfaces have been more widely described. Such coatings are typically applied to hard tissue implants, orthopedic, and dental, to improve integration (Yao and Webster 2006). It has been shown that osteoblast adhesion and activity on nanostructured titanium dioxide surfaces are improved and these effects have been correlated to specific protein adsorption (Webster et al. 1999, 2000b; Colon et al. 2006). The bioactivity of titanium dioxide layers can be further improved through the immobilization of specific peptides (Balasundaram et al. 2008). Positive effects on osteoblast adhesion have also been shown for nanostructured titania-polymer composites (Kay et al. 2002).

3.2.4 Zirconium Oxides

Zirconium dioxide (ZrO_2), also known as zirconium (IV) oxide or zirconia, is the most prominent oxide of the transition metal zirconium. Zirconia is a white powder with a density of $5.68\,g \cdot cm^{-3}$ that is bioinert and insoluble in water. Zirconium dioxide is available in different modifications: monoclinic, tetragonal, and cubic. Tetragonal zirconium dioxide features the highest mechanical stability and is therefore the preferred modification for biomedical applications. At room temperature, however, the monoclinic modification of zirconium dioxide is prevalent. The transformation from monoclinic to tetragonal takes place at heating to 2370°C; the cubic phase can be obtained at 2690°C and further heating leads to melting. In order to stabilize the tetragonal modification at room temperature, different metallic oxides, for example, magnesium oxide and yttrium (III) oxide (Y_2O_3), can be introduced (Rahaman et al. 2007), leading to materials like yttria-doped tetragonal zirconia polycrystal (Y-TZP). Doped TZPs have been shown to demonstrate improved strength and fracture toughness (Chevalier and Gremillard 2009). Zirconium dioxide-based ceramics have been used as compounds of implants, for example, prosthetic knee replacements (Rahaman et al. 2007) and for dental restorations (Yang et al. 2009c).

Ultrafine polymer-stabilized nanocrystalline tetragonal zirconium dioxide powders have been synthesized by a microwave-assisted method from an aqueous solution containing $Zr(NO_3)_4$, poly(vinyl alcohol), and NaOH (Liang et al. 2002). Nanocrystalline monoclinic zirconium dioxide powders were obtained by forced hydrolysis of inorganic zirconyl salts (Hu et al. 1998). Monoclinic nano-grained zirconium dioxide coatings fabricated by atmospheric plasma spraying have demonstrated promising properties for application as coatings for metallic orthopedic implants (Wang et al. 2010). Zirconia coatings exhibited good bioactivity and biocompatibility, high bonding strength with titanium alloys, and high stability in an aqueous environment. Plasma-sprayed HA coatings, in contrast, are often characterized by low crystallinity and poor bonding strength on titanium alloys. In cell culture experiments with osteoblasts, zirconium dioxide coatings supported cell attachment and adhesion, and enhanced cell proliferation could be observed. Nano-grained coatings have also been fabricated from yttria-stabilized zirconia (Racek et al. 2006).

Another application of zirconia is to stabilize HA. Composites that contained low amounts of zirconia and were processed at low sintering temperatures to maintain HA crystallinity revealed higher surface roughness, smaller grain size, and increased osteoblast adhesion (Evis et al. 2006).

3.2.5 Glasses

Bioactive glasses are amorphous solids that are classified as ceramic materials in the biomaterial literature (Vallet-Regi 2001). These glass ceramics are described as surface reactive materials and have been shown to exhibit good biocompatibility. A bioactive glass generally consists of formers and modifiers. The forming materials typically are silicon dioxide and phosphorus pentoxide, whereas modifiers include calcium oxide and sodium oxide. Various other substances such as K_2O, MgO, CaF_2, Al_2O_3, B_2O_3, and Fe_2O_3 can be introduced to create materials with specific properties.

Bioglass® is an FDA-approved material that was developed in 1971 by Larry L. Hench and colleagues and is composed of 46.1 mol% SiO_2, 26.9 mol% CaO, 24.4 mol% Na_2O, and 2.5 mol% P_2O_5 (Hench 1997). One aspect of the bioactivity of these amorphous ceramics is the formation of apatite-like crystals on their surfaces out of initially formed silica hydrogel layers upon contact with body fluids. Some reports claim that bioactive glasses show better performance in bone tissue engineering than HA and that the glasses strongly interact with hard tissues (Xynos et al. 2000). Bioactive glasses have gained attention as promising materials for tissue engineering scaffolds, either as fillers or as coatings of polymer structures or as porous materials themselves (Rezwan et al. 2006). SiO_2–CaO–P_2O_5 ternary bioactive glass ceramic nanoparticles (20 nm in diameter) with different compositions have been prepared via a three-step sol-gel method (Hong et al. 2009). Nanoparticles with low phosphorous and high silicon content exhibited enhanced mineralization capability in simulated body fluid and a higher solubility in phosphate buffered saline.

3.2.6 Other Ceramic Nanobiomaterials

Biomaterial literature includes magnetic oxides, graphite, and pyrolytic carbon in the class of bioceramics (Vallet-Regi 2001). Magnetic nanoparticles are typically used for therapeutic drug, gene, and radionuclide delivery as well as for cancer therapy and as contrast enhancing agents (Pankhurst et al. 2003). These materials are discussed elsewhere in this handbook. Carbon nanostructures, such as nanotubes, are a widely investigated group of nanobiomaterials (Webster et al. 2004) and also discussed in a separate chapter of this handbook. Diamond, a different allotrope of carbon, has lately gained increasing interest as nanobiomaterial. Due to superior mechanical and tribological properties, nanostructured diamond coatings on orthopedic implants are of special interest (Yang et al. 2009a).

3.3 Fabrication of Ceramic Nanoparticles

For the production of nano-sized ceramic biomaterials, many methods have been developed and a selection is highlighted in the following paragraphs. These include milling, precipitation, and emulsion processes as well as the application of ultrasound or microwave irradiation. A general classification of the available techniques can also be done according to whether the nanostructures are obtained by disaggregation of larger particles or by controlled crystal growth from corresponding ions. As calcium phosphate materials are the most frequently used and investigated ceramic biomaterials, most methods have been developed for their fabrication (Dorozhkin 2010b). Consequently, most examples given in this chapter also refer to this class of ceramic nanobiomaterials.

3.3.1 Fabrication of Ceramic Nanoparticles by Disaggregation

Technically, most ceramic nanoparticles can be obtained by milling of larger particles. Suitable disaggregation techniques include the application of ultrasound and milling techniques such as vibro milling and ball milling. Bioapatites with chemical compositions similar to HA can be obtained from different biological sources, such as corals, ivory, teeth, and bone (Roy and Linnehan 1974; Dorozhkin 2010a). Nanocrystalline bovine bioapatite, for example, has been obtained as follows (Ruksudjarlt et al. 2008): bovine bone samples were harvested, carefully cleaned, and boiled several times in distilled water. The deproteinized material was dried at 200°C and afterwards calcined at 800°C. The product was crushed into small pieces, ball-milled for 24 h, and finally vibro-milled to obtain the nanoscaled product.

Another method that includes a disaggregation step is the *controlled hydrolysis* of micron-sized ceramics. For the fabrication of nano-sized HA powder, for example, dicalcium phosphate and calcium carbonate were mixed to achieve a Ca/P ratio of 1.67, poured into a solution of NaOH and stirred at 75°C for 1 h (Shih et al. 2004). The hydrolysis reaction was stopped via cooling with ice water. The aggregates were filtered and washed, and the powder was dried at 60°C. The particles were processed through annealing at 600°C, 800°C, or 1000°C for 4 h.

Through the use of plasma flames, raw ceramic particles can be melted, partially melted or even evaporated. The melted or vaporized material can be quenched or condensed into ultrafine powders by subsequent cooling. Using the *radio frequency thermal plasma method*, nano-sized (10–100 nm) HA powders have been produced (Xu et al. 2004). Coarse HA particles were pre-heated (1000°C) and entered into a plasma torch. The vaporized material was condensed into ultrafine powders and collected.

3.3.2 Controlled Preparation of Ceramic Nanoparticles

Most processes described in the following section are so-called *solution-mediated* fabrication processes with the exception of *mechanochemical processing*. As a common characteristic, nucleation and crystal growth are key steps of these processes. The main parameters that initiate and control these steps differ among fabrication methods. While in *precipitation techniques* solution properties, for example, solution pH and solvent composition, are changed to induce nucleation, a thermal induction is characteristic for *microwave-assisted techniques*. In *mechanochemical processes*, the activation energy for nucleation and growth is delivered by pressure in a mill (Riman et al. 2002).

Controlled precipitation is a straightforward method to prepare nanometer-sized particles from solutions of the constituting ions. Using this method, different ceramic nanoparticles (calcium phosphates, zirconium dioxide, iron oxides, and bioactive glass) have been fabricated. With regard to the different calcium phosphate minerals, the desired Ca/P ratios can be easily adjusted to yield a specific product. Precipitation methods are quite inexpensive and versatile. In addition, key product parameters such as size of the precipitated grains can be controlled. The precipitation can be initiated through different processes, such as pH or temperature adjustment or the addition of a non-solvent. Other protocols involve supersaturated solutions. A classical protocol to obtain nHA starts from individual aqueous solutions of calcium nitrate and diammonium phosphate at pH 11–12 (Wang and Shaw 2009a). Through dropwise addition of the phosphate solution to the calcium solution under stirring, a milky dispersion is obtained. After centrifugation, the slurry is washed and the precipitate sedimented for 6–10 h. The isolated intermediate is dried (90°C), ball-milled, and calcined (300°C) to yield HA nanorods. In comparison to dense HA bodies assembled from micron-sized HA, bodies sintered from the densified nanorods showed enhanced hardness and toughness. For the preparation of bioactive glass nanoceramics, calcium nitrate and tetraethoxysilane were dispersed in a water–ethanol mixture, the pH adjusted with citric acid, dropped into a solution of ammonium dibasic phosphate kept at pH 11, and the precipitated nanoparticles (20–40 nm) were obtained after centrifugation, washing and calcination (Hong et al. 2008, 2009). In order to further reduce the size of particles obtained by a classical precipitation method, a dispersion of the particles can be subjected to a high-intensity ultrasonic field (Jevtic et al. 2009).

Phase-Controlled Methods: Progressing from controlled precipitation out of single-phase systems, a variety of methods strive for better control of size and structure of the prepared nanocrystals through the use of multiple-phase systems during precipitation or growth. Using water-in-oil emulsion, for example, small nano- and micron-sized aqueous, constrained reaction environments for nanocrystal synthesis can be prepared (Dorozhkin 2010b). *Emulsion-based methods* are published for organic composite materials and also for monolithic ceramics. Key parameters during emulsion processes are droplet size of the dispersed phase, temperature, pH of aqueous phase, and mechanical stirring. For the synthesis of nano-sized calcium orthophosphate particles, a water-in-oil microemulsion was prepared from an aqueous calcium hydroxide phase and isooctane using sodium dioctylsulfosuccinate as surfactant (Phillips et al. 2003). After the addition of orthophosphoric acid, the pH was adjusted to 10.5 initiating the precipitation. The nanopowder was obtained after filtration, washing, freeze-drying, and calcination.

Through the combination of a microemulsion-controlled crystal growth and a solid template, lightweight hollow porous shells of calcium carbonate have been fabricated (Walsh and Mann 1995). The process started from a supersaturated aqueous calcium carbonate phase to which magnesium chloride was added to initiate crystal growth. This phase was emulsified in tetradecane as oil phase by means of a cationic surfactant. Polystyrene microspheres were covered with a thin film of the resulting microemulsion and washed with hexane to remove the oil phase and the surfactant. Finally, the polystyrene templates were removed using acetone–ethanol mixtures yielding hollow microshells with a wall thickness of approximately 125 nm.

A method called *liquid–solid–solution* involves a triphasic system for the controlled growth of HA nanorods (Wang et al. 2006a; Wang and Li 2007). The system was assembled from a liquid phase containing linoleic acid in ethanol, sodium linoleate as the solid phase, and an aqueous solution of calcium nitrate (solution phase). Through an ion-exchange process, calcium ions interact with the solid linoleate phase. After the addition of sodium phosphate and thermal treatment in an autoclave, calcium phosphate nanostructures start growing. Along with the reaction and ion exchange process, the linoleic acid can be released and absorbed on the *in situ*–generated HA nanorods with the alkyl chains left outside. These hydrophobic nanocrystals will be separated from the aqueous solution spontaneously and can be collected from the vessel (Figure 3.1A).

The *solvothermal method* for the fabrication of HA nanowires is a process that injects aqueous stock solutions of calcium and phosphate ions into cyclohexane and uses a cationic surfactant as stabilizer (Wang et al. 2006b). As two immiscible phases are combined, this process can strictly be regarded as an emulsion-based technique. For the solvothermal process during which the nanowires are formed, the disperse system is transferred into sealed Teflon containers and subjected to 120°C for 12 h.

Microwave radiation, as applied during hydrothermal *microwave synthesis*, has several advantages over classical chemical processes, such as precipitation, for the fabrication of nanopowders including easy reproducibility, small particle size, narrow particle distribution, high purity, and yield due to fast homogenous nucleation (Siddharthan et al. 2006; Kalita and Verma 2010). Nano-sized HA powders were synthesized from phosphoric acid and calcium hydroxide in a closed-vessel microwave device at 300°C for 30 min. Product characteristics could be controlled by the applied microwave power and Ca/P ratio. At low power and a Ca/P ratio of 1.57, mixed calcium phosphate compounds such as calcium hydroxide, calcium hydrogen phosphate, and HA were yielded, while monophase HA was obtained at higher power and a Ca/P ratio of 1.67 (Han et al. 2006). Bioactive HA nanopowder (5–30 nm) was synthesized from a suspension of calcium nitrate and ethylenediaminetetraacetic acid that was mixed with a sodium hydrogen phosphate solution and adjusted to pH 9 (Kalita and Verma 2010). Following irradiation in a microwave oven, the suspension was filtered and precipitated, and the filtrate was washed and dried in a muffle furnace (200°C, 4 h). Crystalline nHA powder was finally obtained after disaggregation of the dried product.

Hydrothermal methods during which the necessary heat energy is introduced in ways other than microwave irradiation have also been described (Riman et al. 2002). For the fabrication of nHA powders, aqueous solutions of calcium nitrate and diammonium hydrogen phosphate were prepared. The solutions were mixed and different pH values were adjusted for the resulting slurries. The samples were introduced into hydrothermal reactors for 24 h at 50–200°C. The synthesized powders were washed and dried. Depending on the processing parameters stirring and reaction pH, particles with diameters of 20–40 nm as well as nano-sized needles have been fabricated.

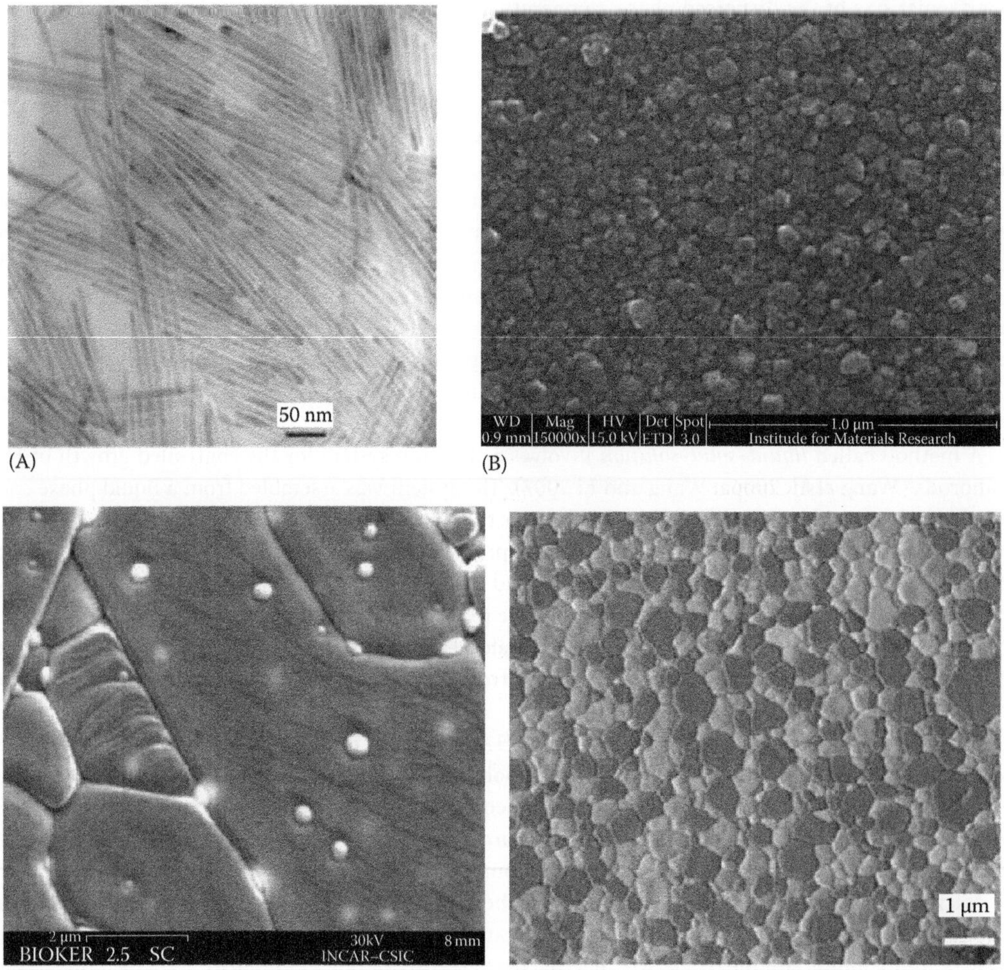

(A) (B) (C) (D)

FIGURE 3.1 Different types of ceramic nanobiomaterials. (A) TEM image of HA nanorods as obtained by liquid–solid solution synthesis. (Wang, X., Zhuang, J., Peng, Q., and Li, Y.: Liquid–solid–solution synthesis of biomedical hydroxyapatite nanorods. *Adv. Mater.* 2006. 18(15). 2031–2034. Copyright Wiley-VCH Verlag GmbH & Co. KGaA. Reproduced with permission.) (B) SEM image of nanocrystalline diamond deposited on a silicon wafer by microwave plasma-enhanced CVD. (Reprinted from *Chem. Phys. Lett.*, 445(4–6), Williams, O.A. et al., Enhanced diamond nucleation on monodispersed nanocrystalline diamond, 255–258, Copyright (2007), with permission from Elsevier.) (C) SEM image of a nano/micro-type composite (intra/inter-type) showing nano-sized zirconia particles embedded in an alumina matrix. (Reprinted with permission from [Chevalier, J. et al., Nanostructured ceramic oxides with a slow crack growth resistance close to covalent materials, *Nano Lett.* 5 (7), 1297–1301]. Copyright [2005] American Chemical Society.) (D) SEM image of a nano/nano-type composite of 30% (w/w) Al_2O_3 and 70% TZ-3Y (ZrO_2 + 3 mol% Y_2O_3). The sample was polished and thermally etched. (Reprinted from *Acta Biomater.*, 6(2), Nevarez-Rascon, A. et al., Al_2O_3(w)-Al_2O_3(n)-ZrO_2 (TZ-3Y)$_n$ multi-scale nanocomposite: An alternative for different dental applications? 563–570, Copyright (2010), with permission from Elsevier.)

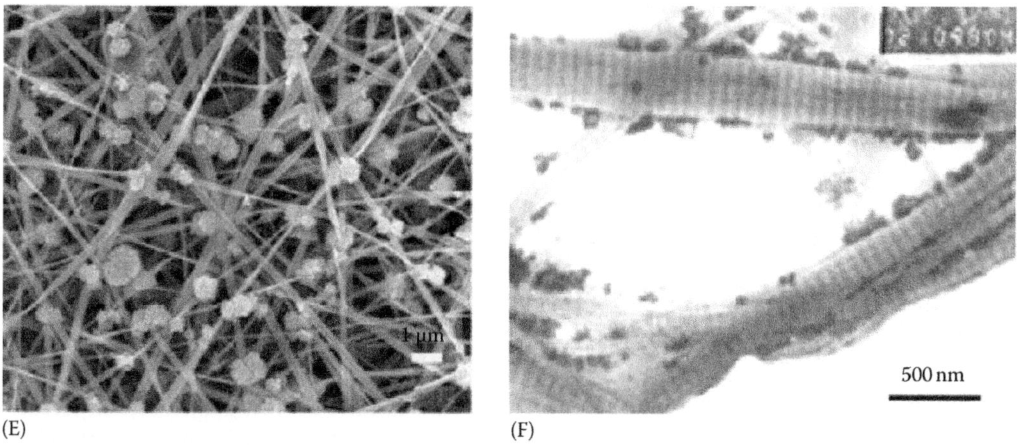

FIGURE 3.1 (continued) (E) SEM image of mineralized bioactive glass nanofibers after 12 h in SBF. (Xia, W. et al., *Nanotechnology*, 18(13), 135601, 2007. Reprinted with permission of IOP Publishing Ltd., U.K.) (F) TEM image of collagen fibers and apatite nanocrystals in an *in situ*-synthesized nanocomposite. (Reprinted from *Mater. Lett.*, 58(27–28), Lin, X. et al., *In situ* synthesis of bone-like apatite/collagen nano-composite at low temperature, 3569–3572, Copyright (2004), with permission from Elsevier.)

Sol–Gel Processes: Compared to classical precipitation techniques, sol–gel combustion is considered advantageous due to its simplicity and shorter fabrication times (Wang and Shaw 2009b). For the fabrication of calcium phosphate nanoparticles, the sol–gel method allows for a molecular-level mixing of the calcium and phosphorous precursors. This way, chemical homogeneity and purity are improved, and the formation temperature of resulting calcium phosphates is decreased in comparison to conventional methods (Han et al. 2004). In a classical sol–gel combustion process to fabricate calcium phosphates, a solution of calcium nitrate and ethyl phosphate was concentrated to form a polymeric gel, which self-combusted into a calcium phosphate precursor powder upon further heating in a hot furnace (Varma et al. 1998). Calcination on heat treatment at 1000°C resulted in β-TCP, HA, or a mixture of the two phases depending on the Ca/P ratio in the gel. Highly pure HA nanopowders were fabricated from a mixture of calcium nitrate and triethyl phosphite in ethanol–water (Wang and Shaw 2009b). The mixture was aged under alkaline conditions and finally gelled by generation of (-Ca-O-P-) oligomers upon heating. A nanopowder with a particle size of approximately 80 nm is obtained via a combustion process after heating the gel with 30°C/min to 350°C for 1 h in a furnace. The process can be traced via TGA measurements and the combustion process was shown to be heating rate dependent. Alternatively, a non-alkoxide sol–gel method, citric acid sol–gel combustion, is available for the preparation of nanocrystalline inorganic powders including HA (Han et al. 2004). The precursors, such as calcium nitrate, citric acid, and diammonium hydrogen phosphate, were dissolved in water and mixed. The solution was acidified and heated under stirring to concentrate the solution and finally form a gel. The gel was dried and calcined to finally yield nHA.

Mechanochemical processing methods have been developed for the fabrication of nano-sized particulates of calcium phosphates, zirconia, and alumina (Yeong et al. 2001; Riman et al. 2002; Dorozhkin 2010b). Typically, lower-energy ball mills or high-energy vibratory and planetary mills are used to introduce the mechanical forces at room temperature. Such processes are relatively straightforward and inexpensive. In early studies that focused on the fabrication of nHA, calcium deficient compositions with low crystallinity were often obtained. Improved routes managed to trigger the formation of single-phase highly crystalline HA from a dry powder mixture of calcium oxide and anhydrous calcium hydrogen phosphate by mechanical activation for more than 20 h. The resulting HA powder exhibits an average particle size of approximately 25 nm (Yeong et al. 2001).

In a process called *mechanical alloying*, a fluorinated HA nanopowder was prepared from a mixture of calcium hydroxide, phosphorous pentoxide, and calcium fluoride (Fathi and Zahrani 2009). The mixture was mechanically alloyed using a high energy planetary ball mill equipped with a zirconia vial and zirconia balls. After 15 h of milling, the final particle size ranged between 35 and 65 nm.

Wet mechanochemical synthesis, a special mechanochemical process and often referred to as mechanochemical-hydrothermal synthesis, introduces an aqueous reaction medium to the milling chamber. For the fabrication of nHA powders with this method, aqueous slurries containing calcium hydroxide, calcium or sodium carbonate, and diammonium hydrogen phosphate were prepared and processed in a multi-ring media mill at 25–35°C for several hours (Riman et al. 2002).

3.4 Applications of Ceramic Nanobiomaterials

Bioceramics, in general, have tremendously evolved over the last five decades of biomaterials research and have become key or even first choice materials in numerous applications including artificial replacements of hips, knees, teeth, tendons, and ligaments. Bioceramics are also used for the repair of periodontal defects, maxillofacial reconstruction, bone augmentation, bone plates, and in spinal fusion (Dorozhkin 2010a,b). Nanoscale bioceramics, in particular, are emerging, because such materials hold promise to overcome long-standing problems associated with these established materials. As a result of the higher surface area of nanostructured materials, the sintering processes are more effective (Dorozhkin 2010b). It has also been hypothesized that nanostructured ceramics have increased bioactivity due to their higher roughness. With a series of dense HA bodies fabricated from well-defined micron- and nano-sized grains, simultaneous improvements in hardness and toughness have been observed in bodies sintered from nHA grains (Wang and Shaw 2009a).

Depending on whether the ceramic nanostructures alone represent the biomaterial or are part of a larger micro- or macrostructure, one can divide ceramic nanobiomaterials into three main groups: (I) nano-sized monolithic ceramic particles, which are used as particulate nanobiomaterial in form of a powder or dispersion, (II) nano-textured biomaterial, in which ceramic nanoparticles are deposited on the surface of a biomaterial construct of larger dimensions, (III) nanocomposites, in which nanoparticulate ceramics are dispersed in the bulk of another biomaterial.

3.4.1 Monolithic Ceramic Nanoparticles

3.4.1.1 Delivery of Therapeutic Agents

Particulate nanoceramics have applications in diagnostics and the delivery of therapeutic agents. Depending on the desired application, ceramic nanoparticles bear advantages over nanostructures made from other materials. Basic properties of the ceramic nanoparticles including size distribution, shape, morphology, and porosity can be directly controlled by the fabrication process (Sahoo and Labhasetwar 2003; Koo et al. 2005). Additionally, the nano-sized ceramic particles are rigid, chemically and physically inert, interact with organic molecules, and are susceptible to specific surface modification. Compared to other materials, ceramic nanoparticles show no pH-induced swelling or change in porosity, are resistant to microbial growth, are stable in physiological aqueous systems, and have the ability to entrap and stabilize drugs. There are some biodegradable ceramics, but most are bioinert, which can be considered critical for delivery applications due to possible accumulation of the particles in the body (Kriven et al. 2004; Medina et al. 2007). Nanoparticulate formulations in general are of special interest for delivery applications because of their submicron size, which is considered a prerequisite for targeting strategies and allows such particles to access almost all areas of the body. Especially for passive drug targeting applications, where access to specific organs or tissues is controlled by particle size, the ability to precisely control nanoparticle size during fabrication and application is essential.

Active drug targeting, which aims at delivering highly potent drugs to specific sites or cell types in order to minimize side effects, requires modification of the nanoparticle surfaces with site-specific structures, such as integrin or other receptor ligands and antibodies (Roy et al. 2003; Torchilin 2006). In combination with the stabilizing effects that some ceramic materials have on sensitive therapeutic molecules, ceramic nanoparticles may open new therapeutic possibilities. To exemplify the spectrum of options ceramic nanoparticles offer for drug delivery strategies, selected applications are described below. The ability to entrap and stabilize proteins has been shown with enzymes (Jain et al. 1998). With 80% efficacy, a peroxidase was encapsulated in mono-disperse hydrated silica nanoparticles by a reverse micelle preparation technique. The enzyme did practically not leach out of the particles over more than a month and remained active showing normal substrate conversion kinetics. It is assumed that substrate can diffuse into the ceramic particle to be converted by the entrapped enzyme. This ability to effectively entrap enzymes and maintain enzyme activity is attractive for therapeutic *in vivo* application, because the risk of allergic or proteolytic reactions of these enzymes is drastically reduced due to their practically zero leachability. Advancing the outlined strategy, a water-insoluble photosensitizing anticancer drug was formulated in silica-based ceramic nanoparticles, passively targeted to the region of therapeutic interest and subsequently, locally activated by irradiation with light. Thereby, the drug remained within the formulation, but the photochemically generated cytotoxic singlet oxygen diffused through the pores of the ceramic nano-matrix and into the tumor tissue (Roy et al. 2003). Aquasomes are self-assembled nanoparticulate delivery vehicles composed of a ceramic core for stability, typically composed of a calcium phosphate or diamond, grafted with a hydrophilic oligomeric polyhydroxy coating, for example, oligosaccharides, to which the therapeutic substances are adsorbed (Cherian et al. 2000; Goyal et al. 2008; Umashankar et al. 2010). Aquasomes are of special interest with regard to peptides and protein delivery, because the hydrophilic glassy coating stabilizes their structure and activity through a water replacement effect (Umashankar et al. 2010). For an insulin delivery formulation, for example, the aquasome formulation was able to reduce the percentage blood glucose more effectively and prolonged compared to a standard plain insulin solution (Cherian et al. 2000). Further applications include the oral delivery of enzymes, the use as hemoglobin-loaded oxygen carriers, and antigen delivery vehicles. In the latter application, benefit and safety of these novel systems remain to be finally determined (Goyal et al. 2008). There is evidence, however, that aquasome formulations have shown better adjuvanticity and effective antigen presentation, because the structural integrity of the proteins is preserved. Applications that involve bioresorbable nanoceramics include gene delivery (Link 2000; Kriven et al. 2004; Ladewig et al. 2010). The ceramic nanoparticles used in such studies are comprised of layered double hydroxides (LDHs), the so-called anionic clays, which consist of cationic brucite ($MgOH_2$)-like layers and exchangeable interlayer anions. By anion exchange, DNA and functional anionic biomolecules can be encapsulated in the inorganic particles. With regard to gene delivery, the interactions of DNA with the LDH nanostructures, similar to other non-viral vectors, neutralize the charges of the DNA and facilitate internalization into cells. Once internalized by endocytosis, the slightly acidic pH in the lysosomes dissolves hydroxide layers of the LDH, and interlayer DNA is replaced by other anions in the cell electrolyte and finally released into the cytosol. A recent study showed that DNA-LDH complexes yield considerable transfection efficiencies when used on adherent cell lines (Ladewig et al. 2010). The transfection of cells in suspension culture, however, was unsuccessful. Transmission electron microscopy (TEM) investigations revealed that, in contrast to smaller anions, plasmid DNA did not become intercalated in the LDHs but was wrapped around the nanoparticles.

3.4.1.2 Diagnostics

For diagnostic application, magnetic iron oxide-based nanoparticles are widely investigated and discussed elsewhere in this handbook. One major application is their use as contrast agents for magnetic resonance imaging (Weissleder et al. 1990; Gupta and Gupta 2005). Magnetic nanoparticles have also been combined with fluorescent moieties (Corr et al. 2008) and quantum dots to obtain magnetic luminescent nanocomposites (Hong et al. 2004).

3.4.1.3 Other Applications of Ceramic Nanoparticles

Calcium phosphate nanoparticles, especially HA and β-TCP, are part of bone and dental cement formulations or of powder mixtures that are processed into tissue engineering scaffolds (Chris Arts et al. 2006; Xu et al. 2008; Dorozhkin 2010b). Due to the favorable interactions of these calcium phosphates with drugs and proteins (Habraken et al. 2007), such matrices can serve as drug delivery systems for a variety of remedies such as antibiotics, antitumor, and anti-inflammatory drugs or growth factors.

3.4.2 Ceramic Nanocoatings

Orthopedic and joint implants have tremendously improved the quality of life for countless individuals. In order to achieve clinical success, implanted materials must form a stable interface with surrounding tissue as well as being compatible with the mechanical properties of natural tissue (Campbell 2003). To date, metals—especially titanium—are the biomaterials of choice for such applications due to their mechanical properties, but the interfacial bonding between the metallic surface and the surrounding bone is poor. Poor interfacial bonding leads to the formation of a non-adherent, fibrous tissue layer, and upon further loosening and movement at the implant-tissue interface, the implant will finally fail. A promising approach to address this problem has been the use of ceramic coatings applied to implant surfaces. Nanoceramic coatings can be achieved by a variety of methods. In the following section, a selection will be discussed with the focus on nanostructured HA coatings (Paital and Dahotre 2009).

3.4.2.1 Coating Methods for Nanostructured HA

HA is probably one of the most applied coatings on implant and prosthesis surfaces, because it resembles the natural inorganic component of bone. Due to its low toughness and brittleness, HA is unsuitable as a load-bearing substrate itself. Coatings of HA on load-bearing substrates, however, especially if they are thin enough, can improve osseointegration and reduce fibrous capsule formation. HA coatings are broadly used in dentistry and orthopedic applications to speed up formation of bone around the device and stabilize the implant in its position (Campbell 2003). As it has been shown for other nano-sized materials, nHA allows for increased osteoblast adhesion and differentiation relative to micrometer-structured surfaces (Catledge et al. 2002). Surface coatings deposited by these processes need to meet the guidelines set by the U.S. Food and Drug Administration (FDA) and the International Organization for Standardization (ISO) (Table 3.2) (Paital and Dahotre 2009). A variety of surface coating processes are known for HA, such as plasma spray deposition, ion beam–assisted deposition (IBAD), electrophoretic deposition (EPD), chemical vapor deposition (CVD), microarc oxidation, magnetron sputtering, sol–gel-derived coatings, and biomineralization (Paital and Dahotre 2009). A selection of these methods is outlined in the following section.

TABLE 3.2 Specifications for HA Coatings by the U.S. Food and Drug Administration (FDA)

Property	Specification
Crystallinity	≥62%
Phase purity	≥95%
Ca/P atomic ratio	1.67–1.76
Tensile strength	>50.8 MPa
Shear strength	>22 MPa

Source: Paital, S.R. and Dahotre, N.B., *Mater. Sci. Eng. R Rep.*, 66(1–3), 1, 2009.

3.4.2.1.1 Plasma Spraying

The most widely commercially used technique for surface coating of implants with HA is plasma spraying (Tang et al. 2010). Due to the extremely high temperature in the plasma flame, almost any coating feedstock material melts and can be coated to a substrate with this method. A typical setup for plasma spraying is given in Figure 3.2A. A plasma gun comprises a chamber with an electrode as cathode and a nozzle as anode. When a plasma-forming gas flows through the chamber, current power is applied to the cathode arching the nozzle (anode) and thereby stripping the gas molecules of their electrons to form a plasma plume. The plasma-forming gas usually consists of argon, helium, nitrogen, or hydrogen. Among these, argon can be easily ionized and provides a stable arc at a low operating voltage (Paital and Dahotre 2009). As the unstable plasma ions recombine back to the gaseous state, a huge amount of thermal energy is released, providing temperatures exceeding 30,000 K in the hottest areas of the

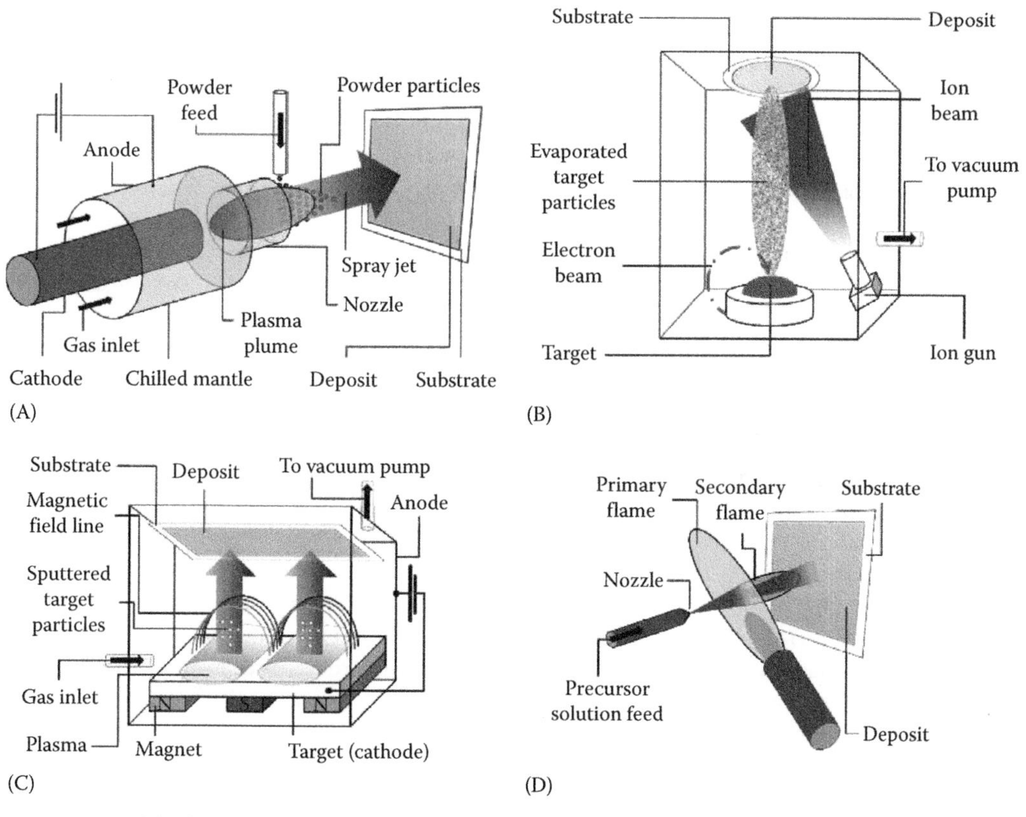

FIGURE 3.2 (A) Plasma spraying: Argon gas is ionized forming a plasma plume in an electric arc between the cathode in the center of the plasma gun and the anodic nozzle. Upon recombination of the plasma ions to the gaseous state, a tremendous amount of energy is released. Into the hot gaseous plume HA powder is fed and propelled toward the substrate. (B) IBAD: The coating material (target) is located in an electron-beam-heated vaporizer. Vaporized target material is deposited on the substrate. Simultaneously, substrate and coating experience bombardment with a high energy ion beam. The process takes place under vacuum conditions. (C) Magnetron sputtering deposition: The argon gas is ionized by an electrical field. The target resides on a series of magnets. The shape of the magnetic field entails circulation of primary and secondary ions close to the target surface due to the Lorentz force and prolongs their flight path. Therefore, argon gas is ionized with high efficiency forming plasma rods. Plasma ions impact into the cathodic target sputtering target molecules that are deposited on the substrate. (D) FACVD: An organic precursor solution is atomized in a nozzle and carried into an oxidizing gas where combustion and pyrolysis take place. By the organic solvent, a secondary flame is formed that contributes to substrate heating and allows for diffusion and good coating adherence to the substrate.

plume. Into the hot gas plume, HA powder is injected, melted, and propelled toward the substrate to be coated, which remains cool. Depending on the particle size of the fed HA powder, HA is coated in a melted or softened particulate form onto the substrate surface. Different physical conditions may coexist especially in larger particles, with an evaporating layer of phosphorous pentoxide and the formation of calcium oxide at the outer layer of the particle resulting in an enrichment of calcium in the coating. In the same particle, a molten layer beneath may solidify as an amorphous coating when deposited due to rapid cooling on the substrate. Furthermore, α-TCP and tetracalcium phosphate (TP) may form, while at lower temperatures, HA remains intact (Dyshlovenko et al. 2004). Therefore, crystalline HA coatings fabricated by plasma spraying contain varying amounts of amorphous HA and impurities of calcium oxide, TCP, and TP (Tang et al. 2010).

Plasma spraying is considered the most efficient and economical method for surface coating with HA for orthopedics and dental implants, but it suffers from low adherence to the substrate, low fracture toughness, and considerable thick coating films of several hundred micrometer. A high thickness of a coating (>100 μm) bears the risk of fatigue failure (Paital and Dahotre 2009).

In order to improve adherence of the plasma-sprayed coatings to the substrate, a mixture of a titanium alloy and HA was deposited on the surface of a titanium alloy implant (Khor et al. 2000). Another recommendation is to pre-coat the substrate with a titanium layer to roughen the surface and to use high plasma power and suitable gas mixtures to ensure melting of HA particles in addition to a subsequent heat treatment at 700°C for 1 h to increase the crystallinity and improve the *in vivo* performance (Tsui et al. 1998).

While low fracture toughness is an inherent shortcoming of HA, a reduction in coating thickness is suitable to positively affect mechanical stability of the coating-implant interface. Therefore, alternative methods focus on thinner coatings (Paital and Dahotre 2009).

3.4.2.1.2 Ion Beam–Assisted Deposition

With this technique, a target material is evaporated usually employing an electron beam or resistive heating and deposited onto the substrate. A reactive ion beam that focuses on the substrate enables surface chemical reaction with the substrate materials and within the coated film (Figure 3.2B). Bond strength between coating and substrate is enhanced by atomic bonding, densification of the coating, and lower thermal stresses as compared to plasma spraying (Bai et al. 2009). Therefore, IBAD is suitable to improve adherence of HA to the substrate surface (Paital and Dahotre 2009).

IBAD coating results in coating thickness in the nanometer range. Coatings show improved mechanical properties meaning good adherence and low tendency for delamination (Rabiei et al. 2006). Crystallinity and phase composition of calcium phosphate are controlled by energy input. This is realized by either heat treatment of the substrate during the coating process (Rabiei et al. 2006; Bai et al. 2009) or by energy input of the ion beam (Hamdi and Ide-Ektessabi 2003). Additionally, the presence of water vapor is discussed to improve phase purity (Bai et al. 2009).

Layering in the structure of the coating can be controlled by ion beam current, control of substrate temperature during coating and post-heat treatment. It has been shown that by manipulating the substrate temperature during the deposition process to avoid time-consuming post-heat treatment, nanocrystalline calcium phosphate forms the layer closest to the substrate surface, and crystallinity decreased with increasing distance to the substrate interface (Bai et al. 2009). This was considered advantageous, because the amorphous layer may dissolve in contact to bone and provide ions for osseointegration while the remaining nanocrystalline layer allows for optimal interaction with adhering osteoblasts.

3.4.2.1.3 Electrophoretic Deposition

An electric field is applied to a conductive substrate in a nonaqueous suspension of charged particles. Depending on the charge of the particles, the substrate acts as anode or cathode, on which the oppositely charged particles are deposited, and the process is called anodic or cathodic eletrophoretic deposition, respectively (Paital and Dahotre 2009). Advantages of this method are the ability to coat irregular surfaces homogeneously in variable thickness and that the coatings adhere strongly. Complex compositions and layered coatings can be easily realized with this comparatively simple setup. HA coatings prepared by EPD typically undergo a sintering step subsequent to EPD. HA coatings densify with increasing sintering temperatures, which may improve the strength of the coating, but on the other hand decrease the osteoconductive potential of the coating that is known to correlate with microporosity (Wang et al. 2002a).

3.4.2.1.4 *Magnetron Sputtering Deposition*

Magnetron sputtering deposition is a physical vapor deposition method to fabricate coatings in a vacuum environment similar to ion beam sputtering (Figure 3.2C) (Paital and Dahotre 2009). Sputtering involves bombardment of a target with high-energy particles to eject atoms or molecules from the target surface. In order to preserve sufficient ion energy for bombardment and to allow for deposition of the ejected molecules on a substrate, vacuum is required. A specially shaped magnetic field is applied to the sputtering target to enhance the effectiveness of the high-energy particle bombardment. HA films generated by magnetron sputtering are comparably thin and tightly bound on the substrates. Crystallinity depends on the substrate temperature during the deposition process or thereafter. In a study that describes the fabrication of HA-coated titanium alloys, it was found that heat treatment (>300°C) either during the sputtering process or as posttreatment was necessary to create crystalline HA layers on the surface (Nelea et al. 2003).

3.4.2.1.5 *Sol–Gel-Derived Coatings*

Similar to sol–gel processes described for nanoparticle fabrication, sol–gel-derived coatings rely on the formation of a sol phase consisting of appropriate calcium and phosphate containing precursor solutions. Upon sol aging, ultrafine HA particles precipitate *in vitro* and start to gel the suspension. The aged sol is homogeneously subjected to a substrate, for example, by dip coating (Nguyen et al. 2004). The sol layer on the substrate finally forms the gel upon drying and becomes annealed onto the substrate. For example, an HA sol was prepared by mixing triethyl phosphate and calcium nitrate in an ethanol–water mixture (Li et al. 2005). The solution was stirred for 1 h and subsequently aged at ambient conditions for 5 days. Substrates were subsequently spin coated, dried at 80°C for 2 h, and finally heat treated at 550°C for 2 h. Sol aging is accelerated by the addition of ammonium hydroxide that works as an acceptor for protons released during HA formation keeping polymerization in progress during aging (Kim et al. 2005). The resulting coatings are thin and show good adherence on metallic surfaces. This technique allows for the incorporation of a variety of organic and inorganic compounds (Ganguli 1993). Processes normally deal with substrate temperatures between 200°C and 600°C for sintering and are comparatively simple, economic, and effective. Depending on the concentration of the suspended particles in the sol and its viscosity, porous structures of a substrate may be preserved or only partly covered (Nguyen et al. 2004; Li et al. 2005).

3.4.2.1.6 *Chemical Vapor Deposition (CVD)*

This method uses a nozzle to atomize a solution of precursor salts that react in the gaseous phase under energy input that is provided by heat, light, or plasma (Choy 2003). Depending on gas temperature, either intermediate species are decomposed forming homogeneous solid products that are deposited onto the substrate or are adsorbed onto the heated substrate and may react heterogeneously with the components of the substrate and form deposits. Higher temperatures that favor homogeneous deposits result in only low adherence to substrates and provide a method for nanoparticle generation. Heterogeneous deposits, on the other hand, diffuse on the substrate surface and form crystallization centers on the substrate. For dense coatings, reaction conditions are tailored to favor heterogeneous reactions, whereas a combination of homogeneous and heterogeneous reactions results in porous coatings. CVD has been described as a non-line-of-sight deposition method that can be used for the deposition of homogenous coatings on complex structure surfaces.

In flame-assisted CVD (FACVD), a subtype of CVD, precursors are dissolved in an organic solvent (Choy 2003). The solution is atomized through a nozzle and carried into a flame by an oxidizing gas. The precursors undergo combustion and pyrolysis in the flame and form the deposit. In addition, a secondary flame is formed by the burning organic solvent that heats the substrate during the coating process and promotes diffusion processes within the forming coating (Figure 3.2D) (Trommer et al. 2007). Substrate temperature was maintained at 500°C during the process. Calcium acetate and ammonium phosphate in combination with nitric acid served as precursors and ethanol as organic solvent. While acetate and ammonium ions are decomposed in the flame, calcium ions are free to react with phosphate. The formation of desired coating components can be controlled through the initial ratio of calcium to phosphate precursors. With this process, for example, HA has been deposited on stainless steel.

3.4.2.1.7 Biomineralization

Biomineralization is a coating process that is induced by soaking a substrate in simulated body fluid (SBF) at physiological temperature and pH. In contrast to many other methods, it is possible to homogeneously coat porous materials with this technique (Habibovic et al. 2002). Calcium phosphate coatings precipitate on substrates by time without the involvement of any cellular activity (Kokubo 1998; Li and Ducheyne 1998). Density and crystallinity of calcium phosphate coatings, which determine the rate of degradation, can be controlled by pH, volume and ionic strength of SBF (Qu and Wei 2008a,b). The original process takes 7–28 days to coat a substrate with a reasonable thickness (5–30 μm) (Li and Ducheyne 1998; Yu et al. 2009). Habibovic et al. introduced a method for accelerated HA mineralization by increasing the ionic concentration of SBF through a transient decrease in pH caused by carbon dioxide addition (Barrere et al. 2002; Habibovic et al. 2002). Upon degassing of the solution, the pH slowly increased and induced rapid precipitation of HA providing coatings within 24 h. The process involved two steps. At first, a thin amorphous layer of calcium phosphate containing multiple nucleation seeds is introduced. In a second step, HA crystals are grown from the crystallization seeds.

Within the process of precipitation, bioactive molecules, such as growth factors, can be co-precipitated within calcium phosphate crystals onto the surface of metallic implant materials (Yu et al. 2009). In a study that used bovine serum albumin (BSA) as a model protein, BSA incorporation into octocalcium phosphate crystals was shown to be highly efficient, if substrate surface area relative to SBF volume was increased.

3.4.2.2 Nanoceramic Coatings Other Than HA

A major challenge for prostheses at articulate surfaces is the necessity for low friction and high wear resistance. It has been shown that wear resistance of ceramics, such as zirconia (Kumar et al. 1991) and diamond coatings, is considerably better than that of metallic surfaces. Especially diamond coatings on metal surfaces have been intensely investigated during the last decade, because such coatings showed low friction and high wear resistance in addition to chemical resistance, high fracture toughness, and bonding strength (Drory et al. 1991; Catledge et al. 2002; Yang et al. 2009a).

3.4.2.2.1 Nanostructured Diamond Coating

Nanostructured diamond coatings are usually generated by CVD or variations of this technique (Catledge et al. 2002), such as microwave-assisted CVD (Yang et al. 2009a,b). In an example of the latter method, substrates were pre-coated with a dispersion of diamond nanopowder in methanol. The precoated substrate was treated with a mixture of methane, hydrogen, and argon gas at high pressure and 800°C for 2 h in order to generate adhesion to the substrate. Diamond, a crystal of tetrahedrally bonded carbon atoms (sp^3), is believed to develop from C_2 dimers that result from collision of acetylene with argon. Hydrogen gas acts as inhibitor for secondary nucleation, and leads to crystal growth and allows for grain size control (Gruen 1999). Surface chemistry and contact angle can be further modified by the addition of oxygen/helium or pure hydrogen plasma during sample cooling (Clem et al. 2008). Figure 3.1B depicts the surface structure of a nanocrystalline diamond film deposited by microwave plasma-enhanced CVD (Williams et al. 2007).

Cell adhesion to nanocrystalline diamond coatings was found to depend on surface chemistry and surface topography, because both parameters likely influence the adsorption of proteins to such surfaces (Yang et al. 2009a,b). A higher number of osteoblasts adhered to nanocrystalline diamond consisting of small crystallites that formed aggregates with mean grain sizes of 30–100 nm compared to submicron crystalline diamond with grain sizes of 100–600 nm. Cells were also more spread and showed higher proliferation and mineralization on nanocrystalline diamond (Yang et al. 2009b). In this context, a correlation between contact angles and surface roughness was considered a relevant factor. Surface roughness, a parameter defined as ratio between geometrical area determined by AFM and projected area, is lower for nano- than for submicron-crystalline diamond. As adsorption and bioactivity of adhesion proteins such as fibronectin and vitronectin are lower on substrates with higher contact angles, the authors discuss this correlation as a factor that contributes to improved osteoblast adhesion (Yang et al. 2008). Moreover, submicron-sized diamond crystals may offer a limited number of adhesion points for osteoblasts. In another study, reduced osteoblast adhesion was shown on surfaces with adhesion points in distances larger than 73 nm (Arnold et al. 2004).

3.4.2.2.2 Alumina Coatings

Alumina offers excellent biocompatibility, strength, and fracture resistance and has therefore been used as prosthesis material for several decades (Vallet-Regi 2001). In addition, alumina shows very high wear resistance at articulate surfaces. As a bioinert material, alumina shows almost no interaction with tissues; especially smooth surfaces do not osseointegrate (Griss et al. 1975). Pores in the micrometer range, however, have been shown to improve integration (Schreiner et al. 2002), and alumina of grain sizes smaller than 100 nm showed increased osteoblast proliferation and differentiation compared to conventional alumina (Webster et al. 2000a).

In another study, the challenge of cementless implant design is addressed (Karlsson et al. 2003; Briggs et al. 2004). HA is known to be bioactive, but the material itself and the bonding to metallic surfaces are weak. Alumina coatings can be deposited onto metallic implants by electron beam evaporation at 300°C. The resulting layer on the implant provides high-bonding mechanical strength. By subsequent anodization, a nanoporous layer of alumina has been formed and is tested *in vitro* and *in vivo* with promising outcome.

3.4.2.2.3 Titania Coatings

Titanium implants have been shown to be susceptible to modification after treatment with H_2O_2/HCl, leading to titania gel formation on the surface that is transformed to anatase (bioactive crystalline titania) by heat treatment at 400°C for 1 h (Wang et al. 2001). This nanostructured titania layer was stable in contact with buffer and allowed for biomineralization within 2 days. It is hypothesized that both the crystal structure and free negative charges promoted biomineralization.

3.4.2.3 Fabrication of Topography

With the objective to generate hierarchical structures and to investigate relevant topographical structures for cells and tissue regeneration, a plethora of studies have been performed. Since there is a cell-specific reaction to topography, these studies also intended to specifically guide osteoblasts to the surface. Some of these studies, especially those involving nano-sized ceramics and micrometer scale patterns, will be highlighted in the following.

In order to guide cell alignment on the surface of an implant, topographical features have been created on silicon wafer model surfaces by photolithography and subsequent coating with nanoceramics. In order to learn about the cooperation between micro- and nanoscale features, the generated micrometer scale structures are tested for effects on cells. Tan and Saltzman, for example, chemically introduced carboxyl groups onto the surface of microstructured silicon wafers with the objective to accelerated HA synthesis by biomineralization (Tan and Saltzman 2004). MG63 cells alignment was directed along the ridges (4 μm height, 10 μm spacing) with and without HA coating. In a similar approach, Lu and Leng investigated the effect of microgrooves of different width (8 and 24 μm) on osteoblasts and myoblasts (Lu and Leng 2009). Smooth HA layers were generated on structured wafers by magnetron sputtering and both cell types were found to align along 8 μm grooves, but only myoblasts showed alignment along 24 μm grooves. Nanocrystalline diamond was coated on microstructured silicon wafers using a microwave plasma enhanced CVD technique (Grausova et al. 2009). Surface hydrophilicity was enhanced by oxygen-containing plasma. Osteoblasts adhered, proliferated, and differentiated better on nanocrystalline diamond surfaces as compared to polystyrene controls.

Beside these techniques for silicon wafer surface modification, techniques exist that can introduce topographical features onto metal surfaces. Laser ablation is a technique that allows for defined removal of materials from surfaces in order to create surface topography (Peruzzi et al. 2004). Defined topography can be realized on a point-to-point basis with a moving beam or through a template, the latter with higher ablation velocities (Norton et al. 2006). Template-assisted electrohydrodynamic atomization spraying can also be used to generate microstructured surfaces (Li et al. 2008a,b). A nanosized HA dispersion in ethanol was syringed through a small needle and deposited onto a wafer on a grounded metal plate. High voltage between the needle and the grounded plate generated a cone-like jet spraying the suspension onto the substrate. To create a surface pattern, a gold template was placed on top of the titanium substrate. In this study, 15 μm wide lines were generated on titanium substrates. Heating of the substrate to 80°C, the boiling point of ethanol, resulted in fast drying and narrow lines (Li et al. 2008a).

In template-assisted ion beam sputtering, a grid is used in front of the substrate in order to generate patterns. Grain size on the patterned surface can be controlled by the deposition rate. Puckett et al. (2008), for example, generated linear micron-sized features on titanium substrates. Ion beam-sputtered parts of the surface were covered with a nanostructured layer of anatase titanium dioxide, while the untreated surface consisted of rutile titanium dioxide. Other approaches to generate micro- and nanostructured titania include acid and oxidative etching (Vetrone et al. 2009) and (sand-) blasting (Zinger et al. 2005).

Bacterial adhesion was compared on different nanostructured titania surfaces (Puckett et al. 2010). Nano-rough surfaces of anatase that were generated by electron beam evaporation showed lower bacterial adhesion as compared to nanotubular and nanotextured amorphous titania surfaces fabricated by anodization.

3.4.2.4 Biological Effects of Nanocoatings

For most nanostructured surfaces, an improved interaction with osteoblasts has been observed (Webster et al. 1999). It is discussed that nanostructured surfaces in comparison to microstructured ones positively affect the adhesion of osteoblasts, while fibroblast adhesion is reduced. This way, improved osseointegration is mediated and fibrous encapsulation of implants is suppressed (Webster et al. 2000b). The role of serum proteins in improved osteoblast adhesion and function has been demonstrated (Webster et al. 2000a,b). Among adhesion-mediating serum proteins, fibronectin and vitronectin are the most prominent ones. In order to determine the impact of these adhesion proteins, osteoblast adhesion to vitronectin and fibronectin pre-coated nanoparticles was investigated (Webster et al. 2000b). On alumina nanoparticles, vitronectin and fibronectin pre-coating led to increased osteoblast adhesion to small nanoparticles compared to 167 nm particles, suggesting that both proteins are involved in the size-dependent effects. When the amount of adsorbed proteins to alumina was determined, significant size-dependent differences were only found for vitronectin. Similar effects have also been found on HA particles. An improved availability of vitronectin adhesion sites for osteoblasts by conformational differences after adsorption to small nanoparticles was discussed to explain the phenomenon. Nuffer and

Siegel also found increased adhesion of osteoblasts, but decreased fibroblast adhesion to small spherical silica nanoparticles (20 nm) compared to 100 nm particles (Nuffer and Siegel 2010). As possible explanation, different conformations of pre-adsorbed fibronectin on small nanoparticles compared to 100 nm particles were discussed. In another study, it was reported that β-sheet structure of the protein was lost on the 100 nm particles as compared to fibronectin in solution and fibronectin pre-adsorbed to small nanoparticles (Ballard et al. 2005). No particle size-dependent effect, however, was found for the conformation of pre-adsorbed vitronectin on spherical silica particles of different diameter. In another approach, it was shown that fibronectin adsorption correlated with surface nanoroughness and surface energy (Khang et al. 2007). In this study, surface nanoroughness was controlled by the amount of carbon nanotubes dispersed in a polymer, and the contribution of surface nanostructure and surface chemistry on protein adsorption was determined. The authors found that 30% of fibronectin adsorption depended on the nanostructure, whereas 70% were controlled by surface chemistry. Moreover, cell adhesion was shown to correlate with surface energy and wettability. Studies on self-assembled monolayers revealed that albumin, the most abundant protein in serum, irreversibly adsorbs to hydrophobic surfaces while displacement with vitronectin and fibronectin is possible for hydrophilic surfaces (Arima and Iwata 2007).

Taken together, protein interactions with nanostructured biomaterials influence cell adhesion and proliferation and have positive effects on osteoblast functions. These favorable effects are partially mediated by the amount and conformation of adsorbed proteins.

3.4.3 Nanocomposites

Nanocomposites are solid dispersions of a nanoparticulate phase within another biomaterial. As bulk materials, ceramics, synthetic and natural polymers, and structural proteins, such as collagen, have been used. Nanocomposites comprising ceramic nanoparticles are typically inspired by the hierarchical nanoscale structures of mineralized tissues and designed to mechanically reinforce a material that is more elastic but less hard than a ceramic (Šupová 2009). Due to the high surface area of ceramic nanoparticles, effective interactions between the particulate filler and the bulk material can be achieved.

Ceramic/ceramic nanocomposites are structurally divided into nano/nano-type and nano/micro-type or nano/macro-type composites (Niihara 1991; Komarneni 1992; Sternitzke 1997; Choi and Awaji 2005). Nano/nano-type composites are obtained by sintering mixtures of two or more nanoparticulate ceramics. Correspondingly, nano/micro- or nano/macro-type composites are sintered from mixtures of the nanoparticulate ceramic and larger particles of the bulk material. These types of nanocomposites can be further classified as inter(granular)-type, intra(granular)-type, and intra(granular)/inter(granular)-type composites. In the intra(granular)-type, the nanoparticulate phase is dispersed mainly within the matrix grains, while in the inter(granular)-type, the nanoparticles are predominantly located at the grain boundaries.

Ceramic/inorganic nanocomposites represent the most researched type of nanocomposite biomaterials for biomedical applications toward the regeneration of bone (Murugan and Ramakrishna 2005; Christenson et al. 2007). Especially nHA/collagen composites have been investigated in a plethora of scientific studies, because this composition most closely resembles the mineralized matrix of bone (Cui et al. 2007). A large number of studies have also been published on composites of nHA with other natural or synthetic polymers or combinations of other nano-sized ceramic components with polymers. Table 3.3 highlights selected *in vivo* studies investigating different ceramic nanocomposites. Most nanocomposites are composed of an organic bulk component, either a natural or a synthetic polymer. The predominant application is in bone regeneration. Ectopic bone formation has been observed for composites with nHA that were seeded with adipose tissue-derived stromal cells (Lin et al. 2007) or loaded with bone morphogenetic protein-2 (BMP-2) (Sotome et al. 2004).

TABLE 3.3 Selected *In Vivo* Studies with Ceramic Nanocomposites

Composition			Implantation Study			
Nano-Sized Ceramic	Bulk	Application	Animal	Site	Duration (Weeks)	Reference
Biocompatibility testing						
HA	Collagen	Bone regeneration	Rat	Subcutaneous (back)	24	Kikuchi et al. (2004a)
HA	PLGA	Bone regeneration	Rabbit	Intramuscular	20	Zhang et al. (2009)
HA	Poly(propylene fumarate)	Bone regeneration	Rabbit	Intramuscular	12	Jayabalan et al. (2010)
HA	Chitosan/ carboxymethyl cellulose	Bone regeneration	Rat	Intramuscular	4	Jiang et al. (2009)
HA	Amino acid (nanoconjugates)	Bone regeneration	Mouse	Subcutaneous	3	Babister et al. (2009)
Hybrid alumoxane	Poly(propylene fumarate)	Bone regeneration	Goat	Subcutaneous (back)	12	Mistry et al. (2010)
Ce-TZP	Alumina	Bone regeneration	Rat	Intramuscular	24	Tanaka et al. (2002)
Alumina	Zirconia	Endoprosthesis	Rat	Stifle	2	Roualdes et al. (2010)
Ectopic implantation						
HA	Collagen/alginate (BMP-2 absorbed)	Bone regeneration	Rat	Intramuscular	5	Sotome et al. (2004)
HA fibers	β-TCP (with stromal cells)	Bone regeneration	Mouse	Subcutaneous	8	Lin et al. (2007)
Orthotopic implantation						
HA	Collagen	Bone regeneration	Beagle	Tibia	12	Kikuchi et al. (2001)
HA	Collagen, cross-linked	Bone regeneration	Rabbit	Tibia	4	Kikuchi et al. (2004b)
HA	Collagen	Bone regeneration	Rat	Tibia	4	Kikuchi et al. (2004a)
HA	Collagen	Bone regeneration	Beagle	Tibia	12	Kikuchi et al. (2004a)
HA	Collagen	Bone regeneration	Beagle	Tibia	24	Itoh et al. (2005)
HA	Collagen/ poly(lactide)	Bone regeneration	Rabbit	Forelimbs	16	Liao et al. (2004)
HA	Collagen/alginate (BMP-2 absorbed)	Bone regeneration	Rat	Femur	5	Sotome et al. (2004)
HA	Chitosan	Bone regeneration	Rabbit	Fibula	12	Kong et al. (2007)
HA	PLGA	Bone regeneration	Rabbit	Forelimbs	8	Zhang et al. (2009)
HA	PLGA	Cartilage regeneration	Rat	Stifle	12	Li et al. (2009)
HA	Poly(propylene fumarate)	Bone regeneration	Rabbit	Femur	48	Jayabalan et al. (2010)
HA	Polyamide	Bone regeneration	Rabbit	Mandible	12	Wang et al. (2007)

TABLE 3.3 (continued) Selected *In Vivo* Studies with Ceramic Nanocomposites

Composition			Implantation Study			
Nano-Sized Ceramic	Bulk	Application	Animal	Site	Duration (Weeks)	Reference
HA fibers	β-TCP (with stromal cells)	Bone regeneration	Rat	Cranium	24	Lin et al. (2007)
HA	Alumina-coated zirconia	Dental implant	Rabbit	Tibia	6	Kong et al. (2002)
TCP	PLGA	Bone regeneration	Rabbit	Cranium	4	Schneider et al. (2009)

3.4.3.1 Fabrication of Ceramic Nanocomposites

As a consequence of the chemical and structural diversity of the ceramic nanocomposites available, it is impossible to provide a comprehensive overview on all utilized fabrication techniques. The following section strives to highlight important techniques in reference to the techniques described for the individual ceramic nanoparticles and to illustrate how different bulk materials can be introduced.

For the fabrication of ceramic/ceramic nanocomposites, mechanochemical processes have been described. For the synthesis of calcium phosphate/titanium particles, commercially available powders of calcium dihydrogen phosphate and titania were mixed and treated in a planetary mill for 15 h (Silva et al. 2007). To avoid excessive heating, the milling was performed in 60 min steps with 10 min pauses. The resulting particles ranged between 20–60 nm in diameter. An intra-type nano-zirconia/alumina composite has been fabricated from alumina powder and zirconium alkoxide (Chevalier et al. 2005). The alumina powder together with a small amount of zirconia is suspended in ethanol, and a zirconium alkoxide solution is added dropwise. The dispersion was dried under stirring at 70°C and the resulting powder was heat-treated at 850°C for 2 h, milled, and finally sintered at 1600°C for 2 h. The resulting intra/inter-type composites contained nano-sized zirconia particles located in between micron-sized alumina grains (Figure 3.1C).

A wet-mechanochemical synthesis route has been described for ceramic/polymer nanocomposites (Wang et al. 2002b). A calcium hydroxide dispersion containing silk fibroin and ammonium polyacrylic acid was prepared. After the addition of phosphoric acid, the mixture was stirred for 1 h and transferred to a multi-ring mill where it was treated for 3 h at 1250 rpm. The milled composite was dried in vacuum, and the dispersed HA nanorods were found to measure 20–30 nm in length and 8–10 nm in width. The following ceramic/ceramic nanocomposite was fabricated by a hydrothermal method (Pushpakanth et al. 2008). In order to improve the load sharing and stress distribution of nHA, *in situ* nanostructured high-strength HA–titanium dioxide was fabricated by microwave-assisted co-precipitation. Minerals were dissolved from cancellous bone with acids and an aqueous solution of titanyl dichloride was added. The pH was then adjusted to 10 and the resulting slurry was placed into a microwave oven and irradiated for 5 min. The precipitate was centrifuged, washed, and dried in a vacuum oven at 100°C to obtain composite nanorods of good chemical and structural uniformity.

Derived from classical controlled precipitation protocols, co-precipitation techniques have become popular for the synthesis of ceramic/polymer nanocomposites. Consequently, parameters that influence the morphology of inorganic/organic composites fabricated via such processes include properties of the co-precipitated polymer, choice of solvent, pH, precipitation time, and temperature (Rusu et al. 2005). For the fabrication of nHA/chitosan composites, for example, an aqueous chitosan solution was prepared with acetic acid and calcium chloride and sodium dihydrogen phosphate were added. The pH was adjusted to 11 and after 24 h the resulting gelatinous dispersion was filtered, washed, and dried to give a rigid material. In a similar process, a nHA/collagen composite was fabricated (Zhang et al. 2003). Collagen, sodium dihydrogen phosphate, and calcium chloride were dissolved in an acidic aqueous solution. Via pH adjustment to 7, co-precipitation is initiated and the composite was obtained by centrifugation, washed and lyophilized. Within the composite, self-assembled collagen nanofibrils were found on which nHA crystals grew along the longitudinal axes of the fibrils.

In all examples mentioned above, the nanoceramic component of the different composites was prepared during composite fabrication. An immense amount of studies investigates ceramic nanocomposites that are fabricated from a ceramic nanopowder that is somehow dispersed into the bulk component and processed further. Ceramic/ceramic nanocomposites, for example, were fabricated from different nano-sized ceramics by mixing and sintering to improve hardness and fracture toughness in dental applications (Nevarez-Rascon et al. 2010). Nano-sized ceramics (Al_2O_3, MgO, Al_2O_3 whiskers, and Y_2O_3–ZrO_2) were dispersed in ethanol and intensely mixed. To the dried mixture magnesium oxide was added to inhibit grain growth of the alumina powder during sintering. The powder was uniaxially pressed at 50 MPa into disks, and the disks were placed into alumina crucibles with zirconia and alumina bed powders and sintered (1500°C, 2 h) (Figure 3.1D). Further routes to obtain ceramic/ceramic nanocomposites include conventional powder processing, sol–gel processing, and polymer processing (Sternitzke 1997; Chevalier and Gremillard 2009). Following the classical powder-processing protocol, the raw powders—ultrafine particles of the matrix-forming ceramic and the nanoparticulate ceramic—are mixed and micro-milled. Wet-mixing is preferred using ultrasound, ball mills, or attrition mills. After drying, the powder mixture is densified, for example, by hot-pressing above 1500°C in a controlled atmosphere. During sol–gel processing, a sol is initially prepared from the nanoparticulate component dispersed in a solution or slurry of matrix precursors. Upon hydrothermal processing, the sol is transferred into a gel, which is turned into an ultrafine powder after drying and calcination. The powder is finally densified by hot-pressing to yield a solid composite with defined density and mechanical properties. As mentioned before, dispersion problems and agglomeration of the ultra-fine powder are critical challenges of these processes.

In approaches toward the fabrication of ceramic/polymer nanocomposites from nano-sized ceramic particles, an effective and homogeneous dispersion of the nano-sized ceramic component is also of critical importance, especially when the bulk materials are hydrophobic. A study that incorporated degradable calcium phosphate particles in a degradable hydrogel matrix showed that a co-precipitation technique for composite fabrication resulted in a much higher degree of dispersion of the ceramic crystals inside the resulting gels compared to a physical mixing strategy (Leeuwenburgh et al. 2007). The physical mixing of ceramic nanoparticles with a solution of the bulk polymer, however, is one of the most popular methods to fabricate ceramic/polymer nanocomposites. In a study that fabricated nHA/poly(lactide-*co*-glycolide) (PLGA) composite nanospheres, HA, which was synthesized by a homogeneous precipitation method in an ultrasonic field from calcium nitrate and ammonium dihydrogen phosphate, was mixed and dispersed within a solution of the polymer in acetone (Jevtic et al. 2009). Polymer precipitation was initiated through the dropwise addition of the non-solvent ethanol, while the reaction vessel was kept in an ultrasonic field and under temperature control. The particulate composites were stabilized through the addition of a polymeric stabilizer, centrifuged, and air-dried. The morphology of the composite particles was highly regular.

Another strategy to improve the dispersion of ceramic nanoparticles within a polymer matrix includes the use of surface-modified nanoparticles. An alumoxane/biodegradable cross-linked polymer nanocomposite was fabricated using surface-modified nanoparticles (Horch et al. 2004). The hybrid alumoxane nanoparticles modified with a long carbon chain and a reactive double bond were dispersed in the pre-polymer mixture and chemically integrated upon cross-copolymerization. The hybrid nanocomposite showed improved dispersion compared to unmodified nanocomposites and significantly improved mechanical properties.

With regard to tissue engineering application, the composite materials often need to be processed into macroporous constructs. Many techniques are available to achieve such structures. Tissue engineering scaffolds from the above mentioned alumoxane/polymer nanocomposites were prepared by a classical salt leaching technique (Mistry et al. 2009). Modified alumoxane nanoparticles were dispersed in pre-polymer mixture and sieved salt particles were added. The dispersion was packed in a mold and photo-cross-copolymerized. The cross-linked blocks were submerged in distilled water to wash out the salt and yield macroporous nanocomposite scaffolds. Besides leaching techniques, thermally induced phase separation is a fairly common method for the fabrication of porous scaffolds for tissue engineering (Liu and Webster 2007). This method typically employs a solvent/non-solvent mixture or a poor solvent to process the matrix material. At ambient or elevated temperature, the matrix polymer is soluble in the mixture. Once the temperature is decreased, the system starts to phase separate into a polymer-solvent phase and a non-solvent phase. The metastable partially phase-separated system is rapidly frozen and both solvent and non-solvent are removed by lyophilization generating a porous matrix. Utilizing this technique, bioactive glass/poly(L-lactide) (PLLA) nanocomposite scaffolds were fabricated (Hong et al. 2009). Bioactive glass powder was homogeneously dispersed in dioxane in an ultrasonic field, and PLLA was added to the solution. After lyophilization for 1 week, porous composite scaffolds were obtained. Ceramic/polymer nanocomposites were also processed into nanofiber meshes by electrospinning (Pham et al. 2006; Nie and Wang 2007).

Another important strategy to generate bioactive, nanostructured calcium phosphate within tissue engineering scaffolds is biomineralization (Bonzani et al. 2006; Kretlow and Mikos 2007; LeGeros 2008; Ma 2008; Palmer et al. 2008). Biomineralization strategies, typically involving the immersion of a polymeric construct in SBF, can be applied to both hydrogel systems and macroporous solids. It has been demonstrated that certain functional groups can promote and regulate crystal growth on the substrate surface (Kretlow and Mikos 2007). Carboxylic acid and hydroxy group that have been generated on poly(hydroxyl esters) surfaces by controlled hydrolysis, for example, have been shown to regulate calcium binding to the polymer surface and heterogeneous mineral growth (Murphy and Mooney 2002). The composition of the mineral grown in SBF solutions was a carbonate apatite similar to vertebrate bone mineral. Crystal size and morphology were predominantly controlled by the mineralization media and not by polymer surface characteristics. A template-driven mineralization technique has also been demonstrated for a poly(2-hydroxyethyl methacrylate) hydrogel scaffold (Song et al. 2003). By a gradual increase in pH controlled by thermal decomposition of urea, carboxylic acid groups were exposed on the surface by ester hydrolysis. These anionic groups promoted high-affinity nucleation and growth of calcium phosphate on the surface along with extensive calcification and the formation of robust surface mineral layers. Biomimetic mineralization processes have also been employed to deposit calcium phosphates on electrospun bioactive glass nanofibers (Figure 3.1E) (Xia et al. 2007) and silica on electrospun polymer fiber meshes (Patel et al. 2009). There are also reports on the fabrication of nano-sized ceramic/ceramic composites by mineralization. Anionic functional groups were chemically introduced in single-walled carbon nanotubes (SWNTs), and HA was shown to nucleate and crystallize on the nanotube surface (Zhao et al. 2005).

3.5 Characterization of Ceramic Nanobiomaterials

The interactions between a nanobiomaterial and biological fluids, cells, or complete biological systems are complex and dependent on many parameters. Especially for nano-sized substrates, the bulk properties of the material become less significant, and commonly known property-response relations have most often to be renewed (Jones and Grainger 2009). As a consequence of the high surface-to-volume ratio, substrate properties are strongly determined by surface state and morphological parameters in contrast to micro- and macro-dimensioned substrates. For some ceramic materials that are considered well biocompatible in the micro- and macroscale, certain cytotoxic effects have been reported for murine fibroblasts and macrophages (Yamamoto et al. 2004). The observed cytotoxic effects of titanium dioxide, aluminum oxide, zirconium dioxide, silicon nitride, and silicon carbide nanoparticles are suggested to be based neither on chemical properties nor on size, but on the total volume of particles and their shape. This so-called mechanical toxicity is supported by findings that spiked nanoparticles bear a higher cytotoxicity than spherical nanoparticles. Such effects were more pronounced in macrophages than in fibroblasts, likely as a result of the phagocytic process. As another example, the toxicity mechanism and long-time health effects of certain silica-based materials such as crocidolite asbestos, although not fully understood yet, depend on properties including respirability or ability to enter the lung, durability due to insolubility and lack of clearance by macrophages, fibrous geometry, aspect ratio, and surface properties associated with the generation of reactive oxygen or nitrogen species (Hillegass et al. 2010). Due to the high surface area of nanomaterials, surface contaminations, which may significantly alter the original surface chemistry and pattern and furthermore the biological response, are another challenge (Grainger and Castner 2008). Surface science provides many tools that can be used for the characterization of surfaces and interfacial conditions of nanobiomaterials. However, significant effort is devoted toward the development of new methods and the adaption of known methods to meet the special requirements of nanoscale analytics. In the next paragraphs, a selection of important morphological, surface, and bulk characterization techniques is given and their applications in ceramic nanotechnology are highlighted.

3.5.1 Particle Size and Morphology

Transmission electron microscope (TEM) is often employed to determine size and morphology of ceramic nanostructures (Roy et al. 2003; Bhattarai et al. 2007; Wang and Shaw 2009b). Usually, TEM samples are prepared from dried nanoparticle suspensions. Alternatively, an electron transparent lamella is cut out of the ceramic sample by ion milling (for details on the milling process, refer to section Bulk Characterization), lifted out of the milling trench, and introduced to TEM optics (Kooi et al. 2003). High resolution TEM (HRTEM) is able to image the crystallographic structure of a sample at resolutions in the sub-Ångström range (O'Keefe et al. 2005). This setup was utilized to explore the mineralization state of collagen fibrils with nanocrystals from HA (Zhang et al. 2003) or basal-plane stacking faults in ceramic titanium silicon carbide crystals (Kooi et al. 2003). In a study that compared nHA powders synthesized by an ethanol-based and a water-based sol–gel technique (Kalita et al. 2007), HRTEM analysis revealed that particle size of powders synthesized via the ethanol-based and the water-based method was 20–50 and 5–10 nm in diameter, respectively.

Scanning electron microscopy (SEM) can be used to image size and morphology of specimens, which are impassable for electrons. With this microscopic technique, different material properties can be assessed depending on the recorded signal, such as secondary electrons, backscattered electrons, Auger electrons, and X-ray bremsstrahlung. Secondary electrons, which are emitted from the uppermost nanometers of the sample surface, are used to detect surface topography and to visualize surface structures. From the electron micrograph, the dimension of nanostructures can be derived (Zhang et al. 2003; Dulgar-Tulloch et al. 2009). For PLGA/HA composite nanospheres, for example, effects of the fabrication process on nanosphere morphology have been comparatively assessed by SEM image analysis (Jevtic et al. 2009).

3.5.2 Surface Characterization

Atomic force microscopy (AFM) is a versatile tool in surface characterization and provides image information on surface structures and patterns as well as data on surface functionalization and reactivity. In the classical operating mode, the cantilever-mounted tip scans the sample surface registering differences in height in order to create a topographical image of the sample surface. With this method, nanometer and micron-scale diamond film surface topographies that showed significant effects on osteoblast functions were imaged (Yang et al. 2009b). To date, variants of the classical AFM method are available, in which the probing tip is modified. Lateral force microscopy (LFM) (Liu et al. 1994; Crossley et al. 1999), chemical force microscopy (CFM) (Frisbie et al. 1994), or friction force microscopy (FFM) (Overney et al. 1992) are such techniques that are used to explore the chemical quality of a surface in AFM contact mode. In order to detect tip-to-surface interactions that depend on the chemical quality of the surface, certain functional groups as well as entire molecules including proteins are covalently bound to the tip. A stronger interaction is represented by a stronger torsion of the tip-bearing cantilever. AFM can also provide information toward the explanation of biological effects observed on nanostructured ceramic surfaces, because it can be used to qualitatively determine protein structures that became adsorbed to the surface (Liu and Webster 2007). AFM was utilized to examine the HA structure at the interface to citrate, in which the carboxylic acid groups have chemical similarities to the residues of osteocalcin that interact with bone minerals (Jiang et al. 2008). In another example of how AFM is utilized, the mineral deposits on dip-coated PLGA and PLGA/collagen nanofibers as well as the nanotexture of the fibers have been visualized by AFM (Ngiam et al. 2009). The most common application for AFM is to determine surface roughness as shown for dense HA and β-TCP substrates (dos Santos et al. 2008).

X-ray photoelectron spectroscopy (XPS) or *electron spectroscopy for chemical analysis* (ESCA) allows for the qualitative and quantitative determination of the surface elemental composition including its chemical state. In XPS, the sample is irradiated with a focused X-ray beam in ultra-high vacuum (UHV) causing a photoelectric effect. Surface electrons are knocked out of their orbitals and their kinetic energies are measured with an electron energy analyzer. The final spectrum is achieved by plotting the electron counts against the calculated electron binding energies. From these spectra, the chemical surface composition can be identified, because each element is represented by a characteristic set of binding energy peaks. Analysis of signal intensity then provides quantitative information. With a sampling depth of less than 10 nm (depending on the angle between analyzer and sample), XPS acquires surface data that is practically unimpaired by signals from the bulk phase. Consequently, surface contaminations can be easily detected in XPS spectra (Chen et al. 2008). With regard to ceramic nanobiomaterials, the formation of HA layers on bioactive titanium in SBF was investigated using XPS (Takadama et al. 2001).

Secondary ion mass spectrometry (SIMS) has a sampling depth of 1–2 nm, which is even lower than that of XPS. However, SIMS is a destructive method, during which molecules are sputtered from the sample surface by a focused primary ion beam in UHV. One distinguishes between dynamic SIMS, which enables compositional analysis from the top layer into deeper layers due to the utilization of a high ion dose beam, and static SIMS (McPhail 2006). Static SIMS utilizes low ion doses to ensure that only the uppermost sample layer is analyzed. Usually, static SIMS is coupled with a time-of-flight analyzer (ToF-SIMS) and is a versatile method to determine the elemental and chemical composition of a surface. It has also been utilized to identify and characterize surface-adsorbed proteins (Tidwell et al. 2001). In a specific example, the interface between strontium-containing HA and cancellous as well as cortical bone was monitored by ToF-SIMS over six months in a rabbit model (Ni et al. 2006). The study revealed that higher concentrations of calcium, phosphor, sodium, and oxygen were found on the interface with cancellous bone indicating different dissolution rates of the ceramic due to the type of adjacent bone.

Energy-dispersive X-ray spectroscopy (EDX or EDS) is another technique that allows for elemental analysis of a surface. The sample surface is irradiated by a SEM electron beam and emitted X-rays are detected. The energy spectra of the emitted X-rays can be correlated to specific elements. This method has been used to characterize nanostructured ceramic coatings due to their elemental composition (Wang et al. 2009) and to analyze minerals deposited during biomineralization of titanium implants (Serro and Saramago 2003).

Auger electron spectroscopy (AES) is another technique in which the sample surface is excited by a SEM electron beam. Electron gaps created in lower orbital shells are filled with electrons from a higher orbital, and the corresponding transition energy is emitted. The peculiarity of the so-called Auger effect is that the emitted energy is absorbed by another electron in the same atom. This excited Auger electron is then emitted from the atom and traceable in UHV. The electron energy patterns are unique for each element providing information about the elemental surface composition. With regard to ceramic materials, it should be considered that most ceramics are isolators and sample charging may occur when exposed to high energy beams. In order to perform AES analysis with the high spatial resolution of a few nanometers, a conductive layer has to be introduced to the reverse side of a thinned sample (Yu and Jin 2001). AES was applied to determine the surface composition of a titanium dioxide coating on a titanium alloy substrate before and after base treatment (Zhao et al. 2006).

Attenuated total reflectance Fourier transform infrared (ATR-FTIR) *spectroscopy* is a powerful tool for surface characterization because one can detect changes in chemical structure or chemical environment as expressed in frequency shifts and changes in relative band intensities. The functional principle of ATR-FTIR is based on so-called evanescent waves. Whenever a light beam hits a phase boundary at an angle of incidence that is larger than the critical angle, total reflection occurs. A certain small amount of the beam energy, however, passes the phase boundary as an evanescent wave, which penetrates the adjacent phase over a distance depending on the refractive index of the material and the wavelength of the reflected beam. The phenomena of total reflection in combination with the formation of evanescent waves occur when the light-conducting phase is of higher optical density than the adjacent phase. Therefore, the crystals utilized in ATR-FTIR spectroscopy are selected to have a higher refractive index than the sample material. In an ATR-FTIR analysis, the attenuation of the irradiated intensity is determined. This loss in energy in the IR-wavelength range is characteristic for the quality and state of the specific atom groups in the sample. In a specific example, the formation of an apatite coating on alginate/chitosan microparticles was confirmed by ATR-FTIR spectroscopy through characteristic signals of phosphate and carbonate groups (Lee et al. 2009). When ATR-FTIR is utilized for surface analysis, it has to be considered that the sampling depth of this method is about 1000 nm and not exclusively limited to the surface layer (Liu and Webster 2007).

3.5.3 Bulk Characterization

Thermogravimetric analysis (TGA) acquires changes in sample mass depending on temperature and time. Measured weight loss, gain, and/or fluctuations at distinct temperatures can be characteristic for a specimen and its composition. Alterations, usually mass loss, in the thermogravimetric curve of a substance are attributed to impurities, chemical modification, degradation, oxidation, or pyrolysis. Magnetic iron oxide nanoparticles stabilized with modified chitosan were examined utilizing TGA to estimate the amount of bound chitosan from the percentage weight loss by pyrolysis of the organic component (Bhattarai et al. 2007). With regard to ceramic materials, TGA is mainly utilized to test for thermal stability of the sample (Kalita and Verma 2010).

Focused ion beam (FIB) *tomography* is a technique of great potential to visualize the inner chemical composition and structure of a material (Möbus and Inkson 2007; Uchic et al. 2007; Munroe 2009). In a so-called milling process, the ion beam ablates nanometer layers of a sample. In a single-beam FIB device, the ion beam is used to both mill the sample and sputter ions from the newly exposed surfaces for analysis. This requires movement of the sample in between the milling and the sputtering step. In combination with SIMS (FIB-SIMS), the elemental composition of a sample can be analyzed layer by layer. With such a setup, a spatial resolution of approximately 50 nm in lateral direction, 5 nm in depth, and a few micrometers in sectional direction can be achieved (Tomiyasu et al. 1998). Dual-beam devices, which use individual beams for milling and detection and therefore do not require the sample to be moved during analysis, provide the highest spatial resolution. In such dual-beam devices, a SEM image of the freshly exposed surface can be recorded in a frequency of roughly 25 slices per hour. From these images, a three-dimensional reconstruction of the sample can be done, which allows for a quantification of pore and grain volumes. A resolution of up to 6 nm in lateral direction, 7 nm in depth, and 17 nm in sectional direction has been achieved (Holzer et al. 2004), and sample volumes larger than 1000 μm^3 can be analyzed (Uchic et al. 2007). Recently, rod-shaped nanoparticles of HA were produced by a hydrothermal synthesis technique and investigated for their infiltration into dentinal tubules of etched human molars (Earl et al. 2009). Information on the depth of infiltration was obtained from sections of dentine prepared using FIB milling (FIB-SEM).

X-ray diffraction (XRD) is a non-destructive technique to determine the crystallographic orientation and structure of a sample. It is based on the scattering of an X-ray beam due to the molecular quality of the target. There are two different modes of data acquisition: small angle X-ray scattering (SAXS) and wide angle X-ray scattering (WAXS) (Cancedda et al. 2007). According to Bragg's law, larger plane spacings in a crystal lattice entail smaller scattering angles. SAXS is used to determine size, shape, and orientation of mesoscopic structures that are relatively large. WAXS is used to determine the distance of molecule or atom layers arranged in an orderly pattern such as in crystalline solids. This way, WAXS provides information about the quality of crystal unit cells that means the crystal inner structure. The simultaneous collection of SAXS and WAXS information with a small area microbeam entails acquisition of highly resolved data, because each volume is described with both acquisition modes at the same time (Guagliardi et al. 2009). Quantitative XRD has been shown to be a more accurate tool than wet chemistry to identify the Ca/P ratio of calcium phosphate apatites that strongly affects the chemical and biological properties of these materials (Raynaud et al. 2001, 2002; Han et al. 2006). Apatite nanostructures have also been identified using this technique in *in situ*–generated nanocomposites with collagen (Lin et al. 2004) (Figure 3.1F). In another application example, biomimetic collagen/nanoapatite composite scaffolds for tissue engineering were prepared by a precipitation method and analyzed (Liu 2008). The crystalline phase was identified as nHA of lower crystallinity than that of a rabbit ulna. It is discussed that crystal growth and final crystallinity are influenced by the orientation of the collagen fibers. With regard to the identification of optimal sintering temperatures and conditions, XRD can also be used to study phase evolution/transformation in calcium phosphate and titanium dioxide samples (Kalita et al. 2007, 2008). In another study, the *in situ* transformation of anhydrous dicalcium phosphate cement into HA was monitored over 24 h by XRD (Hsu et al. 2009).

X-ray computed tomography (CT or μCT) has been applied in mesoscale physics to image porous media and to derive mechanical properties of the specimen from this data (Sakellariou et al. 2004). In addition, pore size and structure of ceramic bodies can be investigated (Ritman 2004; Cancedda et al. 2007). Due to the mesoscale resolution of this technique, a visualization of nanostructures is practically not possible. However, certain applications of μCT with nanocomposites have been described. For a composite of nano-sized calcium phosphate crystals and an injectable hydrogel matrix prepared by precipitation of the mineral in presence of the hydrogel precursors, μCT analysis demonstrated the absence of aggregates in the micrometer range indicating a high degree of dispersion (Leeuwenburgh et al. 2007).

3.5.4 Cytocompatibility, Biocompatibility, and Toxicity

The prominent class of calcium phosphate nanobiomaterials is most likely the least debated with regard to biocompatibility and environmental safety because of the physiological chemistry of these minerals and the high abundance of such nanomaterials in organisms. Reliable and predictive models to determine the acute and chronic biosafety of nanomaterials in general, however, have not yet been developed to a satisfactory level. A common scientific consensus on how to responsibly address urgent questions regarding potential health and environment risks of nanomaterials has not yet been agreed on (Webster 2008; Grainger 2009). *In vitro* cell cultures provide fairly straightforward and cost-effective methods to screen the toxicity of nanomaterials in early stages of product development. Such *in vitro* methods include cell culture assays for cytotoxicity (altered metabolism, decreased growth, lytic or apoptotic cell death), cell stress, proliferation, genotoxicity, altered gene expression, and assays for cell-based production of reactive oxygen species (Jones and Grainger 2009; Hillegass et al. 2010). Following such screening test, which cannot imitate a complementary *in vivo* system, small mammalian models can help to assess possible toxicities and biodistribution of nanomaterials in humans (Fischer and Chan 2007). Furthermore, quick, cheap, and facile models, such as the zebrafish, have been described to conservatively assess toxicity of nanomaterials (Fako and Furgeson 2009). Using this assay, nanocrystalline zinc oxide, nanocrystalline titanium dioxide, and nanocrystalline alumina were tested. Only nanocrystalline zinc oxide showed visible signs of toxicity indicated by a delayed hatching rate and development of zebrafish embryos and larvae as well as tissue damage and decreased survival.

3.6 Conclusion and Perspective

Ceramics have grown to an important class of biomaterials, especially for the fabrication of orthopedic and dental implants and in strategies aiming at the regeneration of bone and teeth (Vallet-Regi 2001, 2006, 2008; Salinas and Vallet-Regi 2007; Chevalier and Gremillard 2009; Dorozhkin 2010a). With ceramic materials being available as nano-sized structures, the opportunities to utilize this class of materials toward a desired mechanical or biological effect have significantly broadened. In the form of nano-grained coatings, ceramic nanobiomaterials improve integration of metal or metal alloy implants with the surrounding hard tissue and have potential to significantly prolong implant lifetimes. In ceramic nanocomposites, the unique chemical composition, mechanical properties, and specific interactions with proteins that are attributed to the bioactive properties of ceramic nanostructures can be embedded in a composite biomaterial with improved bulk properties determined by the bulk material. In both applications, nano-grained ceramic coatings and ceramic nanocomposites, the bulk properties of the resulting biomaterial or medical device are determined by a non-ceramic material in most cases. The most prominent nanoceramic to date is nHA, but a key interest toward the fabrication and use of nanocrystalline cHA can be identified. Another strategy to generate nanocrystalline calcium phosphate that closely resembles bioapatite involves controlled biomineralization processes and holds promise to generate coatings and composites with improved biocompatibility and bioactivity. Such developments have to be accompanied by ongoing analytical efforts to better understand the chemistry and structure of biological apatites and how these parameters control the apatites' physical and biological properties. Rising interest is also devoted to nanocrystalline diamond. Exciting mechanical improvement of articular prostheses by nano-diamond coatings combining ultra-low friction and biomimetic nanostructure has been described. In bone regeneration strategies, it will be necessary to optimize structure, bioactivity, and remodelability, as well as degradative properties of ceramic nanocomposite scaffolds. At the same time, scientific proof of nanoceramic safety and compatibility is a vital prerequisite for patient compliance and a successful regulatory process.

Acknowledgment

The authors thankfully acknowledge financial support by the German Research Foundation (Deutsche Forschungsgemeinschaft, DFG, TRR 67, A1).

References

Arima Y. and Iwata H. 2007. Effect of wettability and surface functional groups on protein adsorption and cell adhesion using well-defined mixed self-assembled monolayers. *Biomaterials* 28 (20): 3074–3082.

Arnold M., Cavalcanti-Adam E.A., Glass R., Blummel J., Eck W., Kantlehner M., Kessler H., and Spatz J.P. 2004. Activation of integrin function by nanopatterned adhesive interfaces. *Chemphyschem* 5 (3): 383–388.

Babister J.C., Hails L.A., Oreffo R.O.C., Davis S.A., and Mann S. 2009. The effect of pre-coating human bone marrow stromal cells with hydroxyapatite/amino acid nanoconjugates on osteogenesis. *Biomaterials* 30 (18): 3174–3182.

Bai X., Sandukas S., Appleford M.R., Ong J.L., and Rabiei A. 2009. Deposition and investigation of functionally graded calcium phosphate coatings on titanium. *Acta Biomater.* 5 (9): 3563–3572.

Balasundaram G., Yao C., and Webster T.J. 2008. TiO_2 nanotubes functionalized with regions of bone morphogenetic protein-2 increases osteoblast adhesion. *J. Biomed. Mater. Res. A* 84A (2): 447–453.

Ballard J.D., Acqua-Bellavitis L.M., Bizios R., and Siegel R.W. 2005. Nanoparticle-decorated surfaces for the study of cell-protein-substrate interactions. *Nanoscale Materials Science in Biology and Medicine Materials Research Society Symposium Proceedings*, Fall 2004, Boston, MA, pp. 339–344.

Barrere F., Van Blitterswijk C.A., de Groot K., and Layrolle P. 2002. Nucleation of biomimetic Ca-P coatings on Ti6Al4V from a SBF x 5 solution: Influence of magnesium. *Biomaterials* 23 (10): 2211–2220.

Bhattarai S.R., Bahadur K.C.R., Aryal S., Khil M.S., and Kim H.Y. 2007. N-acylated chitosan stabilized iron oxide nanoparticles as a novel nano-matrix and ceramic modification. *Carbohydr. Polym.* 69 (3): 467–477.

Bigi A., Cojazzi G., Panzavolta S., Ripamonti A., Roveri N., Romanello M., Noris Suarez K., and Moro L. 1997. Chemical and structural characterization of the mineral phase from cortical and trabecular bone. *J. Inorg. Biochem.* 68 (1): 45–51.

Bonzani I.C., George J.H., and Stevens M.M. 2006. Novel materials for bone and cartilage regeneration. *Curr. Opin. Chem. Biol.* 10 (6): 568–575.

Bow J.S., Liou S.C., and Chen S.Y. 2004. Structural characterization of room-temperature synthesized nano-sized [beta]-tricalcium phosphate. *Biomaterials* 25 (16): 3155–3161.

Briggs E.P., Walpole A.R., Wilshaw P.R., Karlsson M., and Palsgard E. 2004. Formation of highly adherent nano-porous alumina on Ti-based substrates: A novel bone implant coating. *J. Mater. Sci. Mater. Med.* 15 (9): 1021–1029.

Callender R.L., Harlan C.J., Shapiro N.M., Jones C.D., Callahan D.L., Wiesner M.R., MacQueen D.B., Cook R., and Barron A.R. 1997. Aqueous synthesis of water-soluble alumoxanes: Environmentally benign precursors to alumina and aluminum-based ceramics. *Chem. Mater.* 9 (11): 2418–2433.

Campbell A.A. 2003. Bioceramics for implant coatings. *Mater. Today* 6 (11): 26–30.

Cancedda R., Cedola A., Giuliani A., Komlev V., Lagomarsino S., Mastrogiacomo M., Peyrin F., and Rustichelli F. 2007. Bulk and interface investigations of scaffolds and tissue-engineered bones by x-ray microtomography and x-ray microdiffraction. *Biomaterials* 28 (15): 2505–2524.

Catledge S.A., Fries M.D., Vohra Y.K., Lacefield W.R., Lemons J.E., Woodard S., and Venugopalan R. 2002. Nanostructured ceramics for biomedical implants. *J. Nanosci. Nanotechnol.* 2 (3–4): 293–312.

Chen D., Jordan E.H., Gell M., and Wei M. 2008. Apatite formation on alkaline-treated dense TiO_2 coatings deposited using the solution precursor plasma spray process. *Acta Biomater.* 4 (3): 553–559.

Cherian A.K., Rana A.C., and Jain S.K. 2000. Self-assembled carbohydrate-stabilized ceramic nanoparticles for the parenteral delivery of insulin. *Drug Dev. Ind. Pharm.* 26 (4): 459–463.

Chevalier J., Deville S., Fantozzi G., Bartolomé J.F., Pecharroman C., Moya J.S., Diaz L.A., and Torrecillas R. 2005. Nanostructured ceramic oxides with a slow crack growth resistance close to covalent materials. *Nano Lett.* 5 (7): 1297–1301.

Chevalier J. and Gremillard L. 2009. Ceramics for medical applications: A picture for the next 20 years. *J. Eur. Ceram. Soc.* 29 (7): 1245–1255.

Choi S.M. and Awaji H. 2005. Nanocomposites—A new material design concept. *Sci. Technol. Adv. Mater.* 6 (1): 2–10.

Choy K.L. 2003. Chemical vapour deposition of coatings. *Prog. Mater. Sci.* 48 (2): 57–170.

Chris Arts J.J., Verdonschot N., Schreurs B.W., and Buma P. 2006. The use of a bioresorbable nano-crystalline hydroxyapatite paste in acetabular bone impaction grafting. *Biomaterials* 27 (7): 1110–1118.

Christenson E.M., Anseth K.S., van den Beucken L.J.J.P., Chan C.K., Ercan B., Jansen J.A., Laurencin C.T., Li W.J., Murugan R., Nair L.S., Ramakrishna S., Tuan R.S., Webster T.J., and Mikos A.G. 2007. Nanobiomaterial applications in orthopedics. *J. Orthop. Res.* 25 (1): 11–22.

Clem W.C., Chowdhury S., Catledge S.A., Weimer J.J., Shaikh F.M., Hennessy K.M., Konovalov V.V., Hill M.R., Waterfeld A., Bellis S.L., and Vohra Y.K. 2008. Mesenchymal stem cell interaction with ultra-smooth nanostructured diamond for wear-resistant orthopaedic implants. *Biomaterials* 29 (24–25): 3461–3468.

Colon G., Ward B.C., and Webster T.J. 2006. Increased osteoblast and decreased *Staphylococcus epidermidis* functions on nanophase ZnO and TiO_2. *J. Biomed. Mater. Res. A* 78A (3): 595–604.

Corr S.A., Rakovich Y., and Gun'ko Y.K. 2008. Multifunctional magnetic-fluorescent nanocomposites for biomedical applications. *Nanoscale Res. Lett.* 3 (3): 87–104.

Crossley A., Kisi E.H., Summers J.W.B., and Myhra S. 1999. Ultra-low friction for a layered carbide-derived ceramic, Ti3SiC2, investigated by lateral force microscopy (LFM). *J. Phys. D Appl. Phys.* 32 (6): 632–638.

Cui F.Z., Li Y., and Ge J. 2007. Self-assembly of mineralized collagen composites. *Mater. Sci. Eng. R Rep.* 57 (1–6): 1–27.

Dorozhkin S.V. 2010a. Bioceramics of calcium orthophosphates. *Biomaterials* 31 (7): 1465–1485.

Dorozhkin S.V. 2010b. Nanosized and nanocrystalline calcium orthophosphates. *Acta Biomater.* 6 (3): 715–734.

dos Santos E., Farina M., Soares G., and Anselme K. 2008. Surface energy of hydroxyapatite and b-tricalcium phosphate ceramics driving serum protein adsorption and osteoblast adhesion. *J. Mater. Sci. Mater. Med.* 19 (6): 2307–2316.

Drory M.D., Gardinier C.F., and Speck J.S. 1991. Fracture-toughness of chemically vapor-deposited diamond. *J. Am. Ceram. Soc.* 74 (12): 3148–3150.

Drotleff S., Lungwitz U., Breunig M., Dennis A., Blunk T., Tessmar J., and Gopferich A. 2004. Biomimetic polymers in pharmaceutical and biomedical sciences. *Eur. J. Pharm. Biopharm.* 58 (2): 385–407.

Dulgar-Tulloch A.J., Bizios R., and Siegel R.W. 2009. Human mesenchymal stem cell adhesion and proliferation in response to ceramic chemistry and nanoscale topography. *J. Biomed. Mater. Res. A* 90 (2): 586–594.

Dyshlovenko S., Pateyron B., Pawlowski L., and Murano D. 2004. Numerical simulation of hydroxyapatite powder behaviour in plasma jet. *Surf. Coat. Technol.* 179 (1): 110–117.

Earl J.S., Wood D.J., and Milne S.J. 2009. Nanoparticles for dentine tubule infiltration: An in vitro study. *J. Nanosci. Nanotechnol.* 9 (11): 6668–6674.

Evis Z., Sato M., and Webster T.J. 2006. Increased osteoblast adhesion on nanograined hydroxyapatite and partially stabilized zirconia composites. *J. Biomed. Mater. Res. A* 78 (3): 500–507.

Fako V.E. and Furgeson D.Y. 2009. Zebrafish as a correlative and predictive model for assessing biomaterial nanotoxicity. *Adv. Drug Deliv. Rev.* 61 (6): 478–486.

Fathi M.H., and Zahrani E.M. 2009. Fabrication and characterization of fluoridated hydroxyapatite nanopowders via mechanical alloying. *J. Alloy Compd.* 475 (1–2): 408–414.

Fischer H.C. and Chan W.C. 2007. Nanotoxicity: The growing need for in vivo study. *Curr. Opin. Biotechnol.* 18 (6): 565–571.

Fratzl P., Gupta H.S., Paschalis E.P., and Roschger P. 2004. Structure and mechanical quality of the collagen-mineral nano-composite in bone. *J. Mater. Chem.* 14 (14): 2115–2123.

Frieden E. 1972. The chemical elements of life. *Sci. Am.* 227 (1): 52–60.

Frisbie C.D., Rozsnyai L.F., Noy A., Wrighton M.S., and Lieber C.M. 1994. Functional group imaging by chemical force microscopy. *Science* 265 (5181): 2071–2074.

Ganguli D. 1993. Sol-gel processing—A versatile concept for special glasses and ceramics. *Bull. Mater. Sci.* 16 (6): 523–531.

Georgiou G. and Knowles J.C. 2001. Glass reinforced hydroxyapatite for hard tissue surgery—Part 1: Mechanical properties. *Biomaterials* 22 (20): 2811–2815.

Gotman I. 1997. Characteristics of metals used in implants. *J. Endourol.* 11 (6): 383–389.

Goyal A.K., Khatri K., Mishra N., Mehta A., Vaidya B., Tiwari S., and Vyas S.P. 2008. Aquasomes—A nanoparticulate approach for the delivery of antigen. *Drug Dev. Ind. Pharm.* 34 (12): 1297–1305.

Grainger D.W. 2009. Nanotoxicity assessment: All small talk? *Adv. Drug Deliv. Rev.* 61 (6): 419–421.

Grainger D.W. and Castner D.G. 2008. Nanobiomaterials and nanoanalysis: Opportunities for improving the science to benefit biomedical technologies. *Adv. Mater.* 20 (5): 867–877.

Grausova L., Bacakova L., Kromka A., Potocky S., Vanecek M., Nesladek M., and Lisa V. 2009. Nanodiamond as promising material for bone tissue engineering. *J. Nanosci. Nanotechnol.* 9 (6): 3524–3534.

Griss P., Heimke G., Andrianwerburg H.V., Krempien B., Reipa S., Lauterbach H.J., and Hartung H.J. 1975. Morphological and biomechanical aspects of Al_2O_3 ceramic joint replacement—Experimental results and design considerations for human endoprostheses. *J. Biomed. Mater. Res.* 9 (4): 177–188.

Gruen D.M. 1999. Nanocrystalline diamond films. *Annu. Rev. Mater. Sci.* 29: 211–259.

Guagliardi A., Giannini C., Cedola A., Mastrogiacomo M., Ladisa M., and Cancedda R. 2009. Toward the x-ray microdiffraction imaging of bone and tissue-engineered bone. *Tissue Eng. Part B Rev.* 15 (4): 423–442.

Gunatillake P.A. and Adhikari R. 2003. Biodegradable synthetic polymers for tissue engineering. *Eur. Cell Mater.* 5: 1–16.

Gupta A.K. and Gupta M. 2005. Synthesis and surface engineering of iron oxide nanoparticles for biomedical applications. *Biomaterials* 26 (18): 3995–4021.

Habibovic P., Barrere F., Van Blitterswijk C., de Groot K., and Layrolle P. 2002. Biomimetic hydroxyapatite coating on metal implants. *J. Am. Ceram. Soc.* 85 (3): 517–522.

Habibovic P. and de Groot K. 2007. Osteoinductive biomaterials—Properties and relevance in bone repair. *J. Tissue Eng. Regen. Med.* 1 (1): 25–32.

Habraken W.J.E.M., Wolke J.G.C., and Jansen J.A. 2007. Ceramic composites as matrices and scaffolds for drug delivery in tissue engineering. *Adv. Drug Deliv. Rev.* 59 (4–5): 234–248.

Hakim L.F., George S.M., and Weimer A.W. 2005. Conformal nanocoating of zirconia nanoparticles by atomic layer deposition in a fluidized bed reactor. *Nanotechnology* 16 (7): S375–S381.

Hamdi M. and Ide-Ektessabi A. 2003. Preparation of hydroxyapatite layer by ion beam assisted simultaneous vapor deposition. *Surf. Coat. Technol.* 163–164: 362–367.

Han J.K., Song H.Y., Saito F., and Lee B.T. 2006. Synthesis of high purity nano-sized hydroxyapatite powder by microwave-hydrothermal method. *Mater. Chem. Phys.* 99 (2–3): 235–239.

Han Y., Li S., Wang X., and Chen X. 2004. Synthesis and sintering of nanocrystalline hydroxyapatite powders by citric acid sol-gel combustion method. *Mater. Res. Bull.* 39 (1): 25–32.

Hanawa T., Asami K., and Asaoka K. 1998. Repassivation of titanium and surface oxide film regenerated in simulated bioliquid. *J. Biomed. Mater. Res.* 40 (4): 530–538.

Hench L.L. 1997. Sol-gel materials for bioceramic applications. *Curr. Opin. Solid State Mater. Sci.* 2 (5): 604–610.

Hench L.L. 1980. Biomaterials. *Science* 208 (4446): 826–831.

Hench L.L. 1998. Bioceramics. *J. Am. Ceram. Soc.* 81 (7): 1705–1727.

Hillegass J.M., Shukla A., Lathrop S.A., MacPherson M.B., Fukagawa N.K., and Mossman B.T. 2010. Assessing nanotoxicity in cells *in vitro. WIREs Nanomed. Nanobiotechnol.* 2: 219–231.

Holzer L., Indunyi F., Gasser P.H., Münch B., and Wegmann M. 2004. Three-dimensional analysis of porous $BaTiO_3$ ceramics using FIB nanotomography. *J. Microsc.* 216 (1): 84–95.

Hong X., Li J., Wang M.J., Xu J.J., Guo W., Li J.H., Bai Y.B., and Li T.J. 2004. Fabrication of magnetic luminescent nanocomposites by a layer-by-layer self-assembly approach. *Chem. Mater.* 16 (21): 4022–4027.

Hong Z., Reis R.L., and Mano J.F. 2008. Preparation and in vitro characterization of scaffolds of poly(L-lactic acid) containing bioactive glass ceramic nanoparticles. *Acta Biomater.* 4 (5): 1297–1306.

Hong Z., Reis R.L., and Mano J.F. 2009. Preparation and in vitro characterization of novel bioactive glass ceramic nanoparticles. *J. Biomed. Mater. Res. A* 88 (2): 304–313.

Horch R.A., Shahid N., Mistry A.S., Timmer M.D., Mikos A.G., and Barron A.R. 2004. Nanoreinforcement of poly(propylene fumarate)-based networks with surface modified alumoxane nanoparticles for bone tissue engineering. *Biomacromolecules* 5 (5): 1990–1998.

Hsu H.C., Tuan W.H., and Lee H.Y. 2009. In-situ observation on the transformation of calcium phosphate cement into hydroxyapatite. *Mater. Sci. Eng. C Biomim. Mater. Sens. Syst.* 29 (3): 950–954.

Hu M.Z.C., Harris M.T., and Byers C.H. 1998. Nucleation and growth for synthesis of nanometric zirconia particles by forced hydrolysis. *J. Colloid Interface Sci.* 198 (1): 87–99.

Itoh S., Kikuchi M., Koyama Y., Matumoto H.N., Takakuda K., Shinomiya K., and Tanaka J. 2005. Development of a novel biomaterial, hydroxyapatite/collagen (HAp/Col) composite for medical use. *Biomed Mater. Eng.* 15 (1–2): 29–41.

Jain T.K., Roy I., De T.K., and Maitra A. 1998. Nanometer silica particles encapsulating active compounds: A novel ceramic drug carrier. *J. Am. Chem. Soc.* 120 (43): 11092–11095.

Jayabalan M., Shalumon K., Mitha M., Ganesan K., and Epple M. 2010. Effect of hydroxyapatite on the biodegradation and biomechanical stability of polyester nanocomposites for orthopaedic applications. *Acta Biomater.* 6 (3): 763–775.

Jevtic M., Radulovic A., Ignjatovic N., Mitric M., and Uskokovic D. 2009. Controlled assembly of poly(D,L-lactide-co-glycolide)/hydroxyapatite core-shell nanospheres under ultrasonic irradiation. *Acta Biomater.* 5 (1): 208–218.

Jiang L., Li Y., and Xiong C. 2009. Preparation and biological properties of a novel composite scaffold of nano-hydroxyapatite/chitosan/carboxymethyl cellulose for bone tissue engineering. *J. Biomed. Sci.* 16: 65.

Jiang W., Pan H., Cai Y., Tao J., Liu P., Xu X., and Tang R. 2008. Atomic force microscopy reveals hydroxyapatite-citrate interfacial structure at the atomic level. *Langmuir* 24 (21): 12446–12451.

Jones C.F. and Grainger D.W. 2009. In vitro assessments of nanomaterial toxicity. *Adv. Drug Deliv. Rev.* 61 (6): 438–456.

Kalita S.J., Bhardwaj A., and Bhatt H.A. 2007. Nanocrystalline calcium phosphate ceramics in biomedical engineering. *Mater. Sci. Eng. C Biomim. Mater. Sens. Syst.* 27 (3): 441–449.

Kalita S.J., Qiu S., and Verma S. 2008. A quantitative study of the calcination and sintering of nanocrystalline titanium dioxide and its flexural strength properties. *Mater. Chem. Phys.* 109 (2–3): 392–398.

Kalita S.J. and Verma S. 2010. Nanocrystalline hydroxyapatite bioceramic using microwave radiation: Synthesis and characterization. *Mater. Sci. Eng. C Biomim. Mater. Sens. Syst.* 30 (2): 295–303.

Karch J., Birringer R., and Gleiter H. 1987. Ceramics ductile at low temperature. *Nature* 330 (6148): 556–558.

Karlsson M., Palsgard E., Wilshaw P.R., and Di Silvio L. 2003. Initial in vitro interaction of osteoblasts with nano-porous alumina. *Biomaterials* 24 (18): 3039–3046.

Karthikeyan J., Berndt C.C., Tikkanen J., Reddy S., and Herman H. 1997. Plasma spray synthesis of nanomaterial powders and deposits. *Mater. Sci. Eng.* 238 (2): 275–286.

Kay S., Thapa A., Haberstroh K.M., and Webster T.J. 2002. Nanostructured polymer/nanophase ceramic composites enhance osteoblast and chondrocyte adhesion. *Tissue Eng.* 8 (5): 753–761.

Khang D., Kim S.Y., Liu-Snyder P., Palmore G.T.R., Durbin S.M., and Webster T.J. 2007. Enhanced fibronectin adsorption on carbon nanotube/poly(carbonate) urethane: Independent role of surface nano-roughness and associated surface energy. *Biomaterials* 28 (32): 4756–4768.

Khor K.A., Dong Z.L., Quek C.H., and Cheang P. 2000. Microstructure investigation of plasma sprayed HA/Ti6Al4V composites by TEM. *Mater. Sci. Eng.* 281 (1–2): 221–228.

Kikuchi M., Ikoma T., Itoh S., Matsumoto H.N., Koyama Y., Takakuda K., Shinomiya K., and Tanaka J. 2004a. Biomimetic synthesis of bone-like nanocomposites using the self-organization mechanism of hydroxyapatite and collagen. *Compos. Sci. Technol.* 64 (6): 819–825.

Kikuchi M., Itoh S., Ichinose S., Shinomiya K., and Tanaka J. 2001. Self-organization mechanism in a bone-like hydroxyapatite/collagen nanocomposite synthesized in vitro and its biological reaction in vivo. *Biomaterials* 22 (13): 1705–1711.

Kikuchi M., Matsumoto H.N., Yamada T., Koyama Y., Takakuda K., and Tanaka J. 2004b. Glutaraldehyde cross-linked hydroxyapatite/collagen self-organized nanocomposites. *Biomaterials* 25 (1): 63–69.

Kim H.W., Kim H.E., Kim H.W., and Knowles J.C. 2005. Improvement of hydroxyapatite sol-gel coating on titanium with ammonium hydroxide addition. *J. Am. Ceram. Soc.* 88 (1): 154–159.

Kim J.U. and O'Shaughnessy B. 2002. Morphology selection of nanoparticle dispersions by polymer media. *Phys. Rev. Lett.* 89 (23): 238301-1–238301-4.

Kokubo T. 1998. Apatite formation on surfaces of ceramics, metals and polymers in body environment. *Acta Mater.* 46 (7): 2519–2527.

Komarneni S. 1992. Nanocomposites. *J. Mater. Chem.* 2 (12): 1219–1230.

Kong L., Ao Q., Wang A., Gong K., Wang X., Lu G., Gong Y., Zhao N., and Zhang X. 2007. Preparation and characterization of a multilayer biomimetic scaffold for bone tissue engineering. *J. Biomater. Appl.* 22 (3): 223–239.

Kong Y.M., Kim D.H., Kim H.E., Heo S.J., and Koak J.Y. 2002. Hydroxyapatite-based composite for dental implants: An in vivo removal torque experiment. *J. Biomed. Mater. Res.* 63 (6): 714–721.

Koo O.M., Rubinstein I., and Onyuksel H. 2005. Role of nanotechnology in targeted drug delivery and imaging: A concise review. *Nanomedicine* 1 (3): 193–212.

Kooi B.J., Poppen R.J., Carvalho N.J.M., Hosson J.T., and Barsoum M.W. 2003. Ti3SiC2: A damage tolerant ceramic studied with nano-indentations and transmission electron microscopy. *Acta Mater.* 51 (10): 2859–2872.

Kretlow J.D. and Mikos A.G. 2007. Review: Mineralization of synthetic polymer scaffolds for bone tissue engineering. *Tissue Eng.* 13 (5): 927–938.

Kriven W.M., Kwak S.Y., Wallig M.A., and Choy J.H. 2004. Bio-resorbable nanoceramics for gene and drug delivery. *MRS Bull.* 29 (1): 33–37.

Kumar P., Oka M., Ikeuchi K., Shimizu K., Yamamuro T., Okumura H., and Kotoura Y. 1991. Low wear rate of uhmwpe against zirconia ceramic (Y-Psz) in comparison to alumina ceramic and sus 316l alloy. *J. Biomed. Mater. Res.* 25 (7): 813–828.

Ladewig K., Niebert M., Xu Z.P., Gray P.P., and Lu G.Q. 2010. Controlled preparation of layered double hydroxide nanoparticles and their application as gene delivery vehicles. *Appl. Clay Sci.* 48 (1–2): 280–289.

Landi E., Celotti G., Logroscino G., and Tampieri A. 2003. Carbonated hydroxyapatite as bone substitute. *J. Eur. Ceram. Soc.* 23 (15): 2931–2937.

Landry C.C., Pappe N., Mason M.R., Apblett A.W., Tyler A.N., Macinnes A.N., and Barron A.R. 1995. From minerals to materials—Synthesis of alumoxanes from the reaction of boehmite with carboxylic-acids. *J. Mater. Chem.* 5 (2): 331–341.

Lee M., Li W., Siu R.K., Whang J., Zhang X., Soo C., Ting K., and Wu B.M. 2009. Biomimetic apatite-coated alginate/chitosan microparticles as osteogenic protein carriers. *Biomaterials* 30 (30): 6094–6101.

Lee Y.I., Lee J.H., Hong S.H., and Kim D.Y. 2003. Preparation of nanostructured TiO_2 ceramics by spark plasma sintering. *Mater. Res. Bull.* 38: 925–930.

Leeuwenburgh S.C.G., Jansen J.A., and Mikos A.G. 2007. Functionalization of oligo(poly(ethylene glycol) fumarate) hydrogels with finely dispersed calcium phosphate nanocrystals for bone-substituting purposes. *J. Biomater. Sci. Polym. Ed.* 18 (12): 1547–1564.

LeGeros R.Z. 2002. Properties of osteoconductive biomaterials: Calcium phosphates. *Clin. Orthop. Relat. Res.* 395: 81–98.

LeGeros R.Z. 2008. Calcium phosphate-based osteoinductive materials. *Chem. Rev.* 108 (11): 4742–4753.

Li P.J. and Ducheyne P. 1998. Quasi-biological apatite film induced by titanium in a simulated body fluid. *J. Biomed. Mater. Res.* 41 (3): 341–348.

Li X., Huang J., and Edirisinghe M. 2008a. Development of nano-hydroxyapatite coating by electrohydrodynamic atomization spraying. *J. Mater. Sci. Mater. Med.* 19 (4): 1545–1551.

Li X., Huang J., and Edirisinghe M.J. 2008b. Novel patterning of nano-bioceramics: Template-assisted electrohydrodynamic atomization spraying. *J. R. Soc. Interface* 5 (19): 253–257.

Li L.H., Kim H.W., Lee S.H., Kong Y.M., and Kim H.E. 2005. Biocompatibility of titanium implants modified by microarc oxidation and hydroxyapatite coating. *J. Biomed. Mater. Res. A* 73A (1): 48–54.

Li H.D., Zheng Q., Xiao Y.X., Feng J., Shi Z.L., and Pan Z.J. 2009. Rat cartilage repair using nanophase PLGA/HA composite and mesenchymal stem cells. *J. Bioact. Compat. Polym.* 24 (1): 83–99.

Liang J., Deng Z., Jiang X., Li F., and Li Y. 2002. Photoluminescence of tetragonal ZrO_2 nanoparticles synthesized by microwave irradiation. *Inorg. Chem.* 41 (14): 3602–3604.

Liao S.S., Cui F.Z., Zhang W., and Feng Q.L. 2004. Hierarchically biomimetic bone scaffold materials: Nano-HA/collagen/PLA composite. *J. Biomed. Mater. Res.* 69 (2): 158–165.

Lin X., Li X., Fan H., Wen X., Lu J., and Zhang X. 2004. In situ synthesis of bone-like apatite/collagen nano-composite at low temperature. *Mater. Lett.* 58 (27–28): 3569–3572.

Lin Y., Wang T., Wu L., Jing W., Chen X., Li Z., Liu L., Tang W., Zheng X., and Tian W. 2007. Ectopic and in situ bone formation of adipose tissue-derived stromal cells in biphasic calcium phosphate nanocomposite. *J. Biomed. Mater. Res. A* 81A (4): 900–910.

Link A. 2000. Inorganic layered double hydroxides as nonviral vectors. *Angew. Chem. Int. Ed.* 39 (22): 4042–4045.

Liu C.Z. 2008. Biomimetic synthesis of collagen/nano-hydroxyapitate scaffold for tissue engineering. *J. Bionic Eng.* 5 (Suppl 1): 1–8.

Liu S.M., Gan L.M., Liu L.H., Zhang W.D., and Zeng H.C. 2002. Synthesis of single-crystalline TiO_2 nanotubes. *Chem. Mater.* 14 (3): 1391–1397.

Liu X., Ren X., Deng X., Huo Y., Xie J., Huang H., Jiao Z., Wu M., Liu Y., and Wen T. 2010. A protein interaction network for the analysis of the neuronal differentiation of neural stem cells in response to titanium dioxide nanoparticles. *Biomaterials* 31 (11): 3063–3070.

Liu H. and Webster T.J. 2007. Nanomedicine for implants: A review of studies and necessary experimental tools. *Biomaterials* 28 (2): 354–369.

Liu Y., Wu T., and Evans D.F. 1994. Lateral force microscopy study on the shear properties of self-assembled monolayers of dialkylammonium surfactant on mica. *Langmuir* 10 (7): 2241–2245.

Lu X. and Leng Y. 2009. Comparison of the osteoblast and myoblast behavior on hydroxyapatite microgrooves. *J. Biomed. Mater. Res. B* 90B (1): 438–445.

Ma P.X. 2008. Biomimetic materials for tissue engineering. *Adv. Drug Deliv. Rev.* 60 (2): 184–198.

McPhail D. 2006. Applications of secondary ion mass spectrometry (SIMS) in materials science. *J. Mater. Sci.* 41 (3): 873–903.

Medina C., Santos-Martinez M.J., Radomski A., Corrigan O.I., and Radomski M.W. 2007. Nanoparticles: Pharmacological and toxicological significance. *Br. J. Pharmacol.* 150 (5): 552–558.

Mistry A.S., Cheng S.H., Yeh T., Christenson E., Jansen J.A., and Mikos A.G. 2009. Fabrication and in vitro degradation of porous fumarate-based polymer/alumoxane nanocomposite scaffolds for bone tissue engineering. *J. Biomed. Mater. Res. A* 89 (1): 68–79.

Mistry A.S., Pham Q.P., Schouten C., Yeh T., Christenson E.M., Mikos A.G., and Jansen J.A. 2010. In vivo bone biocompatibility and degradation of porous fumarate-based polymer/alumoxane nanocomposites for bone tissue engineering. *J. Biomed. Mater. Res. A* 92 (2): 451–462.

Möbus G. and Inkson B.J. 2007. Nanoscale tomography in materials science. *Mater. Today* 10 (12): 18–25.

Munroe P.R. 2009. The application of focused ion beam microscopy in the material sciences. *Mater. Charact.* 60 (1): 2–13.

Murphy W.L. and Mooney D.J. 2002. Bioinspired growth of crystalline carbonate apatite on biodegradable polymer substrata. *J. Am. Chem. Soc.* 124 (9): 1910–1917.

Murugan R. and Ramakrishna S. 2005. Development of nanocomposites for bone grafting. *Compos. Sci. Technol.* 65 (15–16): 2385–2406.

Murugan R. and Ramakrishna S. 2006. Nanophase biomaterials for tissue engineering. In *Tissue, Cell and Organ Engineering*, ed. C.S.S.R. Kumar, pp. 216–256. Weinheim, Germany: Wiley-VCH.

Nelea V., Morosanu C., Iliescu M., and Mihailescu I.N. 2003. Microstructure and mechanical properties of hydroxyapatite thin films grown by RF magnetron sputtering. *Surf. Coat. Technol.* 173 (2–3): 315–322.

Nevarez-Rascon A., Aguilar-Elguezabal A., Orrantia E., and Bocanegra-Bernal M.H. 2010. Al_2O_3(w)-Al_2O_3(n)-ZrO_2 $(TZ\text{-}3Y)_n$ multi-scale nanocomposite: An alternative for different dental applications? *Acta Biomater.* 6 (2): 563–570.

Ngiam M., Liao S., Patil A.J., Cheng Z., Chan C.K., and Ramakrishna S. 2009. The fabrication of nano-hydroxyapatite on PLGA and PLGA/collagen nanofibrous composite scaffolds and their effects in osteoblastic behavior for bone tissue engineering. *Bone* 45 (1): 4–16.

Nguyen H.Q., Deporter D.A., Pilliar R.M., Valiquette N., and Yakubovich R. 2004. The effect of sol-gel-formed calcium phosphate coatings on bone ingrowth and osteoconductivity of porous-surfaced Ti alloy implants. *Biomaterials* 25 (5): 865–876.

Ni G.X., Lu W.W., Xu B., Chiu K.Y., Yang C., Li Z.Y., Lam W.M., and Luk K.D.K. 2006. Interfacial behaviour of strontium-containing hydroxyapatite cement with cancellous and cortical bone. *Biomaterials* 27 (29): 5127–5133.

Nie H. and Wang C.H. 2007. Fabrication and characterization of PLGA/HAp composite scaffolds for delivery of BMP-2 plasmid DNA. *J. Control Release* 120 (1–2): 111–121.

Niihara K. 1991. New design concept of structural ceramics. Ceramic nanocomposites. *J. Ceram. Soc. Jpn.* 99 (1154): 974–982.

Norton J., Malik K.R., Darr J.A., and Rehman I. 2006. Recent developments in processing and surface modification of hydroxyapatite. *Adv. Appl. Ceram.* 105 (3): 113–139.

Nuffer J.H. and Siegel R.W. 2010. Nanostructure–biomolecule interactions: Implications for tissue regeneration and nanomedicine. *Tissue Eng. Part A* 16 (2): 423–430.

O'Keefe M.A., Allard L.F., and Blom D.A. 2005. HRTEM imaging of atoms at sub-Angstrom resolution. *J. Electron. Microsc. (Tokyo)* 54 (3): 169–180.

Overney R.M., Meyer E., Frommer J., Brodbeck D., Lüthi R., Howald L., Güntherodt H.J., Fujihira M., Takano H., and Gotoh Y. 1992. Friction measurements on phase-separated thin-films with a modified atomic force microscope. *Nature* 359 (6391): 133–135.

Paital S.R. and Dahotre N.B. 2009. Calcium phosphate coatings for bio-implant applications: Materials, performance factors, and methodologies. *Mater. Sci. Eng. R Rep.* 66 (1–3): 1–70.

Palmer L.C., Newcomb C.J., Kaltz S.R., Spoerke E.D., and Stupp S.I. 2008. Biomimetic systems for hydroxyapatite mineralization inspired by bone and enamel. *Chem. Rev.* 108 (11): 4754–4783.

Pang Y.X. and Bao X. 2003. Influence of temperature, ripening time and calcination on the morphology and crystallinity of hydroxyapatite nanoparticles. *J. Eur. Ceram. Soc.* 23 (10): 1697–1704.

Pankhurst Q.A., Connolly J., Jones S.K., and Dobson J. 2003. Applications of magnetic nanoparticles in biomedicine. *J Phys. D Appl. Phys.* 36 (13): R167.

Patel P.A., Eckart J., Advincula M.C., Goldberg A.J., and Mather P.T. 2009. Rapid synthesis of polymer-silica hybrid nanofibers by biomimetic mineralization. *Polymer* 50 (5): 1214–1222.

Peruzzi M., Pedarnig J.D., Sturm H., Huber N., and Bauerle D. 2004. F-2-laser ablation and micro-patterning of GaPO4. *Europhys. Lett.* 65 (5): 652–657.

Pham Q.P., Sharma U., and Mikos A.G. 2006. Electrospinning of polymeric nanofibers for tissue engineering applications: A review. *Tissue Eng.* 12 (5): 1197–1211.

Phillips M.J., Darr J.A., Luklinska Z.B., and Rehman I. 2003. Synthesis and characterization of nanobiomaterials with potential osteological applications. *J. Mater. Sci. Mater. Med.* 14 (10): 875–882.

Pratsinis S.E. 1998. Flame aerosol synthesis of ceramic powders. *Prog. Energy Combust. Sci.* 24 (3): 197–219.

Puckett S., Pareta R., and Webster T.J. 2008. Nano rough micron patterned titanium for directing osteoblast morphology and adhesion. *Int. J. Nanomedicine* 3 (2): 229–241.

Puckett S.D., Taylor E., Raimondo T., and Webster T.J. 2010. The relationship between the nanostructure of titanium surfaces and bacterial attachment. *Biomaterials* 31 (4): 706–713.

Pushpakanth S., Srinivasan B., Sreedhar B., and Sastry T.P. 2008. An in situ approach to prepare nanorods of titania-hydroxyapatite (TiO_2-HAp) nanocomposite by microwave hydrothermal technique. *Mater. Chem. Phys.* 107 (2–3): 492–498.

Qu H.B. and Wei M. 2008a. Improvement of bonding strength between biomimetic apatite coating and substrate. *J. Biomed. Mater. Res. B* 84B (2): 436–443.

Qu H.B. and Wei M. 2008b. The effect of temperature and initial pH on biomimetic apatite coating. *J. Biomed. Mater. Res. B* 87B (1): 204–212.

Rabiei A., Thomas B., Jin C., Narayan R., Cuomo J., Yang Y., and Ong J.L. 2006. A study on functionally graded HA coatings processed using ion beam assisted deposition with in situ heat treatment. *Surf. Coat. Technol.* 200 (20–21): 6111–6116.

Racek O., Berndt C.C., Guru D.N., and Heberlein J. 2006. Nanostructured and conventional YSZ coatings deposited using APS and TTPR techniques. *Surf. Coat. Technol.* 201 (1–2): 338–346.

Rahaman M.N., Yao A., Bal B.S., Garino J.P., and Ries M.D. 2007. Ceramics for prosthetic hip and knee joint replacement. *J. Am. Ceram. Soc.* 90 (7): 1965–1988.

Raynaud S., Champion E., Bernache-Assollant D., and Laval J.P. 2001. Determination of calcium/phosphorus atomic ratio of calcium phosphate apatites using x-ray diffractometry. *J. Am. Ceram. Soc.* 84 (2): 359–366.

Raynaud S., Champion E., Bernache-Assollant D., and Thomas P. 2002. Calcium phosphate apatites with variable Ca/P atomic ratio I. Synthesis, characterisation and thermal stability of powders. *Biomaterials* 23 (4): 1065–1072.

Rezwan K., Chen Q.Z., Blaker J.J., and Boccaccini A.R. 2006. Biodegradable and bioactive porous polymer/inorganic composite scaffolds for bone tissue engineering. *Biomaterials* 27 (18): 3413–3431.

Riman R.E., Suchanek W.L., Byrappa K., Chen C.W., Shuk P., and Oakes C.S. 2002. Solution synthesis of hydroxyapatite designer particulates. *Solid State Ionics* 151 (1–4): 393–402.

Ritman E.L. 2004. Micro-computed tomography—Current status and developments. *Annu. Rev. Biomed. Eng.* 6: 185–208.

Roualdes O., Duclos M.E., Gutknecht D., Frappart L., Chevalier J., and Hartmann D.J. 2010. In vitro and in vivo evaluation of an alumina-zirconia composite for arthroplasty applications. *Biomaterials* 31 (8): 2043–2054.

Roy D.M. and Linnehan S.K. 1974. Hydroxyapatite formed from coral skeletal carbonate by hydrothermal exchange. *Nature* 247 (5438): 220–222.

Roy I., Ohulchanskyy T.Y., Pudavar H.E., Bergey E.J., Oseroff A.R., Morgan J., Dougherty T.J., and Prasad P.N. 2003. Ceramic-based nanoparticles entrapping water-insoluble photosensitizing anticancer drugs: A novel drug-carrier system for photodynamic therapy. *J. Am. Chem. Soc.* 125 (26): 7860–7865.

Ruksudjarlt A., Pengpat K., Rujijanagul G., and Tunkasiri T. 2008. Synthesis and characterization of nanocrystalline hydroxyapatite from natural bovine bone. *Curr. Appl. Phys.* 8 (3–4): 270–272.

Rusu V.M., Ng C.H., Wilke M., Tiersch B., Fratzl P., and Peter M.G. 2005. Size-controlled hydroxyapatite nanoparticles as self-organized organic–inorganic composite materials. *Biomaterials* 26 (26): 5414–5426.

Sahoo S.K. and Labhasetwar V. 2003. Nanotech approaches to drug delivery and imaging. *Drug Discov. Today* 8 (24): 1112–1120.

Sakellariou A., Sawkins T.J., Senden T.J., and Limaye A. 2004. X-ray tomography for mesoscale physics applications. *Physica A* 339 (1–2): 152–158.

Salinas A.J. and Vallet-Regi M. 2007. Evolution of ceramics with medical applications. *Z. Anorg. Allg. Chem.* 633 (11–12): 1762–1773.

Santos J.D., Silva P.L., Knowles J.C., Talal S., and Monteiro F.J. 1996. Reinforcement of hydroxyapatite by adding P_2O_5–CaO glasses with Na_2O, K_2O and MgO. *J. Mater. Sci. Mater. Med.* 7 (3): 187–189.

Schneider O.D., Weber F., Brunner T.J., Loher S., Ehrbar M., Schmidlin P.R., and Stark W.J. 2009. In vivo and in vitro evaluation of flexible, cottonwool-like nanocomposites as bone substitute material for complex defects. *Acta Biomater.* 5 (5): 1775–1784.

Schreiner U., Schroeder-Boersch H., Schwarz M., and Scheller G. 2002. Surface modification of bioinert ceramics enhances osseointegration in an animal model. *Biomed. Tech.* 47 (6): 164–168.

Serro A.P. and Saramago B. 2003. Influence of sterilization on the mineralization of titanium implants induced by incubation in various biological model fluids. *Biomaterials* 24 (26): 4749–4760.

Shih W.J., Chen Y.F., Wang M.C., and Hon M.H. 2004. Crystal growth and morphology of the nano-sized hydroxyapatite powders synthesized from $CaHPO_4$ center dot 2H(2)O and $CaCO_3$ by hydrolysis method. *J. Cryst. Growth* 270 (1–2): 211–218.

Siddharthan A., Seshadri S.K., and Kumar T.S.S. 2006. Influence of microwave power on nanosized hydroxyapatite particles. *Scripta Mater.* 55 (2): 175–178.

Silva C.C., Graca M.P.F., Valente M.A., and Sombra A.S.B. 2007. Crystallite size study of nanocrystalline hydroxyapatite and ceramic system with titanium oxide obtained by dry ball milling. *J. Mater. Sci.* 42 (11): 3851–3855.

Song J., Saiz E., and Bertozzi C.R. 2003. A new approach to mineralization of biocompatible hydrogel scaffolds: An efficient process toward 3-dimensional bonelike composites. *J. Am. Chem. Soc.* 125 (5): 1236–1243.

Sotome S., Uemura T., Kikuchi M., Chen J., Itoh S., Tanaka J., Tateishi T., and Shinomiya K. 2004. Synthesis and in vivo evaluation of a novel hydroxyapatite/collagen-alginate as a bone filler and a drug delivery carrier of bone morphogenetic protein. *Mater. Sci. Eng. C* 24 (3): 341–347.

Sternitzke M. 1997. Structural ceramic nanocomposites. *J. Eur. Ceram. Soc.* 17 (9): 1061–1082.

Šupová M. 2009. Problem of hydroxyapatite dispersion in polymer matrices: A review. *J. Mater. Sci. Mater. Med.* 20 (6): 1201–1213.

Takadama H., Kim H.M., Kokubo T., and Nakamura T. 2001. An x-ray photoelectron spectroscopy study of the process of apatite formation on bioactive titanium metal. *J. Biomed. Mater. Res.* 55 (2): 185–193.

Tan J. and Saltzman W.M. 2004. Biomaterials with hierarchically defined micro- and nanoscale structure. *Biomaterials* 25 (17): 3593–3601.

Tanaka K., Tamura J., Kawanabe K., Nawa M., Oka M., Uchida M., Kokubo T., and Nakamura T. 2002. Ce-TZP/Al_2O_3 nanocomposite as a bearing material in total joint replacement. *J. Biomed. Mater. Res.* 63 (3): 262–270.

Tang Q., Brooks R., Rushton N., and Best S. 2010. Production and characterization of HA and SiHA coatings. *J. Mater. Sci. Mater. Med.* 21 (1): 173–181.

Tidwell C.D., Castner D.G., Golledge S.L., Ratner B.D., Meyer K., Hagenhoff B., and Benninghoven A. 2001. Static time-of-flight secondary ion mass spectrometry and x-ray photoelectron spectroscopy characterization of adsorbed albumin and fibronectin films. *Surf. Interface Anal.* 31 (8): 724–733.

Tomiyasu B., Fukuju I., Komatsubara H., Owari M., and Nihei Y. 1998. High spatial resolution 3D analysis of materials using gallium focused ion beam secondary ion mass spectrometry (FIB SIMS). *Nucl. Instrum. Methods Phys. Res. B* 136–138: 1028–1033.

Torchilin VP. 2006. Multifunctional nanocarriers. *Adv. Drug Deliv. Rev.* 58 (14): 1532–1555.

Trommer R.M., Santos L.A., and Bergmann C.P. 2007. Alternative technique for hydroxyapatite coatings. *Surf. Coat. Technol.* 201 (24): 9587–9593.

Tsui Y.C., Doyle C., and Clyne T.W. 1998. Plasma sprayed hydroxyapatite coatings on titanium substrates Part 2: Optimisation of coating properties. *Biomaterials* 19 (22): 2031–2043.

Uchic M.D., Holzer L., Inkson B.J., Principe E.L., and Munroe P. 2007. Three-dimensional microstructural characterization using focused ion beam tomography. *MRS Bull.* 32 (5): 408–415.

Umashankar M.S., Sachdeva R.K., and Gulati M. 2010. Aquasomes: A promising carrier for peptides and protein delivery. *Nanomedicine* 6 (3): 419–426.

Vallet-Regi M. 2001. Ceramics for medical applications. *J. Chem. Soc. Dalton Trans.* (2): 97–108.

Vallet-Regi M. 2006. Revisiting ceramics for medical applications. *Dalton Trans.* (44): 5211–5220.

Vallet-Regi M. 2008. Bioceramics: Where do we come from and which are the future expectations. *Key Eng. Mater.* 377: 1–18.

Varma H.K., Kalkura S.N., and Sivakumar R. 1998. Polymeric precursor route for the preparation of calcium phosphate compounds. *Ceram. Int.* 24 (6): 467–470.

Veerapandian M., and Yun K. 2009. The state of the art in biomaterials as nanobiopharmaceuticals. *Dig. J. Nanomater. Bios.* 4 (2): 243–262.

Vetrone F., Variola F., de Oliveira P.T., Zalzal S.F., Yi J.H., Sam J., Bombonato-Prado K.F. et al. 2009. Nanoscale oxidative patterning of metallic surfaces to modulate cell activity and fate. *Nano Lett.* 9 (2): 659–665.

Vogelson C.T. and Barron A.R. 2001. Particle size control and dependence on solution pH of carboxylate-alumoxane nanoparticles. *J. Non-Cryst. Solids* 290 (2–3): 216–223.

Walsh D. and Mann S. 1995. Fabrication of hollow porous shells of calcium carbonate from self-organizing media. *Nature* 377 (6547): 320–323.

Wang X. and Li Y.D. 2007. Monodisperse nanocrystals: General synthesis, assembly, and their applications. *Chem. Commun.* 28: 2901–2910.

Wang X.X., Hayakawa S., Tsuru K., and Osaka A. 2001. A comparative study of in vitro apatite deposition on heat-, H_2O_2-, and NaOH-treated titanium surfaces. *J. Biomed. Mater. Res.* 54 (2): 172–178.

Wang Y.J., Lai C., Wei K., Chen X., Ding Y., and Wang Z.L. 2006b. Investigations on the formation mechanism of hydroxyapatite synthesized by the solvothermal method. *Nanotechnology* 17 (17): 4405–4412.

Wang H., Li Y., Zuo Y., Li J., Ma S., and Cheng L. 2007. Biocompatibility and osteogenesis of biomimetic nano-hydroxyapatite/polyamide composite scaffolds for bone tissue engineering. *Biomaterials* 28 (22): 3338–3348.

Wang C., Ma J., Cheng W., and Zhang R.F. 2002a. Thick hydroxyapatite coatings by electrophoretic deposition. *Mater. Lett.* 57 (1): 99–105.

Wang G., Meng F., Ding C., Chu P.K., and Liu X. 2010. Microstructure, bioactivity and osteoblast behavior of monoclinic zirconia coating with nanostructured surface. *Acta Biomater.* 6 (3): 990–1000.

Wang L., Nemoto R., and Senna M. 2002b. Microstructure and chemical states of hydroxyapatite/silk fibroin nanocomposites synthesized via a wet-mechanochemical route. *J. Nanopart Res.* 4 (6): 535–540.

Wang J. and Shaw L.L. 2009a. Nanocrystalline hydroxyapatite with simultaneous enhancements in hardness and toughness. *Biomaterials* 30 (34): 6565–6572.

Wang J. and Shaw L.L. 2009b. Synthesis of high purity hydroxyapatite nanopowder via sol–gel combustion process. *J. Mater. Sci. Mater. Med.* 20 (6): 1223–1227.

Wang D., Tian Z., Shen L., Liu Z., and Huang Y. 2009. Preparation and characterization of nanostructured Al_2O_3-13wt.%TiO_2 ceramic coatings by plasma spraying. *Rare Met.* 28 (5): 465–470.

Wang X., Zhuang J., Peng Q., and Li Y. 2006a. Liquid–solid–solution synthesis of biomedical hydroxyapatite nanorods. *Adv. Mater.* 18 (15): 2031–2034.

Webster T.J. 2008. NanoTox: Hysteria or scientific studies? *Int. J. Nanomedicine* 3 (2): i–ii.

Webster T.J., Ergun C., Doremus R.H., Siegel R.W., and Bizios R. 2000a. Enhanced functions of osteoblasts on nanophase ceramics. *Biomaterials* 21 (17): 1803–1810.

Webster T.J., Ergun C., Doremus R.H., Siegel R.W., and Bizios R. 2000b. Specific proteins mediate enhanced osteoblast adhesion on nanophase ceramics. *J. Biomed. Mater. Res.* 51 (3): 475–483.

Webster T.J., Ergun C., Doremus R.H., Siegel R.W., and Bizios R. 2001a. Enhanced osteoclast-like cell functions on nanophase ceramics. *Biomaterials* 22 (11): 1327–1333.

Webster T.J., Schadler L.S., Siegel R.W., and Bizios R. 2001b. Mechanisms of enhanced osteoblast adhesion on nanophase alumina involve vitronectin. *Tissue Eng.* 7 (3): 291–301.

Webster T.J., Siegel R.W., and Bizios R. 1999. Osteoblast adhesion on nanophase ceramics. *Biomaterials* 20 (13): 1221–1227.

Webster T.J., Waid M.C., McKenzie J.L., Price R.L., and Ejiofor J.U. 2004. Nano-biotechnology: Carbon nanofibres as improved neural and orthopaedic implants. *Nanotechnology* 15 (1): 48–54.

Weissleder R., Elizondo G., Wittenberg J., Rabito C.A., Bengele H.H., and Josephson L. 1990. Ultrasmall superparamagnetic ironoxide: Characterization of a new class of contrast agents for MR imaging. *Radiology* 175 (2): 489–493.

Williams O.A., Douhéret O., Daenen M., Haenen K., Osawa E., and Takahashi M. 2007. Enhanced diamond nucleation on monodispersed nanocrystalline diamond. *Chem. Phys. Lett.* 445 (4–6): 255–258.

Wopenka B. and Pasteris J.D. 2005. A mineralogical perspective on the apatite in bone. *Mater. Sci. Eng. C Biomim. Mater. Sens. Syst.* 25 (2): 131–143.

Xia W., Zhang D., and Chang J. 2007. Fabrication and in vitro biomineralization of bioactive glass (BG) nanofibres. *Nanotechnology* 18 (13): 135601.

Xu J.L., Khor K.A., Dong Z.L., Gu Y.W., Kumar R., and Cheang P.. 2004. Preparation and characterization of nano-sized hydroxyapatite powders produced in a radio frequency (rf) thermal plasma. *Mater. Sci. Eng.* 374 (1–2): 101–108.

Xu H.H.K., Weir M.D., and Simon C.G. 2008. Injectable and strong nano-apatite scaffolds for cell/growth factor delivery and bone regeneration. *Dent. Mater.* 24 (9): 1212–1222.

Xynos I.D., Hukkanen M.V.J., Batten J.J., Buttery L.D., Hench L.L., and Polak J.M. 2000. Bioglass (R) 45S5 stimulates osteoblast turnover and enhances bone formation in vitro: Implications and applications for bone tissue engineering. *Calcif. Tissue Int.* 67 (4): 321–329.

Yamamoto A., Honma R., Sumita M., and Hanawa T. 2004. Cytotoxicity evaluation of ceramic particles of different sizes and shapes. *J. Biomed. Mater. Res. A* 68A (2): 244–256.

Yang L., Sheldon B.W., and Webster T.J. 2008. Topographical evolution of nanocrystalline diamond and its effect on osteoblast interactions. In *Material Research Society Proceedings*, Fall 2007, Boston, MA, pp. 219–226.

Yang L., Sheldon B.W., and Webster T.J. 2009a. Orthopedic nano diamond coatings: Control of surface properties and their impact on osteoblast adhesion and proliferation. *J. Biomed. Mater. Res. A* 91 (2): 548–556.

Yang L., Sheldon B.W., and Webster T.J. 2009b. The impact of diamond nanocrystallinity on osteoblast functions. *Biomaterials* 30 (20): 3458–3465.

Yang S.F., Yang L.Q., Jin Z.H., Guo T.W., Wang L., and Liu H.C. 2009c. New nano-sized Al2O3-BN coating 3Y-TZP ceramic composites for CAD/CAM-produced all-ceramic dental restorations. Part I. Fabrication of powders. *Nanomedicine* 5 (2): 232–239.

Yao C. and Webster T.J. 2006. Anodization: A promising nano-modification technique of titanium implants for orthopedic applications. *J. Nanosci. Nanotechnol.* 6 (9–10): 2682–2692.

Yeong K.C.B., Wang J., and Ng S.C. 2001. Mechanochemical synthesis of nanocrystalline hydroxyapatite from CaO and CaHPO4. *Biomaterials* 22 (20): 2705–2712.

Yu L. and Jin D. 2001. AES and SAM microanalysis of structure ceramics by thinning and coating the backside. *Surf. Interface Anal.* 31 (4): 338–342.

Yu X.H., Qu H.B., Knecht D.A., and Wei M. 2009. Incorporation of bovine serum albumin into biomimetic coatings on titanium with high loading efficacy and its release behavior. *J. Mater. Sci. Mater. Med.* 20 (1): 287–294.

Zhang P., Hong Z., Yu T., Chen X., and Jing X. 2009. In vivo mineralization and osteogenesis of nanocomposite scaffold of poly (lactide-co-glycolide) and hydroxyapatite surface-grafted with poly(L-lactide). *Biomaterials* 30 (1): 58–70.

Zhang W., Liao S.S., and Cui F.Z. 2003. Hierarchical self-assembly of nano-fibrils in mineralized collagen. *Chem. Mater.* 15 (16): 3221–3226.

Zhang L. and Webster T.J. 2009. Nanotechnology and nanomaterials: Promises for improved tissue regeneration. *Nano Today* 4 (1): 66–80.

Zhao B., Hu H., Mandal S.K., and Haddon R.C. 2005. A bone mimic based on the self-assembly of hydroxyapatite on chemically functionalized single-walled carbon nanotubes. *Chem. Mater.* 17 (12): 3235–3241.

Zhao X., Liu X., Ding C., and Chu P.K. 2006. In vitro bioactivity of plasma-sprayed TiO_2 coating after sodium hydroxide treatment. *Surf. Coat. Technol.* 200 (18–19): 5487–5492.

Zinger O., Zhao G., Schwartz Z., Simpson J., Wieland M., Landolt D., and Boyan B. 2005. Differential regulation of osteoblasts by substrate microstructural features. *Biomaterials* 26 (14): 1837–1847.

4

Synthesis, Properties, Characterization, and Processing of Polymeric Nanobiomaterials for Biomedical Applications

Theoni K. Georgiou
University of Hull

Polymeric materials represent the largest class of biomaterials. Many types of polymers are widely used in biomedical applications that include dental, soft tissue, orthopedic, cardiovascular implants, contact lenses, artificial skin, artificial pancreas, and drug and gene delivery. In this chapter, the synthesis, properties, characterization, processing, and common bioapplications of polymers will be considered. Even though polymeric nanobiomaterials as a term is not explicitly used or discussed in this chapter, the tools, techniques, and biomedical applications discussed are equally relevant in the development of polymeric nanobiomaterials.

Polymers as the Greek origin of the word dictates (polymers—poly = πολύς = many, mers = μέρη = parts) are long-chain molecules that are composed of a large number of small repeating units (monomers). Because polymers are so much bigger compared to the repeating units, their characteristics are much more complex than those of the repeating units. These characteristics determine their properties and consequentially their properties affect their biomedical applications.

4.1 Structural Characteristics of Polymers: Shape and Size

4.1.1 Molecular Architecture

The most obvious characteristics of a polymer species are the shape of its molecules and its skeletal structure, which is also called molecular architecture (Gedde 1995; Young and Lovell 2002; Temenoff and Mikos 2008). A polymer can be linear, branched, ladder-like, star, or cross-linked (Allcock and Lampe 1990; Gedde 1995; Young and Lovell 2002; Temenoff and Mikos 2008), as shown in Figure 4.1.

Linear polymers are usually randomly coiled, unless they exhibit a strong tendency to crystallize (Remmp and Merrill 1991). In that case, they form crystallites of various sizes that are connected by parts of chains that have not undergone crystallization, since, even in the most favorable cases, polymers will never be crystallized to 100% (Remmp and Merrill 1991).

The molecular architecture is important for many properties (Remmp and Merrill 1991; Gedde 1995; Young and Lovell 2002). Short-chain branches reduce crystallinity while long-chain branches have profound effects on rheological properties. Ladder polymers have high strength and thermal stability (Gedde 1995). Hyperbranched polymers (branched and star polymers) have different viscosity than that of their linear analogue with the same molecular weight (Gedde 1995). Cross-linked macromolecules (also called polymeric networks) cannot be dissolved in a solvent but they may have the ability to swell or absorb liquids and other molecules depending on the cross-link density and hydrophobicity/hydrophilicity (Patrickios and Georgiou 2003). Thermoset polymers are highly cross-linked macromolecules, in fact so tightly cross-linked that they cannot be swollen nor melted, and they retain the shape of the molds in which they are manufactured (Remmp and Merrill 1991).

4.1.2 Homopolymers and Copolymers

A homopolymer is a macromolecule that contains only one single type of repeat unit in its chain (Gedde 1995; Young and Lovell 2002; Temenoff and Mikos 2008). The chemical structure of a polymer is usually represented by that of the repeat unit enclosed by brackets. Thus, the hypothetical homopolymer

~A—A—A—A—A—A—A~ is represented by $\{A\}_n$ where n is the number of repeat units linked together.

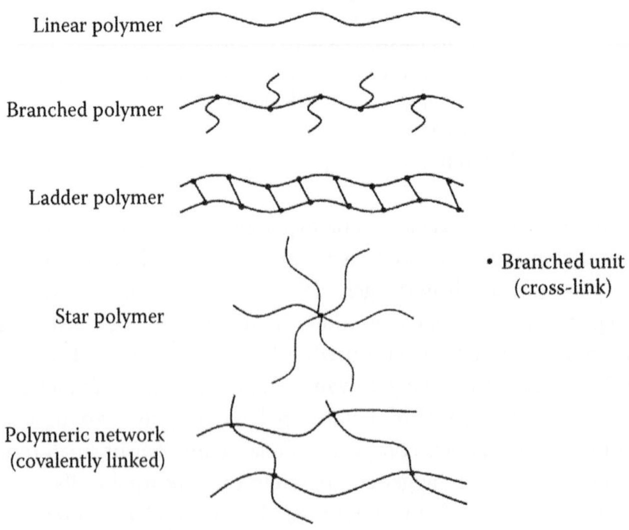

FIGURE 4.1 Schematic representation of structures of polymers with different molecular architecture.

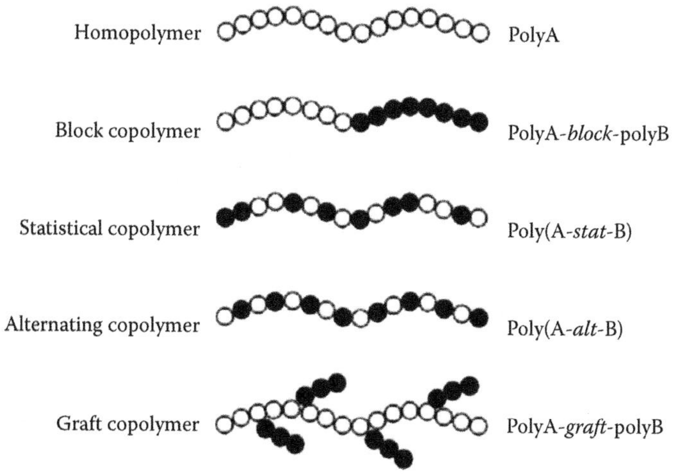

FIGURE 4.2 Homopolymer, different categories of copolymers, and their nomenclature. The different colors represent the different repeating units.

The naming of polymers is often an area of difficulty. The International Union of Pure and Applied Chemistry (IUPAC) recommended a system of nomenclature based on the structure of the monomer or repeat unit (Hiemenz 1984). A polymer is named poly**x**, where **x** is the name of the monomer or the repeat unit, for example, polystyrene. If **x** is more than one word, then parentheses are used, for example, poly(methyl methacrylate) (Gedde 1995). However, many synthetic polymers of commercial importance like Nylon® and Kevlar® are often widely known by trade names (Gedde 1995).

A copolymer consists of two or more repeating units (A, B, etc.) (Gedde 1995; Young and Lovell 2002; and Temenoff and Mikos 2008). There are several categories of copolymers, each being characterized by a particular arrangement of the repeat units along the polymer chain, as shown in Figure 4.2. Block copolymers (polyA-*block*-polyB or A_x-*b*-B_z, where x and z are the degree of polymerizations of A and B, respectively) are polymers in which the repeat units exist in long sequences or blocks of the same type (Young and Lovell 2002). Statistical copolymers, poly(A-*stat*-B), are copolymers in which the sequential distribution of repeat units obeys known statistical law (e.g., Markovian) (Gedde 1995). Random copolymers, poly(A-*ran*-B), are a special type of statistical copolymer in which the distribution of repeat units is truly random (in older textbooks and scientific papers, the term random is often used to describe both random and nonrandom statistical copolymers). Alternating copolymers have only two types of repeat units and these are arranged alternately along the polymer chain (Remmp and Merrill 1991; Young and Lovell 2002). Finally, graft copolymers are branched polymers in which the branches have a different chemical structure to that of the main chain. In the simplest form, they consist of a main homopolymer chain with branches of a different homopolymer. It should be noted that a different nomenclature is used for an unspecified copolymer—poly(A-*co*-B)—and that the copolymers shown in Figure 4.2 consist of only two repeat units. Copolymers can comprise more than two repeat units and this increases the number of different possible ways of distributing the repeat units within the polymer chain and, thus, the molecular architecture. In Figure 4.3, an example of the different architectures triblock copolymers (consisting of three different types of repeating units) can have is shown, while diblock copolymers can only have one (Figure 4.2).

4.1.3 Tacticity: Stereoisomerism

Polymers are capable of assuming many conformations through rotation of valence bonds (Callister 2003; Abramson et al. 2004). Thus, different stereoisomers can be observed. Stereoisomerism denotes the situation in which atoms are linked together in the same order but differ in their spatial arrangement

FIGURE 4.3 Different architecture of triblock copolymers. The different colors represent the different repeating units.

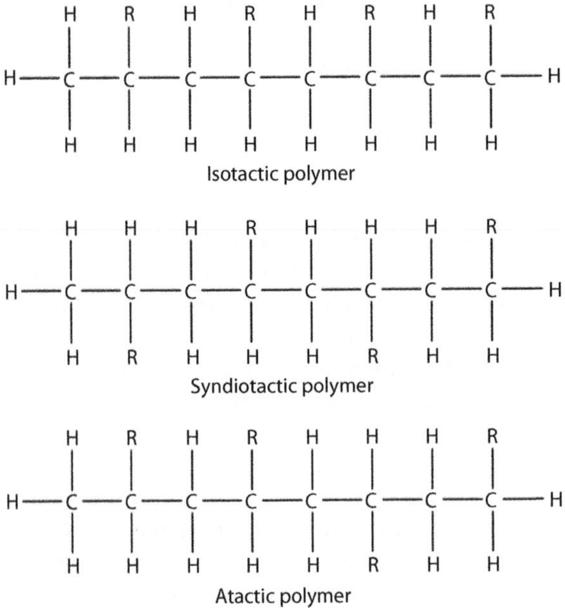

FIGURE 4.4 Schematic of the stereoisomers.

(Callister 2003), as shown in Figure 4.4. For one stereoisomer, all the R groups are situated on the same side of the chain; this is called an isotactic configuration. In a syndiotactic configuration, the R groups alternate the side of the chain, and when they have a random positioning, the term atactic configuration is used.

4.1.4 Molecular Weight: Definition and Distribution

The most important characteristic of polymers that influences all their properties is the molecular weight. Unlike for small molecules, for polymers, there is more than one definition for the molecular weight. This occurs from the fact that when synthesizing a polymer it is usually produced with a distribution of molecular weights; the number of "mers" (structural units) that is defined as the degree of polymerization differs for each polymer (Remmp and Merrill 1991). Therefore, the term molecular weight (or degree of polymerization) cannot be used and the term *average* molecular weight (or *average* degree of polymerization) is introduced (Remmp and Merrill 1991).

The *number average degree of polymerization* P_n is defined as follows:

$$P_n = \frac{\sum_{i=1}^{i=\infty} in_i}{\sum_{i=1}^{i=\infty} n_i} = \sum_{i=1}^{i=\infty} ix_i \tag{4.1}$$

where

n_i is the number of molecules with i monomer units

x_i is the mole fraction of molecules with i monomer units in the chain

If M_i is the molecular weight of this species, the *number average molecular weight* is expressed as

$$M_n = \frac{\sum_{i=1}^{i=\infty} n_i M_i}{\sum_{i=1}^{i=\infty} n_i} = \sum_{i=1}^{i=\infty} x_i M_i = m_0 \sum_{i=1}^{i=\infty} ix_i \tag{4.2}$$

where

m_0 is the molecular weight of a repeat unit, hence assumed constant

M_n is the total weight of polymer divided by the total number of polymer molecules in the sample

The *weight average molecular weight* M_w is defined as the sum of the products of the molecular weight of each fraction multiplied by its weight fraction.

$$M_w = \sum_{i=1}^{i=\infty} w_i M_i \tag{4.3}$$

In terms of the number of molecules, the weight average molecular weight can be expressed as

$$M_w = \frac{\sum_{i=1}^{i=\infty} n_i M_i^2}{\sum_{i=1}^{i=\infty} n_i M_i} \tag{4.4}$$

The ratio of the weight average with the number average molecular weight, M_w/M_n, which by definition should be greater or equal to one, is referred to as the polydispersity index (PDI) and provides important information about the width of the molecular distribution of a polymer sample (Remmp and Merrill 1991; Young and Lovell 2002; Abramson et al. 2004). For an ideal, monodisperse polymer, the value of PDI is one, while the less ideal the polymer sample is (contains polymers with a wide range of molecular weights) the higher the value of the PDI.

4.2 Synthesis

Polymerization reactions can be, in simple terms, classified into two main types: step-growth polymerizations and chain-growth polymerizations (Gedde 1995; Young and Lovell 2002; Abramson et al. 2004). In step-growth polymerization, the polymer chains grow stepwise by reactions that occur between two molecular species, while in chain-growth polymerizations, the polymer chains grow only by reaction of monomer with the reactive end-group on the growing chains (Young and Lovell 2002).

A typical example of step-growth polymerization, also called condensation polymerization, is the synthesis of Nylon 6,6 (shown in Figure 4.5) (Gedde 1995; Abramson et al. 2004). Two monomers react

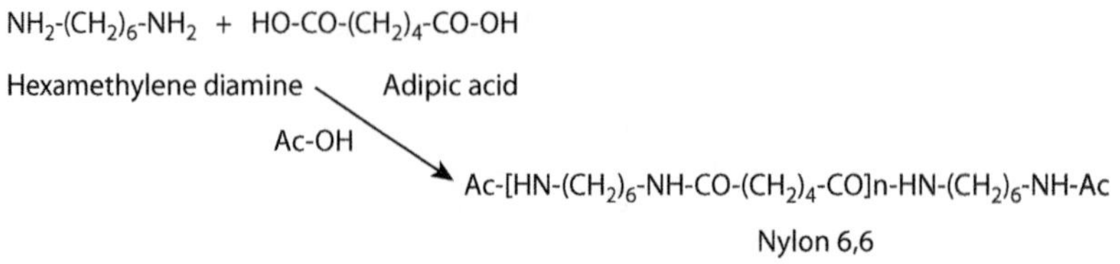

FIGURE 4.5 Nylon 6,6 synthesis by condensation polymerization.

to form a covalent bond, usually with elimination of a small molecule such as water, hydrochloric acid, methanol, or carbon dioxide. Step-growth polymerization is involved in the formation of polyesters and polyamides. Different techniques are available for obtaining a high yield and high molar mass (Gedde 1995). Moreover, polymers with different molecular architectures can be made using monomers of different functionality—trifunctional monomers yield branched and ultimately cross-linked polymers (Gedde 1995). Common biomaterials prepared with this polymerization method are nylon and polyurethanes (Remmp and Merrill 1991; Gedde 1995; Abramson et al. 2004).

Chain-growth polymerization with the exception of ring opening polymerization involves the polymerization of unsaturated monomers (Gedde 1995; Abramson et al. 2004). It usually requires an initial reaction between the monomer and an initiator to start the growth of the chain and thus involves several consecutive stages: initiation, propagation, and termination (Remmp and Merrill 1991; Gedde 1995; Abramson et al. 2004). Each chain is individually initiated and grows until its growth is terminated. The initiators can be free radical, cations, anions, or stereospecific catalysts. The initiator opens the double bond of the monomer, creating another initiation site on the opposite side of the monomer bond for continuing growth. Rapid chain growth ensues during the propagation step until the reaction is terminated by reaction with a radical or a molecule, depending on the polymerization technique. Chain-growth polymerization can be divided into several subgroups depending on the mechanism: radical, anionic, cationic, or coordination polymerization (Remmp and Merrill 1991; Abramson et al. 2004). Commonly used biomaterials that are prepared by step-growth polymerizations are polymethacrylates like poly(methyl methacrylate), PMMA, poly(2-hydroxyl ethyl methacrylate), PHEMA, and poly[2-(dimethylamino)ethyl methacrylate], PDMAEMA, shown in Figure 4.6. It should be noted that some polymerization techniques called "living" or "controlled" polymerization techniques enable the synthesis of well-defined polymers with narrow molecular weight distributions, and the synthesis of polymers with different architectures like block copolymers and star polymers (Webster 1991; Matyjaszewski and Müller 2006). Examples of these techniques include the conventional "living" anionic polymerization (Szwarc 1956; Szwarc et al. 1956; Hadjichristidis et al. 2001), group transfer polymerization (GTP) (Webster et al. 1983; Webster 2000, 2004), ring-opening polymerization (Hashimoto 2000), quasi-living carbocationic polymerization (Kennedy and Iván 1992), and more recently developed polymerization techniques like reversible addition–fragmentation chain transfer (RAFT) polymerization (Moad et al. 2006) and atom transfer radical polymerization (ATRP) (Patten and Matyjaszewski 1998).

4.3 Properties

4.3.1 Crystallinity

Polymers can be divided into fully amorphous and semicrystalline (Gedde 1995; Callister 2003; Abramson et al. 2004). The fully amorphous polymers show no sharp crystalline Bragg reflection in the x-ray diffractograms taken at any temperature (Allcock and Lampe 1990). The reason why these polymers are unable to crystallize is commonly their irregular chain structure and their small side groups (Gedde 1995; Abramson et al. 2004). Atactic polymers, statistical copolymers, and highly branched polymers belong to this class of polymers (Gedde 1995). The semicrystalline polymers show crystalline Bragg reflections superimposed on an amorphous background because they always consist of

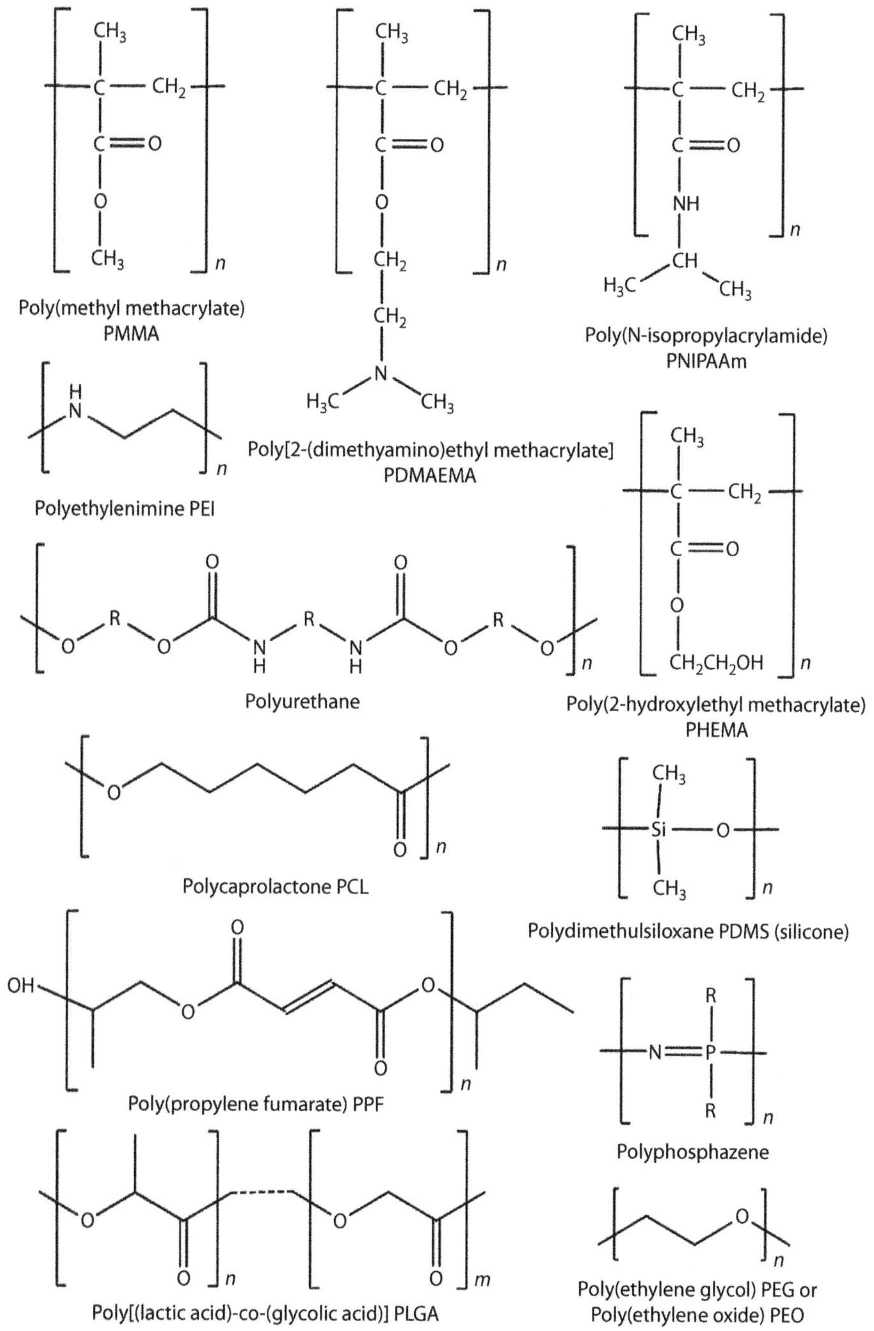

FIGURE 4.6 Chemical structures of common polymeric biomaterials.

two components differing in the degree of order: a component composed of crystals and an amorphous component (Gedde 1995; Callister 2003). The degree of crystallinity can be as high as 90% for certain low molecular weight polyethylenes and as low as 5% for polyvinylchloride (Gedde 1995).

To some extent the physical properties of polymeric materials are influenced by the degree of crystallinity. The presence of crystallites in the polymer usually leads to enhanced mechanical properties, unique thermal behavior, and increased fatigue strength (Callister 2003; Abramson et al. 2004).

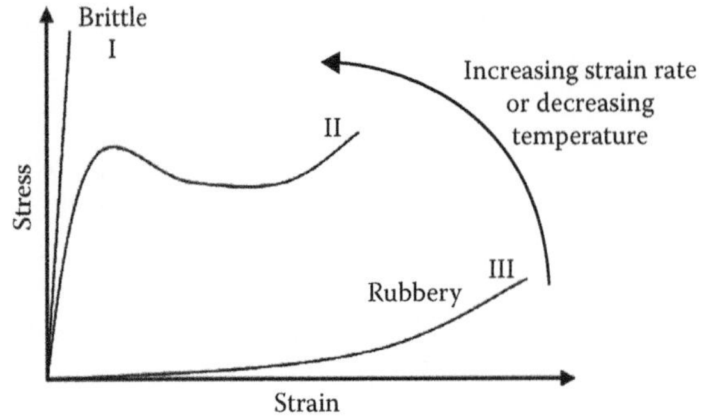

FIGURE 4.7 The stress–strain behavior of (I) brittle, (II) plastic, and (III) highly elastic polymeric materials.

4.3.2 Mechanical Properties

The tensile properties of polymers can be characterized by their stress–strain response and their deformation behavior (Callister 2003; Abramson et al. 2004). Three typically different types of stress–strain behavior are found in polymeric materials (Figure 4.7). Curve I illustrates the stress–strain character of a brittle polymer; Curve II is typical for a plastic material and the initial deformation is elastic, which is followed by yielding and a region of plastic deformation; and Curve III illustrates a totally elastic polymer—a class of polymers that are called elastomers, which have a rubber-like elasticity (Callister 2003; Abramson et al. 2004).

The mechanical property is influenced by the freedom of motion of the polymer chain. The freedom of motion is retained at a local level while a network structure resulting from chemical cross-links and/or chain entanglements prevents large-scale movements or flow. Rubbery polymers tend to exhibit a lower modulus, or stiffness and extensibilities of several hundred percent (Callister 2003; Abramson et al. 2004). Glassy and semicrystalline polymers have higher moduli and lower extensibilities (Callister 2003; Abramson et al. 2004).

The ultimate mechanical properties of polymeric materials at large deformations are important in selecting particular polymers for biomedical applications (Abramson et al. 2004). For example, a rigid, strong material is more suitable for a hip implant, whereas a flexible, less strong material would be sufficient for a vascular graft. Furthermore, the ultimate strength of polymer (the stress at or near failure) is also very important since failure for many biomaterials is catastrophic. Finally, the fatigue behavior of polymers is also important in evaluating materials for applications where dynamic stress is applied, for example, cardiovascular implants that must be able to withstand many cycles of pulsating motion.

4.3.3 Thermal Properties

Unlike small molecules, most polymers exhibit another transition upon decreasing the temperature. The temperature point that this transition happens is called the glass transition temperature, T_g. The glass transition occurs in amorphous (or glassy) and semicrystalline polymers and is due to a reduction of motion of large segments of molecules with decreasing temperature (Callister 2003). The long segments of the polymer before the T_g have enough thermal energy to randomly move, but after the T_g the segment motion ceases (Abramson et al. 2004).

Upon cooling a polymer, it gradually transforms from a liquid to a rubbery material, and finally to a rigid solid (Callister 2003). The latter change, from rubbery to solid, corresponds to the T_g (Callister 2003) and it is different for every polymer (Abramson et al. 2004). In addition, abrupt changes in other physical properties accompany this glass transition, for example, stiffness, heat capacity, and coefficient of thermal expansion (Abramson et al. 2004). Even so, the glass transition is not considered a true thermodynamic phase transition like melting of a crystal (Gedde 1995) and it also takes place over a range of temperatures, usually in a 5°C–10°C temperature span (Callister 2003; Abramson et al. 2004). The T_g, therefore, is an important parameter that has to be taken into consideration for the polymers' applications. For most biomedical applications, depending on the temperature applied, polymers that are within their rubbery region are targeted (Abramson et al. 2004).

The T_g is affected by the molecular weight and the degree of branching of a polymer. Specifically, the increase of the molecular weight tends to raise the T_g (Callister 2003), while a small amount of branching will tend to lower the T_g (Callister 2003). On the other hand, a high density of branches reduces chain mobility and elevates the T_g (Callister 2003). It has been observed that when some amorphous polymers are cross-linked, the T_g is elevated since cross-links restrict molecular motion (Callister 2003). Since the cross-links inhibit flow at all temperatures, chemically cross-linked polymers do not display flow behavior and, thus, cannot be melt processed like linear polymers (Abramson et al. 2004). Instead these materials are processed as reactive liquids or high-molecular-weight amorphous gums that are cross-linked during molding to give the desired product (Abramson et al. 2004). In this way, some polymers can be machined to be formed into useful shapes like, for example, PHEMA (Figure 4.6), the polymeric material used for soft lenses (Abramson et al. 2004).

The T_g is also affected by the polymers' composition and architecture. A copolymer can exhibit two different T_gs or one, depending on its composition and architecture. In particular, a random copolymer will exhibit a T_g that approximates the weighted average of the T_g values of the two homopolymers (Abramson et al. 2004). Block copolymers of sufficient size and incompatible block types will exhibit two individual transitions, each one characteristic of the homopolymer of one of the component blocks (in addition to other thermal transitions), but slightly shifted, owing to incomplete phase separation (Abramson et al. 2004). Even segmented copolymer networks, polymer networks with one type of repeating unit are placed in different segments (blocks) (Patrickios and Georgiou 2003), may exhibit two individual transitions depending on the cross-linking density and the size (Guan et al. 2000).

4.3.4 Aqueous Solution Properties

In solution, the properties of the polymers depend on their compatibility with the solvent; if they are thermodynamically compatible with the solvent. The Flory–Huggins theory describes the thermodynamics of polymer solutions and provides a useful parameter that describes the compatibility of the polymer chain in the solvent, the Flory–Huggins interaction parameter, χ.

This parameter χ is proportional to the square of the difference of the Hildebrand solubility parameters of the solvent, δ_1 and the polymer, δ_2; $\chi \propto (\delta_1 - \delta_2)^2$ (Gedde 1995; Young and Lovell 2002; Rubinstein and Colby 2003) and depending on its value the solvent is considered either a good or a bad solvent for the polymer. In particular, when $\chi > ½$ the solvent is a poor solvent for the polymer and the polymer is precipitated (phase separation) or adopts a collapsed conformation so it will interact as little as possible with the solvent. When $\chi < ½$ the polymer is in a "good" solvent and the polymer chain is extended (Gedde 1995; Rubinstein and Colby 2003). Finally, when $\chi = ½$ the borderline between the good and poor solvent conditions apply and these conditions are called theta, θ (Gedde 1995; Rubinstein and Colby 2003). At θ conditions the polymer has no preference in interacting with itself or the solvent and it adopts a random coil conformation (Gedde 1995; Rubinstein and Colby 2003). The χ parameter is also affected by the temperature as is illustrated in the following equation:

$$\chi = \frac{1}{2} - \frac{C}{\theta}\left(1 - \frac{\theta}{T}\right) \tag{4.5}$$

where

C is a constant

θ is a number, a temperature characteristic for each polymer

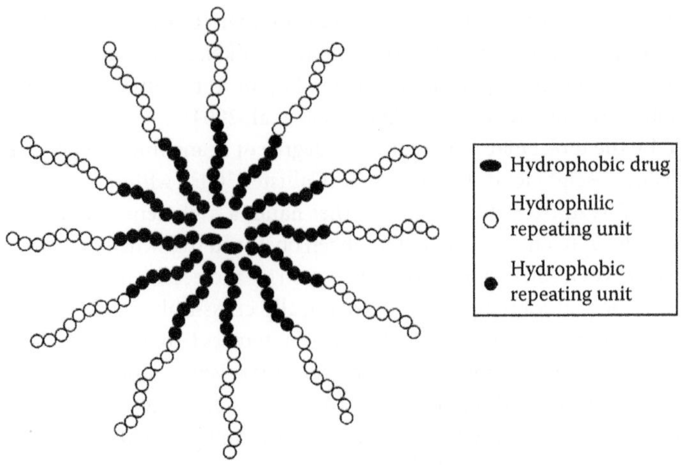

FIGURE 4.8 Drug encapsulation by a micelle. The micelle is formed by amphiphilic block copolymers.

When the temperature T is equal to θ, it is called the theta temperature or Flory temperature and θ conditions apply (Gedde 1995), while depending on the polymer above θ temperature, the polymer phase separates or is soluble in the solvent. In particular, some polymers exhibit an upper critical solution temperature (UCST), while some other polymers exhibit a lower critical solution temperature (LCST) (Gedde 1995; Young and Lovell 2002; Rubinstein and Colby 2003). This temperature limit is important for many applications since many common polymeric biomaterials are based on polymers that exhibit LCST, like poly(ethylene glycol), PEG, also called poly(ethylene oxide), and poly(*N*-isopropylacrylamide), PNIPAAm (Figure 4.6).

For copolymers, the interaction parameters and the compatibility of the polymer and of a segment of the polymer with a solvent are very important. For block copolymers or segmented polymers these parameters become crucial. If the block copolymer is in a solvent that is specific; the solvent will interact more with one block of the copolymer than the other. This will force the block copolymer to form aggregates or micelles. Specifically, if one block copolymer is amphiphilic, it consists of a block that is hydrophilic (from Greek, it means "friend" with the water) and a block that is hydrophobic (from Greek, it means "fears" the water), then the block copolymer is self-assembled in water in such a manner than the hydrophobic part is in contact with the water as little as possible. Since the hydrophobic block of the polymer is not thermodynamically compatible with the solvent that is water, the water compels the polymer chains in solution to form micelles (Hadjichristidis et al. 2003). Micelles are of great importance in biomedical applications since these functional nanomaterials are used for drug and gene delivery. In Figure 4.8 a schematic representation of a micelle (formed by amphiphilic block copolymers) that encapsulates a hydrophobic drug is shown.

4.3.5 Degradation

Many polymers that are commonly used in biomedical applications are degradable. The degradation of the polymer has a crucial rule for the materials applicability. In particular, the material should have the appropriate mechanical properties for the indicated application, and the variation in mechanical properties with degradation should be compatible with the healing or regeneration process (Nair and Laurencin 2007). Other properties that a degradable polymer should have in order to be bioapplicable are biocompatibility, degradation time that matches the healing or regeneration process, degradation products that are nontoxic, and the ability to get metabolized and cleared from the body (Nair and Laurencin 2007). The most common degradable functional groups that biodegradable polymers bear are esters, orthoesters, anhydrides, carbonates, amides, β-amino esters, and urethanes (Nair and Laurencin 2007). The degradation of the polymer or the polymeric material can be studied by monitoring the weight loss (if it is a cross-linked polymer material such as a scaffold for tissue engineering), the molecular weight of the polymer (with techniques that determine the molecular weight), or the breaking of specific chemical bonds (with techniques that analyze the chemical structure of the polymer).

4.4 Characterization

4.4.1 Molecular Weight Analysis

The most important characterization technique of polymers is gel permeation chromatography (GPC), also called size exclusion chromatography (SEC). With this technique, the molecular weight and the molecular weight distribution of a polymer can be determined (Hiemenz 1984; Allcock and Lampe 1990; Remmp and Merrill 1991; Young and Lovell 2002; Callister 2003; Abramson et al. 2004; Temenoff and Mikos 2008). A typical GPC setup is shown in Figure 4.9. A pump pumps the solvent from the solvent reservoir to the collector flask, while it passes through the set of columns and the detector. The sample is injected in the injection port and then it passes through the set of columns. In the columns, the polymer molecules are separated in terms of their size. The smaller polymer molecules enter the smaller pores of the columns and delay, while the bigger polymer molecules do not, and elute faster (Figure 4.9). After passing through the column, the polymer solution passes through the detector (Allcock and Lampe 1990) and then it is collected in the collector flask. Common detectors include a differential defractometer, absorption spectrophotometric detection (such as ultraviolet and infrared), light scattering photometer, and viscometer (Rubinstein and Colby 2003). The most common one is the differential defractometer.

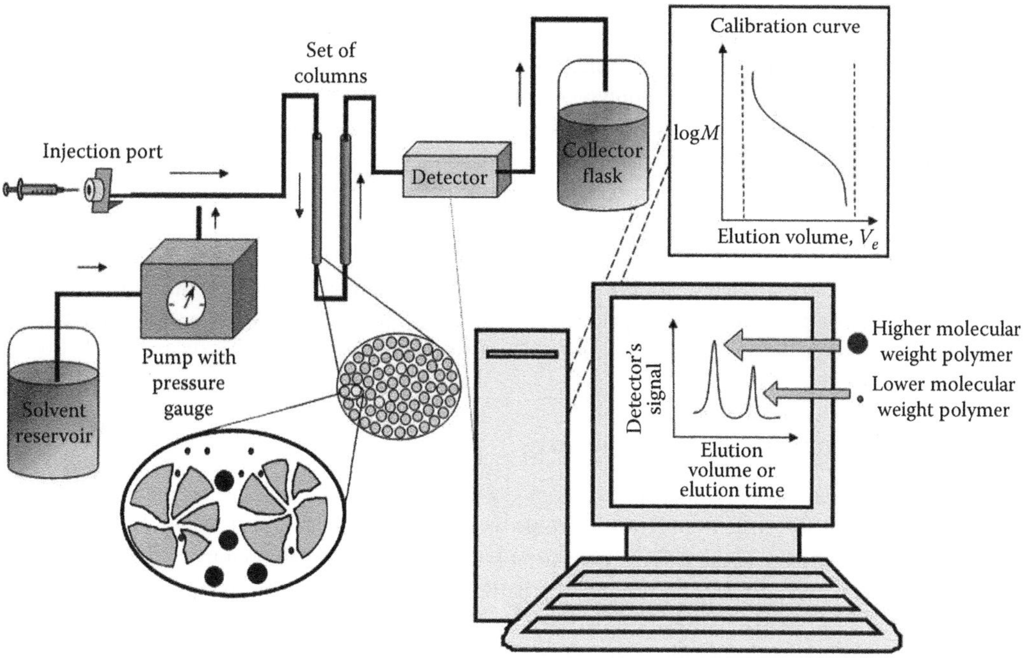

FIGURE 4.9 A typical GPC setup. The sample is injected at the injection port and it passes through the column where the molecules are separated in terms of their size. The bigger molecules do not enter the pores of the polymer beads that the columns are packed with, while the smaller molecules enter the pores and delay, thus eluting later than the bigger molecules.

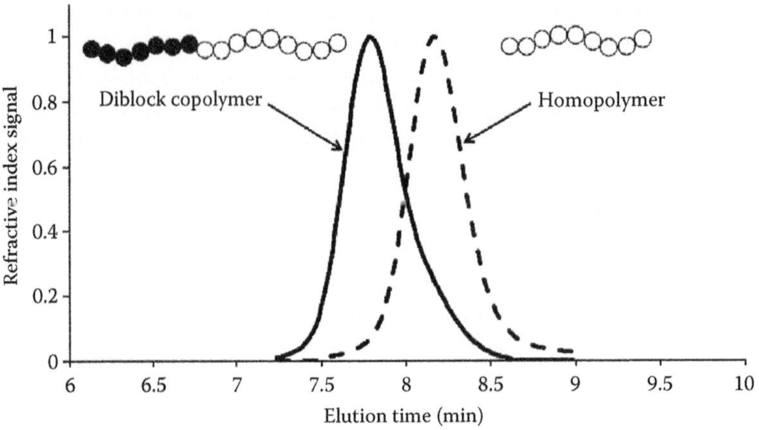

FIGURE 4.10 A typical GPC chromatogram of a diblock copolymer and its precursor.

The detector monitors the concentration of the polymer that is eluted and the chromatograph obtained is a plot of concentration against elution volume, which provides a qualitative indication of the molecular weight distribution. In order to convert a GPC chromatogram into a molecular weight distribution (M_w/M_n) and also calculate the average molecular weights, it is necessary to know the relationship between the molecular weight and the elution volume, V_e. A calibration curve is usually obtained with the use of polystyrene or PMMA standards for GPC systems in organic solvents and PEG standards for aqueous GPC systems.

It should be noted that it is very common to present the chromatograph with respect to the elution time and not the elution volume. What is important to remember is that the higher the molecular weight the smaller the elution volume (or the shorter the elution time) (Allcock and Lampe 1990; Remmp and Merrill 1991). A typical GPC chromatogram is shown in Figure 4.10. The first peak from the right corresponds to the precursor to a diblock copolymer, a homopolymer, while the peak on the left corresponds to the diblock copolymer. The fact that the peak of the diblock copolymer is at a shorter time confirms that the molecular weight of the diblock copolymer is of course bigger than the molecular weight of its precursor. Other useful information that can be obtained from the chromatogram is the full conversion of the homopolymer to the diblock copolymer due the lack of any extra peak in addition to the diblock copolymer curve.

There are other conventional techniques that also determine the molecular weight of polymers but not the molecular weight distribution. Specifically, with static light scattering (SLS) and osmometry, the M_w and M_n can also be determined, respectively (Abramson et al. 2004).

4.4.2 Determination of the Structure

Nuclear magnetic resonance (NMR) spectroscopy is commonly used for determination and confirmation of the chemical structure of polymers (Drobny et al. 2003; Abramson et al. 2004). NMR can provide both qualitative and quantitative information with respect to the comonomer composition and the stereochemical configuration of the polymeric molecules (Drobny et al. 2003). This is due to the fact that there is a proportional relation between the observed peak intensity in the NMR spectrum and the number of nuclei that produce the signal. Both conventional solution and solid-state (particularly for nonsoluble materials) NMR techniques are used for the characterization of polymeric materials (Mathur and Scranton 1996; Drobny et al. 2003; Abramson et al. 2004; Zhang et al. 2005). Many types of nuclei can be observed, but the most frequently used for polymers are proton, ^{1}H NMR and carbon-13, ^{13}C NMR (Drobny et al. 2003).

^{1}H NMR is widely used in order to provide information on the monomeric species used in the preparation of polymers (confirm the chemical structure), the average composition (for copolymers), tacticity, and configuration of polymeric chain (Mathur and Scranton 1996; Drobny et al. 2003; Abramson et al. 2004; Zhang et al. 2005). These studies are done in solution and a disadvantage is that polymer spectra are frequently poorly resolved with broad overlapping lines (Drobny et al. 2003). On the other hand, ^{13}C NMR is more revealing than ^{1}H NMR in polymer work because of the inherently wider spectra separation of the carbon chemical shifts that makes these spectra more interpretable (Drobny et al. 2003; Zhang et al. 2005). NMR can also be used to study micellar solutions and investigate the phenomena within micelles (Drobny et al. 2003), thus provide important information for the biomedical applications of micelles. Solid-state NMR, which is not as conventional, is very useful, since it provides information about secondary structure of polymers, proteins, and peptides (Mathur and Scranton 1996).

Infrared (IR) absorption spectroscopy is also used to provide information on the chemical, structural, and conformational aspects of polymeric chains (Abramson et al. 2004; Kasaal 2008). In IR spectroscopy, absorption of energies corresponding to transitions between vibrational or rotational energy states gives rise to characteristic patterns (Drobny et al. 2003). These characteristic patterns can be translated into qualitative and quantitative information regarding the presence of functional groups, thus identifying the monomer types and their concentration within the polymer chain (Drobny et al. 2003). IR spectroscopy is often used to monitor the degradation and modification of polymeric biomaterials like chitosan and polyurethane (Griesser 1991; Kasaal 2008).

Wide-angle x-ray scattering (WAXS) is a technique useful for providing the local structure of semicrystalline polymeric solid or polymeric networks (Gedde 1995; Callister 2003; Abramson et al. 2004; Matyjaszewski and Müller 2006). Under appropriate conditions, crystalline materials diffract x-rays, giving rise to spots or rings, and these, according to Bragg's laws, can be interpreted as interplanar spacings. By using the appropriate model to fit the data, the crystalline chain conformation and atomic placements can be inferred, for example, if the chain is extended or it has the form of a helix (Abramson et al. 2004). WAXS is used for assessing structure with repeating distances typically less than 1 nm, whereas small angle x-ray scattering (SAXS) is useful for assessing bigger structures (Gedde 1995).

In particular, SAXS is used to determine the structure of many multiphase materials (Gedde 1995; Abramson et al. 2004) and it has been applied to study polymer systems for more than 30 years (Hadjichristidis et al. 2003). This technique requires an electron density difference to be present between two components (Abramson et al. 2004). It has been widely applied to morphological studies of copolymers and ionomers since it can provide information about the molecular weight, overall size, and internal structure of individual micelles (Abramson et al. 2004; Nair and Laurencin 2007). It can probe features of 1–100 nm in size. With appropriate modeling of the data, SAXS can provide detailed structural information like the dimensions of a micellar core (Abramson et al. 2004; Nair and Laurencin 2007).

Small angle neutron scattering (SANS) is in a way very similar to SANS and it is also used to provide information about the dimensions of nanophases of polymer samples of 1–100 nm. This technique is based on scattering neutrons and it also requires the two components to have different scattering densities. It is very common for deuterated analogues of the solvent or one part of the polymer to be used (Drobny et al. 2003). It is commonly used to characterize polymers (especially block copolymers), proteins, DNA, polymeric networks (Seymour et al. 1998; Harada and Kataoka 2006; Melnichenko and Wignall 2007), and their interactions (complexes) (Galant et al. 2005; Melnichenko and Wignall 2007; Horkay and Hammouda 2008). The latter makes SANS a very useful technique to characterize polymer-DNA complexes that are used in gene delivery. Unfortunately, however, SANS instruments are only available at a few places around the world.

4.4.3 Mechanical and Thermal Properties Studies

Dynamic mechanical analysis (DMA) can be used to study the mechanical properties and the deformation behavior of polymers (Abramson et al. 2004; Menard 2008). It can be simply described as applying an oscillating force to the sample and analyzing the material's response to that force (Menard 2008).

Therefore, properties like the tendency to flow (viscosity) and the stiffness (modulus) can be calculated from the phase lag and the sample recovery, respectively (Menard 2008). These properties are often described as the ability to lose energy as heat (damping) and the ability to recover from deformation (elasticity) (Menard 2008), which are of great importance for the biomedical applicability of polymers. What is usually measured with DMA is the sample modulus, which, of course, depends on the temperature, for example, glass at low temperatures has a high modulus, while a rubber at high temperatures has a low modulus. The T_g of the polymers can also be determined by DMA (Abramson et al. 2004).

Differential scanning calorimetry (DSC) is another method that provides information about the thermal properties of the polymers. Specifically by DSC, the crystallization temperature, T_c, the melting temperature, T_m as well as the T_g of a polymer can be determined (Abramson et al. 2004; Kasaal 2008). DSC can also provide useful information about the degradation of a material since many polymeric materials used for bioapplications can be thermolyzed.

4.4.4 Surface Characterization

The surface characteristics of polymeric biomaterials are critically important since it is the surface of the material that will be in contact with the body, and the surface properties and composition are different from the bulk (Abramson et al. 2004). Atomic force microscopy (AFM) and scanning electron microscopy (SEM) are commonly used to characterize the surface of a polymeric biomaterial. SEM provides images of surfaces by focusing an electron beam on it, while AFM provides images of surfaces by applying force on it (Abramson et al. 2004).

Specifically, in AFM a sharp tip attached to a cantilever is scanned across a surface. As the tip moves over the material's surface, changes in surface topography change the interatomic attractive or repulsive forces between the surface and the tip (Dee et al. 2002). The height adjustments or changes in interatomic force are recorded and used to construct images of surface topography (Dee et al. 2002). The resolution of AFM depends on the size of the tip (Dee et al. 2002)—the sharper the tip, the better the resolution. Under the proper conditions, images showing individual atoms can be obtained. Thus, a major feature of AFM is the ability to acquire three-dimensional images with Å or nm lever resolution (Dee et al. 2002). One of AFM's advantages is that imaging can be conducted without scanning, coating, or other preparation and under physiological conditions (Dee et al. 2002).

In SEM, an electron beam is scanned across the sample's surface (Dee et al. 2002; Abramson et al. 2004). The primary electrons penetrate the surface and transfer energy to the material (Dee et al. 2002). In this way, sufficient energy is transferred to the sample and thus electrons (secondary electrons) are emitted from the sample. The intensity of the secondary electrons primarily depends on the topography of the surface (Dee et al. 2002); thus, by scanning the electron beam across the surface and determining the current generated from secondary electrons, images of the surface are obtained (Dee et al. 2002; Abramson et al. 2004). Some chemical information can be obtained from SEM but it is not specific (Dee et al. 2002); brighter and darker images reflect higher atomic number and lower atomic number, respectively (Dee et al. 2002). The disadvantages of SEM is that nonconductive samples, like most polymers and biological materials, must be coated with a conductive film, and that is conducted in a high-vacuum environment, which prevents biological samples from being investigated in their native state (Dee et al. 2002).

X-ray photoelectron spectroscopy (XPS) also known as electron spectroscopy for chemical analysis is based on the process of photoemission and provides chemical information about the surface (identification of the elements of the surface, determination of approximate atomic concentrations, and information about the chemical bonding) (Dee et al. 2002; Abramson et al. 2004).

Finally, contact angle measurements are used to characterize polymeric materials (Dee et al. 2002; Abramson et al. 2004) and are significant, since the adhesion of a number of cells types, including bacteria, granulocytes, and erythrocytes, has been shown, under certain conditions to correlate with solid–vapor surface tension (Abramson et al. 2004).

4.4.5 Characterization in Solution

In previous sections of this chapter, some techniques that characterize the bulk phase of polymers as well as their size in solution were described (SANS and SAXS). There are other techniques that can analyze the polymers in solution. SLS as mentioned before can give information about the M_w of the polymer. It can also provide information about the size of the polymer—specifically, the radius of gyration (R_g) of a polymer (Hiemenz 1984; Allcock and Lampe 1990; Gedde 1995; Young and Lovell 2002; Hadjichristidis et al. 2003). R_g is the root mean square distance of every point of the macromolecular chain from its center of mass. In order to obtain this information, a Zimm plot must be made that follows the equation

$$\frac{KC}{R_\theta} = \left(\frac{1}{M_w} + 2A_2C + \cdots\right)\left(1 + \frac{16\pi^2}{3\lambda^2}R_g^2\sin^2\frac{\theta}{2} + \cdots\right) \tag{4.6}$$

where

λ is the wavelength of the laser of the equipment
θ is the angle at which the detector is located with respect to the transmitted beam
A_2 is the second virial coefficient (a measure of solvent–solute interactions)
K is the material constant
R_θ is the Rayleigh ratio (contains information about the refractive index of the material)
C is the concentration of the polymer solution

However, in order to obtain a Zimm plot (an example of which is shown in Figure 4.11), light scattering measurements of polymer solution of different concentrations at different angles must be made, and that requires a considerable amount of sample. Moreover, these measurements can often prove to be time consuming and tricky.

It should be noted that unlike GPC, SLS does not require a calibration curve in order to determine the molecular weight of the polymer. However, the refractive index of the polymer that is being analyzed must be known in order to obtain the Zimm plot.

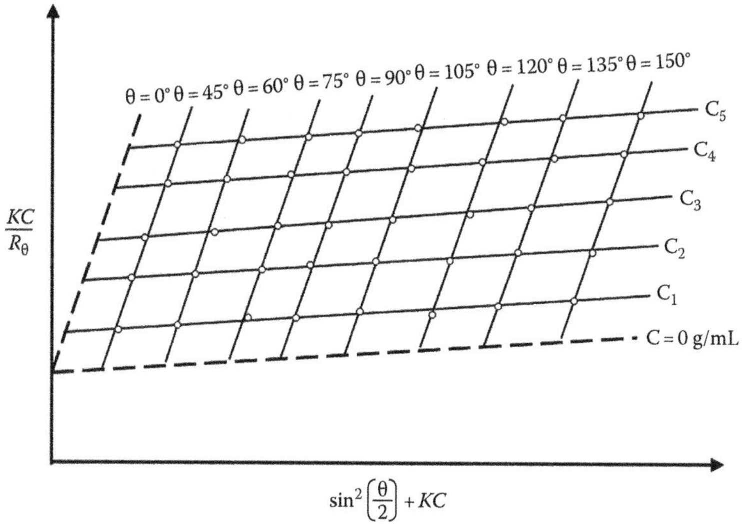

FIGURE 4.11 Zimm plot obtained from SLS measurements. A number of solution of varying concentrations are measured at different angles and the data are extrapolated to zero concentration and angle to determine the molecular weight and the radius of the polymer.

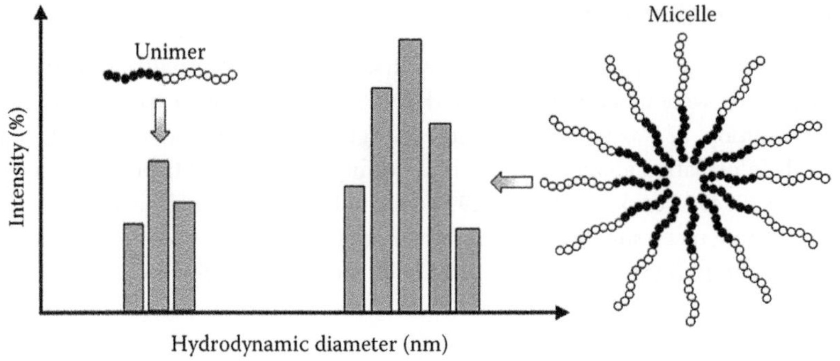

FIGURE 4.12 Dynamic light scattering results: an example of a histogram of the distribution of the hydrodynamic diameter of a block copolymer in aqueous solution. The distribution is bimodal with population being that of micelles formed by block copolymers and the other population being the unimers of the block copolymer.

Dynamic light scattering (DLS) also called quasi-elastic light scattering or photon correlation spectroscopy provides information about the hydrodynamic radius, R_h of the polymer in solution (Hiemenz 1984; Gedde 1995; Young and Lovell 2002; Hadjichristidis et al. 2003). DLS measures the correlation of the scattering intensity. From the correlation graph obtained, the diffusion coefficient (D) can be determined. Consecutively, D can be related to the hydrodynamic radius through the Stokes–Einstein relation

$$R_h = \frac{k_B T}{6\pi\eta D} \tag{4.7}$$

where

k_B is the Boltzmann constant
D is the diffusion coefficient
T is the temperature
η is the solvent viscosity

It is important to point out that the Stokes–Einstein equation assumes that the sample has a spherical shape. DLS is commonly used to determine the size of micelle and aggregates in solution (Hadjichristidis et al. 2003; Harada and Kataoka 2006), an example of which is shown in Figure 4.12. A bimodal distribution of the hydrodynamic diameter of a block copolymer in aqueous solution is shown that corresponds to the unimers (block copolymer) and the micelles. Moreover, DLS is significant for the determination of the size of drug-polymer and DNA-polymer complexes. Another measurement that is useful for the characterization of the polymer complexes is zeta potential measurement, to determine the charge of complexes.

4.5 Processing

Depending on their biomedical application, polymers may need to be processed to produce the right material. Specifically, in order for a polymer to be employed in a medical device, the polymeric material must be manipulated physically, thermally, or mechanically into the desired shape (Allcock and Lampe 1990; Abramson et al. 2004). Polymers can be fabricated into shaped objects by casting, compression molding, injected molding, blow molding (to make hollow objects), thermofusion and thermoforming, and rotational molding (Allcock and Lampe 1990). They can also be expanded and then be stabilized in the expanded structure, for example, to make polyurethane foams (Allcock and Lampe 1990).

In addition, polymers can be coated on a surface using dipping, calendar coating, electrostatic coating, knife coating, roll coating, fluidized-bed coating and powder molding, and radiation-cured coatings (Allcock and Lampe 1990). Moreover, they can be fabricated into sheets or fibers by wet spinning, dry spinning, or electrospinning (Allcock and Lampe 1990; Abramson et al. 2004; Pham et al. 2006; Yoon and Fisher 2007; Sill and von Recum 2008).

For tissue engineering that is one of the most common biomedical applications of polymers, there are two basic strategies of polymeric scaffold fabrication: prefabrication and *in situ* fabrication (Yoon and Fisher 2007). Prefabrication structures are cured before implantation and are often preferred since the polymeric scaffolds are formed outside the body allowing the removal of cytotoxic and nonbio-compatible component prior to implantation (Yoon and Fisher 2007). However, the scaffold may not properly fit in a tissue defect site causing gaps between the engineered graft and the host tissue, leading to undesirable results (Yoon and Fisher 2007). Therefore, in situ scaffold are also being investigated, which involves curing of a polymeric matrix within the tissue defect itself (Yoon and Fisher 2007), like injectable polymeric gels (Kretlow et al. 2007; Klouda and Mikos 2008). This strategy has two main advantages: the deformability of an *in situ* fabricated matrix creates an interface between the scaffold and the surrounding tissue, facilitating tissue integration, and it allows minimally invasive surgery techniques to be used since it may require a little as a narrow path for injection of the liquid scaffold (Yoon and Fisher 2007).

In terms of fabricating the polymeric scaffold, two methods are used: polymer entanglement and polymer cross-linking. Entanglement usually involves intertwining long, linear polymer chains to form a loosely bound polymer network (Yoon and Fisher 2007), a physical gel (not a covalent linked gel) (Patrickios and Georgiou 2003) (see Figure 4.13), while cross-linking involves the formation of covalent or ionic bond between individual polymer chains (Yoon and Fisher 2007). Caution should be made when referring to a polymeric network since it can be chemically cross-linked or be a physical gel where no chemical cross-links exist, only physical entanglements of the polymer chains (shown in Figure 4.13). These physical gels will solubilize in a solvent if given enough space and time to unravel, unlike the covalent linked networks.

The first fabrication method of a polymeric scaffold, polymer entanglement, is simple, allowing the polymer to be molded into a bulk material using hear, pressure, or both. However, the material often lacks mechanical strength, something that the second method has as its advantage. Chemical cross-linking enhances the mechanical strength. However, with the cross-linking method, a radical or ion is needed to promote cross-linking along with an initiator, such as heat, light, chemical accelerant, or time while leading to increased cytotoxicity, especially if the cross-linking takes place in situ (Yoon and Fisher 2007).

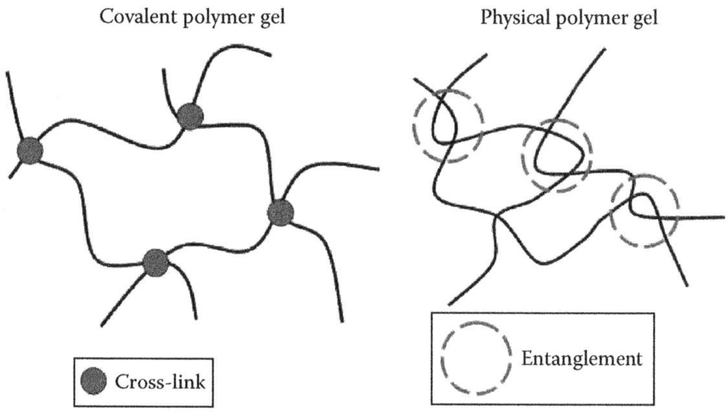

FIGURE 4.13 A covalently linked polymeric network (gel) and a physical gel.

A successful in situ fabrication of a polymer is PMMA when it is used for dental or bone cement (Abramson et al. 2004). The final polymerization step is carried out once the precursors (monomer or low molecular weight prepolymers) are in a casting or a mold device, yielding a solid, shaped end product (Abramson et al. 2004).

4.6 Bioapplications

4.6.1 Polymeric Materials for Tissue Engineering

Tissue engineering as it was defined by Langer and Vacanti is "an interdisciplinary field that applies the principles of engineering and the life sciences towards the development of biological substitutes that restore or improve tissue function" (Langer and Vacanti 1993). It aims to regenerate or replace biological damaged or diseased tissue or generate replacement organs for a wide range of medical conditions such as heart diseases, diabetes, cirrohosis, osteoarthritis, spinal cord injury, and disfiguration (Langer and Vacanti 1993; Calafiore 2001; Matthew 2001; Grigorescu and Hunkelerm 2003; Hoerstup et al. 2004; Salem and Leong 2005; Kretlow et al. 2007; Reddi 2007; Vert 2007; Yoon and Fisher 2007; Klouda and Mikos 2008; Place et al. 2009).

A typical scaffold is a biocompatible polymer in a porous configuration in the desired geometry for the engineered tissue, often modified to facilitate selective adhesion, while in some cases it is selective for a specific circulating cell population (Matthew 2001; Hoerstup et al. 2004). The first phase is the *in vitro* formation of a tissue construct by placing the cells and scaffold in an environment with growth media (in a *bioreactor*), in which the cells proliferate and elaborate extracellular matrix (Hoerstup et al. 2004; Place et al. 2009). In the second phase, the construct is implanted *in vivo* and remodeled to recapitulate the normal functional architecture of an organ or tissue (Hoerstup et al. 2004; Place et al. 2009). The key processes occurring during the *in vitro* and *in vivo* phase of tissue formation and maturation are (1) cell proliferation, sorting, and differentiation; (2) extracellular matrix production and organization; (3) degradation of the scaffold (for most applications); and (4) remodeling and potential growth of the tissue (Hoerstup et al. 2004). In Figure 4.14, a general paradigm of tissue engineering is illustrated.

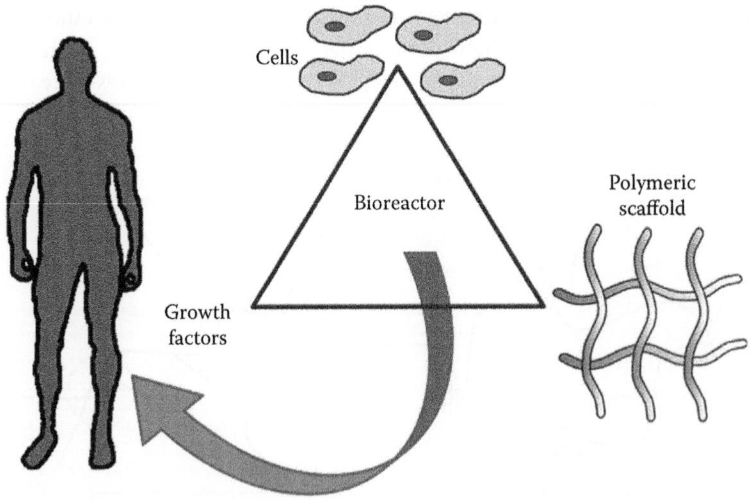

FIGURE 4.14 A general paradigm of tissue engineering is illustrated: (1) *in vitro* formation of a tissue construct by placing the cells and scaffold in an environment with growth media (in a *bioreactor*), in which the cells proliferate and elaborate extracellular matrix; (2) the construct is implanted *in vivo* and remodeled to recapitulate the normal functional architecture of an organ or tissue.

An ideal tissue engineering polymeric scaffold should combine many properties in order to provide a metabolically and mechanically supportive environment to facilitate tissue growth:

1. The first essential criterion is biocompatibility (Matthew 2001; Hoerstup et al. 2004; Place et al. 2009). Some important factors that determine biocompatibility, such as the chemistry, structure, and morphology, can be affected by polymer synthesis, scaffold processing, and sterilization.
2. It must also have a porous structure that will allow cellular ingrowth (Vert 2007). Depending on the application some scaffolds may require to have a porous morphology to help orient cells or surface properties to facilitate selective cell adhesion and/or migration, while in some other applications inhibition of cell adhesion is required (Matthew 2001; Hoerstup et al. 2004). Porosity, pore size, and pore structure are important factors to be considered with respect to nutrient supply to transplanted or regenerated cells (Hoerstup et al. 2004; Salem and Leong 2005; Vert 2007), while the hydrophobicity/hydrophilicity of the polymer should be consider for the enhancement or inhibition of cell adhesion and for the scaffolds wettability.
3. It should be mechanically strong enough to support the structural integrity of the implant. The mechanical stability needed depends on the application and may also vary with time. The mechanical stability is affected by many factors like the polymer chemistry (composition, structure, morphology, molecular weight, and molecular weight distribution) and the scaffold structure (density, shape, size, mass, pore size, and pore structure).
4. Similarly, depending on the application, it could be essential for the polymeric scaffold to be degradable. For most applications, a biodegradable polymeric scaffold that will allow its gradual and orderly replacement with functional tissue is needed (Matthew 2001; Grigorescu and Hunkelerm 2003; Hoerstup et al. 2004; Salem and Leong 2005). The degradation, similarly to the mechanical stability, can be affected by many factors like the polymers chemistry, the scaffold structure, the pH and the ionic strength of the medium, enzymes, and the type and density of the cultured cells (Hoerstup et al. 2004).
5. Finally, since the polymeric scaffolds should be designed to mimic the body, it is desirable for them to have chemically modifiable functional groups onto which sugars, proteins, or peptides can be attached (Matthew 2001; Hoerstup et al. 2004). Moreover, in many citations the scaffold may be required to be modified in order to release, in a controlled manner, tissue-specific growth factors to enhance the process of organ or tissue regeneration (Salem and Leong 2005; Place et al. 2009).

Natural materials like polypeptides (collagen, gelatin, and silk) and polysaccharides (alginate, agarose, chitosan, and hylauronic acid) are commonly used to fabricate tissue engineering scaffolds (Mathur and Scranton 1996; Seymour et al. 1998; Galant et al. 2005). Natural polymers are biocompatible and enzymatically biodegradable and their main advantage is that they contain biofunctional molecules that aid attachment, proliferation, and differentiation of cells. Their disadvantage arises from the fact that they are enzymatically degradable such that their degradation cannot be easily controlled *in vivo* and may not be desirable, depending on the application (Matthew 2001; Yoon and Fisher 2007). Furthermore, natural polymers are often weak in terms of mechanical strength, but cross-linking these polymers has been shown to enhance their structural stability (Yoon and Fisher 2007).

On the other hand synthetic polymers have the advantage that they can be easily moderated to change their structural stability, depending on the application. In general, it is easier to tailor the mechanical and chemical properties of synthetic polymers (Yoon and Fisher 2007; Place et al. 2009). Furthermore, since many synthetic polymers undergo hydrolytic degradation, a scaffold's degradation rate should not vary significantly between hosts (Yoon and Fisher 2007) and should be easier to control. Finally, synthetic polymers must be nontoxic, readily available, and relatively inexpensive to produce, and in many cases should be able to be processed under mild conditions that are compatible with cells (Place et al. 2009). A significant disadvantage for using synthetic polymers is that some degrade into unfavorable products, often acids that can change the local pH and result in adverse responses (Yoon and Fisher 2007).

Synthetic polymers used in tissue engineering are usually polyesters, but polyanhydrites, polycarbonates, and polyphosphazenes are also used (Matthew 2001; Salem and Leong 2005; Yoon and Fisher 2007; Place et al. 2009). Polyesters include poly(α-hydroxy acids) like poly(lactic acid) (PLLA) poly(glycolic acid) (PGA) and their copolymer poly[(lactic acid)-co-{glycolic acid)] (PLGA) that are the most widely used synthetic polymers in tissue engineering (Matthew 2001; Hoerstup et al. 2004; Klouda and Mikos 2008; Place et al. 2009). Other polyester, used in tissue engineering are poly(ε-cabrolactone) (PCL), poly(propylene fumarate) (PPF), and poly(orthoesters) (Matthew 2001; Hoerstup et al. 2004; Yoon and Fisher 2007; Klouda and Mikos 2008). Moreover, PMMA, polyanhydrides, polyphosphates, polyphosphazenes, polycarbonates, and polyurathenes have also been used (Matthew 2001; Hoerstup et al. 2004; Yoon and Fisher 2007; Klouda and Mikos 2008; Place et al. 2009). Most of these polymers with the exception of the poly(α-hydroxy acids) are considered to be hydrophobic. The most common hydrophilic component of tissue engineering scaffolds is PEG. PEG is a hydrophilic, FDA approved, biocompatible polymer, which is mainly used in hydrogels (water absorbing polymeric networks) due to its ability to imbibe water (Grigorescu and Hunkelerm 2003). Furthermore, thanks to its protein repellent effect, it can be useful as a noninterfering background upon which specific biological cues can be built up (Hoerstup et al. 2004). Other hydrophilic polymers like poly(vinyl alcohol) (PVA), poly(acrylic acid) (PAA), and PHEMA have also been applied (Hoerstup et al. 2004).

4.6.2 Polymeric Gene Delivery Systems

Gene delivery is a term used when referring to the delivery of genetic material like DNA and siRNA into cells (also called transfection) (Merdan et al. 2002; Jordan 2003; Langer 2005; Wong et al. 2007; Luten et al. 2008). Gene delivery is essential in gene therapy that aims to treat or cure many diseases (Geddes and Alton 1998; Hersh and Stopeck 1998; Langer 2005), in tissue engineering and is also used to study gene function.

A gene delivery vector is essential in order to carry the hydrophilic, negatively charged DNA through the hydrophobic and negatively charged cell membrane. The first vectors used for gene delivery were viruses, but due to their disadvantages—limitations on the size of DNA that they can carry, their high production cost, and, most importantly, their safety risks (immunogenicity and potential oncogenicity)—nonviral vectors have been developed (Merdan et al. 2002; Jordan 2003; Mrsny 2005; Wong et al. 2007; Luten et al. 2008). Nonviral vectors are divided into two main categories: lipid- and polymer-based, with the polymeric nonviral vectors having the advantage that their properties are easier to customize.

The main steps of polymeric gene delivery (see Figure 4.15) are (Merdan et al. 2002; Jordan 2003; Wong et al. 2007; Luten et al. 2008) as follows: (1) DNA/polymer complexation. The cationic polymer neutralizes the charged phosphate backbone of DNA to prevent charge repulsion with the negatively charged cell membrane and condenses the bulky structure of the DNA to form nanosize complexes. (2) DNA/polymer complex (also called polyplex) passes through the cell membrane. The complex is transported into the cell, through the cell membrane, by a nonspecific or receptor-mediated endocytosis. (3) The complex enters the cytoplasm usually in an endosome (depending on the cell type and the type of entry). The complex is later released from the endosome into the cytoplasm. (4) Cytosolic transport to the nucleus. The complex or the DNA, if it is already released from the complex, passes through the cytoplasm close to the nucleus. (5) The transfer of the genetic material into the nucleus where it is free to be encoded into a therapeutic protein or be inserted into the genome.

The most important property that a polymeric vector should have is to be nontoxic (biocompatible). It is also desirable to be biodegradable. If the biodegradability of the polymer is modulated correctly, with respect to the application, it could decrease the toxicity of the vector and also help the DNA release from the complex into the cytoplasm. A polymer vector must be able to condensate the genetic material. This is usually done through electrostatic interactions by using cationic polymeric vectors. However, studies on noncondensing polymeric systems have also been done (Kabanov et al. 2005; Nicolaou et al. 2005). Polymeric vectors with permanent cationic charges are not preferred since they will condensate the DNA so strongly that it will prevent its release into the cell. Thus, ionizable cationic polymer vectors

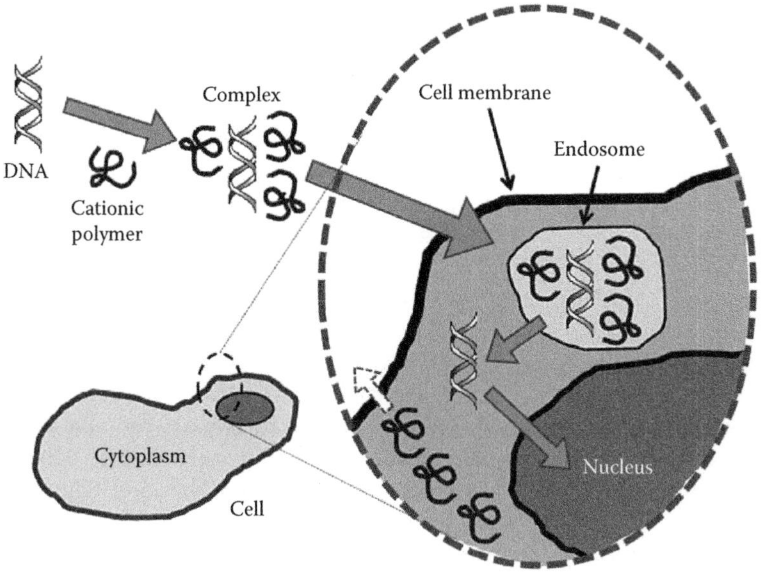

FIGURE 4.15 Main steps of gene delivery using a cationic polymer vector: formation of DNA/polymer complex, transfer of the complex through the cell membrane, DNA release from the endosome into the cytoplasm and DNA transfer into the nucleus.

are used, usually with a pK between 5 and 7. Finally, the polymeric vector should be hydrophilic in order to be mobile in the aqueous medium and the body; the vector may be composed of both hydrophobic and hydrophilic components and be stabilized in an aqueous solution by forming micelles or/and aggregates.

One of the first polymers used as a nonviral gene delivery agent poly(ethylenimine), PEI (chemical formula shown in Figure 4.6) (Boussif et al. 1995; Godbey et al. 1999a,b; Godbey and Mikos 2001; Orgis 2005). PEI has a high positive charged density; every third atom of PEI is a protonatable amino nitrogen atom (Boussif et al. 1995; Orgis 2005) and, thus, condensates the DNA effectively and delivers it into cells. It has been used for both *in vitro* and *in vivo* applications (Orgis 2005) and it is often used as the golden standard to which many novel synthetic polymer vectors are compared. In fact, since 1995 that PEI was first trialed in transfection (Boussif et al. 1995), over 800 publications (as of June 2009) have appeared that use PEI-based polymers as transfection agents. Many of these investigations use copolymers (Merdan et al. 2002; Wong et al. 2007) and degradable (Luten et al. 2008) PEI-based polymers.

PDMAEMA is also one of the first polymers to be investigated that is still commonly used in gene delivery since it has shown great potential (Verbaan et al. 2005). Several in vitro studies on DMAEMA and its derivatives—copolymers (van de Wetering et al. 1997, 1998, 1999a,b, 2000; van Dijk-Wolthuis et al. 1999; Georgiou et al. 2004; Georgiou et al. 2005; Verbaan et al. 2005; Georgiou et al. 2006; Xu et al. 2009) and degradable (Luten et al. 2003; de Wolf et al. 2007; Luten et al. 2008) DMAEMA-based polymers—have been reported in the literature. In vivo evaluation of PDMAEMA-based polymers has also been made (van de Wetering 1999b; Verbaan et al. 2005). PDMAEMA homopolymers like PEI homopolymers are also often used to compare the newly synthesized novel polymer vectors that are being investigated.

PEI and PDMAEMA homopolymers are nondegradable polymers. Other common nondegradable polymers that were used in gene delivery are chitosans (Borchard and Bivas-Benita 2005), poly(L-lysine)s (Lee and Kim 2005), cyclodextrin-containing polymers (Pun and Davis 2005), and dendrimers (Tang et al. 1996; Hudde et al. 1999; Cloninger 2002; Kubasiak and Tomalia 2005; Luten et al. 2008). The latter are spherical, highly branched polymers prepared either by divergent (starting from a central core molecule) or by convergent (starting with what will become the periphery of the molecule building inwards) synthesis strategies (Cloninger 2002; Merdan et al. 2002). The degree of branching is expressed in the generation of the dendrimer (Merdan et al. 2002). Commonly used dendrimers are poly(amidoamines), one of which is commercially available, called SuperFect®. Interestingly, the gene delivery performance is improved with the "activated" dendrimers that are, in fact, fractured dendrimers that show 50-fold enhanced transfection levels compared to the intact polymer (Tang et al. 1996; Cloninger 2002; Merdan et al. 2002). This is may be attributed to a better binding of the DNA, better ability of the polymer to complex the DNA due to the increased flexibility of the fractured polymer (Tang et al. 1996). It should be mentioned that SuperFect is also commonly used as a standard to compare to in gene delivery studies of novel cationic polymers (Georgiou et al. 2004, 2005, 2006).

Degradable polymers used for delivery of genetic material into cells are often polyesters (Lim et al. 2005; Lynn et al. 2005; Luten et al. 2008), especially poly(β-amino ester)s (Lynn et al. 2005), polysaccharides (Azzam and Domb 2005), polyurathenes (Luten et al. 2008), phosphor-containing polymers (Luten et al. 2008), and derivatives of the cationic nondegradable polymers (Luten et al. 2008). The potential advantage of biodegradable carriers as compared to their nondegradable counterparts is their reduced toxicity (provided their degradation products are nontoxic) and the avoidance of accumulation of the polymer in the cells after repeated administration (Luten et al. 2008). Furthermore, the degradation of the polymer can be used as a tool to release the plasmid DNA into the cytoplasm (Luten et al. 2008).

Other important points to consider when engineering a polymeric gene delivery vector, besides its toxicity and chemical structure, is the molecular weight, the molecular structure, and the composition of the polymer. The effect of the molecular weight on the transfection efficiency of the polymer has been studied with contradicting results (Godbey et al. 1999b; Georgiou et al. 2004; de Wolf et al. 2007), probably due to the range of the molecular weight tried, and the difference of the polymers' chemical structure and cell types used. What can be safely concluded from these studies is that by increasing the polymer's molecular weight, its toxicity is also increased (Georgiou et al. 2004; de Wolf et al. 2007). Polymers of different molecular structure such as linear (van de Wetering et al. 1997, 1998, 1999a,b, 2000; van Dijk-Wolthuis et al. 1999; Godbey et al. 1999a,b; Godbey and Mikos 2001; Verbaan et al. 2005), branched, stars (Georgiou et al. 2004, 2005, 2006; Xu et al. 2009), and dendrimers (Tang et al. 1996; Hudde et al. 1999; Cloninger 2002; Kubasiak and Tomalia 2005) have also been studied and, as mentioned before, the molecular structure has shown an important effect on the vectors' ability to transfer genes into cells (Tang et al. 1996; Merdan et al. 2002; Georgiou et al. 2004). Moreover, the introduction of a second monomer, a comonomer, into the polymer's structure influences the polymers transfection efficiency and cytotoxicity (van de Wetering et al. 1998, 1999b, 2000; van Dijk-Wolthuis et al. 1999; Georgiou et al. 2005, 2006; Verbaan et al. 2005; Xu et al. 2009). In general, copolymers with PEG-containing groups have reduced toxicity compared to their homopolymer counterparts (van de Wetering et al. 1998, 2000; Georgiou et al. 2005). Finally, note that, in general, direct comparison of different published studies should be avoided since transfection protocols, reported genes used, molecular weights, and polydispersities of the polymer and cell types used may vary.

It should be stated that the human body is a very complex environment. So naturally this was taken into account and many studies that aim at a specific organ or a specific type of cells, like cancer cells, have been reported (Hersh and Stopeck 1998; Kircheis and Wagner 2001; Ouyang et al. 2001; Merdan et al. 2002; Kinsey et al. 2005; Mrsny 2005). Commonly, studies target cancer cells (Hersh and Stopeck 1998; Merdan et al. 2002; Mrsny 2005) or aim to deliver genetic material into the lung (Kinsey et al. 2005) or liver (Ouyang et al. 2001). In order to achieve this, a targeting moiety, enabling uptake into a specific cell type is incorporated onto the polymer (Merdan et al. 2002).

4.6.3 Polymers for Drug Delivery

The selective and controlled delivery of drugs to malignant cells is essential for a successful treatment. There are many factors that influence the delivery of the drug to the intended target (Bae and Kwon 1998; Yokoyama 1998; Barrat et al. 2001; Hadjichristidis et al. 2003; Harada and Kataoka 2006; Qiu and

Bae 2006; Kabanov and Gendelman 2007; Liu et al. 2009). Specifically, many drugs encounter solubility and stability problems when administered into the body because they are hydrophobic while the body and the blood stream in particular consist mostly of water (Yokoyama 1998; Hadjichristidis et al. 2003; Harada and Kataoka 2006). Moreover, factors like the drug's absorption, distribution, and elimination influence its delivery to the target site (Barrat et al. 2001). Thus, drug delivery systems, also called controlled released systems, have been designed in order to deliver the drug to a specific site, at a specific time scheme, and in a specific release pattern (Bae and Kwon 1998; Barrat et al. 2001). Extensive research has been done on polymer-based drug delivery systems since polymers are easy to modulate and modify to encapsulate the drug, to be target specific, and to be stimuli-responsive to release the drug.

Important characteristics that a polymer-based drug delivery system should have is as follows: to be biocompatible, to have small and uniform size (in the nanometer scale <200 nm), to have long circulation in bloodstream, to be easily sterilized (usually by filtration), not to have long-term accumulation, to be applicable to various drugs, to be able to encapsulate a high drug content but maintain water solubility, to have the right microenvironment for drug preservation (not inhibit the drug's activity), and to release the drug in a target- and time-controlled manner (Bae and Kwon 1998; Yokoyama 1998; Kabanov and Gendelman 2007).

Many polymeric drug delivery systems have been fabricated in the last few decades (Bae and Kwon 1998; Yang and Robinson 1998; Yokoyama 1998; Barrat et al. 2001; Brown 2001; Domb et al. 2001; Haverdings et al. 2001; Felt et al. 2001; Kratz et al. 2001; Maysinger et al. 2001; Michniak and El-Kattan 2001; Woolfson et al. 2001; Worakul and Robinson 2001; Harada and Kataoka 2006; Qiu and Bae 2006; Kabanov and Gendelman 2007; Rapoport 2007; Liu et al. 2009); biodegradable polymeric systems (Domb et al. 2001), for a specific organ or tissue (Felt et al. 2001; Haverdings et al. 2001; Maysinger et al. 2001; Kabanov and Gendelman 2007), for a specific disease (Kratz et al. 2001; Rapoport 2007), for drug delivery via a specific route (Michniak and El-Kattan 2001; Woolfson et al. 2001; Worakul and Robinson 2001), and for specific drug such as insulin (Brown 2001).

It is not very easy to name the most common polymers used for drug delivery. Most polymeric drug delivery systems form micelles or are a part of nanoparticles (Kabanov and Gendelman 2007; Rapoport 2007). As mentioned before, a micelle is formed by amphiphilic block copolymer (Figure 4.8). The inner part of the micelle, called the micelle core, is composed of the hydrophobic core that is usually poly(propylene glycol), poly(D,L-lactide), PCL, while the outer part of the micelle, called the micelle shell is composed of hydrophilic block, which is often PEG (Harada and Kataoka 2006; Kabanov and Gendelman 2007; Rapoport 2007). On the other hand, nanoparticles are often composed of insoluble polymer(s). During their formulation drug is captured within the precipitating polymer, forming nanoparticles, and then released upon degradation of a polymer in the biological environment (Kabanov and Gendelman 2007).

In order for the drug to be released from the polymeric micelle or nanoparticle, often a stimuli response of the polymeric material is required (Bae and Kwon 1998; Liu et al. 2009). Thus, polymeric biomaterials that are responsive to temperature, pH, ionic strength, enzymatic conversion, or magnetic field have been developed (Bae and Kwon 1998; Liu et al. 2009). Alternatively, drug release can be initiated by polymer degradation, thus eliminating the need to remove the scaffold after drug release (Liu et al. 2009).

4.6.4 Other Biomedical Applications of Polymers

Tissue engineering, gene, and drug delivery are just three of a number of biomedical applications of polymers. Just to name these applications in order to demonstrate this wide range of usages, polymers are used in diagnostics (imaging) (Borovetz et al. 2004; Kim et al. 2007; Wolinsky and Grinstaff 2008; Khemtong et al. 2009) in cardiovascular devices and heart valves (Allcock and Lampe 1990; Bhuvaneshwar et al. 2001; El-Zaim and Heggers 2001; Abramson et al. 2004; Borovetz et al. 2004; Venkatraman et al. 2008; Kidane et al. 2009), in surgery and as orthopedic implants (El-Zaim and Heggers 2001; Rokkanen 2001;

Tomita et al. 2001; Borovetz et al. 2004; Tran et al. 2009), as contact lenses (Allcock and Lampe 1990; Abramson et al. 2004), in dressings for burns and wounds (Sheridan et al. 2001; Borovetz et al. 2004), in dental applications (Bascones et al. 2001), and for cosmetic implants (El-Zaim and Heggers 2001).

4.7 Conclusions

Polymers represent a broad family of materials, which are cost-effective and easy to modulate, and have properties that make them useful in a variety of biomedical applications. In this chapter, the main synthetic methodologies, properties, characterization and processing methods, and examples of biomedical applications of polymers were summarized.

References

Abramson, S., H. Alexander, S. Best, J. C. Bokros, J. B. Brunski, A. Colas, S. L. Cooper et al., 2004. Classes of materials used in medicine. In *Biomaterials Science: An Introduction to Materials in Medicine*, B. D. Ratner, A. S. Hoffman, F. J. Schoen, and J. E. Lemons, eds. San Diego, CA: Elsevier Academic Press.

Allcock, H. R. and F. W. Lampe. 1990. *Contemporary Polymer Chemistry*. Englewood Cliffs, NJ: Prentice Hall.

Azzam, T. and A. J. Domb. 2005. Cationic polysaccharides for gene delivery. In *Polymeric Gene Delivery: Principles and Applications*, M. M. Amiji, ed. Boca Raton, FL: Taylor & Francis.

Bae, Y. H. and I. C. Kwon. 1998. Stimuli-sensitive polymers for modulated drug release. In *Biorelated Polymers and Gels: Controlled Release and Applications in Biomedical Engineering*, T. Okano, ed. San Diego, CA: Academic Press.

Barrat, D., G. Couarraze, P. Couvreur, C. Dubernet, E. Fattalm, E. Gref, D. Labarre, P. Legrand, G. Ponchel, and C. Vouthier. 2001. Polymeric micro- and nanoparticles as drug carriers. In *Polymeric Biomaterials*, S. Dumitriu, ed. Boca Raton, FL: Taylor & Francis.

Bascones, A., J. M. Vega, N. Olmo, K. Turnay, J. H. Gavilanes, and M. A. Lizarbe. 2001. Dental and maxillofacial surgery applications of polymers. In *Polymeric Biomaterials*, S. Dumitriu, ed. Boca Raton, FL: Taylor & Francis.

Bhuvaneshwar, G. S., A. V. Ramani, and K. B. Chandran. 2001. Polymeric occluders in titlting disc heart valve prostheses. In *Polymeric Biomaterials*, S. Dumitriu, ed. Boca Raton, FL: Taylor & Francis.

Borchard, G. and M. Bivas-Benita. 2005. Gene delivery using chitosan and chitosan derivatives. In *Polymeric Gene Delivery: Principles and Applications*, M. M. Amiji, ed. Boca Raton, FL: Taylor & Francis.

Borovetz, H. S., J. F. Burke, T. M. S. Chang, A. Colas, A. N. Cranin, J. Curtis, C. J. Gemmell et al., 2004. Application of materials in medicine, biology, and artificial organs. In *Biomaterials Science: An Introduction to Materials in Medicine*, B. D. Ratner, A. S. Hoffman, F. J. Schoen, and J. E. Lemons, eds. San Diego, CA: Elsevier Academic Press.

Boussif, O., F. Lezoualch, M. A. Zanta, M. D. Mergny, D. Scherman, B. Demeneix, and J. P. Behr. 1995. A versatile vector for gene and oligonucleotide transfer cells in culture and in-vivo—Polyethylenimine. *Proceedings of the National Academy of Sciences of the United States of America* 92 (16):7297–7301.

Brown, L. R. 2001. Glucose-mediated insulin delivery from implantable polymers. In *Polymeric Biomaterials*, S. Dumitriu, ed. Boca Raton, FL: Taylor & Francis.

Calafiore, R. 2001. Bioartificial pancreas. In *Polymeric Biomaterials*, S. Dumitriu, ed. Boca Raton, FL: Taylor & Francis.

Callister, W. D. 2003. *Materials Science and Engineering: An Introduction*. Danvers, MA: John Wiley & Sons, Inc.

Cloninger, M. J. 2002. Biological applications of dendrimers. *Current Opinion in Chemical Biology* 6 (6):742–748.

de Wolf, H. K., M. de Raad, C. Snel, M. J. van Steenbergen, Mham Fens, G. Storm, and W. E. Hennink. 2007. Biodegradable poly(2-dimethylamino ethylamino) phosphazene for in vivo gene delivery to tumor cells. Effect of polymer molecular weight. *Pharmaceutical Research* 24 (8):1572–1580.

Dee, K. C., D. A. Puleo, and R. Bizios. 2002. *An Introduction to Tissue-Biomaterial Interactions*. Hoboken, NJ: John Wiley & Sons, Inc.

Domb, A. J., N. Kumar, T. Sheskin, A. Bentolila, J. Slager, and D. Teomim. 2001. Biodegradable polymers as drug carrier systems. In *Polymeric Biomaterials*, S. Dumitriu, ed. Boca Raton, FL: Taylor & Francis.

Drobny, G. P., J. R. Long, T. Karlsson, W. Shaw, J. Popham, N. Oyler, P. Bower et al., 2003. Structural studies of biomaterials using double-quantum solid-state NMR spectroscopy. *Annual Review of Physical Chemistry* 54:531–571.

El-Zaim, H. S. and J. P. Heggers. 2001. Silicones for pharmaceutical and biomedical applications. In *Polymeric Biomaterials*, S. Dumitriu, ed. Boca Raton, FL: Taylor & Francis.

Felt, O., S. Einmahl, P. Furrer, V. Baeyens, and R. Gurny. 2001. Polymer systems for ophthalmic drug delivery. In *Polymeric Biomaterials*, S. Dumitriu, ed. Boca Raton, FL: Taylor & Francis.

Galant, C., C. Amiel, and A. Loic. 2005. Ternary complex formation in aqueous solution between a beta-cyclodextrin polymer, a cationic surfactant and DNA. *Macromolecular Bioscience* 5 (11):1057–1065.

Gedde, U. W. 1995. *Polymer Physics*. London, U.K.: Chapman & Hall.

Geddes, D. and E. W. F. W. Alton. 1998. Cystic fibrosis clinical trials. In *Self-Assembling Complexes for Gene Delivery*, A. V. Kabanov, P. L. Flegner, and L. W. Seymour, eds. West Sussex: John Willey & Sons Ltd.

Georgiou, T. K., L. A. Phylactou, and C. S. Patrickios. 2006. Synthesis, characterization, and evaluation as transfection reagents of ampholytic star copolymers: Effect of star architecture. *Biomacromolecules* 7 (12):3505–3512.

Georgiou, T. K., M. Vamvakaki, C. S. Patrickios, E. N. Yamasaki, and L. A. Phylactou. 2004. Nanoscopic cationic methacrylate star homopolymers: Synthesis by group transfer polymerization, characterization and evaluation as transfection reagents. *Biomacromolecules* 5 (6):2221–2229.

Georgiou, T. K., M. Vamvakaki, L. A. Phylactou, and C. S. Patrickios. 2005. Synthesis, characterization, and evaluation as transfection reagents of double-hydrophilic star copolymers: Effect of star architecture. *Biomacromolecules* 6 (6):2990–2997.

Godbey, W. T. and A. G. Mikos. 2001. Recent progress in gene delivery using non-viral transfer complexes. *Journal of Controlled Release* 72 (1–3):115–125.

Godbey, W. T., K. K. Wu, and A. G. Mikos. 1999a. Poly(ethylenimine) and its role in gene delivery. *Journal of Controlled Release* 60 (2–3):149–160.

Godbey, W. T., K. K. Wu, and A. G. Mikos. 1999b. Size matters: Molecular weight affects the efficiency of poly(ethylenimine) as a gene delivery vehicle. *Journal of Biomedical Materials Research* 45 (3):268–275.

Griesser, H. J. 1991. Degradation of polyurethanes in biomedical applications—A review. *Polymer Degradation and Stability* 33 (3):329–354.

Grigorescu, G. and D. Hunkelerm. 2003. Cell encapsulation: Generalities, methods, applications and bio-artificial pancreas case study. In *Synthetic Polymers for Biotechnology and Medicine*, R. Freitag, ed. Georgetown, DC: Eurekah.com/Landes Bioscience.

Guan, Y., W. C. Zhang, G. X. Wan, and Y. X. Peng. 2000. Polytetrahydrofuran amphiphilic networks. I. Synthesis and characterization of polytetrahydrofuran acrylate ditelechelic and polyacrylamide-*l*-potytetrahydrofuran networks. *Journal of Polymer Science Part A: Polymer Chemistry* 38 (20):3812–3820.

Hadjichristidis, N., S. Pispas, and G. A. Floudas. 2003. *Block Copolymers: Synthetic Strategies, Physical Properties, and Applications*. Hoboken, NJ: John Wiley and Sons, Inc.

Hadjichristidis, N., M. Pitsikalis, S. Pispas, and H. Iatrou. 2001. Polymers with complex architecture by living anionic polymerization. *Chemical Reviews* 101 (12):3747–3792.

Harada, A. and K. Kataoka. 2006. Supramolecular assemblies of block copolymers in aqueous media as nanocontainers relevant to biological applications. *Progress in Polymer Science* 31 (11):949–982.

Hashimoto, K. 2000. Ring-opening polymerization of lactams. Living anionic polymerization and its applications. *Progress in Polymer Science* 25 (10):1411–1462.

Haverdings, R. F. G., R. J. Kok, M. Haas, F. Moolenaarm, D. de Zeeuw, and D. K. F. Meijer. 2001. Drug targeting to the kidney: The low-molecular-weight protein approach. In *Polymeric Biomaterials*, S. Dumitriu, ed. Boca Raton, FL: Taylor & Francis.

Hersh, E. M. and A. T. Stopeck. 1998. Cancer gene therapy using nonviral vectors: Precrinical and clinical observations. In *Self-Assembling Complexes for Gene Delivery*, A. V. Kabanov, P. L. Flegner, and L. W. Seymour, eds. West Sussex: John Willey & Sons Ltd.

Hiemenz, P. C. 1984. *Polymer Chemistry: The Basic Concepts*. New York: Marcel Dekker, Inc.

Hoerstup, S. P., L. Lu, M. J. Lysaght, A. G. Mikos, D. Rein, F. J. Schoen, J. S. Temenoff, J. K. Tessmar, and J. P. Vacanti. 2004. Tissue engineering. In *Biomaterials Science: An Introduction to Materials in Medicine*, B. D. Ratner, A. S. Hoffman, F. J. Schoen, and J. E. Lemons, eds. San Diego, CA: Elsevier Academic Press.

Horkay, F. and B. Hammouda. 2008. Small-angle neutron scattering from typical synthetic and biopolymer solutions. *Colloid and Polymer Science* 286 (6–7):611–620.

Hudde, T., S. A. Rayner, R. M. Comer, M. Weber, J. D. Isaacs, H. Waldmann, D. P. F. Larkin, and A. J. T. George. 1999. Activated polyamidoamine dendrimers, a non-viral vector for gene transfer to the corneal endothelium. *Gene Therapy* 6 (5):939–943.

Jordan, M. 2003. Synthetic and semisynthetic polymers as vehicles for in vitro gene delivery into cultured mammalian cells. In *Synthetic Polymers for Biotechnology and Medicine*, R. Freitag, ed. Georgetown, DC: Eurekah.com/Landes Bioscience.

Kabanov, A. V. and H. E. Gendelman. 2007. Nanomedicine in the diagnosis and therapy of neuro degenerative disorders. *Progress in Polymer Science* 32 (8–9):1054–1082.

Kabanov, A. V., S. Sriabibhatla, and V. Y. Alakhov. 2005. Pluronic® block copolymers for nonviral gene delivery. In *Polymeric Gene Delivery: Principles and Applications*, M. M. Amiji, ed. Boca Raton, FL: Taylor & Francis.

Kasaal, M. R. 2008. A review of several reported procedures to determine the degree of N-acetylation for chitin and chitosan using infrared spectroscopy. *Carbohydrate Polymers* 71 (4):497–508.

Kennedy, J. P. and B. Iván. 1992. *Designed Polymers by Carbocationic Macromolecular Engineering: Theory and Practice*. New York: Hanser Publishers.

Khemtong, C., C. W. Kessinger, and J. M. Gao. 2009. Polymeric nanomedicine for cancer MR imaging and drug delivery. *Chemical Communications* (24):3497–3510.

Kidane, A. G., G. Burriesci, P. Cornejo, A. Dooley, S. Sarkar, P. Bonhoeffer, M. Edirisinghe, and A. M. Seifalian. 2009. Current developments and future prospects for heart valve replacement therapy. *Journal of Biomedical Materials Research Part B—Applied Biomaterials* 88B (1):290–303.

Kim, J. H., K. Park, H. Y. Nam, S. Lee, K. Kim, and I. C. Kwon. 2007. Polymers for bioimaging. *Progress in Polymer Science* 32 (8–9):1031–1053.

Kinsey, B. M., C. L. Densmore, and F. M. Orson. 2005. Gene delivery to the lungs. In *Polymeric Gene Delivery: Principles and Applications*, M. M. Amiji, ed. Boca Raton, FL: Taylor & Francis.

Kircheis, R. and E. Wagner. 2001. Transferrin receptor-targeted gene delivery systems. In *Polymeric Biomaterials*, S. Dumitriu, ed. Boca Raton, FL: Taylor & Francis.

Klouda, L. and A. G. Mikos. 2008. Thermoresponsive hydrogels in biomedical applications. *European Journal of Pharmaceutics and Biopharmaceutics* 68 (1):34–45.

Kratz, F., A. Warnecke, K. Riebessel, and P. C. A. Rodrigues. 2001. Anticancer drug conjugates with macromolecular carriers. In *Polymeric Biomaterials*, S. Dumitriu, ed. Boca Raton, FL: Taylor & Francis.

Kretlow, J. D., L. Klouda, and A. G. Mikos. 2007. Injectable matrices and scaffolds for drug delivery in tissue engineering. *Advanced Drug Delivery Reviews* 59 (4–5):263–273.

Kubasiak, L. A. and D. A. Tomalia. 2005. Cationic dendrimes as gene transfection vectors: Dendri-poly(amidoamines) and Dendri(propylenimines). In *Polymeric Gene Delivery: Principles and Applications*, M. M. Amiji, ed. Boca Raton, FL: Taylor & Francis.

Langer, R. 2005. Introduction. In *Polymeric Gene Delivery: Principles and Applications*, M. M. Amiji, ed. Boca Raton, FL: Taylor & Francis.

Langer, R. and J. P. Vacanti. 1993. Tissue engineering. *Science* 260 (5110):920–926.

Lee, M. and Kim. S. W. 2005. Poly(L-Lysine) and copolymers for gene delivery. In *Polymeric Gene Delivery: Principles and Applications*, M. M. Amiji, ed. Boca Raton, FL: Taylor & Francis.

Lim, Y.-B., Y. Lee, and J.-S. Park. 2005. Cationic polyesters as biodegradable polymer gene delivery carriers. In *Polymeric Gene Delivery: Principles and Applications*, M. M. Amiji, ed. Boca Raton, FL: Taylor & Francis.

Liu, S., R. Maheshwari, and K. L. Kiick. 2009. Polymer-based therapeutics. *Macromolecules* 42 (1):3–13.

Luten, J., C. F. van Nostruin, S. C. De Smedt, and W. E. Hennink. 2008. Biodegradable polymers as non-viral carriers for plasmid DNA delivery. *Journal of Controlled Release* 126 (2):97–110.

Luten, J., J. H. van Steenis, R. van Someren, J. Kemmink, N. M. E. Schuurmans-Nieuwenbroek, G. A. Koning, D. J. A. Crommelin, C. F. van Nostrum, and W. E. Hennink. 2003. Water-soluble biodegradable cationic polyphosphazenes for gene delivery. *Journal of Controlled Release* 89 (3):483–497.

Lynn, D. M., D. G. Anderson, A. Akinc, and R. Langer. 2005. Degradable poly(β-amino ester)s for gene delivery. In *Polymeric Gene Delivery: Principles and Applications*, M. M. Amiji, ed. Boca Raton, FL: Taylor & Francis.

Mathur, A. M. and A. B. Scranton. 1996. Characterization of hydrogels using nuclear magnetic resonance spectroscopy. *Biomaterials* 17 (6):547–557.

Matthew, H. W. T. 2001. Polymers for tissue engineering scaffolds. In *Polymeric Biomaterials*, S. Dumitriu, ed. Boca Raton, FL: Taylor & Francis.

Matyjaszewski, K. and A. H. E. Müller. 2006. 50 Years of living polymerization. *Progress in Polymer Science* 31 (12):1039–1040.

Maysinger, D., R. Savic, J. Tam, C. Allen, and A. Eisenberg. 2001. Recent developments in drug delivery to the nervous system. In *Polymeric Biomaterials*, S. Dumitriu, ed. Boca Raton, FL: Taylor & Francis.

Melnichenko, Y. B. and G. D. Wignall. 2007. Small-angle neutron scattering in materials science: Recent practical applications. *Journal of Applied Physics* 102 (2).

Menard, K. P. 2008. *Dynamic Mechanical Analysis: A Practical Introduction*. Boca Raton, FL: CRC Press.

Merdan, T., J. Kopecek, and T. Kissel. 2002. Prospects for cationic polymers in gene and oligonucleotide therapy against cancer. *Advanced Drug Delivery Reviews* 54 (5):715–758.

Michniak, B. B. and A. El-Kattan. 2001. Transdermal delivery of drugs. In *Polymeric Biomaterials*, S. Dumitriu, ed. Boca Raton, FL: Taylor & Francis.

Moad, G., E. Rizzardo, and S. H. Thang. 2006. Living radical polymerization by the RAFT process—A first update. *Australian Journal of Chemistry* 59 (10):669–692.

Mrsny, R. J. 2005. Tissue- and cell-specific targeting for the delivery of genetic information. In *Polymeric Gene Delivery: Principles and Applications*, M. M. Amiji, ed. Boca Raton, FL: Taylor & Francis.

Nair, L. S. and C. T. Laurencin. 2007. Biodegradable polymers as biomaterials. *Progress in Polymer Science* 32 (8–9):762–798.

Nicolaou, M., P. Chang, and M. J. Newman. 2005. Use of poly(N-vinyl pyrrolidone) with noncondensed plasmid DNA formulations for gene delivery and vaccines. In *Polymeric Gene Delivery: Principles and Applications*, M. M. Amiji, ed. Boca Raton, FL: Taylor & Francis.

Orgis, M. 2005. Gene delivery using polyethylenimine and copolymers. In *Polymeric Gene Delivery: Principles and Applications*, M. M. Amiji, ed. Boca Raton, FL: Taylor & Francis.

Ouyang, E. C., G. Y. Wu, and C. H. Wu. 2001. Biocompatible polymers in liver-targeted gene delivery systems. In *Polymeric Biomaterials*, S. Dumitriu, ed. Boca Raton, FL: Taylor & Francis.

Patrickios, C. S. and T. K. Georgiou. 2003. Covalent amphiphilic polymer networks. *Current Opinion in Colloid & Interface Science* 8 (1):76–85.

Patten, T. E. and K. Matyjaszewski. 1998. Atom transfer radical polymerization and the synthesis of polymeric materials. *Advanced Materials* 10 (12):901–915.

Pham, Q. P., U. Sharma, and A. G. Mikos. 2006. Electrospinning of polymeric nanofibers for tissue engineering applications: A review. *Tissue Engineering* 12 (5):1197–1211.

Place, E. S., J. H. George, C. K. Williams, and M. M. Stevens. 2009. Synthetic polymer scaffolds for tissue engineering. *Chemical Society Reviews* 38 (4):1139–1151.

Pun, S. H. and Davis. M. E. 2005. Cyclodextrin-containing polymer for gene delivery. In *Polymeric Gene Delivery: Principles and Applications*, M. M. Amiji, ed. Boca Raton, FL: Taylor & Francis.

Qiu, L. Y. and Y. H. Bae. 2006. Polymer architecture and drug delivery. *Pharmaceutical Research* 23 (1):1–30.

Rapoport, N. 2007. Physical stimuli-responsive polymeric micelles for anti-cancer drug delivery. *Progress in Polymer Science* 32 (8–9):962–990.

Reddi, A. H. 2007. Growth factors and morphogens: Signals for tissue engineering. In *Tissue Engineering*, J. P. Fisher, A. G. Mikos, and J. D. Bronzino, eds. Boca Raton, FL: Taylor & Francis.

Remmp, P. and E. W. Merrill. 1991. *Polymer Synthesis*. New York: Hüthing u. Wepf.

Rokkanen, P. U. 2001. Bioabsorbable polymers for medical application with an emphasis on orthopedic surgery. In *Polymeric Biomaterials*, S. Dumitriu, ed. Boca Raton, FL: Taylor & Francis.

Rubinstein, M. and R. H. Colby. 2003. *Polymer Physics*. Oxford: Oxford University Press.

Salem, A. K. and K. W. Leong. 2005. Polymeric scaffolds for delivery and regenerative medicine. In *Scaffolding in Tissue Engineering*, P. X. Ma. and J. H. Elisseeff, eds. Boca Raton, FL: Taylor & Francis.

Seymour, L. W., K. Kataoka, and Kabanov. A. V. 1998. Cationic block copolymers as self-assembling vectors for gene delivery. In *Self-Assembling Complexes for Gene Delivery: From Laboratory to Clinical Trial*, A. V. Kabanov, P. L. Felgner, and L. W. Seymour, eds. West Sussex: John Wiley & Sons, Ltd.

Sheridan, R. L., J. R. Morgan, and R. Mohammad. 2001. Biomaterials in burns and wound dressings. In *Polymeric Biomaterials*, S. Dumitriu, ed. Boca Raton, FL: Taylor & Francis.

Sill, T. J. and H. A. von Recum. 2008. Electro spinning: Applications in drug delivery and tissue engineering. *Biomaterials* 29 (13):1989–2006.

Szwarc, M. 1956. Living polymers. *Nature* 178 (4543):1168–1169.

Szwarc, M., M. Levy, and R. Milkovich. 1956. Polymerization initiated by electron transfer to monomer. A new method of formation of block polymers. *Journal of the American Chemical Society* 78 (11):2656–2657.

Tang, M. X., C. T. Redemann, and F. C. Szoka. 1996. In vitro gene delivery by degraded polyamidoamine dendrimers. *Bioconjugate Chemistry* 7 (6):703–714.

Temenoff, J. S. and A. G. Mikos. 2008. *Biomaterials: The Intersection of Biology and Materials Science*. Upper Saddle River, NJ: Pearson Education, Inc.

Tomita, N., K. Nagata, and H. Fujita. 2001. Polymers for artificial joints. In *Polymeric Biomaterials*, S. Dumitriu, ed. Boca Raton, FL: Taylor & Francis.

Tran, P. A., L. Sarin, R. H. Hurt, and T. J. Webster. 2009. Opportunities for nanotechnology-enabled bioactive bone implants. *Journal of Materials Chemistry* 19 (18):2653–2659.

van de Wetering, P., J. Y. Cherng, H. Talsma, D. J. A. Crommelin, and W. E. Hennink. 1998. 2-(Dimethylamino)ethyl Methacrylate based (Co)polymers as gene transfer agents. *Journal of Controlled Release* 53 (1–3):145–153.

van de Wetering, P., J. Y. Cherng, H. Talsma, and W. E. Hennink. 1997. Relation between transfection efficiency and cytotoxicity of poly(2-(dimethylamino)ethyl methacrylate)/plasmid complexes. *Journal of Controlled Release* 49 (1):59–69.

van de Wetering, P., E. E. Moret, N. M. E. Schuurmans-Nieuwenbroek, M. J. van Steenbergen, and W. E. Hennink. 1999a. Structure-activity relationships of water-soluble cationic methacrylate/methacrylamide polymers for nonviral gene delivery. *Bioconjugate Chemistry* 10 (4):589–597.

van de Wetering, P., N. M. E. Schuurmans-Nieuwenbroek, W. E. Hennink, and G. Storm. 1999b. Comparative transfection studies of human ovarian carcinoma cells in vitro, ex vivo and in vivo with poly(2-(dimethylamino)ethyl methacrylate)-based polyplexes. *Journal of Gene Medicine* 1 (3):156–165.

van de Wetering, P., N. M. E. Schuurmans-Nieuwenbroek, M. J. van Steenbergen, D. J. A. Crommelin, and W. E. Hennink. 2000. Copolymers of 2-(dimethylamino)ethyl methacrylate with ethoxytriethylene glycol methacrylate or N-vinyl-pyrrolidone as gene transfer agents. *Journal of Controlled Release* 64 (1-3):193–203.

van Dijk-Wolthuis, W. N. E., P. van de Wetering, W. L. J. Hinrichs, L. J. F. Hofmeyer, R. M. J. Liskamp, D. J. A. Crommelin, and W. E. Hennink. 1999. A versatile method for the conjugation of proteins and peptides to poly 2-(dimethylamino)ethyl methacrylate. *Bioconjugate Chemistry* 10 (4):687–692.

Venkatraman, S., F. Boey, and L. L. Lao. 2008. Implanted cardiovascular polymers: Natural, synthetic and bio-inspired. *Progress in Polymer Science* 33 (9):853–874.

Verbaan, F. J., Crommelin, D. J. A, W. E. Hennink, and G. Storm. 2005. Poly(2-(dimethylamino)ethyl methacrylate)-based polymers for the delivery of genes in vitro and in vivo. In *Polymeric Gene Delivery: Principles and Applications*, M. M. Amiji, ed. Boca Raton, FL: Taylor & Francis.

Vert, M. 2007. Polymeric biomaterials: Strategies of the past vs. strategies of the future. *Progress in Polymer Science* 32 (8–9):755–761.

Webster, O. W. 1991. Living polymerization methods. *Science* 251 (4996):887–893.

Webster, O. W. 2000. The discovery and commercialization of group transfer polymerization. *Journal of Polymer Science Part A—Polymer Chemistry* 38 (16):2855–2860.

Webster, O. W. 2004. Group transfer polymerization: Mechanism and comparison with other methods for controlled polymerization of acrylic monomers. In *New Synthetic Methods*. Berlin, Germany: Springer-Verlag.

Webster, O. W., W. R. Hertler, D. Y. Sogah, W. B. Farnham, and T. V. Rajanbabu. 1983. Group-transfer polymerization. 1. A new concept for addition polymerization organo-silicon initiators. *Journal of the American Chemical Society* 105 (17):5706–5708.

Wolinsky, J. B. and M. W. Grinstaff. 2008. Therapeutic and diagnostic applications of dendrimers for cancer treatment. *Advanced Drug Delivery Reviews* 60 (9):1037–1055.

Wong, S. Y., J. M. Pelet, and D. Putnam. 2007. Polymer systems for gene delivery-past, present, and future. *Progress in Polymer Science* 32 (8–9):799–837.

Woolfson, A. D., R. K. Malcolm, P. A. McCarron, and D. S. Jones. 2001. Bioadhesive drug delivery systems. In *Polymeric Biomaterials*, S. Dumitriu, ed. Boca Raton, FL: Taylor & Francis.

Worakul, N. and J. R. Robinson. 2001. Drug delivery via mucosal routes. In *Polymeric Biomaterials*, S. Dumitriu, ed. Boca Raton, FL: Taylor & Francis.

Xu, F. J., Z. X. Zhang, Y. Ping, J. Li, E. T. Kang, and K. G. Neoh. 2009. Star-shaped cationic polymers by atom transfer radical polymerization from beta-cyclodextrin cores for nonviral gene delivery. *Biomacromolecules* 10 (2):285–293.

Yang, X. and J. R. Robinson. 1998. Bioadhesion in mucosal drug delivery. In *Biorelated Polymers and Gels: Controlled Release and Applications in Biomedical Engineering*, T. Okano, ed. San Diego, CA: Academic Press.

Yokoyama, M. 1998. Novel passive targetable drug delivery with polymeric micelles. In *Biorelated Polymers and Gels: Controlled Release and Applications in Biomedical Engineering*, T. Okano, ed. San Diego, CA: Academic Press.

Yoon, D. M. and J. P. Fisher. 2007. Polymeric scaffolds for tissue engineering applications. In *Tissue Engineering*, J. P. Fisher, A. G. Mikos, and J. D. Bronzino, eds. Boca Raton, FL: Taylor & Francis.

Young, R. J. and P. A. Lovell. 2002. *Introduction to Polymers*. Cheltenham, U.K.: Nelson Thornes Ltd.

Zhang, Q., M. Wang, and K. L. Wooley. 2005. Nanoscopic confinement of semi-crystalline polymers. *Current Organic Chemistry* 9 (11):1053–1066.

5

Carbon-Based Nanomedicine

Michael L. Matson
Rice University

Jeyarama S. Ananta
Rice University

Lon J. Wilson
Rice University

Medical nanotechnology (nanomedicine) is one of the newest fields of medical research. In 2006, the U.S. National Institute of Health Roadmap for Medical Research in Nanomedicine defined nanomedicine as, "an offshoot of nanotechnology, [which] refers to highly specific medical interventions at the molecular scale for curing disease or repairing damaged tissues, such as bone, muscle, or nerve." Earlier, in 2004, the European Medical Research Councils (EMRC) wrote in their Forward Look report that nanomedicine is, "the science and technology of diagnosing, treating, and preventing disease and traumatic injury, of relieving pain, and of preserving and improving human health, using molecular tools and molecular knowledge of the human body." Nanomedicine is the conduit through which nanotechnology encounters biology and medicine—an encounter that will undoubtedly result in great innovation and discovery.

This chapter explores the synthesis, purification, and functionalization of fullerenes, carbon nanotubes, and nanodiamonds (NDs) for nanomedicine applications. Each section includes a brief overview of recent research as well as definitions and relevant applications.

5.1 Introduction to Fullerenes

Following observations of unusual XRD patterns in carbon fibers in 1969, a scientist at the Atomic Energy Research Establishment in Harwell, United Kingdom, proposed a new carbon allotrope with 60 carbon atoms in the shape of a truncated icosahedron (similar to the structure of an American soccer ball) (Thrower 1999). Around this same time, Professor Osawa of the Toyohashi University of Technology suggested that an icosahedral C_{60} molecule could, in fact, be chemically stable based on simulations of the bowl-shaped corannulene molecule (Osawa 1970). Unfortunately, these two observations went largely unnoticed by the scientific community since the former was never published and the latter was published only in Japanese. It was not until 1985 when Professors Curl, Kroto, and Smalley published the results of their famous laser ablation experiment that the existence (and stability) of the

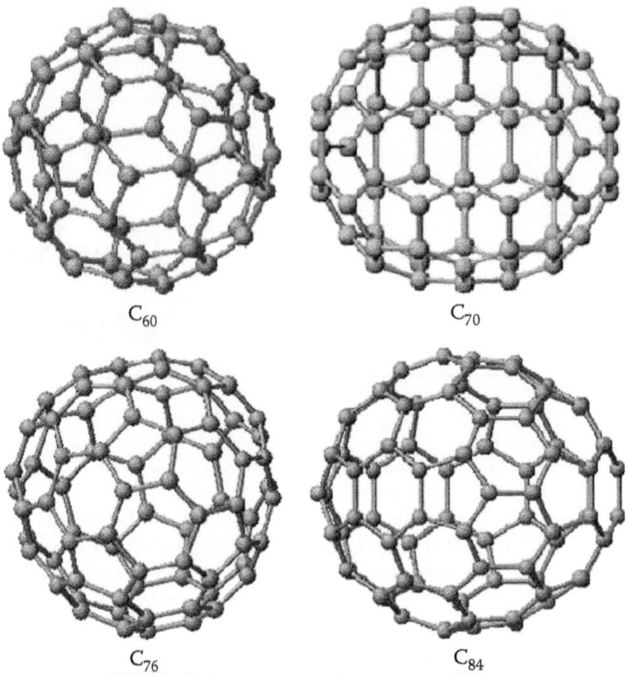

FIGURE 5.1 Ball-and-stick models of some representative fullerene molecules.

C_{60} fullerene molecule in the gas phase was first established (Kroto et al. 1985). Named "buckminsterfullerene" after the American architect who popularized the similarly shaped geodesic dome in the 1960s (Tarnai 1993), the discovery of the C_{60} molecule would later earn the three the 1996 Nobel Prize in Chemistry. While the term "buckminsterfullerene" ("buckyball" for short) is traditionally reserved for the C_{60} molecule alone, the term "fullerenes" is now used to describe an entire class of closed cage molecules consisting of only carbon atoms (Dresselhaus et al. 1996). Perhaps the only rule that guides fullerene classification is Euler's theorem, which simply states that for a closed structure to be purely composed of pentagons and hexagons, it must contain 12 pentagons. This limits the smallest fullerene to a C_{20} structure and renders C_{60} the smallest stable fullerene, as it is the smallest fullerene that does not contain two pentagons side by side in its structure (Zhang et al. 2007). The structures of a few representative fullerenes are shown in Figure 5.1. Fullerenes have several potential applications, but applications in nanomedicine are currently receiving special attention, as discussed in the following.

5.1.1 Synthesis of Fullerenes

In 1974, while completing postdoctoral studies under Professor Donald Levy, Dr. Richard Smalley and coworkers at the University of Chicago reported the use of a continuously operated supersonic molecular beam that allowed a high-resolution study of free particles (Smalley et al. 1974) The rotational cooling that occurred during supersonic expansion upon leaving the nozzle provided the ultracold, collisionless environment required for high-resolution measurements of molecular electronic states. Years later, Smalley and coworkers at Rice University reported the use of a *pulsed* supersonic nozzle apparatus (Liverman et al. 1979), which offered additional advantages of both higher intensity and an absence of condensation effects, and later combined this *pulsed* nozzle capability with a pulsed high-power laser vaporization technique (Dietz et al. 1981). Concurrently, astrophysicists were collaborating with spectroscopists to identify infrared (IR) emission from carbon clusters streaming out of red giant carbon stars (Dresselhaus et al. 1996). This prompted a collaboration between Smalley and Professor

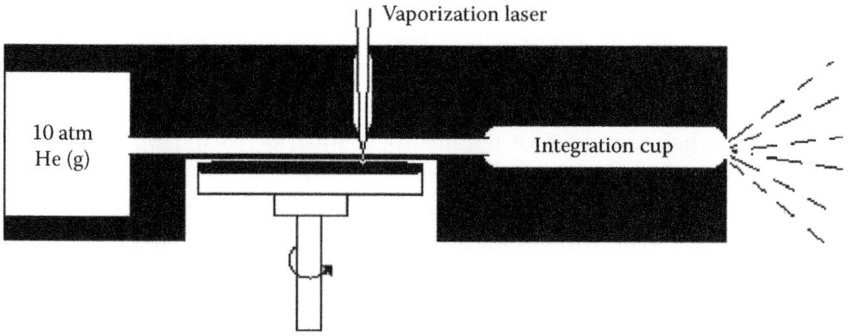

FIGURE 5.2 Original schematic diagram of the pulsed supersonic nozzle used to produce fullerenes. (Adapted from Kroto, H.W. et al., *Nature*, 318(6042), 162, 1985.)

Harry Kroto of Sussex University in England with the goal of using laser irradiation of a graphite target to attempt to recreate the atmosphere in which carbon was nucleating in the observed red stars (Kroto et al. 1985; Dresselhaus et al. 1996).

The original synthesis of C_{60} fullerene involved the use of a pulsed neodymium-doped yttrium aluminum garnet (Nd:YAG) laser focused on a solid disk of graphite into a high-density helium gas flow (Kroto et al. 1985), as shown in Figure 5.2. Upon opening the pulsing valve, the subsequent laser pulse that ablated the surface of the graphite disk in turn created carbon plasma in the He flow. While constrained in the He flow, the carbon plasma can cluster prior to supersonic expansion and cooling into a vacuum. Almost immediately after discovering this synthesis, it was determined that lanthanum (La) atoms could be entrapped inside of fullerenes by doping the graphite target with La (Heath et al. 1985). Unfortunately, these methods were capable of producing only microscopic amounts of fullerenes (mostly C_{60} and C_{70}). It should also be noted that, following their discovery, the fullerenes' presence in nature was discovered both in naturally occurring carbon-rich rocks like shungite of Shunga, Russia (Buseck et al. 1992) and following powerful geological events like wild fires along the K–T boundary in New Zealand (Heymann et al. 1994), lightning strikes (Daly et al. 1993), and meteor strikes (Becker et al. 1994).

It was not until 1990, however, that macroscopic quantities of fullerenes were finally generated in a laboratory environment using ohmic heating (Kratschmer et al. 1990). Two graphite rods, one sharpened to a conical point and continually touching a flat-ended graphite rod by the force of a spring, were connected to copper electrodes in an inert helium environment with a minimum pressure of 25 torr (Diederich et al. 1991). After applying a voltage, resistive heating between the rods produced carbon soot with gram quantities of fullerenes embedded within. This soot is commercially available as produced, containing 7% fullerenes by weight (fullerene composition: C_{60} 76%, C_{70} 22%, with the remainder being higher-order fullerenes). The Krätschmer arc reactor was eventually modified to utilize a gravity feed as opposed to a spring to ensure contact between the carbon rods (Koch et al. 1991). The Wudl reactor, also known as a "contact-arc" apparatus (Dresselhaus et al. 1996), is commonly used for its ease of construction and relatively low cost. A traditional AC arc welder is used as a power supply, and a thin carbon rod (6 mm) is placed atop a thick rod (12 mm) allowing gravity to maintain contact. Operating the apparatus consumes the thin carbon rod and renders fullerene-enriched soot (yields of ~4% extractable C_{60}/C_{70}) (Dresselhaus et al. 1996).

The primary alternative to resistive heating in the laboratory is the arc-discharge technique. As this apparatus is discussed in more detail in the subsequent section on carbon nanotubes, it is sufficient here to simply define this technique as a carbon-plasma arc that forms between two horizontal carbon electrodes separated by a 4 mm fixed gap (Parker et al. 1991). While there are reported yields of 44% fullerene production, it is important to note that one-third of this yield is giant fullerenes (C_{84} or larger).

Finally, though inappropriate for laboratory-scale fullerene production, the combustion synthesis method developed by Dr. Michael Alford at TDA Research Inc. in Wheat Ridge, Colorado, is used at the world's largest fullerene production facility (40 tons per year) (Alford et al. 2008). This method utilizes sooting flames with acetylene, toluene, or benzene as hydrocarbon feedstocks. Polycyclic aromatic feedstocks, such as naphthalene and 1,2,3,4-tetrahydronaphthalene (Bachmann et al. 1994; Alford et al. 2008), have also proven to produce more soot. While resistive heating, electric arcs, and flames are the predominant methods used for fullerene production, it is worth mentioning that fullerenes have also been synthesized with solar furnaces (Chibante et al. 1993), hydrocarbon pyrolysis (Taylor et al. 1993), inductive heating (Peters and Jansen 1992), and coalescence reactions (Yeretzian et al. 1992). Since fullerenes can be found in all forms of soot, even from burning wood and lighting candles (Murr and Garza 2009), it should come as no surprise that there are dozens of other synthetic routes beyond the scope of this chapter.

5.1.2 Purification of Fullerenes

Once fullerene-enriched soot is obtained, isolation of the fullerenes is traditionally accomplished through either sublimation or solvent extraction. For sublimation, the soot is placed at the end of an evacuated quartz tube with a temperature gradient. As the fullerenes possess different sublimation temperatures (350°C for C_{60} and 460°C for C_{70}) (Dresselhaus et al. 1996), higher fullerenes will deposit closer to the soot, while C_{60} will condense on the quartz walls furthest from the soot. One of the benefits of this technique is that there is no contamination of the soot with organic solvents that are potentially toxic. More commonly, however, organic solvent extraction is used for fullerene isolation. Fullerenes have the highest solubility in benzenes, napthalenes, and carbon disulfide, with toluene being favored for its lower cost and lower toxicity (Sivaraman et al. 1992). The use of a Soxhlet extractor is commonly employed (Parker et al. 1991, 1992; Khemani et al. 1992; Hernadi et al. 2004). The extractor can be operated for several hours, allowing the boiling solvent in the round bottom flask to become concentrated with fullerenes while leaving the carbonaceous soot behind in a thimble.

Once extracted, the final purification step is traditionally liquid chromatography (LC). Running a sample of various fullerenes dissolved in solvent (mobile phase) through a column packed with a solid (stationary phase) produces different fullerene fractions off the column since the fullerenes with the lowest molecular weight pass through the stationary phase faster than larger fullerenes. Fractions of different fullerenes dissolved in toluene can actually be identified by color alone since C_{60} has a purple hue, while C_{70} is orange. Most commonly, alumina is used as a stationary phase with a 95/5 hexane/toluene mixture used as a mobile phase (Taylor et al. 1990, 1991). Instead of constantly reintroducing collected solvent for the mobile phase, a modified Soxhlet extractor has been used to automatically recycle the mobile phase. Other stationary phases include activated carbon and polystyrene gel, using chlorobenzene and toluene as eluents, respectively (Komatsu et al. 2004).

5.1.3 Functionalization of Fullerenes

Fullerenes have been described in the literature as a "double-edged sword," being potentially harmful at high concentrations while highly beneficial at low concentrations (Nielsen et al. 2008). While orally administered C_{60} does not induce observable cytotoxicity in rats up to doses of 226 $\mu g/cm^2$ as most of the fullerene will not be absorbed (Jia et al. 2005), it has been suggested that the carbon cage of C_{60} might cause oxidative damage to cell membranes through reactive oxygen species (ROS) formation (Sayes et al. 2005). However, more recent studies have shown aqueous C_{60} suspensions not only have no acute or subacute toxicity in rodents, but that they also act as powerful protective agents against free-radical damage to the liver (Gharbi et al. 2005; Cai et al. 2009). It is possible that the cytotoxicity observed earlier in large fullerene doses was due to improper organic solvent removal prior to animal injection (Gharbi et al. 2005). As unfunctionalized fullerenes are not soluble in water (Sayes et al. 2004), it is not surprising that most fullerene molecules being proposed for medical applications are functionalized. Fullerene chemistry ranges from nucleophilic- and cyclo-additions to hydrogenation and reduction reactions (Hirsch and Brettreich 2005). The rest of this section discusses the role of functionalized C_{60} as diagnostic agents, delivery agents, antioxidants, and finally their function as antimicrobial and antiviral agents.

Fullerenes have been designed for numerous diagnostic imaging platforms, including magnetic resonance imaging (MRI) (Bolskar et al. 2003; Sitharaman and Wilson 2007), x-ray (Wharton and Wilson 2002), and radioimaging (Shultz et al. 2010). The bulk of fullerenes used for MRI contrast agents are endohedral metallofullerenes, which encapsulate gadoliunium ions (Gd^{3+}) known as gadofullerenes and abbreviated as $Gd@C_{2n}$ (Sitharaman et al. 2007). When water-solubilized with hydroxyl or malonate substituents, gadofullerenes aggregate in solution trapping pockets of water molecules (Laus et al. 2007). The magnetic resonance relaxation of these water molecules inside these aggregates renders relaxivity rates (a measure of the effectiveness of an MRI contrast agent) significantly greater than current clinically available Gd-based agents. Moreover, depending on the water-solubilizing functional group selected, the gadofullerenes show a strong pH dependency on proton relaxation rates (Toth et al. 2005). A trimetallic metallofullerene, $Gd_3N@C_{80}$, has also been proposed and studied as an MRI contrast agent (Zhang et al. 2010). Water-solubilized holmium (Ho) metallofullerenes have been proposed as x-ray imaging agents utilizing the scattering effects of the internalized Ho^{3+} ions (Wilson et al. 1999). Moreover, these Ho@C82 molecules can be activated with high-flux neutron irradiation, rendering $^{165}Ho@C_{82}$, a β^-/γ emitter capable of γ-imaging for in vivo biodistribution studies (Cagle et al. 1996).

As delivery agents, functionalized fullerenes have been proposed for the delivery of both drugs and genes. Amphiphilic fullerene monomers (AF-1 or "buckysomes"), C_{60} modified with dendritic moieties, and fatty acid side chains (Brettreich et al. 2000) are attractive due to their potential for vesicle-like self assembly, their ability to encapsulate high payloads of therapeutic molecules, and their tissue specificity when coupled to targeting ligands (i.e., antibodies) (Partha et al. 2007). A fullerene–paclitaxel conjugate has been designed to release paclitaxel via enzymatic hydrolysis following aerosol liposome delivery as a slow-release drug for lung cancer therapy (Zakharian and Wilson 2003). Octa-amino derivatized C_{60} and dodeca-amino derivatized C_{60} molecules have been developed as DNA/gene-delivery vectors (Sitharaman et al. 2008). Using Hirsch–Bengel chemistry to functionalize C_{60}, these transfection agents have the ability to increase cellular DNA uptake. A tissue-vectored bisphosphonate fullerene, $C_{60}(OH)_{16}AMBP$, has been developed as an osteoporosis drug and shown to be an effective inhibitor of hydroxyapatite (HAP) crystal growth in vitro (Gonzalez et al. 2002). A series of bis (bisphosphonate) fullerenes have also been synthesized and proposed as drugs against osteoporosis (Mirakyan et al. 2003).

Fullerenes are also desirable for their ability to scavenge free radicals. Free radicals have been associated with the onset and progression of fatal neurogenerative diseases, like amyotrophic lateral sclerosis (ALS, also referred to as Lou Gehrig's disease) and Parkinson's disease. Therefore, the ability of fullerenes to react with or catalytically dissociate free radicals has sparked additional research as both free radical scavengers and neuroprotective agents (Lin et al. 1999). In this regard, fullerenes have, in fact, come to be regarded as "free radical sponge[s]" (Krusic et al. 1991). The versatility of fullerenes allows for the creation of water-soluble derivatives that have demonstrated neuroprotective effects in animal models (Dugan et al. 1997). Two polyhydroxylated C_{60} derivatives, $C_{60}(OH)_{12}$ and $C_{60}(OH)_{18-20}O_{3-7}$, have shown the ability to quench the superoxide radical prior to conversion to the more harmful hydroxyl radical (Dugan et al. 1996). In addition, fullerene derivatives, acting as free radical scavengers, have demonstrated an ability to reduce injury on the ischemia reperfusion intestine (Lai et al. 2000), as well as an ability to protect against the apoptosis of cell types (Hsu et al. 1998; Bisaglia et al. 2000). Moreover, when given a diet including the C_3 (e,e,e-$C_{60}(C(COOH_2))_3$) compound, mice had an 11% longer average life span than the control group! (Quick et al. 2008)

While not absorbed when ingested, fullerenes are rapidly absorbed and distributed throughout the body when injected. To date, there have been five main areas with demonstrated potential applications for fullerenes: free radical scavenging (discussed above), antihuman immunodeficiency virus (HIV) activity, DNA photocleavage, and antimicrobial activity.

C_{60} derivatives have demonstrated inhibition of HIV periodontitis (HIV-P), an aggressive, painful form of periodontal disease with all the characteristics of HIV plus soft tissue ulceration/necrosis and rapid destruction of the bone (Friedman et al. 1993). Other studies have demonstrated that C_{60} functionalized with polyethylene glycol (C_{60}-PEG), in specific doses, may work as a potential cancer therapy (Tabata et al. 1997) and water-soluble fullerene derivatives have been incorporated into a cancer (melanoma)-targeting antibody (Ashcroft et al. 2006b). One study demonstrated a 19% reduction in cell number within 2 weeks after exposure with ultraviolet (UV) light and dendritic C_{60} monoadduct and malonic acid C_{60} trisadduct (Rancan et al. 2002). Lastly, antimicrobial activity has been observed with various fullerene derivatives; specifically, water-soluble carboxyfullerenes have been shown to inhibit *Escherichia coli*-induced meningitis, notably reducing damage, and thus, offering potential as a therapeutic agent for bacterial meningitis (Tsao et al. 1999).

5.2 Introduction to Single-Walled Carbon Nanotubes

The discovery of tubular forms of carbon on the electrodes of the arc-discharge apparatus used for fullerene synthesis has spurred a plethora of studies based on carbon nanotubes (Iijima 1991; Ebbesen and Ajayan 1992; Iijima and Ichihashi 1993). Carbon nanotubes are a prime example of pseudo one-dimensional structures with a high aspect ratio and a large surface area. Their unique structure provides them with interesting optical, mechanical, electrical, and thermal properties (de Heer et al. 1995; Saito 1997; Wong et al. 1998; Ajayan 1999; Dai 2002). Of particular interest are structures that resemble a single sheet of graphene rolled up seamlessly into a hollow cylindrical tube that are called single-walled carbon nanotubes (SWNTs), as shown in Figure 5.3. The first detailed synthesis of SWNTs was reported in 1993[4], and since then, SWNT research has grown many fold. It is beyond the scope of this short review to cover all aspects of SWNT synthesis and applications; however, we attempt here to give the reader a brief general overview of this new class of carbon nanostructures, with special attention given to potential biological and medical applications.

5.2.1 Synthesis of SWNTs

Three of the most widely used methods for the synthesis of SWNTs are (1) electric arc-discharge, (2) laser ablation or vaporization of a carbon target, and (3) chemical vapor deposition (CVD). Each of the three methods is unique in its own way, yielding SWNTs of differing dimensions and chiralities.

The first reported technique for the synthesis of SWNTs was by electric arc-discharge (Bethune et al. 1993; Iijima and Ichihashi 1993). This is the same process that was used for the first synthesis of

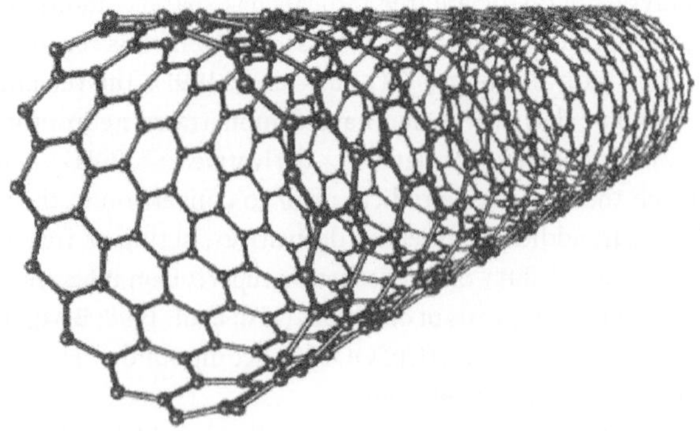

FIGURE 5.3 Ball-and-stick models of a SWNT.

fullerenes and multiwalled nanotubes (MWNTs) (Iijima 1991). MWNTs have multiple concentric layers of rolled up graphene. It was observed that the presence of a small amount of metal particle (Fe, Co, etc.) at the cathode resulted in the synthesis of SWNTs, along with MWNTs and fullerenes. Whereas fullerenes are formed at the cathode, SWNTs are produced mostly in the vapor phase. The vaporization of metal atoms into the gas phase stabilized the elongated fullerene structures resulting in SWNT production (Bethune et al. 1993). The very low yield of SWNTs (fullerenes and MWNTs were formed in excess) and the extensive amorphous carbon coating of SWNTs were initially seen as major drawbacks of the electric-arc process. However, advancements including the use of mixed-metal catalysts have resulted in significant improvement in the yield and purity of SWNTs produced by this process (Journet et al. 1997). The SWNTs produced have a diameter of ~1.2 nm.

In the laser ablation process, a pulsed laser vaporizes a carbon target mixed with metal catalyst particles (Guo et al. 1995). This method resulted in a relatively high yield of SWNTs (~70%), and the use of mixed-metal catalysts such as Co–Ni, Co–Pt, Ni–Y, and Ni–CO results in higher yields of SWNTs compared to their single metal counterparts such as Fe or Co (Kong et al. 1998). Similar to the arc-discharge technique, SWNTs produced by this process are produced in the vapor phase. The SWNTs produced by this method predominantly exist in the form of bundles with very little amorphous carbon coating. The relative cleanliness (less amorphous carbon coating) of the process compared to the arc-discharge technique is attributed to the bundling that reduces the amount of surface available for amorphous carbon coating (Thess et al. 1996; Kong et al. 1998). The laser-ablation technique produces SWNTs with a diameter of ~1.4 nm.

Both the arc-discharge technique and the laser-vaporization technique have been improved to produce high-quality SWNTs. However, neither of these methods are suitable for mass production. This led to the development of a new technique that makes use of disproportionation of a carbon feed stock over preformed metal catalyst particles (Dai et al. 1996; Peigney et al. 1997). A variety of metal catalysts (Fe, Co) and carbon feed stock gases (CH_4, CO, etc.) have been used for the synthesis of SWNTs by CVD (Cassell et al. 1999). The size and composition of the catalyst particle plays an important role in the synthetic process (Harutyunyan 2009), and the reduction of the catalyst particle size from 10–100 to 1–5 nm results in a very high yield of SWNTs. The CVD process produces SWNTs with a range of diameters (1–5 nm).

In spite of the developments made in the production of SWNTs in bulk quantities, all of the methods are "batch methods" yielding only milligram to gram quantities of SWNTs in a few hours. To produce SWNTs in much higher quantities, a continuous-flow synthetic method has been proposed (Tibbetts et al. 1994; Cheng et al. 1998a,b). These schemes involve the principle of introducing metal catalyst particles into a carbon feedstock flow. Many groups have successfully used organometallic precursors as a source of catalyst particles, especially metallocenes and metal carbonyls, to produce SWNTs in bulk quantities. The best results are obtained when iron pentacarbonyl ($Fe(CO)_5$) is used as the catalyst and carbon monoxide as the feed stock (Nikolaev et al. 1999; Bronikowski et al. 2001). High-pressure carbon monoxide, or the HiPco process as it is known, produces large quantities of SWNTs in very high yield (>90%) and purity (no amorphous carbon coating). The HiPco process also produces SWNTs with significantly smaller diameters (0.7–1.1 nm) compared to other SWNT production techniques.

5.2.2 Purification of SWNTs

As produced SWNTs, regardless of the method of production, are contaminated with impurities. The impurities include metal catalyst nanoparticles and carbonaceous impurities such as amorphous carbon, fullerenes, and graphitic shells covering the metal catalyst particles. To utilize the unique properties of SWNTs, it is necessary to obtain them in their purest form. Especially for biomedical applications, the presence of carbonaceous impurities and metal catalyst particles will result in undesirable sample inhomogenity and potential metal-mediated toxicity (Donaldson et al. 2006; Kolosnjaj-Tabi et al. 2010). Initially, hydrothermal treatment and surfactant-assisted microfiltration were proposed for the purification of SWNTs (Tohji et al. 1996; Bandow et al. 1997). However, these techniques are suitable for purification only on the microscale. Many other techniques (oxidation, chemical functionalization, microwave-assisted heating, etc.) have been proposed for the purification of SWNTs on a larger scale. Each of these techniques results in different degrees of purification and modification of the SWNT surface.

Microfiltration and chromatographic separation are nondestructive methods proposed for the purification of SWNTs (Niyogi et al. 2001; Zhao et al. 2001; Chattopadhyay et al. 2002; Farkas et al. 2002). The microfiltration technique is based on particle sizes. As produced SWNTs, they are dispersed in a cationic surfactant and passed through a microfilter. The majority of the fullerenes, other carbon impurities, and metal catalyst particles flow through the filters, whereas SWNTs are retained on the filter. However, this method suffers from the necessity of successive filtration steps and depends on the initial purity of the sample. Chromatographic techniques have been also proposed for the purification of SWNTs based on their size (Niyogi et al. 2001; Zhao et al. 2001). Properly dispersed (aqueous or organic) or solvated SWNTs are passed through columns. The carbonaceous impurities, which have larger diameters than SWNTs, pass through the column more quickly. Similarly, SWNTs of different lengths are found to have different elution times, resulting in a size-based separation. The chromatographic techniques offer better size-based purification than microfiltration; however, they still suffer from a cumbersome procedure and only microscale purification.

Ultrasmall inorganic nanoparticles have been shown to purify SWNTs in a nondestructive manner (Mizoguti et al. 2000; Zhang et al. 2000; Thien-Nga et al. 2002). It has been found that gold nanoparticles selectively bind to carbonaceous impurities of SWNTs dispersed in a cationic surfactant. The selective binding of gold particles increases the chemisorption of O_2, thereby oxidizing the impurities at a lower temperature than the SWNTs by themselves. Similarly, ZrO_2 and $CaCO_3$ nanoparticles have been shown to effectively remove metal catalyst impurities from SWNTs (Zhang et al. 2000).

The purification of large batches of SWNTs was first reported by oxidation in strong acids (3–5 M nitric acid) (Rinzler et al. 1998). Since then, oxidative methods have been preferred over other methods of purification owing to their simplicity and the ability to purify SWNTs on a larger scale (Ivanov et al. 1995; Dujardin et al. 1998; Hu et al. 2003; Zhang et al. 2003; Furtado et al. 2004). The oxidation process also incorporates functional groups such as carboxylic acids, which can be exploited to further functionalize purified SWNTs (Liu et al. 1998). Generally, oxidative treatments are classified either as liquid-phase oxidation using strong acids (HNO_3, HCl, $KMnO_4/H_2SO_4$, H_2O_2, etc.) or as gas-phase oxidation (air, O_2, Cl_2, etc.). The oxidizing agents diffuse through the graphitic layer surrounding the metal catalyst particles and convert them into their corresponding metal oxides. The metal oxides possesses a larger volume than their metal counterparts that results in the disruption of the carbon sheath surrounding the metal particles and the leaching out of metal catalyst impurities (Chiang et al. 2001).

Liquid-phase oxidation methods involve sonication and/or refluxing SWNT samples with strong oxidizing acids, such HNO_3, a 3:1 mixture of H_2SO_4/HNO_3, or $KMnO_4$ in H_2SO_4. Many reports have been published with different temperatures, reaction times, and acid conditions (Liu et al. 1998). Along with the metal catalyst particles, the oxidation process also removes a majority of the carbonaceous impurities. The relative chemical inertness of SWNT materials permits much harsher oxidation conditions than are tolerated by other carbonaceous materials such as fullerenes, graphitic shells, and amorphous carbon. However, the relatively unstable five-membered rings at the SWNT ends react readily with the oxidizing agents (Park et al. 2006), which results in shortening of SWNT lengths with most oxidation methods. The methods also result in the production of some sidewall defects. In order to better purify SWNT samples, liquid purification procedures are generally combined with other oxidation techniques such as wet-air oxidation (Chiang et al. 2001).

Gas-phase oxidative procedures are based on the principle that carbonaceous species are more reactive toward gases than SWNTs themselves (Nagasawa et al. 2000; Zimmerman et al. 2000; Sen et al. 2003; Park et al. 2006). When SWNTs are exposed to oxidizing gases, such as air, O_2, Cl_2, H_2O, and

HCl, the amorphous carbon and graphitic shells covering the metal catalyst particles are etched away. The naked metal particles can then be removed by washing with acid. Most of these oxidative processes result in very low SWNT yields (20%–50%). This problem is partly due to the fact that SWNT synthetic procedures generally produce SWNTs containing a large amount of carbonaceous impurities and metal catalyst particles.

Care must be taken about the severity of the oxidizing agents on SWNTs purification. For example, smaller diameter SWNTs, such as SWNTs produced by the HiPco process, have greater structural strain because of their smaller diameter (Park et al. 2006). Thus, when they are exposed to harsh conditions similar to those used on larger diameter SWNTs, HiPco SWNTs can be completely destroyed. The major impurities in the HiPco process are the metal catalyst particles with very little amorphous carbon (>90% SWNTs). Generally, HiPco SWNTs are purified by wet-air oxidation at 180°C–300°C, followed by treatment with HCl. Recently, a purification protocol involving liquid Br_2 has been shown to be very effective in removing metal catalytic impurities from HiPco SWNTs without compromising their sidewall structure (Mackeyev et al. 2007). The major disadvantage of oxidative methods is that the oxidation process not only removes the metal catalyst and amorphous carbon impurities but also disrupts the physical and electronic structure of SWNTs. Hence, care must be taken to strike a balance between the purification and preservation of physical and electronic structures. Recently, microwave-assisted purification has been proposed for the removal of metal catalyst particles from SWNTs in a nondestructive manner (Martínez et al. 2002).

5.2.3 Characterization of SWNTs

The purity of the SWNTs and their degree of functionalization can be evaluated using analytical techniques such as Raman spectroscopy, UV–Vis spectroscopy, near infrared (NIR) spectroscopy, thermal gravimetric analysis (TGA), and electron microscopy techniques (transmission electron microscopy [TEM], scanning electron microscopy [SEM], etc.).

SWNTs have three signature peaks in the Raman spectrum that can be used for characterization: (1) the radial breathing mode (RBM) (from 150 to 300 cm^{-1}, depending on the diameter of the SWNT) gives information about the diameter and packing, (2) the tangential mode (G-band; from 1515 to 1590 cm^{-1}) gives information about the sp^2-hybridized carbons and can be used to judge purity, and (3) the disorder mode (D-band; from 1280 to 1320 cm^{-1}) is a measure of the sidewall defects, amorphous carbon, and the degree of functionalization, etc. (Rao et al. 1998; Park et al. 2006; Dresselhaus et al. 2010).

NIR spectroscopy has also been used to characterize SWNT materials (Bachilo et al. 2002; O'Connell et al. 2002). Metallic and semiconducting SWNTs can be separately identified using this technique (Ghosh et al. 2010). The technique can also be used to check for functionalization since functionalized SWNTs do not fluoresce in the NIR region as a result of the disruption of the native electronic structure. Similarly, UV–Vis spectroscopy is also useful for characterizing SWNT materials due to their unique electronic transitions between Van Hove singularities (Itkis et al. 2003; Sen et al. 2003).

TGA is one of the most widely used methods to check for the purity and the extent of functionalization since carbonaceous materials and organic substituents decompose at a lower temperature than SWNTs. However, TGA cannot differentiate between different forms of carbonaceous materials such as amorphous carbon and organic functional groups. Hence, the quantification of functional groups by TGA depends on the pre-functionalization purity of SWNT materials and the use of other characterization methods as well. Other techniques such as x-ray photoelectron spectroscopy (XPS) and nuclear magnetic resonance (NMR) spectroscopy can also be used to identify functional groups covalently attached to SWNTs (Ashcroft et al. 2006a). Microscopy techniques such as TEM, SEM, and atomic force microscopy (AFM) have been widely used for the visualization of SWNT materials to study their structural properties, as shown in Figure 5.4. However, electron microscopy techniques use a small, localized fraction of the sample, and hence, such measurements have to be repeated multiple times at different sampling sites to generalize the observation.

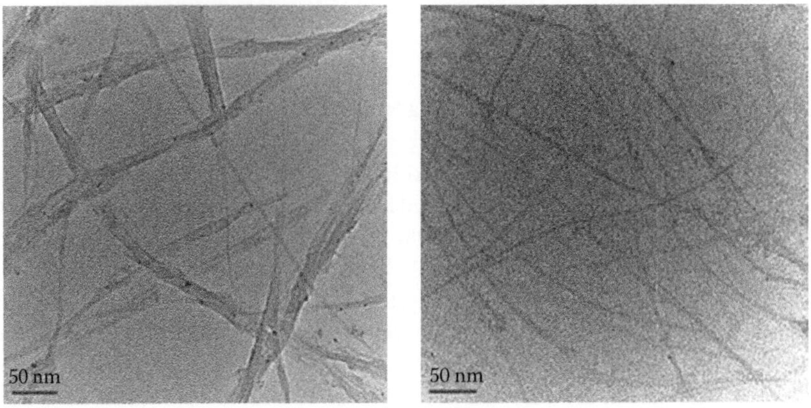

(a) (b)

FIGURE 5.4 TEM images of (a) bundled full-length HiPco SWNTs and (b) debundled full-length HiPco SWNTs. (Reproduced from Liu, Z., et al., *Angewandte Chemie International Edition*, 46(12), 2023, 2007.)

5.2.4 Functionalization of SWNTs

Low-solubility and strong π–π interactions between unfunctionalized SWNTs, make it difficult to manipulate them in their native form. For biological and biomedical applications, SWNTs need to be functionalized to make them "water soluble" and biocompatible. The functionalization of SWNTs can be classified mainly into two types: (1) covalent functionalization (i.e., attachment of functional moieties on the SWNT sidewalls or at the ends) and (2) noncovalent functionalization (i.e., a supramolecular assembly of functional moieties around the SWNTs). In addition, a third important mode of noncovalent functionalization involves the encapsulation of small molecules and ions within the hollow interior of SWNTs.

Most of the covalent functionalization techniques for SWNTs are based on addition reactions such as the Bingel reaction, 1,3-dipolar cycloaddition, free radical addition, fluorination, alkyl diazonium addition, azomethine ylide addition, etc., and coupling reactions with the carboxylic acid groups results from purification processes on SWNT surfaces with amine and alcohol groups.

A number of biologically interesting molecules have been covalently attached to SWNTs either through carbodiimide or acid-chloride-assisted amidation/esterification reactions with carboxylic acid groups. For example, biotin functionalization of SWNTs has been achieved through an EDC-assisted amidation reaction (Huang et al. 2002; Lin et al. 2004; Shi et al. 2004; Asuri et al. 2007). Similarly, BSA has been successfully attached to SWNT surfaces via a carbodiimide-assisted coupling reaction. Upon functionalization, BSA retained 90% of its biological activity. Enzymes such as horseradish peroxidase and soybean peroxidase have also been covalently conjugated to SWNTs through EDC-assisted coupling reactions using *N*-hydroxysuccinamide (NHS) as a linker. All of the enzymes covalently conjugated to SWNTs retained their biological activity.

In addition to coupling reactions with carboxylic acid groups on the sidewalls and ends, the extended π-conjugated sidewalls have also been covalently functionalized for a variety of applications. Current functionalization strategies involve polar, pericyclic, or radical reactions to incorporate carbon–carbon bonds or carbon–heteroatom bonds. The fullerene-like unsaturated π-conjugated structure of SWNTs also allows diverse functionalization approaches including halogenation, hydrogenation, cycloaddition, radical addition, ozonolysis, and electrophilic and nucleophilic addition reactions. A schematic of some of the widely used sidewall functionalization techniques used for SWNTs are presented in Figure 5.5.

Halogenation using elemental fluorine is well established for SWNTs (Mickelson et al. 1998; Touhara and Okino 2000; Kawasaki et al. 2004). The actual mechanism of fluorination is not clear with both

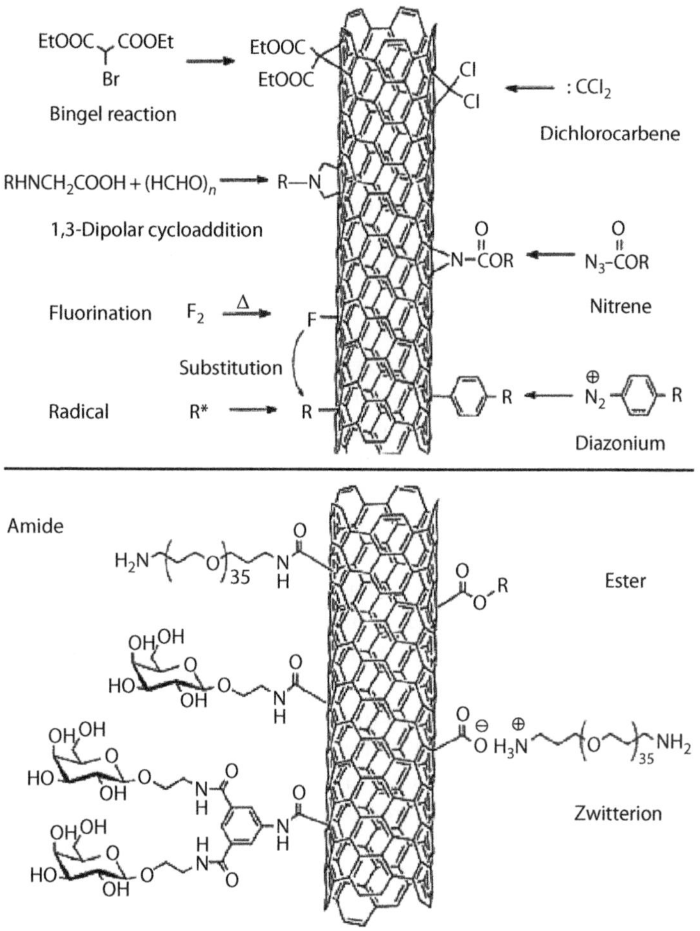

FIGURE 5.5 Schematic of some sidewall functionalization methods for SWNTs. (Reproduced from Liu, K.K. et al., *Biomaterials*, 30(26), 4249, 2009.)

1,2-addition and 1,4-addition pathways proposed with a very small energy difference between them (Boul et al. 1999; Kudin et al. 2001). Fluorinated SWNTs are moderately soluble in alcohol (~1 mg/mL). In addition, the C–F bonds of fluorinated SWNTs can be further substituted with other organic functionalities using Grignard reagents or organolithium agents (alkyl groups) (Boul et al. 1999; Saini et al. 2003), by nucleophilic substitution reactions (diamines and diols) (Zhang et al. 2004), and free radical addition. Further functionalization of these groups can be employed to attach molecules for biological applications. Functionalization involving other halogens such as chlorine and bromine has also been reported (Unger et al. 2002).

Cycloaddition reactions to the SWNT sidewalls are perhaps the most used functionalization methods. Carbenes and nitrenes have been added to the SWNTs via a [2 + 1] cycloaddition mechanism (Chen et al. 1998a,b; Holzinger et al. 2001; Hirsch 2002; Kamaras et al. 2003). A variety of organic functional groups such as alkyl chains, crown ethers, and dendrimers have been attached using [2 + 1] cycloaddition reactions and the functionalized SWNTs can then be further modified to incorporate biomolecules such as DNA (Holzinger et al. 2003; Moghaddam et al. 2003). In another study, 1,3-dipolar cycloaddition of azomethine ylides have been used to functionalize SWNTs (Pantarotto et al. 2003a; Tagmatarchis and Prato 2004). The versatility of this functionalization strategy allows the preparation of SWNT materials for various applications including biological applications where SWNTs are functionalized with amino acids, peptides, and nucleic acids (Georgakilas et al. 2002a; Pantarotto et al. 2003a,b). The azomethine ylide reaction has been also proposed for purification of as-produced SWNTs (Georgakilas et al. 2002b). Similar to fullerenes, the [2 + 1] cyclopropanation (Bingel addition) reaction has also been reported for SWNTs with diethylbromomalonate being used as the carbene precursor (Coleman et al. 2003; Worsley et al. 2004). Microwave-assisted Diels-Alder cycloaddition of o-quinodimethane has also been reported as a way to derivatize SWNTs (Mackeyev et al. 2009), and recently a rhodium-catalyzed [2 + 1] cyclopropanation reaction has been shown to efficiently functionalize SWNTs with amino acids and small peptides (Mackeyev et al. 2009).

A radical addition of organic functionalities to SWNTs was first reported using aryl diazonium salts in the organic media (Bahr et al. 2001; Dyke et al. 2004). The formation of the reactive aryl radical is attributed to the electron transfer between aryl diazonium salts and SWNTs. Interestingly, water-soluble diazonium salts selectively reacted with metallic SWNTs (Strano et al. 2003). It is also reported that the surfactant-assisted dispersion (noncovalent) of SWNTs yielded a highly functionalized material (Dyke and Tour 2003). Oxidative coupling of amine groups have also been reported for individual SWNTs (Kooi et al. 2002), and the amino groups on the SWNT surface have been used as a grafting site for nucleic acids, using a heterobifunctional linker and thiol-modified DNA (Lee et al. 2004). The DNA retained its activity after functionalization. Similarly, glucose oxidase has been grafted onto SWNTs using electrochemical means and its activity tested (Zhang et al. 2005). Radical functionalization of SWNTs has also been successfully demonstrated for both thermal and photochemical routes (Peng et al. 2003; Ying et al. 2003). Reductive intercalation of lithium ions onto the SWNT surface (Billups reaction) has also been shown to be a successful method for functionalizing SWNTs with a variety of organic groups (Liang et al. 2004).

Though ozonolysis has been proposed as a purification method, it can also be used to attach carboxylic acid, ester, aldehyde, and alcohol groups to the SWNT surface (Banerjee and Wong 2002, 2004). SWNTs have also been functionalized by a mechanochemical method (Pan et al. 2002). Simple solid-phase milling of SWNTs with potassium hydroxide resulted in hydroxyl-functionalized SWNTs with high water solubility (~3 mg/mL). Finally, a more detailed review of functionalization techniques for SWNTs can be found in Tasis et al. (2006).

Noncovalent functionalization generally involves a hydrophobic interaction between the aliphatic chain of a dispersion agent and the SWNTs or a π–π interaction between the dispersion agent and SWNT surface. Noncovalent interactions do not disrupt the electronic structure of SWNTs and are preferred for biosensing applications. However, a strong interaction can alter the electronic properties of SWNTs due to "surface doping" effects. Noncovalent dispersion techniques are simple with most of them involving ultrasonication of SWNTs in the dispersion medium, followed by ultracentrifugation. Many different surfactants such as SDS, SDBS, CTAB, and benzyl alkonium chloride have been successfully used to disperse SWNTs in the aqueous medium. As the dispersibility of SWNTs in surfactants is relatively low, polymers such as PEG and polyethylene oxide-polypropylene oxide (Pluronic) have been proposed as an alternative dispersion medium (Moore et al. 2003). Pluronic-suspended SWNTs are especially water soluble and biocompatible (Fernando et al. 2004).

Biologically relevant molecules, such as proteins and peptides have been noncovalently attached to SWNTs (Dieckmann et al. 2003; Nepal and Geckeler 2007; Poenitzsch et al. 2007). It has been shown that common proteins such as lysozyme, histone, hemoglobin, and bovine serum albumin (BSA) are good dispersing agents (Nepal and Geckeler 2007). In fact, proteins have been shown to preferentially coat metallic SWNTs over semiconducting ones, thus suggesting proteins as a potential sorting method for as-prepared SWNTs. Similar to proteins and peptides, ss-DNA disperses SWNTs quite efficiently (Zheng et al. 2003). The bases of the DNA strand π-stack with the SWNT surface and the sugar-phosphate backbone of DNA forms a hydrophilic end. Double-stranded DNA has been reported to also disperse SWNTs efficiently (Zheng et al. 2003; Heller et al. 2006). Short interfering (siRNA) has been attached to SWNTs through noncovalent functionalization using an amine terminated surfactant

(Liu et al. 2007). The nucleic-acid functionalized SWNTs have been actively studied for gene transfection, biosensor, and drug-delivery applications. However, noncovalent functionalization suffers from the lack of control over a large sample volume, and it is unlikely that the noncovalent functionalization of SWNTs will survive in vivo conditions for medical applications.

Finally, SWNTs have been shown to internalize small molecules and ions under proper conditions. The formation of "peapods" or SWNTs with internalized C_{60} molecules, can be achieved by heating SWNTs covered with a thin film of fullerenes to 650°C (Berber et al. 2002; Kataura et al. 2002). In fact, when internalized within a SWNT, fullerenes are more densely packed than fullerenes in the bulk crystal (Yoon et al. 2005). SWNTs have also been shown to encapsulate Zn(II)-diphenylporphyrin (Zn-DPP) when sonicated together in toluene for 1 h (Kataura et al. 2002). Cutting SWNTs via fluorination followed by pyrolysis at 1000°C renders ultrashort SWNTs (US-tubes) with lengths ranging from 20 to 80 nm and a significantly higher degree of sidewall defect (Gu et al. 2002; Ashcroft et al. 2006a). Through these sidewall defects, Gd^{3+} ions (Sitharaman et al. 2005; Hartman et al. 2008), molecular iodine (Ashcroft et al. 2007), and $^{211}AtCl$ molecules readily internalize within the US-tubes (Hartman et al. 2007).

5.3 Nanodiamonds

As this chapter is entitled *Carbon-Based Nanomedicine*, it is important to comment on a fledging field of carbon nanotechnology, which for classified (military) reasons (Shenderova et al. 2002), has only recently become public: NDs. In 1963, investigators in the USSR discovered that the detonation of oxygen-deficient explosives, such as trinitrotoluene (TNT), in an inert medium renders soot containing up to 80% NDs by weight (Shenderova et al. 2002; Danilenko 2004). These 5 nm particles are typically covered with graphite and amorphous carbon, and a lack of purity initially contributed to the slow development of ND research. Acid treatment is currently the most effective method of purification, rendering a carbonaceous material comprised of 90%–97% ND (Xing and Dai 2009). The resulting ND sizes are within the translational range between macromolecules and crystalline solids.

While there has only been limited research in the areas of biocompatibility and cytotoxicity, the results have been generally favorable (Schrand et al. 2007a; Vial et al. 2008; Liu et al. 2009). In one study, NDs demonstrated minimal cytotoxicity effects on cell viability as measured by mitochondria integrity and luminescent ATP production (Schrand et al. 2007b). In addition, another study used Chinese hamster ovary cells to demonstrate the lack of cytotoxicity after 72 h of exposure (Vial et al. 2008). The results of studies such as these are suggestive of ND biocompatibility; however, there is still a great deal to be learned about long-term exposure as well as possible in vivo toxicity, before medical applications can be seriously considered for NDs.

Pristine NDs, ranging in size from 5 nm to μm, exhibit photoluminescence (Chung et al. 2007), a property most widely used for biological imaging as shown in Figure 5.6 (Yu et al. 2005; Vaijayanthimala and Chang 2009). Studies have shown that strong energy beam treatments can increase the luminescence of 100 nm NDs 100-fold (Yu et al. 2005). This is likely due to the surface delocalization of π electrons or color vacancy centers inside the nanoparticles. One recent study by Wee and coworkers observed two-photon-excited fluorescence in the N–V centers after proton irradiation (Wee et al. 2007; Chang et al. 2008). This effect is attributed to the high density of defects ($4.5 \pm 1.1 \times 10^{18}$ centers/cm^3) in NDs. Two photon-excited fluorescence imaging may have significant applications for in vivo imaging because of the low background and high signal-to-noise associated with the technique. NDs present an attractive alternative to quantum dots (QDs) (Fu et al. 2007; Lim et al. 2009). Specifically, it has been demonstrated that NDs can be used to track single particles or a single molecule within a living cell. Finally, a recent report has documented the use of ND as a scaffold for a multicentered MRI contrast agent (Manus et al. 2010).

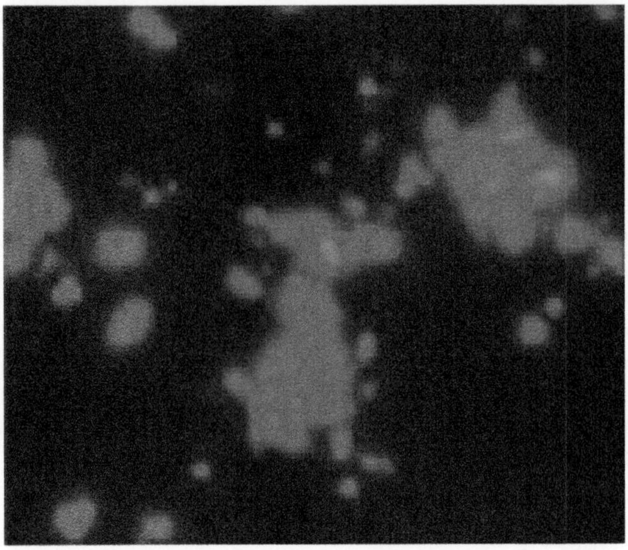

FIGURE 5.6 **(See color insert.)** Epifluorescence image of NDs. (Reproduced from Yu, S.-J. et al., *J. Am. Chem. Soc.*, 127(50), 17604, 2005.)

5.4 Conclusion

The fields of fullerene and carbon nanotube science are relatively mature, having been widely studied for nearly 20 years; by comparison, the field of ND research is far less developed. While many technological applications can be envisioned for fullerene and carbon nanotube materials, medical applications seem especially compelling given their physical properties: (1) an ordered (hollow) structure for containment of medical agents, (2) an exterior carbon surface for facile biological functionalization, (3) an inherent resistance to biological metabolism, and (4) a lipophilic nature producing enhanced cellular uptake with low cytotoxicity. This combination of properties, unique to these carbon nanostructures alone, holds untold promise for advancements in the fields of nanobiology and nanomedicine.

Acknowledgment

The authors wish to gratefully acknowledge the support of the Robert A. Welch Foundation (Grant C-0627) over the years, which contributed to the development of this review.

References

Ajayan, P. M. 1999. Nanotubes from carbon. *Chemical Reviews* 99 (7):1787–1800.

Alford, J. M., C. Bernal, M. Cates, and M. D. Diener. 2008. Fullerene production in sooting flames from 1,2,3,4-tetrahydronaphthalene. *Carbon* 46 (12):1623–1625.

Ashcroft, J. M., K. B. Hartman, K. R. Kissell, Y. Mackeyev, S. Pheasant, S. Young, P. A. W. Van der Heide, A. G. Mikos, and L. J. Wilson. 2007. Single-molecule I-2@US-tube nanocapsules: A new X-ray contrast-agent design. *Advanced Materials* 19 (4):573–576.

Ashcroft, J. M., K. B. Hartman, Y. Mackeyev, C. Hofmann, S. Pheasant, L. B. Alemany, and L. J. Wilson. 2006a. Functionalization of individual ultra-short single-walled carbon nanotubes. *Nanotechnology* 17 (20):5033–5037.

Ashcroft, J. M., D. A. Tsyboulski, K. B. Hartman, T. Y. Zakharian, J. W. Marks, R. B. Weisman, M. G. Rosenblum, and L. J. Wilson. 2006b. Fullerene (C-60) immunoconjugates: Interaction of water-soluble C-60 derivatives with the murine anti-gp240 melanoma antibody. *Chemical Communications* (28):3004–3006.

Asuri, P., S. S. Bale, R. C. Pangule, D. A. Shah, R. S. Kane, and J. S. Dordick. 2007. Structure, function, and stability of enzymes covalently attached to single-walled carbon nanotubes. *Langmuir* 23 (24):12318–12321.

Bachilo, S. M., M. S. Strano, C. Kittrell, R. H. Hauge, R. E. Smalley, and R. B. Weisman. 2002. Structure-assigned optical spectra of single-walled carbon nanotubes. *Science* 298 (5602):2361–2366.

Bachmann, M., J. Griesheimer, and K. H. Homann. 1994. The formation of C-60 and its precursors in naphthalene flames. *Chemical Physics Letters* 223 (5–6):506–510.

Bahr, J. L., J. Yang, D. V. Kosynkin, M. J. Bronikowski, R. E. Smalley, and J. M. Tour. 2001. Functionalization of carbon nanotubes by electrochemical reduction of aryl diazonium salts: A bucky paper electrode. *Journal of the American Chemical Society* 123 (27):6536–6542.

Bandow, S., A. M. Rao, K. A. Williams, A. Thess, R. E. Smalley, and P. C. Eklund. 1997. Purification of single-wall carbon nanotubes by microfiltration. *The Journal of Physical Chemistry B* 101 (44):8839–8842.

Banerjee, S. and S. S. Wong. 2002. Rational sidewall functionalization and purification of single-walled carbon nanotubes by solution-phase ozonolysis. *The Journal of Physical Chemistry B* 106 (47):12144–12151.

Banerjee, S. and S. S. Wong. 2004. Demonstration of diameter-selective reactivity in the sidewall ozonation of SWNTs by resonance Raman spectroscopy. *Nano Letters* 4 (8):1445–1450.

Becker, L., J. L. Bada, R. E. Winans, J. E. Hunt, T. E. Bunch, and B. M. French. 1994. Fullerenes in the 1.85-billion-year-old sudbury impact structure. *Science* 265 (5172):642–645.

Berber, S., Y.-K. Kwon, and D. Tománek. 2002. Microscopic formation mechanism of nanotube peapods. *Physical Review Letters* 88 (18):185502.

Bethune, D. S., C. H. Klang, M. S. de Vries, G. Gorman, R. Savoy, J. Vazquez, and R. Beyers. 1993. Cobalt-catalysed growth of carbon nanotubes with single-atomic-layer walls. *Nature* 363 (6430):605–607.

Bisaglia, M., B. Natalini, R. Pellicciari, E. Straface, W. Malorni, D. Monti, C. Franceschi, and G. Schettini. 2000. C-3-fullero-tris-methanodicarboxylic acid protects cerebellar granule cells from apoptosis. *Journal of Neurochemistry* 74 (3):1197–1204.

Bolskar, R. D., A. F. Benedetto, L. O. Husebo, R. E. Price, E. F. Jackson, S. Wallace, L. J. Wilson, and J. M. Alford. 2003. First soluble M@C-60 derivatives provide enhanced access to metallofullerenes and permit in vivo evaluation of Gd@C-60[C(COOH)(2)](10) as a MRI contrast agent. *Journal of the American Chemical Society* 125 (18):5471–5478.

Boul, P. J., J. Liu, E. T. Mickelson, C. B. Huffman, L. M. Ericson, I. W. Chiang, K. A. Smith et al. 1999. Reversible sidewall functionalization of buckytubes. *Chemical Physics Letters* 310 (3–4):367–372.

Brettreich, M, S. Burghardt, C. Bottcher, T. Bayerl, S. Bayerl, and A. Hirsch. 2000. Globular amphiphiles: Membrane-forming hexaadducts of C(60). *Angewandte Chemie International Edition* 39:1845–1848.

Bronikowski, M. J., P. A. Willis, D. T. Colbert, K. A. Smith, and R. E. Smalley. 2001. Gas-phase production of carbon single-walled nanotubes from carbon monoxide via the HiPco process: A parametric study. *Journal of Vacuum Science and Technology, A: Vacuum, Surfaces, and Films* 19 (4):1800–1805.

Buseck, P. R., S. J. Tsipursky, and R. Hettich. 1992. Fullerenes from the geological environment. *Science* 257 (5067):215–217.

Cagle, D. W., T. P. Thrash, M. Alford, L. P. F. Chibante, G. J. Ehrhardt, and L. J. Wilson. 1996. Synthesis, characterization, and neutron activation of holmium metallofullerenes. *Journal of the American Chemical Society* 118 (34):8043–8047.

Cai, X., J. Hao, X. Zhang, B. Yu, J. Ren, C. Luo, Q. Li et al. 2009. The polyhydroxylated fullerene derivative C60(OH)24 protects mice from ionizing-radiation-induced immune and mitochondrial dysfunction. *Toxicology and Applied Pharmacology* 243 (1):27–34.

Cassell, A. M., J. A. Raymakers, J. Kong, and H. Dai. 1999. Large scale CVD synthesis of single-walled carbon nanotubes. *The Journal of Physical Chemistry B* 103 (31):6484–6492.

Chang, Y. R., H. Y. Lee, K. Chen, C. C. Chang, D. S. Tsai, C. C. Fu, T. S. Lim et al. 2008. Mass production and dynamic imaging of fluorescent nanodiamonds. *Nature Nanotechnology* 3 (5):284–288.

Chattopadhyay, D., S. Lastella, S. Kim, and F. Papadimitrakopoulos. 2002. Length separation of Zwitterion-functionalized single wall carbon nanotubes by GPC. *Journal of the American Chemical Society* 124 (5):728–729.

Chen, Y., R. C. Haddon, S. Fang, A. M. Rao, W. H. Lee, E. C. Dickey, E. A. Grulke, J. C. Pendergrass, A. Chavan, B. E. Haley, and R. E. Smalley. 1998a. Chemical attachment of organic functional groups to single-walled carbon nanotube material. *Journal of Materials Research* 13 (9):2423–2431.

Chen, J., M. A. Hamon, H. Hu, Y. Chen, A. M. Rao, P. C. Eklund, and R. C. Haddon. 1998b. Solution properties of single-walled carbon nanotubes. *Science* 282 (5386):95–98.

Cheng, H. M., F. Li, G. Su, H. Y. Pan, L. L. He, X. Sun, and M. S. Dresselhaus. 1998. Large-scale and low-cost synthesis of single-walled carbon nanotubes by the catalytic pyrolysis of hydrocarbons. *Applied Physics Letters* 72:3282

Cheng, H. M., F. Li, X. Sun, S. D. M. Brown, M. A. Pimenta, A. Marucci, G. Dresselhaus, and M. S. Dresselhaus. 1998. Bulk morphology and diameter distribution of single-walled carbon nanotubes synthesized by catalytic decomposition of hydrocarbons. *Chemical Physics Letters* 289 (5–6):602–610.

Chiang, I. W., B. E. Brinson, A. Y. Huang, P. A. Willis, M. J. Bronikowski, J. L. Margrave, R. E. Smalley, and R. H. Hauge. 2001. Purification and characterization of single-wall carbon nanotubes (SWNTs) obtained from the gas-phase decomposition of CO (HiPco process). *The Journal of Physical Chemistry B* 105 (35):8297–8301.

Chibante, L. P. F., A. Thess, J. M. Alford, M. D. Diener, and R. E. Smalley. 1993. Solar generation of the fullerenes. *Journal of Physical Chemistry* 97 (34):8696–8700.

Chung, P. H., E. Perevedentseva, and C. L. Cheng. 2007. The particle size-dependent photoluminescence of nanodiamonds. *Surface Science* 601 (18):3866–3870.

Coleman, K. S., S. R. Bailey, S. Fogden, and M. L. H. Green. 2003. Functionalization of single-walled carbon nanotubes via the Bingel reaction. *Journal of the American Chemical Society* 125 (29):8722–8723.

Dai, H. 2002. Carbon nanotubes: Synthesis, integration, and properties. *Accounts of Chemical Research* 35 (12):1035–1044.

Dai, H., A. G. Rinzler, P. Nikolaev, A. Thess, D. T. Colbert, and R. E. Smalley. 1996. Single-wall nanotubes produced by metal-catalyzed disproportionation of carbon monoxide. *Chemical Physics Letters* 260:471–475.

Daly, T. K., P. R. Buseck, P. Williams, and C. F. Lewis. 1993. Fullerenes from a fulgurite. *Science* 259 (5101):1599–1601.

Danilenko, V. 2004. On the history of the discovery of nanodiamond synthesis. *Physics of the Solid State* 46 (4):595–599.

de Heer, W. A., A. Châtelain, and D. Ugarte. 1995. A carbon nanotube field-emission electron source. *Science* 270 (5239):1179–1180.

Dieckmann, G. R., A. B. Dalton, P. A. Johnson, J. Razal, J. Chen, G. M. Giordano, E. Muñoz, I. H. Musselman, R. H. Baughman, and R. K. Draper. 2003. Controlled assembly of carbon nanotubes by designed amphiphilic peptide helices. *Journal of the American Chemical Society* 125 (7):1770–1777.

Diederich, F., R. Ettl, Y. Rubin, R. L. Whetten, R. Beck, M. Alvarez, S. Anz et al. 1991. The higher fullerenes—Isolation and characterization of C76, C84, C90, C94, and C70o, an oxide of D5h-C70. *Science* 252 (5005):548–551.

Dietz, T. G., M. A. Duncan, D. E. Powers, and R. E. Smalley. 1981. Laser production of supersonic metal cluster beams. *Journal of Chemical Physics* 74 (11):6511–6512.

Donaldson, K., R. Aitken, L. Tran, V. Stone, R. Duffin, G. Forrest, and A. Alexander. 2006. Carbon nanotubes: A review of their properties in relation to pulmonary toxicology and workplace safety. *Toxicological Sciences* 92 (1):5–22.

Dresselhaus, M. S., G. Dresselhaus, and P. C. Eklund. 1996. *Science of Fullerenes and Carbon Nanotubes.* San Diego, EUA: Academic Press.

Dresselhaus, M. S., A. Jorio, M. Hofmann, G. Dresselhaus, and R. Saito. 2010. Perspectives on carbon nanotubes and graphene Raman spectroscopy. *Nano Letters* 10 (3):751–758.

Dugan, L. L., J. K. Gabrielsen, S. P. Yu, T. S. Lin, and D. W. Choi. 1996. Buckminsterfullerenol free radical scavengers reduce excitotoxic and apoptotic death of cultured cortical neurons. *Neurobiology of Disease* 3 (2):129–135.

Dugan, L. L., D. M. Turetsky, C. Du, D. Lobner, M. Wheeler, C. R. Almli, C. K. F. Shen, T. Y. Luh, D. W. Choi, and T. S. Lin. 1997. Carboxyfullerenes as neuroprotective agents. *Proceedings of the National Academy of Sciences of the United States of America* 94 (17):9434–9439.

Dujardin, E., T. W. Ebbesen, A. Krishnan, and M. M. J. Treacy. 1998. Purification of single-shell nanotubes. *Advanced Materials* 10 (8):611–613.

Dyke, C. A., M. P. Stewart, F. Maya, and J. M. Tour. 2004. Diazonium-based functionalization of carbon nanotubes: XPS and GC-MS analysis and mechanistic implications. *Synlett* 2004 (1):155–160.

Dyke, C. A. and J. M. Tour. 2003. Unbundled and highly functionalized carbon nanotubes from aqueous reactions. *Nano Letters* 3 (9):1215–1218.

Ebbesen, T. W. and P. M. Ajayan. 1992. Large-scale synthesis of carbon nanotubes. *Nature* 358 (6383):220–222.

Farkas, E., M. E. Anderson, Z. Chen, and A. G. Rinzler. 2002. Length sorting cut single wall carbon nanotubes by high performance liquid chromatography. *Chemical Physics Letters* 363 (1–2):111–116.

Fernando, K. A. S., Y. Lin, W. Wang, S. Kumar, B. Zhou, S.-Y. Xie, L. T. Cureton, and Y.-P. Sun. 2004. Diminished band-gap transitions of single-walled carbon nanotubes in complexation with aromatic molecules. *Journal of the American Chemical Society* 126 (33):10234–10235.

Friedman, S. H., D. L. DeCamp, R. P. Sijbesma, G. Srdanov, F. Wudl, and G. L. Kenyon. 1993. Inhibition of the HIV-1 protease by fullerene derivatives: Model building studies and experimental verification. *Journal of the American Chemical Society* 115 (15):6506–6509.

Fu, C. C., H. Y. Lee, K. Chen, T. S. Lim, H. Y. Wu, P. K. Lin, P. K. Wei, P. H. Tsao, H. C. Chang, and W. Fann. 2007. Characterization and application of single fluorescent nanodiamonds as cellular biomarkers. *Proceedings of the National Academy of Sciences of the United States of America* 104 (3):727–732.

Furtado, C. A., U. J. Kim, H. R. Gutierrez, L. Pan, E. C. Dickey, and P. C. Eklund. 2004. Debundling and dissolution of single-walled carbon nanotubes in amide solvents. *Journal of the American Chemical Society* 126 (19):6095–6105.

Georgakilas, V., K. Kordatos, M. Prato, D. M. Guldi, M. Holzinger, and A. Hirsch. 2002a. Organic functionalization of carbon nanotubes. *Journal of the American Chemical Society* 124 (5):760–761.

Georgakilas, V., N. Tagmatarchis, D. Pantarotto, A. Bianco, J.-P. Briand, and M. Prato. 2002b. Amino acid functionalisation of water soluble carbon nanotubes. *Chemical Communications* (24):3050–3051.

Gharbi, N., M. Pressac, M. Hadchouel, H. Szwarc, S. R. Wilson, and F. Moussa. 2005. [60]Fullerene is a powerful antioxidant in vivo with no acute or subacute toxicity. *Nano Letters* 5 (12):2578–2585.

Ghosh, S. S. M. Bachilo, and R. B. Weisman. 2010. Advanced sorting of single-walled carbon nanotubes by nonlinear density-gradient ultracentrifugation. *Nature Nanotechnology* 5 (6):443–450.

Gonzalez, K. A., L. J. Wilson, W. Wu, and G. H. Nancollas. 2002. Synthesis and in vitro characterization of a tissue-selective fullerene: Vectoring C60(OH)16AMBP to mineralized bone. *Bioorganic & Medicinal Chemistry* 10 (6):1991–1997.

Gu, Z., H. Peng, R. H. Hauge, R. E. Smalley, and J. L. Margrave. 2002. Cutting single-wall carbon nanotubes through fluorination. *Nano Letters* 2 (9):1009–1013.

Guo, T., P. Nikolaev, A. Thess, D. T. Colbert, and R. E. Smalley. 1995. Catalytic growth of single-walled manotubes by laser vaporization. *Chemical Physics Letters* 243 (1–2):49–54.

Hartman, K. B., D. K. Hamlin, D. S. Wilbur, and L. J. Wilson. 2007. (AtCl)-At-211@US-tube nanocapsules: A new concept in radiotherapeutic-agent design. *Small* 3 (9):1496–1499.

Hartman, K. B., S. Laus, R. D. Bolskar, R. Muthupillai, L. Helm, E. Toth, A. E. Merbach, and L. J. Wilson. 2008. Gadonanotubes as ultrasensitive pH-smart probes for magnetic resonance imaging. *Nano Letters* 8 (2):415–419.

Harutyunyan, A. R. 2009. The catalyst for growing single-walled carbon nanotubes by catalytic chemical vapor deposition method. *Journal of Nanoscience and Nanotechnology* 9:2480–2495.

Heath, J. R., S. C. Obrien, Q. Zhang, Y. Liu, R. F. Curl, H. W. Kroto, F. K. Tittel, and R. E. Smalley. 1985. Lanthanum complexes of spheroidal carbon shells. *Journal of the American Chemical Society* 107 (25):7779–7780.

Heller, D. A., E. S. Jeng, T.-K. Yeung, B. M. Martinez, A. E. Moll, J. B. Gastala, and M. S. Strano. 2006. Optical detection of DNA conformational polymorphism on single-walled carbon nanotubes. *Science* 311 (5760):508–511.

Hernadi, K., A. Gaspar, J. W. Seo, M. Hammida, A. Demortier, L. Forro, J. B. Nagy, and I. Kiricsi. 2004. Catalytic carbon nanotube and fullerene synthesis under reduced pressure in a batch reactor. *Carbon* 42 (8–9):1599–1607.

Heymann, D., W. S. Wolbach, L. P. F. Chibante, R. R. Brooks, and R. E. Smalley. 1994. Search for Extractable fullerenes in clays from the cretaceous-tertiary boundary of the woodside creek and flaxbourne River sites, New-Zealand. *Geochimica Et Cosmochimica Acta* 58 (16):3531–3534.

Hirsch, A. 2002. Functionalization of single-walled carbon nanotubes. *Angewandte Chemie International Edition* 41 (11):1853–1859.

Hirsch, A. and M. Brettreich. 2005. *Fullerenes: Chemistry and Reactions*, Vol. 1. Weinheim: Wiley-VCH.

Holzinger, M., J. Abraham, P. Whelan, R. Graupner, L. Ley, F. Hennrich, M. Kappes, and A. Hirsch. 2003. Functionalization of single-walled carbon nanotubes with (R-)oxycarbonyl nitrenes. *Journal of the American Chemical Society* 125 (28):8566–8580.

Holzinger, M., O. Vostrowsky, A. Hirsch, F. Hennrich, M. Kappes, R. Weiss, and F. Jellen. 2001. Sidewall functionalization of carbon nanotubes13. *Angewandte Chemie International Edition* 40 (21):4002–4005.

Hsu, S. C., C. C. Wu, T. Y. Luh, C. K. Chou, S. H. Han, and M. Z. Lai. 1998. Apoptotic signal of Fas is not mediated by ceramide. *Blood* 91 (8):2658–2663.

Hu, H., B. Zhao, M. E. Itkis, and R. C. Haddon. 2003. Nitric acid purification of single-walled carbon nanotubes. *The Journal of Physical Chemistry B* 107 (50):13838–13842.

Huang, W., S. Taylor, K. Fu, Y. Lin, D. Zhang, T. W. Hanks, A. M. Rao, and Y.-P. Sun. 2002. Attaching proteins to carbon nanotubes via diimide-activated amidation. *Nano Letters* 2 (4):311–314.

Iijima, S. 1991. Helical microtubules of graphitic carbon. *Nature* 354:56–58.

Iijima, S. and T. Ichihashi. 1993. Single-shell carbon nanotubes of 1-nm diameter. *Nature* 363 (6430):603–605.

Itkis, M. E., D. E. Perea, S. Niyogi, S. M. Rickard, M. A. Hamon, H. Hu, B. Zhao, and R. C. Haddon. 2003. Purity evaluation of As-prepared single-walled carbon nanotube soot by use of solution-phase near-ir spectroscopy. *Nano Letters* 3 (3):309–314.

Ivanov, V., A. Fonseca, J. B. Nagy, A. Lucas, P. Lambin, D. Bernaerts, and X. B. Zhang. 1995. Catalytic production and purification of nanotubules having fullerene-scale diameters. *Carbon* 33 (12):1727–1738.

Jia, G., H. F. Wang, L. Yan, X. Wang, R. J. Pei, T. Yan, Y. L. Zhao, and X. B. Guo. 2005. Cytotoxicity of carbon nanomaterials: Single-wall nanotube, multi-wall nanotube, and fullerene. *Environmental Science & Technology* 39 (5):1378–1383.

Journet, C., W. K. Maser, P. Bernier, A. Loiseau, M. L. delaChapelle, S. Lefrant, P. Deniard, R. Lee, and J. E. Fischer. 1997. Large-scale production of single-walled carbon nanotubes by the electric-arc technique. *Nature* 388 (6644):756–758.

Kamaras, K., M. E. Itkis, H. Hu, B. Zhao, and R. C. Haddon. 2003. Covalent bond formation to a carbon nanotube metal. *Science* 301 (5639):1501.

Kataura, H., Y. Maniwa, M. Abe, A. Fujiwara, T. Kodama, K. Kikuchi, H. Imahori, Y. Misaki, S. Suzuki, and Y. Achiba. 2002. Optical properties of fullerene and non-fullerene peapods. *Applied Physics A: Materials Science & Processing* 74 (3):349–354.

Kawasaki, S., K. Komatsu, F. Okino, H. Touhara, and H. Kataura. 2004. Fluorination of open- and closed-end single-walled carbon nanotubes. *Physical Chemistry Chemical Physics* 6:1769–1772.

Khemani, K. C., M. Prato, and F. Wudl. 1992. A simple soxhlet chromatographic method for the isolation of pure C-60 and C-70. *Journal of Organic Chemistry* 57 (11):3254–3256.

Koch, A. S., K. C. Khemani, and F. Wudl. 1991. Preparation of fullerenes with a simple benchtop reactor. *Journal of Organic Chemistry* 56 (14):4543–4545.

Kolosnjaj-Tabi, J., K. B. Hartman, S. Boudjemaa, J. S. Ananta, G. Morgant, H. Szwarc, L. J. Wilson, and F. Moussa. 2010. In vivo behavior of large doses of ultrashort and full-length single-walled carbon nanotubes after oral and intraperitoneal administration to Swiss mice. *ACS Nano* 4 (3):1481–1492.

Komatsu, N., T. Ohe, and K. Matsushige. 2004. A highly improved method for purification of fullerenes applicable to large-scale production. *Carbon* 42 (1):163–167.

Kong, J., A. M. Cassell, and H. Dai. 1998. Chemical vapor deposition of methane for single-walled carbon nanotubes. *Chemical Physics Letters* 292 (4–6):567–574.

Kooi, S. E., U. Schlecht, M. Burghard, and K. Kern. 2002. Electrochemical modification of single carbon nanotubes13. *Angewandte Chemie International Edition* 41 (8):1353–1355.

Kratschmer, W., L. D. Lamb, K. Fostiropoulos, and D. R. Huffman. 1990. Solid C-60—A new form of carbon. *Nature* 347 (6291):354–358.

Kroto, H. W., J. R. Heath, S. C. Obrien, R. F. Curl, and R. E. Smalley. 1985. C-60—Buckminsterfullerene. *Nature* 318 (6042):162–163.

Krusic, P. J., E. Wasserman, P. N. Keizer, J. R. Morton, and K. F. Preston. 1991. Radical reactions of C60. *Science* 254 (5035):1183–1185.

Kudin, K. N., H. F. Bettinger, and G. E. Scuseria. 2001. Fluorinated single-wall carbon nanotubes. *Physical Review B* 63 (4):045413.

Lai, H. S., W. J. Chen, and L. Y. Chiang. 2000. Free radical scavenging activity of fullerenol on the ischemia-reperfusion intestine in dogs. *World Journal of Surgery* 24 (4):450–454.

Laus, S., B. Sitharaman, E. Toth, R. D. Bolskar, L. Helm, L. J. Wilson, and A. E. Merbach. 2007. Understanding paramagnetic relaxation phenomena for water-soluble gadofullerenes. *Journal of Physical Chemistry C* 111 (15):5633–5639.

Lee, C.-S., S. E. Baker, M. S. Marcus, W. Yang, M. A. Eriksson, and R. J. Hamers. 2004. Electrically addressable biomolecular functionalization of carbon nanotube and carbon nanofiber electrodes. *Nano Letters* 4 (9):1713–1716.

Liang, F., A. K. Sadana, A. Peera, J. Chattopadhyay, Z. Gu, R. H. Hauge, and W. E. Billups. 2004. A convenient route to functionalized carbon nanotubes. *Nano Letters* 4 (7):1257–1260.

Lim, T. S., C. C. Fu, K. C. Lee, H. Y. Lee, K. Chen, W. F. Cheng, W. W. Pai, H. C. Chang, and W. Fann. 2009. Fluorescence enhancement and lifetime modification of single nanodiamonds near a nanocrystalline silver surface. *Physical Chemistry Chemical Physics* 11 (10):1508–1514.

Lin, A. M. Y., B. Y. Chyi, S. D. Wang, H. H. Yu, P. P. Kanakamma, T. Y. Luh, C. K. Chou, and L. T. Ho. 1999. Carboxyfullerene prevents iron-induced oxidative stress in rat brain. *Journal of Neurochemistry* 72 (4):1634–1640.

Lin, Y., L. F. Allard, and Y.-P. Sun. 2004. Protein-affinity of single-walled carbon nanotubes in water. *The Journal of Physical Chemistry B* 108 (12):3760–3764.

Liu, J., A. G. Rinzler, H. Dai, J. H. Hafner, R. K. Bradley, P. J. Boul, A. Lu et al. 1998. Fullerene pipes. *Science* 280 (5367):1253–1256.

Liu, K. K., C. C. Wang, C. L. Cheng, and J. I. Chao. 2009. Endocytic carboxylated nanodiamond for the labeling and tracking of cell division and differentiation in cancer and stem cells. *Biomaterials* 30 (26):4249–4259.

Liu, Z., M. Winters, M. Holodniy, and H. Dai. 2007. siRNA delivery into human t cells and primary cells with carbon-nanotube transporters13. *Angewandte Chemie International Edition* 46 (12):2023–2027.

Liverman, M. G., S. M. Beck, D. L. Monts, and R. E. Smalley. 1979. Fluorescence excitation spectrum of the 1au (Npi-])]-1ag (O-O) band of oxalyl fluoride in a pulsed supersonic free jet. *Journal of Chemical Physics* 70 (1):192–198.

Mackeyev, Y., S. Bachilo, K. B. Hartman, and L. J. Wilson. 2007. The purification of HiPco SWCNTs with liquid bromine at room temperature. *Carbon* 45 (5):1013–1017.

Mackeyev, Y., K. B. Hartman, J. S. Ananta, A. V. Lee, and L. J. Wilson. 2009. Catalytic synthesis of amino acid and peptide derivatized gadonanotubes. *Journal of the American Chemical Society* 131 (24):8342–8343.

Manus, L. M., D. J. Mastarone, E. A. Waters, X. Q. Zhang, E. A. Schultz-Sikma, K. W. MacRenaris, D. Ho, and T. J. Meade. 2010. Gd(III)-nanodiamond conjugates for MRI contrast enhancement. *Nano Letters* 10 (2):484–489.

Martínez, M. T., M. A. Callejas, A. M. Benito, W. K. Maser, M. Cochet, J. M. Andrés, J. Schreiber, O. Chauvet, and J. L. G. Fierro. 2002. Microwave single walled carbon nanotubes purification. *Chemical Communications (Cambridge, England)* (9):1000–1001.

Mickelson, E. T., C. B. Huffman, A. G. Rinzler, R. E. Smalley, R. H. Hauge, and J. L. Margrave. 1998. Fluorination of single-wall carbon nanotubes. *Chemical Physics Letters* 296 (1–2):188–194.

Mirakyan, A. L., L. J. Wilson, and M. P. Cubbage. 2003. Fullerene(C60)-vancomycin conjugates as improved antibiotics. *Abstracts of Papers of the American Chemical Society* 225:U182.

Mizoguti, E., F. Nihey, M. Yudasaka, S. Iijima, T. Ichihashi, and K. Nakamura. 2000. Purification of single-wall carbon nanotubes by using ultrafine gold particles. *Chemical Physics Letters* 321 (3–4):297–301.

Moghaddam, M. J., S. Taylor, M. Gao, S. Huang, L. Dai, and M. J. McCall. 2003. Highly efficient binding of DNA on the sidewalls and tips of carbon nanotubes using photochemistry. *Nano Letters* 4 (1):89–93.

Moore, V. C., M. S. Strano, E. H. Haroz, R. H. Hauge, R. E. Smalley, J. Schmidt, and Y. Talmon. 2003. Individually suspended single-walled carbon nanotubes in various surfactants. *Nano Letters* 3 (10):1379–1382.

Murr, L. E. and K. M. Garza. 2009. Natural and anthropogenic environmental nanoparticulates: Their microstructural characterization and respiratory health implications. *Atmospheric Environment* 43 (17):2683–2692.

Nagasawa, S., M. Yudasaka, K. Hirahara, T. Ichihashi, and S. Iijima. 2000. Effect of oxidation on single-wall carbon nanotubes. *Chemical Physics Letters* 328 (4–6):374–380.

Nepal, D. and K. E. Geckeler. 2007. Proteins and carbon nanotubes: Close encounter in water. *Small* 3 (7):1259–1265.

Nielsen, G. D., M. Roursgaard, K. A. Jensen, S. Seier Poulsen, and S. T. Larsen. 2008. *In vivo* biology and toxicology of fullerenes and their derivatives. *Basic & Clinical Pharmacology & Toxicology* 103 (3):197–208.

Nikolaev, P., M. J. Bronikowski, R. K. Bradley, F. Rohmund, D. T. Colbert, K. A. Smith, and R. E. Smalley. 1999. Gas-phase catalytic growth of single-walled carbon nanotubes from carbon monoxide. *Chemical Physics Letters* 313 (1–2):91–97.

Niyogi, S., H. Hu, M. A. Hamon, P. Bhowmik, B. Zhao, S. M. Rozenzhak, J. Chen, M. E. Itkis, M. S. Meier, and R. C. Haddon. 2001. Chromatographic purification of soluble single-walled carbon nanotubes (s-SWNTs). *Journal of the American Chemical Society* 123 (4):733–734.

O'Connell, M. J., S. M. Bachilo, C. B. Huffman, V. C. Moore, M. S. Strano, E. H. Haroz, K. L. Rialon et al. 2002. Band gap fluorescence from individual single-walled carbon nanotubes. *Science* 297 (5581):593–596.

Osawa, E. 1970. *Kagaku (Kyoto)* 25:854.

Pan, H., L. Liu, Z.-X. Guo, L. Dai, F. Zhang, D. Zhu, R. Czerw, and D. L. Carroll. 2002. Carbon nanotubols from mechanochemical reaction. *Nano Letters* 3 (1):29–32.

Pantarotto, D., C. D. Partidos, R. Graff, J. Hoebeke, J.-P. Briand, M. Prato, and A. Bianco. 2003a. Synthesis, structural characterization, and immunological properties of carbon nanotubes functionalized with peptides. *Journal of the American Chemical Society* 125 (20):6160–6164.

Pantarotto, D., C. D. Partidos, J. Hoebeke, F. Brown, E. Kramer, J.-P. Briand, S. Muller, M. Prato, and A. Bianco. 2003b. Immunization with peptide-functionalized carbon nanotubes enhances virus-specific neutralizing antibody responses. *Chemistry & Biology* 10 (10):961–966.

Park, T.-J., S. Banerjee, T. Hemraj-Benny, and S. S. Wong. 2006. Purification strategies and purity visualization techniques for single-walled carbon nanotubes. *Journal of Materials Chemistry* 16.

Parker, D. H., K. Chatterjee, P. Wurz, K. R. Lykke, M. J. Pellin, and L. M. Stock. 1992. Fullerenes and giant fullerenes—Synthesis, separation, and mass-spectrometric characterization. *Carbon* 30 (8):1167–1182.

Parker, D. H., P. Wurz, K. Chatterjee, K. R. Lykke, J. E. Hunt, M. J. Pellin, J. C. Hemminger, D. M. Gruen, and L. M. Stock. 1991. High-yield synthesis, separation, and mass-spectrometric characterization of fullerenes C60 to C266. *Journal of the American Chemical Society* 113 (20):7499–7503.

Partha, R., M. Lackey, A. Hirsch, S W. Casscells, and J. Conyers. 2007. Self assembly of amphiphilic C60 fullerene derivatives into nanoscale supramolecular structures. *Journal of Nanobiotechnology* 5 (1):6.

Peigney, A., Ch. Laurent, F. Dobigeon, and A. Rousset. 1997. Carbon nanotubes grown in-situ by a novel catalytic method. *Journal of Materials Research* 12 (3):613–615.

Peng, H., P. Reverdy, V. N. Khabashesku, and J. L. Margrave. 2003. Sidewall functionalization of single-walled carbon nanotubes with organic peroxides. *Chemical Communications* (3):362–363.

Peters, G. and M. Jansen. 1992. A new fullerene synthesis. *Angewandte Chemie-International Edition in English* 31 (2):223–224.

Poenitzsch, V. Z., D. C. Winters, H. Xie, G. R. Dieckmann, A. B. Dalton, and I. H. Musselman. 2007. Effect of electron-donating and electron-withdrawing groups on peptide/single-walled carbon nanotube interactions. *Journal of the American Chemical Society* 129 (47):14724–14732.

Quick, K. L., S. S. Ali, R. Arch, C. Xiong, D. Wozniak, and L. L. Dugan. 2008. A carboxyfullerene SOD mimetic improves cognition and extends the lifespan of mice. *Neurobiology of Aging* 29 (1):117–128.

Rancan, F., S. Rosan, F. Boehm, A. Cantrell, M. Brellreich, H. Schoenberger, A. Hirsch, and F. Moussa. 2002. Cytotoxicity and photocytotoxicity of a dendritic C(60) mono-adduct and a malonic acid C(60) tris-adduct on Jurkat cells. *Journal of Photochemistry and Photobiology. B, Biology* 67 (3):157–162.

Rao, A. M., S. Bandow, E. Richter, and P. C. Eklund. 1998. Raman spectroscopy of pristine and doped single wall carbon nanotubes. *Thin Solid Films* 331 (1–2):141–147.

Rinzler, A. G., J. Liu, H. Dai, P. Nikolaev, C. B. Huffman, F. J. Rodriguez-Macias, P. J. Boul et al. 1998. Large-scale purification of single-wall carbon nanotubes: Process, product, and characterization. *Applied Physics A: Materials Science & Processing* 67 (1):29–37.

Saini, R. K., I. W. Chiang, H. Peng, R. E. Smalley, W. E. Billups, R. H. Hauge, and J. L. Margrave. 2003. Covalent sidewall functionalization of single wall carbon nanotubes. *Journal of the American Chemical Society* 125 (12):3617–3621.

Saito, S. 1997. Carbon nanotubes for next-generation electronics devices. *Science* 278 (5335):77–78.

Sayes, C. M., J. D. Fortner, W. Guo, D. Lyon, A. M. Boyd, K. D. Ausman, Y. J. Tao et al. 2004. The differential cytotoxicity of water-soluble fullerenes. *Nano Letters* 4 (10):1881–1887.

Sayes, C. M., A. M. Gobin, K. D. Ausman, J. Mendez, J. L. West, and V. L. Colvin. 2005. Nano-C-60 cytotoxicity is due to lipid peroxidation. *Biomaterials* 26 (36):7587–7595.

Schrand, A. M., H. J. Huang, C. Carlson, J. J. Schlager, E. Osawa, S. M. Hussain, and L. M. Dai. 2007a. Are diamond nanoparticles cytotoxic? *Journal of Physical Chemistry B* 111 (1):2–7.

Schrand, A. M., L. Dai, J. J. Schlager, S. M. Hussain, and E. Osawa. 2007b. Differential biocompatibility of carbon nanotubes and nanodiamonds. *Diamond and Related Materials* 16 (12):2118–2123.

Sen, R., S. M. Rickard, M. E. Itkis, and R. C. Haddon. 2003. Controlled purification of single-walled carbon nanotube films by use of selective oxidation and near-IR spectroscopy. *Chemistry of Materials* 15 (22):4273–4279.

Shenderova, O. A., V. V. Zhirnov, and D. W. Brenner. 2002. Carbon nanostructures. *Critical Reviews in Solid State and Materials Sciences* 27 (3):227–356.

Shi, K., N. W., T. C. Jessop, P. A. Wender, and H. Dai. 2004. Nanotube molecular transporters: Internalization of carbon nanotube protein conjugates into mammalian cells. *Journal of the American Chemical Society* 126 (22):6850–6851.

Shultz, M. D., J. C. Duchamp, J. D. Wilson, C. Y. Shu, J. C. Ge, J. Y. Zhang, H. W. Gibson et al. 2010. Encapsulation of a radiolabeled cluster inside a fullerene cage (LuxLu(3-x)N)-Lu-177@C-80: An interleukin-13-conjugated radiolabeled metallofullerene platform. *Journal of the American Chemical Society* 132 (14):4980–4981.

Sitharaman, B., K. R. Kissell, K. B. Hartman, L. A. Tran, A. Baikalov, I. Rusakova, Y. Sun et al. 2005. Superparamagnetic gadonanotubes are high-performance MRI contrast agents. *Chemical Communications* (31):3915–3917.

Sitharaman, B., L. A. Tran, Q. P. Pham, R. D. Bolskar, R. Muthupillai, S. D. Flamm, A. G. Mikos, and L. J. Wilson. 2007. Gadofullerenes as nanoscale magnetic labels for cellular MRI. *Contrast Media & Molecular Imaging* 2 (3):139–146.

Sitharaman, B. and L. J. Wilson. 2007. Gadofullerenes and gadonanotubes: A new paradigm for high-performance magnetic resonance imaging contrast agent probes. *Journal of Biomedical Nanotechnology* 3 (4):342–352.

Sitharaman, B., T. Y. Zakharian, A. Saraf, P. Misra, J. Ashcroft, S. Pan, Q. P. Pham, A. G. Mikos, L. J. Wilson, and D. A. Engler. 2008. Water-soluble fullerene (C-60) derivatives as nonviral gene-delivery vectors. *Molecular Pharmaceutics* 5 (4):567–578.

Sivaraman, N., R. Dhamodaran, I. Kaliappan, T. G. Srinivasan, P. R. Vasudeva Rao, and C. K. Mathews. 1992. Solubility of C60 in organic solvents. *The Journal of Organic Chemistry* 57 (22):6077–6079.

Smalley, R. E., B. L. Ramakrishna, D. H. Levy, and L. Wharton. 1974. Laser spectroscopy of supersonic molecular-beams—Application to NO2 spectrum. *Journal of Chemical Physics* 61 (10):4363–4364.

Strano, M. S., C. A. Dyke, M. L. Usrey, P. W. Barone, M. J. Allen, H. Shan, C. Kittrell, R. H. Hauge, J. M. Tour, and R. E. Smalley. 2003. Electronic structure control of single-walled carbon nanotube functionalization. *Science* 301 (5639):1519–1522.

Tabata, Y., Y. Murakami, and Y. Ikada. 1997. Photodynamic effect of polyethylene glycol-modified fullerene on tumor. *Japanese Journal of Cancer Research* 88 (11):1108–1116.

Tagmatarchis, N. and M. Prato. 2004. Functionalization of carbon nanotubes via 1,3-dipolar cycloadditions. *Journal of Materials Chemistry* 14 (4):437–439.

Tarnai, T. 1993. Geodesic domes and fullerenes. *Philosophical Transactions of the Royal Society A* 343:145–154.

Tasis, D., N. Tagmatarchis, A. Bianco, and M. Prato. 2006. Chemistry of carbon nanotubes. *Chemical Reviews* 106 (3):1105–1136.

Taylor, R., J. P. Hare, A. K. Abdulsada, and H. W. Kroto. 1990. Isolation, separation and characterization of the fullerenes C-60 and C-70—The 3rd form of carbon. *Journal of the Chemical Society—Chemical Communications* (20):1423–1424.

Taylor, R., G. J. Langley, H. W. Kroto, and D. R. M. Walton. 1993. Formation of C60 by pyrolysis of naphthalene. *Nature* 366 (6457):728–731.

Taylor, R., J. P. Parsons, A. G. Avent, S. P. Rannard, T. J. Dennis, J. P. Hare, H. W. Kroto, and D. R. M. Walton. 1991. Degradation of C60 by light. *Nature* 351 (6324):277–277.

Thess, A., R. Lee, P. Nikolaev, H. Dai, P. Petit, J. Robert, C. Xu et al. 1996. Crystalline ropes of metallic carbon nanotubes. *Science* 273 (5274):483–487.

Thien-Nga, L., K. Hernadi, E. Ljubovic, S. Garaj, and L. Forro. 2002. Mechanical purification of single-walled carbon nanotube bundles from catalytic particles. *Nano Letters* 2 (12):1349–1352.

Thrower, P. A. 1999. Editorial. *Carbon* 37 (11):1677–1678.

Tibbetts, G. G., C. A. Bernardo, D. W. Gorkiewicz, and R. L. Alig. 1994. Role of sulfur in the production of carbon fibers in the vapor phase. *Carbon* 32 (4):569–576.

Tohji, K., T. Goto, H. Takahashi, Y. Shinoda, N. Shimizu, B. Jeyadevan, I. Matsuoka, Y. Saito, A. Kasuya, Te. Ohsuna, K. Hiraga, and Y. Nishina. 1996. Purifying single-walled nanotubes. *Nature* 383 (6602):679–679.

Toth, E., R. D. Bolskar, A. Borel, G. Gonzalez, L. Helm, A. E. Merbach, B. Sitharaman, and L. J. Wilson. 2005. Water-soluble gadofullerenes: Toward high-relaxivity, pH-responsive MRI contrast agents. *Journal of the American Chemical Society* 127 (2):799–805.

Touhara, H. and F. Okino. 2000. Property control of carbon materials by fluorination. *Carbon* 38 (2):241–267.

Tsao, N., P. P. Kanakamma, T.-Y. Luh, C.-K. Chou, and H.-Y. Lei. 1999. Inhibition of *Escherichia coli*-induced meningitis by carboxyfullerence. *Antimicrobial Agents and Chemotherapy* 43 (9):2273–2277.

Unger, E., A. Graham, F. Kreupl, M. Liebau, and W. Hoenlein. 2002. Electrochemical functionalization of multi-walled carbon nanotubes for solvation and purification. *Current Applied Physics* 2 (2):107–111.

Vaijayanthimala, V. and H.-C. Chang. 2009. Functionalized fluorescent nanodiamonds for biomedical applications. *Nanomedicine* 4 (1):47–55.

Vial, S., C. Mansuy, S. Sagan, T. Irinopoulou, F. Burlina, J. P. Boudou, G. Chassaing, and S. Lavielle. 2008. Peptide-grafted nanodiamonds: Preparation, cytotoxicity and uptake in cells. *Chembiochem* 9 (13):2113–2119.

Wee, T. L., Y. K. Tzeng, C. C. Han, H. C. Chang, W. Fann, J. H. Hsu, K. M. Chen, and Y. C. Yull. 2007. Two-photon excited fluorescence of nitrogen-vacancy centers in proton-irradiated type ib diamond. *Journal of Physical Chemistry A* 111 (38):9379–9386.

Wharton, T. and L. J. Wilson. 2002. Highly-iodinated fullerene as a contrast agent for X-ray imaging. *Bioorganic & Medicinal Chemistry* 10 (11):3545–3554.

Wilson, L. J., D. W. Cagle, T. P. Thrash, S. J. Kennel, S. Mirzadeh, J. M. Alford, and G. J. Ehrhardt. 1999. Metallofullerene drug design. *Coordination Chemistry Reviews* 192:199–207.

Wong, S. S., E. Joselevich, A. T. Woolley, C. L. Cheung, and C. M. Lieber. 1998. Covalently functionalized nanotubes as nanometre-sized probes in chemistry and biology. *Nature* 394 (6688):52–55.

Worsley, K. A., K. R. Moonoosawmy, and P. Kruse. 2004. Long-range periodicity in carbon nanotube sidewall functionalization. *Nano Letters* 4 (8):1541–1546.

Xing, Y. and L. M. Dai. 2009. Nanodiamonds for nanomedicine. *Nanomedicine* 4 (2):207–218.

Yeretzian, C., K. Hansen, F. Diederich, and R. L. Whetten. 1992. Coalescence reactions of fullerenes. *Nature* 359 (6390):44–47.

Ying, Y., R. K. Saini, F. Liang, A. K. Sadana, and W. E. Billups. 2003. Functionalization of carbon nanotubes by free radicals. *Organic Letters* 5 (9):1471–1473.

Yoon, M., S. Berber, and D. Tománek. 2005. Energetics and packing of fullerenes in nanotube peapods. *Physical Review B* 71 (15):155406.

Yu, S.-J., M.-W. Kang, H.-C. Chang, K.-M. Chen, and Y.-C. Yu. 2005. Bright fluorescent nanodiamonds: No photobleaching and low cytotoxicity. *Journal of the American Chemical Society* 127 (50):17604–17605.

Zakharian, T. Y. and L. J. Wilson. 2003. Design and synthesis of a paclitaxel-fullerene conjugate. *Abstracts of Papers of the American Chemical Society* 225:U196.

Zhang, C. J., W. X. Sun, and Z. X. Cao. 2007. Most stable structure of fullerene[20] and its novel activity toward addition of alkene: A theoretical study. *Journal of Chemical Physics* 126 (14):144306.

Zhang, J. F., P. P. Fatouros, C. Y. Shu, J. Reid, L. S. Owens, T. Cai, H. W. Gibson, G. L. Long, F. D. Corwin, Z. J. Chen, and H. C. Dorn. 2010. High relaxivity trimetallic nitride (Gd3N) metallofullerene MRI contrast agents with optimized functionality. *Bioconjugate Chemistry* 21 (4):610–615.

Zhang, J., H. Zou, Q. Qing, Y. Yang, Q. Li, Z. Liu, X. Guo, and Z. Du. 2003. Effect of Chemical oxidation on the structure of single-walled carbon nanotubes. *The Journal of Physical Chemistry B* 107 (16):3712–3718.

Zhang, L., V. U. Kiny, H. Peng, J. Zhu, R. F. M. Lobo, J. L. Margrave, and V. N. Khabashesku. 2004. Sidewall functionalization of single-walled carbon nanotubes with hydroxyl group-terminated moieties. *Chemistry of Materials* 16 (11):2055–2061.

Zhang, M., M. Yudasaka, F. Nihey, and S. Iijima. 2000. Effect of ultrafine gold particles and cationic surfactant on burning as-grown single-wall carbon nanotubes. *Chemical Physics Letters* 328 (4–6):350–354.

Zhang, Y., Y. Shen, J. Li, L. Niu, S. Dong, and A. Ivaska. 2005. Electrochemical functionalization of single-walled carbon nanotubes in large quantities at a room-temperature ionic liquid supported three-dimensional network electrode. *Langmuir* 21 (11):4797–4800.

Zhao, B., H. Hu, S. Niyogi, M. E. Itkis, M. A. Hamon, P. Bhowmik, M. S. Meier, and R. C. Haddon. 2001. Chromatographic purification and properties of soluble single-walled carbon nanotubes. *Journal of the American Chemical Society* 123 (47):11673–11677.

Zheng, M., A. Jagota, E. D. Semke, B. A. Diner, R. S. McLean, S. R. Lustig, R. E. Richardson, and N. G. Tassi. 2003a. DNA-assisted dispersion and separation of carbon nanotubes. *Nature Materials* 2 (5):338–342.

Zheng, M., A. Jagota, M. S. Strano, A. P. Santos, P. Barone, S. G. Chou, B. A. Diner, et al. 2003b. Structure-based carbon nanotube sorting by sequence-dependent DNA assembly. *Science* 302 (5650):1545–1548.

Zimmerman, J. L., R. K. Bradley, C. B. Huffman, R. H. Hauge, and J. L. Margrave. 2000. Gas-phase purification of single-wall carbon nanotubes. *Chemistry of Materials* 12 (5):1361–1366.

6

Synthetic and Toxicological Characteristics of Silica Nanomaterials for Imaging and Drug Delivery Applications

Heather Herd
University of Utah

Hamidreza Ghandehari
University of Utah

6.1 Introduction

Silica is the second most prevalent element and as such is found in the many living systems. This element can be found naturally or synthetically and is classified either as crystalline or amorphous. Crystalline materials are by far the most common, as much of the natural element is comprised in quartz, the main component in several rock types and sand. While silica is still considered to be nonessential to sustain life, it does appear to play an important role in maintaining homeostasis and the health of many living organisms (Richmond and Sussman 2003, Martin 2007). It is incorporated in various supplemental plant fertilizers to help maintain growth, mineral nutrition, and ward off fungal diseases (Epstein 1994, Richmond and Sussman 2003). Interestingly, several plant and marine life forms also include cellular pathways that are able to take natural elemental forms of silica and process it into an alternative organic form (Crookes-Goodsodson et al. 2008, Perry 2009). The organic form has proven to play a principle role in plant and animal life, assisting in structural and developmental characteristics in a variety of fashions, such as strengthening of cell walls, bone, and cartilage (Epstein 1994, Martin 2007). Individuals and animals who lack organic silica as a dietary supplement have shown increased risk for cardiovascular disease, osteoporosis, and Alzheimer's (Sahin et al. 2006, Gillette et al. 2007). Much of this is linked to the ability of silica to interact chemically with native metallic ions such as aluminum and iron (Depasse and Warlus 1976, Gillette et al. 2007, Slowing et al. 2009). It however remains unclear at what pivotal concentration level and chemical composition silica switches from a beneficial to detrimental state.

Silica exposure has been associated with autoimmune disease, and crystalline silica has been classified by the International Agency for Research on Cancer as a class one carcinogen (Wilbourn 1997). Additionally, numerous toxicological studies have been performed that provide evidence that the crystalline form induces upregulation of inflammatory and oxidative stress agents, such as cytokines, chemokines, reactive oxygen species (ROS), reactive nitrogen species (RNS), and nitric oxide (Lenz et al. 1992, Fubini and Hubbard 2003, Rimal et al. 2005, Ovrevik et al. 2006, Kleinman et al. 2008). Most of these studies are associated with silicosis and lung cancer in mine workers (Carter and Driscoll 2001, Hnizdo and Vallyathan 2003, Rimal et al. 2005, Cocco et al. 2007, Lacasse et al. 2009).

The chemical and structural properties and abundance of silica provide unique and inexpensive synthetic alternatives for product development. In recent years, the industrial world has seen a drastic increase in the production of products and processes that utilize several forms of silica (Cameron et al. 2007, Chuankrerkkul et al. 2008). Silica can now be found in many cosmetics, foods, and electronics. Due to its attractive synthetic properties, it has become an ideal candidate for biomedical applications including but not limited to sensors (Knopp et al. 2009), drug and gene delivery systems (Slowing et al. 2008), and imaging contrast agents (Sharma et al. 2006). While amorphous silica does not seem to present the same oral or inhalation risks as crystalline silica, the implication of introducing such a material via alternate routes of administration remains unknown. Thus, extensive toxicological studies are needed to understand the environmental and health impacts of silica nanoparticles.

This chapter will attempt to address current amorphous silica nanoparticle synthesis, preparation, characterization, and subsequent toxicological evaluation. Emphasis will be placed on the use of these nanomaterials for biomedical drug delivery and imaging applications.

6.2 Amorphous Silica Nanoparticle Synthesis

The unique inorganic chemical properties of silica provide an exceptional platform from which to build drug delivery and imaging systems. Synthetic sol–gel and polymerization methods provide a simple way to produce these nanomaterials on a large scale. The advantages of silica nanoparticles include the following:

1. Relative chemical and thermal stability
2. Synthetic control over size and size distribution
3. Potential to induce alterations in geometry

4. Ease of surface modification
5. Ability to control encapsulation of molecules of interest
6. Economic affordability and ease of scale-up

These advantages will be individually discussed with a focus on their implications on toxicity and their subsequent potential for drug delivery and imaging applications.

6.2.1 Relative Chemical and Thermal Stability

Silica nanoparticles are generally synthesized via aqueous polymerization of silicic acids or through the Stober method that involves the utilization of silicon alkoxides and their subsequent hydrolysis and condensation (Stober et al. 1968, Mizutani et al. 1998). The two produce two very different core materials, with variations in density. Polymerization tends to permit for classically uniform particles, by allowing for full hydrolysis of monomer repeat units (He et al. 2009). Stober synthesis or silicon alkoxide nanoparticle formation, however, is generated through cluster aggregation, preventing full hydrolysis and thus a reduction in uniformity (Figure 6.1, Stober et al. 1968, Iler 1979). However, with the development of modified Stober methods or sol–gel chemistries, one is able to more carefully control synthetic procedures (Brinker and Scherer 1990). Once formed, both of these processes lead to a very stable, relatively inert colloidal solution, which cannot be disrupted or degraded without further synthetic modification (He et al. 2009). This property is highly sought after in both drug delivery and imaging applications, as it provides a protective environment for the encapsulated material and assists in the reduction of systemic side effects in the human body. Current alternative nanoparticle carriers such as liposomes and micelles have the potential to dissociate in vivo and release potentially toxic materials into circulation (Ostro 1987, Kwon and Kataoka 1995). Additionally, the native surface functionality of these particles is a terminal hydroxyl group, which provides a relatively hydrophilic surface, a property that is known to reduce systemic opsonization and increase circulation times (Carrstensen et al. 1992). It remains largely unknown if the mechanism of toxicity of bare silica nanoparticles is due to hydroxyl functional interactions with the physiological environment or other physiochemical interactions. As will be shown later, the masking of these hydroxyl groups does appear to reduce toxicity and hydroxyl groups facilitate functionalization (Figure 6.2).

6.2.2 Synthetic Control over Size and Size Distribution

Both polymerization and the Stober method allow for excellent control over monodispersity and size of spherical nanoparticles. By simply controlling the reagent concentrations and reaction conditions, one can create a wide range of differing spherical nanoparticle sizes with a polydispersity index within 5% of the total synthesized particle (Figure 6.1, Bogush et al. 1988, He et al. 2009). This provides an advantage over traditional drug delivery and imaging contrast agent systems such as random copolymers, liposomal and micellular constructs, which can be polydisperse (Ostro 1987, Kwon and Kataoka 1995). This leads to increased error in loading and dosing, as variability in size ranges can increase or

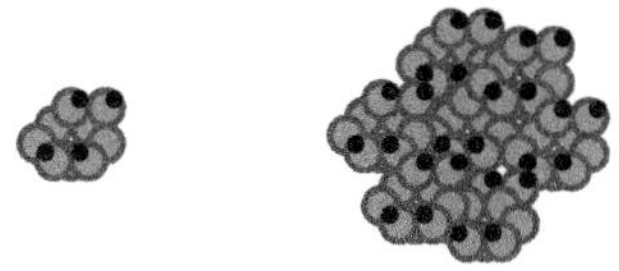

FIGURE 6.1 In the Stober method, silica undergoes spontaneous cluster formation, which then can be used to produce particles. (Iler, R.K.: *The Chemistry of Silica*. 1979. Copyright Wiley-VCH Verlag GmbH & Co. KGaA. Adapted with permission.)

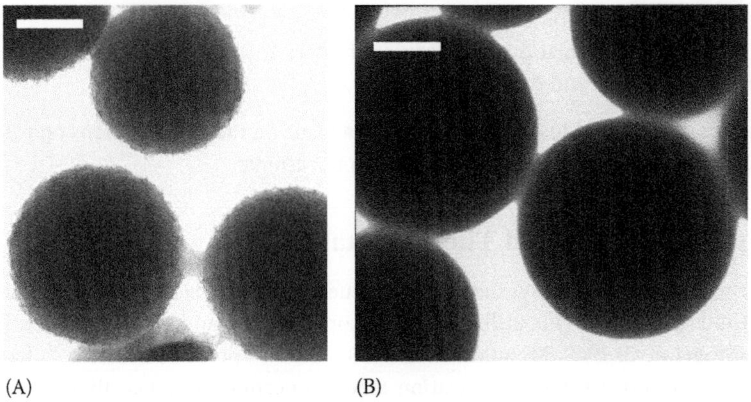

FIGURE 6.2 TEM images illustrating uniformly synthesized porous (A, scale bar 100 nm) and nonporous silica nanoparticles (B, scale bar 100 nm) produced via the Stober and modified Stober methods. (Unpublished data from authors.)

decrease encapsulation. Variation in size distribution can also influence cellular uptake, toxicity, and biodistribution. This is extremely important as the literature outlines nanoparticle size-dependent effects on toxicity, biodistribution, and uptake (Limbach et al. 2005, Cho et al. 2009, Waters et al. 2009). Smaller nanoparticles generally have increased uptake and toxicity, mostly contributing to their increased surface area or exposure to cell surfaces (Figure 6.3, Clift et al. 2008, Waters et al. 2009). The ability to synthesize silica nanoparticles with defined size and size distribution allows systematic correlation of these parameters with cellular uptake, toxicity, and biodistribution, which in turn enables development of silica nanoparticles for safe and effective biomedical applications. Biodistribution and clearance routes have threshold size ranges that prevent or allow for particle accumulation. Thus, this characteristic will be important in engineering or designing drug delivery and imaging systems, as changes in size distribution could significantly alter targeting strategies and dosing mechanisms.

Systematic evaluation is not limited to controlled size and monodisperse systems. Investigators have also explored other synthetic routes to create more interesting nanoparticles. Beck et al. (1992) created one of the first most commonly used mesoporous silica nanoparticles, with the addition of a surfactant, cetryltrimethylammonium bromide (CTAB), using the Stober method. This surfactant facilitates the creation of micelles, which are coated by an initial silica crystal formation, and assists in creating larger gaps during synthetic aggregation hydrolysis by inducing hydrophobic interactions. Similar processes have shown to create mesoporous nanoparticles via polymerization and acid catalysts (Naik and Sokolov 2007, Kobler et al. 2008, He et al. 2009). Alterations in surfactant and polymerization chemistry have allowed for the development of structurally different pores, significantly altering small molecule diffusion patterns (Brohede et al. 2008, Stromme et al. 2009). The control over these diffusion patterns is key to being able to manipulate drug or contrast agent release. By following this process with acid extraction or calcination, the surfactant is effectively removed, leaving the amorphous material with large pores or holes (Huh et al. 2003, Nandiyanto et al. 2009). It is interesting to note that toxicity profiles of porous silica nanomaterials have not shown a significant difference when compared with their bare silica counter parts, and some in vitro studies have even proven to have a reduction in hemolytic capacity of porous materials (Hudson et al. 2008, Slowing et al. 2009). This reduction could potentially be due to a decreased number of hydroxyl groups exposed to erythrocytes in circulation. Such a modification provides additional versatility in silica nanoparticles for delivery applications, as it allows for variations in functionalization and molecular encapsulation within the pores, which have been used potentially for subsequent drug release studies (Li et al. 2004, Slowing et al. 2008).

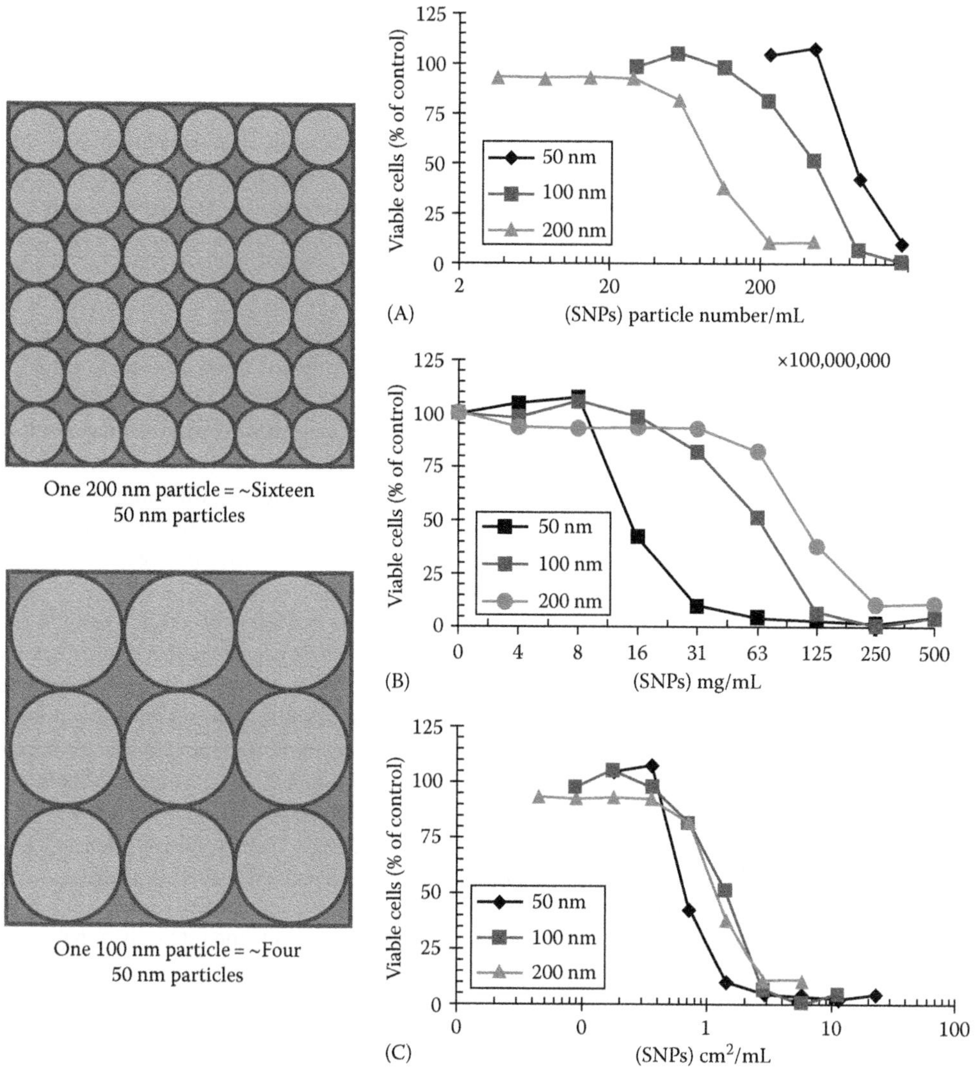

FIGURE 6.3 Plain silica nanoparticle toxicity is dependent on particle–cell surface contact area. The figure illustrates that at equal surface area, smaller particle size yield greater particle number (i.e., for every one 200 nm particle, there exist sixteen 50 nm particles and for every one 100 nm particle there exist four 50 nm particles) and higher toxicity on RAW 264.7 macrophages. Fifty nanometer particles induced the highest toxicity, followed by 100 nm and finally 200 nm. Note that if particles are graphed via particle number, larger particles remain more toxic due to larger cell surface contact area. (A) The toxicity profile when plotted against particle number; a larger particle facilitates higher toxicity due to larger contact area. (B) Toxicity profile when plotted against mass. (C) The toxicity profile when plotted against the surface area; all particles maintain similar toxicity profiles. (Modified data from Malugin, A. et al., Submitted, 2011.)

6.2.3 Potential to Induce Alterations in Geometry

Mesoporous silica synthetic characteristics helped to introduce the production of more diverse geometries. Modifications in surfactants, solvents, catalysts, salts, etc. can significantly alter the structural characteristics of the nanoparticles (Huh et al. 2003). Sol–gel solution phase chemistries can be altered and generate variations in geometries with a change, as little as, an adjustment of the surfactant properties (Trewyn et al. 2007). The surfactants introduce a unique alteration in the interfacial chemistry, causing significant modifications in formation of the solid crystal structure. These changes in the crystal structure facilitate changes in the morphological characteristics. For example, helical mesoporous silica nanorods have been produced utilizing CTAB and cosurfacant hexanol (Zhang et al. 2006). A variety of different morphologies have also emerged utilizing CTAB and the addition of different functional silanes, which provide additional functional alterations in the interfacial chemistry (Huh et al. 2003). These examples illustrate exciting synthetic results because one can easily manipulate sol–gel principles to introduce dramatically different geometries. These geometries have proven to drastically alter toxicity and uptake (Trewyn et al. 2008).

Once the discovery had been made that surfactants could alter crystal and chemical bulk structural formations, investigators started experimenting with differences in nucleation of the same sol–gel chemistry. Instead of utilizing the base alkoxide or acid, other inorganic or surface materials were used. Nuraje et al. (2007) produced such an example of creating variations in geometry would be present in silica geometric creations utilizing interfacial chemistries. At the interface of an organic and aqueous phase, where the aqueous phase contains catalyzing ions, one can introduce a silicon alkoxide and subsequently nucleate off that interface (Figure 6.4, Nuraje et al. 2007). This effectively creates a face for the geometric nanoparticle, and one can alter the shapes obtained by altering the catalyzing ions and organic phase. An aqueous sodium hydroxide (NaOH) phase with an organic butanol phase for example will generate squares, while an aqueous hydrochloric acid (HCl) phase with an organic chloroform phase will generate triangles (Nuraje et al. 2007, Figure 6.4).

Similarly with utilization of sol–gel chemistries and a template nucleation surface, synthetic silica schemes have been developed that provide unique tubular geometries. A template synthesis normally involves the introduction and subsequent nucleation of silica on the surface of a porous alumina membrane. After the silica has coated the inner layer of the membrane, the aluminum is generally dissolved away with phosphoric acid, leaving behind tubular or rodlike structures. One can alter the thickness of the tube wall by controlling the addition of silica to the reaction (Son et al. 2006, Nan et al. 2008).

The addition of another nanoparticle made out of differing materials, such as gold, polystyrene, or other polymers, can also create a nucleation surface (Caruso et al. 1998, Obare et al. 2001, Lu et al. 2004). The silica nanoparticle can grow off of this surface and one can subsequently remove the nanoparticle via high temperature burning or desolvation principles. Investigators have utilized these hollow silica spheres and rods to encapsulate a variety of different materials (Chen et al. 2004, Li et al. 2004).

As stated earlier, the development of materials with alterations in geometry could potentially influence toxicity, cellular uptake, and biodistribution profiles. Recent studies with silica nanotubes in

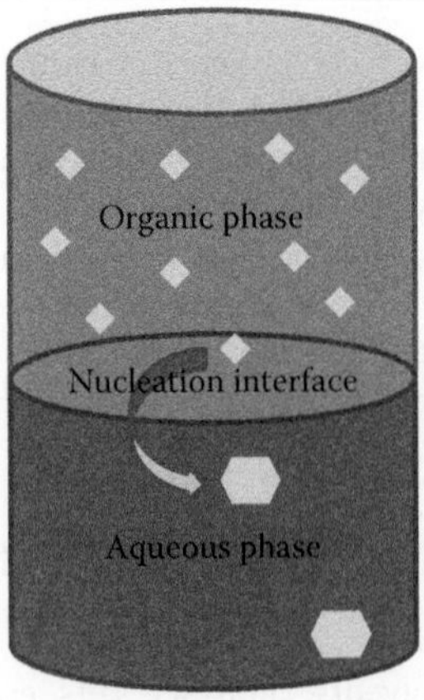

FIGURE 6.4 Silica nanoparticle interfacial chemistry utilizes an organic and aqueous phase. The separation between the two phases produces a nucleation interface, where a silica nanoparticle can form and precipitate into the aqueous phase. (Diagram modified from Nuraje, N. et al., *New J. Chem.*, 31, 1895, 2007.)

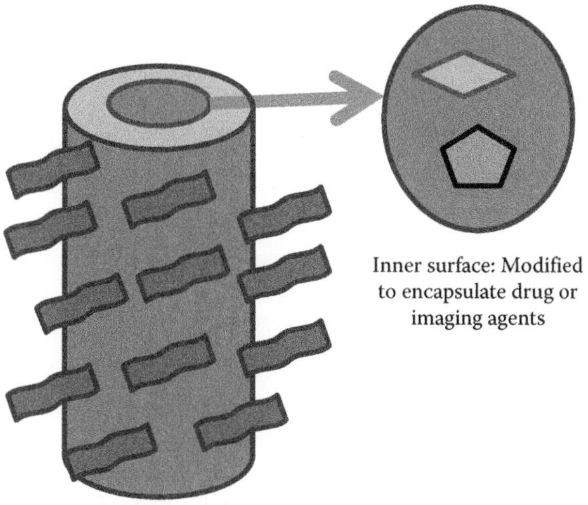

FIGURE 6.5 Diagram representing tubular silica nanoparticles with potential differentiation of inner and outer functional groups. (Modified from Nan, A. et al., *Nano Lett.*, 8, 2150, 2008.)

MDA-MB231cells (human breast cancer epithelial cells) suggest that alterations in nanoparticle surface properties, aspect ratio, or size can influence cellular uptake (Nan et al. 2008, Trewyn et al. 2008, Mitragotri and Lahann 2009, Nelson et al. 2009). Sol–gel mesoporous materials also show significant changes in cellular uptake, as mesoporous tubular-like structures were not uptaken as significantly as mesoporous spherical structures (Trewyn et al. 2008).

The use of various geometries of silica in biomedical applications proves to be an exciting prospect. First, they introduce the ability to potentially change cellular uptake, biodistribution, and toxicity profiles. Second, they provide unique functionalization capabilities where the inner and outer surface can facilitate differential functionalization. As one changes the geometry of the nanoparticle, it provides the potential to present several surfaces, each with the possibility of a different surface characteristic. For example, a silica nanotube (SNT) presents an inner and outer surface (Figure 6.5). One is able to functionalize the inner surface of SNTs so that it incorporates hydrophobic agents, while the outer surface is modified with hydrophilic biomolecular agents (Son et al. 2006, Nan et al. 2008). This characteristic will be helpful in drug delivery and imaging systems, as many prospective deliverable payloads are hydrophobic and in order to increase circulation, statistical site accumulation, and biocompatibility the construct that is delivered needs to have hydrophilic surface properties.

6.2.4 Ease in Surface Modification

Silica nanoparticles have a simple unique surface covered by hydroxyl functional groups. This surface can be easily modified via traditional silane chemistry. Silane chemistry has become an industry standard and is available commercially. The wide array provides the ability to functionalize the surface of these nanoconstructs with a broad variety of materials. This initiates surfaces that have much different characteristics than their respective silica core. These characteristics can be exploited and increase silica nanoconstruct potential for engineering drug delivery and imaging systems.

Functionalization can be as simple as small molecular weight functional groups attached via a silane, such as an amine or a carboxyl, which alters the charge density of particles. Charge density has proven to be extremely important to toxicity, distribution, and uptake in silica nanoparticle constructs

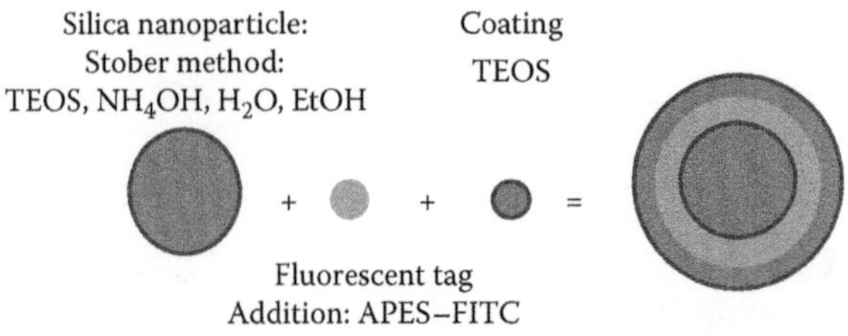

FIGURE 6.6 Schematic of fluorescent encapsulation and labeling of silica nanoparticles. *Abbreviations*: tetraethoxysilane (TEOS), ammonium hydroxide (NH_4OH), water (H_2O), ethanol (EtOH), 3-aminopropyltriethoxysilane (APES), and fluorescein isothiocyanate (FITC). (Modified from Blaaderen, A.V. and Vrij, A., *Langmuir*, 8, 2921, 1992.)

(Clift et al. 2008, He et al. 2008). Positively charged silica nanoparticles are uptaken by mesenchymal stem cells via endocytosis to a greater extent than plain silica nanoparticles while maintaining low toxicity (Chung et al. 2007). While these functional groups can alter toxicity and biodistribution profiles, they also provide the option for introduction of more complex surface chemical synthesis.

Blaaderen and Vrij (1992) have developed a novel fluorescent silica tagging method. By utilizing the spontaneous addition reaction of the amine group of a 3-(aminopropyl)triethoxysilane (APES) and the thioisocyanate group of a fluorescein isothiocyanate (FITC), they were able to create a fluorescent silane group that could be easily coupled to the surface of a silica nanomaterial. This group then proceeded to coat the silica nanoparticle surface with a tetraethoxysilane (TEOS), which provided the FITC with a protective coat that significantly reduced photobleaching (Figure 6.6, Blaaderen and Vrij 1992). Subsequently, research groups have attached other fluorescent molecules to the surface of the nanoparticles and proceeded to assess and compare the potential toxicity of these constructs to their bare silica counterparts (Jin et al. 2007, Kumar et al. 2008). It is important to note that these groups have found that fluorescent particles have similar toxicity profiles to their bare counterparts. The development of this construct was important to translation of these nanoparticles to drug delivery and imaging systems. As it is essential that detection methods that aide in devolving uptake, toxicity, and transport mechanisms should not affect the properties of the construct.

These functional surface modifications have also provided the ability to increase circulation times by reducing protein absorption and opsonization with the addition of hydrophilic surface groups, such as polyethylene glycol (PEG) and other polymers (Huh et al. 2003, Xu et al. 2003, Guo et al. 2005, Thierry et al. 2008, van Schooneveld et al. 2008, Wang et al. 2009b). Stayton et al. (2009) conducted an extensive study investigating the effects of specific protein adsorption on 13.3 nm silica nanoparticles in A549 cells (human lung carcinoma cell line). The analysis included hemoglobin, albumin, histone, and pre-aggregated and complete medium. Protein adsorption differed little from protein to protein. Those cultures incubated with single proteins tended to form particle aggregates, while bare particles had a higher degree of uptake. Additionally, particles adsorbed with histone tended to have a reduced zeta potential, which appeared to correlate with faster uptake. The investigators also created particles with a cadmium surface modification, which appear to reduce toxicity and uptake (Stayton et al. 2009). Furthermore, polymer coatings have shown to increase circulation and significantly reduced particle toxicity on certain cell lines (Clift et al. 2008, He et al. 2008). It is important to note that the proteins adsorbed on the surface of silica nanoparticles can have a significant effect on directed cellular uptake and distribution (Cedervall et al. 2007, Dutta et al. 2007, Aggarwal et al. 2009). The protein itself could potentially interact with cell surfaces and thus initiate adverse events. For example, Chen and Mikecz (2005) suggested that nucleoplasmic protein aggregation, significantly impaired nuclear function, leading to the inhibition of proliferation, transcription, and replication. Surface modification can alter significantly the interactions and thus the protein association with the nanoparticle surface (Karlsson and Carlsson 2005). Thus it will be important to pay attention to protein adsorption onto silica nanoparticles in the context of their biocompatibility, biodistribution, and cellular uptake.

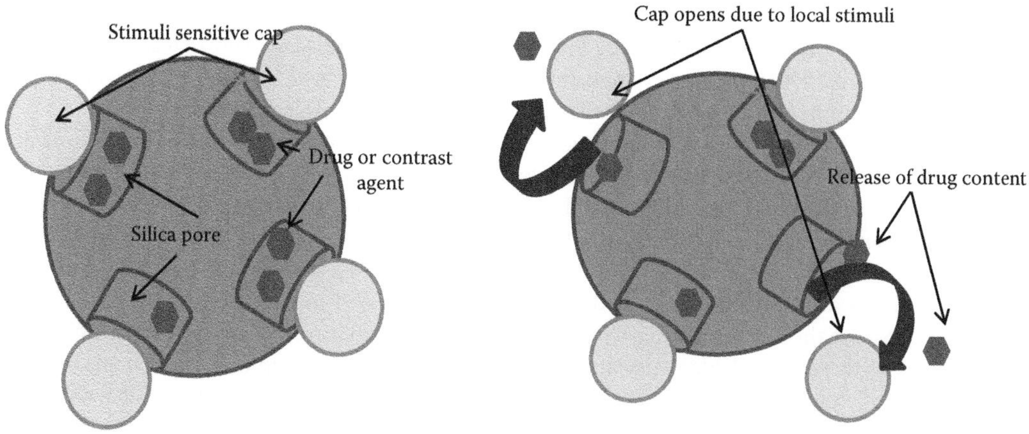

FIGURE 6.7 Diagram representing the ability for porous silica nanoparticles to encapsulate a particular payload (contrast agent or drug) that can be released dependent on environmental stimuli. (Modified from Slowing, I.I. et al., *Adv. Drug Deliv. Rev.*, 60 (11), 1278, 2008.)

Targeting motifs have also been displayed on the surface of these particles to aid in active localized targeting to tumor sites and reduction in nonspecific uptake. Kumar et al. (2008) created fluorescently labeled particles with transferrin surface modification, which showed cell internalization within Mia-PaCa (pancreatic cancer cell line) and without modification showed no internalization. Additionally, Liong et al. (2008) utilized a folic acid–targeting motif and saw a significant increase in uptake of multimodial silica nanoparticles in PANC-1 (pancreatic cancer cell line).

Surface modification can also play a key role in the development of stimuli-sensitive materials. Stimuli sensitivity is a unique property that could provide an additional versatility to drug delivery systems. For example, porous silica nanoparticles can be utilized to trap drug molecules or imaging agents that are usually systemically toxic (Liong et al. 2008, Sharma et al. 2008, Tasciotti et al. 2008). These can be capped with materials such as cadmium sulfate, gold nanoparticles, dendrimers, or polymeric supports until they reach the intended site. The polymeric support can then initiate release depending on changes in the local environment such as pH or reactive species (Figure 6.7, Lai et al. 2003, Radu et al. 2004b, Gruenhagen et al. 2005, Nguyen et al. 2005, Nguyen et al. 2006, Klichko 2008, Aznar et al. 2009). For example, Nguyen et al. have developed such a nanovalve for mesoporous nanomaterials where the valve is opened and closed via oxidation and reduction reactions. Additionally these materials can incorporate enzymatic identification markers to facilitate cleavage of the polymeric support (Klichko 2008, Patel et al. 2008). These stimuli-sensitive nanoparticles have also utilized mechanical processes such as magnetic fields to open magnetically capped porous materials (Angelos et al. 2008). Stimuli-responsive silica composites can be used in delivery of bioactive and imaging agents, or theranostics, where release and/or imaging at the target site is desired.

6.2.5 Ability to Control Encapsulation of Molecules of Interest

In addition to exceptional surface modification characteristics, one is able to utilize the silica sol–gel chemical properties to dope and control encapsulation of other molecular agents. Similar to the synthesis of hollow spheres and rods, it is possible to stimulate nucleation off of other materials and encapsulate them within silica constructs. Such examples are iron oxide, quantum dots, and gold (Insin et al. 2008, Liong et al. 2008, Wang and Shantz 2008, Burns et al. 2009, Knopp et al. 2009). These doped materials provide useful alternative methods of detection such as fluorescence for confocal imaging or magnetic dipole capabilities for magnetic resonance imaging (MRI), but still retain the benefits of silica properties. Initial toxicity profiles on these materials have also proven to be similar to bare silica constructs; however, it is important to note that some of these materials have shown leaching from the silica core (Lai et al. 2003). In the development of these materials, it will be important to ensure that the silica constructs not retain toxicity but also any encapsulated leachable material.

In addition to the encapsulation of imaging agents within these systems, therapeutic molecules have also been incorporated (Radin et al. 2001, Slowing et al. 2008, Brevet et al. 2009, Lee et al. 2009).

Composite materials have also been formed utilizing silica (Satishkumar et al. 2008). These composite materials have provided a unique capability of composite silica degradation, something that traditional silica nanoparticles do not offer. Park et al. (2009) developed a complex of several 3–5 nm silica particles within a dextran coating. This composite created 130–180 nm constructs that could then be degraded to release the 3–5 nm particles that can subsequently be cleared (Park et al. 2009). With the development of these degradable materials, the utilization of silica in drug delivery systems appears to be significantly more promising, as clearance mechanisms of current silica constructs have still not been devolved.

Additionally, these composite materials have encapsulated proteins and immobilized enzymes within the silica nanoparticle. Investigators have done so by pre-hydrolyzing the silicon alkoxide and then simply adjusting and maintaining a stable pH range for which the protein retains stability. This process has provided effective doping capabilities without altering the conformation or shape upon release, which is extremely important for retaining function. This material provides an interesting approach to drug delivery as it helps to reduce degradation potential in circulation (Qhobosheane et al. 2001, Lei et al. 2002). These degradation profiles are promising as they facilitate evidence that inorganic materials could be developed with the ability to specify doping and enhanced control over therapeutic release rates.

6.2.6 Economic Affordability and Ease of Scale-Up

Silica nanoparticle systems can be generated and purified with ease in large quantities at low cost. This is essential in clinical translation of these materials. Pharmaceutical and biomedical corporations are looking to invest in materials that are easy to produce and have the capability to present both an enhanced therapeutic benefit, as well as economic turnaround. Even with the ease in the synthetic chemistry and purification, some additional factors such as characterization and sterilization remain.

6.3 Characterization of Silica Nanoparticles

Once silica nanoparticles are synthesized, it is essential to effectively characterize the constructs. Size, charge, chemical core, and surface composition all need to be validated and verified in order to successfully assess and reevaluate the efficacy and the impact on delivery of these constructs. Presented here are only a few methods of characterization to illustrate each technique's pros and cons. Thus the key point is that validation must be made through multiple sources.

6.3.1 Size

Size characterization is essential because as previously mentioned factors such as cellular uptake, biodistribution, and toxicity are dependent on size and surface area.

6.3.1.1 Dynamic Light Scattering or Photon Correlation Spectroscopy

In this method, a shining light source is used at a solution containing a colloidal dispersion of the sample. The particles in the sample undergo Brownian motion, which creates interference in light penetration and introduces scattering of light. The light that is scattered is detected and translated into a velocity, which can be back calculated into a size. The measurement itself does not serve as a complete characterization of size, as it only provides the polydispersity over the total sample and measures the hydrodynamic radius rather than the actual particle radius. It is however a less expensive option to assess and quantify the size and polydispersity of a nanoparticle sample. The method is also helpful in the identification of aggregation of particles due to surface modification or charge instability on the surface (Berne and Pecora 2000).

6.3.1.2 Transmission Electron Microscopy

TEM is a highly sensitive method that utilizes a beam of electrons emitted and passed through a dried sample on a copper grid. An image is generated via the detection of the electrons through the sample. These images can then be manually or electronically measured to assess the actual size of the particles. Due to the sensitivity of the instrument, particles in the single nanometer range can be detected. It is important to note that once particles are dried, most nonrigid surface modifications such as polymers or peptides are not displayed, due to collapse and nonexistent interaction with a solvent system. Thus, the actual size of the particle could potentially be much different than the representative image. Additionally, it is extremely difficult to ascertain surface topography or visualize porosity of dense materials due to electron penetration (Williams and Barry Carter 2009).

6.3.1.3 Scanning Electron Microscopy

SEM utilizes a beam of electrons and detects the secondary or backscattered electrons of a metallic coated sample, introduced after sample drying. Due to the detection method, it is difficult to provide effective resolution for particles much below 50 nm; however, surface topography can be resolved. Similar to TEM, soft organic modifications or materials are undetectable (Goldstein et al. 2003).

6.3.1.4 Atomic Force Microscopy

AFM utilizes a dragging probe to probe the surface of a material and detects small vibrations, due to surface topography, utilizing a piezo electrode. This method can be performed in solution and thus can provide size information on both rigid and nonrigid materials. AFMs have been utilized to detect and image at the molecular scale, asserting topographical changes within 5 nm (McLean and Sauer 1997). However, changes in instrumentation setup and sample preparation can significantly affect resolution (Gross et al. 2009).

6.3.2 Charge

As stated earlier, charge density plays an important role in directing uptake, distribution, and toxicity. By controlling charge, it is possible to direct the biological fate and toxicity of silica nanoparticles.

6.3.2.1 Zeta Potential

This is a fast, efficient measurement of the electric potential of a colloidal suspension. By passing an electric field through the solution, one is able to detect the charge at the interfacial double layer, which can then be compared to a point in the bulk solution calculating a relative charge solution value. This number is extremely useful in the identification of the stability of the solution; generally zeta potential values between −20 and +20 do not have enough electrostatic repulsion to be stable colloids, and thus they aggregate to create solution stability. Additionally, it can be a useful measurement to provide charge interaction comparison between particles and cell surfaces. It is important to note that this value is highly dependent on pH and salt concentration, since it measures the ions in solution. Thus, when taking zeta potential measurements, one must note that physiological environments differ significantly from measured values, so the number it provides is relative (Hunter 1988).

6.3.2.2 Surface Composition

Cell-cell-mediated interactions via viable, nonviable, native, and foreign materials are generally through conformation arrangement or chemical identification. This also appears to be the case in nanoparticle interactions, and thus surface composition will be key in characterization.

6.3.2.3 Thermogravimetric Analysis

This is measurement based on weight loss due to increases in temperature and effectively burning off attached surface materials (i.e., polymers). Modified silica samples are heated after a thorough drying process and data is extracted and compared to bare silica nanomaterials. Correlations can be drawn between weight loss and average amount of material located on the surface. This measurement provides evidence that compound(s) exist on the surface; however, correlative information is limited to melting temperature of the sample, and thus it is difficult to discern the actual material (Coats and Redfern 1963).

6.3.2.4 Nuclear Magnetic Resonance

This measurement is obtained through the induction of a magnetic field on an object. This field provides spin ratios that can then be correlated back to chemical structural characteristics and subsequent molecular identification. The method relies on heavy training and can be extremely time-consuming based on the surface modification, but it however provides some of the most accurate and quantifiable information (Engelhardt and Michel 1987).

6.3.2.5 Infrared Spectroscopy

This measurement is obtained by emitting infrared band on a sample and recording the absorption of stretching and bending frequencies. This also can be correlated back to chemical structural characteristics and subsequent molecular identification yet limited in quantification (Ferrari et al. 2004).

6.3.2.6 Traditional Fluorescent Labeling or Chemical Substitution Reactions

One can tag the surface modification or utilize a reagent that reacts with the surface modification to create a product that can then be measured via absorbance or fluorescence. The generation of a calibration curve can provide quantification for surface modification (Corrie et al. 2006).

6.4 Silica Nanoparticle Preparation for Biological Evaluation

Following synthesis and characterization, it is essential to effectively sterilize the particles in order to validate and verify the safety and efficacy of the constructs. Again this proves to be a difficult task for inorganic nanoparticles as many traditional sterilization techniques are inefficient, ineffective, or impractical. Without sterilization, drug delivery and imaging devices could potentially be contaminated, leading to toxic side effects not due to the construct itself but the contamination within the solution the construct exists. For a more comprehensive review detailing experiments evaluating the effects of sterilization on nanoparticles, the reader is directed elsewhere (Franca et al. 2010). The following briefly outlines the pros and cons of traditional sterilization techniques.

6.4.1 Heat and Autoclaving

Traditional silica inorganic particles can be heated at extremely high temperatures (~500°C). These temperatures are sufficient to both kill bacteria and burn away endotoxins. However, the temperature induced does have the potential to detrimentally degrade any surface modifications or molecular encapsulations. A lower temperature alternative is autoclavation; however, most natural or mimics of natural materials still degrade at these temperatures or pressures.

6.4.2 Filtration

Filtration through traditional 200 nm pore filters are sufficient to remove most bacteria; however, endotoxins still pose a significant problem. Additionally, 200 nm pores will only work for particles much smaller than the pore size range and many nonspherical or geometric nanomaterials might be caught up and remain within pores themselves.

6.4.3 Gamma Irradiation

Ionizing radiation is emitted on samples to destroy bacteria and alter endotoxin formation effectively reducing inflammatory response. However, irradiating samples can ionize the sample or destroy surface modifications, which is not desirable.

6.4.4 Ethanol

Soaking materials within ethanol effectively destroys bacteria, but endotoxins remain in solution. Sterilization is one of the determining factors behind utilization of any nanotechnology-based construct for clinical applications. Each sterilization technique provides both sufficient advantages and disadvantages, while a combination of the techniques may prove to be ideal. It is important to note that sterilization techniques could potentially alter size and agglomeration, so it will be key to ensuring technical composition is maintained through this process.

6.5 Toxicity of Silica Nanoparticles

Limited toxicity profiles exist on amorphous silica nanoparticles. However, for use of silica nanoparticles in a clinical setting, it is imperative that their toxicity in vitro and in vivo is carefully evaluated. As previously discussed, size, shape, surface modification, and composition have proven to effect toxicity, distribution, and uptake of silica nanoparticle constructs. Outlined here is a brief literature overview of how these characteristics affect silica nanoparticle interaction with in vitro and in vivo environments. By paying close attention to outcomes, this review will also attempt to address modifications or utilization of particular characteristics to obtain better engineered silica nanoparticles for biomedical applications.

There are two distinct modes of cell death: apoptosis (programmed cell death) and necrosis (premature cell death). Cytotoxicity has been indirectly linked to markers of these modes of cell death and their subsequent initiation events. Such an example is caspase-3. This is an important marker of apoptosis and it has been shown to be upregulated in macrophage cell lines following silica nanoparticle treatment (Park and Park 2009). Additionally, other modes of cell death can be initiated by other cytotoxic cellular events. The current initiation modes of cell death remain uncertain. It is however certain that, if silica nanoconstructs are instigating these events, in order to engineer safe and effective constructs, causation and subsequent elimination of these events need to be addressed.

It is important to note that bare silica nanoconstruct toxicity is cell type dependent, and thus the route of administration influences alterations in toxicity as well. Epithelial cells show very little to no cytotoxic effects when treated with silica nanoparticles (Figure 6.8, Lanone et al. 2009, Malugin et al. submitted, 2011). However, fibroblast cell lines and those cell lines with longer population doubling times or lower metabolic rates have been shown to have a substantial increase in susceptibility to toxic effects (Chang et al. 2007). Cells with phagocytic activity such as macrophages and to some degree endothelial cells appear to be the most effected cell types (Hamilton et al. 2008). Thus it will be important to play close attention to the cell type(s) many of these assays are performed on and the mechanisms that induce toxic susceptibility (Figure 6.8, Malugin et al. submitted, 2011). This can ensure that surface or material modifications are investigated to avoid cytotoxic mechanisms that are induced by processes like phagocytic activity. If toxicity has been identified or suggested, it is crucial to determine the causation of cellular toxicity.

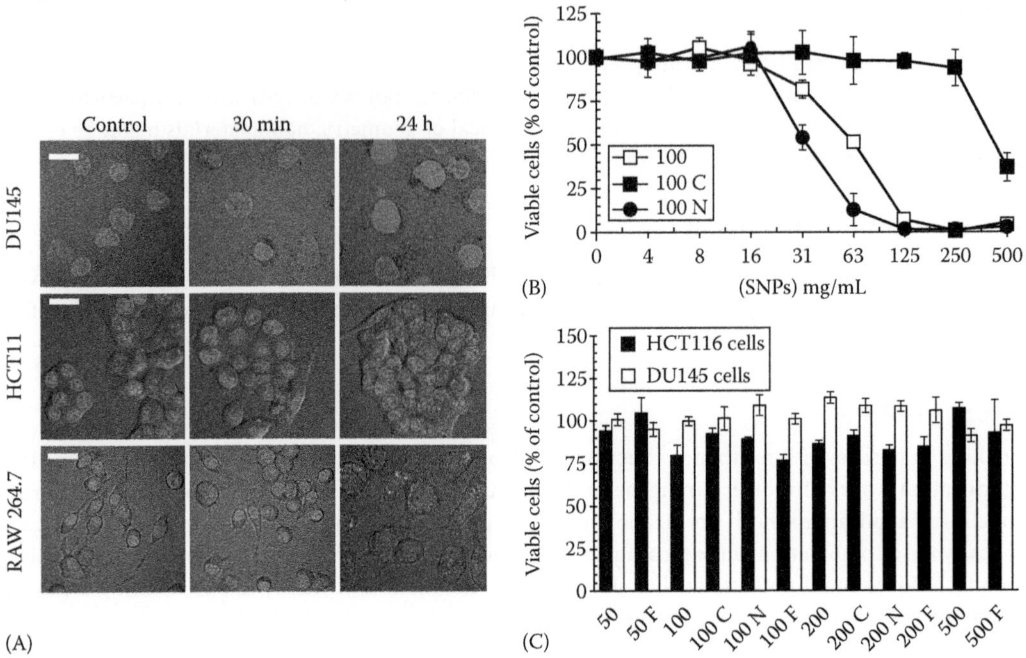

FIGURE 6.8 **(See color insert.)** Influence of cell type and surface modification on toxicity. As shown, silica nanoparticle treatment does not induce HCT116 or DU145 epithelial cell toxicity, while it does induce RAW 264.7 macrophage toxicity. IC_{50} values in RAW 264.7 cells are heavily dependent on surface modification, as amine (N) modified particles are more toxic than unmodified and carboxyl modified particles (C). (A) Confocal microscopy image of DU145, HCT11 and RAW264.7 cell nucleus stained with DRAQ5 and particles labeled with FITC. These images illustrate the relative uptake of these particles. RAW264.7 shows a significant increase in relative uptake. (B) RAW 264.7 WST-8 proliferation assay, 100 nm plain, carboxyl and amine-modified particles. (C) HCT116 and DU145 WST-8 proliferation assay, 100 nm plain, carboxyl- and amine-modified particles, little to no toxicity was observed. (Modified from Malugin, A. et al., Submitted, 2011.)

6.5.1 Toxicity via Cellular Internalization

Prior to the discussion of potential toxicity mechanisms, it will be important to address how nanoparticles can interact with cells and the most probable modes of internalization. Silica nanoparticles display on their surface approximately five hydroxyl functional groups per nanometer (Rahman et al. 2009). The very nature of the particles is foreign to the human body and as such phagocytosis facilitates another uptake mechanism. Cells can also be internalized via caveolin and clathrin-mediated endocytosis, macropinocytosis, or pinocyotsis (Figure 6.9). Each of these mechanisms generally has a size- and surface-dependent threshold.

The most probable mechanistic route of internalization is via endocytosis followed by encapsulation within lysosomal compartments (Xing et al. 2005, Sun et al. 2008, Malugin et al. submitted, 2011, Figure 6.10). It is important to note that the acidic pH of the lysosomal compartment is not sufficient to facilitate degradation of these particles. Thus after internalization these constructs have the potential to do irreversible cell damage if they are released from cellular compartments or internalized into important functional compartments such as the mitochondria. Some literature sources have started to devolve potential silica nanoparticle escape routes as well as cellular compartment recycling. It has been suggested that silica nanoparticles remain within cellular compartments long enough to be released from late-stage lysosomes into the cytoplasm. This has been confirmed via fluorescent microscopy and the delocalization of silica nanoparticles with lysosomal or endocytic compartments (Huang et al. 2005).

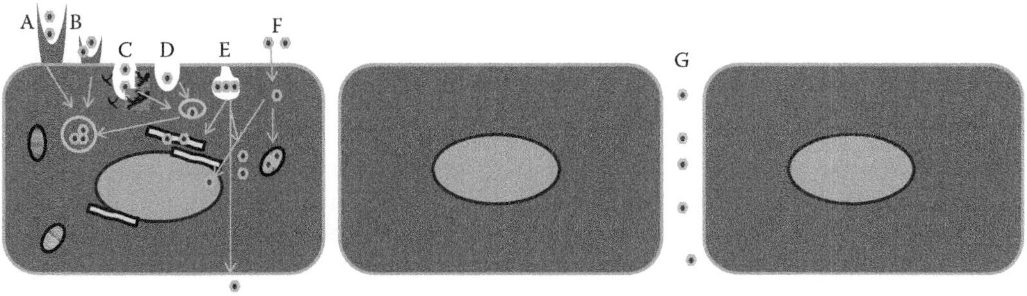

FIGURE 6.9 **(See color insert.)** Mechanisms of nanoparticle uptake: (A) phagocytosis, (B) macropinocytosis, (C) clathrin-mediated endocytosis, (D) clathrin- and caveolae-independent endocytosis, (E) caveolae-mediated endocytosis, (F) transmembrane transport, and (G) paracellular transport.

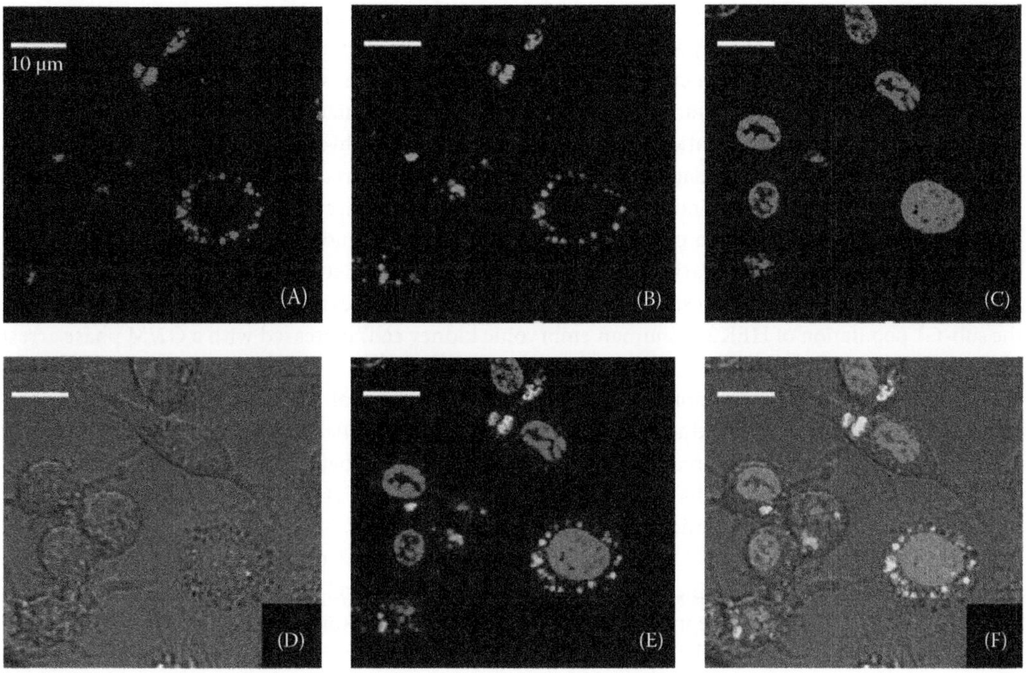

FIGURE 6.10 **(See color insert.)** Confocal images of silica nanoparticle uptake in RAW 264.7 macrophage. Silica nanoparticles are colocalized in lysosomal compartments (yellow color in figures E and F). (A) Lysosomes are stained with lysotracker. (B) Particles are stained with FITC. (C) Nucleus is stained with DRAQ5. (D) Transmitted image of RAW 264.7 cells. (E) Fluorescence overlay. (F) Transmitted and fluorescence overlay. (Modified data from Malugin, A. et al., Submitted, 2011.)

Increased cytoplasmic presence can increase cellular compartmental encapsulation and damage; one such example is the mitochondria. Chang et al. (2007) compared the results of (4,5-dimethylthiazol-2-Yl)-2,5-diphenyltetrazolium bromide (MTT) mitochondrial function assay of ~80 nm bare silica to chitosan silica nanoparticles on WS1, CCD-966sk (human skin adherent fibroblasts), MRC-5 (human lung adherent fibroblast), A549, MKN-28 (human cancer gastric epithelial), and HT-29 (human caner colon epithelial) cells. The results indicate that cancer mitochondrial function was more resilient to silica nanoparticle treatment. Fibroblast cell lines were susceptible to functional damage and that chitosan modification significantly reduced toxicity. Choi et al. (2009), Julien et al. (2009), and Malugin et al. (submitted, 2011) found similar toxicity results.

6.5.2 Toxicity via Particle–Cell Interaction

Cellular internalization however is not the only possible mechanistic route of toxicity. Silica nanoparticle cell interaction could potentially be sufficient to induce inflammatory or coagulation cascades, triggering signaling pathways and subsequent damage. Some investigators believe that nonspecific silica nanoparticle interaction with cells is sufficient to create membrane damage. Lactate dehydrogenase (LDH) or membrane integrity assays have supported this hypothesis (Chang et al. 2007, Yu et al. 2009). It however remains unclear what initiates this damage. Also the influence of surface functional groups, charge, hydrophobicity, and contaminants needs to be examined in detail.

The hydroxyl groups on the surface of the silica nanoparticle can also react with native species, such as iron in the Fenton reaction (Figure 6.11) or cellular receptors. Both modes create radicals that induce oxidative stress and ROS (Dalal et al. 1990). Increase in ROS levels can cause oxidation of deoxyribonucleic acid (DNA), protein, and lipids; induce mitochondrial dysfunction; and significantly alter the genes related to the inflammatory and apoptotic response. Certain markers play a key role in indication of oxidative stress and suggest functional cell damage, such as increases in malondialdehyde (MDA) and thiobarbituric acid reactive substance (TBARS) and decreases in glutathione (GSH). MDA and TBARS indicate amplified lipid peroxidation and GSH is a ubiquitous sulfhydryl-containing molecule that is responsible for maintaining oxidative homeostasis and does so by reducing cellular oxidation species within the cell. Oxidative stress induction via bare silica nanoparticles has been shown to be a potential mechanism of cellular toxicity in macrophage, embryonic kidney, and bronchoalveolar carcinoma cell lines (Lin et al. 2006, Park and Park 2009, Wang et al. 2009a). Wang et al. have linked a potential apoptotic cell death mechanism as a result of oxidative stress with a flow cytometric analysis by proving that the sub-G1 population of HEK293 (human embryonic kidney cell) increased with a G2/M phase arrest (Wang et al. 2003).

Nanoparticle cellular interaction can also stimulate the release of primary and secondary inflammatory mediators, such as cytokines, chemokines, nitric oxide (NO), and expression of inflammatory genes (Park and Park 2009). Increases in interleukins (such as IL-1beta, IL-6, IL-8, IL-10, etc.), tumor necrosis factor alpha (TNF-alpha), transforming growth factor (TGF), monocyte chemoattractant proteins (MCP-1), macrophage and inflammatory proteins (MIP-1, MIP-2, etc.), and their respective messenger ribonucleic acid (mRNA) have been shown to be upregulated in cell lines in environments post in vivo and in vitro with silica nanoparticle treatment (Driscoll 2000, Ovrevik et al. 2006, Park and Park 2009). Many of these represent inflammatory signaling factors that induce cellular migration, proliferation, differentiation, and apoptosis within the biological environment. Direct correlations can be drawn between increases in these inflammatory mediators and increased levels of inflammatory or immunological cells (T, B, and NK cells) in the local environment after silica nanoparticle treatment (Cho et al. 2007, Park and Park 2009). All of these events are hallmark indications of larger developing inflammatory events. Additionally, inflammatory events along with direct particle interaction with cell environments can adversely enhance cellular activation and subsequently induce thromobogenicity (Nemmar et al. 2004). If the induction of these mediators is indeed due to surface interaction, covering the surface or hydroxyl groups of the particle with a biocompatible material could potentially provide a safe alternative.

$$Fe^{3+} + O_2^{\bullet -} \rightarrow Fe^{2+} + O_2$$

$$Fe^{2+} + H_2O_2 \rightarrow Fe^{3+} + OH^{\bullet} + OH^{-}$$

FIGURE 6.11 Fenton reaction demonstrating the ability of OH groups to interact with native metallic elements producing free radical or ROS species.

Circulation of bare and modified particles can induce additional interactions with other circulating blood cells and investigators have begun to study these implications. As stated previously, evidence suggests that bare silica nanoparticles cause the lysis of erythrocytes (Dalal et al. 1990). Modes or mechanisms of cell death that have been proposed include the induction of reactive oxygen species or surface electrostatic binding interactions with tetra-alkyl ammonium groups (Depasse and Warlus 1976). It has been shown that the density of hydroxyl groups on the surface of silica nanoparticles is directly correlated with the rate of hemolysis (Dalal et al. 1990, Murashov et al. 2006). Slowing et al. (2009) reported that mesoporous silica nanoparticles maintain safe levels of biocompatibility, with low hemolysis activity. However contradictory these reports appear to be, it is interesting to note the impact of surface density and charge on the safety of particular constructs. When designing drug delivery or imaging systems, it is important to attempt to minimize hemolysis via surface, bulk, or geometric modifications to reduce side effects.

6.5.3 Toxicity via a Combination of Particle Internalization and Cellular Interaction

Genotoxicity, chromosomal aberrations, and mutagenicity are also extremely important factors to consider both in the long-term health of the patient and in development of these particles. Waters et al. (2009) performed a transcriptional analysis on a variety of different size ranges of particles, utilizing microarrays on RAW 264.7 (murine macrophage cell line). This provided evidence of adverse amplification effects on the expression of genes that are implicated in inflammatory and stress inducing events, such as chemokines and cytokines. Surface area was directly correlated with the level of gene expression abnormalities, which suggests inflammatory stimulation rather than genotoxicity induction. Similarly, Jin et al. (2008) showed that luminescent DNA nanomaterials presented little DNA damage, which did not necessarily correlate with cytotoxicity levels within A549 cells, suggesting cell death occurs through other mechanisms. Similarly Barnes et al. (2008) reported that 3T3-L1 (fibroblast like cells) incubated with silica nanoparticles produced no genotoxicity evidence within a reproducible comet assay that assesses breaks in both single- and double-stranded DNA. These results provide a positive outlook for the application of these materials in drug and imaging systems.

6.5.4 Toxicity via Biodistribution and Clearance Mechanisms

Due to silica's inherent inability to degrade, clearance mechanisms and organ accumulation are extremely important to evaluate, to ensure the safety of these constructs. Borchardt et al. (1994) investigated the biodistribution of a variety of different surface-modified spherical nanoparticles and found that increasing the hydrophilicity of the coatings led to increased intestinal delivery and lower uptake via liver and spleen, while increasing the chain length and attaching a butyltrichlorosilane increased muscle accumulation and bone marrow delivery, respectively. These investigators have proposed that alteration in biodistribution is due to both the hydrophilicity and steric inability for proteins or opsins to adsorb to the surface of the silica constructs. Additionally, the study introduced the aspect that bare silica has a much slower liver absorption than most common drug carriers. It will be important to keep these results in mind in development of silica nanoparticles, as this study suggests that size, geometry, and surface modification could potentially provide directed targeting and initiate elimination in toxicity.

An important issue is clearance of nanoparticles. Ideally, particles should be excreted via the kidney so local liver accumulation does not occur. Liver clearance however is acceptable if particles are cleared through hepatobiliary mechanisms. Nevertheless, toxicity due to prolonged particle accumulation can still pose a significant threat. Burns et al. (2009) were able to develop labeled silica nanoparticles of diameters 3.3 and 6.0 nm and showed complete clearance within 48 h. He et al. (2008) demonstrated that 45 nm silica nanoparticles with free hydroxyl, carboxyl, and PEG modifications were cleared mainly through liver excretion with a high circulation rate for PEGylated particles. Additional data pointed to removal via renal routes in addition to hepatobiliary mechanisms, suggesting that particles are statistically able to accumulate renally and with modification. Hudson et al. (2008) examined rat subcutaneous biodistribution and clearance of 180 nm–4 μm silica nanoparticles at 4 days, and at 2 and 3 months. Significant particle accumulation was found in the subcutaneous nodule at 4 days. However, 2- and 3-month time points showed little to no accumulation. Following intravenous injection into mouse models doses above 1 mg per animal caused lethality.

6.6 Applications of Silica Nanoparticles in Drug Delivery and Imaging

6.6.1 Drug Delivery

Drug delivery is defined as the ability to effectively attach or encapsulate a therapeutic payload, deliver the respective payload to a site of interest, and finally release it at an appropriate therapeutic release rate. Silica nanoconstructs as outlined within the chapter have three distinct advantages over traditional delivery constructs: (1) the construct itself provides protective stability with which one can both protect and encapsulate therapeutics; (2) the physiochemical characteristics of the construct can be manipulated to facilitate controlled release rates; and (3) alterations in physiochemical characteristics and surface attachment can facilitate targetability and effective biocompatibility.

6.6.1.1 Protective Stability with Which One Can Both Protect and Encapsulate Therapeutics

Silica nanoparticles provide a unique thermally and chemically stable nondegradable environment, which can potentially encapsulate therapeutics. Compared to drug delivery systems such as micelles, liposomes, and water-soluble polymeric carriers, silica nanoparticles have some advantages. Unlike silica nanoparticles, polymeric carriers may lack masking capabilities, while micelles and liposomes suffer from inherent potential dissociative properties due to critical concentrations and diminished physiochemical interactions (Kwon and Kataoka 1995). Silica nanoparticles provide a means by which to encapsulate hydrophobic molecules that would normally be insoluble and thus undeliverable. Such an example was produced by Liong et al. (2008) where a mesoporous multimodal imaging phosphonate coated silica nanoparticle with a folic acid targeting motif was used to help deliver loaded hydrophobic chemotherapeutics, specifically camptothecin and paclitaxel. A significant cellular reduction in viability at 20 μg/mL and enhanced cellular uptake with the addition of a folic acid motif were demonstrated. Additionally, silica nanoconstructs can provide a platform to stabilize bioactive agents. As stated earlier in the chapter, enzymes have been linked to the surface of silica to help maintain activity and facilitate a reduction in degradation. Similarly lipids have been stabilized on the surface of these materials. Silica nanoparticle–lipid emulsions were created for oral delivery where the silica component provided protection against lipase digestion in the gastrointestinal tract and a 15-fold increase in digestion was observed compared to lipid emulsions alone (Tan et al. 2010).

It is important to note, however, that silica stability also comes with inherent drawbacks, including but not limited to lack of degradation and aggregation. The key question in the design of novel silica systems for drug delivery will be how they will be cleared after systemic administration. As reported earlier, investigators are looking at degradation chemistries that can help solve this elimination problem. Additionally, because these particles are colloidal suspensions, there could potentially be modes of aggregation or agglomeration. These could counteract alterations in physiochemical characteristics that facilitate directed targeting such as size or geometry.

6.6.1.2 Manipulation of the Construct to Facilitate Controlled Release Rates

While encapsulating the materials is the first step, release and maintenance of effective levels of therapeutics are key to a successful delivery. Mesoporous silica constructs provide a means by which to manipulate drug release rates. Unlike traditional polymers that rely on cleavage of therapeutics or complete disassociation via hydrolysis or enzymatic mechanisms of micelles or liposomes, mesoporous materials rely entirely on physiochemical interactions and their effects on basic diffusion or mass transport phenomena. These porous materials can have interconnected large networks of pores that can create different diffusion length paths and thus alter diffusion release rates. For example, Trewyn et al. (2004) and Stromme et al. (2009) utilized both spherical and rod like mesoporous silica nanoparticles. The spherical particle had ordered aligned pores while the rodlike particle had a tortuous array of pores. These different constructs were loaded with an antibacterial agent that facilitates the destruction of both gram-positive and gram-negative bacteria. The altered pore structures significantly altered the release rates of the drug in in vitro bacterial cultures, as the spheres had an enhanced delivery at 48 h after inoculation. This research suggested that pore geometry influences release rates where more interconnected networks could allow for longer diffusion paths and thus potentially longer release kinetics (Trewyn et al. 2004). Additionally, Brohede et al. (2008) studied the difference between cylindrical and spherical pores. It was found that diffusion through spherical pores was enhanced, suggesting that the interconnected pores of cylindrical particles facilitate longer diffusion paths.

6.6.1.3 Physiochemical Characteristics and Surface Attachment Can Facilitate Targetability and Effective Biocompatibility

Surface functionality is one of the most important aspects in enhancing biocompatibility, targeting, and release of drugs from silica nanoconstructs. With surface functionalization, one can potentially reduce or eliminate unwanted toxicity and influence cellular uptake. For example, silica nanoconstructs have been modified with cationic residues to facilitate nucleic acid adsorption. These constructs have proven to be effective in increasing transfection efficiency of nucleic acids both in vivo and in vitro (Kneuer et al. 2000a,b, Sameti et al. 2003, Bharali et al. 2005, Roy et al. 2005, Shang et al. 2007). Additionally, systemic side effects may be potentially avoided with the addition of specific functionalities. Mal et al. (2003) have utilized coumarin to coat the pores of the mesoporous material and UV light emission at 310 nm to release cholestane and phenanthrene. Several groups have utilized redox alternatives, such as gold and poly(amido amine) dendrimers (PAMAM) to stimulate pore opening capabilities (Lai et al. 2003, Radu et al. 2004a, Gruenhagen et al. 2005). Extensive work has been done with spherical mesoporous nanoparticles, where the attachment of a variety of different types of stimuli-sensitive materials that will only initiate release upon specified stimuli within certain physiological environments (Trewyn et al. 2008). For example, a cadmium sulfide (CdS) cap was utilized to encapsulate neurotransmitters within mesoporous silica nanoparticles, and, as a result of a reducing environment, disulfide bonds were cleaved and the drug was released into circulation (Lai et al. 2003).

Active targeting, through the attachment of specified ligand on the surface of the particles, has the potential to enhance the delivery of silica nanoparticles. As illustrated earlier, the attachment of specific ligands (transferrin and folic acid) and their subsequent interactions with pancreatic cell receptors have facilitated preferential uptake (Kumar et al. 2008 and Liong et al. 2008). Additionally, as stated earlier in the text, Nan et al. (2008) and Trewyn et al. (2008) demonstrated that geometry can also provide preferential uptake. One can extrapolate this in vitro data to translation, as more work is done to understand how these active targeting characteristics facilitate preferential uptake that reduce systemic side effects. Ultimately, combinations of silica with other materials can assist in the creation of an ideal synthetic system that combines appropriate biocompatibility, diffusivity, encapsulation, and release.

6.6.2 Imaging Contrast Agents

Noninvasive methods utilizing contrast agents can effectively aide in earlier detection of disease states and result in better prognosis for patients. The contrast agents need to preferentially differentiate abnormal tissue environments from normal physiology, thus enhancing the local signal. As such, successful new imaging agents are defined by their capability to enhance sensitivity and resolution. Some challenges in toxicity, degradation, and stability limit the use of traditional signal enhancing imaging constructs. However, if the constructs are concealed in a protective environment, their human use can be facilitated. Silica nanoparticles show promise in the development of contrast agent—doped particles, such as those utilized in MRI and optical imaging applications. Silica nanoparticles provide a platform that can aid in developing these modalities by encapsulating specialized imaging materials, thereby protecting both the material and the body, and by enhancing the signal yield.

6.6.2.1 Encapsulating Specialized Imaging Materials

Silica nanoparticle synthetic doping provides an ideal platform by which contrast agents can be doped into particles. Current contrast agents such as gadolinium (MRI) and quantum dots (optical) have an inherent toxicity when placed in physiological environments. However, if the material is encapsulated within an environment that prevents physiological contact, imaging agents can be developed with improved safety and signal intensity. As outlined earlier, gadolinium, iron oxide, quantum dots, gold, along with several different types of materials have been successfully incorporated or doped within silica constructs (Kim et al. 2007, Insin et al. 2008, Liong et al. 2008, Wang et al. 2008, Burns et al. 2009, Knopp et al. 2009). Liong et al. (2008) doped iron oxide constructs within silica nanoparticles and produced localized passive effective tumor targeting in mouse models. This technology has also been utilized to track single stem cells to aid in the identification of the distribution of such cells after injection. Chung et al. (2007) showed that uptake in mesenchymal stem cells does not alter proliferation, function, or differentiation and they were able to subsequently image and track single cell migration through systemic circulation. Multimodal particles utilizing a combination of doped gadolinium and gold nanoparticles provided both MRI and photoacoustic imaging (optical) modalities. These systems were studied via uptake in J 774 macrophage cells to illustrate the proof of concept. However, as stated earlier, occasional reports have suggested that the materials can leak from silica materials (Lai et al. 2003). As a consequence, better conjugation or incorporation methods are needed.

Is it important to note that in addition to being able to protect the body from potentially toxic materials, silica nanoparticles can actually enhance and protect the contrast agent. For example, as stated earlier, the incorporation of fluorophores within these materials have protected them from degradation and reduced photobleaching (Blaaderen and Vrij 1992, Jin et al. 2007, Kumar et al. 2008). Additionally, the incorporation of MRI agents, such as gadolinium, within these materials could also reduce the potential dissociation of the agent from the carrier.

6.6.2.2 Enhancing the Signal Yield

The very premise of contrast agent is to enhance localized signal. Silica nanoparticles have two properties that enhance signal generation. The first is the capability to incorporate multiple contrast agents within one particle. The second is the ability to introduce changes in physiochemical characteristics and surface functionality that provide enhanced delivery potential. By delivering contrast agents in high payloads to a particular site of interest one can enhance the signal significantly. For example Kim et al. (2007) developed silica nanoparticles containing gadolinium and a luminescent particle, with an attached arginine–glycine–aspartic acid (RGD) functionality. They showed little or no uptake in HT-29 (human colon cancer) cells when incubated with particles without RGD and a significant amount of uptake when incubated with particles with RGD. Additionally, Bickford et al. (2009) developed a combination of gold and silica to produce a particle with near-infrared light scattering capabilities and attached human epidermal growth factor-2 (Her-2) targeting ligand. This nanoparticle proved efficacious in its ability to both target and image three separate cancer cell lines that over expressed Her-2.

6.6.3 Theranostics

Dual modalities are useful in that both treatments and diagnostics can be delivered simultaneously. This aids not only in time and cost reduction but also in the identification of localization of each construct. These materials have the same advantages and limitations outlined in drug delivery and imaging agent sections. The following will highlight a few of the examples that are surfacing in the literature with the utilization of silica as a theranostic platform. Lee et al. (2009) combined doxorubicin with iron oxide and dye doped in mesoporous silica nanoparticles. Following subcutaneous injection passive targeting of these particles led to delivery and successful terminal deoxynucleotidyl transferase dUTP nick end labeling (TUNNEL) staining or appearance of apoptotic cells within the tumor site. Similarly, Park et al. (2009) were able to create a luminescent degradable silica nano-composite that incorporated doxorubicin. The relatively nontoxic construct facilitated chemotherapeutic release, allowed for optical imaging, and assessment of the location of the particles, and subsequently degraded, which allowed for clearance of the particles.

The above examples provide evidence of the potential of silica nanoconstructs as delivery systems and the potential for future applications as drug delivery systems, imaging agents, or theranostics.

6.7 Unresolved Issues and Future Directions

As outlined throughout this chapter, silica nanoparticles provide an opportunity for drug delivery and imaging. However, it remains unclear what exactly the implications are of introducing an inorganic material in a biological environment. A key challenge is limited knowledge about mechanisms of toxicity of silica nanoparticles. To facilitate clinical translation it will be important to define how silica interacts with the biological environment. As with any biomaterial, biocompatibility remains the primary concern. The material should elicit an appropriate response without adverse effects, locally or systemically. Taking this into consideration, the traditional definition of toxicity is only part of this larger picture. As highlighted throughout the chapter, these materials do not seem to produce much initial or immediate adverse effects, but in vitro and in vivo data are not without concerns, as the long-term effects still remain unclear.

The initiation of inflammatory or thrombogenic events could potentially possess the greatest risks. It is worth noting that this will be important to understanding the total picture of these materials, as their crystalline counterparts, which are considered highly toxic, usually do not induce their effects for months to years. So, as one looks to the future of silica as a biomedical material, it will be key to maintaining safety, efficacy, and degradability while still affording the robust physiochemical properties. To translate these materials to clinical applications, it is necessary to define a mode of transport, clearance, and treatment.

6.7.1 Defined Mode of Transport

Drug delivery and imaging systems are defined by their capacity to accumulate locally within diseased tissue. The utilization of SNPs to target these sites via alterations in geometry, size, surface modifications, including targeting ligands are still dependent on one key factor, which is the ability to be transported via the blood stream. This transport process is not by any means simplistic or easy to effectively define. Silica nanomaterials must circulate long enough to allow targeting mechanisms to be effective. Thus, they must inherently escape traditional biological removal mechanisms, such as the innate immune system, while still not interfering with the local environment. What is essential here is that a surface or structural modification be made so that these materials evade phagocytic uptake, protein adsorption, and cellular activation. Overcoming hemolytic concerns is imperative to effective maintenance of circulation and affording accumulation at the treatment site. Studies will need to focus on altering the surface chemistry via polymeric, protein, or mesoporous modification that will aid in this endeavor.

6.7.2 Defined Mode of Clearance

As has already been stated, it is impossible to create a material that will elicit no adverse biological response. However, one can create a material that provides limits to the induction of such response. One can do so by developing a material that provides quick access, treatment, and subsequent removal.

The previous section reviewed how to provide access with limited activation via surface modification. Here the focus will be on how to modify these materials so that they can be quickly cleared from the system. This can be done via the development of a degradation chemistry or utilization of small particles that can be excreted readily through renal or hepatobiliary mechanisms. In engineering and developing the chemical degradation of these systems, it will be helpful to consider and exploit native biological environments, such as acidic lysosomal compartments. After the system has performed its function, it should be removed so that the biological environment can return to its functioning hemostatic balance.

6.7.3 Defined Mode of Treatment

This is probably the most difficult and promising aspect of the design. One must carefully investigate the cellular uptake and biodistribution profiles of these systems. Rather than tailoring silica nanoparticles to a disease state or mechanism, it might be potentially useful to utilize their inherent properties as advantageous. For example, since silica nanoparticles appear to be able to escape lysosomal compartments could potentially help to deliver drugs to the cytoplasm of a cell (Huang et al. 2005). Or the fact that they appear to be uptaken preferentially in certain cell types and completely ignored by others could potentially be utilized as a delivery advantage rather than a disadvantage (Malugin et al. submitted, 2011). Additionally, dosing mechanisms of nanoparticle systems may also need to be altered, as it is clear that surface area and particle number have significantly different outcomes in toxicity profiles.

6.8 Concluding Remarks

Silica nanoparticles show great promise for biomedical applications. However, if these materials are to be clinically translated, there is a great need for in depth systematic evaluations of the implications of size, geometry, surface, and core composition. It will be important to delineate and introduce possible solutions for modes of clearance, evasion of phagocytic activity, reduction in inflammatory mediator expression, elimination of thrombogenicity, etc. Each characteristic poses a unique solution for these potential problems.

Acknowledgments

This research was supported by the NIH Grant R01DE19050, NSF-NIRT-ID 0835342, and the Utah Science Technology and Research (USTAR) Initiative.

Abbreviations

3T3-L1	fibroblast like cell line
A549	human lung carcinoma cell line
AFM	atomic force microscopy
APES	3-(aminopropyl)triethoxysilane
CCD-966sk	human skin adherent fibroblast cell line
CdS	cadmium sulfide
CTAB	cetryltrimethylammonium bromide
DLS	dynamic light scattering
DNA	deoxyribonucleic acid
DRAQ5	1,5-bis{[2-(di-methylamino) ethyl]amino}-4,8-dihydroxyanthracene-9,10-dione
DU145	prostate carcinoma epithelial cell line
EtOH	ethanol

FITC	fluorescein isothiocyanate
GSH	glutathione
H_2O	water
HCL	hydrochloric acid
HCT116	colorectal carcinoma epithelial cell line
HEK293	human embryonic kidney cell line
Her-2	human epidermal growth factor-2
HT-29	human cancer colon epithelial cell line
IL-1 beta	interleukin-1 beta
IL-10	interleukin-10
IL-6	interleukin-6
IL-8	interleukin-68
IR	infrared spectroscopy
J 774	macrophage cell line
LDH	lactate dehydrogenase
MCP-1	monocyte chemoattractant protein
MDA	malondialdehyde
MDA-MB231	human breast cancer epithelial cell line
Mia-PaCa	pancreatic cancer cell line
MIP-1	macrophage inflammatory protein-1
MIP-2	macrophage inflammatory protein-2
MKN-28	human cancer gastric epithelial cell line monosodium salt
MRC-5	human lung adherent fibroblast cell line
MRI	magnetic resonance imaging
mRNA	messenger ribonucleic acid
MTT	(4,5-dimethylthiazol-2-Yl)-2,5-diphenyltetrazolium bromide
NaOH	sodium hydroxide
NH_4OH	ammonium hydroxide
NMR	nuclear magnetic resonance
NO	nitric oxide
OH	hydroxyl
PAMAM	poly(amido amine) dendrimers
PANC-1	pancreatic cancer cell line
PEG	polyethylene glycol
RAW 264.7	murine macrophage cell line
RGD	arginine–glycine–aspartic acid
RNS	reactive nitrogen species
ROS	reactive oxygen species
SEM	scanning electron microscopy
SNT	silica nanotube
TBARS	thiobarbituric acid reactive substance
TEM	transmission electron microscopy
TEOS	tetraethoxysilane
TGA	thermogravimetric analysis
TGF	transforming growth factor
TNF-alpha	tumor necrosis factor alpha
TUNNEL	terminal deoxynucleotidyl transferase dUTP nick end labeling
WS1	human skin adherent fibroblast cell line
WST-8	(2-(2-methoxy-4-nitrophenyl)-3-(4-nitrophenyl)-5-(2,4-disulfophenyl)-2H-tetrazolium

References

Aggarwal, P., J. B. Hall, C. B. McLeland, M. A. Dobrovolskaia, and S. E. McNeil. 2009. Nanoparticle interaction with plasma proteins as it relates to particle biodistribution, biocompatibility and therapeutic efficacy. *Adv Drug Deliv Rev* 61 (6):428–437.

Angelos, S., M. Liong, E. Choi, and J. I. Zink. 2008. Mesoporous silicate materials as substrates for molecular machines and drug delivery. *Chem Eng J* 137:4–13.

Aznar, E., M. D. Marcos, R. Martinez-Manez, F. Sancenon, J. Soto, P. Amoros, and C. Guillem. 2009. pH- and photo-switched release of guest molecules from mesoporous silica supports. *J Am Chem Soc* 131 (19):6833–6843.

Barnes, C. A., A. Elsaesser, J. Arkusz, A. Smok, J. Palus, A. Lesniak, A. Salvati et al. 2008. Reproducible comet assay of amorphous silica nanoparticles detects no genotoxicity. *Nano Lett* 8 (9):3069–3074.

Beck, J. S., J. C. Vartuli, W. J. Roth, M. E. Leonowicz, C. T. Kresge, K. D. Schmitt, C. T. W. Chu, D. H. Olson, and E. W. Sheppard. 1992. A new family of mesoporous molecular sieves prepared with liquid crystal templates. *J Amer Chem Soc* 114:10834–10843.

Berne, B. J. and R. Pecora. 2000. *Dynamic Light Scattering: With Applications to Chemistry, Biology, and Physics*. Mineola, NY: Dover Publications, Inc.

Bharali, D. J., I. Klejbor, E. K. Stachowiak, P. Dutta, I. Roy, N. Kaur, E. J. Bergey, P. N. Prasad, and M. K. Stachowiak. 2005. Organically modified silica nanoparticles: A nonviral vector for in vivo gene delivery and expression in the brain. *Proc Natl Acad Sci U S A* 102 (32):11539–11544.

Bickford, L. R., G. Agollah, R. Drezek, and T. K. Yu. 2009. Silica-gold nanoshells as potential intraoperative molecular probes for HER2-overexpression in ex vivo breast tissue using near-infrared reflectance confocal microscopy. *Breast Cancer Res Treat.* 120 (3):547–555.

Blaaderen, A. V. and A. Vrij. 1992. Synthesis and characterization of colloidal dispersions of fluorescent silica spheres. *Langmuir* 8:2921–2931.

Bogush, G. H., M. A. Tracy, and C. F. Zukoski. 1988. Preparation of monodisperse silica particles: Control of size and mass fraction. *J Non Cryst Solids* 104:95–106.

Borchardt, G., S. Brandriss, J. Kreuter, and S. Margel. 1994. Body distribution of 75Se-radiolabeled silica nanoparticles covalently coated with omega-functionalized surfactants after intravenous injection in rats. *J Drug Target* 2 (1):61–77.

Brevet, D., M. Gary-Bobo, L. Raehm, S. Richeter, O. Hocine, K. Amro, B. Loock et al. 2009. Mannose-targeted mesoporous silica nanoparticles for photodynamic therapy. *Chem Commun* (12):1475–1477.

Brinker, C. J. and G. W. Scherer. 1990. *Sol-Gel Science: The Physics and Chemistry of Sol-Gel Processing.* San Diego, CA: Academic Press.

Brohede, U., R. Atluri, A. E. Garcia-Bennett, and M. Stromme. 2008. Sustained release from mesoporous nanoparticles: Evaluation of structural properties associated with release rate. *Curr Drug Deliv* 5 (3):177–185.

Burns, A. A., J. Vider, H. Ow, E. Herz, O. Penate-Medina, M. Baumgart, S. M. Larson, U. Wiesner, and M. Bradbury. 2009. Fluorescent silica nanoparticles with efficient urinary excretion for nanomedicine. *Nano Lett* 9 (1):442–448.

Cameron, N. M. de S., and M. E. Mitchell. 2007. *Nanoscale: Issues and Perspectives for the Nano Century.* Hoboken, NJ: John Wiley & Sons Inc.

Carrstensen, H., R. H. Muller, and B. W. Muller. 1992. Particle size, surface hydrophobicity and interaction with serum of parenteral fat emulsions and model drug carriers as parameters related to RES uptake. *Clin Nutr* 11 (5):289–297.

Carter, J. M. and K. E. Driscoll. 2001. The role of inflammation, oxidative stress, and proliferation in silica-induced lung disease: A species comparison. *J Environ Pathol Toxicol Oncol* 20 (Suppl 1):33–43.

Caruso, F., R. A. Caruso, and H. Möhwald. 1998. Nanoengineering of inorganic and hybrid hollow spheres by colloidal templating. *Science* 282 (5391):1111–1114.

Cedervall, T., I. Lynch, S. Lindman, T. Berggard, E. Thulin, H. Nilsson, K. A. Dawson, and S. Linse. 2007. Understanding the nanoparticle-protein corona using methods to quantify exchange rates and affinities of proteins for nanoparticles. *Proc Natl Acad Sci U S A* 104 (7):2050–2055.

Chang, J. S., K. L. Chang, D. F. Hwang, and Z. L. Kong. 2007. In vitro cytotoxicitiy of silica nanoparticles at high concentrations strongly depends on the metabolic activity type of the cell line. *Environ Sci Technol* 41 (6):2064–2068.

Chen, J.-F., H. M. Ding, J. X. Wang, and L. Shao. 2004. Preparation and characterization of porous hollow silica nanoparticles for drug delivery application. *Biomaterials* 25 (4):723–727.

Chen, M. and A. V. Mikecz. 2005. Formation of nucleoplasmic protein aggregates impairs nuclear function in response to SiO_2 nanoparticles. *Exp Cell Res* 305 (1):51–62.

Cho, M., W. S. Cho, M. Choi, S. J. Kim, B. S. Han, S. H. Kim, H. O. Kim, Y. Y. Sheen, and J. Jeong. 2009. The impact of size on tissue distribution and elimination by single intravenous injection of silica nanoparticles. *Toxicol Lett* 189 (3):177–183.

Cho, W. S., M. Choi, B. S. Han, M. Cho, J. Oh, K. Park, S. J. Kim, S. H. Kim, and J. Jeong. 2007. Inflammatory mediators induced by intratracheal instillation of ultrafine amorphous silica particles. *Toxicol Lett* 175 (1–3):24–33.

Choi, S. J., J. M. Oh, and J. H. Choy. 2009. Toxicological effects of inorganic nanoparticles on human lung cancer A549 cells. *J Inorg Biochem* 103 (3):463–471.

Chuankrerkkul, N. and S. Sangsuk. 2008. Current status of nanotechnology consumer products and nanosafety issues *J Met Mater Miner* 18 (1):75–79.

Chung, T. H., S. H. Wu, M. Yao, C. W. Lu, Y. S. Lin, Y. Hung, C. Y. Mou, Y. C. Chen, and D. M. Huang. 2007. The effect of surface charge on the uptake and biological function of mesoporous silica nanoparticles in 3T3-L1 cells and human mesenchymal stem cells. *Biomaterials* 28 (19):2959–2966.

Clift, M. J., B. Rothen-Rutishauser, D. M. Brown, R. Duffin, K. Donaldson, L. Proudfoot, K. Guy, and V. Stone. 2008. The impact of different nanoparticle surface chemistry and size on uptake and toxicity in a murine macrophage cell line. *Toxicol Appl Pharmacol* 232 (3):418–427.

Coats, A. W. and Redfern, J. P. 1963. Thermogravimetric analysis. A review. *Analyst* 88 (1053):906–924.

Cocco, P., M. Dosemeci, and C. Rice. 2007. Lung cancer among silica-exposed workers: The quest for truth between chance and necessity. *Med Lav* 98 (1):3–17.

Corrie, S. R., G. A. Lawrie, and M. Trau. 2006. Quantitative analysis and characterization of biofunctionalized fluorescent silica particles. *Langmuir* 22 (6):2731–2737.

Crookes-Goodson, W. J., J. M. Slocik, and R. R. Naik. 2008. Bio-directed synthesis and assembly of nanomaterials. *Chem Soc Rev* 37 (11):2403–2412.

Dalal, N. S., X. L. Shi, and V. Vallyathan. 1990. Role of free radicals in the mechanisms of hemolysis and lipid peroxidation by silica: Comparative ESR and cytotoxicity studies. *J Toxicol Environ Health* 29 (3):307–316.

Depasse, J. and J. Warlus. 1976. Relation between the toxicity of silica and its affinity for tetraalkylammonium groups. Comparison between SiO_2 and TiO_2. *J Colloid Interface Sci* 56 (3):618–621.

Driscoll, K. E. 2000. TNFalpha and MIP-2: Role in particle-induced inflammation and regulation by oxidative stress. *Toxicol Lett* 112–113:177–183.

Dutta, D., S. K. Sundaram, J. G. Teeguarden, B. J. Riley, L. S. Fifield, J. M. Jacobs, S. R. Addleman, G. A. Kaysen, B. M. Moudgil, and T. J. Weber. 2007. Adsorbed proteins influence the biological activity and molecular targeting of nanomaterials. *Toxicol Sci* 100 (1):303–315.

Engelhardt, G. and D. Michel. 1987. *High-Resolution Solid-State NMR of Silicates and Zeolites*. New York: John Wiley & Sons.

Epstein, E. 1994. The anomaly of silicon in plant biology. *Proc Natl Acad Sci U S A* 91:11–17.

Ferrari, M., L. Mottola, and V. Quaresima. 2004. Principles, techniques, and limitations of near infrared spectroscopy. *Can J Appl Physiol* 29 (4):463–487.

Franca, A., B. Pelaz, M. Moros, C. Sanchez-Espinel, A. Hernandez, C. Fernandez-Lopez, V. Grazu et al. 2010. Sterilization matters: Consequences of different sterilization techniques on gold nanoparticles. *Small* 6 (1):89–95.

Fubini, B. and A. Hubbard. 2003. Reactive oxygen species (ROS) and reactive nitrogen species (RNS) generation by silica in inflammation and fibrosis. *Free Radic Biol Med* 34 (12):1507–1516.

Gillette Guyonnet, S., S. Andrieu, and B. Vellas. 2007. The potential influence of silica present in drinking water on Alzheimer's disease and associated disorders. *J Nutr Health Aging* 11 (2):119–124.

Goldstein, J., D. E. Newbury, D. C. Joy, P. Echlin, C. E. Lyman, E. Lifshin, and L. Sawyer. 2003. *Scanning Electron Microscopy and X-Ray Microanalysis*. 3rd edn. New York: Kluwer Academic/Plenum Publishers.

Gross, L., F. Mohn, N. Moll, P. Liljeroth, and G. Meyer. 2009. The chemical structure of a molecule resolved by atomic force microscopy. *Science* 325:1110–1114.

Gruenhagen, J. A., C. Y. Lai, D. R. Radu, V. S. Lin, and E. S. Yeung. 2005. Real-time imaging of tunable adenosine 5-triphosphate release from an MCM-41-type mesoporous silica nanosphere-based delivery system. *Appl Spectrosc* 59 (4):424–431.

Guo, Z. X., W. F. Liu, Y. Li, and J. Yu. 2005. Grafting of poly(ethylene glycol)s onto nanometer silica surface by a one-step procedure. *J Macromol Sci Part A— Pure Appl Chem* 42:221–230.

Hamilton, R. F., Jr., S. A. Thakur, and A. Holian. 2008. Silica binding and toxicity in alveolar macrophages. *Free Radic Biol Med* 44 (7):1246–1258.

He, Q., X. Cui, F. Cui, L. Guo, and J. Shi. 2009. Size-controlled synthesis of monodispersed mesoporous silica nano-spheres under a neutral condition. *Microporous Mesoporous Mater* 117:609–616.

He, X., H. Nie, K. Wang, W. Tan, X. Wu, and P. Zhang. 2008. In vivo study of biodistribution and urinary excretion of surface-modified silica nanoparticles. *Anal Chem* 80 (24):9597–9603.

Hnizdo, E. and V. Vallyathan. 2003. Chronic obstructive pulmonary disease due to occupational exposure to silica dust: A review of epidemiological and pathological evidence. *Occup Environ Med* 60 (4):237–243.

Huang, D. M., Y. Hung, B. S. Ko, S. C. Hsu, W. H. Chen, C. L. Chien, C. P. Tsai et al. 2005. Highly efficient cellular labeling of mesoporous nanoparticles in human mesenchymal stem cells: Implication for stem cell tracking. *Faseb J* 19 (14):2014–2016.

Hudson, S. P., R. F. Padera, R. Langer, and D. S. Kohane. 2008. The biocompatibility of mesoporous silicates. *Biomaterials* 29 (30):4045–4055.

Huh, S., J. W. Wiench, J. Yoo, M. Pruski, and V. S.-Y. Lin. 2003. Organic functionalization and morphology control of mesoporous silicas via a co-condensation synthesis method. *Chem. Mater.* 15:4247–4256.

Hunter, R. J. 1988. *Zeta Potential in Colloid Science: Principles and Applications*. London, U.K.: Academic Press.

Iler, R. K. 1979. *The Chemistry of Silica*. New York: John Wiley & Sons, Inc.

Insin, N., J. B. Tracy, H. Lee, J. P. Zimmer, R. M. Westervelt, and M. G. Bawendi. 2008. Incorporation of iron oxide nanoparticles and quantum dots into silica microspheres. *ACS Nano* 2 (2):197–202.

Jin, H., D. A. Heller, and M. S. Strano. 2008. Single-particle tracking of endocytosis and exocytosis of single-walled carbon nanotubes in NIH-3T3 cells. *Nano Lett* 8 (6):1577–1585.

Jin, Y., S. Kannan, M. Wu, and J. X. Zhao. 2007. Toxicity of luminescent silica nanoparticles to living cells. *Chem Res Toxicol* 20 (8):1126–1133.

Julien, D. C., C. C. Richardson, M. F. Beaux, 2nd, D. N. McIlroy, and R. A. Hill. 2009. In vitro proliferating cell models to study cytotoxicity of silica nanowires. *Nanomedicine* 6 (1):84–91.

Karlsson, M. and U. Carlsson. 2005. Protein adsorption orientation in the light of fluorescent probes: Mapping of the interaction between site-directed labeled human carbonic anhydrase II and silica nanoparticles. *Biophys J* 88 (5):3536–3544.

Kim, J. S., W. J. Rieter, K. M. Taylor, H. An, W. Lin, and W. Lin. 2007. Self-assembled hybrid nanoparticles for cancer-specific multimodal imaging. *J Am Chem Soc* 129 (29):8962–8963.

Kleinman, M. T., J. A. Araujo, A. Nel, C. Sioutas, A. Campbell, P. Q. Cong, H. Li, and S. C. Bondy. 2008. Inhaled ultrafine particulate matter affects CNS inflammatory processes and may act via MAP kinase signaling pathways. *Toxicol Lett* 178 (2):127–130.

Klichko, Y. 2008. Mesostructured silica for optical functionality, nanomachines, and drug delivery. *J Am Ceram Soc* 92 (1, Suppl.):S2–S10.

Kneuer, C., M. Sameti, U. Bakowsky, T. Schiestel, H. Schirra, H. Schmidt, and C. M. Lehr. 2000a. A non-viral DNA delivery system based on surface modified silica-nanoparticles can efficiently transfect cells in vitro. *Bioconjugate Chem* 11 (6):926–932.

Kneuer, C., M. Sameti, E. G. Haltner, T. Schiestel, H. Schirra, H. Schmidt, and C. M. Lehr. 2000b. Silica nanoparticles modified with aminosilanes as carriers for plasmid DNA. *Int J Pharm* 196:257–261.

Knopp, D., D. Tang, and R. Niessner. 2009. Review: Bioanalytical applications of biomolecule-functionalized nanometer-sized doped silica particles. *Anal Chim Acta* 647 (1):14–30.

Kobler, J., J. Moller, and T. Bein. 2008. Colloidal suspensions of functionalized mesoporous silcia nanoparticles. *ACS Nano* 2 (4):791–799.

Kumar, R., I. Roy, T. Y. Ohulchanskyy, L. N. Goswami, A. C. Bonoiu, E. J. Bergey, K. M. Tramposch, A. Maitra, and P. N. Prasad. 2008. Covalently dye-linked, surface-controlled, and bioconjugated organically modified silica nanoparticles as targeted probes for optical imaging. *ACS Nano* 2 (3):449–456.

Kumar, R., I. Roy, T. Y. Ohulchanskkky, L. A. Vathy, E. J. Bergey, M. Sajjad, and P. N. Prasad. 2010. In vivo biodistribution and clearance studies using multimodal organically modified silica nanoparticles. *ACS Nano* 4(2):699–708.

Kwon, G. S. and K. Kataoka. 1995. Block copolymer micelles as long circulating drug vehicles. *Adv Drug Deliv Rev* 16 (2–3):295–309.

Lacasse, Y., S. Martin, D. Gagne, and L. Lakhal. 2009. Dose-response meta-analysis of silica and lung cancer. *Cancer Causes Control* 20 (6):925–933.

Lai, C. Y., B. G. Trewyn, D. M. Jeftinija, K. Jeftinija, S. Xu, S. Jeftinija, and V. S. Lin. 2003. A mesoporous silica nanosphere-based carrier system with chemically removable CdS nanoparticle caps for stimuli-responsive controlled release of neurotransmitters and drug molecules. *J Am Chem Soc* 125 (15):4451–4459.

Lanone, S., F. Rogerieux, J. Geys, A. Dupont, E. Maillot-Marechal, J. Boczkowski, G. Lacroix, and P. Hoet. 2009. Comparative toxicity of 24 manufactured nanoparticles in human alveolar epithelial and macrophage cell lines. *Part Fibre Toxicol* 6:14.

Lee, J. E., N. Lee, H. Kim, J. Kim, S. H. Choi, J. H. Kim, T. Kim et al. 2009. Uniform mesoporous dye-doped silica nanoparticles decorated with multiple magnetite nanocrystals for simultaneous enhanced magnetic resonance imaging, fluorescence imaging, and drug delivery. *J Am Chem Soc* 132 (2):552–557.

Lei, C., Y. Shin, J. Liu, and E. J. Ackerman. 2002. Entrapping enzyme in a functionalized nanoporous support. *J Am Chem Soc* 124 (38):11242–11243. 2007; Synergetic effects of nanoporous support and urea on enzyme activity. *Nano Lett* 7 (4):1050–1053.

Lenz, A. G., F. Krombach, and K. L. Maier. 1992. Oxidative stress in vivo and in vitro: Modulation by quartz dust and hyperbaric atmosphere. *Free Radic Biol Med* 12 (1):1–10.

Leroueil, P. R., S. Hong, A. Mecke, J. R. Baker, Jr., B. G. Orr, and M. M. Banaszak Holl. 2007. Nanoparticle interaction with biological membranes: Does nanotechnology present a Janus face? *Acc Chem Res* 40 (5):335–342.

Li, Z. Z., L. X. Wen, L. Shao, and J. F. Chen. 2004. Fabrication of porous hollow silica nanoparticles and their applications in drug release control. *J Control Release* 98 (2):245–254.

Limbach, L. K., Y. Li, R. N. Grass, T. J. Brunner, M. A. Hintermann, M. Muller, D. Gunther, and W. J. Stark. 2005. Oxide nanoparticle uptake in human lung fibroblasts: Effects of particle size, agglomeration, and diffusion at low concentrations. *Environ Sci Technol* 39 (23):9370–9376.

Lin, W., Y. W. Huang, X. D. Zhou, and Y. Ma. 2006. In vitro toxicity of silica nanoparticles in human lung cancer cells. *Toxicol Appl Pharmacol* 217 (3):252–259.

Liong, M., J. Lu, M. Kovochich, T. Xia, S. G. Ruehm, A. E. Nel, F. Tamanoi, and J. I. Zink. 2008. Multifunctional inorganic nanoparticles for imaging, targeting, and drug delivery. *ACS Nano* 2 (5):889–896.

Lu, Y., J. McLellan, and Y. Xia. 2004. Synthesis and crystallization of hybrid spherical colloids composed of polystyrene cores and silica shells. *Langmuir* 20 (8):3464–3470.

Mal, N. K., M. Fujiwara, and Y. Tanaka. 2003. Photocontrolled reversible release of guest molecules from coumarin-modified mesoporous silica. *Nature* 421 (6921):350–353.

Malugin, A., H. L. Herd, and H. Ghandehari. 2010. Differential toxicity of anionic silica nanoparticles towards phagocytic and epithelial cells. Submitted, 2011.

Martin, K. R. 2007. The chemistry of silica and its potential health benefits. *J Nutr Health Aging* 11 (2):94–97.

McLean, S. and B. B. Sauer. 1997. Tapping-mode AFM studies using phase detection for resolution of nanophases in segmented polyurethanes and other block copolymers. *Macromolecules* 30 (26):8314–8317.

Mitragotri, S. and J. Lahann. 2009. Physical approaches to biomaterial design. *Nat Mater* 8 (1):15–23.

Mizutani, T., H. Nagase, N. Fujiwara, and H. Ogoshi. 1998. Silicic acid polymerization catalyzed by amines and polyamines. *Bull Chem Soc Jpn* 71 (8):2017–2022.

Murashov, V., M. Harper, and E. Demchuk. 2006. Impact of silanol surface density on the toxicity of silica aerosols measured by erythrocyte haemolysis. *J Occup Environ Hyg* 3 (12):718–723.

Naik, S. P. and I. Sokolov. 2007. Room temperature synthesis of nanoporous silica spheres and their formation mechanism. *Solid State Commun* 144:437–440.

Nan, A., X. Bai, S. J. Son, S. B. Lee, and H. Ghandehari. 2008. Cellular uptake and cytotoxicity of silica nanotubes *Nano Lett* 8 (8):2150–2154.

Nandiyanto, A. B. D., S. G. Kim, F. Iskandar, and K. Okuyama. 2009. Synthesis of spherical mesoporous silica nanoparticles with nanometer-size controllable pores and outer diameters. *Microporous Mesoporous Mater* 120 (3):447–453.

Nelson, S. M., T. Mahmoud, M. Beaux, 2nd, P. Shapiro, D. N. McIlroy, and D. L. Stenkamp. 2009. Toxic and teratogenic silica nanowires in developing vertebrate embryos. *Nanomedicine* 6 (1):93–102.

Nemmar, A., M. F. Hoylaerts, P. H. Hoet, and B. Nemery. 2004. Possible mechanisms of the cardiovascular effects of inhaled particles: Systemic translocation and prothrombotic effects. *Toxicol Lett* 149 (1–3):243–253.

Nguyen, T. D., K. C. Leung, M. Liong, C. D. Pentecost, J. F. Stoddart, and J. I. Zink. 2006. Construction of a pH-driven supramolecular nanovalve. *Org Lett* 8 (15):3363–3366.

Nguyen, T. D., H. R. Tseng, P. C. Celestre, A. H. Flood, Y. Liu, J. F. Stoddart, and J. I. Zink. 2005. A reversible molecular valve. *Proc Natl Acad Sci U S A* 102 (29):10029–10034.

Nuraje, N., K. Sub, and H. Matsui. 2007. Catalytic growth of silica nanoparticles in controlled shapes at planar liquid/liquid interfaces. *New J Chem* 31:1895–1898.

Obare, S. O., N. R. Jana, and C. J. Murphy. 2001. Preparation of polystyrene- and silica-coated gold nanorods and their use as templates for the synthesis of hollow nanotubes. *Nano Lett* 1 (11):601–603.

Ostro, M. J. 1987. *Liposomes: From Biophysics to Therapeutics*. New York: Marcel Dekker Inc.

Ovrevik, J., M. Refsnes, E. Namork, R. Becher, D. Sandnes, P. E. Schwarze, and M. Lag. 2006. Mechanisms of silica-induced IL-8 release from A549 cells: Initial kinase-activation does not require EGFR activation or particle uptake. *Toxicology* 227 (1–2):105–116.

Park, J. H., L. Gu, G. von Maltzahn, E. Ruoslahti, S. N. Bhatia, and M. J. Sailor. 2009. Biodegradable luminescent porous silicon nanoparticles for in vivo applications. *Nat Mater* 8 (4):331–336.

Park, E. J. and K. Park. 2009. Oxidative stress and pro-inflammatory responses induced by silica nanoparticles in vivo and in vitro. *Toxicol Lett* 184 (1):18–25.

Patel, K., S. Angelos, W. R. Dichtel, A. Coskun, Y. W. Yang, J. I. Zink, and J. F. Stoddart. 2008. Enzyme-responsive snap-top covered silica nanocontainers. *J Am Chem Soc* 130 (8):2382–2383.

Perry, C. C. 2009. An overview of silica in biology: Its chemistry and recent technological advances. *Prog Mol Subcell Biol* 47:295–313.

Qhobosheane, M., S. Santra, P. Zhang, and W. Tan. 2001. Biochemically functionalized silica nanoparticles. *Analyst* 126 (8):1274–1278.

Radin, S., P. Ducheyne, T. Kamplain, and B. H. Tan. 2001. Silica sol–gel for the controlled release of antibiotics. I. Synthesis, characterization, and in vitro release. *J Biomed Mater Res* 57 (2):313–320.

Radu, D. R., C. Y. Lai, K. Jeftinija, E. W. Rowe, S. Jeftinija, and V. S. Lin. 2004a. A polyamidoamine dendrimer-capped mesoporous silica nanosphere-based gene transfection reagent. *J Am Chem Soc* 126 (41):13216–13217.

Radu, D. R., C. Y. Lai, J. W. Wiench, M. Pruski, and V. S. Lin. 2004b. Gatekeeping layer effect: A poly(lactic acid)-coated mesoporous silica nanosphere-based fluorescence probe for detection of amino-containing neurotransmitters. *J Am Chem Soc* 126 (6):1640–1641.

Rahman, I. A., P. Vejayakumaran, C. S. Sipaut, J. Ismail, and C. K. Chee. 2009. Size-dependent physiochemical and optical properties of silica nanoparticles. *Mater Chem Phys* 114:328–332.

Richmond, K. E. and M. Sussman. 2003. Got silicon? The non-essential beneficial plant nutrient. *Curr Opin Plant Biol* 6 (3):268–272.

Rimal, B., A. K. Greenberg, and W. N. Rom. 2005. Basic pathogenetic mechanisms in silicosis: Current understanding. *Curr Opin Pulm Med* 11 (2):169–173.

Roy, I., T. Y. Ohulchanskyy, D. J. Bharali, H. E. Pudavar, R. A. Mistretta, N. Kaur, and P. N. Prasad. 2005. Optical tracking of organically modified silica nanoparticles as DNA carriers: A nonviral, nanomedicine approach for gene delivery. *Proc Natl Acad Sci U S A* 102 (2):279–284.

Sahin, K., M. Onderci, N. Sahin, T. A. Balci, M. F. Gursu, V. Juturu, and O. Kucuk. 2006. Dietary arginine silicate inositol complex improves bone mineralization in quail. *Poult Sci* 85 (3):486–492.

Sameti, M., G. Bohr, M. N. Ravi Kumar, C. Kneuer, U. Bakowsky, M. Nacken, H. Schmidt, and C. M. Lehr. 2003. Stabilisation by freeze-drying of cationically modified silica nanoparticles for gene delivery. *Int J Pharm* 266 (1–2):51–60.

Satishkumar, B. C., S. K. Doorn, G. A. Baker, and A. M. Dattelbaum. 2008. Fluorescent single walled carbon nanotube/silica composite materials. *ACS Nano* 2 (11):2283–2290.

Shang, W., J. H. Nuffer, J. S. Dordick, and R. W. Siegel. 2007. Unfolding of ribonuclease A on silica nanoparticle surfaces. *Nano Lett* 7 (7):1991–1995.

Sharma, P., S. C. Brown, N. Bengtsson, Q. Zhang, G. A. Walter, S. R. Grobmyer, S. Santra, H. Jiang, E. W. Scott, and B. M. Moudgil. 2008. Gold-speckled multimodal nanoparticles for noninvasive bioimaging. *Chem Mater* 20 (19):6087–6094.

Sharma, P., S. Brown, G. Walter, S. Santra, and B. Moudgil. 2006. Nanoparticles for bioimaging. *Adv Colloid Interface Sci* 123–126:471–485.

Slowing, II, J. L. Vivero-Escoto, C. W. Wu, and V. S. Lin. 2008. Mesoporous silica nanoparticles as controlled release drug delivery and gene transfection carriers. *Adv Drug Deliv Rev* 60 (11):1278–1288.

Slowing, II, C. W. Wu, J. L. Vivero-Escoto, and V. S. Lin. 2009. Mesoporous silica nanoparticles for reducing hemolytic activity towards mammalian red blood cells. *Small* 5 (1):57–62.

Son, S. J., X. Bai, A. Nan, H. Ghandehari, and S. B. Lee. 2006. Template synthesis of multifunctional nanotubes for controlled release. *J Control Release* 114 (2):143–152.

Stayton, I., J. Winiarz, K. Shannon, and Y. Ma. 2009. Study of uptake and loss of silica nanoparticles in living human lung epithelial cells at single cell level. *Anal Bioanal Chem* 394 (6):1595–1608.

Stöber, W., A. Fink, and E. Bohn. 1968. Controlled growth of monodisperse silica spheres in the micro size range. *J Colloid Interface Sci* 26:62–69.

Stromme, M., U. Brohede, R. Atluri, and A. E. Garcia-Bennett. 2009. Mesoporous silica-based nanomaterials for drug delivery: Evaluation of structural properties associated with release rate. *Wiley Interdiscip Rev Nanomed Nanobiotechnol* 1 (1):140–148.

Sun, W., N. Fang, B. G. Trewyn, M. Tsunoda, Slowing, II, V. S. Lin, and E. S. Yeung. 2008. Endocytosis of a single mesoporous silica nanoparticle into a human lung cancer cell observed by differential interference contrast microscopy. *Anal Bioanal Chem* 391 (6):2119–2125.

Tan, A., S. Simovic, A. K. Davey, T. Rades, B. J. Boyd, and C. A. Prestidge. 2010. Silica nanoparticles to control the lipase-mediated digestion of lipid-based oral delivery systems. *Mol Pharm* 7 (2):522–532.

Tasciotti, E., X. Liu, R. Bhavane, K. Plant, A. D. Leonard, B. K. Price, M. M. Cheng et al. 2008. Mesoporous silicon particles as a multistage delivery system for imaging and therapeutic applications. *Nat Nanotechnol* 3 (3):151–157.

Thierry, B., L. Zimmer, S. McNiven, K. Finnie, C. Barbe, and H. J. Griesser. 2008. Electrostatic self-assembly of PEG copolymers onto porous silica nanoparticles. *Langmuir* 24:8143–8150.

Trewyn, B. G., J. A. Nieweg, Y. Zhao, and V. S.-Y. Lin. 2008. Biocompatible mesoporous silica nanoparticles with different morphologies for animal cell membrane penetration. *Chem Eng J* 137 (1):23–29.

Trewyn, B. G., I. I. Slowing, S. Giri, H.-T. Chen, and V. S.-Y. Lin. 2007. Synthesis and functionalization of a mesoporous silica nanoparticle based on the sol–gel process and applications in controlled release. *Acc Chem Res* 40:846–853.

Trewyn, B. G., C. M. Whitman, and V. S. Lin. 2004. Morphological control of room-temperature ionic liquid templated mesoporous silica nanoparticles for controlled release of antibacterial agents. *Nano Lett* 4 (11):2139–2143.

van Schooneveld, M. M., E. Vucic, R. Koole, Y. Zhou, J. Stocks, D. P. Cormode, C. Y. Tang et al. 2008. Improved biocompatibility and pharmacokinetics of silica nanoparticles by means of a lipid coating: A multimodality investigation. *Nano Lett* 8 (8):2517–2525.

Verma, A., O. Uzun, Y. Hu, Y. Hu, H. S. Han, N. Watson, S. Chen, D. J. Irvine, and F. Stellacci. 2008. Surface-structure-regulated cell-membrane penetration by monolayer-protected nanoparticles. *Nat Mater* 7 (7):588–595.

Wang, F., F. Gao, M. Lan, H. Yuan, Y. Huang, and J. Liu. 2009a. Oxidative stress contributes to silica nanoparticle-induced cytotoxicity in human embryonic kidney cells. *Toxicol In Vitro* 23 (5):808–815.

Wang, W., B. Gu, L. Liang, and W. Hamilton. 2003. Fabrication of two and three-dimensional silica nanocolloidal particle arrays. *J Phys Chem* 107 (15):3400–3404.

Wang, L., A. Reis, A. Seifert, T. Philippi, S. Ernst, M. Jia, and W. R. Thiel. 2009b. A simple procedure for the covalent grafting of triphenylphosphine ligands on silica: Application in the palladium catalyzed Suzuki reaction. *Dalton Trans* (17):3315–3320.

Wang, Q. and D. F. Shantz. 2008. Ordered mesoporous silica-based inorganic nanocomposites. *J Solid State Chem* 181:1659–1669.

Waters, K. M., L. M. Masiello, R. C. Zangar, B. J. Tarasevich, N. J. Karin, R. D. Quesenberry, S. Bandyopadhyay, J. G. Teeguarden, J. G. Pounds, and B. D. Thrall. 2009. Macrophage responses to silica nanoparticles are highly conserved across particle sizes. *Toxicol Sci* 107 (2):553–569.

Wilbourn, J. D., D. B. McGregor, C. Partensky, and J. M. Rice. 1997. IARC reevaluates silica and related substances. *Environ Health Perspect* 105 (7):756–759.

Williams, D. B. and C. Barry Carter. 2009. *Transmission Electron Microscopy: A Textbook for Materials Science.* 2nd edn. New York: Plenum Publishing Corporation.

Xing, X., X. He, J. Peng, K. Wang, and W. Tan. 2005. Uptake of silica-coated nanoparticles by HeLa cells. *J Nanosci Nanotechnol* 5 (10):1688–1693.

Xu, H., F. Yan, E. E. Monson, and R. Kopelman. 2003. Room-temperature preparation and characterization of poly (ethylene glycol)-coated silica nanoparticles for biomedical applications. *J Biomed Mater Res A* 66 (4):870–879.

Yu, K. O., C. M. Grabinski, A. M. Schrand, R. C. Murdock, W. Wang, B. Gu, J. J. Schlager, and S. M. Hussain. 2009. Toxicity of amorphous silica nanoparticles in mouse ketatinocytes. *J Nanopart Res* 11(1): 15–24.

Zhang, Q., F. Lu, C. Li, Y. Wang, and W. Huilin. 2006. An efficient synthesis of helical mesoporous silica nanorods. *Chem Lett* 35 (2):190–191.

7

Peptide-Based Self-Assembled Nanofibers for Biomedical Applications

Joel M. Anderson
University of Alabama at Birmingham

Meenakshi Kushwaha
University of Alabama at Birmingham

Dong Jin Lim
University of Alabama at Birmingham

Ho-Wook Jun
University of Alabama at Birmingham

7.1 Introduction

The most promising paradigm for regenerative medicine is to engineer a nanostructured environment that mimics the complex hierarchical order and self-assembled formation of native tissue, as opposed to trying to adopt traditional materials to a biomedical need. This approach is emphasized by the ongoing research of biomimetic peptide scaffolds that employ a bottom-up tissue engineering strategy. To capture the self-assembling complexity required, bioactive scaffolds need to emulate the intrinsic nanoscale properties of the desired tissue and surrounding extracellular matrix (ECM). The ECM is a viscoelastic three-dimensional (3D) network consisting of nanofibrillar proteins and polysaccharides that self-assembles into complex supramolecular structures and serves as a critical component in the development and maturation of tissues (Patrick et al. 1998; Hubbell 2003; Daley et al. 2008). In particular, cell–ECM interactions can be recapitulated to directly regulate cell behaviors, such as cell proliferation, growth, survival, polarity, morphology, migration, and differentiation (Kleinman et al. 2003). With this approach, self-assembled nanomaterial constructs can be tailored to precise tissue regenerative needs and other biomedical applications by incorporating specific functional peptide moieties, such as cellular adhesive ligands and enzyme-mediated degradable sites. Furthermore, versatile biomimetic microenvironments can be created by introducing hybrid functionality into the self-assembling peptide (SAP) scaffolds. To this end, peptide-based biomaterials can potentially be combined with other polymers, metals, nanotubes, or growth factors to improve mechanical stability, direct cellular responses, guide degradation, or deliver therapeutic drugs in controlled manners. Overall, this chapter offers only a glimpse into the vast amount of knowledge available. Thus, only a few peptide-based biomaterials have been selected for a more in-depth examination. The focus is on self-assembling nanofibrous structures that have the potential to mediate biological activity and other biomedical applications by closely following the principles of naturally derived phospholipids, molecules critical for the structural stability of membranes in biological systems (Shimizu et al. 2005). The proceeding discussion first starts by describing the self-assembly process of phospholipids, followed by synthetic peptide-based biomaterials developed to mimic this natural self-assembly mechanism.

7.2 Self-Assembly of Phospholipids

Phospholipids occur in nature as an important class of biomolecules. Structurally, they consist of three components—a polar head, one or more hydrophobic tails, and a backbone linking the two parts. Given the versatility of the head and tail regions, lipids are classified based on their backbone. The amphiphilicity of lipids drives their self-assembly in solution. They can assume several shapes based on their structure, concentration, and temperature. Common lamellar and non-lamellar self-assembled structures are shown in Figure 7.1 (Collier and Messersmith 2001). Lamellar bilayers are formed when the hydrophobic alkyl chains are too bulky to fit within a circular micelle, otherwise a non-lamellar conformation is assumed.

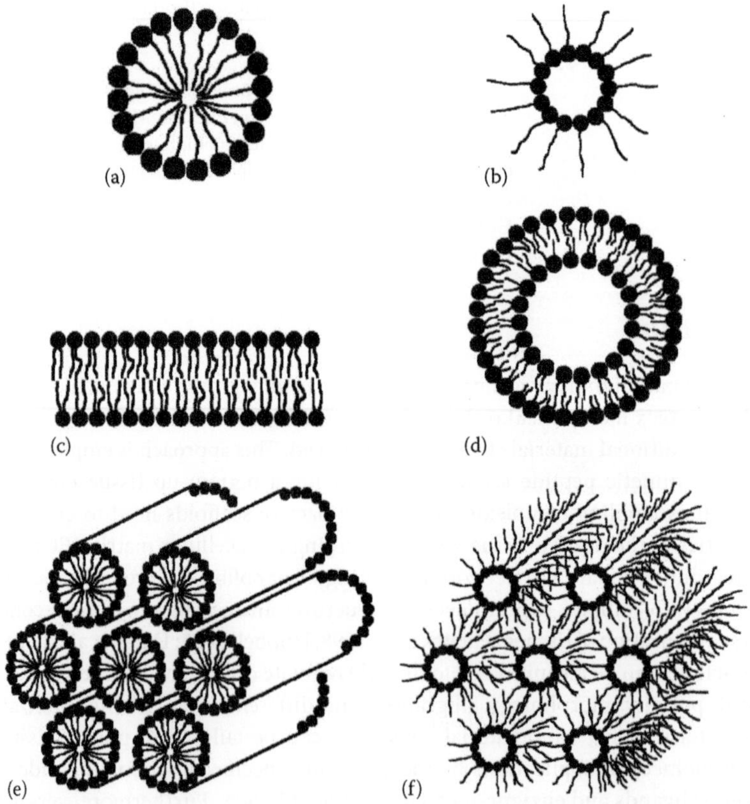

FIGURE 7.1 Common lamellar and non-lamellar self-assembled structures of lipids: (a) micelle, (b) inverse micelle, (c) lamellar bilayer, (d) bilayer vesicle, (e) hexagonal, and (f) inverse hexagonal. (Reprinted from Collier, J.H. and Messersmith, P.B., *Annu. Rev. Mater. Res.*, 31, 237, 2001. With permission. © 2001 by Annual Reviews www.annualreviews.org.)

Due to their inherent biocompatibility and capacity to form self-assembled compartment or layered structures, phospholipids have emerged as attractive candidates for biomedical applications, such as vesicles for drug delivery, tubule and ribbon structures to make scaffolds for tissue engineering, and monolayer or bilayer membrane-like materials for biocompatible coatings of medical devices and implants (Schnur 1993; von Segesser et al. 1993; Gregoriadis 1995). Among the potential functions, considerable efforts have been devoted to developing stimulus-responsive lipid vesicles for site-specific controlled drug delivery. Various stimuli methods for the lipid vesicles or liposomes have been investigated, including temperature, light, and pH change (Thompson et al. 1996; Gerasimov et al. 1999). Messersmith and coworkers have investigated the temperature-responsive approach to engineer therapeutic delivery systems, as chemical reactions between the encapsulated species and the extravesicular species were driven by ambient change in temperature (Messersmith and Starke 1998; Sanborn et al. 2002; Pederson et al. 2003; Burke et al. 2007). The entrapped substances were released from the liposomes at the melting transition temperature (T_m) of the lipid chains. Exploring this temperature sensitive mechanism, the permeability of the bilayers was found to be significantly enhanced at the melting transition temperature, T_m. The bilayers maintained a liquid state below their T_m value but shifted to a gelatinous phase as the temperature increased above the T_m. Both of these states expressed low permeability. However, the permeability of the lipid bilayers was several order of magnitudes higher during the transition between phases, owing to the existence of defect-rich interfacial regions between coexisting gel and fluid domains (Figure 7.2) (Honger et al. 1997).

Applying this methodology to clinical treatments, saturated phosphatidylcholines have been used to make vesicles with a T_m near 37°C to facilitate release of calcium upon injection into the body cavity. The exact T_m value can be determined by the chain length of the phosphatidylcholines. Hence, by selecting miscible phospholipids of appropriate chain length, the T_m of the bilayer has been tailored to fall within 23°C–41°C. This strategy has been used to elicit calcium phosphate mineralization, as the thermally responsive liposomes can be activated to release calcium that reacts with extravesicular phosphate when subjected to physiologic temperatures (Messersmith and Starke 1998).

The liposome strategy has also been used to trigger rapid in situ formation of polymer hydrogels in response to thermal or photochemical stimuli. In this approach, liposomes reacted with $CaCl_2$ were combined with 16 amino acid SAPs called FEK16. (SAPs are discussed in more detail in the following section.) This composite system was created by dispersing liposomes in a low viscosity solution of FEK16, allowing for the long-term maintenance of a fluid state at ambient temperatures. Upon heat activation after exposure to body temperature (37°C) or infrared light (NIR excitation, 800 nm), $CaCl_2$ is released from the liposomes within the composite system (Thompson et al. 1996). The released $CaCl_2$ triggers salt-dependent self-assembly of FEK16, thereby giving rise to a polymer hydrogel. Thus, these suspensions can be stored as stable fluid precursors at room temperature, but they rapidly form polymer hydrogels when induced at physiological conditions (Messersmith and Starke 1998; Collier and Messersmith 2001; Collier et al. 2001; Westhaus and Messersmith 2001).

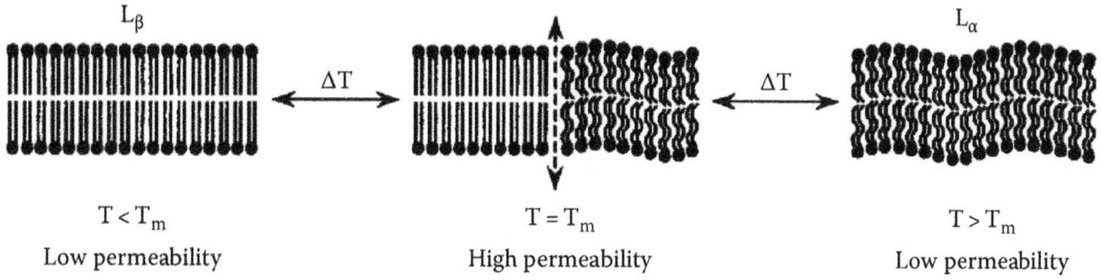

FIGURE 7.2 Schematic illustration of effect of temperature on phospholipid bilayer. (Reprinted from Collier, J.H. and Messersmith, P.B., *Annu. Rev. Mater. Res.*, 31, 237, 2001. With permission. © 2001 by Annual Reviews www.annualreviews.org.)

7.3 Self-Assembling Peptides

Over the past decade Zhang and coworkers have significantly contributed to the field of biologically inspired SAPs that follow some of the same principles as natural phospholipid self-assembly. EAK16-II was the first member of this family and was discovered in the yeast protein, Zuotin (Zhang et al. 1992). Since this initial discovery, a number of peptides have been added to the group (Table 7.1) (Zhang 2002). These SAPs are characterized by an alternating sequence of hydrophobic and hydrophilic residues, as the hydrophilic residues alternate in turn between a positive and a negative charge. Self-assembly is spontaneous, and the peptides are held together by various ionic and nonionic, hydrophobic, and van der Waals interactions (Whitesides et al. 1991; Zhang 2002). Four different types of SAPs have been investigated, differing in their charge distribution and resulting in secondary and tertiary self-assembled structures.

7.3.1 Type I Self-Assembling Peptides

Type I SAPs are characterized by the presence of both a hydrophobic and hydrophilic composite face, which leads to β-sheet formations in aqueous solution. They are also termed as "molecular lego" structures due to their striking similarity to Lego bricks, as they have "pegs and holes" and can only assemble into particular structures at the molecular level. Variations can be made within the peptide sequence to increase the size of the "pegs" and the "holes," producing such sequences as RARADADA and RARARADADADA. Due to their hydrophilic surfaces, SAPs are known to form complementary ionic bonds consisting of regular repeating peptide blocks. The ionic bond arrangements can follow various patterns and serve as the basis for classifying the SAPs into different electrically charged groups (Table 7.1). For example, the molecules within the Type I class have positively (+) and negatively (−) charged amino acids repeating as + − + − + −. Similarly, Type II class SAPs will have amino acids arranged as + + − − + + − − and so on (Zhang et al. 1993).

The alternating charge within SAPs drives the nanofibrous self-assembly when subjected to the right stimuli. Self-assembly of these SAPs into nanofibers can be triggered by exposing the peptides to physiological media or monovalent alkaline cations. This creates a bulk mesh of individually assembled fibers that are typically about 10–20 nm in diameter and up to a few microns in length, as determined by scanning electron microscopy (SEM) and atomic force microscopy (AFM). Pores are prevalent throughout the interwoven fibrillar meshwork and are usually on the order of 50–200 nm, which is the same scale as many vital biomolecules. This size scale is conducive for diffusing biological molecules, along with the subsequent creation of a concentration gradient for programmable drug delivery applications. The density of the nanofibers assembled can be easily controlled, depending on the employed concentration of the peptide solution. Overall, the nanofiber meshwork is very strong as peptide self-assembly is stable over a wide range of pH values, temperatures, and denaturing agents (e.g., urea guanidium hydrochloride) (Zhang et al. 1995; Leon et al. 1998; Caplan et al. 2000; Holmes et al. 2000).

In order to understand the self-assembly of these SAPs, a proposed model for complementary molecular pairing between positively charged lysines and negatively charged glutamates has been described by Zhang et al. (1995). It was found that replacing the charged residues with other amino acids of similar charges does not significantly affect the self-assembly process, as neither the replacement of a positively charged lysine with a positively charged arginine nor a negatively charged glutamate with a negatively charged aspartate had any bearing on the assembled structures. However, replacing the amino acids with the residues of the opposite charge prevented self-assembly. For example, self-assembly cannot occur after substituting a positively charged lysine with a negatively charged glutamate, even though β-sheet structures can still form when exposed to cations (Caplan et al. 2000). Furthermore, enhancing the hydrophobicity of the peptides by replacing an alanine with more hydrophobic residues (e.g., leucine, isoleucine, phenylalanine) helps to promote faster self-assembly and results in improved mechanical strength (Leon et al. 1998). These self-assembled nanofibers become fragmented when subjected to sonication but are able to reassemble after removing the disruptive forces. The kinetics of reassembly

TABLE 7.1 List of Self-Assembling Peptides Studied

Name	Sequence (N→C)	Ionic Modulus	Structure
RADA16-I	+ − + − + − + − n-RADARADARADARADA-c	I	Beta
RGDA16-I	+ − + − + − + − n-RADARGDARADARGDA-c	I	r.c.
RADA8-I	+ − + − n-RADARADA-c	I	r.c.
RAD16-II	+ + − − + + − − n-RARADADARARADADA-c	II	Beta
RAD8-II	+ + − − n-RARADADA-c	II	r.c.
EAKA16-I	− + − + − + − + n-AEAKAEAKAEAKAEAK-c	I	Beta
EAKA8-I	− + − + n-AEAKAEAK-c	I	r.c.
RAEA16-I	+ − + − + − + − n-RAEARAEARAEARAEA-c	I	Beta
RAEA8-I	+ − + − n-RAEARAEA-c	I	r.c.
KADA16-I	+ − + − + − + − n-KADAKADAKADAKADA-c	I	Beta
KADA8-I	+ − + − n-KADAKADA-c	I	r.c.
EAH16-II	− − + + − − + + n-AEAEAHAHAEAEAHAH-c	II	Beta
EAH8-II	− − + + n-AEAEAHAH-c	II	r.c.
EFK16-II	− − + + − − + + n-FEFEFKFKFEFEFKFK-c	II	Beta
EFK12-I	− + − + − + n-FEFKFEFKFEFK-c	I	Beta
EFK8-II	− + − + n-FEFKFEFK-c	I	Beta
ELK16-II	− − + + − − + + n-LELELKLKLELELKLK-c	II	Beta
ELK8-II	− − + + n-LELELKLK-c	II	Beta
EAK16-II	− − + + − − + + n-AEAEAKAKAEAEAKAK-c	II	Beta
EAK12	− − − − + + n-AEAEAEAEAKAK-c	IV/II	Beta/alpha
EAK8-II	− − + + n-AEAEAKAK-c	II	r.c.
KAE16-IV	+ + + + − − − − n-KAKAKAKAEAEAEAEA-c	IV	Beta
EAK16-IV	− − − − + + + + n-AEAEAEAEAKAKAKAK-c	IV	Beta

(continued)

TABLE 7.1 (continued) List of Self-Assembling Peptides Studied

Name	Sequence (N→C)	Ionic Modulus	Structure
KLD12-I	+ − + − + − n-KLDLKLDLKLDL-c	I	Beta
KLE12-I	+ − + − + − n-KLELKLELKLEL-c	I	Beta
RAD16-IV	+ + + + − − − − n-RARARARADADADADA-c	IV	Beta
DAR16-IV	− − − − + + + + n-ADADADADARARARAR-c	IV	Beta/alpha
DAR16-IV[a]	− − − − + + + + n-DADADADARARARARA-c	IV	Beta/alpha
DAR32-IV	− − − − + + + + n-(ADADADADARARARAR)-c	IV	Beta/alpha
EHK16	+ − + − + + + + + − + − + + + + n-HEHEHKHKHEHEHKHK-c	N/A	r.c.
EHK8-I	+ − + − + + + + n-HEHEHKHK-c	N/A	r.c.
VE20[a] (NaCl)	− − − − − − − − − − n-VEVEVEVEVEVEVEVEVEVE-c	N/A	Beta
RF20[a] (NaCl)	+ + + + + + + + + + n-RFRFRFRFRFRFRFRFRFRF-c	N/A	Beta

Source: Adapted from *Biotechnol. Adv.*, 20(5–6), Zhang, S., Emerging biological materials through molecular self-assembly, 321–339, Copyright (2002), with permission from Elsevier.

Beta, β sheet; alpha-helix; r.c., random coil; N/A, not applicable. The numbers that follow the name denote the length of the peptides.

[a] Both VE20 and RF20 are in β-sheet form when they are incubated in solution containing NaCl.

heavily depend on time and are best explained by the sliding diffusion model (Yokoi et al. 2005). On the charged face of the peptide, both positive and negative charges are packed together through intermolecular ionic interactions in a checkerboard pattern. When fragments of the nanofibers first meet, the hydrophobic sides may not fit perfectly together, creating gaps in the formations. However, nonspecific hydrophobic interactions permit each nanofiber to slide along the cylindrical axis in either direction, which minimizes the exposure of hydrophobic residues and eventually seals any gaps.

These SAPs have been extensively tested with different types of mammalian cells to evaluate cellular attachment and growth behaviors. Zhang et al. have systematically studied the adhesion and differentiation behavior of neural stem cells (NSCs) on RAD16-I and compared the results to several naturally derived materials, including collagen, fibronectin, and synthetic polymers (e.g., poly(lactic acid), poly(lactic-co-glycolic acid)). The RADA16-I scaffold was found to support NSC survival and elicit differentiation to a similar degree as other synthetic biomaterials (Gelain et al. 2007). The follow-up evaluation investigated the ability of SAPs to encapsulate cells and ensure viability. Specifically, Zhang et al. explored the utility of these nanofibers to provide a suitable growth environment for endothelial cells (Davis et al. 2005). In this study, 1% RAD16-II peptides were injected into the left heart ventricle of adult mice and established as 3D microenvironments. The recipient hearts were excised at different time points, and hematoxylin and eosin staining of the fixed sections distinguished the synthetic microenvironment from the surrounding tissues. The sectioned peptide scaffolds were found to be populated with both endothelial and smooth muscle cells within 2 weeks. The implanted peptide matrix also recruited α-sacromeric actin-positive cells that are responsible for developing monocytes, as indicated by positive staining for the NKX2.5. transcription factor. Conversely, Matrigel was injected as the control, and minimal penetration of endothelial cells was observed, along with no evidence of putative monocyte precursors. In addition, several therapeutic studies with Type I SAPs have been conducted with other in vivo animal models. Primary rat hippocampal neurons were shown to form fully functional synapses on the peptide scaffolds, indicating the support of neurite outgrowth and active synaptic transmission (Holmes et al. 2000). The peptide scaffolds have been injected into the optical nerve area of severed mice brains and found to promote healing, as the incision was sealed after 2 days (Ellis-Behnke et al. 2006). Bovine chondrocytes seeded within the peptide hydrogel have been shown to deposit cartilage-like ECM within 4 weeks, signifying potential application as a treatment for cartilage tissue repair (Kisiday et al. 2002).

The resourcefulness of SAPs has also been expanded to incorporate cell-specific sequences that promote cell–cell and cell–tissue interactions. Many different designer SAPs inscribed with biologically inspired peptide sequences have been synthesized by Zhang et al. Two examples of biologically functionalized SAPs created by Zhang et al. include RADA-16-Bone Marrow Homing Peptide-I (BMHP1) and RADA16-BMHP2. Both the BMHP1 and BMHP2 SAPs contain cell-specific signals for osteogenic tissue. The addition of these functional motifs did not interfere with the self-assembly of RAD-16, indicating that all SAPs have the potential to be functionalized without any adverse effects. Also, the density of the presented peptide sequences in RADA-16 can be easily controlled as different ratios can be incorporated into the self-assembled nanofiber scaffolds. For cell encapsulation with these biologically functionalized SAPs, the peptide scaffolds were found to significantly enhance adult mouse neuronal stem cell survival after 7 days without external growth factors added to the cell culture media (Gelain et al. 2006). The level of induced neuronal differentiation promoted by these designer scaffolds was very similar to the Matrigel positive control.

7.3.2 Type II Self-Assembling Peptides

Type II SAPs depart from the long-held assumption that assembled secondary peptide structures maintain long-term stability. These peptides have also been described as "molecular switches" because of their predisposition to abruptly transform their secondary molecular structure from α-helix to β-sheet form in response to temperature or pH changes. Similar to Type I SAPs, they have distinct hydrophobic and hydrophilic faces that promote β-sheet assembly (Zhang et al. 1993). A distinguishing feature of SAPs in this class is the clustering of negatively charged residues (e.g., aspartic acid, glutamic acid) toward the N terminus, while positively charged amino acids cluster (e.g., arginine, lysine) toward the C terminus. This distribution of charge balances the C→N dipole moment and facilitates formation of α-helices (Aurora and Rose 1998).

The transition from β-sheets to α-helices and vice versa is usually abrupt and depends on ambient temperature or pH. For example, a 16 residue self-complementary oligopeptide developed by Zhang et al. called DAR16-IV has a β-sheet structure at room temperature that is 5 nm long, but its self-assembled formation undergoes an abrupt structural transition when heated to 60°C to form a stable α-helix with a 2.3 nm length (Zhang and Rich 1997). The temperature at which this transition takes place depends on the peptide sequence. Peptides with more stable β-sheet structures will have to be heated to higher temperatures to transform into α-helices. Once formed, it takes weeks for the α-helices to revert back to the initial β-sheet form. Similarly, adjusting the pH can induce structural transformations, depending on the charge expressed by the constituent amino acids (Altman et al. 2000).

Notably, the slow conversion of α-helix to β-sheet form is similar to the conformational changes implicated in neurological disorders like Alzheimer's disease (Zhang and Rich 1997). These findings offer many potential applications for Type II SAPs, including the study of protein–protein interactions and protein foldings in normal physiology or diseased states, such as Parkinson's or Alzheimer's disorders. These peptides could also be used for other biomedical applications, such as designing peptide biosensors that rapidly respond to in vivo or in vitro ambient pH or temperature changes. Biosensors with such features could potentially be developed as personalized diagnostic devices.

7.3.3 Type III Self-Assembling Peptides

The Type III peptides are designed to self-assemble onto surfaces rather than among themselves, functioning as "molecular paint" or "molecular Velcro" (Zhang 2008). They can be used to form monolayers onto different surfaces, providing recognition or interactive sites that promote the attachment of specific cell types or other molecules. In general, this SAP class has three components, namely, a ligand, an anchor group to facilitate binding to different surfaces, and a linker to connect the ligand and the anchor (Zhang et al. 1999). The ligand can be varied according to the target cell or molecule desired. The peptide anchor displays a specific chemical group that is designed to react with the desired surface and create a self-assembled coating. Besides serving as a connector, the linker also provides versatile mechanical control, as it can be modified to provide flexibility or stiffness depending on the choice of amino acids. For example, incorporating a glycine backbone sequence into the linker segment provides a more flexible structure, as opposed to a chain of valine that imparts a stiff connection (Mrksich et al. 1996).

While investigating Type III SAPs, Zhang et al. designed the following peptide assemblies: RADSRADS and RADSRADSRADS, which were linked to cysteine anchors at the C-terminus by connection sequences of 3–5 alanines (Prieto et al. 1993). Both of these SAPs were self-assembled as monolayers onto micropatterned gold-coated surfaces via the thiol groups in the cysteine anchors. Mammalian cells were seeded on these patterned monolayer surfaces and they aligned in a well-defined manner, reflecting the presence or absence of cell adhesion motifs (Zhang et al. 1999). This simple system can be used to address many questions regarding specific cell–cell and cell–tissue biological interactions. Furthermore, using the cell-responsive ligand as a molecular hook, "intelligent" diagnostic devices can be developed for surface molecular detection.

7.3.4 Type IV Self-Assembling Peptides

This last class of SAPs is designed to mimic the properties of polymeric and lipid surfactant molecules. For this class, the peptide structure is amphiphilic, as the leading head group is composed of at least one charged amino acid followed by a string of six identical hydrophobic amino acids (e.g., alanine, valine) to form the hydrophobic tail. Both the cationic and anionic amphiphiles self-assemble into tubular morphologies at neutral pH. Specifically, it is proposed that the peptides first assemble into bilayers to sequester the hydrophobic tails from an aqueous solution, followed by formation of higher order structures, such as tubular or vesicle formations, that are facilitated by hydrogen bonding between adjacent units (Santoso et al. 2002). Interestingly, when the anionic SAPs are folded into the aggregates, they do not seem to resemble the typical β-sheet or α-helix assemblies. Instead, the folding displays an unusual confirmation of an unknown nature. For the cationic systems, the pH of the surrounding environment is critical because if the pH exceeds the pI value of the head group segment, the assembled tubes collapse into membranous sheets, no longer producing well-defined nanostructures. The anionic systems have only been studied at neutral pH and possess a charged head group in all cases (Maltzahn et al. 2003). By incorporating molecular recognition sites, these vesicles and tubes can be utilized for drug delivery to specific cell types. Additionally, the peptide backbone can be modified to include reactive amino acids that facilitate coupling onto other nanosurfaces for fabrication of devices at the nanoscale.

To conclude, taking cues from the ubiquitous self-assembly found in natural systems, scientists have developed designer peptide-based biomimetic systems. The bottom-up approach in these nanofibrillar peptide assemblies offers numerous potential functionalities, such as molecular switches, Velcro, etc. Thus, the development of these SAPs paves the way for building supramolecular structures with a highly controllable hierarchy that are very attractive for tissue regeneration applications.

7.4 Peptide Amphiphiles

Several synthetic peptide-based fibrillar hydrogels have been investigated as potential ECM mimics, which vary in structure to include diblock co-polypeptide amphiphiles, oligopeptides, or peptide amphiphiles (PAs). Among them, the PAs made from hydrocarbon alkyl chains attached to hydrophilic peptide segments have been known to self-assemble into nanofiber networks similar to the fibrillar mesh-like structure of the ECM. Several distinguishing characteristics of these nanofibrous PA scaffolds peaked interest in studying them as a true ECM microenvironment for tissue engineering. In general, these PAs are very versatile molecules, as their composition can be self-assembled to allow for the concurrent control of nanostructure and biological functionality. Broad utility exists for these peptide-based biomaterials; they can be easily adapted due to amino acid interchangeability and have the potential to inscribe various biologically active sequences. Stupp's laboratory was one of the first to provide valuable research into these PAs, investigating the ability of the molecules to form higher order nanostructures coined as "one-dimensional assemblies" (Hartgerink et al. 2002). They have been designated as 1D because the nanostructure possesses a single dimension that is much longer than the other two, typically showing a 100- to 1000-fold increase (Palmer et al. 2007). Overall, these PAs serve as a synthetic biomaterial designed to interact with cells and proteins in a specific, controllable manner to provide regenerative medicine alternatives. This holds great promise because of the inherent ability of these self-assembling PA systems to direct nanoscopic architecture and alignment, while separately being able to integrate biological functionality.

In 2001, Hartgerink et al. originally investigated these PAs for bone regeneration scaffolding, creating a composite scaffold that combined the organic and inorganic bone phases at the lowest hierarchical level (Hartgerink et al. 2001). This was an innovative system that was designed for these organic PAs to promote nucleation of inorganic hydroxyapatite (HA) on the surfaces of the fibers. The PA structure designed for this study consisted of three functionally distinct peptide regions as shown in Figure 7.3. Four consecutive cysteine amino acids were inscribed next to the hydrophobic core and their inclusion resulted in disulfide bonding between adjacent peptides to stabilize the supramolecular structure. Phosphoserine was also incorporated to provide the proper environment for biomineralization because the phosphorylated amino acid is abundantly common in non-collagenous bone matrix proteins and is believed to interact with HA (Mai et al. 2008). Finally, an Arg-Gly-Asp (RGD) peptide sequence was included as the last region. The RGD motif is a general cell recognition site commonly found in ECM molecules, such as fibronectin and laminin (Hersel et al. 2003). By incorporating this bioactive sequence into the exposed outer domain, these PAs presented a general cell adhesion ligand to promote cell adhesion and growth. This designed PA was induced to self-assemble by lowering the pH, creating a cross-linked fiber network. Successful biomineralization was observed on this PA template, as HA crystals were preferentially aligned down the long fiber axis. Early follow-up studies further expanded the utility of these PAs beyond applicability as a bone tissue regeneration scaffold. In particular, the versatility of these PAs was demonstrated by modifying the molecular structure with different alkyl tail lengths and amino acid compositions, and comprehensively investigating different self-assembly methods (Hartgerink et al. 2002). These variant PAs were all able to self-assemble into 1D nanostructures using several induction methods, such as lowering pH, addition of divalent ions, and drying onto surfaces. These more expansive material characterizations of PAs demonstrated that the biomaterial is tolerant to most chemical modifications, as a vast array of peptide ligands can be incorporated into the molecule. Therefore, this nanofibrous peptide-based system has vast potential for both biological and nonbiological applications, which are subsequently described later on in this chapter.

After the introduction of these PAs as a self-assembling biomaterial, numerous investigations have been conducted and documented in the literature. In general, the structural composition of these PAs consisted of a hydrophilic peptide segment, containing a varying amount of amino acids (6–15 residues), coupled via an amide bond to a hydrophobic alkyl chain that usually fluctuated in length from 10 to 22 carbon atoms (Beniash et al. 2005). The self-assembled configuration of these molecules

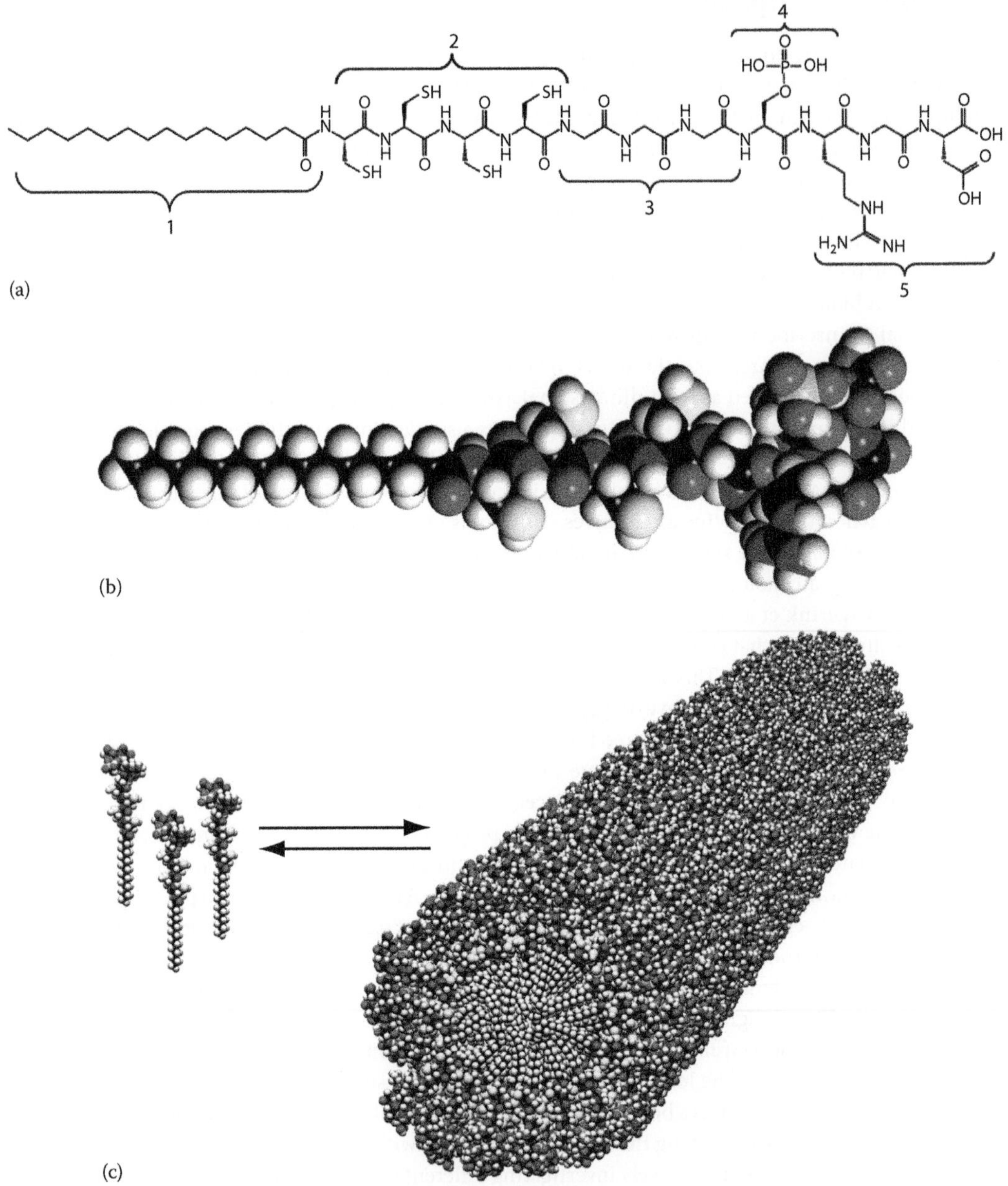

FIGURE 7.3 (a) Chemical structure of the PA, highlighting five key structural features. Region 1 is a long alkyl tail that conveys hydrophobic character to the molecule and, when combined with the peptide region, makes the molecule amphiphilic. Region 2 is composed of four consecutive cysteine residues that when oxidized may form disulfide bonds to polymerize the self-assembled structure. Region 3 is a flexible linker region of three glycine residues to provide the hydrophilic head group flexibility from the more rigid cross-linked region. Region 4 is a single phosphorylated serine residue that is designed to interact strongly with calcium ions and help direct mineralization of HA. Region 5 displays the cell adhesion ligand RGD. (b) Molecular model of the PA showing the overall conical shape of the molecule going from the narrow hydrophobic tail to the bulkier peptide region. (c) Schematic showing the self-assembly of PA molecules into a cylindrical micelle. (From [Hartgerink, J.D., Beniash, E., and Stupp, S.I., Self-assembly and mineralization of peptide-amphiphile nanofibers, *Science*, 294(5547), 1684–1688, 2001]. Reprinted with permission of AAAS.)

mimics native phospholipids and other biological membrane-forming structures (Tovar et al. 2005). In the self-assembled arrangement, the hydrophobic alkyl tails comprise the core, while the hydrophilic peptide segments form a shielding outer surface. Standard solid phase chemistry is typically used for synthesizing the peptide sequences. Single-tailed PAs with only one ionic peptide segment are most commonly studied (Hartgerink et al. 2001). However, PAs with multiple or branched peptide architecture have also been used in past research (Guler et al. 2006; Harrington et al. 2006; Storrie et al. 2007). The branched PAs provide another means for diversifying the nanostructure. In particular, different densities of epitopes can be maintained to control the receptor clustering and signal accessibility (Storrie et al. 2007). This is potentially relevant for studying cell–matrix interactions with the PA because it has been shown that the ligand density affects cellular attachment, spreading, and migration (Massia and Hubbell 1991; Hubbell et al. 1992).

7.4.1 Controlling the Self-Assembly Process of Single-Tailed Peptide Amphiphiles

Overall, these PAs are advantageous because of their ability to self-assemble into sheets, spheres, rod-like fibers, disks, or channels, depending on the shape, charge, and environment (Israelachvili et al. 1977). PA self-assembly creates an intricate nanomatrix environment that is driven by the hydrophobic nature of the covalently attached alkyl tail and primarily stabilized by hydrogen bonding between the adjacent peptides, along with further support provided by electrostatic attraction, ionic bridging, van der Waals forces, and molecular geometry in relation to amphiphilic packing (Hartgerink et al. 2002; Claussen et al. 2003; Stendahl et al. 2006; Jiang et al. 2007). The self-assembly mechanism is initialized by screening the charged groups within these PAs by adjusting the pH or adding soluble metal ions, which results in high-aspect-ratio nanofibers via hydrophobic collapse (Tovar et al. 2005; Palmer et al. 2008). The amphiphilic character of the molecule provides thermodynamic incentive for the assembled formations to maintain peptide shielding and reduce entropically unfavorable interactions between the alkyl tails, especially in an aqueous environment (Stendahl et al. 2006). The formed "one-dimensional assemblies" are typically presented as cylindrical micelle nanostructures because the conical shape of the hydrophilic peptide segment is relatively bulkier than its narrow hydrophobic tail and the employed peptide sequences have a strong β-sheet disposition (Hartgerink et al. 2001, 2002) The β-sheets form parallel to the long axes and are packed radially within the nanofibers, as the hydrophilic peptide segments extend outward toward the surface (Figure 7.4) (Jiang et al. 2007).

The self-assembled PA nanostructures are able to form robust non-covalent cross-links between the fibers, resulting in an interwoven network that gives rise to a macroscopic, self-supporting gel. Rheological characterization has been performed on these PAs to confirm self-supporting gelation, indicating that most of the deformation energy was recovered during elastic stretching rather than being lost as heat during viscous sliding (Stendahl et al. 2006). The gelation process can be induced over a wide PA concentration range, as stable gels can be assembled at concentrations as low as 0.25% by weight (Beniash et al. 2005). Additionally, the gelation kinetics can be controlled without altering any inscribed bioactive epitopes, as described by Niece et al. (2008). Without modifying the outer bioactive peptide domain, they demonstrated that increasing the hydrophobic character of PAs by incorporating specific residues into the peptide core accelerated self-assembly, but the self-assembling formation was suppressed by including more hydrophilic or bulky peptides. Within the self-assembled gels, the morphology of the high-aspect-ratio nanofibers has been well documented as shown in Figure 7.5 under several different high magnification modalities, including transmission electron microscopy (TEM), SEM, and AFM (Palmer et al. 2008). The general nanomatrix observed was a network of cylindrical nanofibers, ranging from 6 to 10 nm in diameter, depending on the length of the self-assembling molecules that form them (Beniash et al. 2005). In principle, there is no limitation on how far each nanofiber can extend along the long axis because of the high potential for orthogonal β-sheet linking between the PA molecules (Jiang et al. 2007). Typically, however, the nanofibers only achieve a length up to several microns

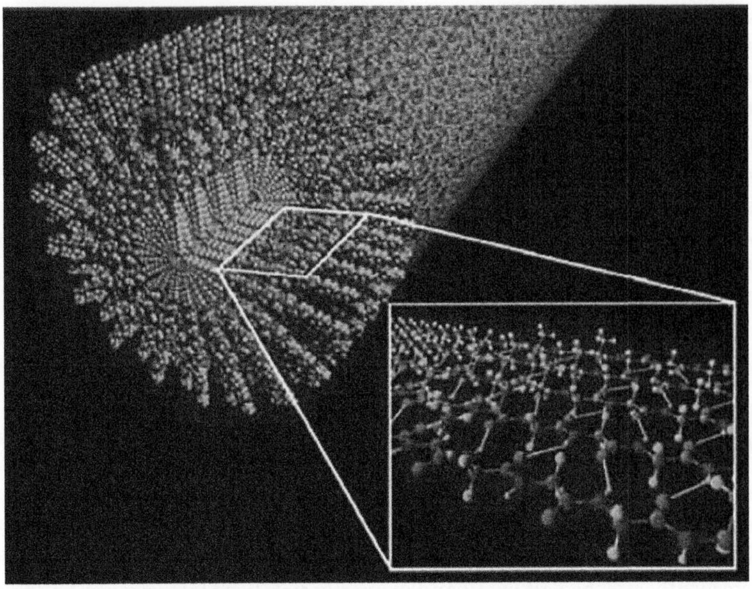

FIGURE 7.4 **(See color insert.)** Schematic representation of β-sheets within PA nanofibers. As depicted in the inset, β-sheets are oriented parallel to the long axis of the nanofibers (inter-β-strand hydrogen bonds are represented as yellow lines; carbon, oxygen, hydrogen, and nitrogen atoms are colored grey, red, light blue, and blue, respectively). (From Jiang, H., Guler, M.O., and Stupp, S.I., The internal structure of self-assembled peptide amphiphiles nanofibers, *Soft Matter*, 3, 454–462, 2007. Reproduced by permission of The Royal Society of Chemistry.)

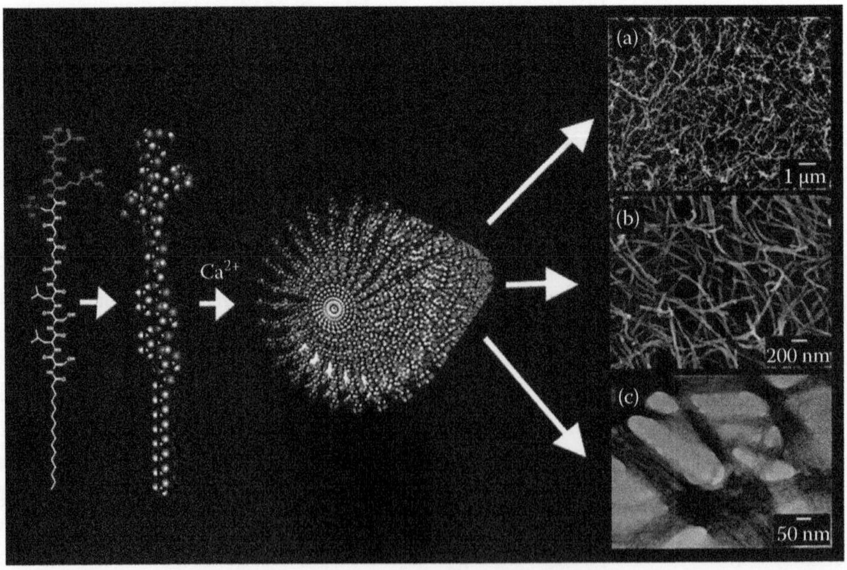

FIGURE 7.5 Schematic illustration of the RGD-PA and the self-assembly into a nanofiber. The low magnification (a) and high magnification (b) scanning electron micrographs and the transmission electron micrograph (c) show fibrous bundles, made up of PA nanofibers approximately 5–7 nm in diameter. The scanning electron micrographs were taken of a critical point dried PA gel, while the transmission electron micrograph was taken of nanofibers dried on a TEM grid and stained with phosphotungstic acid. (Reprinted with permission from [Palmer, L.C., Newcomb, C.J., Kaltz, S.R., Spoerke, E.D., and Stupp, S.I., Biomimetic systems for hydroxyapatite mineralization inspired by bone and enamel, *Chem. Rev.*, 108(11), 4754–4783]. Copyright [2008] American Chemical Society.)

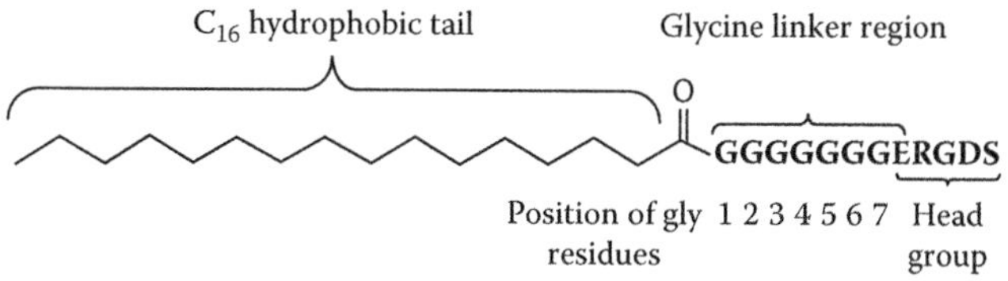

FIGURE 7.6 Schematic representation of PA molecule includes three distinct regions: a hydrophobic alkyl tail, a glycine containing region, and a charged head group. (Reprinted with permission from [Paramonov, S.E., Jun, H.W., and Hartgerink, J.D., Modulation of peptide-amphiphile nanofibers via phospholipid inclusions, *Biomacromolecules*, 7(1), 24–26]. Copyright [2006] American Chemical Society.)

(Hartgerink et al. 2001). Furthermore, because of the ionic nature of the molecule, self-assembly can be reversibly induced by increasing the pH of the PA nanomatrix (Hartgerink et al. 2001, 2002; Guler et al. 2005). This reversibility provides another beneficial mechanism for these peptide-based biomaterial to respond to the local environment by assembling, disassembling, or changing shape, especially as a self-assembled 3D gel (Hartgerink 2004).

From earlier studies, it was revealed that these single-tailed PAs became nanofibers as a result of the hydrophobic interactions between aliphatic carbon chains (Hartgerink et al. 2001; Paramonov et al. 2006a,b). It was also reported that the β-sheet formations between the peptide segments stabilize the nanofibers and electrostatic interactions between the peptide secondary structures influence their stability (Behanna et al. 2005; Paramonov et al. 2006a; Stendahl et al. 2006). Recently, after a more in-depth analysis of the internal peptide region, Paramonov et al. showed that the amino acids closest to the core of the nanofibers form the most critical β-sheet hydrogen bonds needed to achieve higher order assemblies (Paramonov et al. 2006b). Any disruption occurring at these core hydrogen bonds eliminates the ability of these PAs to form elongated, cylindrical nanostructures. The basic structure for all PAs used in this study by Paramonov et al. is depicted in Figure 7.6.

To determine the exact role of hydrogen bonding in the self-assembly process, a series of PAs were prepared, consisting of 19 *N*-methylated variants listed in Table 7.2. After preparing these *N*-methylated PAs, the ability of each PA to self-assemble into fibers (indicated "F") and produce a self-supporting gel (indicated "Gel" or "wGel") was observed. Within the *N*-methylated PA variants, two groups were synthesized to elucidate the relative importance of specific hydrogen bonding locations for nanofiber formation. The first series (PAs 2–8) *N*-methylated a single glycine at position 7 in PA 2 and then progressively added more *N*-methyl groups, moving toward position 1 until all seven linker glycines were methylated. The second series (PAs 9–19) reversed the order of methylation, and a few select variants (PAs 9–14) only contained one *N*-methylated glycine at each position in the glycine linker region. Overall, the introduction of methylated glycine residues lowered the storage modulus values, resulting in weaker gels. Interestingly, the elimination of one hydrogen bond in the core region (methylating a glycine between glycine positions 1–4) disrupted the gel formation, while its absence could be tolerated in the periphery (methylating a glycine between glycine positions 5–7). These findings rationalize that the amino acids further away from the core of the PA nanofiber are less restricted in their conformation and only play a minor role in stabilizing the nanostructure and corresponding macroscopic gel. Thus, there is greater freedom for incorporating bioactive moieties, such as cell adhesive ligands and degradable sites, onto the end of the PA to control cellular behaviors. However, consideration must be given to the resulting assemblies in the hydrophilic peptide region, as unfavorable β-sheet conformations due to random protein folding could reduce the availability of bioactive moieties in the peptide backbone (Paramonov et al. 2006a).

Molecular simulation of PA nanofiber formation has also been investigated to further understand the molecular interactions taking place during self-assembly because a detailed explanation had not been fully realized from experimental characterizations. Velichko et al. used a course-grained model to simulate PA self-assembly (Velichko et al. 2008). This allowed for a simplified simulation that did

TABLE 7.2 Summary of *N*-Methylated Peptide Amphiphiles Prepared

	Glycine Position								
PA	1	2	3	4	5	6	7	Nanostructure	Rheology
1	G	G	G	G	G	G	G	F	Gel
2	G	G	G	G	G	G	^{NMe}G	F	Gel
3	G	G	G	G	G	^{NMe}G	^{NMe}G	F	Gel
4	G	G	G	G	^{NMe}G	^{NMe}G	^{NMe}G	F	wGel
5	G	G	G	^{NMe}G	^{NMe}G	^{NMe}G	^{NMe}G	–	–
6	G	G	^{NMe}G	^{NMe}G	^{NMe}G	^{NMe}G	^{NMe}G	–	–
7	G	^{NMe}G	^{NMe}G	^{NMe}G	^{NMe}G	^{NMe}G	^{NMe}G	–	–
8	^{NMe}G	^{NMe}G	^{NMe}G	^{NMe}G	^{NMe}G	^{NMe}G	^{NMe}G	–	–
9	^{NMe}G	G	G	G	G	G	G	F	–
10	G	^{NMe}G	G	G	G	G	G	F	–
11	G	G	^{NMe}G	G	G	G	G	F	–
12	G	G	G	^{NMe}G	G	G	G	F	–
13	G	G	G	G	^{NMe}G	G	G	F	Gel
14	G	G	G	G	G	^{NMe}G	G	F	Gel
15	^{NMe}G	^{NMe}G	G	G	G	G	G	–	–
16	^{NMe}G	^{NMe}G	^{NMe}G	G	G	G	G	–	–
17	^{NMe}G	^{NMe}G	^{NMe}G	^{NMe}G	G	G	G	–	–
18	^{NMe}G	^{NMe}G	^{NMe}G	^{NMe}G	^{NMe}G	G	G	–	–
19	^{NMe}G	^{NMe}G	^{NMe}G	^{NMe}G	^{NMe}G	^{NMe}G	G	–	–

Source: Adapted with permission from [Paramonov, S.E., Jun, H.W., and Hartgerink, J.D., Modulation of peptide-amphiphile nanofibers via phospholipid inclusions, *Biomacromolecules*, 7(1), 24–26]. Copyright [2006] American Chemical Society.

"F" indicates that the nanofibers were the dominant nanostructure present as observed by vitreous ice cryo-TEM; "–" means no fibers were present, and the sample was principally composed of spherical micelles and amorphous aggregates. For the column indicating rheology, options are Gel or wGel (weak gel), or "–" meaning no gel was formed.

not account for any specific chemical structures of PAs; instead, the amphiphilic molecule was divided into three general regions—hydrophobic, peptide, and epitope head group. The theoretical simulations determined that PA self-assembly into cylindrical nanofibers followed an open association model, indicating that initial hydrogen bonding into β-sheet formations first occurs before reorganization into extended cylindrical nanofibers. This shows that the molecular structure and the balance of dominant intermolecular forces (i.e., hydrogen bonding and hydrophobicity) are the most important factors for achieving an equilibrium state during the PA self-assembly process.

Finally, the environmental conditions must be considered in the PA self-assembly process, especially for biological applications that require a physiological environment to ensure viability. Based heavily on initial factors, such as pH and salinity, PAs can assemble into cylindrical or spherical micelles, an intermediate structure between the two, or not form at all. Thus, Tsonchev et al. built a semiquantitative pH/salinity phase diagram shown in Figure 7.7 to better illustrate how all of these parameters and their corresponding interactions direct PA self-assembly (Tsonchev et al. 2008). The model constructed was based on the competition between electrostatic and hydrophobic forces using both theoretical modeling and experimental data. In general, a disposition toward higher hydrogen bonding favors cylindrical PA formations, but this can be negated with increased salinity. Several more self-assembly generalizations were also formulated based on the pH/salinity interactions investigated by Tsonchev et al. PA molecules tend to stay bundled as cylindrical nanofibers over the pH range of 2–4 at low salt concentrations. The nanostructures become disassembled as the pH is lowered to 0 because of increased protonation breaking apart the hydrogen bonding. However, spherical assemblies are still possible at these highly acidic conditions if the salinity

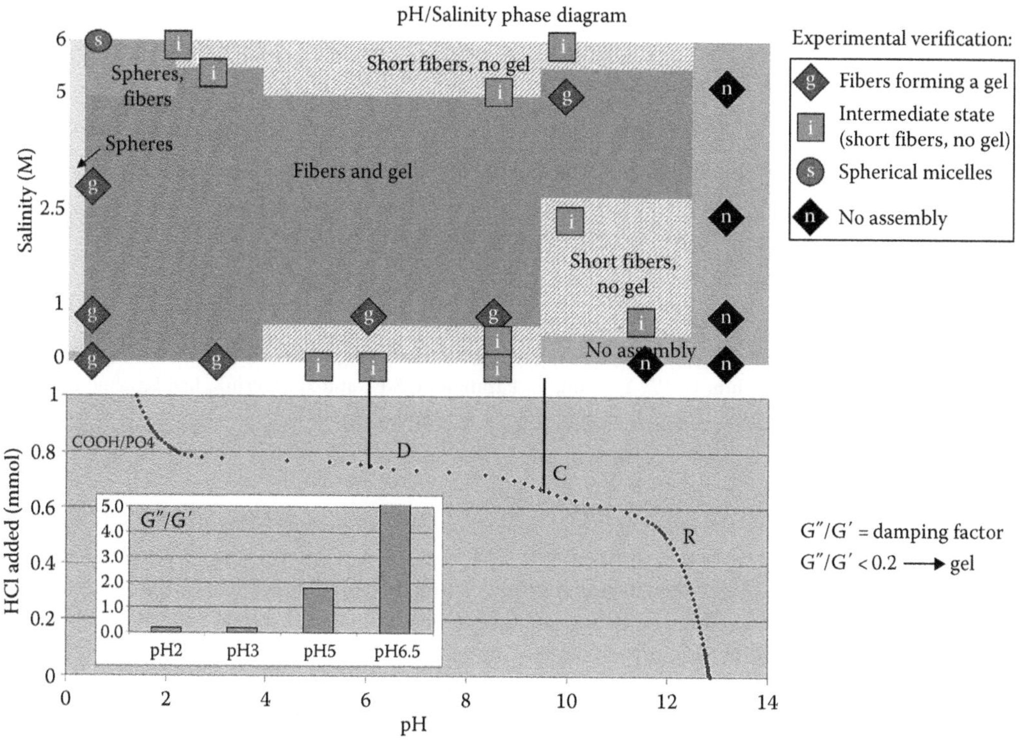

FIGURE 7.7 pH/salinity phase diagram of the self-assembling PAs. The experimental verification points (derived from a combination of visual examination, rheology, TEM) are shown with the figures containing the letters g, i, s, and n, as explained above. Under the diagram, we have shown the titration curve of the amphiphile, where the titration inflection points corresponding to the transition lines of the diagram are shown. The inset shows the dependence of the damping factor G″/G′ on pH, with G″/G′ < 0.2 corresponding to a gel. This provides us with a quantitative picture of the degree of gelation, which is correlated with the length and stability of the fibers. (Reprinted with permission from [Tsonchev, S., Niece, K.L., Schatz, G.C., Ratner, M.A., and Stupp, S.I., Phase diagram for assembly of biologically-active peptide amphiphiles, *J. Phys. Chem. B*, 112(2), 441–447]. Copyright [2008] American Chemical Society.)

is increased enough to sufficiently screen the charges. Conversely, increasing the pH with a negligible salt concentration, increases the electrostatic repulsion between the PA molecules, but the hydrogen bonding, albeit weaker, remains strong enough to still produce cylindrical nanofibers capable of forming gels. For a pH above 9, though, the electrostatic repulsion becomes too high to tolerate PA self-assembly, unless the salinity is greatly increased to screen the repelling forces and allow for cylindrical formations. Altogether, self-assembly of these PA molecules is capable of transitioning across several phases based on the pH and salinity, but for tissue regenerative applications to succeed, a physiologically relevant microenvironment is a required necessity. Hence, this highly controllable PA self-assembly system within the targeted neutral pH range offers numerous opportunities for regenerative treatments based on cell encapsulation within the nanomatrix to direct biological responses.

7.4.2 Incorporation of Enzyme-Mediated Degradation Sites into Single-Tailed Peptide Amphiphiles

These initial molecular characterization studies provided insight into the self-assembly of single-tailed PAs and laid the necessary groundwork for future investigations oriented toward tissue engineering applications. Progressing with this approach, the ultimate goal is to develop an ECM-mimicking

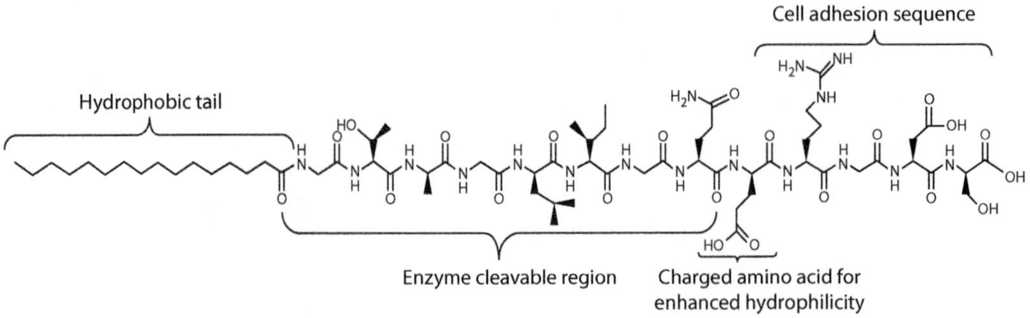

FIGURE 7.8 The chemical structure of enzyme-degradable PA illustrating some of the important considerations for biomaterial design. (From Jun, H.-W., Yuwono, V., Paramonov, S.E., and Hartgerink, J.D: Enzyme-mediated degradation of peptide-amphiphile nanofiber networks. *Adv. Mater.* 2005. 17(21). 2612–2617. 2005. Copyright Wiley-VCH Verlag GmbH & Co. KGaA.)

biomaterial, capturing both the chemical and biological complexity needed. By including degradation sites and cell adhesive ligands isolated from the ECM, one can potentially control cellular behaviors with this PA nanomatrix. Moreover, degradation by cell-mediated enzymes allows the cells to create pathways for migration. Therefore, an ideal ECM-mimicking biomaterial should have all of these vital characteristics. In this regard, an approach demonstrated by Jun et al. is of particular interest (Jun et al. 2005). In this study, single-tailed PAs were utilized to create cell-responsive PA nanofiber networks that simulate several essential properties of the ECM, including self-assembling of nanofibers, presence of cell-adhesive ligands, and cell-mediated degradation.

As shown in Figure 7.8, this degradable PA molecule consisted of three regions: cell-mediated enzyme sensitive peptide sequence (GTAGLIGQ), calcium binding sites via a glutamic acid residue and C-terminal carboxylic acid, and cell adhesive ligand (RGDS). The featured matrix metalloproteinase-2 (MMP-2) specific cleavage site allows for cell-mediated degradation of the PA nanofiber network, thereby enabling a pathway for cellular migration and remodeling. To test the efficiency of the incorporated degradable sequence, PAs were prepared as self-assembled disk-shaped gels and incubated in Type IV collagenase. After 1 week, these PA gels lost 50% of their original weight (Figure 7.9a), and by week 3, egg-shaped fibrillar aggregates that associated into multistranded twisted ribbons were observed (Figure 7.9c). This indicated that an accumulation of defects within these PA nanofibers eventually broke the assemblies into small fragments that diffused out of the gel and decreased the stability.

Finally, rat maxillary incision pulp cells, which play an important role in dentin mineralization and dental tissue development, were encapsulated in these self-assembled nanofibers to assess the ability of these degradable PAs to support cell adhesion and proliferation. The RGDS ligand inscribed within these PAs used for cell encapsulation was varied at different densities to investigate bioactive signal availability. Although the encapsulated cells exhibited a round morphology within the nanofibrous gel after 1 day for all conditions, the cells grown in gels presenting at least 50% of the RGDS ligand eventually formed dense cell colonies throughout the gel, which became fully spread after 4 days (Figure 7.10a and c). However, cells exposed to less than 50% of the RGDS ligands remained spherical throughout (Figure 7.10b and d). This signifies the ability of the encapsulated cells to enzymatically make migratory pathway and track along the adhesive ligands through the network, remodeling the PA nanomatrix. By fully integrating an enzyme-sensitive degradable site, these cell-responsive PA nanofibers offer a more sophisticated biomaterial for mimicking native ECM. The use of single-tailed PAs in this manner provides a large step forward in the development of next-generation biomaterials capable of manipulating cell adhesion, migration, proliferation, and differentiation.

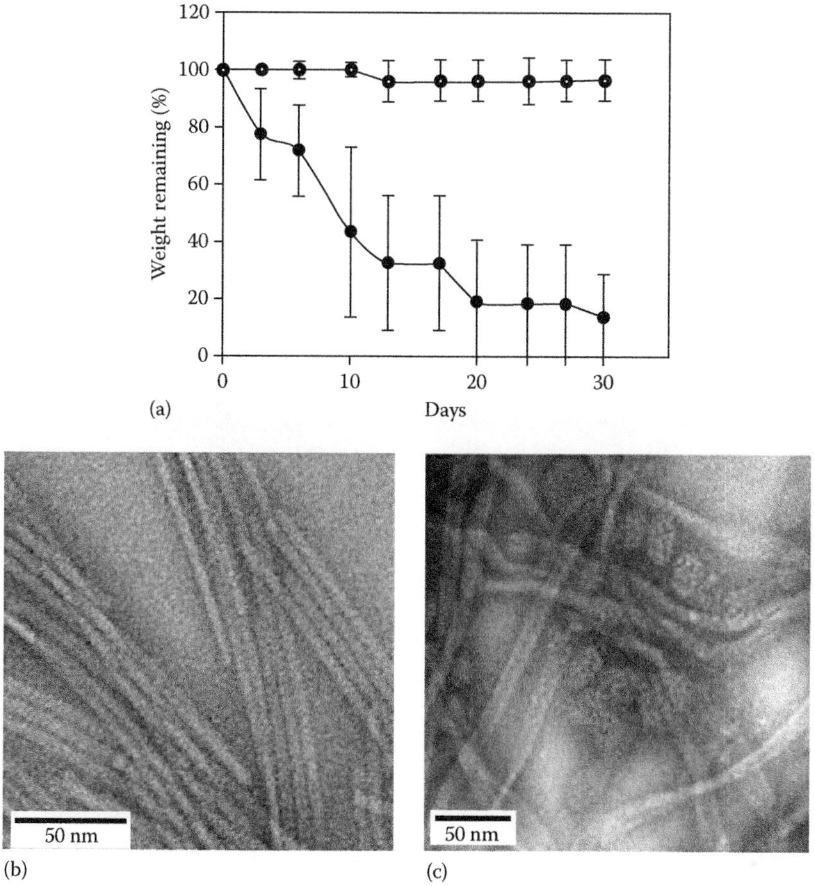

FIGURE 7.9 Proteolytic degradation of nanofiber network. (a) Weight of nanofiber gel remaining after incubation with Type IV collagenase (filled circles) compared to control (open circles), (b) TEM image of nanofiber network before proteolytic degradation, and (c) TEM image after 3 weeks of incubation with Type IV collagenase. (From Jun, H.-W., Yuwono, V., Paramonov, S.E., and Hartgerink, J.D: Enzyme-mediated degradation of peptide-amphiphile nanofiber networks. *Adv. Mater.* 2005. 17(21). 2612–2617. 2005. Copyright Wiley-VCH Verlag GmbH & Co. KGaA.)

7.4.3 Modifications of Single-Tailed Peptide Amphiphiles

The hydrophobic core of these PAs provides a potential region that can be adapted to create synthetic ECM biomaterials directed toward a specific biomedical need. Exploring this versatility, phospholipids have been utilized to modulate the mechanical properties and secondary peptide structures of the self-assembling hydrogels (Paramonov et al. 2006b). Specifically, 1-palmitoyl-2-hydroxy-sn-glycerol-3-phosphocholine was chosen as the phospholipid and was conjugated with PAs as depicted in Figure 7.11. Rheologically, it was demonstrated that these modulated PAs were capable of forming self-supporting gels with lipid inclusions up to 20% molarity. Within this range, normal nanofiber formation was observed with an average diameter of 10.8 ± 0.8 nm; however, increasing the lipid inclusion up to 20% molarity destabilized the hydrogels. Overall, the maximum storage modulus occurred at 5% molarity, indicating the optimal hydrogen bonding and molecular packing of these PA molecules constituting the nanofibers. The ability to introduce small hydrophobic molecules, such as phospholipids, into PAs further expands the resourcefulness of this molecule, especially as a drug delivery system.

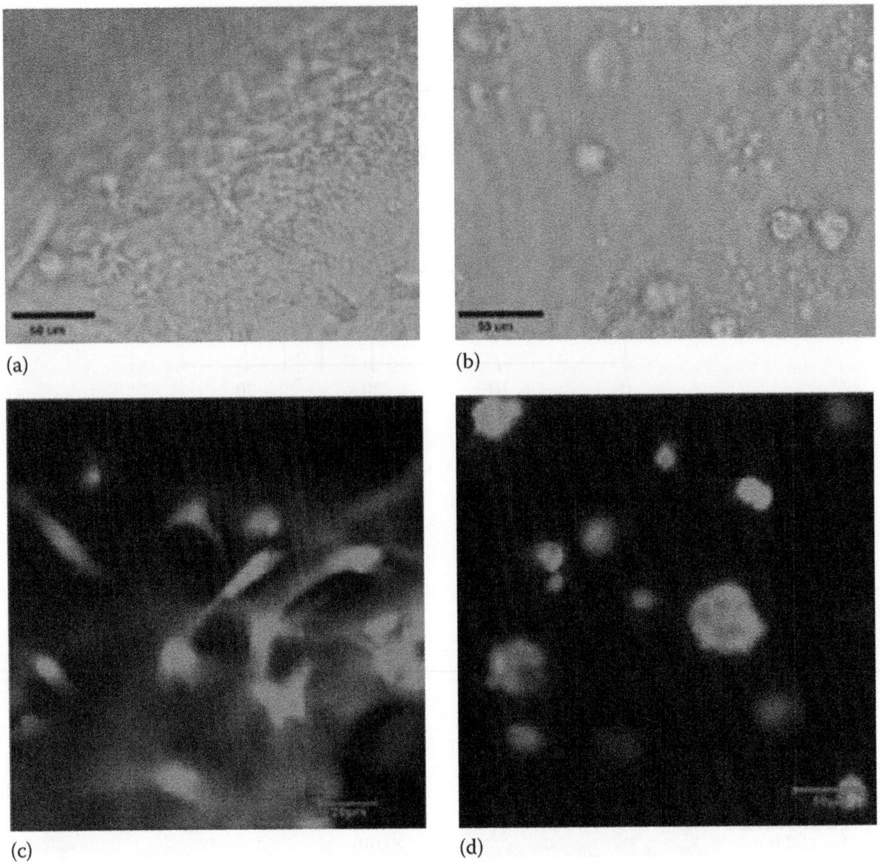

FIGURE 7.10 **(See color insert.)** Response of nanofiber network to encapsulated cells. Rat maxillary incision pulp cells were encapsulated in nanofiber networks with different densities of adhesive ligands, (a) 100% RGDS or (b) RDGS. Confocal laser scanning microscopy images of cells in nanofiber networks fluorescently observed encapsulation of cells in PAs with (c) 100% RGDS or (d) RDGS. Cells were stained with fluorescent dyes (calcein AM and ethidium homodimer-1). All images were taken after 4 days of incubation. (From Jun, H.-W., Yuwono, V., Paramonov, S.E., and Hartgerink, J.D: Enzyme-mediated degradation of peptide-amphiphile nanofiber networks, *Adv. Mater.* 2005, 17(21), 2612–2617, 2005. Copyright Wiley-VCH Verlag GmbH & Co. KGaA.)

In a follow-up study with the enzyme-sensitive PAs described earlier, Jun et al. demonstrated that the selection and combination of specific peptide sequences could be tailored to create a diverse range of ECM-like gels to better control bioactivity, degradability, and mechanical properties (Jun et al. 2008). Investigating this tunable system, three different peptide combinations were used to synthesize three individual PA molecules: a MMP-2 only PA (GTAGLIGQES; PA1), degradable PA containing the cell adhesive RGDS (GTAGLIGQERGDS; PA2), and a scrambled RDGS control (GTAGLIGQERDGS; PA3). All three PAs included the cleavage site (GTAGLIGQ) and a hydrophobic tail composed of palmitic acid. Inducing nanofiber self-assembly with calcium ions ($M_r = 2$, $M_r = [Ca^{2+}]/[PA]$), PA1, which has three lesser amino acids, formed shorter nanofibers with an average length of 500 nm, while PA2, containing the RGDS motif, self-assembled into long nanofibers with a length of several microns. If PA1 and PA2 were mixed at a 1:1 molar ratio, nanofibers with an intermediate length were obtained. Interestingly, the length of the nanofibers affected the mechanical properties of the PA networks. Based on rheometry, the storage modulus (G′) of PA2 only amounted to a fraction of the value for PA1 (Figure 7.12a). However, the nanofiber network modulated to 25% of PA1 and 75% of PA2 showed a storage modulus six times

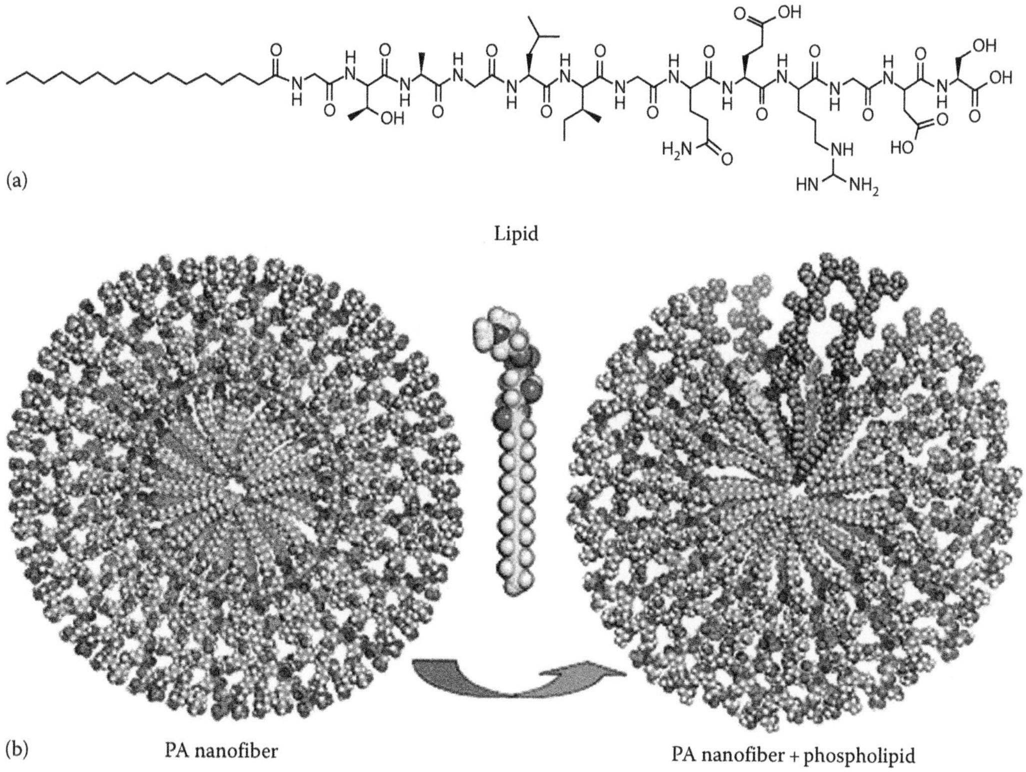

FIGURE 7.11 **(See color insert.)** (a) Chemical structure of the PA and (b) cross section of a PA fiber and a PA fiber containing 6.25 mol% of lipid (yellow). Highlighted in pink are the PA molecules situated adjacent to the lipid molecules. (Reprinted with permission from [Paramonov, S.E., Jun, H.W., and Hartgerink, J.D., Modulation of peptide-amphiphile nanofibers via phospholipid inclusions, *Biomacromolecules*, 7(1), 24–26]. Copyright [2006] American Chemical Society.)

higher than that of PA2 alone, whereas the storage modulus of the 75% PA1 and 25% PA2 composite was 60-fold higher than pure PA2. When the different PA ratio mixtures were incubated with Type IV collagenase, the incubation time needed to achieve a 50% weight reduction for PA2, a 50:50 mixture of PA1/PA2, and PA1 were approximately 1, 2, and 4 weeks, respectively (Figure 7.12b). Therefore, altering the length of the nanofibers also changes the degradation kinetics of PA gels. This follow-up study vividly shows that the nanofiber length is another influential factor in the self-assembly process of PAs into viscoelastic networks. Clearly, these PAs have a promising future as a biomaterial for biomedical applications because of their versatility to form pseudo ECM-like nanostructures with controllable mechanical properties and degradability within the microenvironment.

7.4.4 Biomedical Applications for Peptide Amphiphiles

Potential tissue regenerative applications with these PA molecules are far reaching, as numerous examples are present in the literature and range from directed biological response via cell encapsulation to hybrid scaffolding dual functionality to growth factor delivery. Tissue-specific deviations of this PA nanomatrix have served as bioactive scaffolds for many cell types, including neural progenitor cells, mouse calvarial pre-osteoblastic (MC3T3-E1) cells, primary enamel organ epithelial cells, and pancreatic islets (Silva et al. 2004; Beniash et al. 2005; Huang et al. 2008; Sargeant et al. 2008; Stendahl et al. 2008). In all cases, this PA encapsulation approach has proven to be biocompatible. Specifically, Beniash et al. demonstrated that cell encapsulation within this PA nanomatrix does not deter cell proliferation or

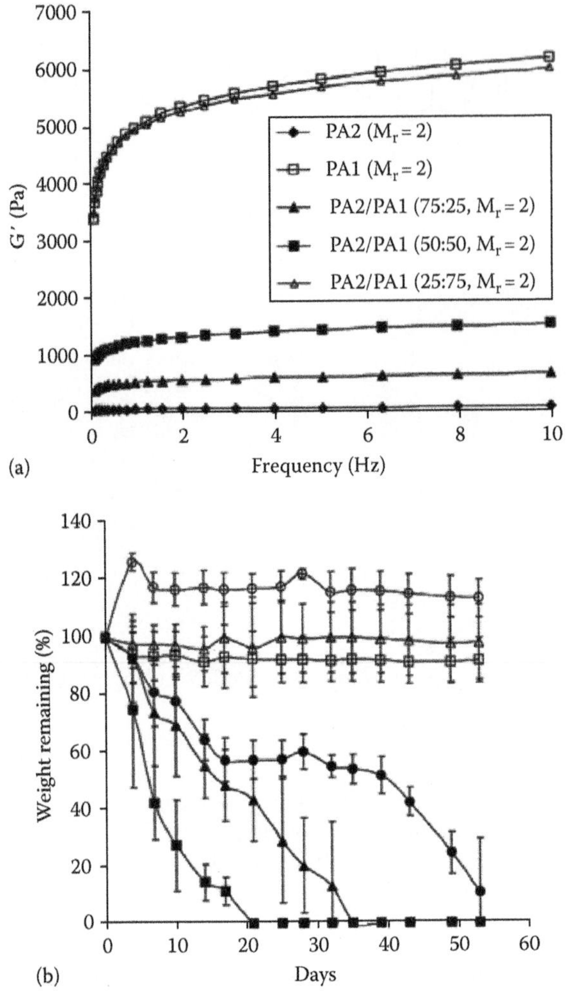

FIGURE 7.12 (a) Viscoelastic properties of the nanofiber network blends with different molar ratios. Storage modulus (G′) of nanofiber networks of PA1, PA2 blends at $M_r = 2$. (b) Enzymatic degradation of nanofiber networks of PA1 (circles), PA1/PA2 (50:50, triangles), and PA2 (squares). Incubation with Type IV collagenase (dark symbols) or buffer (open symbols). (From Jun, H.W., Paramonov, S.E., Dong, H., Forraz, N., McGuckin, C., and Hartgerink, J.D., Tuning the mechanical and bioresponsive properties of peptide-amphiphile nanofiber networks, *J. Biomater. Sci. Polym. Ed.*, 19(5), 665–676. 2008.)

motility, and surprisingly, that the entrapped cells were able to internalize the surrounding nanofibers by endocytosis (Beniash et al. 2005).

Progressing into directed cellular signaling, the PA biomaterial has also been investigated as an instructive scaffold for proliferation and differentiation along specific tissue lineages. For example, neural progenitor cells have been encapsulated within a self-assembled PA nanomatrix expressing the IKVAV epitope in the outer peptide domain (Silva et al. 2004). The IKVAV peptide sequence was isolated from laminin and has shown the ability to promote neurite growth (Wheeler et al. 1999; Kam et al. 2001; Yeung et al. 2001). The presentation of this bioactive epitope within this PA fibrous network was found to selectively enhance differentiation of the progenitor cells into neurons with minimal astrocyte development, as confirmed by neurite outgrowth morphology and positive β-tubulin immunohistological staining. In a similar study, PAs have effectively been used as instructive cell encapsulating scaffolds for enamel formation and long-term tooth regeneration. Specifically, Huang et al. employed a branched PA molecule displaying the RGD peptide signal and co-cultured ameloblast-like cells and primary enamel organ epithelial cells (Huang et al. 2008). This cell-encapsulated PA gel was implanted in embryonic mouse incisors and found to promote proliferation and enhanced expression of differentiation markers, while still maintaining long-term viability. Therefore, these peptide-based biomimetic scaffolds for cell entrapment provide the necessary bioactivity and physical properties desired for regenerative medicine treatments.

Expanding the scope and utility further, these easily tunable PAs can be effortlessly functionalized to create a wide range of hybrid scaffolds and therapeutic delivery vehicles. One such beneficial aspect of this dual functionality is the potential to add bioactive signals to relatively inert surfaces, while still maintaining the physical properties of the original material. Using this approach, PAs have been combined with metal orthopedic implants and carbon nanotubes (Arnold et al. 2005; Sargeant et al. 2008). Sargeant et al. was able to create a hybrid bone implant material consisting of Ti-6Al-4V foam that integrated self-assembled PAs throughout the interconnected pores (Sargeant et al. 2008). These incorporated PAs served as a mineralization template for HA and enhanced bone ingrowth from the surrounding tissue, thus allowing for improved implant fixation, osseointegration, and long-term stability. PAs have also been non-covalently functionalized with carbon nanotubes to provide lacking bioactivity. The carbon nanotube is a hydrophobic material that displays an extremely high length-to-diameter ratio and extraordinary strength (Zheng et al. 2004). Its potential functionality canvases a wide range of applications, such as mechanical, electronic, optical, sensing, and biological (O'Connell et al. 2002; Chen et al. 2003; Dalton et al. 2003; Javey et al. 2003; Kam et al. 2004). Arnold et al. was able to successfully encapsulate carbon nanotubes with several different PAs, as the hydrophobic alkyl tails were strongly attracted to the hydrophobic nanotube surfaces (Arnold et al. 2005). Both of these examples are just one of many endless possibilities for introducing a directed biological response onto other inert materials through the self-assembled formations of PAs.

Finally, the versatility of PAs provides opportunities for developing therapeutic delivery systems, as the peptide segment can serve as a binding construct for growth factors or imaging contrast agents (Bull et al. 2005; Rajangam et al. 2006; Stendahl et al. 2008). For example, heparin-binding sequences have been inscribed into PA molecules by Rajangam et al. (Rajangam et al. 2006). Heparin, itself, is a biological molecule with a strong binding affinity for angiogenic growth factors (Tanihara et al. 2001; Ishihara et al. 2003). Thus, heparin was attracted to specific binding regions in the designed PA, which served as an intermediate for delivering vascular endothelial growth factor (VEGF) and fibroblast growth factor 2 (FGF-2). In vivo studies with these heparin-binding PAs and attached growth factors were found to stimulate significant new blood vessel formation in a rat cornea angiogenesis model, (Rajangam et al. 2006) and the validity of the binding sequence was confirmed by stable heparin interactions and the resulting biological response (Rajangam et al. 2008). These heparin-binding PAs have also been exploited to deliver encapsulated isologous islets and angiogenic growth factors (i.e., VEGF, FGF-2) into a diabetic mouse omentum (Stendahl et al. 2008). The transplantation resulted in increased neovascularization and improved islet engraftment, thereby enhancing the normoglycemia rate in the mice recipients. Besides drug delivery, PAs have been modified to uptake magnetic resonance imaging (MRI) contrast agents, such as Gd(III) (Bull et al. 2005). The magnetic resonance contrast agent was covalently linked to a specific binding sequence isolated from tetraacetic acid and inscribed within PA molecules. By conjugating Gd(III) to these PAs, the relaxivity of the agent was increased, which produced significantly better imaging contrast for longer in vivo observations. This allows for improved image sensitivity and potential cell tracking within these peptide-based scaffolds.

7.5 Conclusions

The potential applications for these peptide-based biomaterials translate across many different fields of biomedical research. This chapter has only highlighted the opportunities presented by two types of peptide molecules—SAPs and PAs. Both work in principle to synthetically capture the self-assembled formations of phospholipids naturally observed under physiological conditions. However, the differences lie in the basic structure, as SAPs are purely peptide based and PAs contain added hydrocarbon tails. Each has its own merits as a self-assembling biomaterial due to the versatility within the internal amino acid composition and ease of conjugating with other biomaterials to diversify functionality. Many different self-assembled configurations are possible based on the endowed physical properties, but the focus here is on nanofibrous assemblies, which present a complex nanostructured environment in the mold of native tissue formations at the most basic level. For both biomaterials, the self-assembly into nanofibers is driven by the inherent amphiphilic nature that results in the outer hydrophilic peptides shielding the inner hydrophobic core in a thermodynamically efficient arrangement. As discussed, this is a highly controllable process that can be directed by molecular and environmental factors, such as charge, hydrophobicity, pH, and salinity. Altogether, these factors work in concert to create cell-responsive peptide-based nanofibers with great promise for biomedical applications in regenerative medicine. Many such examples have been presented, encompassing tissue-engineered scaffold to encapsulate cells and direct cellular responses, therapeutic drug delivery, bioactive implant coatings, and diagnostic biosensors. By having the capacity to concurrently control the nanostructure and biological complexity, the future of peptide-based self-assembling nanofibers as a biomaterial is full of endless possibilities.

References

Altman, M., P. Lee, A. Rich, and S. Zhang. 2000. Conformational behavior of ionic self-complementary peptides. *Protein Sci* 9 (6):1095–1105.

Arnold, M. S., M. O. Guler, M. C. Hersam, and S. I. Stupp. 2005. Encapsulation of carbon nanotubes by self-assembling peptide amphiphiles. *Langmuir* 21 (10):4705–4709.

Aurora, R. and G. D. Rose. 1998. Helix capping. *Protein Sci* 7 (1):21–38.

Behanna, H. A., J. J. Donners, A. C. Gordon, and S. I. Stupp. 2005. Coassembly of amphiphiles with opposite peptide polarities into nanofibers. *J Am Chem Soc* 127 (4):1193–1200.

Beniash, E., J. D. Hartgerink, H. Storrie, J. C. Stendahl, and S. I. Stupp. 2005. Self-assembling peptide amphiphile nanofiber matrices for cell entrapment. *Acta Biomater* 1 (4):387–397.

Bull, S. R., M. O. Guler, R. E. Bras, T. J. Meade, and S. I. Stupp. 2005. Self-assembled peptide amphiphile nanofibers conjugated to MRI contrast agents. *Nano Lett* 5 (1):1–4.

Burke, S. A., M. Ritter-Jones, B. P. Lee, and P. B. Messersmith. 2007. Thermal gelation and tissue adhesion of biomimetic hydrogels. *Biomed Mater* 2 (4):203–210.

Caplan, M. R., P. N. Moore, S. Zhang, R. D. Kamm, and D. A. Lauffenburger. 2000. Self-assembly of a beta-sheet protein governed by relief of electrostatic repulsion relative to van der Waals attraction. *Biomacromolecules* 1 (4):627–631.

Chen, R. J., S. Bangsaruntip, K. A. Drouvalakis, N. W. Kam, M. Shim, Y. Li, W. Kim, P. J. Utz, and H. Dai. 2003. Noncovalent functionalization of carbon nanotubes for highly specific electronic biosensors. *Proc Natl Acad Sci USA* 100 (9):4984–4989.

Claussen, R. C., B. M. Rabatic, and S. I. Stupp. 2003. Aqueous self-assembly of unsymmetric Peptide bolaamphiphiles into nanofibers with hydrophilic cores and surfaces. *J Am Chem Soc* 125 (42): 12680–12681.

Collier, J. H., B. H. Hu, J. W. Ruberti, J. Zhang, P. Shum, D. H. Thompson, and P. B. Messersmith. 2001. Thermally and photochemically triggered self-assembly of peptide hydrogels. *J Am Chem Soc* 123 (38):9463–9464.

Collier, J. H. and P. B. Messersmith. 2001. Phospholipid strategies in biomineralization and biomaterials research. *Annu Rev Mater Res* 31:237.

Daley, W. P., S. B. Peters, and M. Larsen. 2008. Extracellular matrix dynamics in development and regenerative medicine. *J Cell Sci* 121 (Pt 3):255–264.

Dalton, A. B., S. Collins, E. Munoz, J. M. Razal, V. H. Ebron, J. P. Ferraris, J. N. Coleman, B. G. Kim, and R. H. Baughman. 2003. Super-tough carbon-nanotube fibres. *Nature* 423 (6941):703.

Davis, M. E., J. P. Motion, D. A. Narmoneva, T. Takahashi, D. Hakuno, R. D. Kamm, S. Zhang, and R. T. Lee. 2005. Injectable self-assembling peptide nanofibers create intramyocardial microenvironments for endothelial cells. *Circulation* 111 (4):442–450.

Ellis-Behnke, R. G., Y. X. Liang, S. W. You, D. K. Tay, S. Zhang, K. F. So, and G. E. Schneider. 2006. Nano neuro knitting: Peptide nanofiber scaffold for brain repair and axon regeneration with functional return of vision. *Proc Natl Acad Sci USA* 103 (13):5054–5059.

Gelain, F., D. Bottai, A. Vescovi, and S. Zhang. 2006. Designer self-assembling peptide nanofiber scaffolds for adult mouse neural stem cell 3-dimensional cultures. *PLoS ONE* 1:e119.

Gelain, F., A. Lomander, A. L. Vescovi, and S. Zhang. 2007. Systematic studies of a self-assembling peptide nanofiber scaffold with other scaffolds. *J Nanosci Nanotechnol* 7 (2):424–434.

Gerasimov, O. V., J. A. Boomer, M. M. Qualls, and D. H. Thompson. 1999. Cytosolic drug delivery using pH- and light-sensitive liposomes. *Adv Drug Deliv Rev* 38 (3):317–338.

Gregoriadis, G. 1995. Engineering liposomes for drug delivery: Progress and problems. *Trends Biotechnol* 13 (12):527–537.

Guler, M. O., L. Hsu, S. Soukasene, D. A. Harrington, J. F. Hulvat, and S. I. Stupp. 2006. Presentation of RGDS epitopes on self-assembled nanofibers of branched peptide amphiphiles. *Biomacromolecules* 7 (6):1855–1863.

Guler, M. O., S. Soukasene, J. F. Hulvat, and S. I. Stupp. 2005. Presentation and recognition of biotin on nanofibers formed by branched peptide amphiphiles. *Nano Lett* 5 (2):249–252.

Harrington, D. A., E. Y. Cheng, M. O. Guler, L. K. Lee, J. L. Donovan, R. C. Claussen, and S. I. Stupp. 2006. Branched peptide-amphiphiles as self-assembling coatings for tissue engineering scaffolds. *J Biomed Mater Res A* 78 (1):157–167.

Hartgerink, J. D. 2004. Covalent capture: A natural complement to self-assembly. *Curr Opin Chem Biol* 8 (6):604–609.

Hartgerink, J. D., E. Beniash, and S. I. Stupp. 2001. Self-assembly and mineralization of peptide-amphiphile nanofibers. *Science* 294 (5547):1684–1688.

Hartgerink, J. D., E. Beniash, and S. I. Stupp. 2002. Peptide-amphiphile nanofibers: A versatile scaffold for the preparation of self-assembling materials. *Proc Natl Acad Sci USA* 99 (8):5133–5138.

Hersel, U., C. Dahmen, and H. Kessler. 2003. RGD modified polymers: Biomaterials for stimulated cell adhesion and beyond. *Biomaterials* 24 (24):4385–4415.

Holmes, T. C., S. de Lacalle, X. Su, G. Liu, A. Rich, and S. Zhang. 2000. Extensive neurite outgrowth and active synapse formation on self-assembling peptide scaffolds. *Proc Natl Acad Sci USA* 97 (12):6728–6733.

Honger, T., K. Jorgensen, D. Stokes, R. L. Biltonen, and O. G. Mouritsen. 1997. Phospholipase A2 activity and physical properties of lipid-bilayer substrates. *Methods Enzymol* 286:168–190.

Huang, Z., T. D. Sargeant, J. F. Hulvat, A. Mata, P. Bringas, Jr., C. Y. Koh, S. I. Stupp, and M. L. Snead. 2008. Bioactive nanofibers instruct cells to proliferate and differentiate during enamel regeneration. *J Bone Miner Res* 23 (12):1995–2006.

Hubbell, J. A. 2003. Materials as morphogenetic guides in tissue engineering. *Curr Opin Biotechnol* 14 (5):551–558.

Hubbell, J. A., S. P. Massia, and P. D. Drumheller. 1992. Surface-grafted cell-binding peptides in tissue engineering of the vascular graft. *Ann N Y Acad Sci* 665:253–258.

Ishihara, M., K. Obara, T. Ishizuka, M. Fujita, M. Sato, K. Masuoka, Y. Saito et al. 2003. Controlled release of fibroblast growth factors and heparin from photocrosslinked chitosan hydrogels and subsequent effect on in vivo vascularization. *J Biomed Mater Res A* 64 (3):551–559.

Israelachvili, J. N., D. J. Mitchell, and B. W. Ninham. 1977. Theory of self-assembly of lipid bilayers and vesicles. *Biochim Biophys Acta* 470 (2):185–201.

Javey, A., J. Guo, Q. Wang, M. Lundstrom, and H. Dai. 2003. Ballistic carbon nanotube field-effect transistors. *Nature* 424 (6949):654–657.

Jiang, H., M. O. Guler, and S. I. Stupp. 2007. The internal structure of self-assembled peptide amphiphiles nanofibers. *Soft Matter* 3:454–462.

Jun, H. W., S. E. Paramonov, H. Dong, N. Forraz, C. McGuckin, and J. D. Hartgerink. 2008. Tuning the mechanical and bioresponsive properties of peptide-amphiphile nanofiber networks. *J Biomater Sci Polym Ed* 19 (5):665–676.

Jun, H.-W., V. Yuwono, S. E. Paramonov, and J. D. Hartgerink. 2005. Enzyme-mediated degradation of peptide-amphiphile nanofiber networks. *Adv Mater* 17 (21):2612–2617.

Kam, N. W. S., T. C. Jessop, P. A. Wender, and H. Dai. 2004. Nanotube molecular transporters: Internalization of carbon nanotube-protein conjugates into Mammalian cells. *J Am Chem Soc* 126 (22):6850–6851.

Kam, L., W. Shain, J. N. Turner, and R. Bizios. 2001. Axonal outgrowth of hippocampal neurons on micro-scale networks of polylysine-conjugated laminin. *Biomaterials* 22 (10):1049–1054.

Kisiday, J., M. Jin, B. Kurz, H. Hung, C. Semino, S. Zhang, and A. J. Grodzinsky. 2002. Self-assembling peptide hydrogel fosters chondrocyte extracellular matrix production and cell division: Implications for cartilage tissue repair. *Proc Natl Acad Sci USA* 99 (15):9996–10001.

Kleinman, H. K., D. Philp, and M. P. Hoffman. 2003. Role of the extracellular matrix in morphogenesis. *Curr Opin Biotechnol* 14 (5):526–532.

Leon, E. J., N. Verma, S. Zhang, D. A. Lauffenburger, and R. D. Kamm. 1998. Mechanical properties of a self-assembling oligopeptide matrix. *J Biomater Sci Polym Ed* 9 (3):297–312.

Mai, R., R. Lux, P. Proff, G. Lauer, W. Pradel, H. Leonhardt, A. Reinstorf et al. 2008. O-phospho-L-serine: A modulator of bone healing in calcium-phosphate cements. *Biomed Tech (Berl)* 53 (5):229–233.

Maltzahn, G. von, S. Vauthey, S. Santoso, and S. Zhang. 2003. Positively charged surfactant like peptides self assemble into nanostructures. *Langmuir* 19:4332.

Massia, S. P. and J. A. Hubbell. 1991. An RGD spacing of 440 nm is sufficient for integrin alpha V beta 3-mediated fibroblast spreading and 140 nm for focal contact and stress fiber formation. *J Cell Biol* 114 (5):1089–1100.

Messersmith, P. B. and S. Starke. 1998. Thermally triggered calcium phosphate formation from calcium-loaded liposomes. *Chem Mater* 10:117–124.

Mrksich, M., C. S. Chen, Y. Xia, L. E. Dike, D. E. Ingber, and G. M. Whitesides. 1996. Controlling cell attachment on contoured surfaces with self-assembled monolayers of alkanethiolates on gold. *Proc Natl Acad Sci USA* 93 (20):10775–10778.

Niece, K. L., C. Czeisler, V. Sahni, V. Tysseling-Mattiace, E. T. Pashuck, J. A. Kessler, and S. I. Stupp. 2008. Modification of gelation kinetics in bioactive peptide amphiphiles. *Biomaterials* 29 (34): 4501–4509.

O'Connell, M. J., S. M. Bachilo, C. B. Huffman, V. C. Moore, M. S. Strano, E. H. Haroz, K. L. Rialon et al. 2002. Band gap fluorescence from individual single-walled carbon nanotubes. *Science* 297 (5581):593–596.

Palmer, L. C., C. J. Newcomb, S. R. Kaltz, E. D. Spoerke, and S. I. Stupp. 2008. Biomimetic systems for hydroxyapatite mineralization inspired by bone and enamel. *Chem Rev* 108 (11):4754–4783.

Palmer, L. C., Y. S. Velichko, M. O. de la Cruz, and S. I. Stupp. 2007. Supramolecular self-assembly codes for functional structures. *Philos Transact A Math Phys Eng Sci* 365 (1855):1417–1433.

Paramonov, S. E., H. W. Jun, and J. D. Hartgerink. 2006a. Modulation of peptide-amphiphile nanofibers via phospholipid inclusions. *Biomacromolecules* 7 (1):24–26.

Paramonov, S. E., H. W. Jun, and J. D. Hartgerink. 2006b. Self-assembly of peptide-amphiphile nanofibers: The roles of hydrogen bonding and amphiphilic packing. *J Am Chem Soc* 128 (22):7291–7298.

Patrick, Jr., C. W., A. G. Mikos, L. V. McIntire, and R. S. Langer. 1998. Prospectus of tissue engineering. In *Frontiers in Tissue Engineering*. Oxford: Pergamon.

Pederson, A. W., J. W. Ruberti, and P. B. Messersmith. 2003. Thermal assembly of a biomimetic mineral/collagen composite. *Biomaterials* 24 (26):4881–4890.

Prieto, A. L., G. M. Edelman, and K. L. Crossin. 1993. Multiple integrins mediate cell attachment to cytotactin/tenascin. *Proc Natl Acad Sci USA* 90 (21):10154–10158.

Rajangam, K., M. S. Arnold, M. A. Rocco, and S. I. Stupp. 2008. Peptide amphiphile nanostructure-heparin interactions and their relationship to bioactivity. *Biomaterials* 29 (23):3298–3305.

Rajangam, K., H. A. Behanna, M. J. Hui, X. Han, J. F. Hulvat, J. W. Lomasney, and S. I. Stupp. 2006. Heparin binding nanostructures to promote growth of blood vessels. *Nano Lett* 6 (9):2086–2090.

Sanborn, T. J., P. B. Messersmith, and A. E. Barron. 2002. In situ crosslinking of a biomimetic peptide-PEG hydrogel via thermally triggered activation of factor XIII. *Biomaterials* 23 (13):2703–2710.

Santoso, S., W. Hwang, H. Hartman, and S. Zhang. 2002. Self-assembly of surfactant-like peptides with variable glycine tails to form nanotubes and nanovesicles. *Nano Lett* 2 (7):687–691.

Sargeant, T. D., M. O. Guler, S. M. Oppenheimer, A. Mata, R. L. Satcher, D. C. Dunand, and S. I. Stupp. 2008. Hybrid bone implants: Self-assembly of peptide amphiphile nanofibers within porous titanium. *Biomaterials* 29 (2):161–171.

Schnur, J. M. 1993. Lipid tubules: A paradigm for molecularly engineered structures. *Science* 262 (5140): 1669–1676.

Shimizu, T., M. Masuda, and H. Minamikawa. 2005. Supramolecular nanotube architectures based on amphiphilic molecules. *Chem Rev* 105 (4):1401–1443.

Silva, G. A., C. Czeisler, K. L. Niece, E. Beniash, D. A. Harrington, J. A. Kessler, and S. I. Stupp. 2004. Selective differentiation of neural progenitor cells by high-epitope density nanofibers. *Science* 303 (5662): 1352–1355.

Stendahl, J. C., M. S. Rao, M.O. Guler, and S. I. Stupp. 2006. Intermolecular forces in the self-assembly of peptide amphiphile nanofibers. In *Advanced Functional Materials*. Wiley-VCH Verlag GmbH & Co. KgaA: Weinheim.

Stendahl, J. C., L. J. Wang, L. W. Chow, D. B. Kaufman, and S. I. Stupp. 2008. Growth factor delivery from self-assembling nanofibers to facilitate islet transplantation. *Transplantation* 86 (3):478–481.

Storrie, H., M. O. Guler, S. N. Abu-Amara, T. Volberg, M. Rao, B. Geiger, and S. I. Stupp. 2007. Supramolecular crafting of cell adhesion. *Biomaterials* 28 (31):4608–4618.

Tanihara, M., Y. Suzuki, E. Yamamoto, A. Noguchi, and Y. Mizushima. 2001. Sustained release of basic fibroblast growth factor and angiogenesis in a novel covalently crosslinked gel of heparin and alginate. *J Biomed Mater Res* 56 (2):216–221.

Thompson, D. H., O. V. Gerasimov, J. J. Wheeler, Y. Rui, and V. C. Anderson. 1996. Triggerable plasmalogen liposomes: Improvement of system efficiency. *Biochim Biophys Acta* 1279 (1):25–34.

Tovar, J. D., R. C. Claussen, and S. I. Stupp. 2005. Probing the interior of peptide amphiphile supramolecular aggregates. *J Am Chem Soc* 127 (20):7337–7345.

Tsonchev, S., K. L. Niece, G. C. Schatz, M. A. Ratner, and S. I. Stupp. 2008. Phase diagram for assembly of biologically-active peptide amphiphiles. *J Phys Chem B* 112 (2):441–447.

Velichko, Y. S., S. I. Stupp, and M. O. de la Cruz. 2008. Molecular simulation study of peptide amphiphile self-assembly. *J Phys Chem B* 112 (8):2326–2334.

von Segesser, L. K., A. Olah, B. Leskosek, and M. Turina. 1993. Coagulation patterns in bovine left heart bypass with phospholipid versus heparin surface coating. *ASAIO J* 39 (1):43–46.

Westhaus, E. and P. B. Messersmith. 2001. Controlled release of calcium from lipid vesicles: Adaption of a biological strategy for rapid gelation of polysaccharide and protein hydrogels. *Biomaterials* 22:453–462.

Wheeler, B. C., J. M. Corey, G. J. Brewer, and D. W. Branch. 1999. Microcontact printing for precise control of nerve cell growth in culture. *J Biomech Eng* 121 (1):73–78.

Whitesides, G. M., J. P. Mathias, and C. T. Seto. 1991. Molecular self-assembly and nanochemistry: A chemical strategy for the synthesis of nanostructures. *Science* 254 (5036):1312–1319.

Yeung, C. K., L. Lauer, A. Offenhausser, and W. Knoll. 2001. Modulation of the growth and guidance of rat brain stem neurons using patterned extracellular matrix proteins. *Neurosci Lett* 301(2):147–150.

Yokoi, H., T. Kinoshita, and S. Zhang. 2005. Dynamic reassembly of peptide RADA16 nanofiber scaffold. *Proc Natl Acad Sci USA* 102 (24):8414–8419.

Zhang, S. 2002. Emerging biological materials through molecular self-assembly. *Biotechnol Adv* 20 (5–6): 321–339.

Zhang, S. 2008. Designer self-assembling peptide nanofiber scaffolds for study of 3-d cell biology and beyond. *Adv Cancer Res* 99:335–362.

Zhang, S., T. C. Holmes, C. M. DiPersio, R. O. Hynes, X. Su, and A. Rich. 1995. Self-complementary oligopeptide matrices support mammalian cell attachment. *Biomaterials* 16 (18):1385–1393.

Zhang, S., T. Holmes, C. Lockshin, and A. Rich. 1993. Spontaneous assembly of a self-complementary oligopeptide to form a stable macroscopic membrane. *Proc Natl Acad Sci USA* 90 (8):3334–3338.

Zhang, S., C. Lockshin, A. Herbert, E. Winter, and A. Rich. 1992. Zuotin, a putative Z-DNA binding protein in Saccharomyces cerevisiae. *EMBO J* 11 (10):3787–3796.

Zhang, S. and A. Rich. 1997. Direct conversion of an oligopeptide from a beta-sheet to an alpha-helix: A model for amyloid formation. *Proc Natl Acad Sci USA* 94 (1):23–28.

Zhang, S., L. Yan, M. Altman, M. Lassle, H. Nugent, F. Frankel, D. A. Lauffenburger, G. M. Whitesides, and A. Rich. 1999. Biological surface engineering: A simple system for cell pattern formation. *Biomaterials* 20 (13):1213–1220.

Zheng, L. X., M. J. O'Connell, S. K. Doorn, X. Z. Liao, Y. H. Zhao, E. A. Akhadov, M. A. Hoffbauer et al. 2004. Ultralong single-wall carbon nanotubes. *Nat Mater* 3 (10):673–676.

8

Electrostatically Self-Assembled Nanomaterials

Helmut Strey
Stony Brook University

Interaction between macroions of opposite charge is a recurring concept in biological self-assembly (Holm et al. 2001). Examples include complexation of DNA with histone proteins inside cell nuclei (Darnell et al. 1990), protamine-induced DNA condensation inside sperm heads (Bloomfield 1996; Podgornik et al. 1998; Strey et al. 1998), and the formation of arterial plaque—a complex between positively charged low-density lipoprotein and negatively charged polysaccharides (Camejo et al. 1985; Camejo et al. 1993). Over the last few years great strides have been made to understand polyelectrolyte-mediated interaction between like-charged objects. The corresponding experimental and theoretical efforts were recently reviewed by Cleasson and Podgornik (Claesson et al. 2005; Podgornik and Licer 2006).

While purely electrostatic interactions undoubtedly play a role during complexation, counterion release is believed to be the major driving force for the self-assembly process in these and other highly charged systems [10–16]. Prior to complexation, the polyelectrolyte and surfactant counterions are restricted to regions close to the surfaces of both the surfactant micelles and the polyelectrolyte chains, a phenomenon known as Manning condensation (Manning 1979). Upon adsorption of a polyelectrolyte chain to the surface of an oppositely charged object (lipid bilayer, surfactant micelle, or polyelectrolyte layer), the counterions bound to both species are released into the bulk solution, which gives rise to a significant increase in the overall entropy of the system (see Figure 8.1). Other possible contributions to the free energy include the conformational entropy of the polyelectrolyte, the elastic energy of the surface, and hydrophobic and steric interactions between the polyelectrolyte chain and the surface. In this chapter, we will review strategies to employ electrostatically driven self-assembly for making long-range ordered nanostructured materials for applications in filtration, catalysis, and drug delivery (Figure 8.2).

Throughout the past 2 decades, researchers have recognized the potential of electrostatically driven self-assembly as a facile structure-forming tool, and significant efforts have been made to elucidate assembly mechanisms, as well as to characterize the wide variety of observed structures.

Here we will discuss three major applications of electrostatically self-assembled nanomaterials: (1) layer-by-layer deposition, (2) polyelectrolyte–surfactant complexes, (3) DNA–lipid complexes, and (4) DNA–cationic polymer complexes.

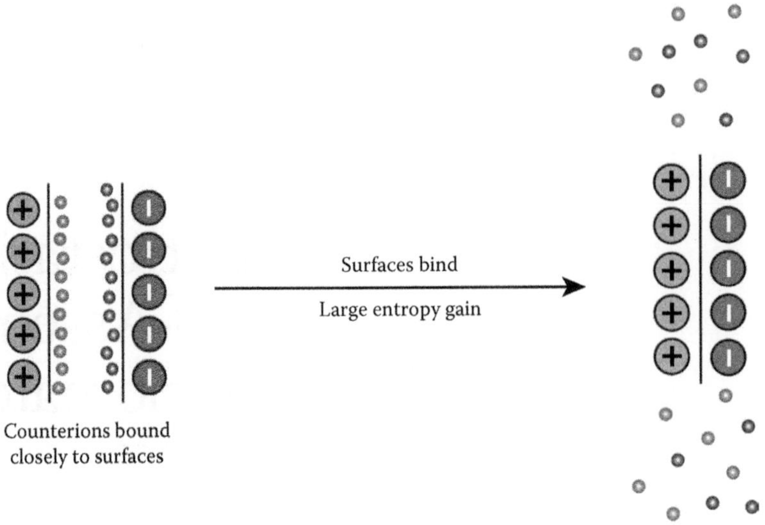

FIGURE 8.1 Counterion release mechanism.

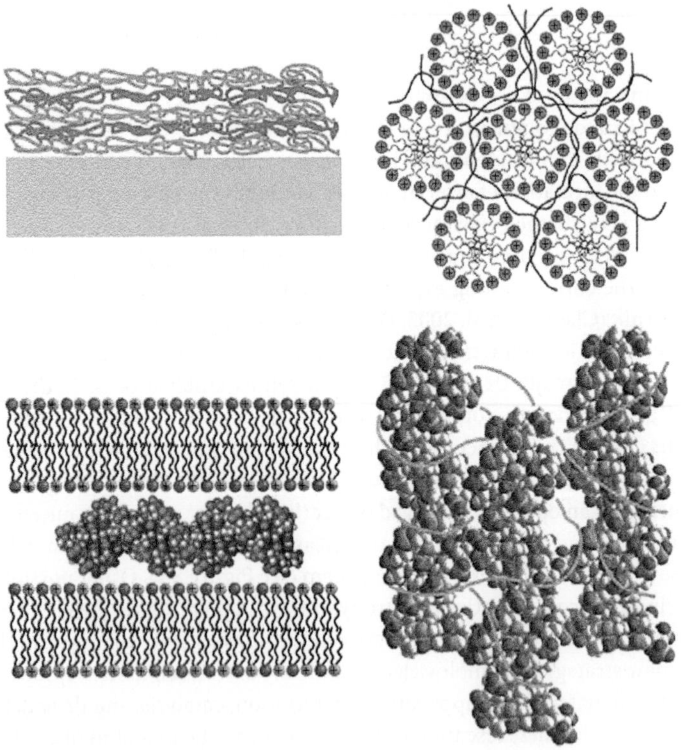

FIGURE 8.2 Examples of electrostatically self-assembled nanomaterials: (upper left) layer-by-layer deposition, (upper right) polyelectrolyte–surfactant complexes, (lower left) DNA–lipid complexes, (lower right) DNA–polycation complexes.

8.1 Layer-by-Layer Deposition

This technique relies on the fact that when polyelectrolytes bind to an oppositely charged surface under low to moderate salt concentrations, the surface overcharges, which leads to the reversal of the surface charge. This reversal allows repeated layering of many (up to several hundreds) polyelectrolyte layers that can be assembled in a regular fashion (Decher 1997). Under the right conditions, these layers exhibit long-range order and can be used for reflective coatings. Usually, the layering is achieved by bathing the surface in oppositely charged polyelectrolyte solutions using a robotic system. Such surface treatments take a long time and are tedious and can therefore not easily be applied commercially. To remedy these shortcomings, alternative coating strategies have been pursued such as alternate spraying of oppositely charged polyelectrolytes to build up multilayers (Schlenoff et al. 2000). Recent efforts and progress in layer-by-layer deposition have been reviewed in Hammond (2004).

8.2 Polyelectrolyte–Surfactant Complexes

Complexes between polyelectrolytes and oppositely charged, small-molecule surfactants provide especially good examples of electrostatic self-assembly, and as such, have attracted a significant amount of attention, both theoretically (Wallin and Linse 1998; Diamant and Andelman 1999, 2000; Hansson 2001) and experimentally (Thalberg and Lindman 1991; Ober and Wegner 1997). Complexes formed between flexible, highly charged polyelectrolytes and oppositely charged surfactants at stoichiometric charge ratios have been of particular experimental interest, since they often form water-insoluble complexes possessing long-range nanoscopic order (Antonietti et al. 1994).

To illustrate the general behavior of these systems, we will briefly review the phase diagram of a model system: cetyltrimethylammonium chloride, copolymers of poly (acrylic acid), and poly (acrylamide) at different salt concentrations (Leonard and Strey 2003). The phase diagram is summarized in Figure 8.3. In this system three long-range ordered structures are found: Pm3n cubic (micellar), hexagonal (cylindrical micelles), and lamellar (bilayers). As the density of the system increases (this can be achieved by decreasing salt concentration, increasing polyelectrolyte charge density, increasing osmotic pressure), the phase structure shifts toward phases that can pack denser (Pm3n turns into hexagonal; hexagonal turns into lamellar). The phase diagram can therefore be explained by a balance between spontaneous

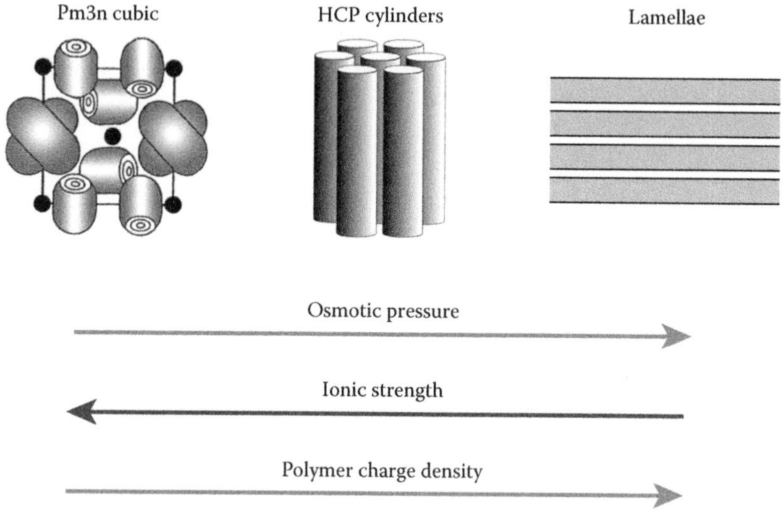

FIGURE 8.3 Generic phase diagram of polyelectrolyte–surfactant complexes.

curvature of the surfactant and close packing constrains. The surfactant tends to form spherical micelles and therefore favors Pm3n cubic phases, whereas packing constrains favor lamellar phases.

Because they exhibit such rich phase behavior, stoichiometric polyelectrolyte–surfactant complexes may serve as attractive precursors for nanostructured materials for potential drug delivery and molecular separation applications. Self-assembled colloidal systems have long been used as templates for creating mesoporous and nanoporous materials (Beck et al. 1992; Monnier et al. 1993; McGrath et al. 1997; Zhao et al. 1998). Common approaches to these materials have involved the self-assembly of surfactants or block copolymers in the presence of various silica species, followed by the sintering of the silica phase. Since the temperatures used for the sintering process often exceed 500°C, the material comprising the organic template completely decomposes, leaving a rigid nanoporous matrix. In such systems, pore sizes tend to be monodisperse as they are dictated by the self-assembled amphiphile structure. Materials possessing monodisperse nanopores are attractive because they may allow for improved size-based selectivity and filtering capabilities over materials with broad distributions of pore sizes.

The amphiphilic templating approach described above has also been used to create non-siliceous mesostructured and nanostructured materials [52–55], but has been met with much more limited success. Removal of amphiphilic templates from non-siliceous, soft matter systems has proven to be especially challenging, since these materials are much less chemically and structurally stable than aluminosilicates. However, the potential of non-siliceous nanostructured materials is extremely high, especially in the area of biocompatible materials for drug delivery and molecular filtration (Figure 8.3).

Self-assembled, polyelectrolyte–surfactant complexes may provide a route to such materials. Crosslinkable, biocompatible polyelectrolytes (e.g., charged polysaccharides) may provide the means to "lock in" the desired structure prior to removal of the surfactant template. The resulting hydrogel should be biocompatible and would possess relatively monodisperse nanopores. This proposed three-step process—complexation, matrix crosslinking, and surfactant removal—is illustrated schematically in Figure 8.4.

There is a great deal of current interest in the use of biocompatible hydrogels in the biomedical and pharmaceutical industries. These types of gels are already being adapted for many applications in areas including drug delivery, wound healing, and bio-separation. It is important to understand that for any material to be considered for implantation within the body, it must meet a number of important criteria. Firstly, it must be biocompatible, meaning that it does not produce a toxic or immunological response in living systems. Secondly, it must not encourage the absorption of nonspecific proteins to its surface. Thirdly, it must be biodegradable, meaning that it will break down in the body over time. There is also a great deal of new interest in materials that are not only biocompatible but also bioactive. Bioactive materials are not only accepted by the body as harmless but they work with the body to perform certain functions. Increasingly, the biomedical industry will be moving toward the use of more and more biologically active materials for in vivo applications.

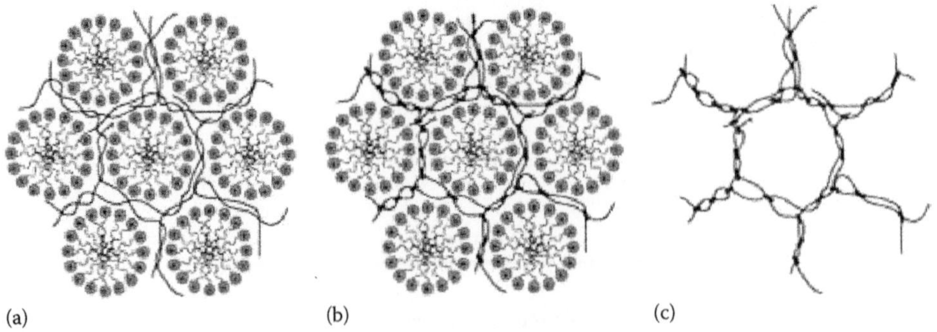

FIGURE 8.4 Proposed approach for producing a nanostructured hydrogel: (a) self-assembly, (b) polymer crosslinking by chemical or physical means, (c) surfactant removal, leaving a nanoporous polymer matrix.

One potential use of polyelectrolyte–surfactant complexes could be in the creation of nanostructured hydrogels. Because their structure is mediated by the surfactant phase, the surfactant can be used as temporary scaffolding, giving form to the polysaccharide phase. The surfactant phase will determine the phase structure of the complex, and the polyelectrolyte phase can then be crosslinked, in theory, locking in the structure of the hydrogel. If the surfactant phase were then removed, the result would be a porous nanostructured hydrogel with a very narrow distribution of pore sizes. Structured gels of this nature could have a number of potential applications ranging from drug delivery to biomedical scaffolds to filtration applications. Because each pore would be of the same diameter and geometry, these materials would be perfect for separations based on particle size. These pores could also potentially be used as wells for the capture of contaminants in water or the surfaces of the pores could be modified and used as sites for chemical reactions due to their high surface area.

8.3 DNA–Lipid Complexes

DNA–lipid complexes have been fascinating not only because of their use in nonviral gene therapy (Felgner et al. 1987) but also because they form an unusual 2D smectic liquid crystalline phase (Lasic et al. 1997; Radler et al. 1997; Salditt et al. 1997). X-ray scattering experiments revealed that complexes of cationic lipid/neutral lipid and DNA form lamellar phases as indicated in Figure 8.2. In addition to the lamellar x-ray peaks, the experiments showed peaks that are associated with the DNA–DNA spacing, which meant that the DNA between the lipid layers exhibits long-range orientational order (smectic). Counterion release in these systems has been discussed in Wagner et al. (1997).

The group around Safinya has controlled the structure of DNA–lipid complexes by manipulating the spontaneous curvature of the lipid phase. Koltover et al. (1998) showed that by introducing 1,2-Dioleoyl-sn-Glycero-3-Phosphoethanolamine (DOPE), instead of the usual 1,2-Dioleoyl-sn-Glycero-3-Phosphocholine (DOPC), as helper lipid, an inverted hexagonal DNA-lipid phase is formed. There is also strong evidence that gene transfection efficiencies are linked to the structure and size of DNA–lipid assemblies.

8.4 DNA–Polycation Complexes

When DNA is mixed with polycations, a complex is formed at low salt concentrations. Depending on the strength of the interaction and charge density of the polycation, hexagonal, cholesteric, or nematic phases can be formed (DeRouchey et al. 2005).

In general, the phase diagram of DNA-polycation follows the same sequence as DNA in monovalent salt concentration as the concentration of DNA is increased: isotropic, cholesteric, hexagonal (Podgornik et al. 1998; Strey et al. 1998). The existence of a cholesteric phase indicates that DNA–DNA interactions are chiral, which causes a twisting of the nematic long-range order. The DNA cholesteric pitch, as a function of density and salt, has been extensively studied (Stanley et al. 2005) and theoretically modeled (Kornyshev et al. 2007; Cherstvy 2008).

Because of the importance of these systems in nonviral gene therapy applications, many naturally occurring and synthetic biodegradable polycations have been investigated (DeRouchey et al. 2005; Eliyahu et al. 2005; Wong et al. 2007; Midoux et al. 2008; Stanley and Strey 2008).

In addition, several attempts have been reported to employ electrostatically self-assembled DNA complexes as biosensors (Skuridin et al. 1996; Yevdokimov et al. 1997; Yevdokimov 2000; Stanley and Strey 2008). In particular, Stanley (Stanley and Strey 2008) showed that DNA complexed by a ABA-triblock copolymer of two short charged blocks of poly (lysine) and a functional block of elastin forms a material that inherits properties of both the DNA liquid crystalline matrix as well as of the functional mid-block. In this case, the temperature dependent properties of the elastin control the cholesteric pitch of the DNA phase.

8.5 Conclusion

In this review, we presented several examples of ordered materials that were self-assembled by electrostatic interactions. From these and other examples, some more general rules of self-assembly can be derived:

1. The hierarchy of forces determines the structure.
   In several of our cases, building blocks are self-assembled by stronger forces. In the case of lipid, lipid molecules are assembled into lipid membranes by hydrophobic interactions. The double helix of DNA is formed by hydrogen bonds. When assembling a material one has to take care not to overwhelm the interactions responsible for the stability of the desired building blocks. This is why electrostatic interactions are more versatile than other interactions. By changing salt concentration and charge density, the range and strength of the interactions can be controlled.
2. The stiffest building block dominates the structure.
   This rule applies to many self-assembled systems. In our examples, the stiffness (or persistence length) of the components goes as (from stiff to soft): lipid assemblies–DNA–surfactant–polyelectrolyte. When mixing polyelectrolyte with surfactant, the surfactant determines the phase structure (Pm3n, hexagonal, lamellar). When mixing DNA with polycations, DNA dominates that structure (cholesteric, hexagonal). On the other hand, in mixtures of DNA and lipid, lipid exhibits a much larger persistence length, and therefore, forces the DNA into the plane.

In summary, we demonstrated the versatility of electrostatic interactions in self-assembled systems. At this time, only a very small subset of possibilities for molecular self-assembly have been explored and applications of self-assembled materials are beginning to emerge in tissue engineering, drug delivery, energy, catalysis, and biomolecular separation.

References

Antonietti, M., J. Conrad, and A. Thünemann. 1994. Polyelectrolyte–surfactant complexes: A new type of solid, mesomorphous material. *Macromolecules* 27:6007–6011.

Beck, J.S., J.C. Vartulli, W.J. Roth, M.E. Leonowicz, C.T. Kresge, K.D. Schmitt, C.T.-W. Chu et al. 1992. A new family of mesoporous molecular sieves prepared with liquid crystal templates. *Journal of the American Chemical Society* 114:10834–10843.

Bloomfield, V.A. 1996. DNA condensation. *Current Opinion in Structural Biology* 6:334–341.

Camejo, G., G. Fager, B. Rosengren, E. Hurtcamejo, and G. Bondjers. 1993. Binding of low-density lipoproteins by proteoglycans synthesized by proliferating and quiescent human arterial smooth-muscle cells. *Journal of Biological Chemistry* 268 (19):14131–14137.

Camejo, G., A. Lopez, F. Lopez, and J. Quinones. 1985. Interaction of low-density lipoproteins with arterial proteoglycans—The role of charge and sialic-acid content. *Atherosclerosis* 55 (1):93–105.

Cherstvy, A.G. 2008. DNA cholesteric phases: The role of DNA molecular chirality and DNA–DNA electrostatic interactions. *Journal of Physical Chemistry B* 112 (40):12585–12595.

Claesson, P.M., E. Poptoshev, E. Blomberg, and A. Dedinaite. 2005. Polyelectrolyte-mediated surface interactions. *Advances in Colloid and Interface Science* 114–115:173–187.

Darnell, J., H. Lodish, and D. Baltimore. 1990. *Molecular Cell Biology*. 2nd edn. New York: Scientific American Books.

Decher, G. 1997. Fuzzy nanoassemblies: Toward layered polymeric multicomposites. *Science* 277 (5330): 1232–1237.

DeRouchey, J., R.R. Netz, and J.O. Radler. 2005. Structural investigations of DNA-polycation complexes. *The European Physical Journal E, Soft Matter* 16 (1):17–28.

Diamant, H. and D. Andelman. 1999. Onset of self-assembly in polymer-surfactant systems. *Europhysics Letters* 48 (2):170–176.

Diamant, H. and D. Andelman. 2000. Self-assembly in mixtures of polymers and small associating molecules. *Macromolecules* 33:8050–8061.

Eliyahu, H., Y. Barenholz, and A.J. Domb. 2005. Polymers for DNA delivery. *Molecules* 10 (1):34–64.

Felgner, P.L., T.R. Gadek, M. Holm, R. Roman, H.W. Chan, M. Wenz, J.P. Northrop, G.M. Ringold, and M. Danielsen. 1987. Lipofection: A highly efficient, lipid-mediated DNA-transfection procedure. *Proceedings of the National Academy Science United States America* 84:7413–7417.

Hammond, P.T. 2004. Form and function in multilayer assembly: New applications at the nanoscale. *Advanced Materials* 16 (15):1271–1293.

Hansson, P. 2001. Self-assembly of ionic surfactants in polyelectrolyte solutions: A model for mixtures of opposite charge. *Langmuir* 17:4167–4180.

Holm, C., P. Kékicheff, and R. Podgornik, eds. 2001. *Electrostatic Effects in Soft Matter and Biophysics.* Dordrecht: Kluwer Academic Publishers.

Koltover, I., T. Salditt, J.O. Radler, and C.R. Safinya. 1998. An inverted hexagonal phase of cationic liposome-DNA complexes related to DNA release and delivery. *Science* 281 (5373):78–81.

Kornyshev, A.A., D.J. Lee, S. Leikin, and A. Wynveen. 2007. Structure and interactions of biological helices. *Reviews of Modern Physics* 79 (3):943–996.

Lasic, D.D., H. Strey, R. Podgornik, M.C.A.R. Stuart, and P.M. Frederik. 1997. DNA-cationic liposome complexes: Structure and structure-activity. *Journal of American Chemical Society* 119:832–833.

Leonard, M.J. and H.H. Strey. 2003. Phase diagrams of stoichiometric polyelectrolyte-surfactant complexes. *Macromolecules* 36 (25):9549–9558.

Manning, G.S. 1979. Counterion binding in polyelectrolyte theory. *Accounts of Chemical Research* 12 (12):443–449.

McGrath, K.M., D.M. Dabbs, N. Yao, I.A. Aksay, and S.M. Gruner. 1997. Formation of silicate L3 phase with continuously adjustable pore size. *Science* 277:552–556.

Midoux, P., G. Breuzard, J.P. Gomez, and C. Pichon. 2008. Polymer-based gene delivery: A current review on the uptake and intracellular trafficking of polyplexes. *Current Gene Therapy* 8 (5):335–352.

Monnier, A., F. Schüth, Q. Huo, D. Kumar, D. Margolese, R.S. Maxwell, G.D. Stucky et al. 1993. Cooperative formation of inorganic–organic interfaces in the synthesis of silicate mesostructures. *Science* 261:1299–1303.

Ober, C.K. and G. Wegner. 1997. Polyelectrolyte-surfactant complexes in the solid state: Facile building blocks for self-organized materials. *Advanced Materials* 9:17–31.

Podgornik, R. and M. Licer. 2006. Polyelectrolyte bridging interactions between charged macromolecules. *Current Opinion in Colloid & Interface Science* 11 (5):273–279.

Podgornik, R., H.H. Strey, and V.A. Parsegian. 1998. Colloidal DNA. *Current Opinion in Colloid & Interface Science* 3 (5):534–539.

Radler, J.O., I. Koltover, T. Salditt, and C.R. Safinya. 1997. Structure of DNA-cationic liposome complexes: DNA intercalation in multilamellar membranes in distinct interhelical packing regimes. *Science* 275 (5301):810–814.

Salditt, T., I. Koltover, J.O. Radler, and C.R. Safinya. 1997. Two-dimensional smectic ordering of linear DNA chains in self-assembled DNA-cationic liposome mixtures. *Physical Review Letters* 79 (13):2582–2585.

Schlenoff, J.B., S.T. Dubas, and T. Farhat. 2000. Sprayed polyelectrolyte multilayers. *Langmuir* 16 (26):9968–9969.

Skuridin, S.G., Y.M. Yevdokimov, V.S. Efimov, J.M. Hall, and A.P. Turner. 1996. A new approach for creating double-stranded DNA biosensors. *Biosensors and Bioelectronics* 11 (9):903–911.

Stanley, C.B., H. Hong, and H.H. Strey. 2005. DNA cholesteric pitch as a function of density and ionic strength. *Biophysical Journal* 89 (4):2552–2557.

Stanley, C.B. and H.H. Strey. 2008. Electrostatically driven self-assembly of hybrid elastin-DNA liquid crystals. *Soft Matter* 4 (2):241–244.

Strey, H.H., R. Podgornik, D.C. Rau, and V.A. Parsegian. 1998. DNA–DNA interactions. *Current Opinion in Structural Biology* 8 (3):309–313.

Thalberg, K. and B. Lindman. 1991. Gel formation in aqueous systems of a polyanion and an oppositely charged surfactant. *Langmuir* 7:277–283.

Wagner, K., E. Keyes, T.W. Kephart, and G. Edwards. 1997. Analytical Debye–Hückel model for electrostatic potentials around dissolved DNA. *Biophysical Journal* 73:21–30.

Wallin, T. and P. Linse. 1998. Polyelectrolyte-induced micellization of charged surfactantts. Calculations based on a self-consistent field model. *Langmuir* 14:2940–2949.

Wong, S.Y., J.M. Pelet, and D. Putnam. 2007. Polymer systems for gene delivery—past, present, and future. *Progress in Polymer Science* 32 (8–9):799–837.

Yevdokimov, Y.M. 2000. Double-stranded DNA liquid-crystalline dispersions as biosensing units. *Biochemical Society Transactions* 28 (2):77–81.

Yevdokimov, Y.M., V.I. Salyanov, L.V. Buligin, A.T. Dembo, E. Gedig, F. Spener, and M. Palumbo. 1997. Liquid-crystalline structure of nucleic acids: Effect of antracycline drugs and copper ions. *Journal of Biomolecular Structure and Dynamics* 15 (1):97–105.

Zhao, D., J. Feng, Q. Huo, N. Melosh, G.H. Fredrickson, B.F. Chmelka, and G.D. Stucky. 1998. Triblock copolymer synthesis of mesoporous silica with periodic 50 to 300 angstrom pores. *Science* 279:548–552.

9

Peptide-Based Nanomaterials for siRNA Delivery: Design, Evaluation, and Challenges

Seong Loong Lo
Institute of Bioengineering and Nanotechnology

Yukti Choudhury
Institute of Bioengineering and Nanotechnology

Shu Wang
Institute of Bioengineering and Nanotechnology

and

National University of Singapore

9.1 Introduction

The use of small interfering RNA (siRNA) holds great promises for the development of gene-specific therapeutics. Similar to plasmid DNA transfection, cellular uptake of naked siRNA is difficult due to the lack of transport mechanisms to move negatively charged, large nucleic acids across the cell membrane. Double-stranded siRNA is more stable than single-stranded RNA, but unmodified siRNA is still prone to enzymatic degradation by nucleases in serum and can be destroyed within minutes (Layzer et al. 2004; Morrissey et al. 2005a). Also, unmodified siRNA might trigger Toll-like receptor 7 pathway and induce nonspecific activation of the immune system (Hornung et al. 2005; Judge et al. 2005). Therefore, efficient siRNA delivery systems have to be developed to protect siRNA from rapid degradation in serum, enhance cellular internalization, and reduce immunostimulatory activity. Different strategies have been employed to enhance siRNA delivery efficiency, including physical methods, chemical modification of siRNA, and non-covalent encapsulation of siRNA with lipids, polymers, or peptides.

Peptides are very attractive nanomaterials with significant potential for biomedical applications. The development of peptide nanomaterials is based on the biochemical understanding that the active sites of protein molecules, such as enzymes, receptor ligands, and antibodies, usually involve only 5–20 amino acid residues (Sparrow et al. 1998). With the rapid advances in structural biology and high-throughput genomics and proteomics, the identification of peptide motifs associated with biological functions has been drastically accelerated (Saito et al. 2007). Thus, peptide materials offer a highly attractive feature of incorporating various natural or synthetic sequences with biological activities, for example, cell targeting domain and nuclear targeting domain. Peptides are relatively easy to be synthesized in large scale and can be characterized with well-established chemistry and instrumental operation. As biomaterials, peptides are generally less toxic and have low immunogenicity compared to high molecular weight (MW) polymers (Fabre and Collins 2006) and undergo degradation in the body to naturally occurring compounds. Different nanometric structures can arise from peptide-cargo or interpeptide interactions of electrostatic, hydrophobic, or aromatic nature. A good example is nanometric complex formed by peptides and nucleic acids. In this chapter, published studies using peptides to encapsulate siRNA noncovalently into nanoparticles will be reviewed and discussed in detail. A short summary of commonly used methods to deliver siRNA will also be presented.

9.2 RNA Interference

RNA interference (RNAi) is both an intrinsic and nearly universal mechanism of gene expression regulation and a means to control specific gene expression extrinsically. RNAi hinges on simple yet specific Watson–Crick base-pairing between small RNA and messenger RNA (mRNA), resulting in reduced gene expression at the posttranscriptional level. With its generally inhibitory effect on gene function, RNAi has greatly impacted the area of functional genomics and very rapidly entered the arena of therapeutic development in disease settings.

Double-stranded RNA (dsRNA) of various origins and lengths are the initiators of RNAi. They are processed into short dsRNAs, 21–28 nucleotides (nt), depending on species (Hutvagner and Zamore 2002), which then affect the sequence-specific degradation of complementary single-stranded RNAs. dsRNA can be of viral origin (converted from single-stranded form by RNA-dependent RNA polymerases), overlapping transcripts from repetitive sequences such as transposons (Waterhouse et al. 2001) or artificially introduced long dsRNAs. MicroRNAs (miRNAs) that form dsRNA hairpins via intramolecular complementarity are endogenous initiators of RNAi (Bartel 2004).

The use of short dsRNAs or siRNAs in mammalian cells as a direct mediator of RNAi is the method of choice for specific gene silencing in mammalian cells (Caplen et al. 2001; Elbashir et al. 2001). This overcomes the major hurdle in the use of long dsRNA-mediated RNAi that appears to cause nonspecific degradation of mRNAs and/or general toxicity in vertebrates, including zebrafish and mammalian cells (Tuschl et al. 1999; Caplen et al. 2000; Oates et al. 2000; Zhao et al. 2001). The reason for this toxicity is understood to be a dsRNA-induced interferon response (Manche et al. 1992; Kumar and Carmichael 1998; Stark et al. 1998). The discovery of endogenous miRNAs also led to the development of tools for the intracellular expression of RNAi triggers that mimics miRNAs in the form of short-hairpin RNAs (shRNA) (Paddison et al. 2004).

The canonical RNAi pathway begins with the excision of short dsRNA fragments from long dsRNA in the cytoplasm by the multidomain RNaseIII endonuclease Dicer. The dsRNA products of the Dicer activity are siRNA duplexes about 19–25 nt in length and have 5′ phosphates and 2-nucleotide 3′ overhangs (Bernstein et al. 2001; Elbashir et al. 2001; Macrae et al. 2006). Endogenous miRNAs are transcribed in the nucleus as primary structures that are cleaved by a nuclear RNaseIII enzyme Drosha to ~70 nt precursor miRNAs (pre-miRNA). The pre-miRNAs are exported to the cytoplasm and are processed by Dicer to produce an miRNA duplex similar to the siRNA duplex (Zeng and Cullen 2004). In practice, synthetic siRNAs are designed to resemble Dicer cleavage products of 21–22 nt length duplexes with 2-nucleotide 3′ overhangs. The miRNA or siRNA duplex generated by Dicer is loaded into the RNA-induced silencing complex (RISC) by RISC-loading complex (RLC), a trimeric complex including Dicer (Maniataki and Mourelatos 2005). Synthetic siRNAs are most likely directly loaded to RISC, presumably independent of Dicer (part of RLC), at least in vitro in mammalian cells (Macrae et al. 2006; Carthew and Sontheimer 2009). The single-stranded siRNA guide strand, once loaded into RISC, guides RISC to mRNA targets that are perfectly complementary, orchestrating a sequence-specific degradation of targets. Although siRNAs typically function to cleave target mRNA through perfect complementarity, mismatches in the siRNA/target duplex often result in cleavage or translational repression of unintended targets (Carthew and Sontheimer 2009). This aspect of siRNA function is nearly identical to that of miRNAs and is dictated largely by sequence complementarity between 7-nucleotide "seed" region at the positions 2–8 of antisense strand of siRNA and the target. This is summed up as the "off-target" effect of siRNAs (Jackson et al. 2003; Saxena et al. 2003). Chemical modification of siRNAs to limit such effects has been explored, and 2′-*O*-Me modifications and DNA substitutions have been demonstrated to be effective (Jackson et al. 2006; Ui-Tei et al. 2008). A summary of the RNAi pathway is illustrated in Figure 9.1.

The mechanistic understanding of RNAi has promoted rational siRNA design for experimental use. Thus, the following aspects should be taken into account: (1) structural requirements of siRNA duplex, having length of 19–21 nt, absence of 5′ overhangs, GC content of ~50%, and appropriate thermodynamics of duplex for guide strand selection; (2) target mRNA accessibility; and (3) "seed" region choice to minimize off-target effects (Reynolds et al. 2004). Interestingly, it has been shown that RISC programming is more efficient when a longer siRNA (27 mers) is used, as they are incorporated into the Dicer-processing step and linked to RISC activation (Gregory et al. 2005; Kim et al. 2005; Siolas et al. 2005).

9.3 siRNA Delivery

Long-term stable gene silencing can be established with the use of viral delivery vectors for shRNA. Like miRNAs, shRNAs form hairpin structures and are Dicer substrates. The typical design uses small inverted repeats (19–29 nt) expressed from an RNA Pol III promoter that transcribes self-complementary shRNAs (Paddison et al. 2004). These are exported out to the cytoplasm and processed by Dicer.

One of the limitations associated with the use of shRNA is its inevitable interference with the endogenous miRNA pathway, given its nuclear phase. A study using a deno-associated virus, AAV/shRNA vectors for silencing luciferase transgene reported liver toxicity in ~50% of animals that were administered these shRNA vectors (Grimm et al. 2006). The saturation of the miRNA pathway, particularly nuclear export, resulted in downregulation of endogenous miRNAs, manifesting in outward toxicity, although optimizing shRNA dose and sequence may avoid the oversaturation. No similar toxicity and disruption of miRNA pathway was reported when synthetic siRNA duplexes were systemically administered to animals (John et al. 2007). In this sense, direct use of siRNA displayed a better safety profile even though silencing effect of siRNA might be transient.

9.3.1 Physical Methods

Physical methods such as electroporation and hydrodynamic injection are the most direct methods to introduce foreign substance into cells. Electroporation utilizes externally applied electrical field to increase the permeability of cell membrane, create pores, and allow extracellular material to diffuse into cytoplasm. Electroporation has been demonstrated as a useful tool to facilitate in vitro siRNA delivery into primary cells and difficult-to-transfect cells such as human primary fibroblasts, human umbilical vein endothelial cells (HUVEC), and neuroblastoma cell lines (Jordan et al. 2008). The potential of electroporation for in vivo siRNA delivery has been reported in the rat brain (Akaneya et al. 2005), rats and mice muscle tissue (Kishida et al. 2004; Kong et al. 2004; Golzio et al. 2005; Takayama et al. 2009), and tumor xenograft (Takahashi et al. 2005; Takei et al. 2008). However, the use of electroporation is usually limited due to high rate of cell mortality caused by high-voltage pulses and the availability of other more effective delivery systems.

Another physical method is hydrodynamic injection, which involves the delivery of samples into tissue by intravascular injection of a relatively large volume of samples with high hydrostatic pressure (Liu et al. 1999). In fact, the first demonstration of RNAi in mammals such as mice was using siRNA delivered by hydrodynamic injection (Lewis et al. 2002; McCaffrey et al. 2002). While hydrodynamic injection is usually performed to deliver siRNA into the liver (Lewis et al. 2002; McCaffrey et al. 2002; Giladi et al. 2003; Klein et al. 2003; Song et al. 2003; Xu et al. 2005; Morrissey et al. 2005b), the uptake

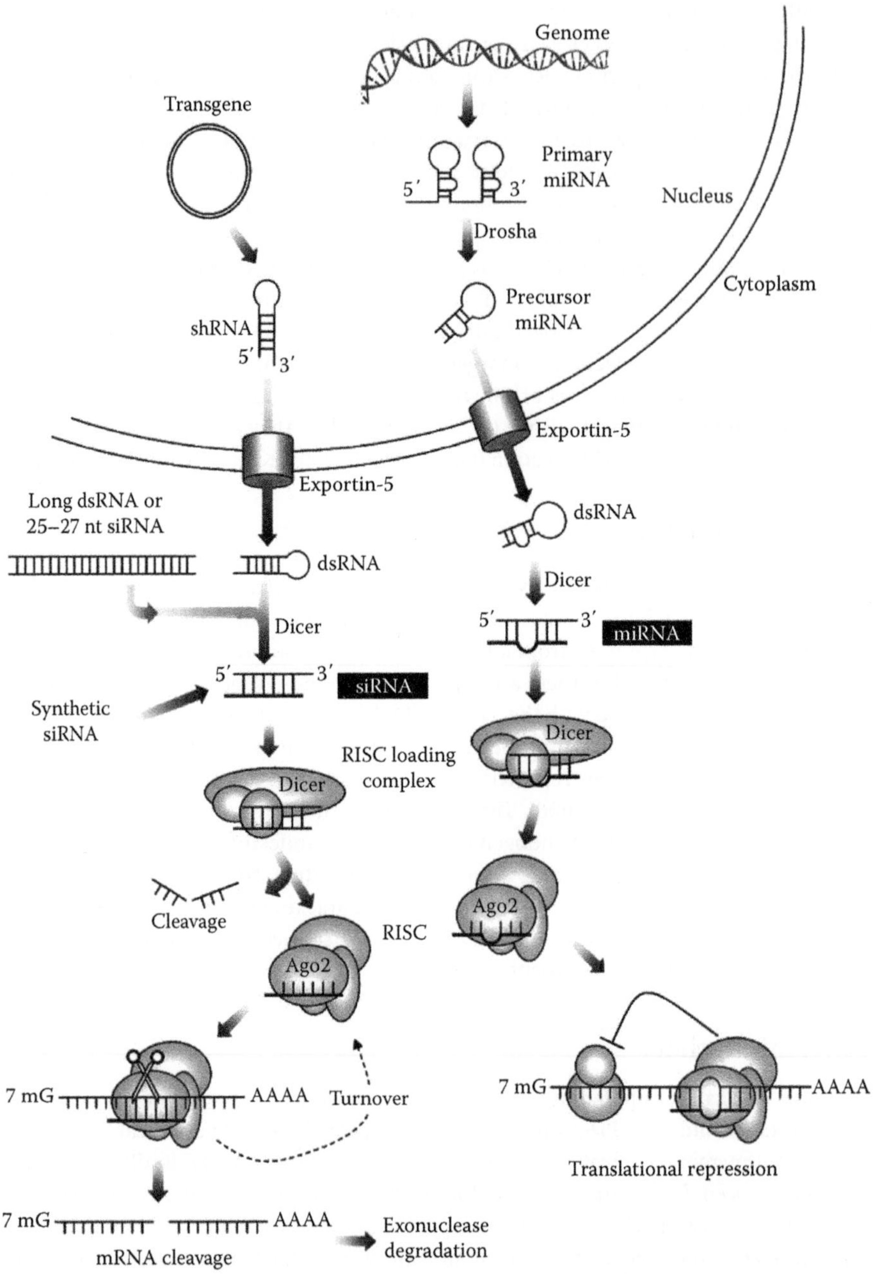

FIGURE 9.1 The mechanism of RNAi. When exogenously introduced, the triggers of RNAi can be DNA-encoded shRNA, long dsRNA (Dicer-substrates), or chemically synthesized siRNA. In the nucleus, shRNAs are transcribed as hairpin-structures from a transgene and exported to the cytoplasm. The endogenous RNAi pathway begins with the transcription of primary miRNA structure, which undergoes nuclear processing by the enzyme Drosha into pre-miRNA. The pre-miRNA is exported to the cytoplasm. In the cytoplasm, the enzyme Dicer mediates processing of longer forms of dsRNAs (long dsRNA, shRNA, pre-miRNA) into 21–23 nt duplex siRNA or miRNA, with typical features such as 3′-overhangs and 5′-phosphate. The duplexes are loaded into RISC via an intermediate RLC. The passenger strand (—) is unwound and cleaved. The single guide strand (—) then directs target gene silencing based on sequence complementarity. Typically, the outcome of siRNA activity is target degradation (full complementarity). The outcome of miRNA activity could be translational repression (partial complementarity).

of siRNA by other organs such as the kidney (Hamar et al. 2004), pancreas (Bradley et al. 2005), and lung (Tompkins et al. 2004) were also possible. Hydrodynamic injection of siRNA is a relatively efficient method to deliver siRNA, but this method has limited application in humans due to the possible complications caused by rapid injection of large volume of fluid into the blood vessel.

9.3.2 Chemical Modification of siRNA

Ideally, chemical modifications of siRNA increase stability in serum, reduce immunostimulatory activity, increase silencing ability, and enhance cellular uptake. There are three commonly used chemical modifications for siRNA, including phosphodiester modification, such as phosphorothioate or boranophosphonate, at siRNA 3′-end (Braasch et al. 2004; Hall et al. 2004), modification of 2′-base sugar, such as 2′-*O*-methyl or 2′-deoxy-2′-fluoro, of selected nucleotides (Chiu and Rana 2003; Layzer et al. 2004), and the use of locked nucleic acid (LNA) in siRNA strand (Braasch et al. 2003; Elmen et al. 2005). Although these methods have been demonstrated to protect siRNA against nuclease degradation and reduce off-target effects, cytotoxicity and reduced gene-silencing activity associated with modifications were observed in some reports (de Fougerolles et al. 2007; de Paula et al. 2007; Rana 2007).

Besides chemically modifying nucleotide structure, covalent conjugation of biologically functional molecules is an attractive method to enhance siRNA delivery. The conjugation is normally done at the sense strand as it is less likely to affect the silencing effect of the siRNA. Cholesterol conjugation has previously been used to enhance delivery of antisense oligonucleotide and nucleic acid into liver cells (Biessen et al. 1999; Cheng et al. 2006). Injection of cholesterol-conjugated siRNA into mouse was able to silence in vivo mRNA expression of apolipoprotein B required for transport of cholesterol (Soutschek et al. 2004), and a mutated gene related to Huntington's disease (DiFiglia et al. 2007). Another example of conjugation is to exploit peptides derived from cell-penetrating peptide (CPP) or protein transduction domain (PTD). CPPs are known to be capable of delivering different cargoes into cells efficiently (Stewart et al. 2008). Different CPPs, such as penetratin, transportan, or Tat, have been conjugated to siRNAs through reducible disulfide bond linkages (Chiu et al. 2004; Davidson et al. 2004; Muratovska and Eccles 2004; Moschos et al. 2007). In earlier reports, enhanced cellular uptake of the conjugated siRNA and successful inhibition of reporter gene were observed. However, because no purification after conjugation reaction between CPP and siRNA was done (Chiu et al. 2004; Davidson et al. 2004; Muratovska and Eccles 2004), the observed RNAi activity could be due to the complex formed by non-covalent encapsulation of siRNA with excess cationic peptide in the reaction mixture (Meade and Dowdy 2007). This assumption was confirmed by later findings showing that purified CPP-conjugated siRNA did not increase cellular uptake and distribution in vitro (Moschos et al. 2007; Meade and Dowdy 2008).

9.3.3 Non-Covalent Encapsulation of siRNA

Difficulties in large-scale purification and characterization of siRNA covalently conjugated with functional molecules might raise concerns over the quality of siRNA-based therapeutics. Hence, non-covalent encapsulation of siRNA with cationic molecules could be a better alternative for siRNA delivery. One of the most commonly used encapsulation reagents is cationic lipid. Cationic lipid is an amphiphilic molecule composed of cationic hydrophilic amine groups and hydrophobic side chains. Cationic lipids are assembled into bilayer-structured liposomes and can interact with nucleic acids to form a complex termed lipoplex. A great variety of lipid-based products are commercially available for siRNA delivery, for example, Lipofectamine™ 2000 (Invitrogen), DharmaFECT™ set (Dharmacon), and siPORT™ *NeoFX*™ (Ambion). Despite their popularity and excellent in vivo results (Sioud and Sorensen 2003; Sorensen et al. 2003; Flynn et al. 2004; Ma et al. 2005), toxicity of some cationic lipids remains a safety issue for in vivo application (Ma et al. 2005; Lv et al. 2006; Akhtar and Benter 2007).

Many polymers originally developed for plasmid DNA delivery are also possible encapsulation reagents of siRNA, for instance, polyethylenimine (Urban-Klein et al. 2005; Werth et al. 2006;

Zintchenko et al. 2008), chitosan (Howard et al. 2006; Katas and Alpar 2006), and cyclodextrin polymers (Hu-Lieskovan et al. 2005; Bartlett et al. 2007; Heidel et al. 2007; Bartlett and Davis 2008). Polyethylene-glycol (PEG) modified lipid (Zimmermann et al. 2006), stearyl octa-arginine (Tonges et al. 2006), and cholesteryl nona-arginine (Kim et al. 2006) have also been demonstrated to encapsulate and deliver siRNA efficiently in vivo.

9.4 Peptide-Based siRNA Delivery Vectors

Over the past few decades, different peptides have been designed and used for plasmid DNA delivery (Mann et al. 2008). Although both plasmid DNA and siRNA are double-stranded nucleic acids in nature, there are some differences that are worthy of being taken into account in designing peptide-based vectors. First of all, siRNA is more susceptible to hydrolysis by serum nucleases than DNA because of the hydroxyl group in the 2′ position of the pentose ring in RNA backbone. Secondly, in terms of size, plasmid DNA is often several kilo base pairs whereas siRNA is 19–21 base pairs. Thirdly, plasmid DNA has to be delivered into the nucleus where the functional gene can be expressed, whereas siRNAs usually induce degradation of target mRNAs in the cytosol, except for rare cases in which transcriptional gene silencing is involved in the nucleus (Kawasaki and Taira 2004; Morris et al. 2004). Therefore, peptides that are efficient in plasmid DNA delivery might not work similarly for siRNA. Keeping this in mind, some of the strategies developed for plasmid DNA delivery provide good starting points for new peptide vector design for siRNA delivery. Depending upon their design, peptide-based vectors involved in siRNA delivery can be categorized as (1) amphipathic peptides, (2) CPPs, (3) histidine-rich branched peptides, and (4) peptide-based reducible polymers (PRPs) (Table 9.1). In the following section, we highlight and discuss key features of each design.

9.4.1 Amphipathic Peptides

An amphipathic peptide is a peptide that possesses both hydrophobic (nonpolar) and hydrophilic (polar) properties. The amphipathicity characteristic could originate from either a primary structure that contains both hydrophobic domain and hydrophilic domain or a secondary structure that allows hydrophobic and hydrophilic amino acid residues to be positioned on the opposite side of peptide conformation (Fernandez-Carneado et al. 2004).

9.4.1.1 Primary Amphipathic Peptide

The most well-characterized primary amphipathic peptides are MPG-based peptides. These are a series of peptides composed of a hydrophobic domain derived from glycine-rich region of the membrane fusion sequence of HIV gp41 (Gallaher 1987; Rafalski et al. 1990) and a hydrophilic domain derived from the nuclear localization signal (NLS) of SV40 large T-antigen (Kalderon et al. 1984; Dingwall and Laskey 1992). The first generation of the MPG-based peptides, MPG-W, was designed to study intracellular localization of the peptides with different chemical modification (Vidal et al. 1996). Based on the MPG-W sequence, the MPG-mNLS peptide that contains a mutated NLS sequence was developed to deliver oligonucleotides (Morris et al. 1997). To derive the MPG-mNLS peptide, the following changes were made: (1) phenylalanine (F) residue at position 7 in the hydrophobic domain was restored; (2) the short linker that connects the hydrophobic and hydrophilic domains was changed to tryptophan-serine-glutamine (WSQ), allowing sensitive monitoring and quantification of interaction of peptide with oligonucleotides by measuring fluorescence quenching; (3) in hydrophilic domain, the proline (P) residue from the NLS of SV40 large T-antigen was included and the second lysine (K) residue was mutated to serine (S) residue; and (4) N- and C-termini were modified with acetyl group and cysteamide group, respectively, to improve the ability to cross the membrane (Mery et al. 1993). The MPG-mNLS peptide protects oligonucleotides from DNase degradation, suggesting that the peptide interacts with oligonucleotides strongly. At both 37°C and 4°C, peptide/nucleotide complexes prepared at molar ratio of 20,

TABLE 9.1 List of Peptides Tested for siRNA Delivery

Amphipathic Peptides	
Primary amphipathic peptides	
MPG-W	X-GALFLGWLGAAGSTMGA-R-KKKRKV-Cya-X′
MPG-mNLS	Ac-GALFLGFLGAAGSTMGA-WSQ-PKSKRKV-Cya
MPG-NLS	Ac-GALFLGFLGAAGSTMGA-WSQ-PKKKRKV-Cya
MPGα-NLS	Ac-GALFLAFLAAALSLMGL-WSQ-PKKKRKV-Cya
MPGα-mNLS	Ac-GALFLAFLAAALSLMGL-WSQ-PKSKRKV-Cya
Secondary amphipathic peptides	
KALA	WEAKLAKALAKALAKHLAKALAKALKACEA
CADY	Ac-GLWRALWRLLRSLWRLLWRA-Cya
Classical Cell-Penetrating Peptides	
Penetratin-based peptide	
EB1	LIRLWSHLIHIWFQNRRLKWKKK-amide
Oligoarginine-based peptide	
R_9	RRRRRRRRR
RVG	YTIWMPENPRPGTPCDIFTNSRGKRASNG
RV-Mat	MNLLRKIVKNRRDEDTQKSSPASAPLDDG
RVG-9r	YTIWMPENPRPGTPCDIFTNSRGKRASNG -GGG-rrrrrrrrr
RV-Mat-9r	MNLLRKIVKNRRDEDTQKSSPASAPLDDG -GGG-rrrrrrrrr
Novel peptide	
POD	$GGG(ARKKAAKA)_4$
Histidine-Rich Branched Peptides	
H^3K8b	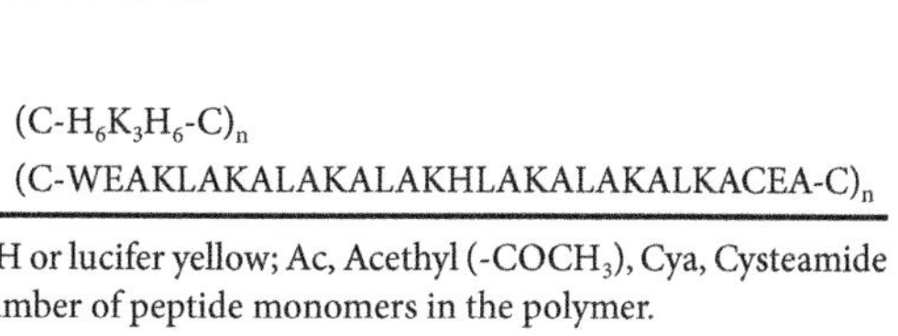
$H^3K(+H)4b$	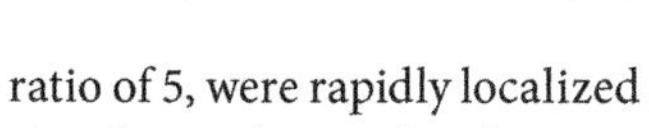
Peptide-Based Reducible Polymers	
HIS6-RPC	$(C\text{-}H_6K_3H_6\text{-}C)_n$
cl-KALA	$(C\text{-WEAKLAKALAKALAKHLAKALAKALKACEA-}C)_n$

X—H, Ac, or methoxy coumarin; X′—H or lucifer yellow; Ac, Acethyl ($-COCH_3$), Cya, Cysteamide ($-NH-CH_2-CH_2-SH$); r, D-arginine; n, number of peptide monomers in the polymer.

corresponding to an amine/phosphate (N/P) ratio of 5, were rapidly localized in the nucleus of fibroblast cells (HS-68 and NIH-3T3) within 1 h, indicating that endosomal pathway was not involved in the internalization of peptide-mediated delivery. Even though both MPG-W and MPG-mNLS peptides adopted similar β-sheet structure in phospholipid solution (Chaloin et al. 1998; Vidal et al. 1998), alteration of the short linker to WSQ changed the localization pattern from membrane-associated to nucleus-associated localization. The authors reasoned that an additional arginine (R) residue in the MPG-W peptide possibly blocked the nuclear translocation property of NLS as suggested previously (Whitley et al. 1995).

In view of the success of oligonucleotide delivery, the potential use of the MPG-mNLS peptide was extended to plasmid DNA delivery (Morris et al. 1999). The MPG-mNLS peptide could bind to plasmid DNA through electrostatic interaction. Formation of peptide cage around DNA, rather than just charge neutralization, was hypothesized since large excess of the peptides was required for complex formation. The MPG-mNLS peptide was capable of efficiently delivering a plasmid vector expressing luciferase gene into various types of cell lines at an N/P ratio of 10, without exhibiting any cytotoxic effects. After delivering a plasmid DNA expressing antisense full-length cDNA human cdc25c (cell division cycle 25 homolog C) into late G1 phase human fibroblast (HS-68), efficient inhibition (70%) of entry into mitosis was observed, suggesting the loss of cdc25C protein that plays a key role in the regulation of cell division.

Following the discovery of RNAi, the possibility of using MPG-mNLS for siRNA delivery was explored (Simeoni et al. 2003). The authors first compared the difference between an MPG peptide with wild-type NLS (MPG-NLS) and MPG-mNLS in nuclear targeting ability. Plasmid DNA delivery efficiency of the MPG-mNLS peptide in HS-68, as measured by luciferase activity, was only about one-third of that offered by the MPG-NLS peptide. When fluorescently labeled siRNA was delivered by the two peptides, the delivery efficiencies were similar. However, siRNA localized to the nucleus when delivered by MPG-NLS but remained mostly in the cytoplasm when delivered by MPG-mNLS, although it has been previously demonstrated that oligonucleotide delivered by the MPG-mNLS peptide localized in the nucleus (Mery et al. 1993). The authors claimed that the cytoplasmic localization of siRNA was due to the reduced nuclear-targeting ability of the MPG-mNLS peptide. The discrepancy in intracellular distribution pattern could arise from the re-localization of free siRNAs released from the complex. The MPG-mNLS/siRNA complex might first localize in the nucleus, followed by early disassembly of the complex due to a weaker binding of the mutated NLS to siRNA. The free siRNA will then be actively excluded from the nucleus by Exportin-5 (Ohrt et al. 2006), resulting in cytoplasmic localization. Moreover, since the MPG-mNLS peptide was not fluorescently labeled, there is no evidence to demonstrate that the MPG-mNLS peptide was co-localized with siRNA cargo in the cytoplasm. Nonetheless, despite the uncertain effect of NLS mutation, MPG-mNLS-mediated siRNA delivery was effective in silencing target gene expression. In HeLa and COS7 cells pre-transfected with a luciferase plasmid, MPG-mNLS peptide/siRNA against luciferase complexes at an N/P ratio of 10 reduced the luciferase activities by 90% and 95%, respectively. This silencing effect was comparable with that of commercial lipid-based delivery vector Oligofectamine™. Northern blot analysis showed that siRNA against glyceraldehyde 3-phosphate dehydrogenase (GAPDH) delivered by the MPG-mNLS peptide was able to reduce 80% of protein expression in HS-68.

Based on results from conformational analysis (Deshayes et al. 2004a,b; 2006a,b), a cellular entry mechanism mediated by the MPG-mNLS and MPG-NLS peptides was proposed. The model consists of five steps: (1) formation of peptide/cargo complex, (2) electrostatic interaction between cellular membrane component and hydrophilic cationic domain of the peptides, (3) transient formation of β-sheet-related transmembrane channels leading to insertion of the peptide/cargo complex, (4) internalization of peptide/cargo complex, and (5) translocation to the nucleus.

A derivative of original MPG-NLS peptide, MPGα-NLS, was also used in a study that established techniques to analyze peptide-mediated siRNA internalization along with its biological effects (Veldhoen et al. 2006). The MPGα-NLS peptide, with a partial α-helical structure resulting from five mutations in hydrophobic domain, was originally designed to study the entry mechanism of the MPG-NLS peptide (Deshayes et al. 2004a). To test whether the MPGα-NLS peptide could mediate siRNA delivery, Veldhoen et al. first established two cell lines stably expressing firefly luciferase, HeLa-TetOff Luc (HTOL) and a derivative of human urinary bladder carcinoma cells (ECV304 GL3). These two cell lines were transfected with 50 nM of siRNA against luciferase complexed with either Lipofectamine 2000 (LF2000, 10 μg/mL) or MPGα-NLS (at an N/P ratio of 15). The MPGα-NLS/siRNA complexes reduced luciferase activity by 80%–90%, a silencing level similar to that offered by the LF2000/siRNA complexes. However, the apparent value of half maximal inhibition (IC50) of the MPGα-NLS/siRNA complexes was ~0.8 nM, which was about 20–40 times higher than that of the LF2000/siRNA complexes. The authors suggested that MPGα-NLS might be less efficient due to a lower rate of siRNA internalization. To quantify the amount of intracellular siRNAs, a sensitive liquid hybridization method developed previously (Overhoff et al. 2004) was adapted. Briefly, 4 h after transfection with peptide/siRNA complexes, the cells were treated with heparin (15 U/mL) three times to remove extracellularly bound complexes. The cells were incubated for another 24 h, followed by cell detachment with trypsin. Cellular RNA was extracted, hybridized with ^{32}P-labeled sense-strand at 95°C for 10 min, followed by 1 h incubation at 37°C. The samples were resolved by polyacrylamide gel electrophoresis (PAGE) and blotted onto membrane for quantification. The authors discovered that to obtain 50% inhibition of maximum luciferase activity, 10,000 siRNA molecules were required in MPGα-NLS-mediated delivery. This value was 30-fold higher than the case of LF2000-mediated delivery. Based on microscopic observation of cellular distribution of fluorescently labeled siRNA and effects of various inhibitors/effectors of endocytosis, the authors concluded that accumulation in endosomes is the bottleneck of siRNA delivery mediated by the MPGα-NLS peptide. As opposed to what was reported in the case of MPG-mNLS (Simeoni et al. 2003), similar cellular localization and luciferase activity were observed when cells were transfected with siRNA complexed with either MPGα-NLS or MPGα-mNLS. This suggests that the mutation of lysine (K) residue in NLS might have a minor effect on nuclear-targeting ability of the peptide.

9.4.1.2 Secondary Amphipathic Peptide

As an essential protein for influenza viral entry, hemagglutinin (HA) mediates a low pH-dependant membrane fusion through the exposure of its amphipathic α-helix (Skehel et al. 1982). Based on the idea of mimicking the α-helical structure associated with the membrane fusion feature of viral proteins, several secondary amphipathic peptides have been designed to function as a membrane destabilization agent or a DNA delivery vector (Fernandez-Carneado et al. 2004). Among them, KALA peptide is known to undergo conformational change from pH 5.0–7.5 (Wyman et al. 1997) and has been chosen to condense and deliver siRNA conjugated with PEG via a disulfide linkage (Lee et al. 2007). The studies suggested that the self-assembled polyelectrolyte complex micelle has an inner core with KALA peptide-condensed siRNA, surrounded by protective PEG corona. The average size of KALA/siRNA-PEG complexes was around 200 nm at an N/P ratio of 6, a size favorable for cellular uptake. Vascular endothelial growth factor (VEGF) expression in KALA/siRNA-PEG complex-transfected prostate carcinoma (PC-3) cells was reduced to 20% of that in the untransfected control. Although cytotoxicity of the KALA peptides was not observed in this study, it was previously reported that the KALA peptides exhibited hemolytic activity at a physiological pH due to nonspecific membrane destabilization (Wyman et al. 1997). Further cell viability tests with a prolonged incubation time and an increased amount of the KALA peptides should be performed to evaluate the cytotoxicity.

CADY peptide is another secondary amphipathic peptide tested for siRNA delivery. The peptide was designed based on two amphipathic peptides called JTS1 (Gottschalk et al. 1996) and ppTG1 (Rittner et al. 2002). To improve interaction with siRNA and cell membrane, arginine (R) and tryptophan (W) were included in the CADY peptide (Crombez et al. 2009). Similar to MPG-based peptides, N- and C-termini of the CADY peptide were modified by acetyl group and cysteamide group, respectively. The CADY peptide formed complex with siRNAs and protected them from serum nuclease degradation with increasing molar ratio of peptide to siRNA, most significantly at a molar ratio of 80. Flow cytometry analysis showed that cellular uptake of fluorescently labeled siRNA was improved by increasing the molar ratio. To investigate CADY peptide-mediated siRNA delivery, functional siRNAs targeting GAPDH or p53 were used to monitor silencing effects. After transfection of CADY/siRNA-GAPDH complexes at a molar ratio of 40, the silencing effect was >80% after 24 h in human osteosarcoma cells (U2OS) and primary HUVEC. When the CADY peptide was used to deliver siRNA against p53 into U2OS cells at a molar ratio of 40, the inhibitory effect was maintained for at least 5 days, with 97% and 60% knockdown at day 2 and 5, respectively. The pretreatment of cells with different inhibitors of the endocytosis pathway had no significant effect in the cellular uptake and silencing effect mediated by the CADY/siRNA complexes. Unlike the JTS1 peptide, the CADY peptide adopts α-helical structure independent of pH. Hence, it was hypothesized that the cellular uptake of the CADY peptide was due to the direct interaction of aromatic tryptophan (W) residues with cell membrane components.

9.4.2 Classical Cell-Penetrating Peptides

Classical CPPs are usually rich in arginine (R) and lysine (K) that are highly positively charged. Some CPPs, like penetratin (Derossi et al. 1994) and Tat (Vives et al. 1997), are directly derived from natural proteins, while others have a designed sequence, such as oligoarginine (Mitchell et al. 2000) and transportan (Pooga et al. 1998). Many different cargoes have been successfully delivered into cells by CPPs, mainly by covalent conjugation and also after non-covalent complex formation (Stewart et al. 2008).

9.4.2.1 Penetratin-Based Peptide

Although classical CPPs are able to cross cell membrane effectively, endosomal escape of cargos carried by classical CPPs remains as the limiting step for efficient intracellular delivery. To furnish the penetratin peptide with endosomolytic property, an analogue peptide called EB1 was designed, in which certain amino acids were replaced with histidine to adopt α-helical structure upon protonation in acidic endosomes (Lundberg et al. 2007). Six amino acid residues were also added at N-terminus to provide length required to span endosomal membrane. Both EB1 and penetratin peptides contain seven cationic amino acid residues, but ethidium bromide exclusion assay showed that EB1 had better siRNA-binding efficiency than penetratin. This indicates that hydrophobic interactions might be involved in EB1 peptide/siRNA complex formation. However, no conformational studies were performed to support the assumption.

The cellular uptake of fluorescently labeled siRNA delivered by EB1 or penetratin was compared in HeLa cells. In agreement with ethidium bromide exclusion assay, siRNA delivered by EB1 was enhanced by at least 2.5-fold, depending on the molar ratio used for complex formation. When delivering siRNA against luciferase into HeLa cells transiently transfected with luciferase plasmid, no silencing effect was observed for penetratin. By contrast, luciferase activity was reduced by 45% by EB1 peptide-mediated siRNA delivery. This silencing effect was similar to that provided by the MPG-mNLS peptide, a control included in the study. The authors hypothesized that endosomal pH change would induce conformational change of the EB1 peptide, although the cellular uptake pathway of EB1 peptide/siRNA complexes was not investigated. Moreover, no experiment was performed using inhibitor of vacuolar proton pump, such as bafilomycin A1 (Bowman et al. 1988), to confirm the pH-dependant conformational change.

9.4.2.2 Oligoarginine-Based Peptide

DNA/RNA-binding protein domains are often found to be rich in arginine residues (Tan and Frankel 1995), suggesting that oligoarginine peptides could be a potential nucleic acid carrier. It was also discovered that to achieve efficient internalization, the guanidinium group and the number of arginine residues are more important than the presence of positive charge or backbone structure (Mitchell et al. 2000; Futaki et al. 2002; Rothbard et al. 2002). Accordingly, a nona-arginine (R_9) peptide was demonstrated to have the capability to deliver siRNA into mammalian cells. Gel retardation assay revealed that the R_9 peptides could form complex with siRNA at low N/P ratios. When human gastric carcinoma cells stably expressing enhanced green fluorescent protein (GC-EGFP) were exposed to R_9 peptide/siRNA against EGFP complexes, the EGFP intensity was reduced to 57% of the untransfected cell control after 48 h incubation. The distribution of fluorescently labeled siRNA was found to be cytoplasmic.

To extend the potential use of CPPs, cell-targeting sequence can be fused with CPPs for cell type-specific delivery. In a particular study, a short peptide sequence derived from rabies virus glycoprotein (RVG) was fused with nona-D-arginine to deliver siRNA specifically to neuronal cells with nicotinic acetylcholine receptor (AchR) (Kumar et al. 2007). The authors first investigated the in vitro binding specificity of the RVG peptide and confirmed that the peptide bound only to AchR-expressing Neuro2a cells but not to receptor-negative HeLa cells. The snake-venom toxin α-bungarotoxin (BTX), which specifically binds to AchR (Lentz 1990), inhibited the in vitro RVG peptide binding in a dose-dependent manner. After tail vein injection of biotinylated RVG peptides, primary neuronal cells in mice brains were stained positive for the peptide, indicating that the RVG peptides were able to cross the blood brain barrier (BBB). To use the RVG peptide for siRNA delivery, a chimeric peptide (RVG-9r) was designed by addition of a spacer and nona-D-arginine at C-terminus of the RVG peptide. A peptide derived from rabies virus matrix protein, RVG-Mat-9r, was used to serve as a control peptide. Both RVG-9r and RVG-Mat-9r peptides bound to siRNA at a molar ratio of 10:1, but only the RVG-9r peptide could deliver fluorescently labeled siRNA into Neuro2a cells. In Neuro2a cells stably expressing green fluorescence protein (GFP), RVG-9r/siRNA targeting GFP complexes silenced GFP expression up to 70%, similar to that offered by Lipofectamine 2000 transfection. When challenged by serum nucleases, the RVG-9r peptide partially protected siRNA for up to 8 h, indicating the potential of the peptide for siRNA transvascular delivery to brain cells. Indeed, after intravenous injection of RVG-9r/FITC-siRNA complexes into mice, fluorescein isothiocynate (FITC) fluorescence was detected in the brain, but not in the liver or spleen. To test brain-specific gene silencing, RVG-9r/siRNA against GFP complexes were injected into GFP transgenic mice for three consecutive days. Two days after final injection, GFP expression in brain was reduced 30% while the expression in the liver or spleen was not affected. Furthermore, treatment with multiple intravenous injections of RVG-9r/antiviral siRNA complexes improved the survival of mice challenged by fatal Japanese encephalitis virus (JEV). The presence of antiviral siRNA in the brain tissue was confirmed by northern blot analysis. This remarkable study is the first one reporting a peptide-based approach for noninvasive siRNA delivery into mammalian brain without significant toxicity. Use of chemically modified siRNAs or liposomal nanoparticles decorated with the RVG-9r peptide might enhance complex stability in blood circulation and make significant impact for future targeted brain delivery. It is also possible to replace the RVG sequence with other cell-targeting sequences to achieve tissue or cell type-specific delivery in other organs.

9.4.2.3 Peptide for Ocular Delivery

To improve delivery of small molecules into ocular tissues, a novel peptide with a sequence of $GGG(ARKKAAKA)_4$ and named as peptide for ocular delivery (POD) was designed (Johnson et al. 2008). Lissamine-conjugated POD peptide (L-POD) was taken up by human embryonic retinal (HER) 911 cells with cytoplasmic distribution. Cellular uptake of L-POD was inhibited by chondroitin sulfate and heparan sulfate, suggesting that cell-surface proteoglycans were involved in binding and internalization. The L-POD peptide could penetrate retinal or ocular tissues through delivery into subretinal space or topical application, respectively. An N-terminus cysteinyl POD peptide (C-POD) was able to deliver plasmid DNA and streptavidin-coated quantum dots into HER 911 cells. When HER 911 cells were co-transfected with plasmid encoding EGFP and C-POD/siRNA against EGFP complexes, the percentage of EGFP-positive cells was reduced twofold when compared to that of the control without siRNA delivery. Unfortunately, the authors did not attempt any in vivo siRNA delivery. Nevertheless, the C-POD peptide inhibited bacterial growth on Lysogeny Broth (LB) agar plate in a concentration-dependant manner, a property useful for treatment of eye infection.

9.4.3 Histidine-Rich Branched Peptides

Cationic peptides bind electrostatically to nucleic acids to form nanoparticles, but they are incapable of mediating the endosomal escape of the delivered cargos into the cytoplasm. Thus, endosomolytic agents, such as chloroquine, are often used to enhance in vitro transfection efficiency (Wattiaux et al. 2000). The use of these agents for in vivo application might not be a feasible approach due to possible toxicity. To overcome the problem, histidine residues are commonly incorporated into the design of peptide-based vectors. pH-dependant liposome fusion in the presence of poly(L-histidine) was found to correlate with the protonation of imidazole group of histidine residues (Wang and Huang 1984; Uster and Deamer 1985). Linear peptides containing multiple histidine residues ($pK_a = 6.0$) were also demonstrated to enhance plasmid DNA transfection efficiency by buffering acidic endosomes (Midoux et al. 1998; Kichler et al. 2003; Lo and Wang 2008). Another interesting histidine-rich peptide design is a series of branched peptides consisting of histidine-rich branches emanating from an uncharged lysine core (Chen et al. 2001; Chen et al. 2002). Depending on the degree of branching, these branched peptides could enhance transfection efficiency in combination with cationic liposomes. To work without liposomes, the design of these branched peptides was improved by increasing histidine contents in the branches (Leng and Mixson 2005a). In cell transfection with luciferase plasmid DNA, the branched peptides with histidine-rich tail were more effective than their counterparts without the tail. One of the designs, H^2K4bT, could deliver gene into cells more effectively than two commercial transfection agents, Lipofectamine and SuperFect. After replacing the histidine-rich tail of the H^2K4bT peptide with other peptide sequences, the transfection efficiencies were reduced significantly, indicating that histidine-rich tail is essential for endosomal escape.

The above branched peptides, although effective in gene delivery, were unable to deliver siRNA effectively (Leng et al. 2005). To study the effect of the number of terminal branches and the histidine contents in the branches on siRNA delivery, siRNAs targeting β-galactosidase were complexed with different peptides and delivered into mouse endothelial cells stably expressing β-galactosidase (SVR-bag4). The authors discovered that the H^3K4b peptide with fewer lysine residues in the branches was more effective than the H^2K4b peptide with more lysine residues in the branches, suggesting that strong binding between siRNA and branched peptides might not be favorable for efficient siRNA delivery. A peptide with eight terminal branches, H^3K8b, was able to reduce β-gal activity to 80% of the untransfected control. Addition of integrin-binding ligand arginine-glycine-aspartic acid (RGD) to the H^3K8b peptide (H^3K8 + RGD) further increased the silencing effect by 20% when an optimal weight/weight ratio of 4:1 was used for in vitro siRNA delivery. However, the in vitro results could not translate to in vivo application (Leng and Mixson 2005b). In an attempt to deliver siRNA targeting Raf-1 (activated substrate of oncogenic Ras) intratumorally to inhibit subcutaneous tumor growth in adult nude mice, the H^3K8b peptide was less efficient than the H^3K4b peptide with the lowest lysine:histidine ratio. Since the H^3K4b peptide was easier and cheaper to be synthesized, the authors recently focused on the modification of H^3K4b for systemic delivery of siRNAs (Leng et al. 2008). A new branched peptide $H^3K(+H)4b$, with one additional histidine residue in each of the branches of H^3K4b, was designed to improve endosomal escape. Nanoparticles of around 230 nm arose from mixing the $H^3K(+H)4b$ peptide with siRNA at weight/weight ratio of 4:1. In comparison to the earlier design of the H^3K8b peptide, the $H^3K(+H)4b$ peptide was slightly more efficient (10%–20%) in inhibiting growth of cancer cell lines by delivering siRNA targeting Raf-1. Systemically delivered $H^3K(+H)4b$/fluorescently labeled siRNA complexes into adult nude mice were observed inside tumor xenograft and other tissues, with the greatest accumulation in the kidney. Seven injections of $H^3K(+H)4b$/siRNA against Raf-1 complexes significantly reduced tumor growth, confirmed by histological and immunochemical studies. Even though no toxicity was observed in other organs, the $H^3K(+H)4b$ peptide was moderately toxic in in vitro studies. Further evaluation of the toxicity of these peptide/siRNA complexes might be required. Alternatively, cell-targeting sequence or PEG could be used to address the biocompatibility issue of the $H^3K(+H)4b$ peptide.

9.4.4 Peptide-Based Reducible Polymers

A PRP is prepared by connecting cysteine-containing peptide monomer through formation of interpeptide disulfide bonds, either by auto-oxidation or chemical oxidation. Cargos carried by PRP will be released in the cytoplasm due to the cleavage of disulfide bonds by reductive glutathione species in the cellular environment. Findings from previous studies suggest that the environment in the endosomal and lysosomal compartments does not permit efficient cleavage of disulfide bonds (Feener et al. 1990; Austin et al. 2005; Yang et al. 2006). Therefore, endosomolytic properties would be an essential requirement in the design of PRP monomers. After incorporating histidine residues in the peptide design, it has been proven that the resulting polymers could mediate endosomal escape without chloroquine (McKenzie et al. 2000a,b; Read et al. 2005; Manickam and Oupicky 2006; Lo and Wang 2008).

A particular reducible polycation (RFC) with 12 histidines (HIS6-RPC, MW of 113 kDa) obtained by oxidization with dimethyl sulfoside (DMSO) is a highly versatile nucleic acid delivery vector (Read et al. 2005). HIS6-RPC mediated high levels of transfection with DNA or mRNA, encoding GFP at weight/weight (HIS6-RPC/nucleic acid) ratio of 40. HIS6-RPC/siRNA complexes formed at weight/weight ratio of 24 also successfully suppressed expression of EGFP in a human prostatic cell line (PC-3) transiently expressing EGFP. While plasmid DNA requires large MW PRPs for efficient condensation, low molecular weight (LMW) PRPs might be an appropriate alternative for siRNA delivery. Hence, a series of HIS6-RPC with different MW ranging from 38 to 162 kDa were synthesized by varying the time of oxidative polymerization (Stevenson et al. 2008). HIS6-RPC 162 kDa was not able to retard siRNA mobility in agarose gel, whereas LMW HIS6-RPCs (38, 44, 80, and 114 kDa) retained siRNA at weight/weight ratio of 10. HIS6-RPC 80 kDa formed 90 nm nanoparticles with siRNA, approximately half the size of those formed with HIS6-RPC 162 kDa. Consistent with these observations, only siRNA delivered by LMW HIS6-RPCs could provide up to 30% decrease in fluorescence intensity in human hepatocyte carcinoma cells (HepG2) stably expressing EGFP. LMW HIS6-RPCs' terminally incorporated cell-targeting domain, derived from the circumsporozoite protein of malaria parasite, improved cell-specific siRNA delivery into hepatocytes. However, HIS6-RPCs were generally less efficient than commercially available lipid-based vectors such as N-[1-(2,3-dioleoyloxy)propyl]-N, N, N-trimethylammonium methyl-sulfate (DOTAP). Interestingly, a primary miRNA (pri-miRNA-23a), delivered by a similar histidine-rich polymer consisting of nuclear localization sequence, was processed into mature miRNA in HeLa cells, indicating successful nuclear-targeted delivery (Rahbek et al. 2008).

A KALA PRP (cl-KALA) was developed recently for siRNA delivery (Mok and Park 2008). Considering that individual KALA peptide with fusogenic activities would only be produced from cl-KALA after cleavage of disulfide bonds intracellularly, it has been speculated that the cytotoxicity of the KALA peptide associated with nonspecific membrane destabilization would be reduced. However, the reduction of gene expression in GFP over-expressing cells transfected by cl-KALA/siRNA complexes was less than 25%, even when an N/P ratio of as high as 64 was used. This inefficient silencing activity could be related to reduced endosomal escape. Unlike histidine-rich peptides that can buffer acidic endosomes, a significant amount of the KALA peptides might be required to perturb endosomal membrane for efficient cargo release, since it was previously demonstrated that the cleavage of disulfide bonds in endosomes is not efficient. On the other hand, the polydispersity of the polymers could be another contributing factor that affects the delivery efficiency.

9.5 Methods for Evaluating Peptide-Based Vectors

In this section, important experimental methods that are useful in evaluating the efficiency of peptide-based vectors for siRNA delivery will be discussed. Although some of the methods are not designed or performed using siRNA as a cargo, it is always possible to adapt the protocols for this purpose.

The most commonly used method to investigate the interaction between siRNA and peptide is electrophoretic mobility shift assay (EMSA), also known as gel retardation assay. Mobility of siRNA through an agarose gel will reduce after its binding to the peptide. This is visible after staining with a nucleic acid stain like ethidium bromide. Using the same method, it is also possible to find out whether peptide binding could prevent siRNA degradation after being subjected to nuclease treatment. The rate of binding and binding constants can be estimated using fluorescence resonance energy transfer (FRET) and fluorescence correlation spectroscopy (FCS) methods (Ayame et al. 2008). Both methods will require fluorescent labeling of siRNAs and peptides. Hence, additional experiments might be necessary to show that the binding of peptide to siRNA will not be affected by the attachment of fluorophores. Other possible methods to study interactions between siRNA and peptides include size-exclusion chromatography (SEC) (Morris et al. 2001), fluorescence titration (Morris et al. 1997), UV-vis absorbance spectroscopy (Law et al. 2008), and circular dichroism (Law et al. 2008).

The major cellular uptake pathways in eukaryotic cells are phagocytosis, macropinocytosis, clathrin-mediated endocytosis, and caveole-mediated endocytosis. Surface charge and size of peptide/siRNA complexes are two parameters that affect cellular uptake. The surface charge of the complexes can be measured as zeta potential. Complexes with large positive zeta potential are usually more favorable for cellular uptake due to interaction with cellular membrane and a less possibility to aggregation. A good

TABLE 9.2 List of Different Inhibitors or Effectors That Are Commonly Used to Study the Cellular Uptake Pathway of Peptide/siRNA Complexes

Treatment	Effects
4°C	Inhibits of energy-dependent processes
Chloroquine, monensin	Inhibits acidification of endosomes, enhances endosomal escape
Cytochalasin B	Disrupts microfilaments, inhibits macropinocytosis
Amiloride	Inhibits Na^+/H^+ exchange required for macropinocytosis
Filipin complex, nystatin	Sterol-binding agent, disrupts caveolar structure and function
Okadaic acid	Activates/inducts caveolin-mediated/dependent endocytosis
Sucrose, ikarugamycin	Inhibits clathrin-mediated endocytosis
Wortmannin	PI(3)K-inhibitor, inhibits endosome fusion
Bafilomycin A1	Inhibits vacuolar ATPase, prevents acidification of early endosomes

estimation of particle size can be achieved by measuring hydrodynamic diameter using dynamic light scattering method. Particle size can be further confirmed by microscopic method such as atomic force microscopy (AFM), scanning electron microscopy (SEM), and transmission electron microscopy (TEM).

Study on the cellular uptake of peptide/siRNA complexes often relies upon fluorescence imaging or flow cytometry analysis using peptides and siRNAs labeled by different fluorophores. Importantly, only living cells, but not fixed cells, should be used in these studies. Fixation procedures can result in artifacts, leading to overestimation of cellular uptake, especially for membrane-bound CPP, that are internalized upon fixation (Richard et al. 2003). When using living cells for localization studies, extra washing steps with trypsin or heparin are important to remove the extracellularly bound complex (Richard et al. 2003; Veldhoen et al. 2006). Additionally, confocal laser scanning microscopy should be routinely used to demonstrate the intracellular localization of the complexes. Combined with these methods, it is possible to investigate the cellular uptake pathway of peptide/siRNA complexes by using different inhibitors or effectors listed in Table 9.2 (Simeoni et al. 2003; Veldhoen et al. 2006; Ayame et al. 2008). Sensitive methods developed for quantification of siRNA delivered by peptides, such as liquid hybridization method and competitive quantitative PCR (qPCR) (Veldhoen et al. 2006; Liu et al. 2009), should be very useful in optimizing peptide design and complex formulation.

Following studies of cellular uptake, the next step is to investigate silencing effects in the cells transfected with peptide/siRNA complexes. The cells usually carry different siRNA-targeted reporter genes expressed transiently or stably, for example, luciferase, GFP, or β-galactosidase. Housekeeping genes like GAPDH are convenient targets useful to siRNA silencing study (Simeoni et al. 2003). At the protein level, expression of reporter genes can be confirmed by Western blot and quantified by available ELISA methods. At the mRNA level, Northern blot and qPCR can quantify the extent of silencing of the target gene. When a dose–response curve is plotted based on reduction of expression levels, the apparent value of maximal inhibition (IC_{max}) and half maximal inhibition (IC_{50}) can be identified to compare the effectiveness of peptide/siRNA complexes with different compositions (Veldhoen et al. 2006). With the safety profile of a delivery vector being of utmost concern, in vitro and in vivo toxicity of individual components of the complexes should also be evaluated.

9.6 Conclusion

With increasingly extensive application of siRNA in disease treatment, the bottleneck for developing siRNA-based therapeutics will be efficient delivery. As the use of physical methods and chemically modified siRNAs might be limited, a delivery vector that can encapsulate siRNA molecules non-covalently appears to be more desirable. An ideal siRNA delivery vector should fulfill the following criteria: (1) synthesized easily in large scale; (2) relatively stable for a long shelf life; (3) biodegradable and biocompatible; (4) low toxicity, as well as low immunogenicity that allows multiple administrations; (5) flexibility to be functionalized for tissue or cell type-specific delivery; (5) high drug-loading capacity; (6) protection of siRNA from nuclease degradation; and (7) ability to release siRNA efficiently within target cells.

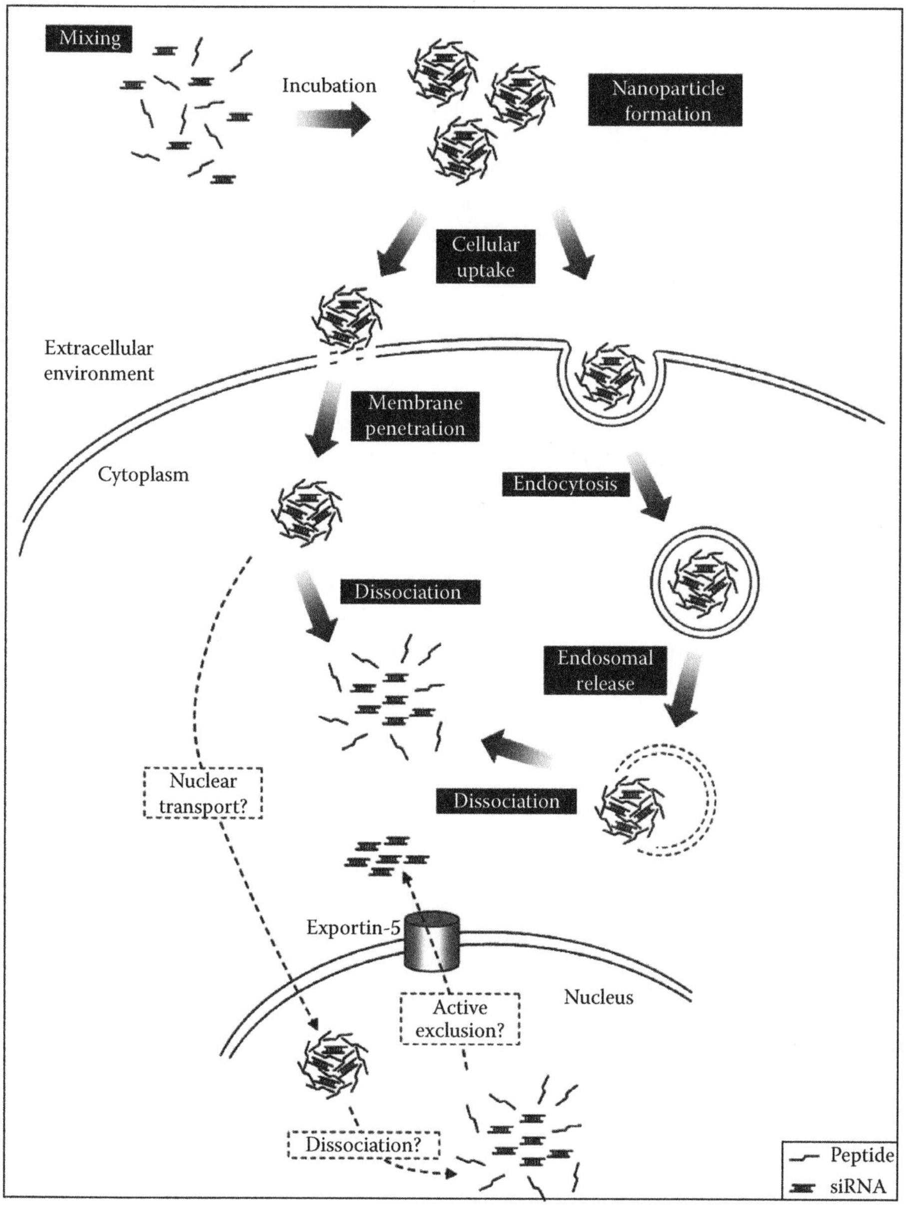

FIGURE 9.2 Schematic diagram on delivery of siRNA by peptide-based vectors. Peptide/siRNA complexes of nanometer size are usually formed by incubating a mixture of peptide and siRNA. These complexes will be internalized by cells either through direct membrane penetration or endocytosis. In the cytoplasm, the complexes will dissociate and the released siRNAs will then silence gene expression. If the peptide contains nuclear localization sequence, it is possible that the complexes are transported into the nucleus and free siRNAs can be actively excluded from the nucleus into the cytoplasm.

Taking these into consideration, peptides are emerging as attractive nanomaterials for siRNA delivery. In fact, a wide range of peptides with different properties have been tested and utilized in in vitro studies, as reviewed here. The delivery of siRNA using peptide-based vectors is graphically summarized in the schematic diagram in Figure 9.2.

Before the applications of peptide-based vectors can be translated from in vitro to in vivo and from the bench to the bedside, several concerns need to be addressed. First, it is highly unlikely to have only one single entry pathway for different peptides and peptide/siRNA complexes, due to differences in peptide sequence, conformation, surface charge, and particle size. In order to have better understanding on cellular entry mechanisms, the characterization of relevant biophysical properties of the complexes under various conditions is desirable. Second, silencing effect on a reporter gene system like luciferase or GFP expression might be insufficient to illustrate the transfection efficiency of peptides. Sensitive quantification methods to quantify siRNA should be employed to enable comparison of the amount of internalized siRNA mediated by different vectors. Third, 10% serum-containing medium has been widely used to study the stability of peptide/siRNA complexes. Nonetheless, it is arguable whether this trustfully reflects in vivo condition, where many components are present in blood stream with higher concentrations that can potentially result in inactivating effects on the complexes in a dose-dependent manner. Intravenous administration of the complexes into animal models might be essential to justify whether tested peptides will be able to protect and deliver siRNA efficiently in the body. Fourth, the immunogenicity of peptide, siRNA, and complexes should be investigated carefully when moving toward in vivo application. A combined use of siRNA with high specificity and peptide with cell-targeting sequence might minimize the possibility of stimulating the immune system and lower the dosage to be used for therapeutic purpose.

In conclusion, recent progress provides increasing evidence that peptides are promising nanomaterials for siRNA delivery. With inputs from different research disciplines from molecular biology, genetics, chemistry, physics, materials sciences to nanoscience, we will have a much better understanding of siRNA mechanism, peptide chemistry, biophysical properties of peptide/siRNA complex, pharmacodynamics, and pharmacokinetics, thereby accelerating the development of peptide-based vectors for in vivo therapeutic applications of siRNAs.

References

Akaneya, Y., B. Jiang, and T. Tsumoto. 2005. RNAi-induced gene silencing by local electroporation in targeting brain region. *J Neurophysiol* 93 (1):594–602.

Akhtar, S. and I. Benter. 2007. Toxicogenomics of non-viral drug delivery systems for RNAi: Potential impact on siRNA-mediated gene silencing activity and specificity. *Adv Drug Deliv Rev* 59 (2–3):164–182.

Austin, C. D., X. Wen, L. Gazzard, C. Nelson, R. H. Scheller, and S. J. Scales. 2005. Oxidizing potential of endosomes and lysosomes limits intracellular cleavage of disulfide-based antibody-drug conjugates. *Proc Natl Acad Sci USA* 102 (50):17987–17992.

Ayame, H., N. Morimoto, and K. Akiyoshi. 2008. Self-assembled cationic nanogels for intracellular protein delivery. *Bioconjug Chem* 19 (4):882–890.

Bartel, D. P. 2004. MicroRNAs: Genomics, biogenesis, mechanism, and function. *Cell* 116 (2):281–297.

Bartlett, D. W. and M. E. Davis. 2008. Impact of tumor-specific targeting and dosing schedule on tumor growth inhibition after intravenous administration of siRNA-containing nanoparticles. *Biotechnol Bioeng* 99 (4):975–985.

Bartlett, D. W., H. Su, I. J. Hildebrandt, W. A. Weber, and M. E. Davis. 2007. Impact of tumor-specific targeting on the biodistribution and efficacy of siRNA nanoparticles measured by multimodality in vivo imaging. *Proc Natl Acad Sci USA* 104 (39):15549–15554.

Bernstein, E., A. A. Caudy, S. M. Hammond, and G. J. Hannon. 2001. Role for a bidentate ribonuclease in the initiation step of RNA interference. *Nature* 409 (6818):363–366.

Biessen, E. A., H. Vietsch, E. T. Rump, K. Fluiter, J. Kuiper, M. K. Bijsterbosch, and T. J. van Berkel. 1999. Targeted delivery of oligodeoxynucleotides to parenchymal liver cells in vivo. *Biochem J* 340 (Pt 3):783–792.

Bowman, E. J., A. Siebers, and K. Altendorf. 1988. Bafilomycins: A class of inhibitors of membrane ATPases from microorganisms, animal cells, and plant cells. *Proc Natl Acad Sci USA* 85 (21):7972–7976.

Braasch, D. A., S. Jensen, Y. Liu, K. Kaur, K. Arar, M. A. White, and D. R. Corey. 2003. RNA interference in mammalian cells by chemically-modified RNA. *Biochemistry* 42 (26):7967–7975.

Braasch, D. A., Z. Paroo, A. Constantinescu, G. Ren, O. K. Oz, R. P. Mason, and D. R. Corey. 2004. Biodistribution of phosphodiester and phosphorothioate siRNA. *Bioorg Med Chem Lett* 14 (5):1139–1143.

Bradley, S. P., C. Rastellini, M. A. da Costa, T. F. Kowalik, A. B. Bloomenthal, M. Brown, L. Cicalese, G. P. Basadonna, and M. E. Uknis. 2005. Gene silencing in the endocrine pancreas mediated by short-interfering RNA. *Pancreas* 31 (4):373–379.

Caplen, N. J., J. Fleenor, A. Fire, and R. A. Morgan. 2000. dsRNA-mediated gene silencing in cultured Drosophila cells: A tissue culture model for the analysis of RNA interference. *Gene* 252 (1–2):95–105.

Caplen, N. J., S. Parrish, F. Imani, A. Fire, and R. A. Morgan. 2001. Specific inhibition of gene expression by small double-stranded RNAs in invertebrate and vertebrate systems. *Proc Natl Acad Sci USA* 98 (17):9742–9747.

Carthew, R. W. and E. J. Sontheimer. 2009. Origins and mechanisms of miRNAs and siRNAs. *Cell* 136 (4):642–655.

Chaloin, L., P. Vidal, P. Lory, J. Mery, N. Lautredou, G. Divita, and F. Heitz. 1998. Design of carrier peptide-oligonucleotide conjugates with rapid membrane translocation and nuclear localization properties. *Biochem Biophys Res Commun* 243 (2):601–608.

Chen, Q. R., L. Zhang, P. W. Luther, and A. J. Mixson. 2002. Optimal transfection with the HK polymer depends on its degree of branching and the pH of endocytic vesicles. *Nucleic Acids Res* 30 (6):1338–1345.

Chen, Q. R., L. Zhang, S. A. Stass, and A. J. Mixson. 2001. Branched co-polymers of histidine and lysine are efficient carriers of plasmids. *Nucleic Acids Res* 29 (6):1334–1340.

Cheng, K., Z. Ye, R. V. Guntaka, and R. I. Mahato. 2006. Enhanced hepatic uptake and bioactivity of type alpha1(I) collagen gene promoter-specific triplex-forming oligonucleotides after conjugation with cholesterol. *J Pharmacol Exp Ther* 317 (2):797–805.

Chiu, Y. L., A. Ali, C. Y. Chu, H. Cao, and T. M. Rana. 2004. Visualizing a correlation between siRNA localization, cellular uptake, and RNAi in living cells. *Chem Biol* 11 (8):1165–1175.

Chiu, Y. L. and T. M. Rana. 2003. siRNA function in RNAi: A chemical modification analysis. *RNA* 9 (9):1034–1048.

Crombez, L., G. Aldrian-Herrada, K. Konate, Q. N. Nguyen, G. K. McMaster, R. Brasseur, F. Heitz, and G. Divita. 2009. A new potent secondary amphipathic cell-penetrating peptide for siRNA delivery into mammalian cells. *Mol Ther* 17 (1):95–103.

Davidson, T. J., S. Harel, V. A. Arboleda, G. F. Prunell, M. L. Shelanski, L. A. Greene, and C. M. Troy. 2004. Highly efficient small interfering RNA delivery to primary mammalian neurons induces MicroRNA-like effects before mRNA degradation. *J Neurosci* 24 (45):10040–100406.

de Fougerolles, A., H. P. Vornlocher, J. Maraganore, and J. Lieberman. 2007. Interfering with disease: A progress report on siRNA-based therapeutics. *Nat Rev Drug Discov* 6 (6):443–453.

de Paula, D., M. V. Bentley, and R. I. Mahato. 2007. Hydrophobization and bioconjugation for enhanced siRNA delivery and targeting. *RNA* 13 (4):431–456.

Derossi, D., A. H. Joliot, G. Chassaing, and A. Prochiantz. 1994. The third helix of the Antennapedia homeodomain translocates through biological membranes. *J Biol Chem* 269 (14):10444–10450.

Deshayes, S., S. Gerbal-Chaloin, M. C. Morris, G. Aldrian-Herrada, P. Charnet, G. Divita, and F. Heitz. 2004a. On the mechanism of non-endosomial peptide-mediated cellular delivery of nucleic acids. *Biochim Biophys Acta* 1667 (2):141–147.

Deshayes, S., T. Plenat, G. Aldrian-Herrada, G. Divita, C. Le Grimellec, and F. Heitz. 2004b. Primary amphipathic cell-penetrating peptides: Structural requirements and interactions with model membranes. *Biochemistry* 43 (24):7698–7706.

Deshayes, S., M. C. Morris, G. Divita, and F. Heitz. 2006a. Interactions of amphipathic carrier peptides with membrane components in relation with their ability to deliver therapeutics. *J Pept Sci* 12 (12):758–765.

Deshayes, S., T. Plenat, P. Charnet, G. Divita, G. Molle, and F. Heitz. 2006b. Formation of transmembrane ionic channels of primary amphipathic cell-penetrating peptides. Consequences on the mechanism of cell penetration. *Biochim Biophys Acta* 1758 (11):1846–1851.

DiFiglia, M., M. Sena-Esteves, K. Chase, E. Sapp, E. Pfister, M. Sass, J. Yoder et al. 2007. Therapeutic silencing of mutant huntingtin with siRNA attenuates striatal and cortical neuropathology and behavioral deficits. *Proc Natl Acad Sci USA* 104 (43):17204–17209.

Dingwall, C. and R. Laskey. 1992. The nuclear membrane. *Science* 258 (5084):942–947.

Elbashir, S. M., J. Harborth, W. Lendeckel, A. Yalcin, K. Weber, and T. Tuschl. 2001. Duplexes of 21-nucleotide RNAs mediate RNA interference in cultured mammalian cells. *Nature* 411 (6836):494–498.

Elmen, J., H. Thonberg, K. Ljungberg, M. Frieden, M. Westergaard, Y. Xu, B. Wahren et al. 2005. Locked nucleic acid (LNA) mediated improvements in siRNA stability and functionality. *Nucleic Acids Res* 33 (1):439–447.

Fabre, J. W. and L. Collins. 2006. Synthetic peptides as non-viral DNA vectors. *Curr Gene Ther* 6 (4):459–480.

Feener, E. P., W. C. Shen, and H. J. Ryser. 1990. Cleavage of disulfide bonds in endocytosed macromolecules. A processing not associated with lysosomes or endosomes. *J Biol Chem* 265 (31):18780–18785.

Fernandez-Carneado, J., M. J. Kogan, S. Pujals, and E. Giralt. 2004. Amphipathic peptides and drug delivery. *Biopolymers* 76 (2):196–203.

Flynn, M. A., D. G. Casey, S. M. Todryk, and B. P. Mahon. 2004. Efficient delivery of small interfering RNA for inhibition of IL-12p40 expression in vivo. *J Inflamm (Lond)* 1 (1):4.

Futaki, S., I. Nakase, T. Suzuki, Z. Youjun, and Y. Sugiura. 2002. Translocation of branched-chain arginine peptides through cell membranes: Flexibility in the spatial disposition of positive charges in membrane-permeable peptides. *Biochemistry* 41 (25):7925–7930.

Gallaher, W. R. 1987. Detection of a fusion peptide sequence in the transmembrane protein of human immunodeficiency virus. *Cell* 50 (3):327–328.

Giladi, H., M. Ketzinel-Gilad, L. Rivkin, Y. Felig, O. Nussbaum, and E. Galun. 2003. Small interfering RNA inhibits hepatitis B virus replication in mice. *Mol Ther* 8 (5):769–776.

Golzio, M., L. Mazzolini, P. Moller, M. P. Rols, and J. Teissie. 2005. Inhibition of gene expression in mice muscle by in vivo electrically mediated siRNA delivery. *Gene Ther* 12 (3):246–251.

Gottschalk, S., J. T. Sparrow, J. Hauer, M. P. Mims, F. E. Leland, S. L. Woo, and L. C. Smith. 1996. A novel DNA-peptide complex for efficient gene transfer and expression in mammalian cells. *Gene Ther* 3 (5):448–457.

Gregory, R. I., T. P. Chendrimada, N. Cooch, and R. Shiekhattar. 2005. Human RISC couples microRNA biogenesis and posttranscriptional gene silencing. *Cell* 123 (4):631–640.

Grimm, D., K. L. Streetz, C. L. Jopling, T. A. Storm, K. Pandey, C. R. Davis, P. Marion, F. Salazar, and M. A. Kay. 2006. Fatality in mice due to oversaturation of cellular microRNA/short hairpin RNA pathways. *Nature* 441 (7092):537–541.

Hall, A. H., J. Wan, E. E. Shaughnessy, B. Ramsay Shaw, and K. A. Alexander. 2004. RNA interference using boranophosphate siRNAs: Structure–activity relationships. *Nucleic Acids Res* 32 (20):5991–6000.

Hamar, P., E. Song, G. Kokeny, A. Chen, N. Ouyang, and J. Lieberman. 2004. Small interfering RNA targeting Fas protects mice against renal ischemia–reperfusion injury. *Proc Natl Acad Sci USA* 101 (41):14883–14888.

Heidel, J. D., Z. Yu, J. Y. Liu, S. M. Rele, Y. Liang, R. K. Zeidan, D. J. Kornbrust, and M. E. Davis. 2007. Administration in non-human primates of escalating intravenous doses of targeted nanoparticles containing ribonucleotide reductase subunit M2 siRNA. *Proc Natl Acad Sci USA* 104 (14):5715–5721.

Hornung, V., M. Guenthner-Biller, C. Bourquin, A. Ablasser, M. Schlee, S. Uematsu, A. Noronha et al. 2005. Sequence-specific potent induction of IFN-alpha by short interfering RNA in plasmacytoid dendritic cells through TLR7. *Nat Med* 11 (3):263–270.

Howard, K. A., U. L. Rahbek, X. Liu, C. K. Damgaard, S. Z. Glud, M. O. Andersen, M. B. Hovgaard et al. 2006. RNA interference in vitro and in vivo using a novel chitosan/siRNA nanoparticle system. *Mol Ther* 14 (4):476–484.

Hu-Lieskovan, S., J. D. Heidel, D. W. Bartlett, M. E. Davis, and T. J. Triche. 2005. Sequence-specific knockdown of EWS-FLI1 by targeted, nonviral delivery of small interfering RNA inhibits tumor growth in a murine model of metastatic Ewing's sarcoma. *Cancer Res* 65 (19):8984–8992.

Hutvagner, G. and P. D. Zamore. 2002. RNAi: Nature abhors a double-strand. *Curr Opin Genet Dev* 12 (2):225–232.

Jackson, A. L., S. R. Bartz, J. Schelter, S. V. Kobayashi, J. Burchard, M. Mao, B. Li, G. Cavet, and P. S. Linsley. 2003. Expression profiling reveals off-target gene regulation by RNAi. *Nat Biotechnol* 21 (6):635–637.

Jackson, A. L., J. Burchard, D. Leake, A. Reynolds, J. Schelter, J. Guo, J. M. Johnson et al. 2006. Position-specific chemical modification of siRNAs reduces "off-target" transcript silencing. *RNA* 12 (7):1197–1205.

John, M., R. Constien, A. Akinc, M. Goldberg, Y. A. Moon, M. Spranger, P. Hadwiger et al. 2007. Effective RNAi-mediated gene silencing without interruption of the endogenous microRNA pathway. *Nature* 449 (7163):745–747.

Johnson, L. N., S. M. Cashman, and R. Kumar-Singh. 2008. Cell-penetrating peptide for enhanced delivery of nucleic acids and drugs to ocular tissues including retina and cornea. *Mol Ther* 16 (1):107–114.

Jordan, E. T., M. Collins, J. Terefe, L. Ugozzoli, and T. Rubio. 2008. Optimizing electroporation conditions in primary and other difficult-to-transfect cells. *J Biomol Tech* 19 (5):328–334.

Judge, A. D., V. Sood, J. R. Shaw, D. Fang, K. McClintock, and I. MacLachlan. 2005. Sequence-dependent stimulation of the mammalian innate immune response by synthetic siRNA. *Nat Biotechnol* 23 (4):457–462.

Kalderon, D., W. D. Richardson, A. F. Markham, and A. E. Smith. 1984. Sequence requirements for nuclear location of simian virus 40 large-T antigen. *Nature* 311 (5981):33–38.

Katas, H. and H. O. Alpar. 2006. Development and characterisation of chitosan nanoparticles for siRNA delivery. *J Control Release* 115 (2):216–225.

Kawasaki, H. and K. Taira. 2004. Induction of DNA methylation and gene silencing by short interfering RNAs in human cells. *Nature* 431 (7005):211–217.

Kichler, A., C. Leborgne, J. Marz, O. Danos, and B. Bechinger. 2003. Histidine-rich amphipathic peptide antibiotics promote efficient delivery of DNA into mammalian cells. *Proc Natl Acad Sci USA* 100 (4):1564–1568.

Kim, D. H., M. A. Behlke, S. D. Rose, M. S. Chang, S. Choi, and J. J. Rossi. 2005. Synthetic dsRNA Dicer substrates enhance RNAi potency and efficacy. *Nat Biotechnol* 23 (2):222–226.

Kim, W. J., L. V. Christensen, S. Jo, J. W. Yockman, J. H. Jeong, Y. H. Kim, and S. W. Kim. 2006. Cholesteryl oligoarginine delivering vascular endothelial growth factor siRNA effectively inhibits tumor growth in colon adenocarcinoma. *Mol Ther* 14 (3):343–350.

Kishida, T., H. Asada, S. Gojo, S. Ohashi, M. Shin-Ya, K. Yasutomi, R. Terauchi et al. 2004. Sequence-specific gene silencing in murine muscle induced by electroporation-mediated transfer of short interfering RNA. *J Gene Med* 6 (1):105–110.

Klein, C., C. T. Bock, H. Wedemeyer, T. Wustefeld, S. Locarnini, H. P. Dienes, S. Kubicka, M. P. Manns, and C. Trautwein. 2003. Inhibition of hepatitis B virus replication in vivo by nucleoside analogues and siRNA. *Gastroenterology* 125 (1):9–18.

Kong, X. C., P. Barzaghi, and M. A. Ruegg. 2004. Inhibition of synapse assembly in mammalian muscle in vivo by RNA interference. *EMBO Rep* 5 (2):183–188.

Kumar, M. and G. G. Carmichael. 1998. Antisense RNA: Function and fate of duplex RNA in cells of higher eukaryotes. *Microbiol Mol Biol Rev* 62 (4):1415–1434.

Kumar, P., H. Wu, J. L. McBride, K. E. Jung, M. H. Kim, B. L. Davidson, S. K. Lee, P. Shankar, and N. Manjunath. 2007. Transvascular delivery of small interfering RNA to the central nervous system. *Nature* 448 (7149):39–43.

Law, M., M. Jafari, and P. Chen. 2008. Physicochemical characterization of siRNA-peptide complexes. *Biotechnol Prog* 24 (4):957–963.

Layzer, J. M., A. P. McCaffrey, A. K. Tanner, Z. Huang, M. A. Kay, and B. A. Sullenger. 2004. In vivo activity of nuclease-resistant siRNAs. *RNA* 10 (5):766–771.

Lee, S. H., S. H. Kim, and T. G. Park. 2007. Intracellular siRNA delivery system using polyelectrolyte complex micelles prepared from VEGF siRNA-PEG conjugate and cationic fusogenic peptide. *Biochem Biophys Res Commun* 357 (2):511–516.

Leng, Q. and A. J. Mixson. 2005a. Modified branched peptides with a histidine-rich tail enhance in vitro gene transfection. *Nucleic Acids Res* 33 (4):e40.

Leng, Q. and A. J. Mixson. 2005b. Small interfering RNA targeting Raf-1 inhibits tumor growth in vitro and in vivo. *Cancer Gene Ther* 12 (8):682–690.

Leng, Q., P. Scaria, P. Lu, M. C. Woodle, and A. J. Mixson. 2008. Systemic delivery of HK Raf-1 siRNA polyplexes inhibits MDA-MB-435 xenografts. *Cancer Gene Ther* 15 (8):485–495.

Leng, Q., P. Scaria, J. Zhu, N. Ambulos, P. Campbell, and A. J. Mixson. 2005. Highly branched HK peptides are effective carriers of siRNA. *J Gene Med* 7 (7):977–986.

Lentz, T. L. 1990. Rabies virus binding to an acetylcholine receptor alpha-subunit peptide. *J Mol Recognit* 3 (2):82–88.

Lewis, D. L., J. E. Hagstrom, A. G. Loomis, J. A. Wolff, and H. Herweijer. 2002. Efficient delivery of siRNA for inhibition of gene expression in postnatal mice. *Nat Genet* 32 (1):107–108.

Liu, F., Y. Song, and D. Liu. 1999. Hydrodynamics-based transfection in animals by systemic administration of plasmid DNA. *Gene Ther* 6 (7):1258–1266.

Liu, W. L., M. Stevenson, L. W. Seymour, and K. D. Fisher. 2009. Quantification of siRNA using competitive qPCR. *Nucleic Acids Res* 37 (1):e4.

Lo, S. L. and S. Wang. 2008. An endosomolytic Tat peptide produced by incorporation of histidine and cysteine residues as a nonviral vector for DNA transfection. *Biomaterials* 29 (15):2408–2414.

Lundberg, P., S. El-Andaloussi, T. Sutlu, H. Johansson, and U. Langel. 2007. Delivery of short interfering RNA using endosomolytic cell-penetrating peptides. *FASEB J* 21 (11):2664–2671.

Lv, H., S. Zhang, B. Wang, S. Cui, and J. Yan. 2006. Toxicity of cationic lipids and cationic polymers in gene delivery. *J Control Release* 114 (1):100–109.

Ma, Z., J. Li, F. He, A. Wilson, B. Pitt, and S. Li. 2005. Cationic lipids enhance siRNA-mediated interferon response in mice. *Biochem Biophys Res Commun* 330 (3):755–759.

Macrae, I. J., K. Zhou, F. Li, A. Repic, A. N. Brooks, W. Z. Cande, P. D. Adams, and J. A. Doudna. 2006. Structural basis for double-stranded RNA processing by Dicer. *Science* 311 (5758):195–198.

Manche, L., S. R. Green, C. Schmedt, and M. B. Mathews. 1992. Interactions between double-stranded RNA regulators and the protein kinase DAI. *Mol Cell Biol* 12 (11):5238–5248.

Maniataki, E. and Z. Mourelatos. 2005. A human, ATP-independent, RISC assembly machine fueled by pre-miRNA. *Genes Dev* 19 (24):2979–2990.

Manickam, D. S. and D. Oupicky. 2006. Multiblock reducible copolypeptides containing histidine-rich and nuclear localization sequences for gene delivery. *Bioconjug Chem* 17 (6):1395–1403.

Mann, A., G. Thakur, V. Shukla, and M. Ganguli. 2008. Peptides in DNA delivery: Current insights and future directions. *Drug Discov Today* 13 (3–4):152–160.

McCaffrey, A. P., L. Meuse, T. T. Pham, D. S. Conklin, G. J. Hannon, and M. A. Kay. 2002. RNA interference in adult mice. *Nature* 418 (6893):38–39.

McKenzie, D. L., K. Y. Kwok, and K. G. Rice. 2000a. A potent new class of reductively activated peptide gene delivery agents. *J Biol Chem* 275 (14):9970–9977.

McKenzie, D. L., E. Smiley, K. Y. Kwok, and K. G. Rice. 2000b. Low molecular weight disulfide cross-linking peptides as nonviral gene delivery carriers. *Bioconjug Chem* 11 (6):901–909.

Meade, B. R. and S. F. Dowdy. 2007. Exogenous siRNA delivery using peptide transduction domains/cell penetrating peptides. *Adv Drug Deliv Rev* 59 (2–3):134–140.

Meade, B. R. and S. F. Dowdy. 2008. Enhancing the cellular uptake of siRNA duplexes following noncovalent packaging with protein transduction domain peptides. *Adv Drug Deliv Rev* 60 (4–5):530–536.

Mery, J., C. Granier, M. Juin, and J. Brugidou. 1993. Disulfide linkage to polyacrylic resin for automated Fmoc peptide synthesis. Immunochemical applications of peptide resins and mercaptoamide peptides. *Int J Pept Protein Res* 42 (1):44–52.

Midoux, P., A. Kichler, V. Boutin, J. C. Maurizot, and M. Monsigny. 1998. Membrane permeabilization and efficient gene transfer by a peptide containing several histidines. *Bioconjug Chem* 9 (2):260–267.

Mitchell, D. J., D. T. Kim, L. Steinman, C. G. Fathman, and J. B. Rothbard. 2000. Polyarginine enters cells more efficiently than other polycationic homopolymers. *J Pept Res* 56 (5):318–325.

Mok, H. and T. G. Park. 2008. Self-crosslinked and reducible fusogenic peptides for intracellular delivery of siRNA. *Biopolymers* 89 (10):881–888.

Morris, K. V., S. W. Chan, S. E. Jacobsen, and D. J. Looney. 2004. Small interfering RNA-induced transcriptional gene silencing in human cells. *Science* 305 (5688):1289–1292.

Morris, M. C., L. Chaloin, J. Mery, F. Heitz, and G. Divita. 1999. A novel potent strategy for gene delivery using a single peptide vector as a carrier. *Nucleic Acids Res* 27 (17):3510–3517.

Morris, M. C., J. Depollier, J. Mery, F. Heitz, and G. Divita. 2001. A peptide carrier for the delivery of biologically active proteins into mammalian cells. *Nat Biotechnol* 19 (12):1173–1176.

Morris, M. C., P. Vidal, L. Chaloin, F. Heitz, and G. Divita. 1997. A new peptide vector for efficient delivery of oligonucleotides into mammalian cells. *Nucleic Acids Res* 25 (14):2730–2736.

Morrissey, D. V., K. Blanchard, L. Shaw, K. Jensen, J. A. Lockridge, B. Dickinson, J. A. McSwiggen et al. 2005a. Activity of stabilized short interfering RNA in a mouse model of hepatitis B virus replication. *Hepatology* 41 (6):1349–1356.

Morrissey, D. V., J. A. Lockridge, L. Shaw, K. Blanchard, K. Jensen, W. Breen, K. Hartsough et al. 2005b. Potent and persistent in vivo anti-HBV activity of chemically modified siRNAs. *Nat Biotechnol* 23 (8):1002–1007.

Moschos, S. A., S. W. Jones, M. M. Perry, A. E. Williams, J. S. Erjefalt, J. J. Turner, P. J. Barnes, B. S. Sproat, M. J. Gait, and M. A. Lindsay. 2007. Lung delivery studies using siRNA conjugated to TAT(48–60) and penetratin reveal peptide induced reduction in gene expression and induction of innate immunity. *Bioconjug Chem* 18 (5):1450–1459.

Muratovska, A. and M. R. Eccles. 2004. Conjugate for efficient delivery of short interfering RNA (siRNA) into mammalian cells. *FEBS Lett* 558 (1–3):63–68.

Oates, A. C., A. E. Bruce, and R. K. Ho. 2000. Too much interference: Injection of double-stranded RNA has nonspecific effects in the zebrafish embryo. *Dev Biol* 224 (1):20–28.

Ohrt, T., D. Merkle, K. Birkenfeld, C. J. Echeverri, and P. Schwille. 2006. In situ fluorescence analysis demonstrates active siRNA exclusion from the nucleus by Exportin 5. *Nucleic Acids Res* 34 (5):1369–1380.

Overhoff, M., W. Wunsche, and G. Sczakiel. 2004. Quantitative detection of siRNA and single-stranded oligonucleotides: Relationship between uptake and biological activity of siRNA. *Nucleic Acids Res* 32 (21):e170.

Paddison, P. J., A. A. Caudy, R. Sachidanandam, and G. J. Hannon. 2004. Short hairpin activated gene silencing in mammalian cells. *Methods Mol Biol* 265:85–100.

Pooga, M., M. Hallbrink, M. Zorko, and U. Langel. 1998. Cell penetration by transportan. *FASEB J* 12 (1):67–77.

Rafalski, M., J. D. Lear, and W. F. DeGrado. 1990. Phospholipid interactions of synthetic peptides representing the N-terminus of HIV gp41. *Biochemistry* 29 (34):7917–7922.

Rahbek, U. L., K. A. Howard, D. Oupicky, D. S. Manickam, M. Dong, A. F. Nielsen, T. B. Hansen, F. Besenbacher, and J. Kjems. 2008. Intracellular siRNA and precursor miRNA trafficking using bioresponsive copolypeptides. *J Gene Med* 10 (1):81–93.

Rana, T. M. 2007. Illuminating the silence: Understanding the structure and function of small RNAs. *Nat Rev Mol Cell Biol* 8 (1):23–36.

Read, M. L., S. Singh, Z. Ahmed, M. Stevenson, S. S. Briggs, D. Oupicky, L. B. Barrett et al. 2005. A versatile reducible polycation-based system for efficient delivery of a broad range of nucleic acids. *Nucleic Acids Res* 33 (9):e86.

Reynolds, A., D. Leake, Q. Boese, S. Scaringe, W. S. Marshall, and A. Khvorova. 2004. Rational siRNA design for RNA interference. *Nat Biotechnol* 22 (3):326–330.

Richard, J. P., K. Melikov, E. Vives, C. Ramos, B. Verbeure, M. J. Gait, L. V. Chernomordik, and B. Lebleu. 2003. Cell-penetrating peptides. A reevaluation of the mechanism of cellular uptake. *J Biol Chem* 278 (1):585–590.

Rittner, K., A. Benavente, A. Bompard-Sorlet, F. Heitz, G. Divita, R. Brasseur, and E. Jacobs. 2002. New basic membrane-destabilizing peptides for plasmid-based gene delivery in vitro and in vivo. *Mol Ther* 5 (2):104–114.

Rothbard, J. B., E. Kreider, C. L. VanDeusen, L. Wright, B. L. Wylie, and P. A. Wender. 2002. Arginine-rich molecular transporters for drug delivery: Role of backbone spacing in cellular uptake. *J Med Chem* 45 (17):3612–3618.

Saito, H., T. Minamisawa, and K. Shiba. 2007. Motif programming: A microgene-based method for creating synthetic proteins containing multiple functional motifs. *Nucleic Acids Res* 35 (6):e38.

Saxena, S., Z. O. Jonsson, and A. Dutta. 2003. Small RNAs with imperfect match to endogenous mRNA repress translation: Implications for off-target activity of small inhibitory RNA in mammalian cells. *J Biol Chem* 278 (45):44312–44319.

Simeoni, F., M. C. Morris, F. Heitz, and G. Divita. 2003. Insight into the mechanism of the peptide-based gene delivery system MPG: Implications for delivery of siRNA into mammalian cells. *Nucleic Acids Res* 31 (11):2717–2724.

Siolas, D., C. Lerner, J. Burchard, W. Ge, P. S. Linsley, P. J. Paddison, G. J. Hannon, and M. A. Cleary. 2005. Synthetic shRNAs as potent RNAi triggers. *Nat Biotechnol* 23 (2):227–231.

Sioud, M. and D. R. Sorensen. 2003. Cationic liposome-mediated delivery of siRNAs in adult mice. *Biochem Biophys Res Commun* 312 (4):1220–1225.

Skehel, J. J., P. M. Bayley, E. B. Brown, S. R. Martin, M. D. Waterfield, J. M. White, I. A. Wilson, and D. C. Wiley. 1982. Changes in the conformation of influenza virus hemagglutinin at the pH optimum of virus-mediated membrane fusion. *Proc Natl Acad Sci USA* 79 (4):968–972.

Song, E., S. K. Lee, J. Wang, N. Ince, N. Ouyang, J. Min, J. Chen, P. Shankar, and J. Lieberman. 2003. RNA interference targeting Fas protects mice from fulminant hepatitis. *Nat Med* 9 (3):347–351.

Sorensen, D. R., M. Leirdal, and M. Sioud. 2003. Gene silencing by systemic delivery of synthetic siRNAs in adult mice. *J Mol Biol* 327 (4):761–766.

Soutschek, J., A. Akinc, B. Bramlage, K. Charisse, R. Constien, M. Donoghue, S. Elbashir et al. 2004. Therapeutic silencing of an endogenous gene by systemic administration of modified siRNAs. *Nature* 432 (7014):173–178.

Sparrow, J. T., V. V. Edwards, C. Tung, M. J. Logan, M. S. Wadhwa, J. Duguid, and L. C. Smith. 1998. Synthetic peptide-based DNA complexes for nonviral gene delivery. *Adv Drug Deliv Rev* 30 (1–3):115–131.

Stark, G. R., I. M. Kerr, B. R. Williams, R. H. Silverman, and R. D. Schreiber. 1998. How cells respond to interferons. *Annu Rev Biochem* 67:227–264.

Stevenson, M., V. Ramos-Perez, S. Singh, M. Soliman, J. A. Preece, S. S. Briggs, M. L. Read, and L. W. Seymour. 2008. Delivery of siRNA mediated by histidine-containing reducible polycations. *J Control Release* 130 (1):46–56.

Stewart, K. M., K. L. Horton, and S. O. Kelley. 2008. Cell-penetrating peptides as delivery vehicles for biology and medicine. *Org Biomol Chem* 6 (13):2242–2255.

Takahashi, Y., M. Nishikawa, N. Kobayashi, and Y. Takakura. 2005. Gene silencing in primary and metastatic tumors by small interfering RNA delivery in mice: Quantitative analysis using melanoma cells expressing firefly and sea pansy luciferases. *J Control Release* 105 (3):332–343.

Takayama, K., A. Suzuki, T. Manaka, S. Taguchi, Y. Hashimoto, Y. Imai, S. Wakitani, and K. Takaoka. 2009. RNA interference for noggin enhances the biological activity of bone morphogenetic proteins in vivo and in vitro. *J Bone Miner Metab* 27 (4):402–411.

Takei, Y., T. Nemoto, P. Mu, T. Fujishima, T. Ishimoto, Y. Hayakawa, Y. Yuzawa, S. Matsuo, T. Muramatsu, and K. Kadomatsu. 2008. In vivo silencing of a molecular target by short interfering RNA electroporation: Tumor vascularization correlates to delivery efficiency. *Mol Cancer Ther* 7 (1):211–221.

Tan, R. and A. D. Frankel. 1995. Structural variety of arginine-rich RNA-binding peptides. *Proc Natl Acad Sci USA* 92 (12):5282–5286.

Tompkins, S. M., C. Y. Lo, T. M. Tumpey, and S. L. Epstein. 2004. Protection against lethal influenza virus challenge by RNA interference in vivo. *Proc Natl Acad Sci USA* 101 (23):8682–8686.

Tonges, L., P. Lingor, R. Egle, G. P. Dietz, A. Fahr, and M. Bahr. 2006. Stearylated octaarginine and artificial virus-like particles for transfection of siRNA into primary rat neurons. *RNA* 12 (7):1431–1438.

Tuschl, T., P. D. Zamore, R. Lehmann, D. P. Bartel, and P. A. Sharp. 1999. Targeted mRNA degradation by double-stranded RNA in vitro. *Genes Dev* 13 (24):3191–3197.

Ui-Tei, K., Y. Naito, S. Zenno, K. Nishi, K. Yamato, F. Takahashi, A. Juni, and K. Saigo. 2008. Functional dissection of siRNA sequence by systematic DNA substitution: Modified siRNA with a DNA seed arm is a powerful tool for mammalian gene silencing with significantly reduced off-target effect. *Nucleic Acids Res* 36 (7):2136–2151.

Urban-Klein, B., S. Werth, S. Abuharbeid, F. Czubayko, and A. Aigner. 2005. RNAi-mediated gene-targeting through systemic application of polyethylenimine (PEI)-complexed siRNA in vivo. *Gene Ther* 12 (5):461–466.

Uster, P. S. and D. W. Deamer. 1985. pH-dependent fusion of liposomes using titratable polycations. *Biochemistry* 24 (1):1–8.

Veldhoen, S., S. D. Laufer, A. Trampe, and T. Restle. 2006. Cellular delivery of small interfering RNA by a non-covalently attached cell-penetrating peptide: Quantitative analysis of uptake and biological effect. *Nucleic Acids Res* 34 (22):6561–6573.

Vidal, P., L. Chaloin, A. Heitz, N. Van Mau, J. Mery, G. Divita, and F. Heitz. 1998. Interactions of primary amphipathic vector peptides with membranes. Conformational consequences and influence on cellular localization. *J Membr Biol* 162 (3):259–264.

Vidal, P., L. Chaloin, J. Mery, N. Lamb, N. Lautredou, R. Bennes, and F. Heitz. 1996. Solid-phase synthesis and cellular localization of a C- and/or N-terminal labelled peptide. *J Pept Sci* 2 (2):125–133.

Vives, E., P. Brodin, and B. Lebleu. 1997. A truncated HIV-1 Tat protein basic domain rapidly translocates through the plasma membrane and accumulates in the cell nucleus. *J Biol Chem* 272 (25):16010–16017.

Wang, C. Y. and L. Huang. 1984. Polyhistidine mediates an acid-dependent fusion of negatively charged liposomes. *Biochemistry* 23 (19):4409–4416.

Waterhouse, P. M., M. B. Wang, and T. Lough. 2001. Gene silencing as an adaptive defence against viruses. *Nature* 411 (6839):834–842.

Wattiaux, R., N. Laurent, S. Wattiaux-De Coninck, and M. Jadot. 2000. Endosomes, lysosomes: Their implication in gene transfer. *Adv Drug Deliv Rev* 41 (2):201–208.

Werth, S., B. Urban-Klein, L. Dai, S. Hobel, M. Grzelinski, U. Bakowsky, F. Czubayko, and A. Aigner. 2006. A low molecular weight fraction of polyethylenimine (PEI) displays increased transfection efficiency of DNA and siRNA in fresh or lyophilized complexes. *J Control Release* 112 (2):257–270.

Whitley, P., G. Gafvelin, and G. von Heijne. 1995. SecA-independent translocation of the periplasmic N-terminal tail of an Escherichia coli inner membrane protein. Position-specific effects on translocation of positively charged residues and construction of a protein with a C-terminal translocation signal. *J Biol Chem* 270 (50):29831–29835.

Wyman, T. B., F. Nicol, O. Zelphati, P. V. Scaria, C. Plank, and F. C. Szoka, Jr. 1997. Design, synthesis, and characterization of a cationic peptide that binds to nucleic acids and permeabilizes bilayers. *Biochemistry* 36 (10):3008–3017.

Xu, J., L. Li, Z. Qian, J. Hong, S. Shen, and W. Huang. 2005. Reduction of PTP1B by RNAi upregulates the activity of insulin controlled fatty acid synthase promoter. *Biochem Biophys Res Commun* 329 (2):538–543.

Yang, J., H. Chen, I. R. Vlahov, J. X. Cheng, and P. S. Low. 2006. Evaluation of disulfide reduction during receptor-mediated endocytosis by using FRET imaging. *Proc Natl Acad Sci USA* 103 (37):13872–13877.

Zeng, Y. and B. R. Cullen. 2004. Structural requirements for pre-microRNA binding and nuclear export by Exportin 5. *Nucleic Acids Res* 32 (16):4776–4785.

Zhao, Z., Y. Cao, M. Li, and A. Meng. 2001. Double-stranded RNA injection produces nonspecific defects in zebrafish. *Dev Biol* 229 (1):215–223.

Zimmermann, T. S., A. C. Lee, A. Akinc, B. Bramlage, D. Bumcrot, M. N. Fedoruk, J. Harborth et al. 2006. RNAi-mediated gene silencing in non-human primates. *Nature* 441 (7089):111–114.

Zintchenko, A., A. Philipp, A. Dehshahri, and E. Wagner. 2008. Simple modifications of branched PEI lead to highly efficient siRNA carriers with low toxicity. *Bioconjug Chem* 19 (7):1448–1455.

10

Nucleic Acid Nanobiomaterials

Bin Wang
Marshall University

In the past 2 decades, there has been an incredible growth in "nano"-related research. Nano refers to a world that is on the nanometer-length scale (<100 nm) in at least one dimension. Nanoscale building blocks are widely existent in the biological world and include molecules such as nucleic acids, proteins, and peptides. There are also a variety of nonbiological nanomaterials that exist, among which the most widely investigated for use in nanotechnology-based projects include semiconductor (e.g., quantum dots such as CdS, CdSe, and CdTe), silica, metallic (e.g., gold, silver, and platinum), magnetic (e.g., iron oxides), synthesized polymeric (e.g., poly(L-lactide), poly(hydroxybutyrate-co-hydroxyvalerate), and poly(lactic acid-co-glycolic acid)), and carbon-based (e.g., carbon nanotube, and fullerene derivatives) materials. This chapter focuses on the nucleic acid-based nanobiomaterials. However, it is worth mentioning that the development of novel materials to integrate biological with nonbiological nanomaterials will have enormous potential in applications such as biosensors, nanoelectronics, nanorobotics, smart materials, drug delivery, and gene therapy.

Structural nucleic acid nanotechnology, which refers to the application of deoxyribonucleic acid (DNA) and ribonucleic acid (RNA) nanomaterials to construct highly ordered nanostructures and/or well-controlled nanomechanical devices, was founded by Nad Seeman's research group at New York University. Although still in its infancy, structural nucleic acid technology has attracted quite a bit of attention since the 1990s (Seeman 1998, 2002, 2003; Guo 2005a,b; Jaeger and Chworos 2006; Triberis and Dimakogianni 2009). This chapter reviews some important nanoarchitectures and nanodevices constructed using DNA and RNA nanobiomaterials. The content is divided into four sections: Section 10.1 demonstrates the design and construction of DNA-based static nanoarchitectures, Section 10.2 describes RNA-based nanostructures and their advantage compared to DNA counterparts, Section 10.3 describes mobile nucleic acid nanodevices and their potential applications, and Section 10.4 offers concluding remarks and commentary on the future direction of nano-based research.

10.1 DNA-Based Static Nanoarchitectures

DNA is a nucleic acid comprising double-helical biopolymer chains of nucleotides. Each single nucleotide contains a sugar backbone, a phosphate group, and one of four of the following bases: cytosine, guanine, adenine, and thymine. According to the Watson–Crick base-pairing model, cytosine forms three hydrogen bonds with guanine, while adenine forms two hydrogen bonds with thymine. The complementary sequences make an antiparallel double-stranded DNA molecule. The diameter of a DNA double helix chain is about 2 nm; one nucleotide unit is about 0.33 nm long. The persistence length (a mechanical measurement of stiffness of long polymers) of the DNA duplex strand in buffers of moderate salt concentration is 45–50 nm (Hagerman 1988). Within cells, DNA functions as a genetic storage material; the DNA-encoded genes can be expressed to functional proteins through the itermediate RNA carriers.

Due to its structural robustness and programmability, DNA has been investigated as a potential building block for nanomaterial science for more than 2 decades. The theoretical basis of structural DNA nanotechnology is the self-assembly of the DNA double helix to form Watson–Crick base-pairing complementary secondary structures. Two double-stranded DNA molecules with complementary overhangs (termed "sticky ends") can cohere with each other and be ligated to form a B-DNA (right-handed double helix) with predictable structures (see Figure 10.1A and B). Because linear DNA is not a good candidate for nanostructure building blocks, asymmetrically branched DNA molecules with sticky ends were synthesized to construct nanoarchitectures (see Figure 10.1C) (Seeman and Kallenbach 1994; Constantinou and West 2004; Liu and West 2004). Figure 10.1D shows an example of the sticky-ended assembly of branched DNA molecules. The branched DNA

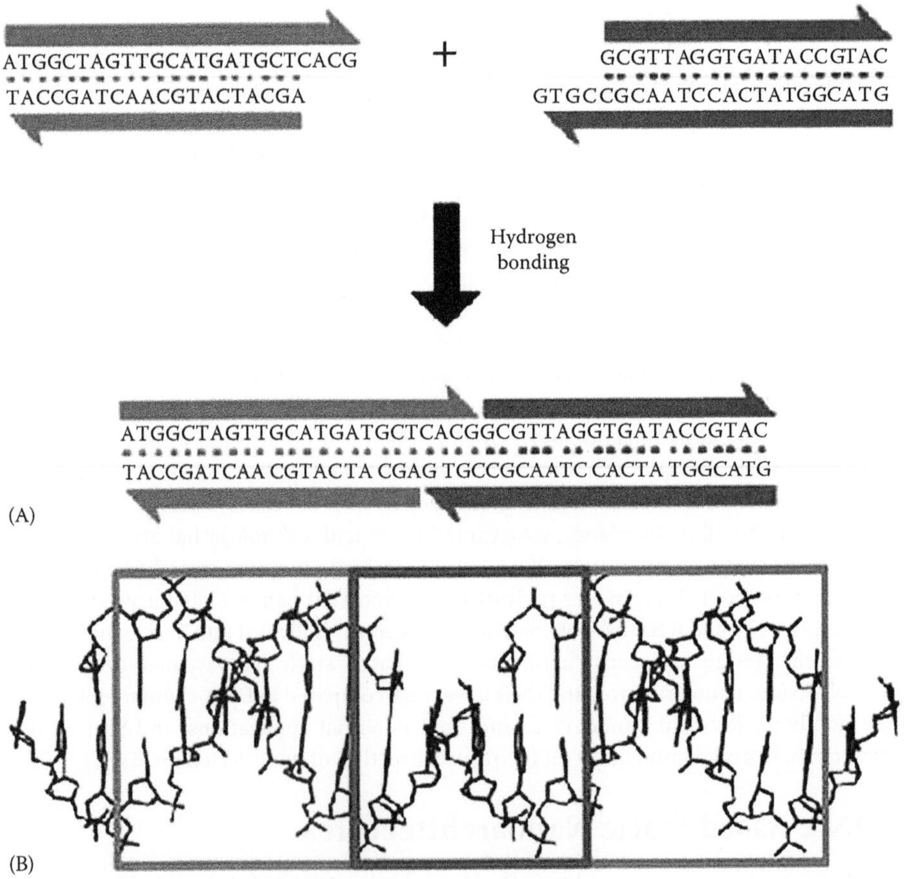

FIGURE 10.1 Assembly of branched DNA molecules with sticky-end cohesion: (A) the bonding of two molecules of DNA through the cohesion of their sticky ends; (B) a portion of the crystal structure of the DNA double helix formed by sticky-end cohesion;

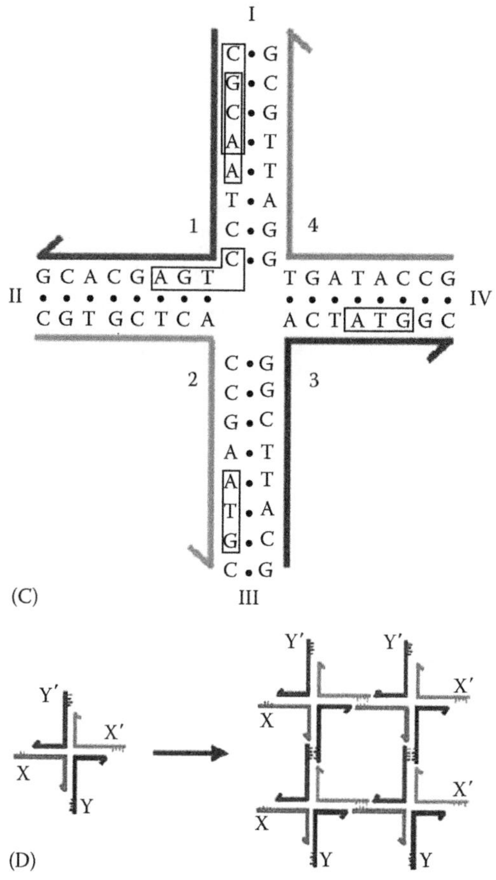

FIGURE 10.1 (continued) (C) an asymmetrical branched DNA junction; sequences in boxes such as CGCA, GCAA, and CTGA are unique; and (D) the assembly of branched DNA molecules into a quadriplex with four sticky ends. (Adapted from Seeman, N.C., *Chem. Biol.*, 10(12), 1151, 2003a. With permission.)

contains four arms with sticky ends complementary to each other. The X end is complementary to and sticks to the X′ end, and the Y end is complementary to and sticks to the Y′ end, thus allowing the self-assembly of four branched DNA molecules into a single quadrilateral structure. Moreover, an infinite number of branched DNA molecules can attach by assembling to the unpaired sticky ends to the outside of the quadrilateral structure (Seeman 1982; Seeman and Kallenbach 1983; Seeman 2003). The stable and immobile DNA branched junctions permit the design and construction of desired nanostructures (Wang et al. 1991; Zhang et al. 1993; Du and Seeman 1994; Seeman and Kallenbach 1994).

The first stick polyhedral object (a cube; see Figure 10.2A) constructed by multiply connected DNA was pioneered by Seeman's research group in 1991 (Chen and Seeman 1991a,b; Seeman 1991). The cube was formed by the catenation of single-stranded cyclic oligonucleotides to branched DNA molecules, as illustrated in Figure 10.3. Briefly, the construction started with the cyclization of two 80-nucleotide single-stranded DNA molecules, followed by the association of branched DNA strands with the DNA rings to form two squares, each with four sticky ends. When the sticky ends are ligated (i.e., ligate C–C′ and D–D′) (see Figure 10.3), a tricyclic belt is formed. When the remaining outer sticky ends are ligated (i.e., ligate A–A′ and B–B′) (see Figure 10.3), the belt is closed to form a cube (Chen and Seeman 1991b).

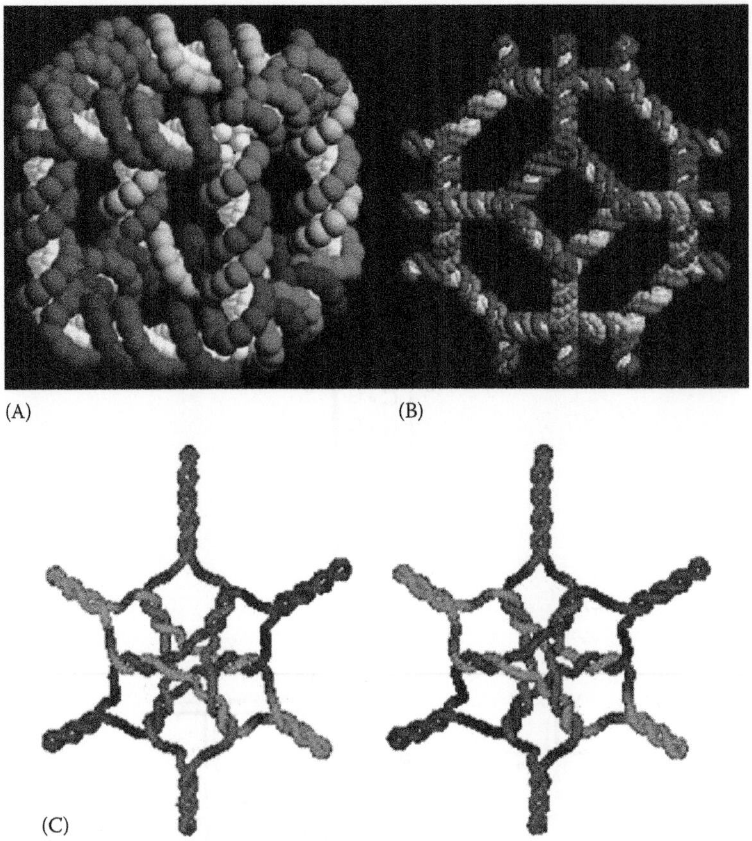

FIGURE 10.2 Topologies of (A) a DNA cube, (B) a DNA-truncated octahedron, and (C) Borromean rings in stereo view. A right-handed three-arm junction is in front, and a left-handed three-arm junction is at the rear. (Adapted from Seeman, N.C., *Chem. Biol.*, 10(12), 1151, 2003a. With permission.)

In 1994, Seeman's group constructed a more complicated truncated octahedron in which the DNA self-assembly occurred on a solid support (see Figure 10.2B) (Zhang and Seeman 1994). In 1997, an intriguing structure called Borromean rings was constructed by the same research group (see Figure 10.2C). In this structure, three interlocked rings were assembled using B-DNA and Z-DNA (a left-handed double helix) with opposite helical senses (Mao et al. 1997).

In addition to sticky-end cohesion, inspired by the DNA reciprocal exchange mechanism (see Figure 10.4A), Seeman et al. used "crossovers" to generate more rigid DNA motifs (Seeman 2001). Double-crossovers (DX) refer to two parallel helices with the same or opposite polarity that interchange their single strands at two crossover points, where cohesion comes from pairing of alternate half turns in an inter-wrapped double helix (see Figure 10.4B) (Fu and Seeman 1993; Li et al. 1996). The DX molecule is approximately twice as stiff as a linear DNA duplex (Sa-Ardyen et al. 2003). As an extension of the DX motif, triple-crossovers (TX) combine a DX molecule with another DNA double helix (LaBean et al. 2000). The paranemic crossover (PX) motif contains a crossover at every possible position, and the two helices have the same polarity (Zhang et al. 2002). The difference between a JX_2 motif and a PX motif is that reciprocal exchange in JX_2 is omitted at two adjacent juxtapositions (see Figure 10.4B) (Seeman 2001, 2003). Using these rigid motifs, Seeman et al. built a series of two-dimensional DNA arrays, including the DX array, TX array, and DNA parallelogram array (Winfree et al. 1998; Mao et al. 1999; Yan et al. 2002; Spink et al. 2009).

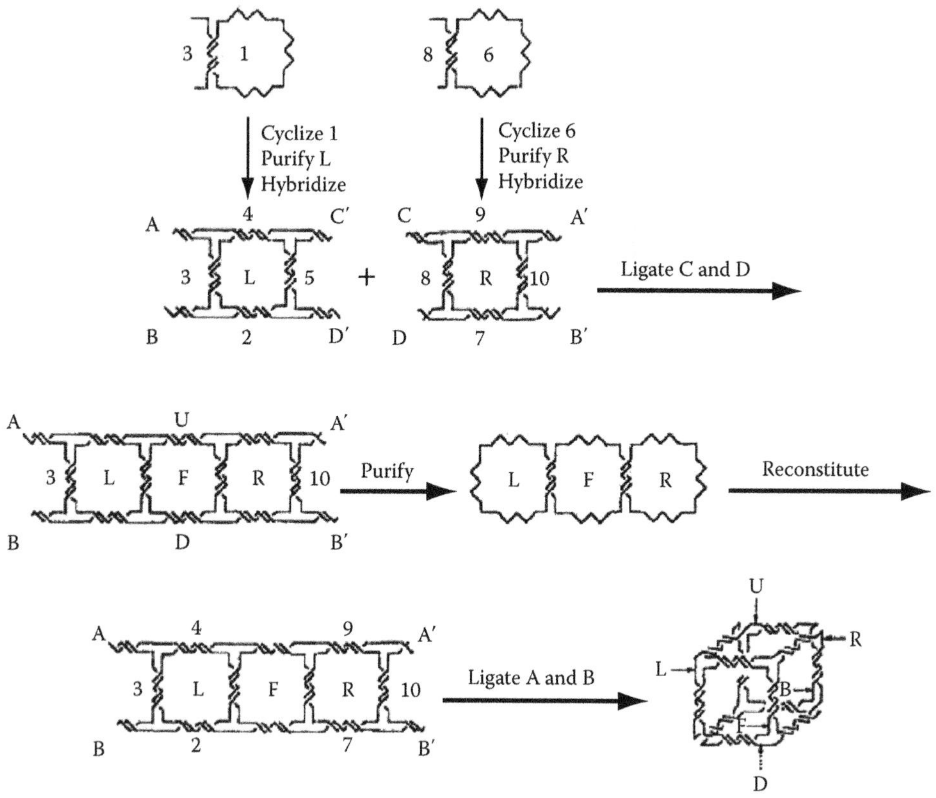

FIGURE 10.3 Steps followed to synthesize a DNA cube nanostructure. The numbers 1–10 represent different DNA strands. L, R, F, B, U, and D represent left, right, front, back, up, and down faces of the cube, respectively. A, A′, B, B′, C, C′, D, and D′ are pairs of complementary sticky ends (A sticks to A′, B sticks to B′, C sticks to C′, and D sticks to D′). (Adapted from Chen, J.H. and Seeman, N.C., *Nature*, 350(6319), 631, 1991. With permission.)

The initial success of structural DNA nanotechnology encouraged more researchers to pursue research in this field. Turberfield et al. constructed a shape-persistent DNA tetrahedron and a trigonal bypyramid (Goodman et al. 2005; Erben et al. 2006, 2007). Joyce et al. made an octahedron by the assembly of PX motifs (Shih et al. 2004). Rothemund et al. demonstrated the generation of "DNA origami" by starting with a long single-stranded DNA folded into accordion pleats and adding several short strands to staple the long DNA strand in place, a method that can be used to construct a wide variety of desired shapes (Rothemund 2006). Yan et al. described a DNA nanoarray using TX complexes (Liu et al. 2005); LaBean et al. constructed DNA nanotubes with a diameter of approximately 25 nm, self-assembled from TX tiles (Liu et al. 2004); and Winfree et al. made DNA nanotubes using DX tiles (Rothemund et al. 2004). Kiedrowski et al. introduced branched "trisoligonucleotidyls" (triple-stranded DNA) to resemble the topology of cyclobutadiene and acetylene (Scheffler et al. 1999).

In addition to Watson–Crick base-pairing interactions, other interactions such as G-quadruplexes and $C \cdot C^+$ base pairing have been exploited to build nanostructures (Alberti et al. 2006). When DNA molecules are rich in cytidine, they form four-stranded quadruplexes by base pairing. These quadruplexes are made of two parallel duplexes that contain hemi-protonated $C \cdot C^+$ pairs intercalated to become what is known as an i-motif or i-DNA (see Figure 10.5A) (Gehring et al. 1993). In humans, cytosine-rich telomeric repeats such as $(CCCTAA)_4$ can form i-motifs (Phan and Mergny 2002).

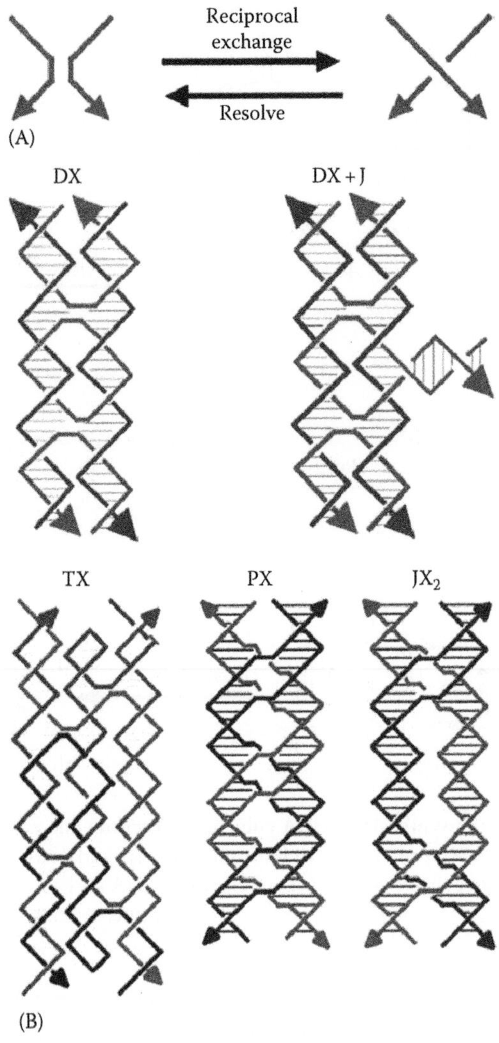

FIGURE 10.4 Generation of DNA motifs: (A) the process of reciprocal exchange and (B) topologies of several DNA motifs. (Adapted from Seeman, N.C., *Chem. Biol.*, 10(12), 1151, 2003a. With permission.)

DNA molecules rich in guanine often form four-stranded structures called G-quadruplexes (also termed G-tetrads or G4-DNA) (Henderson et al. 1987; Williamson et al. 1989; Jin et al. 1990). Four adjacent guanines are held together by eight hydrogen bonds to form a tetrad plane (see Figure 10.5B). A monovalent cation (e.g., K^+ or Na^+) binds to the center between two planar tetrads, thus interacting with eight guanines to stabilize quartet stacking. G-quadruplexes are exceptionally rigid and can self-assemble to form G-wires (Alberti et al. 2006). For example, the G-wire $G_4T_2G_4$ is a rod with a length of approximately 1 μm and a diameter of 2.5 nm (Marsh and Henderson 1994).

There are many potential applications of DNA nanostructures. Ned Seeman's motivation for pioneering structural DNA nanotechnology was born of his frustration with biomolecule crystallization. He wanted to build a DNA host lattice to use as a scaffold for macromolecule crystallization (Seeman 2003, 2008). One potential application of such DNA nanostructures is in the field of nanoelectronics. It is anticipated that DNA scaffolds could be used to arrange nanoparticles and/or carbon nanotubes into arrays that would form functional hardware or other entities. As novel DNA materials are designed and created, the applications of such nanostructures will continue to grow.

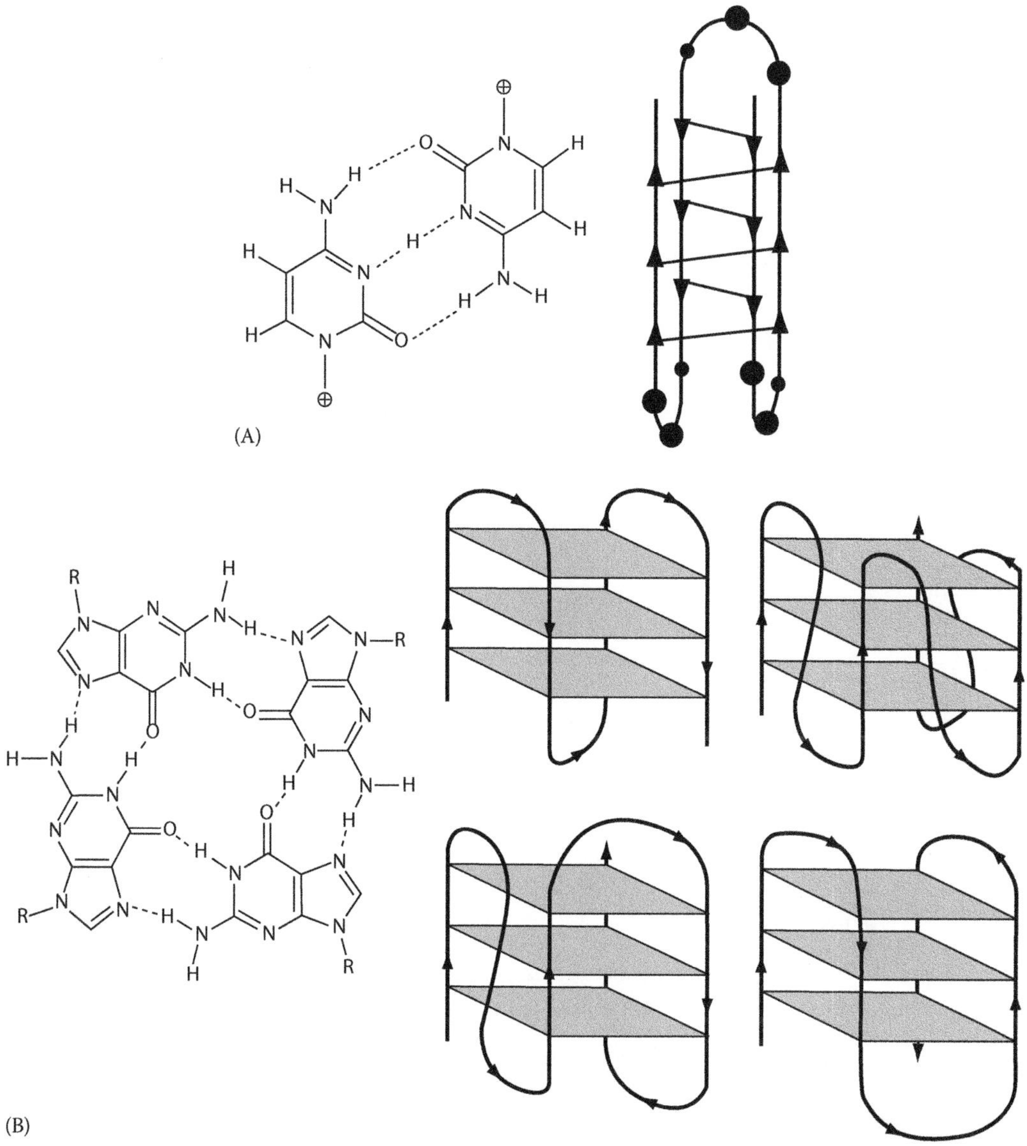

FIGURE 10.5 DNA interactions other than conventional Watson–Crick base pairing: (A) C·C⁺ base pairing and a schematic drawing of i-DNA (i-motif) and (B) a G-quartet and a schematic drawing of a G-quadruplex with four possible conformations. (Adapted from Alberti, P. et al., *Org. Biomol. Chem.*, 4(18), 3383, 2006. With permission.)

10.2 RNA-Based Static Nanostructures

There are three major differences between DNA and RNA. First, the sugar backbone of RNA is ribose rather than deoxyribose. Second, rather than containing thymine, RNA includes the base uracil, which pairs with adenine by way of two hydrogen bonds. Third, RNA is usually single-stranded, while DNA is generally double-stranded. RNA is not merely the intermediate between DNA and protein but has two major functions. It can both encode genetic information and act as a biocatalyst (enzyme) (Gesteland et al. 2005; Holbrook 2005; Leontis et al. 2006).

Much as with structural DNA nanotechnology, RNA has recently emerged as an important nano-architecture material. RNA molecules are easy to generate, either by solid-phase chemical synthesis or by a template-driven enzymatic approach that involves in vitro transcription of PCR-generated DNA templates using T7 RNA polymerase. Distinct from the double-stranded DNA molecules, where self-assembling nanostructures mainly rely on base-pairing between complementary strands to form selective secondary structures (Phan et al. 2006), single-stranded RNA molecules are more mobile and have the capability to fold into complex tertiary structures (Batey et al. 1999; Ferre-D'Amare and Doudna 1999; Doudna 2000). Tertiary RNA structures are formed when secondary structural elements associate

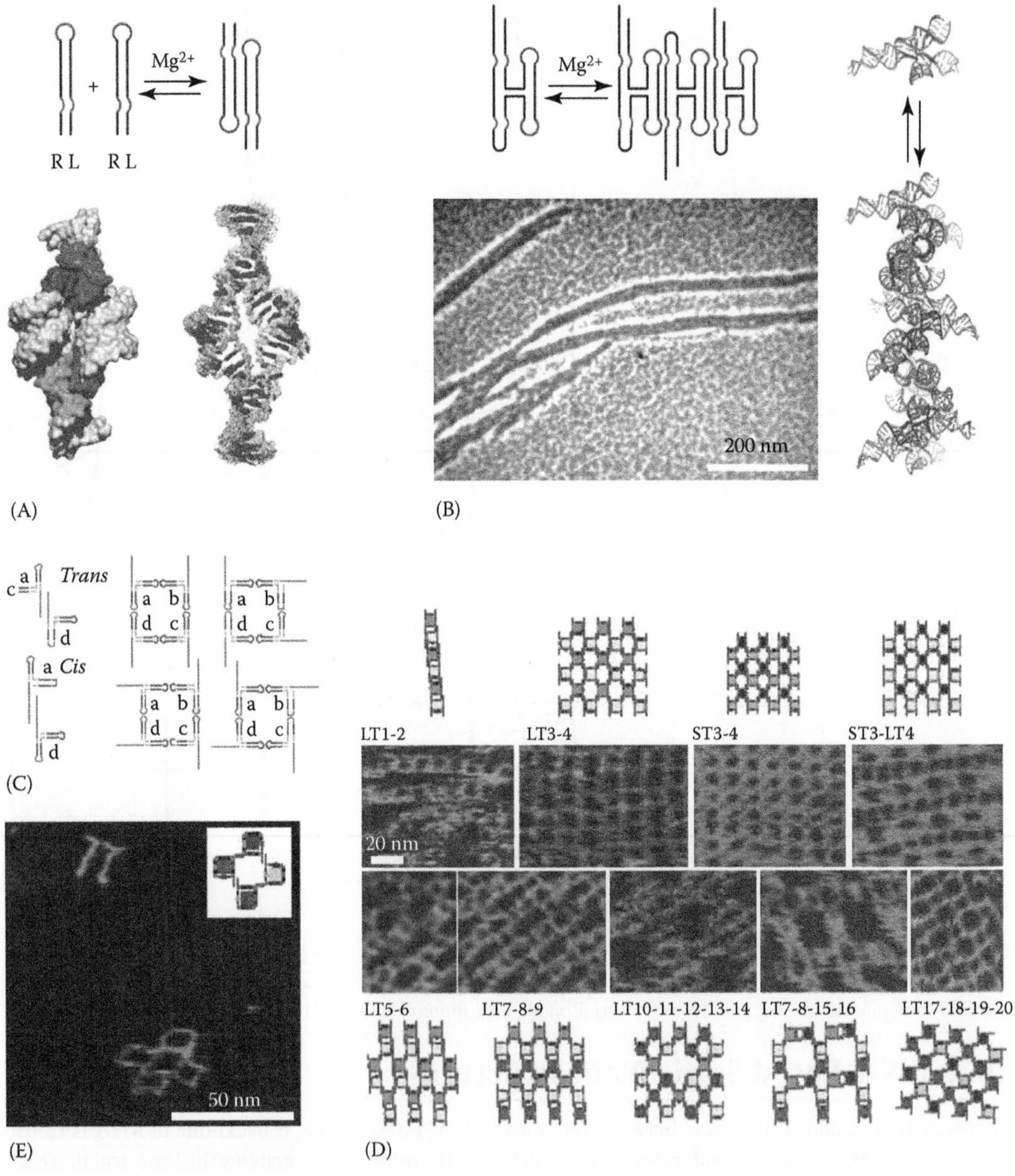

FIGURE 10.6 (See color insert.) RNA nanoarchitectures: (A) Top: the schematic diagram of receptor-loop (RL) building blocks that form a dimeric tectoRNA. Bottom: the theoretically predicted 3D model of the dimer (left) and the actual NMR-determined structure (right); (B) Top: the schematic drawing of a one-dimensional filament formed by H-shaped tectoRNAs. Bottom: TEM image of RNA filaments; (C) the schematic drawing of RNA tectosquare nanostructures; (D) and (E) two-dimensional RNA nanoarchitectures constructed from tectosquare building blocks. (Adapted from Jaeger, L. and Chworos, A., *Curr. Opin. Struct. Biol.*, 16(4), 531, 2006. With permission.)

through hydrogen bonding that involves flexible (i.e., single-stranded) loop/bulge regions, van der Waals contacts, π stacking, or metal coordination (Doudna 1997; Doudna and Doherty 1997; Strobel and Doudna 1997; Ferre-D'Amare and Doudna 1999; Doudna 2000; Paliy et al. 2009).

Many RNA nanoarchitectures have been constructed within the past 10 years (Hendrix et al. 2005). Examples of such structures include tectoRNA (Westhof et al. 1996; Jaeger and Leontis 2000; Jaeger et al. 2001), "kissing-loop" RNA (Hansma et al. 2003), and RNA jigsaw puzzles (Chworos et al. 2004) constructed by Jeager's group and RNA LEGO from Harada's group (Horiya et al. 2002).

In 2000–2001, Jaeger and Leontis pioneered a tectoRNA nanostructure (Jaeger and Leontis 2000; Jaeger et al. 2001). In their words, "RNA-tectonics refers to the modular character of natural RNA molecules, which can be decomposed and reassembled to create new nanoscale molecular objects." They used an 11-nucelotide motif receptor to bind to 5′-GAAA-3′ tetraloops. A one-dimensional nanostructure was formed by tertiary interactions between a loop and its receptor. Figure 10.6A (top) shows the schematic drawing of the hairpin tetraloop (L) and tetraloop receptor (R), forming a dimeric tectoRNA structure in the presence of Mg^{2+}. The theoretically predicted three-dimensional structure of this RNA dimer (Figure 10.6A, bottom left) was revealed to be an excellent model of the actual NMR-confirmed tectoRNA structure (Figure 10.6A, bottom right) (Davis et al. 2005). Figure 10.6B shows that loop-receptor motifs can form an H-shaped structure, which can then self-assemble into a directional one-dimensional RNA filament (Jaeger and Chworos 2006; Nasalean et al. 2006).

In 2003, Jaeger's group constructed a "kissing-loop" RNA nanostructure using 230-nucleotide-long RNA fragments from the Moloney murine leukaemia virus (MoMLV) genome (Hansma et al. 2003). AFM images showed that two RNA molecules dimerized through base-pairing in the hairpin loop region (nucleotides 210–219, 5-GCUGGCCAGC-3′) to form a "kissing-loop" RNA. Harada's research group assembled RNA structures using an approach similar to that for "kissing-loop" RNAs; they called their RNA motifs "RNA LEGO" (Horiya et al. 2003).

In 2004, Jaeger's group constructed RNA jigsaw puzzle units that they named tectosquares (Chworos et al. 2004). The tectosquare was formed by four tectoRNA units via kissing-loop interactions; each tectoRNA contained two hairpin loops connected by a 90° right angle motif (see Figure 10.6C). A sticky 3′ end was attached to each tectoRNA to allow self-assembly to generate higher-order aggregates. As shown in Figure 10.6D and E, the square-shaped RNA building blocks with 3′ sticky tails could be carefully engineered to form a variety of two-dimensional nanoarchitectures (Chworos et al. 2004; Severcan et al. 2009).

The RNA nanoarchitectures discussed above demonstrate that both artificial and natural RNA motifs can be programmed to self-assemble into a variety of complex large molecular architectures; and that the topology, addressability, and directionality of such structures can be well-controlled. The ability of RNA to fold into rigid tertiary structures provides potential modules for supramolecular engineering, which is expected to have far-reaching applications in nanodiagnostics, pathogen detection, and drug/gene delivery (Jain 2008).

10.3 Nucleic Acid-Based Mobile Nanomachines

Sections 10.1 and 10.2 discuss the static assembly of nucleic acid nanostructures. Based on those initial assemblies, dynamically assembled nucleic acid objects with the ability to perform mechanical movements in response to environmental stimuli have been developed (Liu and Liu 2009). The mobile nucleic acid nanomechanical device, often called a nanomachine, was, once again, pioneered by Seeman's research group. As shown in Figure 10.7A, the first DNA nanodevice demonstrated a branch point migration in a cruciform junction, which was induced by adding an ethidium intercalator to unwind a circular DNA molecule (Yang et al. 1998). The second DNA nanodevice was capable of undergoing a B–Z transition that is not dependent on the presence of a specific sequence (Mao et al. 1999). As shown in Figure 10.7B, the device on top is a B-DNA comprising two DX molecules connected by a 20-base-pair segment of proto-Z DNA that forms a shaft at the center (yellow color). When $Co(NH_3)_6^{3+}$ is added to form a high ionic strength solution, the two DX molecules change their relative positions, and the B–Z transition takes place.

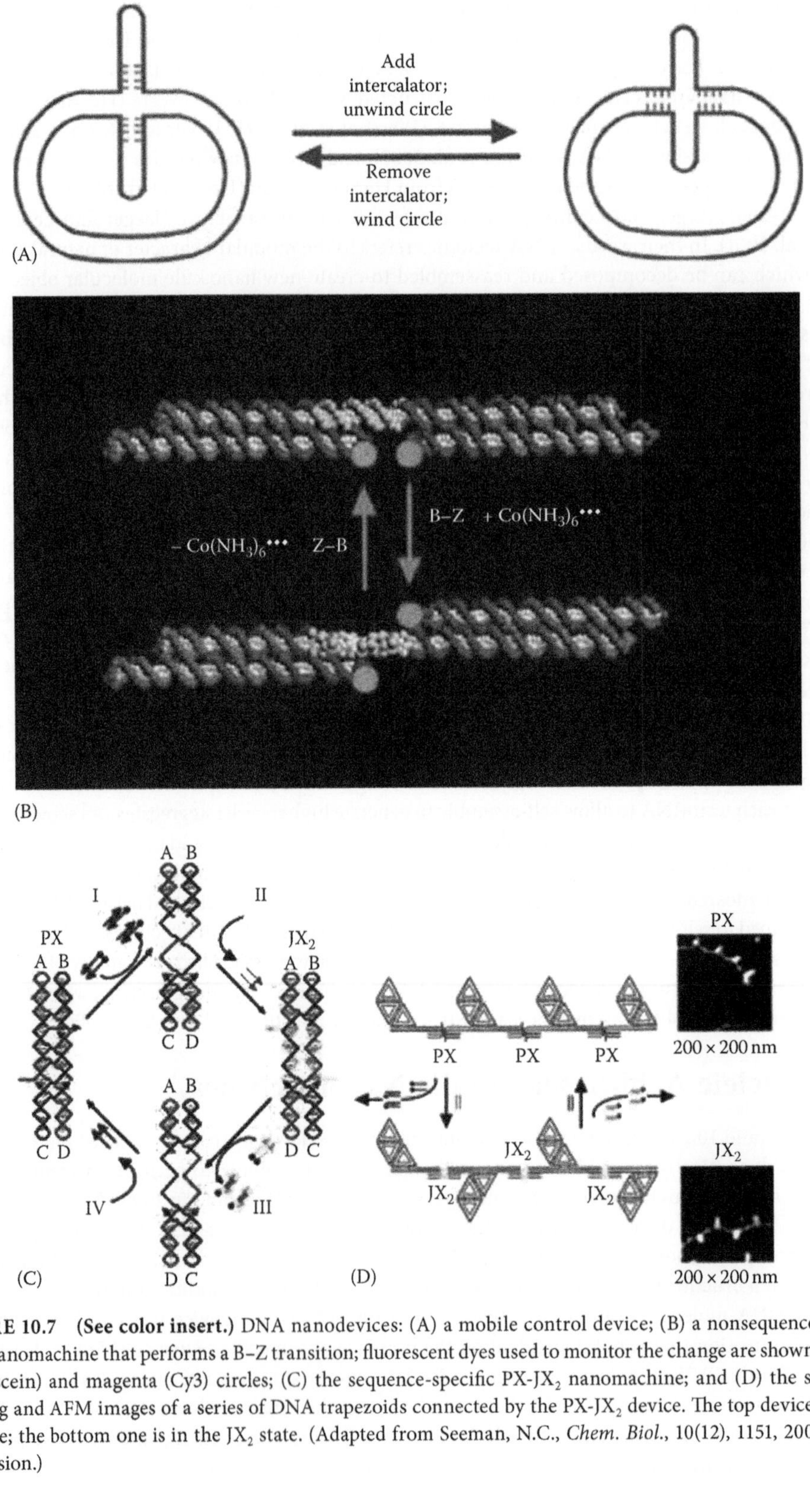

FIGURE 10.7 **(See color insert.)** DNA nanodevices: (A) a mobile control device; (B) a nonsequence-specific DNA nanomachine that performs a B–Z transition; fluorescent dyes used to monitor the change are shown in green (fluorescein) and magenta (Cy3) circles; (C) the sequence-specific PX-JX_2 nanomachine; and (D) the schematic drawing and AFM images of a series of DNA trapezoids connected by the PX-JX_2 device. The top device is in the PX state; the bottom one is in the JX_2 state. (Adapted from Seeman, N.C., *Chem. Biol.*, 10(12), 1151, 2003a. With permission.)

A sequence-specific DNA nanodevice was constructed based on the PX-JX_2 conversion. (Note: PX and JX_2 motifs differ by a half-turn rotation). As shown in Figures 10.7C and D, starting with the PX state, the removal of the strands shown in green leaves a less-structured frame, whereas the addition of the strands shown in yellow converts the frame to the JX_2 state. Removal of the yellow strands again leaves a less-structured frame, and the addition of the green strands recovers the PX structure (Yan et al. 2002; Seeman 2003; Zhong and Seeman 2006).

In 2002–2003, Tan's and Mergny's research groups first designed quadruplex-based nanomachines that can perform extension–contraction movement (Li and Tan 2002; Alberti and Mergny 2003). This design involved the reversible folding of a single 21-base oligonucleotide into a G-quadruplex. As shown in Figure 10.8, fueling by the addition of a complementary single-strand DNA (C-fuel) converted the closed G-quadruplex into an open duplex, whereas the addition of a complementary G-fuel (a different single-stranded DNA) removed the C-fuel and allowed G-quadruplex to refold. This nanomachine

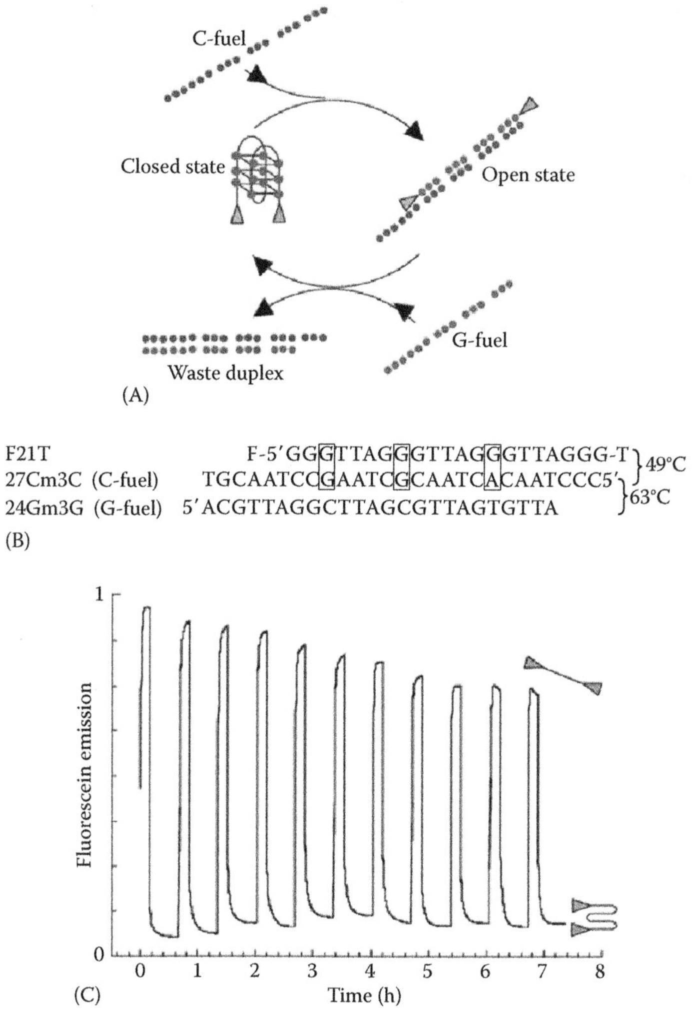

FIGURE 10.8 G-quadruplex nanodevice: (A) the working mechanism of the device; (B) the sequences of the G-quadruplex, C-fuel, and G-fuel, and the T_m of the two duplexes; and (C) G-quadruplex–duplex conformational interconversions monitored by FRET. The device was cycled in a 10 mM sodium cacodylate buffer at pH 7.2, with 0.1 M KCl and 20 mM $MgCl_2$, at 45°C. The two fluorescent dyes used were fluorescein and tetramethylrhodamine. (Adapted from Alberti, P. et al., *Org. Biomol. Chem.*, 4(18), 3383, 2006. With permission.)

generates a DNA duplex waste molecule with each cycle. In 2004, Simmel's group applied this quadruplex–duplex interconversion model to the real world: they designed a G-quadruplex aptamer that binds to the human blood-clotting factor, α-thrombin (Dittmer et al. 2004). The addition of a specific "opening" DNA strand converted the G-quadruplex into a duplex that cannot bind to thrombin, whereas the addition of a "removal" DNA strand that is complementary to the "opening" strand recovered the G-quadruplex structure, which is again able to bind to thrombin. Fahlman et al. constructed a nanopinching device (Fahlman et al. 2003). In this design, G·G mismatches within the DNA double helix bind to Sr^{2+} to form a G-quadruplex, which converted the DNA double helix into a "pinning" structure. The addition of EDTA to intercalate the Sr^{2+} metal ion removed the pinched structure and restored the DNA double helix.

In addition to the G-quadruplex, the i-motif has also been used to construct DNA nanomachines. Balasubramanian's research group designed a proton-fueled, pH-driven nanomachine (Liu and Balasubramanian 2003). A 21-nucleotide single-stranded oligonucleotide was used to form a "closed" i-motif at pH 5.0. Raising the pH to 8.0 resulted in the unwinding of the i-motif and the formation of an "open" duplex structure in the presence of a complementary DNA strand. Lowering the pH recovered the i-motif. The waste product of this nanomachine is salt and water rather than DNA duplexes (see Figure 10.9) (Liu and Balasubramanian 2003; Alberti et al. 2006; Liu and Liu 2009).

DNA-based nanodevices that mimic biological molecular motors have also been designed. In 2004, two research groups pioneered DNA walking devices. Seeman's group designed a biped DNA walker using TX molecules that can perform linear unidirectional walking on a three-step track (Sherman and Seeman 2004). Pierce's group demonstrated a simpler DNA walker that comprises a 20-base-pair helix with two single-stranded legs (each 23 nucleotides long) (Shin and Pierce 2004). The track was designed with four single-stranded branches (each 20 nucleotides long) (see Figure 10.10) placed on the same side of the track with approximately 5 nm between each branch. As shown in Figure 10.10, A1 and A2 DNA strands helped anchor two legs of the DNA walker to the first and second track branches, respectively, by forming helixes with both the leg and the branch. The first walker leg was then released from the first track branch by using a D1 strand that is complementary to the A1 strand. A DNA duplex is released as waste, and the first walker leg was free for the next step.

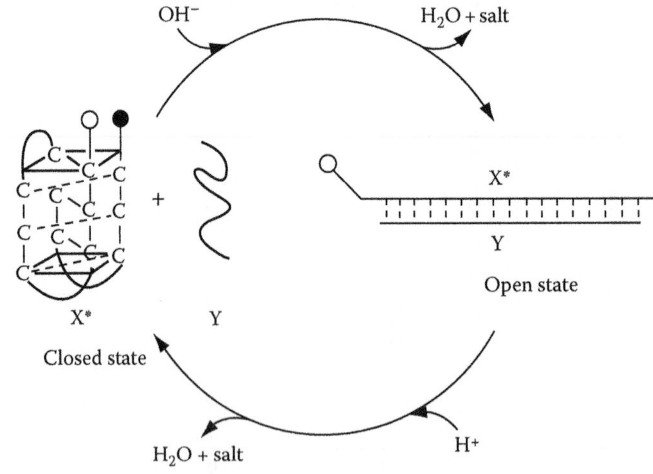

X: 5′CCCTAACCCTAACCCTAACCC 3′

X* 5′ Rhodamine Green-CCCTAACCCTAACCCTAACCC-Dabcyl 3′

Y: 3′ GATTGTGATTGTGATTG 5′

FIGURE 10.9 Working mechanism of an i-motif nanodevice. (Adapted from Liu, D. and Balasubramanian, S., *Angew. Chem. Int. Ed. Engl.*, 42(46), 5734, 2003. With permission.)

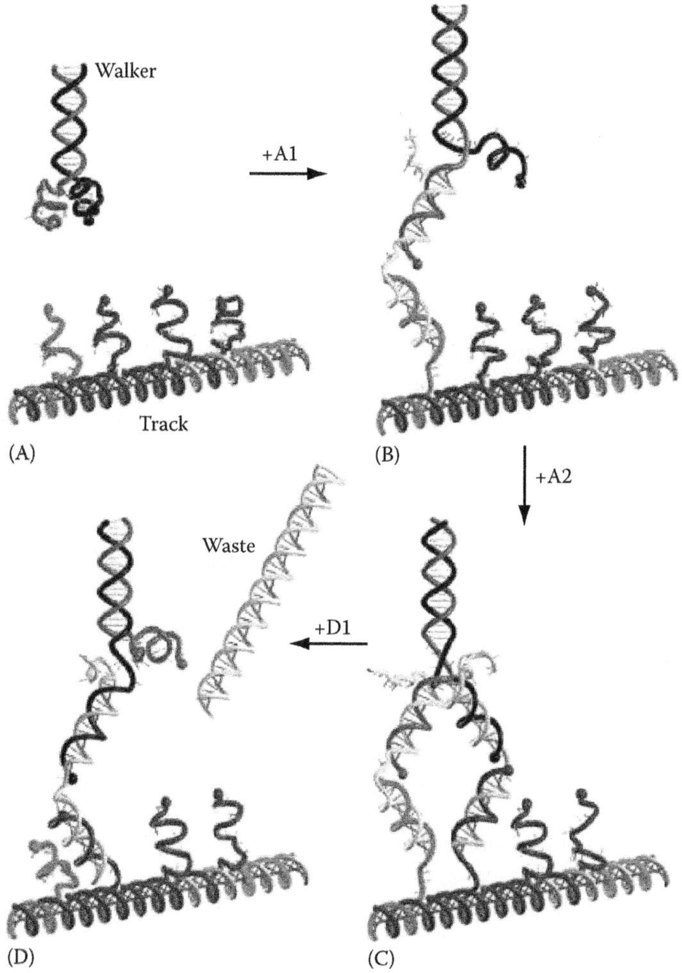

FIGURE 10.10 Illustration of a DNA walker that moves along a track: (A) unbound walker, (B) walker attached to the first track branch, (C) walker attached to the first and second track branches, and (D) walker released from the first track branch with a DNA duplex waste. (Adapted from Shin, J.S. and Pierce, N.A., *J. Am. Chem. Soc.*, 126(35), 10834, 2004. With permission.)

Guo's research group used synthetic nucleic acids and purified recombinant proteins to construct a nanomotor to mimic the naturally occurring bacteriophage Phi29 DNA-packaging motor (Guo 2005a,b; Guo and Lee 2007). Bacteriophage Phi29 of *Bacillus subtilis* is a well-studied virus that packages its genome DNA into a procapsid during maturation. The DNA encapsidation process requires the aid of a Phi29-encoded RNA called packing RNA (pRNA), which is 120 nucleotides long (Guo et al. 1998). Six pRNA molecules together form a hexameric ring via "hand-in-hand" interaction (Chen et al. 1999; Trottier et al. 2000; Hoeprich and Guo 2002). When fueled by ATP hydrolysis, the RNA strands kick against the axle in succession to turn the nanomotor (Guo 2002a,b).

In addition to the constructions mentioned above, a large number of intriguing nucleic acid-based nanodevices have been developed in the past few decades, demonstrating that nucleic acids are an important material in nanotechnology. It is quite likely that in the near future, robust and highly efficient DNA nanomachines that produce fewer by-products and are capable of directionally controllable movement will be developed. The construction of RNA nanodevices will also continue to increase. Their potential applications in biomedical fields may include building nanorobotics that can perform movement at the nanometer scale in order to regulate gene expression, deliver drugs, and remove cellular wastes (Jain 2008). They may also be used to build thermometers, biosensors, tweezers, and computers (Mao et al. 2000; Benenson et al. 2001; Braich et al. 2002; Saghatelian et al. 2003; Miyoshi et al. 2006; Kossoy et al. 2007). The future of nucleic acid-based nanotechnology is fascinating.

10.4 Conclusions and Outlook

To date, negatively charged DNA and RNA molecules have been exploited as structural building blocks for static nanoarchitectures and mobile nanomachines. The success of using nucleic acids as nanobiomaterials is expected to expedite the investigation and development of their variants and/or analogs, such as neutral polyamide nucleic acids (PNA), as building blocks in order to accommodate different environments. In addition, as mentioned in the introduction, the construction of novel nanostructures and nanodevices by combining DNA or RNA with nonbiological nanomaterials such as metallic nanoparticles or carbon nanotubes will greatly facilitate the development of nanorobotics, nanoelectronics, and biosensors.

References

Alberti, P., A. Bourdoncle, B. Sacca, L. Lacroix, and J. L. Mergny. 2006. DNA nanomachines and nanostructures involving quadruplexes. *Org. Biomol. Chem.* 4 (18):3383–3391.

Alberti, P. and J. L. Mergny. 2003. DNA duplex–quadruplex exchange as the basis for a nanomolecular machine. *Proc. Natl. Acad. Sci. U.S.A.* 100 (4):1569–1573.

Alberti, P., P. Schmitt, C. H. Nguyen, C. Rivalle, M. Hoarau, D. S. Grierson, and J. L. Mergny. 2002. Benzoindoloquinolines interact with DNA tetraplexes and inhibit telomerase. *Bioorg. Med. Chem. Lett.* 12 (7):1071–1074.

Batey, R. T., R. P. Rambo, and J. A. Doudna. 1999. Tertiary motifs in RNA structure and folding. *Angew. Chem. Int. Ed. Engl.* 38 (16):2326–2343.

Benenson, Y., T. Paz-Elizur, R. Adar, E. Keinan, Z. Livneh, and E. Shapiro. 2001. Programmable and autonomous computing machine made of biomolecules. *Nature* 414 (6862):430–434.

Braich, R. S., N. Chelyapov, C. Johnson, P. W. Rothemund, and L. Adleman. 2002. Solution of a 20-variable 3-SAT problem on a DNA computer. *Science* 296 (5567):499–502.

Chen, J. H. and N. C. Seeman. 1991a. The electrophoretic properties of a DNA cube and its substructure catenanes. *Electrophoresis* 12 (9):607–611.

Chen, J. H. and N. C. Seeman. 1991b. Synthesis from DNA of a molecule with the connectivity of a cube. *Nature* 350 (6319):631–633.

Chen, C., C. Zhang, and P. Guo. 1999. Sequence requirement for hand-in-hand interaction in formation of RNA dimers and hexamers to gear phi29 DNA translocation motor. *RNA* 5 (6):805–818.

Chworos, A., I. Severcan, A. Y. Koyfman, P. Weinkam, E. Oroudjev, H. G. Hansma, and L. Jaeger. 2004. Building programmable jigsaw puzzles with RNA. *Science* 306 (5704):2068–2072.

Constantinou, A. and S. C. West. 2004. Holliday junction branch migration and resolution assays. *Methods Mol. Biol.* 262:239–253.

Davis, J. H., M. Tonelli, L. G. Scott, L. Jaeger, J. R. Williamson, and S. E. Butcher. 2005. RNA helical packing in solution: NMR structure of a 30 kDa GAAA tetraloop-receptor complex. *J. Mol. Biol.* 351 (2):371–382.

Dittmer, W. U., A. Reuter, and F. C. Simmel. 2004. A DNA-based machine that can cyclically bind and release thrombin. *Angew. Chem. Int. Ed. Engl.* 43 (27):3550–3553.

Doudna, J. A. 1997. RNA structure. A molecular contortionist. *Nature* 388 (6645):830–831.

Doudna, J. A. 2000. Structural genomics of RNA. *Nat. Struct. Biol.* 7 (Suppl):954–956.

Doudna, J. A. and E. A. Doherty. 1997. Emerging themes in RNA folding. *Fold. Des.* 2 (5):R65–R70.

Du, S. M. and N. C. Seeman. 1994. The construction of a trefoil knot from a DNA branched junction motif. *Biopolymers* 34 (1):31–37.

Erben, C. M., R. P. Goodman, and A. J. Turberfield. 2006. Single-molecule protein encapsulation in a rigid DNA cage. *Angew. Chem. Int. Ed. Engl.* 45 (44):7414–7417.

Erben, C. M., R. P. Goodman, and A. J. Turberfield. 2007. A self-assembled DNA bipyramid. *J. Am. Chem. Soc.* 129 (22):6992–6993.

Fahlman, R. P., M. Hsing, C. S. Sporer-Tuhten, and D. Sen. 2003. Duplex pinching: A structural switch suitable for contractile DNA nanoconstructions. *Nano Lett.* 3:1073–1078.

Ferre-D'Amare, A. R. and J. A. Doudna. 1999. RNA folds: Insights from recent crystal structures. *Annu. Rev. Biophys. Biomol. Struct.* 28:57–73.

Fu, T. J. and N. C. Seeman. 1993. DNA double-crossover molecules. *Biochemistry* 32 (13):3211–3220.

Gehring, K., J. L. Leroy, and M. Gueron. 1993. A tetrameric DNA structure with protonated cytosine-cytosine base pairs. *Nature* 363 (6429):561–565.

Gesteland, R. F., T. R. Cech, and J. F. Atkins. 2005. *The RNA World.* 3rd edn. New York: Cold Spring Harbor Laboratory Press.

Goodman, R. P., I. A. Schaap, C. F. Tardin, C. M. Erben, R. M. Berry, C. F. Schmidt, and A. J. Turberfield. 2005. Rapid chiral assembly of rigid DNA building blocks for molecular nanofabrication. *Science* 310 (5754):1661–1665.

Guo, P. 2002a. Structure and function of phi29 hexameric RNA that drives the viral DNA packaging motor: Review. *Prog. Nucleic Acid Res. Mol. Biol.* 72:415–472.

Guo, P. X. 2002b. Methods for structural and functional analysis of an RNA hexamer of bacterial virus phi29 DNA packaging motor. *Sheng Wu Hua Xue Yu Sheng Wu Wu Li Xue Bao (Shanghai)* 34 (5):533–543.

Guo, P. 2005a. Bacterial virus phi29 DNA-packaging motor and its potential applications in gene therapy and nanotechnology. *Methods Mol. Biol.* 300:285–324.

Guo, P. 2005b. RNA nanotechnology: Engineering, assembly and applications in detection, gene delivery and therapy. *J. Nanosci. Nanotechnol.* 5 (12):1964–1982.

Guo, P. and T. J. Lee. 2007. Viral nanomotors for packaging of dsDNA and dsRNA. *Mol. Microbiol.* 64 (4):886–903.

Guo, P., C. Zhang, C. Chen, K. Garver, and M. Trottier. 1998. Inter-RNA interaction of phage phi29 pRNA to form a hexameric complex for viral DNA transportation. *Mol. Cell.* 2 (1):149–155.

Hagerman, P. J. 1988. Flexibility of DNA. *Annu. Rev. Biophys. Biophys. Chem.* 17:265–286.

Hansma, H. G., E. Oroudjev, S. Baudrey, and L. Jaeger. 2003. TectoRNA and 'kissing-loop' RNA: Atomic force microscopy of self-assembling RNA structures. *J. Microsc.* 212 (Pt 3):273–279.

Henderson, E., C. C. Hardin, S. K. Walk, I. Tinoco, Jr., and E. H. Blackburn. 1987. Telomeric DNA oligonucleotides form novel intramolecular structures containing guanine–guanine base pairs. *Cell* 51 (6):899–908.

Hendrix, D. K., S. E. Brenner, and S. R. Holbrook. 2005. RNA structural motifs: Building blocks of a modular biomolecule. *Q. Rev. Biophys.* 38 (3):221–243.

Hoeprich, S. and P. Guo. 2002. Computer modeling of three-dimensional structure of DNA-packaging RNA (pRNA) monomer, dimer, and hexamer of Phi29 DNA packaging motor. *J. Biol. Chem.* 277 (23):20794–20803.

Holbrook, S. R. 2005. RNA structure: The long and the short of it. *Curr. Opin. Struct. Biol.* 15 (3):302–308.

Horiya, S., X. Li, G. Kawai, R. Saito, A. Katoh, K. Kobayashi, and K. Harada. 2002. RNA LEGO: Magnesium-dependent assembly of RNA building blocks through loop–loop interactions. *Nucleic Acids Res. Suppl.* (2):41–42.

Horiya, S., X. Li, G. Kawai, R. Saito, A. Katoh, K. Kobayashi, and K. Harada. 2003. RNA LEGO: Magnesium-dependent formation of specific RNA assemblies through kissing interactions. *Chem. Biol.* 10 (7):645–654.

Jaeger, L. and A. Chworos. 2006. The architectonics of programmable RNA and DNA nanostructures. *Curr. Opin. Struct. Biol.* 16 (4):531–543.

Jaeger, L. and N. B. Leontis. 2000. Tecto-RNA: One-dimensional self-assembly through tertiary interactions. *Angew. Chem. Int. Ed. Engl.* 39 (14):2521–2524.
Jaeger, L., E. Westhof, and N. B. Leontis. 2001. TectoRNA: Modular assembly units for the construction of RNA nano-objects. *Nucleic Acids Res.* 29 (2):455–463.
Jain, K. K. 2008. *The Handbook of Nanomedicine.* Totowa: Humana Press.
Jin, R. Z., K. J. Breslauer, R. A. Jones, and B. L. Gaffney. 1990. Tetraplex formation of a guanine-containing nonameric DNA fragment. *Science* 250 (4980):543–546.
Kossoy, E., N. Lavid, M. Soreni-Harari, Y. Shoham, and E. Keinan. 2007. A programmable biomolecular computing machine with bacterial phenotype output. *Chembiochem* 8 (11):1255–1260.
LaBean, T., H. Yan, J. Kopatsch, F. Liu, E. Winfree, J. H. Reif, and N. C. Seeman. 2000. The construction, analysis, ligation and self-assembly of DNA triple crossover complexes. *J. Am. Chem. Soc.* 122:1848–1860.
Leontis, N. B., A. Lescoute, and E. Westhof. 2006. The building blocks and motifs of RNA architecture. *Curr. Opin. Struct. Biol.* 16 (3):279–287.
Li, J. J. and W. Tan. 2002. A single DNA molecule nanomotor. *Nano Lett.* 2 (4):315–318.
Li, X., X. Yang, J. Qi, and N. C. Seeman. 1996. Antiparallel DNA double crossover molecules as components for nanoconstruction. *J. Am. Chem. Soc.* 118:6131–6140.
Liu, D. and S. Balasubramanian. 2003. A proton-fuelled DNA nanomachine. *Angew. Chem. Int. Ed. Engl.* 42 (46):5734–5736.
Liu, H. and D. Liu. 2009. DNA nanomachines and their functional evolution. *Chem. Commun. (Camb)* (19):2625–2636.
Liu, Y. and S. C. West. 2004. Happy Hollidays: 40th anniversary of the Holliday junction. *Nat. Rev. Mol. Cell. Biol.* 5 (11):937–944.
Liu, Y., Y. G. Ke, and H. Yan. 2005. Self-assembly of symmetirc finite-size DNA nanoarrays. *J. Am. Chem. Soc.* 127:17140–17141.
Liu, D., S. H. Park, J. H. Reif, and T. H. LaBean. 2004. DNA nanotubes self-assembled from triple-crossover tiles as templates for conductive nanowires. *Proc. Natl. Acad. Sci. U.S.A.* 101:717–722.
Mao, C., T. H. LaBean, J. H. Relf, and N. C. Seeman. 2000. Logical computation using algorithmic self-assembly of DNA triple-crossover molecules. *Nature* 407 (6803):493–496.
Mao, C., W. Sun, and N. C. Seeman. 1997. Assembly of Borromean rings from DNA. *Nature* 386 (6621):137–138.
Mao, C., W. Sun, Z. Shen, and N. C. Seeman. 1999. A nanomechanical device based on the B-Z transition of DNA. *Nature* 397 (6715):144–146.
Marsh, T. C. and E. Henderson. 1994. G-wires: Self-assembly of a telomeric oligonucleotide, d(GGGGTTGGGG), into large superstructures. *Biochemistry* 33 (35):10718–10724.
Miyoshi, D., M. Inoue, and N. Sugimoto. 2006. DNA logic gates based on structural polymorphism of telomere DNA molecules responding to chemical input signals. *Angew. Chem. Int. Ed. Engl.* 45 (46):7716–7719.
Nasalean, L., S. Baudrey, N. B. Leontis, and L. Jaeger. 2006. Controlling RNA self-assembly to form filaments. *Nucleic Acids Res.* 34 (5):1381–1392.
Paliy, M., R. Melnik, and B. A. Shapiro. 2009. Molecular dynamics study of the RNA ring nanostructure: A phenomenon of self-stabilization. *Phys. Biol.* 6 (4):46003.
Phan, A. T., V. Kuryavyi, and D. J. Patel. 2006. DNA architecture: From G to Z. *Curr. Opin. Struct. Biol.* 16 (3):288–298.
Phan, A. T. and J. L. Mergny. 2002. Human telomeric DNA: G-quadruplex, i-motif and Watson-Crick double helix. *Nucleic Acids Res.* 30 (21):4618–4625.
Rothemund, P. W. 2006. Folding DNA to create nanoscale shapes and patterns. *Nature* 440 (7082):297–302.
Rothemund, P. W., A. Ekani-Nkodo, N. Papadakis, A. Kumar, D. K. Fygenson, and E. Winfree. 2004. Design and characterization of programmable DNA nanotubes. *J. Am. Chem. Soc.* 126 (50):16344–16352.

Sa-Ardyen, P., A. V. Vologodskii, and N. C. Seeman. 2003. The flexibility of DNA double crossover molecules. *Biophys. J.* 84 (6):3829–3837.

Saghatelian, A., N. H. Volcker, K. M. Guckian, V. S. Lin, and M. R. Ghadiri. 2003. DNA-based photonic logic gates: AND, NAND, and INHIBIT. *J. Am. Chem. Soc.* 125 (2):346–347.

Scheffler, M., A. Dorenbeck, S. Jordan, M. Wusterfeld, and G. von Kiedrowski. 1999. Self-assembly of trisoligo-nucleotidyls: The case for nano-acetylene and nano-cyclobutadiene. *Angew. Chem. Int. Ed.* 38:3312–3315.

Seeman, N. C. 1982. Nucleic acid junctions and lattices. *J. Theor. Biol.* 99 (2):237–247.

Seeman, N. C. 1991. Construction of three-dimensional stick figures from branched DNA. *DNA Cell. Biol.* 10 (7):475–486.

Seeman, N. C. 1998. DNA nanotechnology: Novel DNA constructions. *Annu. Rev. Biophys. Biomol. Struct.* 27:225–248.

Seeman, N. C. 2001. DNA nicks and nodes and nanotechnology. *Nano Lett.* 1:22–26.

Seeman, N. C. 2002. Key experimental approaches in DNA nanotechnology. *Curr. Protoc. Nucleic Acid Chem.* Chapter 12:Unit 12 1.

Seeman, N. C. 2003a. At the crossroads of chemistry, biology, and materials: Structural DNA nanotechnology. *Chem. Biol.* 10 (12):1151–1159.

Seeman, N. C. 2003b. Biochemistry and structural DNA nanotechnology: An evolving symbiotic relationship. *Biochemistry* 42 (24):7259–7269.

Seeman, N. C. 2003c. DNA in a material world. *Nature* 421 (6921):427–431.

Seeman, N. 2008. A conversation with prof. Ned Seeman: Founder of DNA nanotechnology. Interview by Paul S. Weiss. *ACS Nano* 2 (6):1089–1096.

Seeman, N. C. and N. R. Kallenbach. 1983. Design of immobile nucleic acid junctions. *Biophys. J.* 44 (2):201–209.

Seeman, N. C. and N. R. Kallenbach. 1994. DNA branched junctions. *Annu. Rev. Biophys. Biomol. Struct.* 23:53–86.

Severcan, I., C. Geary, E. Verzemnieks, A. Chworos, and L. Jaeger. 2009. Square-shaped RNA particles from different RNA folds. *Nano Lett.* 9 (3):1270–1277.

Sherman, W. B. and N. C. Seeman. 2004. A precisely controlled DNA biped walking device. *Nano Lett.* 4:1203–1207.

Shih, W. M., J. D. Quispe, and G. F. Joyce. 2004. A 1.7-kilobase single-stranded DNA that folds into a nanoscale octahedron. *Nature* 427 (6975):618–621.

Shin, J. S. and N. A. Pierce. 2004. A synthetic DNA walker for molecular transport. *J. Am. Chem. Soc.* 126 (35):10834–10835.

Spink, C. H., L. Ding, Q. Yang, R. D. Sheardy, and N. C. Seeman. 2009. Thermodynamics of forming a parallel DNA crossover. *Biophys. J.* 97 (2):528–538.

Strobel, S. A. and J. A. Doudna. 1997. RNA seeing double: Close-packing of helices in RNA tertiary structure. *Trends Biochem. Sci.* 22 (7):262–266.

Triberis, G. P. and M. Dimakogianni. 2009. DNA in the material world: Electrical properties and nano-applications. *Recent Pat. Nanotechnol.* 3 (2):135–153.

Trottier, M., Y. Mat-Arip, C. Zhang, C. Chen, S. Sheng, Z. Shao, and P. Guo. 2000. Probing the structure of monomers and dimers of the bacterial virus phi29 hexamer RNA complex by chemical modification. *RNA* 6 (9):1257–1266.

Wang, Y. L., J. E. Mueller, B. Kemper, and N. C. Seeman. 1991. Assembly and characterization of five-arm and six-arm DNA branched junctions. *Biochemistry* 30 (23):5667–5674.

Westhof, E., B. Masquida, and L. Jaeger. 1996. RNA tectonics: Towards RNA design. *Fold. Des.* 1 (4):R78–R88.

Williamson, J. R., M. K. Raghuraman, and T. R. Cech. 1989. Monovalent cation-induced structure of telomeric DNA: The G-quartet model. *Cell* 59 (5):871–880.

Winfree, E., F. Liu, L. A. Wenzler, and N. C. Seeman. 1998. Design and self-assembly of two-dimensional DNA crystals. *Nature* 394 (6693):539–544.

Yan, H., X. Zhang, Z. Shen, and N. C. Seeman. 2002. A robust DNA mechanical device controlled by hybridization topology. *Nature* 415 (6867):62–65.

Yang, X., A. V. Vologodskii, B. Liu, B. Kemper, and N. C. Seeman. 1998. Torsional control of double-stranded DNA branch migration. *Biopolymers* 45 (1):69–83.

Zhang, Y. W. and N. C. Seeman. 1994. Construction of a DNA-truncated octahedron. *J. Am. Chem. Soc.* 116:1661–1669.

Zhang, S., T. J. Fu, and N. C. Seeman. 1993. Symmetric immobile DNA branched junctions. *Biochemistry* 32 (32):8062–8067.

Zhang, X., H. Yan, Z. Shen, and N. C. Seeman. 2002. Paranemic cohesion of topologically-closed DNA molecules. *J. Am. Chem. Soc.* 124 (44):12940–12941.

Zhong, H. and N. C. Seeman. 2006. RNA used to control a DNA rotary nanomachine. *Nano Lett.* 6 (12):2899–2903.

11

Emerging Technologies in Nanomedicine

F. Kyle Satterstrom
Harvard University

Marjan Rafat
Harvard University

Jin-Oh You
Harvard University

Debra T. Auguste
Harvard University

Nanoscale biomaterials are designed to enhance a drug's therapeutic effects while minimizing the toxicity that would be caused by delivery of the drug alone. Materials used to encapsulate drugs are often referred to as "nanoparticles" or "nanocarriers" because of their size and function. In this chapter, we discuss nanocarrier functions such as shielding the drug from the immune system, targeting the drug to a specific location in the body, controlling the release of a drug, or reacting to a particular microenvironment to trigger drug release (Table 11.1). We discuss these functions in the context of three material systems commonly used in drug delivery—liposomes, polymers, and metals. We then discuss multifunctional nanocarriers, which are at the forefront of current drug delivery technology.

First synthesized in 1965, liposomes are one of the most well-known nanocarriers in drug delivery. Liposomes are spherical vesicles formed by lipid bilayers, the same material that composes a cell's membrane in the body (Bangham et al. 1965). Liposomes enclose small volumes of an aqueous solution and restrict transport of large polar molecules. Liposomes are made in the laboratory from mixtures of lipids and other amphiphilic molecules (e.g., cholesterol). Methods for constructing liposomes include "reverse-phase evaporation," in which an aqueous solution is added to lipid solubilized in an organic solvent followed by evaporation of the solvent. To encapsulate a hydrophilic drug, the drug is simply dissolved in the aqueous phase before addition to the organic phase. Liposome size can be controlled through methods such as "extrusion," in which liposomes are forced through a filter using a pressurized vessel (Szoka and Papahadjopoulos 1978). In drug delivery, liposomes have been employed in a variety of applications, including delivery of dextran to erythrocytes (Shangguan et al. 1998) and delivery of DNA to ovarian cancer cells (Shangguan et al. 2000). Current pharmaceuticals on the market that utilize liposome technology include the antifungal agent Amphotericin B and the anticancer agents Doxil and Evacet, both of which contain the chemotherapeutic drug doxorubicin.

In addition to liposomes, polymeric drug delivery vehicles have also been designed to release therapeutic molecules in a controlled and sustained manner. Important early work in this field was done by Langer and Folkman in the 1970s (Langer and Folkman 1976). Polymers' mechanical properties, such as degradation rate, can be tuned by altering their composition and molecular weight. Biodegradable

TABLE 11.1 Outline of Chapter

Topic	References
Example materials	
Liposomes	Bangham et al. (1965); Szoka and Papahadjopoulos (1978); Shangguan et al. (1998); Shangguan et al. (2000); Auguste et al. (2006)
Polymers	Chasin and Langer (1990); Shive and Anderson (1997); Lu and Chen (2004); Astete and Sabliov (2006); You and Auguste (2008)
Metallic nanoparticles	Wang et al. (2001); Loo et al. (2005); Chertok et al. (2008); Kamei et al. (2009)
Immune evasion	
Immune system removes particles from circulation	Bazile et al. (1992); Chonn et al. (1992); Liu et al. (1995); Ahl et al. (1997)
PEG helps prevent recognition	Blume and Cevc (1990); Gabizon et al. (1994); Beduaddo and Huang (1995); Harris et al. (2001); Reddy et al. (2002); Auguste et al. (2003); Savic et al. (2003); Cheng et al. (2007)
Targeting	
Passive (i.e., EPR)	Matsumura and Maeda (1986); Barratt (2000)
Antibodies	Martin and Papahadjopoulos (1982); Baselga et al. (1998); McLaughlin et al. (1998)
Ligands (e.g., folic acid)	Pan et al. (2003); Saul et al. (2006)
Peptides (e.g., RGD)	Singh et al. (2009)
Magnetism	Chertok et al. (2008)
Imaging	
MRI	Wang et al. (2001); Lanza et al. (2002); Corot et al. (2006); Sun et al. (2008)
Ultrasound	Lanza et al. (2000); Blomley et al. (2001); Liang and Blomley (2003); Pitt et al. (2004); Crowder et al. (2005)
Release	
Degradation	Shive and Anderson (1997); Wu and Wang (2001); Panyam and Labhasetwar (2003)
Cation-mediated	Felgner et al. (1987); Xu and Szoka (1996); Bivas-Benita et al. (2004); Landen et al. (2005); Zhang et al. (2006); Sato et al. (2008); Kim et al. (2009)
pH sensitivity	Kuhn et al. (1950); Felt et al. (1998); Fischel-Ghodsian et al. (1988); Jarvinen et al. (1998); Lowman et al. (1999); Lim, Yeom, and Park (2000); You et al. (2001); Li et al. (2002); Jones et al. (2003); Dai et al. (2004); Napoli et al. (2004); Heffernan and Murthy (2005); Mahkam (2005); Auguste et al. (2006); Kong et al. (2007); Auguste et al. (2008); You and Auguste (2008); Qi et al. (2009)
Photo-induced	Lai et al. (2003); Vivero-Escoto et al. (2009); Volodkin et al. (2009)
Ultrasonic disruption	Nelson et al. (2002); Pitt et al. (2004); Husseini and Pitt (2008)
Multifunctionality	
Examples of multifunctional delivery systems	Nasongkla et al. (2006); Weng et al. (2008); Guo et al. (2009); Lu et al. (2009); Mikhaylova et al. (2009); Tai et al. (2009); Wang et al. (2009)
Nanoparticle toxicity	
Nanomaterial toxic effects	Oberdorster et al. (1994); Li et al. (2003); Brown et al. (2004); Derfus et al. (2004); Gupta and Curtis (2004); Lam et al. (2004); The Royal Society & The Royal Academy of Engineering (2004); Risom et al. (2005); Nel et al. (2006); Sayes et al. (2006); Buzea et al. (2007); United States Environmental Protection Agency (2007); United States Food and Drug Administration (2007); Apopa et al. (2009)

polymers are used for controlled release purposes. One such polymer, poly(D,L lactide-co-glycolide) (PLGA), is a hydrophobic, biodegradable, and biocompatible aliphatic polyester (Shive and Anderson 1997; Astete and Sabliov 2006). PLGA typically breaks down by hydrolysis into the nontoxic degradation products—lactic acid and glycolic acid. Poly(ε-caprolactone) (PCL), also an aliphatic polyester, is another highly used biodegradable polymer that is used in sutures and FDA-approved implantable devices

(Chasin and Langer 1990; Lu and Chen 2004). PCL is degraded by hydrolysis more slowly than PLGA under physiological conditions and is suitable for longer-term controlled release (Chasin and Langer 1990). These types of biodegradable polymeric materials are valuable in designing therapeutic release systems because of their biocompatibility, tunable properties, and controlled release capabilities. Tables 11.2 and 11.3 catalogue the chemical structures of widely used polymers and lipids in drug delivery.

TABLE 11.2 Biocompatible Polymers in Drug Delivery

Polymer	Structure	References
Poly(D,L lactide-co-glycolide) (PLGA)	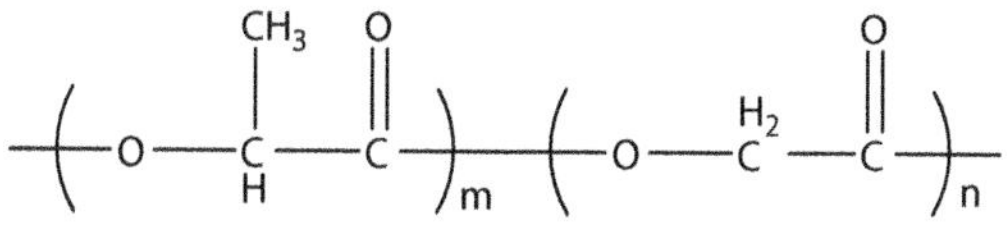	Cheng et al. (2007); Chan et al. (2009); Singh et al. (2009); Wang et al. (2009)
Poly(ethylene glycol) (PEG)	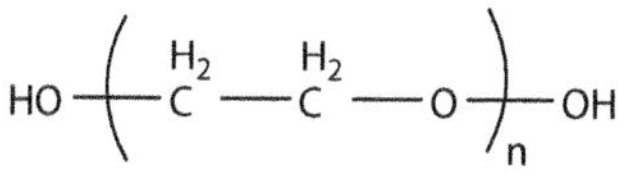	Blume and Cevc (1990); Gabizon et al. (1994); Beduaddo and Huang (1995); Gabizon and Papahadjopoulos (1988); Jarvinen et al. (1998); Reddy et al. (2002); Auguste et al. (2003); Savic et al. (2003); Gupta and Curtis (2004); Auguste et al. (2006); Cheng et al. (2007); Wang et al. (2009)
Poly(ethyleneimine) (PEI)	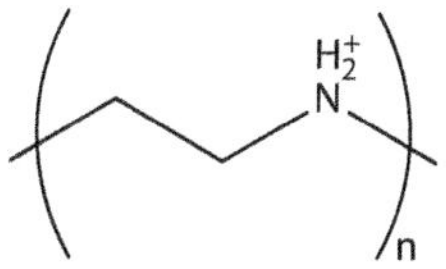	Bivas-Benita et al. (2004); Lungwitz et al. (2005)
Poly(methacrylic acid) (PMAA)	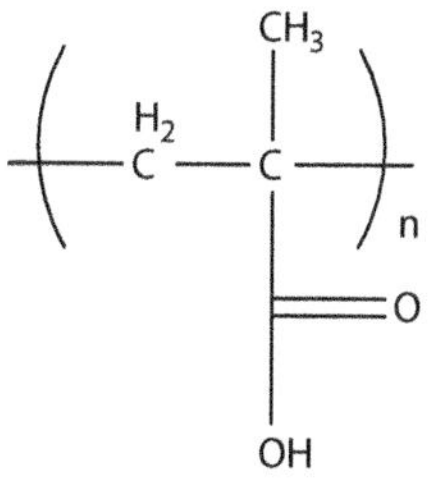	Jones et al. (2003); Guo et al. (2009)
Poly(acrylic acid) (PAA)	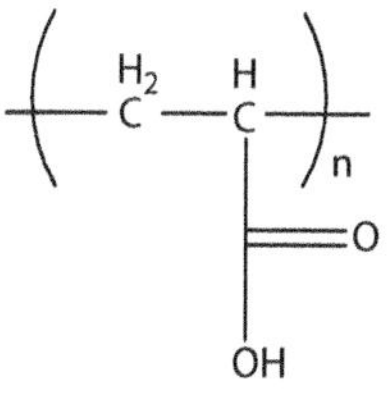	Jarvinen et al. (1998)
Poly(*N,N*-dimethylaminoethyl methacrylate) (DMAEMA)	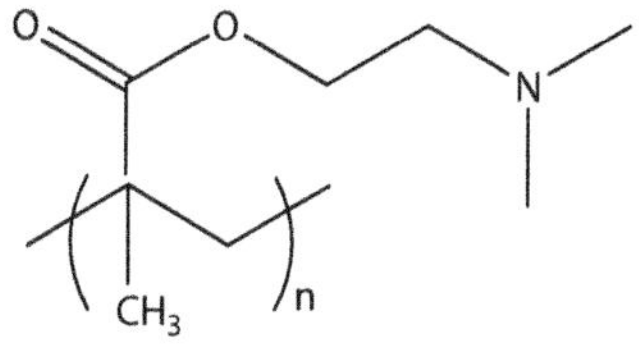	You and Auguste (2008)
Poly(2-hydroxyethyl methacrylate) (HEMA)	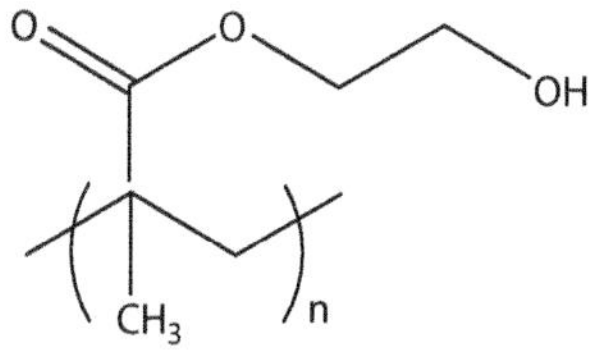	You and Auguste (2008)

TABLE 11.3 Lipids in Drug Delivery

Lipid	Structure[a]	References
Phosphatidylethanolamine (PE)	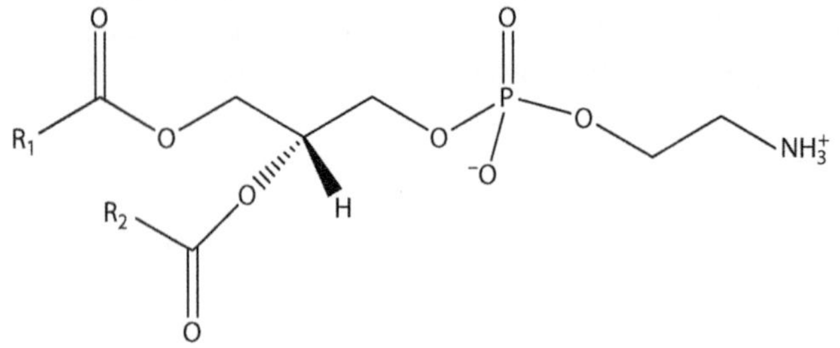	Felgner et al. (1987)
Phosphatidylcholine (PC)	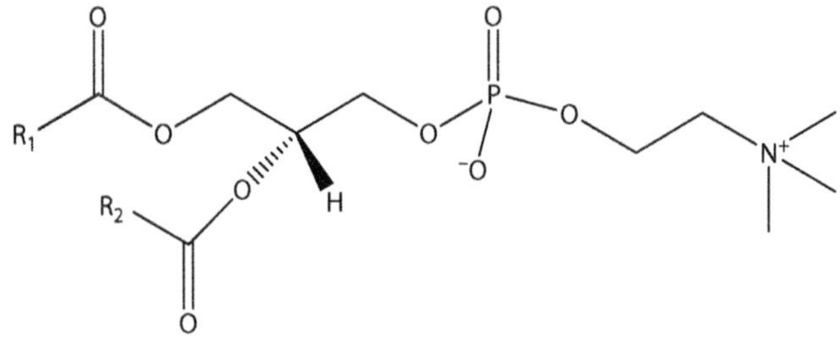	Auguste et al. (2006); Tai et al. (2009)
Phosphatidylglycerol (PG)	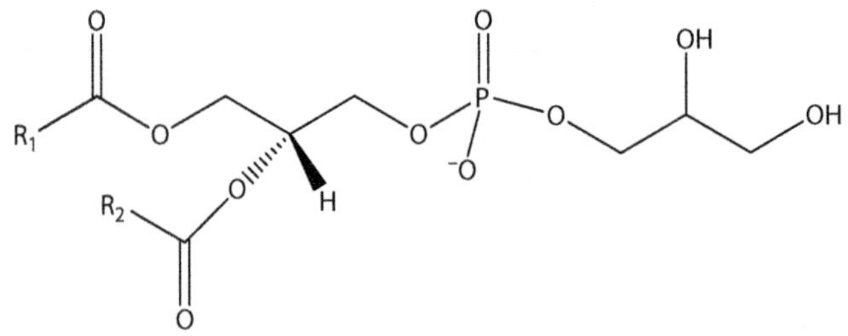	Auguste et al. (2006)
N-[1-(2,3-dioleyloxy) propyl]-*N,N,N*-trimethylammonium chloride (DOTMA)	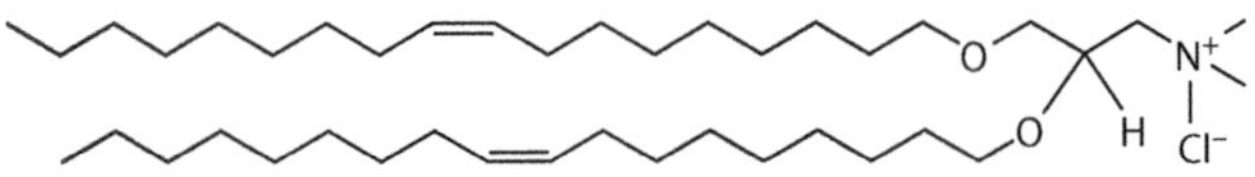	Felgner et al. (1987)
Dimethylammonium propane (DAP)	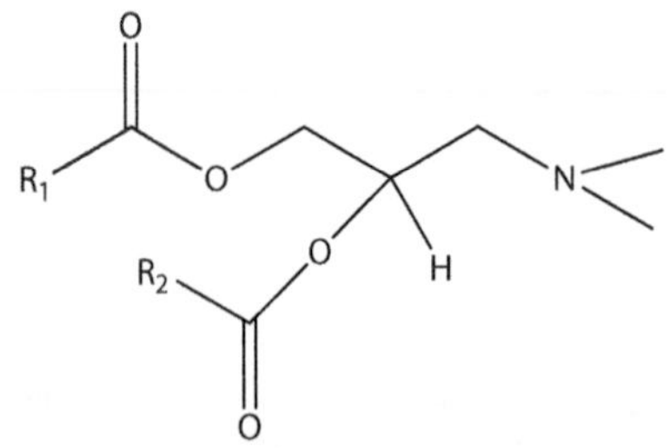	Auguste et al. (2006, 2008)

[a] R groups represent the carbon backbone of the lipid fatty acid and are variable depending on the lipid.

Both liposomes and polymeric nanocarriers may be designed to alter the biodistribution of drugs within the body. Nanoparticles less than 250 nm in size passively localize to sites of tumors and inflammation (Matsumura and Maeda 1986; Maeda et al. 2009). Active targeting may also be achieved by utilizing molecules that bind to receptors of specific cells (Martin and Papahadjopoulos 1982; Lee and Low 1995). Additionally, nanocarriers can trigger release based on changes in their microenvironment, such as a change in pH, ionic strength, electric field, or temperature (Galaev and Mattiasson 1999; de Las Heras Alarcon et al. 2005), For example, liposomes incorporating a pH-dependent lipid (dimethylammonium propane [DAP]) have been shown to be useful in delivery of short interfering RNA (siRNA) (Auguste et al. 2008). In addition, liposomes may be formulated with disruptive proteins that can destabilize the liposomal envelope (Mastrobattista et al. 2002). Polymeric nanocarriers comprised of poly(*N*,*N*-dimethylaminoethyl methacrylate) (DMAEMA) and poly(2-hydroxyethyl methacrylate) (HEMA) allow pH-sensitive release of paclitaxel due to swelling in low pH environments (You and Auguste 2008). Such alteration of a drug's biodistribution has the benefits of increasing the amount of drug at the desired location, delivering less drug overall to reduce costs and minimizing toxic effects.

Finally, metallic nanoparticles composed of gold and iron oxide have also been explored in drug delivery methods. Among many favorable properties, gold and iron oxide can easily be functionalized with biological molecules and are biocompatible (Kamei et al. 2009). Both can be visualized by noninvasive imaging techniques such as magnetic resonance imaging (MRI) (iron oxide) or near-infrared (IR) imaging (gold) (Wang et al. 2001; Loo et al. 2005). Additionally, iron oxide nanoparticles can be manipulated by a magnetic field to trap them within a target area (Chertok et al. 2008). Using these types of materials in drug delivery is beneficial not only for targeting specific areas in the body but also for monitoring the nanocarrier's location in the body.

The following discussion introduces several such functions to consider when designing a nanocarrier for drug delivery, including the ability to evade uptake by immune cells, target an area in the body, monitor noninvasively, and release a therapeutic in a controlled manner.

11.1 Immune Evasion

A major goal for any effective nanocarrier is to circulate long enough to reach the location in the body where the drug is needed. While the nanocarrier is traveling in circulation prior to being taken up by tissue, plasma proteins called opsonins will adsorb onto it. This process, called opsonization, is an element of the complement system that helps the immune system to recognize foreign agents. The activation of the complement system leads to the uptake and degradation of nanocarriers by the mononuclear phagocytic system (MPS), which recognizes these coated carriers and rapidly removes them from circulation (Bazile et al. 1992; Liu et al. 1995; Ahl et al. 1997). This occurrence is problematic since rapid elimination from circulation will alter the biodistribution and increase drug concentration in the kidney, liver, and spleen (organs responsible for clearance), which increases the overall amount of drug required to reach a given concentration at the therapeutic site of interest and may cause adverse side effects.

For this reason, much of the development of nanocarriers has focused on their modification to avoid removal by the immune system, allowing them to circulate for longer periods of time and increasing the probability that any given carrier will reach the site of interest. Early research showed that bare liposomes would bind proteins in vivo, an interaction correlated with removal by the immune system (Chonn et al. 1992). In attempts to evade uptake by phagocytes and increase circulation time, liposomes have been coated with many different molecules, including polymers and carbohydrates (Senior 1987; Gabizon and Papahadjopoulos 1988; Sunamoto et al. 1992). To date, the most successful and popular coating has been poly(ethylene glycol) (PEG). PEG is an amphiphilic polymer used for its "stealth" properties. Employing PEG in delivery vehicles enables evasion of the immune response by decreasing protein adsorption and opsonization, allowing for an increase in bioavailability and circulation time in the bloodstream (Harris et al. 2001).

One of the most effective PEG formulations for liposome carriers uses a lipid attached to a 5 kDa PEG chain, forming distearoylphosphatidylethanolamine-PEG5k (DSPE-PEG5k) (Beduaddo and Huang 1995). Blume and Cevc (1990) showed that a layer of 10 mol% DSPE-PEG5k on liposomes increased circulation half-life in mice from 0.47 to 8.4 h. Studies in humans found that levels of the chemotherapeutic agent doxorubicin in tumor effusions from cancer patients were several times larger when delivered via PEG-modified liposomes rather than with free doxorubicin, indicating that more drug had reached its target (Gabizon et al. 1994). Patients who received the drug in PEG-modified liposomes also showed lower levels of drug metabolites in their excretions, showing that less drug had been cleared from the bloodstream through excretion pathways (Gabizon et al. 1994). PEG has been broadly used as a method for shielding nanocarriers from the immune system, and it has recently been shown that incorporating hydrophobically modified PEG polymers on liposomes can inhibit liposome–protein interactions more effectively than the conventional, singly anchored PEGylated lipids (Auguste et al. 2003). Because of its use inhibiting liposome–protein interactions, PEG is also the subject of debate over whether it inhibits liposome–cell interactions. Some researchers believe that PEG can protect nanocarriers from MPS uptake without preventing cellular uptake (Savic et al. 2003), while others have found that the addition of PEG to liposomal nanocarriers has inhibited cell entry (Reddy et al. 2002).

PLGA nanocarriers have also utilized PEG to increase circulation time in vivo. For example, PLGA-block-PEG copolymers have been used for delivery to prostate tumors (Cheng et al. 2007). PLGA nanocarriers, composed of a 50:50 monomer ratio and intrinsic viscosity of 0.20 dL/g, were loaded with the chemotherapeutic docetaxel by the "nanoprecipitation" method (Fonseca et al. 2002). The nanoparticles were modified using 1-ethyl-3(3-dimethylaminopropyl) carbodiimide (EDC) hydrochloride and *N*-hydroxysuccinimide (NHS) (which conjugate carboxyl-containing molecules to amine-containing molecules) to attach targeting molecules called aptamers to carboxylated PEG. Aptamers are short nucleic acid chains (~20 bp) that have the ability to target cells. These nanoparticles enhanced delivery to prostate tumors in an in vivo mouse model. Thus, the inclusion of PEG on the surface of the nanoparticles provided not only the ability to evade the immune system but also the ability to conjugate molecules that could target a specific cell type.

PLGA-lecithin-PEG core-shell nanoparticles have been used in controlled drug delivery, an approach that combined polymeric and lipid-based materials (Chan et al. 2009). PLGA, with a 50:50 monomer ratio and intrinsic viscosity of 0.71–0.92 dL/g, was loaded with docetaxel in an organic phase and then precipitated in an aqueous solution containing carboxylated DSPE-PEG2k and lecithin at 65°C to allow for self-assembly of the particles. The drug remained protected in the hydrophobic PLGA core by the lecithin monolayer, allowing drug release to be controlled and manipulated by changing the lipid density of the particle. The outer carboxylated PEG shell provided electrostatic and steric stability due to the length of the PEG chains and the charge repulsion of the negatively charged carboxyl groups, and it also extended circulation time in vivo.

Immune recognition of a delivered therapeutic is a challenging obstacle when designing a nanocarrier. Cationic particles, for example, are easily recognized by the immune system and are known to be cytotoxic (Filion and Phillips 1997). As we have shown, PEG is widely used for decreasing adsorption of plasma proteins, which would result in an immune response. PEG increases the circulation time of administered drugs, allowing time for the nanocarriers to target specific locations in the body.

11.2 Targeting

Targeting a drug to a specific cell type is highly desirable in many therapeutic treatments. Many different strategies have been adopted to improve the distribution of drug in the body, including controlling the size of the carrier. Tumor endothelium is discontinuous, with gaps of approximately 250 nm (Barratt 2000). Blood vessels in tumors are therefore leaky compared to blood vessels elsewhere in the body. Thus, when liposomes and other nanoparticles less than 250 nm in size are introduced into the body, they passively accumulate in tumors, as well as the liver, spleen, and bone marrow, which also have discontinuous endothelium. This phenomenon has been termed the "enhanced permeability and retention effect" (EPR effect) and was first reported in 1986 (Matsumura and Maeda 1986).

Nanocarriers may also be actively targeted to specific locations in the body through chemical moieties. Antibodies, perhaps the best example of this idea, were first conjugated to liposomes for targeting purposes in the early 1980s (Martin and Papahadjopoulos 1982). Since then, they have become widespread in drug delivery research and have made their way into commercial products. The anticancer drugs, trastuzumab (Baselga et al. 1998) and rituximab (McLaughlin et al. 1998), for example, are antibodies. Trastuzumab (sold under the brand name Herceptin) is an anti-HER2 antibody used in the treatment of breast cancers, which express human epidermal growth factor receptor 2 (HER2), while rituximab is an antibody that targets the B-cell marker CD20 and is used in the treatment of B-cell lymphomas and leukemias.

Nanocarriers can also target certain cancers by displaying the ligand for any surface receptor molecules overexpressed by the cancer cells. For example, many cancers overexpress the membrane-bound folate receptor. Thus, the incorporation of folic acid into the lipid bilayer helps target liposomes to the cancer cells and facilitate endocytosis (Lee and Low 1995). In one study, the addition of 0.5 mol% folic acid-conjugated lipid to a liposome preparation encapsulating doxorubicin increased murine survival by 31% compared to a nontargeted liposomal control (Pan et al. 2003). Some researchers have added multiple targeting components to liposome formulations; for example, Saul et al. synthesized liposomal nanocarriers displaying both folic acid and a monoclonal antibody against epidermal growth factor receptor (EGFR), leading to enhanced targeting of the human KB cancer cell line, which overexpresses both the folate receptor and EGFR (Saul et al. 2006).

In another instance of taking advantage of a specific binding pair, peptides such as the arginine-glycine-aspartate (RGD) sequence have also been used for targeting purposes. The RGD peptide selectively binds the $\alpha_v\beta_3$-integrin, a membrane protein through which cells normally attach to the extracellular matrix. Both liposomes and PLGA nanoparticles can be surface functionalized with RGD peptides. In one study, transferrin and an RGD peptide were conjugated to the surface of PLGA, a combination that facilitated endocytosis and enhanced delivery to the neovascular eye for treatment of age-related macular degeneration (Singh et al. 2009).

Nonchemical means such as magnetism have also been employed for targeting. One example method uses a magnetic field to control the localization of particles made from magnetically responsive materials including iron oxide (Fe_3O_4). Nanoparticles with a core of iron oxide and a shell of starch have been used to target delivery to brain tumors using an electromagnet (Chertok et al. 2008). These particles displayed superparamagnetic behavior, meaning that they did not have a magnetic moment in the absence of an external magnetic field and thus did not aggregate beyond the targeted region. The particles were injected intravenously, and they were retained in the tumors even after the magnetic field was removed.

11.3 Imaging

Once a drug delivery vehicle is administered into the body, it is helpful to monitor its location noninvasively. One extremely useful detection method is MRI. The technique is based on the principle that protons align due to an applied magnetic field and can then be perturbed by a transverse radiofrequency to return to their original state, a process called relaxation (Sun et al. 2008). Iron oxide nanoparticles enhance proton spin-spin relaxation and have a large magnetic moment, allowing them to be used as "contrast agents" for MRI detection and suggesting a reason for their inclusion in engineered nanocarriers (Wang et al. 2001; Corot et al. 2006).

Targeted perfluorocarbon nanoparticles have also been used for MRI detection. One such method incorporated biotinylated phosphatidylethanolamine (PE) into the outer lipid monolayer of a perfluorocarbon microemulsion (Lanza et al. 2002). The microemulsion included lecithin and cholesterol as surfactants and gadolinium diethylene-triamine-pentaacetic acid-bis-oleate (Gd-DTPA-BOA) as a contrast agent. After emulsification, tissue factor, a protein involved in the blood coagulation cascade, was conjugated on the surface to target vascular smooth muscle cells in vitro. The study concluded that targeted iron oxide nanoparticles could provide a useful method for visualizing drug delivery in vivo.

Another widely used noninvasive imaging method is ultrasound. This technique utilizes low-intensity and high-frequency sound waves to create a 2D image (Liang and Blomley 2003; Pitt et al. 2004). Ultrasound contrast agents such as gas-filled bubbles can be introduced to obtain an improved image. These "microbubbles" can be injected intravenously to detect a material with different acoustic properties from the tissue (Blomley et al. 2001). Perfluorocarbon nanoparticles have also been used as ultrasound-imaging contrast agents (Lanza et al. 2000; Crowder et al. 2005). Though they have poor acoustic reflectivity, they can be strongly detected when bound to specific targets.

11.4 Release

11.4.1 Material-Dependent Release

Once a nanocarrier has evaded the immune system long enough to reach its therapeutic target, its next function is to release its drug contents. In the case of polymers, this may be as simple as undergoing degradation. Polymers' degradation rates depend on their chemical architecture, which can be engineered so that the drug is released in a controlled manner. For example, PLGA breaks down by hydrolysis, leaving nontoxic degradation products, which can be removed by the Krebs cycle (Panyam and Labhasetwar 2003). The degradation rate is tuned based on the number of lactic acid or glycolic acid subunits: increasing the glycolide ratio will increase the degradation rate (Wu and Wang 2001). After hydrolysis, the degradation products produce a low-pH environment, which has been a challenge for protein delivery. Other factors influencing polymer degradation rates include molecular weight and porosity, with low molecular weight and high porosity corresponding to faster degradation (Shive and Anderson 1997).

Release of a drug may also be triggered by response of the nanocarrier to its microenvironment. When a liposome is endocytosed by a cell, it becomes sequestered in an endosome, preventing it from delivering its cargo to the cytoplasm. Successful delivery requires the liposome to fuse with or destabilize the endosomal membrane, thereby releasing its contents. If cationic lipids are used in the liposome, the endosomal membrane is disrupted and transfection efficiency is increased. This is because the DNA-cationic lipid complexes destabilize the endosomal membrane after endocytosis, leading to a rearrangement in which anionic lipids in the membrane shift position to neutralize the charge of the cationic lipids. This results in the release of DNA from the cationic lipids into the cytoplasm (Xu and Szoka 1996). The use of cationic lipids to facilitate transfection was reported by Felgner et al. (1987), who used dioleoyl phosphatidylethanolamine (DOPE) and the cationic lipid *N*-[1-(2,3-dioleyloxy)propyl]-*N*,*N*,*N*-trimethylammonium chloride (DOTMA).

Cationic liposomes have also been used to deliver siRNA in RNA interference (RNAi) (Landen et al. 2005; Zhang et al. 2006). RNAi technology regulates protein expression by degrading specific mRNA sequences (Elbashir et al. 2001). Delivery of siRNA can be difficult since RNAs are charged, small, susceptible to nucleases, and require delivery in high numbers to single cells. Sato et al. (2008) used cationic liposomes coupled to vitamin A to deliver siRNA against a collagen-specific chaperone to the liver of cirrhotic rats. The vitamin A promoted liposome uptake by hepatic stellate cells, which play an important role in the fibrogenesis typical of liver cirrhosis. The siRNA reduced collagen secretion and thus the fibrogenesis conducted by the cells. In another study, Kim et al. (2009) enhanced siRNA delivery to livers with the hepatitis C virus by encapsulating the RNA in cationic liposomes that were also complexed with apolipoprotein A-I, a ligand for high-density lipoprotein (HDL) receptors on hepatocytes. The delivered siRNA knocked down expression of its target gene by 95%. This success is not without a significant drawback, however—cationic liposomes have proved to be toxic (Song et al. 1997; Dokka et al. 2000).

Polyethyleneimine (PEI), a positively charged polymer, has also been used for gene delivery. PEI forms complexes with negatively charged DNA, which condenses the nucleic acid for protection from enzymatic hydrolysis (Lungwitz et al. 2005). Its high transfection efficiency is associated with its ability to buffer against the endosomal pH, coined the "proton sponge effect" (Boussif et al. 1995). PEI has been used for gene delivery to the pulmonary epithelium as an alternative to viral gene carriers (Bivas-Benita et al. 2004). Unfortunately, like the use of cationic liposomes, the use of PEI has been limited due to cellular toxicity (Chollet et al. 2002).

Because of the toxicity of cationic materials, researchers have looked at making drug delivery vehicles responsive to internal microenvironments. This has been achieved by engineering vehicles incorporating stimuli-sensitive polymers, which show a change in their properties upon changes in environmental conditions (either physical or chemical) such as temperature (Lam et al. 2004; Ramanan et al. 2006; Dromi et al. 2007), pH (Li et al. 2002; Kong et al. 2007), light (Suzuki and Tanaka 1990; Gerasimov et al. 1999), ionic strength (Demanuele and Staniforth 1992), and electric (Kwon et al. 1991) or magnetic fields

(Alexiou et al. 2000). Among these, pH-sensitive polymers have received a large amount of attention in the fields of drug delivery (Lowman et al. 1999; Dai et al. 2004; Mahkam 2005), gene delivery (Lim et al. 2000), and insulin delivery (Napoli et al. 2004). pH-sensitive polymers generally have ionizable groups of weak acids or weak bases with a pK_a value of 3–10. Weak acids and weak bases, such as carboxyl acids, phosphoric acid, and amino groups, exhibit a change in ionization state with pH change.

Systems able to alter their character based on local pH changes have been investigated. For example, pH-sensitive liposomes may shield themselves from the immune system with PEG and still take advantage of the membrane disruption abilities of cationic liposomes when they reach their target. Auguste et al. (2006) formulated liposomes with variable surface charge by varying the composition of a zwitterionic lipid (phosphatidylcholine) and an anionic lipid (phosphatidylglycerol). To induce a pH-dependent liposome surface charge, a titratable lipid (DAP) with a pK_a of 6.7 was also incorporated. The liposome was then shielded with polycation-block-PEG polymers to confer protection from immune recognition while in circulation. The protective polymers remained electrostatically bound to the liposomes at the bloodstream pH of 7.4. Upon endocytosis and a local shift to the endosomal pH of 5.5, the polymer was released from the liposome's surface, allowing the liposome to interact with the endosomal membrane and lead to membrane disruption (Figure 11.1). Recently, siRNA has been delivered using this pH-sensitive liposomal technology (Auguste et al. 2008).

Many pH-sensitive polymers have also been investigated, dating back to Eisenberg et al. in 1950 (Kuhn et al. 1950), generally with the goal of using physiological pH changes to facilitate swelling and thus control release of embedded drug molecules. Biocompatible and biodegradable materials have been developed, but few conventional pH-sensitive polymers have been used because of their limited sensitivity near the pH of blood (7.4). Moreover, common pH-sensitive polymers, such as natural polymers (alginate [You et al. 2001] and chitosan [Felt et al. 1998]) and synthesized polymers (based on poly(acrylic acid) [Jarvinen et al. 1998] and poly(methacrylic acid) [PMAA] [Jones et al. 2003]), swell at high pH due to ionizable functional groups on the backbone or side chain of the polymer. Thus, these polymers are not responsive under most physiological conditions.

New technologies have recently been introduced that are responsive under conditions that are more physiologically relevant. In 2005, Heffernan and Murthy (2005) developed a pH-sensitive biodegradable nanocarrier using polyketal nanoparticles. The ketal linkage in the particles allows acid-catalyzed hydrolysis into low molecular weight, water-soluble compounds. Thus, the degradation of the particle is accelerated under acidic conditions. In contrast, You and Auguste (2008) synthesized new pH-sensitive nanoparticles that focused on swelling for drug release (Figure 11.2). The pH-sensitive particles are comprised of DMAEMA and HEMA. DMAEMA is a pH-responsive material, which has a tertiary amine

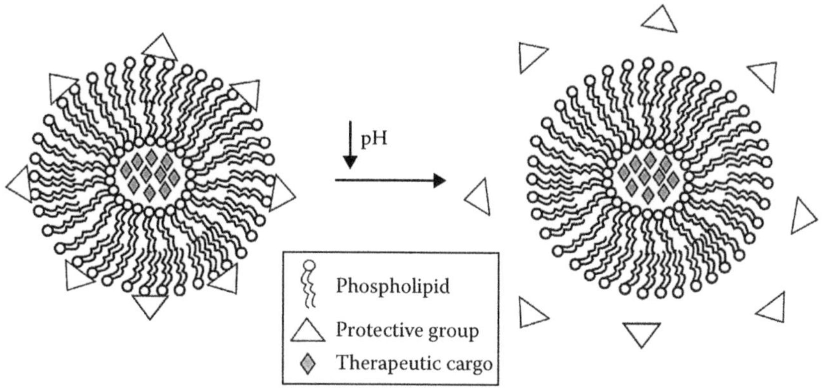

FIGURE 11.1 Release of protective, polycationic polymers from pH-sensitive liposomes in response to lowered pH. (From Auguste, D.T. et al., *J. Control. Release*, 130(3), 266, 2008.)

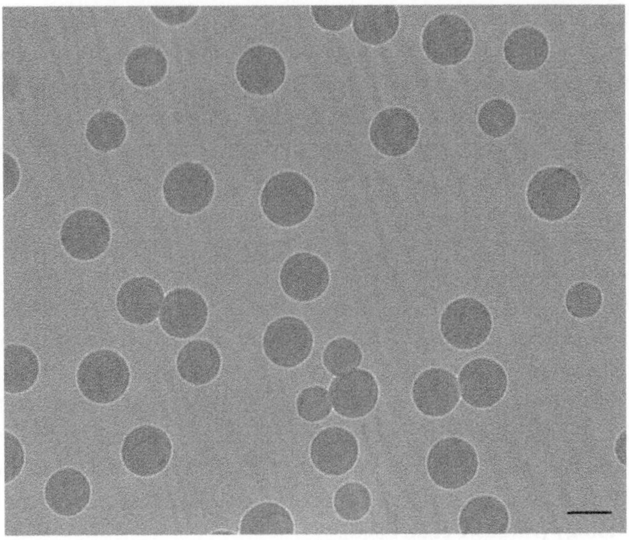

FIGURE 11.2 Transmission electron microscope image of DMAEMA/HEMA nanoparticles used for drug delivery. Scale bar is 200 nm. (From You, J.O. and Auguste, D.T., *Biomaterials*, 29(12), 1950, 2008.)

functional group with a pK_a of 7.5. Monodisperse pH-sensitive DMAEMA/HEMA nanocarriers have been successfully synthesized for pH-triggered paclitaxel delivery. The particles were 230 nm in diameter and showed a high volume swelling ratio at low pH, low crosslinking density, and high content of DMAEMA. Due to nanoparticle swelling, paclitaxel was released quickly at low pH and low crosslinker concentration (Figure 11.3).

Other stimuli-sensitive delivery systems have been used to deliver proteins such as insulin, with the goal of sensing local glucose levels and responding by releasing the proper amount of insulin. Langer and colleagues introduced an ethylene/vinyl acetate copolymer system, incorporating trilysyl insulin and glucose oxidase. The glucose oxidase reacted with glucose to produce gluconic acid, which lowered

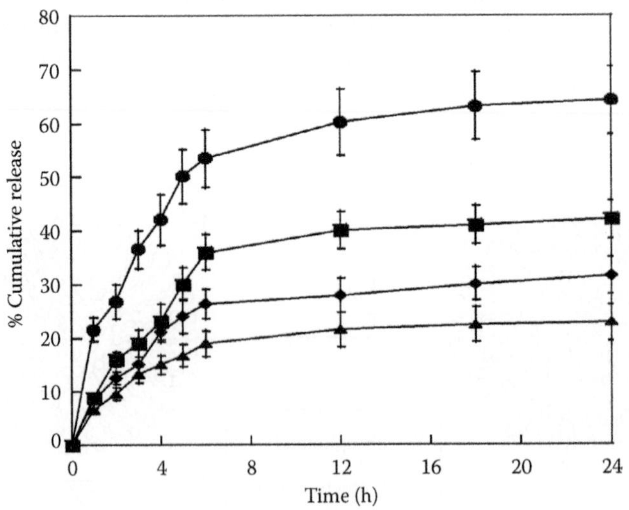

FIGURE 11.3 Controlled release of paclitaxel from DMAEMA/HEMA nanoparticles. Paclitaxel-loaded 10/90 (mol/mol) DMAEMA/HEMA nanoparticles crosslinked with 3 mol% TEGDMA were monitored for paclitaxel release (a) in medium at pH 6.8 (●), 7.0 (■), 7.2 (◆), and 7.4 (▲). The error is the standard deviation of the mean, where n = 3. (From You, J.O. and Auguste, D.T., *Biomaterials*, 29(12), 1950, 2008.)

the pH of the local microenvironment. Decreasing pH increased insulin solubility, resulting in a greater release of insulin when greater concentrations of glucose were present (Fischel-Ghodsian et al. 1988). Recently, a multilayer shell system for insulin delivery has been developed, which contains both glucose oxidase and catalase, and operates on much the same principle (Qi et al. 2009).

11.4.2 Externally Mediated Release

A drug can also be released from its nanocarrier in response to external cues. Light, for example, has been used to trigger drug release through the cleaving of photolabile chemical groups. One design uses mesoporous 100 nm silica nanospheres capped with 5 nm gold nanospheres functionalized with a cationic photoreactive linker (Lai et al. 2003; Vivero-Escoto et al. 2009). Drug compounds can be encapsulated within the mesoporous nanosphere, which is capped using the electrostatic attraction between the positively charged linker and the negatively charged silica particles. Photoirradiation using ultraviolet (UV) light for 10 min at 0.49 mW/cm^2 cleaves the photolabile linker, causing uncapping of the silica because of charge repulsion between the gold and silica particles. This allows for control of drug release. The system can be used for shuttling hydrophobic compounds into cells without the initial burst release that is commonly experienced with controlled release systems.

Light has been used to trigger release from liposomes as well. Volodkin et al. (2009) used near-IR light to trigger release from gold nanoparticle-conjugated liposomes. The gold nanoparticles absorbed applied laser light, heating up and disrupting the lipid bilayer, allowing the release of liposomal contents. Use of light is hindered by its penetration depth, but new techniques are being developed that may allow broader use of light-responsive delivery.

Drug release can also be facilitated by ultrasonic disruption (Pitt et al. 2004). Ultrasonic waves transmit pressure waves above human hearing at 20 kHz and produce a physical force. Because ultrasound can be focused and targeted to specific areas of the body, collapse cavitation has been explored in drug release (Husseini and Pitt 2008). In one example, doxorubicin was encapsulated within polymeric nanomicelles composed of 10% pluronic 105 and *N,N*-diethylacrylamide for targeting tumors in rats in vivo (Nelson et al. 2002). Applying low-frequency ultrasound at specific tumor sites caused the micelles to release the drug, significantly decreasing tumor volumes. Like the use of light, ultrasound application does not damage the surrounding tissue, making it an attractive area for further drug delivery research.

11.5 Multifunctionality

Emerging drug delivery technologies often combine the functionalities described above in a quest to create the optimal nanocarrier. This section discusses the utilization of immune evasion, targeting, controlled release, and/or imaging in one delivery vehicle to maximize the therapeutic potential of a drug (Table 11.4).

One common application of multifunctionality is the combination of immune evasion with active targeting functionalities. For example, a PLGA-doxorubicin nanocarrier has utilized the properties of PEG with RGD targeting (Wang et al. 2009). The nanoparticle was PEGylated to decrease immune recognition and also modified with RGD peptide for targeting $\alpha_v\beta_3$-integrin-expressing cancer cells such as the B16F10 mouse melanoma, DU145 human prostate carcinoma, and MDA-MB-231 human mammary carcinoma cell lines. The nanoparticles were more easily taken up by MDA-MB-231 and B16F10 cell lines, which overexpress the integrin, rather than the DU145 cell line, which has the lowest integrin expression level.

Immune evasion and noninvasive imaging have been combined in one nanocarrier for another simple instance of multifunctionality. Mikhaylova et al. (2009) studied the effects of adding MRI contrast agents to their liposome-based siRNA delivery system. The liposomes included both PEGylated lipids to avoid immune recognition and cationic lipids to aid delivery of the siRNA against COX-2. The MRI

TABLE 11.4 Multifunctional Drug Delivery

Authors	Particle Type	Payload	Immune Evasion Method	Targeting Method	Imaging Method	Release Method	References
Wang et al. (2009)	PLGA	Doxorubicin	PEG	RGD	Doxorubicin is intrinsically fluorescent	Degradation	Wang et al. (2009)
Mikhaylova et al. (2009)	Liposomes	siRNA	PEG	N/A	MRI via encapsulation of contrast agents	Use of cationic liposomes	Mikhaylova et al. (2009)
Guo et al. (2009)	Iron oxide	Doxorubicin	N/A	Magnetism possible	MRI possible via iron oxide particles	pH-sensitive	Guo et al. (2009)
Weng et al. (2008)	Liposomes	Doxorubicin	PEG	Anti-Her2 antibodies	Conjugation to quantum dots	N/A	Weng et al. (2008)
Nasongkla et al. (2006)	MAL-PEG-PLA and MPEG-PLA micelles	Doxorubicin	PEG	RGD	MRI via attached iron oxide particles	pH-sensitive	Nasongkla et al. (2006)
Tai et al. (2009)	Liposomes	Carboxyl-fluorescein	N/A	N/A	MRI suggested but not performed	Heat-induced via encapsulated iron oxide particles	Tai et al. (2009)
Lu et al. (2009)	Hollow gold nanospheres	The particles themselves	PEG	α-melanocyte-stimulating hormone analog	Surface plasmon resonance	Heat-induced activation of particles	Lu et al. (2009)

contrast agents Magnevist and Feridex were also loaded into the liposomes, and results in a cell culture system showed that the gadolinium-based Magnevist adversely affected the specificity of the siRNA's effects, whereas the iron-based Feridex did not. The liposomes were found in tumors within 30 min of tail injection into mice, but after several hours, they were largely aggregated in the lungs, kidney, stomach, and liver. It is possible that adding a targeting moiety such as an antibody to this system would increase tumor specificity.

Another versatile delivery system involves combining noninvasive imaging, active targeting, and pH-sensitive therapeutic release. In one example, a modified iron oxide particle was used to analyze the controlled release of doxorubicin (Guo et al. 2009). A mesoporous iron oxide core, enabling both imaging using MRI and also targeting using magnetism, was synthesized using a solvothermal method and functionalized with PMAA for drug loading. The mesoporous structure allowed a drug-loading capacity twice that of similar nonporous nanoparticles, and the pH-sensitive PMAA exhibited faster release of the drug at lower pH.

In addition to metallic nanoparticles, liposomes have also been designed to include threefold multifunctionality. Weng et al. (2008) designed a liposome drug delivery system capable of immune evasion, noninvasive imaging, and active targeting using antibodies. The liposomes displayed tumor-targeting antibodies as well as quantum dots for fluorescence visualization, and both functional groups were conjugated to the liposome via PEG to shield it from the host immune system. In a cell culture model system, liposomes with anti-HER2 antibody were taken up by a HER2-displaying cell line much more readily than untargeted liposomes. Quantum dot circulation time in an athymic mouse model was also greatly increased by conjugation to this liposome system, bringing the plasma half-life from less than 10 min (for bare quantum dots) to almost 3 h (for the quantum dot-conjugated liposomes).

Adding even more functional layers, polymeric micelle nanocarriers have been designed to include noninvasive imaging, active targeting using peptides, immune evasion, and pH-sensitive release. In one such system, amphiphilic block copolymers of maleimide-terminated PEG-block-poly(D,L lactide) (MAL-PEG-PLA) and methoxy-terminated PEG-block-PLA (mPEG-PLA) were used to form pH-sensitive micelles 100 nm in size (Nasongkla et al. 2006). The hydrophilic shell of the micelles was surface-functionalized with cyclic integrin-binding RGD peptide to specifically target $\alpha_v\beta_3$-integrins on SLK human sarcoma endothelial cells. Superparamagnetic iron oxide particles were then encapsulated, along with the hydrophobic drug of interest, within the hydrophobic core of the micelles. The clustering of the iron oxide particles within the micelle core enabled ultrasensitive MRI.

In addition to uses in targeting and as MRI contrast agents, iron oxide particles are also said to be "thermally sensitive" because they can generate heat through vibrations resulting from an applied alternating magnetic field. Dextran-coated iron oxide nanoparticles have been encapsulated within thermosensitive dipalmitoylphosphatidylcholine/cholesterol (DPPC/Chol) liposomes using a thin-film hydration and extrusion method (Tai et al. 2009). By encapsulating iron oxide nanoparticles within these liposomes, the delivery vehicle can be targeted to a specific site using magnetism, and then heating can be induced to release the encapsulated drug without adversely affecting the temperature of the larger-scale environment.

Gold nanoparticles can also be used in photothermal ablation therapies (Schwartzberg et al. 2006). These particles can also be monitored by noninvasive [^{18}F] fluorodeoxyglucose positron emission topography (PET) scans. PEGylated hollow gold nanospheres (40 nm) have been conjugated to the agonist of a receptor overexpressed by melanoma cells (Lu et al. 2009). The nanoparticles were selectively delivered and endocytosed by murine melanoma cells in vivo, allowing the tumor site to be irradiated with a near-IR-region laser for 1 min at a low output power of 0.5 W/cm^2. It was found that the treated tumors exhibited favorable necrotic responses. Although the biodistribution and long-term fate of the nanoparticles have yet to be analyzed, these multifunctional gold nanoparticles combined immune evasion, noninvasive imaging, targeted delivery, and photothermal sensitivity for a promising clinical treatment concept.

Overall, multifunctional nanoparticles have been useful for delivering therapeutics to their intended targets while concurrently minimizing cytotoxicity and maximizing effectiveness and efficiency. Using nanoparticles in drug delivery systems enables release in the correct area and at the intended time, allowing for lower drug loading concentrations and thereby minimizing the potential hazards of the payload. The examples described above illustrate how different levels of complexity can be added to nanocarriers for more effective treatment strategies than can be achieved using only one function. PEG is widely used to evade protein adsorption and immune recognition. Active targeting directs the vehicle to the intended site of the body—either through biological moieties or an external force—and adding controlled release properties based on triggers from the environment or degradation of the material aids in maximizing the efficiency of the therapeutic. Finally, being able to image noninvasively is beneficial for monitoring the drug in vivo. Emerging methods employing these combinations of techniques will aid in the discovery of new treatment modalities. Further research should involve monitoring their biodistribution, analyzing cellular uptake, and understanding their toxic effects, which are all important concerns before moving forward with a treatment strategy.

11.6 Nanoparticle Toxicity

While maximizing the therapeutic effect of the delivered drug, an ideal nanocarrier should not introduce toxicity on its own. Yet because many nanoparticle preparations contain materials in forms not previously studied, the in vivo fate of such nanoparticles is not well understood. To date, this issue has received higher-profile attention in the fields of environmental and human health risk assessment than it has in nanomedicine, with governmental bodies such as the U.S. Environmental Protection Agency issuing reports on the potential risks and benefits of nanotechnology (The Royal Society & The Royal Academy of Engineering 2004; United States Environmental Protection Agency 2007; United States Food and Drug Administration 2007). In contrast to the intravenously delivered systems discussed in this chapter, these reports generally take an approach rooted in traditional risk assessment and consider environmental and occupational exposures via inhalation, ingestion, and transdermal pathways. However, their concern with toxicity is as relevant to nanomedicine as it is to the environment.

Predicting the toxicity of a given nanomaterial is complicated by many factors. For simple substances, the toxicity of larger-sized formulations may not be indicative of nanomaterial toxicity. Inhalation studies have shown that nanoparticles less than approximately 100 nm in diameter are more toxic on a mass basis than larger particles, suggesting that surface area may be a better metric for comparison (Oberdorster et al. 1994). Other factors such as charge, reactivity, and shape are important as well. For example, CdSe quantum dots are toxic in a cell culture model when not coated to prevent release of Cd^{2+} ions (Derfus et al. 2004), and carbon is more toxic when inhaled by mice as carbon nanotubes than as carbon black nanoparticles (Lam et al. 2004). The unifying factor underlying these toxicity data may be the particles' ability to generate reactive oxygen species (ROS) (Nel et al. 2006; Sayes et al. 2006).

Indeed, many nanoparticles are thought to cause cellular toxicity via oxidative stress (Brown et al. 2004; Nel et al. 2006; Buzea et al. 2007). The ROS responsible for this stress may be generated in response to physical damage of the cell, especially the mitochondria, or by metals on the nanocarrier able to catalyze free radical reactions (Li et al. 2003; Risom et al. 2005). ROS are also associated with inflammation, as well as the normal cellular functions of respiration and metabolism (Buzea et al. 2007). Once generated, the ROS can damage cells by reacting with lipids, proteins, DNA, and other molecules, which may interfere with gene transcription, signaling functions, and other cellular activities (Brown et al. 2004).

The particulars of this process may be highly dependent on the formulation of the nanocarrier and the types of cells it encounters. In one study, PEG-coated iron oxide (Fe_3O_4) particles were introduced to hTERT-BJ1 primary human dermal fibroblasts where it was found that cell attachment, morphology, cytoskeletal organization, and viability were not affected significantly by the particles (Gupta and Curtis 2004). In a different study, bare Fe_2O_3 particles incubated with human microvascular endothelial cells (HMVECs) caused an increase in permeability of the cell layer as a consequence of ROS production

(Apopa et al. 2009). As evidenced by these cases, researchers should evaluate the toxicities of specific nanoparticle formulations on a case-by-case basis in order to ascertain the effects of the materials. Understanding how the particles interact with cells and the level of their cytotoxic effects is crucial for assessing and moving forward with in vivo treatment strategies (Linkov et al. 2008).

11.7 Summary

Liposomes, polymers, and other nanocarrier technologies have progressed significantly in the past few decades. Current nanocarriers extend circulation time by helping the drug evade the immune system. They also target the drug to specific therapeutic sites of interest, enable noninvasive monitoring of drug location, and control release of the drug. The combination of these roles in recent multifunctional systems allows for the creation of more powerful delivery vehicles, leading to a favorable alteration of drug biodistribution. In therapeutic applications, this technology has the potential to reduce side effects felt by the patient as well as cost, as less drug is needed.

As research continues into novel polymers and stimuli-sensitive delivery systems, nanocarriers are likely to offer greater accuracy in targeting and more control of release profiles. Further research into the mechanisms of toxicity may lead to the development of nanocarriers that minimize ROS production. The delivery technologies likely to emerge from this research will be exciting steps forward for the field.

References

Ahl, P. L., S. K. Bhatia, P. Meers, P. Roberts, R. Stevens, R. Dause, W. R. Perkins, and A. S. Janoff. 1997. Enhancement of the in vivo circulation lifetime of L-alpha-distearoylphosphatidylcholine liposomes: Importance of liposomal aggregation versus complement opsonization. *Biochim. Biophys. Acta—Biomembranes* 1329 (2):370–382.

Alexiou, C., W. Arnold, R. J. Klein, F. G. Parak, P. Hulin, C. Bergemann, W. Erhardt, S. Wagenpfeil, and A. S. Lubbe. 2000. Locoregional cancer treatment with magnetic drug targeting. *Cancer Res.* 60 (23):6641–6648.

Apopa, P. L., Y. Qian, R. Shao, N. L. Guo, D. Schwegler-Berry, M. Pacurari, D. Porter, X. Shi, V. Vallyathan, V. Castranova, and D. C. Flynn. 2009. Iron oxide nanoparticles induce human microvascular endothelial cell permeability through reactive oxygen species production and microtubule remodeling. *Part. Fibre Toxicol.* 6:1.

Astete, C. E. and C. M. Sabliov. 2006. Synthesis and characterization of PLGA nanoparticles. *J. Biomater. Sci. Polym. Ed.* 17 (3):247–289.

Auguste, D. T., S. P. Armes, K. R. Brzezinska, T. J. Deming, J. Kohn, and R. K. Prud'homme. 2006. pH triggered release of protective poly(ethylene glycol)-b-polycation copolymers from liposomes. *Biomaterials* 27 (12):2599–2608.

Auguste, D. T., K. Furman, A. Wong, J. Fuller, S. P. Armes, T. J. Deming, and R. Langer. 2008. Triggered release of siRNA from poly(ethylene glycol)-protected, pH-dependent liposomes. *J. Control. Release* 130 (3):266–274.

Auguste, D. T., R. K. Prud'homme, P. L. Ahl, P. Meers, and J. Kohn. 2003. Association of hydrophobically-modified poly(ethylene glycol) with fusogenic liposomes. *Biochim. Biophys. Acta—Biomembranes* 1616 (2):184–195.

Bangham, A. D., M. M. Standish, and J. C. Watkins. 1965. Diffusion of univalent ions across lamellae of swollen phospholipids. *J. Mol. Biol.* 13 (1):238–252.

Barratt, G. M. 2000. Therapeutic applications of colloidal drug carriers. *Pharm. Sci. Technol. Today* 3 (5):163–171.

Baselga, J., L. Norton, J. Albanell, Y. M. Kim, and J. Mendelsohn. 1998. Recombinant humanized anti-HER2 antibody (Herceptin (TM)) enhances the antitumor activity of paclitaxel and doxorubicin against HER2/neu overexpressing human breast cancer xenografts. *Cancer Res.* 58 (13):2825–2831.

Bazile, D. V., C. Ropert, P. Huve, T. Verrecchia, M. Marlard, A. Frydman, M. Veillard, and G. Spenlehauer. 1992. Body distribution of fully biodegradable [14C]-poly(lactic acid) nanoparticles coated with albumin after parenteral administration to rats. *Biomaterials* 13 (15):1093–1102.

Beduaddo, F. K. and L. Huang. 1995. Interaction of PEG-phospholipid conjugates with phospholipid—Implications in liposomal drug-delivery. *Adv. Drug Deliv. Rev.* 16 (2–3):235–247.

Bivas-Benita, M., S. Romeijn, H. E. Junginger, and G. Borchard. 2004. PLGA-PEI nanoparticles for gene delivery to pulmonary epithelium. *Eur. J. Pharm. Biopharm.* 58 (1):1–6.

Blomley, M. J., J. C. Cooke, E. C. Unger, M. J. Monaghan, and D. O. Cosgrove. 2001. Microbubble contrast agents: A new era in ultrasound. *BMJ* 322 (7296):1222–1225.

Blume, G. and G. Cevc. 1990. Liposomes for the sustained drug release in vivo. *Biochim. Biophys. Acta* 1029 (1):91–97.

Boussif, O., F. Lezoualch, M. A. Zanta, M. D. Mergny, D. Scherman, B. Demeneix, and J. P. Behr. 1995. A versatile vector for gene and oligonucleotide transfer into cells in culture and in-vivo—Polyethylenimine. *Proc. Natl. Acad. Sci. U.S.A.* 92 (16):7297–7301.

Brown, D. M., K. Donaldson, P. J. Borm, R. P. Schins, M. Dehnhardt, P. Gilmour, L. A. Jimenez, and V. Stone. 2004. Calcium and ROS-mediated activation of transcription factors and TNF-alpha cytokine gene expression in macrophages exposed to ultrafine particles. *Am. J. Physiol. Lung Cell. Mol. Physiol.* 286 (2):L344–L353.

Buzea, C., I. I. Pacheco, and K. Robbie. 2007. Nanomaterials and nanoparticles: Sources and toxicity. *Biointerphases* 2 (4):MR17–MR71.

Chan, J. M., L. Zhang, K. P. Yuet, G. Liao, J. W. Rhee, R. Langer, and O. C. Farokhzad. 2009. PLGA-lecithin-PEG core-shell nanoparticles for controlled drug delivery. *Biomaterials* 30 (8):1627–1634.

Chasin, M. and R. Langer. 1990. *Biodegradable Polymers as Drug Delivery Systems*, Informa Health Care. Marcel Dekker, New York.

Cheng, J., B. A. Teply, I. Sherifi, J. Sung, G. Luther, F. X. Gu, E. Levy-Nissenbaum, A. F. Radovic-Moreno, R. Langer, and O. C. Farokhzad. 2007. Formulation of functionalized PLGA-PEG nanoparticles for in vivo targeted drug delivery. *Biomaterials* 28 (5):869–876.

Chertok, B., B. A. Moffat, A. E. David, F. Yu, C. Bergemann, B. D. Ross, and V. C. Yang. 2008. Iron oxide nanoparticles as a drug delivery vehicle for MRI monitored magnetic targeting of brain tumors. *Biomaterials* 29 (4):487–496.

Chollet, P., M. C. Favrot, A. Hurbin, and J. L. Coll. 2002. Side-effects of a systemic injection of linear polyethylenimine-DNA complexes. *J. Gene Med.* 4 (1):84–91.

Chonn, A., S. C. Semple, and P. R. Cullis. 1992. Association of blood proteins with large unilamellar liposomes in vivo—Relation to circulation lifetimes. *J. Biol. Chem.* 267 (26):18759–18765.

Corot, C., P. Robert, J. M. Idee, and M. Port. 2006. Recent advances in iron oxide nanocrystal technology for medical imaging. *Adv. Drug Deliv. Rev.* 58 (14):1471–1504.

Crowder, K. C., M. S. Hughes, J. N. Marsh, A. M. Barbieri, R. W. Fuhrhop, G. M. Lanza, and S. A. Wickline. 2005. Sonic activation of molecularly-targeted nanoparticles accelerates transmembrane lipid delivery to cancer cells through contact-mediated mechanisms: Implications for enhanced local drug delivery. *Ultrasound Med. Biol.* 31 (12):1693–1700.

Dai, J. D., T. Nagai, X. Q. Wang, T. Zhang, M. Meng, and Q. Zhang. 2004. pH-sensitive nanoparticles for improving the oral bioavailability of cyclosporine A. *Int. J. Pharm.* 280 (1–2):229–240.

de Las Heras Alarcon, C., S. Pennadam, and C. Alexander. 2005. Stimuli responsive polymers for biomedical applications. *Chem. Soc. Rev.* 34 (3):276–285.

Demanuele, A. and J. N. Staniforth. 1992. An electrically modulated drug delivery device. 2. Effect of ionic-strength, drug concentration, and temperature. *Pharm. Res.* 9 (2):215–219.

Derfus, A. M., W. C. W. Chan, and S. N. Bhatia. 2004. Probing the cytotoxicity of semiconductor quantum dots. *Nano Lett.* 4 (1):11–18.

Dokka, S., D. Toledo, X. G. Shi, V. Castranova, and Y. Rojanasakul. 2000. Oxygen radical-mediated pulmonary toxicity induced by some cationic liposomes. *Pharm. Res.* 17 (5):521–525.

Dromi, S., V. Frenkel, A. Luk, B. Traughber, M. Angstadt, M. Bur, J. Poff et al. 2007. Pulsed-high intensity focused ultrasound and low temperature sensitive liposomes for enhanced targeted drug delivery and antitumor effect. *Clin. Cancer Res.* 13 (9):2722–2727.

Elbashir, S. M., J. Harborth, W. Lendeckel, A. Yalcin, K. Weber, and T. Tuschl. 2001. Duplexes of 21-nucleotide RNAs mediate RNA interference in cultured mammalian cells. *Nature* 411 (6836):494–498.

Felgner, P. L., T. R. Gadek, M. Holm, R. Roman, H. W. Chan, M. Wenz, J. P. Northrop, G. M. Ringold, and M. Danielsen. 1987. Lipofection—A highly efficient, lipid-mediated DNA-transfection procedure. *Proc. Natl. Acad. Sci. U.S.A.* 84 (21):7413–7417.

Felt, O., P. Buri, and R. Gurny. 1998. Chitosan: A unique polysaccharide for drug delivery. Drug development and industrial pharmacy. *Drug Dev. Ind. Pharm.* 24 (11):979–993.

Filion, M. C. and N. C. Phillips. 1997. Toxicity and immunomodulatory activity of liposomal vectors formulated with cationic lipids toward immune effector cells. *Biochim. Biophys. Acta—Biomembranes* 1329 (2):345–356.

Fischel-Ghodsian, F., L. Brown, E. Mathiowitz, D. Brandenburg, and R. Langer. 1988. Enzymatically controlled drug delivery. *Proc. Natl. Acad. Sci. U.S.A.* 85 (7):2403–2406.

Fonseca, C., S. Simoes, and R. Gaspar. 2002. Paclitaxel-loaded PLGA nanoparticles: Preparation, physicochemical characterization and in vitro anti-tumoral activity. *J. Control. Release* 83 (2):273–286.

Gabizon, A., R. Catane, B. Uziely, B. Kaufman, T. Safra, R. Cohen, F. Martin, A. Huang, and Y. Barenholz. 1994. Prolonged circulation time and enhanced accumulation in malignant exudates of doxorubicin encapsulated in polyethylene-glycol coated liposomes. *Cancer Res.* 54 (4):987–992.

Gabizon, A. and D. Papahadjopoulos. 1988. Liposome formulations with prolonged circulation time in blood and enhanced uptake by tumors. *Proc. Natl. Acad. Sci. U.S.A.* 85 (18):6949–6953.

Galaev, I. Y. and B. Mattiasson. 1999. 'Smart' polymers and what they could do in biotechnology and medicine. *Trends Biotechnol.* 17 (8):335–340.

Gerasimov, O. V., J. A. Boomer, M. M. Qualls, and D. H. Thompson. 1999. Cytosolic drug delivery using pH- and light-sensitive liposomes. *Adv. Drug Deliv. Rev.* 38 (3):317–338.

Guo, S., D. Li, L. Zhang, J. Li, and E. Wang. 2009. Monodisperse mesoporous superparamagnetic single-crystal magnetite nanoparticles for drug delivery. *Biomaterials* 30 (10):1881–1889.

Gupta, A. K. and A. S. Curtis. 2004. Surface modified superparamagnetic nanoparticles for drug delivery: Interaction studies with human fibroblasts in culture. *J. Mater. Sci. Mater. Med.* 15 (4):493–496.

Harris, J. M., N. E. Martin, and M. Modi. 2001. Pegylation: A novel process for modifying pharmacokinetics. *Clin. Pharmacokinet.* 40 (7):539–551.

Heffernan, M. J. and N. Murthy. 2005. Polyketal nanoparticles: A new pH-sensitive biodegradable drug delivery vehicle. *Bioconjugate Chem.* 16 (6):1340–1342.

Husseini, G. A. and W. G. Pitt. 2008. Micelles and nanoparticles for ultrasonic drug and gene delivery. *Adv. Drug Deliv. Rev.* 60 (10):1137–1152.

Jarvinen, K., S. Akerman, B. Svarfvar, T. Tarvainen, P. Viinikka, and P. Paronen. 1998. Drug release from pH and ionic strength responsive poly(acrylic acid) grafted poly(vinylidenefluoride) membrane bags in vitro. *Pharm. Res.* 15 (5):802–805.

Jones, M. C., M. Ranger, and J. C. Leroux. 2003. pH-sensitive unimolecular polymeric micelles: Synthesis of a novel drug carrier. *Bioconjugate Chem.* 14 (4):774–781.

Kamei, K., Y. Mukai, H. Kojima, T. Yoshikawa, M. Yoshikawa, G. Kiyohara, T. A. Yamamoto, Y. Yoshioka, N. Okada, S. Seino, and S. Nakagawa. 2009. Direct cell entry of gold/iron-oxide magnetic nanoparticles in adenovirus mediated gene delivery. *Biomaterials* 30 (9):1809–1814.

Kim, S. I., D. Shin, H. Lee, B. Y. Ahn, Y. Yoon, and M. Kim. 2009. Targeted delivery of siRNA against hepatitis C virus by apolipoprotein A-I-bound cationic liposomes. *J. Hepatol.* 50 (3):479–488.

Kong, S. D., A. Luong, G. Manorek, S. B. Howell, and J. Yang. 2007. Acidic hydrolysis of N-ethoxybenzylimidazoles (NEBIs): Potential applications as pH-sensitive linkers for drug delivery. *Bioconjugate Chem.* 18 (2):293–296.

Kuhn, W., B. Hargitay, A. Katchalsky, and H. Eisenberg. 1950. Reversible dilation and contraction by changing the state of ionization of high-polymer acid networks. *Nature* 165 (4196):514–516.

Kwon, I. C., Y. H. Bae, and S. W. Kim. 1991. Electrically erodible polymer gel for controlled release of drugs. *Nature* 354 (6351):291–293.

Lai, C. Y., B. G. Trewyn, D. M. Jeftinija, K. Jeftinija, S. Xu, S. Jeftinija, and V. S. Lin. 2003. A mesoporous silica nanosphere-based carrier system with chemically removable CdS nanoparticle caps for stimuli-responsive controlled release of neurotransmitters and drug molecules. *J. Am. Chem. Soc.* 125 (15):4451–4459.

Lam, C.-W., J. T. James, R. McCluskey, and R. L. Hunter. 2004. Pulmonary toxicity of single-wall carbon nanotubes in mice 7 and 90 days after intratracheal instillation *Toxicol. Sci.* 77:126–134.

Landen, C. N., A. Chavez-Reyes, C. Bucana, R. Schmandt, M. T. Deavers, G. Lopez-Berestein, and A. K. Sood. 2005. Therapeutic EphA2 gene targeting in vivo using neutral liposomal small interfering RNA delivery. *Cancer Res.* 65 (15):6910–6918.

Langer, R. and J. Folkman. 1976. Polymers for the sustained release of proteins and other macromolecules. *Nature* 263 (5580):797–800.

Lanza, G. M., D. R. Abendschein, C. S. Hall, M. J. Scott, D. E. Scherrer, A. Houseman, J. G. Miller, and S. A. Wickline. 2000. In vivo molecular imaging of stretch-induced tissue factor in carotid arteries with ligand-targeted nanoparticles. *J. Am. Soc. Echocardiogr.* 13 (6):608–614.

Lanza, G. M., X. Yu, P. M. Winter, D. R. Abendschein, K. K. Karukstis, M. J. Scott, L. K. Chinen, R. W. Fuhrhop, D. E. Scherrer, and S. A. Wickline. 2002. Targeted antiproliferative drug delivery to vascular smooth muscle cells with a magnetic resonance imaging nanoparticle contrast agent: Implications for rational therapy of restenosis. *Circulation* 106 (22):2842–2847.

Lee, R. J. and P. S. Low. 1995. Folate-mediated tumor-cell targeting of liposome-entrapped doxorubicin in-vitro. *Biochim. Biophys. Acta—Biomembranes* 1233 (2):134–144.

Li, F., W. G. Liu, and K. D. Yao. 2002. Preparation of oxidized glucose-crosslinked N-alkylated chitosan membrane and in vitro studies of pH-sensitive drug delivery behaviour. *Biomaterials* 23 (2):343–347.

Li, N., C. Sioutas, A. Cho, D. Schmitz, C. Misra, J. Sempf, M. Y. Wang, T. Oberley, J. Froines, and A. Nel. 2003. Ultrafine particulate pollutants induce oxidative stress and mitochondrial damage. *Environ. Health Perspect.* 111 (4):455–460.

Liang, H. D. and M. J. Blomley. 2003. The role of ultrasound in molecular imaging. *Br. J. Radiol.* 76 (Suppl 2):S140–S150.

Lim, D. W., Y. I. Yeom, and T. G. Park. 2000. Poly(DMAEMA-NVP)-b-PEG-galactose as gene delivery vector for hepatocytes. *Bioconjugate Chem.* 11 (5):688–695.

Linkov, I., F. K. Satterstrom, and L. M. Corey. 2008. Nanotoxicology and nanomedicine: Making hard decisions. *Nanomedicine* 4 (2):167–171.

Liu, D. X., F. Liu, and Y. K. Song. 1995. Recognition and clearance of liposomes containing phosphatidylserine are mediated by serum opsonin. *Biochim. Biophys. Acta—Biomembranes* 1235 (1):140–146.

Loo, C., A. Lowery, N. Halas, J. West, and R. Drezek. 2005. Immunotargeted nanoshells for integrated cancer imaging and therapy. *Nano Lett.* 5 (4):709–711.

Lowman, A. M., M. Morishita, M. Kajita, T. Nagai, and N. A. Peppas. 1999. Oral delivery of insulin using pH-responsive complexation gels. *J. Pharm. Sci.* 88 (9):933–937.

Lu, Y. and S. C. Chen. 2004. Micro and nano-fabrication of biodegradable polymers for drug delivery. *Adv. Drug Deliv. Rev.* 56 (11):1621–1633.

Lu, W., C. Xiong, G. Zhang, Q. Huang, R. Zhang, J. Z. Zhang, and C. Li. 2009. Targeted photothermal ablation of murine melanomas with melanocyte-stimulating hormone analog-conjugated hollow gold nanospheres. *Clin. Cancer Res.* 15 (3):876–886.

Lungwitz, U., M. Breunig, T. Blunk, and A. Gopferich. 2005. Polyethylenimine-based non-viral gene delivery systems. *Eur. J. Pharm. Biopharm.* 60 (2):247–266.

Maeda, H., G. Y. Bharate, and J. Daruwalla. 2009. Polymeric drugs for efficient tumor-targeted drug delivery based on EPR-effect. *Eur. J. Pharm. Biopharm.* 71 (3):409–419.

Mahkam, M. 2005. Using pH-sensitive hydrogels containing cubane as a crosslinking agent for oral delivery of insulin. *J. Biomed. Mater. Res. B* 75B (1):108–112.

Martin, F. J. and D. Papahadjopoulos. 1982. Irreversible coupling of immunoglobulin fragments to preformed vesicles—An improved method for liposome targeting. *J. Biol. Chem.* 257 (1):286–288.

Mastrobattista, E., G. A. Koning, L. van Bloois, A. C. S. Filipe, W. Jiskoot, and G. Storm. 2002. Functional characterization of an endosome-disruptive peptide and its application in cytosolic delivery of immunoliposome-entrapped proteins. *J. Biol. Chem.* 277 (30):27135–27143.

Matsumura, Y. and H. Maeda. 1986. A new concept for macromolecular therapeutics in cancer-chemotherapy—Mechanism of tumoritropic accumulation of proteins and the antitumor agent SMANCS. *Cancer Res.* 46 (12):6387–6392.

McLaughlin, P., A. J. Grillo-Lopez, B. K. Link, R. Levy, M. S. Czuczman, M. E. Williams, M. R. Heyman et al. 1998. Rituximab chimeric anti-CD20 monoclonal antibody therapy for relapsed indolent lymphoma: Half of patients respond to a four-dose treatment program. *J. Clin. Oncol.* 16 (8):2825–2833.

Mikhaylova, M., I. Stasinopoulos, Y. Kato, D. Artemov, and Z. M. Bhujwalla. 2009. Imaging of cationic multifunctional liposome-mediated delivery of COX-2 siRNA. *Cancer Gene Ther.* 16 (3):217–226.

Napoli, A., M. J. Boerakker, N. Tirelli, R. J. M. Nolte, N. A. J. M. Sommerdijk, and J. A. Hubbell. 2004. Glucose-oxidase based self-destructing polymeric vesicles. *Langmuir* 20 (9):3487–3491.

Nasongkla, N., E. Bey, J. Ren, H. Ai, C. Khemtong, J. S. Guthi, S. F. Chin, A. D. Sherry, D. A. Boothman, and J. Gao. 2006. Multifunctional polymeric micelles as cancer-targeted, MRI-ultrasensitive drug delivery systems. *Nano Lett.* 6 (11):2427–2430.

Nel, A., T. Xia, L. Madler, and N. Li. 2006. Toxic potential of materials at the nanolevel. *Science* 311 (5761):622–627.

Nelson, J. L., B. L. Roeder, J. C. Carmen, F. Roloff, and W. G. Pitt. 2002. Ultrasonically activated chemotherapeutic drug delivery in a rat model. *Cancer Res.* 62 (24):7280–7283.

Oberdorster, G., J. Ferin, and B. E. Lehnert. 1994. Correlation between particle-size, in-vivo particle persistence, and lung injury. *Environ. Health Perspect.* 102 (Suppl 5):173–179.

Pan, X. Q., H. Q. Wang, and R. J. Lee. 2003. Antitumor activity of folate receptor-targeted liposomal doxorubicin in a KB oral carcinoma murine xenograft model. *Pharm. Res.* 20 (3):417–422.

Panyam, J. and V. Labhasetwar. 2003. Biodegradable nanoparticles for drug and gene delivery to cells and tissue. *Adv. Drug Deliv. Rev.* 55 (3):329–347.

Pitt, W. G., G. A. Husseini, and B. J. Staples. 2004. Ultrasonic drug delivery—A general review. *Expert Opin. Drug Deliv.* 1 (1):37–56.

Qi, W., X. H. Yan, J. B. Fei, A. H. Wang, Y. Cui, and J. B. Li. 2009. Triggered release of insulin from glucose-sensitive enzyme multilayer shells. *Biomaterials* 30 (14):2799–2806.

Ramanan, R. M. K., P. Chellamuthu, L. P. Tang, and K. T. Nguyen. 2006. Development of a temperature-sensitive composite hydrogel for drug delivery applications. *Biotechnol. Prog.* 22 (1):118–125.

Reddy, J. A., C. Abburi, H. Hofland, S. J. Howard, I. Vlahov, P. Wils, and C. Leamon. 2002. Folate-targeted, cationic liposome-mediated gene transfer into disseminated peritoneal tumors. *Gene Ther.* 9 (22):1542–1550.

Risom, L., P. Moller, and S. Loft. 2005. Oxidative stress-induced DNA damage by particulate air pollution. *Mutat. Res.* 592 (1–2):119-137.

Sato, Y., K. Murase, J. Kato, M. Kobune, T. Sato, Y. Kawano, R. Takimoto, K. Takada, K. Miyanishi, T. Matsunaga, T. Takayama, and Y. Niitsu. 2008. Resolution of liver cirrhosis using vitamin A-coupled liposomes to deliver siRNA against a collagen-specific chaperone. *Nat. Biotechnol.* 26 (4):431–442.

Saul, J. M., A. V. Annapragada, and R. V. Bellamkonda. 2006. A dual-ligand approach for enhancing targeting selectivity of therapeutic nanocarriers. *J. Control. Release* 114 (3):277–287.

Savic, R., L. B. Luo, A. Eisenberg, and D. Maysinger. 2003. Micellar nanocontainers distribute to defined cytoplasmic organelles. *Science* 300 (5619):615–618.

Sayes, C. M., R. Wahi, P. A. Kurian, Y. P. Liu, J. L. West, K. D. Ausman, D. B. Warheit, and V. L. Colvin. 2006. Correlating nanoscale titania structure with toxicity: A cytotoxicity and inflammatory response study with human dermal fibroblasts and human lung epithelial cells. *Toxicol. Sci.* 92 (1):174–185.

Schwartzberg, A. M., T. Y. Olson, C. E. Talley, and J. Z. Zhang. 2006. Synthesis, characterization, and tunable optical properties of hollow gold nanospheres. *J. Phys. Chem. B* 110 (40):19935–19944.

Senior, J. H. 1987. Fate and behavior of liposomes in vivo—A review of controlling factors. *CRC Crit. Rev. Ther. Drug Carrier Syst.* 3 (2):123–193.

Shangguan, T., D. Cabral-Lilly, U. Purandare, N. Godin, P. Ahl, A. Janoff, and P. Meers. 2000. A novel N-acyl phosphatidylethanolamine-containing delivery vehicle for spermine-condensed plasmid DNA. *Gene Ther.* 7 (9):769–783.

Shangguan, T., C. C. Pak, S. Ali, A. S. Janoff, and P. Meers. 1998. Cation-dependent fusogenicity of N-acyl phosphatidylethanolamine. *Biochim. Biophys. Acta—Biomembranes* 1368 (2):171–183.

Shive, M. S. and J. M. Anderson. 1997. Biodegradation and biocompatibility of PLA and PLGA microspheres. *Adv. Drug Deliv. Rev.* 28 (1):5–24.

Singh, S. R., H. E. Grossniklaus, S. J. Kang, H. F. Edelhauser, B. K. Ambati, and U. B. Kompella. 2009. Intravenous transferrin, RGD peptide and dual-targeted nanoparticles enhance anti-VEGF intraceptor gene delivery to laser-induced CNV. *Gene Ther.* 16 (5):645–659.

Song, Y. K., F. Liu, S. Y. Chu, and D. X. Liu. 1997. Characterization of cationic liposome-mediated gene transfer in vivo by intravenous administration. *Hum. Gene Ther.* 8 (13):1585–1594.

Sun, C., J. S. Lee, and M. Zhang. 2008. Magnetic nanoparticles in MR imaging and drug delivery. *Adv. Drug Deliv. Rev.* 60 (11):1252–1265.

Sunamoto, J., T. Sato, T. Taguchi, and H. Hamazaki. 1992. Naturally-occurring polysaccharide derivatives which behave as an artificial cell-wall on an artificial cell liposome. *Macromolecules* 25 (21):5665–5670.

Suzuki, A. and T. Tanaka. 1990. Phase-transition in polymer gels induced by visible-light. *Nature* 346 (6282):345–347.

Szoka, F. and D. Papahadjopoulos. 1978. Procedure for preparation of liposomes with large internal aqueous space and high capture by reverse-phase evaporation. *Proc. Natl. Acad. Sci. U.S.A.* 75 (9):4194–4198.

Tai, L.-A., P.-J. Tsai, Y.-C. Wang, Y.-J. Wang, L.-W. Lo, and C.-S. Yang. 2009. Thermosensitive liposomes entrapping iron oxide nanoparticles for controllable drug release. *Nanotechnology* 20:135101–135109.

The Royal Society & The Royal Academy of Engineering. 2004. *Nanoscience and Nanotechnologies: Opportunities and Uncertainties*. Latimer Trend Ltd, Plymouth, U.K.

United States Environmental Protection Agency. 2007. *Nanotechnology White Paper*. Washington, DC.

United States Food and Drug Administration. 2007. *Nanotechnology*. Rockville, MD.

Vivero-Escoto, J. L., I. I. Slowing, C. W. Wu, and V. S. Lin. 2009. Photoinduced intracellular controlled release drug delivery in human cells by gold-capped mesoporous silica nanosphere. *J. Am. Chem. Soc.* 131 (10):3462–3463.

Volodkin, D. V., A. G. Skirtach, and H. Mohwald. 2009. Near-IR remote release from assemblies of liposomes and nanoparticles. *Angew. Chem. Int. Ed.* 48 (10):1807–1809.

Wang, Z., W. K. Chui, and P. C. Ho. 2009. Design of a multifunctional PLGA nanoparticulate drug delivery system: Evaluation of its physicochemical properties and anticancer activity to malignant cancer cells. *Pharm. Res.* 26 (5):1162–1171.

Wang, Y. X., S. M. Hussain, and G. P. Krestin. 2001. Superparamagnetic iron oxide contrast agents: Physicochemical characteristics and applications in MR imaging. *Eur. Radiol.* 11 (11):2319–2331.

Weng, K. C., C. O. Noble, B. Papahadjopoulos-Sternberg, F. F. Chen, D. C. Drummond, D. B. Kirpotin, D. H. Wang, Y. K. Hom, B. Hann, and J. W. Park. 2008. Targeted tumor cell internalization and imaging of multifunctional quantum dot-conjugated immunoliposomes in vitro and in vivo. *Nano Lett.* 8 (9):2851–2857.

Wu, X. S. and N. Wang. 2001. Synthesis, characterization, biodegradation, and drug delivery application of biodegradable lactic/glycolic acid polymers. Part II: Biodegradation. *J. Biomater. Sci. Polym. Ed.* 12 (1):21–34.

Xu, Y. H. and F. C. Szoka. 1996. Mechanism of DNA release from cationic liposome/DNA complexes used in cell transfection. *Biochemistry* 35 (18):5616–5623.

You, J. O. and D. T. Auguste. 2008. Feedback-regulated paclitaxel delivery based on poly(N,N-dimethylaminoethyl methacrylate-co-2-hydroxyethyl methacrylate) nanoparticles. *Biomaterials* 29 (12):1950–1957.

You, J. O., S. B. Park, H. Y. Park, S. Haam, C.H. Chung, and W. S. Kim. 2001. Preparation of regular sized Ca-alginate microspheres using membrane emulsification method. *J. Microencapsul.* 18 (4):521–532.

Zhang, C. L., N. Tang, X. J. Liu, W. Liang, W. Xu, and V. P. Torchilin. 2006. siRNA-containing liposomes modified with polyarginine effectively silence the targeted gene. *J. Control. Release* 112 (2):229–239.

12

Nanomaterials for Therapeutic Drug Delivery

Dinesh Jegadeesan
University of Toronto

M. Eswaramoorthy
Jawaharlal Nehru Centre for Advanced Scientific Research

12.1 Introduction to Therapeutic Drug Delivery

The use of nanomaterials for the therapeutic delivery of drugs is by itself an established concept. The analogy could be traced back to the Bronze Age stories of Trojan War[1] during which the Greeks, in a desperate attempt to enter the city of Troy, built a huge wooden structure resembling a horse and secretly loaded it with armed soldiers in its interior. The *Trojan Horse*, mistaken by its artistic exterior, was given a free access into Troy, which eventually led to its capture. Since then, the term *Trojan Horse* refers to a stratagem that causes the target to invite a foe into its protected circle. In the context of drug delivery, nanomaterials are often used as Trojan horses that are loaded with drugs or genes or proteins that are aimed at destroying/repairing an infected cell. However, the major challenge lies in their safe and specific delivery to the infected cells without affecting the healthy neighboring cells. In addition to this, there are various natural defense barriers in our body, which consider any such attempts of delivery as foreign invasion and trigger hypersensitive responses.

On the other hand, viruses and bacteria have evolved to become the most sophisticated nanocarriers, which are known to surpass the natural defense barriers to deliver their genetic material inside the cells. A few illustrations would substantiate the point. Our body possesses certain types of cells called macrophages, which freely circulate in the bloodstream. Their main role is to recognize and engulf foreign particles (e.g., pathogens) that circulate in the blood. However, in certain cases like mycobacterium (which causes tuberculosis) is evolved to replicate and thrive specifically inside such macrophages. For example, lysosomes, which are membrane-bound organelles, contain acid hydrolases that are highly deleterious to most organic molecules. Viruses, like influenza, are extremely efficient in taking advantage of the acidity in endosomes. Influenza virus has developed surface proteins that remain inactive until placed in an acidic environment where the protonation of the carboxylic acid residues induces a conformational change from a random coil structure to an α-helical structure.[2] The conformational change leads to the extension of several hydrophobic regions capable of inserting themselves into lipid bilayers and mediating the transport of viral contents across the membrane.[3] Though the strategies adopted by viruses and bacteria may be exemplary, their complete emulation to nano drug delivery systems is tedious and less practical. Using viruses poses high risk factors of ethical concern because of which developing nonviral carriers based on synthetic materials have become an essential alternative. Nevertheless, the lessons learnt from viruses and bacteria prescribe that an efficient carrier capable of protecting the cargo (drugs) from the harsh acidic and enzymatic environment and still specifically targeting the infected tissues without getting engulfed by the macrophages (and other types of cells) is essential to realize enhanced therapeutic results.

12.2 Therapeutic Drug Delivery—Challenges Involved

The inefficient delivery of drugs and biomolecules to a specific location may be attributed to the following reasons: (a) the hostile environmental conditions present in blood circulation and inside the cells, (b) nature of the drug (may be highly hydrophobic or highly cytotoxic or lack of selectivity for target tissues), and (c) poor pharmacokinetics (rapid clearance).

A number of materials based on lipids, polymers, inorganic materials have been developed in the past in order to have better pharmacokinetics and biodistribution of the drugs compared to the free drugs. The search for such materials in the nanometer regime is also logical because of many desirable properties that come with the reduced size like high surface-to-volume ratio, access to multiple functionalities, tunability of properties, and controlled synthesis. But there are stringent requirements that need to be fulfilled by any carrier for successful commercialization.

Some of the desired features of a drug carrier are[4] (a) high loading efficiency of more than 80 wt%, (b) biodegradability, (c) small size (<5 μm diameter), and (d) slow clearance from the blood. Flexibility in the release kinetics is also important in choosing a carrier. For example, in the treatment of diseases like diabetes, a lot of thrust is being put on systems where there can be a gradual release of the drugs. On the other hand, a burst release of large amount of drugs is desirable in cancer therapy. pH, light, magnetic field, radiofrequency are some of the external stimuli, which can be used to noninvasively trigger the release mechanisms. Further challenges that can be addressed by research include the search for the best delivery system, which can retain the potency of a particular drug molecule, the stability of the drug during incorporation procedures, the advantages of using a drug delivery system conjugates over a free drug; carrier toxicity, integrated diagnosis and therapy, etc.

12.3 Drug Delivery Systems—Classification

There are a wide range of drug delivery systems. The classification of the drug delivery systems as shown in Figure 12.1 is based on the type of the material composition of the carrier and their size. The blue region includes materials in the nanometer regime (including size lesser than 500 nm in diameter). The red region includes submicron and millimeter-sized particles. Some excellent reviews on various types of drug carriers can be found in the references.[5,6] An overview of each type of nanocarriers is given in the following sections. Table 12.1 shows the various stages of development of different types of nanocarriers.

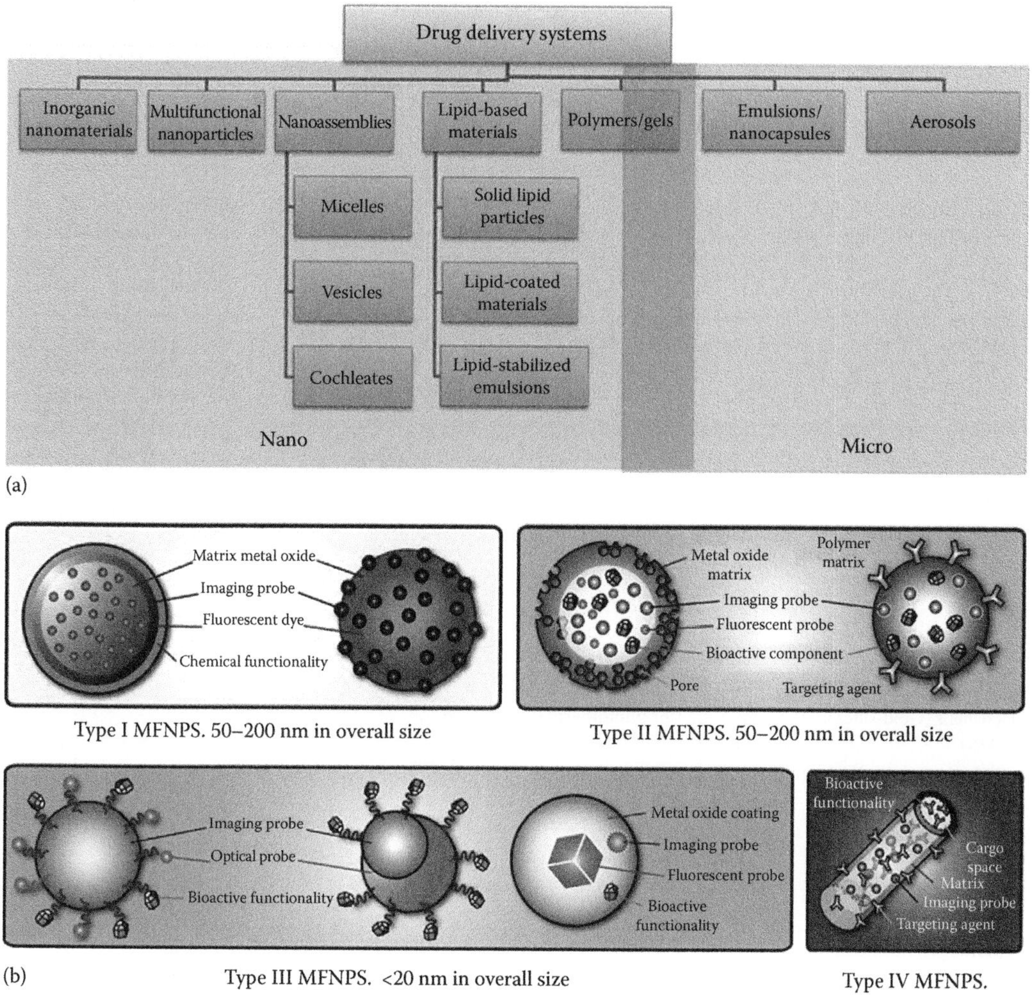

FIGURE 12.1 (a) Classification of drug delivery vehicles. (b) Categorization of multifunctional nanoparticles. (From Suh, W.H. et al., *Nano Today*, 4, 27, 2009.)

12.3.1 Polymers and Hydrogels

Polymers are the most commonly explored materials for constructing nanoparticle-based drug carriers. One of the earliest reports of polymer nanoparticle use has been described by Couvreur et al. in 1979, explaining the adsorption of anticancer drug to polyalkylcyanoacrylate onto the polymer and its release and tissue distribution.[7,8] Polymeric nanoparticles can be derived from synthetic polymers like poly(lactic acid) and poly(lactic co-glycolic acid). Natural polymers include cellulose, starch, chitosan, or collagen, which may be used without any chemical modification. Advances in polymer chemistry have led to the synthesis of a number of varieties of polymer nanoparticles with interesting applications.[9,10] Synthetic dendrimers are another class of organic nanoparticles that have been used in targeted drug delivery. Dendrimers can be synthesized in multiple ways.[11] Owing to their large number of surface groups, dendrimers have the ability to create multivalent interactions.[12] Dendritic structures may also be engineered to encapsulate certain hydrophobic drugs like indomethacin.[13,14] The rich knowledge in polymer chemistry can be used to overcome some of its current drawbacks on uncontrolled release of the drugs.[15]

TABLE 12.1 Stages of Development of Various Types of Nanocarriers

Type of Carrier and Mean Diameter (nm)	Drug Entrapped or Linked	Current Stage of Development	Type of Cancer (for Clinical Trials)
Polymer–drug conjugates (6–15)	Doxorubicin, paclitaxel, camptothecin, platinate, TN P-470	Twelve products under clinical trials (Phases I–III) and in vivo	Various tumors
Liposomes (both PEG and non-PEG coated) (85–100)	Lurtotecan, platinum compounds, annamycin	Several products in clinical trials (Phases I–III) and in vivo	Solid tumors, renal cell carcinoma, mesothelioma, ovarian and acute lymphoblastic leukaemia
Polymeric nanoparticles (50–200)	Doxorubicin, paclitaxel, platinum-based drugs, docetaxel	Several products are in clinical trials (Phases I–III) and in vivo	Adenocarcinoma of the esophagus, metastatic breast cancer and acute lymphoblastic leukemia
Polymersomes (~100)	Doxorubicin, paclitaxel	In vivo	
Micelles (lipid based and polymeric) (5–100)	Doxorubicin	Clinical trials (Phase I)	Metastatic or recurrent solid tumors refractory to conventional chemotherapy
	Paclitaxel	Clinical trials (Phase I)	Pancreatic, bile duct, gastric and, colonic cancers
	Platinum-based drugs (carboplatin/cisplatin), camptothecin, tamoxifen, epirubicin	In vivo and in vitro	
Nanoshells (Gold-silica) (~130)	No drug (for photothermal therapy)	In vivo	
Gold nanoparticles (10–40)	No drug (for photothermal ablation)	In vivo	
Nanocages (30–40)	No drug chemistry, structural analysis	In vitro	
Dendrimers (~5)	Methotrexate	In vitro/in vivo	
Immuno-PEG-liposomes (100)	Doxorubicin	Clinical trials (Phase I)	Metastatic stomach cancer
Immunoliposomes (100–150)	Doxorubicin, platinum-based drugs, vinblastin, vincristin, topotecan, paclitaxel	In vivo	
Immunotoxins, Immunopolymers, and fusion proteins (3–15)	Various drugs, toxins	Clinical trials (Phases I–III)	Various types of cancer

Source: Modified from Torchilin, V.P., Ed., *Nanoparticulates as Drug Carriers*, Imperial College Press, London, U.K., 2006.

Hydrogels are three-dimensional networks composed of hydrophilic polymer chains. They have the ability to swell in water without dissolving. The type of cross-linking between the polymer chains can be chemical (covalent bonds) or physical (hydrogen bonds or hydrophobic interactions). The high water content in these materials makes them highly biocompatible. There are natural hydrogels such as DNA, proteins, or synthetic (e.g., poly(2-hydroxyethyl methacrylate), poly(*N*-isopropylacrylamide)) or a biohybrid.[16,17] The release mechanism can be induced by temperature or pH. Temperature-controlled release is due to the competition between hydrogen bonding and hydrophobic interactions. At lower temperatures, the hydrogen bonding between polar groups of the polymer is predominant causing the polymer to swell in water. At higher temperatures, the hydrophobic interactions takeover leading to its shrinkage.[18,19] Glucose-sensitive hydrogels can release insulin in a controlled fashion in response to the demand.[20]

12.3.2 Lipid-Based Materials

Lipids are esters of glycerol and fatty acids and therefore are natural surfactants. Their association with the living systems is also well established. The use of solid lipids or conjugated lipids as nanoparticles or in the form of coatings is therefore considered as biocompatible.[6] However, making the right choice of lipids is important to avoid any side effects. Liposomes, which are formed by the assembly of lipids, have been used in cancer therapy for a long time.[21] Lipids can also be used to stabilize microemulsions or gas bubbles in water when the drugs are hydrophobic.[22] Lipid-coated microbubbles can also be used as blood substitute in oxygen delivery. They are also used as contrast agents in ultrasound imaging.[22] Lipoproteins are another preferred class of drug carriers[23] as they do not trigger any immunological response and can be used to carry hydrophobic and cytotoxic agents without affecting the native structure of the lipoprotein.[24] As most of the cancerous cells possess higher low-density lipoprotein receptors on their cell surface, using them may be a great advantage in cancer therapy.

Solid lipid particles[25–27] are currently tested as nanocarriers for both in vitro and in vivo applications. They are usually prepared by various procedures like high pressure homogenization,[28] microemulsion,[29] and nanoprecipitation. These partially crystalline lipid particles are composed of a mixture of glycerides with different fatty acids possessing various chain length and degree of unsaturation. The rearrangement of the less stable α-crystalline structure to a more stable and ordered β-crystalline structure on standing[30] leads to an expulsion of the active substances into the amorphous regions.[31,28] A second type of lipid particles, called multiple lipid particles, is obtained by mixing liquid lipids with solid lipids when preparing the nanoparticles. However, the polymorphism of these lipid matrices and possible crystal rearrangements must to be controlled to improve the stability of these structures (gelification problems[32]). Moreover, the release of the active molecules incorporated into these solid nanoparticles is not always well controlled, which limits their applications. Nanocapsules are also recently developed materials wherein a lipidic core is surrounded by a shell of polyethylene glycol.[33] Many lipid-based nanocarriers have already reached the market.[9]

12.3.3 Nanoassemblies

Self-assembly of biocompatible molecules into regular supramolecular structures has an important implication. The molecular aggregates can range from simple sphere shape as in case of micelles to more complex structures like bilayers, vesicles, and other liquid crystalline phases. Though a number of factors like concentration, temperature, pH, electrolytes influence the formation of a specific type of aggregate structure, there has been a rationale to predict them based on the chemical structure of the amphiphilic molecule. The controlling parameter in the formation of molecular aggregates is called critical packing parameter (CPP), which is given by the term v/al, where "v" is the volume of the hydrophobic portion of the molecule, "a" is the interfacial area, and "l" is the chain length of the hydrophobic tail. A summary of the structures to be expected from molecules falling into various categories is given in Table 12.2. In Figure 12.2, cartoon representation of various nanoassemblies is presented.

12.3.3.1 Micelles

Biodegradable amphiphilic block copolymers assemble into spherical micelles approximately 10–80 nm in diameter with core-shell architecture.[34] A micelle consists of a hydrophobic core for loading the drug and a hydrophilic shell that provides steric hindrance against aggregation, opsonization, and protein binding. Micelles can be formed by the self-assembly of surfactant molecules, which remain dynamic by continuous exchange of molecules between the assemblies and bulk solution. Micelles can also form by the self-assembly of the block copolymers (polymers containing both hydrophobic and hydrophilic units). These structures are much less dynamic than that formed by surfactant molecules, and hence, they have better mechanical stability. FDA has approved the use of hydrophilic groups like poly(ethylene glycol) (PEG) and poly(ethylene oxide) (PEO) in micelles for drug delivery. Different

TABLE 12.2 Expected Aggregate Characteristics in Relation to Surfactant Critical Packing Parameter, *v*/al

Critical Packing Parameter (CPP)	General Surfactant Type	Expected Aggregate Structure
<0.33	Simple surfactants with single chains and relatively large head groups	Spherical or ellipsoidal micelles
0.33–0.5	Simple surfactants with relatively small head groups, or ionics in the presence of large amounts of electrolyte	Cylindrical or rod-shaped micelles
0.5–1.0	Double-chain surfactants with large head groups and flexible chains	Vesicles and flexible bilayers
1.0	Double-chain surfactants with small head groups or rigid immobile chains	Planar extended bilayers
>1.0	Double-chain surfactants with small head groups, very large and bulky hydrophobic groups	Reversed or inverted micelles

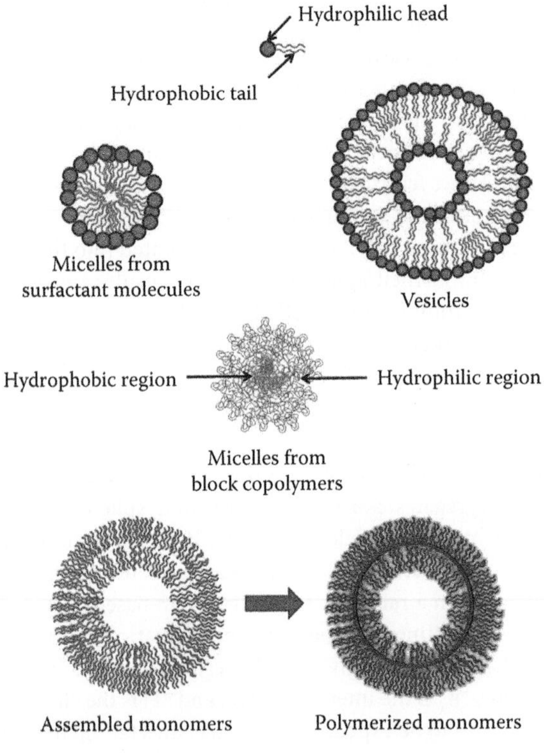

FIGURE 12.2 Cartoon representation of nanoassemblies.

types of hydrophobic groups have been developed over a period of time. They are poly(L-amino acids), biodegradable polyesters of poly(glycolic acid), poly(D-lactic acid), poly(D,L-lactic acid), copolymers of lactide/glycolide and poly(ε-caprolactone), phospholipids and poly(propylene oxide). The choice of hydrophobic group is dictated by the stability of the micelle and drug-compatibility.

12.3.3.2 Vesicles—Polymerosomes, Niosomes, Discomes, and Lipososmes

The spontaneous organization of amphiphilic molecules in a solvent creates compartments encapsulating a definite volume within it.[35] These structures besides being biomimetic (like a membrane-bound cell or a cell organelle) are also valuable for the delivery of drugs, genes, and proteins. Unlike polymeric micelles, the synthesis of vesicles needs input of energy.[36] Polymeric vesicles may be formed from the self-assembly of the polymers or from the polymerization of the assembled monomers. Such vesicles are generally termed as polymerosomes. The thickness of the vesicle is determined by the degree of polymerization of the hydrophobic groups. And the extra thick membranes provide the vesicles with exceptional stability to mechanical stress.[37,38] For example, the polymerosomes produced from Poly(ethylene oxide)-*block*-poly[3-(trimethoxysilyl)propyl methacrylate] in water, methanol, and triethylamine mixtures are stable up to 1 year.[39] They are also less leaky,[40] thermostable,[41] and can be isolated as dry powders.[42] Some new interesting properties also emerge when polymers form vesicles. For example, poly(L-lysine) possesses reduced cytotoxicity when it forms vesicles.[43]

Vesicle formation in nonionic surfactant molecules (Niosomes) is observed when a CPP of the surfactants is between 0.5 and 10.[1644,45] Niosomes are 30–120 μm in size and often they are surface stabilized using minor quantities of ionic molecules. Other variations, like polyhedral vesicles and giant vesicles (discomes) are formed when hexadecyl diglycerol ether, Solulan C24, and cholesterol are added. Of these, discomes are thermoresponsive due to possible gel-liquid phase transition at 37°C, which makes them suitable for temperature-controlled release of drugs. The vesicles that are formed due to the self-assembly of phospholipids are termed as liposomes. Liposomes are very identical in composition to the natural cell membranes.

12.3.3.3 Cochleates

The negatively charged phospholipid vesicles bind with divalent cations like Ca^{2+} to form precipitates called cochleates. These are nanometer-sized cigar-like rolled structures.[46–49] The formation of cochleate is easier from small unilamellar vesicles, while in case of multi-lamellar vesicles, the Ca^{2+} first disrupts the membrane architecture and then reorganizes into cochleate.

12.3.4 Inorganic Nanomaterials

Inorganic nanocarriers are of many types. In particular, a variety of metals, metal oxides, and semiconducting nanoparticles offer optical and magnetic properties that can be tuned with their size. Many of the recent applications that aim at simultaneous diagnosis and therapy are most likely to use inorganic nanomaterials. However, they also have quite a few serious limitations, the first being their toxicity and the second is the inevitable surface modification, which can increase the processing cost. For more details, reviews can be referred.[50]

12.3.4.1 Metal Nanoparticles

Gold nanoparticles are preferred materials in biomedical applications because of their biocompatibility and optical properties.[51] The plasmon absorption of the gold nanoparticles can be tuned from visible region (spherical nanoparticles) to near infrared (IR) (nanorods/nanocages) by tuning its shape.[49] Accordingly, their applications can also vary from optical imaging to photothermal therapy. The high affinity of Au with sulfur is well exploited in their surface functionalization. Also bimetallic nanorods of Ni/Au are also of interest because of the added magnetic functionality by Ni.[52] Besides gold nanomaterials, there have been interests on gadolinium nanoparticles, which are a potential agent for neutron capture therapy in cancer.[53]

12.3.4.2 Carbon Nanomaterials

Carbon nanomaterials for drug delivery applications mainly include fullerenes and carbon nanotubes (single and multiwalled). Considerable amount of work has been done to utilize them as nanocarriers for drug delivery.[54–56] The inert surface of these materials has posed challenges in terms of surface modifications to make them water soluble, biocompatible, and fluorescent. But despite all these, a number of recent reports establish that carbon nanotubes are toxic.[57] More recently glucose-derived functionalized carbon spheres[58] seem to present hopes as efficient nanocarriers. They have been shown to be nuclear targeting and nontoxic. But more detailed studies on their mechanism of entry and other possible applications are awaited.

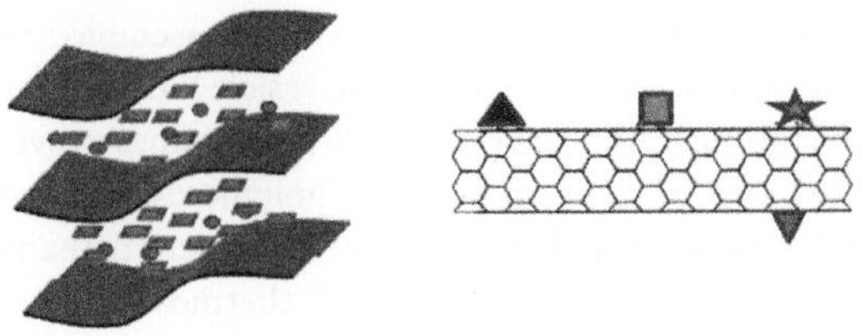

FIGURE 12.3 Cartoon representation of LDH and carbon nanotubes.

12.3.4.3 Layered Double Hydroxides

The use of layered double hydroxides (LDH) as nanocarriers is explained in Section 12.8. Figure 12.3 shows the cartoon representation of carbon nanotubes and LDH.

12.3.4.4 Oxide Nanoparticles

Silicon dioxide (amorphous silica)[59] and iron oxide nanoparticles (Fe_3O_4)[60] are the most commonly used oxide nanomaterials for drug delivery applications. Silica nanomaterials are not associated with any special properties except for the versatility in its surface functionalization. Mesoporous silica has a higher surface area and pore volume than their solid counterparts and, therefore, has higher drug carrying capacity. Traditionally believed to be biocompatible with an ability to dissolve by hydrolysis in acid and alkaline medium, these materials pose new challenges due to the recent reports of their impact on cellular respiration.[61] Iron oxide, on the other hand, has been proved to be very biocompatible especially when coated with polyethylene glycol. These nanoparticles are superparamagnetic and therefore used as MRI contrast agents. Besides, by specifically targeting them to tumors, one can expect to have an integrated imaging and therapy.[62]

12.3.5 Multifunctional Nanoparticles

In recent years, there is a steep rise in the use of multifunctional nanoparticles for biomedical applications. These materials could be hybrid of organic–inorganic or two different organic or polymer composites, but the necessary criteria being multiple functions like stimuliresponsive, optical, magnetic, tissue specificity, biocompatibility, etc. Recently, Suh et al.[63] have classified these materials into four types based on their size and structure. In brief, type I particles are spherical and nonporous, sub 200 nm particles (e.g., SiO_2 spheres loaded with Gd^{3+}), type II particles are spherical and porous particles in the size range of 50–200 nm (e.g., mesoporous materials, which incorporate fluorescent or magnetic nanoparticles). Type III nanoparticles lie in sub 20 nm regime and include a wide variety of core-shell spherical particles that can integrate many properties ($CdSe@Fe_3O_4$). Most of the nanoparticles are surface modified to have more functions like biocompatibility or tissue specificity. Type IV particles are those with high aspect ratio like tubes and rods. Metal-organic framework nanorods coated with SiO_2 are classified under this category.[64]

12.3.6 Emulsions and Nanocapsules

Emulsions are heterogenous dispersions of two immiscible liquids such as oil in water (O/W) or water in oil (W/O). Without surfactant molecules, they are susceptible for rapid degradation by coalescence or flocculation leading to phase separation.[65] The use of micro- and nanoemulsions are becoming increasingly common in drug delivery systems. Microemulsion is used to denote a thermodynamically stable, fluid, transparent (or translucent) dispersion of oil and water, stabilized by an interfacial film of amphiphilic molecules.[66] The striking difference between a conventional emulsion (1–10 μm) and the microemulsion (200 nm–1 μm) is that the latter does not need any mechanical input for their formation as they are thermodynamically more stable. On the other hand, nanoemulsions (20–200 nm) are at best kinetically more stable. Wretlind[67] developed the first intravenously injectable O/W emulsion as a source of nutrients for patients unable to feed themselves orally or unable to metabolize food. Since then, many hydrophobic drugs like diazepam,[68] antibiotics,[69] proteins,[70] and anticancer drugs[71] have been administered using this method. The drawback in emulsion system is the use of high concentration of surfactant, which leads to toxicity and embolism.[72]

12.3.7 Others

Other carriers usually used are not strictly classified in the nanometer regime. For example, many natural carriers like cells, cell ghosts, viruses, bacteria have been shown to be effective in certain cases of drug delivery.[6] Besides being in the nanometer to micrometer size ranges, they all suffer from the drawback of immunogenicity and rapid clearance from the circulation. Aerosols, which are tiny droplets of liquid dispersed in a jet of air, are used effectively in the treatment of respiratory diseases.

12.4 Drug Loading and Release

Following types of drug loadings are common in nanocarrier-based delivery systems. The type of loading procedure is adopted based on the nature of the nanocarrier used, nature of the cargo (its stability and chemical/physical properties), and, finally, the desired release profile. Figure 12.4 shows schematically the types of drug loading.

Surface linking of the cargo is either done by physical adsorption or by chemical covalent linking/complexing. In case of physical adsorption, the hydrophobic or electrostatic interactions between the

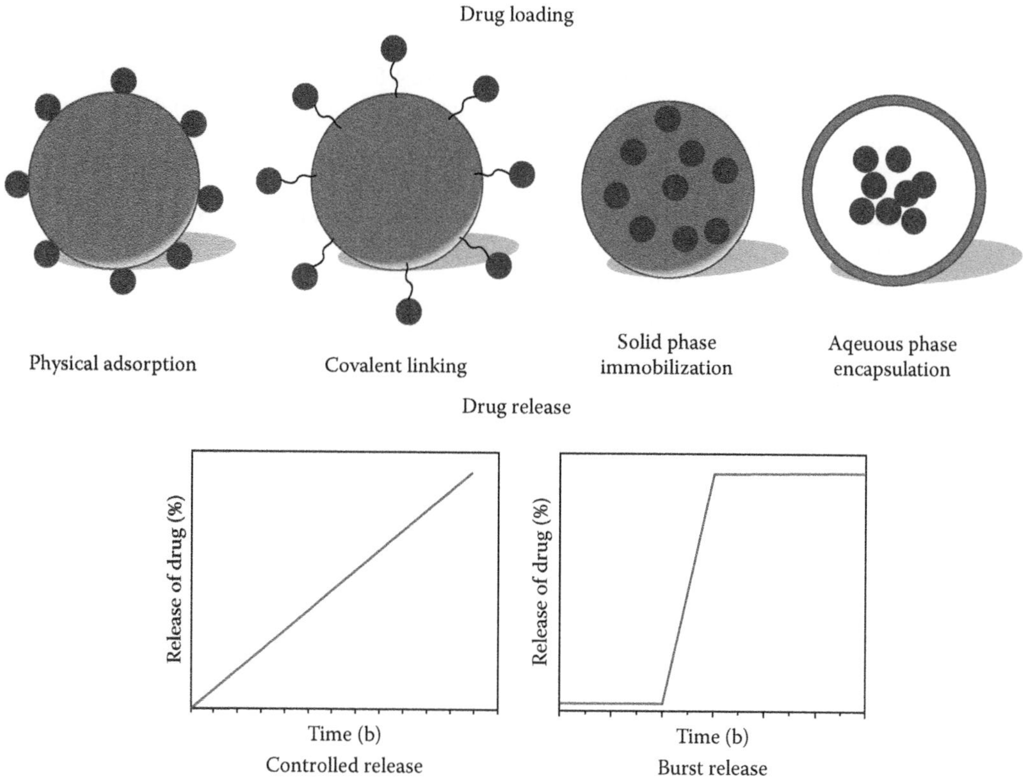

FIGURE 12.4 Cartoon representation of various forms of drug loading and release profiles. Blue spheres represent the nanocarriers and the red spheres represent the drugs (do not scale).

drug and the nanocarrier surface are exploited. On the other hand, the chemical linking involves suitable procedures involving click chemistry, amide chemistry, Au–S bond interaction, sulfhydryl chemistry, silane chemistry or more specific interactions involving biotin-streptavidin.[73] The basic necessity is that the drug molecule should be stable throughout the loading procedure without affecting its pharmacophore or in other words, nanocarrier-drug complex must remain therapeutically efficient. Linking of the drugs to the nanoparticle surface does not ensure safety to the cargo from the enzymes and other acidic conditions. Usually, cargos like proteins are preferred for such loading as they are naturally biocompatible during circulation. Often, the surface of the nanoparticles is modified to provide targeted delivery and enhanced circulation times.

Nanoassemblies such as vesicles, micelles, etc., usually possess a large encapsulated aqueous volume in which a desired drug molecule can be dissolved. Since the cargo in this case is in native aqueous environment, small sensitive molecules can remain active. Liposomes also afford effective loading of small hydrophobic molecules. Cochleates can encapsulate a variety of molecules of all shapes and sizes (hydrophobic or charged) with more preference toward hydrophobic molecules. As calcium induces dehydration of the interbilayer domains, the amount of water in this region is low; therefore, small hydrophilic molecules will not be suitable for cochleate structures.[74]

Solid-phase immobilization of drugs is an alternative strategy in which crystallized or lyophilized protein molecules are loaded as suspensions within the solid core of organic, hydrophobic nanoparticles.[75] The loading of DNA/RNA in precipitated calcium phosphate or carbonate nanoparticles, clay sheets, porous silica also belong to this category.[76]

In the context of drug release mechanisms, three types have been identified.[6] In the first case, the release of the drug from the carrier is slow. This can occur naturally by the gradual hydrolysis of the nanocarrier. Such releases maintain a sustained high local concentration of drugs. The second type of release involves a sudden burst of release of drugs, which is more preferred in the cancer therapy. Such release mechanisms can be externally actuated using magnetic field, IR rays, or radiofrequency. A plot of release against the time is graphically shown in Figure 12.4 for the first two types. The third type involves no actual release of cargo. In this case the substrate (H_2O_2, glucose, oxygen) diffuses into the nanovehicles containing enzymes where the reaction occurs.

12.5 Behavior of Nanoparticles In Vivo: Implications on Its Journey

The study of nanoparticles in the living systems is an important topic with immense relevance to their role as drug carriers. Their behavior is the outcome of the interaction of the nanoparticles/nanoparticles–drug conjugates with the proteins, electrolytes, and circulating cells present in the body fluid (blood, lymph, and interstitial fluid). The dynamics of the nanoparticles in the body fluid flow is a subject of recent study.[77] This topic assumes importance as, irrespective of the mode of administration of the nanoparticles to the body (oral or intravenous), the interaction between nanoparticles and fluid flow is inevitable. The fluid dynamics involved as the particles flow through the bloodstream also plays an important role in deciding the pharmacokinetics on the nanocarrier. It is also mandatory to know the fate of the drug carrier in terms of their removal, accumulation, and toxicity. For convenience, the term nanoparticle conjugates is used to refer to nanoparticles conjugated to drugs, fluorescent tags, and target ligands.

12.5.1 Nanoparticle Flow and Shape Effect

The flow of nanoparticles could be either convective or diffusive. The former occurs in the regions of high fluid velocity (e.g., arteries, veins), while the latter occurs in the regions of low fluid velocity (e.g., lymphatics, interstitial fluids, and tumor vasculature). In physical terms, the following situations could be considered: (i) particle flow in rapidly flowing bloodstream can cause aggregation or deposition, (ii) effect of shear forces on the particles that adhere to the walls of the capillaries. Both of these situations can vary as a function of particle size, and the complications arise because the sources of polydispersity in nanoparticle size are many and usually uncontrolled. For example, nanoparticles used usually come in a range of sizes, and when conjugated with the drugs, the polydispersity in size is further enhanced. In case of interaction of the nanoparticles with the blood constituents (adsorption of albumin, IgG, and fibrinogen from the blood on to the hydrophobic surface of the particle or adsorption of the nanoparticles on the erythrocytes) the dynamics of flow is decided by the size of the entire complex. All these can result in segregation during flow or differential migration leading to concentration differences and nonuniform distribution of the nanoparticles and their drug conjugates. Nanoparticles that adhere to the walls of the tissue can be removed by the blood flow. Factors that affect this phenomenon are flexibility of the nanoparticles and the velocity of the blood. While the former is an intrinsic property of the nanoparticle, the latter varies with the type of the vessel (12 mm s^{-1} in venules and 30 mm s^{-1} in arterioles for a similar diameter of 50 μm),[78] diameter of the vessel (in arterioles, the blood velocity can decrease to 5 mm s^{-1} for a diameter of 10 μm), and also along the diameter with a maximum velocity at the centre of the tube. The flexibility of the nanoparticle concerned plays a crucial role at the bifurcations and pores in the blood vessels. The narrowest blood vessel could be on the order of a few micrometers, which is big enough to permit the free flow of nanosystems. However, after extravasation, the passage of nanoparticles shifts to cellular networks (interstitial spaces) where the fluid flow is static. The diffusion of particles in this regime is analogous to particle movement in porous networks. A diagrammatic representation of the areas of flow of nanoparticles is given below. The works of Alexander T. Florence and others may be referred for more information on the nanoparticle flow through the blood vessels.[79–81] The fate of nanoparticles in flowing blood, their adhesion, extravasation, and permeation into the target tissue depends on a number of factors like particle diameter, surface ligand density, orientation, shape, capillary diameter, bifurcations, rugosity, and viscosity.

Nanosystems can be prepared in a variety of shapes like rods, tubes, toroids, discs, cups, etc. In some cases, the shape can present advantages of enhanced cargo loading or even unique magnetic or optical properties.[82] Therefore, the study of effect of shape on the in vivo flow of nanoparticles has been the interest of many researchers.[83–85] A recent work, which was inspired by the fact that a number of infectious viruses are also filamentous, reiterates the shape effect on enhanced blood circulation.[86] Synthetically designed copolymers of PEG and polyethylethylene or polycaprolactone form filamentous micelles, which have been shown to retain in the circulation for a longer period than their spherical counterparts. They evade the macrophages that circulate in the blood by using the blood flow to relax the filaments from coiled to expanded structure. This results in the micelles having minimal contact with the circulating macrophages. Ironically, there are some of the naturally occurring deadly viruses like Ebola and Marburg, which though being filamentous are suspected to interact most strongly with phagocytic cells causing hemorrhagic fevers. But in these cases, more complex signaling reactions must be involved besides the shape factor alone.

12.5.2 Receptor Effect

Most often the surface of the nanosystems are modified with special ligands, which can specifically bind to the receptors present on a tissue. This modification is done in addition to the primary cargo, which the nanosystems are expected to deliver. The attachment to the receptors can occur both in static and dynamic flow conditions of the body fluid, depending on the location of the target tissue. Both the situations are complex and many factors like diffusion, convection, geometrical interception, migration under gravity, and tangential interaction play a role in the successful attachment.[87,88]

12.5.3 EPR Effect

The EPR effect or the enhanced permeation and retention effect is a well-known concept in pharmacokinetics in the context of cancer therapy. A number of reviews are available on this topic.[89,90] Tumor tissues

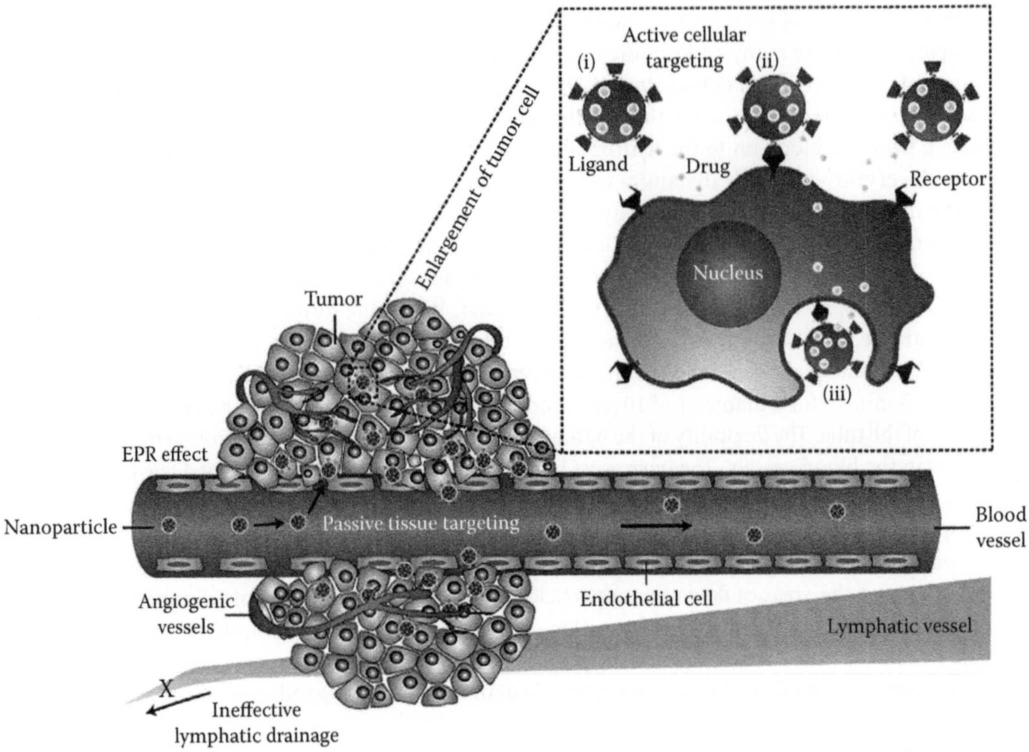

FIGURE 12.5 Schematic representation of different mechanisms by which nanocarriers can deliver drugs to tumors. Polymeric nanoparticles are shown as representative nanocarriers (circles). Passive tissue targeting is achieved by extravasation of nanoparticles through increased permeability of the tumor vasculature and ineffective lymphatic drainage (EPR effect). Active cellular targeting (inset) can be achieved by functionalizing the surface of nanoparticles with ligands that promote cell-specific recognition and binding. The nanoparticles can (i) release their contents in close proximity to the target cells, (ii) attach to the membrane of the cell and act as an extracellular sustained-release drug depot, or (iii) internalize into the cell. (From Peer, D. et al., *Nat. Nanotechnol.*, 2, 751, 2007. With permission.)

are physiologically different from their normal healthy counterparts. The vasculature is poor, often constricted with slow and leaky blood flow. All these lead to enhanced permeation of delivery systems from the blood vessels into the tissue surroundings and be retained there. The phenomenon has been used to the advantage for effective delivery of the drugs to tumors. Figure 12.5 shows a schematic diagram of various ways of delivering drugs to the tumor.

12.6 Mechanism of Cellular Uptake and Design Strategies

The uptake of extracellular materials into the cells has been a subject of interest for cell biologists for more than a century. Knowledge of the cellular uptake mechanisms helps in better designing of nano drug delivery vehicles with high uptake efficiency, better cargo protection, and even organelle-specific delivery of nanoparticles. There are a few reviews available on this topic.[91,92] In cells, the uptake of nutrients and the communications within the cell are mediated by the cell plasma membrane. Endocytosis is a process by which a cell transports a particle from outside the cell using a vesicle derived from the invagination and pinched-off pieces of plasma membrane. The process results in the formation of particles encapsulated inside the vesicles called endosomes. Endosomes become lysosomes after a brief interaction with golgi complex during which digestive enzymes are imported. The pH inside lysosomes is 4.5, while the surrounding cytoplasm remains alkaline. As such

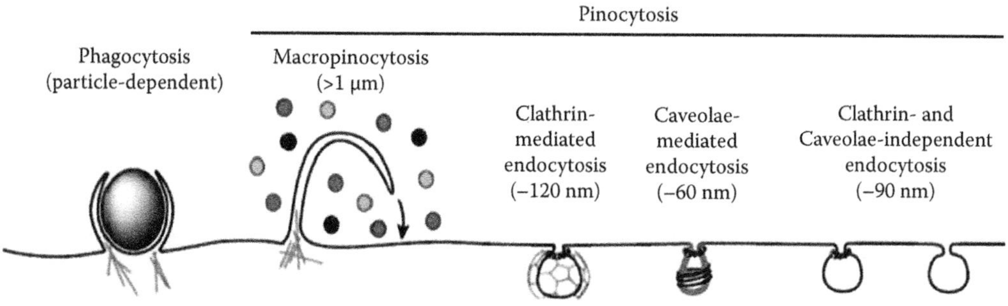

FIGURE 12.6 Various pathways of endocytosis. (From Conner, S.D. and Schmid, S.L., *Nature*, 422, 37, 2003. With permission.)

endocytosis can be divided into two broad categories phagocytosis (cell eating) or pinocytosis (cell drinking). Figure 12.6 summarizes the various processes involved.

The type of endocytosis that will occur depends on the size of the particle and their surface ligands. Experimentally, it is possible to differentiate whether a mechanism is by endocytosis or non-endocytosis by studying the extent of uptake of the materials by the cells at lower temperatures (4 instead of 37°C) and also in ATP-depleted environment by adding sodium azide.[93] Phagocytosis is conducted by specialized cells like monocytes, neutrophils, and macrophages to clear large pathogens like bacteria, yeast, or cell debris. It is a highly regulated process triggered by the antibodies present on the surface of the foreign body. Pinocytosis is more relevant to the uptake mechanism of nanomaterials. The distinction between clathrin-mediated or caveolae-mediated is experimentally identified by performing uptake experiments in K^+-depleted medium, which inhibits clathrin formation. Control experiments under similar conditions using transferrin are performed to establish clathrin-mediated pathways. Caveolae-mediated pathways are usually active in the uptake of lipid-based materials. When the carrier size is less than 100 nm, permeation across endothelial and epithelial barriers is possible via transcellular and pericellular pathways.[94] Sub-micron carriers are less likely to pass through intercellular junctions in the linings. However, they can pass through the organs with fenestrated endothelium having large, micron-sized openings such as in liver and spleen. However, even carriers as large as 500 nm have been known to be internalized via receptor-mediated endocytosis. Studies concerning the mechanism of uptake of nanomaterials can be found in the references.[93]

12.7 Targeted Drug Delivery

Most often the drugs intended to be delivered to treat a disease are cytotoxins. Thus, it is necessary to ensure that the toxic drugs are carefully delivered only to the infected region of the body without affecting the surrounding healthy tissues. Targeted drug delivery falls into two broad categories: (a) passive targeting and (b) active targeting. In the former case, the situation is such that the cells that are to be targeted migrate toward the drug-carrying vehicles. Though these types of systems are not strictly in the nanometer regime, they are widely used in the delivery of cells like neutrophils, macrophages, dendritic cells for vaccination purposes.[95] Passive targeting can also exploit EPR effect in the tumor vasculature. The second type is active targeting, which involves rational design of nanosytems with suitable surface engineering performed with acceptable chemical linking strategies to specifically target the cell receptors of a target tissue. Usually, polymer-based materials are versatile in active targeting as they can be designed to have suitable functional groups that can be used as chemical handles to hold the receptors.[96,97] Furthermore, the targeting operates at two levels; firstly the targeting of tissue/system in order to enrich the concentration of the carriers at the infected site. The right choice of administration of the nanocarriers will avoid its unnecessary clearance or hypersensitive reactions of the body and effectively deliver them to the targeted tissue. After the nanocarriers are administered via an appropriate route, suitable ligands present onto the surface of the nanocarriers can lead to high cellular uptake, and, in some cases, the subcellular components are also targeted to bring about the intended results. However, the successful implementation of this strategy is greatly limited due to the presence of a number of naturally occurring barriers in the body.

12.7.1 Barriers to Carriers

The body consists of a number of barriers that serve to protect its interior from a variety of external invaders and toxins, for example, skin. In a similar way, many systems like nervous systems or gastrointestinal system have their own custom-designed borders that physiologically differentiate them from the other systems. Barriers could be physiological or biochemical or chemical depending on the nature of resistance they offer. It is essential to overcome the natural barriers to achieve successful targeting.

The system linings are usually made up of layers of epithelial cells, which are attached to each other presenting physiological barriers. The origin of physiological barriers is the intercellular junctions between the cells, which are rich in protein strands. The junctions could be as long as 80 nm with a main function to prevent the entry of molecules larger than 1.1 nm. The entry of particles or drugs between the cells through the intercellular spaces is called paracellular pathway. In some cases, the particles may be too large to cross the cell junctions, in which case, the transcellular pathway (pass through the cells, rather than passing between the cells) is known to occur (Figure 12.7).

The most apical region of the cells is composed of tight junctions (Figure 12.8). The pores present in this region restrict the motion of the particles based on their size and charge. Immediately below this region is a region of cell adhesion whose initial formation actually facilitates the formation of

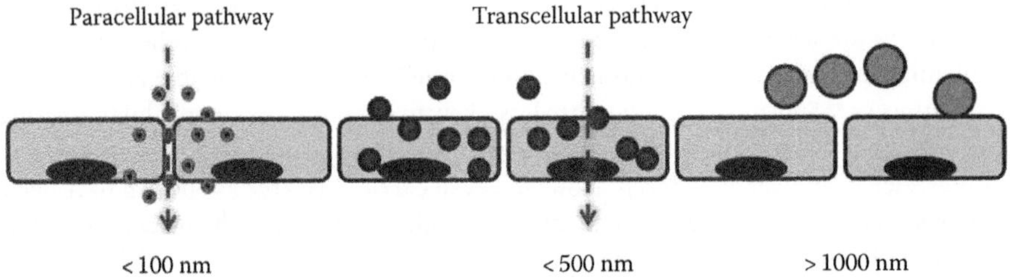

FIGURE 12.7 Scheme showing the size-dependent transport of particles through vascular endothelium. Particles of size less than 100 nm can be either endocytosed or travel between the cells. Particles of size 500 nm cannot travel between the cells. Particles of size larger than 100 nm are predominantly retained in the lumen of the vasculature.

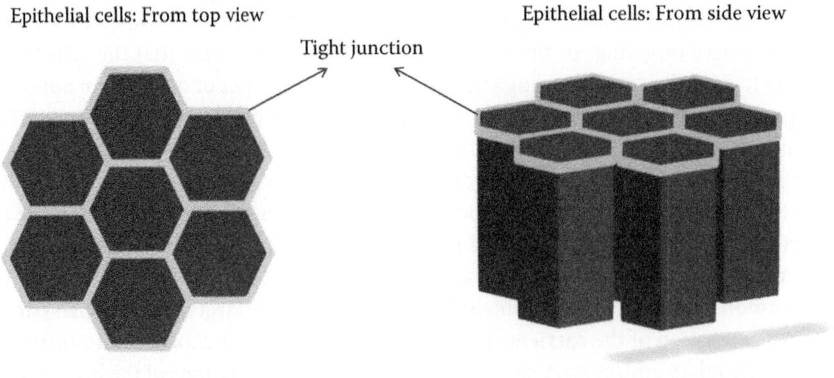

FIGURE 12.8 Cartoon showing the apical location of tight junctions in endothelial cells.

tight junctions. Resistance for paracellular pathway increases as the number of strands in this region increases. The last region of the paracellular pathway is called desmosomes, which has been found to be less critical. The transcellular pathway is mediated by specialized processes like endocytosis, which are described in Section 12.6. Certain systems have specialized secretions on their surface, for example, HCl, proteolytic pepsins, trypsin, carboxypeptidases, etc. in the gastrointestinal tract, which will be major biochemical barriers to a peptide-based carrier. Certain specialized cells may also present enzymes and proteins on their surface, which are classified as biochemical barriers. In order to target a nanocarrier to a particular system, it is essential to design them to traverse these specialized layers of cells and their secretions.

12.7.2 System Targeting

Cells like macrophages, macrophage precursors, specialized endothelial cells lining the sinusoids of spleen, liver, and bone marrow and the reticular cells of the lymphatic tissue (macrophages) and bone marrow (fibroblasts) form the vital reticuloendothelial system (RES). These ranges of cells freely circulate in the bloodstream and are capable of phagocytosis (ability to take up and sequester inert particles and vital dyes). The stationary members of RES are found in pulmonary alveoli, liver, spleen, and joints. They have functions related to inflammatory and immune responses. Most nanocarriers that are in the blood circulation/lymphatic system are targeted by the cells of RES. It is rather straightforward if the delivery is intended for the RES. Periportal and midzonal Kupffer cells of the liver, marginal zone and red pulp macrophages of the spleen are actively involved in scavenging the injected nanoparticles. The role of bone marrow is highly specific. A number of factors like surface charge, hydrophilicity, and particle size have been noticed to influence the in vivo organ distribution. The effect of zeta potential of a particle on the phagocytosis and biodistribution has been reported.[98] The mechanism of uptake is usually by opsonization. When the surface of the particle is coated with certain protein ligands called "opsonin" (the process is called opsonization), they become targets for phagocytes. Specific targeting to RES organs can be achieved by suitable surface modifications. For example, surface modifications using block copolymer polyethylene oxide/polypropylene oxide specifically targets the particles to the liver.[99] Presence of poly(organo phosphazenes) usually direct the particles to the bone marrow.[100] By introducing terminal amine groups in polyethylene oxide, sequestration to spleen was increased.[101] Mannose-based polymeric delivery systems have been specifically used in the treatment of leishmaniasis as they are recognized by the mannosyl fucosyl receptors present on the surface of the macrophages.[102] Adsorption of proteins like IgG, transferrin, or fibronectin under in vitro conditions can be performed to see the probability of uptake via opsonization. More related works on this topic can be found elsewhere.[103–105]

Excepting the RES-targeted drug delivery, higher circulation time in blood is required for many other systems. In such cases, grafting the surface of the nanocarriers with large molecular weight hydrophilic polymers (neutral or negative) like polyethylene glycol extends the circulation time.[4,106] The hydrophilic molecules usually minimize interaction with the surface of the circulating cells and proteins and thus evade the RES. The use of cationic nanocarriers like cationic liposomes in DNA delivery is a common practice. Though the positive charge gives a good electrostatic interaction with the anionic glycocalyx covering the endothelium, it also attracts many negatively charged blood components leading to emboli formation/serum inhibition.[107] A brief overview of various systems and nanocarriers used to target them is described in Table 12.3.

A special mention is required about the delivery of drugs to the central nervous system. The major component of the central nervous system is brain and disorder in its function can give rise to a wide range of diseases like Alzheimer's disease, stroke, HIV infection, and the most aggressive human cancers like malignant gliomas. Difficulty in brain delivery arises because of the presence of tight junctions in the endothelium, which restrict the paracellular pathway and the limited presence of transcytotic activities.[108]

TABLE 12.3 Summary of Nanocarriers in Targeted Delivery to Various Systems of the Human Body

System	Components	Diseases/Disorders	Nanocarriers	Challenges Involved
Reticuloendothelial	Liver, spleen, bone marrow	AIDS, pulmonary tuberculosis, typhoid, leishmaniasis, trypanosomiasis, ulcerative Colitis	Nanoparticles of albumin, PEG, PLA, PLGA, poloxamine 908 Poly(organo phosphazene)	Targeted delivery
Cardiovascular	Heart, blood vessels	Heart attacks and strokes	PEG-modified immunoliposomes	Targeted delivery
Nervous	Brain, spinal cord	Parkinson's disease, Alzheimer's disease, cancer, other psychological disorders	Solid lipid nanoparticles, functionalized amorphous carbon, poloxamer, polysorbate, cremaphor	Blood–brain barrier
Pulmonary	Lungs, pulmonary vasculature	Asthma	Lipid-coated microbubbles	Precipitation onto oropharynx and trachea
Gastrointestinal	Oral cavity, pharynx, esophagus, intestines, rectum, and anus	Candidiosis, diabetes, inflammatory bowel diseases, ulcers, Crohn's disease, immunosuppressions	PLGA, poly(styrene-co maleic acid), PVA, polyamino acids, dextran, starch, chitosan, lectin	Mucus secretion and digestive enzymes
Ocular	Sclera, cornea, iris and ciliary body	Iritis, corneal ulcer, keratoconus, scleritis, Fuch's dystropy, diabetic retinopathy, macular edema, glaucoma, floaters	Polyacrylic- and polysaccharide-based nanoparticles	Poor bioavailability
Lymphatics (secondary vascular system)	Lymphatic vessels and lymph nodes	Cancer, AIDS, elephantiasis, tuberculosis, anthrax, pneumonia	Polymer spheres coated with poloxamer, PEG-coated liposomes, albumin-stabilized N2 bubbles, PLGA, PEO-based micelles	Enhanced clearance

Source: Modified from Torchilin, V.P., Ed., *Nanoparticulates as Drug Carriers*, Imperial College Press, London, U.K., 2006.

Recent studies show that when glucose-derived functionalized carbon spheres (as large as 400 nm) were injected into mice, its localization was observed in spleen and liver with a significant amount crossing the blood–brain barrier (BBB).[58] A detailed study is essential to conclude the mechanism of uptake across BBB. This important discovery is yet to be completely harnessed in the field of neural drug delivery. The mechanism of the delivery of drugs using other nanoparticles is not elucidated completely. However, following suggestions have been put forth as mechanism.[109,110]

An increased retention of the nanoparticles in the brain blood capillaries combined with an adsorption to the capillary walls. This could create a higher concentration gradient that would enhance the transport across the endothelial cell layer, and as a result, the delivery to the brain. The polysorbate 80 used as the coating agent could inhibit the influx system, especially P-glycoprotein (Pgp). A general toxic effect on the brain vasculature leading to the permeabilization of the brain blood vessel endothelial cells. A general surfactant effect characterized by a solubilization of the endothelial cell membrane lipids that would lead to membrane fluidization and enhanced drug permeability through BBB. The nanoparticles could lead to an opening of the tight junctions between the endothelial cells. The drugs could then permeate through the tight junctions in free, or together with the nanoparticles, in bound form. The nanoparticles may be endocytosed by the endothelial cells, followed by the release of drugs within these cells and the delivery to the brain. The nanoparticles with bound drugs could be transcytosed through the endothelial cell layer.

12.7.3 Organelle Targeting

After reaching a desired tissue, the next step involves the entry of the nanoparticles inside the cells. The mechanisms involved in cellular uptake is described in the next section but here, an overview of targeting the subcellular components is presented. The fluid phase of the cytoplasm has roughly the same viscosity as the water. However, the presence of dense cytoskeleton fibers is a major hindrance to the free movement of the particle. It has been shown that particles bigger than 26 nm in size cannot diffuse freely in the cytoplasm. Cytoskeletons also interact with negatively charged drugs and cationic polymers. If cytoplasm is the ultimate destiny of nanocarriers, then there are no further barriers beyond this. However, in many cases, targeting must be specified to further organelles within the cell like mitochondria,[111,112] endoplasmic reticulum,[113–116] etc. Specific release of the drug in the cytoplasm is also possible by exploiting the redox state. For example, the extracellular components are strongly oxidizing (stabilize disulfide bonds), while the intracellular compartment is strongly reducing due to the presence of glutathione. Thus, drugs attached to a polymer via a disulfide bond can be selectively released in the cytoplasm.

12.7.3.1 Nuclear Targeting

Nucleus is an important organelle of the cell where most genetic material is densely packed. Gene delivery has a lot of significance in nucleus targeting as the latter is described as the hub of transcriptional activities. Nuclear targeting is generally believed to occur during mitotic division during which the nuclear membrane disintegrates and reforms. Besides mitosis, there is little known about the mechanism of entry of polymers like polyethyleneimine[117,118] or amorphous carbon.[58] In Figure 12.9, the confocal fluorescent image and time-dependent nuclear targeting of amorphous carbon spheres are shown. Nanoparticles when conjugated to viral peptide-based nuclear localization sequences can be delivered

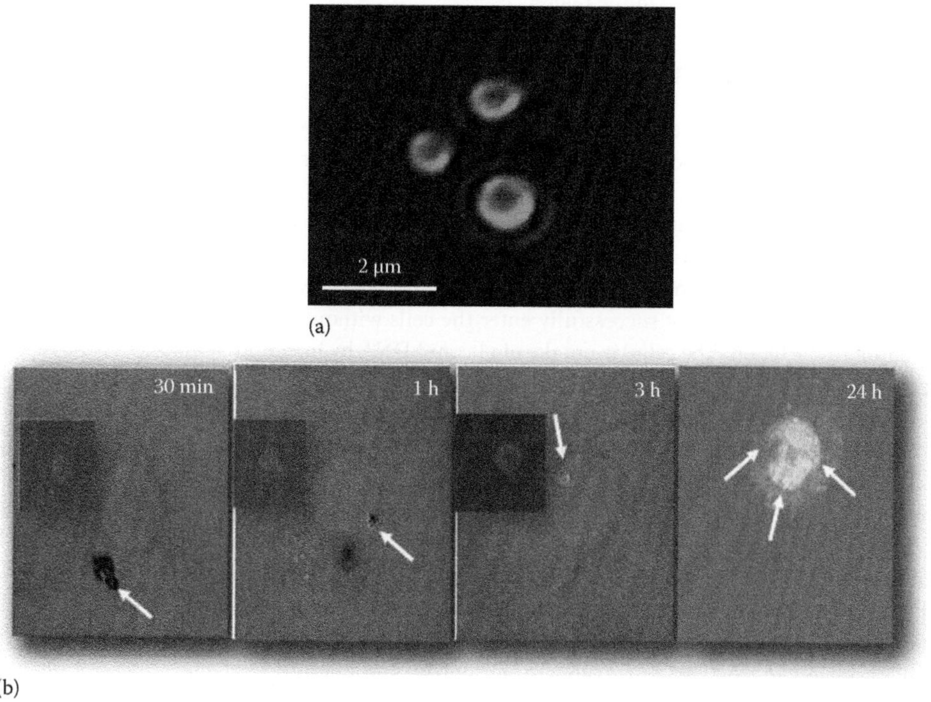

FIGURE 12.9 **(See color insert.)** Glucose-derived fluorescent carbon nanospheres. (a) Confocal laser scanning image. (b) Nuclear entry into the HeLa cells is indicated by white arrow marks. (From Selvi, R. et al., *Nano Lett.*, 8, 3182, 2008.)

into the nucleus.[119–121] These sequences are known to bind to karyopherin proteins, which facilitate the trafficking and binding to the nuclear membrane.[122]

12.7.3.2 Mitochondrial Targeting

Mitochondria play a very important role as the powerhouse of the cell. They are also responsible for regulated cell death, modulation of intracellular calcium concentration, and damaging reactive oxygen species. A large number of diseases are associated with the malfunctioning of mitochondria. Diseases like diabetes, cardiomyopathy, infertility, migraine, blindness, deafness, stroke, age-related neurodegenerative diseases, and neuromuscular diseases make the targeting of mitochondria pharmacologically important.[123–126] Mitochondria targeting nanocarriers are usually vesicles assembled from dicationic quinolinium derivatives. The symmetrical molecules with two delocalized charge centers separated by a hydrophobic chain spontaneously self-assemble to form vesicles called bolasomes/DQAsomes.[127] Their intrinsic mitochondriotropism has relation to their response to the mitochondrial membrane potential.

12.7.3.3 Lysosomal Bypass

Lysosomes are very deleterious to the stability of the payload of any drug carrier because of their acidity. This has remained the rate-limiting step in the delivery of genes or any drugs. There are a few virally inspired synthetic peptides, which when linked to the nanocarriers can disrupt the lysosomal membrane at its acidic pH.[128] Cells pretreated with chloroquine were found to have lysosomes with increased pH.[129] However, the clinical applicability of this method was greatly hindered by the toxicity of the chloroquine. Further developments lead to polymer nanocarriers with pH-buffering ability such as polyamidoamine dendrimers[130] and polyethyleneimine.[131] The mechanism in which such nanoparticles operate is described in terms of proton sponge hypothesis, which suggest an increased influx of protons and chloride ions leading to lysosomal burst rather than the neutralization of pH. Further information on this hypothesis can be found in the references.[132–136]

12.8 Delivery of Nucleic Acids

A controlled introduction of genetic sequences into mammalian cells has become an essential tool in the analyses and regulation of diseases that are both genetic and acquired in origin. The delivery of DNA or RNA into eukaryotic cells is called transfection. Therapeutic significance can occur in case of a stable transfection where the integration of the transfected nucleic acids with the host chromosome can result in passing on the vital information to next generations. Introduction of small interfering RNA (siRNA) can selectively turn off the production of specific proteins (gene silencing or antisense therapy).[137]

However, DNA itself cannot successfully enter the cells without a suitable vector due to enzymatic degradation in the blood. The cellular uptake of plasmid DNA by injection in muscles is less than 1% of the total dose. The current methods of gene transfers and their disadvantages are given in the Table 12.4. The information is adapted from a review by Sokolova and Epple.[138]

Nanoparticle-mediated gene delivery involves micelles, liposomes, cationic polymers, and inorganic particles. Cationic compounds hold the DNA by electrostatic adsorption. Nanoparticles made up of polylysine,[139,140] polyethyleneimine,[141–144] are known to enter cells with lower transfection efficiency. However, both these polymers are toxic and cannot be used. Liposomes made up of cationic and neutral lipids like dioleoyltrimethylammonium chloride and dioleoylphosphatidylethanolamine are successfully used for gene delivery. Among the inorganic nanomaterials, a wide range of options are available primarily due to rich covalent chemistry. Metallic nanoparticles,[145–147] carbon nanotubes,[148] iron oxide nanoparticles,[149–152] silica nanoparticles,[153–155] calcium phosphate, and quantum dots have successfully demonstrated their use in gene delivery.[156–159] Another set of materials known as LDH[160–162] deserve a special mention as they have a unique mechanism of delivering any negatively charged molecules inside the cell. LDH are positively charged inorganic sheets, which are held together in lamellar fashion by counter anions. Their desired feature is that the anions can be exchanged with the negatively charged molecules

TABLE 12.4 Summary of Various Transfection Methods with Their Advantages and Disadvantages

Transfection Method	Advantages	Disadvantages
Viral methods	Highly efficient	Immunogenicity, carcinogenicity, inflammation
Electroporation	Easy to perform, efficient	Optimized for every cell line required, large amount of DNA is necessary
Microinjection	Exact direction of nucleic acid into a single cell	Slow, cannot be scaled-up
Gene gun	Good for genetic vaccination	Shallow penetration of DNA into the tissues
Recombinant proteins	High biocompatibility	Expensive
Cationic polymers	Easy preparation	Toxic
Polymeric nanoparticles	Easy preparation, functionalization	Limited efficiency, some are toxic
Inorganic nanoparticles	Easy preparation, functionalization	Limited efficiency, some are toxic

Source: Modified from Sokolova, V. and Epple, M., *Angew. Chem. Int. Ed.*, 47, 1382, 2007.

like DNA, RNA, etc. LDH–DNA complexes have been shown to enter the cells and release DNA into the cytoplasm in a controlled fashion. In the process, LDH also dissolves into its ions causing no toxic effects.

12.9 Toxicity

There are twofold risk aspects involved with the nanocarriers. Firstly, the nanomaterials (especially inorganic materials) are highly reactive compared to the bulk—a fact that though well known is still not completely studied in biological systems. Secondly, nanoparticles have gained increased access to different types of tissues and cells. As the use of nanomaterials in various consumer products including pharmaceuticals has been increasing, the study of their interaction with the human body, their long-term effects on accumulation becomes highly relevant. As these nanoparticles are designed deliberately to interact with the cellular systems, it is important to ensure that both the naked particles and their drug conjugates will undergo biodegradation inside cellular systems. Moreover, it is also important to study the responses of the cell to the degraded products of the nanoparticles. By definition, the term nanotoxicity is now commonly used to describe cellular injuries produced by various nanoparticles.[163] The in vitro assays (toxicity assays in cell cultures) in general are less expensive and less stringent. To get unambiguous results of nanotoxicity assays of nanoparticles, it is necessary to monitor more than one cellular response using various nanoparticle concentrations and incubation times. Table 12.5 summarizes various assays that can be performed to ascertain the toxic effects of the nanoparticles. Besides, it is also advisable to test the toxicity in more than one cell line. Following cell lines are usually used in the studies of nanoparticle cytotoxicity:[164] human dermal fibroblasts (HDF), human liver carcinoma (HepG2), guinea pig alveolar macrophages, neuronal human astrocytes (NHA), monocyte-derived macrophages, human monocyte macrophages, human epidermal keratinocytes (HEK), human umbilical vein endothelial cells (HUVEC), 3T3 cells, immortalized human epidermal keratinocytes (HaCaT), mouse peritoneal macrophage-like cells (J774.1A), human embryonic kidney (HEK293), human promyelocytic leukemia cells (HL60), Jurkat T cells, HeLa cells, lung carcinoma cells (A549, H1299), rat alveolar macrophage cells (NR8383), mesothelioma cells (MSTO-211H), human skin fibroblasts (HSF42), human embryonic lung fibroblasts (IMR-90), Sprague-Dawley rat peritoneal macrophages, murine alveolar macrophages (RAW267.9), human acute monocytic leukemia cells (THP-1), Wister male rats T lymphocytes, human osteoblastic line (hFOB1.19), red blood cells, and human breast carcinoma xenograft cells.

Various types of in vitro cytotoxicity assays are tabulated. Many commercially available cytotoxicity kits photometrically probe the mitochondrial activity of the cell. In the viable cells, tetrazolium-based salts are usually reduced by mitochondrial dehydrogenases to formazan. More recently, there have been works related to the toxic effects of many potential nanocarriers like carbon nanotubes, fullerenes,

TABLE 12.5 Cytotoxicity Assays Performed on Nanoparticles

Cell Response	Assay	Description
1. Cellular and nuclear morphology	—	Visual inspection of cell and nuclear morphology using light microscopy.
2. Plasma membrane integrity: The integrity of plasma membrane is vital in maintaining the ion homeostasis over the surrounding media.	(a) Neutral red assay	Cationic dyes like neutral red or toluylene red can cross the plasma membrane by diffusion and gets accumulated in lysosome. Altered cell membrane can leak out the dye.
	(b) Trypan blue assay	An azo dye permeable into cells with compromised plasma membrane. Dead cells are stained blue while the live cells remain colorless.
	(c) Calceinacetoxymethyl (calcein AM) and ethidium bromide assay	Calcein AM is an electrically neutral esterified molecule. Live cells produce esterases, which convert them into green fluorescent molecule. Ethidium bromide permeates and binds to the nucleic acids of the dead cells only giving a red fluorescence.
	(d) Lactate dehydrogenase (LDH assay)	Dead cells leak out enzymes like LDH, which oxidize lactate to pyruvate that promotes the conversion of tetrazolium salts to formazan.
3. Mitochondrial activity: Active mitochondria produce dehydrogenases, which can cleave tetrazolium ring to formazan derivatives.	(a) MTT assay	3-(4,5-Dimethylthiazol-2-yl)-2,5-diphenyl tetrazolium bromide.
	(b) MTS assay	3-(4,5-Dimethylthiazol-2-yl)-5-(3-carboxymethoxyphenyl)-2-(4-sulfophenyl)-2H-tetrazolium.
	(c) WST-1 or WST-8 assays	2-(4-Iodophenyl)-3-(4-nitrophenyl)-5- (2,4-disulfophenyl)-2H-tetrazolium.
	(d) Resazurin or Alamar blue assay	Non-fluorescent resazurin acts as an electron acceptor for enzymes like NADP and FADH to form highly fluorescent resorufin.
	(e) Tetramethylrhodamin ethyl ester (TMRE assay)	Quantitative marker for the activity of mitochondria, which requires no metabolic action of the cell.
4. Oxidative stress	(a) Glutathione assay (GSH assay)	GSH is a major antioxidant that is oxidized to glutathione disulfide (GSSG) in the presence of reactive oxygen species. GSH/GSSG ratio is maintained high in viable cells by the enzyme reductases. Level of GSH can be monitored using Ellman's reagent.
	(b) Thiobarbituric acid assay (TBA assay)	Malondialdehyde is a toxic byproduct formed during the lipid peroxidation of plasma membrane. TBA reacts in acidic pH to form pink-colored chromagen.
5. Inflammation	Enzyme-linked immunosorbent assay (ELISA)	Protein signals of inflammatory responses include IL-1β, IL-6, TNF-α, and IL-8.
6. Genotoxicity: Extent of DNA damage is estimated as a mark of the potential genotoxicity of the nanoparticles.	—	DNA intercalating dyes are used to estimate the cellular DNA content to determine the proportion of apoptotic cells, e.g., propidium iodide.

silica-based materials, metal nanoparticles, superparamagnetic iron oxides, and quantum dots.[165–173] However, more recently, the validity of these experiments were questioned with the carbon-based nanomaterials as they were found to interact with the water insoluble formazan crystals. It is therefore advised to cross check the cytotoxicity of carbon nanomaterials with water soluble modification assays.[167] Formation of peroxy radical, which is responsible for membrane degradation and superoxide radical formation in case of some fullerenes, is measured using specific dyes like 1,1,3,3-tetraethoxypropane and iodophenol, respectively. More sophisticated measurements of cytotoxicity have involved monitoring the micromotility of the cells using impedance spectroscopy.[165] Inhibition of cellular respiration has also been observed in case of morphologically different mesoporous silica.[61]

However, in vivo toxicological studies are more relevant and closer to reality than the in vitro results. It is necessary to have detailed characterization on the biodistribution, kinetics of clearance, and other toxicological data for every potential nanocarrier. A number of reviews deal with the subject of nanotoxicity,[172–180] but in this section, discussion will be restricted to nanocarriers used in biomedical applications like drug delivery and diagnostic imaging.

12.10 Recent Advances and Future Prospects

There is a growing quest in the treatment of cancer therapy using nanomaterials as drug carrier. Recent reports emphasize more importance to the targeted delivery of nanomaterials to the tumors. An integrated approach of using nanomaterials for both diagnostic imaging and therapy is also being widely worked out. Nanomaterials like dendrimers, carbon materials, which have more than one functionality like magnetic fluorescence are attached with suitable ligands specific to surface receptors on the cancer cells. While fluorescence, scattering, magnetic resonance properties of the nanomaterials are used for the diagnostic imaging of the tumorous tissues, magnetic hyperthermia and photothermal therapy are used as therapeutic tools. However, the toxicity of nanomaterials still continues to remain as a challenge that hinders their fast development. Though, a number of reports exclusively deal with the cellular toxicity of the nanomaterials, future works must deal with their in vivo fate. More efforts must also be spent to suitably modify the synthetic procedures to induce biocompatibility, biodegradability, and nontoxicity.

12.11 Conclusions

It must be admitted that developing an efficient drug delivery technology is as complex and expensive as the drug discovery itself. Nevertheless, all innovative and successful therapeutic strategies and devices have resulted in the convergence of many technologies. The development of inhaled insulin is one such classic example where combined efforts from a vast array of experts ranging from biology to vacuum industry were involved. In that perspective, nanocarriers, which are also a product of interdisciplinary research, seem to present bright hopes in the field of drug delivery with enhanced circulation time and control over release options. Proper engineering of the surfaces can offer targeted drug delivery both at the levels of tissue and organelle. They can improve the therapeutic index of the already existing drug molecules with their well-established therapeutic profiles. This certainly removes enormous efforts and risks associated in the search of molecules with better therapeutic indices. Besides, the current generation therapeutic molecules are likely to be proteins, peptides, oligonucleotides, plasmids, etc., whose successful delivery is ambiguous without the use of nanocarriers. Again, though we have a wide range of nanomaterials at our disposal, toxicity needs to be addressed more rigorously than ever.

For sure, the awesome intricacies and mechanics of the human body would only force us to spend more time and money on the nanocarrier research than predicted earlier. As John S. Patton rightly said in his historical perspective,[175] "If there is one thing that we should have learned by now about medicine, it is that human biology does not quickly, predictably or inexpensively yield its secrets." The efforts of the scientists working in this field therefore would be not to let the spirits of *nano* die off as any other buzzword but to develop it into a more realistic technology affordable to common man.

References

1. http://en.wikipedia.org/wiki/Trojan_War (2009).
2. Y. G. Yu, D. S. King, and Y. K. Shin. 1994. Insertion of a coiled-coil peptide from influenza virus hemagglutinin into membranes. *Science* 266: 274–276.
3. T. Shangguan, D. P. Siegel, J. D. Lear, P. H. Axelsen, D. Alford, and J. Bentz. 1998. Morphological changes and fusogenic activity of influenza virus hemagglutinin. *Biophys. J.* 74(1): 54–62.
4. R. Gref, Y. Minamatake, M. T. Peracchia, V. Trubetskoy, V. Torchilin, and R. Langer. 1994. Biodegradable long-circulating polymeric nanospheres. *Science* 263: 1600–1603.
5. E. Soussan, S. Cassel, M. Blanzat, and I. R. Lattes. 2009. Drug delivery by soft matter: Matrix and vesicular carriers. *Angew. Chem. Int. Ed.* 48: 274–288.
6. V. P. Torchilin (Ed). 2006. *Nanoparticulates as Drug Carriers*. Imperial College Press, London, UK.
7. P. Couvreur, B. Kante, M. Roland, and P. Speiser. 1979. Adsorption of anti-neoplastic drugs to polyalkylcyanoacrylate nanoparticles and their release in calf serum. *J. Pharm. Sci.* 68(12): 1521–1524.
8. P. Couvreur. B. Kante, V. Lenaerts, V. Scailteur, M. Roland, and P. Speiser. 1980. Tissue distribution of anti-tumor drugs associated with polyalkylcyanoacrylate nanoparticles. *J. Pharm. Sci.* 69(2): 199–202.
9. D. Peer, J. M. Karp, S. Hong, O. C. Farokhzad, R. Margalit, and R. Langer. 2007. Nanocarriers as an emerging platform for cancer therapy. *Nat. Nanotechnol.* 2(12): 751–760.
10. D. A. LaVan, T. McGuire, and R. Langer. 2003. Small-scale systems for in vivo drug delivery. *Nat. Biotechnol.* 21: 1184–1191.
11. D. A. Tomalia, H. Baker, M. Hall, G. Kallos, S. Martin, J. Roeck, J. Ryder, and P. Smith. 1985. A new class of polymers: Starburst-dendritic macromolecules. *Polym. J.* 17(1): 117–132.
12. M. Mammen, S. K. Choi, and G. M. Whitesides. 1998. Polyvalent interactions in biological systems: Implications for design and use of multivalent ligands and inhibitors. *Angew. Chem. Int. Ed.* 37: 2754–2794.
13. M. Liu and J. M. J. Frèchet.1999. *Polym. Mater. Sci. Eng.* 80: 167.
14. M. Liu, K. Kono, and J. M. J. Frèchet. 2000.Water-soluble dendritic unimolecular micelles: Their potential as drug delivery agents. *J. Control. Release* 65: 121–131.
15. J. Z. Hilt and M. E. Byrne. 2004. Configurational biomimesis in drug delivery: Molecular imprinting of biologically significant molecules. *Adv. Drug Del. Rev.* 56: 1599–1620.
16. N. A. Peppas, J. Z. Hilt, A. Khademhosseini, and R. Langer. 2006. Hydrogels in biology and medicine: From molecular principles to bionanotechnology. *Adv. Mater.* 18(11): 1345–1360.
17. K. Letchford and H. Burt. 2007. A review of the formation and classification of amphiphilic block copolymer nanoparticulate structures: Micelles, nanospheres, nanocapsules and polymersomes. *Eur. J. Pharm. Biopharm.* 65: 259–269.
18. Y. H. Bae, T. Okano, and S. W. Kim. 1991. "On-Off" thermocontrol of solute transport. II solute release from thermosensitive hydrogels. *Pharm. Res.* 8: 624–628.
19. Y. H. Bae, T. Okano, and S. W. Kim. 1991. "On-Off" thermocontrol of solute transport. I. Temperature dependent swelling of N-isoprpylacrylamide networks modified with hydrophobic components in water. *Pharm. Res.* 8: 531–537.
20. S. H. Yuk, S. H. Cho, and S. H. Lee. 1997. pH/temperature-responsive polymer composed of poly(N,N-dimethylaminoethyl methacrylate-*co*-ethylacrylamide). *Macromolecules* 30: 6856–6859.
21. T. M. Allen, T. Mehra, C. Hansen, and Y. C. Chin. 1992. Stealth liposomes: An improved sustained release system for 1-β-D-arabinofuranosylcytosine. *Cancer Res.* 52(9): 2431–2439.
22. K. Ferrara, R. Pollard, and M. Borden. 2007. Ultrasound microbubble contrast agents: Fundamentals and application to gene and drug delivery. *Ann. Rev. Biomed. Eng.* 9: 415–447.
23. N. S. Chung and K. M. Vasan. 2004. Potential role of low density lipoprotein receptor family as mediators of cellular drug uptake. *Adv. Drug Del. Rev.* 56: 1315–1334.

24. A. J. Domb. 1993. Lipospheres for controlled delivery of substances. US patent No. 5188837.
25. A. J. Domb. 1995. Long acting injectable oxytetracycline-liposphere formulation. *Int. J. Pharm.* 124: 271–278.
26. A. J. Domb. 1993. Liposphere parenteral delivery system. *Proc. Int. Symp. Control. Rel. Bioact. Mater.* 20: 346–347.
27. A. Illing, T. Unruh, and M. H. J. Koch. 2004. Investigation on particle self-assembly in solid lipid based colloidal drug carrier systems. *Pharm. Res.* 21: 592–597.
28. K. Jores. 2007. Lipid nanodispersions as drug carrier systems—A physicochemical characterisation. Thesis University of Halle.
29. A. Illing and T. Unruh. 2004. Investigations on flow behavior of dispersions of solid triglyceride nanoparticles. *Int. J. Pharm.* 284: 123–131.
30. E. B. Souto et al. 2004. Evaluation of physical stability of SLN and NLC before and after incorporation into hydrogel formulations. *Eur. J. Pharm. Biopharm.* 58: 83–90.
31. H. Bunjes, K. Westesen, and M. H. J. Koch. 1996. Crystallization tendency and polymorphic transitions in triglyceride nanoparticles. *Int. J. Pharm.* 129: 159–173.
32. W. Mehnert and K. Mäder. 2001. Solid lipid nanoparticles: Production, characterization and applications. *Adv. Drug Del. Rev.* 47: 165–196.
33. V. P. Torchilin. 2004. Targeted polymeric micelles for delivery of poorly soluble drugs. *Cell Mol. Life Sci.* 61: 2549–2559.
34. V. P. Torchilin. 2001. Structure and design of polymeric surfactant-based drug delivery systems. *J. Control. Release* 73: 137–172.
35. M. Antonietti and S. Forster. 2003. Vesicles and liposomes: A self-assembly principle beyond lipids. *Adv. Mater.* 15: 1323–1333.
36. B. Discher, Y. Y. Won, D. S. Ege, J. C. M. Lee, F. S. Bates, D. E. Discher, and D. A. Hammer. 1999. Polymerosomes: Tough vesicles made from diblock copolymers. *Science* 284: 1143–1146.
37. J. Z. Du and Y. M. Chen. 2004. Preparation of organic/inorganic hybrid hollow particles based on gelation of polymer vesicles. *Macromolecules* 37: 5710–5716.
38. I. Cho and Y. D. Kim. 1997. Synthesis and properties of tocopherol-containing polymeric vesicle system. *Macromol. Symp.* 118: 631–640.
39. I. Cho and Y. D. Kim.1998. Formation of stable polymeric vesicles by tocopherol containing amphiphiles. *Macromol. Rapid Commun.* 19: 27–30.
40. I. Cho and K. C. Chung. 1988. Cholesterol containing polymeric vesicles—Synthesis, characterization and separation as a solid powder. *Macromolecules* 21: 565–571.
41. M. D. Brown, A. Schatzlein, A. Brownline, V. Jack, W. Wang, L. Tetley, A. I. Gray, and I. F. Uchegbu. 2000. Preliminary characterization of novel amino acid based polymeric vesicles as gene and drug delivery agents. *Bioconjug. Chem.* 11(6): 880–891.
42. I. F. Uchegbu and A. T. Florence. 1995. Nonionic surfactant vesicles (niosomes)—Physical and pharmaceutical chemistry. *Adv. Coll. Inter. Sci.* 58: 1–55.
43. I. F. Uchegbu and S. P. Vyas. 1998. Nonionic surfactant vesicles (niosomes) in drug delivery. *Int. J. Pharm.* 172: 33–70.
44. L. Zarif. 2002. Elongated supramolecular assemblies in drug delivery. *J. Control. Release Rev.* 81: 7–23.
45. L. Zarif, J. R. Graybill, D. Perlin, and R. J. Mannino. 2000. Cochleates: New lipid-based drug delivery system. *J. Liposome Res.* 10(4): 523–538.
46. D. Papahadjopoulos, W. J. Wail, K. Jacobson et al. 1975. Cochleate lipid cylinders: Formation by fusion of unilamellar lipid vesicles. *Biochim. Biophys. Acta* 394: 483–491.
47. D. Papahadjopoulos. 1978. Large unilamellar vesicles and the method of preparing the same. US patent No. 4078052.
48. Z. P. Xu, H. Zeng, G. Qing Lu, and A. B. Yu. 2006. Inorganic nanoparticles as carriers for efficient cellular delivery. *Chem. Eng. Sci.* 61(3): 1027–1040.

49. P. K. Jain, X. Huang, I. H. El Sayed, and M. A. El Sayed. 2008. Nobel metals on the nanoscale: Optical and photothermal properties and some applications in imaging, sensing, biology and medicine. *Acc. Chem. Res.* 41: 1578–1586.
50. M. Hu, J. Chen, Z. L. Li, L. Au, G. V. Hartland, X. Li, M. Marquez, and Y. Xia. 2006. Gold nanostructures: Engineering their plasmonic properties for biomedical applications. *Chem. Soc. Rev.* 35(11): 1084–1094.
51. S. Eustis and M. A. El Sayed. 2006. Why gold nanoparticles are more precious than pretty gold: Noble metal surface plasmon resonance and its enhancement of the radiative and non radiative properties of nanocrystals of different shapes. *Chem. Soc. Rev.* 35(3): 209–217.
52. A. K. Salem, P. C. Searson, and K. W. Leong. 2003. Multifunctional nanorods for gene delivery. *Nature Mat.* 2: 668–671.
53. J. L. Bridot, D. Dayde, C. Rivière, C. Mandon, C. Billotey, S. Lerondel, R. Sabattier et al. 2009. Hybrid gadolinium oxide nanoparticles combining imaging and therapy. *J. Mat. Chem.* 19(16): 2328–2335.
54. A. Bianco and M. Prato. 2003. Can carbon nanotubes be considered useful tools for biological applications? *Adv. Mat.* 15: 1765–1768.
55. A. Bianco, K. Kostarelos, and M. Prato. 2005. Applications of carbon nanotubes in drug delivery. *Curr. Opin. Chem. Bio.* 9: 674–679.
56. N. W. S. Kam and H. Dai. 2005. Carbon nanotubes as intracellular protein transporters: Generality and biological functionality. *J. Amer. Chem. Soc.* 127: 6021–6026.
57. G. Jia, H. Wang, L. Yan, X. Wang R. Pei, T. Yan, Y. Zhao, and X. Guo. 2005. Cytotoxicity of carbon nanomaterials: Single wall nanotube, multiwall nanotube and fullerene. *Environ. Sci. Technol.* 39(5): 1378–1383.
58. R. Selvi, D. Jagadeesan, B. S. Suma, G. Nagashankar, M. Arif, K. Balasubrahmanyam, M. Eswaramoorthy, and T.K. Kundu. 2008. Intrinsically fluorescent carbon nanospheres as a nuclear targeting vector: Delivery of membrane-impermeable molecule to modulate gene expression in vivo. *Nano Lett.* 8(10): 3182–3188.
59. C. Barbe, J. Barlett, L. Kong, K. Finnie, H. Q. Lin, M. Larkin, S. Calleja, A. Bush, and G. Calleja. 2004. Silica particles: A novel drug delivery system. *Adv. Mat.* 16(21): 1959–1966.
60. A. Petri-Fink, M. Chastellain, L. Juillerat-Jeanneret, A. Ferrari, and H. Hofmann. 2005. Development of functionalized superparamagnetic iron oxide nanoparticles for interaction with human cancer cells. *Biomaterials* 26(15): 2685–2694.
61. Z. Tao, M. P. Morrow, T. Asefa, K. K. Sharma, C. Duncan, A. Anan, H. S. Penefsky, J. Goodisman, and A.K. Souid. 2008. Mesoporous silica nanoparticles inhibit cellular respiration. *Nano Lett.* 8(5): 1517–1526.
62. C. Loo, A. Lowrey, N. Halas, J. West, and R. Drezek. 2005. Immunotargeted nanoshells for integrated cancer imaging and therapy. *Nano Lett.* 5(4): 709–711.
63. W. H. Suh et al. 2009. Multifunctional nanosystems at the interface of physical and life sciences. *Nano Today* 4: 27–36.
64. S. Pal, D. Jagadeesan, K. L. Gurunatha, M. Eswaramoorthy, and T. K. Maji. 2008. Construction of bi-functional inorganic–organic hybrid nanocomposites. *J. Mat. Chem.* 18(45): 5448–5451.
65. S. Fukushima, S. Kishimoto, Y. Takeuchi, and M. Fukushima. 2000. Preparation and evaluation of o/w type emulsions containing antitumor prostaglandin. *Adv. Drug Del. Rev.* 45(1): 65–75.
66. I. Danielsson and B. Lindman. 1981. The definition of a microemulsion. *Coll. Surf.* 3: 391–392.
67. A. Wretlind. 1999. The development of fat emulsions. *Nutrition* 15: 641–645.
68. R. Jeppsson and S. Ljungberg. 1975. *Acta Pharmacol. Toxicol.* 36: 312.
69. G. M. Tejado, S. Bouttier, J. Fourniat, J. L. Grossiord, J. P. Marty, M. Seiller. 2005. Release of antiseptics from the aqueous compartments of a w/o/w multiple emulsions. *Int. J. Pharm.* 288(1): 63–72.
70. F. Cournarie, M. P. Savelli, W. Rosilio, F. Bretez, C. Vauthier, J. L Grossiord, M. Seiller. 2004. Insulin-loaded W/O/W multiple emulsions: Comparison of the performances of systems prepared with medium-chain-triglycerides and fish oil. *Eur. J. Pharm. Biopharm.* 58(3): 477–482.

71. S. Higashi and T. Setoguchi. 2000. Hepatic arterial injection chemotherapy for hepatocellular carcinoma with epirubicin aqueous solution as numerous vesicles in iodinated poppy-seed oil microdroplets: Clinical application of water-in-oil-in-water emulsion prepared using a membrane emulsification technique. *Adv. Drug Del. Rev.* 45: 57–64.
72. B. D. Spiess. 2009. Perfluorocarbon emulsions as a promising technology: A review of tissue and vascular gas dynamics. *J. Appl. Physiol.* 106: 1444–1452.
73. E. Estephan, M. B. Saab, C. Larroque et al. 2009. Peptides for functionalization of InP semiconductors. *J. Coll. Int. Sci.* 337: 358–363; C. J. Murphy, A. M. Gole, J. W. Stone et al. 2008. Gold nanoparticles in biology: Beyond toxicity to cellular imaging. *Acc. Chem. Res.* 41: 1721–1730.
74. A. Portis, C. Newton, W. Pangborn, and D. Papahajopoulous. 1979. Studies on the mechanism of membrane fusion: Evidence for an intermembrane Ca^{+2}-phospholipid complex, synergism with Mg^{+2}, and inhibition by spectrin. *Biochemistry* 18(5): 780–790.
75. L. Zhang, S. Qiao, Y. Jin, L. Cheng, Z. Yan, and G. Q. Lu. 2008. Hydrophobic functional group initiated helical mesostructured silica for controlled drug release. *Adv. Funct. Mat.* 18(23): 3834–3842.
76. J. J. Green, R. Langer, and D. G. Anderson. 2008. A Combinatorial polymer library approach yields insight into the nonviral gene delivery. *Acc. Chem. Res.* 41: 749–759.
77. Z. Adamczyk, B. Siwek, K. Jaszczolt, and P. Weroński. 2004. Deposition of latex particles on heterogenous surfaces. *Coll. Surf. A Physicochem. Eng. Asp.* 249(1–3): 95–98.
78. R. K. Jain. 2001. Delivery of molecular medicine to solid tumors: Lessons from in vitro imaging and gene expression and function. *J. Control. Release* 74: 7–25.
79. M. Sugihara-Seki and R. Skalak. 1997. Asymmetric flows of spherical particles in a cylindrical tube. *Biorheology*. 34: 155–159.
80. H. Wang and R. Skalak. 1969. Viscous flow in a cylindrical tube containing a line of spherical particles. *J Fluid Mech.* 38: 75–96.
81. B. Nasseri and A. T. Florence. 2003. Microtubules formed by capillary extrusion and fusion of surfactant vesicles. *Int. J. Pharm.* 266: 91–98.
82. D. Jagadeesan, U. Mansoori, P. Mandal, A. Sundaresan, and M. Eswaramoorthy. 2008. Hollow spheres to nanocups: Tuning the morphology and the magnetic properties of single crystalline α-Fe_2O_3 nanostructures. *Angew. Chem. Int. Ed.* 47: 7685–88; J. B. Lassiter, M. W. Knight, N. A. Mirin, and N. J. Halas. 2009. Reshaping the plasmonic properties of a single nanoparticle. 2009. *Nano Lett.* 9(12): 4326–4332.
83. W. R. Bowen and A. Mongruel. 1998. Calculation and collective diffusion coefficient of electrostatically stabilised colloidal particles. *Coll. Surf. A.* 138: 161–172.
84. I. F. Uchegbu et al. 1997. Polyhedral non ionic surfactant vesicles. *J Pharm. Pharmacol.* 49: 606–610.
85. R. Vasanthi and S. Battacharya. 2005. Anisotropic diffusion of spheroids in liquids: Slow orientational relaxation of the oblates. *J. Chem. Phys.* 116: 1092–1096.
86. Y. Geng, P. Dalhaimer, S. Cai, R. Tsai, M. Tewari, T. Minko, and D. E. Discher. 2007. Shape effects of filaments versus spherical particles in flow and drug delivery. *Nat. Nanotech.* 2(2): 249–255.
87. Z. Adamczyk, B. Siwek, K. Jaszczólt, and P. Weroński. 2004. Deposition of latex particles at heterogenous surfaces. *Coll. Surf. A Physicochem. Eng. Asp.* 249(1–3): 95–98.
88. D. B. Warheit, B. R. Laurence, K. L. Reed, D. H. Roach, G. A. M. Reynolds, and T. R. Webb. 2004. Comparative pulmonary toxicity assessment of single-wall carbon nanotubes in rats. *Toxicol Sci.* 77(1): 117–125.
89. H. Maeda, J. Wu, T. Sawa, Y. Matsumura, and K. Hori. 2000. Tumor vascular permeability and the EPR effect in macromolecular therapeutics: A review. *J Control. Release* 65: 271–284.
90. H. Maeda, T. Sawa, and T. Konno. 2001. Mechanism of tumor-targeted delivery of macromolecular drugs, including the EPR effect in solid tumor and clinical overview of the prototype polymeric drug SMANCS. *J. Control. Release* 74: 47–61.
91. M. Marsh and H. T. McMahon. 1999. The structural era of endocytosis. *Science* 285: 215–219; S. D. Conner and S. L. Schmid. 2003. Regulated portals of entry into the cell. *Nature* 422: 37–44.

92. K. Kostarelos, L. Lacerda, G. Pastorin, W. Wu, S. Wieckowski, J. Luangsivilay, S. Godefroy, D. Pantarotto, J. P. Briand, S. Muller, M. Prato, and A. Bianco. 2007. Cellular uptake of functionalised carbon nanotubes is independent of functional group and cell type. *Nat. Nanotech.* 2(2): 108–113.
93. N. W. S. Kam, Z. Liu, and H. Dai. 2006. Carbon nanotubes as intracellular transporters for proteins and DNA: An investigation of the uptake mechanism and pathway. *Angew. Chem. Int. Ed.* 45: 577–581.
94. S. Muro, M. Koval, and V. Muzykantov. 2004. Endothelial endocytotic pathways: Gates for vascular drug delivery. *Curr. Vasc. Pharmacol.* 2: 281–299.
95. M. L. Hedley, J. Curley, and R. Urban. 1998. Microspheres containing plasmid-encoded antigens elicit cytotoxic T-cell responses. *Nat Med.* 4: 365–368.
96. G. Hermanson. 1996. *Bioconjugate Techniques*. Academic Press Inc., San Diego, CA, pp. 570–592.
97. T. Merdan, J. Kopecek, and T. Kissel. 2002. Prospects for cationic polymers in gene and oligonucleotide therapy against cancer. *Adv. Drug Del. Rev.* 54(5): 715–758.
98. M. Roser, D. Fischer, and T. Kissel. 1998. Surface-modified biodegradable albumin nano and microspheres II. Effect of surface charges on in vitro phagocytosis and biodistribution in rats. *Eur. J. Pharm. Biopharm.* 46: 255–263.
99. T. I. Armstrong, S. M. Moghimi, S. S. Davis, and L. Illum. 1997. Activation of mononuclear phagocyte system by poloxamine 908: Its implications for targeted drug delivery. *Pharm Res.* 14: 1629–1633.
100. J. Vandorpe, E. Schaht, S. Dunn, A. Hawley, S. Stolnik, S. S. Davis, M. C. Garnett, M. C. Davies, L. Illum. 1997. Long-circulating biodegradable poly(phosphazene) nanoparticles surface modified with poly(phosphazene) and poly(ethyleneoxide) copolymer. *Biomaterials* 18(17): 1147–1152.
101. J. C. Neal et al. 1998. Modification of the copolymers polaxomer 407 and poloxamine 908 can affect the physical and biological properties of surface modified nanospheres. *Pharm Res.* 15: 324–328.
102. S. Medda, P. Jaisankar, R. K. Manna et al. 2003. Phospholipid microspheres: A novel delivery mode for targeting antileishmanial agent in experimental leishmaniasis. *J. Drug Targ.* 11: 123–128.
103. S. M. Moghimi, A. E. Hawley, N. M. Christy, T. Gray, L. Illum, and S. S. Davis. 1994. Surface engineered nanospheres with enhanced drainage into lymphatics and uptake by macrophages of the regional lymph nodes. *FEBS Lett.* 344(1): 25–30.
104. A. E. Hawley, L. Illum, and S. S. Davis. 1997. Lymph node localization of biodegradable nanospheres surface modified with poloxamer and poloxamine block copolymers. *FEBS Lett.* 400(3): 319–323.
105. A. E. Hawley, L. Illum, and S. S. Davis. 1997. Preparation of biodegradable surface engineered PLGA nanospheres with enhanced lymphatic drainage and lymph node uptake. *Pharm Res.* 14: 657–661.
106. J. M. Harris. 1992. *Poly(ethyleneglycol) Chemistry: Biotechnical and Biomedical Applications*. Plenum Press: New York.
107. A. Benigni, S. Tomasoni, and G. Remuzzi. 2002. Impediments to successful gene transfer to the kidney in the context of transplantation and how to overcome them. *Kidney Int.* 61: 115–119.
108. M. Brightman. 1992. Ultrastructure of the brain endothelium. M.W.B. Bradbury (ed.) *Physiology and the Pharmacology of the Blood Brain Barrier. Handbook of Experimental Pharmacology*. Springer: Berlin, Heidelberg, vol. 103, pp. 1–22.
109. J. Kreuter. 2001. Nanoparticle systems for brain delivery of drugs. *Adv. Drug Del.* 47: 65–81.
110. J. Kreuter. 2002. Transport of drugs across the blood brain barrier by the nanoparticles. *Curr. Med. Chem. Cent. Nerv. Sys. Agents* 2: 241–249.
111. A. Szewczyk and L. Wojtczak. 2002. *Pharmacol. Rev.* 54: 101–127.
112. G. G. D'Souza, S. V. Boddapati, and V. Weissig. 2005. Mitochondrial leader sequence-plasmid DNA conjugates delivered into mammalian cells by DQAsomes co-localize with mitochondria. *Mitochondrion* 5: 352–358.
113. F. Rodriguez, L. L. An, S. Harkins et al. 1998. DNA immunization with minigenes: Low frequency of memory cytotoxic T lymphocytes and inefficient antiviral protection are rectified by ubiquitination. *J. Virol.* 72: 5174–5181.

114. F. Rodriguez, J. Zhang, and J. L. Whitton. 1997. DNA immunization: Ubiquitination of a viral protein enhances cytotoxic T-lymphocyte induction and antiviral protection but abrogates antibody induction. *J. Virol.* 71: 8497–8503.
115. G. Delogu, A. Howard, F. M. Collins, and S. L. Morris. 2000. DNA vaccination against tuberculosis: Expression of a ubiquitin-conjugated tuberculosis protein enhances antimycobacterial immunity. *Infect. Immun.* 68: 3097–3102.
116. T. Tobery and R. F. Siliciano. 1999. Cutting edge: Induction of enhanced CTL-dependent protective immunity in vivo by N-end rule targeting of a model tumor antigen. *J. Immunol.* 162: 639–642.
117. S. Brunner, E. Furtbauer, T. Sauer, M. Kursa, and E. Wagner. 2002. Overcoming the nuclear barrier: Cell cycle independent nonviral gene transfer with linear polyethylenimine or electroporation. *Mol. Ther.* 5(1): 80–86.
118. W. T. Godbey, K. K. Wu, and A. G. Mikos. 1999. Tracking the intracellular path of poly(ethylenimine)/DNA complexes for gene delivery. *Proc. Natl. Acad. Sci. USA* 96: 5177–5181.
119. L. J. Branden, A. J. Mohamed, and C. I. Smith, 1999. Peptide nucleic acid–nuclear localization signal fusion that mediates nuclear transport of DNA. *Nat. Biotechnol.* 17: 784–787.
120. M. A. Zanta, P. Belguise-Valladier, and J. P. Behr. 1999. Gene delivery: A single nuclear localization signal peptide is sufficient to carry DNA to the cell nucleus. *Proc. Natl. Acad. Sci. USA* 96: 91–96.
121. C. K. Chan and D. A. Jans. 1999. Enhancement of polylysine-mediated transfection by nuclear localization sequences (NLSs): Polylysine does not function as an NLS. *Hum. Gene Ther.*10: 1695–1702.
122. C. W. Pouton. 1998. Nuclear import of polypeptides, polynucleotides and supramolecular complexes. *Adv. Drug Del. Rev.* 34: 51–64.
123. R. A. Smith, C. M. Porteous, A. M. Gane, and M. P. Murphy. 2003. Delivery of bioactive molecules to mitochondria in vivo. *Proc. Natl. Acad. Sci. USA* 100(9): 5407–5412.
124. M. P. Murphy and R. A. Smith. 2000. Drug delivery to mitochondria: The key to mitochondrial medicine. *Adv. Drug Del. Rev.* 41: 235–250.
125. A. Muratovska, R. N. Lightowlers, R. W. Taylor, J. A. Wilce, and M. P. Murphy. 2001. Targeting large molecules to mitochondria. *Adv. Drug Del. Rev.* 49(1–2): 189–198.
126. A. Szewczyk and L. Wojtczak. 2002. Mitochondria as a pharmacological target. *Pharmacol. Rev.* 54: 101–127.
127. A. Weissig, J. Lasch, G. Erdos et al. 1998. DQAsomes: A novel potential drug and gene delivery system made from dequalinium. *Pharm. Res.* 15: 334–337.
128. N. K. Subbarao, R. A. Parente, F. C. Szoka, Jr, L. Nadasdi, and K. Pongracz. 1987. The pH dependent bilayer destabilization by an amphipathic molecule. *Biochemistry* 26(11): 2964–2972.
129. E. Wagner, C. Plank, K. Zatloukal, M. Cotten, and M. L. Bimstiel. 1992. Influenza virus hemagglutinin HA-2 N-terminal fusogenic peptides augment gene transfer by transferrin-polylysine-DNA complexes: Toward a synthetic virus-like gene-transfer vehicle. *Proc. Natl. Acad. Sci. USA* 89(17): 7934–7938.
130. J. Haensler and F. C. Szoka, Jr. 1993. Polyamidoamine cascade polymers mediate transfection of cells in culture. *Bioconjugate Chem.* 4: 372–379.
131. O. Boussif, F. Lezoualc'h, M. A. Zanta, M. D. Mergny, D. Scherman, B. Demeneix, and J. P. Behr. 1995. A versatile vector for gene and oligonucleotide transfer into cells in culture and in vivo: Polyethylenimine. *Proc. Natl. Acad. Sci. USA* 92(16): 7297–7301.
132. T. Merdan, K. Kunath, D. Fischer et al. 2002. Intracellular processing of —poly(ethylene imine)/ribozyme complexes can be observed in living cells using confocal laser scanning microscopy and inhibitor experiments. *Pharm. Res.* 19: 140–146.
133. A. Akinc, M. Thomas, A. M. Klibanov, and R. Langer. 2005. Exploring polyethylenimine—mediated DNA transfection and the proton sponge hypothesis. *J. Gene Med.* 7: 657–663.
134. M. X. Tang, C. T. Redemann, and F. C. Szoka, Jr. 1996. In vitro gene delivery by dragged polyamidoamine dendrimers. *Bioconjugate Chem.* 7: 703–714.

135. W. T. Godbey, M. A. Barry, P. Saggau, K. K. Wu, and A. G. Mikos. 2000. Poly(ethyleneimine) mediated transfection: A new paradigm for gene delivery. *J. Biomed. Mater. Res.* 51(3): 321–328.
136. B. Neukamm, A. Weimann, S. Wu, M. Danevad, C. Lang, and R. Geßner. 2006. Novel two stage screening procedure leads to the identification of a new class of transfection enhancers. *J. Gene Med.* 8(6): 745–753.
137. C. C. Mello and D. Conte Jr. 2004. Revealing the world of RNA interference. *Nature* 431: 338–342.
138. V. Sokolova and M. Epple. 2007. Inorganic nanoparticles as carriers of nucleic acids into cells. *Angew. Chem. Int. Ed.* 47: 1382–1395.
139. E. Wagner, C. Plank, K. Zatloukal, M. Cotten, and M. L. Bimstiel. 1992. Influenza virus hemagglutinin HA-2 N-terminal fusogenic peptides augment gene transfer by transferrin-polylysine-DNA complexes: Toward a synthetic virus-like gene-transfer vehicle. *Proc. Natl. Acad. Sci. USA* 89(17): 7934–7938.
140. M. A. Wolfert, P. R. Dash, O. Nazarova, and D. Oupicky. 1999. Polyelectrolyte vectors for gene delivery: Influence of cationic polymer on biophysical properties of complexes formed with DNA. *Bioconjugate Chem.* 10: 993–1004.
141. R. Kircheis, L. Wightman, and E. Wagner. 2001. Design and gene delivery activity of modified polyethylenimines. *Adv. Drug Del. Rev.* 53: 341–358.
142. S. Kazuyoshi and S. W. Kim. 2002. Multipulse drug permeation across a membrane driven by a chemical pH-oscillator. *J. Control. Release* 79: 271–281.
143. C. Rudolph, U. Schillinger, and C. Plank. 2002. Nonviral gene delivery to the lung with copolymer-protected and transferrin-modified polyethylenimine. *Biochim. Biophys. Acta* 1573: 75–83.
144. D. G. Anderson, D. M. Lynn, and R. Langer. 2003. Semi-automated synthesis and screening of a large library of degradable cationic polymers for gene delivery. *Angew. Chem. Int. Ed.* 42: 3153–3158.
145. G. Schmid. 1992. Large clusters and colloids: Metals in the embryonic state. *Chem. Rev.* 92: 1709–1727.
146. M. Oishi, J. Nakaogami, T. Ishii, and Y. Nagasaki. 2006. Smart PEGylated gold nanoparticals for the cytoplasmic delivery of siRNA to induce enhanced gene silencing. *Chem. Lett.* 35: 1046–1047.
147. N. L. Rosi, D. A. Giljohann, C. S. Thaxton, A. K. R. Lytton-Jean, M. S. Han, and C. A. Mirkin. 2006. Oligonucleotide-modified gold nanoparticles for intracellular gene regulation. *Science* 312: 1027–1030.
148. Z. Liu, M. Winters, M. Holodniy, and H. Dai. 2007. siRNA Delivery into Human T Cells and primary cells with carbon-nanotube transporters. *Angew. Chem. Int. Ed.* 46: 2023–2027.
149. P. Gould. 2004. Nanoparticles probe biosystems. *Mater. Today* 7: 36–43.
150. I. J. Bruce, J. Taylor, M. Todd, M. J. Davies, E. Borioni, C. Sangregorio, and T. Sen. 2004. Synthesis, characterisation and application of silica-magnetite nanocomposites. *J. Magn. Magn. Mater.* 284: 145–160.
151. A. Campo, T. Sen, J. P. Lellouche, and I. J. Bruce. 2005. Multifunctional magnetite and silica–magnetite nanoparticles: Synthesis, surface activation and applications in life sciences. *J. Magn. Magn. Mater.* 293: 33–40.
152. N. Morishita, H. Nakagami, R. Morishita et al. 2005. Magnetic nanoparticles with surface modification enhanced gene delivery of HVJ-E vector. *Biochem. Biophys. Res. Commun.* 334: 1121–1126.
153. D. Luo and W. M. Saltzman. 2000. Enhancement of transfection by physical concentration of DNA at the cell surface. *Nat. Biotechnol.* 18: 893–895.
154. R. A. Gemeinhart, D. Luo, and W. M. Saltzman. 2005. Cellular fate of a modular DNA delivery system mediated by silica nanoparticles. *Biotechnol. Prog.* 21: 532–537.
155. H. Shen, J. Tan, and W. M. Saltzman. 2004. Surface-mediated gene transfer from nanocomposites of controlled texture. *Nat. Mater.* 3: 569–574.
156. F. L. Graham and A. J. van der Eb. 1973. A new technique for the assay of infectivity of human adenovirus 5 DNA. *Virology* 52: 456–467.
157. A. Maitra. 2005. Calcium phosphate nanoparticles: Second-generation nonviral vectors in gene therapy. *Expert Rev. Mol. Diagn.* 5: 893–905.

158. Y. Kakizawa and K. Kataoka. 2002. Block copolymer self-assembly into monodispersive nanoparticles with hybrid core of antisense DNA and calcium phosphate. *Langmuir* 18: 4539–4543.
159. D. Olton, J. Li, M. E. Wilson. T. Rogers, J. Close, L. Huang, P. N. Kumta, and C. Sfeir. 2007. Nanostructured calcium phosphates (NanoCaPs) for non-viral gene delivery: Influence of the synthesis parameters on transfection efficiency. *Biomaterials* 28(6): 1267–1279.
160. S. Aisawa, H. Hirahara, K. Ishiyama, W. Ogasawara, Y. Umetsu, and E. Narita. 2003. Sugar–anionic clay composite materials: Intercalation of pentoses in layered double hydroxide. *J Solid State Chem.* 174(2): 342–348.
161. B. Li, J. He, D. G. Evans, and X. Duan. 2004. Inorganic layered double hydroxides as a drug delivery system—intercalation and in vitro release of fenbufen. *Appl. Clay Sci.* 27: 199–207.
162. J. H. Choy, S. Y. Kwak, Y. J. Jeong, and J. S. Park. 2000. Inorganic layered double hydroxides as non viral vectors. *Angew. Chem. Int. Ed.* 39: 4041–4045.
163. H. M. Kipen and D. L. Laskin. 2005. Smaller is not always better: Nanotechnology yields nanotoxicology. *Am. J. Physiol: Lung Cell. Mol. Physiol.* 289: L696–L697.
164. N. Lewinski, V. Colvin, and R. Drezek. 2008. Cytotoxicity of nanoparticles. *Small* 4: 26–49.
165. M. Tarantola, D. Schneider, E. Sunnick, H. Adam, S. Pierrat, C. Rosman, V. Breus, C. Sonnichsen, T. Basche, J. Wegener, and A. Janshoff. 2009. Cytotoxicity of metal and semiconductor nanoparticles indicated by cellular micromotility. *ACS Nano* 3(1): 213–222.
166. Y. Pan, S. Neuss, A. Leifert, M. Fischler, F. Wen, U. Simon, G. Schmid, W. Brandau, and W. Jahnen-Dechent. 2007. Size-dependent toxicity of gold nanoparticles. *Small* 3(11): 1941–1949.
167. A. Magrez, S. Kasas. V. Salicio, N. Pasquier, J. W. Seo, M. Celio, S. Catsicas, B. Schwaller, and L. Forró. 2006. Cellular toxicity of carbon based nanomaterials. *Nano Lett.* 6(6): 1121–1125.
168. C. M. Sayes, A. M. Marchione, K. L. Reed, and D. B. Warheit. 2007. Comparative pulmonary toxicity assessments of C60 water suspensions in rats: Few differences in fullerene toxicity in vivo in contrast to in vitro profiles. *Nano Lett.* 7(8): 2399–2406.
169. A. Nan, X. Bai, S. J. Son, S. B. Lee, and H. Ghandehari. 2008. Cellular uptake and cytotoxicity of silica nanotubes. *Nano Lett.* 8(8): 2150–2154.
170. L. Ding, J. Stilwell, T. Zhang, O. Elboudwarej, H. Jiang, J. P. Selegue, P. A. Cooke, J. W. Gray, and F. F. Chen. 2005. Molecular characterisation of the cytotoxic mechanism of multiwall carbon nanotubes and nano onions on human skin fibroblasts. *Nano Lett.* 5(12): 2448–2464.
171. J. Miyawaki, M. Yudasaka, T. Azami Y. Kubo, and S. Iijima. 2008. Toxicity of single walled carbon nanohorns. *ACS Nano* 2(2): 213–226.
172. A. M. Derfus, W. C. W. Chan, and S. N. Bhatia. 2004. Probing the cytotoxicity of semiconductor quantum dots. *Nano Lett.* 4(1): 11–18.
173. C. M. Sayes, J. D. Fortner, W. Guo, D. Lyon, A. M. Boyd, K. D. Ausman, Y. J. Tao et al. 2004. The differential cytotoxicity of water soluble fullerenes. *Nano Lett.* 4(10): 1881–1887; G. Bhabra, A. Sood, B. Fisher, L. Cartwright, M. Saunders, W. H. Evans, A. Surprenant et al. 2009. Nanoparticles can cause DNA damage across a cellular barrier. *Nature Nanotech.* 4: 876–883.
174. J. M. Worle-Knirsch, K. Pulskamp, and H. F. Krug. 2006. Oops they did it again! Carbon nanotubes hoax scientists in viability assays. *Nano Lett.* 6(6): 1261–1268.
175. J. S. Patton. 2006. A historical perspective on convergence technology. *Nature Biotech.* 24(3): 280–281.

13

Nanobiomaterials for Nonviral Gene Delivery

Xiujuan Zhang
Tulane University

Daniel A. Balazs
Tulane University

W.T. Godbey
Tulane University

13.1 Introduction

Since its first clinical trial in 1990 (Anderson 1992), the field of gene therapy has grown exponentially and drawn more and more attention from the fields of biotechnology, pharmaceutical research, and medicine (Park et al. 2006; Ragusa et al. 2007). Despite the added attention, only a minor percentage of clinical trials utilizing gene delivery have reached phase III, which indicates the level of difficulty associated with the methods and the strict restrictions to such promising and at the same time risky treatments (Ragusa et al. 2007).

One important obstacle to gene therapy is the delivery of therapeutic polynucleotides past the plasma membrane and into the cells of interest. The delivery of naked DNA has yielded little success due mainly to limited cellular uptake; its use is only feasible in select tissues such as the skeletal muscle (Giannoukakis et al. 1999; Liu et al. 2001a). Because of this limitation in uptake, the development of efficient and safe delivery systems has been an essential component of gene therapy research. A good gene delivery vector should be able to effectively compact and protect DNA, bypass the immune system of the host, traverse the plasma membrane (typically through endocytosis), disrupt the endosomal membrane, and deliver the DNA into the nucleus (Mahato 1999). (In gene therapy, not only could DNA be delivered, but small interfering RNA [siRNA] molecules could be used instead for silencing the expression of specific host genes. The use of RNA entails different cellular goals versus DNA delivery. However, since this chapter is focused on gene delivery, siRNA delivery will not be discussed further.)

Gene delivery systems can be divided into two general categories: viral transduction systems and nonviral transfection systems. Although viral gene delivery vectors have an established history of efficacy, the nonviral delivery systems have merits in that there is limited-to-no induction of immune responses, there is virtually no limitation on the size of the genes that can be delivered, and the cost of production is relatively low (Lee and Kim 2005). For nonviral vectors, the size of the delivery complexes depends on the molecular weight of the vector, the ratio between the vector nitrogens and the DNA phosphates (termed the N:P ratio), and the salt concentration of the buffer solution. For example, complexes of poly(L-lysine) (PLL)/DNA with N:P ratios greater than 0.5 form either 25–50 nm toroids or 40–80 nm rods (Kwoh et al. 1999) and complexes of poly(ethylenimine) (PEI)/DNA (using 25 kDa PEI) at N:P = 2.3:1 are homogenous 40–60 nm toroids (Tang and Szoka 1997). (The same group also reported that the size of PEI/DNA complexes ranged from 90 to 130 nm when dynamic light scattering was used, as opposed to the 40–60 nm seen when electron microscopy was used [Tang and Szoka 1997].) In general, the sizes of the complexes formed by the nonviral gene delivery vector and DNA fit into the nanoparticle category, which is usually defined as particles having diameters less than 100 nm. Using this definition, viral gene delivery vectors can also be considered nanoparticles due to the fact that viruses typically have a maximum dimension between 10 and 100 nm (www.nano.gov). In this chapter, we will focus upon the development of nonviral vectors used for gene therapy.

Apart from the polycations, another class of nonviral gene delivery materials is the cationic lipid. A solution of cationic lipids, often formed with neutral helper lipids, can be mixed with DNA to form a positively charged complex termed a lipoplex (Wasungu and Hoekstra 2006). Well-characterized and widely used commercial reagents for cationic lipid transfection include *N*-[1-(2,3-dioleyloxy)propyl]-*N*,*N*,*N*-trimethylammonium chloride (DOTMA) (Felgner et al. 1987), [1,2-bis(oleoyloxy)-3-(trimethylammonio) propane] (DOTAP) (Leventis and Silvius 1990), 3β[*N*-(*N*′,*N*′-dimethylaminoethane)-carbamoyl]cholesterol (DC-Chol) (Gao and Huang 1991), and dioctadecylamidoglycylspermine (DOGS) (Behr et al. 1989). Dioleoylphosphatidylethanolamine (DOPE), a neutral lipid, is often used in conjunction with cationic lipids because of its membrane destabilizing effects at low pH, which aid in endolysosomal escape (Farhood et al. 1995).

Many cationic lipid compounds have been formulated since the advent of DOTMA (Behr 1994; Farhood et al. 1994). Each lipid has different structural aspects that confer distinct characteristics to the lipid/DNA complex, which affect association with and uptake into the cell. However, the basic structure of cationic lipids mimics the chemical and physical attributes of biological lipids (Maurer et al. 1999). Cationic lipids used for gene delivery have a hydrophobic region that is often in the form of two fatty acid tails linked by a (glycerol) backbone molecule to a cationic head-group. The positive charge facilitates spontaneous electrostatic interaction with DNA as well as binding of the resulting lipoplexes to the negatively charged components of the cell membrane prior to cellular uptake (Nicolau and Papahadjopulos 1998). The use of a cation is a recurring theme for virtually all chemically mediated gene delivery vectors, including polymers, lipids, and nondegradable nanoparticles.

13.2 Cationic Polymers

DNA, when combined with sufficient amounts of cationic polymers, will condense into discrete entities known as polyplexes (Vuorimaa et al. 2008). The polyplexes are compact nanoparticles formed through electrostatic interactions between the positive charges of amines and the negative charges of DNA phosphates. The strength of DNA binding to the polymers is related to the N:P ratio. For example, in theory, the DNA could be completely complexed by the 25 kDa branched PEI at N:P ratios of 1.0, but agarose gel electrophoresis data imply that complete complexation does not occur below N:P = 2.0 (Zanta et al. 1997). Moreover, the surface charge concentrations (zeta potentials) of the polycation/DNA complexes, which can range from strongly negative to strongly positive (Erbacher et al. 1999), is also a function of the N:P ratio. For the branched 25 kDa PEI, the zeta potential is about −50 mV at N:P = 2, around 0 mV at N:P = 3, and around 20 mV at N:P = 10 (Zanta et al. 1997; Erbacher et al. 1999; Godbey et al. 1999a).

Positive zeta potentials indicate an excess of positive charges on the surface of the nanoparticles, which aids in their attachment to the negatively charged exterior of the plasma membrane, facilitating the internalization of the nanoparticles.

The most common cationic polymers used as nonviral gene delivery vectors include chitosan, PLL, PEI, poly(amido amine) (PAMAM) dendrimers, and select polypeptides.

13.2.1 Chitosan-Based Gene Delivery Systems

Chitosan is obtained by the alkaline deacetylation of chitin, which is the second most abundant polysaccharide in nature (Synowiecki and Al-Khateeb 2003). The main commercial sources of chitosan are the crustacean shell wastes of crabs, shrimps, and lobsters (Koping-Hoggard et al. 2001; Hejazi and Amiji 2003). Chitosan is a polysaccharide composed of randomly distributed β-(1-4)-linked D-glucosamines and *N*-acetyl-D-glucosamines, with different molecular weights (50–200 kDa), degrees of deacetylation (40%–98%), and viscosities (Illum 1998). Chitosan is biodegradable, biocompatible, and nontoxic at low molecular weights (10–50 kDa) (Lee et al. 2001). However, chitosan has shown concentration-dependent cytoxicity in B16F10 cells *in vitro* (Carreño-Gómez and Duncan 1997). It has been suggested that the toxicity of chitosan is perhaps due to impurities in the chitosan polymers (Koping-Hoggard et al. 2001).

Chitosan is a robust vector with kinetics that differ from those of other polycations. *In vitro* transfections of Hela cells yielded positive results even in the presence of 10% serum, and gene expression increased over time to be 10 times more efficient than PEI over 96 h (Erbacher et al. 1998). In separate trials using 293 cells (Koping-Hoggard et al. 2001), it was found that the onset of gene expression using chitosan was slower than when PEI was used, but gene expression increased over time. After 120–144 h, chitosan yielded a maximal transgene expression in 293 cells that was lower, but statistically comparable, to that yielded by PEI. The group also reported that the transfection efficiency of chitosan in HT-1080 and Caco-2 cell lines was found to be much lower than that of PEI, which suggested that the transfection ability of chitosan was cell line dependent (Koping-Hoggard et al. 2001).

Because of the adhesive and transport properties of chitosan in the gastrointestinal (GI) tract, this polymer has also been used to form chitosan/DNA polyplexes for oral gene therapy applications (Roy et al. 1999; Chew et al. 2003). Chitosan/pCMVArah2 (Arah2 is the dominant anaphylaxis-inducing antigen in mice sensitized to peanuts) has been administered into a strain of mouse from the Jackson Laboratory (AKR/J) mice as an oral immunization method for modulating peanut antigen-induced murine anaphylactic responses (Roy et al. 1999). Chitosan/pDer p 1 (Der p 1 is a major triggering factor for mite allergy) has been investigated for its potential as an oral vaccination against mite allergy with promising results (Chew et al. 2003). These two studies show that chitosan possesses characteristics suitable for oral vaccination and could potentially be used for oral gene delivery in general.

Derivatives of chitosan have been developed in an attempt to target specific cell types and improve transfection efficiencies. Examples of such modifications include conjugation with folate (Mansouri et al. 2006; Chan et al. 2007), thiolation (Lee et al. 2007), and glycolation (Yoo et al. 2005). These modifications reportedly did not affect the ability of chitosan to condense and compact DNA, but they did produce enhanced gene expression in the targeted cells.

13.2.2 PLL-Based Gene Delivery Systems

PLL is a well-known polycation that has been widely studied as a nonviral gene delivery vector since the first reported formation of PLL/DNA complexes (Laemmli 1975). PLL is a polypeptide of the essential amino acid L-lysine that can be produced via bacterial fermentation (Shima 1977). At physiological pH, each repeating unit of PLL carries a positive charge on the ε-amine of the side chain, a property that has been exploited to allow PLL to condense plasmid DNA to varying degrees depending upon salt concentration (Gonsho et al. 1994). However, aggregation and precipitation of PLL/DNA complexes have also been found to be dependent upon salt concentration (Liu et al. 2001b).

To combat aggregation, the surfactant dextran has been used to increase the solubility and stability of PLL/DNA complexes without considerably hindering the electrostatic interaction between PLL and DNA (Ferdous et al. 1998).

Although the structure of PLL appears to be suitable for gene delivery, unmodified versions of this polymer are associated with low transfection efficiencies and cytotoxicity. Partially gluconoylated PLL, by reaction with β-gluconolactone, has improved transfection efficiencies in HepG2 cells over those of unmodified PLL (Erbacher et al. 1997). The acylation with β-gluconolactone partially blocked the ε-amino groups of the PLL, and as a result the electrostatic interactions between the PLL and DNA were decreased, promoting dissociation between the plasmid and the carrier (Erbacher et al. 1997). Another modification to PLL has been the conjugation of poly(ethylene glycol) (PEG) at the ε-amino group, with the result of reduced toxicity and improved transfection efficiency (30-fold) in HepG2 cells (Choi et al. 1998a). Glycosylation has also been used in an attempt to decrease cytotoxicity (Boussif et al. 1999), and while improved cell survival was achieved with this approach, transfection efficiencies were low and remained comparable to unmodified PLL (Boussif et al. 1999).

Just as with other nonviral gene delivery vehicles, PLL had been conjugated with various ligands in an attempt to improve the specific cellular uptake of the targeted cells or tissues while at the same time reducing side effects to neighboring tissues. For example, lactose has been attached to PLL to target the asialoglycoprotein of hepatocytes (Midoux et al. 1993; Choi et al. 1998b). Folate has also been conjugated to PLL in an attempt to target the folate-overexpressing cancer cells (Cho et al. 2005).

13.2.3 PEI-Based Gene Delivery Systems

Branched PEI is produced by the acid-catalyzed polymerization of aziridine (Dick and Ham 1970). PEI has been employed in industry for years in processes such as shampoo manufacturing, paper production, and water purification. However, PEI was not, until 1995, introduced as a "versatile vector" for gene delivery (Boussif et al. 1995; Godbey et al. 2000). At that time, Boussif et al. (1995) used PEI as the vector to deliver plasmids coding for luciferase into various cell types including 3T3, HepG2, Cos-7, and chicken embryonic hypothalamic neurons, and into newborn mice. PEI has been a popular cationic gene delivery vehicle because of its relatively high transfection efficiency in a variety of cell lines (Boussif et al. 1996; Godbey et al. 2000; Oh et al. 2007; Yao et al. 2007; Ye et al. 2007). Like PLL, PEI/DNA complexes are prone to aggregate (Tang and Szoka 1997; Sharma et al. 2005), and overdosage of PEI is toxic to the cells. Many investigations have focused on the PEGylation of PEI in an attempt to reduce aggregation and the cytotoxicity (Ogris et al. 1999; Lee et al. 2001; Petersen et al. 2002; Shi et al. 2003). It has been shown that PEI cytotoxicity in 3T3 fibroblasts can be modulated by the degree of PEGylation, independent of the molecular weight of the PEG used (Petersen et al. 2002). This is consistent with observations that an overabundance of cations in gene delivery complexes is deleterious to cells.

In order for the transfecting complexes to reach cells of interest, many researchers have studied the conjugation of PEI with targeting moieties. An $\alpha_v\beta_3/\alpha_v\beta_5$ integrin-binding the tri-peptide Argening-Glycine-Aspartate (RGD) peptide, ACDCRGDCFC, has been conjugated into PEI via a PEG spacer to target angiogenic human dermal microvascular endothelial cells (HDMEC) (Suh et al. 2002). The conjugated PEI/DNA complexes showed approximately five times greater transfection efficiencies in vascular endothelial growth factor (VEGF)-stimulated angiogenic HDMEC versus unconjugated PEI/DNA complexes. However, the conjugated PEI/DNA complexes showed much lower transfection efficiencies in angiostatic HDMEC than unconjugated PEI/DNA complexes, which suggested that the RGD conjugated PEI was highly selective toward angiogenic endothelial cells. The human epidermal growth factor receptor-2 (HER-2) has also been used for targeted gene delivery, in this case to target breast cancer cells (Chiu et al. 2004). The HER-2-conjugated PEI showed enhanced transfection efficiencies in HER-2 overexpressing human breast adenocarcinoma cells (Sk-Br3) as compared to unmodified PEI, but not in breast cancer cells expressing low levels of HER-2 (MDA-MB-231). Other molecules conjugated with PEI for targeting include anti-CD3 (Kircheis et al. 1997), transferrin (Ogris et al. 1998), and folate (Benns et al. 2001).

The linear form of PEI (L-PEI) is also being used for gene delivery. L-PEI also arises from cationic polymerization, but from a 2-substituted 2-oxazoline monomer (instead of aziridine). The polymerization product, for example, linear poly(*N*-formalethylenimine) is then hydrolyzed to yield L-PEI. The linear form of PEI can also be obtained by the same process as that used to obtain branched PEI, but the reaction must take place at a relatively low temperature (reviewed and described in Tomalia and Killat 1985).

Linear PEI has been widely studied as a gene delivery vector, especially in the transfection of the lung (Goula et al. 1998; Goula et al. 2000; Uduehi et al. 2001a,b). It has been reported that L-PEI/DNA complexes can pass the capillary barrier in the lung to reach and transfect other pulmonary cell types with minimal-to-no toxicity (Goula et al. 1998; Goula et al. 2000). However, as separate investigations have shown, L-PEI-mediated gene delivery to rat lungs is associated with the moderate impairment of lung function (Uduehi et al. 2001a,b).

13.2.4 Dendrimer-Based Gene Delivery Systems

The first article using the term "dendrimer" was written by Tomalia et al. (1985) in which the preparation of PAMAM dendrimers was described in detail. At the same time though, Newkome et al. (1985), independently reported the synthesis of similar macromolecules that they termed arborols. "Dendrimer" is the term that is in general usage today.

Dendrimers are branched polymers that are synthesized in a stepwise fashion to control both monodispersity and the exact number of branching layers, or "generations" (Fréchet and Tomalia 2001). Dendrimers can be synthesized by either divergent or convergent methods (Tomalia and Fréchet 2002). For the divergent method, the dendrimer grows in a stepwise fashion outwards from a multifunctional core molecule (Figure 13.1). Slight structural defects can occur in larger molecules, especially at higher generation numbers. For the convergent method, the dendrimer is constructed beginning with the end groups and progressing inwards (Figure 13.1). Defective structures can be more readily separated with this method (Tomalia et al. 1985). Unlike hyperbranched polymers, dendrimers are polymerized in a tightly controlled, stepwise fashion to produce monodisperse sets of macromolecules. Because of their

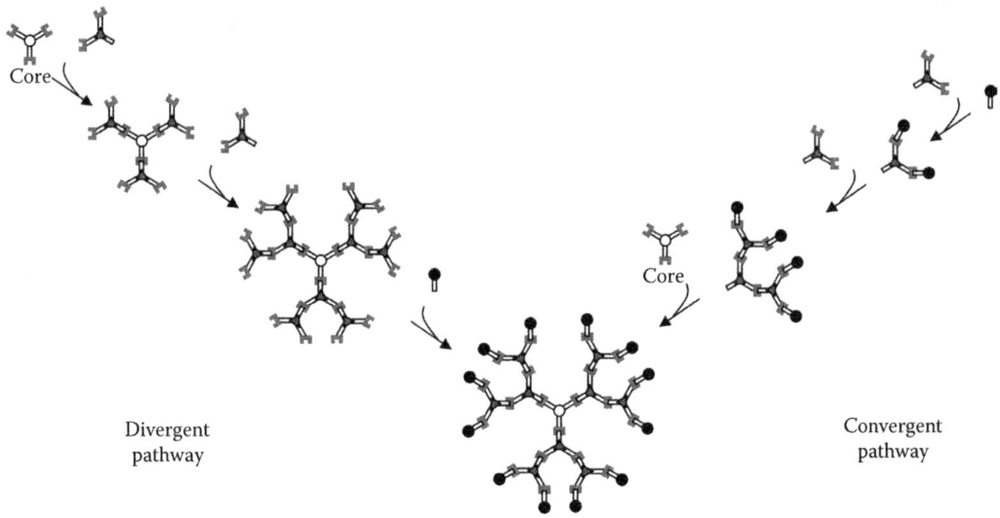

FIGURE 13.1 Dendrimers can be constructed via divergent or convergent pathways. In the divergent pathway (left), the dendrimer is extended outward from a multifunctional core molecule, often ending with a functionalized terminal group. The convergent method (right) begins at the outer ends and polymerization extends toward what will be the interior of the dendrimer, ending with the addition of the core molecule.

globular shapes and the presence of internal cavities, dendrimers could be used to encapsulate guest molecules. Dendrimer/DNA complexes are often called dendriplexes to preserve terminology that is analogous with lipoplexes and polyplexes. PAMAM, and PAMAM derivatives have been investigated as gene delivery vectors both *in vitro* and *in vivo* (Bielinska et al. 1996; Kukowska-Latallo et al. 2000; Kim et al. 2004; Huang et al. 2007). A commercially available dendrimeric formulation is sold under the name SuperFect™ (Qiagen, Valencia, CA). This reagent consists of activated dendrimers with a defined spherical architecture. The low pK_as of the amines (3.9 and 6.9) afford the dendrimer the potential to buffer pH changes during the acidification of the endosome (Klajnert and Bryszewska 2001), which may contribute to the favorable transfection efficiencies associated with SuperFect/DNA complexes.

13.2.5 Peptide-Based Gene Delivery Systems

Poly(L-lysine) (PLL), as introduced earlier, was one of the first studied cationic peptides used to mediate gene delivery. However, as the length of PLL increases, so does the cytotoxicity. Moreover, the polydispersity of PLL complicates modifications with ligands, making the chemical synthesis of PLL conjugates hard to control (Martin and Rice 2007). To solve these problems, the synthesis of oligolysines was investigated and reported (Gottschalk et al. 1996; Wadhwa et al. 1997; Adami and Rice 1999; McKenzie et al. 2000; Yang et al. 2001; Kwok et al. 2003). A peptide with the sequence $YKAK_nWK$ ($4 \leq n \leq 12$) was designed to determine the minimum lysine required for DNA binding and gene expression (Gottschalk et al. 1996), and the results showed that an eight-lysine cluster was sufficient to condense DNA. Efficient gene expression in a variety of cell lines was also shown when an additional endosome-disruptive peptide was coupled to the lysine-containing sequence. Another peptide, CWK_n (n = 3, 6, 8, 13, 16, 18, 26, and 36) was used to transfect HepG2 and Cos-7 cells (Wadhwa et al. 1997). Reports from this investigation showed a 40-fold reduction in particle size and a 1000-fold amplification in transfection efficiency for CWK_{18}/DNA condensates relative to K_{19}. Although oligopeptides containing lysine showed efficient gene transfection *in vitro*, they were not stable enough for *in vivo* transfection due to their low affinity for DNA. Glutaraldehyde (Adami and Rice 1999; Yang et al. 2001) and disulfide bonds (McKenzie et al. 2000; Kwok et al. 2003) have been used to cross-link the peptides in an attempt to stabilize the complexes, and results showed that the cross-linkage stabilized the peptide/DNA condensates and enhanced the metabolic stability of the carried DNA.

Another class of peptides that have been used for gene delivery was created with the goal of inducing endosome disruption. One of the oldest and most well-studied endosome-disruptive peptides is the INF peptide, which was derived from the amino-terminal sequence of the influenza virus hemagglutinin HA-2 (Wagner et al. 1992; Plank et al. 1994). The peptide has membrane perturbation activity that is triggered by an acidic environment. Other endosome-disruptive peptides include trans-activating transcriptional activator (TAT, a 86 amino acid peptide derived from the human immunodeficiency virus-1 [HIV-1]) (Frankel et al. 1988; Ruben et al. 1989), penetratin (a 16-mer peptide derived from the third α-helix of the homeodomain of Antennapedia) (Derossi et al. 1994), and transportan (a 27 amino acid peptide, containing the peptide sequence from the amino terminus of the neuropeptide galani) (Pooga et al. 1998). Both TAT and penetratin are arginine-rich peptides, and transportan consists of amphipathic helical peptides (Gupta et al. 2005). Synthetic amphipathic peptides have also been designed for endosome disruption, and include such molecules as model amphipathic peptide (MAP, an 18-mer peptide) (Oehlke et al. 1998), and GALA (a 30 amino acid peptide, containing a repeating unit of glutamic acid [G]-alanine [A]-leucine [L]-alanine [A]) (Subbarao et al. 1987).

13.3 Cationic Lipids

A cationic lipid is an amphipathic molecule that consists of a hydrophobic region and a hydrophilic region (Wasungu and Hoekstra 2006). The hydrophobic region usually consists of one or two hydrophobic fatty acid chains linked via a glycerol backbone to a cationic head group that can vary based on how the lipid is being administered (*in vitro* or *in vivo*), and what cell lines are being transfected (Wasungu and Hoekstra 2006). A cholesterol-like moiety is sometimes used with or instead of a fatty acid. The hydrophobic group of the molecule allows for self-assembly into micelles or liposomes of varying morphologies, depending on the lipid being used (Tranchant et al. 2004).

There are a number of structures that are known to appear during polynucleotide compaction in the liposome. Each structure is formed in the most energetically favorable conformation based upon the characteristics of the lipids used in the system (Israelachvili 1991). The structure-packing parameter suggests what shape the amphiphile will take depending on the ratio of size variables. The packing parameter is

$$P = \frac{v}{al_c}$$

where

v is the volume of the hydrocarbon
a is the effective area of the head group
l_c is the length of the lipid tail

This correlation predicts a range of structures according to the following conditions (Israelachvili 1991; Hsu et al. 2005) (Figure 13.2):

$P < 1/3 \rightarrow$ spherical micelle
$1/3 \leq P < 1/2 \rightarrow$ cylindrical micelle
$1/2 \leq P < 1 \rightarrow$ flexible bilayers, vesicles
$P = 1 \rightarrow$ planar bilayers
$P > 1 \rightarrow$ inverted micelles (hexagonal (H_{II}) phase)

Between 8 and 18 carbons commonly constitute the hydrocarbon tails of lipids used for gene delivery. The tails are typically saturated, but a single double bond is occasionally seen. The combination of hydrocarbon chains attached to glycerol can be symmetric or asymmetric. It has been shown that asymmetric lipids with both a shorter saturated carbon chain lipid and a long unsaturated carbon chain produce relatively high transfection efficiencies as compared to symmetric cationic lipids (Ferrari et al. 2002).

Hydrophobic tails are not the only liposomal features that play a role in effective gene delivery—ionizable head groups are also involved. Some examples are the multivalent cationic lipids 2,3-dioleylo xy-*N*-[2(sperminecarboxamido)ethyl]-*N*,*N*-dimethyl-l-propanaminium trifluoroacetate (DOSPA) and DOGS (covered in Section 13.3.2), both of which have a functionalized spermine head group that confers the ability to act as a buffer, such as in the case where there is an influx of protons into a maturing endosome/endolysosome (Remy et al. 1994). The buffering could extend the amount of time needed to activate acid hydrolases, and could explain why some multivalent cationic lipids can exhibit higher transfection efficiencies versus their monovalent counterparts (Behr et al. 1989; Uchida et al. 2002).

13.3.1 Monovalent Cationic Lipids

13.3.1.1 DOTMA

DOTMA, was one of the first synthesized and commercially available cationic lipids used for gene delivery (Figure 13.3). Its structure consists of two unsaturated oleoyl chains (C18:1Δ^9), bound by an ether bond to the three-carbon skeleton of a glycerol, with a quarternary amine as the cationic head group (Felgner et al. 1987). As compared to other methods of gene transfer used in the late 1980s, DOTMA proved to facilitate up to 100-fold more efficient gene delivery than the use of diethylaminoethyl (DEAE)-dextran coprecipitation or calcium phosphate (Felgner et al. 1987). The ability to entrap DNA or RNA

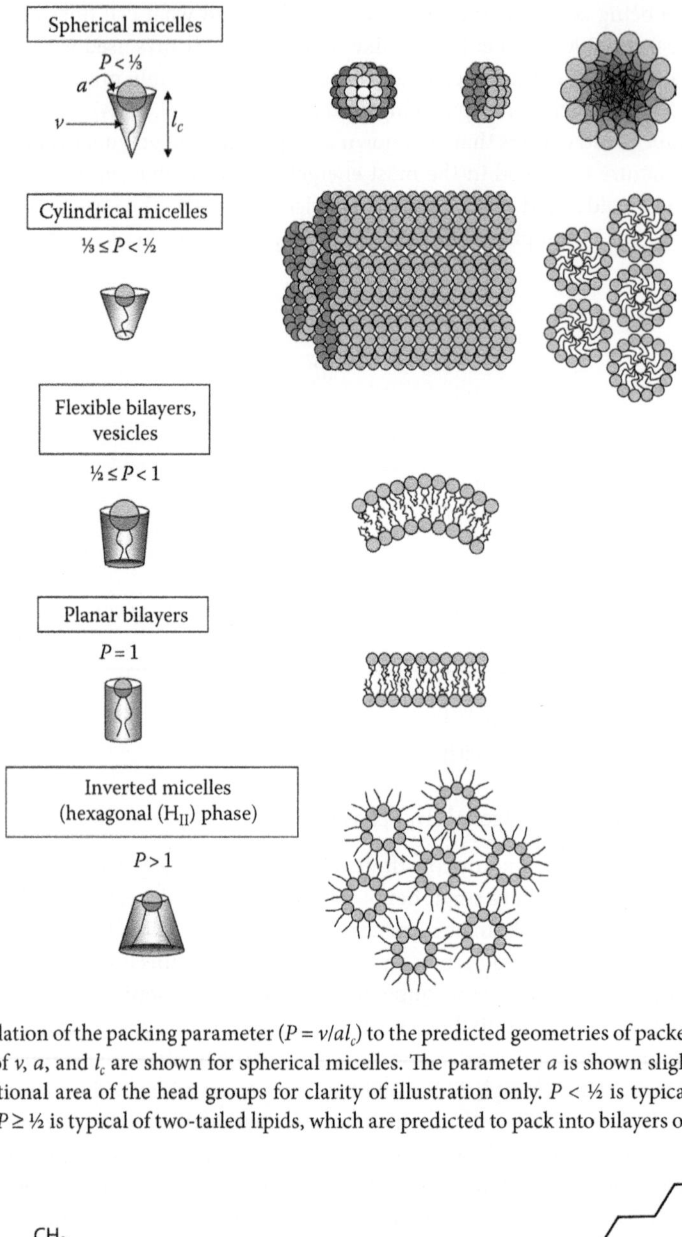

FIGURE 13.2 Relation of the packing parameter ($P = v/al_c$) to the predicted geometries of packed lipids. The physical interpretation of v, a, and l_c are shown for spherical micelles. The parameter a is shown slightly larger than the maximal cross-sectional area of the head groups for clarity of illustration only. $P < ½$ is typically seen for single-tailed lipids, while $P \geq ½$ is typical of two-tailed lipids, which are predicted to pack into bilayers or inverted micelles.

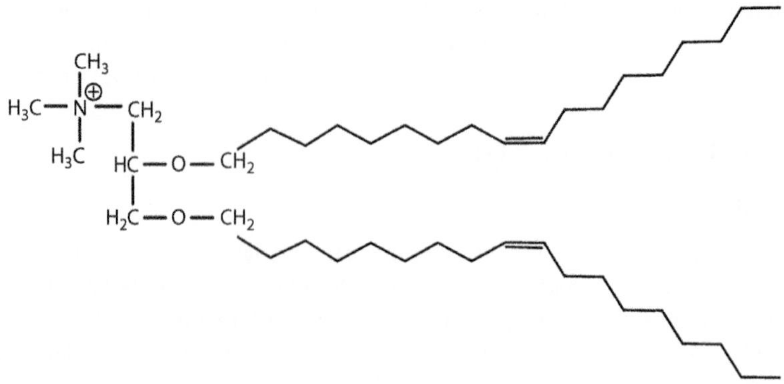

FIGURE 13.3 DOTMA.

in a liposome in a relatively simple fashion, with effective gene delivery, significantly influenced and improved the potential of nonviral agents for gene therapy (Felgner et al. 1987; Malone et al. 1989). The initial success of *in vitro* transfection of multiple cell lines with DOTMA sparked a number of attempts to improve the lipid formulation and resulted in the creation of many effective formulations including such notable lipids as DOTAP (Leventis and Silvius 1990) (see Section 13.3.1.2) and DC-Chol (Gao and Huang 1991) (Section 13.3.1.3).

Commercialization of DOTMA as Lipofectin® involved its coupling with DOPE (Section 13.3.3.1) in a 1/1 ratio due to the ability of DOPE to increase transfection efficiencies. Once commercialized, improvements in Lipofectin were desired, motivating others to add functional groups to the DOTMA. Many alterations made in the four major moieties of DOTMA (head group, linker, linkage bonds, and hydrocarbon chains) have reflected widespread efforts to reduce toxicity and increase transfection efficiencies (Leventis and Silvius 1990; Ren et al. 2000). Felgner et al. (1994) also experimented with novel lipid formulations by altering DOTMA and DOPE to obtain a more robust understanding of the mechanism of biological action. The structural changes included different combinations of side chains and alkyl attachments to the head groups as well as the replacement of a methyl group on the quarternary amine of DOTMA with a hydroxyl. Their report suggested that compounds with such a hydroxyl modification display improved transfection efficiencies over DOTMA. The stabilization of the bilayer vesicles was purported to occur as a result of the hydroxyl group remaining in contact with the aqueous layer surrounding the liposome. Compounds lacking this moiety were hypothesized to become entrenched in the aliphatic region, thus destabilizing the membrane. It was also indicated that the aliphatic chain length had a large effect on the efficacy of lipid vectors. As the lengths of the saturated chains were increased in the DOTMA analogs, transfection efficiencies decreased. This was thought to be due to increased bilayer stiffness, which may have prevented efficient fluid interactions with the endosomal membrane to thus hamper the release of the liposomes or pDNA from the endosomal compartments.

13.3.1.2 DOTAP

DOTAP, was first synthesized by Leventis and Silvius in 1990. The molecule consists of a quarternary amine head group coupled to a glycerol backbone with two oleoyl chains. The only differences between this molecule and DOTMA are that ester bonds link the chains to the backbone rather than ether bonds. It was originally hypothesized that ester bonds, which are hydrolysable, could render the lipid biodegradable and reduce cytotoxicity (Figure 13.4).

The use of 100% DOTAP for gene delivery is inefficient due to the density of positive charges on the liposome surface, which possibly prevents counterion exchange (Zuidam and Barenholz 1998). DOTAP is completely protonated at pH 7.4 (which is not the case for all other cationic lipids) (Zuidam and

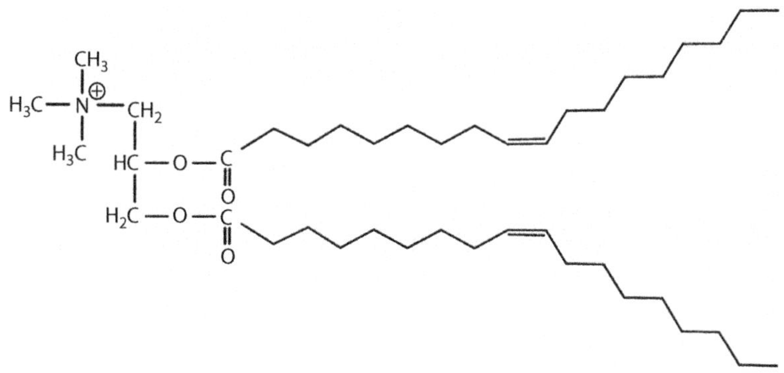

FIGURE 13.4 DOTAP.

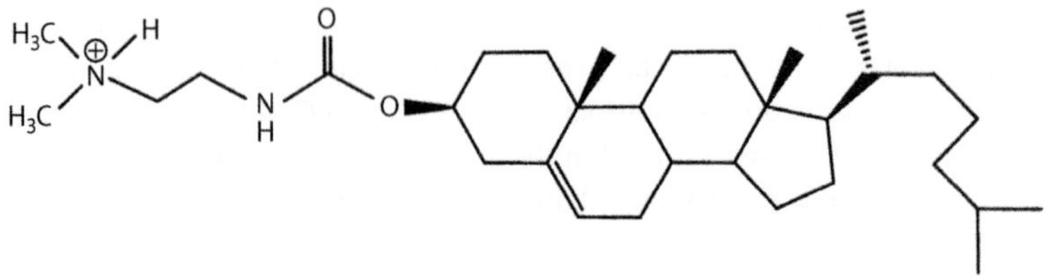

FIGURE 13.5 DC-Chol.

Barenholz 1998), so it is possible that more energy is required to separate the DNA from the lipoplex for successful transfection (Zabner et al. 1995). Thus, for DOTAP to be more effective in gene delivery, it should be combined with a helper lipid, as seems to be the case for most cationic lipid formulations.

High temperature and long incubation times have been used to create lipoplexes that exhibit resistance to serum interaction (Yang and Huang 1998). Interestingly, this approach was only observed to affect monovalent cationic lipids such as DOTMA, DOTAP, or DC-Chol, as opposed to multivalent cationic lipids. The specific reasons for this phenomenon remain unclear. In fact, the specific mechanism behind serum inactivation of lipoplexes in general is as yet unexplained. Several hypotheses have been offered as to the mechanism, including the prevention of lipoplex binding to cell membranes by serum proteins (Yang and Huang 1997), the prevention of structural complex maturation by serum proteins binding to cationic charges on the lipoplexes (Yang and Huang 1998), and the disparity of endocytosis pathways—which have varying kinetics—that are used for lipoplex endocytosis, with the method of endocytosis being regulated by the size of the lipoplexes or aggregates of lipoplexes plus serum proteins (Marchini et al. 2009).

13.3.1.3 DC-Chol

3β[*N*-(*N*′,*N*′-dimethylaminoethane)-carbamoyl]cholesterol, or DC-Chol, was first synthesized by Gao and Huang (1991). DC-Chol contains a cholesterol moiety attached by an ester bond to a hydrolysable dimethylethylenediamine. Cholesterol was reportedly chosen for its biocompatibility and the stability it imparts to lipid membranes, an idea which was supported by desirable transfection efficiencies with reduced cytotoxicities in many cell lines (Gao and Huang 1991) (Figure 13.5).

In contrast to cationic liposomes containing fully charged quarternary amines (e.g., DOTMA and DOTAP), DC-Chol in a 1:1 lipid ratio with DOPE, contains a tertiary amine that is charged on 50% of the liposome surface at pH 7.4 (Zuidam and Barenholz 1997). This feature is thought to reduce the aggregation of lipoplexes leading to a higher transgene expression (Ajmani and Hughes 1999). The reduction in overall lipoplex charge can also aid in DNA dissociation during gene delivery (Zuidam and Barenholz 1998), which has been proven to be necessary for successful transfection (Zabner et al. 1995).

13.3.2 Multivalent Cationic Lipids

13.3.2.1 DOSPA

DOSPA is another cationic lipid synthesized as a derivative of DOTMA. The structure is similar to DOTMA except for a spermine group that is bound via a peptide bond to the hydrophobic chains. This cationic lipid, used with the neutral helper lipid DOPE at a 3:1 ratio, is commercially available as the transfection reagent Lipofectamine® (Figure 13.6).

In general, the addition of the spermine functional group allows for a more efficient packing of DNA in terms of liposome size. The efficient condensation is possibly due to the many ammonium groups in spermine. It has been shown that spermine can interact via hydrogen bonds with the bases of DNA in such a way as to be attracted on one strand and wind around the major groove to interact with complementary bases of the opposite strand (Jain et al. 1989).

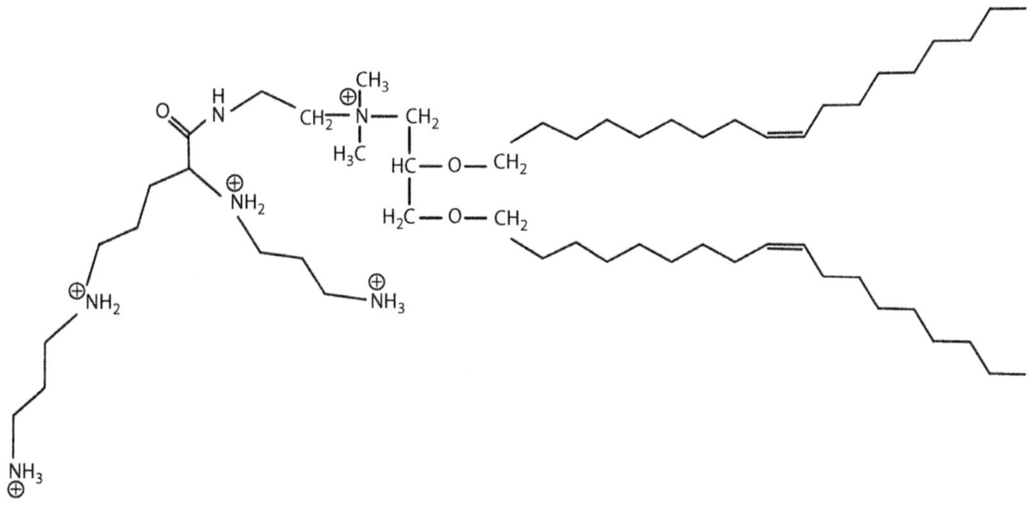

FIGURE 13.6 DOSPA.

13.3.2.2 DOGS

DOGS has a structure similar to DOSPA; both molecules have a multivalent spermine head group and two 18-carbon alkyl chains. However, the chains in DOGS are saturated, are linked to the head group through a peptide bond, and lack a quarternary amine. DOGS is commercially available under the name Transfectam®. This lipid has been used to transfect many cell lines with relatively low levels of toxicity (Behr et al. 1989) (Figure 13.7).

Much like the multivalent cationic lipid DOSPA, DOGS is very efficient at binding and packing DNA, a result of the spermine head group that so closely associates with DNA (Behr et al. 1989). Characterization of the head group of DOGS was determined to facilitate not only the efficient condensation of DNA but also the buffering of the endosomal compartment, which was thought to protect the delivered DNA from degradation by pH-sensitive nucleases (Remy et al. 1994). DOGS is a multifaceted molecule in terms of buffering capacity. At pH values lower than 4.6, all the amino groups in the spermine are protonated, while at pH = 8 only two are purportedly ionized, which promotes arrangement into a lamellar structure (Boukhnikachvili et al. 1997). The packing ability of DOGS is due in part to the dynamics of the large head group molecule and the length of long unsaturated carbon chains.

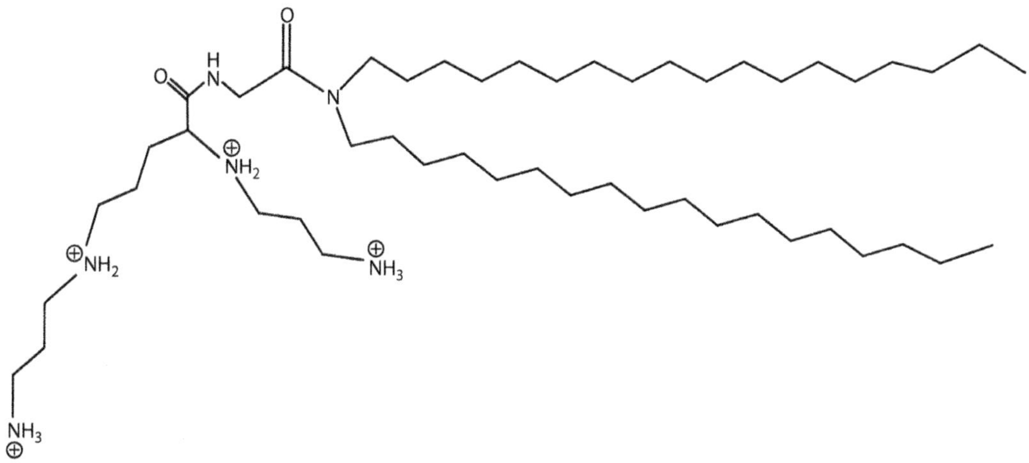

FIGURE 13.7 DOGS.

13.3.3 Improvements to Cationic Lipids for Increased Transfection Efficiency

13.3.3.1 Use of DOPE and DOPC as Neutral Helper Lipids

Most liposomal formulations used for gene delivery consist of a combination of charged lipids and neutral helper lipids. The neutral helper lipids used are often DOPE, which is the most widely used neutral helper lipid, or dioleoylphosphatidylcholine (DOPC). Results have shown that the use of DOPE versus DOPC as the helper lipid yields higher transfection efficiencies in many cell types (Farhood et al. 1994; Simoes et al. 1998), thought to be due to a conformational shift to an inverted hexagonal packing structure that is imparted by DOPE at low pH (Figure 13.2). In contrast to the creation of repeated layers of DNA/lipids as is the case in lamellar packing, the inverted hexagonal packing structure is similar to that of a honeycomb of tubular structures that condense DNA inside the tubes through electrostatic interactions. The tubes aggregate due to van der Waals interactions between the lipid tails that spread out to encircle each tube. Fusion and destabilization of the lipoplexes during transfection are thought to occur due to the exposure of the endosomal membrane to invasive hydrocarbon chains (Chesnoy and Huang 2000). It has been shown that a hexagonal conformation allows for efficient escape of complexed DNA from endosomal vesicles via destabilization of the vesicle membrane (Hui et al. 1996; Zuhorn et al. 2005). With the lysosomotropic agent chloroquine inhibiting the activity of DOPE-containing lipoplexes, it is reasonable to assume that the membrane-destabilizing hexagonal conformation associated with DOPE is brought about at acidic pH (Legendre and Szoka 1992; Farhood et al. 1995) (Figure 13.8).

In DOTAP-mediated DNA-binding studies, it was discovered that liposomes—formulated without DOPE—would not effectively complex with DNA to neutralize it until a 2:1 N:P ratio was reached

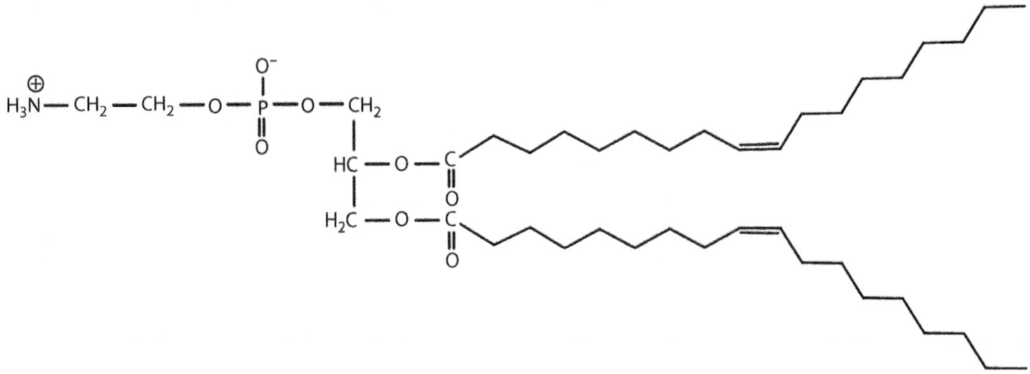

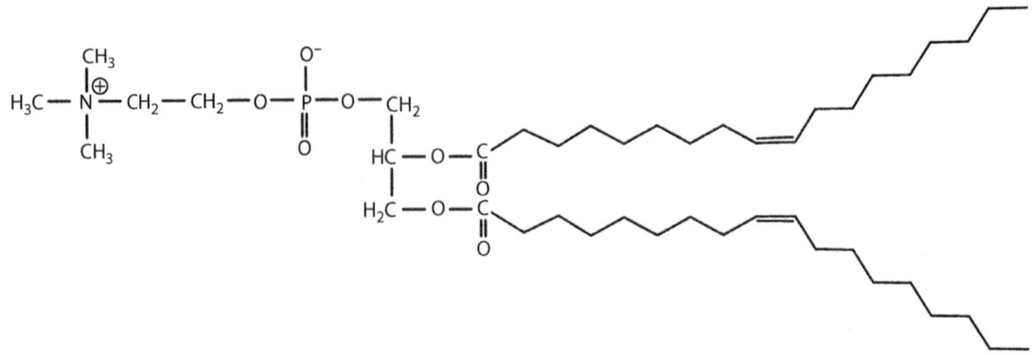

FIGURE 13.8 DOPE (top) and DOPC.

due to an inability to displace counterions bound to the cationic lipid head groups (Zuidam and Barenholz 1998). In contrast, complexes with a 1:1 ratio of DOTAP/DOPE continuously neutralized and complexed with the negatively charged DNA at all charge ratios. This is possibly due to salt bridges more easily forming between the positively charged head groups of the cationic lipids and the phosphate groups of DOPE moieties. This association would force the primary amine of DOPE to stabilize itself in the plane of the liposome surface and allow for more close interactions with the negatively charged phosphate of the DNA. DOPE could also facilitate counterion release from the positively charged lipid head group, thus lowering the energy required for binding DNA (Zuidam and Barenholz 1998). Circular dichromism has been used to indicate that the use of DOPE as a helper lipid allows for much closer contact and packing of DNA helices (Zuidam and Barenholz 1998).

DC-Chol and other cholesterol derivatives have been incorporated into the lipoplex assembly for increased transfection efficiency *in vivo* (Bennett et al. 1995; Hong et al. 1997). Galactosylated cholesterol derivatives have been shown to lower cytotoxicity levels and improve transfection efficiencies in human hepatoma cells (Hep G2), likely due to the affinity of cellular receptors for galactosylated ligands (Kawakami et al. 1998). This result indicates that lipoplexes can be formulated for cell-specific uptake through the addition of specific ligands.

13.3.3.2 Poly(ethylene) Glycol

Recent improvements in lipofection have involved liposomal targeting and facilitated protection from degradation *in vivo*, both due to surface modifications with PEG. PEG presents many attractive qualities as a liposomal coating, such as availability in a variety of molecular weights, lack of toxicity, ready excretion by the kidneys, and ease of application (Metselaar et al. 2003). Methods of modifying liposomal surfaces with PEG include physical adsorption to the surface of the complex and covalent attachment onto premade liposomes (Immordino et al. 2006).

It has been shown by Kim et al. (2003), that PEGylated lipoplexes yield increased transfection efficiencies in the presence of serum as compared to liposomal transfection methods lacking surface attachments. Additionally, the PEGylated lipoplexes display improved stabilities and longer circulation times in blood. It is thought that the PEG forms a steric barrier around the lipoplexes, which stifles clearance due to reduced macrophage uptake (Immordino et al. 2006), and may allow the liposome to overcome aggregation problems through mutually repulsive interactions between the PEG molecules (Needham et al. 1992). These characteristics increase bioavailability, facilitating higher transfection efficiencies due to improved tissue distribution and larger available concentrations (Decastro et al. 2006).

Because of decreased immune responses and increased circulation times associated with PEG-modified liposomes, they are sometimes referred to as "stealth liposomes." However, such liposomes lack specificity with regard to cellular targeting. Notably, Shi et al. (2002) found that PEGylation inhibited endocytosis of the lipoplexes in a fashion that was dependent upon the mole percentage of PEG on the liposome as well as the identity of certain functional groups that were conjugated to the lipoplexes. Additionally, upon incorporation into the cell, PEG worked to deter proper complex dissociation by stabilizing a lamellar phase of DNA packing. As a result of these findings, a need has arisen for the creation of novel PEG-containing liposomes whereby the attached PEG is removed following endocytosis via a hydrolysable connecting molecule.

Alternative formulations utilizing PEG and other polymers are being produced with the aim of creating steric protection. The goals of such a system include biocompatibility, a flexible structure, and solubility in physiological systems (Immordino et al. 2006). A report by Metselaar et al. (2003) on L-amino-acid-based polymers found an extended circulation time and reduced clearance by macrophages at levels similar to PEG-modified lipids. These oligopeptides are attractive alternatives to PEG due to advantages such as increased biodegradability and favorable pharmokinetics when lower concentration doses are used.

Liposomes can also be coupled to targeting moieties through the use of PEG to impart attraction to affected tissues for optimal routing and transfection. Targeting ligands are selected based upon specific target cell receptors. The target cells can be normal or transformed tumor cells. Examples are transferrin

(Ishida et al. 2001), a popular ligand for the delivery of anticancer drugs to solid tumors *in vivo*, and haloperidol (Mukherjee et al. 2005), a ligand that associates with sigma receptors that are overexpressed in many types of cancer.

13.4 Nondegradable Nanoparticles

13.4.1 Magnetic Nanoparticles

One limitation of gene therapy is that often only low amounts of DNA reach cells of interest. The principle behind magnetofection is the use of magnetic nanoparticles, attached to gene delivery complexes, which are guided via an applied magnetic field to specific tissues, organs, or cells. Goals of magnetofection are to reduce the total amount of DNA used, to decrease the time needed for complexes to reach the targeted areas, and to enhance the percentage of cells that express the delivered genes. Figure 13.9 is a pictorial description of magnetofection in an *in vitro* setting.

Magnetic nanoparticles used for gene delivery have been based on a concept established in 1978 by Widder et al., where magnetic micro- and nanoparticles were used for drug delivery (Widder et al. 1978). The use of magnetic microparticles for gene delivery was demonstrated 22 years later *in vitro* in C12S cells and *in vivo* in mice using adeno-associated viruses linked to magnetic microspheres via heparin (Dobson 2006). The use of magnetic nanoparticles linked to nonviral vectors (PEI, Lipofectamine, DOTAP-Cholesterol, PLL) was described shortly thereafter in 2002 (Scherer et al. 2002). Since these initial studies, this technique, now termed "magnetofection," has been drawing more and more attention after successful transfections were demonstrated in various additional types of cells.

The magnetic nanoparticles used for gene or drug delivery are usually designed as follows: a magnetic core is coated by a protective layer, which can be further functionalized covalently or non-covalently with therapeutic agents such as carrier/DNA complexes (or other drugs). A schematic design for a magnetic nanoparticle is shown in Figure 13.10. The magnetic core can be made from a wide variety of materials having a range of magnetic properties. The most widely used cores are superparamagnetic iron oxide nanoparticles (IONPs), especially those using magnetite (Fe_3O_4) and maghemite (γ-Fe_2O_3). These materials are currently used as contrast agents in magnetic resonance imaging (MRI) and their pharmacokinetics and toxicities have been extensively studied (Dobson 2006; Ragusa et al. 2007).

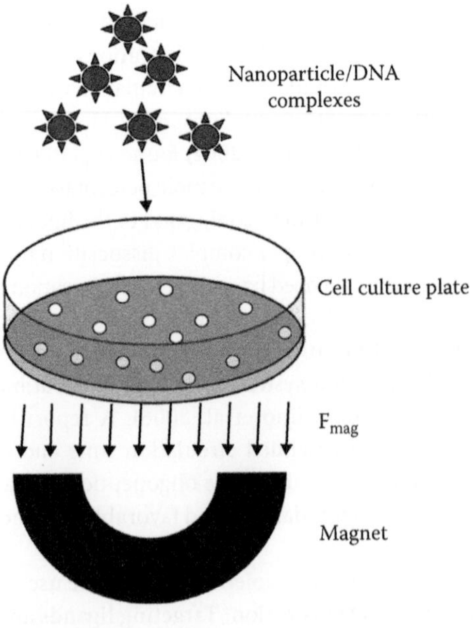

FIGURE 13.9 Schematic representation of magnetofection *in vitro*.

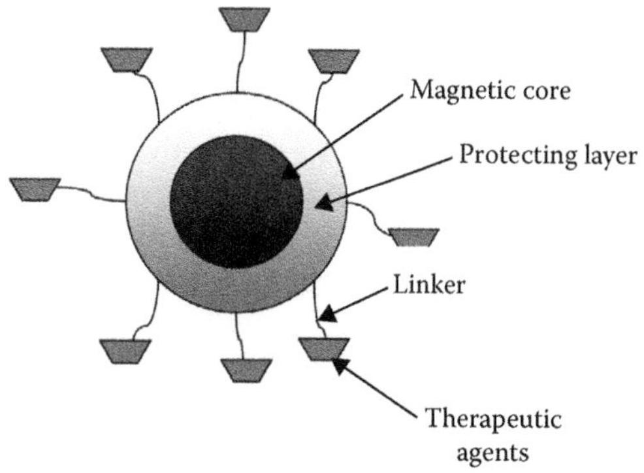

FIGURE 13.10 Schematic design of magnetic nanoparticles.

The magnetic nanoparticle core is fairly reactive, and is usually coated to prevent corrosion. The coating could also prevent the leaking of potentially toxic components into the body when applied *in vivo*. The coating materials can be silica; gold; natural polymers such as dextran; or synthetic polymers such as PEI, PLL, PEG, and polyvinyl alcohol (PVA) (McBain et al. 2008). Organic linkers are used to alter surface properties to suit various therapeutic agents. Commonly used organic linkers include amines, thiols, and aldehydes (McBain et al. 2008). However, in gene delivery, the most suitable coating surfaces for the magnetic particles are those that are strongly cationic because of the negatively charged DNA molecules that are to be delivered. Although organic linkers provide a good approach for the attachment of biomolecules, this approach is not always necessary because DNA could attach to positively charged coating polymers such as PEI or PLL through electrostatic interactions.

As mentioned earlier, the first investigation of magnetic nanoparticles linked to nonviral vectors such as PEI, PLL, and lipofectamine indicated that magnetofection could improve vector targeting and efficacy (Scherer et al. 2002). In 2004, the mechanism of magnetic INOPs coated with PEI entering cells was studied (Huth et al. 2004). The investigation showed that cellular uptake of the coated INOPs was similar to that of PEI polyplexes. It seemed that the magnetic field did not aid in the uptake process, it just quickened the gathering of vector/DNA complexes near the cell surface, which increased the chance of internalization of the complexes. In 2006, Kamau et al. (2006) coupled DNA to PEI-coated INOPs, and then the vector/DNA complexes were delivered to different cell lines (Hela, 293T, Cos-7) that were subsequently exposed to permanent or pulsating magnetic fields. The transfection efficiencies were found to be much higher than in cells not exposed to magnetic fields, with cellular uptake observed 5 min after exposure to the magnetic fields.

At present, magnetofection has been applied to transfect a number of cell types such as primary lung epithelial cells (Gersting et al. 2004) and human umbilical vein endothelial cells (HUVEC) (Krotz et al. 2003b), which are known to be resistant to transfection. Moreover, magnetofection has been used to deliver antisense oligonucleotides (Krotz et al. 2003a) and siRNA (Schillinger et al. 2005) to downregulate gene expression.

The main advantage of magnetofection is rapid gene delivery, taking just a few minutes as compared to traditional transfection methods that can take hours. Moreover, *in vivo*, the magnetic field could not only enhance transfection but also target therapeutic genes to specific organs, which could save considerable amounts of transfection materials. Although magnetofection is relatively new, optimization of parameters such as those associated with the magnetic field have shown great promise for the advancement of this technique.

13.4.2 Gold Nanoparticles

Gold nanoparticles (GNPs or AuNPs) have been widely used for their optical properties, and have recently drawn attention as potential candidates as gene delivery vectors. The size of GNPs can range from a few to several hundred nanometers, and the synthesis processes in aqueous solutions as well as in organic solvents are very well established (Sperling et al. 2008). GNPs can be made very small to provide a high surface-to-volume ratio, and the surfaces of GNPs can be easily modified by conjugation with various ligands (Sperling et al. 2008). Due to the high binding affinity of thiol mioeties to gold surfaces, thiol-modified ligands are frequently used for binding to the GNP surface via Au-sulfur bonds. Such modifications are usually called monolayer-protected clusters (MPCs) or mixed monolayer protected clusters (MMPCs). Since the gold core is essentially inert, GNPs are regarded as biocompatible in cell culture experiments. To date, there has been no acute cytotoxicity reported for GNP use.

The Rotello group has investigated the interaction between cationic GNPs and DNA, showing that tetraalkylammonium ligands conjugated to GNPs could completely inhibit *in vitro* DNA tanscription by T7 RNA polymerase (McIntosh et al. 2001). Branched PEI (2 kDa)-conjugated GNPs have been used as gene delivery vectors to transfect Cos-7 (monkey kidney) cells, with transfection efficiencies varying with the PEI:gold ratio in the hybrid GNP-PEI. The hybrid GNP-PEI conjugate was about 12 times more efficient than PEI alone at the optimized ratio (Thomas and Klibanov 2003). β-Cyclodextrin was also attached to the oligo(ethylenediamino)-modified GNPs (OEA-CD-NP), and it was found that the modified GNPs could effectively bind and concentrate DNA (Wang et al. 2007). The data demonstrated that OEA-CD-NP could deliver plasmid DNA to breast cancer cells (MCF-7).

13.4.3 Silica Nanoparticles

Silica is a major component of sand and glass, and has been studied extensively in material science and engineering due to the variety of chemical and physical modifications possible. Pure silica nanoparticles without surface modifications cannot condense and deliver DNA. To address this issue, silica particles are often functionalized aminosilanes (Luo and Saltzman 2006; Ragusa et al. 2007). *N*-(6-aminohexyl)-3-aminopropyltrimethoxysilane and *N*-(2-aminoethyl)-3 aminopropyltrimethoxysilane have been used to modify silica, and the resulting surface-functionalized silica can condense and deliver DNA to Cos-1 cells with low toxicity (Kneuer et al. 2000). Amino-hexyl-aminopropyltrimethoxysilane (AHAPS) has also been used to functionalize silica, yielding a material that successfully transfected mouse lung *in vivo* (Ravi Kumar et al. 2004).

Amino-functionalized, organically modified silica (ORMOSIL) has been used for *in vivo* gene delivery in mouse brain (Bharali et al. 2005). Intraventricular injection of ORMOSIL/pEGFP complexes showed effective transfection and expression of enhanced green fluorescent protein (EGFP) in neuron-like cells in the periventricular brain regions and the subventricular zone. Moreover, the transfection of ORMOSIL/FGFR-1 (nucleus targeting fibroblast growth factor receptor type 1) caused cells to withdraw from the cell cycle, which resulted in neuronal differentiation. These studies provided groundwork for the application of ORMOSIL nanoparticles to *in vivo* gene transfer of the central nervous system.

13.5 Problems Associated with Nanobiomaterials for Nonviral Gene Delivery

As stated in the beginning of this chapter, nonviral gene delivery agents have the merits of relatively low induction of immune responses, virtually no limitation to the size of delivered genetic material, and a relatively low cost of production. However, there are some drawbacks associated with nonviral gene delivery vectors including their low transfection efficiencies and cytotoxicity.

In order to achieve effective gene expression, investigators must address a series of issues including the intrinsic stability of the vector/DNA complexes to be constructed; how the complexes will fare with respect to plasma membranes, endolysosomes, and nuclear envelopes; and cytosolic transport (Wiethoff and Middaugh 2003).

DNA complexed with cationic polymers or lipids can easily form large self-aggregates, which can be recognized and cleared by macrophages. Moreover, interactions between positively charged complexes and negatively charged molecules in the extracellular milieu, such as serum albumin, can inhibit the delivery of genetic cargo (Zelphati et al. 1998). It has been hypothesized that cationic complexes are coated nonspecifically by negatively charged proteins, leading to reduced binding to the negatively charged surfaces of cells. Another hypothesis is that the cationic complexes dissociate due to interactions with anionic serum components that act to pull cationic molecules away from the carried DNA via electrostatic attractions (Zelphati et al. 1998).

The cellular uptake of cationic vector/DNA complexes is mediated by nonspecific binding between positive charges on the exteriors of the complexes and negatively charged components of the extracellular portion of the plasma membrane, with the complexes internalized via endocytosis. Endosomal escape is one of the major barriers to efficient gene delivery (Zabner et al. 1995). For complexes utilizing cationic lipids, the mechanism of endosomal escape seems to involve lipid–lipid interactions between the membranes of the endosomes and lipoplexes, leading to membrane disruption and DNA release into the cytoplasm (Xu and Szoka 1996). However, the exact mechanism of the process has not yet been defined. For complexes utilizing cationic polymers, the mechanism involved following endocytosis is even less clear. Several hypotheses have been proposed. One indicates that endosomal membrane disruption is caused by direct interaction between the negatively charged endosomal membrane and the cationic polymers (Zhang and Smith 2000). Another hypothesis is that of the "proton sponge" (Boussif et al. 1995), which states that as protons are pumped into the endolysosome by vesicular ATPases, the pH of the endolysosome is buffered by polyplex amines, preventing the activation of acid hydrolases. In the meantime, chloride ions enter the endolysosome via chloride channels to relieve the developing ionic gradient. Water molecules then enter the endolysosome to relieve the ensuing osmotic gradient, and the endolysosome will swell to the point that its membrane becomes leaky, allowing the endocytosed polyplexes to be released into the cytoplasm. Still other research indicates a lack of interaction between certain types of polyplexes and lysosomes, with DNA being delivered into nuclei still attached to their delivery vehicles (Godbey et al. 1999b, 2000; Akinc and Langer 2002).

Cytotoxicity is another obstacle to the use of nonviral gene delivery vectors for gene therapy. For cationic lipids, cytotoxicity is associated with the cationic nature of the head groups. For example, quaternary amine headgroups are more toxic than groups that employ tertiary amines. Moreover, the type of linker bond that is used to join the polar and nonpolar regions of the lipids also plays a role in cytotoxicity, which is partially due to the end products created by cationic lipid degradation (Lv et al. 2006). For cationic polymers, cytotoxicity is interrelated to polymer structure. As an example, the cytotoxicity of PEI is related to factors, such as molecular weight, degree of branching, zeta potential, and particle size (Kircheis et al. 1999; Kunath et al. 2003).

Despite the fact that nonviral vectors are used for their reduced immunogenic properties, there are still vectors that do stimulate portions of the immune system. These vectors and their antigenic properties must be characterized before optimal gene delivery vehicles can be created and used in the clinic. While DOTAP has been widely used *in vitro*, this vector—and many other cationic liposomes—have been found to stimulate the immune system activation markers CD80 and CD86 despite a lack of pro-inflammatory cytokine secretion (Yan et al. 2007). Another report has indicated that the increased activation of these markers is correlated with unsaturated or shortened saturated hydrocarbon chains in comparison to a lower level of activation by liposomes containing lipids with longer or saturated acyl chains (Vangasseri et al. 2006). A recently designed shorter phospholipid, DiC14-amidine, contains hydrocarbon chains with 14 saturated carbons. DiC14-amidine was found to not only stimulate CD80/86 marker proteins, but also to activate production and secretion of pro-inflammatory cytokines (Tanaka et al. 2008). It is necessary to keep the immunogenic characteristics of the structure of DiC14-amidine and other cationic lipids in mind when designing new and ideally effective vectors that will not act as immune system agonists.

The complement system is a major aspect of the nonspecific immune system that must remain inactivated for efficient gene delivery. All of the cationic vectors examined, including DOTAP, DC-Chol/DOPE,

DOGS/DOPE, DOTMA/DOPE, and cationic polymer vectors, activate the complement cascade, with the lipopolyamines (DOGS) acting as one of the most potent activators (Plank et al. 1996). Of the cationic polymers, long-chain polylysines activate the complement system very strongly. Attenuation of these polymers to a length between 19 and 28 segments decreases complement activation by orders of magnitude (Plank et al. 1996). Other data by Plank and Szoka indicate that cationic peptides can be very active in gene delivery and hardly activate the complement system if their lengths are under 10 repeats (Plank et al. 1996). Cationic polymers complexed with DNA such that they are electrically neutral do not activate the complement system. The trend of activation seems to be dependent upon charge ratio or charge density: uncomplexed multivalent cationic liposomes activate complement to a greater degree than uncomplexed monovalent cationic liposomes. However, when complexed with DNA, neither multivalent nor monovalent liposomes activate to the same extent as the uncomplexed forms of the lipids (Plank et al. 1996).

13.6 Conclusion

The use of nanobiomaterials for gene delivery presents great potential for future medical applications both *in vitro* and *in vivo*. In terms of transfection efficiency, nonviral gene delivery vectors are still inferior to their viral counterparts. However, they are associated with advantages in that there is limited induction of immune responses, there is virtually no limitation on the size of the genes that can be delivered, and the cost of production is relatively low. Although nonviral gene delivery vectors have been widely investigated, considerable improvements are still needed to meet the requirements associated with clinical use. Nevertheless, the rapid development of nanotechnology should help to realize this aim sooner.

References

Adami, R. C. and K. G. Rice (1999). Metabolic stability of glutaraldehyde cross-linked peptide DNA condensates. *J Pharm Sci* 88(8): 739–746.

Ajmani, P. S. and J. A. Hughes (1999). 3Beta [N-(N′,N′-dimethylaminoethane)-carbamoyl] cholesterol (DC-Chol)-mediated gene delivery to primary rat neurons: Characterization and mechanism. *Neurochem Res* 24(5): 699–703.

Akinc, A. and R. Langer (2002). Measuring the pH environment of DNA delivered using nonviral vectors: Implications for lysosomal trafficking. *Biotechnol Bioeng* 78(5): 503–508.

Anderson, W. F. (1992). Human gene therapy. *Science* 256(5058): 808–813.

Behr, J. P. (1994). Gene transfer with synthetic cationic amphiphiles: Prospects for gene therapy. *Bioconjug Chem* 5(5): 382–389.

Behr, J. P., B. Demeneix et al. (1989). Efficient gene transfer into mammalian primary endocrine cells with lipopolyamine-coated DNA. *Proc Natl Acad Sci USA* 86(18): 6982–6986.

Bennett, M. J., M. H. Nantz et al. (1995). Cholesterol enhances cationic liposome-mediated DNA transfection of human respiratory epithelial cells. *Biosci Rep* 15(1): 47–53.

Benns, J. M., A. Maheshwari et al. (2001). Folate-PEG-folate-graft-polyethylenimine-based gene delivery. *J Drug Target* 9(2): 123–139.

Bharali, D. J., I. Klejbor et al. (2005). Organically modified silica nanoparticles: A nonviral vector for in vivo gene delivery and expression in the brain. *Proc Natl Acad Sci USA* 102(32): 11539–11544.

Bielinska, A., J. F. Kukowska-Latallo et al. (1996). Regulation of in vitro gene expression using antisense oligonucleotides or antisense expression plasmids transfected using starburst PAMAM dendrimers. *Nucleic Acids Res* 24(11): 2176–2182.

Boukhnikachvili, T., O. Aguerre-Chariol et al. (1997). Structure of in-serum transfecting DNA-cationic lipid complexes. *FEBS Lett* 409(2): 188–194.

Boussif, O., T. Delair et al. (1999). Synthesis of polyallylamine derivatives and their use as gene transfer vectors in vitro. *Bioconjug Chem* 10(5): 877–883.

Boussif, O., F. Lezoualc'h et al. (1995). A versatile vector for gene and oligonucleotide transfer into cells in culture and in vivo: Polyethylenimine. *Proc Natl Acad Sci USA* 92(16): 7297–7301.

Boussif, O., M. A. Zanta et al. (1996). Optimized galenics improve in vitro gene transfer with cationic molecules up to 1000-fold. *Gene Ther* 3(12): 1074–1080.

Carreño-Gómez, B. and R. Duncan (1997). Evaluation of the biological properties of soluble chitosan and chitosan microspheres. *Int J Pharm* 148: 231–240.

Chan, P., M. Kurisawa et al. (2007). Synthesis and characterization of chitosan-g-poly(ethylene glycol)-folate as a non-viral carrier for tumor-targeted gene delivery. *Biomaterials* 28(3): 540–549.

Chesnoy, S. and L. Huang (2000). Structure and function of lipid-DNA complexes for gene delivery. *Annu Rev Biophys Biomol Struct* 29: 27–47.

Chew, J. L., C. B. Wolfowicz et al. (2003). Chitosan nanoparticles containing plasmid DNA encoding house dust mite allergen, Der p 1 for oral vaccination in mice. *Vaccine* 21(21–22): 2720–2729.

Chiu, S. J., N. T. Ueno et al. (2004). Tumor-targeted gene delivery via anti-HER2 antibody (trastuzumab, Herceptin) conjugated polyethylenimine. *J Control Release* 97(2): 357–369.

Cho, K. C., S. H. Kim et al. (2005). Folate receptor-mediated gene delivery using folate-poly(ethylene glycol)-poly(L-lysine) conjugate. *Macromol Biosci* 5(6): 512–519.

Choi, Y. H., F. Liu et al. (1998a). Polyethylene glycol-grafted poly-L-lysine as polymeric gene carrier. *J Control Release* 54(1): 39–48.

Choi, Y. H., F. Liu et al. (1998b). Lactose-poly(ethylene glycol)-grafted poly-L-lysine as hepatoma cell-targeted gene carrier. *Bioconjug Chem* 9(6): 708–718.

Decastro, M., Y. Saijoh et al. (2006). Optimized cationic lipid-based gene delivery reagents for use in developing vertebrate embryos. *Dev Dyn* 235(8): 2210–2219.

Derossi, D., A. H. Joliot et al. (1994). The third helix of the Antennapedia homeodomain translocates through biological membranes. *J Biol Chem* 269(14): 10444–10450.

Dick, C. R. and G. E. Ham (1970). Characterization of polyethylenimine. *J Macromol Sci Chem* A, 4: 1301–1314.

Dobson, J. (2006). Gene therapy progress and prospects: Magnetic nanoparticle-based gene delivery. *Gene Ther* 13(4): 283–287.

Erbacher, P., T. Bettinger et al. (1999). Transfection and physical properties of various saccharide, poly(ethylene glycol), and antibody-derivatized polyethylenimines (PEI). *J Gene Med* 1(3): 210–222.

Erbacher, P., A. C. Roche et al. (1997). The reduction of the positive charges of polylysine by partial gluconoylation increases the transfection efficiency of polylysine/DNA complexes. *Biochim Biophys Acta* 1324(1): 27–36.

Erbacher, P., S. Zou et al. (1998). Chitosan-based vector/DNA complexes for gene delivery: Biophysical characteristics and transfection ability. *Pharm Res* 15(9): 1332–1339.

Farhood, H., X. Gao et al. (1994). Cationic liposomes for direct gene transfer in therapy of cancer and other diseases. *Ann N Y Acad Sci* 716: 23–34; discussion 34–35.

Farhood, H., N. Serbina et al. (1995). The role of dioleoyl phosphatidylethanolamine in cationic liposome mediated gene transfer. *Biochim Biophys Acta* 1235(2): 289–295.

Felgner, J. H., R. Kumar et al. (1994). Enhanced gene delivery and mechanism studies with a novel series of cationic lipid formulations. *J Biol Chem* 269(4): 2550–2561.

Felgner, P. L., T. R. Gadek et al. (1987). Lipofection: A highly efficient, lipid-mediated DNA-transfection procedure. *Proc Natl Acad Sci USA* 84(21): 7413–7417.

Ferdous, A., H. Watanabe et al. (1998). Comb-type copolymer: Stabilization of triplex DNA and possible application in antigene strategy. *J Pharm Sci* 87(11): 1400–1405.

Ferrari, M. E., D. Rusalov et al. (2002). Synergy between cationic lipid and co-lipid determines the macroscopic structure and transfection activity of lipoplexes. *Nucleic Acids Res* 30(8): 1808–1816.

Frankel, A. D., D. S. Bredt et al. (1988). Tat protein from human immunodeficiency virus forms a metal-linked dimer. *Science* 240(4848): 70–73.

Fréchet, J. M. J. and D. A. Tomalia., Eds. (2001). *Dendrimers and Dendritic Polymers*. New York, John Wiley & Sons.

Gao, X. and L. Huang (1991). A novel cationic liposome reagent for efficient transfection of mammalian cells. *Biochem Biophys Res Commun* 179(1): 280–285.

Gersting, S. W., U. Schillinger et al. (2004). Gene delivery to respiratory epithelial cells by magnetofection. *J Gene Med* 6(8): 913–922.

Giannoukakis, N., A. Thomson et al. (1999). Gene therapy in transplantation. *Gene Ther* 6(9): 1499–1511.

Godbey, W. T., M. A. Barry et al. (2000). Poly(ethylenimine)-mediated transfection: A new paradigm for gene delivery. *J Biomed Mater Res* 51(3): 321–328.

Godbey, W. T., K. K. Wu et al. (1999a). Improved packing of poly(ethylenimine)/DNA complexes increases transfection efficiency. *Gene Ther* 6(8): 1380–1388.

Godbey, W. T., K. K. Wu et al. (1999b). Tracking the intracellular path of poly(ethylenimine)/DNA complexes for gene delivery. *Proc Natl Acad Sci USA* 96(9): 5177–5181.

Gonsho, A., K. Irie et al. (1994). Tissue-targeting ability of saccharide-poly(L-lysine) conjugates. *Biol Pharm Bull* 17(2): 275–282.

Gottschalk, S., J. T. Sparrow et al. (1996). A novel DNA-peptide complex for efficient gene transfer and expression in mammalian cells. *Gene Ther* 3(5): 448–457.

Goula, D., N. Becker et al. (2000). Rapid crossing of the pulmonary endothelial barrier by polyethylenimine/DNA complexes. *Gene Ther* 7(6): 499–504.

Goula, D., C. Benoist et al. (1998). Polyethylenimine-based intravenous delivery of transgenes to mouse lung. *Gene Ther* 5(9): 1291–1295.

Gupta, B., T. S. Levchenko et al. (2005). Intracellular delivery of large molecules and small particles by cell-penetrating proteins and peptides. *Adv Drug Deliv Rev* 57(4): 637–651.

Hejazi, R. and M. Amiji (2003). Chitosan-based gastrointestinal delivery systems. *J Control Release* 89(2): 151–165.

Hong, K., W. Zheng et al. (1997). Stabilization of cationic liposome-plasmid DNA complexes by polyamines and poly(ethylene glycol)-phospholipid conjugates for efficient in vivo gene delivery. *FEBS Lett* 400(2): 233–237.

Hsu, W. L., H. L. Chen et al. (2005). Mesomorphic complexes of DNA with the mixtures of a cationic surfactant and a neutral lipid. *Langmuir* 21(21): 9426–9431.

Huang, R. Q., Y. H. Qu et al. (2007). Efficient gene delivery targeted to the brain using a transferrin-conjugated polyethyleneglycol-modified polyamidoamine dendrimer. *Faseb J* 21(4): 1117–1125.

Hui, S. W., M. Langner et al. (1996). The role of helper lipids in cationic liposome-mediated gene transfer. *Biophys J* 71(2): 590–599.

Huth, S., J. Lausier et al. (2004). Insights into the mechanism of magnetofection using PEI-based magnetofectins for gene transfer. *J Gene Med* 6(8): 923–936.

Illum, L. (1998). Chitosan and its use as a pharmaceutical excipient. *Pharm Res* 15(9): 1326–1331.

Immordino, M. L., F. Dosio et al. (2006). Stealth liposomes: Review of the basic science, rationale, and clinical applications, existing and potential. *Int J Nanomed* 1(3): 297–315.

Ishida, O., K. Maruyama et al. (2001). Liposomes bearing polyethyleneglycol-coupled transferrin with intracellular targeting property to the solid tumors in vivo. *Pharm Res* 18(7): 1042–1048.

Israelachvili, J. N. (1991). *Intermolecular and Surface Forces*. San Diego, Academic Press Inc.

Jain, S., G. Zon et al. (1989). Base only binding of spermine in the deep groove of the A-DNA octamer d(GTGTACAC). *Biochemistry* 28(6): 2360–2364.

Kamau, S. W., P. O. Hassa et al. (2006). Enhancement of the efficiency of non-viral gene delivery by application of pulsed magnetic field. *Nucleic Acids Res* 34(5): e40.

Kawakami, S., F. Yamashita et al. (1998). Asialoglycoprotein receptor-mediated gene transfer using novel galactosylated cationic liposomes. *Biochem Biophys Res Commun* 252(1): 78–83.

Kim, J. K., S. H. Choi et al. (2003). Enhancement of polyethylene glycol (PEG)-modified cationic liposome-mediated gene deliveries: Effects on serum stability and transfection efficiency. *J Pharm Pharmacol* 55(4): 453–460.

Kim, T. I., H. J. Seo et al. (2004). PAMAM-PEG-PAMAM: Novel triblock copolymer as a biocompatible and efficient gene delivery carrier. *Biomacromolecules* 5(6): 2487–2492.

Kircheis, R., A. Kichler et al. (1997). Coupling of cell-binding ligands to polyethylenimine for targeted gene delivery. *Gene Ther* 4(5): 409–418.

Kircheis, R., S. Schuller et al. (1999). Polycation-based DNA complexes for tumor-targeted gene delivery in vivo. *J Gene Med* 1(2): 111–120.

Klajnert, B. and M. Bryszewska (2001). Dendrimers: Properties and applications. *Acta Biochim Pol* 48(1): 199–208.

Kneuer, C., M. Sameti et al. (2000). A nonviral DNA delivery system based on surface modified silica-nanoparticles can efficiently transfect cells in vitro. *Bioconjug Chem* 11(6): 926–932.

Koping-Hoggard, M., I. Tubulekas et al. (2001). Chitosan as a nonviral gene delivery system. Structure-property relationships and characteristics compared with polyethylenimine in vitro and after lung administration in vivo. *Gene Ther* 8(14): 1108–1121.

Krotz, F., C. de Wit et al. (2003a). Magnetofection—A highly efficient tool for antisense oligonucleotide delivery in vitro and in vivo. *Mol Ther* 7(5 Pt 1): 700–710.

Krotz, F., H. Y. Sohn et al. (2003b). Magnetofection potentiates gene delivery to cultured endothelial cells. *J Vasc Res* 40(5): 425–434.

Kukowska-Latallo, J. F., E. Raczka et al. (2000). Intravascular and endobronchial DNA delivery to murine lung tissue using a novel, nonviral vector. *Hum Gene Ther* 11(10): 1385–1395.

Kunath, K., A. von Harpe et al. (2003). Low-molecular-weight polyethylenimine as a non-viral vector for DNA delivery: Comparison of physicochemical properties, transfection efficiency and in vivo distribution with high-molecular-weight polyethylenimine. *J Control Release* 89(1): 113–125.

Kwoh, D. Y., C. C. Coffin et al. (1999). Stabilization of poly-L-lysine/DNA polyplexes for in vivo gene delivery to the liver. *Biochim Biophys Acta* 1444(2): 171–190.

Kwok, K. Y., Y. Park et al. (2003). In vivo gene transfer using sulfhydryl cross-linked PEG-peptide/glycopeptide DNA co-condensates. *J Pharm Sci* 92(6): 1174–1185.

Laemmli, U. K. (1975). Characterization of DNA condensates induced by poly(ethylene oxide) and polylysine. *Proc Natl Acad Sci USA* 72(11): 4288–4292.

Lee, D., W. Zhang et al. (2007). Thiolated chitosan/DNA nanocomplexes exhibit enhanced and sustained gene delivery. *Pharm Res* 24(1): 157–167.

Lee, H., J. H. Jeong et al. (2001). A new gene delivery formulation of polyethylenimine/DNA complexes coated with PEG conjugated fusogenic peptide. *J Control Release* 76(1–2): 183–192.

Lee, M. and S. W. Kim (2005). Polyethylene glycol-conjugated copolymers for plasmid DNA delivery. *Pharm Res* 22(1): 1–10.

Lee, M., J. W. Nah et al. (2001). Water-soluble and low molecular weight chitosan-based plasmid DNA delivery. *Pharm Res* 18(4): 427–431.

Legendre, J. Y. and F. C. Szoka, Jr. (1992). Delivery of plasmid DNA into mammalian cell lines using pH-sensitive liposomes: Comparison with cationic liposomes. *Pharm Res* 9(10): 1235–1242.

Leventis, R. and J. R. Silvius (1990). Interactions of mammalian cells with lipid dispersions containing novel metabolizable cationic amphiphiles. *Biochim Biophys Acta* 1023(1): 124–132.

Liu, F., K. W. Liang et al. (2001a). Systemic administration of naked DNA: Gene transfer to skeletal muscle. *Mol Interv* 1(3): 168–172.

Liu, G., M. Molas et al. (2001b). Biological properties of poly-L-lysine-DNA complexes generated by cooperative binding of the polycation. *J Biol Chem* 276(37): 34379–34387.

Luo, D. and W. M. Saltzman (2006). Nonviral gene delivery: Thinking of silica. *Gene Ther* 13(7): 585–586.

Lv, H., S. Zhang et al. (2006). Toxicity of cationic lipids and cationic polymers in gene delivery. *J Control Release* 114(1): 100–109.

Mahato, R. I. (1999). Non-viral peptide-based approaches to gene delivery. *J Drug Target* 7(4): 249–268.

Malone, R. W., P. L. Felgner et al. (1989). Cationic liposome-mediated RNA transfection. *Proc Natl Acad Sci USA* 86(16): 6077–6081.

Mansouri, S., Y. Cuie et al. (2006). Characterization of folate-chitosan-DNA nanoparticles for gene therapy. *Biomaterials* 27(9): 2060–2065.

Marchini, C., M. Montani et al. (2009). Structural stability and increase in size rationalize the efficiency of lipoplexes in serum. *Langmuir* 25(5): 3013–3021.

Martin, M. E. and K. G. Rice (2007). Peptide-guided gene delivery. *Aaps J* 9(1): E18–E29.

Maurer, N., A. Mori et al. (1999). Lipid-based systems for the intracellular delivery of genetic drugs. *Mol Membr Biol* 16(1): 129–140.

McBain, S. C., H. H. Yiu et al. (2008). Magnetic nanoparticles for gene and drug delivery. *Int J Nanomed* 3(2): 169–180.

McIntosh, C. M., E. A. Esposito et al. (2001). Inhibition of DNA transcription using cationic mixed monolayer protected gold clusters. *J Am Chem Soc* 123(31): 7626–7629.

McKenzie, D. L., K. Y. Kwok et al. (2000). A potent new class of reductively activated peptide gene delivery agents. *J Biol Chem* 275(14): 9970–9977.

Metselaar, J. M., P. Bruin et al. (2003). A novel family of L-amino acid-based biodegradable polymer-lipid conjugates for the development of long-circulating liposomes with effective drug-targeting capacity. *Bioconjug Chem* 14(6): 1156–1164.

Midoux, P., C. Mendes et al. (1993). Specific gene transfer mediated by lactosylated poly-L-lysine into hepatoma cells. *Nucleic Acids Res* 21(4): 871–878.

Mukherjee, A., T. K. Prasad et al. (2005). Haloperidol-associated stealth liposomes: A potent carrier for delivering genes to human breast cancer cells. *J Biol Chem* 280(16): 15619–15627.

Needham, D., T. J. McIntosh et al. (1992). Repulsive interactions and mechanical stability of polymer-grafted lipid membranes. *Biochim Biophys Acta* 1108(1): 40–48.

Newkome, G. R., Z. Yao et al. (1985). Micelles. Part 1. Cascade molecules: A new approach to micelles. A [27]-arborol. *J Org Chem* 50(11): 2003–2004.

Nicolau, C. and D. Papahadjopulos. (1998). Gene therapy: Liposomes and gene delivery—A perspective. In *Medical Applications of Liposomes*. D. Papahadjopulos and D. D. Lasic. Eds. Amsterdam, Elsevier: pp. 347–352.

Oehlke, J., A. Scheller et al. (1998). Cellular uptake of an alpha-helical amphipathic model peptide with the potential to deliver polar compounds into the cell interior non-endocytically. *Biochim Biophys Acta* 1414(1–2): 127–139.

Ogris, M., S. Brunner et al. (1999). PEGylated DNA/transferrin-PEI complexes: Reduced interaction with blood components, extended circulation in blood and potential for systemic gene delivery. *Gene Ther* 6(4): 595–605.

Ogris, M., P. Steinlein et al. (1998). The size of DNA/transferrin-PEI complexes is an important factor for gene expression in cultured cells. *Gene Ther* 5(10): 1425–1433.

Oh, S., G. E. Pluhar et al. (2007). Efficacy of nonviral gene transfer in the canine brain. *J Neurosurg* 107(1): 136–144.

Park, T. G., J. H. Jeong et al. (2006). Current status of polymeric gene delivery systems. *Adv Drug Deliv Rev* 58(4): 467–486.

Petersen, H., P. M. Fechner et al. (2002). Polyethylenimine-graft-poly(ethylene glycol) copolymers: Influence of copolymer block structure on DNA complexation and biological activities as gene delivery system. *Bioconjug Chem* 13(4): 845–854.

Plank, C., K. Mechtler et al. (1996). Activation of the complement system by synthetic DNA complexes: A potential barrier for intravenous gene delivery. *Hum Gene Ther* 7(12): 1437–1446.

Plank, C., B. Oberhauser et al. (1994). The influence of endosome-disruptive peptides on gene transfer using synthetic virus-like gene transfer systems. *J Biol Chem* 269(17): 12918–12924.

Pooga, M., M. Hallbrink et al. (1998). Cell penetration by transportan. *FASEB J* 12(1): 67–77.

Ragusa, A., I. Garcia et al. (2007). Nanoparticles as nonviral gene delivery vectors. *IEEE Trans Nanobiosci* 6(4): 319–330.

Ravi Kumar, M. N., M. Sameti et al. (2004). Cationic silica nanoparticles as gene carriers: Synthesis, characterization and transfection efficiency in vitro and in vivo. *J Nanosci Nanotechnol* 4(7): 876–881.

Remy, J. S., C. Sirlin et al. (1994). Gene transfer with a series of lipophilic DNA-binding molecules. *Bioconjug Chem* 5(6): 647–654.

Ren, T., Y. K. Song et al. (2000). Structural basis of DOTMA for its high intravenous transfection activity in mouse. *Gene Ther* 7(9): 764–768.

Roy, K., H. Q. Mao et al. (1999). Oral gene delivery with chitosan—DNA nanoparticles generates immunologic protection in a murine model of peanut allergy. *Nat Med* 5(4): 387–391.

Ruben, S., A. Perkins et al. (1989). Structural and functional characterization of human immunodeficiency virus tat protein. *J Virol* 63(1): 1–8.

Scherer, F., M. Anton et al. (2002). Magnetofection: Enhancing and targeting gene delivery by magnetic force in vitro and in vivo. *Gene Ther* 9(2): 102–109.

Schillinger, U., T. Brill et al. (2005). Advances in magnetofection—Magnetically guided nucleic acid delivery. *J Magn Magn Mater* 293(1): 501–508.

Sharma, V. K., M. Thomas et al. (2005). Mechanistic studies on aggregation of polyethylenimine-DNA complexes and its prevention. *Biotechnol Bioeng* 90(5): 614–620.

Shi, F., L. Wasungu et al. (2002). Interference of poly(ethylene glycol)-lipid analogues with cationic-lipid-mediated delivery of oligonucleotides: Role of lipid exchangeability and non-lamellar transitions. *Biochem J* 366(Pt 1): 333–341.

Shi, L., G. P. Tang et al. (2003). Repeated intrathecal administration of plasmid DNA complexed with polyethylene glycol-grafted polyethylenimine led to prolonged transgene expression in the spinal cord. *Gene Ther* 10(14): 1179–1188.

Shima, S. and H. Sakai, (1977). Polylysine produced by Streptomyces. *Agric Biol Chem* 41: 1807–1809.

Simoes, S., V. Slepushkin et al. (1998). Gene delivery by negatively charged ternary complexes of DNA, cationic liposomes and transferrin or fusigenic peptides. *Gene Ther* 5(7): 955–964.

Sperling, R. A., P. Rivera Gil et al. (2008). Biological applications of gold nanoparticles. *Chem Soc Rev* 37(9): 1896–1908.

Subbarao, N. K., R. A. Parente et al. (1987). pH-dependent bilayer destabilization by an amphipathic peptide. *Biochemistry* 26(11): 2964–2972.

Suh, W., S. O. Han et al. (2002). An angiogenic, endothelial-cell-targeted polymeric gene carrier. *Mol Ther* 6(5): 664–672.

Synowiecki, J. and N. A. Al-Khateeb (2003). Production, properties, and some new applications of chitin and its derivatives. *Crit Rev Food Sci Nutr* 43(2): 145–171.

Tanaka, T., A. Legat et al. (2008). DiC14-amidine cationic liposomes stimulate myeloid dendritic cells through Toll-like receptor 4. *Eur J Immunol* 38(5): 1351–1357.

Tang, M. X. and F. C. Szoka (1997). The influence of polymer structure on the interactions of cationic polymers with DNA and morphology of the resulting complexes. *Gene Ther* 4(8): 823–832.

Thomas, M. and A. M. Klibanov (2003). Conjugation to gold nanoparticles enhances polyethylenimine's transfer of plasmid DNA into mammalian cells. *Proc Natl Acad Sci USA* 100(16): 9138–9143.

Tomalia, D. A., H. Baker et al. (1985). A new class of polymers: Starburst-dendritic macromolecules. *Polym J* 17(1): 117–132.

Tomalia, D. A. and J. M. J. Fréchet (2002). Discovery of dendrimers and dendritic polymers: A brief historical perspective. *J Polym Sci Part A: Polym Chem* 40(16): 2719–2728.

Tomalia, D. A. and G. R. Killat (1985). Encyclopedia of Polymer Science and Engineering, Wiley, New York, 1, 680.

Tranchant, I., B. Thompson et al. (2004). Physicochemical optimisation of plasmid delivery by cationic lipids. *J Gene Med* 6 (Suppl 1): S24–S35.

Uchida, E., H. Mizuguchi et al. (2002). Comparison of the efficiency and safety of non-viral vector-mediated gene transfer into a wide range of human cells. *Biol Pharm Bull* 25(7): 891–897.

Uduehi, A. N., U. Stammberger et al. (2001a). Efficiency of non-viral gene delivery systems to rat lungs. *Eur J Cardiothorac Surg* 20(1): 159–163.

Uduehi, A. N., U. Stammberger et al. (2001b). Effects of linear polyethylenimine and polyethylenimine/DNA on lung function after airway instillation to rat lungs. *Mol Ther* 4(1): 52–57.

Vangasseri, D. P., Z. Cui et al. (2006). Immunostimulation of dendritic cells by cationic liposomes. *Mol Membr Biol* 23(5): 385–395.

Vuorimaa, E., A. Urtti et al. (2008). Time-resolved fluorescence spectroscopy reveals functional differences of cationic polymer-DNA complexes. *J Am Chem Soc* 130(35): 11695–11700.

Wadhwa, M. S., W. T. Collard et al. (1997). Peptide-mediated gene delivery: Influence of peptide structure on gene expression. *Bioconjug Chem* 8(1): 81–88.

Wagner, E., C. Plank et al. (1992). Influenza virus hemagglutinin HA-2 N-terminal fusogenic peptides augment gene transfer by transferrin-polylysine-DNA complexes: Toward a synthetic virus-like gene-transfer vehicle. *Proc Natl Acad Sci USA* 89(17): 7934–7938.

Wang, H., Y. Chen et al. (2007). Synthesis of oligo(ethylenediamino)-beta-cyclodextrin modified gold nanoparticle as a DNA concentrator. *Mol Pharm* 4(2): 189–198.

Wasungu, L. and D. Hoekstra (2006). Cationic lipids, lipoplexes and intracellular delivery of genes. *J Control Release* 116(2): 255–264.

Widder, K. J., A. E. Senyel et al. (1978). Magnetic microspheres: A model system of site specific drug delivery in vivo. *Proc Soc Exp Biol Med* 158(2): 141–146.

Wiethoff, C. M. and C. R. Middaugh (2003). Barriers to nonviral gene delivery. *J Pharm Sci* 92(2): 203–217.

Xu, Y. and F. C. Szoka, Jr. (1996). Mechanism of DNA release from cationic liposome/DNA complexes used in cell transfection. *Biochemistry* 35(18): 5616–5623.

Yan, W., W. Chen et al. (2007). Mechanism of adjuvant activity of cationic liposome: Phosphorylation of a MAP kinase, ERK and induction of chemokines. *Mol Immunol* 44(15): 3672–3681.

Yang, J. P. and L. Huang (1997). Overcoming the inhibitory effect of serum on lipofection by increasing the charge ratio of cationic liposome to DNA. *Gene Ther* 4(9): 950–960.

Yang, J. P. and L. Huang (1998). Time-dependent maturation of cationic liposome-DNA complex for serum resistance. *Gene Ther* 5(3): 380–387.

Yang, Y., Y. Park et al. (2001). Cross-linked low molecular weight glycopeptide-mediated gene delivery: Relationship between DNA metabolic stability and the level of transient gene expression in vivo. *J Pharm Sci* 90(12): 2010–2022.

Yao, Y. H., Y. B. Liu et al. (2007). Synthesis and screening of polyethylenimine as a carrier for gene transfer into cultured human tumor cells. *Ai Zheng* 26(7): 790–794.

Ye, L., H. Haider et al. (2007). Nonviral vector-based gene transfection of primary human skeletal myoblasts. *Exp Biol Med (Maywood)* 232(11): 1477–1487.

Yoo, H. S., J. E. Lee et al. (2005). Self-assembled nanoparticles containing hydrophobically modified glycol chitosan for gene delivery. *J Control Release* 103(1): 235–243.

Zabner, J., A. J. Fasbender et al. (1995). Cellular and molecular barriers to gene transfer by a cationic lipid. *J Biol Chem* 270(32): 18997–19007.

Zanta, M. A., O. Boussif et al. (1997). In vitro gene delivery to hepatocytes with galactosylated polyethylenimine. *Bioconjug Chem* 8(6): 839–844.

Zelphati, O., L. S. Uyechi et al. (1998). Effect of serum components on the physico-chemical properties of cationic lipid/oligonucleotide complexes and on their interactions with cells. *Biochim Biophys Acta* 1390(2): 119–133.

Zhang, Z. Y. and B. D. Smith (2000). High-generation polycationic dendrimers are unusually effective at disrupting anionic vesicles: Membrane bending model. *Bioconjug Chem* 11(6): 805–814.

Zuhorn, I. S., U. Bakowsky et al. (2005). Nonbilayer phase of lipoplex-membrane mixture determines endosomal escape of genetic cargo and transfection efficiency. *Mol Ther* 11(5): 801–810.

Zuidam, N. J. and Y. Barenholz (1997). Electrostatic parameters of cationic liposomes commonly used for gene delivery as determined by 4-heptadecyl-7-hydroxycoumarin. *Biochim Biophys Acta* 1329(2): 211–222.

Zuidam, N. J. and Y. Barenholz (1998). Electrostatic and structural properties of complexes involving plasmid DNA and cationic lipids commonly used for gene delivery. *Biochim Biophys Acta* 1368(1): 115–128.

14

Nanobiomaterials for Cancer-Targeting Therapy

Mingji Jin
Chinese Academy of Medical Sciences

Zhonggao Gao
Chinese Academy of Medical Sciences

14.1 Introduction

Amongst various cancer therapies, chemotherapy is one of the major treatment modalities along with debulking surgery (Gonzalez-Angulo et al. 2007; Matei 2007). However, cancer chemotherapy is often complicated by toxic side effects of anticancer drugs. Despite advances in the development of new antitumor drugs, these drugs still cause significant serious side effects (Lee et al. 2008).

However, the main problem of cancer therapy today is that the treatment is unable to focus on the tumor tissue only. While we are dealing with the tumor tissue, we are also killing the normal tissue, especially the cells in good condition (for instance marrow cells or endothelial cells of small intestine), thereby causing serious side effects. Hence, the ideal anticancer drugs may have the specificity that can get right into the tumor tissue. Also, some factors like unstable environment and high dose of drugs restrict the use of anticancer drugs. Nanoparticles can be modified or combined with some special materials to achieve the action of targeting.

Nanotechnology is an emerging branch of science dealing with cancer (Sandhiya et al. 2009). The nanoparticles used in drug delivery include liposomes, polymers, micelles, dendrimers, quantum dots, nanoshells, gold nanoparticles, paramagnetic nanoparticles, and carbon nanotubes (Mitra et al. 2006). Cancer cells have certain receptors on the surface, which can be made use of in targeting delivery of nanoparticles by attaching a monoclonal antibody or cell surface receptor ligand (Hobbs et al. 1998). Tumors are distinct from normal cells as their endothelial cells possess wide fenestrations ranging from 200 nm to 1.2 μm. The large pore size allows passage of nanoparticles into the extravascular spaces (Rawatet al. 2006). As shown in Figure 14.1, normal blood vessel is smooth and regularly growing (A–B), but the tumor blood vessels are in twists and turns and irregular (C through E). The outside of the tumor blood vessel is rough and has many indefinite pores (F–G). Hashizume et al. (2000) found that the pore size of the blood vessel is different in different kinds of tumors. Nanoparticles easily leak

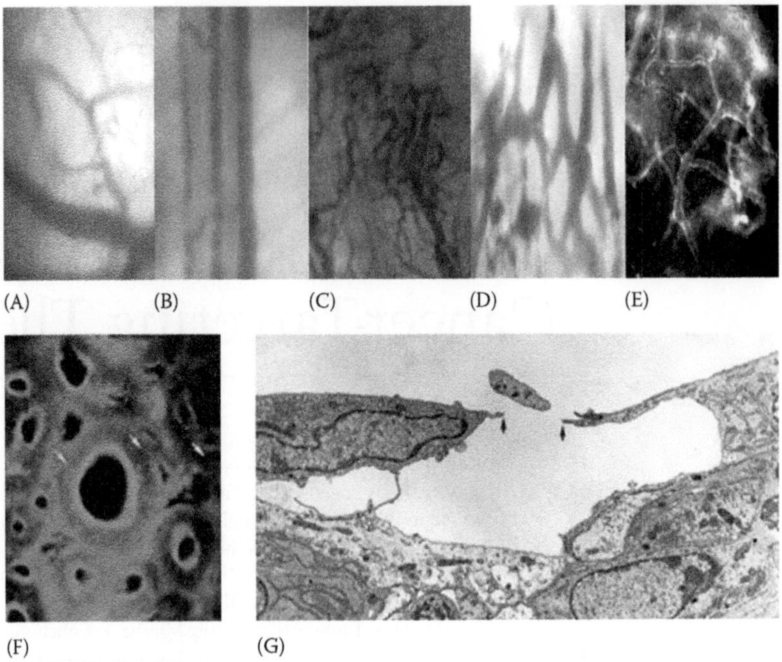

FIGURE 14.1 (**See color insert.**) Normal blood vessel (A–B); tumor blood vessel (C–E); the pore in the tumor blood vessel wall (F); Hematoxylin and Eosin (HE) staining of pores in tumor blood vessel wall (G).

into these deviant structures of tumor blood vessels. Most nanoparticles could passively accumulated in the tumor tissue and avoid entering into normal tissues, thereby significantly reducing the side effects and targeting the tumor. Triggered release of a drug at tumor sites while maintaining a minimal release rate during circulation may result in a very desirable property in carrier design for tumor chemotherapy (Gao et al. 2005).

We developed some modalities of tumor chemotherapy aimed at circumventing side effects of the treatment via drug targeting to tumors. In this chapter, we introduce ultrasound-enhanced tumor-targeting nanoparticles and tumor pH targeting nanotechnology.

14.2 Ultrasound-Triggered Drug Targeting to Tumor

Present-day chemotherapy is associated with severe side effects caused by the drug-induced action on healthy tissues. In addition, in the process of chemotherapy, cells often become resistant to active chemotherapeutic agents (Rapoport et al. 2009). In the course of our previous work, we have made significant progress in site-specific tumor chemotherapy by ultrasound activated drug-loaded polymeric micelles. The micelles passively accumulate in the tumor interstitium presumably via the enhanced penetration and retention effect (EPR) (Gao et al. 2004). The biodistribution of micelle-encapsulated drug closely followed that of micelles, with a significant degree of tumor targeting (Gao et al. 2005). The use of nanomedicine methods opens new lines of attack on the problem of systemic toxicity of anticancer drugs.

14.2.1 Ultrasound in Drug Delivery

Ultrasound is a type of transmission of pressure waves; its frequencies are above human hearing or above 20,000 Hz. These ultrasonic waves can be reflected, refracted (bent), focused, and absorbed. They are actual movement of molecules when the medium is compressed at high pressure and expanded at low pressure, and thus, ultrasound can act physically upon biomolecules and cells. Ultrasonic waves are absorbed relatively little by water, flesh, and other tissues. Therefore, ultrasound can "see" into the body (e.g., diagnostic ultrasound) and can be used to transmit energy into the body at precise locations. This safe, noninvasive, and painless transmission of energy into the body is the key to ultrasonic-activated drug delivery (Pitt et al. 2004). Clinical ultrasound units being manufactured typically deliver ultrasound frequencies of 1 and 3 MHz with duty cycles ranging from 20% to 100%. Duty cycles less than 100% are usually termed pulsed ultrasound, while a 100% duty cycle is referred to as continuous ultrasound. In medicine, ultrasound is used as either a diagnostic or a therapeutic modality. The main advantage of ultrasound is its noninvasive nature; the transducer is placed in contact with a water-based gel or water layer on the skin, and no insertion or surgery is required (Rapoport 2006).

All energy-based tumor treatment modalities, including thermal treatments, require prior imaging. The field of thermal tumor therapy has advanced rapidly, particularly for high-intensity focused ultrasound (HIFU), which causes tumor ablation by coagulative necrosis of tumor tissue. Some other clinical applications of thermal therapy such as adjuvant hyperthermia with radiation therapy, drug and gene delivery, or coagulation of tumor blood vessels can be envisioned (Haar and Coussios 2007). For most systems, magnetic resonance imaging (MRI) remains the imaging modality of choice due to the possibility of combining tumor imaging with MRI thermometry, which would eventually allow development of feedback treatment control systems. However, the possibility for combining diagnostic and therapeutic ultrasound for tumor imaging and treatment appears more attractive, for reasons of simplicity, time-, and cost-effectiveness. Dual-modality ultrasound-imaging-treatment systems are expected to allow precisely controlled, ultrasound-enhanced tumor chemotherapy in a clinical environment. However, ablative techniques have a number of problems related to the precise control of heat deposition. Patient motion and breathing during the treatment are also problematic. Currently, long treatment time, incomplete treatment of large targets, a nonuniform thermal dose distribution, and unintended normal tissue damage continue to impede broad penetration of HIFU therapies into clinical practice. There is a therapeutic technique that utilizes nonthermal mechanisms of ultrasound interaction for targeted drug delivery and tumor imaging. This technique may rely on the same instruments that are used for HIFU, but driven at substantially lower ultrasound energies. In addition, thermal treatment and ultrasound-mediated drug delivery may complement each other because drug delivery works well in the highly perfused regions of a tumor (generally the periphery), while HIFU can destroy a poorly perfused tumor core. Again, tumor imaging prior to treatment is a crucial element of the proposed technique. Contrast-enhanced ultrasonic tumor imaging is still in its infancy. Development of ultrasound contrast agents and the concomitant use of harmonic ultrasound imaging have made possible imaging of liver (Wilson et al. 2000), breast (Kedar et al. 1996), and prostate tumors (Halpern et al. 2001). Also, ultrasound-stimulated acoustic emission of microbubbles has been used for color Doppler imaging of liver metastases (Marcil and Goulet 2002). Polymeric ultrasound contrast agents with targeting potential have been explored by the Wheatley group (El-Sherif et al. 2004; Lathia et al. 2004). Molecularly targeted microbubbles have been successfully used for the ultrasonic imaging of angiogenesis (Takalkar et al. 2004; Rychak et al. 2006). Bubble targeting using ultrasound radiation force has also been developed (Dayton and Ferrara 2002; Bloch et al. 2004, 2005). Combining contrast-enhanced ultrasound tumor imaging with targeted drug delivery is a challenging task. We have developed novel ultrasound-sensitive multifunctional nanoparticles composed of nanoscale polymeric micelles that function as drug carriers and nano- or microscale echogenic bubbles that combine the properties of drug carriers, enhancers of the ultrasound-mediated drug delivery, and long-lasting ultrasound contrast agents (Rapoport et al. 2007). Drug carrying, tumor targeting, and retention in the tumor volume are functions of the micelles and/or nanobubbles. Ultrasound contrast properties are provided by the microbubbles formed in a tumor volume by the coalescence of nanobubbles. The technology of ultrasound requires order of magnitude lower ultrasound energy, reducing the risk of coagulative necrosis of healthy tissues associated with HIFU.

14.2.2 Polymeric Micelles for Combining Ultrasonic Tumor Imaging

Polymeric micelles are nanoparticles formed by self-assembly of amphiphilic block copolymers; they have a size between 10 and 100 nm and core-shell structure; they are most commonly spherical in shape (Allen et al. 1999). The polymeric micelles are formed by hydrophobic–hydrophilic block copolymers at concentrations that are above their critical micelle concentration (CMC). Below the CMC, block copolymer molecules exist in solution in the form of individual molecules (unimers) or their loose, water-penetrated aggregates. At concentrations above the CMC, copolymer molecules self-assemble into dense micelles with hydrophobic cores and a hydrophilic corona. The size of micelles (~10–50 nm) and their surface properties provide for a high drug-loading capacity and long circulation time in the vascular system, which makes them attractive drug carriers (Yu et al. 1998; Rappoprt et al. 2004; Rapoport 2006). Ultrasonic irradiation of the tumor triggered drug release from micelles and transiently altered cell membrane permeability, resulting in effective intracellular drug uptake by the tumor cells (Rapoport et al. 2007). The method of drug targeting to tumors is based on drug encapsulation within polymeric micelles followed by localized triggering of drug release and intracellular uptake induced by ultrasound focused at the tumor site. A rationale behind this approach is that drug encapsulation in micelles decreases the systemic concentration of free drug, diminishes intracellular drug uptake by normal cells, and provides for passive drug targeting to the tumor via the EPR effect. Drugs are encapsulated in nanoparticles (Figure 14.2) that cannot penetrate blood vessels of healthy tissues; however, they penetrate through inter-endothelial gaps and accumulate in a tumor. Liposomes, micelles, nanoemulsions, and nanobubbles are examples of such nanocarriers of drugs. The dosage forms shown in Figure 14.2 comprise polymeric micelles (small circles), nanodroplets (stars), and microbubbles (large circles); the micelles are formed by biodegradable block copolymers (e.g.,poly(ethylene glycol)-*block*-poly(L-lactide) (PEG-PLLA) or poly(ethylene glycol)-*block*-poly(caprolactone) (PEG-PCL)); the droplets and bubbles are formed by perfluorocarbon (e.g., perfluoropentane (PFP) and are stabilized by the same (or another) block copolymer(Rapoport et al. 2009). Drug targeting to the tumors reduces unwanted drug interactions with healthy tissues. Upon accumulation of

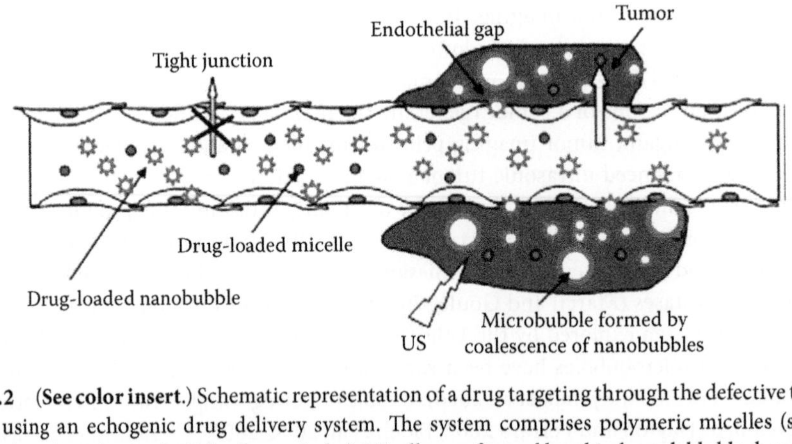

FIGURE 14.2 (**See color insert.**) Schematic representation of a drug targeting through the defective tumor microvasculature using an echogenic drug delivery system. The system comprises polymeric micelles (small circles), nanobubbles (stars), and microbubbles (large circles). Micelles are formed by a biodegradable block copolymer (e.g., PEG-PLLA or PEG-PCL); bubbles are formed by perfluorocarbon (e.g., PFP) stabilized by the same biodegradable block copolymer. Lipophilic drug (e.g., doxorubicin) is localized in the micelle cores and the walls of nano/microbubbles. Tight junctions between endothelial cells in blood vessels of normal tissues do not allow extravasation of drug-loaded micelles or nano/microbubbles (indicated by cross). In contrast, tumors are characterized by defective vasculature with large gaps between the endothelial cells, which allow extravasation of drug-loaded micelles and small nanobubbles, resulting in their accumulation in the tumor interstitium. On accumulation in the tumor tissue, small nanobubbles coalesce into larger, highly echogenic microbubbles that release their drug load in response to therapeutic ultrasound. (From Rapoport, N. et al., *J. Natl. Cancer Inst.*, 99, 1095, 2007; Gao, Z. et al., *Ultrasonics*, 48, 260, 2008.)

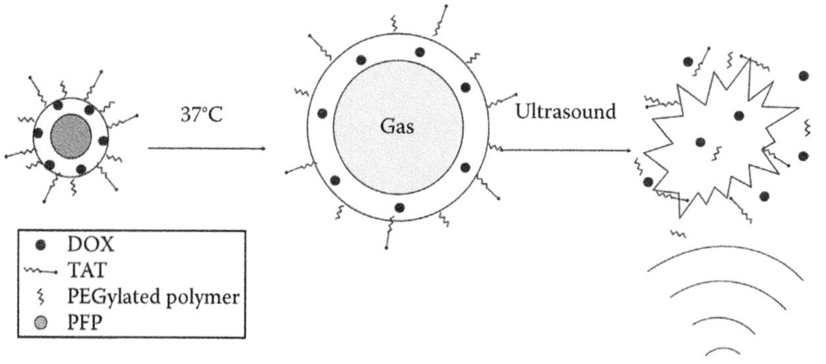

FIGURE 14.3 The figure shows the procedure of liquid droplets changing into microbubbles when the temperature increases from 25°C to 37°C and bubble collapse by the application of ultrasound.

micelles in the tumor interstitium, effective intracellular drug uptake by the tumor cells can be induced by using ultrasonic irradiation. In vitro, ultrasound triggered drug release from micelles substantially enhanced intracellular uptake of both released and encapsulated drug. This was observed for various cell lines suggesting that the ultrasound effect on intracellular drug uptake was a general phenomenon.

A new class of multifunctional nanoparticles that combine properties of polymeric drug carriers, ultrasound imaging contrast agents, and enhancers of ultrasound-mediated drug delivery has been developed. At room temperature, the developed systems comprise perfluorocarbon nanodroplets stabilized by the walls made of biodegradable block copolymers. Upon heating to physiological temperatures, the nanodroplets convert into nano/microbubbles. The phase state of the systems and bubble size may be controlled by the copolymer/perfluorocarbon volume ratio. Upon intravenous injections, a long-lasting, strong, and selective ultrasound contrast is observed in the tumor volume indicating nanobubble extravasation through the defective tumor microvasculature, suggesting their coalescence into larger, highly echogenic microbubbles in the tumor tissue. Under the action of tumor-directed ultrasound, microbubbles cavitate and collapse resulting in a release of the encapsulated drug and dramatically enhanced intracellular drug uptake by the tumor cells. This effect is tumor selective; no accumulation of echogenic microbubbles is observed in other organs (Gao et al. 2008).

In our previous work, we developed a kind of nanoparticles which was made from PLLA-PEG. The different molecular weight of co-polymers PLLA-PEG were synthesized for engineering these nanoparticles. Nanoparticles with the size less than 1 μm were encapsulated with anticancer drugs and PFP. Because PFP's boiling point was 29°C, the nanoparticles were stable at room temperature, but, when the temperature came up to the human normal physiological temperature (37°C), PFP evaporated into a gaseous form and the nanoparticles became nano/microbubbles. When we injected these nanoparticles into mice, after irradiation with ultrasound imaging to observe the tumor tissues, the nanoparticles were found to have strong, long-lasting imaging of microbubbles. When we used 1–3 MHz ultrasonic to irradiate the tumor tissue, the bubbles were cavitated and released drugs; in the meanwhile, the cytomembrane was pierced and could enhance drug release, killing the tumor cells effectively (Figure 14.3).

14.2.3 Evaluation of Ultrasound-Triggered Nano/Microbubbles' Tumor-Targeting Modality

14.2.3.1 Nanodroplet/Microbubble Conversion upon Heating to Physiological Temperature

Upon heating to physiological temperatures, nanodroplets are converted into larger nanobubbles due to the vaporization of PFP inside the bubble walls (Figure 14.4). The initial nanodroplets or nanobubbles were not resolved at the 100× magnification that was the maximum available for our fluorescence microscopy system (Figure 14.5A). In addition to reversible nanodroplet/nanobubble conversion,

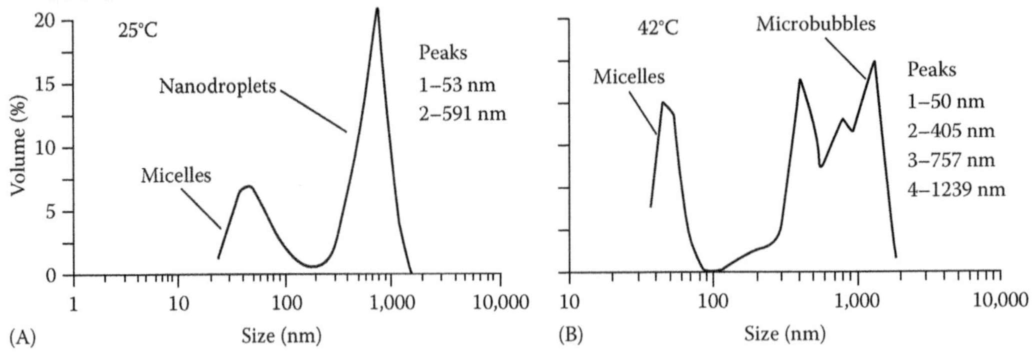

FIGURE 14.4 Nanodroplet/microbubble conversion on heating of the 1% PEG-PLLA/0.5% PFP system to 42°C for 5 min; the nanodroplets (691 nm) were converted into microbubbles (1.24 μm). The effect was reversible; upon cooling to room temperature, the initial size distribution shown in panel (A) was restored. (B) the particle size distribution was changed after heating to 42°C. (From Gao, Z. et al., *Ultrasonics*, 48, 260, 2008.)

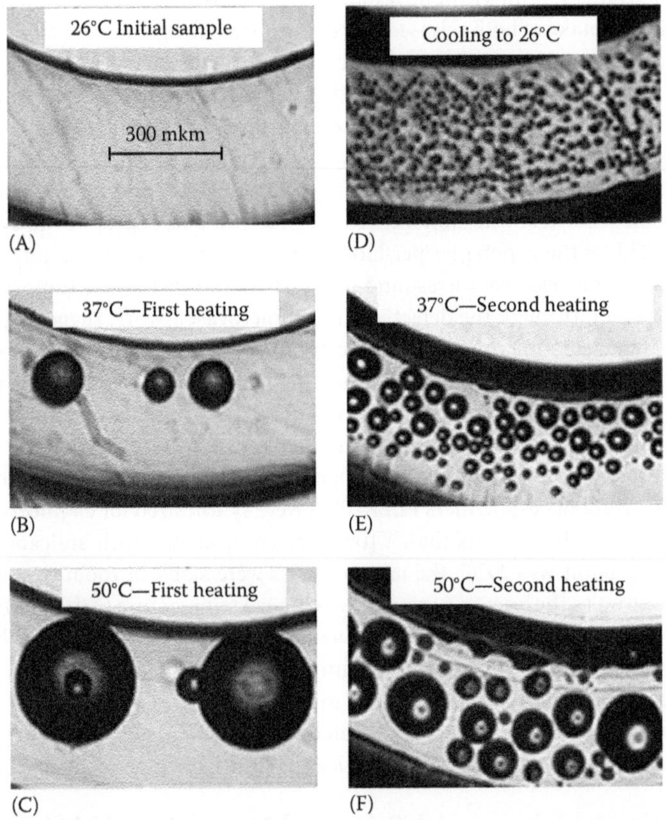

FIGURE 14.5 Optical images of a 0.75 mg/mL Dox/0.5% PEG-PLLA/2% PFP formulation placed in a closed plastic capillary (internal diameter 340 mm) of a snake mixer slide (XXS). (A) At 26°C, nanodroplets of the initial 0.5% PEG-PLLA/2% PFP formulation were not resolved at the highest available magnification (×100). Upon heating to 37°C (B) and 50°C (C), larger bubbles grew by attraction and coalescence of small bubbles. (D) After cooling back to room temperature, the initial sample structure was not restored; a large number of small microdroplets were formed through disintegration of large microbubbles and PFP condensation. (E and F) During a second heating step to 37°C and 50°C, respectively, the growth of large microbubbles through the attraction and coalescence with small microbubbles was manifested by a progressive decrease in the number of small microbubbles. (From Gao, Z. et al., *Ultrasonics*, 48, 260, 2008.)

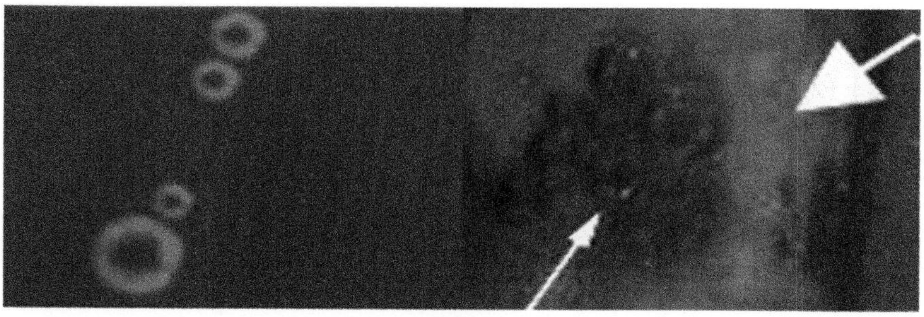

(A) (B)

FIGURE 14.6 **(See color insert.)** Fluorescence images of the 0.75 mg/mL Dox/0.5% PEG-PLLA/2% PFP formulation placed in a plastic capillary (internal diameter 340 mm) of a snake mixer slide. (A) The formulation was heated to 46°C to generate visible microbubbles; the Dox-derived fluorescence (red) of the microbubbles was localized in the bubble walls formed by the bubble-stabilizing copolymer. (B) The MDA MB231 cells incubated with Dox-loaded microbubbles at 37°C did not fluoresce during 30 min incubation (data not shown). Under the action of ultrasound, the cells (shown by a thick arrow) acquired strong fluorescence, while the fluorescence of microbubbles (indicated by thin arrow) was substantially decreased. Unfocused ultrasound of 3 MHz was applied to the slide through the Aquasonic coupling gel for 150 s at a 2 W/cm^2 nominal SATA power density and a 20% duty cycle (the actual power density experienced by the sample may be different from the nominal because the sonication took place in the ultrasound near field). The data indicate that without ultrasound Dox was strongly retained in microbubble walls; however, under the ultrasonic action Dox was released from the bubbles and internalized by the cells.

a small number of microbubbles were formed irreversibly at 37°C through coalescence of nanobubbles (Figure 14.5B). Sample overheating to hyperthermia temperatures (44°C or higher) resulted in a significant coalescence of the nanobubbles into larger microbubbles whose size and number increased with time (Figure 14.5C); this effect was irreversible, as shown in the sequence of photographs presented in Figure 14.5C through F. Large microbubbles, formed under hyperthermic conditions may potentially be useful for localized occlusion of tumor blood vessels.

14.2.3.2 Drug Release in the Cell

At room temperature, the nanoparticles in the formulation were not resolved by fluorescence microscopy, when heated to 46°C, the drug (doxorubicin, Dox) was localized in the bubble wall formed by the stabilizing copolymer, as indicated by fluorescence imaging (Figure 14.6A). When bubbles were incubated with cells at 37°C without the application of ultrasound, neither decrease in bubble fluorescence nor increase in cell fluorescence was observed, thus indicating strong drug retention by the microbubble wall. The application of ultrasound resulted in a decrease in bubble fluorescence and increase in cell fluorescence (Figure 14.6B through D), indicating drug transfer from the bubbles to the cells (Gao et al. 2008).

14.2.3.3 Nano/Microbubbles as Enhancers of Ultrasonic Drug Delivery

For a wide variety of tumor samples and ultrasound parameters, the ultrasound-mediated intracellular drug uptake was substantially enhanced in the presence of microbubbles compared to that of micelles; the effect of the microbubbles was statistically significant. A typical example is shown in Figure 14.7 (Rapoport et al. 2007).

14.2.3.4 Nano/Microbubbles as Ultrasound Contrast Agents

Intratumoral injections of nanobubbles produced long-lasting hyperechoic "dots" in the ultrasound scans of the MDA MB231 breast cancer or A2780 ovarian carcinoma tumors grown subcutaneously in nu/nu mice (Figure 14.8). The generation of hyperechoic sites in the tumors was also observed

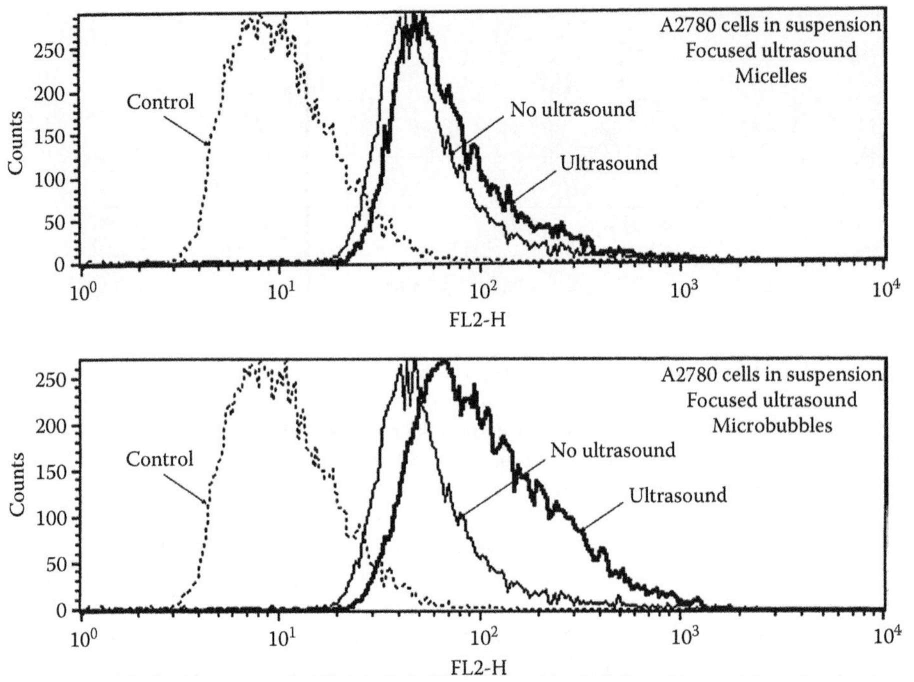

FIGURE 14.7 Effect of microbubbles and ultrasound on the intracellular Dox uptake by A2780 ovarian carcinoma cells in suspension as demonstrated by fluorescence histograms (cell fluorescence on the x-axis; corresponding cell number on the y-axis); Dox at a concentration of 0.75 mg/mL was encapsulated in either 0.5% PEG-PLLA micelles (upper graph) or 0.5% PEG-PLLA/2% PFP microbubbles; microbubble concentration was 5×10^8 mL^{-1}. Continuous wave focused 1.1 MHz ultrasound was applied for 30 s at a peak pressure of 2.9 MPa. The effect of ultrasound on intracellular Dox uptake was much greater for the microbubbles than for the micelles.

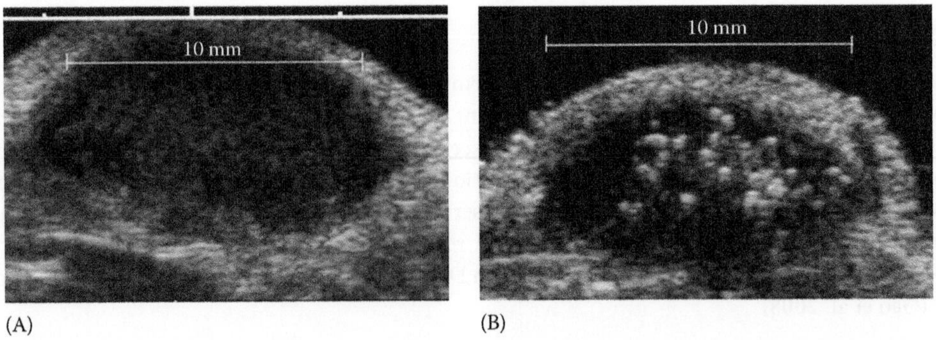

FIGURE 14.8 B-mode ultrasound images of MDA MB231 human breast cancer tumor: (A) before the intratumoral contrast injection and (B) 4 h after the intratumoral injection of 100 ml 0.5% PEG-PLLA/2% PFP formulation. Images were taken using a 14 MHz linear Acuson Sequoia 512 transducer. Strong ultrasound contrast in the tumor persisted for several days. (From Rapoport, N. et al., *J. Natl. Cancer Inst.*, 99, 1095, 2007.)

after the intravenous injections of the nanobubbles (Figure 14.9). Because the functional cutoff size of the endothelial gaps in the tumor capillaries does not exceed 1 μm (Campbell 2006), the observation of echoes in the tumors of intravenously injected mice suggested that nanobubbles were extravasated into the tumor tissue. Because high echogenicity could not be expected from the nanoscaled bubbles at 14 MHz, the observation of the echoes shown in Figure 14.9 suggested that the injected nanobubbles coalesced into highly echogenic microbubbles inside the tumor tissue.

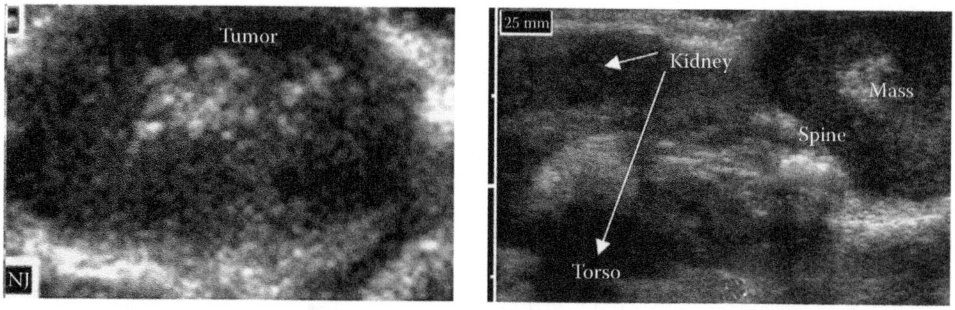

(A) (B)

FIGURE 14.9 B-mode ultrasound images of a MDA MB 231 breast cancer bearing nu/nu mouse following intravenous contrast injection: (A) tumor image taken 4.5 h after the intravenous injection of 100 ml 0.5% PEG-PLLA/2% PFP formulation and (B) trans-torso image of the same mouse showing a tumor (designated as "mass"), kidneys, and spine. The images show that echogenic microbubbles were accumulated in the tumor but not in the kidney. (From Rapoport, N. et al., *J. Natl. Cancer Inst.*, 99, 1095, 2007.)

No hyperechoic entities were observed in the kidney or liver of intravenously injected mice suggesting that the nanobubble extravasation was tumor selective.

14.2.4 Conclusions

Multifunctional nano/microbubble formulations have been developed for combining ultrasonic tumor imaging and ultrasound-enhanced chemotherapeutic treatment. The polymeric copolymer-stabilized PFP nanoemulsion systems undergo nanodroplet/microbubble conversion in vivo, accumulate locally in tumor tissue and coalesce into larger, highly echogenic microbubbles, which provide for long-lasting ultrasound contrast in the tumor and allow on-demand release of the encapsulated drug under the action of therapeutic ultrasound. This technique offers prospects for treating multi-drug-resistant tumors that fail conventional chemotherapy, are expected to have good forehand in cancer treatment.

14.3 pH-Sensitive Polymeric Micelle Targeting to Tumor

14.3.1 Acid Tumor pH

The difference in pH between solid tumors and normal tissue properties has been long recognized (Lee and Bae 2006). The pH of normal tissues and blood is around pH 7.4, and the intracellular pH of normal cells is around pH 7.2. However, in most tumor cells the pH gradient is reversed, the intracellular pH is higher than the extracellular pH, and the pH outside most tumors is about 7.06 (Tannock and Rotin 1989; Engin et al. 1995; Hobbs et al. 1998; Stubbs et al. 2000). As shown in Figure 14.10, when we determined tumor pH using ^{1}H, ^{19}F, ^{31}P after implanting human cancer cells into mice, the results showed that the average tumor pH was 6.84. The main reason is that cancer cells absorb and metabolize protein much faster than normal cells. During the same time, through protein metabolization, cancer cells can produce 2 numerators' lactic acid ($pK_a = 3.9$) and 2 numerators' ATP; however, the normal cells only produce 1 lactic acid and 1 ATP. The cancer cells produce excess lactic acid so as to reduce pH and become acidified. The difference in pH is small but apparent as a natural signal of solid tumors for triggered release.

As we know, chemotherapy is often complicated by toxic side effects of anticancer drugs. Effective chemotherapy regimens are frequently hindered by the cross-resistance or multidrug resistance of tumor cells to one or more drugs. Developing a pH-sensitive nanotechnology that targets only the tumor tissue can be effective for treatment of drug-sensitive and multidrug-resistant tumor cells (Rapoport 2006).

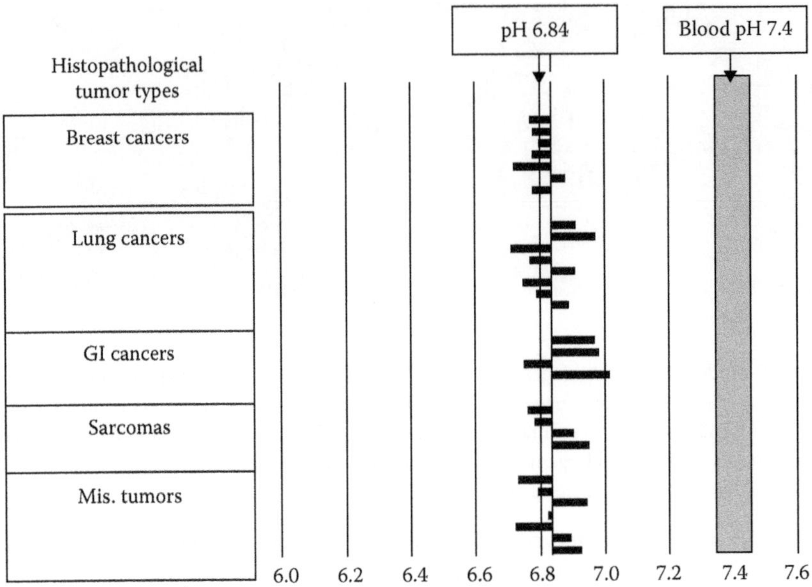

FIGURE 14.10 pH 6.84 is the average pH of a tumor (various kinds of human tumors, 268 samples); the black bars indicate different pH in different tumors and the deviation with the average pH value. (From Volk, T. et al., *Br. J. Cancer*, 68, 492, 1993; Gasparini, G. et al., *Am. J. Clin. Oncol.*, 14, 38, 1991.)

Polymeric micelles have high stability, but they may still dissociate upon dilution after injection and interact with blood components (Yokoyama 1998; Moghimi et al. 2001). Considering the dose–response relationship, it was suggested that future improvement in carrier design, providing a triggerable mechanism for drug release upon reaching the tumor sites, could be the most efficacious delivery strategy (Drummond et al. 1999; Lee and Bae 2006).

14.3.2 Tumor pH Targeting Micelle

Triggered release of a drug from long-circulating vesicles at tumor sites while maintaining a minimal release rate during circulation, which may result in a high local dose in tumor sites and less side effects, is a very desirable property in carrier designs for tumor chemotherapy and has been proven to be effective for treating multidrug-resistant tumors. Approaches to achieve triggering systems include magnetic (Pardoe et al. 2003), thermal (Needham et al. 2000), and ultrasonic activation (Rappoprt et al. 2004) for enhanced drug release from the carriers, which were specifically designed to respond to external signals or energy. The external activation could be useful for the tumors whose location is well identified and which are accessible by external signal or energy sources. In addition, it would be technically difficult to restrict the activation only in tumor regions, and thus the external activation does not distinguish between the blood compartment and tumor cells. Nowadays, many approaches have described drug release by the cleavages of chemical bonds, which conjugate a drug to polymers by endosomal pH (Kataoka et al. 2000). It has been difficult to develop effective pH-sensitive formulations responding to tumor extracellular pH due to the lack of a proper pH-sensitive functional group in the physiological pH range (Sikic et al. 1997). Tumor tissue has an acidic environment, and it's ideal to design polymeric micelles that have long circulation and active transport in the acidic environment of the tumor tissue. Outside of the micelle is PEG, which long circulates in vivo and so can effectively accumulate near the tumor tissues. When the micelles come into the acidic environment, because of the neutralization of positive and negative charges, the positive charge carriers pop up outside the micelle, so that the micelles' active transport can happen.

Cell-penetrating proteins (CPP) are a kind of polypeptide substances that can accelerate cytomembrane transportation. A typical substance is TAT micromolecular polypeptide. TAT has a strong ability to penetrate into the cytomembrane (Torchilin et al. 2003; Torchilin 2008a,b). It was discovered by scientists in 1988 during their research on the project of AIDS virus (HIV) (Green and Loewenstein 1988). The approaches to target various solid tumors by pH include micelle systems with a triggered drug release mechanism, and exposing nonspecific cationic TAT peptide by a shielding/deshielding mechanism or by a pop-up mechanism (Lee et al. 2008). Particular polysulfonamides are negatively charged at blood pH (pH 7.4) and can be neutralized at acidic pH (tumor pH) (Carelle et al. 2002). Poly (methacryloyl sulfadimethoxine)-b-PEG is negatively charged and interacts electrostatically with TAT molecules (shielding) at blood pH. However, charge density on this polymer decreases by decreasing pH. Below pH 6.8, due to destabilized electrostatic interactions, the TAT will be deshielded. The zeta potential measurements on micelle comprising of PLLA-b-PEG-TAT demonstrated the shield/deshielding process. It was shown that the zeta potential is close to zero between pH 8.0 and 6.8, which indicates complete shielding of TAT, and from pH 6.6 to 6.0 it increased to 6.0 mV, which is close to the measured zeta potential for TAT-decorated micelles without masking. When the shielded and unshielded TAT micelles were tested for tumor cell internalization at pHs 7.4 and 6.6 by incubating for an hour, unshielded micelles were internalized into both the cells and their nucleus at pH 7.4 and 6.6. However, the micelle shielded with poly (methacryloyl sulfadimethoxine)-b-PEG was not internalized at pH 7.4, indicating TAT was masked even as it internalized into cells and the nucleus at pH 6.6. This shield/deshielding mechanism suggests that an optimized pH-targeting system with an appropriate sulfonamide polymer is feasible (Kim et al. 2005; Sethuraman and Bae 2007; Oh et al. 2008; Yin et al. 2008). The carriers were modified with acidic sulfonamide and characterized for enhanced drug release and internalization into cells at tumor pH (Na and Bae 2002; Han et al. 2003; Na et al. 2003). These self-assembled nanocarriers or micelles switched their surface properties from hydrophilic to hydrophobic by deionization of sulfonamide group at tumor pH, resulting in distribution and reorganization of the self-assembly structures. It causes enhanced drug release and interactions with cells for internalization (Figure 14.11).

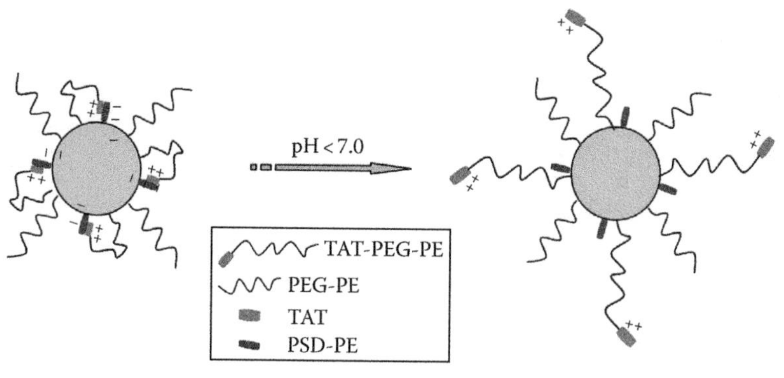

FIGURE 14.11 The process of targeting-ligand's exposure in the acid environment: the carrier system consists of two components, poly (L-lactic acid)-b-PEG-TAT micelles and pH-sensitive poly (methacryloyl sulfadimethoxine)-b-PEG. At normal blood pH, polysulfonamide is negatively charged, and, when mixed with TAT, polysulfonamide shields the TAT by electrostatic interaction. Only PEG is exposed to the outside, which could make the carrier long circulating. When the system experiences a decrease in pH (near tumor), polysulfonamide loses charge and detaches, exposing TAT for interaction with tumor cells. (From Sethuraman, V.A. and Bae, Y.H., *J. Control. Release*, 118, 216, 2007.)

14.3.3 Conclusions

Tumor pH sensitive micelle is a novel anticancer drug delivery system for overcoming limitations of conventional drug delivery system. It increases target drug accumulation at tumor site, brings less drug distribution to normal tissues and organs, and is effective for treatment of multidrug resistance. pH-sensitive micelles have good forehand and provide useful information to the drug treatment of cancer chemotherapy.

14.4 Mouse Dorsal Skinfold Window Chamber Model for Evaluation of Nanoparticle Release from Tumor Blood Vessel

The main cause of cancer treatment failure is the invasion of normal tissues by cancer cells that have migrated from a primary tumor. An important obstacle to understanding cancer invasion has been the inability to acquire detailed, direct observations of the process over time in a living system (Condeelis et al. 2005; Makale 2007). Professor Jain from Harvard University and Professor Dewhirst from Duke University developed the dorsal skinfold window chamber model. It's a kind of model that can be observed under the fluorescence microscope. Dorsal skinfold window chamber model is a known model to visualize the distribution profiles of nanoparticles in the tumor area (Warburton et al. 2004). Chambers made from titanium frames, which are mirror images of each other, will be implanted so as to sandwich the extended double layer of the skin. One layer of the skin will be removed in a circular area of approximately 15 mm in diameter, and the remaining layer, consisting of subcutaneous tissue, epidermis, and the striated skin muscle, will be covered with a cover slip incorporated into one of the frames. The animals will be allowed to recover for hours following the microsurgery after the cells will

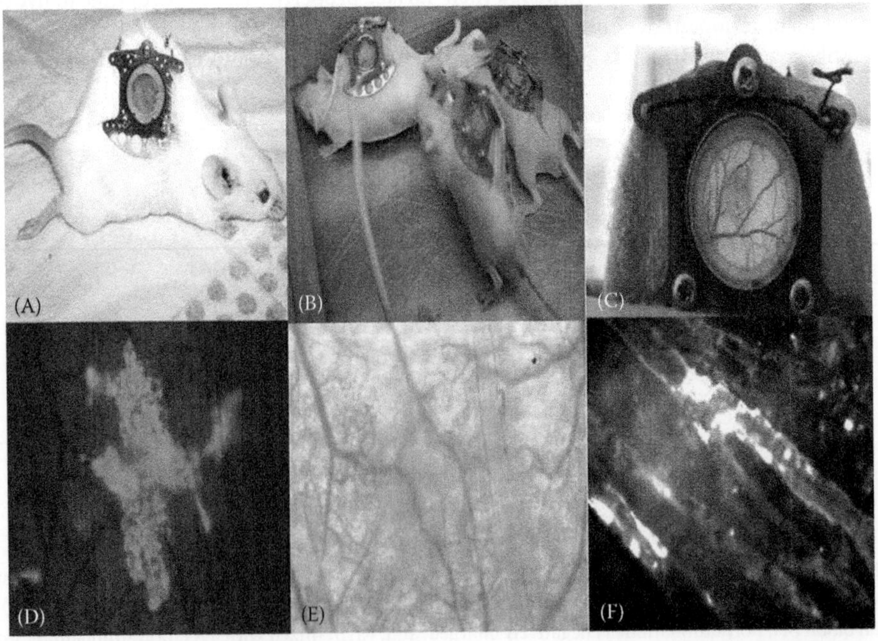

FIGURE 14.12 **(See color insert.)** Fluorescence labeling of drug release from tumor blood vessel dorsal skinfold window chamber model (A through C); after injection of green fluorescent protein (GFP)-transfected 4T1 mammary cancer cell (D: as observed under the fluorescence, E: as observed under the white light); drug release in the blood vessel (F).

be injected. We observe the window chamber under the fluorescence microscope, so the whole processes including tumor vessel's growth in the tissue, drug release from tumor blood vessel, drug controlling action can be directly viewed with the model (Liu et al. 2005). As shown in Figure 14.12, using a special designed microscope, we observed division and reproduction of cancer cells and the process of growth of tumor tissues. Many scientists use this model to reveal malignant tumor's law of development and method of controlling (Jain 1991) and to evaluate drug/gene release in the tumor blood vessel (Shan et al. 2006).

We can implant both tumor tissue and tumor cell using the dorsal skinfold window chamber. Tumor cells grow in the window chamber and can be observed under the microscope. If we inject fluorescence-labeled nanoparticles, we may read the release of nanoparticles into the tumor tissue and calculate strength and area of the fluorescence to evaluate a dynamic process of the nanoparticle's whole circuit. This model is not only very useful in the research of drug-targeting nanomaterial but also considered to be a versatile tool to study the microcirculation in health and disease (Menger et al. 2002).

References

Allen, C., D. Maysinger, and A. Eisenberg. 1999. Nano-engineering block copolymer aggregates for drug delivery. *Colloids Surf B Biointerfaces* 16:3.

Bloch, S. H., P. A. Dayton, and K. W. Ferrara. 2004. Targeted imaging using ultrasound contrast agents. Progess and opportunities for clinical and research applications. *IEEE Eng Med Biol Mag* 23 (5):18–29.

Bloch, S. H., R. E. Short, K. W. Ferrara, and E. R. Wisner. 2005. The effect of size on the acoustic response of polymer-shelled contrast agents. *Ultrasound Med Biol* 31 (3):439–444.

Campbell, R. B. 2006. Tumor physiology and delivery of nanopharmaceuticals. *Anticancer Agents Med Chem* 6 (6):503–512.

Carelle, N., E. Piotto, A. Bellanger, J. Germanaud, A. Thuillier, and D. Khayat. 2002. Changing patient perceptions of the side effects of cancer chemotherapy. *Cancer* 95 (1):155–163.

Condeelis, J., R. H. Singer, and J. E. Seagall. 2005. The great escape: When cancer cells hijack the genes for chemotaxis and motility. *Annu Rev Cell Dev Biol* 21:695.

Dayton, P. A. and K. W. Ferrara. 2002. Targeted imaging using ultrasound. *J Magn Reson Imaging* 16 (4):362–377.

Drummond, D. C., O. Meyer, K. Hong, D. B. Kirpotin, and D. Papahadjopoulos. 1999. Optimizing liposomes for delivery of chemotherapeutic agents to solid tumors. *Pharmacol Rev* 51 (4):691–743.

El-Sherif, D. M., J. D. Lathia, N. T. Le, and M. A. Wheatley. 2004. Ultrasound degradation of novel polymer contrast agents. *J Biomed Mater Res A* 68 (1):71–78.

Engin, K., D. B. Leeper, J. R. Cater, A. J. Thistlethwaite, L. Tupchong, and J. D. McFarlane. 1995. Extracellular pH distribution in human tumors. *Int J Hyperthermia* 11:211.

Gao, Z., H. D. Fain, and N. Ya. Rapoport. 2004. Ultrasound-enhanced tumor targeting of polymeric micellar drug carriers. *Mol Pharm* 1:317.

Gao, Z., H. D. Fain, and N. Rapoport. 2005. Controlled and targeted tumor chemotherapy by micellar-encapsulated drug and ultrasound. *J Control Release* 102:203.

Gao, Z., A. M. Kennedy, D. A. Christensen, and N. Y. Rapoport. 2008. Drug-loaded nano/microbubbles for combining ultrasonography and targeted chemotherapy. *Ultrasonics* 48 (4):260–270.

Gao, Z. G., D. H. Lee, D. I. Kim, and Y. H. Bae. 2005. Doxorubicin loaded pH-sensitive micelle targeting acidic extracellular pH of human ovarian A2780 tumor in mice. *J Drug Targeting* 13:391.

Gasparini, G., S. Dal Fior, G. A. Panizzoni, S. Favretto, and F. Pozza. 1991. Weekly epirubicin versus doxorubicin as second line therapy in advanced breast cancer. A randomized clinical trial. *Am J Clin Oncol* 14:38.

Gonzalez-Angulo, A. M., F. Morales-Vasquez, and G. N. Hortobagyi. 2007. Overview of resistance to systemic therapy in patients with breast cancer. *Adv Exp Med Biol* 608:1.

Green, M. and P. M. Loewenstein. 1988. Autonomous functional domains of chemically synthesized human immunodeficiency virus tat trans-activeator protein *Cell* 55:1179.

Haar, G. T. and C. Coussios. 2007. High intensity focused ultrasound: Past, present and future. *Int J Hyperthermia* 23 (2):85–87.

Halpern, E. J., M. Rosenberg, and L. G. Gomella. 2001. Prostate cancer: Contrast-enhanced us for detection. *Radiology* 219 (1):219–225.

Han, S. K., K. Na, and Y. H. Beae. 2003. Sulfonamide based pH-sensitive polymeric micelles: Physicochemical characteristics and pH-dependent aggregation. *Colloids Surf A Physicochem Eng Aspects* 214:49.

Hashizume, H., P. Baluk, S. Morikawa, and J. W. Mclean. 2000. Openings between defective endothelial cells explain tumor vessel leakiness. *Am J Pathol* 156:1363.

Hobbs, S. K., W. L. Monsky, and Yuan, F. 1998. Regulation of transport pathways in tumor vessels: Role of tumor type and microenvironment. *Proc Natl Acad Sci USA* 95:4607.

Hobbs, S. K., W. L. Monsky, F. Yuan, W. G. Roberts, L. Griffith, V. P. Torchilin, and R. K. Jain. 1998. Regulation of transport pathways in tumor vessels: Role of tumor type and microenvironment. *Proc Natl Acad Sci USA* 95 (8):4607–4612.

Jain, R. K. 1991. Haemodynamic and transport barriers to the treatment of solid tumors. *Int J Radiat Biol* 60:85.

Kataoka, K., T. Matsumoto, M. Yokoyama, T. Okano, Y. Sakurai, S. Fukushima, K. Okamoto, and G. S. Kwon. 2000. Doxorubicin-loaded poly(ethylene glycol)-poly(beta-benzyl-L-aspartate) copolymer micelles: Their pharmaceutical characteristics and biological significance. *J Control Release* 64:143.

Kedar, R. P., D. Cosgrove, V. R. McCready, J. C. Bamber, and E. R. Carter. 1996. Microbubble contrast agent for color Doppler US: Effect on breast masses. Work in progress. *Radiology* 198 (3):679–686.

Kim, G. M., Y. H. Bae, and W. H. Jo. 2005. pH-induced micelle formation of poly(histidine-co-phenylalanine)-block-poly(ethylene glycol) in aqueous media. *Macromol Biosci* 5 (11):1118–1124.

Lathia, J. D., L. Leodore, and M. A. Wheatley. 2004. Polymeric contrast agent with targeting potential. *Ultrasonics* 42 (1–9):763–768.

Lee, E. S. and Y. H. Bae. 2006. Polymeric micelles targeting tumor pH. In *Nanotechnology for Cancer Therapy*, Ed. M. M. Amiji.

Lee, E. S., Z. Gao, and Y. H. Bae. 2008. Recent progress in tumor pH targeting nanotechnology. *J Control Release* 132:164.

Liu, P., A. Zhang, Y. Xu, and L. X. Xu. 2005. Study of non-uniform nanoparticle in liposome extravasation in tumor. *Int J Hyperthermia* 21:259.

Makale, M. 2007. Intravital imaging and invasion. *Methods Enzymol* 426:375.

Marcil, I. and C. Goulet. 2002. [The journal club: From research to practice]. *Infirm Que* 9 (5):46–51.

Matei, D. 2007. Novel agents in ovarian cancer. *Export Opin Investig Drugs* 16:1227.

Menger, M. D., M. W. Laschke, and B. Vollmar. 2002. Viewing the microcirculation through the window: Some twenty years experience with the Hamster Dorsal Skinfold Chamber. *Eur Surg Res* 34:83.

Mitra, A., Nan. A, Line. B. P., and Ghanderhari. H. 2006. Nanocarriers for nuclear imaging and radiotherapy of cancer. *Curr Pharm Des* 12:4729.

Moghimi, S. M., A. C. Hunter, and J. C. Murray. 2001. Long-circulating and target-specific nanoparticles: Theory to practice. *Pharmacol Rev* 53 (2):283–318.

Na, K. and Y. H. Bae. 2002. Self-assembled hydrogel nanoparticles responsive to tumor extracellular pH from pullulan derivative/sulfonamide conjugate: Characterization, aggregation, and adriamycin release in vitro. *Pharm Res* 19 (5):681–688.

Na, K., E. S. Lee, and Y. H. Bae. 2003. Adriamycin loaded pullulan acetate/sulfonamide conjugate nanoparticles responding to tumor pH: pH-dependent cell interaction, internalization and cytotoxicity in vitro. *J Control Release* 87 (1–3):3–13.

Needham, D., G. Anyarambhatla, G. Kong, and M. W. Dewhirst. 2000. A new temperature-sensitive liposome for use with mild hyperthermia: Characterization and testing in a human tumor xenograft model. *Cancer Res* 60:1197.

Oh, K. T., E. S. Lee, D. Kim, and Y. H. Bae. 2008. L-histidine-based pH-sensitive anticancer drug carrier micelle: Reconstitution and brief evaluation of its systemic toxicity. *Int J Pharm* 358 (1–2):177–183.

Pardoe, H., P. R. Clark, T. G. St Pierre, P. Moroz, and S. K. Jones. 2003. A magnetic resonance imaging based method for measurement of tissue iron concentration in liver arterially embolized with ferrimagnetic particles designed for magnetic hyperthermia treatment of tumors. *Magn Reson Imaging* 21:483.

Pitt, W. G., G. A. Husseini, and B. Staples. 2004. Ultrasound drug delivery—A General review. *Expert Opin Drug Deliv* 1:37.

Rapoport, N. 2006. Combined cancer therapy by micellar-encapsulated drugs and ultrasound. In *Nanotechnology for Cancer Therapy*, Eds. M. Amiji, CRC Press: Boca Raton, FL, 2006; pp. 417–437.

Rappoprt, N. Y., D. A. Christensen, H. D. Fain, L. Barrows, and Z. Gao. 2004. Ultrasound-triggered drug targeting of tumors in vitro and in vivo. *Ultrasonics* 42:943.

Rapoport, N., Z. Gao, and A. Kennedy. 2007. Multifunctional nanoparticles for combining ultrasonic tumor imaging and targeted chemotherapy. *J Natl Cancer Inst* 99 (14):1095–1106.

Rapoport, N. Ya., K. H. Nam, Z. G. Gao, and A. Kennedy. 2009. Application of ultrasound for targeted nanotherapy of malignant tumors. *Acoust Phys* 55:594.

Rawat, M., D. Singh, and S. Saraf. 2006. Nanocarriers: Promising vehicle for bioactive drugs. *Biol Pharm Bull* 29:1790.

Rychak, J. J., B. Li, S. T. Acton, A. Leppanen, R. D. Cummings, K. Ley, and A. L. Klibanov. 2006. Selectin ligands promote ultrasound contrast agent adhesion under shear flow. *Mol Pharm* 3 (5):516–524.

Sandhiya, S., S. A. Dkhar, and A. Surendiran. 2009. Emerging trends of nanomedicine—An overview. *Fundamental Clin Pharmacol* 23:263.

Sethuraman, V. A. and Y. H. Bae. 2007. TAT peptide-based micelle system for potential active targeting of anti-cancer agents to acidic solid tumors. *J Control Release* 118 (2):216–224.

Shan, S., C. Flowers, C. D. Peltz, H. Sweet, N. Maurer, E. G. Kwon, A. Krol, F. Yuan, and M. W. Dewhirst. 2006. Preferential extravasation and accumulation of liposomal vincristine in tumor comparing to normal tissue enhances antitumor activity. *Cancer Chemother Pharmacol* 58:245.

Sikic, B. I., G. A. Fisher, B. L. Lum, J. Halsey, L. Beketic-Oreskovic, and G. Chen. 1997. Modulation and prevention of multidrug resistance by inhibitors of P-glycoprotein. *Cancer Chemother Pharmacol* 40:13.

Stubbs, M., R. M. J. McSheehy, J. R. Griffiths, and L. Bashford. 2000. Causes and consequences of tumor acidity and implications for treatment. *Opinion* 6:15.

Takalkar, A. M., A. L. Klibanov, J. J. Rychak, J. R. Lindner, and K. Ley. 2004. Binding and detachment dynamics of microbubbles targeted to P-selectin under controlled shear flow. *J Control Release* 96 (3):473–482.

Tannock, I. F. and D. Rotin. 1989. Acid pH in tumors and its potential for therapeutic exploitation. *Cancer Res* 49:4373.

Torchilin, V. P. 2008a. Tat peptide-mediated intracellular delivery of pharmaceutical nanocarriers. *Adv Drug Deliv Rev* 60:548.

Torchilin, V. P. 2008b. Cell penetrating peptide-modified pharmaceutical nanocarriers for intracellular drug and gene delivery. *Biopolymers* 90:604.

Torchilin, V. P., T. S. Levchenko, R. Rammohan, N. Volodina, B. Papahadjopoulos-Sternberg, and G. G. D'Souza. 2003. Cell transfection in vitro and in vivo with nontoxic TAT peptide-liposome-DNA complexes. *Proc Natl Acad Sci USA* 100:1972.

Volk, T., E. Jahde, H. P. Fortmeyer, K. H. Glusenkamp, and M. F. Rajewsky. 1993. pH in human tumor xenografts: Effect of intravenous administration of glucose. *Br J Cancer* 68:492.

Warburton, C., W. H. Dragowska, and K. Gelmon. 2004. Treatment of HER-2/neu overexpressing breast cancer xenograft models with trastuzumab (herceptin) and gefitinib (ZD1839): Drug combination effects on tumor growth, HER-2/neu and epidermal growth factor receptor expression, and viable hypoxic cell fraction. *Clin Cancer Res* 10:2512.

Wilson, S. R., P. N. Burns, D. Muradali, J. A. Wilson, and X. Lai. 2000. Harmonic hepatic US with microbubble contrast agent: Initial experience showing improved characterization of hemangioma, hepatocellular carcinoma, and metastasis. *Radiology* 215 (1):153–161.

Yin, H., E. S. Lee, D. Kim, K. H. Lee, K. T. Oh, and Y. H. Bae. 2008. Physicochemical characteristics of pH-sensitive poly(L-histidine)-b-poly(ethylene glycol)/poly(L-lactide)-b-poly(ethylene glycol) mixed micelles. *J Control Release* 126 (2):130–138.

Yokoyama, M. 1998. Novel passive targetable drug delivery with polymeric micelles. In *Biorelated Polymers and Gels*, Ed. T. Okano. San Diego, CA: Academic Press, pp. 193–229.

Yu, B. G., T. Okano, K. Kataoka, and G. Kwon. 1998. Polymeric micelles for drug delivery: Solubilization and haemolytic activity of amphotericin B. *J Control Release* 53 (1–3):131–136.

15

Nanobiomaterials for Ocular Applications

Rinti Banerjee
Indian Institute of Technology

15.1 Introduction

The development of suitable biomaterials for replacement of different parts of the eye is challenging due to the stringent criteria that need to be fulfilled by such materials in order to maintain vision and prevent foreign body sensation on implantation. Various nanostructured materials and nanoparticles have been developed in recent years, which have the potential to improve ocular residence times, cellular interactions, and drug delivery in the eye. This chapter describes some of these advances and is organized into two main areas, namely, nanobiomaterials for ocular drug delivery and nanobiomaterials for ocular implants.

15.2 Nanobiomaterials for Ocular Drug Delivery

15.2.1 Anatomical Considerations for Ocular Drug Delivery

The delivery of drugs to the posterior segment of the eye after topical administration is very less. This is because drugs have to cross many anatomical barriers, and the sources of removal at each step are multiple. The drug is first distributed in the tear fluid and then passes through the cornea to the anterior chamber. To reach the retina and adjacent structures, it has to cross through the vitreous cavity.

The precorneal tear film (PTF) is the first barrier encountered by the topically administered drug. It is secreted by lacrimal glands and covers the cornea and conjunctiva in the form of a thin film. Spontaneous blinking of eyelids, volume of tear fluid, and its drainage dynamics influence the ocular bioavailability and residence time of drugs in the PTF. Reflex blinking ensures the uniform spread of the tear film and prevents the cornea from drying. It also helps in the spread of the drug in the PTF and causes its removal through nasolacrimal drainage. In humans, blinking occurs approximately at the rate of 15 times per minute. A mucin layer covers the anterior surface of the conjunctiva and cornea and is secreted by the goblet cells of the conjunctiva. Ocular mucoadhesives interact with this layer to prolong the residence time of drugs.

The cornea may act as a pathway, a barrier, or a reservoir for topically applied drugs. Since the lipid content of the epithelium and the endothelium are much higher than that of the stroma, the transfer of drugs through the cornea is largely determined by phase solubility. The epithelium and endothelium are highly permeable to the substances that are fat soluble whereas the stroma is permeable to substances that are water soluble.

Further routes to the posterior chamber have several anatomical barriers. Lipophilic drugs can partition into the lens from aqueous humor and then diffuse around the cortex to the vitreous body. When the drug molecule reaches the anterior surface of the vitreous, it can further progress to the fundus by diffusion through the vitreous gel or by convection when it is liquefied. The several topical barriers prevent adequate drug levels in the posterior chamber and the presence of the blood retinal barrier prevents systemic drugs from reaching adequate levels in the retinal tissues.

15.2.2 Need for Drug Delivery Systems

The need for novel drug delivery agents exists in ophthalmology due to the poor residence time and pulse kinetics of the conventional formulations like eyedrops. Materials in the form of solid reservoirs, liposomes, nanoparticles, and gels can be used for sustained release of drugs in the eye. This is particularly important for conditions of the posterior segment of the eye as only 1%–2% of the conventional eyedrops actually reach this site.

15.2.3 Materials Used in Conventional Ocular Formulations

To overcome the disadvantages of eyedrops, various ophthalmic drug delivery systems such as hydrogels, and shields have been developed. Polymeric gels used for ophthalmic drug delivery may be either preformed gels or in situ gelling systems. Hydrogels are used in the eye to increase the ocular residence time of the drugs due to their increased viscosity and in some cases the mucoadhesive property. Hyaluronic acid is an example of a biological polymer that has mucoadhesive properties and hence can be used for increasing the ocular residence time of drugs. Since it is a component of aqueous and vitreous humor, its ocular tolerance and safety are not a concern. In a study by Kyyronen et al. (1992), the in vivo release of methylprednisolone in hyaluronic acid gels was 9–12 times lower than the control suspension. The burst release of the control was not observed when using the gel, which showed a slow sustained release of drug over several hours. Synthetic mucoadhesive polymers include water-soluble polymers that are linear chains and water-insoluble polymers that are swellable networks joined by cross-linking agents. The polyanionic polyacrylic acid is one example of a synthetic mucoadhesive drug delivery agent (Le Bourlais et al., 1998). Mucoadhesion occurs due to interactions between the polymer chains and mucin molecules present in the ocular mucus layer.

One disadvantage of these gels is that they interfere with eyelid motion and lead to ocular discomfort and blurring of vision due to their highly viscous nature. Though the lacrimation and blinking mechanisms help in improving the ocular tolerance to the eye, in situ forming gels, which can be instilled in solution form, are preferred. In situ acting gelling systems, as described earlier, are viscous liquids that on exposure to physiological conditions will shift to a gel phase. The poloxamers are polyols with thermal gelling properties whose solution viscosity increases when temperature is raised from its critical temperature of 16°C to the ocular temperature of 33°C–34°C. Cellulose acetophthalate is a solution at pH 4.4, which undergoes coagulation when the pH is raised by tear fluid to 7.4 (Ding, 1998). Gellan gum, an anionic extracellular polysaccharide secreted by *Pseudomonas elodea*, can be administered as an aqueous solution that forms clear gels in the presence of mono or divalent cations present in the tear fluid.

Shields or inserts act as solid reservoirs of drugs over prolonged durations but have low patient tolerance. Ocusert is an insoluble ophthalmic insert that consists of a central reservoir of drug enclosed between two semipermeable membranes that allow the drug to diffuse at a predetermined rate over several days (usually a week). However, after prolonged wear over 12 h, these inserts swell and get partially fragmented. Collagen shields are soluble ophthalmic inserts that are fabricated from porcine scleral tissue, which has a collagen composition similar to that of human cornea. The collagen shields are hydrated before being placed on the eye. On hydration with tear fluids, they form clear thin films of around 0.1 mm in thickness, which conform to the corneal surface and dissolves slowly over a predetermined period usually a couple of days to cause prolonged release of the drug.

15.2.4 Role of Nanoparticles in Ocular Drug Delivery

Nanoparticles have many advantages as potential carriers for drugs in the eyes. These include their ability to penetrate through several anatomical barriers in the eye. Hence, the nanoparticles can reach the posterior parts of the eyes even when applied to the anterior surface. Further, nanoparticles can be made of mucoadhesive polymers, which allow them to stick to the surface of the sclera and increase the ocular residence time of the drugs, withstanding washout by the tear fluid. This allows the drugs to be delivered in more patient compliant schedules. The nanoparticles also act as depots of the drug maintaining a sustained release of the drugs over several hours. The small sizes of the nanoparticles allow them to be directly internalized by endocytosis by corneal endothelial cells. Figure 15.1 depicts the different types of nanoparticles commonly used in ocular drug delivery.

15.3 Liposomes and Solid Lipid Nanoparticles

Liposomes are microscopic vesicles that consist of membrane like lipid bilayers, which surround aqueous compartments. They are biocompatible and biodegradable and can be used for delivery of both hydrophilic and hydrophobic drugs. Liposomes with positive surface charge are relevant for the eye as they provide more stable adsorption to the corneal surface, which is coated with negatively charged mucin (Durrani et al., 1992). However, stearylamine is avoided as it causes irritation to the eye (Taniguchi et al., 1988). Positively charged unilamellar liposomes increased the flux of penicillin G across isolated rabbit corneas more than fourfold. Solid lipid nanoparticles have also been developed as nanocarriers for ocular drug delivery. Cavalli et al. (2002) have developed solid lipid nanoparticles <100 nm in size loaded with tobramycin, which achieved higher bioavailability in the rabbit aqueous humor than the free drug. Attama et al. (2008) have shown increased oentration of diclofenac-loaded solid lipid nanoparticles through human corneal constructs.

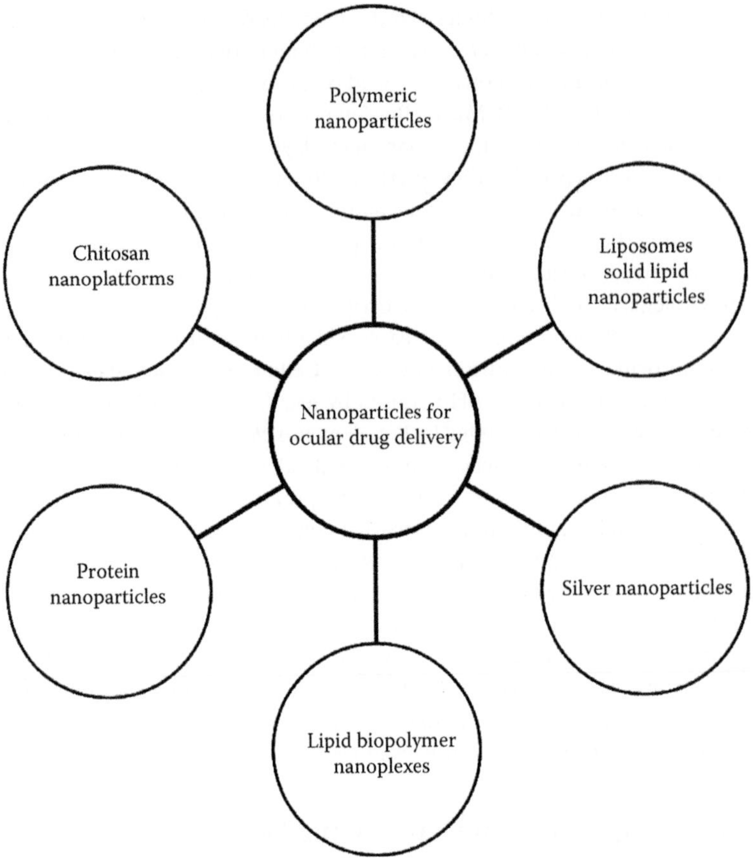

FIGURE 15.1 Overview of nanoparticles used in ocular drug delivery.

15.4 Niosomes

Niosomes are bilayered nonionic surfactant vesicles that encapsulate hydrophilic and hydrophobic drugs and are chemically more stable than liposomes. Niosomes have been found to increase the ocular bioavailability of cyclopentolate (Sahoo et al., 2008).

15.5 Polymeric Nanoparticles

Nanoparticles are colloidal particles of sizes 10 nm to less than 1 μm in diameter and are in the form of either nanocapsules or nanospheres. The small sizes allow increased penetration through the ocular membranes and less of a foreign body sensation on topical administration. Slow sustained release of drugs is obtained by the use of nanoparticulate carriers. Materials like polyalkyl cyanoacrylate, poly-ε-caprolactone, polylactic-*co*-glycolic acid, gelatin, and albumin have been used for the preparation of the nanoparticles. All these are widely studied biodegradable polymers offering good ocular tolerance (Zimmer and Kreuter, 1995). Gupta et al. (in press) developed polylactide-*co*-glycolide nanoparticles loaded with sparfloxacin, which showed higher ocular residence time than the conventional formulation. Recently, Das et al. (in press) synthesized amphotericin B–loaded Eudragit nanoparticles, which were found to be nontoxic to the eyes and can be used for ocular drug delivery. Pignatello et al. (2002) have developed Eudragit nanosuspensions loaded with ibuprofen for ocular delivery of nonsteroidal anti-inflammatory agents. Gupta et al. (2000) have developed polymeric micelles of copolymers

N-isopropylacrylamide, vinyl pyrrolidone, and acrylic acid loaded with ketorolac and found a sustained ocular anti-inflammatory effect as compared to the free drug. Yenice et al. (2008) found that hyaluronic acid coated polycaprolactone nanoparticles delivered several folds higher cyclosporine A levels in the cornea as compared to the free drugs.

15.6 Biopolymeric Nanoparticles

Campos et al. (2001) have developed chitosan nanoparticles loaded with cyclosporine A by ionic gelation. The nanoparticles were 290 nm in diameter and following topical administration in rabbits, achieved therapeutic concentrations in cornea and conjunctiva over 48 h higher than those achieved by the free drug solutions. Yuan et al. (2006) also studied cyclosporine A–loaded nanoparticles using cholesterol modified chitosan self-aggregated nanoparticles <230 nm in diameter, which showed increased retention in precorneal areas and may have the potential for treatment of extraocular conditions like keratoconjunctivitis sicca, which is limited by the lacrimal washout and poor ocular residence of conventional eyedrops.

15.7 Protein Nanoparticles

Merodio et al. (2002) have developed ganciclovir-loaded albumin nnaoparticles by coacervation technique. Intravitreal injection of these nanoparticles were well tolerated and 2 weeks after the injection, a significant amount of the nanoparticles were found in the vitreous humor, in a thin layer overlying the retina and close to the blood aqueous barrier. Such nanoparticles have the potential to treat cytomegalovirus infections that affect the endothelial cells of ocular blood vessels, optic nerve, and the retina. Das et al. (2005) have also developed aspirin-loaded albumin nanoparticles for ocular dug delivery. Similarly, gelatin nanoparticles have also been developed as platforms for ocular delivery using model drugs pilocarpine hydrochloride and hydrocortisone (Vandervoort and Ludwig, 2004). Jain and Banerjee (2008) have compared albumin, gelatin, chitosan, and solid lipid nanoparticles as carriers for ciprofloxacin hydrochloride in the eye. Solid lipid nanoparticles and chitosan nanoparticles were found to have favorable properties of high encapsulation, stability, and sustained release suitable for carriers in ocular drug delivery.

15.8 Combination Nanosystems

Newer delivery systems are being developed that combine the advantages of two or more of the above systems. Polylactide-*co*-glycolide nanoparticles can be suspended in a mucoadhesive polymeric solution or an in situ gelling solution to obtain colloidal particles trapped within a gel (Dillen et al., 2004).

Diebold et al. (2007) developed a novel nanoplatfom that combines the advantages of hydrophilic and hydrophobic carriers. Liposome chitosan nanocomplexes were developed that allow efficient coencapsulation of different types of drugs. Liposome chitosan nanocomplexes were found to have high cellular uptake within conjunctival epithelial cells and were well tolerated in rabbit eyes. The penetration and distribution of the nanostructures could be modulated depending on the compositions and charges.

15.9 Nanoparticles for Ocular Gene Delivery

Efficient delivery of genes to the retina can be promising in the treatment of several retinal diseases like retinitis pigmentosa and proliferative vitreoretinopathy. Viral gene carriers have risks of infection. Intravitreal injection of nonviral nanoparticles are a safe and promising strategy for gene delivery in retinal diseases. The nanoparticles need to be optimized to reduce their binding to the glycosaminoglycans in the vitreous to increase their internalization within the target retinal cells and their escape from the endosome (Sanders et al., 2007; Cai et al., 2008). Ultrapure oligomeric chitosan DNA complexes have been shown to be effective for transfection in corneal cells (Klausner et al., 2010). Chitosan oligomers that are fully deacetylated and have a low molecular weight <9.5 kDa are preferred to higher weight chitosans for ocular gene delivery as they overcome the viscosity-enhancing properties of higher weight chitosans and are more easily dissociated from DNA, allowing more efficient release from the nanoparticles.

15.10 Nanoparticles for Ocular Delivery of Nucleic Acids

Antisense oligonucleotides are complementary to a RNA sequence to which they bind and prevent the translation of a specific protein. Some targets for gene inhibition in the eye include vascular endothelial growth factor (VEGF), transforming growth factor B, fibroblast growth factor, herpes simplex virus, and cytomegalovirus. Oligosense nucleotides are hydrophilic, negatively charged, and require nanocarriers for delivery in the posterior segment of the eye. The nanoparticles are required to allow intracellular delivery and sustained residence time intraocularly, preventing the need for multiple intravitreal injections (Fattal and Bochot, 2006). For example, lipoaminoacids like lipid lysine dendrimers have been found to deliver a phosphodiester oligosense nucleotide, which inhibits VEGF in human retinal cell lines (Wimmer et al., 2002). A 40%–60% reduction in VEGF expression was observed within 24 h.

15.11 Functional Nanoparticles for Ocular Diseases

Diabetes causes severe vascular complications in the eyes including diabetic retinopathy. Advanced glycation products cause increased endothelial permeability of the blood retinal barrier. Sheikpranbabu et al. (2010) have found that silver nanoparticles (without any drugs) acted as potent antipermeability agents and blocked the biological effects of advanced glycation products by targeting the Src signaling pathway and tight junction proteins.

15.12 Nanobiomaterials for Ocular Implants

Nanoparticles are incorporated in many implants. Nanostructured implants allow various improvements over conventional ocular implants with regard to their cellular interactions, physical properties, and ability to be multifunctional. Table 15.1 summarizes some of the recent nanostructured ocular implant strategies.

15.13 Contact Lenses

Contact lenses are a means of vision correction by the use of a lens in intimate contact with the cornea. Contact lenses are used to correct refractive errors of the eyes as an alternative to spectacles. They have the advantages of an increased field of vision and the ability to correct irregularities of the corneal surface and the corresponding astigmatism. Further, they are preferred in cases of

TABLE 15.1 Types of Nanostructured Ocular Implants

Implant	Nanostructure Strategies
Soft contact lenses	Molecular imprinting
	Cyclodextrin linked
	Microemulsion droplets
Intraocular lens	Inorganic nanoparticle incorporated within polymeric matrix
Ocular inserts	Nanoparticle in polymeric membranes
Corneal adhesives	Dendrimers
Vitreous substitutes	Nanoparticles in gels

severe ametropia and in cases where the powers of refraction are widely different in each eye. Recent advances also allow the use of drug-loaded contact lenses for sustained topical release of the drugs. Cosmetic uses of tinted contact lenses are also common.

15.13.1 Anatomical Considerations for Contact Lenses

The anatomical considerations for development of a contact lens are that the lens is to be implanted on the anterior most surface of the eye, overlying the cornea. The cornea is an avascular structure that obtains oxygen from the atmosphere and the tear film for adequate nutrition and for maintenance of its structure. The cornea consists of an outer epithelium, an internal stroma containing keratocytes and an internal endothelial layer. The stroma comprises over 90% of the total thickness of the cornea and is composed mainly of parallel arrays of collagen fibers. This regular arrangement allows the maintenance of corneal transparency under normal fluid conditions. The Bowman's layer, consisting of randomly arranged collagen fibers and proteoglycans, separates the stroma from the corneal epithelium that resides on the outer surface of the eye. The endothelial layer is only single cell thick and is separated from the stroma by the Descemet's membrane. The endothelial cells play an important role in the maintenance of transparency in the corneal stroma both by actively pumping water out via a sodium–potassium adenosine triphosphatase and a coupled bicarbonate pump and by serving as a tight barrier to fluid entry (Geroshi et al., 1995). The ophthalmic compatibility of a contact lens on the eye requires that the lens maintain a stable, continuous tear film for clear vision, sustain normal hydration, and be permeable to oxygen in order to maintain corneal metabolism. Further, the material should be nonirritating and must have excellent surface characteristics.

15.14 Current Materials for Contact Lenses

Historically, though the earliest reference to the concept of contact lenses dates back to Leonardo da Vinci, the first clinical application was that of a glass corneoscleral lens in the 1880s. The first polymeric contact lens was devised of polymethymethacrylate (PMMA) in the 1940s. Today, contact lenses are classified into hard and soft lenses based on the modulus of elasticity of the materials. The PMMA lens is now classified as a hard contact lens. It has good optical properties and high durability but is less well tolerated. The low oxygen permeability limits the long-term wear of these lenses. To avoid corneal anoxia, these lenses need to be small in diameter and need to be designed to float on a precorneal tear film such that tear film exchange during blinking and movement of the lens allows oxygenation of the cornea. Soft contact lenses are either hydrogels or are silicone-based elastomers and are more comfortable to the user. A third category of lenses includes the rigid gas permeable lenses that have superior oxygen permeability while maintaining the mechanical properties similar to hard lenses. This category may be considered as an improved version of the original hard contact lenses.

The first soft contact lens material was polyhydroxyethyl methacrylate (PHEMA), which contained 38% water, had excellent wettability and improved wearer comfort. Since then, several modifications have been made to develop soft contact lenses with superior oxygen permeability. The oxygen permeability of these hydrogels can be increased by increasing the water content or by decreasing the thickness of the lenses. Very thin lenses lead to difficulties in handling. Increasing the water content of the lenses is achieved by copolymerizing HEMA with hydrophilic monomers such as methacrylic acid and vinyl pyrrolidone. Hydrogels with high water content (>70%) have the advantage of improved oxygen permeability but have low tear resistance and an increased tendency to adsorb proteins from the tear fluid leading to spoilation. Actifresh 400, from Hydron, is an example of a conventional hydrogel soft contact lens composed mainly of MMA/NVP having 73% water content and an oxygen permeability of 36 Barrers (Lloyd et al., 2001).

Soft contact lenses based on the silicone elastomer, polydimethysiloxane (PDMS), also can be used as extended wear contact lenses. The material has excellent optical properties and high oxygen permeability (Dk upto 600 Barrers). The main disadvantage of this material is its low surface energy leading to poor tear fluid wetting and an increased tendency to bind lipids from the tear fluid. This can be overcome by surface modification techniques and by the grafting hydrophilic polymers to its surface. The combination of the high oxygen permeability of PDMS with the excellent wettability and patient tolerance of conventional hydrogels can be achieved by the development of silicone-based hydrogels. Silsoft is a PDMS-based contact lens from Bausch & Lomb that has a water content of 0.2% and an oxygen permeability of 340 Barrers. There is a distinct difference in the relation of water content to oxygen permeability in the two types of soft contact lenses. In case of conventional hydrogels, the oxygen permeability increases with the increasing water content. However, in case of silicone hydrogels, the increase in water content is achieved by increasing the portion of the conventional hydrogel monomer to the silicone monomer and this reduces the oxygen permeability of the material (Lloyd et al., 2001).

Daily wear, rigid, gas-permeable lenses have been developed by copolymerization of methyl methacrylate with methacrylate functionalized siloxanes like methacryloxypropyltris (trimethyl siloxy silane) (TRIS). The various properties of these lenses are modulated by the TRIS-to-cross-linker ratio. The disadvantages of decreased wettability with the use of high concentrations of TRIS are overcome by the addition of methacrylic acid. Quantum II lens made of silicone acrylate is a product of Bausch & Lomb and has an oxygen permeability of around 100 Barrers with no water content. Further improvements in rigid gas-permeable lenses have been developed by the use of fluoromethacrylates. The low surface energies of such lenses cause reduced spoliation but have the disadvantage of poor wettability. Some of these lenses are suitable for 7 day extended wear.

15.15 Advances in Contact Lenses due to Nanotechnology

The use of a contact lens for localized sustained delivery of drugs has been achieved successfully recently (Hiratani et al., 2005). Therapeutic soft contact lenses fabricated by the molecular imprinting method have been found to have a drug loading capacity two- to threefold greater than that of the contact lenses made by conventional methods (Hiratani and Alvarez-Lorenzo, 2002). Furthermore, the adsorption affinities for the drug used as template were 9- to 20-fold higher in the case of the imprinted contact lenses (Hiratani and Alvarez-Lorenzo, 2004). Thus, adsorption sites capable of capturing the target drug can be effectively encoded into the polymer network by molecular imprinting and, in consequence, can improve the specific drug loading capacity of the contact lenses. Such imprinted soft contact lenses are able to provide greater and more sustained drug concentrations in tear fluid with lower doses than conventional eyedrops.

Similarly, Santos et al. (2009) developed soft contact lenses functionalized with pendant cyclodextrins for controlled drug delivery over 2 weeks in the lacrimal fluid. Ophthalmic solutions containing cyclodextrins undergo instantaneous decomplexation of drugs when diluted in ocular fluids preventing sustained release. On the contrary, if the cycodextrins are attached to a polymeric network, dilution effects are minimized and controlled release of the drug can be achieved. Such networks have potential as drug-loaded soft contact lenses.

The development of glucose-sensing contact lenses has been envisaged. These will aid in the determination and monitoring of tear fluid glucose levels, which in turn track blood glucose with an approximate 30 min lag time. These disposable colorless contact lenses can be worn by diabetics and changes in the color of their lenses can give an indication of glucose levels in tear fluid and blood (Badugu et al., 2005).

Further, synthetic corneal onlays or implantable contact lenses are also being developed either as part of partial thickness epikeratopathy or as the optical portions of full-thickness keratoprostheses.

15.16 Corneal Adhesives and Nanoparticulate Inserts

Corneal wounds are presently repaired using nylon sutures. Sutures have many disadvantages like additional trauma due to multiple sites of suturing, uneven healing leading to astigmatism, sites of infection, and removal if nonbiodegradable. Dendrimers are highly branched polymers having three structural parts, namely, the central core, the intermediate branching layers, and the peripheral groups.

Dendrimer-based adhesives can offer a minimally invasive technique for efficient corneal wound healing. Dendrimers have been recently explored as adhesives for corneal wound healing posttraumatic injury or postsurgical procedures (Grinstaff, 2007). Photocross-linking or peptide ligation strategies have been used to couple individual dendrimers to form an adhesive.

Jain et al. (2010) have developed nanoparticulate polymeric membranes as biodegradable drug-loaded inserts for ocular drug delivery. The nanoparticles act as depots for the sustained release of drugs and the polymeric matrix acts as an insert and allows the ease of regional application of the nanoparticles on the eye. These inserts were found to sustain drug release over 5 days in simulated ocular conditions and allowed increased penetration of drug into posterior segment of the explanted goat eyeballs. The nanoparticulate inserts are biodegradable and act as platform technologies for delivering both hydrophilic and hydrophobic drugs in the eye. The inserts are mucoadhesive and prevent washout of the drugs by the tear fluid.

15.17 Intraocular Lenses

An intraocular lens is a synthetic lens that is placed within the eye to replace the dysfunctional crystalline human lens after its removal. The commonest condition that requires the use of intraocular lenses is cataract. In cataract, there is opacification of the normally transparent crystalline lens leading to loss of vision. On removal of the cataractous lens, the intraocular lens is placed within the eye as a substitute.

15.18 Anatomical Considerations for Intraocular Lenses

The human ocular lens is situated just behind the iris and the pupil and is attached to the ciliary body via the suspensory ligament. It is transparent and biconvex with an outer capsule consisting of type IV collagen. Inner to the anterior capsule is a single layer of lens epithelial cells. These epithelial cells give rise to the nondividing enucleate lens fiber cells that along with the interstitial material form the main part of the internal lens substance. The high refractive index of the lens is due to the specialized proteins called crystallins secreted by the lens cells.

A special feature of the human lens is its ability to change its shape and hence focal lengths to allow images of both near and far objects to form sharply on the retina. This is achieved by accommodation. During this process, the contraction and relaxation of the ciliary muscles and suspensory ligament control the shape of the lens. During the viewing of distant objects, the lens is flattened by the pull of the suspensory ligaments on the lens margin. By contrast, for viewing near objects, there is contraction of the ciliary muscle causing relaxation of the suspensory ligament and leading to thickening of the lens. These factors must be kept in mind while designing intraocular lenses.

15.18.1 Current Strategies for Intraocular Lenses

In 1950, Harold Ridley first implanted a polymethylmethacrylate (PMMA) intraocular lens (IOL) following removal of a cataractous crystalline lens. The initial procedure entailed total cataractous lens extraction or intracapsular cataract extraction and fixation of an IOL in the anterior chamber, which was supported by the iris. In due course of time, it was realized that several complications are reduced if the posterior capsule of the lens is not removed. Based on these considerations, the procedure was modified to leave the lens capsule in place. This change caused the procedure to be redesignated as extracapsular cataract extraction. Though several other polymeric materials have been developed, PMMA remains one of the standard materials for intraocular lenses even today.

The types of intraocular lenses are based on their position of replacement within the eye as well as the mechanical properties of the lens. According to position, intraocular lenses are classified as anterior chamber IOLs, which are positioned in front of the iris but behind the cornea, and iris clip lenses, which straddle the pupil and posterior chamber IOLs, which are placed behind the iris within or on the lens capsular bag. Based on the mechanical properties of intraocular lenses, they are either rigid or foldable. This classification has implications on the manner of handling of the IOLs. Previously, the lenses were almost always made of rigid material, and an incision of approximately 6 mm (roughly equal to the diameter of the lens) was required to insert the lens into the capsule. Smaller incisions are clinically advantageous, with less patient discomfort and faster recovery from surgery but perhaps more importantly, they reduce astigmatism. These advantages have fostered the development of foldable IOLs. Foldable intraocular lenses can be inserted through smaller incisions of less than 3.5 mm size. These new lenses can be folded into a "taco" shape, to be inserted through a small tube into the eye. After the folded IOL has been pushed completely through a delivery tube introduced into the eye, the IOLs "memory" causes it to spring back, regaining its original shape.

Intraocular lenses consist of two parts an optic and haptics. The optic is the central part of the lens responsible for the refraction through the lens. The haptics are projections from the optic that allow the attachment of the lens within the eye. Though PMMA has the advantages of excellent optical properties and low weight, its low surface energy leads to complications like corneal endothelial damage and post-operative adhesion of inflammatory cells to its surface. This has led to the development of lenses with soft high energy surfaces using *N*-vinyl pyrrolidone and hydroxyethylmethacrylate.

Foldable lenses are made from silicone elastomers, PHEMA hydrogels, and acrylic polymers. A special requirement of these lenses is that they must be easy to insert and should unfold slowly in a controlled manner without any creases. In this respect the acrylic lenses unfold more slowly than the silicone lenses (Kohnen, 1996; Lloyd et al., 2001). Examples of foldable intraocular lenses are Alcon HydroSof made of HEMA having a water content of 38% and a refractive index of 1.44 and Chiron C10UB made of PDMS having <1% water content and a refractive index of 1.41.

Current IOLs have different chemical properties and shapes. The shape of the lens optics differs in an effort to design lenses that require smaller incision sizes and reduce the incidence of posterior capsule opacification. Prolate IOLs have also been developed that are steeper centrally and flatter peripherally to offset the age-induced spherical aberration (Lazzaro, 2005).

Phosphorylcholine-based polymeric coated intraocular lenses have been developed in an effort to reduce protein adsorption, cellular adhesion, and inflammatory changes (Lloyd et al., 1997). Also, heparin covalently bound to PMMA surfaces have been associated with improved compatibility due to lower activation of complement and decreased outgrowth of fibroblasts and macrophages, activation of granulocytes, and adhesion of platelets (Lamson et al., 1989).

15.18.2 Advances in Intraocular Lenses due to Nanotechnology

Nanoparticles have been impregnated within intraocular lenses to produce haptics having high fracture toughness. Further, in the optical portion incorporation of nanosized inorganic fillers of high refractive index can lead to the increase in the global refractive index and the development of thinner intraocular lenses requiring smaller incisions for surgical procedures.

The need for improving standard intraocular lenses to provide accommodation is recognized. Recently, a two-piece lens system has been developed in which the distance between the two lenses is controlled by the pressure exerted by the ciliary muscle on a U-shaped flange connecting the periphery of the two lenses. Relaxation of the ciliary muscle causes the lens to flatten and the focal length of the lens can be dynamically altered (Smith, 1989; Lloyd et al., 2001). Injectable intraocular lenses that consist of viscoelastics have also been developed in which the viscoelastic gel dimensions and refractive properties are modified as the ciliary muscle contracts and relaxes leading to stretching or bulging of the capsular bag.

Alcon Laboratories has developed AcrySof Natural IOL, which mimics the UV and blue-light attenuating properties of human crystalline lens (Ernest, 2004). This IOL contains a covalently bound chromophore that absorbs light in the 400–500 nm range, adding this light protection to that already provided in the UV range.

15.19 Vitreous Substitutes

The vitreous humor is a transparent gel that occupies the posterior two-thirds of the eye. It is damaged in various vitreo-retinal pathologies and requires replacement by a suitable substitute to aid in vision and to support the retina preventing its detachment.

15.20 Anatomical Considerations for Vitreous Substitutes

The vitreous is a gelatinous mass located posterior to the lens and consists mainly of hyaluronic acid and collagen. The vitreous can be considered as a fiber-reinforced biocomposite with the amounts of the nonaqueous components reduced to a minimum compatible with maintaining mechanical stability (Bishop, 2000). The water component of vitreous accounts for 99% of its composition. This ensures that a minimal amount of solid matter comes in the light path.

Both vitreal and retinal pathologies together comprise the posterior segment disorders that require vitreous substitutes. These conditions are interrelated as retinal tears can lead to vitreous displacement and vitreous scarring can cause retinal detachment. When the vitreous gel shrinks due to disease- or age-related changes, it pulls on the retina and leads to a tear, the vitreous then seeps in between the different layers of the retina, and causes their separation and detachment from the eyeball, leading to blindness. Common conditions requiring vitreous replacements are traumatic injuries to the eye, diabetic retinopathy, and age-related macular degeneration. Vitreous substitutes are required to act as short-term tamponade agents providing retinal support during surgical procedures of the posterior segment of the eye, as well as a long-term tamponade to support the retina in cases of retinal detachment and cases of degeneration of vitreous humor.

15.20.1 Current Strategies for Vitreous Replacements

Air was first used as early as 1911 as a vitreous replacement but the major problem with air, in addition to tissue reactions, was the rapid absorption from the vitreous cavity by diffusion across the retina and hence a reduction in the tamponade effect.

Perfluorocarbon gases like perfluoroethane and perfluoropropane expand on intravitreal injection due to diffusion of other gases from the blood stream. Perfluoropropane expands to four times its original volume by the fourth day of injection and is resorbed at a slower rate than air. They are useful in short-term procedures like pneumatic retinopexy. They have disadvantages of requiring postoperative positioning of the patient and causing increases in the intraocular pressure. They are not suited for long-term replacement.

Silicone oil and its derivatives are the commonly used vitreous substitutes for long-term tamponade. The silicone oils in current use are polydimethylsiloxanes with various chain lengths and molecular weights. The viscosity of the oils varies linearly with the chain lengths and molecular weight. The high interfacial surface energy of silicone oil at the tamponade/aqueous/retina interface ensures the closure of the retinal breaks and reduces subretinal leakage. However, the hydrophobic nature of silicone oil leads to a poor contact with the retina and aqueous fluids, which inhibits the total filling of the vitreous cavity, which is required for effective closure of retinal breaks. Also, the persistence of silicone oil leads to life-threatening complications of cataract, glaucoma, and keratopathy, which necessitate its removal after 2–3 months (Nakamura et al., 1991; Ohira et al., 1991).

Perfluorocarbon liquids have also been tried as vitreous substitutes but cannot be used for more than a week due to the irreversible damage caused to the inferior retina. Further, their low interfacial surface tensions make them poor tamponade agents.

Semisynthetic polymers like hyaluronic acid, collagen, their mixtures, and hydroxypropylmethylcellulose have been evaluated as possible vitreous substitutes. Hyaluronic acid and collagen have the advantages of biocompatibility and hydrophilic nature but do not form suitable gels even in several fold higher concentrations than those present in the vitreous. Perhaps cross-linked structures may help in improving the performance of these materials (Nakagawa et al., 1997).

Polyvinylpyrrolidinone (PVP) was the first synthetic polymer to be used experimentally as a vitreous substitute in rabbits in 1954. It was used as an aqueous solution and had a short residence time. However, hydrogels of PVP may be used as vitreous substitutes. They undergo biodegradation within 4 weeks and hence can be used only for short-term purposes. Cross-linked polyvinyl alcohol is a promising candidate as a vitreous substitute. It forms a transparent hydrogel and has superior tamponade properties (Colthurst et al., 2000; Soman and Banerjee, 2003).

15.20.2 Recent Advances in Vitreous Substitutes

The possibility of using smart materials that undergo in situ gelation as vitreous substitutes is promising. The material requires to undergo a sol–gel transition due to physiological triggers like ocular temperature, ionic contents in ocular fluids, or changes in pH. Such a material would be easily injected and would form a gel with low syneresis within the posterior chamber of the eye. Such in situ gelling systems have been developed for drug delivery systems and can be combined to serve both as a vitreous substitutes as well as drug delivery agents. Cross-linked polyacrylic acid derivatives have been used for in situ gel formation triggered by pH changes. Methycellulose is an example of a temperature-triggered in situ gelling system whereas alginates can form gels in the presence of cations. However, to function as a vitreous replacement, the agent will need to be nondegradable or very slowly degradable with a prolonged intravitreal residence time.

A vitreous body prosthetic device, though not a vitreous substitute strictly, can act as a functional replacement of the vitreous temporarily. Such a device comprises a thin-walled inflatable balloon made of silicone rubber. The balloon is provided with means for stabilizing and fixing the balloon within the eye. The balloon has an inflow tube made of silicone rubber and is in fluid-tight communication with the interior of the balloon. The other end of the inflow tube is connected to a bulb through which fluid can be introduced into and removed from the tube. Thus the degree of inflation of the balloon can be controlled in order to force the thin-walled balloon against the retinal surface, leading to short-term functional support of the retina (Joseph, 1990).

15.21 Miscellaneous Applications

Orbital implants are used to replace the eyeball for cosmetic reasons after enucleation of the eye. Hydroxyapatite or porous polyethylene implants are used for this purpose.

Intracorneal lenses are implanted within the central stroma of the cornea and augment the normal corneal function. PMMA intracorneal lenses have good optical properties but disrupt nutrient transport across the cornea. Hydrogel and polysulfone intracorneal lenses have also been developed.

Keratoprostheses are penetrating total replacements of the cornea. Lee et al. (1996) identified a variety of criteria that must be met by an ideal keratoprosthesis. These are as follows: (1) the device should be tightly retained in the cornea to prevent extrusion, (2) it should be easily and completely colonized by corneal epithelial cells on the external surface, and (3) it should suppress downgrowth of such cells into the implant bed. Custom-made prostheses of polytetrafluoroethylene having 50 μm pores have shown a refractive index similar to the natural cornea, good collagen synthesis and have been clinically successful (Legeais et al., 1994).

Glaucoma filtration devices are useful in the drainage of fluid in glaucoma. Galucoma is a condition in which raised intraocular pressure causes progressive optic nerve damage and visual field loss. The devices consist of silicone tubes and plates of silicone, polypropylene, and silicone–PMMA combinations (Lim et al., 1998).

Scleral buckles are materials that are used to indent the sclera bringing the choroid in contact with the retina and thus being useful in retinal support in cases of retinal detachment. The simplest absorbable scleral buckles consist of autogenous tendons and fascia lata from the patient. Another resorbable material commonly used as a scleral buckle is pigskin gelatin, which is resorbed over 3–6 months (Schepens and Acosta, 1991). Nonabsorbable buckles are made of silicone or hydrogels. An advantage of the use of silicones is the formation of a tough fibrous capsule around the implant, which both strengthens the sclera and allows easy removal of the implant in the future if required.

15.22 Clinical Implications and Future Prospects

The ocular toxicity of nanoparticles needs to be clearly evaluated. In a recent study, Prow et al. (2008) evaluated the ocular toxicity and transfection potential of chitosan and synthetic polymeric and magnetic nanoparticles in retinal cells. Intravitreal injection of chitosan nanoparticles were found to show an inflammatory reaction. Subretinal magnetic nanoparticles did not show any inflammation or toxic changes over 7 days and had a high transfection potential. The study indicates that different types of nanoparticles have varying cellular interactions in the eyes and ocular toxicity needs to be evaluated over long-term periods. Many biopolymeric and lipid nanoparticles have been found to be nontoxic when applied in the eye.

Lipimix, a drug-free phospholipid emulsion, has been commercialized to replace the lipid layer of the acrimal fluid after refractive surgeries. Clinical trials of Piloplex (pilocarpine-loaded nanospheres made of polymethylmethacrylate acrylic acid copolymer) showed a reduction in intraocular pressure levels but commercialization was not undertaken due to nonbiodegradability, local toxicity, and difficulty in large-scale manufacture of sterile preparations (Araújo et al., 2009). Pegylated RNA aptamers directed at vascular endothelial growth factor have been approved for the treatment of neovascularization in acute macular degeneration (Santos et al., 2006). The use of nonviral nanoparticle delivery strategies can improve transfection and reduce the risk of repeated intravitreal injections.

Nanobiomaterials can provide many promising strategies for diagnostic and therapeutic improvements in ocular drug delivery and in ocular implants. The nanoparticles can especially be useful to overcome the ocular penetration barriers and to prevent washout by the lacrimal fluid. The importance of such nanoparticles is particularly for posterior segment ocular diseases like retinopathies, acute macular degeneration, and other retinal diseases due to the negligible reach of current modalities of therapy in such conditions. The concepts of nanoparticle-impregnated implants like inserts, corneal bandages, and contact lenses can open up new therapeutic avenues for combined biomaterial applications and drug delivery. Further preclinical and clinical trials are warranted to allow the clinical benefits of nanotechnology to be reaped in ophthalmic practice.

References

Araújo, J., Gonzalez, E., Egea, M. A., Garcia, M. L., and Souto, E. B. (2009) Nanomedicines for ocular NSAIDs: Safety on drug delivery. *Nanomedicine* **5**, 394–401.

Attama, A. A., Reichl, S., and Muller-Goymann, C. C. (2008) Diclofenac sodium delivery to the eye: In vitro evaluation of novel solid lipid nanoparticle formulation using human cornea construct. *Int. J. Pharm.* **355**, 307–313.

Badugu, R., Lakowicz, J. R., and Geddes, C. D. (2005) A glucose-sensing contact lens: From bench top to patient. *Curr. Opin. Biotech.* **16(1)**, 100–107.

Benjamin, W. J. (1996) Down sizing of Dk and Dk/l: The difficulty in using hPa instead of mm Hg. *Int. Contact Lens Clin.* **23**, 188–189.

Bishop, P. (2000) Structural macromolecules and supramolecular organization of the vitreous gel. *Prog. Ret. Eye Res.* **19**, 323–344.

Brennan, N. A. (1988) New technology in contact lens materials. *Trans. Br. Contact Lens Assoc. Ann. Clin. Conf.* **11**, 23–28.

Bruinsma, G. M., van der Mei, H. C., and Busscher, H. J. (2001) Bacterial adhesion to surface hydrophilic and hydrophobic contact lenses. *Biomaterials* **22**, 3217–3224.

Cai, X., Conley, S., and Naash, M. (2008) Nanoparticle applications in ocular gene therapy. *Vision Res.* **48**, 319–324.

Calvo, P., Thomas, C., Alonso, M. J., Vila-Jato, J., and Robinson, J. R. (1994) Study of the mechanism of interaction of poly(ε-caprolactone) nanocapsules with the cornea by confocal laser scanning microscopy. *Int. J. Pharm.* **103**, 283–291.

Campos, A. M. D., Sanchez, A., Alonso, M. J. (2001) Chitosan nanoparticles: A new vehicle for the improvement of the delivery of drugs to the ocular surface. Application to cyclosporin A. *Int. J. Pharm.* **224**, 159–168.

Cavalli, R., Gasco, M. R., Chetoni, P., Burgalassi, S., and Saettone, M. F. (2002) Solid lipid nanoparticles as ocular delivery system for tobramycin. *Int. J. Pharm.* **238**, 241–245.

Colthurst, M., Williams, R., Hiscott, P., and Grierson, I. (2000) Biomaterials used in the posterior segment of the eye. *Biomaterials* **21**, 649–665.

Das, S., Banerjee, R., and Bellare, J. (2005) Aspirin loaded albumin nanoparticles by coacervation: Implications in drug delivery. *Trends Biomater. Artif. Organs* **18(2)**, 203–212.

Das, S., Suresh, P. K., and Desmukh, R. Design of Eudragit RL 100 nanoparticles by nanoprecipitation method for ocular drug delivery. *Nanomedicine.* In press.

Davson, H. (1990) *Physiology of the Eye: The Aqueous Humor and Intraocular Pressure.* London: Macmillan Press, pp. 3–65.

Diebold, Y., Jarrin, M., Saez, V., Carvalho, E. L. S., Orea, M., Calonge, M., Seijo, M., and Alonso, M. J. (2007) Ocular drug delivery by liposome–chitosan nanoparticle complexes (LCS-NP). *Biomaterials* **28**, 1553–1564.

Dillen, K., Weyenberg, W., Vandervoort, J., and Ludwig, A. (2004) The influence of the use of viscosifying agents as dispersion media on the drug release properties from PLGA nanoparticles. *Eur. J. Pharm. Biopharm.* **58**, 539–549.

Ding, S. (1998) Recent developments in ophthalmic drug delivery. *PSTT* **1**, 328–336.

Dufrene, Y. F., Boonaert, C. J. P., and Rouxhet, P. G. (1996) Adhesion of *Azospirillum brasilense*: Role of proteins at the cell–support interface. *Colloids Surf. B* **7**, 113–128.

Durrani, A. M., Davies, N. M., Thomas, M., and Kellaway, I. W. (1992) Pilocarpine bioavailability from a muco-adhesive liposomal ophthalmic drug delivery system. *Int. J. Pharm.* **88**, 409–415.

Ernest, P. H. (2004) Light-transmission-spectrum comparison of foldable intraocular lenses. *J. Cataract Refract. Surg.* **30**, 1755–1758.

Fatt, I. (1996) New physiological paradigms to assess the effect of lens oxygen transmissibility on corneal health. *Contact Lens Assoc. Ophthalmol. J.* **22**, 25–29.

Fattal, E. and Bochot, A. (2006) Ocular delivery of nucleic acids: Antisense oligonucleotides, aptamers and siRNA. *Adv. Drug Deliv. Rev.* **58**, 1203–1223.

Geroshi, D. H., Matsuda, M., and Yee, R. W. (1995) Pump function of the human corneal endothelium. *Ophthalmology* **92**, 1.

Grinstaff, M. W. (2007) Designing hydrogel adhesives for corneal wound repair. *Biomaterials* **28**, 5205–5214.

Gupta, H., Aqil, M., Khar, R. K., Ali, A., Bhatnagar, A., and Mittal, G. Sparfloxacin loaded PLGA nanoparticles for sustained ocular drug delivery. *Nanomedicine*. In press.

Gupta, A. K., Madan, S., Majumdar, D. K., and Maitra, A. (2000) Ketorolac entrapped in polymeric micelles: Preparation, characterisation and ocular anti-inflammatory studies. *Int. J. Pharm.* **209**, 1–14.

Hiratani, H. and Alvarez-Lorenzo, C. (2002) Timolol uptake and release by imprinted soft contact lenses made of N, N-diethylacrylamide and methacrylic acid. *J. Control. Release* **83**, 223–230.

Hiratani, H. and Alvarez-Lorenzo, C. (2004) The nature of backbone monomers determines the performance of imprinted soft contact lenses as timolol drug delivery systems. *Biomaterials* **25**, 1105–1113.

Hiratani, H., Fujiwara, A., Tamiya, Y., Mizutani, Y., and Alvarez-Lorenzo, C. (2005) Ocular release of timolol from molecularly imprinted soft contact lenses. *Biomaterials* **26**, 1293–1298.

Holden, B. A. and Mertz, G. W. (1984) Critical oxygen levels to avoid corneal edema for daily and extended wear contact lenses. *Invest. Ophthalmol. Vis. Sci.* **25**, 1161–1167.

Jain, D. and Banerjee, R. (2008) Comparison of ciprofloxacin hydrochloride loaded protein, lipid and chitosan nanoparticles for drug delivery. *J. Biomed. Mater. Res. B* **86(1)**, 105–112.

Jain, D., Carvalho, E., and Banerjee, R. (2010) Biodegradable hybrid polymeric membranes for ocular drug delivery. *Acta Biomaterialia* **6(4)**, 1370–1379.

Joseph, N. (1990) Vitreous body prosthetic device. U.S. Patent No. 4902492.

Ketelson, H. A., Meadows, D. L., and Stone, R. P. (2005) Dynamic wettability properties of a soft contact lens hydrogel. *Colloids Surf. B* **40**, 1–9.

Klausner, E. A., Zhang, Z., Chapman, R. L., Multack, R. F., and Volin, M. V. (2010) Ultrapure chitosan oligomers as carriers for corneal gene transfer. *Biomaterials* **31**, 1814–1820.

Kohnen, T. (1996) The variety of foldable intraocular lens materials. *J. Cataract Refract. Surg.* **22**, 1255–1257.

Kyyronen, K., Hume, L., Benedetti, L., Urtti, A., Topp, E., and Stella, V. (1992) Methylprednisolone esters of hyaluronic acid: In ophthalmic drug delivery: In vitro and in vivo release studies. *Int. J. Pharm.* **80**, 161–169.

Lamson, R., Selbn, G., Bjiirklund, H., and Fagerholm, P. (1989) Intraocular PMMA lenses modified with surface-immobilized heparin: Evaluation of biocompatibility in vitro and in vivo. *Biomaterials* **10**, 511–516.

Lazzaro, D. R. (2005) What's new in ophthalmic surgery. *J. Am. Coll. Surg.* **200**, 96–102.

Le Bourlais, C., Acar, L., Zia, H., Sado, P. A., Needham, T., and Leverge, R. (1998) Ophthalmic drug delivery systems recent advances. *Progr. Retinal Eye Res.* **17**, 33–58.

Lee, S. D., Hsiue, G. H., Kao, C. Y., and Chang, P. C. T. (1996) Artificial cornea: Surface modification of silicone rubber membrane by graft polymerization of pHEMA via glow discharge. *Biomaterials* **17**, 587–595.

Legeais, J. M., Renard, G., Parel, J. M., Serdarevic, O., Mei Mui, M., and Pouliquen, Y. (1994) Expanded fluorocarbon for keratoprosthesis cellular ingrowth and transparency. *Exp. Eye Res.* **58**, 41–52.

Lim, K. S., Allan, B. D. S., Lloyd, A. W., Muir, A., and Khaw, P. T. (1998) Glaucoma filtration implants: Past, present and future. *Br. J. Ophthalmol.* **82**, 1083–1089.

Lindstrom, R. L. and Doddi, N. (1986) Ultraviolet light absorption in intraocular lenses. *J. Cataract Refract. Surg.* **12**, 285–289.

Lloyd, A. W., Bowers, R. W. J., Dropcova, S., Denyer, S. P., Faragher, R. G. A., Gard, P. R., Hall et al. (1997) In vitro evaluation of novel biomimetic intraocular lens materials. *Invest. Ophthalmol. Vis. Sci.* **38**, 884.

Lloyd, A. W., Faragher, R. G. A., and Denyer, S. P. (2001) Ocular biomaterials and implants. *Biomaterials* **22**, 769–785.

Lundberg, F., Gouda, I., Larm, O., Galin, M. A., and Ljungh, A. (1998) A new model to assess staphylococcal adhesion to intraocular lenses under in vitro flow conditions. *Biomaterials* **19**, 1727–1733.

Mainster, M. A. (1986) The spectra, classification, and rationale of ultraviolet-protective intraocular lenses. *Am. J. Ophthalmol.* **102**, 727–732.

Merodio, M., Irache, J. M., Valamanesh, F., and Mirshahi, M. (2002) Ocular disposition and tolerance of ganciclovir-loaded albumin nanoparticles after intravitreal injection in rats. *Biomaterials* **23**, 1587–1594.

Nakagawa, M., Tanaka, M., and Miyata, T. (1997) Evaluation of collagen gel and hyaluronic acid as vitreous substitutes. *Ophthalmic Res.* **29**, 409–420.

Nakamura, K., Refojo, M., Crabtree, D., Pastor, J., and Leong, F. (1991) Ocular toxicity of low molecular-weight components of silicone and fluorosilicone oils. *Invest. Ophthalmol. Vis. Sci.* **32**, 3007–3020.

Ohira, A., Wilson, C., de Juan Jr., E., Murata, Y., Soji, T., and Oshima, K. (1991) Experimental retinal tolerance to emulsified silicone oil. *Retina* **11**, 259–265.

Pignatello, R., Bucolo, C., Ferrara, P., Maltese, A., Puleo, A., and Puglisi, G. (2002) Eudragit RS100® nanosuspensions for the ophthalmic controlled delivery of ibuprofen. *Eur. J. Pharm. Sci.* **16**, 53–61.

Prow, T. W., Bhutto, I., Kim, S. Y., Grebe, R., Merges, C., McLeod, D. S., Uno et al. (2008) Ocular nanoparticle toxicity and transfection of the retina and retinal pigment epithelium. *Nanomedicine* **4**, 340–349.

Sahoo, S. K., Dilnawaz, F., and Krishnakumar, S. (2008) Nanotechnology in opcular drug delivery. *Drug Discov. Today* **13**, 144–150.

Saika, S. (2004) Relationship between posterior capsule opacification and intraocular lens biocompatibility. *Progr. Retinal Eye Res.* **23**, 283–305.

Sanders, N. N., Peeters, L., Lentacker, I., Demeester, J., and De Smedt, S. C. (2007) Wanted and unwanted properties of surface PEGylated nucleic acid nanoparticles in ocular gene transfer. *J. Control. Release* **122**, 226–235.

Santos, J. F. R., Alvarez-Lorenzo, C., Silva, M., Balsa, L., Couceiro, J., Torres-Labandeira, J. J., and Concheiro, A. (2009) Soft contact lenses functionalized with pendant cyclodextrins for controlled drug delivery. *Biomaterials* **30**, 1348–1355.

Santos, A. L. G. D., Bochot, A., Doyle, A., Tsapis, N., Siepmann, J., Siepmann, F., Schmaler, J., Besnard, M., Behar-Cohen, F., and Fattal, E. (2006) Sustained release of nanosized complexes of polyethylenimine and anti-TGF-beta2 oligonucleotide improves the outcome of glaucoma surgery. *J. Control. Release* **112**, 369–381.

Schepens, C. L. and Acosta, F. (1991) Scleral implants: An historical perspective. *Surv. Ophthalmol.* **35**, 447–453.

Schneider, R. P. and Marshall, K. C. (1994) Retention of the Gram-negative marine bacterium SW8 on surfaces effects of microbial physiology, substratum nature and conditioning films. *Colloids Surf. B* **2**, 387–396.

Sheikpranbabu, S., Kalishwaralal, K., Lee, K., Vaidyanathan, R., Eom, S. H., and Gurunathan, S. (2010) The inhibition of advanced glycation end-products-induced retinal vascular permeability by silver nanoparticles. *Biomaterials* **31**, 2260–2271.

Smith, S. G. (1989) Accommodating intraocular lens and method of implanting and using same. U.S. Patent 50782.

Soman, N. and Banerjee, R. (2003) Artificial vitreous replacements. *Biomed. Mater. Eng.* **13**, 59–74.

Sparrow, J. R., Miller, A. S., and Zhou, J. (2004) Blue light-absorbing intraocular lens and retinal pigment epithelium protection in vitro. *J. Cataract Refract. Surg.* **30**, 873–878.

Stenevi, U., Gwin, T., Harfstrand, A., and Apple, D. (1993) Demonstration of hyaluronic acid binding to corneal endothelial cells in human eye-bank eyes. *Eur. J. Implant Refract. Surg.* **5**, 228–232.

Taniguchi, K., Yamamoto, Y., ltakura, K., Miichi, H., and Hayashi, S. (1988) Assessment of ocular irritability of liposome preparation. *J. Pharmacobiodyn.* **11**, 607–611.

Tighe, B. J. and Ng, C. O. (1979) The mechanical properties of contact lens materials. *Ophthal. Optician* **19**, 394–402.

Tranoudis, I. and Efron, N. (2004a) Tensile properties of soft contact lens materials. *Contact Lens Ant. Eye* **27**, 177–191.

Tranoudis, I. and Efron, N. (2004b) Water properties of soft contact lens materials. *Contact Lens Ant. Eye* **27**, 193–208.

Vandervoort, J. and Ludwig, A. (2004). Preparation and evaluation of drug-loaded gelatin nanoparticles for topical ophthalmic use. *Eur. J. Pharm. Biopharm.* **57**, 251–261.

Wimmer, N., Marano, R. J., Kearns, P. S., Rakoczy, E. P., and Toth, I. (2002) Syntheses of polycationic dendrimers on lipophilic peptide core for complexation and transport of oligonucleotides. *Bioorg. Med. Chem. Lett.* **12**, 2635–2637.

Yenice, I., Mocan, M. C., Palaska, E., Bochot, A., Bilensoy, E., Vural, I., Irkeç, M., and Hincal, A. A. (2008) Hyaluronic acid coated poly-3-caprolactone nanospheres deliver high concentrations of cyclosporine A into the cornea. *Exp. Eye Res.* **87**, 162–167.

Yuan, X., Li, H., and Yuan, Y. (2006) Preparation of cholesterol-modified chitosan self-aggregated nanoparticles for delivery of drugs to ocular surface. *Carbohydr. Polym.* **65**, 337–345.

Zimmer, A. and Kreuter, J. (1995) Microspheres and nanoparticles used in ocular delivery systems. *Adv. Drug Deliv. Rev.* **16**, 61–73.

16

Nucleic Acid Based Nanobiosensing

Nicholas M. Fahrenkopf
University at Albany

Phillip Z. Rice
University at Albany

Nathaniel C. Cady
University at Albany

16.1 Introduction

Nucleic acids are a popular target for biosensors and bioanalytical devices. Unlike proteins, lipids, or carbohydrates, nucleic acids have extremely high information content and are relatively stable across a range of temperature and chemical conditions. Furthermore, nucleic acids contain sequence information that can reveal the identity of an organism or genetic changes that reflect disease status or mutation. For these reasons, thousands of different biosensing strategies have been developed for nucleic acid detection and analysis. Currently, many of these strategies incorporate nanotechnology or nanostructures. Nanoscale devices and materials have unique surface, electronic, optical, and mechanical properties that can increase device sensitivity and reduce analysis time. Therefore, much effort has been placed on using nanotechnology for improvement of nucleic acid biosensing.

This chapter primarily focuses on the use of nanotechnology for nucleic acid biosensing applications and on the strategies used to incorporate nucleic acids onto nanoscale materials and devices. Section 16.2 covers more traditional biosensing strategies on bulk materials and nanofabricated devices. This section also gives an overview of the main transduction mechanisms used for nucleic acid biosensors. Section 16.3 the focus is on nanoscale materials and how they are incorporated into nucleic acid detection systems. Section 16.4 covers current methods for linking nucleic acids to sensor surfaces and nanomaterials. Each section focuses on current research, and discusses both the benefits and pitfalls of various strategies.

16.2 Solid-State Nucleic Acid Biosensors

Solid-state sensors are distinct from nanomaterial-based systems, since they can be produced using integrated circuit (IC) fabrication techniques, including micro- and nanofabrication methodologies. These manufacturing techniques allow for large-scale, high-precision manufacturing methods derived from the semiconductor device manufacturing sector. Most of these solid-state sensing systems rely upon nucleic acid hybridization as a means of detection. In all cases, probe nucleic acids are linked to the sensor surface, which are then available to hybridize with target strands. The additional mass, electrical charge, or post-hybridization modification yields a measurable signal that can be transduced by these sensors. A wide range of transduction mechanisms have been used for solid-state detection, the most prevalent of which are described below.

16.2.1 Optical Detection

16.2.1.1 Fluorescent Detection

Optical detection of nucleic acids via fluorescent dyes has been well developed for quantification, optical localization (for microscopy), and in the form of DNA microarrays (Drummond et al. 2003). While most of these methods are restricted to use in liquid format, DNA microarray technology is a true solid-state detection method. DNA microarrays typically consist of glass slides that have been printed or photolithographically patterned with small spots of probe DNA. One of the leaders in microarray technology, Affymetrix (www.affymetrix.com) utilizes photolithographic patterning to achieve extremely high spot density. This patterning technique builds off of nano/microfabrication techniques to build DNA probes in situ. For each spot, the DNA probe is patterned in place by exposing the region to ultraviolet light (through a photomask) to de-protect functional groups on the probe and add subsequent nucleotides. Using this technique, as many as 10,000 or more spots can be printed on a single slide, making it possible to detect nearly 10,000 targets in a single assay. These slides are then exposed to samples containing fluorescently labeled target DNA. These targets are typically generated as copy DNA (cDNA) from mRNA targets representing two different samples of interest. Each subpopulation of cDNA is labeled with a different color of fluorescent dye, such that they can be differentiated after hybridization. cDNA from both samples are mixed together during the hybridization procedure, in which they compete for hybridization with immobilized DNA. Specificity can be adjusted by changing buffer salt concentrations or by adjusting hybridization temperature. The resulting hybridized array will typically have fluorescent spots with the following characteristics. One group of spots will have only hybridized to target A and will be the color of target A's dye label. A second group of spots will be hybridized to target B and will be the color of target B's label. A third group of spots will have hybridized to both A and B targets and will have a mixed fluorescent signal of both labels. By measuring the relative intensity of the two labels for each individual spot, one can then determine the relative concentration of various targets. A small segment of a typical DNA microarray is shown in Figure 16.1.

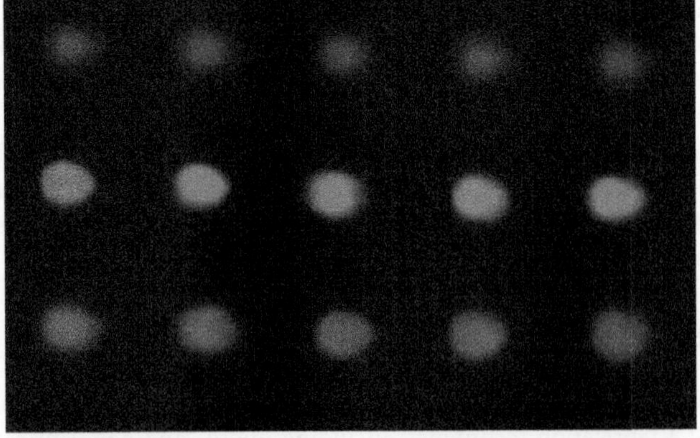

FIGURE 16.1 **(See color insert.)** Small segment of a typical DNA microarray, which has been hybridized with target DNA. The top row of spots hybridized with a red-labeled target, while the second row hybridized to a green-labeled target. The orange spots in the third row correspond to co-hybridization with red and green labeled targets.

In addition to pre-labeling target DNA prior to hybridization, it is also possible to introduce fluorescent labels after hybridization. One popular fluorescent DNA stain, PicoGreen, binds preferentially to double-stranded nucleic acids and undergoes an ~1000-fold increase in fluorescent intensity compared to its unbound form (Homs 2002). Using this type of fluorescent label can significantly simplify the detection procedure, since target DNA does not need to be pre-labeled prior to the experiment. Instead, sample DNA can be exposed to the substrate, hybridized, rinsed, and then stained. Successful hybridization events to target DNA are then indicated by bright fluorescent spots on the slide surface.

In addition to microarray-based detection strategies, a wealth of fluorescence detection strategies focuses on the use of DNA amplification via the polymerase chain reaction (PCR). The PCR is a biochemical technique that essentially produces multiple copies of a particular segment of DNA (Mullis et al. 1986). PCR is essential for many nucleic acid detection technologies, since it offers a way to produce many copies from a small population of target molecules. PCR has been extensively described in a wealth of journal articles, review papers, and textbooks; therefore, a full description of the technique is not provided in this chapter. The use of PCR in fluorescence detection strategies, however, will be discussed below.

PCR can be coupled to fluorescent nucleic acid detection assays via multiple methods. The most common methods are as follows: (1) the 5′ nuclease assay (TaqMan) (Livak 1999), (2) PicoGreen fluorogenic PCR (Cady et al. 2005), and (3) molecular beacons (Tyagi and Kramer 1996). The 5′ nuclease assay (or TaqMan assay) utilizes typical PCR primers, as well as a third oligonucleotide "probe." The probe DNA contains a fluorophore (fluorescent dye) and a quencher molecule on opposing 5′ and 3′ ends of the ssDNA. In this state, when the fluorophore and quencher are in close proximity, the fluorophore will not emit light when excited, but rather will transfer energy to the quencher molecule in a process known as fluorescent resonant energy transfer (FRET). If the probe hybridizes between the two PCR primers, it can be degraded by the 5′–3′ nuclease activity of the Taq DNA polymerase, releasing the fluorophore and quencher separately into solution. This action prevents FRET from occurring, and allows the fluorophore to emit photons (i.e., fluoresce). In this way, during each PCR cycle, additional probes are degraded, yielding a net increase in fluorescence signal. This contrasts from SYBRGreen style assays in which fluorescent dyes interact with newly synthesized dsDNA during each PCR cycle, yielding an increase in fluorescence. Unlike the 5′ nuclease assay, SYBRGreen style assays typically exhibit lower specificity, since formation of any dsDNA (including nonspecific products such as primer dimers) can also yield increased fluorescence.

Fluorescence detection can also be built into the probe sequence itself in the form of molecular beacons. Molecular beacons also utilize FRET and consist of two parts, which are added to opposite ends of the specially engineered probe sequence: a donor fluorophore and an acceptor chromophore (quencher). When in the "off" state, the probe is engineered to fold on itself in a stem loop in order to bring the donor and the quencher close to each other so that the quencher accepts the excitation energy from the donor and therefore does not give off fluorescence (Cady 2006). If the strand is hybridized, and the donor and the quencher are spatially separated, an increase in fluorescence intensity will be observed (Kim et al. 2007; Wang et al. 2009b). An example of a molecular beacon is shown in Figure 16.2. Our group has demonstrated molecular beacon-based sensors using nanoparticle quantum dots as fluorophores and gold nanoparticles as quenchers (Cady et al. 2007). Molecular beacons have also been used in solid-state sensors, where one end of the DNA hairpin is attached to a gold surface and the other end is attached to the fluorophore (Du et al. 2003). In the hairpin configuration, the fluorophore is quenched by the gold surface and cannot fluoresce. When hybridized with target DNA, however, the beacon unfolds, separating the fluorophore from the surface and allowing fluorescence. This technology can potentially be used as a label-free alternative to typical microarray-based detection.

Each of these fluorescence-based detection methods has drawbacks. Pre-tagging target DNA with fluorescent dyes requires additional sample handling steps that can significantly increase the complexity of the detection assay. Staining after hybridization requires the use of an additional, expensive, reagent that makes the hybridized complex sensitive to ambient light and temperature. Molecular beacons are

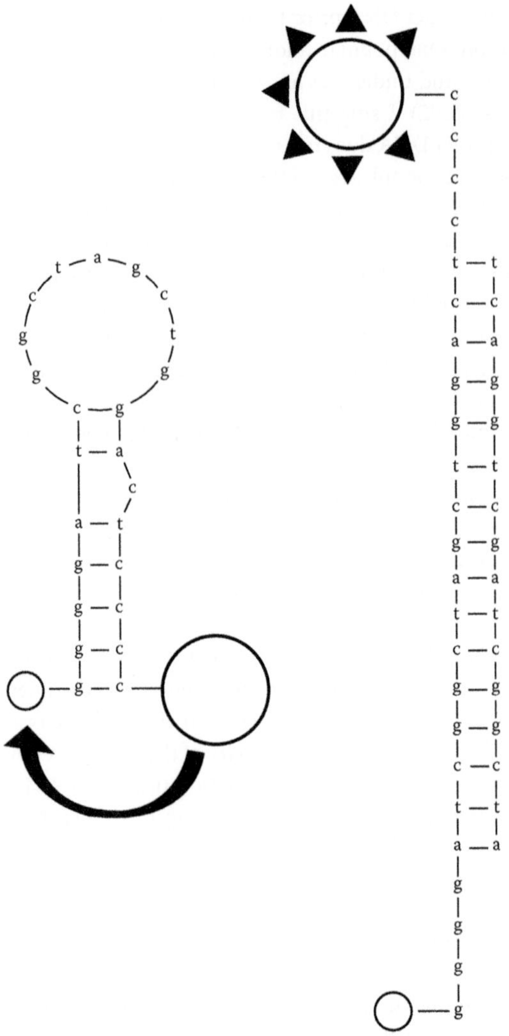

FIGURE 16.2 Example of a molecular beacon is shown. Molecular beacons are typical comprised of DNA stem-loop structures (left) that have a fluorophore (large circle) and quencher (small circle) linked to opposing ends of the DNA strand. Hybridization with target DNA can open the strand, separating the fluorophore from the quencher, which yields an increase in fluorescence (right).

sensitive to small changes in temperature or denaturing solution conditions (presence of denaturants) and therefore must be used in highly controlled environments. These methods all share the common drawback of the optical readout itself. In order to visualize, or digitize the intensity of the fluorescence at each spot complex optical detectors are required. Although fluorescence microscopy can be used for the simplest cases, many groups have developed customized spot readers that can employ complex techniques such as laser scanning confocal microscopy. As the optical analysis grows in complexity, the entire sensor system typically becomes more bulky and less applicable to field or point-of-care usage. To reduce system complexity, however, some groups have sought to use IC manufacturing techniques. These efforts capitalize on a wealth of research and development on solid-state devices that produce narrow-bandwidth light across the visible spectrum and into the UV and IR. In addition, solid-state devices have been developed to detect electromagnetic radiation at different wavelengths (Khanna 2007). By incorporating DNA hybridization onto the surface of these devices, excitation and readout can be achieved in miniaturized fashion. Furthermore, individual devices can now be fabricated on the same size scale as arrayed nucleic acid spots (micron-scale), which could facilitate highly multiplexed devices, without the need for optical magnification or high complexity optical readers.

16.2.1.2 Surface Plasmon Resonance-Based Detection

Surface plasmon resonance (SPR) has been commercially developed as an optical detection technology for label-free biosensing (Tan et al. 2004; Jin et al. 2009). Polarized light passed through a prism excites plasmons in a thin film of gold or silver (Kretschmann configuration) (Tan et al. 2004; Jin et al. 2009). In this configuration, surface plasmons propagate on the upper surface of the metal film, where a nucleic acid probe can be immobilized, and a sample solution can be exposed to the probes. Since the plasmons are extremely sensitive to changes in the interface of the metal/external medium (due to changes in the refractive index), the adsorption of extremely small concentrations of molecules can be detected—both as the probe is immobilized and as the target hybridizes to the probe (Homs 2002; Su et al. 2005; Cady 2006). Traditional SPR yields plots of reflectance (or absorbance) with respect to incident angle or wavelength. As material accumulates on the sensor surface, these peaks will shift in height, width, and/or position. While commercially available SPR sensors can be very accurate and sensitive, they are of limited usefulness for multiplexed detection. For multiplexed applications SPR imaging (SPRI) has been recently developed in a research setting, and to a limited extent in some commercial products. In SPRI the key advancement is the use of imaging lenses and a CCD camera to take 2D measurements across the SPR chip. SPRI allows for multiple probes to be immobilized in an array on the metal surface and therefore multiple sequences to be tested simultaneously. The SPR chip also includes two special mirrors that completely block the CCD (a dark area) or allow total reflection to the CCD (a bright area), each of which is utilized as a control for software normalization of image contrast (Ladd et al. 2008; Piliarik et al. 2009). With these improvements, SPRI systems are nearly as powerful as traditional SPR systems, but with the key advantage of multiplexed detection.

16.2.2 Electrochemical Detection

Electrochemical sensors can be divided into two general categories: direct (or label free) and indirect (labeled). In both cases, the sensor transduces an electrical signal (change in current, capacitance, etc.) in the presence of the biomolecule of interest. Direct electrochemical sensors sense the biomolecule without the need for an additional label or modification. This can be achieved for many DNA or RNA targets since these molecules have an inherent negative charge, due to their phosphate backbone. When this electrical charge is not enough for signal transduction, the immobilized or hybridized DNA can be further modified with electroactive compounds or materials. This "indirect" method often provides increased sensitivity. For most electrochemical sensors probe nucleic acid sequences are immobilized on or near conductive electrodes that register the hybridization of the probe and the target either directly or with the help of a secondary label.

Electrochemical sensors are similar to optical sensors in that they rely on the hybridization of two strands of DNA to cause an increase in signal (Drummond et al. 2003). For direct electrical sensors, an increase in negative charge is detected when going from single stranded (probe only) to double stranded (hybridized to target). Therefore, as a second strand of DNA hybridizes with a probe strand immobilized on an electrode, the addition of charge will influence the electrical characteristics of the electrode (i.e., impedance, capacitance) (Davis et al. 2005; Tosar et al. 2009). Beyond changes in impedance or capacitance, some electrochemical sensors detect the direct oxidation/reduction of the DNA/RNA nitrogenous bases and/or sugar-phosphate backbone (Drummond et al. 2003; Cady 2006; Lee and Hsing 2006; Khanna 2007). Most of the nitrogenous bases (and especially guanine) are electroactive, which can be useful for detecting single strands of nucleic acid or hybridization events (Cady 2006).

One direct detection strategy that is of interest to our research group is the detection of DNA hybridization on transistors. In a typical field-effect transistor (FET) the applied voltage on the gate electrode dictates the amount of current flowing from source to drain electrodes. Our group, among others, has shown that the negative charge from immobilized probe and hybridized target DNA are enough to influence the amount of current flowing from source to drain (Xuan et al. 2005; Zhang and Subramanian 2007; Goncalves et al. 2008; Mohanty and Berry 2008), when this DNA is applied to the FET gate. FETs, however, are only one type of semiconductor device that can be used as a direct electrochemical sensor. High electron mobility transistor (HEMT) devices, such as those consisting of group III/V materials, have similar properties to FETs. For instance, AlGaN/GaN heterostructures form a unique 2D electron gas at their interfacial region. When DNA (or other charged biomolecules) are placed in proximity to this interface, they can be readily detected via changes in electrical current through the device (Kang et al. 2006). Both FET and HEMT devices are fabricated using typical IC fabrication techniques, making them easily integrated with electronic devices. An example of an FET-based sensor is shown in Figure 16.3.

Indirect detection strategies utilize additional steps after hybridization to increase or improve electrical signal transduction. For instance, hybridized nucleic acids can participate in oxidation/reduction (redox) reactions (Cady 2006; Khanna 2007). Redox-based electrical sensors typically employ ruthenium- or cobalt-based metallointercalators, or organics such as Hoechst 33258 and daunomycin (Homs 2002; Drummond et al. 2003). Iron cyanide is a redox indicator that will readily oxidize at an electrode but is screened by both single- and double-stranded DNA (Cady 2006). A neutralizing agent allows the iron cyanide to the surface where the amount of oxidation can be measured. Methylene blue has also been used for electrical nucleic acid detection. Since a perfectly formed double strand of DNA can conduct along its length, current through the strand can oxidize the methylene blue that has been intercalated. However, if mismatches exist, current cannot flow through the strand to oxidize the methylene blue, yielding a negative result (Cady 2006). Depending on the type of indicator that is being

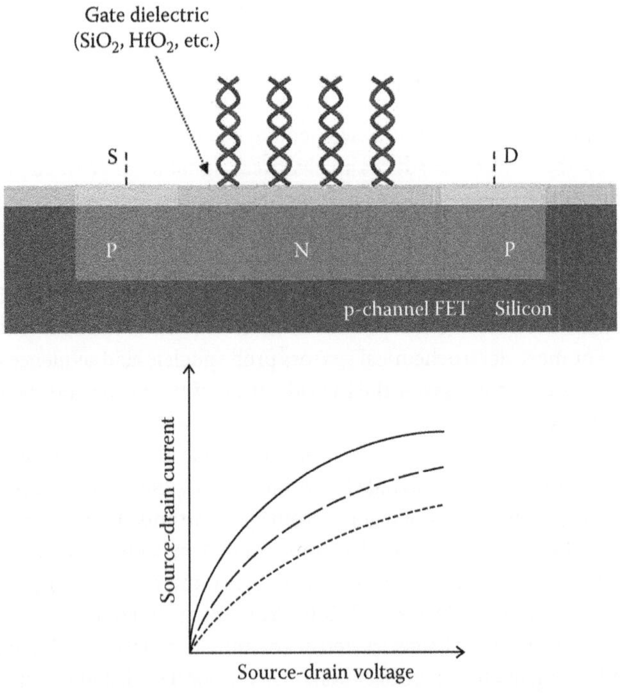

FIGURE 16.3 FET-based biosensor for detection of nucleic acids. Nucleic acids in proximity to the gate region of the FET (top) can cause changes in source to drain current. A typical current–voltage (IV) curve is shown (bottom) representing the initial device (solid line), a device with single-stranded probe DNA (large dotted line), and a device with hybridized target (small dotted line).

used, detection is accomplished by monitoring capacitance, conductivity, or through techniques such as cyclic voltammetry (Homs 2002; Lee and Hsing 2006).

16.2.3 Mass-Based Detection

The hybridization of a probe strand of DNA to its complement can also be detected via mechanical techniques, due to the additional mass deposited on the sensor surface. Most mass-based sensors function through disruptions in the mechanical vibrations or oscillations of a material. Changes to mechanical oscillations can be measured via acoustic waves on the surface of a piezoelectric substrate, such as a surface acoustic waves (SAW) sensor, or through bulk changes in oscillations, such as in a quartz crystal microbalance (QCM) or cantilever-based device. How these mechanical changes can be converted into a measurable signal is discussed in the following sections.

16.2.3.1 Surface Acoustic Wave Detection

Acoustic waves that travel along the surface of a material are called surface acoustic waves. These waves are extremely sensitive to changes to the surface they travel along, which affects their amplitude, wavelength, and/or frequency (Shiokawa and Kondoh 2004). SAW sensors excite an acoustic wave on the surface of a piezoelectric material, which interacts with molecules that have been attached to the surface. When additional mass is added to the surface (which can occur during DNA hybridization events), the acoustic properties of the surface change. These events can then be measured as a shift in resonance or acoustic wave transmission. SAW devices have been developed by multiple groups, and have been used to detect DNA targets (Gronewold et al. 2006). Typically, a piezoelectric material is patterned with electrodes using conventional photolithographic techniques. On one end of the material, an electric field is applied causing a stress and therefore a mechanical wave on the surface of the material. On the opposite side of the device, the resulting wave is converted back into an electrical signal and analyzed for changes in amplitude, phase, or wavelength (Hur et al. 2005). This can be used to detect the hybridization of a target strand of DNA to an immobilized probe strand of DNA. Simple hybridization events can cause measurable changes in the properties of the SAW, which can therefore function as a mass-based nucleic acid sensor (Hur et al. 2005; Gronewold et al. 2006).

16.2.3.2 Quartz Crystal Microbalance-Based Detection

Similar, but distinctly different from SAW sensors, quartz crystal microbalances take advantage of bulk acoustic waves that propagate through a piezoelectric material. In commercially available QCMs, a thin, flat, piezoelectric material is sandwiched between two electrodes. A square or sine wave of alternating polarity is applied to these electrode, which results in the production of acoustic waves. This, in turn, creates standing waves, which turn the entire device into a simple resonator (Homs 2002; Cady 2006). As mass accumulates on the surface of the QCM, the resonant frequency of the material will change, allowing for a direct correlation between frequency shift and mass attached to the sensor surface (Satjapipat et al. 2001; Su et al. 2005). When applied for biological sensing, an immobilized nucleic acid probe can hybridize with its complementary strand in a sample solution. Hybridization to target DNA introduces additional mass, which can be detected by subtle changes in the resonant frequency of the device.

16.2.3.3 Cantilever-Based Detection

Nonpiezoelectric devices can also be used for high sensitivity nucleic acid detection. Microfabricated micro electromechanical systems (MEMS)-based devices can measure subtle changes in mass on a surface without the use of acoustic waves. In these devices, free-standing cantilevers are fabricated, using simple MEMS processing steps (photolithographic patterning and plasma-based etching). Similar to atomic force microscope (AFM) probes, these cantilevers oscillate at specific resonant frequencies. The resonant frequency, however, can be modulated by attachment of a mass to the surface of the cantilever. For nucleic acid sensing applications, probe molecules are immobilized onto the cantilever and then are allowed to hybridize to target DNA (Madou 1997). The additional mass changes the resonant frequency, which can be measured by optical or electrical methods. Optical detection is the most commonly used method in which light is reflected off of the oscillating cantilever and detected by a photodetector or CCD (Madou 1997). As mass accumulates on the cantilever tip, the cantilever can bend (through changes in surface stress), or can undergo changes in resonant frequency, both of which can be detected via light reflected off of the cantilever tip (Cady 2006, 2008). Cantilevers can also be fabricated with electrical actuators (electrodes or semiconductor materials) (Madou 1997; Pawsey et al. 2003).

16.3 Nanomaterials-Based Biosensors

Nanomaterials are materials that are less than 100 nm in size, which often exhibit properties that are different than their bulk materials. This discrepancy in materials properties is due to size quantization and high surface area to volume ratio that is present at these small scales. In addition, some nanomaterials exhibit quantum effects due to their small length scale, yielding unique electrical, magnetic, and optical properties. In our continuing discussion of nucleic acid nanobiosensing, we divide nanomaterials into three main areas: (1) nanoparticles, (2) nanowires, and (3) nanotubes. Using nanomaterials for biosensing applications can result in both increased sensitivity and unique detection and transduction mechanisms. The following sections detail the current state of each nanomaterial as well as some nucleic acid based biosensors that have been developed using each technology.

16.3.1 Nanoparticles

Nanoparticles are small particulates (which are often referred to as colloids) that can be synthesized from an extremely wide range of starting materials. When these materials have unique optical, magnetic, or electrical properties, they become excellent components of biosensors or biosensing assays. Due to their small size, nanoparticles can be directly linked to biological molecules, and can function as a type of "tag" or "handle" during the biosensing process. When used as a "tag," nanoparticles label the biomolecule with a measurable optical, magnetic, or electrical signal, which significantly increases detection sensitivity for that molecule. When used as a "handle," nanoparticles interact with biomolecules to aid in separation, purification, or manipulation. Quantum dots, which are fluorescent nanoparticles, are excellent examples of nanoparticulate "tags." Magnetic nanoparticles, on the other hand, can be used as "handles" for separating biomolecules from complex mixtures, or manipulating biomolecules within a 3D environment. The following paragraphs discuss the utility of various nanoparticles in biosensing assays.

Gold is one of the most widely used nanomaterials for biosensing applications, due to its ease of synthesis, uniform shape, unique optical and electrical properties, and ability to act as a substrate for formation of self-assembled monolayers (SAMs). SAMs, and other surface immobilization methods, provide a direct linkage between biomolecules and a surface. As discussed in the immobilization strategy section below, thiol (-SH) functional groups readily bind to gold surface, providing an extremely strong surface linkage. This thiol-gold interaction is the basis for SAMs in many biosensing applications. Functionalized gold nanoparticles using thiol linkers have been monitored using SAW (Chiu and Gwo 2008), cyclic voltammetry (Zhang et al. 2008b), dynamic light scattering (Zhao et al. 2008), and SPR (Wark et al. 2007). Our group has performed simple SPR experiments to observe DNA interactions with gold nanoparticles that are embedded in yttria-stabilized zirconia (YSZ). An example of these experiments is shown in Figure 16.4. One of the more common analysis techniques used for analyzing DNA immobilization on both gold and silver nanoparticles is the use of surface enhanced Raman spectroscopy (SERS), which causes an enhancement of surface plasmons present on these materials (Braun et al. 2007). DNA immobilization on Ag nanoparticles have also been studied by cyclic voltammetry, following electrodeposition on glassy carbon electrodes (Wu et al. 2006).

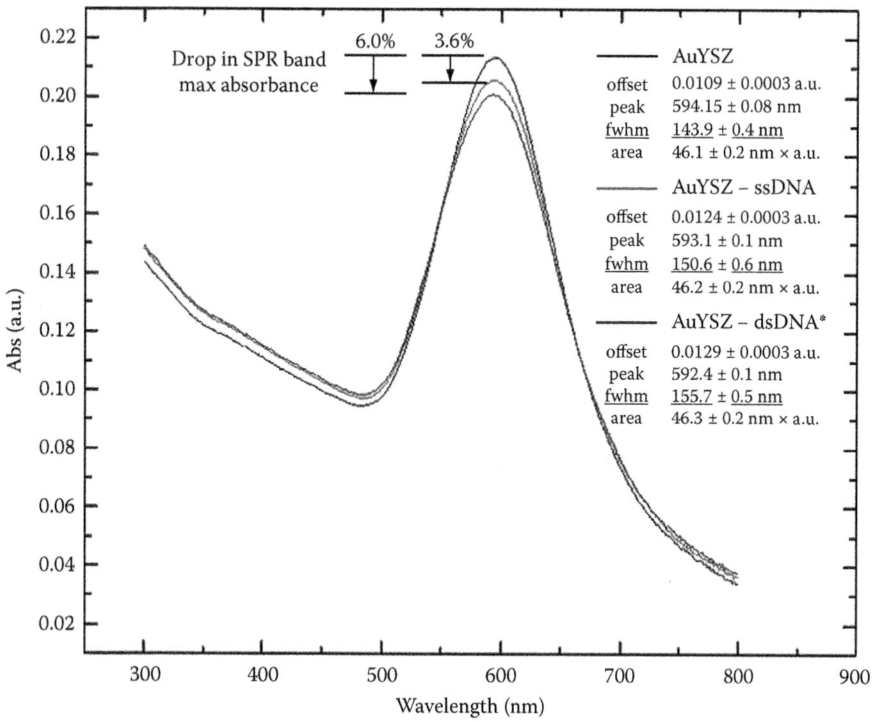

FIGURE 16.4 **(See color insert.)** DNA-gold nanoparticle interactions and its effect on SPR. Single and double-stranded DNA was incubated with gold nanoparticles embedded in YSZ films. Addition of DNA reduced the SPR peak intensity by 3.6% for ssDNA and 6% for dsDNA.

Compound metal oxide semiconductors (CMOS) structures, such as FETs, have been previously discussed for biosensing applications. Metal oxides offer unique properties, such as tunable electronic and magnetic properties, as well as photocatylitic effects. For these reasons, metal oxide nanoparticles are currently of interest for biosensor applications by our research group and others. Numerous metal oxide nanoparticle have been demonstrated for biosensing applications, including: Fe_3O_4, TiO_2, ZnO_2, ZrO_2. Magnetic Fe_3O_4 nanoparticles have been used for immunoassay and DNA detection (Chan et al. 2006), as well as detecting point mutations in DNA (via quartz crystal microbalance) (Pang et al. 2007). These methods typically employ dextran (Chan et al. 2006) or gold-coated (Pang et al. 2007) Fe_3O_4 nanoparticles in order to immobilize the DNA/antibody being studied. Similar studies have been conducted using polyvinyl pyrrolidone (PVP-k30) to increase the dispersion of Fe_3O_4 and TiO_2 nanoparticles as well as help immobilize the DNA onto the particles (Chang and Tzeng 2008). To date, most methods for conjugating DNA to metal oxide nanoparticles call for surface modifications, such as the case with TiO_2 (Kim 2009), or coating the particles in a polymer matrix to incorporate the DNA. Today, few groups have investigated the possibility of binding DNA directly to metal oxide nanoparticles through adsorption of terminal phosphate groups or phosphate backbone along the DNA. Our group has been investigating the differences in binding between short chains of ssDNA and dsDNA, as well as bacteriophage λ DNA binding to TiO_2 nanoparticles through their adsorption to the surface. The authors have found that DNA sequence length can affect DNA-nanoparticle affinity. In addition, we have also found that single-stranded and double-stranded DNA exhibit subtle differences in nanoparticle affinity. These differences in affinity can be exploited for discriminatory assays, as well as for immobilization strategies.

In addition to TiO_2 particles, ZrO_2 particles have been used to develop sensors on gold electrodes (Zhang et al. 2008a) and on multiwalled carbon nanotubes (MWCNTs) (Yang et al. 2007b) for the detection of DNA–DNA hybridization. One of the reasons that TiO_2 and ZrO_2 nanoparticles are of such great

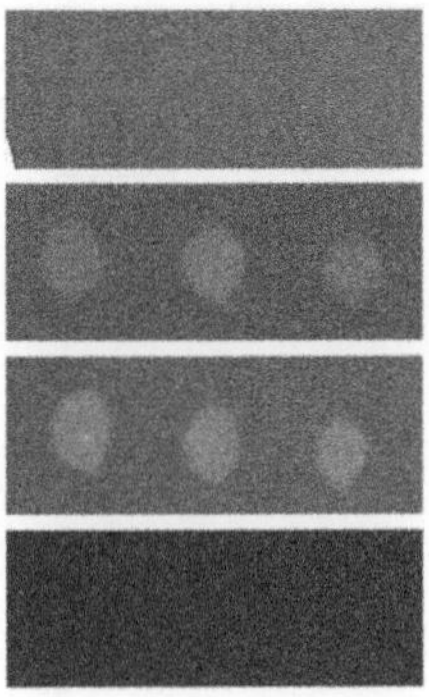

FIGURE 16.5 Phosphate-dependent DNA attachment to hafnium oxide. DNA without a terminal phosphate, with a 5′ phosphate, or with a 3′ phosphate were printed onto hafnium oxide and subsequently hybridized with target DNA. PicoGreen fluorescent dye was used to label hybridized DNA and showed that a terminal (5′ or 3′) phosphate is required for surface immobilization. A control experiment was performed with 5′ phosphorylated DNA, but using a noncomplementary target DNA.

interest is due to the ability for group IIIb metal oxides (Ti, Zr, Hf) to conjugate directly to terminal 5′ phosphate groups of DNA (Xu et al. 2002; Pawsey et al. 2003; Nonglaton et al. 2004). While much of this work has been done with bulk materials, there are some indications that 5′ phosphates of DNA can interact with metal oxide nanoparticles. More typically, biosensing strategies utilize indirect linkage methods to attach DNA to the particles. For instance, ZnO_2 nanoparticles have been used to detect DNA hybridization events (Zhang et al. 2008c,d), but utilizing a chitosan-based linker. Currently work is being performed by the authors to investigate other metal oxides such as those of Zn, Al, and Fe, to better characterize direct DNA–metal oxide interactions. To date, terminal phosphate-dependent DNA-metal oxide interactions have only been observed on bulk materials (Figure 16.5).

A variety of other nanoparticles have been used for the development of DNA biosensors. For example, platinum nanoparticles (Pt) have been used for the detection of DNA and protein recognition events (Polsky et al. 2006). Copper sulfide (CuS) nanoparticles have also been used to detect DNA hybridization events with up to two mismatched bases (Ding et al. 2008). These sensors immobilize ssDNA onto the CuS nanoparticles through thiol modification, while the complementary ssDNA is immobilized onto glass-carbon electrodes using chitosan as a linker. Lead sulfide (PbS) nanoparticles have also been used for high sensitivity electrochemical sensors. Mercaptoacetic acid (MAA) modified PbS nanoparticles have been prepared in literature (Lin et al. 2003) and these particles can be covalently modified using oligonucleotides (Liu et al. 2004a). Along with PbS nanoparticles, Liu et al. (2004a) have demonstrated oligonucleotide immobilization using ZnS, CdS, and CuS particles. Electrochemical detection of DNA hybridization has been demonstrated using complementary strands of ssDNA immobilized on both PbS nanoparticles and the sides of a microwell plate. After hybridization, the PbS particles were dissolved using nitric acid and the Pb^{2+} ions were detected using differential pulse anode stripping voltammetry (DSASV) (Gao et al. 2009). Silica nanoparticles are also commonly used for biosensors since their surface hydroxyl groups can be readily reacted with cross-linkers to provide amine (Chrisey et al. 1996; Lobert et al. 2003), carboxyl (Jang et al. 2006), aldehyde (Zammatteo et al. 2000; Lobert et al. 2003), epoxy (Lamture et al. 1994), and thiol (Cavic et al. 2001) functional groups. These functionalized nanoparticles have all been used as substrates for immobilizing DNA and other biomolecules. Silica-labeled biomolecules have been used for imaging and detection of both DNA and whole cells (Tan et al. 2004).

16.3.2 Nanowires

Nanowires are high aspect ratio nanostructures that have recently been used for a variety of biosensing applications. When positioned between two conductors, a nanowire can act as conductive circuit element, which is susceptible to subtle changes in the electronic properties of the surrounding solution. Therefore, attachment of biological molecules to nanowires can result in changes in conductivity, capacitance, and impedance. In addition, when nanowires have semiconductor properties, they can also be used in simple transistor-like structures. Previously in this chapter, we have discussed FETs and solid-state devices used for biosensing applications. FETs typically detect biological molecules via electronic interactions with the gate portion of the FET, which is located between the source and drain. The charge status, or electrical field associated with the biomolecule can modulate the carrier density (electrons or holes) in the gate region, which in turn modulates the current or on/off voltage of the FET. Unlike planar structures that are fabricated with traditional semiconductor processing techniques, FETs can also be constructed using doped silicon nanowires, or other "1D" nanowires (Cui and Lieber 2001). The term 1D refers to the extremely high aspect ratio of nanowires, having very small widths (100 nm or less) compared to their length (1 μm or more). Nanowire-based FETs operate in a similar manner to traditional FET's, however, the gate or junction material between the source and drain is replaced with a conductive (or semiconductive) nanowire. Nanowire-based FETs can yield increased sensitivity due to the high surface area to volume ratio, and their increased susceptibility to minute changes in electric field (as compared to traditional FETs). There are multiple reviews that describe the detection mechanism and application of nanowire-based sensors (Patolsky and Lieber 2005), as well as methods for fabricating and functionalizing DNA-based nanowires (Qun et al. 2006).

Initial nanowire based biosensors were developed using silicon nanowires due to their similarities with traditional planar FET structures. Silicon nanowires have been used to detect micro-RNAs (miRNA), which play an important role in developmental and cell biology (Zhang et al. 2009). These silicon nanowires were functionalized with peptide nucleic acid (PNA) and could detect single nucleotide differences (mismatches) in the target strand. In addition to silicon nanowires, silica (SiO_2) nanowires have been used for biosensing applications. Similar to silica nanoparticles, the free hydroxyl groups on the surface of these nanowires can be used to immobilize DNA or other biomolecules. Silica has also been used to coat many different nanoparticles including Au and Ag nanospheres as well as metal core nanorods and wires. For example, silica coatings have been applied to Au/Ag striped nanowires, functionalized with thiolated DNA, and used to detect target DNA as well as single base mismatches (Sioss et al. 2007). Using Au/Ag striped nanowires allows one to have visual "barcodes" for each unique strand of DNA attached to the nanowire. Subtle differences in reflectivity between Au and Ag allows for visual confirmation of proper DNA hybridization or detection. Alternatively, solid Au and Ag nanowires have been widely used in other biosensing applications. For example, Ag nanowires have been formed using templated pectin, which has been immobilized through the hybridization of PNA and ssDNA targets (Kong et al. 2008). Along with Ag nanowires, Au nanowires are commonly used with thiolated DNA/PNA immobilized on the surface, and allowed to hybridize with the complementary target (Fang and Kelley 2008). In this particular case, hybridization events were monitored using cyclic voltammetry to measure the electrocatalytic currents of Ru^{3+} and Fe^{3+} ions in solution. In addition to direct electrical measurements, nanowires have also been used for enhancing spectroscopic signals. For instance, SERS is a phenomenon in which Ag and Au nanoparticles and nanowires enhance and amplify the Raman signal, through plasmon-based interactions. Yoon et al. (2008) have shown that single nanowires of Au or Ag cause enhancement in Raman spectra for biosensing applications.

Metal alloys have also been formed into nanowire structures for nucleic acid biosensing. Multisegment CdTe-Au-CdTe striped nanowires have been fabricated and used to detect DNA hybridization or capture of target molecules using thiolated ssDNA by immobilizing the thiolated target on the Au portion of the wire (Wang and Ozkan 2008). These types of heterostructure nanowires have garnered great interest and potential in the electronics industry for the creation of tunneling diodes (Bjork et al. 2002), and could be used for the detection and sensing of biomolecules. Metal oxide nanowires are another commonly employed class of nanowires used for biosensing applications. They can be grown through various catalytic methods and can be functionalized with DNA, using a variety of surface modification strategies. Metal oxide nanowires can be used for FET applications, such as has been performed using In_2O_3 (Curreli et al. 2005). In another approach, nanowires can be synthesized during the detection assay, using DNA as a template for their formation. This has been demonstrated for ZnO nanowires (Lazareck et al. 2006).

Although most nanowire-based biosensors are electrical devices, there are some reports of optoelectronic sensors based upon nanowires. Optoelectronic devices are used to detect and control light and mediate electrical-optical or optical-electrical transduction events. Gallium nitride (GaN) devices are direct band-gap semiconductors that are commonly used in LEDs and nanoelectronics. GaN-based nanowires maintain the same transport properties at bulk GaN, but have improved surface characteristics such as higher surface area and improved charge transfer (Chen et al. 2008). By functionalizing GaN nanowires with single-stranded probe DNA, hybridization to target strands has been demonstrated (Ganguly et al. 2009). Although this demonstration utilized cyclic voltammetry for detection, it is expected that future studies will utilize optical-based detection of DNA using semiconductor nanowires.

16.3.3 Nanotubes

Carbon nanotubes are well-studied nanostructures that form ordered tubelike structures. Carbon nanotubes come in multiple varieties, including single-walled carbon nanotubes (SWCNTs) and multiwalled carbon nanotubes (MWCNTs). The primary difference between these varieties is the number of concentric tubes that form individual structures. SWCNTs have walls that are only one layer of carbon thick. MWCNTs, on the other hand, consist of multiple SWCNTs of varying diameter. Nanotubes, like many other nanomaterials, allow for miniaturization of devices, increased sensitivity for sensors, and reduction of sample. Although nanotubes can be formed from other materials, carbon nanotubes are by far the most studied form of nanotube used for DNA biosensing. Biosensing research using CNTs has resulted in the numerous devices that can probe single cells (Wu et al. 2008), mimic FET devices (Hwang et al. 2008), and be used electrochemical sensors (Li et al. 2009). DNA immobilization on CNT devices is difficult but can be performed through adsorption dialysis (Barone et al. 2005; Jeng et al. 2006), through immobilization with polymeric films (Martinez et al. 2009), as well as through covalent strategies (Zhu et al. 2005).

Polymeric materials have also used for the formation of nanotubes. Polyanaline (PAN) has been used for the formation of nanotubes/nanowires to create dendritic structures (Li et al. 2006). PAN nanotubes have also been used for DNA immobilization on glassy carbon electrodes detection of hybridization events through cyclic voltammetry and impedance spectroscopy (Feng et al. 2008). Using polymers for nanotube formation allows for simple incorporation of a wide variety of surface functional groups. This makes it possible to use many different types of surface immobilization strategies. Interestingly, DNA, which is a naturally occurring polymer, has also been used for nanotube formation (Hou et al. 2005). These "bionanotubes" form spontaneously via hydrogen bonding between complementary strands. Upon heating these DNA nanotubes, the structure denatures, which has proven useful for cargo and drug delivery schemes.

Inorganic nanotubes have also been demonstrated for DNA biosensor applications. These nanotubes can be integrated with solid-state devices for nanoelectronics applications. Silica nanotubes have been synthesized by growing silicon nanowires as temples, and allowing oxidation to form an SiO_2 sheath (Fan et al. 2003). These rigid nanotubes provide a wide variety of applications for microfluidics as well as semiconductor applications. For example, TiO_2 is a well known semiconductor that has photocatalytic properties for solar cell applications (Fujishima et al. 2000). Flexible TiO_2 nanotubes have been fabricated in a similar manner as silica nanotubes mentioned above, and provide a similar support structure for functionalizing the surface with DNA and other biomolecules for sensing applications (Wang et al. 2009a).

16.4 Immobilization Strategies

All the biosensors discussed in the preceding sections require immobilization of nucleic acids onto some component of the sensor for signal transduction to occur. In order to immobilize nucleic acids onto a biosensor, one must consider the substrate in which the nucleic acids will be immobilized and what functional groups are available on the molecule or the surface of interest. We have divided immobilization methods into two general categories, those using unmodified nucleic acids, and those using chemically modified nucleic acids. Both of these categories are discussed in detail below.

16.4.1 Unmodified DNA

Unmodified nucleic acids are those that have not been changed to incorporate novel functional groups. These "naturally occurring" nucleic acids have a 5′ phosphate group, a 3′ hydroxyl group (and a 2′ hydroxyl if RNA), as the remaining phosphate backbone and nitrogenous bases. Unmodified nucleic acids are typified by their bulk negative charge (due to the phosphate backbone). Due to this bulk negative charge, nucleic acids can be immobilized on some surfaces via electrostatic interactions. Many immobilization strategies take advantage of these charge-based interactions. Since DNA is polyanionic it will tightly bind to positively charged surfaces like the natural cationic polymer chitosan (Teles and Fonseca 2008). It has been shown that DNA can be cross-linked onto poly-L-lysine (a positively charged peptide) modified glass slides in order to monitor the expression of genes in a high-capacity microarray system (Schena et al. 1995). This method requires no modification to the DNA as the negatively charge phosphate backbone interacts with the amine groups of the poly-L-lysine. This method has also been repeated with polystyrene, polymethylmethacrylate, and polyhistidine, which is demonstrated by the pH dependence for binding affinity (Allemand et al. 1997). Immobilization has also been shown to occur through vinyl and aminated silanized supports (Allemand et al. 1997). Similar polymer-DNA constructs have been formed through the use of layer-by-layer assembly using DNA and poly-(allylamine hydrochloride) (PAH) (Lang and Liu 1999). The assembly of each layer can be monitored by measuring the UV-visible spectrum as layer upon layer of DNA and PAH is formed (up to 20 layers). Langmuir–Blodgett films have also been employed to conduct DNA hybridization with oligonucleotides, which bind to nucleobase lipid monolayers. These films have been monitored and detected using a QCM-based methods (Ebara et al. 2000). DNA adsorption to TiO_2 modified indium tin oxide (ITO) electrodes through the phosphate backbone of the nucleic acid also been demonstrated and quantified using cyclic voltammetry (Liu et al. 2005). Adsorption of nucleic acids through the phosphate backbone has also been demonstrated using ZrO_2 nanoparticles and confirmed using X-ray photo-electron spectroscopy (XPS) as well as differential pulse and cyclic voltammetry (Liu et al. 2004b). Other research has shown that surface attachment can occur through the nitrogenous bases (A,G,T,C, and U) along the DNA or RNA strand. Hydrophobic bases will physically adsorb to hydrophobic surfaces like pentacene film (Zhang and Subramanian 2007). Alternatively, the amine groups of the bases can react through diimide-activated amidation to covalently bind to the surface of interest, such as aminated MWCNTs (Xu et al. 2009a).

In addition to electrostatic interactions or covalent linkage to naturally occurring functional groups, nucleic acids can also be mixed with monomers to form a polymer matrix with during electrodeposition. DNA has been deposited onto metal electrodes through this method using polyaniline (Davis et al. 2005) and polypyrrole (Tosar et al. 2009). Polypyrrole has also been used to electrochemically deposit DNA onto MWCNTs (Xu et al. 2009a). While sensors have been fabricated using this technique, the nucleic acids are physically incorporated into the polymer matrix, and are not all exposed the surface of the film. This limits availability for hybridization and sensing.

The major drawback to many of these indirect techniques, especially electrostatic interactions, is that at the backbone of the nucleic acid is attached at multiple points along the strand. This limits the flexibility and mobility of the strand, which can make it difficult for the probe sequence to come into contact with the target sequence, and to hybridize in a stable manner. Alternatively, the nucleic acid probe sequences can be immobilized through the terminal hydroxyl or phosphate group (Teles and Fonseca 2008). A surface that has been silanized to include an epoxy functional group can covalently bond to the 5′ phosphate (Mahajan et al. 2006). Multiple groups have developed biosensors that use immobilized DNA that has been physisorbed onto metal surfaces (Sassolas et al. 2008). Much research has shown organic molecules with the phosphate or phosphonate terminal group will selectively immobilize on surfaces such as zirconium (Pawsey et al. 2003; Liu et al. 2004a; Nonglaton et al. 2004), $Al_2(C_4BP)$, Ti (Pawsey et al. 2003), and Hf (Jespersen et al. 2007). Our research group has explored the direct immobilization of DNA through terminal phosphate groups, as shown in Figure 16.5. Our work has shown that DNA can be directly immobilized onto HfO_2 and AlGaN and that immobilization depends greatly on the terminal phosphate (Fahrenkopf et al. 2009; Xu et al. 2009b). The concept of directly immobilizing probe sequences to a surface through naturally occurring terminal groups is novel and represents a new mode of immobilization methods.

16.4.2 Modified DNA

The second general class of modification strategies requires the nucleic acid strand to be modified with additional functional groups. These groups can include, but are not limited to disulfides, amines, thiols, and aldehyde. Most typically, these groups are added to either the 5′ or 3′ terminus of the nucleic acid. Alternatively, individual nucleotides can be modified and incorporated to the nucleic acid during synthesis. These internally modified nucleic acids are less commonly used for immobilization strategies, but are commonly used for fluorophore linkages in fluorescence-based assays. In order to covalently bind modified nucleic acids to a substrate, the surface must have the complementary reactive group, or compatible chemical modification for linkage. Functionalizing silicon, gold, or glass surfaces (among others) offers the ability to integrate nucleic acids with CMOS, SPR, and/or microarray systems.

Gold-thiol linkages have been one of the most widely studied examples of surface attachment due to positive binding energy shift in the Au 4f-core level electrons, causing changes in the electronic properties between the gold and thiol (Büttner et al. 2006). These shifts in electronic properties are useful for self-assembled monolayer (SAM) formation via n-alkyl thiols on gold surfaces (Porter et al. 1987). This well-documented technique has benefited from extensive studies on how n-alkyl chain lengths and terminal groups affect gold-thiol SAM packing (Folkers et al. 1992). SAMs of thiolated ssDNA probes have been prepared by using gold surfaces, analyzed using SPR, and allow antibodies to be specifically oriented on a surface using antibody (Boozer et al. 2006) and G-protein (Jung et al. 2007) modified complementary DNA sequences. Numerous examples of gold interactions using thiol modified DNA can be found in literature, which utilize gold nanoparticle substrates (Hurst et al. 2006; Lu et al. 2008; Nykypanchuk et al. 2008). Thiol modified DNA oligos have been used as probe sequences immobilized on gold surfaces in SPR (Wang et al. 2004), SAW-based sensors (Hur et al. 2005), HEMTs (Kang et al. 2006), and microcantilever resonance biosensor platforms (Su et al. 2003).

Although gold-thiol based linkages are used in many nucleic acid sensors, another popular strategy is to use silanization of metal oxide and silicon surfaces. As discussed previously, alkoxy-silanes are reactive with metal oxides and silicon surfaces. Linkers that contain silanes can also incorporate functional or reactive groups, which can be used for nucleic acid linkage. For example, 3-aminopropyltriethoxysilane (APTES) has been used as a silanization agent, and was subsequently functionalized with sulfo-EMCS ([N-e-maleimidocaproyloxy]sulfosuccinimide ester). Finally, DNA was immobilized through a thioether linkage (Goncalves et al. 2008). In other cases, an alkoxy-silane containing maleimide groups was used to modify a silicon surface, and was then used to covalently bind 5′ thiol modified oligonucleotides (Cabeca et al. 2008). Similarly, thiolated silanes can be used to modify surfaces and covalently bind to disulfide modified DNA through formation of thiolether bonds (Cavic et al. 2001). In addition to planar materials, nanostructures can also be modified with these techniques. For instance, hydroxylated GaN nanowires were modified with 3-methacryloxypropyltrimethoxysilane (MPTS) for thiol modified DNA to immobilize for voltammetric detection (Ganguly et al. 2009).

There are a number of other schemes that utilize alternative organic functional groups such as carboxylic acids, amines, aldehydes, epoxies, esters, and phosphates. Zammatteo et al. (2000) have shown a variety of covalent linkage strategies including carboxylated or phosphorylated DNA to amino-modified glass as well as amine terminated DNA to carboxyl and aldehyde modified glass. They have shown that for the development of DNA microarrays, aldehyde modified surfaces with synthetic amine terminated oligonucleotides provide the most efficient method for attaching DNA onto glass surfaces and have optimized the coupling procedure. Similar studies have been performed in reverse using aminated surfaces with aldehyde terminated ssDNA to form a bond between silane support and nucleic acid (Lobert et al. 2003). Lamture et al. (1994) have also covalently immobilized amine terminated DNA onto epoxide activated silicon thin films to detect DNA hybridization using a charge coupled device. Silica surfaces have also been treated with 2,2,2-trifluroethyl undec-10-enoate (TFEE) to form ester modified surfaces in which amine terminated DNA strands could be covalently coupled to the glass surface (Vong et al. 2009). The advantage of using these types of surface chemistries is that they can be used on a wide range of materials for biosensing, not just silicon, glass, or gold electrodes. For instance, hydrogen-terminated, oxygen-terminated, and fluorine-terminated diamond surfaces were amidated so that carboxylated probes could be immobilized through the amine (Yang et al. 2007a). The 12-phosphonododecanoic acid has also been used to adsorb onto ITO electrodes. The exposed carboxylic acid head group bound with the amine that had been added to the 5′ end of the probe DNA (Moses et al. 2007).

16.5 Summary and Outlook

Nucleic acid biosensing has rapidly evolved from bulk techniques to the incorporation of nanomaterials and unique nanofabrication methods. Nanotechnology offers many advantages over traditional detection technologies, making them more sensitive, faster, and more portable. As nanotechnology continues to develop, we anticipate that additional detection modalities and unique materials properties will be discovered, which may benefit nucleic acid biosensing. At this time, however, we have not seen an explosion of rapid, low-cost, high-sensitivity nanobiosensors on the commercial market. Most of the detection systems described in this chapter are still in the development phase and therefore will not find their way into commercial products for years to come. Several major hurdles, which have not been discussed in this chapter, must be overcome. One of the key challenges for nucleic acid biosensors is the transition from sample to device. Although nucleic acids offer a unique target for detection, they must be isolated and purified from biological specimens before detection can occur. Similarly, sample size can play an important role. In many cases, nanotechnology-based sensors are capable of analyzing very small samples, which may not accurately represent the entire specimen. For these reasons, improved sample handling and analysis techniques must be developed to make these sensors viable. However, the promise of high speed, high sensitivity nucleic acid detection makes this a worthwhile endeavor.

References

Allemand, J. F., D. Bensimon, L. Jullien, A. Bensimon, and V. Croquette. 1997. pH-dependent specific binding and combing of DNA. *Biophysical Journal* 73 (4):2064–2070.

Barone, P. W., S. Baik, D. A. Heller, and M. S. Strano. 2005. Near-infrared optical sensors based on single-walled carbon nanotubes. *Nature Materials* 4 (1):86–92.

Bjork, M. T., B. J. Ohlsson, C. Thelander, A. I. Persson, K. Deppert, L. R. Wallenberg, and L. Samuelson. 2002. Nanowire resonant tunneling diodes. *Applied Physics Letters* 81 (23):4458.

Boozer, C., J. Ladd, S. Chen, and S. Jiang. 2006. DNA-directed protein immobilization for simultaneous detection of multiple analytes by surface plasmon resonance biosensor. *Analytical Chemistry* 78 (5):1515–1519.

Braun, G., S. J. Lee, M. Dante, T.-Q. Nguyen, M. Moskovits, and N. Reich. 2007. Surface-enhanced Raman spectroscopy for DNA detection by nanoparticle assembly onto smooth metal films. *Journal of the American Chemical Society* 129 (20):6378–6379.

Büttner, M., H. Kröger, I. Gerhards, D. Mathys, and P. Oelhafen. 2006. Changes in the electronic structure of gold particles upon thiol adsorption as a function of the mean particle size. *Thin Solid Films* 495 (1–2):180–185.

Cabeca, R., D. M. F. Prazeres, V. Chu, and J. P. Conde. 2008. Electrical and chemical control of surfaces for DNA immobilization and hybridization. *Materials Research Society Symposium Proceedings* 1093:8.

Cady, N. C. 2006. DNA-based biosensors. In *Encyclopedia of Sensors*, eds. C. A. Grimes, E. C. Dickey, and M. V. Pishko, Stevenson Ranch, CA: American Scientific Publishers.

Cady, N. C., A. Gadre, and A. E. Kaloyeros. 2008. Nanobiological sensors. In *Dekker Encyclopedia of Nanoscience and Nanotechnology*, eds. J. A. Schwarz, C. I. Contescu, and K. Putyera, New York: Taylor & Francis.

Cady, N. C., S. Stelick, M. V. Kunnavakkam, and C. A. Batt. 2005. Real-time PCR detection of *Listeria* monocytogenes using an integrated microfluidics platform. *Sensors and Actuators B* 107 (1):332–341.

Cady, N. C., A. D. Strickland, and C. A. Batt. 2007. Optimized linkage and quenching strategies for quantum dot molecular beacons. *Molecular and Cellular Probes* 21:116–124.

Cavic, B. A., M. E. McGovern, R. Nisman, and M. Thompson. 2001. High surface density immobilization of oligonucleotide on silicon. *The Analyst* 126:485–490.

Chan, H. T., Y. Y. Do, P. L. Huang, P. L. Chien, T. S. Chan, R. S. Liu, C. Y. Huang, S. Y. Yang, and H. E. Horng. 2006. Preparation and properties of bio-compatible magnetic Fe_3O_4 nanoparticles. *Journal of Magnetism and Magnetic Materials* 304 (1):e415–e417.

Chang, Ho and W.-C. Tzeng. 2008. A combined conjugation and hybridization technology for different types of DNA and nanoparticles. *Materials Transactions* 49 (6):1467–1473.

Chen, C.-P., A. Ganguly, C.-H. Wang, C.-W. Hsu, S. Chattopadhyay, Y.-K. Hsu, Y.-C. Chang, K.-H. Chen, and L.-C. Chen. 2008. Label-free dual sensing of DNA molecules using GaN nanowires. *Analytical Chemistry* 81 (1):36–42.

Chiu, C.-S. and S. Gwo. 2008. Quantitative surface acoustic wave detection based on colloidal gold nanoparticles and their bioconjugates. *Analytical Chemistry* 80 (9):3318–3326.

Chrisey, L. A., G. U. Lee, and C. E. O'Ferrall. 1996. Covalent attachment of synthetic DNA to self-assembled monolayer films. *Nucleic Acids Research* 24:3031–3039.

Cui, Y. and C. M. Lieber. 2001. Functional nanoscale electronic devices assembled using silicon nanowire building blocks. *Science* 291 (5505):851–853.

Curreli, M., C. Li, Y. Sun, B. Lei, M. A. Gundersen, M. E. Thompson, and C. Zhou. 2005. Selective functionalization of In_2O_3 nanowire mat devices for biosensing applications. *Journal of the American Chemical Society* 127 (19):6922–6923.

Davis, F., A. V. Nabok, and S. P. J. Higson. 2005. Species differentiation by DNA-modified carbon electrodes using an ac impedimetric approach. *Biosensors and Bioelectronics* 20:7.

Ding, C., H. Zhong, and S. Zhang. 2008. Ultrasensitive flow injection chemiluminescence detection of DNA hybridization using nanoCuS tags. *Biosensors and Bioelectronics* 23 (8):1314–1318.

Drummond, T. G., M. G. Hill, and J. K. Barton. 2003. Electrochemical DNA sensors. *Nature Biotechnology* 21 (10):8.

Du, H., M. D. Disney, B. L. Miller, and T. D. Krauss. 2003. Hybridization-based unquenching of DNA hairpins on Au surfaces: Prototypical "molecular beacon" biosensors. *Journal of the American Chemical Society* 125 (14):4012–4013.

Ebara, Y., K. Mizutani, and Y. Okahata. 2000. DNA hybridization at the air–water interface. *Langmuir* 16 (6):2416–2418.

Fahrenkopf, N. M., S. Oktyabrsky, E. Eisenbraun, M. Bergkvist, H. Shi, and N. C. Cady. 2009. Phosphate-dependent DNA immobilization on Hafnium oxide for bio-sensing applications. *Materials Research Society Symposium Proceedings* 1191.

Fan, R., W. Yiying, L. Deyu, Y. Min, M. Arun, and Y. Peidong. 2003. Fabrication of silica nanotube arrays from vertical silicon nanowire templates. *Journal of the American Chemical Society* 125 (18):5254–5255.

Fang, Z. and S. O. Kelley. 2008. Direct electrocatalytic mRNA detection using PNA-nanowire sensors. *Analytical Chemistry* 81 (2):612–617.

Feng, Y., T. Yang, W. Zhang, C. Jiang, and K. Jiao. 2008. Enhanced sensitivity for deoxyribonucleic acid electrochemical impedance sensor: Gold nanoparticle/polyaniline nanotube membranes. *Analytica Chimica Acta* 616 (2):144–151.

Folkers, J. P., P. E. Laibinis, and G. M. Whitesides. 1992. Self-assembled monolayers of alkanethiols on gold: Comparisons of monolayers containing mixtures of short- and long-chain constituents with methyl and hydroxymethyl terminal groups. *Langmuir* 8 (5):1330–1341.

Fujishima, A., T. N. Rao, and D. A. Tryk. 2000. Titanium dioxide photocatalysis. *Journal of Photochemistry and Photobiology C: Photochemistry Reviews* 1 (1):1–21.

Ganguly, A., C.-P. Chen, Y.-T. Lai, C.-C. Kuo, C.-W. Hsu, K.-H. Chen, and L.-C. Chen. 2009. Functionalized GaN nanowire-based electrode for direct label-free voltammetric detection of DNA hybridization. *Journal of Materials Chemistry* 19 (7):928–933.

Gao, H., J. Zhong, P. Qin, C. Lin, W. Sun, and K. Jiao. 2009. Electrochemical DNA hybridization assay for the FMV 35S gene sequence using PbS nanoparticles as a label. *Microchimica Acta* 165 (1–2):173–178.

Goncalves, D., D. M. F. Prazeres, V. Chu, and J. P. Conde. 2008. Detection of DNA and protiens using amorphous silicon ion-sensitive thin-film field effect transistors. *Biosensors and Bioelectronics* 24 (4):6.

Gronewold, T. M. A., A. Baumgartner, E. Quandt, and M. Famulok. 2006. Discrimination of single mutations in cancer-related gene fragments with a surface acoustic wave sensor. *Analytical Chemistry* 78:6.

Homs, M. C. I. 2002. DNA sensors. *Analytical Letters* 35 (12):20.

Hou, S., J. Wang, and C. R. Martin. 2005. Template-synthesized DNA nanotubes. *Journal of the American Chemical Society* 127 (24):8586–8587.

Hur, Y., J. Han, J. Seon, Y. Eugene Pak, and Y. Roh. 2005. Development of an SH-SAW sensor for the detection of DNA hybridization. *Sensors and Actuators A* 120:5.

Hurst, S. J., A. K. R. Lytton-Jean, and C. A. Mirkin. 2006. Maximizing DNA loading on a range of gold nanoparticle sizes. *Analytical Chemistry* 78 (24):8313–8318.

Hwang, J. S., H. T. Kim, H. K. Kim, M. H. Son, J. H. Oh, S. W. Hwang, and D. Ahn. 2008. Electronic transport characteristics of a single wall carbon nanotube field effect transistor wrapped with deoxyribonucleic acid molecules. *Journal of Physics: Conference Series* 109:012015.

Jang, J., S. Ko, and Y. Kim. 2006. Dual-functionalized polymer nanotubes as substrates for molecular-probe and DNA-carrier applications. *Advanced Functional Materials* 16 (6):754–759.

Jeng, E. S., A. E. Moll, A. C. Roy, J. B. Gastala, and M. S. Strano. 2006. Detection of DNA hybridization using the near-infrared band-gap fluorescence of single-walled carbon nanotubes. *Nano Letters* 6 (3):371–375.

Jespersen, M. L., C. E. Inman, G. J. Kearns, E. W. Foster, and J. E. Hutchison. 2007. Alkanephosphonates on hafnium-modified gold: A new class of self-assembled organic monolayers. *Journal of the American Chemical Society* 129 (10):5.

Jin, W., Xiaochen Lin, Shaowu Lv, Ying Zhang, Qinhan Jin, and Ying Mu. 2009. A DNA sensor based on surface plasmon resonance for apoptosis-associated genes detection. *Biosensors and Bioelectronics* 24:3.

Jung, Y., J. M. Lee, H. Jung, and B. H. Chung. 2007. Self-directed and self-oriented immobilization of antibody by protein G–DNA conjugate. *Analytical Chemistry* 79 (17):6534–6541.

Kang, B. S., S. J. Pearton, J. J. Chen, F. Ren, J. W. Johnson, R. J. Therrien, P. Rajagopal, J. C. Roberts, E. L. Piner, and K. J. Linthicum. 2006. Electrical detection of deoxyribonucleic acid hybridization with AlGaN/GaN high electron mobility transistors. *Applied Physics Letters* 89:3.

Khanna, V. K. 2007. Existing and emerging detection technologies for DNA (deoxyribonucleic acid) finger printing, sequencing, bio- and analytical chips: A multidisciplinary development unifying molecular biology, chemical and electronics engineering. *Biotechnology Advances* 25:13.

Kim, W. J. 2010. Method for immobilizing bio-material on titanium dioxide nanoparticles and titanium dioxide nanoparticles immobilized by bio-material. United States patent US 2010/0261244 Al.

Kim, S., L. Chen, S. Lee, G. H. Seong, J. Choo, E. K. Lee, C.-H. Oh, and S. Lee. 2007. Rapid DNA hybridization analysis using a PDMS microfluidic sensor and a molecular beacon. *Analytical Sciences* 23:5.

Kong, J., A. R. Ferhan, X. Chen, L. Zhang, and N. Balasubramanian. 2008. Polysaccharide templated silver nanowire for ultrasensitive electrical detection of nucleic acids. *Analytical Chemistry* 80 (19):7213–7217.

Ladd, J., A. D. Taylor, M. Piliarik, J. Homola, and S. Jiang. 2008. Hybrid surface platform for the simultaneous detection of proteins and DNAs using a surface plasmon resonance imaging sensor. *Analytical Chemistry* 80 (11):5.

Lamture, J. B., K. L. Beatie, B. E. Burke, H. D. Eggero, D. J. Ehrlich, and R. Fowler. 1994. Direct detection of nucleic acid hybridization on the surface of a charge couple device. *Nucleic Acids Research* 22:2121–2125.

Lang, J. and M. Liu. 1999. Layer-by-layer assembly of DNA films and their interactions with dyes. *The Journal of Physical Chemistry B* 103 (51):11393–11397.

Lazareck, A. D., G. S. Cloutier, T.-F. Kuo, J. B. Taft, S. O. Kelley, and J. M. Xu. 2006. DNA-directed synthesis of zinc oxide nanowires on carbon nanotube tips. *Nanotechnology* 17 (10):2661.

Lee, T. M.-H. and I.-M. Hsing. 2006. DNA-based bioanalytical microsystems for handheld device applications. *Analytica Chimica Acta* 556:12.

Li, G., S. Pang, G. Xie, Z. Wang, H. Peng, and Z. Zhang. 2006. Synthesis of radially aligned polyaniline dendrites. *Polymer* 47 (4):1456–1459.

Li, J., W. Zhu, and H. Wang. 2009. Novel magnetic single-walled carbon nanotubes/methylene blue composite amperometric biosensor for DNA determination. *Analytical Letters* 42 (2):366–380.

Lin, Z., S. Cui, H. Zhang, Q. Chen, B. Yang, X. Su, J. Zhang, and Q. Jin. 2003. Studies on quantum dots synthesized in aqueous solution for biological labeling via electrostatic interaction. *Analytical Biochemistry* 319 (2):239–243.

Liu, G., T. M. H. Lee, and J. Wang. 2004a. Nanocrystal-based bioelectronic coding of single nucleotide polymorphisms. *Journal of the American Chemical Society* 127 (1):38–39.

Liu, S.-Q., J.-J. Xu, and H.-Y. Chen. 2004b. A reversible adsorption–desorption interface of DNA based on nano-sized zirconia and its application. *Colloids and Surfaces B: Biointerfaces* 36 (3–4):155–159.

Liu, J., C. Roussel, G. Lagger, P. Tacchini, and H. H. Girault. 2005. Antioxidant sensors based on DNA-modified electrodes. *Analytical Chemistry* 77 (23):7687–7694.

Livak, K. J. 1999. Allelic discrimination using fluorogenic probes and the 5′ nuclease assay. *Genetic Analysis: Biomolecular Engineering* 14:143–149.

Lobert, P. E., D. Bourgeois, R. Pampin, A. Akheyar, L. M. Hagelsieb, D. Flandre, and J. Remacle. 2003. Immobilization of DNA on CMOS compatible materials. *Sensors and Actuators B: Chemical* 92 (1–2):90–97.

Lu, W., Y. Jin, G. Wang, D. Chen, and J. Li. 2008. Enhanced photoelectrochemical method for linear DNA hybridization detection using Au-nanoparticle labeled DNA as probe onto titanium dioxide electrode. *Biosensors and Bioelectronics* 23 (10):1534–1539.

Madou, M. 1997. *Fundamentals of Microfabrication*. 1st edn. Boca Raton, FL: CRC Press.

Mahajan, S., P. Kumar, and K. C. Gupta. 2006. Oligonucleotide microarrays: Immobilization of phosphorylated oligonucleotides on epoxylated surface. *Bioconjugate Chemistry* 17:6.

Martinez, M. T., Y.-C. Tseng, N. Ormategui, I. Loinaz, R. Eritja, and J. Bokor. 2009. Label-free DNA biosensors based on functionalized carbon nanotube field effect transistors. *Nano Letters* 9 (2):530–536.

Mohanty, N. and V. Berry. 2008. Graphene-based single-bacterium resolution biodevice and DNA transistor: Interfacing graphene derivatives with nanoscale and microscale biocomponents. *Nano Letters* 8 (12):7.

Moses, S., S. H. Brewer, S. Kraemer, R. R. Fuierer, L. B. Lowe, C. Agbasi, M. Sauthier, and S. Franzen. 2007. Detection of NSA hybridization in indium tin oxide surfaces. *Sensors and Actuators B* 125:7.

Mullis, K., F. Faloona, S. Scharf, R. Saiki, G. Horn, and H. Erlich. 1986. Specific enzymatic amplification of DNA in vitro: The polymerase chain reaction. *Cold Spring Harbor Symposium Quantitative Biology* 51 (Pt 1):263–273.

Nonglaton, G., I. O. Benitez, I. Guisle, M. Pipelier, J. Leger, D. Dubreuil, C. Tellier, D. R. Talham, and B. Bujoli. 2004. New approach to oligonucleotide microarrays using zirconium phosphonate-modified surfaces. *Journal of the American Chemical Society* 126 (5):1497–1502.

Nykypanchuk, D., M. M. Maye, D. van der Lelie, and O. Gang. 2008. DNA-guided crystallization of colloidal nanoparticles. *Nature* 451 (7178):549–552.

Pang, L.-L., J.-S. Li, J.-H. Jiang, Y. Le, G. L. Shen, and R.-Q. Yu. 2007. A novel detection method for DNA point mutation using QCM based on Fe_3O_4/Au core/shell nanoparticle and DNA ligase reaction. *Sensors and Actuators B: Chemical* 127 (2):311–316.

Patolsky, F. and C. M. Lieber. 2005. Nanowire nanosensors. *Materials Today* 8 (4):20–28.

Pawsey, S., M. McCormick, S. De Paul, R. Graf, Y. S. Lee, L. Reven, and H. W. Spiess. 2003. 1H fast MAS NMR studies of hydrogen-bonding interactions in self-assembled monolayers. *Journal of the American Chemical Society* 125 (14):4174–4184.

Piliarik, M., L. Parova, and J. Homola. 2009. High-throughput SPR sensor for food safety. *Biosensors and Bioelectronics* 24:5.

Polsky, R., R. Gill, L. Kaganovsky, and I. Willner. 2006. Nucleic acid-functionalized Pt nanoparticles: Catalytic labels for the amplified electrochemical detection of biomolecules. *Analytical Chemistry* 78 (7):2268–2271.

Porter, M. D., T. B. Bright, D. L. Allara, and C. E. D. Chidsey. 1987. Spontaneously organized molecular assemblies. 4. Structural characterization of n-alkyl thiol monolayers on gold by optical ellipsometry, infrared spectroscopy, and electrochemistry. *Journal of the American Chemical Society* 109 (12):3559–3568.

Qun, G., C. Chuanding, G. Ravikanth, S. Shivashankar, A. Sathish, D. Kun, and T. H. Donald. 2006. DNA nanowire fabrication. *Nanotechnology* 17 (1):R14.

Sassolas, A., B. D. Leca-Bouvier, and L. J. Blum. 2008. DNA biosensors and microarrays. *Chemical Reviews* 108:31.

Satjapipat, M., R. Sanedrin, and F. Zhou. 2001. Selective desorption of alkanethiols in mixed self-assembled monolayers for subsequent oligonucleotide attachment and DNA hybridization. *Langmuir* 17:7.

Schena, M., D. Shalon, R. W. Davis, and P. O. Brown. 1995. Quantitative monitoring of gene expression patterns with a complementary DNA microarrays. *Science* 270:467–470.

Shiokawa, S. and J. Kondoh. 2004. Surface acoustic wave sensors. *Japanese Journal of Applied Physics* 43 (5B):4.

Sioss, J. A., R. L. Stoermer, M. Y. Sha, and C. D. Keating. 2007. Silica-coated, Au/Ag striped nanowires for bioanalysis. *Langmuir* 23 (22):11334–11341.

Su, M., S. Li, and V. P. Dravid. 2003. Microcantilever resonance-based DNA detection with nanoparticle probes. *Applied Physics Letters* 82 (20):3.

Su, X., Y.-J. Wu, and W. Knoll. 2005. Comparison of surface plasmon resonance spectroscopy and quartz crystal microbalance techniques for studying DNA assembly and hybridization. *Biosensors and Bioelectronics* 21:7.

Tan, W., K. Wang, X. He, X. J. Zhao, T. Drake, L. Wang, and R. P. Bagwe. 2004. Bionanotechnology based on silica nanoparticles. *Medicinal Research Reviews* 24 (5):621–638.

Teles, F. R. R. and L. P. Fonseca. 2008. Trends in DNA Biosensors. *Talanta* 77:18.

Tosar Pablo, J., K. Keel, and J. Laiz. 2009. Two independent label-free detection methods in one electrochemical DNA sensor. *Biosensors and Bioelectronics* 24:6.

Tyagi, S. and F. R. Kramer. 1996. Molecular beacons: Probes that fluoresce upon hybridization. *Nature Biotechnology* 14 (3):303–308.

Vong, T., J. ter Maat, T. A. van Beek, B. van Lagen, M. Giesbers, J. C. M. van Hest, and H. Zuilhof. 2009. Site-specific immobilization of DNA in glass microchannels via photolithography. *Langmuir* 25 (24):13952–13958.

Wang, X. and C. S. Ozkan. 2008. Multisegment nanowire sensors for the detection of DNA molecules. *Nano Letters* 8 (2):398–404.

Wang, D., Y. Liu, C. Wang, F. Zhou, and W. Liu. 2009a. Highly flexible coaxial nanohybrids made from porous TiO_2 nanotubes. *ACS Nano* 3 (5):1249–1257.

Wang, K., Z. Tang, C. J. Yang, Y. Kim, X. Fang, W. Li, Y. Wu et al. 2009b. Molecular engineering of DNA: Molecular beacons. *Angewandte Chemie International Edition* 48:14.

Wang, R., S. Tombelli, M. Minunni, M. M. Spiriti, and M. Mascini. 2004. Immobilisation of DNA probes for the development of SPR-based sensing. *Biosensors and Bioelectronics* 20:967.

Wark, A. W., H. J. Lee, A. J. Qavi, and R. M. Corn. 2007. Nanoparticle-enhanced diffraction gratings for ultrasensitive surface plasmon biosensing. *Analytical Chemistry* 79 (17):6697–6701.

Wu, Y., J. A. Phillips, H. Liu, R. Yang, and W. Tan. 2008. Carbon nanotubes protect DNA strands during cellular delivery. *ACS Nano* 2 (10):2023–2028.

Wu, S., H. Zhao, H. Ju, C. Shi, and J. Zhao. 2006. Electrodeposition of silver-DNA hybrid nanoparticles for electrochemical sensing of hydrogen peroxide and glucose. *Electrochemistry Communications* 8 (8):1197–1203.

Xu, K., J. Huang, Z. Ye, Y. Ying, and Y. Li. 2009a. Recent development of nano-materials used in DNA biosensors. *Sensors* 9 (7):5534–5557.

Xu, X., V. Jindal, F. Shahedipour-Sandvik, M. Bergkvist, and N. C. Cady. 2009b. Direct immobilization and hybridization of DNA on group III nitride semiconductors. *Applied Surface Science* 255 (11):5905–5909.

Xu, X.-H., H. C. Yang, T. E. Mallouk, and A. J. Bard. 2002. Immobilization of DNA on an aluminum(III) alkanebisphosphonate thin film with electrogenerated chemiluminescent detection. *Journal of the American Chemical Society* 116 (18):8386–8387.

Xuan, G., J. Kolodzey, V. Kapoor, and G. Gonye. 2005. Characteristics of field-effect devices with gate oxide modification by DNA. *Applied Physics Letters* 87:3.

Yang, J.-H., K.-S. Song, S. Kuga, and H. Kawarada. 2007a. Direct immobilization of DNA on partially functionalized diamond surface. *Materials Research Society Symposium Proceedings* 950:8.

Yang, Y., Z. Wang, M. Yang, J. Li, F. Zheng, G. Shen, and R. Yu. 2007b. Electrical detection of deoxyribonucleic acid hybridization based on carbon-nanotubes/nano zirconium dioxide/chitosan-modified electrodes. *Analytica Chimica Acta* 584 (2):268–274.

Yoon, I., T. Kang, W. Choi, J. Kim, Y. Yoo, S.-W. Joo, Q.-H. Park, H. Ihee, and B. Kim. 2008. Single nanowire on a film as an efficient SERS-active platform. *Journal of the American Chemical Society* 131 (2):758–762.

Zammatteo, N., L. Jeanmart, S. Hamels, S. Courtois, P. Louette, L. Hevesi, and J. Remacle. 2000. Comparison between different strategies of covalent attachment of DNA to glass surfaces to build DNA microarrays. *Analytical Biochemistry* 280 (1):143–150.

Zhang, Q. and V. Subramanian. 2007. DNA hybridization detection with organic thin film transistors: Toward fast and disposable DNA microarray chips. *Biosensors and Bioelectronics* 22:5.

Zhang, G.-J., J. H. Chua, R.-E. Chee, A. Agarwal, and S. M. Wong. 2009. Label-free direct detection of MiRNAs with silicon nanowire biosensors. *Biosensors and Bioelectronics* 24 (8):2504–2508.

Zhang, W., T. Yang, C. Jiang, and K. Jiao. 2008a. DNA hybridization and phosphinothricin acetyltransferase gene sequence detection based on zirconia/nanogold film modified electrode. *Applied Surface Science* 254 (15):4750–4756.

Zhang, S., J. Xia, and X.i. Li. 2008b. Electrochemical biosensor for detection of adenosine based on structure-switching aptamer and amplification with reporter probe DNA modified Au nanoparticles. *Analytical Chemistry* 80 (22):8382–8388.

Zhang, W., T. Yang, D. M. Huang, and K. Jiao. 2008c. Electrochemical sensing of DNA immobilization and hybridization based on carbon nanotubes/nano zinc oxide/chitosan composite film. *Chinese Chemical Letters* 19 (5):589–591.

Zhang, W., T. Yang, D. Huang, K. Jiao, and G. Li. 2008d. Synergistic effects of nano-ZnO/multi-walled carbon nanotubes/chitosan nanocomposite membrane for the sensitive detection of sequence-specific of PAT gene and PCR amplification of NOS gene. *Journal of Membrane Science* 325 (1):245–251.

Zhao, W., W. Chiuman, J. C. F. Lam, S. A. McManus, W. Chen, Y. Cui, R. Pelton, M. A. Brook, and Y. Li. 2008. DNA aptamer folding on gold nanoparticles: From colloid chemistry to biosensors. *Journal of the American Chemical Society* 130 (11):3610–3618.

Zhu, N., Z. Chang, P. He, and Y. Fang. 2005. Electrochemical DNA biosensors based on platinum nanoparticles combined carbon nanotubes. *Analytica Chimica Acta* 545 (1):21–26.

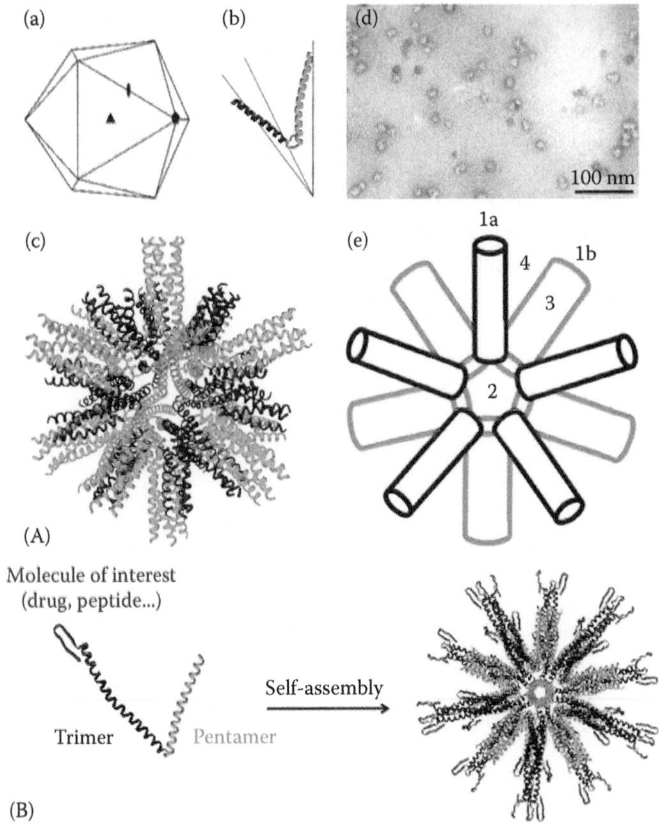

FIGURE 1.3 (A) Design of peptide nanoparticles: (a) icosahedron with the symbols of the three different symmetry elements; (b) monomeric building block composed of a trimeric (blue) and pentameric coiled-coil-α-helix; (c) model of an assembled nanoparticle with icosahedral symmetry; (d) electron micrograph of the peptide nanoparticles functionalized with somatostatin; and (e) schematic diagram of the particle with possible modification sites. (B) Functionalized nanoparticles (right) formed by self-assembly of peptides (right). The trimeric domain can be modified by a ligand or by a drug.

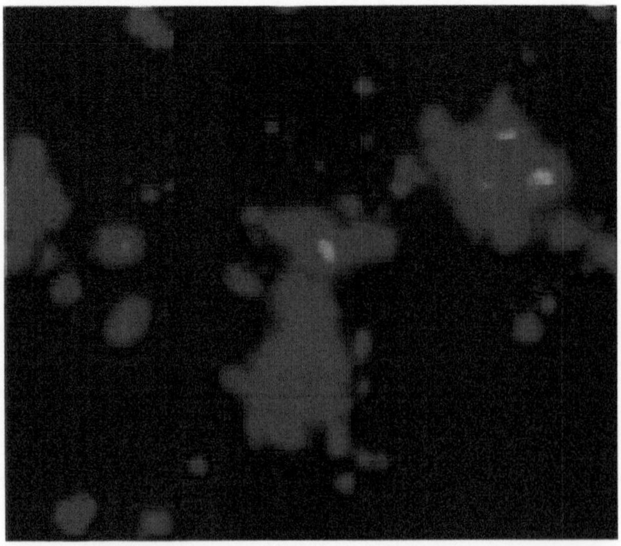

FIGURE 5.6 Epifluorescence image of NDs. (Reproduced from Yu, S.-J. et al., *J. Am. Chem. Soc.*, 127(50), 17604, 2005.)

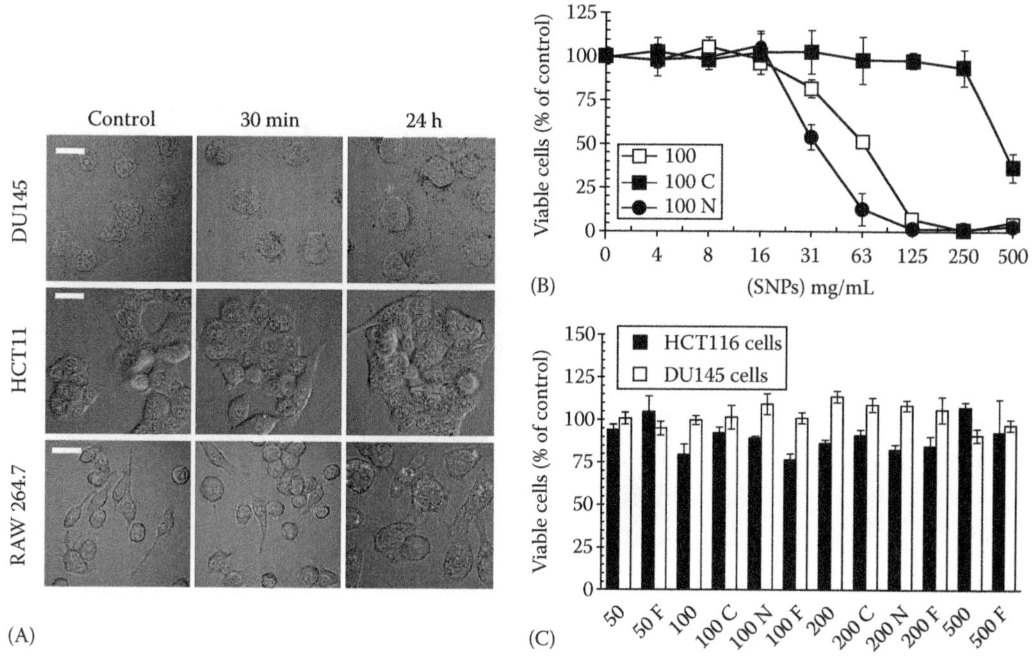

FIGURE 6.8 Influence of cell type and surface modification on toxicity. As shown, silica nanoparticle treatment does not induce HCT116 or DU145 epithelial cell toxicity, while it does induce RAW 264.7 macrophage toxicity. IC_{50} values in RAW 264.7 cells are heavily dependent on surface modification, as amine (N) modified particles are more toxic than unmodified and carboxyl modified particles (C). (A) Confocal microscopy image of DU145, HCT11 and RAW264.7 cell nucleus stained with DRAQ5 and particles labeled with FITC. These images illustrate the relative uptake of these particles. RAW264.7 shows a significant increase in relative uptake. (B) RAW 264.7 WST-8 proliferation assay, 100 nm plain, carboxyl and amine-modified particles. (C) HCT116 and DU145 WST-8 proliferation assay, 100 nm plain, carboxyl- and amine-modified particles, little to no toxicity was observed. (Modified from Malugin, A. et al., Submitted, 2011.)

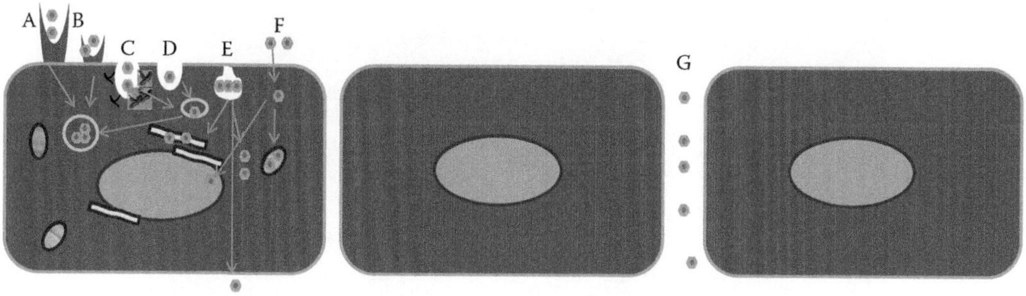

FIGURE 6.9 Mechanisms of nanoparticle uptake: (A) phagocytosis, (B) macropinocytosis, (C) clathrin-mediated endocytosis, (D) clathrin- and caveolae-independent endocytosis, (E) caveolae-mediated endocytosis, (F) transmembrane transport, and (G) paracellular transport.

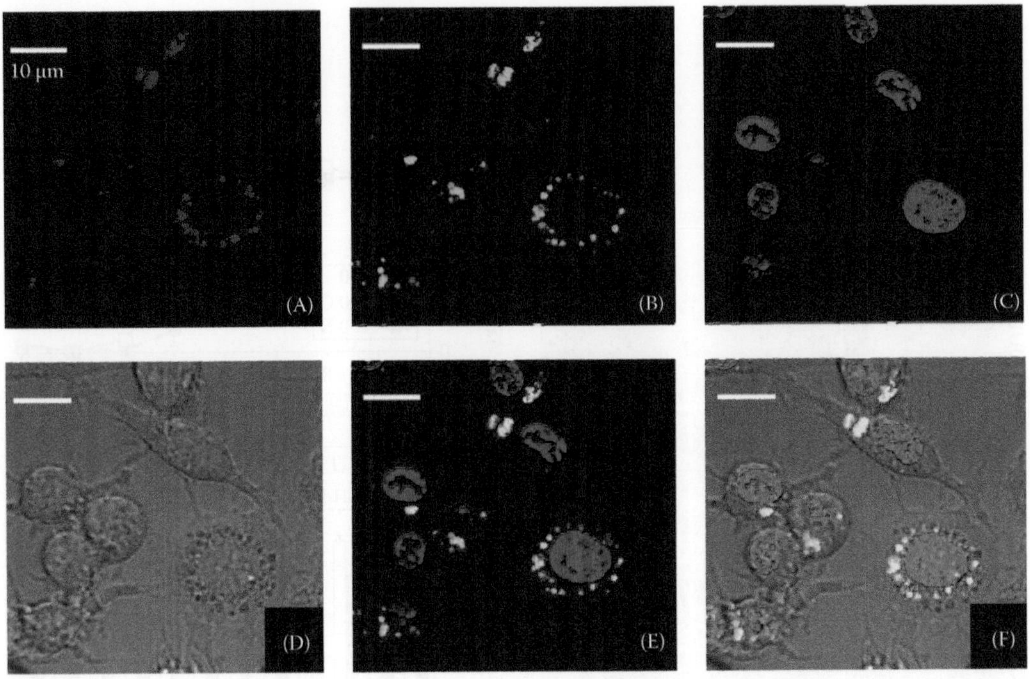

FIGURE 6.10 Confocal images of silica nanoparticle uptake in RAW 264.7 macrophage. Silica nanoparticles are colocalized in lysosomal compartments (yellow color in figures E and F). (A) Lysosomes are stained with lysotracker. (B) Particles are stained with FITC. (C) Nucleus is stained with DRAQ5. (D) Transmitted image of RAW 264.7 cells. (E) Fluorescence overlay. (F) Transmitted and fluorescence overlay. (Modified data from Malugin, A. et al., Submitted, 2011.)

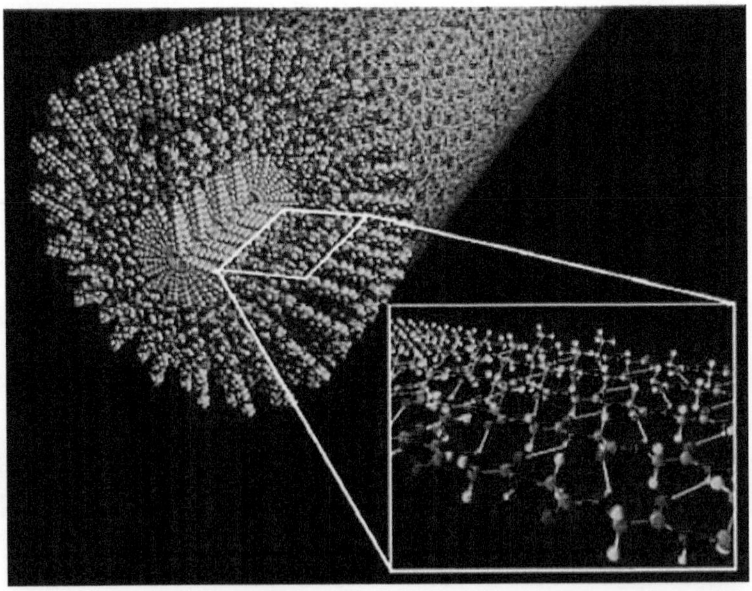

FIGURE 7.4 Schematic representation of β-sheets within PA nanofibers. As depicted in the inset, β-sheets are oriented parallel to the long axis of the nanofibers (inter-β-strand hydrogen bonds are represented as yellow lines; carbon, oxygen, hydrogen, and nitrogen atoms are colored grey, red, light blue, and blue, respectively). (From Jiang, H., Guler, M.O., and Stupp, S.I., The internal structure of self-assembled peptide amphiphiles nanofibers, *Soft Matter*, 3, 454–462, 2007. Reproduced by permission of The Royal Society of Chemistry.)

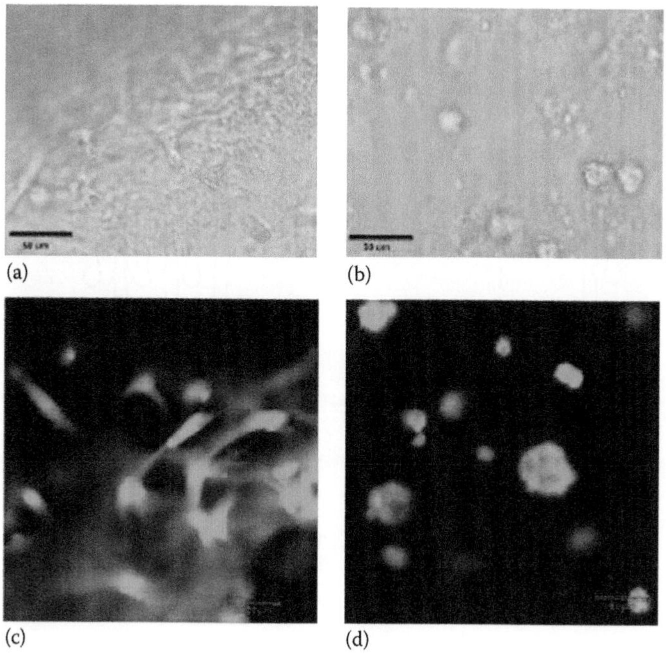

FIGURE 7.10 Response of nanofiber network to encapsulated cells. Rat maxillary incision pulp cells were encapsulated in nanofiber networks with different densities of adhesive ligands, (a) 100% RGDS or (b) RDGS. Confocal laser scanning microscopy images of cells in nanofiber networks fluorescently observed encapsulation of cells in PAs with (c) 100% RGDS or (d) RDGS. Cells were stained with fluorescent dyes (calcein AM and ethidium homodimer-1). All images were taken after 4 days of incubation. (From Jun, H.-W., Yuwono, V., Paramonov, S.E., and Hartgerink, J.D: Enzyme-mediated degradation of peptide-amphiphile nanofiber networks, *Adv. Mater.* 2005, 17(21), 2612–2617, 2005. Copyright Wiley-VCH Verlag GmbH & Co. KGaA.)

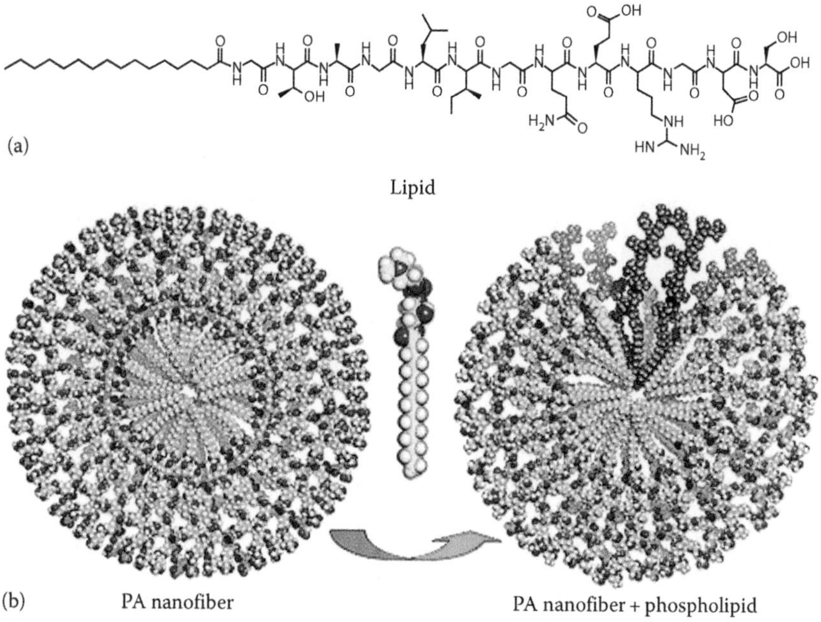

FIGURE 7.11 (a) Chemical structure of the PA and (b) cross section of a PA fiber and a PA fiber containing 6.25 mol% of lipid (yellow). Highlighted in pink are the PA molecules situated adjacent to the lipid molecules. (Reprinted with permission from [Paramonov, S.E., Jun, H.W., and Hartgerink, J.D., Modulation of peptide-amphiphile nanofibers via phospholipid inclusions, *Biomacromolecules*, 7(1), 24–26]. Copyright [2006] American Chemical Society.)

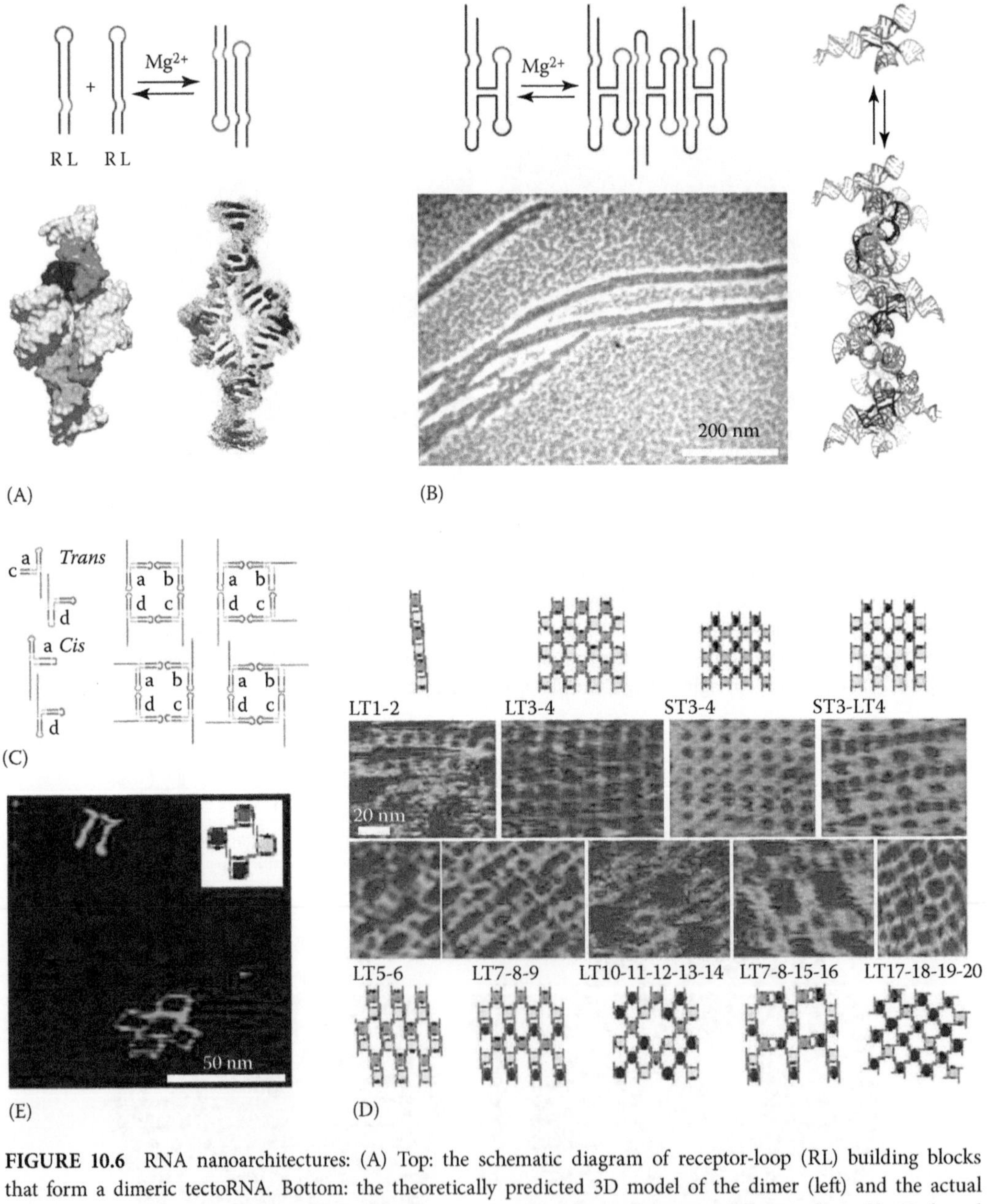

FIGURE 10.6 RNA nanoarchitectures: (A) Top: the schematic diagram of receptor-loop (RL) building blocks that form a dimeric tectoRNA. Bottom: the theoretically predicted 3D model of the dimer (left) and the actual NMR-determined structure (right); (B) Top: the schematic drawing of a one-dimensional filament formed by H-shaped tectoRNAs. Bottom: TEM image of RNA filaments; (C) the schematic drawing of RNA tectosquare nanostructures; (D) and (E) two-dimensional RNA nanoarchitectures constructed from tectosquare building blocks. (Adapted from Jaeger, L. and Chworos, A., *Curr. Opin. Struct. Biol.*, 16(4), 531, 2006. With permission.)

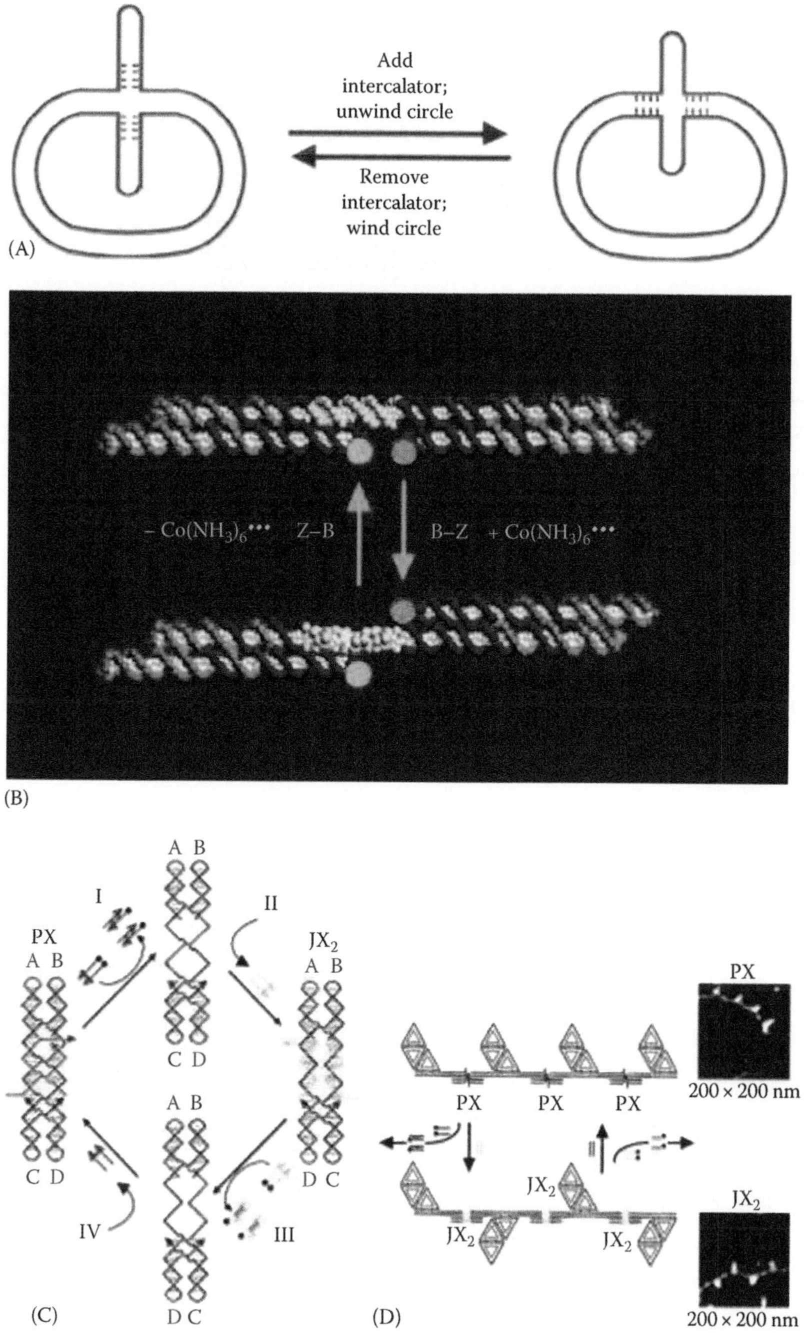

FIGURE 10.7 DNA nanodevices: (A) a mobile control device; (B) a nonsequence-specific DNA nanomachine that performs a B–Z transition; fluorescent dyes used to monitor the change are shown in green (fluorescein) and magenta (Cy3) circles; (C) the sequence-specific PX-JX$_2$ nanomachine; and (D) the schematic drawing and AFM images of a series of DNA trapezoids connected by the PX-JX$_2$ device. The top device is in the PX state; the bottom one is in the JX$_2$ state. (Adapted from Seeman, N.C., *Chem. Biol.*, 10(12), 1151, 2003a. With permission.)

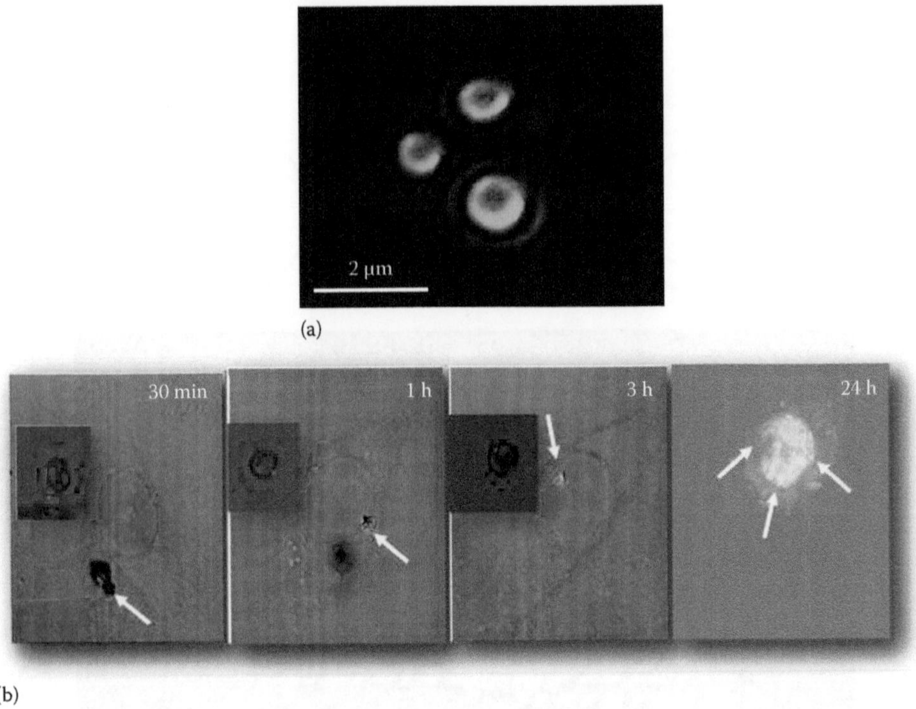

FIGURE 12.9 Glucose-derived fluorescent carbon nanospheres. (a) Confocal laser scanning image. (b) Nuclear entry into the HeLa cells is indicated by white arrow marks. (From Selvi, R. et al., *Nano Lett.*, 8, 3182, 2008.)

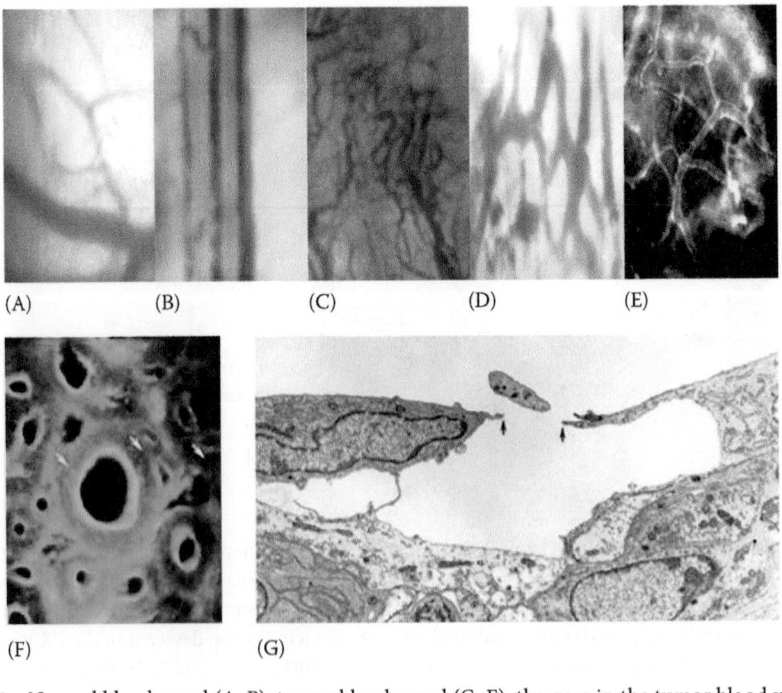

FIGURE 14.1 Normal blood vessel (A–B); tumor blood vessel (C–E); the pore in the tumor blood vessel wall (F); Hematoxylin and Eosin (HE) staining of pores in tumor blood vessel wall (G).

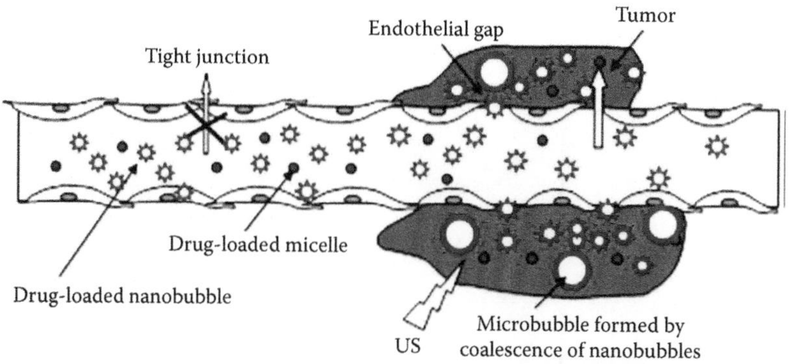

FIGURE 14.2 Schematic representation of a drug targeting through the defective tumor microvasculature using an echogenic drug delivery system. The system comprises polymeric micelles (small circles), nanobubbles (stars), and microbubbles (large circles). Micelles are formed by a biodegradable block copolymer (e.g., PEG-PLLA or PEG-PCL); bubbles are formed by perfluorocarbon (e.g., PFP) stabilized by the same biodegradable block copolymer. Lipophilic drug (e.g., doxorubicin) is localized in the micelle cores and the walls of nano/microbubbles. Tight junctions between endothelial cells in blood vessels of normal tissues do not allow extravasation of drug-loaded micelles or nano/microbubbles (indicated by cross). In contrast, tumors are characterized by defective vasculature with large gaps between the endothelial cells, which allow extravasation of drug-loaded micelles and small nanobubbles, resulting in their accumulation in the tumor interstitium. On accumulation in the tumor tissue, small nanobubbles coalesce into larger, highly echogenic microbubbles that release their drug load in response to therapeutic ultrasound. (From Rapoport, N. et al., *J. Natl. Cancer Inst.*, 99, 1095, 2007; Gao, Z. et al., *Ultrasonics*, 48, 260, 2008.)

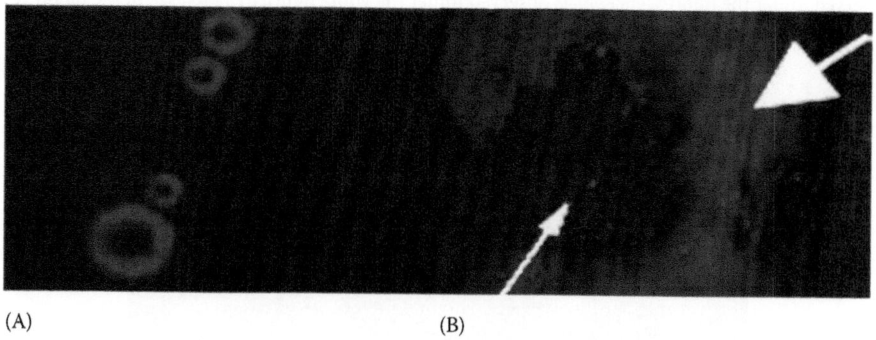

FIGURE 14.6 Fluorescence images of the 0.75 mg/mL Dox/0.5% PEG-PLLA/2% PFP formulation placed in a plastic capillary (internal diameter 340 mm) of a snake mixer slide. (A) The formulation was heated to 46°C to generate visible microbubbles; the Dox-derived fluorescence (red) of the microbubbles was localized in the bubble walls formed by the bubble-stabilizing copolymer. (B) The MDA MB231 cells incubated with Dox-loaded microbubbles at 37°C did not fluoresce during 30 min incubation (data not shown). Under the action of ultrasound, the cells (shown by a thick arrow) acquired strong fluorescence, while the fluorescence of microbubbles (indicated by thin arrow) was substantially decreased. Unfocused ultrasound of 3 MHz was applied to the slide through the Aquasonic coupling gel for 150 s at a 2 W/cm^2 nominal SATA power density and a 20% duty cycle (the actual power density experienced by the sample may be different from the nominal because the sonication took place in the ultrasound near field). The data indicate that without ultrasound Dox was strongly retained in microbubble walls; however, under the ultrasonic action Dox was released from the bubbles and internalized by the cells.

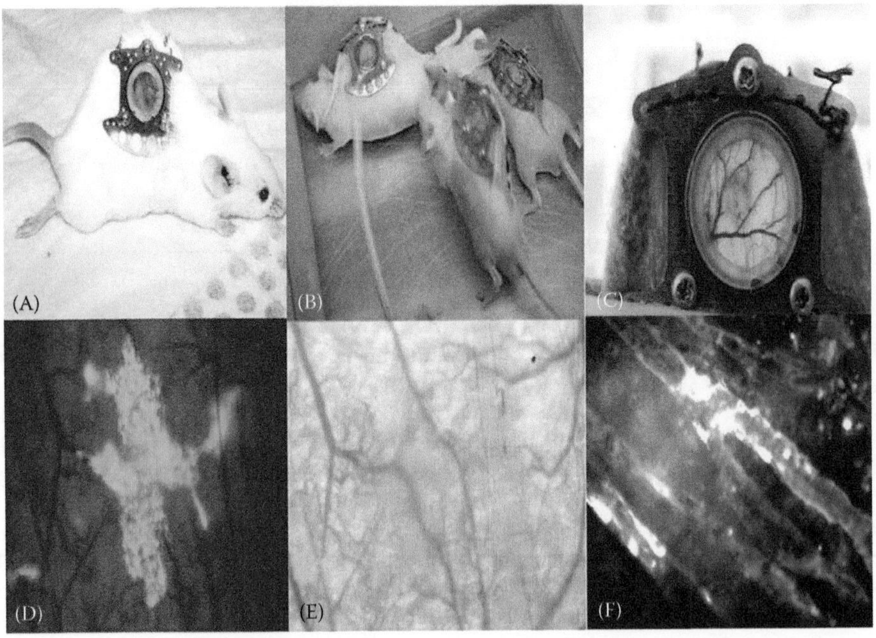

FIGURE 14.12 Fluorescence labeling of drug release from tumor blood vessel dorsal skinfold window chamber model (A through C); after injection of green fluorescent protein (GFP)-transfected 4T1 mammary cancer cell (D: as observed under the fluorescence, E: as observed under the white light); drug release in the blood vessel (F).

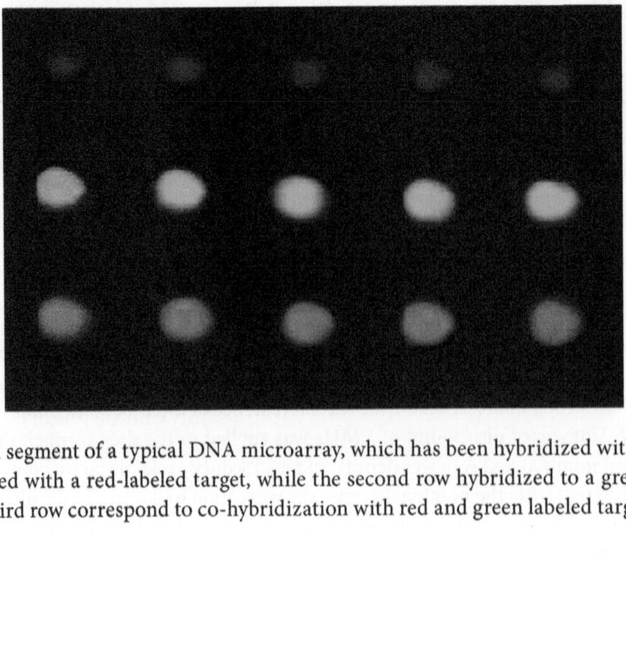

FIGURE 16.1 Small segment of a typical DNA microarray, which has been hybridized with target DNA. The top row of spots hybridized with a red-labeled target, while the second row hybridized to a green-labeled target. The orange spots in the third row correspond to co-hybridization with red and green labeled targets.

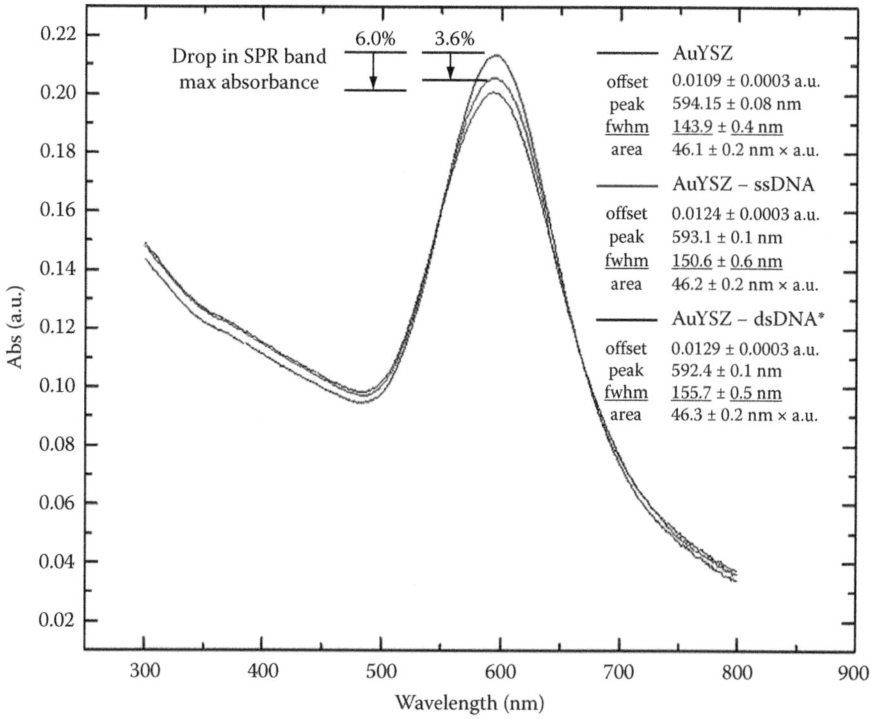

FIGURE 16.4 DNA-gold nanoparticle interactions and its effect on SPR. Single and double-stranded DNA was incubated with gold nanoparticles embedded in YSZ films. Addition of DNA reduced the SPR peak intensity by 3.6% for ssDNA and 6% for dsDNA.

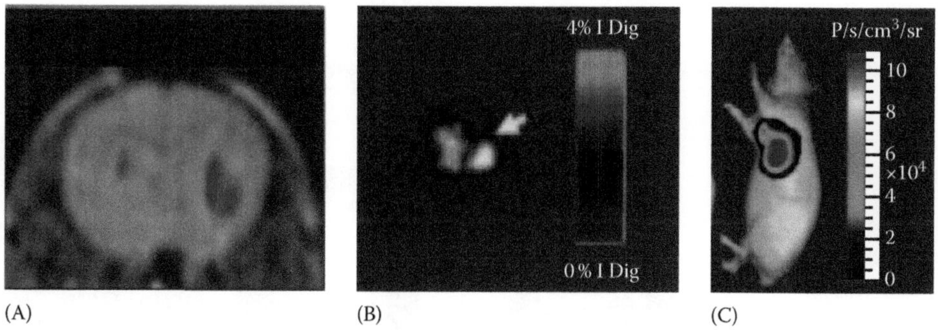

FIGURE 17.3 (A) 9.4 T MR image of a mouse brain 3 days post-engraftment of PFC-labeled neural stem cells, with the ^{19}F signal superimposed on the ^{1}H MR image. The cells labeled with fluorinated emulsions appear as "hot spots." (From Ruiz-Cabello, J. et al., *Magn. Resonance Med.*, 60 (6), 1506, 2008.) (B,C) PET (B) and bioluminescent (C) images of a mouse injected with stem cells carrying thymidine kinase (PET) and luciferase gene 4 weeks post-injection. (Courtesy of Dr. Joseph C. Wu, Stanford University, Stanford, CA.) Cells were visible as "hot spots" on the PET and BL images.

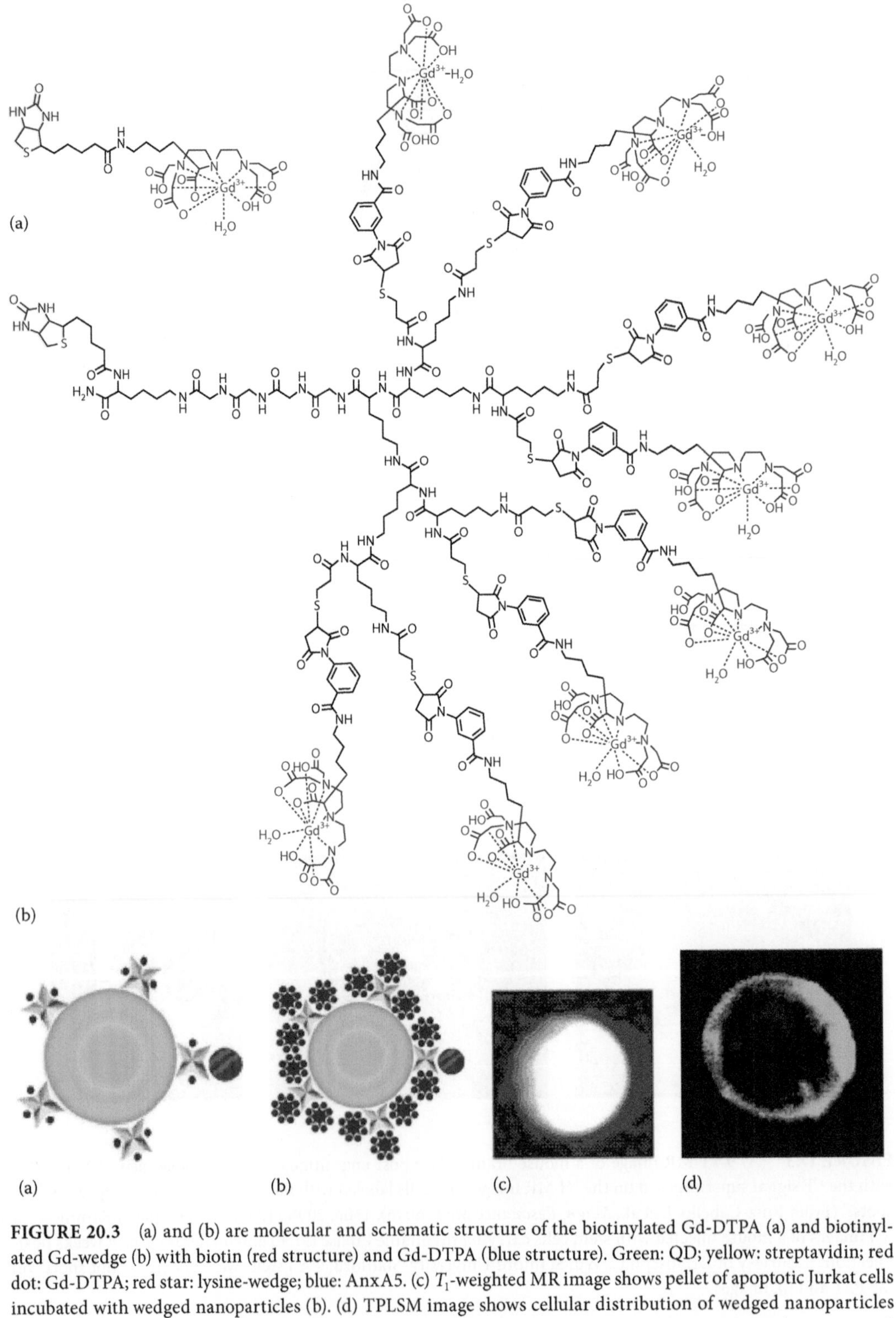

FIGURE 20.3 (a) and (b) are molecular and schematic structure of the biotinylated Gd-DTPA (a) and biotinylated Gd-wedge (b) with biotin (red structure) and Gd-DTPA (blue structure). Green: QD; yellow: streptavidin; red dot: Gd-DTPA; red star: lysine-wedge; blue: AnxA5. (c) T_1-weighted MR image shows pellet of apoptotic Jurkat cells incubated with wedged nanoparticles (b). (d) TPLSM image shows cellular distribution of wedged nanoparticles (b) in late apoptotic Jurkat cells. (Reproduced from Prinzen, L. et al., *Nano Lett.*, 7(1), 93, 2007. With permission.)

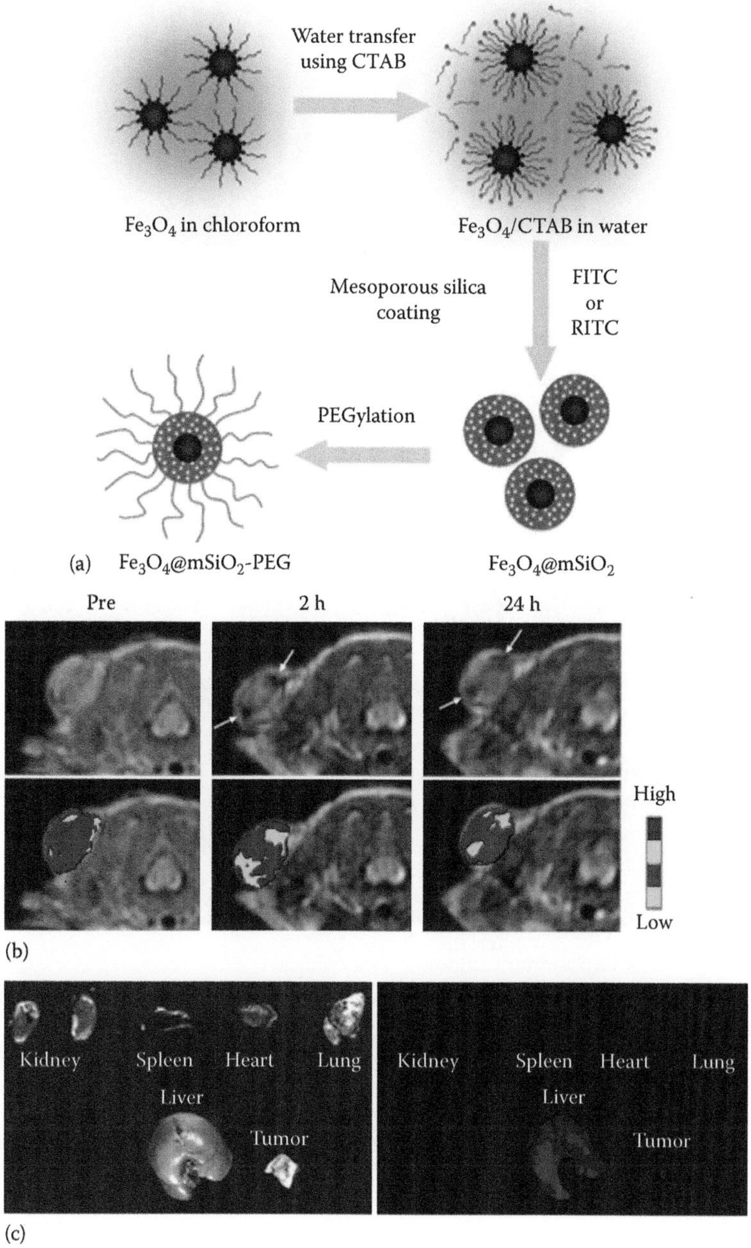

FIGURE 20.5 Multimodality MR/fluorescence imaging of tumor using $Fe_3O_4/mSiO_2$ core/shell nanoparticles in a mouse model. (a) Schematic synthesis of magnetite nanocrystal/mesoporous silica core–shell nanoparticles; (b) In vivo T_2-weighted MR images (upper row) and color maps (lower row) of tumor before and after intravenously injection of probe. A decrease of signal intensity on T_2-weighted MR images was detected at the tumor site (arrows); (c) photographic image and corresponding fluorescence image of several organs and the xenograft tumor 24 h after intravenous injection. (Reproduced from Kim, J. et al., *Angew. Chem. Int. Ed.*, 47(44), 8438, 2008. With permission.)

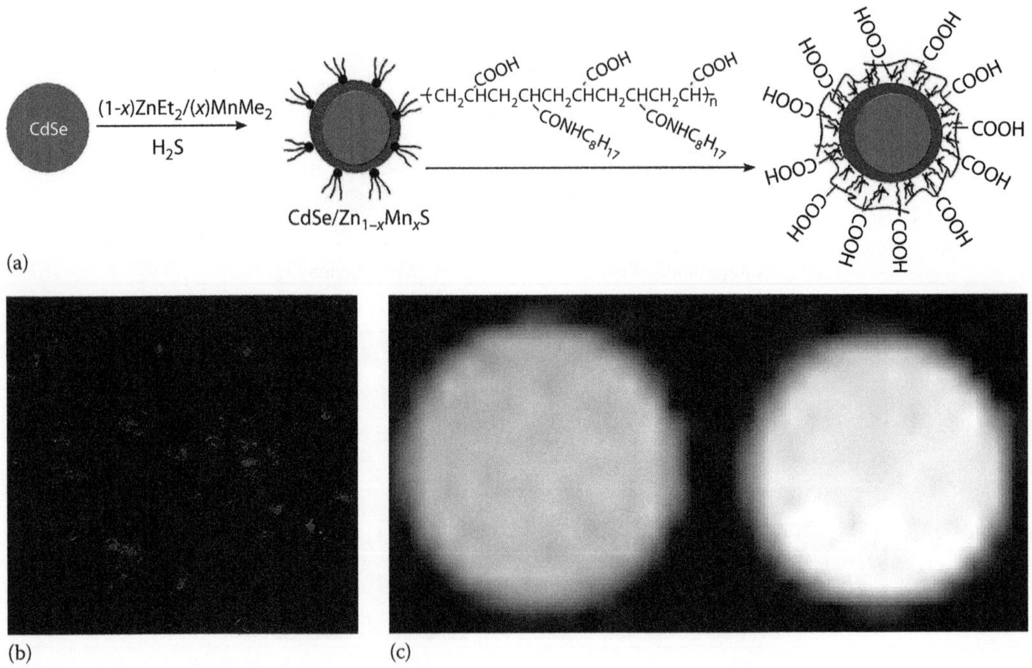

FIGURE 20.6 (a) Synthetic representation of water-soluble core/shell CdSe/$Zn_{1-x}Mn_xS$; (b) cells incubated with CdSe/$Zn_{1-x}Mn_xS$ exhibit strong fluorescence in confocal microscopy; (c) T_1-weighted MR images from tubes containing cell lysates show that lysates of cells incubated with CdSe/$Zn_{1-x}Mn_xS$ (right) show significant contrast enhancement as compared to cells that have not been exposed to CdSe/$Zn_{1-x}Mn_xS$. (Reproduced from Wang, S. et al., *J. Am. Chem. Soc.*, 129(13), 3848, 2007. With permission.)

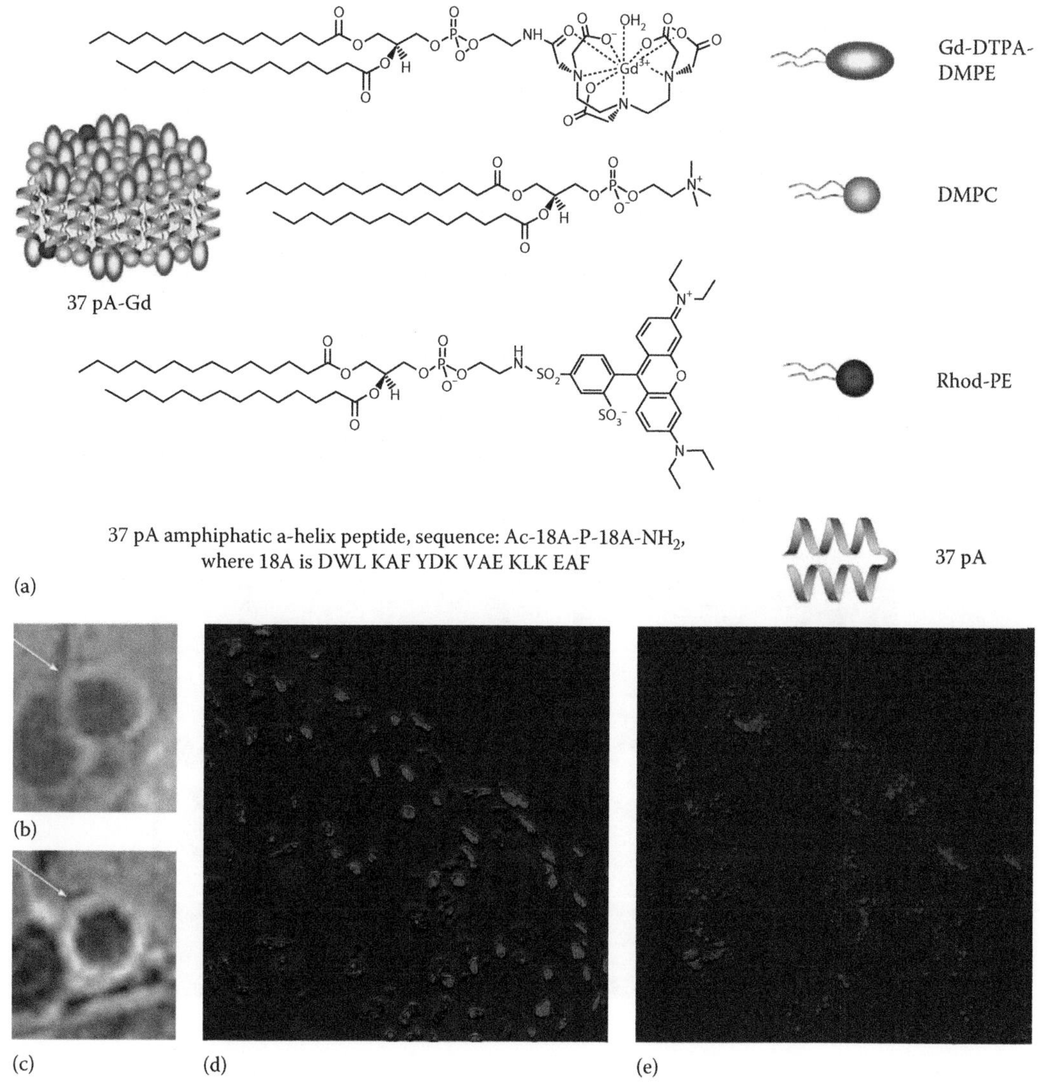

FIGURE 20.8 (a) Schematic depiction of the reconstituted, (bi)layer nanodisk HDL with both magnetic and fluorescent properties; (b) and (c) MR images of the abdominal aorta, indicated by arrows, of an apo E-/- mouse, before (top) and 24 h postinjection (bottom) with the probe; (d) and (e) confocal microscopy images of aortic tissue from an apo E-/- mouse injected with the probe 24 h prior to excision. (d) The nuclei are blue via a DAPI stain and (e) the areas containing the probe are red due to the fluorescence from the rhodamine lipid. (Reproduced from Cormode, D.P. et al., *Small* 4(9), 1437, 2008. With permission.)

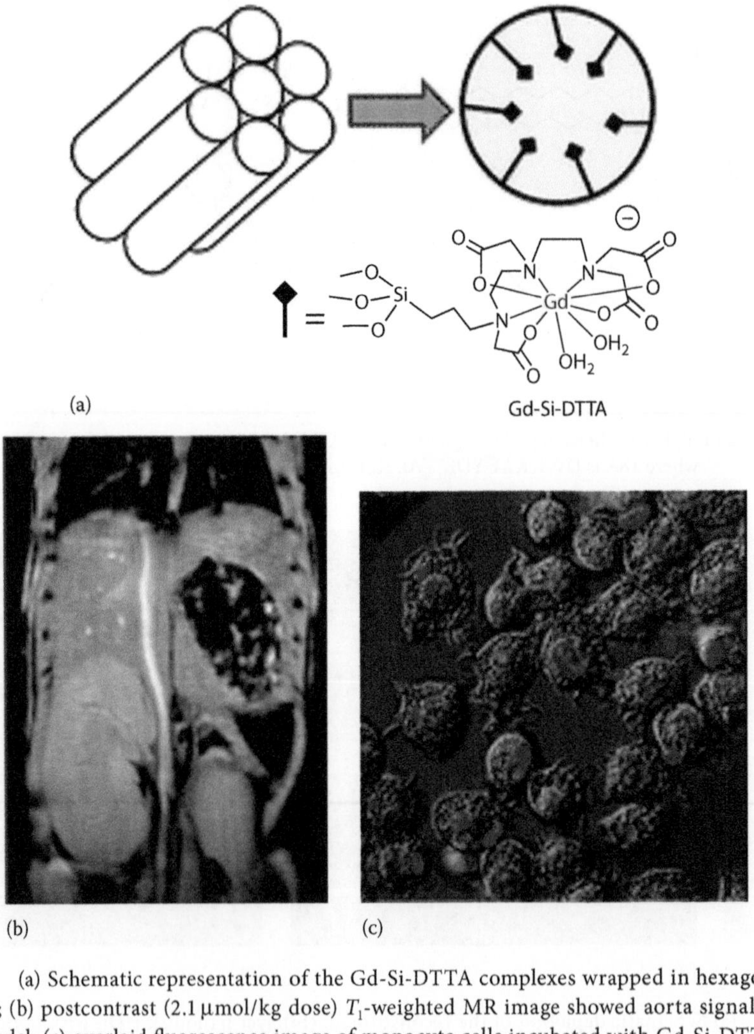

FIGURE 20.9 (a) Schematic representation of the Gd-Si-DTTA complexes wrapped in hexagonally ordered nanochannels; (b) postcontrast (2.1 μmol/kg dose) T_1-weighted MR image showed aorta signal enhancement in a mouse model; (c) overlaid fluorescence image of monocyte cells incubated with Gd-Si-DTTA complexes. (Reproduced from Taylor, K.M.L. et al., *J. Am. Chem. Soc.*, 130(44), 14358, 2008. With permission.)

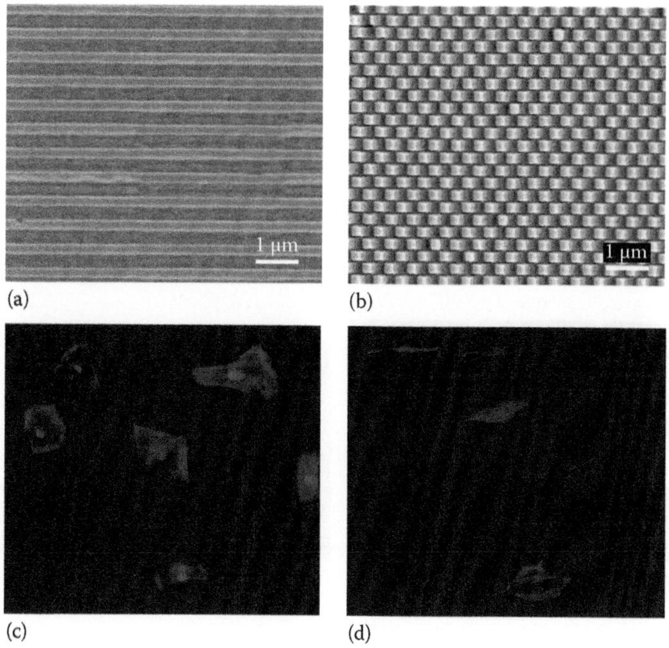

FIGURE 26.5 (a), (b) Nanopatterns with a range of different shapes and dimensions were created with electron beam lithography. Cells were cultured on the substrates and subsequently studied on their morphology (i.e. orientation, area and elongation) by fluorescence microscopy on filamentous actin. (c) Cells aligned to wide grooves (600 nm) and became more elongated, whereas cells spread randomly on small grooves and were more rounded (d).

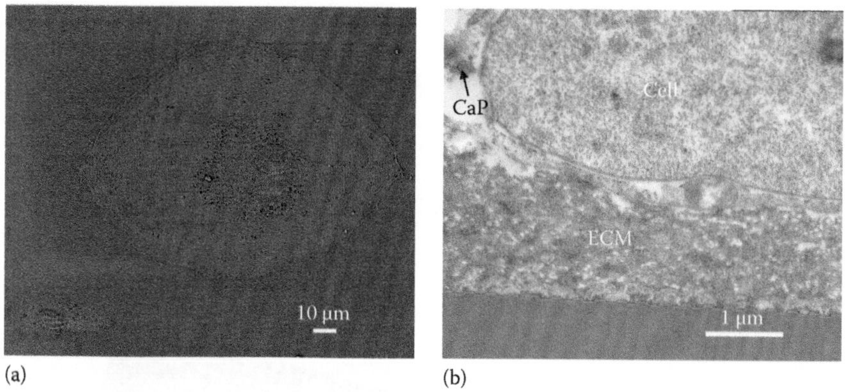

FIGURE 26.7 (a) Osteoblasts align to grooves with a width of 100 nm and a ridge of 100 nm filamentous actin is stained red and the nuclei blue. (b) Bone extracellular matrix (ECM) is deposited inside grooves with a width of 150 nm.

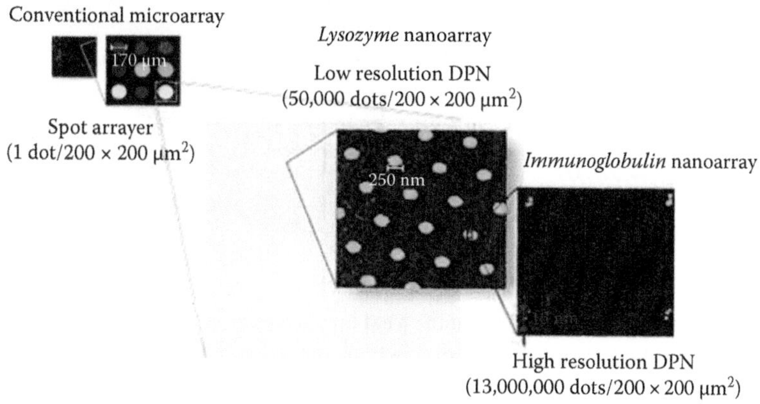

FIGURE 27.1 The relative spot sizes of microarrays and nanoarrays. (From Rosi, N.L. and Mirkin, C.A., *Chem. Rev.*, 105, 1547, 2005. With permission.)

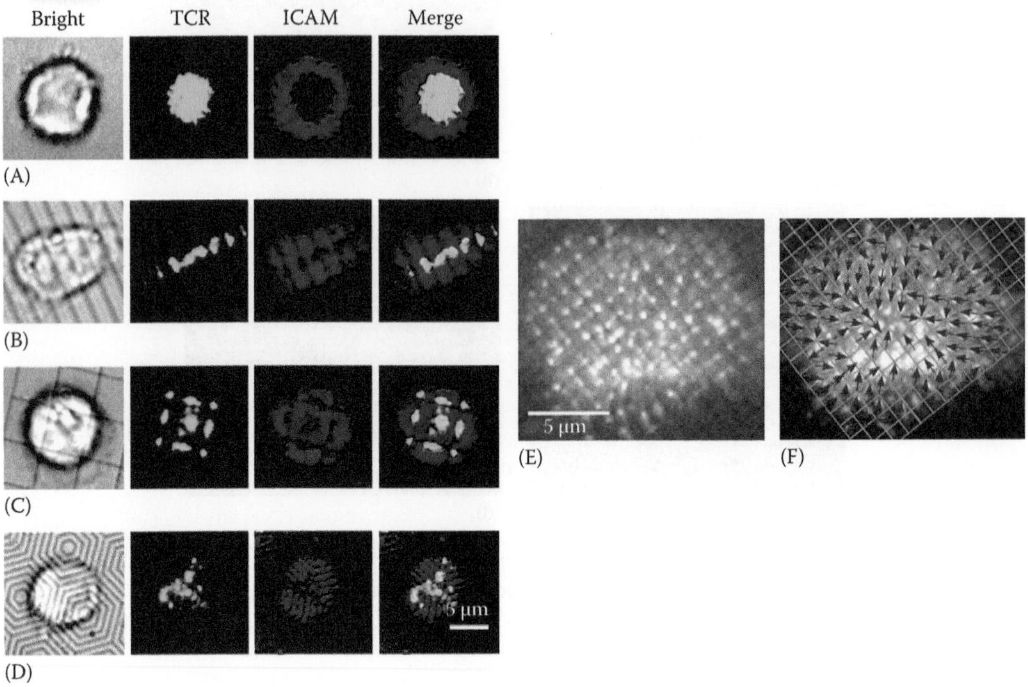

FIGURE 27.2 Lipids membrane deposited on (A) unpatterned chromium, (B) 2 μm lines, (C) 5 μm square grid, (D) hexagonal patterns spaced 1 μm apart. The distribution of TCR in the membrane is shown in (E) and its transport is visualized using red arrows as shown in (F). (From Mossman, K.D. et al., *Science*, 310, 1191, 2005. With permission.)

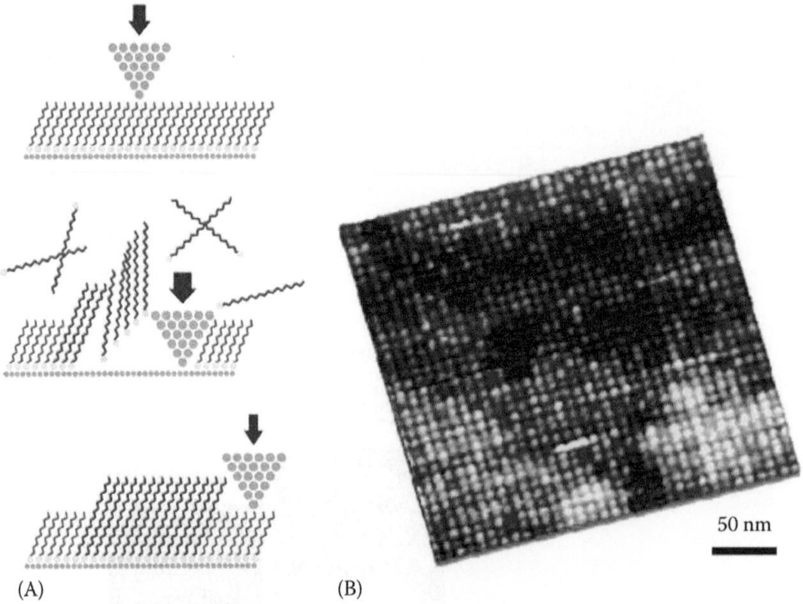

FIGURE 27.4 (A) In nanografting, the force of the AFM tip removes a portion of the surface SAM in the presence of the alternative solution phase SAM, which uses thiol linkages to bind to the exposed gold. (B) A 33 × 33, biotin nanoarray produced via nanografting. (From Liu, M. et al., *Annu. Rev. Phys. Chem.*, 59, 367, 2008. With permission.)

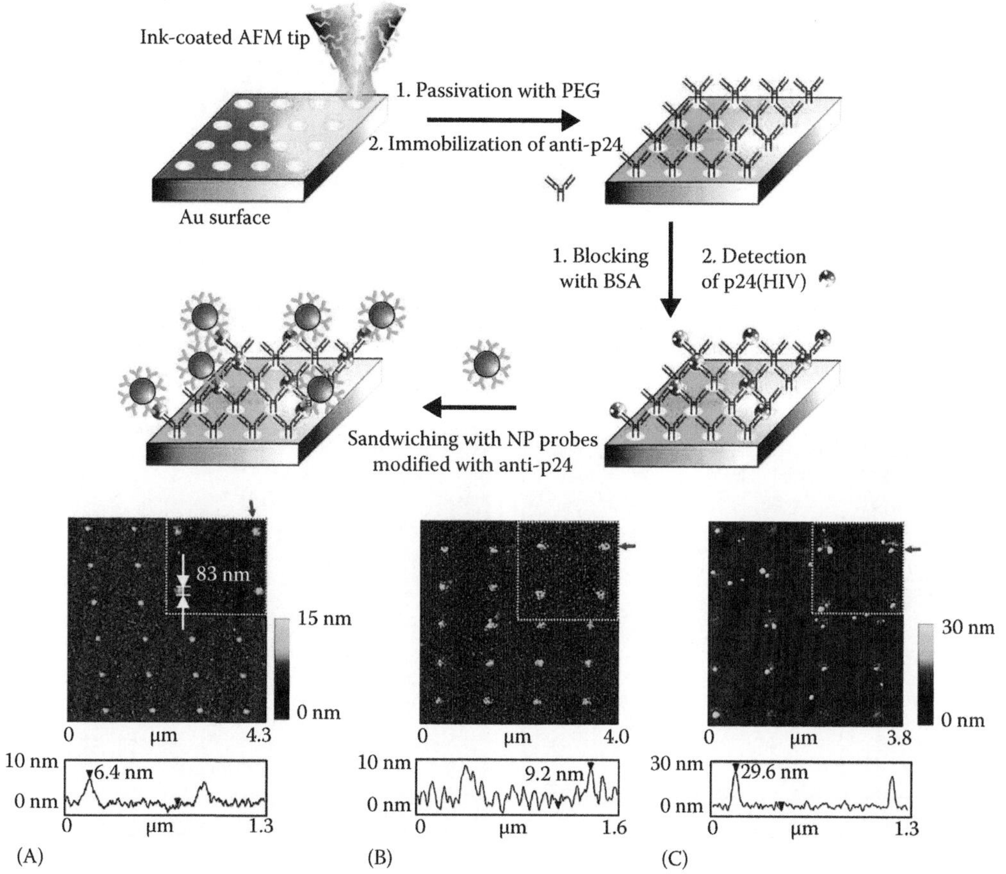

FIGURE 27.6 Schematic of the construction of an HIV-detection nanoarray using DPN. (A) An AFM height profile of the HIV-antibody array, showing an average feature height of 6.4 nm. (B) Upon exposure to HIV, the feature heights rise to 9.2 nm. (C) To amplify the signal, gold nanoparticles functionalized with HIV antibodies were exposed to the nanoarray and bound to the immobilized viruses, increasing the feature height to 29.6 nm. (From Lee, K.B. et al., *Nano Lett.*, 4, 1869, 2004. With permission.)

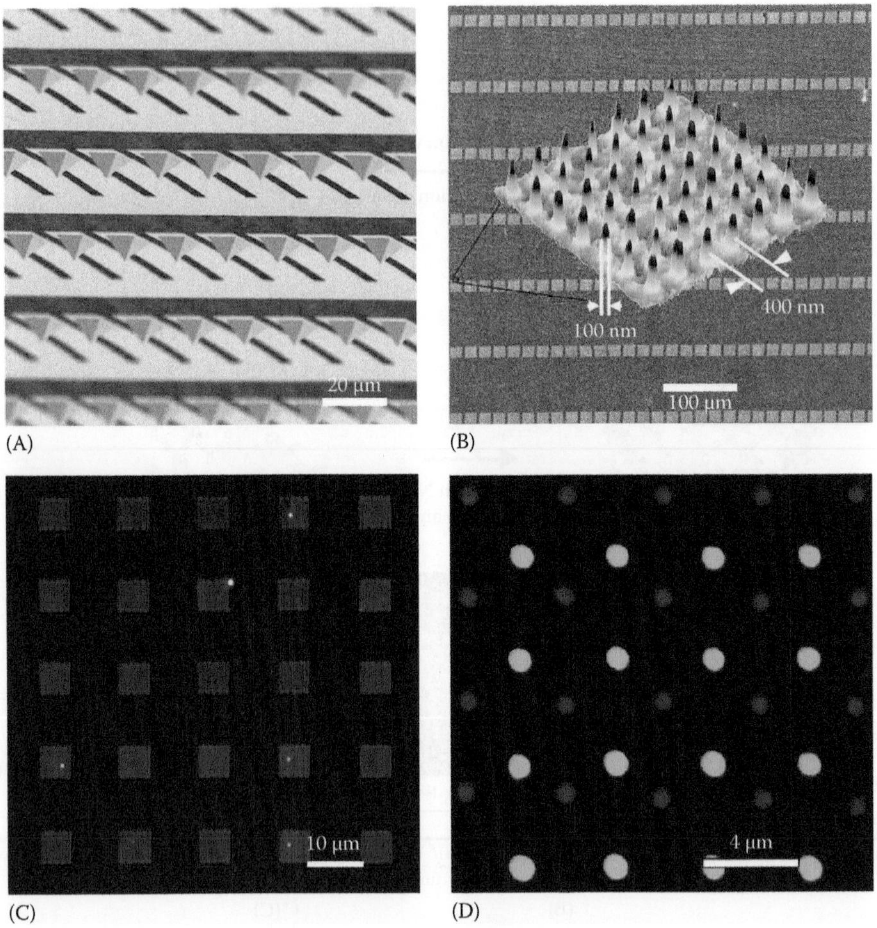

FIGURE 27.9 (A) SEM image of part of a 55,000-pen 2D array. (B) SEM image of part of gold nanostructures created by etching DPN patterns. The inset is an AFM topographic image of part of an array patterned by one single cantilever within the 2D pen array. (With permission from Salaita, K. et al., *Angew. Chem. Int. Ed.*, 45, 7220, 2006.) (C) Fluorescence microscopy of 1 mol% rhodamine-labeled 1,2-dioleoyl-sn-glycero-3-phosphocholine. (With permission from Lenhert, S. et al., *Small*, 3, 71, 2007.) (D) Epifluorescence image of a two sequence oligonucleotide DPN nanoarray on as SiO_x substrate hybridized by fluorophore-labeled sequences (Oregon Green 488-X and Texas Red-X). (From Demers, L.M. et al., *Science*, 296, 1836, 2002. With permission.)

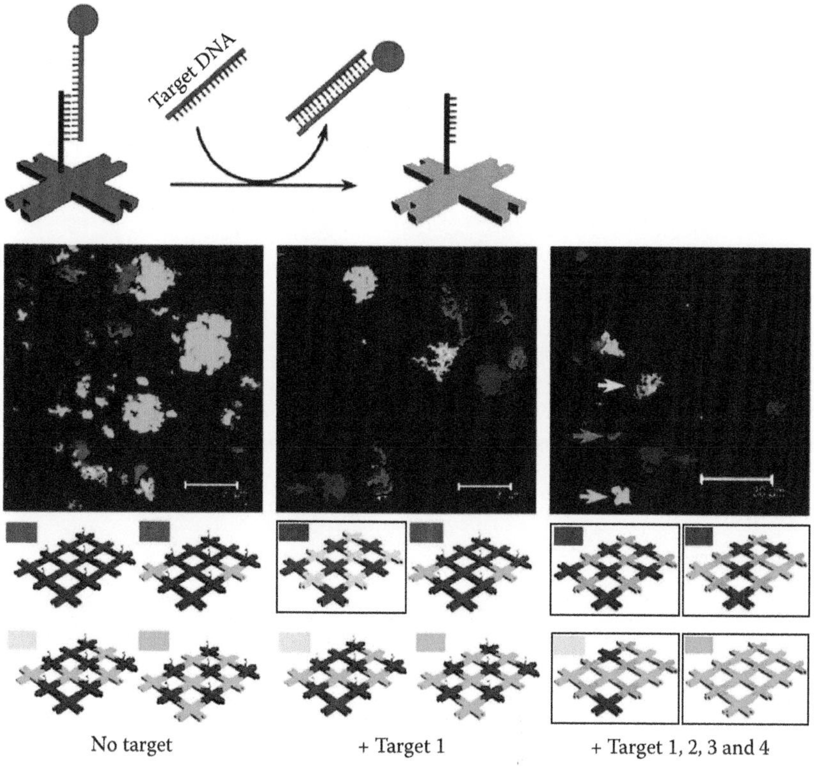

FIGURE 27.10 DNA-tile nanoarray detection upon hybridization with a complementary strand. (From Lin, C.X. et al., *Nano Lett.*, 7, 507, 2007.)

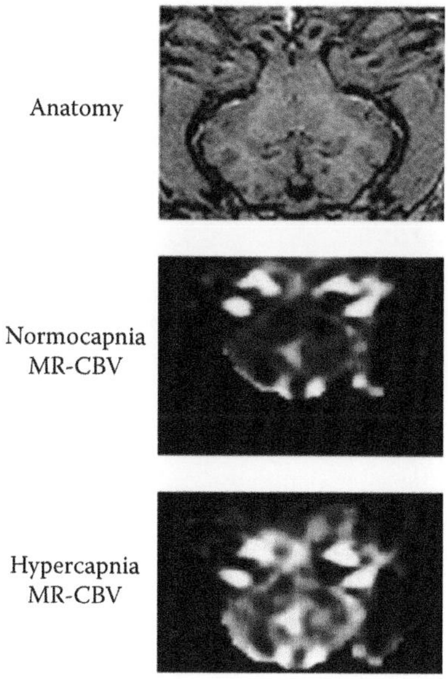

FIGURE 29.3 Cerebral blood volume (CBV) maps obtained in patients after bolus administration of Dy-DTPA-BMA. The maps were obtained by plotting the relative CBV on a pixel by pixel basis within the brain. CBV values were calculated based upon the signal intensity versus time curves obtained during the bolus phase of the CA (technique know as dynamic contrast enhancement [DCE] imaging). Brighter colors (yellow/white) correlate to higher CBV values.

17

Nanobiomaterials for Molecular Imaging

Dian Respati Arifin
The Johns Hopkins University School of Medicine

Jeff Bulte
The Johns Hopkins University School of Medicine

17.1 Introduction to Molecular Imaging

Molecular imaging is an emerging field that can be broadly defined as *in vivo* visualization, characterization, and quantification of biological processes at the cellular and molecular levels (Weissleder and Mahmood 2001; Massoud and Gambhir 2003). Conventional diagnostic imaging mostly measures nonspecific anatomic, physiological, or metabolic changes that mark pathological differentiation of diseased tissues or conditions from healthy tissues. Molecular imaging, on the other hand, aims to probe the cellular and molecular pathways and changes that are the fundaments of diseases.

This field offers a noninvasive tool to investigate the cellular and molecular phenomena inside a living animal, where cellular and molecular studies can be performed in physiological conditions representative of a clinical scenario. Serial studies can be carried out in the same animal without the need to sacrifice animals for *in vitro* analysis by histopathological means, providing continuous and more meaningful data. The development of a noninvasive imaging method to detect and predict the state and progress of diseases has the potential to shift the use of medical imaging from clinical diagnosis to prognosis. Detection of diseases can be achieved at a much earlier time point where it is possible to treat and potentially even cure diseases. In contrast to some current procedures, molecular imaging techniques are noninvasive, therefore significantly reducing patients' discomfort. Likewise, these techniques can be used to evaluate the therapeutic effects, efficacy and/or toxicity of a pharmaceutical agent or a treatment regime.

Molecular imaging typically requires an agent to create signals detectable by the imaging systems. These agents are called contrast agents, molecular probes, tracers, or reporters. Molecular probes can broadly be categorized as either passive or responsive ("smart") probes. Passive tracers facilitate *in vivo* visualization of tissues, cells, or molecules of interest. For smart probes, changes in biochemical or physiological parameters, or the presence of a particular biological species, affect the signals produced by them (Sherry and Woods 2008). Therefore, a "smart" probe reports on the biological activities of the target tissues, cells, and molecules. This chapter aims to summarize the properties, applications, and limitations of different molecular probes used with each imaging modality.

17.2 Magnetic Resonance Imaging

Magnetic resonance imaging (MRI) offers an excellent soft tissue contrast and a spatial resolution close to the size of a single cell. Biological as well as anatomical information can be simultaneously obtained by MRI. MR manipulation is mostly performed on water protons (^1H), taking advantage of the water abundance in human and animal bodies. The most common ^1H MRI protocols are T_1-weighted and T_2 or T_2^*-weighted. MR probes can broadly be categorized as T_2-weighted or negative contrast agents and T_1-weighted or positive contrast agents. Negative agents attenuate MR signals and appear as dark signals or hypointensities. In contrast, positive probes cause enhancement of signals or hyperintensities. Examples of negative and positive contrast MRI are shown in Figure 17.1.

17.2.1 Negative MR Contrast Agents

Superparamagnetic iron oxide-based nanoparticles (SPIOs) are the most common and widely used negative MR contrast agents. SPIOs are composed of nanocrystalline magnetite (Fe_3O_4) or maghemite (γFe_2O_3) cores and typically have a median diameter around 50 nm. Those with a median diameter smaller than 50 nm are called ultrasmall superparamagnetic iron oxide-based nanoparticles (USPIOs). To improve their biocompatibility and colloidal stability, iron oxide nanoparticles are coated with a biocompatible material: dextran (these nanoparticles are called ferumoxides), carboxydextran (ferrixans or ferucarbotrans), or silicone (ferumoxsils). Feridex® (AMAG Pharmaceutical Inc., this product is now discontinued) is an FDA-approved ferumoxides and is being sold in Europe as Endorem™. GastroMARK® (AMAG Pharmaceutical Inc.) and Lumirem® (European brand name) are commercial ferumoxsils, while Resovist® (Bayer Schering Pharma AG) is a ferucarbotran approved for the European market.

Inside the body, iron oxide nanoparticles are taken up by the reticuloendothelial system of the liver (Kupffer cells) and the spleen (macrophages). These nanoparticles biodegrade into their coating and their iron core, with the latter subsequently entering the normal iron metabolic pathway. SPIOs were first formulated for noninvasive MR detection and visualization of hepatic tumors. Hepatic tumor cells do not take up injected SPIOs, whereas Kupffer cells in the normal liver do, resulting in MR-signal differences between healthy and cancerous cells.

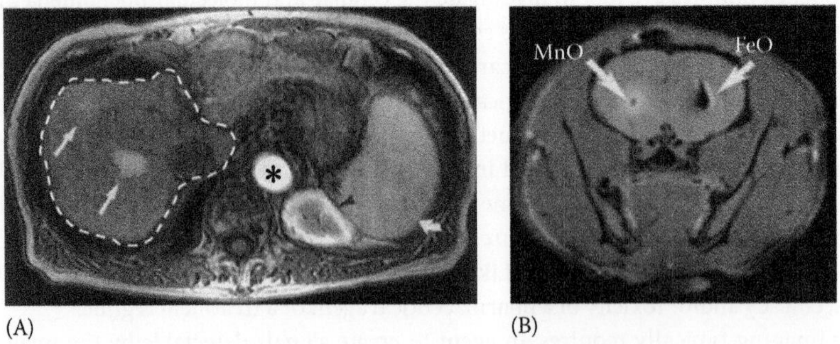

FIGURE 17.1 (A) 1.5T MR image of a patient with Gd-DTPA accumulated in hepatocellular carcinomas (straight arrows, dashed area is liver), appearing as hyperintensities. (From Kanematsu, M. et al., *Radiology*, 225 (2), 407, 2002.) (B) 9.4T MR image of a rat brain injected with manganese oxide and iron oxide nanoparticles-labeled cells that are visible as hyper- and hypointensities, respectively. (From Gilad, A.A., et al. *Magn. Resonance Med.*, 60 (1), 1, 2008.)

(A) (B) (C)

FIGURE 17.2 (A) Microscopic image of an alginate microcapsule containing one human pancreatic islet mixed with Feridex® Scale bar = 150 μm. (B,C) MR images of microcapsules before (B) and after (C) rupture using glass bead agitation. The signal intensity of ruptured capsules is weaker due to the release of nanoparticles. (From Barnett, B.P. et al., *Nature Med.*, 13 (8), 986, 2007.)

Currently, transplantation of therapeutic cells is performed "blindly." Cells may be injected into the wrong site and/or never reach the intended target site. To allay these potential problems, cells can be labeled with SPIOs before engraftment. Hence, the *in vivo* location, migration, and cell–tissue interaction of transplanted cells can be followed and studied in real time by MRI (Figure 17.1B). It is estimated that only 1.4–3 pg of iron per cell is required for the visualization of a single cell using a clinical MRI setup (Walter et al. 2007). Cells can be labeled by prolonged incubation with SPIOs, i.e., via phagocytosis (Bulte and Kraitchman 2004; Kraitchman and Bulte 2009), using an electroporation technique (Walczak et al. 2005) or using polycationic transfection agents, such as poly-L-lysine (Frank et al. 2003) or protamine sulfate (Arbab et al. 2006). USPIOs are taken up by macrophages following intravenous injections, and can thus be used as an endogenous macrophage labeling probe (Bulte and Kraitchman 2004).

A relatively new approach is to encapsulate cells and SPIOs inside immunoprotective alginate microcapsules (Figure 17.2A) (Barnett et al. 2007). An advantage here is that a high concentration of SPIOs can be confined inside the capsules, providing a stronger hypointense MR signal. Ruptured capsules exhibit weaker signals due to the release of nanoparticles (Figure 17.2B and C). However, this strategy only reports on the integrity of the capsules but not on the viability and functionality of the encapsulated cells.

New kinds of negative contrast agents are continuously being discovered and developed. A major improvement on iron oxide nanoparticles is the so-called magnetism-engineered iron oxide (MEIO) nanoprobes or metal-doped Fe_2O_4 nanoparticles (Lee et al. 2007) with enhanced MR-sensitivity (lower concentration offers sufficient MR-contrast) compared to regular SPIOs. The use of ferritin, a natural globular protein complex and the main intracellular iron storage vehicle, has also been explored. Free Fe^{3+} ions are toxic, but inside the ferritin shell, Fe^{3+} ions form crystallites with phosphate and hydroxides. Wild-type ferritin is a poor negative contrast agent. To improve MR-sensitivity, ferritin aggregates were fabricated by the conjugation of ferritin to actin, followed by the polymerization of actin (Bennett et al. 2008). Bouchard et al. (2009) created cobalt nanoparticles coated with bioinert gold that are seven times more sensitive than SPIOs. Cobalt nanoparticles were an unattractive choice due to their oxidation-induced instability and toxicity, but these issues could be circumvented with gold coating.

17.2.2 Positive MR Contrast Agents

Hyperintense MR signals are easier to be detected *in vivo* than hypointense signals due to the endogenous hypointense background in T_2 or T_2^*-weighted images. The interpretation of these hypointensities may be ambiguous and can have a physiological origin, such as iron-containing hemoglobin in blood, or pathological origin, such as blood clots due to trauma. Paramagnetic gadolinium (Gd) is the most effective positive contrast MR agent due to its seven unpaired electrons, but free Gd^{3+} ions are toxic and must be chelated for clinical and biomedical applications. Diethylene triamine pentaacetic acid (DTPA), tetraazacyclododecane tetraacetic acid (DOTA), and their derivatives are examples of clinically used Gd-chelates. A Gd-DTPA enhanced clinical MRI exam is shown in Figure 17.1A.

Gd-chelates can easily be inserted into cells through pinocytosis but their MR sensitivity is much lower than iron oxide nanoparticles. A myriad of strategies has been explored to increase the sensitivity of this contrast agent. Multiple Gd-chelates were bound inside a fullerene cage (Shu et al. 2006; MacFarland et al. 2008) or into a dendrimer (Bryant et al. 1999) or inside a nanotube (Sitharaman and Wilson 2007). Gd-chelates were encapsulated inside vesicles that can be further functionalized (Mulder et al. 2006; Sofou and Sgouros 2008), such as liposomes (Mulder et al. 2006) or porous polymeric shells (Cheng and Tsourkas 2008). Gd-based nanoparticles have also been fabricated, for example, nanoparticles coated or embedded with gadolinium (Gerion et al. 2007; Zhu et al. 2008) and gadolinium oxide (Gd_2O_3) nanocrystals capped with a biopolymer (Engstrom et al. 2006). Multiple Gd^{3+} ions were chelated on the surface of gold nanoparticles to create MR and X-ray compatible agents (Alric et al. 2008). Gd-agents can also be encapsulated inside immunoprotective capsules.

Another strong positive contrast MR agent is manganese ions. Manganese salt ($MnCl_2$) crystals (Aoki et al. 2006), manganese carbonate ($MnCO_3$) (Shapiro and Koretsky 2008), and manganese oxide (MnO, MnO_2, and Mn_3O_4) nanoparticles (Gilad et al. 2008; Shapiro and Koretsky 2008) were formulated for cell labeling applications. Intact particles are detected by T_2^*-weighted MRI. However, they dissociate into manganese ions upon internalization by cells, producing hyperintense signals (Figure 17.1B). Similar to Gd-based agents, the biocompatibility of these compounds is a major issue and currently under investigation. When combined with negative contrast agents, positive contrast agents offer a method of MR "double labeling" of two different cell populations.

17.2.3 Perfluorocarbons for ^{19}F MR Imaging

Perfluorocarbons (PFC) are derived from hydrocarbons by replacing hydrogen atoms with fluorine atoms. These compounds are bioinert, and chemically and thermally stable. Their fluorine nuclei can be detected by ^{19}F MRI. ^{19}F MRI cannot provide anatomic information since an abundance of fluorine is not naturally present in a living body and therefore its use is always combined with ^{1}H MRI. The use of PFCs offers the possibility of "hot spot" MR-imaging, analogous to nuclear imaging (exemplified in Figure 17.3A). PFCs also function as X-ray and ultrasound agents, an advantageous feature, which will be discussed in the later sections. PFCs currently being used for molecular imaging applications are perfluorooctyl bromide (PFOB) and perfluoro-15-crown-5-ether (PFCE). These compounds are liquid

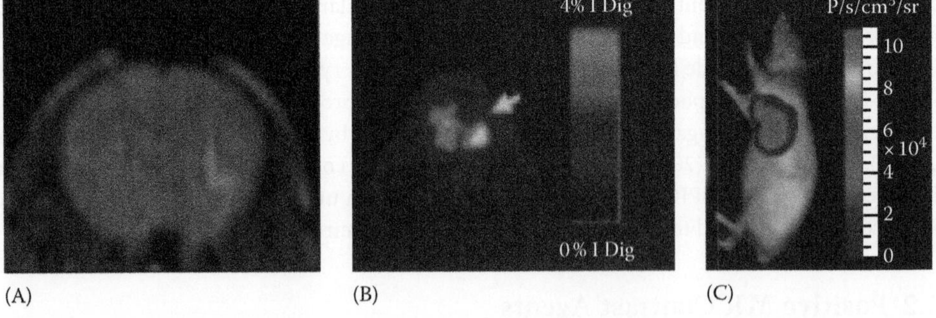

FIGURE 17.3 (See color insert.) (A) 9.4 T MR image of a mouse brain 3 days post-engraftment of PFC-labeled neural stem cells, with the ^{19}F signal superimposed on the ^{1}H MR image. The cells labeled with fluorinated emulsions appear as "hot spots." (From Ruiz-Cabello, J. et al., *Magn. Resonance Med.*, 60 (6), 1506, 2008.) (B,C) PET (B) and bioluminescent (C) images of a mouse injected with stem cells carrying thymidine kinase (PET) and luciferase gene 4 weeks post-injection. (Courtesy of Dr. Joseph C. Wu, Stanford University, Stanford, CA.) Cells were visible as "hot spots" on the PET and BL images.

at room temperature and have to be emulsified in order to be used as a contrast agent. Nanoscale emulsions of PFCs are prepared with surfactants, the most common being phospholipids, to maintain their colloidal stability.

Neural stem cells were directly labeled with cationic PFCE-phospholipid emulsions without the use of transfection agents (Ruiz-Cabello et al. 2008). The authors reported an MR-sensitivity of approximately 140 pmol of PFCE per cell at 9.4 T (Figure 17.3A), and no observable detrimental effects on the viability and proliferation of stem cells. Emulsions of PFOB were co-encapsulated with mesenchymal stem cells inside immunoprotective alginate microcapsules to create ^{19}F MR, X-ray, and ultrasound-trackable capsules (Fu et al. 2009). Oxygen dissolves in PFOB in a linear relationship, hence co-encapsulated PFOB emulsions also function as an oxygen carrier for the encapsulated cells. The surface of PFC emulsions can be conjugated with paramagnetic Gd-chelates (Winter et al. 2007) or with targeting-ligands (Schmieder et al. 2005) to improve MR imaging. To date, PFCs have not been used in clinics but have shown promising results in animal models.

17.2.4 Chemical Exchange Saturation Transfer Imaging

Chemical exchange saturation transfer (CEST) imaging is an emerging MRI technique that detects signals arising from the proton exchange between the imaging target and the surrounding water molecules. Some endogenous species can directly be detected by CEST imaging without the need of contrast agents, for example, the hydroxyl protons of glycogen and the amide protons in brain tissues. CEST imaging has been used for real-time, noninvasive studies of glycogen metabolism in a mouse liver (van Zijl et al. 2007) and ischemic regions in a rat brain (Zhou et al. 2003).

Recent work on exogeneous CEST agents includes diamagnetic peptide-based (DIACEST) agents and paramagnetic CEST (PARACEST) agents. A lysine-rich protein (LRP) reporter (Gilad et al. 2007), a prototype CEST reporter gene, was developed for transfection into cells. Lysine is a peptide rich in exchangeable amide protons and a strong CEST agent. This reporter presents a biocompatible and biodegradable agent that provides a constant endogenous CEST signal even after cell proliferation. Since LRP is strongly pH-dependent, the contrast is lowered by an order of magnitude during ischemia or cell death, offering a noninvasive method to monitor cell viability *in vivo* (Gilad et al. 2007). The CEST potential of a library of prototype polypeptides with exchangeable amine, amide, and hydroxyl protons was further investigated (McMahon et al. 2008). These three exchangeable groups have different chemical shifts and therefore different "colors" can be assigned to each of them, presenting a possibility of multiple labeling and simultaneous imaging of multiple targets.

Paramagnetic CEST (PARACEST) agents are complexes of paramagnetic lanthanides, such as Eu^{3+}, Ho^{3+}, Dy^{3+}, Tb^{3+}, Tm^{3+}, Pr^{3+}, and Yb^{3+}. The exchange rate of these agents increases with increasing temperature, therefore PARACEST agents have the potential for detecting temperature changes (Zhang et al. 2002, 2005) in addition to passive CEST imaging. Moreover Pr- and YbDOTA-tetraamide complexes are sensitive to pH changes (Aime et al. 2002; Terreno et al. 2004). However, lanthanide ions are toxic, therefore their chelated forms must have a high *in vivo* stability and a safe route of clearance from the body. Zhang et al. (2003) predicted that only a few lanthanides may function as PARACEST agents in a low magnetic field. Since clinical MRI is performed at 1.5–3 T magnetic field strength, this will narrow the use of most PARACEST agents to research applications. DIACEST and PARACEST agents can be functionalized for targeting a tissue, cell, or molecule of interest.

A current drawback of CEST agents is its relatively low sensitivity, requiring a concentration in the range of 1–10 mM (Sherry and Woods 2008). Sensitivity can be enhanced by increasing the number of exchangeable groups on a probe or by assembling multiple probes in/on a substrate. Multiple Eu^{3+} complexes were conjugated on the surface of PFC-filled nanoparticles, resulting in a PARACEST and ^{19}F MR agent (Schmieder et al. 2005). Complexes were easily encapsulated in the aqueous cores of liposomes (LIPOCEST) (Aime et al. 2005). Liposomes are an attractive option due to their biocompatibility and potential for surface-functionalization: with targeting ligands, with poly(ethylene glycol) to prolong *in vivo* circulation life, and/or with an extra contrast-generating agent such as Gd chelates (Torchilin 2005; Zheng et al. 2006). Moreover, the exchange of water molecules across the phospholipid membranes can be manipulated by varying the membrane composition (Sherry and Woods 2008). Terreno et al. (2009) improved the sensitivity by creating osmotically shrunken LIPOCESTs. Other inventions include polypeptides rich in exchangeable protons (McMahon et al. 2008) and multiple agents bound to a dendrimer (Snoussi et al. 2003).

17.3 Nuclear Imaging

Positron emission tomography (PET) is a nuclear imaging system that detects pairs of gamma rays emitted by positron-emitting radionuclides. Common positron-emitting radionuclides are ^{18}F, ^{15}O, ^{13}N, and ^{11}C. Less common ones are ^{64}Cu, ^{62}Cu, ^{14}O, ^{124}I, ^{76}Br, ^{68}Ga, and ^{82}Rb. Single photon emission computed tomography (SPECT) utilizes a gamma camera that rotates around the subject and directly measured the gamma rays emitted by the radionuclides. Some SPECT radionuclides are ^{111}In, ^{123}I, ^{131}I and ^{99m}Tc. SPECT agents typically have longer half-lives than PET radionuclides. Most radionuclides are produced in a cyclotron and a few of them by a generator. Radionuclide production sites should be in close proximity to the hospitals since the half-lives of some radionuclides are short, for example, the half-life of ^{18}F is 110 min. ^{99m}Tc is the most clinically used agent since it is a gamma-emitter of low energy, has a moderate half-life (6 h), and a ^{99}Mo/^{99m}Tc generator that is relatively inexpensive.

The most attractive feature of nuclear imaging is the small size of radiotracers compared to the probes of other imaging techniques. Atoms in molecules, proteins, or drugs can be easily substituted with radionuclides without perturbing their native properties. For example, hydrogen atoms can readily be substituted with ^{18}F. Nuclear imaging allows the quantification of injected radiotracers, a technique that can be translated to the quantification of therapeutic agents. This modality has an excellent sensitivity, and agents in a mere nanomolar concentration can be detected. Similar to ^{19}F MRI, nuclear imaging has to be combined with ^{1}H MRI or X-ray imaging to obtain both metabolic and anatomical information. "Hot spot" PET imaging is shown in Figure 17.3B. However, it is unsuitable for repeated and extensive imaging due to the short half-lives of radionuclides. ^{111}In for SPECT and ^{64}Cu for PET have relatively longer half-lives (2.8 days and 12.7 h, respectively) compared to other radioligands. Cells were labeled with these agents via direct incubation. In the case of ^{111}In-labeled cells, serial tracking by SPECT was attainable for only 5–7 days (Kraitchman and Bulte 2009). Radionuclides are potentially radiotoxic, therefore these agents are clinically used in a conservative manner or avoided if possible.

Cells can be transduced with a reporter gene for nuclear imaging applications. Transduced cells produce a particular substance that accumulates and can be detected by the administration of a reporter probe. The main advantage is that only viable cells are detectable, providing a way to track viable versus nonviable cells. Herpes simplex virus 1 thymidine kinase (HSV-Tk) is the best known reporter gene for nuclear imaging (Cao et al. 2006). Prodrugs (ganciclovir, penciclovir, or fialuridine) labeled with radioactive ^{18}F or ^{124}I easily penetrate cell membranes. Inside transduced cells, prodrugs are phosphorylated by the viral thymidine kinase and are trapped. This accumulation of radioactivity allows the monitoring of viable, engrafted cells inside the host. Another example is sodium-iodide symporter (NIS) labeled with radioactive iodine and ^{99m}Tc-pertechnetate (Kang et al. 2005).

To improve the payload of radiotracers, radiocolloids and liposome-based radiotracers have been fabricated. Radiocolloids are colloidal aggregates consisting of or containing radionuclides. An example of these is ^{99m}Tc-sulphur colloids that have been utilized to image lymph nodes in oncology patients and bone marrows in patients with hematological and muscoskeletal diseases (Moffat and Gulec 2007).

Liposomes are biocompatible, can be surface-functionalized, and shield radionuclides from potentially destabilizing external environment. Radiolabeled liposomes were formulated with various methods but the following two yield the best labeling efficiency and *in vivo* stability: (1) the incorporation of radiolabeled amphiphilic chelators into the bilayers of preformed liposomes (Laverman et al. 1999); and (2) the after-loading method: radionuclides chelated with lipophilic molecules are shuttled across the bilayers of liposomes, which encapsulate chelators with higher affinity for the radionuclides, thereby trapping the radionuclides inside the liposomes (Bao et al. 2003). A variation of the first method is the use of polychelating polymers containing a hydrophobic, phospholipid fragment to amplify radionuclide loading (Torchilin 2000; Erdogan et al. 2006). Phospholipids with saturated long acyl chains and cholesterol are chosen as the membrane composition to improve the stability and rigidity of bilayers, therefore reducing the leakage or dissociation of radionuclides. Liposomes are removed by the reticuloendothelical system in the liver and spleen. Hence, radiolabeled liposomes can be used to identify and visualize hepatic and splenic tumors in a fashion similar to iron oxide nanoparticles for MRI. A few clinical studies on radioloabeled liposomes have been completed with promising results.

17.4 Optical Imaging

Optical imaging typically requires an external light source to excite the fluorophores or optical agents. An ideal fluorescence-based probe should have the following properties: high photo- and chemical stability, sufficient biocompatibility, emission in the near-infrared range (650–900 nm where tissue adsorption is low), high quantum yield, and a chemical matrix for multifunctionalization (Bremer et al. 2003; Sharma et al. 2006).

The most commonly used fluorophores for passive imaging are organic fluorescent dyes, such as fluorescein isothiocyanate (FITC) and derivatives of rhodamine. Targeting-ligands can be conjugated to these dyes. Some dyes are currently used in clinical screening (Pierce et al. 2008). However, they undergo rapid photo-bleaching, are detectable only at a superficial tissue depth, and are not well suited for simultaneous multicolor imaging of different target populations. Some dyes have even been reported to adversely affect cell viability and proliferation (Modo 2008). New generations of dyes such as Alexa Fluors® (Molecular Probes) and DyLight Fluors™ (Thermo Fisher Scientific) are more photo-stable, brighter, and less pH-sensitive.

Fluorescent nanoparticles are an attractive alternative since their signal intensity can be stronger than organic dyes. Silica nanoparticles doped with fluorescent dyes suffer from the drawbacks mentioned above (Veiseh et al. 2005). Porous silicon nanoparticles are naturally luminescent but they biodegrade in less than a day and hence are not suitable for extended *in vivo* imaging. On the other hand, these nanoparticles are appealing for the image guided-delivery of therapeutic agents (Canham 1990; Park et al. 2009). The potential of gold nanoparticles for fluorescence imaging is currently being explored (Medley et al. 2008).

Quantum dots (QDs) are comprised of a semiconductor core, capped by a coating of a second semiconductor material, the most common being CdSe/ZnS core/shell. The core/shell structures are coated with silica, biopolymers, amphiphilic, or hydrophilic ligands, for biocompatibility and solubility, followed by conjugation with functional molecules. QDs are brighter and more resistant to photo-bleaching than standard dyes, and have tuneable optical properties for multiplexed imaging and/or *in vivo* imaging of deep tissues (Zimmer et al. 2006; Pierce et al. 2008; Walling et al. 2009). The size of QDs is typically 2–10 nm. The core of QDs is toxic and a careful design is pertinent to ensure their *in vivo* biocompatibility and stability. Moreover, the fluorescence signals of QDs display intermittent on/off behavior, which can complicate measurements (Pierce et al. 2008; Walling et al. 2009).

The following two agents produce signals only when the cells are viable, providing means to monitor dead versus live cells *in vivo*. When the genes encoding for these proteins are inserted into the DNA, proteins will be continuously produced and not diluted by cell proliferation and therefore can be used to study cell lineages *in vivo*. The first one is green fluorescent protein (GFP), extracted from bioluminescent jellyfish *Aequorea Victoria*. A library of mutagenized fluorescent proteins with excitation wavelengths in the 350–630 nm range and emission wavelengths in the 450–650 nm range was developed for multicolor labeling (Hadjantonakis et al. 2003). However, these proteins have a shallow penetration depth since their emission is in the visible light range, which is highly adsorbed by the tissues (Bremer et al. 2003).

Bioluminescence imaging does not require an external source of light. Molecules or cells of interest are engineered to carry the enzyme luciferase. Luciferase genes have been cloned from a variety of sources: fireflies, coral, jellyfish, and bacteria. When the substrate luciferin is introduced, luciferase converts luciferin to oxyluciferin with the simultaneous emission of light. An example of *in vivo* bioluminescence imaging is shown in Figure 17.3C. However, bioluminescent signals are attenuated by tissue thickness due to nonhomogeneous scattering (Modo 2008). The use of this technique is limited to research applications since luciferin is unlikely to be administered to patients (Bremer et al. 2003).

17.5 X-Ray and Ultrasound Imaging

X-ray or computed tomography (CT) and ultrasound imaging systems are traditionally not considered sensitive enough for molecular imaging applications. However, recent development in X-ray and ultrasound probes has opened the possibility of using these modalities for molecular imaging in the near future.

Radiopaque iodine-based agents such as iohexol, iopromide, iomeprol, or iodinated oil droplets were encapsulated inside liposomes (Zheng et al. 2006) or polymeric micelles (Trubetskoy et al. 1997) to increase their sensitivity. Silica nanoparticles carrying electron-dense metal ions such as ruthenium and gadolinium (Santra et al. 2005) are a potential CT-probe. Gold nanoparticles are radiopaque with better sensitivity than iodine-based agents, bioinert, and their fabrication method is simple (Figure 17.4A and B) (Hainfeld et al. 2006). PFCs, particularly PFOB, are radiopaque, bioinert, and can multifunction as an oxygen carrier (Sanchez et al. 1995). Arifin et al. (unpublished results) fabricated a new type of immunoprotective Ba^{2+}-gelled alginate microcapsules with inherent radiopacity. Hence, these capsules can potentially be imaged by CT without co-encapsulation of a contrast agent.

Microbubbles are biocompatible shells containing acoustically trackable gases with PFC gases as the most ideal agent. The typical diameter is 1–3 μm. The shells maintain the colloidal stability of the cores and should accommodate acoustic contraction and expansion of the cores, therefore flexible phospholipids (liposomes) are the best choice. The surface can be further functionalized (Klibanov 2007). Microbubbles can be destroyed *in vivo* with high ultrasound energies, opening the potential for targeted-drug and gene delivery (Mayer et al. 2008). A few microbubbles have been approved for clinical use (Figure 17.4C) (Blomley et al. 2001; Feinstein 2004). Alginate microcapsules were found to have an acoustic property similar to microbubbles and empty microcapsules appeared in an ultrasound image (Arifin et al. 2011). The signals were significantly enhanced with co-encapsulation of contrast agents (Arifin et al. 2011; Fu et al. 2009; Figure 17.5D and E).

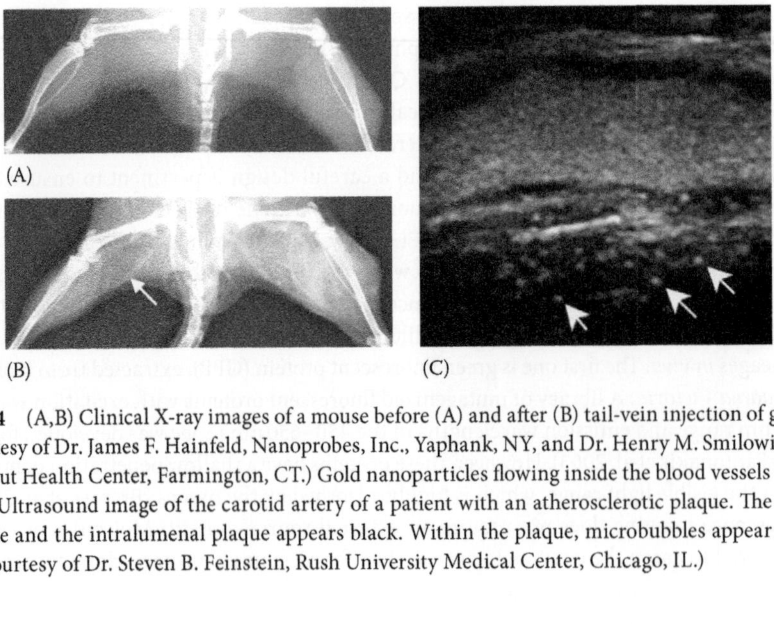

FIGURE 17.4 (A,B) Clinical X-ray images of a mouse before (A) and after (B) tail-vein injection of gold nanoparticles. (Courtesy of Dr. James F. Hainfeld, Nanoprobes, Inc., Yaphank, NY, and Dr. Henry M. Smilowitz, University of Connecticut Health Center, Farmington, CT.) Gold nanoparticles flowing inside the blood vessels appear bright (arrow). (C) Ultrasound image of the carotid artery of a patient with an atherosclerotic plaque. The artery lumen appears white and the intralumenal plaque appears black. Within the plaque, microbubbles appear as white dots (arrows). (Courtesy of Dr. Steven B. Feinstein, Rush University Medical Center, Chicago, IL.)

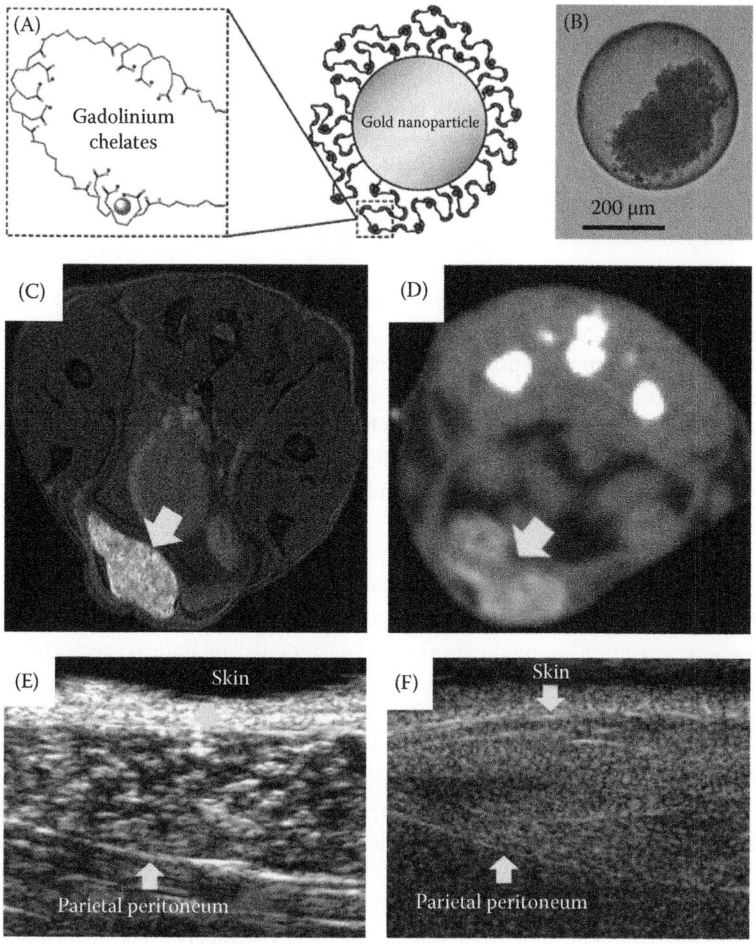

FIGURE 17.5 (A) Schematic of gold nanoparticles with Gd^{3+} ions chelated on the surface (GadoGold) (Alric et al. 2008). (B) GadoGold microcapsule containing one human pancreatic islet. (C–E) 9.4 T positive contrast MRI (C), micro-CT (D), and 40 MHz ultrasound (E) images of GadoGold microcapsules following subcutaneous injection in mouse abdomen. (F) Saline injection as negative U.S. image control. (From Arifin, D.R., et al. Novel trimodal gadolinium-gold microcapsules for simultaneous immunoprotection and positive contrast MRI, x-ray and ultrasound imaging of human pancreatic islet cells. 2011.) Arrows indicate capsules.

17.6 Multimodal Imaging Probes

A multimodal imaging probe combines the strengths and applications of multiple imaging techniques. Moreover, more complete information on a target can be gathered from each imaging modality. PFCs are a natural multimodal imaging probe and are detectable by ^{19}F MR and X-ray imaging. Nanoparticles or nanoshells are a convenient platform for creating a multimodal contrast agent. The core can contain an agent visible to one system, while the surface can be decorated with a second agent visible with the second imaging system. Encapsulation of more than one agent is a possibility as long as a quenching effect does not exist. Liposomes, silica nanoparticles, or a combination of both are popular choices for the shells/cores. Examples of these are silica nanoparticles doped with fluorescent dyes (Veiseh et al. 2005) or quantum dots (Koole et al. 2008) with Gd-chelates on the surface for MRI and optical imaging; gold nanoparticles (GadoGold, Figure 17.5) (Alric et al. 2008) or iodinated liposomes (Zheng et al. 2006) with Gd^{3+} ions incorporated on the surface for X-ray and MR imaging; PFC or iodinated cores surface-decorated with Gd-chelates for ^{19}F and ^{1}H MRI and X-ray imaging (Winter et al. 2007); liposomes encapsulating radionuclides and Gd-chelates for nuclear and MR imaging (Zielhuis et al. 2006).

Santra et al. (2005) fabricated a trimodal imaging probe: silica nanoparticles doped with organometal Ru(bpy) and surface-conjugated with Gd-chelates, for MR, X-ray and optical imaging. An ingenious use of bimodal imaging was demonstrated by Langereis et al. (2009), who created liposomes containing Tm^{3+}-complexes (CEST) and NH_4PF_6 (^{19}F) for image guided-drug delivery. Intact liposomes were trackable only by CEST imaging due to the quenching of the ^{19}F effect. When the liposomes were ruptured, resulting in the release of drugs, the CEST effect vanished and ^{19}F agents were detected.

Cells can be designed to carry more than one probe for the multimodal imaging of engrafted cells. For example, stem cells were engineered to carry thymidine kinase and luciferase genes (Cao et al. 2006) for PET and bioluminescence imaging, respectively (Figure 17.3B and C); cells were "housed" inside immunoprotective microcapsules with a multimodal probe (Arifin et al. 2011; Fu et al. 2009). Encapsulating these probes inside acoustically reflective alginate microcapsules may impart an ultrasound-visibility. The use of Ba^{2+}/alginate microcapsules with an intrinsic radiopacity (Arifin et al. 2011) may amplify or add CT-visibility. GadoGold microcapsules were readily imaged by MR, X-ray, and ultrasound imaging (Figure 17.5) (Arifin et al. 2011), while PFOB-loaded capsules were visible in X-ray, ^{19}F MRI, and ultrasound images (Fu et al. 2009).

17.7 "Smart" Contrast Agents

The signals of "smart" probes are affected by changes in biochemical or physiological parameters (pH, temperature, ion concentration) or by the presence of a particular biological species (enzyme, glucose). Since a myriad of responsive agents is currently under development, this segment aims to present a few examples of "smart" agent designs.

A "smart" probe can differentiate between viable and nonviable cells by (1) direct measurement where only viable cells exhibit signals, for example, thymidine kinase transduced cells (PET or SPECT) (Cao et al. 2006) and GFP- (fluorescence) or luciferase-carrying cells (bioiluminescence) (Bremer et al. 2003); (2) the indirect measurement of pH since dead cells release their acidic cellular contents, for example, cells transfected with LRP (CEST) (Gilad et al. 2007). Urano et al. (2009) developed a probe based on a boron-dipyrromethene fluorophore, which is highly fluorescent only in the protonated form, i.e., in an intracellular acidic environment. This probe was conjugated to a cancer-targeting ligand for internalization by viable cells and hence for the detection of viable cancer cells.

DIACEST agents (Gilad et al. 2007; McMahon et al. 2008) and some PARACEST agents, such as Pr- and YbDOTA-tetraamide complexes (Aime et al. 2002; Terreno et al. 2004), can be used as a pH-sensor. Positive contrast MR-properties of gadofullerene derivatives depend on their aggregation state, which in turn depends on the local pH (Toth et al. 2005). PARACEST agents (Zhang et al. 2002, 2005) and quantum dots (Ngo et al. 2009) show potential for detecting temperature changes.

In the case of ion, protein, or enzyme-sensing probes, interaction with target ions/protein or enzymatic activity alters the pathways for signal production and therefore allows detection of the target. The simplest enzyme-indicator is multiple fluorescent dyes grafted on a biopolymer. The fluorescence is quenched when the dyes are in close vicinity. Enzymes cleave the polymer, releasing dyes which become fluorescent (Klohs et al. 2008). An example of the mechanism of enzyme-detection is depicted in Figure 17.6A. In a Gd-based, Zn^{2+}-responsive agent, the binding of Zn^{2+} ions opens the access of water protons to chelated Gd^{3+} ions, eliciting an MR-positive contrast (Major et al. 2007). Louie et al. (2000) designed a Gd-based agent in which the access of water protons is blocked with a substrate that can be removed by marker enzyme galactosidase. An agent composed of two GFP-variants linked by a kinase-inducible domain (KID) is used to fluorescently detect protein phosphorylation by cAMP-dependent protein kinase A (PKA). The phosphorylation of KID by PKA converts the fluorescence emission of the probe (Nagai et al. 2000). A MR-counterpart of this probe is ferritin-based nanoparticles, which aggregate in the presence of PKA, thereby altering their MRI contrast (Shapiro et al. 2009). Ren et al. (2008) fabricated a glucose-sensor: Eu^{3+}-chelated by a macrocyclic ligand containing a glucose recognition site.

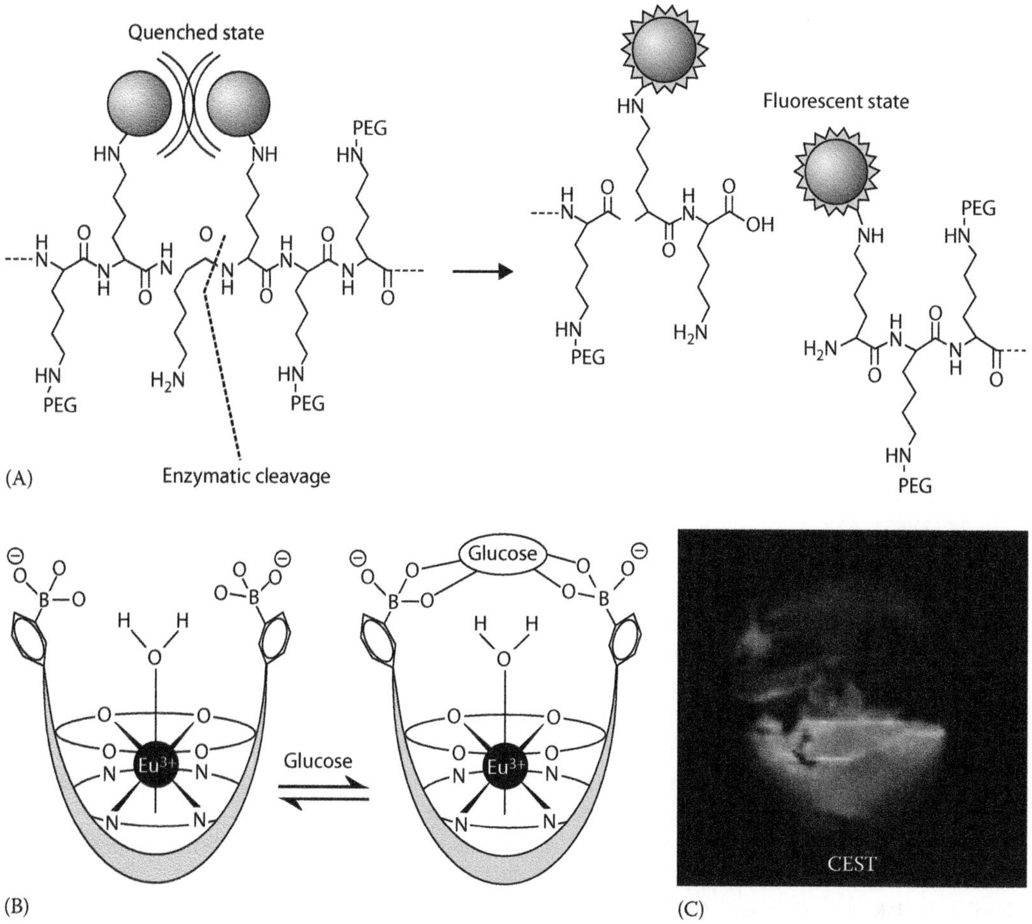

FIGURE 17.6 (A) Multiple fluorescent dyes grafted on a biopolymer as an enzyme-detector. The fluorescence is quenched when the dyes are in close vicinity. Enzymes cleave the polymer, releasing dyes which then become fluorescent. (From Klohs, J. et al., *Basic Res. Cardiol.*, 103 (2), 144, 2008.) (B) The proposed binding mechanism of glucose with Eu^{3+}-based PARACEST agent. (C) *In vitro* CEST images of a mouse liver perfused with PARACEST agents in the absence (top liver) and presence (bottom liver) of glucose. (From Ren, J.M. et al., *Magn. Reson. Med.*, 60 (5), 1047, 2008.)

The binding of glucose decreases the water exchange rate of Eu^{3+} ions with bulk water, changing the CEST contrast (Figure 17.6B and C). Likewise, enzymatic cleavage by caspase-3 (Yoo and Pagel 2006) or the presence of Zn^{2+} ions (Trokowski et al. 2005) causes the CEST effect to disappear.

17.8 Conclusion

Molecular imaging typically requires a contrast agent to facilitate visualization. The key points to consider in designing a contrast agent are biocompatibility, design feasibility, detection sensitivity, and specificity for the application. Although a variety of imaging modalities (MR, PET, SPECT, optical, X-ray, and ultrasound imaging) are available for molecular imaging applications, each system has its limitations and offers different information. Therefore, there is currently a high interest in developing a multimodal probe. Another novel design is a responsive or "smart" agent, which is able to provide information on the biological activities of a target in addition to passive imaging. To date, the majority of molecular probes are in their research and development stage, and their clinical suitability is yet to be determined.

References

Aime, S., A. Barge, D. D. Castelli, F. Fedeli, A. Mortillaro, F. U. Nielsen, and E. Terreno. 2002. Paramagnetic lanthanide(III) complexes as pH-sensitive chemical exchange saturation transfer (CEST) contrast agents for MRI applications. *Magnetic Resonance in Medicine* 47 (4):639–648.

Aime, S., D. D. Castelli, and E. Terreno. 2005. Highly sensitive MRI chemical exchange saturation transfer agents using liposomes. *Angewandte Chemie—International Edition* 44 (34):5513–5515.

Alric, C., J. Taleb, G. Le Duc, C. Mandon, C. Billotey, A. Le Meur-Herland, T. Brochard et al. 2008. Gadolinium chelate coated gold nanoparticles as contrast agents for both X-ray computed tomography and magnetic resonance imaging. *Journal of the American Chemical Society* 130 (18):5908–5915.

Aoki, L., T. Yoshiyuki, K. H. Chuang, A. C. Silva, T. Igarashi, C. Tanaka, R. W. Childs, and A. P. Koretsky. 2006. Cell labeling for magnetic resonance imaging with the T-1 agent manganese chloride. *NMR in Biomedicine* 19 (1):50–59.

Arbab, A. S., V. Frenkel, S. D. Pandit, S. A. Anderson, G. T. Yocum, M. Bur, H. M. Khuu, E. J. Read, and J. A. Frank. 2006. Magnetic resonance imaging and confocal microscopy studies of magnetically labeled endothelial progenitor cells trafficking to sites of tumor angiogenesis. *Stem Cells* 24 (3):671–678.

Arifin, D. R., C. M. Long, A. A. Gilad, S. Manek, E. Call, C. Alric, S. Roux et al. 2011. Novel trimodal gadolinium-gold microcapsules for simultaneous immunoprotection and positive contrast MRI, X-ray and ultrasound imaging of human pancreatic islet cells. *Radiology*, in press.

Bao, A., B. Goins, R. Klipper, G. Negrete, M. Mahindaratne, and W. T. Phillips. 2003. A novel liposome radiolabeling method using Tc-99m-"SNS/S" complexes: *In vitro* and *in vivo* evaluation. *Journal of Pharmaceutical Sciences* 92 (9):1893–1904.

Barnett, B. P., A. Arepally, P. V. Karmarkar, D. Qian, W. D. Gilson, P. Walczak, V. Howland et al. 2007. Magnetic resonance-guided, real-time targeted delivery and imaging of magnetocapsules immunoprotecting pancreatic islet cells. *Nature Medicine* 13 (8):986–991.

Bennett, K. M., E. M. Shapiro, C. H. Sotak, and A. P. Koretsky. 2008. Controlled aggregation of ferritin to modulate MRI relaxivity. *Biophysical Journal* 95 (1):342–351.

Blomley, M. J. K., J. C. Cooke, E. C. Unger, M. J. Monaghan, and D. O. Cosgrove. 2001. Science, medicine, and the future—Microbubble contrast agents: A new era in ultrasound. *British Medical Journal* 322 (7296):1222–1225.

Bouchard, L. S., M. S. Anwar, G. L. Liu, B. Hann, Z. H. Xie, J. W. Gray, X. D. Wang, A. Pines, and F. F. Chen. 2009. Picomolar sensitivity MRI and photoacoustic imaging of cobalt nanoparticles. *Proceedings of the National Academy of Sciences of the United States of America* 106 (11):4085–4089.

Bremer, C., V. Ntziachristos, and R. Weissleder. 2003. Optical-based molecular imaging: Contrast agents and potential medical applications. *European Radiology* 13 (2):231–243.

Bryant, L. H., Jr., M. W. Brechbiel, C. Wu, J. W. Bulte, V. Herynek, and J. A. Frank. 1999. Synthesis and relaxometry of high-generation (G = 5, 7, 9, and 10) PAMAM dendrimer-DOTA-gadolinium chelates. *Journal of Magnetic Resonance Imaging* 9 (2):348–352.

Bulte, J. W. M. and D. L. Kraitchman. 2004. Monitoring cell therapy using iron oxide MR contrast agents. *Current Pharmaceutical Biotechnology* 5 (6):567–584.

Canham, L. T. 1990. Silicon quantum wire array fabrication by electrochemical and chemical dissolution of wafers. *Applied Physics Letters* 57 (10):1046–1048.

Cao, F., S. Lin, X. Y. Xie, P. Ray, M. Patel, X. Z. Zhang, M. Drukker et al. 2006. *In vivo* visualization of embryonic stem cell survival, proliferation, and migration after cardiac delivery. *Circulation* 113 (7):1005–1014.

Cheng, Z. L. and A. Tsourkas. 2008. Paramagnetic porous polymersomes. *Langmuir* 24 (15):8169–8173.

Engstrom, M., A. Klasson, H. Pedersen, C. Vahlberg, P. O. Kall, and K. Uvdal. 2006. High proton relaxivity for gadolinium oxide nanoparticies. *Magnetic Resonance Materials in Physics Biology and Medicine* 19 (4):180–186.

Erdogan, S., A. Roby, R. Sawant, J. Hurley, and V. P. Torchilin. 2006. Gadolinium-loaded polychelating polymer-containing cancer cell-specific immunoliposomes. *Journal of Liposome Research* 16 (1):45–55.

Feinstein, S. B. 2004. The powerful microbubble: From bench to bedside, from intravascular indicator to therapeutic delivery system, and beyond. *American Journal of Physiology—Heart and Circulatory Physiology* 287 (2):H450–H457.

Frank, J. A., B. R. Miller, A. S. Arbab, H. A. Zywicke, E. K. Jordan, B. K. Lewis, L. H. Bryant, Jr., and J. W. Bulte. 2003. Clinically applicable labeling of mammalian and stem cells by combining superparamagnetic iron oxides and transfection agents. *Radiology* 228 (2):480–487.

Fu, Y. L., D. A. Kedziorek, R. Ouwerkerk, V. Crisostomo, W. D. Gilson, N. Azene, A. Arepally et al. 2009. Novel perfluorinated alginate microcapsules for CT and MRI monitoring of mesenchymal stem cell delivery and engraftment tracking. *Journal of the American College of Cardiology* 53 (10):A447–A447.

Gerion, D., J. Herberg, R. Bok, E. Gjersing, E. Ramon, R. Maxwell, J. Kurhanewicz et al. 2007. Paramagnetic silica-coated nanocrystals as an advanced MRI contrast agent. *Journal of Physical Chemistry C* 111 (34):12542–12551.

Gilad, A. A., M. T. McMahon, P. Walczak, P. T. Winnard, V. Raman, H. W. M. van Laarhoven, C. M. Skoglund, J. W. M. Bulte, and P. C. M. van Zijl. 2007. Artificial reporter gene providing MRI contrast based on proton exchange. *Nature Biotechnology* 25 (2):217–219.

Gilad, A. A., P. Walczak, M. T. McMahon, H. Bin Na, J. H. Lee, K. An, T. Hyeon, P. C. M. van Zijl, and J. W. M. Bulte. 2008. MR tracking of transplanted cells with "positive contrast" using manganese oxide nanoparticles. *Magnetic Resonance in Medicine* 60 (1):1–7.

Hadjantonakis, A. K., M. E. Dickinson, S. E. Fraser, and V. E. Papaioannou. 2003. Technicolour transgenics: Imaging tools for functional genomics in the mouse. *Nature Reviews Genetics* 4 (8):613–625.

Hainfeld, J. F., D. N. Slatkin, T. M. Focella, and H. M. Smilowitz. 2006. Gold nanoparticles: A new X-ray contrast agent. *British Journal of Radiology* 79 (939):248–253.

Kanematsu, M., R. C. Semelka, M. Matsuo, H. Kondo, M. Enya, S. Goshima, N. Moriyama, and H. Hoshi. 2002. Gadolinium-enhanced MR imaging of the liver: Optimizing imaging delay for hepatic arterial and portal venous phases—A prospective randomized study in patients with chronic liver damage. *Radiology* 225 (2):407–415.

Kang, J. H., D. S. Lee, J. C. Paeng, J. S. Lee, Y. H. Kim, Y. J. Lee, D. W. Hwang et al. 2005. Development of a sodium/iodide symporter (NIS)-transgenic mouse for imaging of cardiomyocyte-specific reporter gene expression. *Journal of Nuclear Medicine* 46 (3):479–483.

Klibanov, A. L. 2007. Ultrasound molecular imaging with targeted microbubble contrast agents. *Journal of Nuclear Cardiology* 14 (6):876–884.

Klohs, J., A. Wunder, and K. Licha. 2008. Near-infrared fluorescent probes for imaging vascular pathophysiology. *Basic Research in Cardiology* 103 (2):144–151.

Koole, R., M. M. van Schooneveld, J. Hilhorst, K. Castermans, D. P. Cormode, G. J. Strijkers, C. D. Donega et al. 2008. Paramagnetic lipid-coated silica nanoparticles with a fluorescent quantum dot core: A new contrast agent platform for multimodality imaging. *Bioconjugate Chemistry* 19 (12):2471–2479.

Kraitchman, D. L. and J. W. M. Bulte. 2009. In vivo imaging of stem cells and beta cells using direct cell labeling and reporter gene methods. *Arteriosclerosis, Thrombosis and Vascular Biology* 29 (7):1025–1030.

Langereis, S., J. Keupp, J. L. J. van Velthoven, I. H. C. de Roos, D. Burdinski, J. A. Pikkemaat, and H. Grull. 2009. A temperature-sensitive liposomal H-1 CEST and F-19 contrast agent for MR image-guided drug delivery. *Journal of the American Chemical Society* 131 (4):1380–1381.

Laverman, P., E. T. M. Dams, W. J. G. Oyen, G. Storm, E. B. Koenders, R. Prevost, J. W. M. van der Meer, F. H. M. Corstens, and O. C. Boerman. 1999. A novel method to label liposomes with Tc-99m by the hydrazino nicotinyl derivative. *Journal of Nuclear Medicine* 40 (1):192–197.

Lee, J. H., Y. M. Huh, Y. Jun, J. Seo, J. Jang, H. T. Song, S. Kim et al. 2007. Artificially engineered magnetic nanoparticles for ultra-sensitive molecular imaging. *Nature Medicine* 13 (1):95–99.

Louie, A. Y., M. M. Huber, E. T. Ahrens, U. Rothbacher, R. Moats, R. E. Jacobs, S. E. Fraser, and T. J. Meade. 2000. *In vivo* visualization of gene expression using magnetic resonance imaging. *Nature Biotechnology* 18 (3):321–325.

MacFarland, D. K., K. L. Walker, R. P. Lenk, S. R. Wilson, K. Kumar, C. L. Kepley, and J. R. Garbow. 2008. Hydrochalarones: A novel endohedral metallofullerene platform for enhancing magnetic resonance imaging contrast. *Journal of Medicinal Chemistry* 51 (13):3681–3683.

Major, J. L., G. Parigi, C. Luchinat, and T. J. Meade. 2007. The synthesis and *in vitro* testing of a zinc-activated MRI contrast agent. *Proceedings of the National Academy of Sciences of the United States of America* 104 (35):13881–13886.

Massoud, T. F. and S. S. Gambhir. 2003. Molecular imaging in living subjects: Seeing fundamental biological processes in a new light. *Genes & Development* 17 (5):545–580.

Mayer, C. R., N. A. Geis, H. A. Katus, and R. Bekeredjian. 2008. Ultrasound targeted microbubble destruction for drug and gene delivery. *Expert Opinion on Drug Delivery* 5 (10):1121–1138.

McMahon, M. T., A. A. Gilad, M. A. DeLiso, S. D. C. Berman, J. W. M. Bulte, and P. C. M. van Zijl. 2008. New "multicolor" polypeptide diamagnetic chemical exchange saturation transfer (DIACEST) contrast agents for MRI. *Magnetic Resonance in Medicine* 60 (4):803–812.

Medley, C. D., J. E. Smith, Z. Tang, Y. Wu, S. Bamrungsap, and W. H. Tan. 2008. Gold nanoparticle-based colorimetric assay for the direct detection of cancerous cells. *Analytical Chemistry* 80 (4):1067–1072.

Modo, M. 2008. Noninvasive imaging of transplanted cells. *Current Opinion in Organ Transplantation* 13 (6):654–658.

Moffat, F. L. and S. A. Gulec. 2007. Sulphur colloid for imaging lymph nodes and bone marrow. In *Nanoparticles in Biomedical Imaging*, eds. J. W. M. Bulte and M. M. J. Modo. New York: Springer.

Mulder, W. J. M., G. J. Strijkers, G. A. F. van Tilborg, A. W. Griffioen, and K. Nicolay. 2006. Lipid-based nanoparticles for contrast-enhanced MRI and molecular imaging. *NMR in Biomedicine* 19 (1):142–164.

Nagai, Y., M. Miyazaki, R. Aoki, T. Zama, S. Inouye, K. Hirose, M. Iino, and M. Hagiwara. 2000. A fluorescent indicator for visualizing cAMP-induced phosphorylation *in vivo*. *Nature Biotechnology* 18 (3):313–316.

Ngo, C. Y., S. F. Yoon, and S. J. Chua. 2009. Ambient temperature dependence on emission spectrum of InAs quantum dots. *Physica Status Solidi B—Basic Solid State Physics* 246 (4):799–802.

Park, J. H., L. Gu, G. von Maltzahn, E. Ruoslahti, S. N. Bhatia, and M. J. Sailor. 2009. Biodegradable luminescent porous silicon nanoparticles for *in vivo* applications. *Nature Materials* 8 (4):331–336.

Pierce, M. C., D. J. Javier, and R. Richards-Kortum. 2008. Optical contrast agents and imaging systems for detection and diagnosis of cancer. *International Journal of Cancer* 123 (9):1979–1990.

Ren, J. M., R. Trokowski, S. R. Zhang, C. R. Malloy, and A. D. Sherry. 2008. Imaging the tissue distribution of glucose in livers using a PARACEST sensor. *Magnetic Resonance in Medicine* 60 (5):1047–1055.

Ruiz-Cabello, J., P. Walczak, D. A. Kedziorek, V. P. Chacko, A. H. Schmieder, S. A. Wickline, G. M. Lanza, and J. W. M. Bulte. 2008. *In vivo* "hot spot" MR imaging of neural stem cells using fluorinated nanoparticles. *Magnetic Resonance in Medicine* 60 (6):1506–1511.

Sanchez, V., J. Greiner, and J. G. Riess. 1995. Highly concentrated 1,2-bis(perfluoroalkyl)iodoethene emulsions for use as contrast agents for diagnosis. *Journal of Fluorine Chemistry* 73 (2):259–264.

Santra, S., R. P. Bagwe, D. Dutta, J. T. Stanley, G. A. Walter, W. Tan, B. M. Moudgil, and R. A. Mericle. 2005. Synthesis and characterization of fluorescent, radio-opaque, and paramagnetic silica nanoparticles for multimodal bioimaging applications. *Advanced Materials* 17 (18):2165–2169.

Schmieder, A. H., P. M. Winter, S. D. Caruthers, T. D. Harris, T. A. Williams, J. S. Allen, E. K. Lacy et al. 2005. Molecular MR imaging of melanoma angiogenesis with alpha(nu)beta(3)-targeted paramagnetic nanoparticles. *Magnetic Resonance in Medicine* 53 (3):621–627.

Shapiro, E. M. and A. P. Koretsky. 2008. Convertible manganese contrast for molecular and cellular MRI. *Magnetic Resonance in Medicine* 60 (2):265–269.

Shapiro, M. G., J. O. Szablowski, R. Langer, and A. Jasanoff. 2009. Protein nanoparticles engineered to sense kinase activity in MRI. *Journal of the American Chemical Society* 131 (7):2484–2486.

Sharma, P., S. Brown, G. Walter, S. Santra, and B. Moudgil. 2006. Nanoparticles for bioimaging. *Advances in Colloid and Interface Science* 123–126:471–485.

Sherry, A. D. and M. Woods. 2008. Chemical exchange saturation transfer contrast agents for magnetic resonance imaging. *Annual Review of Biomedical Engineering* 10:391–411.

Shu, C. Y., L. H. Gan, C. R. Wang, X. L. Pei, and H. B. Han. 2006. Synthesis and characterization of a new water-soluble endohedral metallofullerene for MRI contrast agents. *Carbon* 44 (3):496–500.

Sitharaman, B. and L. J. Wilson. 2007. Gadofullerenes and gadonanotubes: A new paradigm for high-performance magnetic resonance imaging contrast agent probes. *Journal of Biomedical Nanotechnology* 3 (4):342–352.

Snoussi, K., J. W. M. Bulte, M. Gueron, and P. C. M. van Zijl. 2003. Sensitive CEST agents based on nucleic acid imino proton exchange: Detection of poly(rU) and of a dendrimer-poly(rU) model for nucleic acid delivery and pharmacology. *Magnetic Resonance in Medicine* 49 (6):998–1005.

Sofou, S. and G. Sgouros. 2008. Anti body-targeted liposomes in cancer therapy and imaging. *Expert Opinion on Drug Delivery* 5 (2):189–204.

Terreno, E., D. D. Castelli, G. Cravotto, L. Milone, and S. Aime. 2004. Ln(III)-DOTAMGIY complexes: A versatile series to assess the determinants of the efficacy of paramagnetic chemical exchange saturation transfer agents for magnetic resonance imaging applications. *Investigative Radiology* 39 (4):235–243.

Terreno, E., D. D. Castelli, E. Violante, H. M. H. F. Sanders, N. A. J. M. Sommerdijk, and S. Aime. 2009. Osmotically shrunken LIPOCEST agents: An innovative class of magnetic resonance imaging contrast media based on chemical exchange saturation transfer. *Chemistry—A European Journal* 15 (6):1440–1448.

Torchilin, V. P. 2000. Polymeric contrast agents for medical imaging. *Current Pharmaceutical Biotechnology* 1:183–215.

Torchilin, V. P. 2005. Recent advances with liposomes as pharmaceutical carriers. *Nature Reviews Drug Discovery* 4 (2):145–160.

Toth, E., R. D. Bolskar, A. Borel, G. Gonzalez, L. Helm, A. E. Merbach, B. Sitharaman, and L. J. Wilson. 2005. Water-soluble gadofullerenes: Toward high-relaxivity, pH-responsive MRI contrast agents. *Journal of the American Chemical Society* 127 (2):799–805.

Trokowski, R., J. M. Ren, F. K. Kalman, and A. D. Sherry. 2005. Selective sensing of zinc ions with a PARACEST contrast agent. *Angewandte Chemie—International Edition* 44 (42):6920–6923.

Trubetskoy, V. S., G. S. Gazelle, G. L. Wolf, and V. P. Torchilin. 1997. Block-copolymer of polyethylene glycol and polylysine as a carrier of organic iodine: Design of long-circulating particulate contrast medium for X-ray computed tomography. *Journal of Drug Targeting* 4 (6):381–388.

Urano, Y., D. Asanuma, Y. Hama, Y. Koyama, T. Barrett, M. Kamiya, T. Nagano et al. 2009. Selective molecular imaging of viable cancer cells with pH-activatable fluorescence probes. *Nature Medicine* 15 (1):104–109.

van Zijl, P. C. M., C. K. Jones, J. Ren, C. R. Malloy, and A. D. Sherry. 2007. MR1 detection of glycogen *in vivo* by using chemical exchange saturation transfer imaging (glycoCEST). *Proceedings of the National Academy of Sciences of the United States of America* 104 (11):4359–4364.

Veiseh, O., C. Sun, J. Gunn, N. Kohler, P. Gabikian, D. Lee, N. Bhattarai et al. 2005. Optical and MRI multifunctional nanoprobe for targeting gliomas. *Nano Letters* 5 (6):1003–1008.

Walczak, P., D. A. Kedziorek, A. A. Gilad, S. Lin, and J. W. M. Bulte. 2005. Instant MR labeling of stem cells using magnetoelectroporation. *Magnetic Resonance in Medicine* 54 (4):769–774.

Walling, M. A., J. A. Novak, and J. R. E. Shepard. 2009. Quantum dots for live cell and *in vivo* imaging. *International Journal of Molecular Sciences* 10 (2):441–491.

Walter, G.A., S. Santra, B. Thattaliyath, and S. C. Grant. 2007. (Super)paramagnetic nanoparticles: Applications in noninvasive MR imaging of stem cell transfer. In *Nanoparticles in Biomedical Imaging*, eds. J. W. M. Bulte and M. M. J. Modo. New York: Springer.

Weissleder, R. and U. Mahmood. 2001. Molecular imaging. *Radiology* 219 (2):316–333.

Winter, P. M., S. D. Caruthers, A. H. Schmieder, A.M. Neubauer, G. M. Lanza, and S.A. Wickine. 2007. Molecular MR imaging with paramagnetic perfluorocarbon nanoparticles. In *Nanoparticles in Biomedical Imaging*, eds. J. W. M. Bulte and M. M. J. Modo. New York: Springer.

Yoo, B. and M. D. Pagel. 2006. A PARACEST MRI contrast agent to detect enzyme activity. *Journal of the American Chemical Society* 128 (43):14032–14033.

Zhang, S. R., C. R. Malloy, and A. D. Sherry. 2005. MRI thermometry based on PARACEST agents. *Journal of the American Chemical Society* 127 (50):17572–17573.

Zhang, S. R., M. Merritt, D. E. Woessner, R. E. Lenkinski, and A. D. Sherry. 2003. PARACEST agents: Modulating MRI contrast via water proton exchange. *Accounts of Chemical Research* 36 (10):783–790.

Zhang, S. R., L. Michaudet, S. Burgess, and A. D. Sherry. 2002. The amide protons of an ytterbium(III) dota tetraamide complex act as efficient antennae for transfer of magnetization to bulk water. *Angewandte Chemie—International Edition* 41 (11):1919–1921.

Zheng, J. Z., G. Perkins, A. Kirilova, C. Allen, and D. A. Jaffray. 2006. Multimodal contrast agent for combined computed tomography and magnetic resonance imaging applications. *Investigative Radiology* 41 (3):339–348.

Zhou, J. Y., J. F. Payen, D. A. Wilson, R. J. Traystman, and P. C. M. van Zijl. 2003. Using the amide proton signals of intracellular proteins and peptides to detect pH effects in MRI. *Nature Medicine* 9 (8):1085–1090.

Zhu, D., X. L. Lu, P. A. Hardy, M. Leggas, and M. Jay. 2008. Nanotemplate-engineered nanoparticles containing gadolinium for magnetic resonance imaging of tumors. *Investigative Radiology* 43 (2):129–140.

Zielhuis, S. W., J. H. Seppenwoolde, V. A. P. Mateus, C. J. G. Bakker, G. C. Krijger, G. Storm, B. A. Zonnenberg, A. D. van het Schip, G. A. Koning, and J. F. W. Nijsen. 2006. Lanthanide-loaded liposomes for multimodality imaging and therapy. *Cancer Biotherapy and Radiopharmaceuticals* 21 (5):520–527.

Zimmer, J. P., S. W. Kim, S. Ohnishi, E. Tanaka, J. V. Frangioni, and M. G. Bawendi. 2006. Size series of small indium arsenide-zinc selenide core-shell nanocrystals and their application to *in vivo* imaging. *Journal of the American Chemical Society* 128 (8):2526–2527.

18

Gadolinium-Based Bionanomaterials for Magnetic Resonance Imaging

Lothar Helm
Ecole Polytechnique Fédérale de Lausanne

Eva Toth
Centre National de la Recherche Scientifique

18.1 Introduction

In the last two decades, magnetic resonance imaging (MRI) has evolved into one of the most powerful diagnostic modalities in clinical and biomedical imaging. MRI provides remarkable resolution, with no depth limit. To increase the signal intensity and image contrast, a reduction of the longitudinal or transverse spin relaxation time, T_1 or T_2, respectively, is often induced by the application of suitable paramagnetic or superparamagnetic compounds, called MRI contrast agents. In today's practice, agents that affect mostly the T_1 relaxation time (so-called T_1 agents) represent the majority of the imaging probes used. Nearly all T_1 agents developed so far are compounds based on the gadolinium ion, Gd^{3+} (Aime et al. 2007a; Geraldes and Laurent 2009). Gd^{3+} is the metal of choice since it possesses the highest electron spin ($S = 7/2$) for a paramagnetic cation and a slow electron spin relaxation. The efficiency of the agent can be boosted by increasing the relaxation effect generated per paramagnetic center (expressed by the term relaxivity, r_1) and/or by bringing many of these ions to the target. Based on theoretical considerations, an estimation of $r_1 \sim 100\,\text{mM}^{-1}\,\text{s}^{-1}$ per Gd^{3+} can be deduced as an upper limit of the attainable relaxivity, depending on the magnetic field strength (Caravan et al. 1999; Toth et al. 2001; Caravan 2007). Given its intrinsic low sensitivity, MRI requires a relatively large amount of the relaxation agent to provide detectable contrast enhancement. In order to attain high amounts of Gd^{3+} to be delivered to the target site, various kinds of nanomaterials have been explored in the context of MRI contrast agent research. In addition to increasing the paramagnetic payload per particle, the slow rotation of nano-sized systems also contributes to the increase of their relaxation efficiency per Gd^{3+}. This chapter covers the basic physical-chemistry aspects related to the development of high-efficiency MRI contrast agents, followed by a more detailed description of the different classes of gadolinium-based nano-sized imaging probes, involving both inorganic and organic bionanomaterials.

18.2 Relaxivity of Gd^{3+} Complexes

The relaxation enhancement of solvent nuclear spins around a paramagnetic center is the fundamental physical phenomenon on which contrast agents for MRI are based on (Koenig and Brown 1995; Tweedle and Kumar 1999; Krause 2002). Both the longitudinal relaxation rate, $1/T_1$, and the transverse relaxation rate, $1/T_2$, are accelerated through mainly dipolar interaction with the electron spin of the paramagnetic species. The observed ^{1}H relaxation rate, $1/T_{i,obs}$ ($i = 1,2$), is the sum of a diamagnetic term, $1/T_{i,d}$, corresponding to the relaxation rate measured in the absence of the paramagnetic substance, and a paramagnetic term, $1/T_{i,p}$, which describes the relaxation enhancement caused by the paramagnetic solute. In general, $1/T_{i,p}$ is a linear function of the concentration of the paramagnetic ion, which is, in the case of MRI contrast agents, normally the Gd^{3+} ion. By convention, the concentration is expressed in mmol/L (mM) and the constant of proportionality is called relaxivity, r_1. Deviation from linear concentration dependence can be due to the formation of complexes with macromolecules or formation of aggregates, such as micelles, for example.

$$\frac{1}{T_{i,obs}} = \frac{1}{T_{i,d}} + \frac{1}{T_{i,p}} = \frac{1}{T_{i,d}} + r_i[\mathrm{Gd}^{3+}] \quad i = 1,2 \tag{18.1}$$

The paramagnetic relaxation of the water proton is dominated by the interaction between the proton nuclear spin and the fluctuating local magnetic field generated by the spins of the unpaired electrons of the paramagnetic compound (Banci et al. 1991; Bertini and Luchinat 1996; Bertini et al. 2001). The fluctuation of the magnetic field is caused by rotational and translational diffusion of the water molecules and the complex accommodating the ion. Three types of water molecules are typically distinguished in respect to their interaction with the paramagnetic complex. The first type comprises the "bound" water molecules. These are in general directly bound in the first coordination sphere of the metal ion with a relatively long mean residence time, τ_m. The contribution to relaxivity caused by these water molecules is called inner sphere relaxivity, r_i^{IS}. The second type of water is represented by bulk water molecules that are assumed to undergo free translational diffusion in the vicinity of the complex. The corresponding contribution to r_i is called outer sphere contribution, r_i^{OS}. In some cases, a third type of water is present that is formed of water molecules loosely bound at the surface of the complex (Botta 2000). These molecules are called second sphere water molecules, and the relaxivity generated by them is called second sphere contribution, r_i^{2nd}.

$$r_i = r_i^{IS} + r_i^{OS} + r_i^{2nd} \quad i = 1,2 \tag{18.2}$$

The general theoretical description of relaxation enhancement by paramagnetic centers has been developed since more than 50 years now. The inner sphere relaxivity is characterized by a number of parameters:

q the number of water molecules in the first coordination sphere; it is evident that r_i^{IS} depends linearly on this parameter

τ_m the mean residence time of the water molecule(s) in the first coordination sphere; this time is the inverse of the exchange rate constant, k_{ex}, for the bound water molecules

τ_R the correlation time describing the rotational diffusion of the complex; this correlation time is linked to the rotational diffusion constant by $\tau_R = 6/D_R$
r_{IS} the distance between the spins of the unpaired electrons, *S*, and the nuclear spin, *I*
T_{ie} the longitudinal ($i = 1$) and transverse ($i = 2$) relaxation times of the electron spin

The oldest and most widely used theory, which allows calculating r_i^{IS} from these parameters, has been developed by Solomon, Bloembergen, and Morgan (SBM) (Solomon 1955; Solomon and Bloembergen 1956; Bloembergen 1957; Bloembergen and Morgan 1961). Based on several approximations, it leads to simple analytical equations that allow calculating the inner sphere relaxivity as a function of external parameters like temperature and the applied magnetic field. In view of gadolinium-based contrast agents, the most restrictive approximations concern the treatment of the relaxation of the electron spin and the use of the Redfield relaxation theory (Kowalewski et al. 1985; Kowalewski et al. 2005). More general theoretical descriptions of r_i^{IS} have been developed over the years (Kowalewski et al. 2005; Belorizky et al. 2008). Some of them are universally valid but very difficult to apply in practice because heavy numerical calculations are necessary. Others are limited to special cases like slow motion as observed, for example, with large molecules such as proteins.

The relaxivity induced by second sphere water molecules is difficult to quantify (Botta 2000; Lebduskova et al. 2007; Hermann et al. 2008). It is described by the same theoretical approach as inner sphere relaxivity; however, the definition of the parameters is less precise. Various second sphere binding sites with different distances, r_i^{2nd}, and different mean residence times, τ_m^{2nd}, can be present. The residence times of those water molecules are much shorter compared to τ_m of the water molecule(s) in the first coordination sphere. Good estimations of the second sphere effect on relaxivity are only available on complexes with no inner sphere water molecules ($q = 0$) or from computer simulations. For instance, classical molecular dynamics simulations for the polyamino-phosphonate-based $[Gd(DOTP)]^{5-}$ led to four second sphere water molecules with $\tau_m^{2nd} \sim 56\,ps$ (Borel et al. 2001). An increase of the number of second sphere water molecules as well as their residence time would increase this contribution to the overall relaxivity.

The source for outer sphere relaxivity is in water molecules that diffuse close to the paramagnetic complex (bulk water) (Ayant et al. 1975; Hwang and Freed 1975; Freed 1978). The intermolecular dipolar interaction between the electron spin and the nuclear spins is governed by random translational motion, and analytical equations could be obtained if any attractive or repulsive interaction between the water molecules and the complex is neglected. In this case, the parameters characterizing r_i^{OS} are

D the diffusion constant for relative translational diffusion
d the closest distance of approach of spins *I* and *S*; the characteristic correlation time for the fluctuation is $\tau_d = d^2/D$
T_{ie} the longitudinal ($i = 1$) and transverse ($i = 2$) relaxation times of the electron spin

Outer sphere relaxivity is difficult to modulate. The relative translational diffusion is close to the self-diffusion of water molecules ($2.3 \times 10^{-9}\,m^2\,s^{-1}$) (Holz et al. 2000) and cannot be changed. The closest distance of approach, which is fitted for most chelating ligands to 3.5–3.6 Å, cannot be decreased; it is already relatively close to that of first sphere water molecules (~3.1 Å). Slowing down the relaxation of the electron spin would increase outer sphere relaxivity as it would increase the inner sphere relaxation.

All the parameters mentioned above influence the relaxivity of gadolinium-based MRI contrast agents in a characteristic way. In the following section, we will discuss the inner sphere contribution in more detail.

18.2.1 Inner Sphere Proton Relaxivity

The distance between the spins of the unpaired electrons and the nuclear spin is approximated by the internuclear distance between the gadolinium ion and the water protons. It has been shown by quantum mechanical calculations that this so-called point dipole approximation is valid in this case (Yazyev et al. 2005). For directly bound water molecules, this distance is between 3 and 3.2 Å as determined by

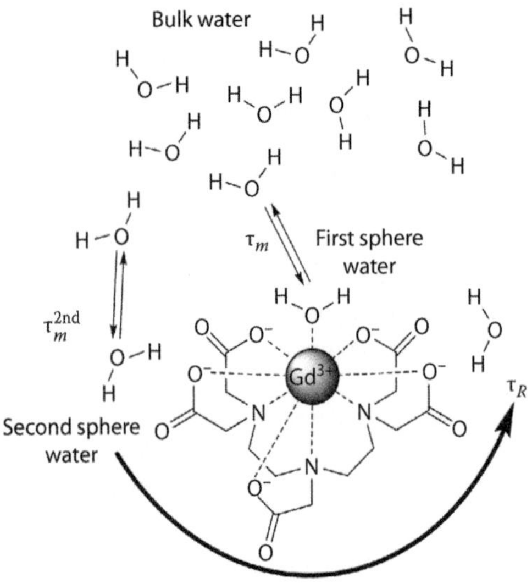

FIGURE 18.1 Gadolinium complex with one inner sphere water molecule and second- and outer sphere water molecules.

pulsed ENDOR (Caravan et al. 2003; Astashkin et al. 2004). A shortening of this distance would lead to a marked increase in relaxivity due to the r_{IS}^{-6} dependence of the dipolar relaxation contribution. It is however difficult to conceive how to achieve this shortening because the triple positive charge of the cation strongly tends to align linearly the electric dipole moment of the water molecules.

The number of water molecules bound in the first coordination sphere scales the inner sphere relaxivity linearly. Lanthanide complexes have in general coordination numbers, CN, of eight or nine. All contrast agents in clinical use have CN = 9 with eight coordination sites occupied by the chelating ligand and one by the oxygen atom of a water molecule (q = 1) (Caravan et al. 1999; Hermann et al. 2008). Increasing q to two can be achieved with ligands with CN = 7. This leads in general to less stable complexes preventing them to be approved by the controlling drug agencies. Another feature encountered with some q = 2 chelates, in particular those based on the macrocyclic ligand DO3A, is the ability to form ternary complexes with bi-dentate anions like CO_3^{2-} and lactate (Dickins et al. 2002; Terreno et al. 2005). On the one hand, this enables in principle these compounds to detect the presence of such anions; on the other hand, it prevents them to be used as general contrast agents due to the disappearance of the inner sphere contribution to relaxivity. Three types of q = 2 complexes have found special interest in recent years and will be discussed below.

The parameter mostly addressed to design complexes with high relaxivity is the rotational correlation time τ_R (Caravan et al. 1999; Toth et al. 2001; Caravan 2006; Caravan et al. 2009). The main reason is that longer correlation times always lead to higher relaxivities at magnetic fields used by clinical MRI scanners ($0.15\,T \leq B \leq 3\,T$). Slowing down the rotation of molecules can be realized by forming large entities by linking the chelating ligand to macromolecules like proteins, by synthesizing dendrimers and polymers loaded with many Gd^{3+}-binding groups, and by assembling micelles and nanoparticles with hundreds and thousands of paramagnetic centers. Rotational correlation times longer than 1 ns lead to a characteristic hump of longitudinal relaxivity r_1 at Larmor frequencies between 10 and 130 MHz (magnetic fields of 0.2 and 3 T) (Figure 18.2). At very high fields (above 9.4 T), the dependence on τ_R becomes inverted and r_1 is proportional to τ_R^{-1} (Helm 2006). The transverse relaxivity, r_2, however continues to increase with the magnetic field strength. The development of MRI scanners working at high fields induced the search for special contrast agents with high longitudinal relaxivities at high fields.

The rigidity of the large molecules constructed is an important feature with respect to their MRI efficiency (Dunand et al. 2001; Nicolle et al. 2002b). The rotational correlation time to be considered

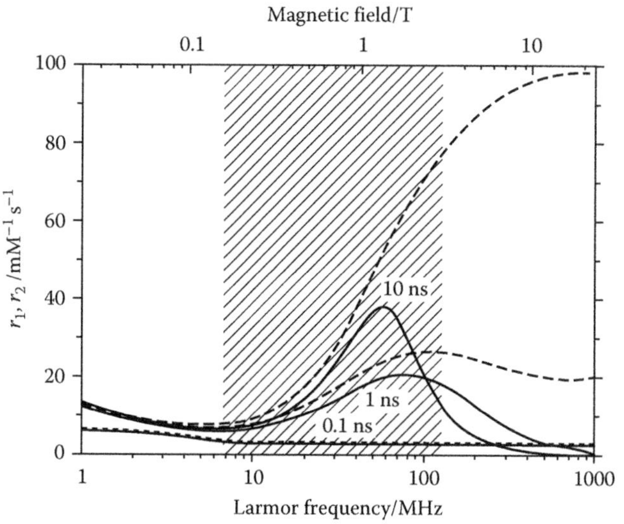

FIGURE 18.2 Simulation of the inner sphere longitudinal (r_i^{IS}, ——) and transverse (r_i^{IS}, - - -) relaxivity as a function of the magnetic field for three values of the rotational correlation time, τ_R, using the Solomon–Bloembergen–Morgan (SBM) model. The field range used in clinical MRI scanners (0.15–3 T) is dashed. The parameters used in the simulation were $r_{IS} = 3.1$ Å, $\tau_v = 10$ ps, $^2_t = 0.1 \times 10^{20}$ s^{-2}, $\tau_m = 20$ ns.

is for the reorientation of the vector connecting the nuclei of Gd^{3+} and inner sphere water hydrogen. Any internal or anisotropic motion will accelerate the reorientation of this vector and therefore influence relaxivity (Woessner 1962; Werbelow and Grant 1977; Kowalewski and Mäler 2006). Different dynamic models for nonrigid molecules have been developed to describe spin relaxation. The number of possibilities together with the lack of detailed information in the case of macromolecules prevents the application of these models. A satisfactory description for the dynamics of contrast agents with internal motion has been found by using the "model-free" approach, developed by Lipari and Szabo in the context of protein dynamics (Lipari and Szabo 1982a,b). It is based on an isotropic overall reorientation of the large particle, which is uncorrelated with faster internal motions. The degree of restriction of the internal motion is described by an order parameter S. In the absence of internal motion $S^2 = 1$ whereas $S^2 = 0$ is found for fully unrestricted internal motion. If $S^2 < 0.8$ the relaxation hump observed at clinical MRI fields is considerably depleted (Figure 18.3).

The mean residence time of water molecules in the first coordination sphere also found considerable advertence in the research to optimize relaxivity of contrast agents (Toth and Merbach 1998; Toth et al. 2001; Caravan et al. 2009). The residence time τ_m must be short to transmit the relaxation effect induced by the paramagnetic center to the bulk solvent. If, however, τ_m is very short, meaning below 10 ns, the contact between the ^{1}H nuclear spin and the electron spin is too short to allow an effective relaxation of the first one leading to a reduced relaxivity. An optimal residence time situated between 10 and 100 ns exists, the exact value of which depends on the magnetic field strength, the rotational correlation time, and the electron spin relaxation (Caravan et al. 2009). The exchange rate of water molecules, $k_{ex} = 1/\tau_m$, can be influenced by modifications of the chelating ligand (Toth et al. 2002, 2004; Helm and Merbach 2005). It has been shown that increasing the steric hindrance on the first sphere water will increase k_{ex} in the case of dissociative activation of the water exchange reaction, which is active in general for $q = 1$ systems. For example, replacing an ethylene bridge between the amines in the backbone of the common diethylenetriamine-pentaacetate (DTPA) chelator by a longer propylene bridge allows the carboxylate oxygens of the ligand to come closer together, restricting the space available for the coordinated water oxygen (Laus et al. 2003). As for the increase of q, care has to be taken not to reduce the thermodynamic stability of the gadolinium complex in optimizing the water exchange.

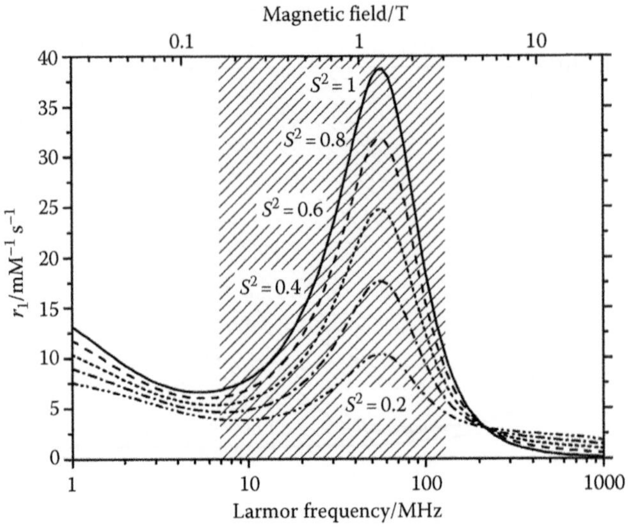

FIGURE 18.3 Relaxivity r_i^{IS} as a function of the magnetic field calculated using the Lipari–Szabo approach for internal motion for five values of the order parameter S^2. The parameters used in the simulation were r_{IS} = 3.1 Å, t_v = 10 ps, $^{2}_{t}$ = 0.1 × 10^{20} s^{-2}, τ_m = 20 ns, the global rotational correlation time τ_R = 10 ns, the local, fast correlation time τ_l = 0.1 ns.

The last factor acting on the relaxivity of inner sphere water molecules is the electron spin relaxation. For Gd^{3+} ions with seven unpaired S electrons, it is governed by the zero-field splitting (ZFS) energy modulated by the rotation and fast distortions of the complex, the latter caused by inharmonic distortions and collisions with solvent molecules. The theoretical description of the electron spin relaxation and its influence on the relaxivity of the nuclear spin is very complex. For a general description, it has to be considered that four distinct relaxation rates exist for transitions between spin states (McLachlan 1964; Strandberg and Westlund 1996; Rast et al. 1999; Aaman and Westlund 2007). Furthermore,

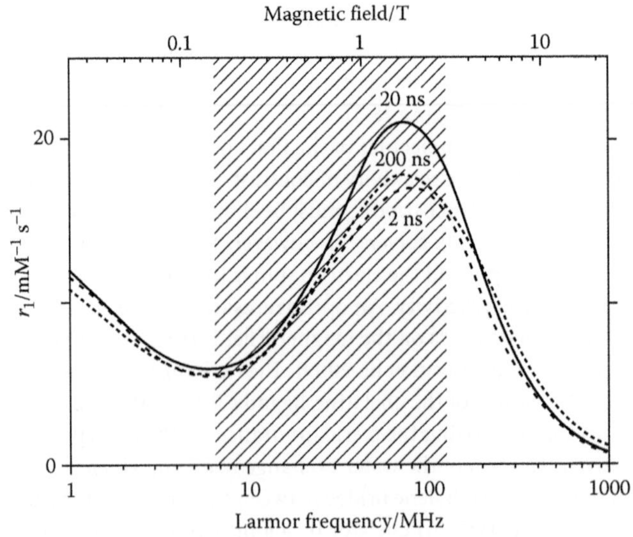

FIGURE 18.4 Relaxivity r_i^{IS} as a function of the magnetic field for three values of the residence time τ_m. of the inner sphere water molecule. The parameters used in the simulation were r_{IS} = 3.1 Å, τ_v = 10 ps, $^{2}_{t}$ = 0.1 × 10^{20} s^{-2}, τ_R = 1 ns.

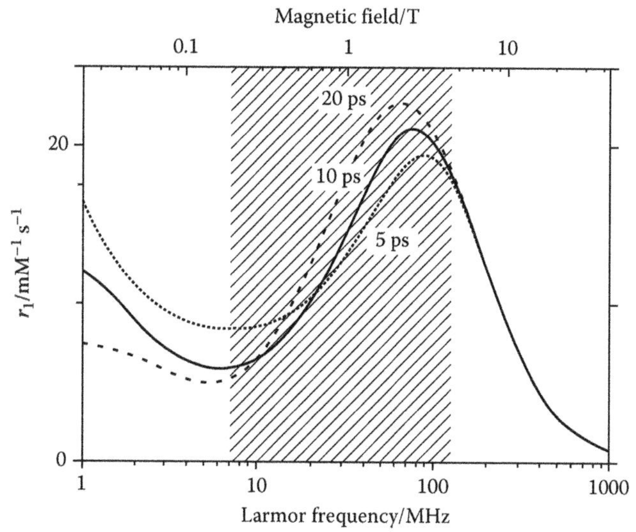

FIGURE 18.5 Relaxivity r_i^{IS} as a function of the magnetic field for three values of the correlation time τ_v, describing the fluctuation of the transient ZFS. The parameters used in the simulation were r_{IS} = 3.1 Å, ^2_t = 0.1 × 10^{20} s^{-2}, τ_R = 1 ns, τ_m = 20 ns.

depending on the magnetic field strength and the amplitude of the ZFS, the perturbation theory as applied by Redfield in his relaxation theory is no longer valid (Rast et al. 2001; Kowalewski et al. 2002; Kowalewski et al. 2005; Fries and Belorizky 2007; Belorizky et al. 2008). Several approaches have been developed leading to solutions valid in the case of fast or slow rotation, high or low magnetic field, or strong or weak ZFS. In all approaches, the ZFS is split in a part with static amplitude, Δ_s, which is modulated by the rotational diffusion (τ_R) and a transient part with amplitude Δ_t that is modulated by a fast process characterized by the correlation time τ_v. Fortunately, at higher magnetic fields (>0.23 T, 10 MHz), the simple analytical description of SBM theory based on the fast fluctuating transient ZFS coincides with more correct, mainly numerical calculations taking also into account the slower varying static ZFS (Belorizky and Fries 2004; Fries and Belorizky 2005). In general terms, water proton relaxivity increases when the electron spin relaxation is slowed down either by lower ZFS amplitude or by slower modulation (longer τ_v) (Figure 18.5).

18.3 Chelating Units for Gadolinium-Based Contrast Agents

MRI contrast agents are in general administered intravenously and distribute through the extracellular and intravascular spaces. The relatively high doses of 0.1–0.3 mmol kg^{-1} of body weight of gadolinium imply that the complexes have to stay intact in the body because both the Gd^{3+} ion and the chelating ligand itself are toxic. The agents approved by the drug agencies are eight-dentate and involve either macrocyclic or acyclic polyaminocarboxylates. The macrocyclic ligands are based on 1,4,7,10-tetraazacyclododecane with four additional binding substituents on the four nitrogen atoms. The three low-molecular-weight compounds used in clinics are [Gd(DOTA)H_2O]$^-$ (Gd-DOTA), [Gd(HP-DO3A)H_2O] (Gd-HP-DO3A), and [Gd(BT-DO3A)H_2O] (Gd-BT-DO3A). The acyclic ligands are based on a diethylenetriamine substituted with at least three acetic acid groups and at least two other groups with coordinating atoms. Examples for complexes with acyclic polyaminocarboxylate complexes are [Gd(DTPA)H_2O]$^{2-}$ (Gd-DTPA), [Gd(DTPA-BMA)H_2O] (Gd-DTPA-BMA), and [Gd(EOB-DTPA)H_2O]$^{2-}$ (Gd-EOB-DTPA).

These basic chelating groups have been derivatized in many different ways either to build dimers, polymers, or other larger units or to attach groups that interact non-covalently among themselves to form, for example, micelles or with other molecules like proteins. These functionalizing groups can be either attached to a carbon of an ethylene bridge or to the CH_2-group of an acetic acid arm. A special group of ligands is formed if one or several carboxylate groups are replaced by phosphorus acid analogues. This replacement often leads to higher second sphere contributions to relaxivity due to the ability of the phosphate group to form hydrogen bonds with water (Kotek et al. 2003). Furthermore, in many of the phosphorus-containing complexes, water exchange is faster than on the typical polyaminocarboxylate compounds (Hermann et al. 2008).

Gadolinium complexes with two inner sphere water molecules can be obtained using ligands with only seven binding sites. Using the macrocyclic DO3A and fixing a nonbinding substituent on the fourth nitrogen lead to such complexes. As already mentioned, formation of ternary complexes is often observed if multidentate anions replace the two adjacent water molecules from the first coordination sphere. Replacing the acetate group on the central nitrogen of the DTPA ligand by a non-coordinating group leads to the DTTA chelator (Moriggi et al. 2008). Gadolinium complexes of DTTA-type ligands have a stability similar to [Gd(DTPA-BMA)H_2O] allowing to use them in studies with animals (Livramento et al. 2006; de Sousa et al. 2008). The two water molecules in the first sphere of Gd^{3+} are probably not adjacent and formation of ternary complexes are absent or weak (Moriggi et al. 2008). The incorporation of a pyridine unit into the ligand skeleton (L^1) has been shown beneficial for the complex stability, while the bishydration and the fast water exchange of the Gd^{3+} chelate have been preserved (Pellegatti et al. 2008).

Another seven-dentate chelating ligand with a different structure is the AAZTA based on the 1,4-diazepine ring system (Scheme 18.2) (Aime et al. 2004; Baranyai et al. 2009). The Gd^{3+} complex of AAZTA shows high stability (Table 18.1) and the water exchange is considerably faster than on Gd-DOTA or Gd-DTPA. As with the DTTA, no ternary complex formation with lactate or phosphate has been observed.

Another family of seven-coordinate chelators is based on the hydroxypyridinonate (HOPO) ligand (Xu et al. 1995; Cohen et al. 2000; O'Sullivan et al. 2003; Thompson et al. 2004; Xu et al. 2004;

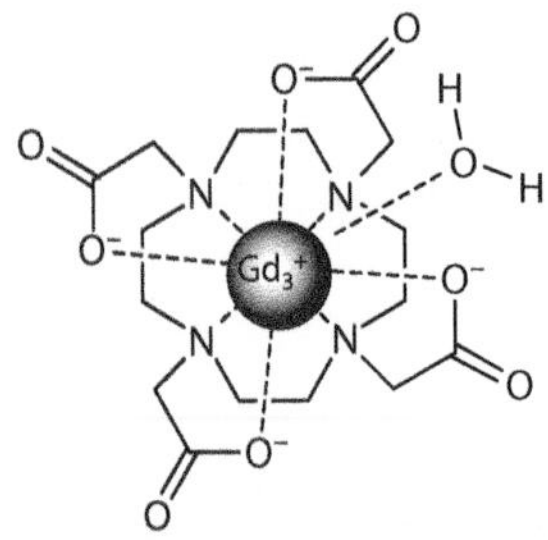

$[Gd(DOTA)H_2O]^-$

Dotarem®

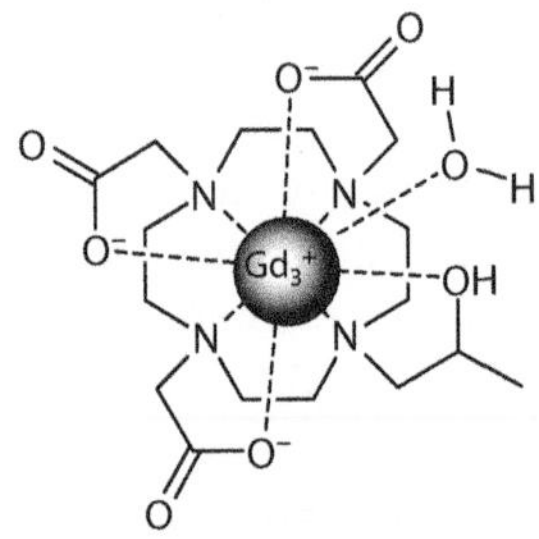

$[Gd(HP\text{-}DO3A)H_2O]^-$

ProHance®

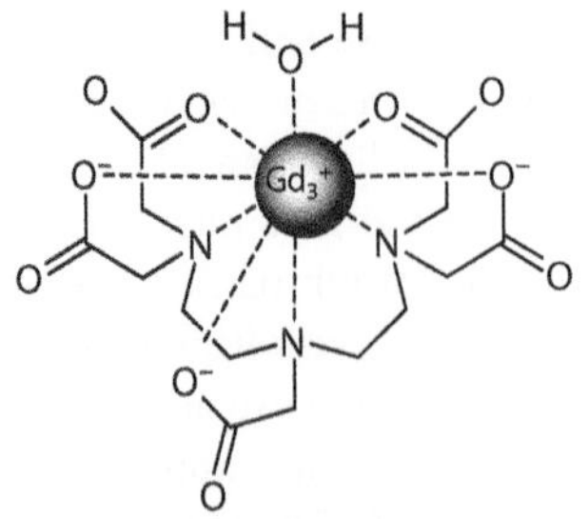

$[Gd(DTPA)H_2O]^{2-}$

Magnevist®

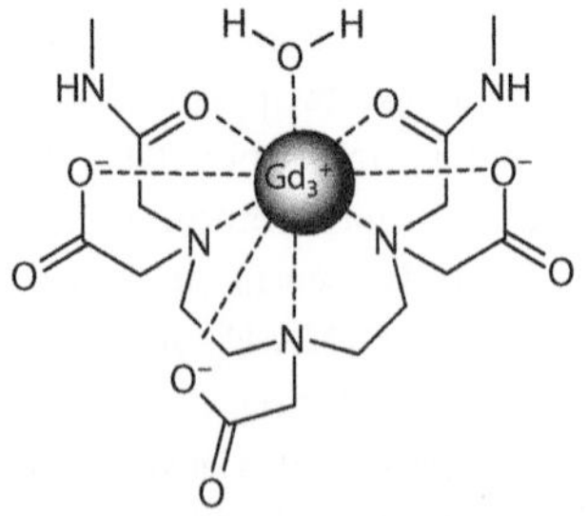

[Gd(DTPA-BMA)H_2O]

Omniscan®

SCHEME 18.1 Clinically used gadolinium-based MRI contrast agents.

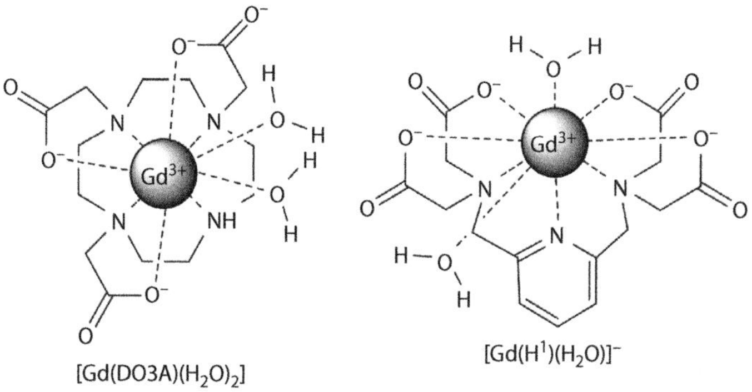

SCHEME 18.2 Gadolinium complexes with $q = 2$ studied as potential MRI contrast agents.

TABLE 18.1 Thermodynamic Stability Constants $\log K_{GdL}^{298}$ and Water Exchange Rates k_{ex}^{298} for Selected Gd^{3+} Complexes

Complex	Thermodynamic Stability Constant ($\log K_{GdL}^{298}$)	References	Water Exchange Rate k_{ex}^{298} ($10^6\ s^{-1}$)	References
$[Gd(DTPA)H_2O]^{2-}$	22.1	Tweedle et al. (1991)	3.3	Powell et al. (1996)
$[Gd(DTPA\text{-}BMA)H_2O]$	16.8	Wedeking et al. (1992)	0.45	Powell et al. (1996)
$[Gd(DTPA\text{-}BMEA)H_2O]^-$	16.84	Imura et al. (1997)	0.39	Toth et al. (1998)
$[Gd(EOB\text{-}DTPA)H_2O]^-$	23.46	Mühler and Weinmann (1995)	1.1	Laurent et al. (2004)
$[Gd(DOTA)H_2O]^-$	25.4	Tweedle et al. (1991)	4.1	Powell et al. (1996)
$[Gd(HP\text{-}DO3A)H_2O]$	22.8	Tweedle et al. (1991)	2.1	Laurent et al. (2006)
$[Gd(BT\text{-}DO3A)H_2O]$	21.8	Tombach and Heindel (2002)	2.7	Laurent et al. (2006)
$[Gd(DTTA\text{-}Me)(H_2O)_2]^-$	18.6	Moriggi et al. (2008)	—	Moriggi et al. (2008)
$[Gd(AAZTA)(H_2O)_2]^-$	20.24	Baranyai et al. (2009)	11.1	Aime et al. (2004); Baranyai et al. (2009)
$[Gd(L^1)(H_2O)_2]^-$	18.6	Pellegatti et al. (2008)	8.3	Pellegatti et al. (2008)
$[Gd(3,2\text{-}HOPO)(H_2O)_2]$	19.2	Doble et al. (2003), Werner et al. (2008)		
$[Gd(bisHOPO\text{-}tam\text{-}Me)(H_2O)_2]$	24.1	Cohen et al. (2000)	60–100	Cohen et al. (2000)

Pierre et al. 2006a; Werner et al. 2008; Datta and Raymond 2009). These complexes typically possess two inner sphere water molecules, which exchange relatively rapidly. Heteropodal complexes that include a terephthalamide (TAM) moiety can also favorably stabilize the coordination of a third water molecule on the Gd(III) center ($q = 3$) (Pierre et al. 2006b). Interestingly, the coordination of the third water molecule has been achieved without destabilizing the complex.

18.3.1 Gadolinium-Based Contrast Agents Based on Carbon Nanomaterials

We have seen that very stable complexes of gadolinium are required as contrast agents because of the toxicity of free Gd^{3+} ions. Another possibility to protect living organisms from the free gadolinium ion is its encapsulation in carbon nanomaterials like fullerenes and nanotubes. Water soluble gadofullerenes show high relaxivities despite the absence of water molecules directly coordinated to the paramagnetic ion. The relaxation enhancement is produced by exchanging protons from OH- or COOH-groups at the surface of the carbon cage. It has been shown that gadofullerenes form aggregates in aqueous solution, which are mainly responsible for the high relaxivities observed. Even higher relaxivities are observed with ultrashort single-walled carbon nanotubes partially filled with gadolinium ions. It has been shown that no Gd^{3+} is released from the tubes. These carbon-based bionanomaterials are presented in detail in Chapter 5.

18.4 "Inorganic" Bionanomaterials for MRI

Besides the nanomaterials based on organic compounds like dendrimers and micelles or carbon, inorganic materials are also used to construct containers and cores to transport many gadolinium ions to the cellular or molecular targets. Materials based on gold nanoparticles, silicon, and gadolinium oxide have been considered so far and will be presented in the following.

18.4.1 Gold Nanoparticles

One possible route to create nano-sized particles involves nanocrystals made of noble metals such as gold or silver. These units can be functionalized at the surface with different ligands, opening multimodal perspectives. An interesting feature of gold nanoparticles is that the content of elements with high atomic number is large, and these substances can be applied in vivo as x-ray contrast agents (Hainfeld et al. 2006). Gold nanoparticles coated at the surface with Gd^{3+} chelating units have been synthesized by reducing $HAuCl_4$ in the presence of dithiolated derivatives of DTPA (DTDTPA) (Debouttiere et al. 2006; Alric et al. 2008; Park et al. 2008). The thiol groups bind to gold atoms at the surface of the nanocrystals and also form disulfide bonds between DTDTPA molecules (Debouttiere et al. 2006). In this way, the gold nanoparticles with a diameter between 2 and 3 nm are covered by an organic shell composed of ~150 DTDTPA units, each being able to bind a gadolinium ion. Even if the relaxivity at 7 T/300 MHz is only slightly higher than for common $[Gd(DTPA)(H_2O)]^{2-}$, 3.9 mM^{-1} s^{-1} compared to 3.0 mM^{-1} s^{-1}, an overall relaxation enhancement per nanoparticle of 585 mM^{-1} s^{-1} can be calculated.

Markedly higher relaxivities have been achieved by functionalizing Au-particles with a small monothiol derivative of DTTA. This led to very rigid nanoparticles with ~50 gadolinium ions rather densely packed at the surface (Moriggi et al. 2009). Because each Gd-DTTA unit binds two water molecules, a very high relaxivity maximum of ~60 mM^{-1} s^{-1} at 30 MHz was found. The relaxation enhancement per particle at the relaxivity maximum is higher than 3000 mM^{-1} s^{-1}.

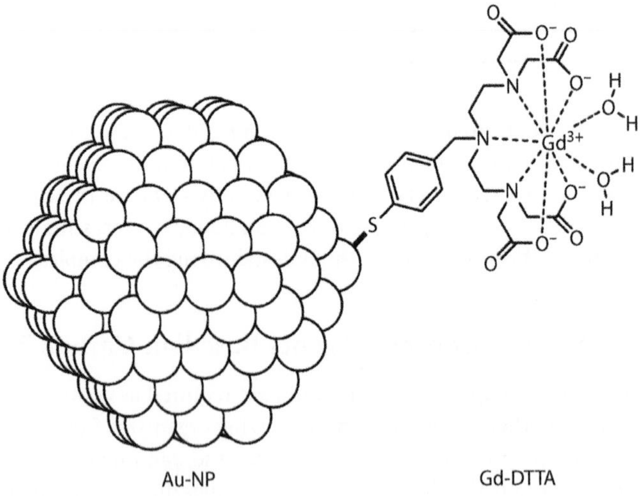

SCHEME 18.3 Gold nanoparticle functionalized with Gd-DTTA.

18.4.2 Silicon-Based Particles

A first group of particles used as MRI contrast agents based on silicon is formed by zeolites. Balkus et al. (Bresinska and Balkus 1994) developed Gadolite® for the imaging of the gastrointestinal tract, a NaY zeolite in which Na^+ is partially exchanged for Gd^{3+}. A study of nonspherical nano-sized GdNaY-zeolite (60–100 nm) showed that the measured relaxivity decreased with increasing load with gadolinium (Platas-Iglesias et al. 2002). A maximum relaxivity of 37 mM^{-1} s^{-1} (at 37°C and 40 MHz) has been obtained with a 1.3% (mass) loading. The results showed that the Gd^{3+} ions are immobilized inside the zeolite cavities and the relaxivities are mainly limited by relatively slow diffusion of water protons from the interior to the bulk water. The decrease of relaxivity with increasing gadolinium loading is explained by a decrease of the number of water molecules per Gd^{3+} available in the cavities (Platas-Iglesias et al. 2002). Upon dealumination of the zeolites, the volume of the cavities can be increased and more uncoordinated water molecules can be accommodated inside (Csajbók et al. 2005). Zeolite NaA is closely related to NaY with a framework that is also made up of sodalite cages, but in a different arrangement than in zeolite Y. For these particles, the supercages share an eight-membered ring with an open diameter of 4.1 Å compared to a 12-membered ring with 7.4 Å for zeolite Y. The importance of diffusion for relaxivity is demonstrated by the dramatic decrease of more than a factor of three in r_1 going from GdNaY to GdNaA (Csajbók et al. 2005).

A second class of silicon-based nanoparticles build as MRI contrast agents are nanocontainers (Tsotsalas et al. 2008) and mesoporous nanoparticles (Lin et al. 2004; Taylor et al. 2008; Tsai et al. 2008). A dual optical and magnetic imaging probe has been synthesized from nanometer-sized zeolite L crystals (Tsotsalas et al. 2008). Zeolite L is a cylindrically shaped, porous aluminosilicate, featuring a one-dimensional channel system oriented with a pore opening of 0.71 nm (Suarez et al. 2007). Molecular dyes have been entrapped in the channels, and the surface has been functionalized with about 370 Gd-DOTA units per zeolite crystal. A relaxivity of ~30 mM^{-1} s^{-1} (at 0.47 T/20 MHz, 298 K) has been measured, which leads to an overall relaxivity of ~11,000 mM^{-1} s^{-1} per particle. The relaxivity is limited by the slow water exchange from the $[Gd(DOTA)H_2O]$-unit.

Mesoporous silica nanoparticles that are functionalized with a green fluorescent dye and paramagnetic gadolinium have been synthesized to track stem cells (Hsiao et al. 2008; Tsai et al. 2008). Both the dye and the gadolinium DTPA-type chelate are loaded inside the pores. A relaxivity of 22 mM^{-1} s^{-1} (at 0.47 T/20 MHz) has been measured. The particles are internalized in stem cells and can be visualized in a 1.5 T MRI scanner in vivo for at least 14 days (Hsiao et al. 2008). Mesoporous silica nanospheres with diameters ranging from 60 to 1100 nm have been grafted with Gd-DTPA chelates on the surface via a siloxane linkage (Rieter et al. 2007; Taylor et al. 2008). Relaxivities measured are high at high magnetic field: r_1 = 28.8 mM^{-1} s^{-1} (3.0 T/128 MHz) and 10.2 mM^{-1} s^{-1} (9.4 T/400 MHz). The relaxation enhancement on a particle basis is 7×10^5 mM^{-1} s^{-1} (3.0 T), which allows imaging of labeled monocyte cells (Taylor et al. 2008).

18.4.3 Gadolinium Oxide Particles

Small and ultrasmall superparamagnetic iron oxide particles (SPIO and USPIO) are now well introduced in clinics as T_2 relaxation agents (Muller et al. 2001; Di Marco et al. 2007; Laurent et al. 2008; Sun et al. 2008). These agents are characterized by a large r_2/r_1 ratio, leading to a negative contrast in the image. In recent years, techniques have been developed to synthesize nano-sized particles of gadolinium oxide, Gd_2O_3, which are paramagnetic at room temperature and not superparamagnetic like Fe_3O_4 or γ-Fe_2O_3. Ultrasmall (5–10 nm) Gd_2O_3 nanocrystals capped with diethylene glycol (DEG) showed considerable T_1 and T_2 relaxation increase at 1.5 T compared to, for example, $[Gd(DTPA)H_2O]^{2-}$ (Engstroem et al. 2006; Fortin et al. 2007). In a spin echo sequence, short T_1 leads to increase and short T_2 to decrease of the magnetic resonance signal. At low concentration ($[Gd^{3+}] \leq 1\,mM$), the small Gd_2O_3 particles create a positive contrast like polyaminocarboxylate complexes of Gd^{3+}. Encapsulating Gd_2O_3 cores within polysiloxane doped with organic fluorophores and coated with PEG leads to multimodal optical and magnetic contrast agents (Bridot et al. 2007). It has been shown that these particles do not accumulate in the lung and liver but in kidneys and the bladder. Depending on the core size of the particles, T_1 relaxation enhancements per mM of particle are as high as 39,800 $mM^{-1}\,s^{-1}$.

18.5 "Organic" Bionanomaterials for MRI

18.5.1 Gd-Containing Lipidic Nanoparticles

Lipid-based colloidal aggregates, such as micelles, liposomes, or microemulsions have been extensively explored as drug carriers, (Lukyanov and Torchilin 2004; Samad et al. 2007) and more recently also in imaging applications (Mulder et al. 2006b; Mulder et al. 2007; Delli Castelli et al. 2008; Aime et al. 2009; Mulder et al. 2009). They are composed of lipids and/or other amphiphilic molecules containing both hydrophilic and hydrophobic moieties that spontaneously assemble to form aggregates in aqueous solutions. Such systems do not only provide improved pharmacokinetics and targeting possibilities of the contrast agent incorporated, but they can carry a very large number of paramagnetic ions and, by their macromolecular size, they also contribute to the increase of relaxivity per Gd. In addition, they offer easy ways to the preparation of multimodal agents. Lipidic aggregates represent therefore a very promising approach to deliver high payloads of gadolinium to specific sites and are widely investigated in molecular imaging applications.

The self-assembly of amphiphilic molecules is governed by hydrophobic interactions between the hydrophobic tails and repulsive interactions between the hydrophilic headgroups in a way that organizes the hydrophobic parts together and projects the hydrophilic moieties face to the water. The nature and the relative size of the hydrophobic chain and the hydrophilic headgroup will determine the structure of the lipidic aggregate. Among possible structures, micelles typically form from amphiphiles with a large headgroup and a single hydrophobic chain, while liposomes can be created from bilayer-forming lipids, typically containing double hydrophobic chains (Figure 18.6). Cholesterol is often added to stabilize the lipid bilayers, or PEGylated lipids are incorporated to increase circulation lifetimes of the liposome. Microemulsions, formed from water, oil, and an amphiphilic molecule, are stable dispersions of oil covered by a lipid monolayer.

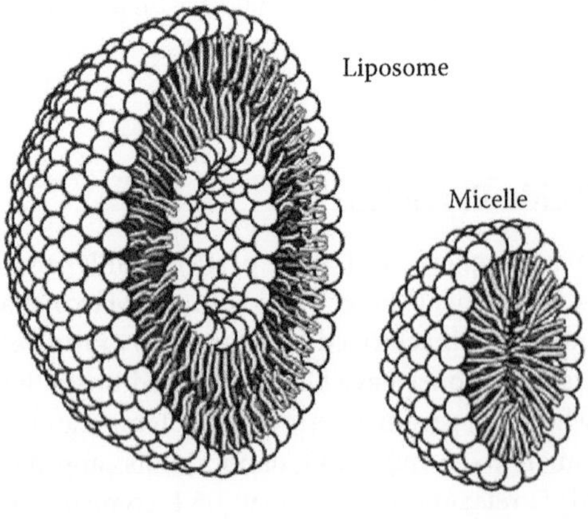

FIGURE 18.6 Schematic representation of liposomes and micelles. (Adapted from http://schools-wikipedia.org/wp/l/Lipid.htm)

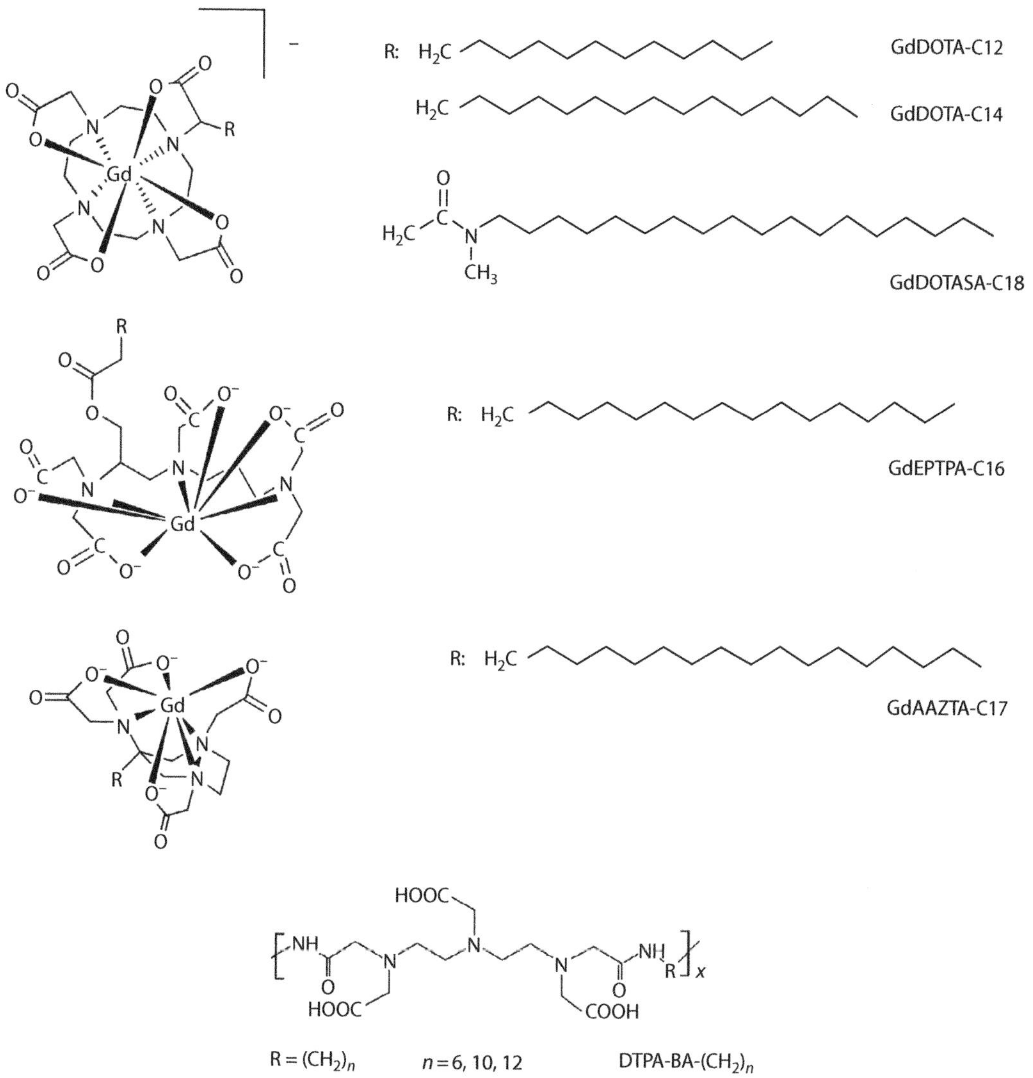

SCHEME 18.4 Examples of micelle-forming Gd^{3+} complexes.

18.5.1.1 Micellar Systems

Paramagnetic micelles of amphiphilic Gd^{3+} complexes were first considered for MRI applications more than 10 years ago (Andre et al. 1999). Indeed, from the synthetic point of view, the self-assembly of amphiphilic Gd^{3+} complexes is likely one of the easiest ways of creating nano-sized systems. Several Gd^{3+} complexes bearing hydrophobic side chains have been synthesized and their relaxometric properties investigated (Scheme 18.4) (Glogard et al. 2000; Nicolle et al. 2002a; Hovland et al. 2003; Accardo et al. 2004; Parac-Vogt et al. 2004; Nasongkla et al. 2006; Torres et al. 2006; Accardo et al. 2007). Micelle-like structures can also form from copolymers that consist of a nonionic hydrophilic polar part (the Gd chelate) and a hydrophobic part (for instance, methylene groups), which behave as typical nonionic polymeric surfactants (Scheme 18.4) (Toth et al. 1998, 1999; Dunand et al. 2001).

A relaxometric method has been established to assess the critical micellar concentration (*cmc*) (Nicolle et al. 2002a). The *cmc* corresponds to the concentration of the amphiphilic molecule above which micelles are present in solution. At concentrations below the *cmc*, no micelles form; thus, the proton relaxation rate measured in the paramagnetic solution is only due to the paramagnetic complex present as free surfactant and is given by Equation 18.3:

$$R_1^{obs} - R_1^d = r_1^{na} \times c_{\mathrm{Gd}} \tag{18.3}$$

where

r_1^{na} represents the relaxivity of the free, nonaggregated Gd^{3+} complex expressed in per millimole per second

c_{Gd} is the analytical Gd^{3+} concentration

R_1^d is the diamagnetic contribution to the longitudinal relaxation rate (the relaxation rate of pure water)

Above the *cmc*, the measured relaxation rate is the sum of two contributions, one due to the complex as free surfactant present at a concentration equal to the *cmc* and the other due to the aggregated complex in the form of micelles.

$$R_1^{obs} - R_1^d = r_1^{na} \times cmc + r_1^a\ (c_{\mathrm{Gd}} - cmc) \tag{18.4}$$

Equation 18.4 could be then written as follows:

$$R_1^{obs} - R_1^d = (r_1^{na} - r_1^a)\ cmc + r_1^a \times c_{\mathrm{Gd}} \tag{18.5}$$

The *cmc* is then determined from the plot of $R_1^{obs} - R_1^d$ as a function of c_{Gd} and by a simultaneous least-squares fit of the two straight lines, whose slopes define r_1^{na} and r_1^a below and above the *cmc*, respectively. As an example, the variation of $R_1^{obs} - R_1^d$ versus the total Gd^{3+} concentration, c_{Gd}, for $[Gd(DOTAC14)(H_2O)]^-$ is shown at 25°C and 60 MHz in Figure 18.7.

^1H relaxivity measurements provide an accurate and convenient tool to determine the *cmc* of a paramagnetic surfactant. Very importantly, this method can be easily used over a wide temperature range, which is not always accessible by other commonly applied techniques. It is evident that the *cmc* is strongly dependent on the length and the nature of the hydrophobic chain, and it was found to vary between 0.06 and 7.2 mM from C18 to C12-linked Gd-DOTA chelates (Scheme 18.4). A longer hydrophobic tail will result in a lower *cmc*.

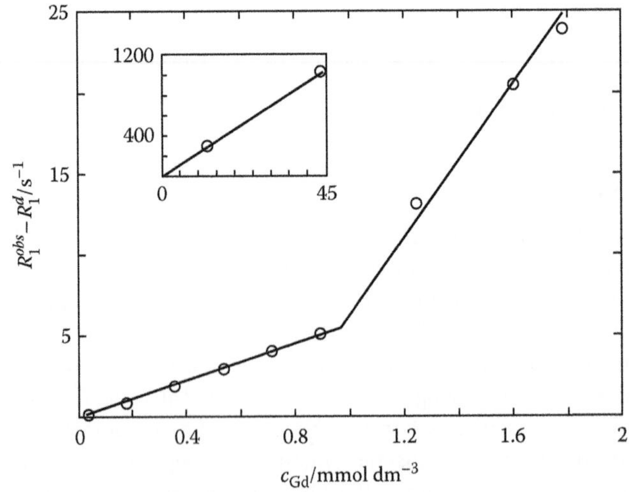

FIGURE 18.7 Variation of the paramagnetic proton relaxation enhancement as a function of the Gd^{3+} concentration in the solution of the amphiphilic complex. $[Gd(DOTAC14)(H_2O)]^-$ at 25°C and 60 MHz. The break point corresponds to the *cmc*.

Water exchange rates have been directly determined on several micellar systems by ^{17}O NMR measurements, and NMRD profiles have been recorded to fully describe the rotational dynamics. As expected, the relaxivities do increase for the micellar aggregates with respect to the monomeric chelates. However, typically they do not exceed $25\,mM^{-1}\,s^{-1}$ for monohydrated complexes in the frequency range 20–60 MHz, where the slower rotation has the greatest effect on the relaxivity. For several Gd-DOTA(Cn) derivatives, the ^{17}O NMR measurements confirmed that the rate of the water exchange is not considerably influenced by the micellar aggregation (Andre et al. 1999; Nicolle et al. 2002a; Torres et al. 2006) as it has also been found for polymeric Gd^{3+}-based surfactants (Kellar et al. 1997).

The relaxivities of these systems are still mainly limited by the relatively fast rotational motions. The rotational dynamics has been investigated in detail for several Gd-based micelles, both by ^{17}O longitudinal relaxation rate and ^{1}H relaxivity measurements (Nicolle et al. 2002a; Torres et al. 2006). These data evidenced an important internal flexibility of the micelles. The rotational dynamics has been described by the Lipari–Szabo model-free approach developed for non-extreme narrowing conditions. In the Lipari–Szabo approach, two kinds of motion are assumed to affect relaxation: a rapid, local motion, which lies in the extreme narrowing limit, and a slower, global motion (Lipari and Szabo 1982a,b). For the micellar systems, this analysis allows the separation of fast local motions of the Gd-containing headgroups from the slower overall motion of the entire micellar entity. Provided they are statistically independent and the global motion is isotropic, the reduced spectral density function can be written as

$$J(\omega) = \left(\frac{S^2\tau_g}{1+\omega^2\tau_g^2} + \frac{(1-S^2)\tau}{1+\omega^2\tau^2} \right) \quad \tau^{-1} = \tau_g^{-1} + \tau_l^{-1} \tag{18.6}$$

where

τ_g is a correlation time for the global motion, common to the whole micellar entity

τ_l is the correlation time for the fast local motion, specific for the individual relaxation axis, which is related to the motion of the Gd^{3+} chelate unit

The generalized order parameter, S^2, is a model-independent measure of the degree of spatial restriction of the local motion: $S^2 = 0$ if the internal motion is isotropic and $S^2 = 1$ if the motion is completely restricted. This analysis provides a phenomenological description of the rotational motion, which can be very useful for comparing systems with similar structure.

The local and global rotational correlation times as well as the generalized order parameters are presented for various micellar systems in Table 18.2. In general, we observe rather low values for the order parameter, S^2, indicating that the rotational motion is mainly modulated by the local rotational correlation time. This considerable flexibility of the micellar assemblies is further underlined by a comparison to Gd^{3+}-based linear copolymers or dendrimers, where higher values of the order parameter, around 0.5, have been reported (Nicolle et al. 2002b). Comparing the local rotational correlation times, τ_l, it seems that both the chain length and the chemical structure (hydrophobicity modified by linking function

TABLE 18.2 Global (τ_g^{298}) and Local (τ_l^{298}) Rotational Correlation Times and Order Parameters (S^2) for Various Micellar Gd^{3+} Complexes as Obtained from ^{17}O and ^{1}H Longitudinal Relaxation Rates Using the Lipari–Szabo Approach

Ligand	DOTA-C12 (Nicolle et al. 2002a)	DOTA-C14 (Nicolle et al. 2002a)	EPTPA-C16 (Torres et al. 2006)	AAZTA-C17 (Gianolio et al. 2007)	DOTASAC18 (Nicolle et al. 2002a)
τ_g^{298}/ps	1600	2220	2100	2540	2810
τ_l^{298}/ps	430	820	330	295	330
S^2	0.23	0.17	0.41	0.4	0.28

between the alkyl chain and the chelate) have an influence. The global correlation times increase continuously with increasing chain length within the series of analogous compounds. This parameter is directly linked to the overall mass of the micelles; thus, the augmentation of τ_g observed here would indicate the increase of the micelle size with the chain length. This could be expected supposing that the structure (shape) of the micelles formed by the various surfactants is similar; thus, a longer hydrophobic side chain implies a larger diameter for the micelle.

It has been demonstrated for different micellar systems that the dipolar interactions between the close paramagnetic centres on the micellar surface can lead to an important acceleration of the electron spin relaxation, which translates to a decrease in proton relaxivity. This limiting effect might vanish at frequencies above 10 MHz, as it was the case for Gadofluorine 8 (Nicolle et al. 2003). However, for GdAAZTA-C17 (Scheme 18.4), which has optimized water exchange, the relaxivities have been limited by such dipolar interactions all over the frequency range studied (0.01–80 MHz) (Gianolio et al. 2007). This undesirable effect of the large number of Gd centres being in close proximity to each other can be circumvented by diluting the paramagnetic metal by a diamagnetic analogue like yttrium. Indeed, for GdAAZTAC17, such dilution (2% Gd and 98% Y) led to an increase of ~40% of the relaxivities. However, in this approach, the Gd payload per micelle is largely reduced.

18.5.1.2 Micelles in Biological Environment: Imaging Applications

Given their size, micelles can remain in the blood pool for an extended time. At the same time, with the micellar approach, one can avoid the problems of slow elimination from the body that often are encountered with polymeric systems. Attaching PEG moieties to the micelle surface can further prolong their circulation half-life without being recognized by plasma proteins and phagocytic cells. Binding interactions between GdAAZTA-C17 micelles and both fatted and defatted HSA have been proved by relaxometric measurements (Gianolio et al. 2007). The slight increase in relaxivity of the micellar GdAAZTA-C17 in the presence of HSA was attributed to the formation of a supramolecular adduct driven by electrostatic interactions between the negatively charged micellar surface and the positive charges of certain protein residues. The formation of the adduct has also been confirmed by dynamic light-scattering measurements. The electrostatic nature of the binding was evidenced by the fact that increasing amounts of added NaCl led to a decrease in relaxivity and that fatty acid binding to the protein did not interfere with the interaction of the micelle. In contrast, no interaction has been found between HSA and GdEPTPA-C16 micelles (Torres et al. 2006). Meding et al. (2007) have used fluorescently and radioactively labeled derivates of Gadofluorine to show that following intravenous injection, the micelles formed by Gadofluorine M break down and the individual molecules bind to albumin. Recently, Bonnet et al. (2010) proposed a method based on amphiphilic luminescent lanthanide complexes and a hydrophobic chromophore incorporated into the micellar structure to follow the fate of micelles in biological media.

The presence of negative surface charges on GdAAZTA-type micelles has been exploited by Aime et al. in a novel approach to cellular labeling by micelles (Gianolio et al. 2008). The micelles have been anchored on the outer cell surface by the help of a cationic polymeric linker, which binds both the negatively charged cell surface and the micelle. Different polycationic species, including protamine, polylysine, polyornithine, and polyarginine, have been successfully tested. For instance, by using 1.1×10^8 polyarginine molecules per cell, it has been found that each cell can load up to approximately 5×10^6 particles. This corresponds to approximately 5×10^9 Gd/cell, well above the threshold for MRI visualization. Furthermore, as the Gd-containing units are anchored on the external surface of cells and not entrapped in endosomes inside the cells, the relaxivity is not quenched increasing the Gd-loading (Terreno et al. 2006) but increases linearly with the quantity of Gd-particles added.

Fayad et al. investigated the ability of targeted immunomicelles to detect macrophages in atherosclerotic plaque using MRI in vitro (Lipinski et al. 2006) and in vivo (Amirbekian et al. 2007). Macrophages play a central role in atherosclerosis and are associated with plaques vulnerable to rupture. Therefore, macrophage scavenger receptor (MSR) was chosen as a target for molecular MRI. MSR-targeted immunomicelles were tested in Apolipoprotein E knockout and WT mice. In vivo MRI revealed that at 24 h postinjection, immunomicelles provided a 79% increase in signal intensity of atherosclerotic aortas in Apolipoprotein E knockout mice compared with only 34% using untargeted micelles and no enhancement using GdDTPA. There was a strong correlation between macrophage content in atherosclerotic plaques and the matched in vivo MRI results as measured by the percent-normalized enhancement ratio. Immunomicelles provided excellent validated in vivo enhancement of atherosclerotic plaques.

In order to achieve active targeting of micelles to specific tumour sites, targeting agents can be coupled to the micelles. Accardo has reported mixed micelles prepared from amphiphilic molecules bearing one or two hydrophobic tails (18 carbon atoms) and the CCK8 peptide or the GdDTPAGlu moiety (Accardo et al. 2004, 2007). The CCK8 peptide is known to display high affinity for cholecystokinin receptors. These receptors, belonging to the G-protein coupled receptors (GPCRs) superfamily, are localized in the cell membrane and are promising targets due to their overexpression in many tumors like pancreatic cancer, small cell lung cancer, colon, and gastric cancers, medullary thyroid carcinomas, astrocitomas, and stromal ovarian tumors.

Gadofluorine is a very particular micellar contrast agent composed of a GdDOTA-type chelate and a strongly hydrophobic perfluoroalkyl chain, which proved to be applicable in diverse fields. Originally, it has been proposed in imaging of lymph nodes (Misselwitz et al. 2004). Gadofluorine M (a Gd-GlyMe-DOTA-perfluorooctyl-mannose-conjugate) accumulates rapidly in the lymph nodes and could be used to detect lymph node metastases after intravenous injection with a minimum effective diagnostic dose of 0.025 mmol/kg. Gadofluorine has also been successfully applied to detect atherosclerotic plaques in hyperlipidemic rabbits (Barkhausen et al. 2003; Meding et al. 2007) and plaque progression (Zheng et al. 2008). More recently, it has been tested as a novel tool to assess nerve outgrowth in vivo (Bendszus et al. 2005). After systemic application, Gadofluorine M selectively accumulated and persisted in nerve fibers undergoing Wallerian degeneration causing bright contrast on T1-weighted MR images. MRI enhancement was present already at 48 h within the entire nerve segments undergoing degeneration and subsequently disappeared from proximal to distal parts in parallel to regrowth of nerve fibers. Finally, in a recent study, Gadofluorine M was used for quantitative characterization of tumors according to their histopathologic grade and for differentiation of benign and low-grade carcinomas from high-grade malignant neoplasms (Raatschen et al. 2006). Due to the tendency of this agent to form large micelles in water and to bind strongly to hydrophobic sites on proteins, patterns of dynamic tumor enhancement could be used to differentiate benign from malignant lesions, to grade the severity of malignancies and to define areas of tumour necrosis.

Micelles have great potential to deliver hydrophobic drugs; consequently, paramagnetic micelles can be used for controlled drug delivery. Gao et al. reported a multifunctional, polymeric micellar system with cancer-targeting ability via $\alpha_v\beta_3$ integrins endowed with efficient MRI reporter characteristics for controlled delivery of the anticancer agent doxorubicin (Nasongkla et al. 2006). In another approach, Shuai et al. used folate functionalized micelles to deliver doxorubicin to hepatic carcinoma cells (Hong et al. 2008). Though these particular systems contained iron oxide nanoparticles inside the micelle core as MRI reporter, similar structures bearing Gd^{3+} complexes can also be imagined.

18.5.1.3 Liposomal Contrast Agents

Liposomes are nano-sized vesicles composed of phospholipid layers encapsulating an aqueous core. Liposomes are highly biocompatible, they can carry hydrophilic (inside the liposome) and hydrophobic or amphiphilic substances (in the lipid bilayer), their preparation is relatively easy, and their pharmacokinetic properties can be widely modulated via the chemical composition of the bilayer. The first liposomes loaded with various paramagnetic ions or chelates were investigated more than 20 years ago, and research still continues to synthesize more efficient and more stable liposomes (Hak et al. 2009). It has been recognized early on that the limited water permeability of the lipid bilayer can considerably reduce (by a factor of 10) the relaxation efficiency of the liposome-entrapped paramagnetic species on the bulk water (Fossheim et al. 1999). Nevertheless, liposomes loaded with Gd^{3+} complexes have been demonstrated to be useful in various imaging applications, such as in visualizing tumors (Pirollo et al. 2006), atherosclerotic plaques (Mulder et al. 2006a), lymph nodes (Wisner et al. 1997), etc. On the other hand, the quenching effect of the liposome encapsulation on the relaxivity could be exploited to create imaging agents that are responsive to pH or temperature. Indeed, these physicochemical parameters can strongly affect the permeability of the liposome bilayer; thus, their variation can be detected via the relaxivity change related to the increased water permeability of the bilayer or to the release of the paramagnetic complex. The use of paramagnetic pH-sensitive liposomes was recently suggested as a new approach for monitoring pathologic changes in pH by MRI. Such liposomes must be stable in blood and selectively release the encapsulated paramagnetic agent when exposed to lower pH in the target tissue. Løkling et al. reported pH-responsiveness of liposomes consisting of phosphatidylethanolamine-based phospholipids and dipalmyitoylglycerosuccinate, encapsulating GdDTPA-BMA (Løkling et al. 2001, 2004). Other types of phospholipidic components can allow for the design of temperature-responsive liposomic agents (Lindner et al. 2005). The transition of these systems from the gel to the liquid state occurs at a well-defined temperature, above which the liposome bilayer is characterized by a greater fluidity and water permeability. Moreover, the liposome might completely release its paramagnetic content as well, which accounts for the large increase observed in relaxivity (Fossheim et al. 2000). Such temperature-sensitive liposomes have also been studied in vivo in guided hyperthermia treatment of cancer in animal models (Frich et al. 2004). Liposomes containing a paramagnetic Mn^{2+} salt or a Gd^{3+} complex and doxorubicin have been tested to detect drug release in general (Viglianti et al. 2004) or to monitor the delivery of the anticancer agent to brain tumors in rodents (Mamot et al. 2004; Saito et al. 2005).

Liposomes can also provide a convenient platform to integrate various imaging probes in the objective of creating bimodal or multimodal imaging agents, which can be further combined with targeting agents. Mulder et al. (2004) introduced a bimodal-targeted liposomal contrast agent for the detection of molecular markers with both MRI and fluorescence microscopy. The liposomes contained Gd–DTPA attached to stearyl chains and a fluorescent lipid for MRI and optical detection, respectively, and were tested in various in vitro and in vivo models (Mulder et al. 2005).

Aime et al. (2006) have proposed a novel approach to design improved responsive liposomes (Terreno et al. 2008a). By encapsulating a high amount of Gd^{3+} chelate in the vesicle, one obtains a nanosystem that simultaneously affects both the longitudinal and the transverse relaxation rate of the surrounding water protons. This allows creating thermosensitive liposomes whose MRI response is no longer dependent on the concentration of the probe. The method is based on the determination of the ratio between the transverse and longitudinal paramagnetic contributions to the observed water proton relaxation rates. Both of these quantities are proportional to the concentration of the paramagnetic agent; thus, their ratio provides a concentration-independent response on the MR images. In order to be effective as a responsive agent, this ratio has to be sensitive to the parameter that one wants to monitor. This responsiveness can typically be related to the differences in the mechanism that governs the transverse and longitudinal relaxation. This concept can be particularly well exploited for temperature measurements since the longitudinal and transverse relaxivities of large, Gd^{3+}-encapsulated liposomes at high magnetic fields have opposite temperature dependency.

Paramagnetic liposomes can act as very efficient T_2 susceptibility agents. The typical transverse relaxivity values have been estimated from 10^6 to $10^9\,mM^{-1}\,s^{-1}$ referred to the liposome concentration for 7 T and 298 K (Terreno et al. 2008b). These values make them comparable in efficacy to the gold standard iron oxide nanoparticles.

Aime et al. have exploited paramagnetic liposomes as CEST agents. Chemical Exchange Saturation Transfer (CEST) agents (Zhang et al. 2001; Aime et al. 2002; Woods et al. 2006) are compounds containing protons in relatively slow chemical exchange with the bulk water protons. These protons are typically situated either on functional groups of the molecule (such as amides, amines, and hydroxyls) or on a very slowly exchanging water molecule of a lanthanide chelate. When the NMR signal of these exchangeable protons is selectively saturated (the populations of the both energy levels are equilibrated and the NMR signal is reduced or disappears), the chemical exchange will also transfer the saturation to the second proton pool (bulk water proton), and, consequently, the intensity of the NMR signal of the second pool is decreased, which can be detected on CEST MR images. The main advantage of CEST agents with respect to traditional Gd-based MRI probes is that the contrast can be generated at will by selectively irradiating the system at the appropriate frequency. It also implies that different CEST agents can be independently visualized in the same region of interest (Terreno et al. 2008b). Furthermore, CEST agents can be particularly well adapted to design responsive probes, and in the last years, they have been widely investigated in this respect. One major drawback of the CEST technique is its low sensitivity. To circumvent this problem, Aime et al. suggested the use of liposomes entrapping a shift reagent that induces a selective shift of the proton resonance of the intraliposomal water with respect to the bulk water (LIPOCEST agents) (Aime et al. 2005). The proton exchange rate between the intra- and extraliposomal water pool falls into the right region to exploit the CEST effect. In addition, the large number of protons inside the liposome (10^7–10^9 per vesicle) provides great sensitivity, pushing the limit of detection to subnanomolar level. Moreover, nonspherical liposomes ensure much larger shifts of the entrapped water protons due to the contribution from the bulk magnetic susceptibility (Aime et al. 2007b; Terreno et al. 2007; Terreno et al. 2008a). Such nonspherical liposomes could be obtained via osmotic shrinkage and by incorporating amphiphilic shift reagents into the liposome bilayer. Moreover, the sign of the intraliposomal water proton shift is dependent on the orientation of the nonspherical liposomes in the magnetic field, itself related to the magnetic anisotropy of the shift reagent incorporated.

18.5.1.4 Lipoproteins

Low-density lipoprotein (LDL) and high-density lipoprotein (HDL) have an important role in the transport of cholesterol. Fayad et al presented an HDL-like nanoparticle contrast agent that selectively targets atherosclerotic plaques (Frias et al. 2004). These HDL-like particles have several advantages: a small size (7–12 nm diameter); protein components that are endogenous, biodegradable, and do not trigger immunoreactions; and particles that are not recognized by the reticuloendothelial system. Furthermore, HDL-like particles are easily reconstituted and can carry a considerable contrast agent payload. The techniques for the preparation of this contrast agent involve isolation and delipidation of normal human HDL to obtain apo-HDL proteins. The apolipoproteins are extracted and reconstituted with phospholipids, including a phospholipid-derivative Gd^{3+}-based contrast agent (monohydrated GdDTPA-BMA [Frias et al. 2004] or bishydrated GdAAZTA [Briley-Saebo et al. 2009]) that becomes incorporated into the reconstituted particle. Reconstituted HDLs can be obtained in large scale, suitable for therapeutic use. The synthesis has to be carried out below the cmc of the lipidic Gd^{3+} complex, otherwise disrupted particles form (Briley-Saebo et al. 2009). Fluorescent dyes or other imaging probes can also be simultaneously integrated into the lipoprotein to create bimodal or multimodal agents (Cormode et al. 2008; Vucic et al. 2009). Gd^{3+}-loaded LDL particles have been used to identify and characterize vulnerable atherosclerotic plaques (Chen et al. 2008; Cormode et al. 2009a,b) or tumor cells (Crich et al. 2007).

18.5.2 Microemulsions

Microemulsions consist of water, oil, and an amphiphilic molecule whose mixture gives rise to a thermodynamically stable solution. Microemulsions based on a perfluorocarbon (PFC) core have been largely exploited by the group of Lanza and Wickline (Lanza et al. 2005; Winter et al. 2005, 2007). PFC nanoparticles are distinctly different from other oil-based emulsions, thanks to the physicochemical properties of fluorine, the most electronegative of all elements. The biocompatibility of liquid fluorocarbons is well documented. Even at large doses, most fluorocarbons are innocuous and physiologically inactive. No toxicity, carcinogenicity, or mutagenicity effects have been reported for pure fluorocarbons. PFC nanoparticles can serve as a platform technology for molecular imaging and targeted drug-delivery applications (Winter et al. 2007). These nanoparticles are approximately 250 nm in diameter and are encapsulated in a phospholipid shell, which provides an ideal surface for the incorporation of targeting ligands, imaging agents, and drugs. PFC emulsion nanoparticles may be functionalized for targeted MR molecular imaging by the surface incorporation of paramagnetic chelates and homing ligands into the outer phospholipid monolayer. With a nominal diameter of 250 nm, PFC nanoparticles present an enormous surface area to transport and concentrate paramagnetic metal to important vascular biomarkers sites. They can deliver 50,000–90,000 gadolinium ions each, and in the case of PFC nanoparticles, all of the paramagnetic ions are presented to the outer aqueous phase for maximum relaxivity. A variety of different epitopes, including alpha(v)beta(3)-integrin, tissue factor, and fibrin, have been imaged using PFC nanoparticles (Winter et al. 2003; Morawski et al. 2004; Lanza et al. 2005; Marsh et al. 2007). This platform is not only effective as a T1-weighted agent but also supports F-19 MR spectroscopy and imaging (Neubauer et al. 2008). Lipophilic drugs can also be incorporated into the outer lipid shell of nanoparticles for targeted delivery (Pan et al. 2008). By combining targeted molecular imaging and localized drug delivery, PFC nanoparticles provide diagnosis and therapy with a single agent.

18.5.3 Dendrimers

Dendrimers have been used for a long time as nanoscale scaffolds for the attachment of Gd^{3+} complexes (Kobayashi and Brechbiel 2005; Longmire et al. 2008). These highly ordered, three-dimensional, branched synthetic polymers have several advantages including low polydispersity and easy control of their size, shape, and loading with Gd chelates (Tomalia et al. 1990). In contrast to many other macromolecular systems, dendrimers in general are rigid and monodisperse and often have a quasi-spherical or spherical three-dimensional structure. They are synthesized in repetitive reaction steps starting from a small core molecule, leading to consecutive generations. The surface groups, whose number is largely increasing with increasing generation, can be used for the conjugation of Gd^{3+} chelates. This method has made possible the accumulation of a very large number of gadolinium chelates within one molecule (up to 1860 reported for a generation 10 PAMAM (polyamidoamine) dendrimer) (Bryant et al. 1999). As for their proton relaxivity, an important issue is the right choice of the linker group between the macromolecule and the Gd^{3+} complex. This has to be rigid enough so that the slow rotation of the rigid dendrimer molecule is transmitted to the surface chelate itself. Since the first report on a dendrimeric Gd^{3+}-based contrast agent, by Wiener et al. (1994), a large number of Gd^{3+}-loaded dendrimers have been synthesized. Most of them were based on polyamidoamine, PAMAM, and dendrimers and were loaded with DOTA- (Toth et al. 1996; Bryant et al. 1999; Rudovsky et al. 2005) or DTPA-type (Wiener et al. 1994; Lebduskova et al. 2006) chelates, often using a *p*-NCS-benzyl linker. For these types of dendrimer-based Gd^{3+} complexes, the relaxivity was increasing with increasing generation before reaching a plateau for the high generation compounds (above generation 7, e.g., for $[Gd(p\text{-NCS-bz-DOTA})]^-$-loaded dendrimers) (Bryant et al. 1999). The NMRD profiles show typical high field peaks around 20 MHz, characteristic of slow rotation. The relaxivities for the high generation dendrimers ($G = 5$–10) often decrease as the temperature decreases, indicating that slow water exchange limits relaxivity. Slow water exchange and fast internal motions were found responsible for the limited relaxivities for Gd(DOTA-monoamide)-functionalized PAMAM dendrimers (Toth et al. 1996), as well as for Gadomer 17, a lysine-based dendrimer with a 1,3,5-tricarboxylic acid core and 24 Gd^{3+} chelates on the surface (Nicolle et al. 2002b). Therefore, dendrimers loaded with Gd^{3+} complexes that possess fast water exchange have also been synthesized, though without spectacular increase in the proton relaxivity. This has been related to limitations by fast internal motions of the Gd^{3+} segments (Laus et al. 2005; Rudovsky et al. 2005; Lebduskova et al. 2006; Rudovsky et al. 2006; Polasek et al. 2009). Such fast motions of the surface groups are particularly important when non-charged chelates are attached on the surface (Polasek et al. 2009). This has been demonstrated in a detailed analysis of the rotational dynamics by using the Lipari–Szabo treatment on a series of generation 4 PAMAM dendrimers and hyperbranched (HB) dendrimeric structures loaded with macrocyclic Gd^{3+} chelates on their surface (Jaszberenyi et al. 2007). The Gd^{3+} was complexed either by the tetraaza-tetracarboxylate DOTA unit (DOTA-*p*Bn) or by the tetraaza-tricarboxylate-monoamide

DO3A-MA chelator (Scheme 18.5). More than twice as high proton relaxivities have been found at frequencies below 200 MHz for the dendrimers loaded with the negatively charged [Gd(DOTA-*p*Bn)(H_2O)]$^-$ in comparison to the dendrimeric complex bearing the neutral [Gd(DO3A-MA)(H_2O)] moieties. The analysis of the field-dependent proton relaxivities in terms of local and global rotational motion allowed concluding that it is almost exclusively the different rotational dynamics that are responsible for the different proton relaxivities. The slower rotation of the [Gd(DOTA-*p*Bn)(H_2O)]$^-$-loaded dendrimers and the consequently higher relaxivity were related to a negative charge of the complex, which creates more rigidity and increases the overall size of the macromolecule as compared to the dendrimer loaded with the neutral [Gd(DO3A-MA)(H2O)] complex. This study also showed that HB dendrimers can be as good macromolecular scaffolds for Gd^{III} complexes with respect to proton relaxivity as the regular PAMAM dendrimers.

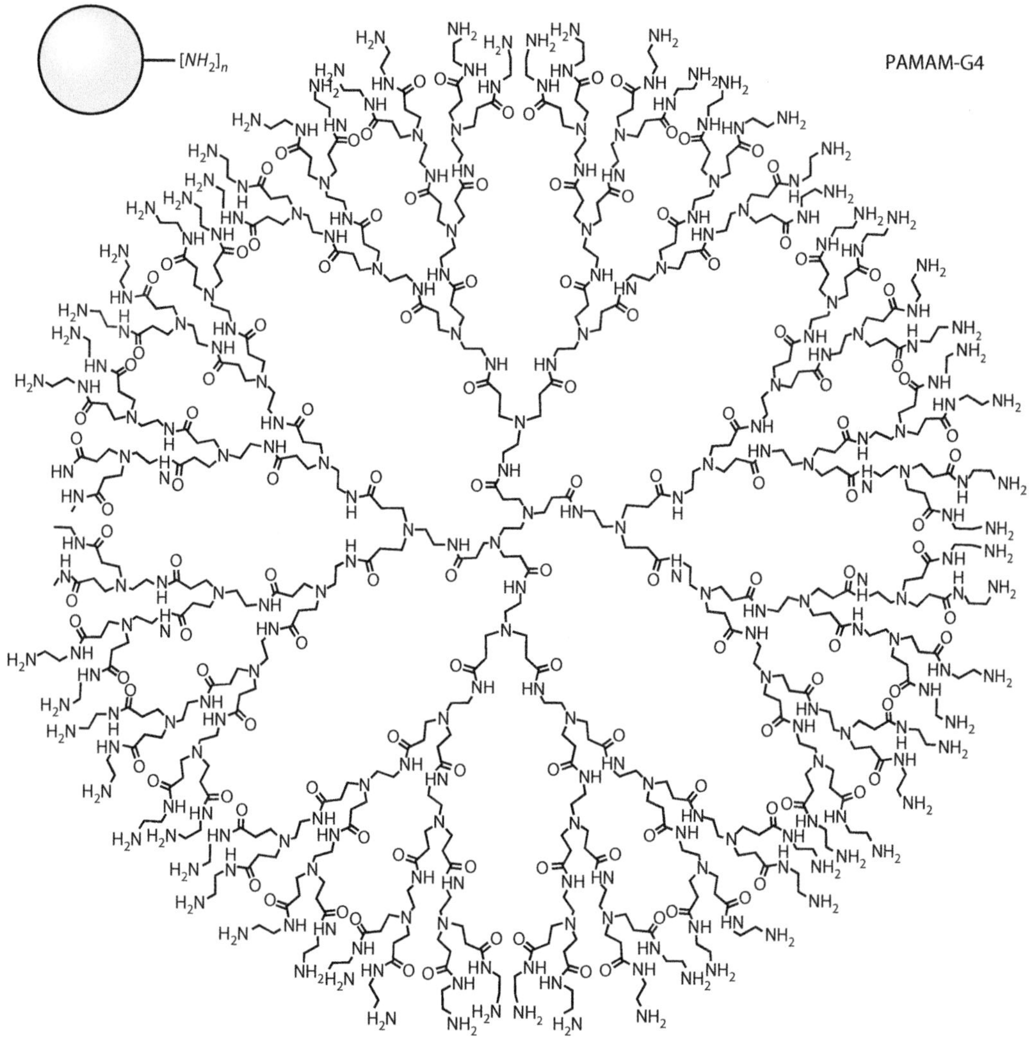

SCHEME 18.5 Dendrimers as macromolecular scaffolds for MRI contrast agents. Top: ethylenediamine-cored polyamido amine, generation 4 (PAMAM G4); bottom left: hyperbranched, ethylenediamine-cored polyethylene imine (HB-PEI); bottom right: hyperbranched, amino-functionalized polyglycerol (HB-PG). Different moieties attached to the respective dendrimers via amide or thiourea bonds.

HB-PEI

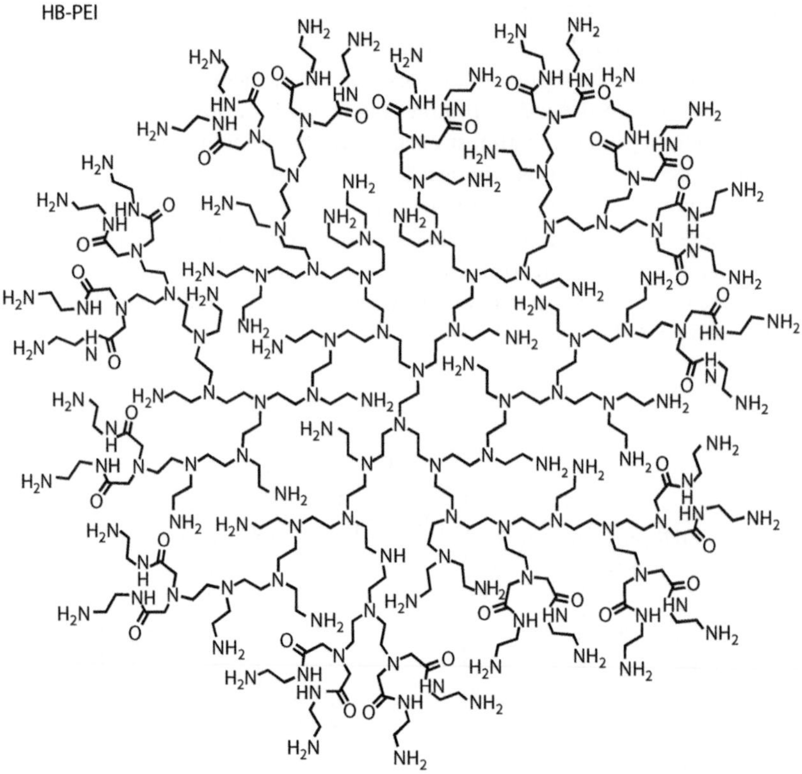

HB-PG

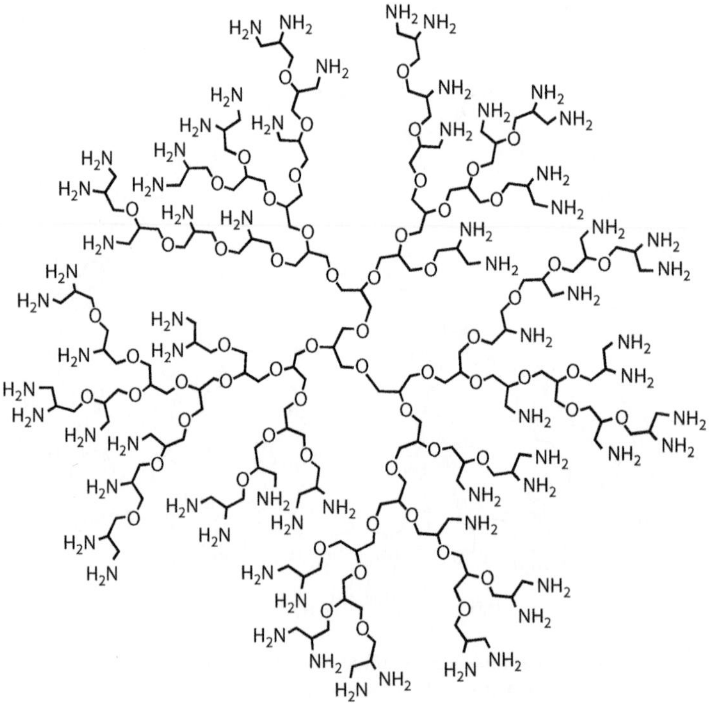

SCHEME 18.5 (continued)

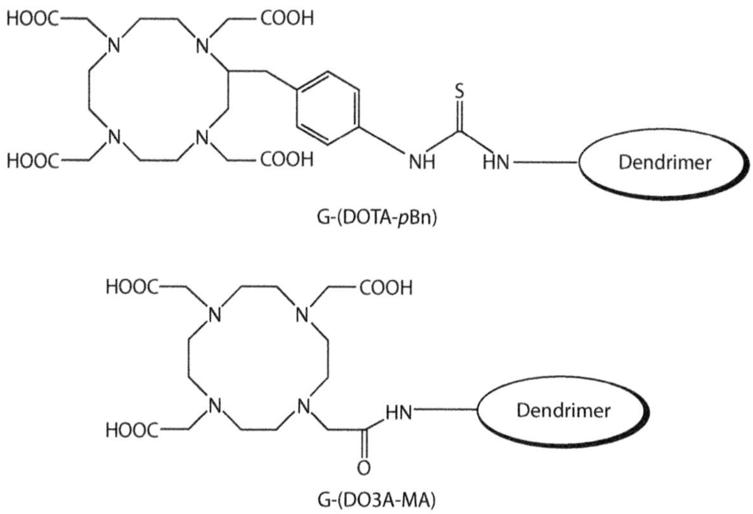

SCHEME 18.5 (continued)

In addition to the relatively high proton relaxivities and the high Gd^{3+} payload achievable, dendrimeric structures possess an extensive flexibility that makes them ideal for molecular-imaging applications (Tomalia et al. 2007; Longmire et al. 2008). They can be modified atom by atom on cores, interiors, and surface groups that allows for a rational manipulation to optimize their physical and biological characteristics (Kobayashi et al. 2003). They can be adapted to localize preferentially to areas or organs of interest for facilitating target-specific imaging. Furthermore, one can take advantage of the numerous binding sites on the surface of a single dendrimer molecule and develop multimodality imaging agents (Talanov et al. 2006; Koyama et al. 2007).

Early applications of dendrimers in molecular imaging concerned "passive" targeting based on variations in dendrimer size (generation number) and charge. It has been recognized early on that small changes in diameter of the dendrimeric agents can substantially modify their pharmacokinetics, vascular permeability, excretion, or recognition. Kobayashi et al. synthesized a series of dendrimeric Gd^{3+} complexes with molecular weights of 29–3850 kDa corresponding to molecular sizes of 3–15 nm to systematically investigate the effect of these parameters on blood retention and tissue perfusion of excretion rate (Kobayashi and Brechbiel 2005). The applicability of dendrimeric contrast agents was first demonstrated in imaging the blood pool (Kobayashi and Brechbiel 2005) and tumor angiogenesis (Barrett et al. 2006). Dendrimers with hydrophobic character were used in hepatic imaging (Kobayashi and Brechbiel 2005). Their usefulness in kidney imaging has also been proven (Dear et al. 2006).

A recent research direction involves the development of biodegradable dendrimers to facilitate the clearance of the agent from the body after MRI examinations. In this perspective, Lu et al. used disulfide groups in the interior of the dendrimer (Lu et al. 2006; Zong et al. 2009). These novel agents have relatively long blood circulation time and can thus act as macromolecular contrast agents for in vivo imaging, then gradually degrade into small Gd^{3+} complexes, which are rapidly excreted via renal filtration. These agents result in effective and prolonged in vivo contrast enhancement in the blood pool and tumor tissue in animal models, yet demonstrate minimal Gd^{3+} tissue retention as the clinically used low-molecular-weight agents.

18.6 Conclusions and Perspectives

In the last decade, a great variety of nano-sized systems containing Gd^{3+} complexes have been explored in the perspective of contrast agent applications in MRI. Nanoparticular MRI probes have several invaluable properties with respect to small molecular weight agents:

1. Due to their slow rotation, they ensure considerably higher molar relaxivities, thus higher MRI efficiency at typical clinical magnetic fields.
2. They allow for the delivery of a large Gd^{3+} payload to the site of interest, which, together with the high molar relaxivity, provides important contrast enhancement.
3. They provide unique possibilities to develop multimodal imaging probes, which integrate, within the same molecular entity, features required for various imaging modalities.
4. They open the way to the development of theranostic agents, which combine diagnostic and therapeutic properties in the objective of directly monitoring the delivery of the therapeutic agent and its effect by imaging techniques.

In the recent years, nano-sized agents have largely contributed to the important advances in molecular MRI, providing a noninvasive, in vivo real-time monitoring of molecular events occurring in cells and organisms. Their modularity endows them with the most versatile properties, which ensure a bright future for nanomaterials in MRI.

Acknowledgments

ET is grateful to the National Cancer Institute, the Cancer Ligue, and the National Research Agency (ANR), France, for financial support. LH acknowledges financial support of the Swiss National Science Foundation and the Swiss State Secretariat for Education and Research (SER). This work was performed in the frame of the EU COST Action D38 "Metal Based Systems for Molecular Imaging Applications."

References

Aaman, K. and P.-O. Westlund. 2007. Direct calculation of $^{1}H_2O$ T_1 NMRD profiles and EPR lineshapes for the electron spin quantum numbers S = 1, 3/2, 2, 5/2, 3, 7/2, based on the stochastic Liouville equation combined with Brownian dynamics simulation. *Phys. Chem. Chem. Phys.* 9 (6):691–700.

Accardo, A., D. Tesauro, G. Morelli, E. Gianolio, S. Aime, M. Vaccaro, G. Mangiapia, L. Paduano, and K. Schillen. 2007. High-relaxivity supramolecular aggregates containing peptides and Gd complexes as contrast agents in MRI. *J. Biol. Inorg. Chem.* 12:267–276.

Accardo, A., D. Tesauro, P. Roscigno, E. Gianolio, L. Paduano, G. D'Errico, C. Pedone, and G. Morelli. 2004. Physicochemical properties of mixed micellar aggregates containing CCK peptides and Gd complexes designed as tumor specific contrast agents in MRI. *J. Am.Chem. Soc.* 124:3097–3107.

Aime, S., Z. Baranyai, E. Gianolo, and E. Terreno. 2007a. Paramagnetic contrast agents. In *Molecular and Cellular MR Imaging*, eds. M. M. J. Modo and J. W. M. Bulte. Boca Raton, FL: CRC Press.

Aime, S., L. Calabi, C. Cavallotti, E. Gianoloio, G. B. Giovenzana, P. Losi, A. Maiocchi, G. Palmisano, and M. Sisti. 2004. [Gd-AAZTA]⁻: A new structural entry for an improved generation of MRI contrast agents. *Inorg. Chem.* 43:7588–7590.

Aime, S., D. D. Castelli, S. G. Crich, E. Gianolio, and E. Terreno. 2009. Pushing the sensitivity envelope of lanthanide-based magnetic resonance imaging (MRI) contrast agents for molecular imaging applications. *Acc. Chem. Res.* 42 (7):822–831.

Aime, S., D. D. Castelli, D. Lawson, and E. Terreno. 2007b. Gd-loaded liposomes as T1, susceptibility, and CEST agents, all in one. *J. Am. Chem. Soc.* 129:2430–2431.

Aime, S., D. D. Castelli, and E. Terreno. 2002. Novel pH-reporter MRI contrast agents. *Angew. Chem. Int. Ed. Engl.* 41 (22):4334–4336.

Aime, S., D. D. Castelli, and E. Terreno. 2005. Highly sensitive MRI chemical exchange saturation transfer agents using liposomes. *Angew. Chem. Int. Ed. Engl.* 44 (34):5513–5515.

Aime, S., F. Fedeli, A. Sanino, and E. Terreno. 2006. A R2/R1 ratiometric procedure for a concentration-independent, pH-responsive, Gd(III)-based MRI agent. *J. Am. Chem. Soc.* 128 (35):11326–11327.

Alric, C., J. Taleb, G. Le Duc, C. Mandon, C. Billotey, A. Le Meur-Herland, T. Brochard et al. 2008. Gadolinium chelate coated gold nanoparticles as contrast agents for both x-ray computed tomography and magnetic resonance imaging. *J. Am. Chem. Soc.* 130 (18):5908–5915.

Amirbekian, V., M. J. Lipinski, K. C. Briley-Saebo, S. Amirbekian, J. G. S. Aguinaldo, D. B. Weinreb, E. Vucic et al. 2007. Detecting and assessing macrophages in vivo to evaluate atherosclerosis noninvasively using molecular MRI. *Proc. Natl. Acad. Sci. U.S.A.* 104 (3):961–966.

Andre, J. P., E. Toth, H. Fischer, A. Seelig, H. Maecke, and A. E. Merbach. 1999. High relaxivity for monomeric Gd(DOTA)-based MRI contrast agents, thanks to micellar self-organization. *Chem. Eur. J.* 5 (10):2977–2983.

Astashkin, A. V., A. M. Raitsimring, and P. Caravan. 2004. Pulsed ENDO study of water coordination to Gd3+ complexes in orientationally disordered systems. *J. Phys. Chem. A* 108:1990–2001.

Ayant, Y., E. Belorizky, J. Alizon, and J. Gallice. 1975. Calculation of the spectral densities for relaxation resulting from random translational motion by magnetic dipolar coupling in liquids. *J. Phys.* 36:991–1004.

Banci, L., I. Bertini, and C. Luchinat. 1991. *Nuclear and Electron Relaxation.* Weinheim, Germany: VCH.

Baranyai, Z., F. Uggeri, G. B. Giovenzana, A. Benyei, E. Brucher, and S. Aime. 2009. Equilibrium and kinetic properties of the lanthanoids (III) and various divalent metal complexes of the heptadentate ligand AAZTA. *Chem. Eur. J.* 15 (7):1696–1705.

Barkhausen, J., W. Ebert, C. Heyer, J. F. Debatin, and H.-J. Weinmann. 2003. Detection of atherosclerotic plaque with Gadofluorine-enhanced magnetic resonance imaging. *Circulation* 108 (5):605–609.

Barrett, T., H. Kobayashi, M. Brechbiel, and P. L. Choyke. 2006. Macromolecular MRI contrast agents for imaging tumor angiogenesis. *Eur. J. Radiol.* 60 (3):353–366.

Belorizky, E. and P. H. Fries. 2004. Simple analytical approximation of the longitudinal electronic relaxation rate of Gd(III) complexes in solutions. *Phys. Chem. Chem. Phys.* 6 (9):2341–2351.

Belorizky, E., P. H. Fries, L. Helm, J. Kowalewski, D. Kruk, R. R. Sharp, and P.-O. Westlund. 2008. Comparison of different methods for calculating the paramagnetic relaxation enhancement of nuclear spins as a function of the magnetic field. *J. Chem. Phys.* 128 (5):052307.

Bendszus, M., C. Wessig, A. Schutz, T. Horn, C. Kleinschnitz, C. Sommer, B. Misselwitz, and G. Stoll. 2005. Assessment of nerve degeneration by gadofluorine M-enhanced magnetic resonance imaging. *Ann. Neurol.* 57 (3):388–395.

Bertini, I. and C. Luchinat. 1996. NMR of paramagentic substances. *Coord. Chem. Rev.* 150:1–296.

Bertini, I., C. Luchinat, and G. Parigi. 2001. *Solution NMR of Paramagnetic Molecules.* Vol. 2. In *Current Methods in Inorganic Chemistry.* Amsterdam, the Netherlands: Elsevier.

Bloembergen, N. 1957. Proton relaxation times in paramagnetic solutions. *J. Chem. Phys.* 27:572–573.

Bloembergen, N. and L. O. Morgan. 1961. Proton relaxation times in paramagnetic solutions. Effects of electron spin relaxation. *J. Chem. Phys.* 34:842–850.

Bonnet, C. S., L. Pellegatti, F. Buron, C. M. Shade, S. Villette, V. Kubicek, G. Guillaumet, F. Suzenet, S. Petoud, and E. Toth. 2010. Hydrophobic chromophore cargo in micellar structures: A different strategy to sensitize lanthanide cations. *Chem. Commun.* 46 (1):124–126.

Borel, A., L. Helm, and A. E. Merbach. 2001. Molecular dynamics simulations of MRI-relevant Gd(III) chelates: Direct access to outer-sphere relaxivity. *Chem. Eur. J.* 7:600–610.

Botta, M. 2000. Second coordination sphere water molecules and relaxivity of gadolinium(III). *Eur. J. Inorg. Chem.* 2000:399–407.

Bresinska, I. and K. J. Balkus, Jr. 1994. Studies of Gd(III)-exchanged Y-type zeolites relevant to magnetic resonance imaging. *J. Phys. Chem.* 98 (49):12898–12994.

Bridot, J.-L., A.-C. Faure, S. Laurent, C. Riviere, C. Billotey, B. Hiba, M. Janier et al. 2007. Hybrid gadolinium oxide nanoparticles: Multimodal contrast agents for in vivo imaging. *J. Am. Chem. Soc.* 129 (16):5076–5084.

Briley-Saebo, K. C., S. Geninatti-Crich, D. P. Cormode, A. Barazza, W. J. M. Mulder, W. Chen, G. B. Giovenzana, E. A. Fisher, S. Aime, and Z. A. Fayad. 2009. High-relaxivity gadolinium-modified high-density lipoproteins as magnetic resonance imaging contrast agents. *J. Phys. Chem. B* 113 (18):6283–6289.

Bryant, L. H., M. W. Brechbiel, C. Wu, J. W. M. Bulte, V. Herynek, and J. A. Frank. 1999. Synthesis and relaxometry of high-generation (G=5,7,9, and 10) PAMAM dendrimer-DOTA-gadolinium chelates. *J. Magn. Reson. Imaging* 9:348–352.

Caravan, P. 2006. Strategies for increasing the sensitivity of gadolinium based MRI contrast agents. *Chem. Soc. Rev.* 35 (6):512–523.

Caravan, P. 2007. Physicochemical principles of MR contrast agents. In *Molecular and Cellular MR Imaging*, eds. M. M. J. Modo and J. W. M. Bulte. Boca Raton, FL: CRC Press LLC.

Caravan, P., A. V. Astashkin, and A. M. Raitsimring. 2003. The gadolinium(III)-water hydrogen distance in MRI contrast agents. *Inorg. Chem.* 42 (13):3972–3974.

Caravan, P., J. J. Ellison, T. J. McMurry, and R. B. Lauffer. 1999. Gadolinium(III) chelates as MRI contrast agents: Structure, dynamics, and applications. *Chem. Rev.* 99 (9):2293–2352.

Caravan, P., C. T. Farrar, L. Frullano, and R. Uppal. 2009. Influence of molecular parameters and increasing magnetic field strength on relaxivity of gadolinium- and manganese-based T_1 contrast agents. *Contrast Media Mol. Imaging* 4 (2):89–100.

Chen, W., E. Vucic, E. Leupold, W. J. M. Mulder, D. P. Cormode, K. C. Briley-Saebo, A. Barazza, E. A. Fisher, M. Dathe, and Z. A. Fayad. 2008. Incorporation of an apoE-derived lipopeptide in high-density lipoprotein MRI contrast agents for enhanced imaging of macrophages in atherosclerosis. *Contrast Media Mol. Imaging* 3 (6):233–242.

Cohen, S. M., J. Xu, E. Radkov, K. N. Raymond, A. Barge, and S. Aime. 2000. Syntheses and relaxation properties of mixed gadolinium hydroxypyridinonate MRI contrast agents. *Inorg. Chem.* 39:5747–5756.

Cormode, D. P., R. Chandrasekar, A. Delshad, K. C. Briley-Saebo, C. Calcagno, A. Barazza, W. J. M. Mulder, E. A. Fisher, and Z. A. Fayad. 2009a. Comparison of synthetic high density lipoprotein (HDL) contrast agents for MR imaging of atherosclerosis. *Bioconjug. Chem.* 20 (5):937–943.

Cormode, D. P., J. C. Frias, Y. Q. Ma, W. Chen, T. Skajaa, K. Briley-Saebo, A. Barazza, K. J. Williams, W. J. M. Mulder, Z. A. Fayad, and E. A. Fisher. 2009b. HDL as a contrast agent for medical imaging. *Clin. Lipidol.* 4 (4):493–500.

Cormode, D. P., T. Skajaa, M. M. van Schooneveld, R. Koole, P. Jarzyna, M. E. Lobatto, C. Calcagno et al. 2008. Nanocrystal core high-density lipoproteins: A multimodality contrast agent platform. *Nano Lett.* 8 (11):3715–3723.

Crich, S. G., S. Lanzardo, D. Alberti, S. Belfiore, A. Ciampa, G. Giovenzana, C. Lovazzano, R. Pagliarin, and S. Aime. 2007. Magnetic resonance imaging detection of tumor cells by targeting low-density lipoprotein receptors with Gd-loaded low-density lipoprotein particles. *Neoplasia* 9:1046–1056.

Csajbók, E., I. Bányai, L. Vander Elst, R. N. Muller, W. Zhou, and J. A. Peters. 2005. Gadolinium(III)-loaded nanoparticulate zeolites as potential high-field MRI contrast agents: Relationship between structure and relaxivity. *Chem. Eur. J.* 11:4799–4807.

Datta, A. and K. N. Raymond. 2009. Gd-hydroxypyridinone (HOPO)-based high-relaxivity magnetic resonance imaging (MRI) contrast agents. *Acc. Chem. Res.* 42:938–947.

de Sousa, P. L., J. B. Livramento, L. Helm, A. E. Merbach, W. Meme, B. T. Doan, J. C. Beloeil et al. In vivo Mill assessment of a novel Gd-III-based contrast agent designed for high magnetic field applications. *Contrast Media Mol. Imaging* 3 (2):78–85.

Dear, J. W., H. Kobayashi, M. W. Brechbiel, and R. A. Star. 2006. Imaging acute renal failure with polyamine dendrimer-based MRI contrast agents. *Nephron Clin. Pract.* 103 (2):C45–C49.

Debouttiere, P.-J., S. Roux, F. Vocanson, C. Billotey, O. Beuf, A. Favre-Reguillon, Y. Lin, S. Pellet-Rostaing, R. Lamartine, P. Perriat, and O. Tillement. 2006. Design of gold nanoparticles for magnetic resonance imaging. *Adv. Funct. Mater.* 16 (18):2330–2339.

Delli Castelli, D., E. Gianolio, S. Geninatti Crich, E. Terreno, and S. Aime. 2008. Metal containing nano-sized systems for MR-molecular imaging applications. *Coord. Chem. Rev.* 252 (21–22):2424–2443.

Di Marco, M., C. Sadun, M. Port, I. Guilbert, P. Couvreur, and C. Dubernet. 2007. Physicochemical characterization of ultrasmall superparamagnetic iron oxide particle (USPIO) for biomedical application as MRI contrast agents. *Int. J. Nanomed.* 2 (4):609–622.

Dickins, R. S., S. Aime, A. Batsanov, A. Beeby, M. Botta, J. I. Bruce, J. A. K. Howard et al. 2002. Structural, luminescence, and NMR studies of the reversible binding of acetate, lactate, citrate, and selected amino acids to chiral diaqua ytterbium, gadolinium, and europium complexes. *J. Am. Chem. Soc.* 124 (43):12697–12705.

Doble, D. M. J., M. Melchior, B. O'Sullivan, C. Siering, J. Xu, V. C. Pierre, and K. N. Raymond. 2003. Toward optimized high-relaxivity MRI agents: The effect of ligand basicity on the thermodynamic stability of hexadentate hydroxypyridonate/catecholate gadolinium(III) complexes. *Inorg. Chem.* 42 (16):4930–4937.

Dunand, F. A., E. Toth, R. Hollister, and A. E. Merbach. 2001. Lipari-Szabo approach as a tool for the analysis of macromolecular gadolinium(III)-based MRI contrast agents illustrated by the $[Gd(EGTA\text{-}BA\text{-}(CH_2)_{12})]_n^{n+}$ polymer. *J. Biol. Inorg. Chem.* 6:247–255.

Engstroem, M., A. Klasson, H. Pedersen, C. Vahlberg, and P.-O. Käll, K. Uvdal. 2006. High proton relaxivity for gadolinium oxide nanoparticles. *Magn. Reson. Mat. Phys. Biol. Med.* 19 (4):180–184.

Fortin, M.-A., R. M. Petoral Jr., F. Soederlind, A. Klasson, M. Engstroem, T. Veres, P.-O. Kaell, and K. Uvdal. 2007. Polythylene glycol-covered ultra-small Gd_2O_3 nanoparticles for positive contrast at 1.5 T magnetic resonance clinical scanning. *Nanotechnology* 18 (39):395501/1–9.

Fossheim, S. L., A. K. Fahlvik, J. Klaveness, and R. N. Muller. 1999. Paramagnetic liposomes as MRI contrast agents: Influence of liposomal properties on the in vitro relaxivity. *Magn. Reson. Imaging* 17:83–89.

Fossheim, S. L., K. A. Il'yasov, J. Hennig, and A. Bjornerud. 2000. Thermosensitive paramagnetic liposomes for temperature control during MR imaging-guided hyperthermia: In vitro feasibility studies. *Acad. Radiol.* 7 (12):1107–15.

Freed, J. H. 1978. Dynamic effect of pair correlation functions on spin relaxation by translational diffusion in liquids. II. Finite jumps and independent T_1 processes. *J. Chem. Phys.* 69:4034–4037.

Frias, J. C., K. J. Williams, E. A. Fischer, and Z. A. Fayad. 2004. Recombinant HDL-like nanoparticles: A specific contrast agent for MRI of atherosclerotic plaques. *J. Am. Chem. Soc.* 126:16316–16317.

Frich, L., A. Bjornerud, S. Fossheim, T. Tillung, and I. Gladhaug. 2004. Experimental application of thermosensitive paramagnetic liposomes for monitoring magnetic resonance imaging guided thermal ablation. *Magn. Reson. Med.* 52 (6):1302–1309.

Fries, P. H. and E. Belorizky. 2005. Electronic relaxation of paramagnetic metal ions and NMR relaxivity in solution: Critical analysis of various approaches and application to a Gd(III)-based contrast agent. *J. Chem. Phys.* 123 (12):124510.

Fries, P. H. and E. Belorizky. 2007. Relaxation theory of the electronic spin of a complexed paramagnetic metal ion in solution beyond the Redfield limit. *J. Chem. Phys.* 126:204503.

Geraldes, C. F. G. C. and S. Laurent. 2009. Classification and basic properties of contrast agents for magnetic resonance imaging. *Contrast Media Mol. Imaging* 4 (1):1–23.

Gianolio, E., G. B. Giovenzana, A. Ciampa, S. Lanzardo, D. Imperio, and S. Aime. 2008. A novel method of cellular labeling: Anchoring MR-imaging reporter particles on the outer cell surface. *ChemMedChem* 3 (1):60–62.

Gianolio, E., G. B. Giovenzana, D. Longo, I. Longo, I. Menegotto, and S. Aime. 2007. Relaxometric and modelling studies of the binding of a lipophilic Gd-AAZTA complex to fatted and defatted human serum albumin. *Chemistry* 13 (20):5785–5797.

Glogard, C., R. Hovland, S. L. Fossheim, A. J. Aasen, and J. Klaveness. 2000. Synthesis and physicochemical characterization of new amphiphilic gadolinum DO3A complexes as contrast agents for MRI. *J. Chem. Soc., Perkin Trans.* 2:1047–1052.

Hainfeld, J. F., D. N. Slatkin, T. M. Focella, and H. M. Smilowitz. 2006. Gold nanoparticles: A new x-ray contrast agent. *Br. J. Radiol.* 79:248–253.

Hak, S., H. M. H. F. Sanders, P. Agrawal, S. Langereis, H. Grull, H. M. Keizer, F. Arena, E. Terreno, G. J. Strijkers, and K. Nicolay. 2009. A high relaxivity Gd(III)DOTA-DSPE-based liposomal contrast agent for magnetic resonance imaging. *Eur. J. Pharm. Biopharm.* 72 (2):397–404.

Helm, L. 2006. Relaxivity in paramagnetic systems: Theory and mechanism. *Prog. NMR Spectrosc.* 49 (1):45–64.

Helm, L. and A. E. Merbach. 2005. Inorganic and bioinorganic solvent exchange mechanisms. *Chem. Rev.* 105 (6):1923–1960.

Hermann, P., J. Kotek, V. Kubicek, and I. Lukes. 2008. Gadolinium(III) complexes as MRI contrast agents: Ligand design and properties of the complexes. *Dalton Trans.* (23):3027–3047.

Holz, M., S. R. Heil, and A. Sacco. 2000. Temperature-dependent self-diffusion coefficients of water and six selected molecular liquids for calibration in accurate 1H NMR PFG measurements. *Phys. Chem. Chem. Phys.* 2 (20):4740–4742.

Hong, G., R. Yuan, B. Liang, J. Shen, X. Yang, and X. Shuai. 2008. Folate-functionalized polymeric micelle as hepatic carcinoma-targeted, MRI-ultrasensitive delivery system of antitumor drugs. *Biomed. Microdevices* 10 (5):693–700.

Hovland, R., C. Glogard, A. J. Aasen, and J. Klaveness. 2003. Preparation and in vitro evaluation of a novel amphiphilic GdPCTA-[12] derivative; a micellar MRI contrast agent. *Org. Biomol. Chem.* 1 (4):644–647.

Hsiao, J.-K., C.-P. Tsai, T.-H. Chung, Y. Hung, M. Yao, H.-M. Liu, C.-Y. Mou, C.-S. Yang, Y.-C. Chen, and D.-M. Huang. 2008. Mesoporous silica nanoparticles as a delivery system of gadolinium for effective human stem cell tracking. *Small* 4 (9):1445–1452.

Hwang, L. P. and J. H. Freed. 1975. Dynamic effects of pair correlation functions on spin relaxation by translational diffusion in liquids. *J. Chem. Phys.* 63:4017–4025.

Imura, H., G. R. Choppin, W. P. Cacheris, L. A. de Learie, T. J. Dunn, and D. H. White. 1997. Thermodynamics and NMR studies of DTPA-bis(methoxyethylamide) and its derivatives. Protonation and complexation with Ln(III). *Inorg. Chim. Acta* 258 (2):227–236.

Jaszberenyi, Z., L. Moriggi, P. Schmidt, C. Weidensteiner, R. Kneuer, A. E. Merbach, L. Helm, and E. Toth. 2007. Physicochemical and MRI characterization of Gd3+-loaded polyamidoamine and hyperbranched dendrimers. *J. Biol. Inorg. Chem.* 12 (3):406–420.

Kellar, K., P. M. Henrichs, R. Hollister, S. H. Koenig, J. Eck, and D. Wei. 1997. High relaxivity linear Gd(DTPA)-polymer conjugates: The role of hydrophobic interactions. *Magn. Reson. Med.* 38:712–716.

Kobayashi, H. and M. W. Brechbiel. 2005. Nano-sized MRI contrast agents with dendrimer cores. *Adv. Drug Deliv. Rev.* 57 (15):2271–2286.

Kobayashi, H., S. Kawamoto, S.-K. Jo, H. L. Bryant, Jr., M. W. Brechbiel, and R. A. Star. 2003. Macromolecular MRI contrast agents with small dendrimers: Pharmacokinetic differences between sizes and cores. *Bioconjug. Chem.* 14 (2):388–94.

Koenig, S. H. and R. D. Brown, III. 1995. Relaxivity of MRI magnetic contrast agents. Concepts and principles. In *Handbook of Metal-Ligand Interactions in Biological Fluids. Bioiniorganic Chemistry*, ed. G. Berthon. New York: Dekker.

Kotek, J., P. Lebduskova, P. Hermann, L. Vander Elst, R. N. Muller, C. F. G. C. Geraldes, T. Maschmeyer, I. Lukes, and J. A. Peters. 2003. Lanthanide(III) complexes of novel mixed carboxylic-phosphorus acid derivatives of diethylenetriamine: A step towards more efficient MRI contrast agents. *Chem. Eur. J.* 9:5899–5915.

Kowalewski, J., D. Kruk, and G. Parigi. 2005. NMR relaxation in solution of paramagnetic complexes: Recent theoretical progress for $S \geq 1$. In *Advances in Inorganic Chemistry*, eds. R. Van Eldik and I. Bertini. San Diego, CA: Elsevier.

Kowalewski, J., C. Luchinat, T. Nilsson, and G. Parigi. 2002. Nuclear spin relaxation in paramagnetic systems: Electron spin relaxation effects under near-redfield limit conditions and beyond. *J. Phys. Chem. A* 106:7376–7382.

Kowalewski, J. and L. Mäler. 2006. *Nuclear Spin Relaxation in Liquids: Theory, Experiments, and Applications*, eds. J. H. Moore and N. D. Spencer. *Series in Chemical Physics*. New York: Taylor & Francis.

Kowalewski, J., L. Nordenskiöld, N. Benetis, and P. Westlund. 1985. Theory of nuclear spin relaxation in paramagnetic systems in solution. *Prog. NMR Spectrosc.* 17:141–185.

Koyama, Y., V. S. Talanov, M. Bernardo, Y. Hama, C. A. S. Regino, M. W. Brechbiel, P.L. Choyke, and H. Kobayashi. 2007. A dendrimer-based nanosized contrast agent dual-labeled for magnetic resonance and optical fluorescence imaging to localize the sentinel lymph node in mice. *J. Magn. Reson. Imaging* 25:866–871.

Krause, W. 2002. *Contrast Agents I., Magnetic Resonance Imaging.* Vol. 221, ed. W. Krause. *Topics in Current Chemistry.* Heidelberg, Germany: Springer.

Lanza, G. M., P. Winter, A. M. Neubauer, S. D. Caruthers, F.D. Hockett, and S.A. Wickline. 2005b. 1H/19F magnetic resonance molecular imaging with perfluorocarbon nanoparticles. *Curr. Top. Dev. Biol.* 70:57–76.

Laurent, S., F. Botteman, L. Vander Elst, and R. N. Muller. 2004. New bifunctional contrast agents: Bisamide derivatives of C-substituted Gd-DTPA. *Eur. J. Inorg. Chem.* 2004 (3):463–168.

Laurent, S., D. Forge, M. Port, A. Roch, C. Robic, L. Vander Elst, and R. N. Muller. 2008. Magnetic iron oxide nanoparticles: Synthesis, stabilization, vectorization, physicochemical characterizations, and biological applications. *Chem. Rev.* 108 (6):2064–2110.

Laurent, S., L. Vander Elst, and R. N. Muller. 2006. Comparative study of the physicochemical properties of six clinical low molecular weight gadolinium contrast agents. *Contrast Media Mol. Imaging* 1:128–137.

Laus, S., R. Ruloff, E. Toth, and A. E. Merbach. 2003. Gd^{III} complexes with fast water exchange and high thermodynamic stability: Potential building blocks for high-relaxivity MRI contrast agents. *Chem. Eur. J.* 9 (15):3555–3566.

Laus, S., A. Sour, R. Ruloff, E. Toth, and A. E. Merbach. 2005. Rotational dynamics account for pH-dependent relaxivities of PAMAM dendrimeric, Gd-based potential MRI contrast agents. *Chem. Eur. J.* 11 (10):3064–3076.

Lebduskova, P., P. Hermann, L. Helm, E. Toth, J. Kotek, K. Binnemans, J. Rudovsky, I. Lukes, and A. E. Merbach. 2007. Gadolinium(III) complexes of mono- and diethyl esters of monophosphonic acid analogue of DOTA as potential MRI contrast agents: Solution structures and relaxometric studies. *Dalton Trans.* (4):493–501.

Lebduskova, P., A. Sour, L. Helm, E. Toth, J. Kotek, I. Lukes, and A. E. Merbach. 2006. Phosphinic derivative of DTPA conjugated to a G5 PAMAM dendrimer: An O-17 and H-1 relaxation study of its Gd(III) complex. *Dalton Trans.* (28):3399–3406.

Lin, Y.-S., Y. Hung, J.-K. Su, R. Lee, C. Chang, M.-L. Lin, and C.-Y. Mou. 2004. Gadolinium(III)-incorporated nanosized mesoporous silica as potential magnetic resonance imaging contrast agents. *J. Phys. Chem. B* 108 (40):15608–15611.

Lindner, L. H., H. M. Reinl, M. Schlemmer, R. Stahl, and M. Peller. 2005. Paramagnetic thermosensitive liposomes for MR-thermometry. *Int. J. Hyperther.* 21 (6):575–588.

Lipari, G. and A. Szabo. 1982a. Model-free approach to the interpretation of nuclear magnetic resonance relaxation in macromolecules. 1. Theory and range of validity. *J. Am. Chem. Soc.* 104:4546–4559.

Lipari, G. and A. Szabo. 1982b. Model-free approach to the interpretation of nuclear magnetic resonance relaxation in macromolecules. 2. Analysis of exerimental results. *J. Am. Chem. Soc.* 104:4559–4570.

Lipinski, M. J., V. Amirbekian, J. C. Frias, J. G. S. Aguinaldo, V. Mani, K. C. Briley-Saebo, V. Fuster, J. T. Fallon, E. A. Fisher, and Z. A. Fayad. 2006. MRI to detect atherosclerosis with gadolinium-containing immunomicelles targeting the macrophage scavenger receptor. *Magn. Reson. Med.* 56 (3):601–610.

Livramento, J. B., C. Weidensteiner, M. I. M. Prata, P. R. Allegrini, C. F. G. C. Geraldes, L. Helm, R. Kneuer, A. E. Merbach, A. C. Santos, P. Schmidt, and E. Toth. 2006. First in vivo MRI assessment of a self-assembled metallostar compound endowed with a remarkable high field relaxivity. *Contrast Media Mol. Imaging* 1 (1):30–39.

Løkling, K. E., S. L. Fossheim, R. Skurtveit, A. Bjornerud, and J. Klaveness. 2001. pH-sensitive paramagnetic liposomes as MRI contrast agents: In vitro feasibility studies. *Magn. Reson. Imaging* 19:731–738.

Løkling, K. E., R. Skurtveit, A. Bjornerud, and S. L. Fossheim. 2004. Novel pH-sensitive paramagnetic liposomes with improved MR properties. *Magn. Reson. Med.* 51:688–696.

Longmire, M., P. L. Choyke, and H. Kobayashi. 2008. Dendrimer-based contrast agents for molecular imaging. *Curr. Top. Med. Chem.* 8:1180–1186.

Lu, Z.-R., A. M. Mohs, Y. Zong, and Y. Feng. 2006. Polydisulfide Gd(III) chelates as biodegradable macromolecular magnetic resonance imaging contrast agents. *Int. J. Nanomed.* 1 (1):31–40.

Lukyanov, A. N. and V. P. Torchilin. 2004. Micelles from lipid derivatives of water-soluble polymers as delivery systems for poorly soluble drugs. *Adv. Drug Deliv. Rev.* 56 (9):1273–1289.

Mamot, C., J. B. Nguyen, M. Pourdehnad, P. Hadaczek, R. Saito, J. R. Bringas, D. C. Drummond et al. 2004. Extensive distribution of liposomes in rodent brains and brain tumors following convection-enhanced delivery. *J. Neurooncol.* 68 (1):1–9.

Marsh, J. N., A. Senpan, G. Hu, M. J. Scott, P. J. Gaffney, S. A. Wickline, and G. M. Lanza. 2007. Fibrin-targeted perfluorocarbon nanoparticles for targeted thrombolysis. *Nanomedicine* 2 (4):533–543.

McLachlan, A. D. 1964. Line widths of electron resonance spectra in solution. *Proc. R. Soc. A* 280:271–288.

Meding, J., M. Urich, K. Licha, M. Reinhardt, B. Misselwitz, Z. A. Fayad, and H.-J. Weinmann. 2007. Magnetic resonance imaging of atherosclerosis by targeting extracellular matrix deposition with Gadofluorine M. *Contrast Media Mol. Imaging* 2 (3):120–129.

Misselwitz, B., J. Platzek, and H.-J. Weinmann. 2004. Early MR lymphography with gadofluorine M in rabbits. *Radiology* 231 (3):682–688.

Morawski, A. M., P. Winter, K. C. Crowder, S. D. Caruthers, R. W. Fuhrhop, M. J. Scott, J. D. Robertson, D. R. Abednschein, G. M. Lanza, and S. A. Wickline. 2004. Targeted nanoparticles for quantitative imaging of sparse molecular epitopes with MRI. *Magn. Reson. Med.* 51:480–486.

Moriggi, L., C. Cannizzo, E. Dumas, C. R. Mayer, A. Ulianov, and L. Helm. 2009. Gold nanoparticles functionalized with gadolinium chelates as high-relaxivity MRI contrast agents. *J. Am. Chem. Soc.* 131:10828–10829.

Moriggi, L., C. Cannizzo, C. Prestinari, F. Berrière, and L. Helm. 2008. Physicochemical properties of the high-field MRI-relevant [Gd(DTTA-Me)$(H_2O)_2$]-complex. *Inorg. Chem.* 47 (18):8357–8366.

Mühler, A. and H.-J. Weinmann. 1995. Biodistribution and excretion of 153Gd-labeled gadolinium ethoxybenzyl diethylenetriamine pentaacetic acid following repeated intravenous administration to rats. *Acad. Radiol.* 2:313–318.

Mulder, W. J. M., K. Douma, G. A. Koning, M. A. van Zandvoort, E. Lutgens, M. J. Daemen, K. Nicolay, and G. J. Strijkers. 2006a. Liposome-enhanced MRI of neointimal lesions in the ApoE-KO mouse. *Magn. Reson. Med.* 55 (5):1170–1174.

Mulder, W. J. M., A. W. Griffioen, G. J. Strijkers, D. P. Cormode, K. Nicolay, and Z. A. Fayad. 2007. Magnetic and fluorescent nanoparticles for multimodality imaging. *Nanomedicine* 2 (3):307–324.

Mulder, W. J. M., G. J. Strijkers, A. W. Griffioen, L. van Bloois, G. Molema, G. Storm, G. A. Koning, and K. Nicolay. 2004. A liposomal system for contrast-enhanced magnetic resonance imaging of molecular targets. *Bioconjug. Chem.* 15:799–806.

Mulder, W. J. M., G. J. Strijkers, J. W. Habets, E. J. W. Bleeker, D. W. J. van der Schaft, G. Storm, G. A. Koning, A. W. Griffioen, and K. Nicolay. 2005. MR molecular imaging and fluorescence microscopy for identification of activated tumor endothelium using a bimodal lipidic nanoparticle. *FASEB J.* 19 (14):2008–2010.

Mulder, W. J. M., G. J. Strijkers, G. A. F. van Tilborg, D. P. Cormode, Z. A. Fayad, and K. Nicolay. 2009. Nanoparticulate assemblies of amphiphiles and diagnostically active materials for multimodality imaging. *Acc. Chem. Res.* 42 (7):904–914.

Mulder, W. J. M., G. J. Strijkers, G. A. F. van Tilborg, A. W. Griffioen, and K. Nicolay. 2006b. Lipid-based nanoparticles for contrast-enhanced MRI and molecular imaging. *NMR Biomed.* 19 (1):142–164.

Muller, R. N., A. Roch, J. M. Colet, A. Ouakssim, and P. Gillis. 2001. Particulate magnetic contrast agents. In *The Chemistry of Contrast Agents in Magnetic Resonance Imaging*, eds. A. E. Merbach and E. Toth. Chichester, U.K.: John Wiley & Sons Ltd.

Nasongkla, N., E. Bey, J. Ren, H. Ai, C. Khemtong, J. S. Guthi, S.-F. Chin, A. D. Sherry, D. A. Boothman, and J. Gao. 2006. Multifunctional polymeric micelles as cancer-targeted, MRI-ultrasensitive drug delivery systems. *Nano Lett.* 6:2427–2430.

Neubauer, A. M., J. Myerson, S. D. Caruthers, F. D. Hockett, P. M. Winter, J. J. Chen, P. J. Gaffney, J. D. Robertson, G. M. Lanza, and S. A. Wickline. 2008. Gadolinium-modulated F-19 signals from perfluorocarbon nanoparticles as a new strategy for molecular imaging. *Magn. Reson. Med.* 60 (5):1066–1072.

Nicolle, G. M., L. Helm, and A. E. Merbach. 2003. 8S paramagnetic centers in molecular assemblies: Possible effect of their proximity on the water proton relaxivity. *Magn. Reson. Chem.* 41:794–799.

Nicolle, G. M., E. Toth, K.-P. Eisenwiener, H. R. Macke, and A. E. Merbach. 2002a. From monomers to micelles: Investigation of the parameters influencing proton relaxivity. *J. Biol. Inorg. Chem.* 7 (7–8):757–769.

Nicolle, G. M., E. T¢th, H. Schmitt-Willich, B. Raduchel, and A. E. Merbach. 2002b. The impact of rigidity and water exchange on the relaxivity of a dendritic MRI contrast agent. *Chem. Eur. J.* 8 (5):1040–1048.

O'Sullivan, B., D. M. J. Doble, M. K. Thompson, C. Siering, J. Xu, M. Botta, S. Aime, and K. N. Raymond. 2003. The effect of ligand scaffold size on the stability of tripodal hydroxypyridonate gadolinium complexes. *Inorg. Chem.* 42:2577–2583.

Pan, D., S. D. Caruthers, G. Hu, A. Senpan, M. J. Scott, P. J. Gaffney, S. A. Wickline, and G. M. Lanza. 2008. Ligand-directed nanobialys as theranostic agent for drug delivery and manganese-based magnetic resonance imaging of vascular targets. *J. Am. Chem. Soc.* 130:9186–9187.

Parac-Vogt, T. N., K. Kimpe, S. Laurent, C. Pirart, L. Vander Elst, R. N. Muller, and K. Binnemans. 2004. Gadolinium DTPA-monoamide complexes incorporated into mixed micelles as possible MRI contrast agents. *Eur. J. Inorg. Chem.* 2004:3538–3543.

Park, J.-A., P. A. N. Reddy, H.-K. Kim, I.-S. Kim, G.-C. Kim, Y. Chang, and T.-J. Kim. 2008. Gold nanoparticles functionalised by Gd-complex of DTPA-bis(amide) conjugate of glutathione as an MRI contrast agent. *Bioorg. Med. Chem. Lett.* 18:6135–6137.

Pellegatti, L., J. Zhang, B. Drahos, S. Villette, F. Suzenet, G. Guillaumet, S. Petoud, and E. Toth. 2008. Pyridine-based lanthanide complexes: Towards bimodal agents operating as near infrared luminescent and MRI reporters. *Chem. Commun.* (48):6591–6593.

Pierre, V. C., M. Botta, S. Aime, and K. N. Raymond. 2006a. Tuning the coordination number of hydroxypyridonate-based gadolinium complexes: Implications for MRI contrast agents. *J. Am. Chem. Soc.* 128 (16):5344–5345.

Pierre, V. C., M. Botta, S. Aime, and K. N. Raymond. 2006b. Substituent effects on Gd(III)-based MRI contrast agents: Optimizing the stability and selectivity of the complex and the number of coordinated water molecules. *Inorg. Chem.* 45 (20):8355–8364.

Pirollo, K. F., J. Dagata, P. Wang, M. Freedman, A. Vladar, S. Fricke, L. Ileva, Q. Zhou, and E. H. Chang. 2006. A tumor-targeted nanodelivery system to improve early MRI detection of cancer. *Mol. Imaging* 5 (1):41–52.

Platas-Iglesias, C., L. Vander Elst, W. Zhou, R. N. Muller, C. F. G. C. Geraldes, T. Maschmeyer, and J. A. Peters. 2002. Zeolite GdNaY nanoparticles with very high relaxivity for application as contrast agents in magnetic resonance imaging. *Chem. Eur. J.* 8 (22):5121–5131.

Polasek, M., M. Sedinova, J. Kotek, L. V. Elst, R. N. Muller, P. Hermann, and I. Lukes. 2009. Pyridine-N-oxide analogues of DOTA and their gadolinium(III) complexes endowed with a fast water exchange on the square-antiprismatic isomer. *Inorg. Chem.* 48 (2):455–465.

Powell, D. H., O. M. Ni Dhubhghaill, D. Pubanz, L. Helm, Y. S. Lebedev, W. Schlaepfer, and A. E. Merbach. 1996. Structural and dynamic parameters obtained from O-17 NMR, EPR, and NMRD studies of monomeric and dimeric Gd^{3+} complexes of interest in magnetic resonance imaging—An integrated and theoretically self consistent approach. *J. Am. Chem. Soc.* 118 (39):9333–9346.

Raatschen, H.-J., R. Swain, D. M. Shames, Y. Fu, Z. Boyd, M. L. Zierhut, M. F. Wendland et al. 2006. MRI tumor characterization using Gd-GlyMe-DOTA-perfluorooctyl-mannose-conjugate (Gadofluorine M), a protein-aviod contrast agent. *Contrast Media Mol. Imaging* 1 (3):113–120.

Rast, S., P. H. Fries, and E. Belorizky. 1999. Theoretical study of electronic relaxation processes in hydrated Gd^{3+} complexes in solutions. *J. Chim. Phys.* 96 (9/10):1543–1550.

Rast, S., P. H. Fries, E. Belorizky, A. Borel, L. Helm, and A. E. Merbach. 2001. A general approach to the electronic spin relaxation of Gd(III) complexes in solutions. Monte Carlo simulations beyond the Redfield limit. *J. Chem. Phys.* 115 (16):7554–7563.

Rieter, W. J., J. S. Kim, K. M. L. Taylor, H. An, W. Lin, T. Tarrant, and W. Lin. 2007. Hybrid silica nanoparticles for multimodal imaging. *Angew. Chem. Int. Ed.* 46 (20):3680–3682.

Rudovsky, J., M. Botta, P. Hermann, K. I. Hardcastle, I. Lukes, and S. Aime. 2006. PAMAM dendrimeric conjugates with a Gd-DOTA phosphinate derivative and their adducts with polyaminoacids: The interplay of global motion, internal rotation, and fast water exchange. *Bioconjug. Chem.* 17 (4):975–987.

Rudovsky, J., P. Hermann, M. Botta, S. Aime, and I. Lukes. 2005. Dendrimeric Gd(III) complex of a monophosphinated DOTA analogue: Optimizing relaxivity by reducing internal motion. *Chem. Commun.* (18):2390–2392.

Saito, R., M. T. Krauze, J. R. Bringas, C. Noble, T. R. McKnight, P. Jackson, M. F. Wendland et al. 2005. Gadolinium-loaded liposomes allow for real-time magnetic resonance imaging of convection-enhanced delivery in the primate brain. *Exp. Neurol.* 196:381–389.

Samad, A., Y. Sultana, and M. Aqil. 2007. Liposomal drug delivery systems: An update review. *Curr. Drug Deliv.* 4 (4):297–305.

Solomon, I. 1955. Relaxation processes in a system of two spins. *Phys. Rev.* 99 (2):559–565.

Solomon, I. and N. Bloembergen. 1956. Nuclear magnetic interactions in the HF molecule. *J. Chem. Phys.* 25:261–266.

Strandberg, E. and P.-O. Westlund. 1996. 1H NMRD profile and ESR lineshape calculation for an isotropic electron spin system with S = 7/2. A generalized modified Solomon-Bloembergen-Morgan theory for nonextreme-narrowing conditions. *J. Magn. Reson. A* 122:179–191.

Suarez, S., A. Devaux, J. Banuelos, O. Bossart, A. Kunzmann, and G. Calzaferri. 2007. Transparent zeolite-polymer hybrid materials with adaptable properties. *Adv. Funct. Mat.* 17 (14):2298–2306.

Sun, C., J. S. H. Lee, and M. Zhang. 2008. Magnetic nanoparticles in MR imaging and drug delivery. *Adv. Drug Deliv. Rev.* 60 (11):1252–1265.

Talanov, V. S., C. A. S. Regino, H. Kobayashi, M. Bernardo, P. L. Choyke, and M. W. Brechbiel. 2006. Dendrimer-based nanoprobe for dual modality magnetic resonance and fluorescence imaging. *Nano Lett.* 6 (7):1459–1463.

Taylor, K. M. L., J. S. Kim, W. J. Rieter, H. An, W. Lin, and W. Lin. 2008. Mesoporous silica nanospheres as highly efficient MRI contrast agents. *J. Am. Chem. Soc.* 130 (7):2154–2155.

Terreno, E., M. Botta, P. Boniforte, C. Bracco, L. Milone, B. Mondino, F. Uggeri, and S. Aime. 2005. A multinuclear NMR relaxometry study of ternary adducts formed between heptadentate Gd^{III} chelates and L-lactate. *Chem. Eur. J.* 11 (19):5531–5537.

Terreno, E., C. Cabella, C. Carrera, D. Delli Castelli, R. Mazzon, S. Rollet, J. Stancanello, M. Visigalli, and S. Aime. 2007. From spherical to osmotically shrunken paramagnetic liposomes: An improved generation of LIPOCEST MRI agents with highly shifted water protons. *Angew. Chem. Int. Ed. Engl.* 46 (6):966–968.

Terreno, E., A. Barge, L. Beltrami, G. Cravotto, D. D. Castelli, F. Fedeli, B. Jebasingh, and S. Aime. 2008a. Highly shifted LIPOCEST agents based on the encapsulation of neutral polynuclear paramagnetic shift reagents. *Chem. Commun.* (5):600–602.

Terreno, E., D. Delli Castelli, C. Cabella, W. Dastru, A. Sanino, J. Stancanello, L. Tei, and S. Aime. 2008b. Paramagnetic liposomes as innovative contrast agents for magnetic resonance (MR) molecular imaging applications. *Chem. Biodivers.* 5 (10):1901–1912.

Terreno, E., D. Delli Castelli, L. Milone, S. Rollet, J. Stancanello, E. Violante, and S. Aime. 2008c. First ex-vivo MRI co-localization of two LIPOCEST agents. *Contrast Media Mol. Imaging* 3 (1):38–43.

Terreno, E., S. Geninatti Crich, S. Belfiore, L. Biancone, C. Cabella, G. Esposito, A. D. Manazza, and S. Aime. 2006. Effect of the intracellular localization of a Gd-based imaging probe on the relaxation enhancement of water protons. *Magn. Reson. Med.* 55:491–497.

Thompson, M. K., D. M. J. Doble, L. S. Tso, S. Barra, M. Botta, S. Aime, and K. N. Raymond. 2004. Hetero-tripodal hydroxypyridonate gadolinium complexes: Syntheses, relaxometric properties, water exchange dynamics, and human serum albumin binding. *Inorg. Chem.* 43 (26):8577–8586.

Tomalia, D. A., A. M. Naylor, and W. A. Goddard. 1990. Starburst dendrimers: Molecular-level control of size, surface chemistry, topology, and flexibility from atoms to macroscopic matter. *Angew. Chem. Int. Ed. Engl.* 29:138–175.

Tomalia, D. A., L. A. Reyna, and S. Svenson. 2007. Dendrimers as multi-purpose nanodevices for oncology drug delivery and diagnostic imaging. *Biochem. Soc. Trans.* 35 (Pt 1):61–67.

Tombach, B. and W. Heindel. 2002. Value of 1.0-M gadolinium chelates: Review of preclinical and clinical data on gadobutrol. *Eur. Radiol.* 12:1550–1556.

Torres, S., J. A. Martins, J. P. Andre, C. F. G. C. Geraldes, A. E. Merbach, and E. Toth. 2006. Supramolecular assembly of an amphiphilic GdIII chelate: Tuning the reorientational correlation time and the water exchange rate. *Chem. Eur. J.* 12:940–948.

Toth, É., F. Connac, K. Adzamli, and A. E. Merbach. 1998. O-17-NMR, EPR and NMRD characterization of [Gd(DTPA-BMEA)(H_2O)]: A neutral MRI contrast agent. *Eur. J. Inorg. Chem.* 12:2017–2021.

Toth, E., L. Helm, K. Kellar, and A. E. Merbach. 1999. Gd(DTPA-bisamide)alkyl copolymers: A hint for the formation of MRI contrast agents with very high relaxivity. *Chem. Eur. J.* 5 (4):1202–1211.

Toth, É., L. Helm, and A. E. Merbach. 2001. Relaxivity of gadolinium(III) complexes: Theory and mechanism. In *The Chemistry of Contrast Agents in Medical Magnetic Resonance Imaging*, eds. A. E. Merbach and E. Toth. Chichester, U.K.: John Wiley & Sons.

Toth, É., L. Helm, and A. E. Merbach. 2002. Relaxivity of MRI contrast agents. In *Magnetic Resonance Contrast Agents*, ed. W. Krause. Heidelberg, Germany: Springer.

Toth, É., L. Helm, and A. E. Merbach. 2004. Metal complexes as MRI contrast enhancement agents. In *Applications of Coordination Chemistry*, ed. by M. Ward. Oxford, U.K.: Elsevier.

Toth, É. and A. E. Merbach. 1998. Water exchange dynamics: The key for high relaxivity contrast agents in medical magnetic resonance imaging. *ACH—Models Chem.* 135 (5):873–884.

Toth, E., D. Pubanz, S. Vauthey, L. Helm, and A. E. Merbach. 1996. The role of water exchange in attaining maximum relaxivities for dendrimeric MRI contrast agents. *Chem. Eur. J.* 2 (12):209–217.

Toth, E., I. van Uffelen, L. Helm, A. E. Merbach, D. Ladd, K. Briley-Saebo, and K. Kellar. 1998. Gadolinium-based linear polymer with temperature-independent proton relaxivities: A unique interplay between the water exchange and rotational contributions. *Magn. Reson. Chem.* 36:S125–S134.

Tsai, C.-P., Y. Hung, Y.-H. Chou, D.-M. Huang, J.-K. Hsiao, C. Chang, Y.-C. Chen, and C.-Y. Mou. 2008. High-contrast paramagnetic fluorescent mesoporous silica nanorods as a multifunctional cell-imaging probe. *Small* 4 (2):186–191.

Tsotsalas, M., M. Busby, E. Gianolio, S. Aime, and L. De Cola. 2008. Functionalized nanocontainers as dual magnetic and optical probes for molecular imaging applications. *Chem. Mater.* 20 (18):5888–5893.

Tweedle, M. F., J. J. Hagan, K. Kumar, S. Mantha, and C. A. Chang. 1991. Reaction of gadolinium chelates with endogenously available ions. *Magn. Reson. Imaging* 9:409–415.

Tweedle, M. F. and K. Kumar. 1999. Magnetic resonance imaging (MRI) contrast agents. In *Metallopharmaceuticals II, Diagnosis and Therapy*, eds. M. J. Clarke and P. J. Sadler. Berlin, Germany: Springer.

Viglianti, B. L., S. A. Abraham, C. R. Michelich, P. S. Yarmolenko, J. R. MacFall, M. B. Bally, and M. W. Dewhirst. 2004. In vivo monitoring of tissue pharmacokinetics of liposome/drug using MRI: Illustration of targeted delivery. *Magn. Reson. Med.* 51 (6):1153–1162.

Vucic, E., H. M. H. F. Sanders, F. Arena, E. Terreno, S. Aime, K. Nicolay, E. Leupold et al. 2009. Well-defined, multifunctional nanostructures of a paramagnetic lipid and a lipopeptide for macrophage imaging. *J. Am. Chem. Soc.* 131:406–407.

Wedeking, P., K. Kumar, and M. F. Tweedle. 1992. Dissociation of gadolinium chelates in mice: Relationship to chemical characteristics. *Magn. Reson. Imaging* 10:641–648.

Werbelow, L. G. and D. M. Grant. 1977. Intramolecular dipolar relaxation in multispin systems. *Adv. Magn. Reson.* 9:190–299.

Werner, E. J., A. Datta, C. J. Jocher, and K. N. Raymond. 2008. High-relaxivity MRI contrast agents: Where coordination chemistry meets medical imaging. *Angew. Chem. Int. Ed.* 47 (45):8568–8580.

Wiener, E. C., M. W. Brechbiel, H. Brothers, R. L. Magin, O. A. Gansow, D. A. Tomalia, and P. C. Lauterbur. 1994. Dendrimer—Based metal chelates: A new class of MRI contrast agents. *Magn. Reson. Med.* 31:1–8.

Winter, P., P. Athey, G. E. Kiefer, G. Gulyas, K. Frank, R. W. Fuhrhop, D. Robertson, S.A. Wickline, and G. M. Lanza. 2005. Improved paramagnetic chelate for molecular imaging with MRI. *J. Magn. Magn. Mater.* 293:540–545.

Winter, P., S. D. Caruthers, X. Yu, S.-K. Song, J. Chen, B. Miller, J. W. M. Bulte, J. D. Robertson, P. J. Gaffney, S. A. Wickline, and G. M. Lanza. 2003. Improved molecular imaging contrast agent for detection of human thrombus. *Magn. Reson. Med.* 50:411–416.

Winter, P. M., K. Cai, S. D. Caruthers, S. A. Wickline, and G. M. Lanza. 2007. Emerging nanomedicine opportunities with perfluorocarbon nanoparticles. *Expert Rev. Med. Devices* 4 (2):137–145.

Wisner, E. R., K. L. Aho-Sharon, M. J. Bennett, S. G. Penn, C. B. Lebrilla, and M. H. Nantz. 1997. A modular lymphographic magnetic resonance imaging contrast agent: Contrast enhancement with DNA transfection potential. *J. Med. Chem.* 40 (25):3992–3996.

Woessner, D. E. 1962. Nuclear spin relaxation in ellipsoids undergoing rotational Brownian motion. *J. Chem. Phys.* 37 (3):647–654.

Woods, M., D. E. Woessner, and A. D. Sherry. 2006. Paramagnetic lanthanide complexes as PARACEST agents for medical imaging. *Chem. Soc. Rev.* 35:500–511.

Xu, J., D. G. Churchill, M. Botta, and K. N. Raymond. 2004. Gadolinium(III) 1,2-hydroxypyridonate-based complexes: Toward MRI contrast agents of high relaxivity. *Inorg. Chem.* 43 (18):5492–5494.

Xu, J., S. J. Franklin, D. W. Whisenhunt, Jr., and K. N. Raymond. 1995. Gadolinium complex of tris[(3-hydroxy-1-methyl- 2-oxo-1,2-didehydropyridine-4-carboxamido)ethyl]-amine: A new class of gadolinium magnetic resonance relaxation agents. *J. Am. Chem. Soc.* 117 (27):7245–7246.

Yazyev, O. V., L. Helm, V. G. Malkin, and O. L. Malkina. 2005. Quantum chemical investigation of hyperfine coupling constants on first coordination sphere water molecule of gadolinium(III) aqua complexes *J. Phys. Chem. A* 109 (48):10997–11005.

Zhang, S., P. Winter, K. Wu, and A. D. Sherry. 2001. A novel europium(III)-based MRI contrast agent. *J. Am. Chem. Soc.* 123 (7):1517–1518.

Zheng, J., E. Ochoa, B. Misselwitz, D. Yang, I. El Naqa, P. K. Woodard, and D. Abendschein. 2008. Targeted contrast agent helps to monitor advanced plaque during progression: A magnetic resonance imaging study in rabbits. *Invest Radiol.* 43 (1):49–55.

Zong, Y. D., X. L. Wang, E. K. Jeong, D. L. Parker, and Z. R. Lu. 2009. Structural effect on degradability and in vivo contrast enhancement of polydisulfide Gd(III) complexes as biodegradable macromolecular MRI contrast agents. *Magn. Reson. Imaging* 27 (4):503–511.

19

Nanostructured Materials for Improved Magnetic Resonance Imaging

Hamsa Jaganathan
Purdue University

Albena Ivanisevic
Purdue University

19.1 Introduction

Magnetic resonance imaging (MRI) is the leading noninvasive for diagnostics in medicine. Among other clinical imaging techniques (i.e., x-ray, computed x-ray tomography, positron emission tomography, and ultrasound), MRI is able to image and provide anatomical information for both hard and soft tissues (Na et al. 2009). Since MRI displays qualitative information, it must produce well-defined images, accounting for differences among all types of tissues. Failure to do so may result in images that cause physicians to misinterpret or misdiagnose diseases. High spatial and contrast resolution are needed to distinguish the location and the nature of tissues in a quality MRI image. Clinical MRI scanners with 1.5 or 3 T magnets provide sufficient signal-to-noise ratio (SNR) to produce images up to micron-scaled spatial resolution (25–100 μm) (Massoud and Gambhir 2003). Since heterogeneous tissues have unique proton densities and relaxation times that generate different magnetic resonance (MR) signals, image contrast is achieved (Na et al. 2009). Detecting small differences in signal intensity, however, is limited by the strength of the scanner magnet and timing parameters set in pulse sequences, resulting in low contrast in MRI images (Massoud and Gambhir 2003). One inexpensive approach to acquire high MRI contrast is to use magnetic agents that can increase the local MR signal from selective tissues. Contrast agents (CAs) selectively shorten proton relaxations of tissues to enhance MR signal differences, improving image contrast. However, since problems with MR signal enhancement exist in current CAs, there is a need to improve the design of MRI CAs.

Typically, an ideal MRI CA needs to (a) have a stable, biocompatible structure, (b) exhibit high magnetization and proton relaxation rates, and (c) demonstrate long blood circulation times in vivo (Bai et al. 2008). The structure of CAs needs to be stable and biocompatible to remain intact and function effectively on in vivo systems. In addition, high magnetization and relaxation rates in CAs are needed to efficiently shorten proton relaxation and to acquire a strong MR signal. Furthermore, prolonged blood circulation time for CAs is necessary to detect a quality MR signal. The CAs in use today are severely limited by some of these requirements.

For example, commercially available gadolinium chelates produce low magnetic fields due to their proton relaxation mechanism. Additionally, the gadolinium ion structure has demonstrated to be toxic in the body (Bhushan 2004). While ferric oxide nanoparticles (NPs), another commercially available CA, are biocompatible and exhibit superparamagnetism, their nanoscaled (5 nm), crystallized structure demonstrates rapid blood circulation time and renal clearance within minutes (Park et al. 2008). The rapid biodistribution of NPs cause the MR signal to quickly fade away in vivo (Bai et al. 2008). The shortcomings in both commercially available CAs can be attributed to the flaws in their structural design.

In order to improve the effectiveness of CA, current research is mainly focused on manipulation of NP diameter to increase magnetism for strong MR signal detection (Okuhata 1999). However, the magnitude of NP diameter is not the principal factor that can cause high magnetism. Other extrinsic properties, such as volume, surface area, and shape, also affect the magnetic property of nanostructures. For example, both zero-dimensional (0D: isotropic in all directions) and one-dimensional (1D: anisotropic in one direction) nanostructures can be magnetically anisotropic, a behavior needed for superparamagnetism (Gossuin et al. 2009). Despite this understanding, 1D nanostructures are not commonly used as CAs due to complex fabrication methods to control shape (Champion and Mitragotri 2006). A feasible fabrication method is needed for the mass production of MRI CAs. The fabrication method must exhibit minimal synthesis energy and require inexpensive materials that can produce stable CAs (Geraldes and Laurent 2009). In this chapter, the utility of 1D, superparamagnetic nanostructures are reviewed as potential imaging agents to improve MRI. Herein, fabrication methods and magnetic and in vivo properties for 1D superparamagnetic nanostructures are reviewed.

19.2 Basic Principles of MRI

MRI offers three-dimensional (3D) anatomical images through a noninvasive technique. It is based on detecting signals from the relaxation of protons from tissues and organs in aqueous environments and reconstructs these signals to obtain3D, gray scale images (Na et al. 2009). In short, when a strong magnetic field is applied to nuclei of protons, they precess, or oscillate, at the Larmor frequency and the proton spins align parallel to the magnetic field, obtaining a net magnetization, M_z (Figure 19.1A). When a radio frequency (RF) pulse is applied, the protons absorb energy and are excited into a higher energy state perpendicular to the magnetic field in the *xy*-plane, altering the net magnetization, M_{xy}. After the duration of the RF pulse, the spins begin to relax to their initial low energy state, M_z. The relaxation varies for different tissues and organs due to different proton densities and biomolecule characteristics.

Longitudinal (T_1) and transverse (T_2) are two types of relaxations observed in MRI. The recovery time of M_z following the RF pulse is called the longitudinal relaxation (T_1) (Figure 19.1B), and the decay time of M_{xy} after the RF pulse is called the transverse relaxation (T_2) (Figure 19.1C). In T_1-weighted images, signals from T_1 relaxation are largely detected and displayed as positive contrast. Conversely, in T_2-weighted images, MRI detects signals from T_2 relaxation and displays as negative contrast (Nakata et al. 2008).

Although MRI is able to display contrast for large tissue volumes, it is difficult to delineate between proximal, small-sized tissues due to overlapping protons that skew the MR signal (Burtea et al. 2008). In addition, low magnetic fields from 1.5 to 3 T and imaging parameters set in pulse sequences are also aspects to generate low contrast in MRI images (Geraldes and Laurent 2009; Na et al. 2009). In recent decades, the use of CAs has significantly improved distinguishability among different tissues in images (Burtea et al. 2008).

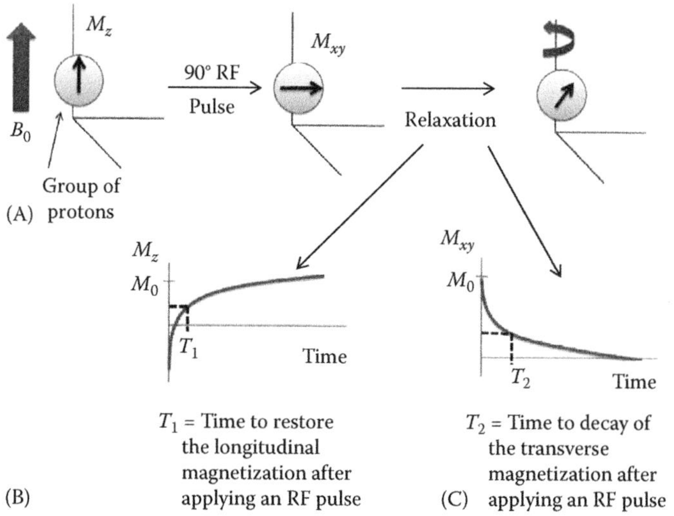

FIGURE 19.1 Graphic representation of the two types of magnetic relaxations. (A) A nuclei of protons are aligned due to an external magnetic field, and after applying a RF pulse, protons tend to relax. (B) Longitudinal relaxation time (T_1). (C) Transverse relaxation time (T_2). (Figure adapted from Cheon, J. and Lee, J.H., *Acc. Chem. Res.*, 41, 1630, 2008.)

19.3 Overview of MRI Contrast Agents

The utility of MRI CAs is governed by intrinsic (composition) and extrinsic (size and shape) physicochemical properties (Thorek et al. 2006). Paramagnetic and superparamagnetic materials are currently the most effective CA compositions and exhibit strong magnetization only in the presence of an external magnetic field (Table 19.1).

First generation agents, such as gadolinium chelates, are paramagnetic and have magnetic moments due to unpaired electron orbits in the outer electron shell. The T_1 of nearby tissue protons tends to

TABLE 19.1 Types of Magnetism and Corresponding MRI Contrast Agent Examples

Type of Magnetism	Definition	Contrast Agent Examples	References
Diamagnetism	All electrons are paired causing weak magnetic interactions	Since 99% of biological tissue is diamagnetic, MR signal from diamagnetic CAs have low detectable changes	Hendee and Ritenour (2002)
Ferromagnetism	Permanently magnetic from domains of atoms having unpaired electrons.	Permanent magnetism can be detrimental to metallic implants and little research is focused on ferromagnetic CA	Pamboucas and Rokas (2008); Bhushan (2004)
Paramagnetism	Molecules contain unpaired electrons and exhibit magnetism in the presence of a strong external magnetic field	Gadolinium chelates: (Gd-DOTA) *Commercially available*	Froehlich (2008)
Super-paramagnetism	Similar to paramagnetism but single crystallites contain single magnetic moments	Ferric oxide NPs (~5 nm diameter): *Commercially available*	Froehlich (2008)

Source: Images adapted from Burtea, C. et al., *Handbook of Experimental Pharmocology*, Springer-Verlag, Berlin, 2008.

shorten due to a direct interaction of electron exchange with gadolinium chelates, resulting in a strong MR signal for T_1-weighted images (Froehlich 2008). However, a direct, intramolecular interaction between the paramagnetic ion and tissues is not ideal to distinguish proximal tissue characteristics. Protons from different, yet proximal tissues may experience similar interactions with chelates, and the magnitude of MR signal becomes equal for both tissues. This limitation may cause indistinguishable contrast in images (Okuhata 1999). In addition to problems arising from the relaxation mechanism, gadolinium products are toxic to the human body (Bhushan 2004). Therefore, paramagnetic CAs can cause health and safety concerns.

Since paramagnetic CAs may cause problems, second generation, superparamagnetic materials were proposed as a potential solution. These materials are superparamagnetic and have greater magnetization (10–1000-fold) than paramagnetic materials due to their single-crystal lattice arrangement. The single-crystal arrangement results in a greater ability to strengthen the local magnetic field and create larger susceptibilities, compared to paramagnetic materials (Froehlich 2008). For example, in the presence of ferric oxide NPs, T_2 relaxation is affected by *inter*molecular interactions between water protons and NPs. Superparamagnetic nanoparticles (SPNs) affect the local magnetic field through their magnetization strength, and this local inhomogeneous magnetic field leads to an accelerated dephasing of surrounding protons. The rapid proton dephasing causes T_2 to shorten and creates a pronounceable change in MR signal. The change in MR signal causes a negative contrast in T_2-weighted images (Hendee and Ritenour 2002). Since SPNs do not directly interact with tissues and rather alter the local field to influence proton relaxation, superparamagnetic materials are highly desired for in vivo systems as opposed to paramagnetic materials (Duguet et al. 2006).

While SPNs do not require direct interaction with tissues, issues arising from extrinsic properties (size and shape) affect the utility of these second generation CAs. SPNs (~5 nm diameter) demonstrate rapid blood circulation time and renal clearance (Thorek et al. 2006; Sun et al. 2008). The rapid biodistribution of NPs cause MR signal to quickly fade away in vivo (Bai et al. 2008). Researchers have attempted to increase NP diameter to acquire high magnetism and a stronger MR signal (Okuhata 1999; Geraldes and Laurent 2009). Large diameter NPs (greater than 150 nm), however, activate the monocyte-phagocyte system and are collected into the spleen within minutes (Sun et al. 2008). Drawbacks from the SPN design result in a need to improve the structural design of CAs.

19.4 Fabrication Methods for 1D Superparamagnetic Nanostructures

The shape of superparamagnetic nanostructures has not been a major consideration in the CA design. One reason for this gap in CA research is due to the complicated, expensive fabrication methods to control geometry in nanoscaled structures (Champion and Mitragotri 2006). However, superparamagnetic particles (0D), which do not exhibit a specific shape, are synthesized by simple fabrication approaches and reviewed frequently in literature (Gupta and Gupta 2005; Thorek et al. 2006). Consequently, by organizing and arranging preformed SPNs into a linear shape, 1D nanostructures can be constructed with geometric control (Figure 19.2).

An ideal CA structure needs to exhibit biocompatibility and stability in order to function effectively in vivo. The biocompatibility is governed by the basic materials utilized to form the nanostructure, while stability of a CA is governed by the fabrication approach. Top-down and bottom-up are two types of approaches to fabricate nanoscaled structures. A top-down approach consists of expensive lithographic techniques that construct nanostructures from bulk material. Due to disadvantages in size resolution and low throughput, lithographic techniques are not optimal to construct 1D, superparamagnetic nanostructures (Martín et al. 2003). High yields of structures at the nanoscale are achieved by bottom-up approaches. Subsequently, this report focuses on three bottom-up fabrication techniques that are able to arrange preformed NPs in a linear fashion (Table 19.2).

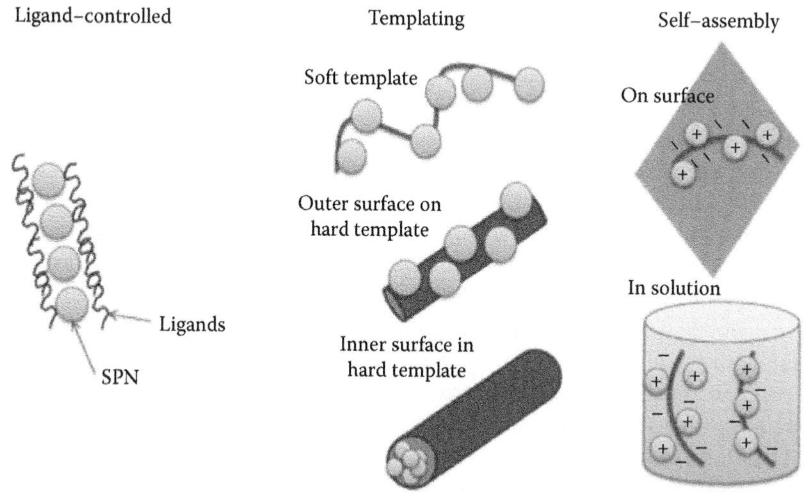

FIGURE 19.2 Methods to construct 1D nanostructures using NPs.

TABLE 19.2 Three Types of Bottom-Up Approaches to Assemble NPs into Linear Nanostructures

Bottom-Up Approach	Description	Control over Dimensions by	Advantages	Disadvantages	References
Ligand-controlled	Use of ligands in reaction to form linear-shaped nanostructure	Ligand concentration and molecular weight Time of ligand addition in reaction	High yield Colloidal stability Inexpensive	No control over diameter size May cause impurities in structure	Feng et al. (2004); Nakata et al. (2008); Park et al. (2008); Rebolledo et al. (2008); Leung et al. (2009); Miyasaka et al. (2009); Nath et al. (2009); Shanmugam et al. (2009)
Templating	Uses a template to direct NPs in a linear arrangement	Length of template Density of template Flexibility of emplate Size of NPs	High yield Feasible synthesis	Template impurities Low colloidal stability	MacLachlan et al. (2000); Byrne et al. (2004); Correa-Duarte et al. (2005); Sitharaman et al. (2005); Son et al. (2005); Choi et al. (2007); Jana et al. (2007); Jung et al. (2007); Bai et al. (2008); Corr et al. (2008a); Corr et al. (2008b); Ersen et al. (2008); Piao et al. (2008); Kanamadi et al. (2009); Zhang and Zhang (2009)
Self-assembly	Caused by a spontaneous reaction that minimizes thermo-dynamic energy in standard conditions that arranges NPs in a linear fashion	Type of interaction between NPs Size of NPs	One step synthesis High yield No external energy or steps	No control over morphology sizes Intrinsic disorder of proximal NPs	Nyamjav and Ivanisevic (2005); Salgueirino-Maceira et al. (2006); Sheparovych et al. (2006); Kinsella and Ivanisevic (2007); Fresnais et al. (2008); Kinsella and Ivanisevic (2008); Liu et al. (2009)

19.4.1 Ligand-Controlled Approach

Ligand-controlled approach uses ligands, or small molecules, to form anisotropic shape in nanostructures. Ligands, such as cystamine (Leung et al. 2009), poly(vinylpyrrolidone) (Feng et al. 2004), and aluminum (Rebolledo et al. 2008) demonstrated to control length of 1D superparamagnetic nanostructures. One biocompatible ligand used in commercially available CA is dextran. Dextran can manipulate anisotropic shape formation in addition to providing a biocompatible outer covering (Nath et al. 2009). For example, Park et al. demonstrated that adding high molecular weight of dextran (20 kDa) to Fe(II) and Fe(III) salts formed a chain of NPs, or "nanoworms," whereas low-molecular-weight dextran formed dispersed, spherical NPs (Park et al. 2008). Although it was unclear to whether the NPs were connected or agglomerated in the dextran coating, the entire size of the "nanoworm" had a diameter of ~5 nm and length of ~50 nm. The time of ligand addition can also affect the structure formation. Nath et al. found that dextran added 60 s after iron oxide nucleation reduced the yield of nanorods. However, when dextran was added 30 s after iron oxide nucleation, monodispersed, superparamagnetic nanorods were formed (Nath et al. 2009).

Under the ligand-controlled approach, nanostructures were produced with high yield. While the approach seems promising, the long-term stability of these ligand-controlled nanostructures is still unclear. In addition, the fabrication method requires precise synthesis steps, such as refluxing at high temperatures and controlling pressure, to form a 1D structure (Tang and Kotov 2005). As ligands can be used to construct 1D nanostructures, another bottom-up approach uses linear templates.

19.4.2 Templating Approach

Templating approach directs the arrangement of NPs along an already linear-shaped nanostructure. Categorized by structural flexibility, templates can be (a) soft, such as biomolecules or linear polymers, or (b) hard, such as carbon nanotubes (CNTs) or anodic alumina.

Soft templates, such as linear polyelectrolytes (Corr et al. 2008a,b) and biomolecules (Byrne et al. 2004), organize NPs into 1D structures by electrostatic or covalent attachments. For example, Corr et al. stabilized Fe_3O_4 NPs to linear poly(sodium-4-styrene) sulfonate by a reduction reaction in solution (Corr et al. 2008b). The polyelectrolyte acted both as a stabilizer for the iron ions and as a template for NP arrangement (Corr et al. 2008a). Monodispersity and high yield were observed for soft template, 1D superparamagnetic nanostructures (Corr et al. 2008b). However, due to its flexible design, a soft template for CAs may cause fragmentation or clumping on in vivo systems, decreasing its structural stability. A thick, biocompatible coating around the structure is needed in order to achieve stability. On the contrary, hard templates may offer high structural stability due to its rigid path to assemble SPNs into a linear structure.

Hard templates, such as carbon (Sitharaman et al. 2005; Ersen et al. 2008) and silica (Bai et al. 2008) nanotubes, can direct SPNs to assemble on the (a) inner or (b) outer surfaces of the tube (Tang and Kotov 2005). For example, magnetite (Fe_3O_4) NPs were arranged on the *inner* surface of silica nanotubes while exposing the outer surface for further functionalization of tissue-targeting biomolecules (Son et al. 2005). One design constraint is the density of nanotube limits the number of SPNs arranged inside, and thereby, control over magnetic properties is restrictive (Ersen et al. 2008). Conversely, superparamagnetic iron platinum (FePt) NPs were covalently bonded on the *outer* surface of CNTs (Tsoufis et al. 2008). After annealing the nanotube surface, transmission electron microscope (TEM) images presented high dispersity of NPs coated on the nanotube surface and no evidence of nanostructure agglomeration.

NPs templated by either hard or soft material to form 1D nanostructures is a potential method for fabrication of linear MRI CAs due to high yields of production. However, main concerns with both templating approaches are (i) observed low colloidal stability, (ii) template impurities disrupting intrinsic properties of arranged NPs, and (iii) long-term issues regarding structural and morphological alterations of the linear template for in vivo systems. Since this fabrication approach encompasses several of these issues, researchers attempted to assemble NPs anisotropically without the use of templates.

19.4.3 Self-Assembly Approach

Self-assembly approach exploits spontaneous reactions among NPs that can arrange into anisotropic nanostructures. NPs can self-organize into a linear chain by driving forces, such as van der Waals, electrostatic, and hydrogen-bond forces, in a manner that minimizes their thermodynamic energy. The approach is feasible and inexpensive approach because it is a one-step process. Assembling either (a) on surface or (b) in solution are two routes for self-assembly fabrication.

Constructing linear structures on surface provides a platform for controlled, linear formation of NP chains. For example, superparamagnetic Fe_3O_4 NPs, coated with citric acid, stacked themselves into a chain-like shape on a plastic surface under 1 T magnetic field (Sheparovych et al. 2006). Cationic polyelectrolyte, poly(2-vinyl-*N*-methylpyridinium iodide), was added to act as a stabilizer in the formation of a linear shape. When the magnetic field was removed, Sheparovych et al. observed that the NP chains detached from the surface but remained intact. The rate of NP injection and time length of applied magnetic field controlled the dimensions of the chains in this specific study. One concern for this design is the charge stability. A CA that does not exhibit a neutral surface charge can cause protein adsorption and cellular toxicity (Sun et al. 2008). Another study demonstrated NPs chain formation on a surface without the use of a magnetic field (Nyamjav and Ivanisevic 2005). After linear DNA molecules were stretched on a silicon oxide surface, cationic ligand-coated ferric oxide NPs organized along the stretched DNA through an electrostatic interaction. Although the length of the chains is determined by the length of the DNA molecule, it may be difficult to detach these linear structures from silicon oxide surfaces and disperse in solution.

In solution, self-assembly approach provides dispersity of linear nanostructures. Similar to ligand-controlled, in solution self-assembly uses small molecules to stabilize linear chains of NPs. Self-assembly, however, is performed in one step under standard conditions (room temperature, atmospheric pressure, etc.). Ligand-controlled requires further fabrication steps, such as refluxing at high temperatures. For example, after adding a solution of poly(trimethylammonium ethylacrylate methylsulfate)-*b*-poly(acrylamide) to polyacrylic acid-coated iron oxide NPs, superparamagnetic nanowires were formed in solution under an applied magnetic field (Fresnais et al. 2008). These nanowires were stable for many months in solution even after the removal of the external magnetic field. In addition, ferric oxide and cobalt ferric oxide 1D nanostructures were formed when cationic-coated SPNs aligned along linear DNA in a buffer solution (Kinsella and Ivanisevic 2008). This study used advantages from soft templating and self-assembly in order to create stable nanostructures. Self-assembly in solution is the most attractive option, among the discussed fabrication methods, due to its one-step synthesis and minimum energy consumption, which are beneficial for mass production.

Although the feasibility aspect of self-assembly approach promises to provide an optimal solution to the problem of reliably fabricating linear MRI CAs, future research needs to focus on controlling the dimensions and morphology of self-assembled, superparamagnetic nanowires. The main concern of any self-assembled nanostructure is the stability and control over shape in order to function in vivo. Also, intrinsic interactions between proximal NPs may cause disorder to the morphology (Tang and Kotov 2005; Lin and Samia 2006), which can then alter the magnetic properties (superparamagnetic to ferromagnetic) of the nanostructure (Salgueirino-Maceira et al. 2006; Nakata et al. 2008). While stability and biocompatibility are important in CA design, high magnetization and relaxation rates are also required to acquire strong MR signals.

19.5 Effects of 1D Superparamagnetic Nanostructures on MRI Properties

The second requirement to achieve the ideal design is to exhibit high magnetization and relaxation rates. To characterize the magnetic properties of CAs, research groups commonly report saturation magnetization and relaxivity. Saturation magnetization (M_s) is the maximum net magnetization exhibited by a NP at high magnetic fields and is dependent on the intrinsic property, such as the degree of crystallization, of a nanostructure (Cheon and Lee 2008). The strength of M_s increases the magnitude of the local magnetic field for surrounding protons and directly affects the MR signal intensity. Since the degree of crystallization in a nanostructure is governed by the method of fabrication, Table 19.3 compares M_s values to the fabrication approach.

Hard templating provides a wide range of M_s values from 2.8 to 95 emu/g, demonstrating that NP clustering on either the inner or outer surfaces may not be reliable for controlling the saturation magnetization in nanostructures. On the contrary, when NPs are aligned either by ligands or on a soft template, a relatively high M_s value (74.2 and 89 emu/g) is obtained. In addition, Kinsella (2008) demonstrated that by combining two fabrication methods (soft templating and self-assembly), M_s was higher (89 emu/g) than the M_s value (30–50 emu/g) retained from soft templating alone.

As a large M_s value is desired in CAs, a high relaxivity, or the rate of relaxation per mole of colloid, is also important. It is dependent on the extrinsic properties (size and shape) of the superparamagnetic nanostructure. Relaxivity of paramagnetic CAs is reported by its longitudinal relaxivity, r_1, and relaxivity of superparamagnetic CAs is reported by its transverse relaxivity, r_2 (Thorek et al. 2006) A large r_2 shortens T_2 faster, which increases MR signal and provides high contrast in T_2-weighted images. To understand the influence of shape in CA, Table 19.4 compares r_2 values for two differently shaped superparamagnetic nanostructures (0D and 1D).

Since a magnetically anisotropic behavior is needed in superparamagnetism (Gossuin et al. 2009), two-dimensional (2D) and 3D magnetic structures are not included as they do not exhibit this type of magnetic behavior (Petit et al. 2003). Table 19.4 is limited to reports that only presented transverse relaxivity information for 1D, superparamagnetic nanostructures. The relaxivity is compared to its analog SPNs made of the same material. While there have been numerous studies with various sizes, materials, and surface functionalizations for SPNs in literature, the table only summarizes relevant NP studies that exhibited similarities in surface functionalization, NP size, or fabrication method. Table 19.4 reports information on the diameter for 0D nanostructures and the aspect ratio (length: diameter) for 1D nanostructures along with their corresponding transverse relaxivity and magnetic field strength. Relaxivity is dependent on the strength of the magnetic field (B_0) due to the direct relationship between B_0 and the rate of proton precession (Jun et al. 2006).

One limitation in Table 19.4 is that not all relaxation rates were not measured at the same magnetic field (B_0). However, an overall, general trend is provided, exhibiting that relaxation rates for 1D

TABLE 19.3 Saturation Magnetization Value for 1D Superparamagnetic Nanostructures

Structural Design	Fabrication Method	M_s (emu/g)	References
Fe_2O_3 nanoworms	Ligand controlled	74.2	Park et al. (2008)
Silica nanotube—Fe_3O_4 NPs inside	Hard template	~2.8	Son et al. (2005)
Carbon nanotube (Fe_3O_4 NPs outside)	Hard template	50	Correa-Duarte et al. (2005)
Fe_2O_3 NPs on gold nanorods	Hard template, multicomponent	65	Gole et al. (2008)
Silica nanotube—Fe_2O_3 NPs inside	Hard template	95	Bai et al. (2008)
Fe_3O_4 NPs on PSS	Soft template	30–50	Corr et al. (2008b)
$CoFe_2O_4$ NPs on DNA	Soft template, self-assembly	89	Kinsella (2008)

TABLE 19.4 Comparison of Relaxation Rates of 0D and 1D Nanostructures

	0D Nanostructures						1D Nanostructures						Summary
Superparamagnetic Material		Surface	Diameter (nm)	r_2	B_0	References		Details	Aspect Ratio	r_2	B_0	References	r_2 Increase from 0D to 1D
Fe_2O_3	A_0	Dextran (ferum-oxide)	4.96	120	1.5	Jung and Jacobs (1995)	A_1	Nano-particles in silica nanotube	8–28	358	3	Bai et al. (2008)	3X
	B_0	Dextran	5	79	4.7	Park et al. (2008)	B_1	Dextran coated Nano-worms	10	116	4.7	Park et al. (2008)	46%
Fe_3O_4	C_0	Dextran (ferum-oxtran)	5–5.84	61–65	1.5	Wang et al. (2001); Josephson et al. (1999)	C_1[a]	Nano-particle chain on PSS	95	89.4–132.4	1.5	Corr et al. (2008a)	46–117%
							D_1[a]	Nano-rods	2.3	300	0.47	Nath et al. (2009)	5X
				$r_1 = 10–20$	1.5	Meledandri et al. (2008)	E_1	Nano-particle chain on DNA	>111	$r_1 = 140$	23	Byrne et al. (2004)	7–14X[c]
Gadolinium	F_0	PEG[b]	3	$r_1 = 8.8–9.4$	1.5	Fortin et al. (2007)	F_1[a]	Nano-capsules using CNT	6.6–8	$r_1 = 164–173$	1	Sitharaman et al. (2005)	18X[c]
Au–Fe_3O_4	G_0	Au NP core/ Fe_3O_4 coated	20	114	3	Xu et al. (2008)	G_1	Au Nano-rods—Fe_3O_4 NPs coated	3.6 (Au)–15 nm dia. (Fe_3O_4)	248	3	Wang et al. (2009)	2X
	H_0	Au NP attached with Fe_3O_4 NP	33	204	1.5	Choi et al. (2008)							21%

Source: Feng, L. et al., *J. Colloid. Interf. Sci.*, 278, 372, 2004.

By convention, NPs are defined to exhibit an aspect ratio (length: diameter) of 1. An aspect ratio greater than 1 and less than 20 is considered a nanorod, while nanowires exhibit aspect ratios greater than 20. Dia. = diameter; r_2 units are $mM^{-1} \cdot s^{-1}$ and B_0 units are in Tesla (T).

[a] Measured temperature at 310 K; all other studies measured at 298 K.

[b] PEG-coated gadolinium NPs exhibited paramagnetism. However, their r_1 value is compared to the r_1 value for superparamagnetic nanocapsules.

[c] r_1 increase from 0D to 1D.

nanostructures are significantly greater than relaxation rates for 0D nanostructures (Table 19.4—summary column). Ferumoxides (maghemite: γFe_2O_3) and ferumoxtrans (magnetite: Fe_3O_4), two commercially available CAs, were set as standard SPNs to compare with 1D nanostructures. For example, maghemite NPs arranged in a silica nanotube (Table 19.4—A_1) exhibited an r_2 three times the r_2 of ferumoxides (Table 19.4—A_0) alone. Additionally, magnetite NPs arranged on linear polystyrene sulfonate (PSS) (Table 19.4—C_1) retained 46% greater relaxation than ferumoxtrans (Table 19.4—C_0) alone. Although it is inconclusive from Table 19.4 to which type of NP arrangement (clustering or alignment) offers a better relaxation rate, the distinct effect in proton relaxation due to shape (from 0D to 1D) can be explained by the following mechanisms.

The outer sphere spin–spin relaxation theory is used to predict the effectiveness of relaxation rates of water protons to nearby SPNs. The theory involves the rotational and transitional diffusion of water molecules in the presence of SPNs (Cheon and Lee 2008). In brief, the outer sphere theory explains that a high relaxation rate is possible when more water molecules are able to interact with the magnetic dipole of the NP (Figure 19.3A) (Nelson and Runge 1996).

When NPs are brought into proximity and clustered, magnetic spin moments are coupled, generating strong local magnetic fields (Figure 19.3B). The coupling effect alters the relaxation rate by providing more dipole–dipole interactions between the water molecules and the NP cluster (Ai et al. 2005). This type of relaxation enhancement may explain the r_2 increase when NPs are clustered inside a linear nanotube (i.e., Table 19.4; A_1 and F_1).

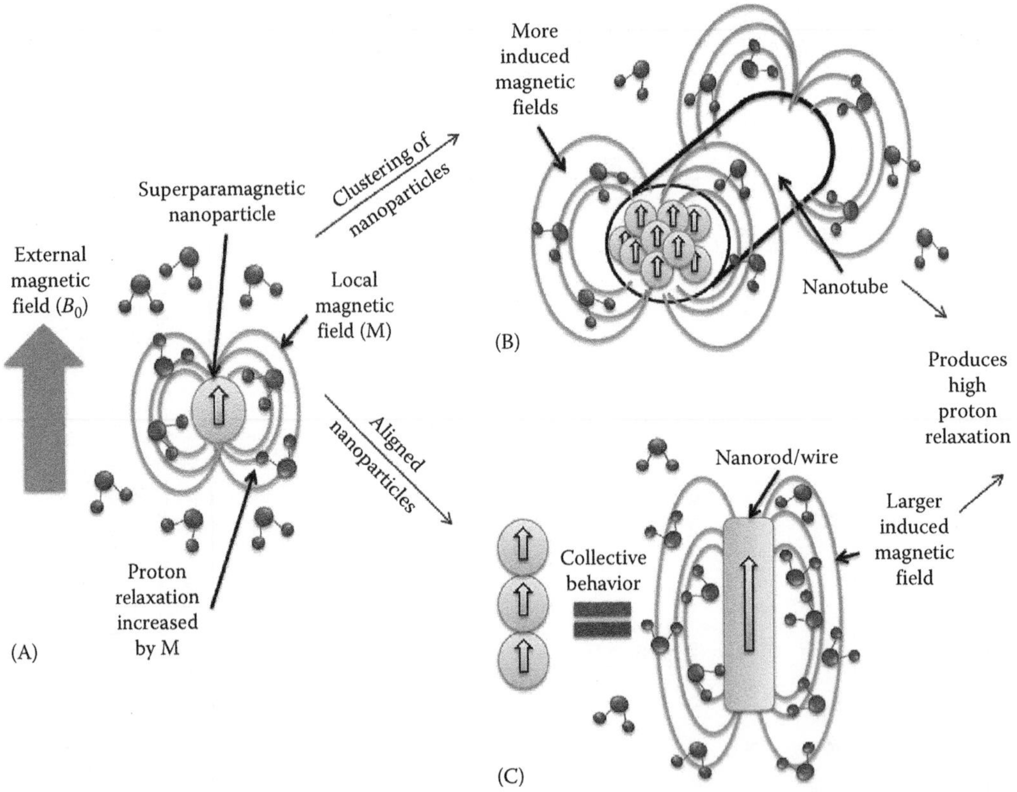

FIGURE 19.3 Mechanism of superparamagnetic CA affecting water protons. (A) Under an external magnetic field, the magnetic dipole of SPN aligns along B_0 and produces a local magnetic field, which increases T_2 relaxation of nearby protons. (B) The clustering of NPs in a linear nanotube induces more local magnetic fields, which increases greater T_2 relaxation of nearby protons. (C) The alignment of NPs in a linear shape produces a larger magnetic field, which increases greater T_2 relaxation of NPs.

In the case where NPs are arranged into a linear chain without aggregation, the magnetic dipole interactions between aligned NPs behave as a magnetic wire (Figure 19.3C) (Petit et al. 2003; Kinsella and Ivanisevic 2007; Lee et al. 2007). The "collective" behavior due to alignment of NPs exhibited an increase in interaction between the NPs and water molecules. Consequently, the interparticle interaction may explain the higher relaxation rates measured for structurally aligned NPs compared to dispersed NPs (i.e., Table 19.4; B_1, C_1, D_1, E_1, and G_1).

An issue not thoroughly documented in this chapter is spin-incoherence relaxation (T_2^*). It is a subset of T_2 that accounts for spatial inhomogeneties of the magnetic field. While T_2^* and other magnetic properties (i.e., blocking temperature, spin life time, coercivity, and susceptibility) strongly influence magnetic behavior in nanoscale structures (Jun et al. 2005), the aspect of saturation magnetization and relaxation rates are the focus to provide an initial understanding about shape dependence in CA effectiveness.

It is also important to note that contrast in T_1- and T_2-weighted images does not solely rely on signal from proton relaxation. Parameters, such as echo time (TE), repetition time (TR), and flip angle, defined in pulse sequences and environmental factors, such as temperature and medium of measurement, also affect contrast intensity. For further information about factors in image contrast, refer to Liang and Lauterbur (2000). Nevertheless, an ideal CA needs to exhibit stability with high magnetism on in vivo systems. The next section evaluates initial studies on in vivo properties for 1D, superparamagnetic nanostructures.

19.6 Effects of 1D Nanostructures on In Vivo Systems

The third requirement to achieve an ideal design is to attain optimal in vivo properties. It is important to understand the in vivo effects of CAs (i.e., biodistribution properties, foreign body response, monocyte-phagocyte interactions, and nonspecific protein adsorption) to address health and safety concerns and improve MR signal detection. Blood circulation time for MRI CA is one of the important in vivo properties to evaluate. Since MRI exhibits low sensitivity (10^{-3}–10^{-5} mol/L) and temporal resolution (minutes to hours) (Massoud and Gambhir 2003), it becomes difficult to detect a strong MR signal from CAs with short blood half-life (seconds to minutes) administered at a minimal concentration (Faraj et al. 2009). A prolonged half-life (hours) in blood provides more time to detect a quality MR signal.

Extrinsic properties, such as (i) size, (ii) charge, (iii) surface coating, and (iv) shape, affect the utility of SPN-based CAs on in vivo systems. Many studies have focused on size, charge, and surface coating effects for SPNs. Particles that are intravenously injected are usually in the size order of 5–150 nm to effectively evade splenic filtration and avoid renal clearance (Neuberger et al. 2005). Additionally, positively charged surfaces of NPs exhibit nonspecific binding to cells, while negatively charged surfaces of NPs exhibit increased liver uptake. The general consensus from literature accepts that a neutral surface on NPs increases blood circulation time (Sun et al. 2008). Surface charge can be controlled by the type of coating. Derivatives of dextran, poly(ethyleneglycol), starch, albumin, silica, and others provide a biocompatible covering for SPNs. Such coatings are necessary to stabilize SPNs in a biological environment at pH 7.4 and also avoid their immediate uptake by the reticulendothelial system (Neuberger et al. 2005). Although size, charge, and surface coating are studied extensively for numerous in vivo applications, there has been little added to the understanding on the effects of shape of the nanostructure (Geng et al. 2007).

CNTs, a popular group of linear nanostructures, have received the most attention than any other linear nanostructures for in vivo applications. Structural characteristics of CNTs have been studied for effects in toxicity (Deng et al. 2007; Koyama et al. 2009), inflammation (Sato et al. 2005), liver accumulation (Yang et al. 2007), and blood clearance times (Liu et al. 2007, 2008; Yang et al. 2008). Specifically, the length of CNTs used turned out to be an important parameter for in vivo systems. The Kostarelos group demonstrated CNTs with lengths ranging from 300 to 1000 nm exhibited a 3 h long blood half-life (Singh et al. 2006). If the ideal MRI CA can circulate as long as CNTs, a high-quality MR signal can be obtained.

In addition to CNTs, gold nanorods are another type of anisotropic nanostructure that researchers have investigated for in vivo, optical applications (Niidome et al. 2006; Eghtedari et al. 2007; Eghtedari et al. 2008; Chen-Wei et al. 2008; Takuro et al. 2008), but very few studies have dealt with the issue of linear, superparamagnetic nanostructures for in vivo MRI applications. Studies from Corr et al. (2008a) and Richard et al. (2008) demonstrated that high image contrast is possible when using linear CAs in vivo but did not report the biodistribution characteristics. Conversely, Faraj et al. observed a 24 h blood clearance after administrating a CNT-based CA (Faraj et al. 2009), and Lacerda et al. explained that water permeability was increased in the hollow nanotube structures, aiding in prolonged blood circulation time (Lacerda et al. 2008). Park et al., however, demonstrated the structure does not need to be hollow to display long blood circulation times. Superparamagnetic nanoworms exhibited a prolonged blood half-life for 15 h with no sign of phagocyte uptake while 0D SPNs were cleared within minutes in rats' bodies (Park et al. 2008, 2009). The overall trend observed from these initial studies is that 1D nanostructures exhibit prolonged blood circulation time when compared to dispersed 0D nanostructures.

19.7 Future Directions

While 1D superparamagnetic nanostructures demonstrate promising aspects for improving CA design, future directions in this research area are to develop a new generation of nanosized agents that exhibit multiple functions. Specifically, imaging agents must contribute to diagnosis and treatment of medical conditions (Moghimi et al. 2005).

Diagnosis by imaging agents primarily relies on their ability to identify abnormal tissues. New generation agents would need to target these specific tissues. The use of different biomolecules has contributed to the function of tissue targeting. Small biomolecules, or ligands, such as amino acids, peptides, sugars, can have unique recognition properties to specific cells and tissues. One example of cellular targeting is the direct binding of a ligand to a membrane protein (Vivès et al. 2008). Therefore, ligands can be attached to the exterior surface of nanostructures and aid to target a specific component, such as membrane proteins, of a cell. Due to the high aspect ratio of 1D nanostructures, the surface area is greater than the surface area for NPs. A higher surface area provides more ligand attachments on to the nanostructure. If more targeting ligands are available for binding, then the nanostructure has a greater ability to locate the tissue by specific targeting.

Treatment by imaging agents relies on their ability to carry and deliver the appropriate drugs. After locating and identifying the abnormal tissue, new generation agents would need to serve as a vehicle that can carry drugs and treat the targeted, abnormal tissue (Sun et al. 2008).

One method to incorporate drug carrying capabilities into the agent is to design a hollow nanostructure (McCarthy and Weissleder 2008). Hollow, 1D nanostructures, such as nanotubes, can provide space in the inner tube to carry drug molecules. Also, 1D nanocapsules can provide area inside for drug molecule storage. One concern using hollow nanostructures is to understand the drug release method from the nanostructure into the tissues and cells. Structures must either be flexible or degradable to release the drugs out into the tissues.

Another method to carry drugs is to include other types of materials in the nanostructure. Multicomponent nanostructures exhibit two or more different materials in which each material serves a specific function (Shubayev et al. 2009). In this type of design, the magnetic NPs would serve as the primary imaging agent, whereas another type of material, such as metallic NPs, would serve to carry the drugs. Specifically, metals like silver and gold are advantageous components to imaging agents. Since gold is a soft acid, it allows for strong binding of thiols and amine functionalities, which permits numerous chemical modifications on nanostructures (Chowdhury and Akaike 2005). Although biomolecules and drugs can attach to surfaces of ferric oxide NPs, the chemistry can be complex and time consuming (Jeong et al. 2007; Na et al. 2009). Instead, the use of metal NPs in the multicomponent design allows for feasible chemistry of ligand and drug attachment for applications in cellular targeting and drug delivery (Karim et al. 2006).

19.8 Conclusion

Nanostructures in the biomedical field exhibit great potential for diagnosis and therapy. Specifically, nanostructured agents, such as ferric oxide NPs, have shown to inexpensively improve contrast in MRI images. Ferric oxide NPs, however, have exhibited undesired biodistribution properties in vivo, and there is a need to improve their utility. One way to improve current imaging agents is to address the structural design. Particularly, 1D, superparamagnetic nanostructures have demonstrated to have high relaxation rates and long blood circulation times. These nanostructures may provide promising improvements for MRI CAs. Moreover, the future for CAs holds to further develop the field of nanomedicine by functioning as both a diagnostic and a therapeutic tool.

References

Ai, H., C. Flask, B. Weinberg, X. Shuai, M. D. Pagel, D. Farrell, J. Duerk, and J. Gao. 2005. Magnetite-loaded polymeric micelles as ultrasensitive magnetic-resonance probes. *Adv. Mater.* 17 (16):1949–1952.

Bai, X., S. J. Son, S. Zhang, W. Liu, E. K. Jordan, J. A. Frank, T. Venkatesan, and S. B. Lee. 2008. Synthesis of superparamagnetic nanotubes as MRI contrast agents and for cell labeling. *Nanomedicine* 3 (2):163–174.

Bhushan, B. 2004. *Springer Handbook of Nanotechnology*. New York: Springer-Verlag.

Burtea, C., S. Laurent, L. V. Elst, and R. N. Muller. 2008. Contrast agents: Magnetic resonance. In *Handbook of Experimental Pharmocology*, W. Semmler and M. Schwaiger, eds. Springer-Verlag: Berlin.

Byrne, S. J., S. A. Corr, Y. K. Gun'ko, J. M. Kelly, D. F. Brougham, and S. Ghosh. 2004. Magnetic nanoparticle assemblies on denatured DNA show unusual magnetic relaxivity and potential applications for MRI. *Chem. Commun.* (22):2560–2561.

Champion, J. A. and S. Mitragotri. 2006. Role of target geometry in phagocytosis. *PNAS* 103 (13):4930–4934.

Chen-Wei, C. W., C. K. Liao, Y. Y. Chen, C. R. C. Wang, A. A. Ding, D. B. Shiehd, and P. C. Li, 2008. In vivo photoacoustic imaging with multiple selective targeting using bioconjugated gold nanorods. *SPIE International Symposium on Biomedical Optics*, San Jose, CA.

Cheon, J. and J. H. Lee. 2008. Synergistically integrated nanoparticles as multimodal probes for nanobiotechnology. *Acc. Chem. Res.* 41 (12):1630–1640.

Choi, J. H., F. T. Nguyen, P. W. Barone, D. A. Heller, A. E. Moll, D. Patel, S. A. Boppart, and M. S. Strano. 2007. Multimodal biomedical imaging with asymmetric single-walled carbon nanotube/iron oxide nanoparticle complexes. *Nano Lett.* 7 (4):861–867.

Choi, S. H., H. B. Na, Y. I. Park, K. An, S. G. Kwon, Y. Jang, M. H. Park et al. 2008. Simple and generalized synthesis of oxide—Metal heterostructured nanoparticles and their applications in multimodal biomedical probes. *J. Am. Chem. Soc.* 130 (46):15573–15580.

Chowdhury, E. H. and T. Akaike. 2005. Bio-functional inorganic materials: An attractive branch of gene-based nano-medicine delivery for 21st century. *Curr Gene Ther.* 5 (6):669–676.

Corr, S. A., S. J. Byrne, R. Tekoriute, C. J. Meledandri, D. F. Brougham, M. Lynch, C. Kerskens, L. O'Dwyer, and Y. K. Gun'ko. 2008a. Linear assemblies of magnetic nanoparticles as MRI contrast agents. *J. Am. Chem. Soc.* 130 (13):4214–4215.

Corr, S. A., Y. K. Gun'ko, R. Tekoriute, C. J. Meledandri, and D. F. Brougham. 2008b. Poly(sodium-4-styrene)sulfonate iron oxide nanocomposite dispersions with controlled magnetic resonance properties. *J. Phys. Chem. C* 112 (35):13324–13327.

Correa-Duarte, M. A., M. Grzelczak, V. Salgueirino-Maceira, M. Giersig, L. M. Liz-Marzan, M. Farle, K. Sierazdki, and R. Diaz. 2005. Alignment of carbon nanotubes under low magnetic fields through attachment of magnetic nanoparticles. *J. Phys. Chem. B* 109 (41):19060–19063.

Deng, X., G. Jia, H. Wang, H. Sun, X. Wang, S. Yang, T. Wang, and Y. Liu. 2007. Translocation and fate of multi-walled carbon nanotubes in vivo. *Carbon* 45 (7):1419–1424.

Duguet, E., S. Vasseur, S. Mornet, and J. M. Devoisselle. 2006. Magnetic nanoparticles and their applications in medicine. *Nanomedicine* 1 (2):157–168.

Eghtedari, M., A. V. Liopo, J. A. Copland, A. A. Oraevsky, and M. Motamedi. 2008. Engineering of hetero-functional gold nanorods for the in vivo molecular targeting of breast cancer cells. *Nano Lett.* 9 (1):287–291.

Eghtedari, M., A. Oraevsky, J. A. Copland, N. A. Kotov, A. Conjusteau, and M. Motamedi. 2007. High sensitivity of in vivo detection of gold nanorods using a laser optoacoustic imaging system. *Nano Lett.* 7 (7):1914–1918.

Ersen, O., S. Begin, M. Houlle, J. Amadou, I. Janowska, J. M. Greneche, C. Crucifix, and C. Pham-Huu. 2008. Microstructural investigation of magnetic $CoFe_2O_4$ nanowires inside carbon nanotubes by electron tomography. *Nano Lett.* 8 (4):1033–1040.

Faraj, A. A., K. Cieslar, G. Lacroix, S. Gaillard, E. Canet-Soulas, and Y. Cremillieux. 2009. In vivo imaging of carbon nanotube biodistribution using magnetic resonance imaging. *Nano Lett.* 9 (3):1023–1027.

Feng, L., L. Jiang, Z. Mai, and D. Zhu. 2004. Polymer-controlled synthesis of Fe_3O_4 single-crystal nanorods. *J. Colloid Interf. Sci.* 278 (2):372–375.

Fortin, M. A., R. M. Petoral Jr., F. Soderlind, A. Klasson, M. Engstrom, T. Veres, P. O. Kall, and K. Uvdal. 2007. Polyethylene glycol-covered ultra-small Gd_2O_3 nanoparticles for positive contrast at 1.5T magnetic resonance clinical scanning. *Nanotechnology* 18:1–9.

Fresnais, J., J. F. Berret, B. Frka-Petesic, O. Sandre, and R. Perzynski. 2008. Electrostatic co-assembly of iron oxide nanoparticles and polymers: Towards the generation of highly persistent superparamagnetic nanorods. *Adv. Mater.* 20 (20):3877–3881.

Froehlich, J. M. 2008. MR contrast agents. In *How Does MRI Work?* Berlin: Springer.

Geng, Y., P. Dalhaimer, S. Cai, R. Tsai, M. Tewari, T. Minko, and D. E. Discher. 2007. Shape effects of filaments versus spherical particles in flow and drug delivery. *Nat. Nanotechnol.* 2:249–255.

Geraldes, C. F. G. C. and S. Laurent. 2009. Classification and basic properties of contrast agents for magnetic resonance imaging. *Contrast Media Mol. I* 4:1–23.

Gole, A., J. W. Stone, W. R. Gemmill, H. C. zur Loye, and C. J. Murphy. 2008. Iron oxide coated gold nanorods: Synthesis, characterization, and magnetic manipulation. *Langmuir* 24 (12):6232–6237.

Gossuin, Y., P. Gillis, A. Hocq, Q. L. Vuong, and A. Roch. 2009. Magnetic resonance relaxation properties of superparamagnetic particles. *Wiley Interdiscip. Rev. Nanomed. Nanobiotechnol.* 1:299–310.

Gupta, A. K. and M. Gupta. 2005. Synthesis and surface engineering of iron oxide nanoparticles for biomedical applications. *Biomaterials* 26 (18):3995–4021.

Hendee, W. R. and E. R. Ritenour. 2002. *Medical Imaging Physics*. New York: John Wiley & Sons.

Jana, S., S. Basu, S. Pande, S. K. Ghosh, and T. Pal. 2007. Shape-selective synthesis, magnetic properties, and catalytic activity of single crystalline MnO_2 nanoparticles. *J. Phys. Chem. C* 111 (44):16272–16277.

Jeong, U., X. Teng, Y. Wang, H. Yang, and Y. Xia. 2007. Superparamagnetic colloids: Controlled synthesis and niche applications. *Adv. Mater.* 19 (1):33–60.

Josephson, L., C. H. Tung, A. Moore, and R. Weissleder. 1999. High-efficiency intracellular magnetic labeling with novel superparamagnetic-tat peptide conjugates. *Bioconjugate Chem.* 10 (2):186–191.

Jun, Y. W., Y. M. Huh, J. S. Choi, J. H. Lee, H. T. Song, K. Kim, S. Yoon et al. 2005. Nanoscale size effect of magnetic nanocrystals and their utilization for cancer diagnosis via magnetic resonance imaging. *J. Am. Chem. Soc.* 127 (16):5732–5733.

Jun, Y. W., J. T. Jang, and J. Cheon. 2006. Magnetic resonance nanoparticle probes for cancer imaging. In *Nanomaterials for Cancer Diagnosis*, C. Kumar, ed. Weinheim, Germany: Wiley-VCH.

Jung, C. W. and P. Jacobs. 1995. Physical and chemical properties of superparamagnetic iron oxide MR contrast agents: Ferumoxides, ferumoxtran, ferumoxsil. *Magn. Reson. Imaging* 13 (5):661–671.

Jung, J. S., J. H. Lim, L. Malkinski, A. Vovk, K. H. Choi, S. L. Oh, Y. R. Kim, and J. H. Jun. 2007. Magnetic properties of structurally confined FePt nanoparticles within mesoporous nanotubes. *J. Magn. Magn. Mater.* 310 (2, Part 3):2361–2363.

Kanamadi, C. M., B. K. Das, C. W. Kim, H. G. Cha, E. S. Ji, D. I. Kang, A. P. Jadhav, J. K. Heo, D. Kim, and Y. S. Kang. 2009. Template assisted growth of cobalt ferrite nanowires. *J. Nanosci. Nanotechnol.* 9:4942–4947.

Karim, S., M. E. Toimil-Molares, F. Maurer, G. Miehe, W. Ensinger, J. Liu, T. W. Cornelius, and R. Neumann. 2006. Synthesis of gold nanowires with controlled crystallographic characteristics. *Appl. Phys. A* 84 (4):403–407.

Kinsella, J. M. 2008. Magnetotransport of one-dimensional chains of $CoFe_2O_4$ nanoparticles ordered along DNA. *J. Phys. Chem. C* 112 (9):3191–3193.

Kinsella, J. M. and A. Ivanisevic. 2007. DNA-templated magnetic nanowires with different compositions: Fabrication and analysis. *Langmuir* 23 (7):3886–3890.

Kinsella, J. M. and A. Ivanisevic. 2008. Fabrication of ordered metallic and magnetic heterostructured DNA-nanoparticle hybrids. *Colloid. Surf. B* 63 (2):269–300.

Koyama, S., Y. A. Kim, T. Hayashi, K. Takeuchi, C. Fujii, N. Kuroiwa, H. Koyama, T. Tsukahara, and M. Endo. 2009. In vivo immunological toxicity in mice of carbon nanotubes with impurities. *Carbon* 47 (5):1365–1372.

Lacerda, L., M. A. Herrero, K. Venner, A. Bianco, M. Prato, and K. Kostarelos. 2008. Carbon-nanotube shape and individualization critical for renal excretion. *Small* 4 (8):1130–1132.

Lee, J. H., Y. M. Huh, Y. W. Jun, J. Wook. Seo, J. T. Jang, H. T. Song, S. Kim et al. 2007. Artificially engineering magnetic nanoparticles for ultra-sensitive molecular imaging. *Nat. Med.* 13 (1):95–99.

Leung, K. C. F., Y. J. Wang, H. Wang, S. Xuan, C. P. Chak, and C. H. K. Cheng. 2009. Biological and magnetic contrast evaluation of shape-selective Mn-Fe nanowires. *IEEE JNL* 8 (2).

Liang, Z. P. and P. C. Lauterbur. 2000. *Principles of Magnetic Resonance Imaging*, M. Akay, ed. New York: IEEE Press.

Lin, X. M. and A. C. S. Samia. 2006. Synthesis, assembly and physical properties of magnetic nanoparticles. *J. Magn. Magn. Mater.* 305 (1):100–109.

Liu, Y., W. Jiang, S. Li, and F. Li. 2009. Electrostatic self-assembly of Fe3O4 nanoparticles on carbon nanotubes. *Appl. Surf. Sci.* 255 (18):7999–8002.

Liu, Z., W. Cai, L. He, N. Nakayama, K. Chen, X. Sun, X. Chen, and H. Dai. 2007. In vivo biodistribution and highly efficient tumour targeting of carbon nanotubes in mice. *Nat. Nanotechnol.* 2:47–52.

Liu, Z., C. Davis, W. Cai, L. He, X. Chen, and H. Dai. 2008. Circulation and long-term fate of functionalized biocompatible single-walled carbon nanotubes in mice probed by Raman Spectroscopy. *PNAS* 105 (5):1410–1415.

MacLachlan, M. J., M. Ginzburg, N. Coombs, N. P. Raju, J. E. Greedan, G. A. Ozin, and I. Manners. 2000. Superparamagnetic ceramic nanocomposites: Synthesis and pyrolysis of ring-opened poly(ferrocenylsilanes) inside periodic mesoporous silica. *J. Am. Chem. Soc.* 122 (16):3878–3891.

Martín, J. I., J. Nogués, K. Liu, J. L. Vicent, and I. K. Schuller. 2003. Ordered magnetic nanostructures: Fabrication and properties. *J. Magn. Magn. Mater.* 256 (1–3):449–501.

Massoud, T. F. and S. S. Gambhir. 2003. Molecular imaging in living systems: Seeing fundamental biological processes in a new light. *Genes Dev.* 17:545–580.

McCarthy, J. R. and R. Weissleder. 2008. Multifunctional magnetic nanoparticles for targeted imaging and therapy. *Adv. Drug Deliv. Rev.* 60 (11):1241–1251.

Meledandri, C. J., J. K. Stolarczyk, S. Ghosh, and D. F Brougham. 2008. Nonaqueous magnetic nanoparticle suspensions with controlled particle size and nuclear magnetic resonance properties. *Langmuir* 24 (24):14159–14165.

Miyasaka, H., M. Julve, M. Yamashita, and R. Clerac. 2009. Slow dynamics of the magnetization in one-dimensional coordination polymers: Single-chain magnets. *Inorg. Chem.* 48 (8):3420–3437.

Moghimi, S. M., A. C. Hunter, and J. C. Murray. 2005. Nanomedicine: Current status and future prospects. *FASEB* 19:311–330.

Na, H. B., I. C. Song, and T. Hyeon. 2009. Inorganic nanoparticles for MRI contrast agents. *Adv. Mater.* 21:1–16.

Nakata, K., Y. Hu, O. Uzun, O. Bakr, and F. Stellacci. 2008. Chains of superparamagnetic nanoparticles. *Adv. Mater.* 20:4294–4299.

Nath, S., C. Kaittanis, V. Ramachandran, N. S. Dalal, and J. M. Perez. 2009. Synthesis, magnetic characterization, and sensing applications of novel dextran-coated iron oxide nanorods. *Chem. Mater* 21 (8):1761–1767.

Nelson, K. L. and V. M. Runge. 1996. Principles of MR contrast. In *Contrast-Enhanced Clinical Magnetic Resonance Imaging*, V. M. Runge, ed. Lexington, KY: University Press of Kentucky.

Neuberger, T., B. Schopf, H. Hofmann, M. Hofmann, and B. Von. Rechenberg. 2005. Superparamagnetic nanoparticles for biomedical applications: Possibilities and limitations of a new drug delivery system. *J. Magn. Magn. Mater.* 293 (1):483–496.

Niidome, T., M. Yamagata, Y. Okamoto, Y. Akiyama, H. Takahashi, T. Kawano, Y. Katayama, and Y. Niidome. 2006. PEG-modified gold nanorods with a stealth character for in vivo applications. *J. Control. Release* 114 (3):343–347.

Nyamjav, D. and A. Ivanisevic. 2005. Templates for DNA-templated Fe_3O_4 nanoparticles. *Biomaterials* 26 (15):2749–2757.

Okuhata, Y. 1999. Delivery of diagnostic agents for magnetic resonance imaging. *Adv. Drug. Deliv. Rev.* 37 (1–3):121–137.

Pamboucas, C. A. and S. G. Rokas. 2008. Clinical safety of cardiovascular magnetic resonance: Cardiovascular devices and contrast agents. *Hellenic J. Cardiol.* 49:352–356.

Park, J. H., G. V. Maltzahn, L. Zhang, A. M. Derfus, D. Simberg, T. J. Harris, E. Ruoslahti, S. N. Bhatia, and M. J. Sailor. 2009. Systematic surface engineering of magnetic nanoworms for in vivo tumor targeting. *Small* 5 (6):694–700.

Park, J. H., G. V. Maltzahn, L. Zhang, M. P. Schwartz, E. Ruoslahti, S. N. Bhatia, and M. J. Sailor. 2008. Magnetic iron oxide nanoworms for tumor targeting and imaging. *Adv. Mater.* 20 (9):1630–1635.

Petit, C., V. Russier, and M. P. Pileni. 2003. Effect of the structure of cobalt nanocrystal organization on the collective magnetic properties. *J. Phys. Chem. B* 107 (38):10333–10336.

Piao, Y., J. Kim, H. B. Na, D. Kim, J. S. Baek, M. K. Ko, J. H. Lee, M. Shokouhimehr, and T. Hyeon. 2008. Wrap-bake-peel process for nanostructural transformation from [beta]-FeOOH nanorods to biocompatible iron oxide nanocapsules. *Nat. Mater.* 7 (3):242–247.

Rebolledo, A. F., O. Bomatí-Miguel, J. F. Marco, and P. Tartaj. 2008. A facile synthetic route for the preparation of superparamagnetic iron oxide nanorods and nanorices with tunable surface functionality. *Adv. Mater.* 20 (9):1760–1765.

Richard, C., B. T. Doan, J. C. Beloeil, M. Bessodes, E. Toth, and D. Scherman. 2008. Noncovalent functionalization of carbon nanotubes with amphiphilic Gd3+ chelates: Toward powerful T1 and T2 MRI contrast agents. *Nano Lett.* 8 (1):232–236.

Salgueirino-Maceira, V., M. A. Correa-Duarte, A. Hucht, and M. Farle. 2006. One-dimensional assemblies of silica-coated cobalt nanoparticles: Magnetic pearl necklaces. *J. Magn. Magn. Mater.* 303 (1):163–166.

Sato, Y., A. Yokoyama, K. Shibata, Y. Akimoto, S. Ogino, Y. Nodasaka, T. Kohgo et al. 2005. Influence of length on cytotototoxicity of multi-walled carbon nanotubes against human acute monocytic leukemia cell line THP-1 in vitro and subcutaneous tissue of rats in vivo. *Mol. Biosyst.* 1:176–182.

Shanmugam, S., T. Nakanishi, and T. Osaka. 2009. Morphology and magnetic properties of iron oxide nanostructures synthesized with biogenic polyamines. *J. Electrochem. Soc.* 156 (7):K121–K127.

Sheparovych, R., Y. Sahoo, M. Motornov, S. Wang, H. Luo, P. N. Prasad, I. Sokolov, and S. Minko. 2006. Polyelectrolyte stabilized nanowires from Fe_3O_4 nanoparticles via magnetic field induced self-assembly. *Chem. Mater.* 18 (3):591–593.

Shubayev, V. I., T. R. P. Ii, and S. Jin. 2009. Magnetic nanoparticles for theragnostics. *Adv. Drug Deliv. Rev.* 61 (6):467–477.

Singh, R., D. Pantarotto, L. Lacerda, G. Pastorin, C. Klumpp, M. Prato, A. Bianco, and K. Kostarelos. 2006. Tissue biodistribution and blood clearance rates of intravenously administered carbon nanotube radiotracers. *PNAS* 103 (9):3357–3362.

Sitharaman, B., K. R. Kissell, K. B. Hartman, L. A. Tran, A. Baikalov, I. Rusakova, Y. Sun et al. 2005. Superparamagnetic gadonanotubes are high-performance MRI contrast agents. *Chem. Commun.* (31):3915–3917.

Son, S. J., J. Reichel, B. He, M. Schuchman, and S. B. Lee. 2005. Magnetic nanotubes for magnetic-field-assisted bioseparation, biointeraction, and drug delivery. *J. Am. Chem. Soc.* 127 (20):7316–7317.

Sun, C., J. S. H. Lee, and M. Zhang. 2008. Magnetic nanoparticles in MR imaging and drug delivery. *Adv. Drug. Deliv. Rev.* 60 (11):1252–1265.

Takuro, N., A. Yasuyuki, S. Kohei, K. Takahito, M. Takeshi, K. Yoshiki, and N. Yasuro. 2008. In vivo monitoring of intravenously injected gold nanorods using near-infrared light. *Small* 4 (7):1001–1007.

Tang, Z. and N. A. Kotov. 2005. One-dimensional assemblies of nanoparticles: Preparation, properties, and promise. *Adv. Mater.* 17 (8):951–962.

Thorek, D. L. J., A. K. Chen, J. Czupryna, and A. Tsourkas. 2006. Superparamagnetic iron oxide nanoparticle probes for molecular imaging. *Ann. Biomed. Eng.* 34 (1):23–38.

Tsoufis, T., A. Tomou, D. Gournis, A. P. Douvalis, I. Panagiotopoulos, B. Kooi, V. Georgakilas, I. Arfaoui, and T. Bakas. 2008. Novel nanohybrids derived from the attachment of FePt nanoparticles on carbon nanotubes. *J. Nanosci. Nanotechnol.* 8 (11):5942–5951.

Vivès, E., J. Schmidt, and A. Pèlegrin. 2008. Cell-penetrating and cell-targeting peptides in drug delivery. *Biochim. Biophys. Acta (BBA)—Rev. Cancer* 1786 (2):126–138.

Wang, C., J. Chen, T. Talavage, and J. Irudayaraj. 2009. Gold nanorod/Fe_3O_4 nanoparticle "nano-pearl-necklaces" for simultaneous targeting, dual-mode imaging, and photothermal ablation of cancer cells. *Angew. Chem. Int. Ed.* 48 (15):2759–2763.

Wang, Y. X. J., S. M. Hussain, and G. P. Krestin. 2001. Superparamagnetic iron oxide contrast agents: Physicochemical characteristics and applications in MR imaging. *Eur. J. Radiol.* 11 (11):2319–2331.

Xu, C., J. Xie, D. Ho, C. Wang, N. Kohler, E. G. Walsh, J. R. Morgan, Y. E. Chin, and S. Sun. 2008. Au-Fe_3O_4 dumbbell nanoparticles as dual-functional probes. *Angew. Chem. Int. Ed.* 47 (1):173–176.

Yang, S. T., K. A. S. Fernando, J. H. Liu, J. Wang, H. F. Sun, Y. Liu, M. Chen et al. 2008. Covalently PEGylated carbon nanotubes with stealth character in vivo. *Small* 4 (7):940–944.

Yang, S. T., W. Guo, Y. Lin, X. Y. Deng, H. F. Wang, H. F. Sun, Y. F. Liu et al. 2007. Biodistribution of pristine single-walled carbon nanotubes in vivo. *J. Phys. Chem. C* 111 (48):17761–17764.

Zhang, L. and Y. Zhang. 2009. Fabrication and magnetic properties of Fe_3O_4 nanowire arrays in different diameters. *J. Magn. Magn. Mater.* 321 (5):L15–L20.

20

Nanobiomaterials for Dual-Mode Molecular Imaging: Advances in Probes for MR/Optical Imaging Applications

Chuqiao Tu
University of California at Davis

Angelique Louie
University of California at Davis

20.1 Introduction

The term "nanotechnology" refers to the fabrication and application of materials in the size range of nanometers; at this size scale, materials can possess unique chemical and physical properties in comparison with their bulk counterparts. In the past two decades, reports related to nanomaterials have grown exponentially and now nanotechnology is an active area of research in many diverse fields ranging from chemistry, physics, material science, and engineering to biology, pharmacology, and medicine. One of the major applications of nanomaterials in the biomedical field is their use as contrast agents for noninvasive molecular imaging (Cormode, Skajaa et al. 2009; Riehemann et al. 2009; Suh et al. 2009). Compared with traditional molecular imaging agents, nanomaterials can contain a high payload of the contrast-generating materials, which greatly improves their detectability in vivo. Furthermore, the surfaces of nanomaterials are readily modified and functionalized to provide biocompatibility, biostability, and specific biomarker targeting.

There have been a number of breakthroughs in the use of nanomaterials as bioimaging contrast agents for detecting and diagnosing various forms of diseases. For example, semiconductor quantum dots (QDs) have been applied in various imaging applications including fixed cell labeling, imaging of live cell dynamics, in situ tissue profiling, biochemical sensing, and in vivo animal imaging in

TABLE 20.1 Features of Some Imaging Techniques

Modality	MRI	CT	PET	Fluorescence Imaging
Energy Type	Radiowaves	X-rays + contrast	γ-rays	Visible and near-infrared light
Resolution	10–100 μm	50 μm	1–2 mm	In vivo, 2–3 mm; in vitro, sub-μm
Sensitivity	mM to μM	Not well characterized	pM	nM to pM
Depth	No limit	No limit	No limit	<1 cm
Acquisition time	Minutes to hours	Minutes	Minutes to hours	Seconds to minutes
Imaging agents	Paramagnetic chelates, para- or superparamagnetic particles	Iodine-based or barium-based molecules	Radionuclide (^{18}F-, ^{64}Cu-, ^{11}C-, ^{13}N-, ^{15}O- or ^{124}I-labelled compounds)	Photoproteins, fluorochromes, quantum dots, gold nanoparticles
Target	Anatomical, physiological, molecular	Anatomical, physiological	Physiological, molecular	Physiological, molecular
Clinical use	Versatile imaging modality with high soft-tissue contrast	Imaging lungs and bones	Versatile imaging modality with many tracers	Rapid screening of molecular events in surface-based diseases

Source: Weissleder, R. and Pittet, M.J., *Nature*, 452(7187), 580, 2008; Koo, V. et al., *Cell. Oncol.*, 28(4), 127, 2006; Cheon, J. and Lee, J.H., *Acc. Chem. Res.* 41(12), 1630, 2008.

many biological systems (Medintz et al. 2005; Michalet et al. 2005; Cheon and Lee 2008; Xing and Rao 2008). Metallic gold nanoparticles have shown powerful potential in biomedical fields and are being actively explored for applications such as localized surface plasmon (LSP) resonance for optical sensing and detection of a variety of biomolecules (Qian et al. 2008; Sperling et al. 2008). Magnetic nanomaterials have been used as magnetic resonance imaging (MRI) contrast agents for liver imaging, cellular labeling, metastatic lymph node imaging, early detection of atherosclerosis, and tumor imaging (Sharma et al. 2007; Laurent et al. 2008). However, to date, most of the nanoscale imaging agents reported are single functionality and are restricted to detection by a single imaging modality, which limits their utility for diagnostic imaging. For in vivo imaging the ideal readout would be quantitative, high resolution, longitudinal (to allow imaging over time), and sensitive to molecular perturbations in the system. No single imaging modality possesses all of the ideal traits, as evident in Table 20.1, where typical molecular imaging modalities are summarized (Koo et al. 2006; Weissleder and Pittet 2008).

There has been a rapid increase in research focused on the development of multimodality systems for noninvasive molecular imaging. Among various possibilities, the combination of MRI and optical imaging is particularly attractive. The two modalities are both nonionizing and they have highly complementary features in anatomical resolution and detection sensitivity with the potential for strong preclinical and clinical synergies. The hybrid has shown potential for significantly increased diagnostic accuracy compared to standalone imaging (Cheon and Lee 2008; Niedre and Ntziachristos 2008; Lee and Chen 2009). To optimize the use of multiple modalities and to register images from each, ideally a single probe detectable by multiple modalities would be employed. A hybrid MR/fluorescence imaging probe, which imparts MR and fluorescence contrast, may be resolved by both MR and optical instrumentation, e.g., using MRI for noninvasive disease detection and optical imaging during subsequent surgical intervention to identify diseased tissues. This approach can be used to merge noninvasive MRI findings with corresponding highly sensitive intraoperative or endoscopic detection of optical signals during follow-up procedures (Niedre and Ntziachristos 2008). Nanomaterials are ideal for the purpose as they provide a particularly useful platform to integrate multiple functionalities into a single entity (Mulder et al. 2007; Cheon and Lee 2008; Corr et al. 2008; Kim et al. 2009). A growing number of particle-based multimodality imaging probes have been developed in recent years.

In this chapter, we summarize the recent progress in the development of nanobiomaterials for dual-modality MR/florescence imaging from the point-of-view of the design, fabrication, and applications of nanomaterials.

20.2 MRI Contrast Agents and Fluorescence Imaging Probes

To understand the basis for multimodal probes, it is helpful to consider the current state of the art for typical single-mode probes used with these modalities. We here briefly provide an overview of probes used in MRI and in optical imaging.

20.2.1 MRI Contrast Agents

As the endogenous contrast difference among various types of tissues can be small, a contrast agent is usually used in MRI to achieve sufficient contrast in the region of interest. Contrast agents are in the form of either paramagnetic metal chelates or magnetic iron oxide nanoparticles that take effect by shortening the relaxation times of bulk water protons. The majority of currently available MRI contrast agents are chelates of gadolinium (Gd(III)), because the lanthanide ion has a large magnetic moment and long electronic relaxation time. However, the effectiveness of small molecular weight MRI contrast agents is limited by fast rotation, rapid excretion by kidney, and short serum half-life, resulting in insufficient accumulation at target sites (Toth et al. 2002; Caravan 2006). Particle-based technologies provide an effective way of increasing relaxivity by slowing down rotational tumbling time and increasing the local concentration of paramagnetic ions. Gd(III) chelates have been coupled to polymers, and incorporated into micelles or silica to form gadolinium nanoparticles, increasing their high relaxivity and extending lifetime in the blood pool; these have been used as contrast agents for MRI of atherosclerosis and other diseases (Toth et al. 2002; Sharma et al. 2007). Magnetic iron oxide nanoparticles are another primary nanoscale contrast agent for MRI due to their advantages such as higher sensitivity and low toxicity of iron. Magnetic iron oxide nanoparticles are already commercially available and are routinely used as contrast agents for targeting organs (liver and spleen) or lymph nodes (Sharma et al. 2006; Laurent et al. 2008).

20.2.2 Fluorescence Imaging Probes

Fluorescence imaging has been one of the most rapidly adapted imaging technologies. The fundamental barriers for in vivo optical imaging are high photon absorption by hemoglobin (Hb) and scattering (Sharma et al. 2006). In the body deoxy- and oxyhemoglobin (HbO_2) are the primary absorbers of visible light, while water and lipids are the primary absorbers of infrared light. However, absorption coefficients are lower in the near-infrared (NIR) region (Weissleder 2001). The highest tissue penetration occurs for red and NIR light of 650–900 nm, and these wavelengths have been used for tomographic imaging based on bulk tissue optical properties. For in vivo use, therefore, exogenously administered fluorophores that fluoresce in the red and NIR region are desired (Sharma et al. 2006).

Organic dyes are the most commonly used fluorophores for fluorescence imaging. However, organic dyes are prone to fast photobleaching (i.e., the loss of fluorescence upon continuous excitation) and thus, are not ideal agents for long-term imaging and tracking studies. Moreover, the emission/excitation of organic dyes is often susceptible to variations in local chemical environment, e.g., pH, interacting ions, etc. (Sharma et al. 2006; Cai et al. 2007). Nanoparticle-based optical contrast agents, particularly QDs, are emerging as a new class of fluorescent contrast agents, which overcome the limits of traditional organic dyes listed above. The emission of QDs is environmentally stable as the production of photons stems from a bandgap process rather than the singlet-singlet transition typically observed in organic dyes (Wang et al. 2007). Typical QDs are core–shell (e.g., CdSe core with a ZnS shell) or core-only (e.g., CdTe) structures functionalized with different coatings to optimize fluorescence quantum yield (QY), decay kinetics, and stability (Resch-Genger et al. 2008). In comparison with classical organic dyes, QDs have many advantages such as higher molar extinction, much greater resistance to photobleaching and chemical degradation, narrower emission spectra, and larger effective Stokes shift. These characteristics make QDs 10–20 times brighter than classical organic dyes under photon-limited in vivo and in vitro conditions, where the light intensities are severely attenuated by tissue scattering and absorption (Bakalova et al. 2007; Cai et al. 2007).

20.3 Multifunctional Nanomaterials for Both MR and Fluorescence Imaging

Most multifunctional nanocomposites for MR/fluorescence imaging are designed and fabricated starting from a magnetic or fluorescent nanoparticle core. A core, i.e., iron oxide nanoparticles or QDs, can be coated with a polymer, an amphiphilic lipid, or a silica shell containing complementary fluorescent or magnetic components, which are usually organic dyes or Gd(III) chelates. In some cases the QDs are directly coated with paramagnetic Gd(III) chelates. Other structures include dual magnetic and fluorescent cores, magnetically doped QDs, and paramagnetic Gd(III) chelates and organic dyes coupled to a polymer, or encapsulated within an amphiphilic lipid or a silica matrix (Mulder et al. 2007; Quarta et al. 2007; Corr et al. 2008). In this chapter, we will provide a brief overview of all of these approaches.

20.3.1 Magnetic or Fluorescent Core–Based MR/Fluorescence Imaging Agents

20.3.1.1 Polymer-Coated MR/Fluorescence Imaging Agents

20.3.1.1.1 Dextran Coating

Coating of the iron oxide particles, including in situ and postsynthesis coatings, is crucial for stabilization and further functionalization of magnetic colloidal ferrofluids. Dextran is the most commonly used coating material for iron oxide particles because of its biocompatibity. In practice, the dextran on the particles is usually chemically cross-linked to generate cross-linked iron oxide nanoparticles (CLIO). CLIO have been widely used for in vivo as well as in vitro MR imaging due to their small hydrodynamic size and stable hydrophilic polymer coating. A common method to introduce multifunctionality to iron oxide nanoparticles is to covalently attach a fluorophore to the surface of the nanoparticles. One of the most widely used multifunctional magnetic and fluorescent nanoparticles was developed by Josephson et al. (2002) who conjugated CLIO with the indocyanine dye, Cy5.5. As shown in Figure 20.1, monocrystalline iron oxide nanoparticle (MION-47) consisting of a 3 nm core of $(Fe_2O_3)_n(Fe_3O_4)_m$ with a layer of 10 kDa dextran was cross-linked with epichlorohydrin to form CLIO-47. CLIO-47 was aminated by reacting with ammonia to generate CLIO-NH_2, which had an average of 62 primary amine groups per particle. The particle had a hydrodynamic size of 38 nm, an r_1 of 21 mM^{-1} s^{-1}, and an r_2 of 62 mM^{-1} s^{-1} (37°C, 0.5 T). Cyanine 5.5 (λ_{em} 692 nm) was dissolved in DMSO and reacted with CLIO-NH_2 to afford a MR/fluorescence imaging probe CLIO-Cy5.5, which had an iron content of 6.5 mg mL^{-1} and an average of 2.5 fluorophores per particle (Weissleder et al. 2005; Jaffer et al. 2006).

The CLIO-Cy5.5 particles have been coupled to many different targeting agents for MR/fluorescence imaging of atherosclerotic plaques (Nahrendorf et al. 2006; Sosnovik et al. 2007), siRNA delivery and silencing in tumors (Medarova et al. 2007), biomarkers of tumors such as $\alpha_v\beta_3$ integrin (Montet et al. 2006), E-selectin (Kang et al. 2006) or uMUC-1 (Medarova, Pham et al. 2006, 2009), and grafted pancreatic islets in type 1 diabetes (Evgenov et al. 2006; Medarova, Evgenov et al. 2006). For example, using the peptide VHPKQHR, CLIO-Cy5.5 can be targeted to vascular cell adhesion molecule 1 (VCAM-1), a biomarker found in the early stage of atherosclerotic plaques. In vitro, the CLIO-Cy5.5 targeted all cell types expressing VCAM-1. In vivo, MR and fluorescence imaging in apo E-/- mice after injection with CLIO-Cy5.5 revealed signal enhancement in the aortic root of mice receiving the nanoparticles

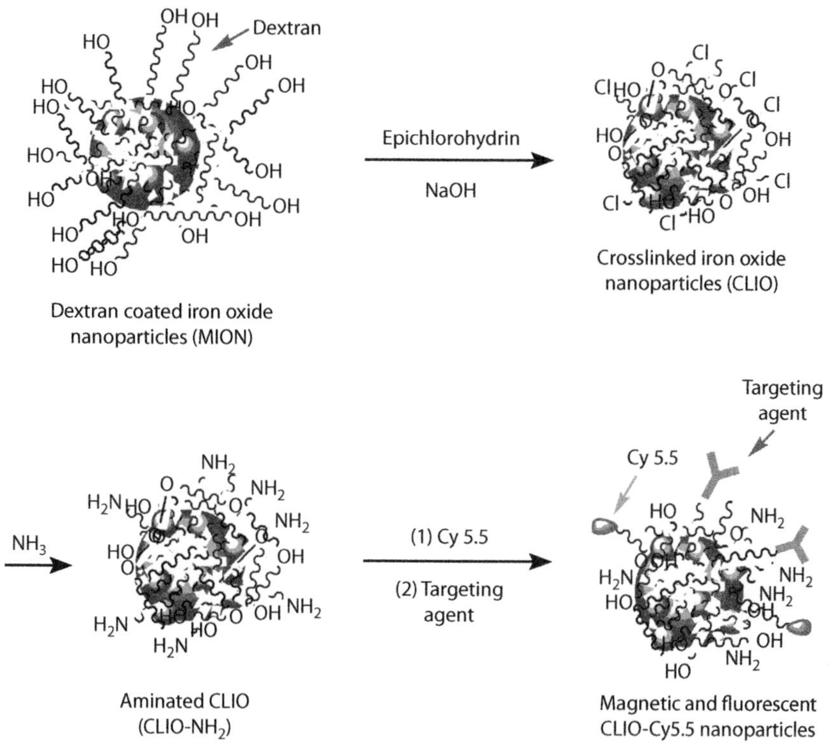

FIGURE 20.1 Schematic. Synthesis of iron oxide nanoparticle-based multifunctional magnetic and fluorescent nanoparticles CLIO-Cy5.5.

(Nahrendorf et al. 2006). The CLIO-Cy5.5 particles were also used for imaging the infarcted myocardium of C57BL/6 mice with MRI and fluorescence molecular tomography (FMT). FMT is an emerging imaging technique that combines advantages of macroscopic visualization with the inherent sensitivity of microscopic fluorescence imaging. The technology allows for 3D imaging of fluorescence biodistribution in whole animals, by accounting for tissue optical heterogeneity, and the nonlinear dependence of fluorescence intensity on depth and optical properties (Ntziachristos 2006). MRI showed myocardial probe accumulation in the anterolateral walls of the infarcted mice but not in the sham-operated mice 48 h after injection of CLIO-Cy5.5. Fluorescence intensity over the heart was also significantly greater in the FMT images of the infarcted mice (Sosnovik et al. 2007).

20.3.1.1.2 Polyethylene Glycol Coating

Water-soluble and biocompatible, polyethylene glycol (PEG) is another frequently used polymer material for iron oxide nanoparticles coating. Jeong and coworkers reported the synthesis of thermally cross-linked superparamagnetic iron oxides with a polymer coating composed of PEG, $Si(OCH_3)_3$, and *N*-hydroxysuccinimide(NHS)-activated carboxylic acid groups. Upon heating at 80°C, trimethoxysilane and NHS ester are spontaneously converted to trihydroxysilane and carboxylic acid for cross-linking and further conjugation with Cy5.5, respectively. Dynamic light scattering (DLS) measurements revealed that the probe had a hydrodynamic mean size of 25.6 ± 2.7 nm with the magnetite (Fe_3O_4) cores ranging from approximately 3–10 nm (Transmission Electron Microscopy [TEM]). The probe was administered to Lewis lung carcinoma tumor allograft mice by intravenous injection. The in vivo T_2-weighted MR performed with a 1.5 T MR scanner and NIR fluorescence imaging revealed that the probes were accumulated at the tumor site via the enhanced permeation and retention effect. The ex vivo NIR fluorescent images of harvested tumor and other major organs indicated that the highest fluorescence intensity was from the tumor (Lee et al. 2007). Hyeon and coworkers coupled $POCl_3$ with PEG to generate PEG-derivatized phosphine oxide ligands (PO-PEG). Then the oleic acid originally used to stabilize monodisperse magnetite nanocrystals was replaced by the phosphine group in PO-PEG. After amination, a fluorescent dye, fluorescein isothiocyanate (FITC), was attached to the nanoparticles. The probes were shown to possess dual-modality properties of T_2-weighted MR and optical imaging in cell studies (Bin Na et al. 2007).

Zhang et al. functionalized the PEG-coated iron oxide nanoparticles with peptide-MHC (major histocompatibility complex) monomers and further coupled the particles with the fluorophore Alexa Fluor 647. The iron oxide nanoparticles were synthesized via a co-precipitation of iron chloride and sodium hydroxide. The PEG-immobilized nanoparticles were generated by sonication of PEG-trifluoroethylester silane and nanoparticle suspension. The primary amine was created on the immobilized PEG chain termini by flooding the nanoparticle suspension with excess ethylenediamine (EDA), as shown in Figure 20.2a (Kohler et al. 2004). The resultant nanoparticles had 26 linear PEG chains, each

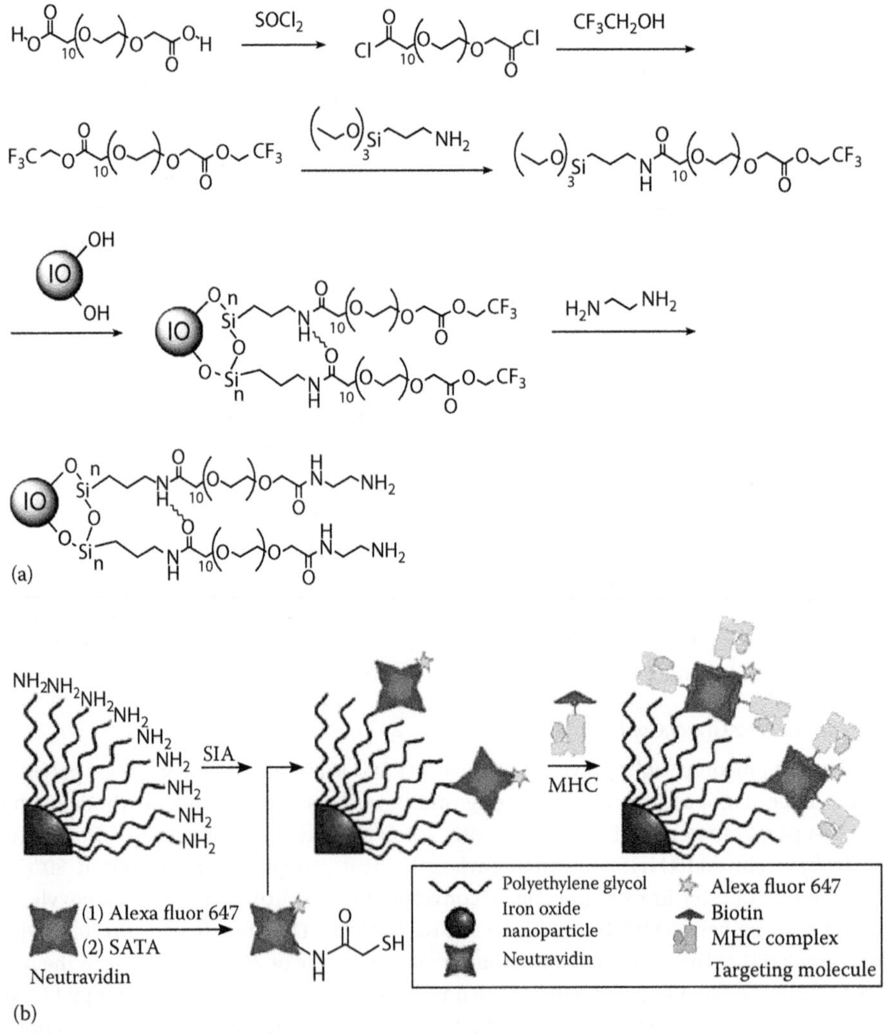

FIGURE 20.2 Schematic illustration of (a) immobilization of aminated PEG on iron oxide nanoparticles and (b) synthesis of florescent iron oxide nanoparticles targeted to CTLs. (Reproduced from Gunn, J. et al., *Small*, 4(6), 712, 2008; Kohler, N. et al., *J. Am. Chem. Soc.*, 126(23), 7206, 2004. With permission.)

bearing a reactive amine attached to the surface of each iron oxide core. On the other hand, the neutravidin was made fluorescent by shaking the protein with Alexa Fluor 647 monosuccinimidyl ester. After reaction with N-succinimidyl-S-acetylthioacetate (SATA), which endows the protein with active sulfhydryl groups, the fluorophore-labeled neutravidin protein was covalently conjugated with the sulfhydryl-reactive, magnetic iron nanoparticle via a thioether linkage. The biotinylated peptide-MHC was then attached to neutravidin, lending the particle targeting specificity for CTL cells expressing T-cell receptors (TCR), as shown in Figure 20.2b. The particle had a core diameter of 10 nm and a hydrodynamic size of 64.8 nm (PEG-coated nanoparticle), which increased to approximately 71.0 nm subsequent to the attachment of neutravidin and peptide-MHC. It was estimated that there were 2.59 dyes per neutravidin and 13.0 neutravidins per nanoparticle (Gunn et al. 2008). The nanoparticles showed significant CTL binding (58.47%), while nontargeting nanoparticles showed only 15% of the CTLs labeled, indicating the selective cell labeling. The T_2 relaxation times for CTL and non-CTL samples were 24 ± 3 and 71 ± 2 ms, respectively. The corresponding MR phantom image shows the CTLs significantly darker than the non-CTL cells, further confirming the uptake specificity (Gunn et al. 2008).

20.3.1.1.3 Chitosan Coating

Recently, chitosan has aroused great interest in biological applications due to its properties of being hydrophilic, biocompatible, biodegradable, and nonantigenic. Chitosan has a positive zeta potential and, thus, can interact with negative domain of cell membranes by nonspecific electrostatic interactions (Weecharangsan et al. 2008). Gu and coworkers coated maghemite (γ-Fe_2O_3) with FITC-labeled chitosan. Compared with naked particles, the zeta potential of FITC-labeled particles possessed higher positive charge at physiological environment (pH 7.4), which favored the association to the negative domain of the cell membrane. The FITC-labeled particle had an average diameter of 13.8 ± 5.3 nm and exhibited an intense and narrow emission spectrum with a peak at 520 nm, similar to that of FITC with a peak at 518 nm. The small redshift (2 nm) resulted from the surrounding environments of the amino groups or the interaction between the dye and the iron oxide nanoparticles. Flow cytometry and MRI (1.5 T) showed that the nanoparticles had a high affinity to SMMC-7721 cancer cells but efficiency of cell labeling was dependent on the incubation time and concentration of nanoparticles. The localization patterns of particles in cells were monitored by fluorescence and TEM. Nanoparticles were internalized within the cell inside late endosomes or lysosomes. No uptake into endoplasmatic reticulum, mitochondria, and structures of the Golgi organ or the nucleus was found (Ge et al. 2009).

20.3.1.2 Streptavidin-Conjugated MR/Fluorescence Imaging Agents

Streptavidin is a tetrameric protein, which shows a very high affinity to biotin. The dissociation constant (K_d) of the biotin-streptavidin complex is on the order of ~10^{-15} mol L^{-1}, ranking among the strongest noncovalent interactions known in nature. After conjugation with biotinylated Annexin A5, the streptavidin-conjugated particle has been used for visualization of apoptosis, or programmed cell death (PCD). Apoptosis or PCD is a process of organized cell suicide. In body, there is a balance between cell proliferation and cell death, but the balance is disturbed in pathological conditions, such as acute myocardial infarctions, heart failure, or unstable atherosclerotic plaques, which show increased PCD, whereas in cancer this balance shifts toward cell proliferation. Annexin A5 is a protein that binds specifically to phosphatidylserine, a phospolipid overexpressed at the outer lipid layer of apoptotic cells, with a K_d in the nanometer range in the presence of Ca^{2+} ions (Hofstra et al. 2000; Dumont et al. 2001).

Prinzen et al. (2007) functionalized streptavidin-conjugated QD525 with biotinylated Annexin A5 and Gd-DTPA (DTPA: diethylene triamine pentaacetic acid), which enabled the particles to be used for visualization of PCD. The earlier version of the nanoparticle consisted of biotinylated AnxA5 coupled to streptavidin-conjugated QDs in a 1:1 ratio. AnxA5 contained a single biotin and the remaining binding sites of streptavidin were saturated with biotinylated Gd(III)-DTPA (Figure 20.3a). The r_1 relaxivity was 17.5 mM^{-1} s^{-1}. There is approximately 10 streptavidin molecules per QD; therefore, the r_1 value per

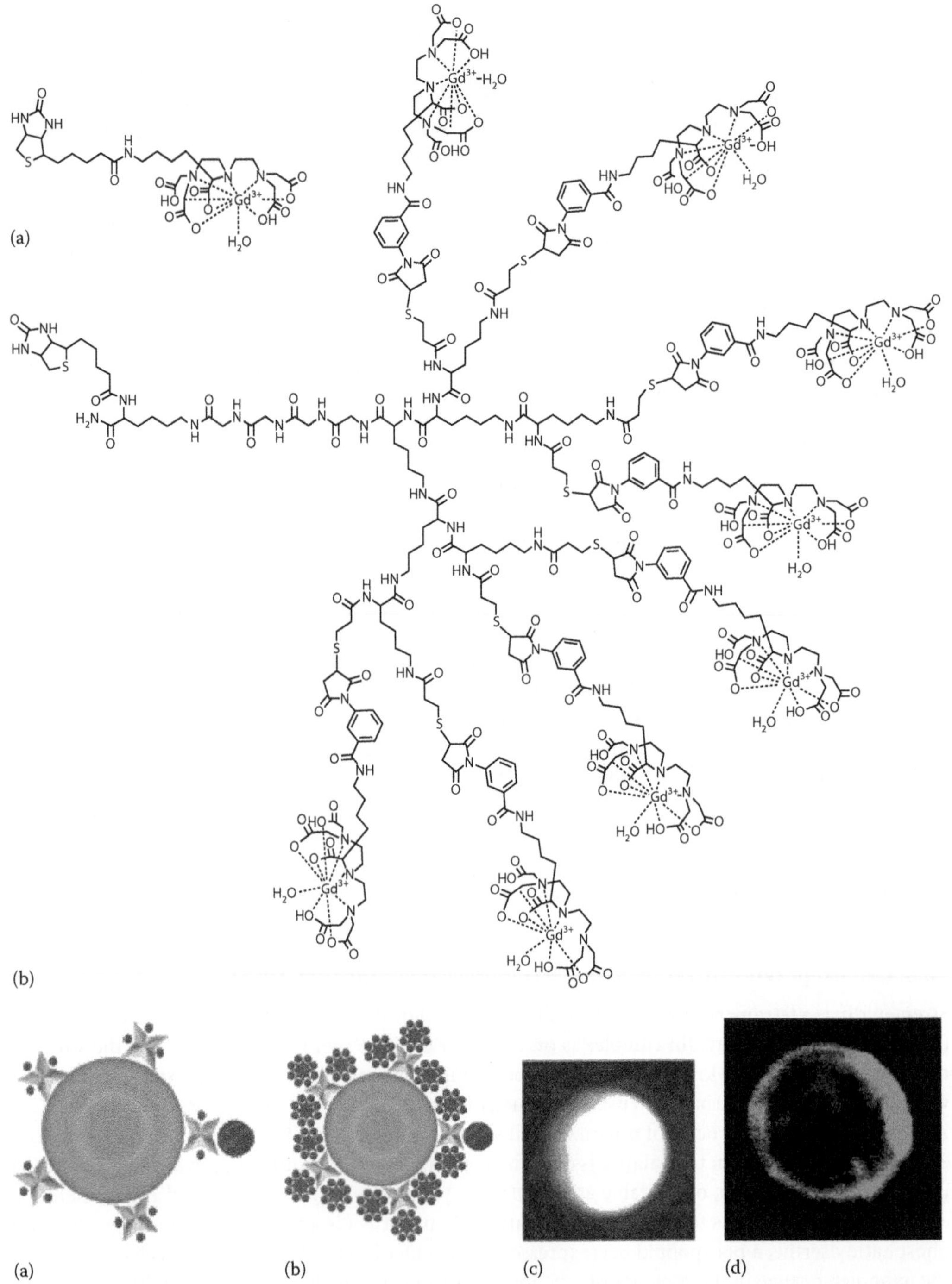

FIGURE 20.3 (See color insert.) (a) and (b) are molecular and schematic structure of the biotinylated Gd-DTPA (a) and biotinylated Gd-wedge (b) with biotin (red structure) and Gd-DTPA (blue structure). Green: QD; yellow: streptavidin; red dot: Gd-DTPA; red star: lysine-wedge; blue: AnxA5. (c) T_1-weighted MR image shows pellet of apoptotic Jurkat cells incubated with wedged nanoparticles (b). (d) TPLSM image shows cellular distribution of wedged nanoparticles (b) in late apoptotic Jurkat cells. (Reproduced from Prinzen, L. et al., *Nano Lett.*, 7(1), 93, 2007. With permission.)

nanoparticle was 420–630 $mM^{-1}\ s^{-1}$. The Gd-DTPA loading is significantly increased by using a lysine dendritic wedge with eight Gd-DTPA complexes attached to the periphery, as shown in Figure 20.3b. Although the r_1 relaxivity of Gd-wedge coupled to avidin was 15.6 $mM^{-1}\ s^{-1}$, the r_1 value per nanoparticle increased to 3,000–4,500 $mM^{-1}\ s^{-1}$. Low-temperature TEM showed that functionalized QDs in suspension were spherical, monodisperse, and had a diameter of 6.7 ± 1.0 nm. By using the probe, the cell death and activated platelets were visualized with both two-photon laser scanning microscopy (TPLSM) and MRI in vitro (Prinzen et al. 2007).

Backes and coworkers functionalized streptavidin-conjugated QDs with paramagnetic dendritic wedge with cyclic Asn-Gly-Arg (cNGR) for the noninvasive assessment of tumor angiogenic activity, the formation of new blood vessels. The tumor-homing capability of cNGR was shown to be threefold higher compared with cRGD, a peptide which was commonly used in recognition of $\alpha_v\beta_3$-integrins overexpressed on the surface of angiogenic endothelial cells (Arap et al. 1998). Intravenous injection of cNGR-QDs in tumor-bearing mice resulted in increased quantitative contrast in the tumor rim but not in tumor core or muscle tissue. Ex vivo TPLSM showed that cNGR-QDs were primarily located on the surface of tumor endothelial cells and to a lesser extent in the vessel lumen, which further supports the MRI results (Ostendorp et al. 2008).

20.3.1.3 Micelle Encapsulated MR/Fluorescence Imaging Agents

A number of approches to lipid coating imaging agents has been reported including synthetic lipid coatings, and coatings that are intended to mimic endogenously occurring lipoproteins.

High-density lipoprotein (HDL), a natural nanosized particle composed of a hydrophobic core of triglycerides and cholesterol esters covered in a monolayer of phospholipids into which apolipoprotein A-I (apoA-I) is embedded, has been used as a platform to construct multimodality contrast agents. Fayad and coworkers created endogenous nanoparticle-inorganic material composites for multimodality MR/fluorescence imaging by replacing the hydrophobic core of HDL with inorganic nanocrystals, such as QDs or iron oxide nanoparticles, and coating them with paramagnetic phospholipids (Gd-DTPA-DMPE) or rhodamine-labeled phospholipids (Cormode, Skajaa et al. 2008). A 10-fold excess of the relevant phospholipids and the hydrophobic nanoparticle were co-dissolved in a chloroform/methanol (9/1) solvent mixture. The solution was added dropwise to hot (80°C), stirred, deionized water. The organic solvents were evaporated immediately, resulting in the swift formation of micellar structures with inorganic nanoparticles lodged in the core of phospholipid aggregates. The HDL coated QD had a core size of 6.5 nm (TEM) and a hydrodynamic size of 12.9 nm (DLS), which are within the normal size range for HDL (7–13 nm). The relaxivities r_1 and r_2 at 60 MHz and 40°C for the QD-HDL were 11.7 and 14.8 $mM^{-1}\ s^{-1}$, respectively. In vitro confocal microscopy and MRI demonstrated that the nanocrystal HDL was avidly taken up by macrophages. QD-HDL particles were injected into apo E-/- mice at a 50 μmol Gd kg^{-1} dose (equivalent to 57 nmol QD per mouse). The abdominal aorta of these mice was MR imaged pre- and 24 h postintravenous administration of the agents (9.4 T). A clear increase in the signal intensity was detected in the aortic wall in the postscans. Analysis of the normalized enhancement ratio of the postscan compared to the prescan revealed a signal enhancement of 69% ± 23%. The localization of the particles in the aorta was confirmed by confocal microscopy and Maestro fluorescence imaging system on sections of aortas excised from the mice 24 h postinjection. The control, PEG-coated QD, was synthesized. Although QD-PEG has the similar parameters to QD-HDL, the uptake of QD-PEG into the aorta wall is much less extensive than QD-HDL (Cormode, Skajaa et al. 2008).

Van Tilborg et al. (2006) entrapped iron oxide particles within PEGylated micelles followed by incorporation of lipids containing PE-carboxyfluorescein into the lipid (bi)layer. The resultant probe had a core size of approximately 5 nm and the total diameter of the conjugate was measured to be approximately 10 nm with DLS. The dual-modality MR/optical imaging contrast agent has a relatively high r_2/r_1 ratio of 12 (159.6/13.3 $mM^{-1}\ s^{-1}$, 20 MHz, room temperature (rt)), which is typical for negative MR contrast agents. They also prepared another probe using Gd(III) chelates in place of iron oxide particles. By incorporation of Gd-DTPA-bis(stearylamide) lipids within the lipid (bi)layer of PEGylated liposomes, they obtained a positive MR contrast agents with a r_2/r_1 ratio of 1.7 (6.8/4.1 mM^{-1} s^{-1}, 20 MHz, rt). After conjugation with annexin A5, both probes were able to detect apoptotic cells with parallel MR/optical imaging. The probes were shown to significantly alter the relaxation rates of apoptotic cell pellets compared to untreated control cells and apoptotic cells that were treated with nonfunctionalized nanoparticles.

Recently, Yang et al. (2009) developed amphiphilic polymer coated fluorescent iron oxide nanoparticles for imaging of pancreatic cancer. Iron oxide nanoparticles (10 nm) were coated with a high-molecular-weight (100 kDa) copolymer with an elaborate ABC triblock structure and a grafted eight-carbon alkyl side chain. In contrast to simple polymers and amphiphilic lipids, this triblock polymer consists of a polybutylacrylate segment (hydrophobic), a polyethylacrylate segment (hydrophobic), a polymethacrylic acid segment (hydrophilic), and a hydrophobic hydrocarbon side chain, allowing the polymer to disperse and encapsulate single hydrophobic nanoparticles via a spontaneous self-assembly process (Gao et al. 2004). Urokinase plasminogen activator (uPA) is a serine protease that regulates matrix degradation, cell invasion, and angiogenesis through interaction with its cellular receptor (uPAR) overexpressed in more than 86% of pancreatic cancer tissues. The binding of uPA to uPAR occurs with high affinity through the amino-terminal fragment (ATF, residues 1–135 amino acids) with a $K_d < 1$ nmol L^{-1}. Conjugation of ATF or Cy5.5-ATF peptides to the nanoparticles, via cross-linking of carboxyl groups of the amphiphilic polymers to amino side groups of the peptides, generated the expected probe for MR/fluorescence imaging of pancreatic cancer. In vivo MRI (3 T) demonstrated that the nanoparticles selectively accumulated within tumors of orthotopically xenografted human pancreatic cancer in nude mice. Ex vivo NIR fluorescence imaging of the tumor and normal tissues confirmed the presence of a strong NIR signal in the tumor but not in most normal tissues except for the kidney and bladder (Yang et al. 2009).

20.3.1.4 Small Molecule–Coated MR/Fluorescence Imaging Agents

Amphiphilic lipid or polymer capping preserves the photophysical properties of the QDs, but may result in particles with a final size three or four times larger than the original nanocrystal diameter. Another strategy to coat QDs is ligand exchange. The naturally bound surface ligands are replaced by bifunctional ligands, which contain surface anchoring moieties to bind to the QD surface and hydrophilic end groups to provide water solubility. This method yields a small final diameter, but may affect the physicochemical and photophysical stability of QDs in buffered solutions (Cai et al. 2007; Hezinger et al. 2008).

Jin et al. (2008) used a natural thiol compound, glutathione (GSH [γ-L-glutamyl-L-cysteinylglycine], reduced form) to replace naturally bound ligands on a QD surface. The GSH coated CdSeTe/CdS QDs retained ca. 60% of the fluorescence intensity after ligand exchange. Then the surface of the QDs were functionalized with DOTA (1,4,7,10-tetraazacyclododecane-1,4,7,10-tetraacetic acid) followed by insertion of Gd^{3+} ions (Figure 20.4). The size of QDs before and after the conjugation reaction was measured by fluorescence correlation spectroscopy, which uses the fluctuation of fluorescence intensity in a tiny excitation volume to determine the diffusion times of fluorescent molecules. The hydrodynamic diameters are calculated to be 7.0 ± 0.4 and 10 ± 0.2 nm for GSH-QDs and Gd(III)-DOTA-QDs, respectively, on an assumption that the QD possesses a spherical body. The fluorescence intensity of the Gd(III)-DOTA-QDs was seven times higher than that of organic dye indocyanine green (ICG) when excited at 695 nm, where the absorbance was set to be the same value. The r_1 and r_2 relaxivity values for Gd(III)-DOTA-QDs were calculated to be 365 mM^{-1} s^{-1} and 6,779 mM^{-1} s^{-1} (11.7 T), respectively, on the basis of an average 77 ± 18 Gd(III)-DOTA complexes per QD particle. To test the feasibility of the Gd(III)-DOTA-QDs as a dual modal contrast agent for in vivo imaging, a phantom (a polyethylene tube with 1.5 mm diameter) containing 5 mL of Gd(III)-DOTA-QDs (10 μM) was embedded into a mouse abdomen for fluorescence imaging and MRI. NIR fluorescence emission was detected from the inside of the mouse abdomen and the location of the phantom in the mouse was determined from the T_1-weighted MR image.

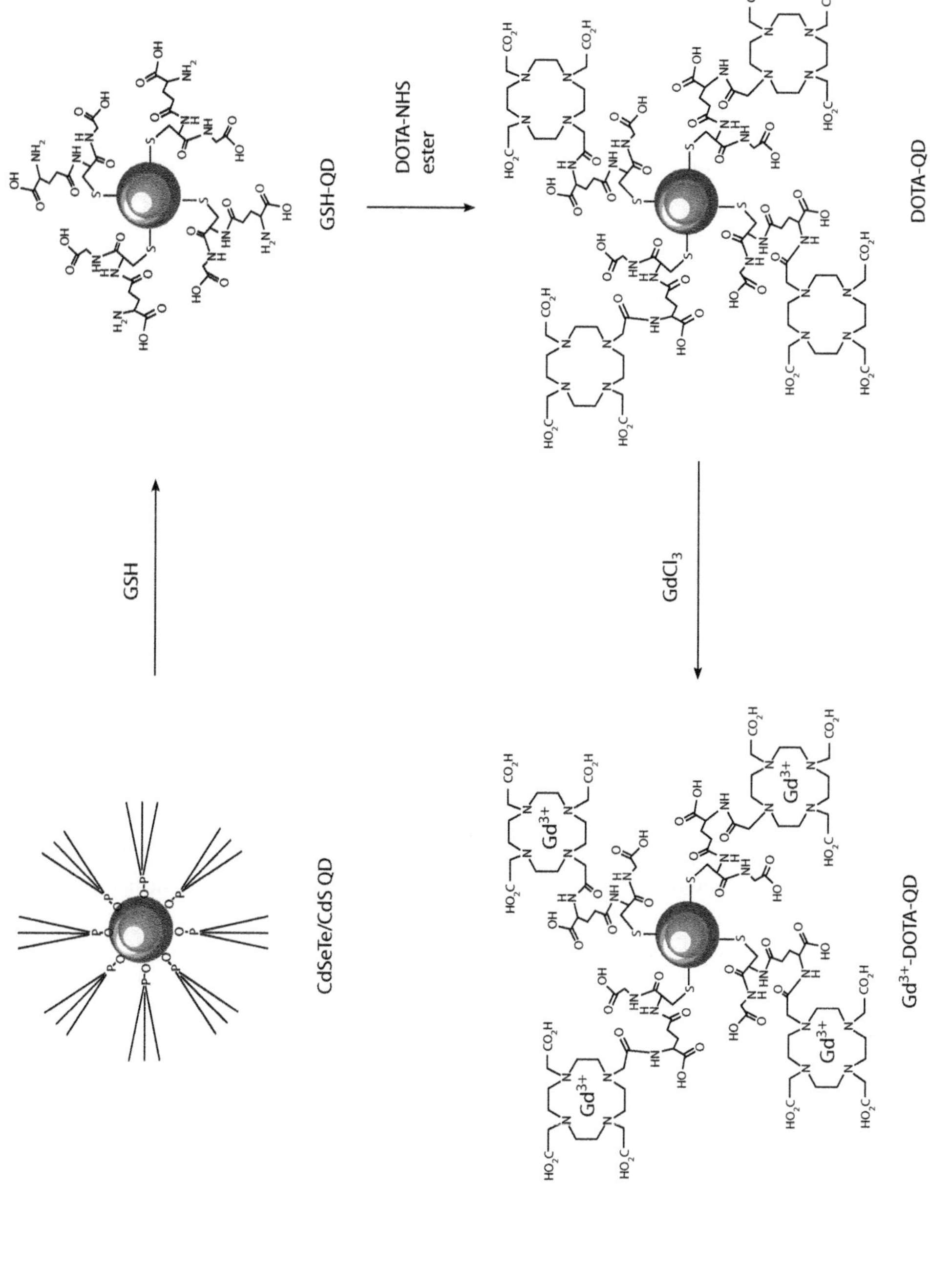

FIGURE 20.4 Schematic representation of Gd-DOTA functionalized CdSeTe/CdS QDs with GSH coating. (Reproduced from Jin, T. et al., *Chem. Commun.*, (44), 5764, 2008. With permission.)

20.3.1.5 Silica Shell–Coated MR/Fluorescence Imaging Agents

A major concern that may restrict the use of QDs in vivo is the possible toxicity associated with cadmium (Derfus et al. 2004). The encapsulation of isolated QDs in silica spheres has been widely studied as the silica coating is biocompatible, biological inert, and is not at risk of desorption, which has been found for surfactant coatings. The silica isolates the QDs from the environment, preventing the possible leakage of cadmium, and activation of immunocompatible cells (Bakalova et al. 2007). Recently, Mulder et al. incorporated the CdSe/CdS/$Cd_{0.5}Zn_{0.5}S$/ZnS core–shell–shell (CSS) QDs (CSS-QDs) in a silica sphere. The silica surface was made hydrophobic, and then it was surrounded by pegylated, paramagnetic (Gd-DTPA), and biofunctional lipids, which resulted in a monodisperse agent possessing both fluorescent and paramagnetic properties. The nanoparticles are made target-specific by the conjugation of multiple RGD-peptides, which are recognized by $\alpha_v\beta_3$-integrins overexpressed on both the surface of angiogenic endothelial cells and tumor cells. The initial QY of the CSS-QDs was 60%, but dropped by half after the silica coating. The decrease of QY of silica-coated QDs was also seen in other cases. Silica is a porous material, therefore, water and/or oxygen can penetrate inside the sphere and cause negative effects on the fluorescent QY (Bakalova et al. 2007). The nanoparticles had a size of 31 nm and a r_1 value of 14.4 mM^{-1} s^{-1} (60 MHz). The estimated maximum number of lipids in a densely packed lipid layer surrounding a 31 nm silica particle is approximately 5,000, using a literature value of 60 A^2/lipid (Petrache et al. 2000). Since half of the lipids are paramagnetic, the r_1 relaxivity per particle was estimated to be 36,000 mM^{-1} s^{-1}. The specific uptake of the nanoparticles by endothelial HUVEC cells in vitro was demonstrated using fluorescence microscopy, quantitative fluorescence imaging, and MRI (Koole et al. 2008). The pharmacokinetics of the lipid-coated and bare silica nanoparticles were studied both in vitro and in vivo in the other paper from the same group. The results showed that the PEG-lipid coating increased the blood circulation half-life time of silica particles by a factor of 10 (165 versus 15 min for lipid-coated or bare silica particles, respectively). Compared to the bare silica particles, the longer circulation time of lipid-coated silica particles resulted in a more favorable tissue distribution profile at organ level, tissue level, and cellular or nanoparticle level by the results obtained from MRI, fluorescence imaging, and TEM (van Schooneveld et al. 2008).

In comparison with conventional nanoparticles with exposed fluorescent dyes, dyes embedded in silica have a higher fluorescence intensity, improved photostability due to the protective layer around encapsulated optical agents, and a higher payload of fluorescent dyes (Tallury et al. 2008). A few magnetic-fluorescent core–shell, or core–satellite hybrid nanoparticles has been reported. After conjugation with targeting agents, the particles exhibited both T_2-weighted MRI contrast enhancement and optical properties in cancer cells (Yoon et al. 2006), human mesenchymal stem cells (Lu et al. 2007), neuroblastoma cells (Lee et al. 2006), or the in vivo imaging (Kim et al. 2008). For example, Hyeon and coworkers synthesized a nanocomposite with a Fe_3O_4 magnetic core encapsulated with fluorescent silica shell for dual-modality imaging. The synthetic route was illustrated in Figure 20.5a. The 15 nm Fe_3O_4 magnetic core was capped with cetyltrimethylammonium bromide (CTAB) which served as the stabilizing surfactant for the transfer of hydrophobic Fe_3O_4 nanocrystals to the aqueous phase and also as the organic template for the formation of mesopores in the following sol-gel reaction. Then, tetraethylorthosilicate (TEOS) and fluorescent dyes FITC or rhodamine B isothiocyanate (RITC, emission @ 577 nm) were added to the reaction solution to yield the expected magnetic fluorescent core–shell nanoparticles with a size of 53 nm. The r_1 and r_2 relaxivity values of the resultant particles were 3.40 and 245 mM^{-1} s^{-1}, respectively (r_2/r_1 = 72.1), indicating that they are primarily suited to be T_2-weighted MRI contrast agents. The particles had well-developed 2.6 nm mesopores, and the surface area and the total pore volume were 481 and 1.07 cm^3 g^{-1}, respectively. Finally, the surface of the particle was modified with PEG to render them biocompatible. DLS measurements revealed that the hydrodynamic size of the probe in PBS solution is approximately 97 nm. The probe exhibited darker fluorescence than the Fe_3O_4 removed particles, indicating that Fe_3O_4 nanocrystals can partially quench the fluorescence of the dye. The nanoparticles were intravenously injected into

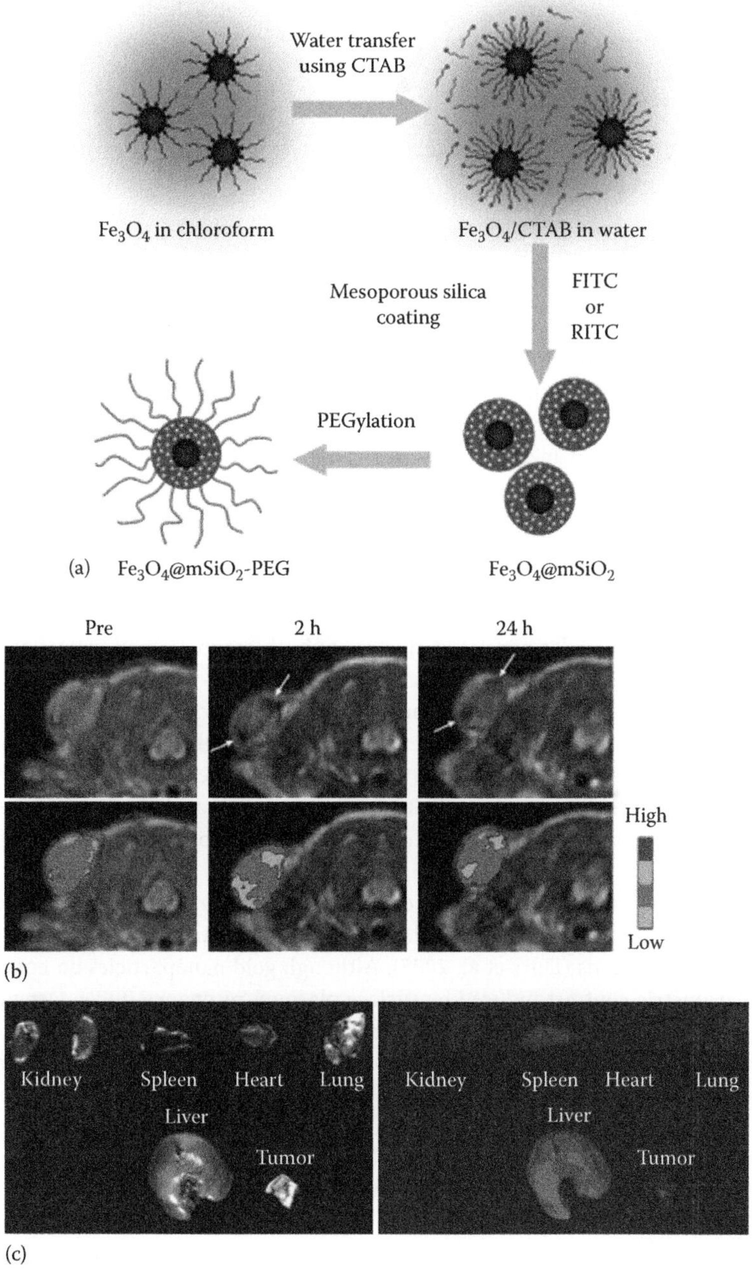

FIGURE 20.5 **(See color insert.)** Multimodality MR/fluorescence imaging of tumor using Fe_3O_4/mSiO_2 core/shell nanoparticles in a mouse model. (a) Schematic synthesis of magnetite nanocrystal/mesoporous silica core–shell nanoparticles; (b) In vivo T_2-weighted MR images (upper row) and color maps (lower row) of tumor before and after intravenously injection of probe. A decrease of signal intensity on T_2-weighted MR images was detected at the tumor site (arrows); (c) photographic image and corresponding fluorescence image of several organs and the xenograft tumor 24 h after intravenous injection. (Reproduced from Kim, J. et al., *Angew. Chem. Int. Ed.*, 47(44), 8438, 2008. With permission.)

the nude mice bearing tumors with diameter of 0.5 cm on their shoulder. T_2-weighted MR images were obtained before, 2 and 24 h after injection. The accumulation of the nanoparticles in tumors was detectable at 2 h after the injection and was still observable at 24 h after injection (Figure 20.5b). The ex vivo fluorescence imaging confirmed the accumulation of the nanoparticles in tumors (Figure 20.5c) (Kim et al. 2008).

20.3.2 Dual Magnetic and Fluorescent Core MR/Fluorescence Imaging Agents

Instead of organic dyes, QDs have also been used as the fluorescent moiety in multifunctional iron oxide nanoparticles. For example, in work by Ying and coworkers CdSe, QDs were grown onto preformed Fe_2O_3 cores at 300°C in the presence of organic surfactants, yielding either heterodimers or a homogeneous dispersion of QDs around the cores, which exhibited superparamagnetism and tunable optical emission properties. The Fe_2O_3–CdSe hybrid nanoparticles were rendered water-soluble and nontoxic by using a silica-coating process, and used for the labeling of live cell membranes (Selvan et al. 2007). Another strategy is to simultaneously encapsulate NIR QDs and iron oxide nanoparticles into either mesoporous silica spheres (Kim et al. 2006) or micelles composed of a PEG-modified phospholipid (Park et al. 2008). In work by Sailor and coworkers, spherical oleic acid coated magnetic nanoparticles (11 nm) and NIR QDs (10–12 nm) were encapsulated simultaneously within micelles composed of a PEG-modified phospholipid. The micellar hybrid nanoparticles were removed from the micellar magnetic nanoparticles, micellar QDs, and empty micelle by-products by magnetic separation and centrifugation. TEM and DLS measurements revealed that the micellar hybrid nanoparticles consist of clusters of both magnetic nanoparticles and QDs and had a hydrodynamic size of 60–70 nm. The magnetic nanoparticles—QDs ratio within the individual micelles—could be adjusted by changing the mass ratio of magnetic nanoparticles to QDs during the synthesis. After conjugation with F3, a peptide known to target cell surface nucleolin in endothelial cells in tumor blood vessels and in tumor cells, the nanoparticles were incubated with MDA-MB-435 human cancer cells. NIR fluorescence and MR imaging demonstrated significant contrast between cells incubated with F3-conjugated nanoparticles and cells incubated with unmodified nanoparticles (Park et al. 2008).

Recently, the photon luminescence effect of gold nanoparticles has attracted much attention because of the particle's excellent anti-photobleaching properties and the chemically inert behavior under physiological conditions (Durr et al. 2007). Although gold nanoparticles do not fluoresce, they can provide colorimetric contrast induced by surface plasmon resonance (SPR). Depending on their size, shape, degree of aggregation, and local environment, gold nanoparticles can appear red, blue, or other colors (Murphy et al. 2008). Sun and coworkers reported the synthesis of dumbbell-shaped Fe_3O_4–Au nanoparticles by decomposing iron pentacarbonyl on the surfaces of Au nanoparticles in the presence of oleic acid and oleylamine. After functionalized by a surfactant exchange reaction, the epidermal growth factor receptor antibody (EGFRA) was linked to the Fe_3O_4 surface through PEG (M_r = 3,000) and dopamine. The 8–20 nm Au–Fe_3O_4 dumbbell nanoparticles exhibit superparamagnetic at room temperature and a plasmonic absorption in PBS at 530 nm, and were demonstrated to be effective in MR/fluorescence imaging of A431 (human epithelial carcinoma cell line) cells (Xu et al. 2008). Recently, Irudayaraj and coworkers synthesized "nano-pearl-necklace" structured nanoparticles comprising a single, amine-modified Au rod decorated with multiple "pearls" of Fe_3O_4 nanoparticles capped with carboxyl groups. The Au_{rod}–Fe_3O_4 nano-pearl-necklaces (Au_{rod}-$(Fe_3O_4)_n$, where $n > 5$) were further stabilized with thiol-modified PEG (M_r = 5,000) and functionalized with Herceptin for imaging and photothermal killing of human breast cancer cells (SK-BR-3 cells). Au_{rod}–$(Fe_3O_4)_n$ had a r_2 value of 248.1 $mM^{-1}\ s^{-1}$ (3 T) and a longitudinal plasmon (LP) band at 785 nm, and were shown to be magnetically and optically active in dual-modality imaging and photothermal ablation of SK-BR-3 breast cancer cells (Wang et al. 2009).

20.3.3 Magnetically Doped QDs

Besides conjugating paramagnetic Gd(III) chelates to the surface of QDs, the other method for adding magnetism to semiconductors is the direct magnetic impurity doping. The magnetic impurities can act as paramagnetic centers in the semiconductor lattice, thus enabling the semiconductor to possess both optical and magnetic properties (Erwin et al. 2005). Manganese (Mn)-doped group II–VI semiconductor nanocrystals have been intensively investigated over the last decade (Yang et al. 2005, 2008). Several semiconductor host materials such as ZnS, CdS, and ZnSe have been used for Mn-doped QDs (QD_{Mn}) with various synthetic routes and surface passivation. Most of studies focus on the technological applications such as spin injectors and magnetic memory elements, or alteration of the optical properties of QDs by Mn dopants, including decay lifetime, emission wavelengths, etc. Although these QDs possess both fluorescent and magnetic properties in organic or aqueous solution, the application of them in multimodality MR/optical imaging is rare (Erwin et al. 2005; Santra, Yang, Stanley et al. 2005; Santra, Yang, Holloway et al. 2005; Yang et al. 2005, 2006, 2008; Wang et al. 2007; Yong 2009).

20.3.3.1 Magnetically Doped Group II–VI Core/Shell–Structured QDs

20.3.3.1.1 Manganese Dopant in the Core of QDs

Santra et al. prepared 3.1 nm size of Mn-doped CdS QDs with ZnS shell (CdS:Mn/ZnS) using a water-in-oil (W/O) microemulsion system. After coated with amine functionalized silica by the hydrolysis and co-condensation reaction of tetraethyl orthosilicate, 3-(aminopropyl) triethoxysilane and 3-(trihydroxysilyl)propyl methylphosphonate in the presence of ammonium hydroxide base, the QDs became highly water-dispersible. The QD_{Mn} had a bright yellow emission when excited with a 366 nm multiband handheld UV light. The QD_{Mn} were also radioopaque under excitation of x-ray, and therefore were expected to generate contrast under a CT scan. The magnetic measurement of the QD_{Mn} was performed by using a superconducting quantum interference device (SQUID) magnetometer. The QD_{Mn} showed a typical hysteresis curve for paramagnetic material. To demonstrate in vivo bioimaging capability, the QD_{Mn} were applied for brain imaging. One of the major barriers for delivery of imaging agents to the brain is the endothelial cell membrane. An effective way for macromolecules to overcome the cellular membrane barrier is conjugation with membrane translocation peptides such as TAT peptide. After attachment of TAT peptides, the QD_{Mn} were intraarterially delivered to a rat brain. A gross fluorescence could be visualized from the whole rat brain using a low power handheld UV lamp. Histological data showed that TAT-conjugated QD_{Mn} migrated beyond the endothelial cell line and reached the brain parenchyma (Santra, Yang, Holloway et al. 2005; Santra, Yang, Stanley et al. 2005).

Yong reported the synthesis of Mn-doped CdTeSe/CdS core/shell QDs emitting in the NIR region with a QY of 15%. A mixture of cadmium oxide and manganese acetylacetonate in TOPO (tri-*n*-octylphosphine oxide) and myristic acid was heated at ~290°C for 35 min under argon flow, then the TOP:Se:Te solution (TOP: tri-*n*-octylphosphine) was injected under vigorous stirring into the hot reaction mixture. After stirring for ~15–20 min at 230°C, an oleylamine-sulfur solution was slowly added to the hot reaction solution and the mixture was left in the reaction pot at 200°C for 2 h. The particles were separated from the surfactant solution by addition of ethanol and centrifugation. The particle sizes were estimated to be in the range of 4–5 nm (TEM). The decay lifetime of QD_{Mn} was estimated to be 70 ns, which is three times larger than that of CdSe/ZnS QDs, suggesting that QD_{Mn} have better potential as real-time and long-term imaging probes in comparison with the undoped QDs. The percentage of Mn in QD_{Mn} was at 3% determined by EDS (energy-dispersive x-ray spectroscopy) measurement. The magnetic measurement of the QD_{Mn} using SQUID indicated the presence of an active paramagnetic component—a typical hysteresis curve for paramagnetic material. A ligand exchange experiment with pyridine was performed to determine the status of Mn atoms in the QD_{Mn}. After pyridine wash, only a 10%–20% decrease of Mn composition was observed, indicating Mn atoms were mostly distributed within the QD matrix because pyridine wash would remove the Mn atoms that were bound to the surface of the QD. Because each lysine molecule contains two amino groups and one carboxyl group, the surface of QD_{Mn}

were functionalized with lysine to make them stably dispersed in aqueous media and followed by conjugation with monoclonal and polyclonal antibodies rendering the QD_{Mn} targeted to pancreatic cancer cells such as Panc-1 and MiaPaCa. After lysine coating, the hydrodynamic diameter of the particle was ~50 nm (DLS). An uptake of the antibody-conjugated, lysine-coated QD_{Mn} was observed in confocal imaging of Panc-1 and MiaPaCa pancreatic cancer cells. The uptake specificity of the QD_{Mn} was verified by control experiments using lysine-coated QD_{Mn} incubated with the Panc-1 cells, which exhibited only minimal uptake. The cytotoxicity of lysine-coated QD_{Mn} was evaluated by MTS assay on Panc-1 and MiaPaCa pancreatic cancer cell lines. After incubation with QD_{Mn} for 24 h, the viability of both cells is over 80% even at particle concentrations as high as 800 μg mL^{-1} (Yong 2009).

In work by Santra and Yong described above, the relaxivity of QD_{Mn} was not reported, probably due to the low relaxivity of QD_{Mn} as the paramagnetic manganese ions were buried in the particles. The mechanism of T_1 relaxation is a through-space dipole-dipole interaction between the unpaired electrons of the paramagnetic metal ion and bulk water molecules that are not "bound" to the metal ion. In images derived from changes in T_1, regions that are associated with a paramagnetic metal ion (nearby water molecules) have higher signal intensity. The mechanism was verified by Yang et al. (2006) who further functionalized the CdS:Mn/ZnS QDs developed by Santra et al. with Gd(III) chelates. The QD_{Mn} were mixed with *n*-(trimethoxysilylpropyl)ethyldiamine, triacetic acid trisodium salt (TSPETE)—a molecule with five reactive coordination sites for the capture of Gd^{3+} ions, and gadolinium(III) acetate. The generated QDs had a r_1 value of 20.5 mM^{-1} s^{-1} and a r_2 value of 151 mM^{-1} s^{-1} at 4.7 T. The r_2/r_1 ratio of ~7.4 indicates that the Gd-QDs may be more effective as a T_2 contrast agent.

20.3.3.1.2 Manganese Dopant in the Shell of QDs

A T_1 contrast agent requires that the paramagnetic metal ion directly interacts with water, suggesting that the paramagnetic ions should be localized at or near the particle surface to achieve the satisfactory contrast enhancement. We developed a method to introduce paramagnetism into QDs by capping luminescent CdSe nanoparticle cores with a paramagnetically doped ZnS surface (Wang et al. 2007). CdSe nanoparticles were synthesized by stirring Se and cadmium acetate in TOPO and hexadecylamine (HAD) under flowing argon at high temperature. To achieve approximately 1.5 or 6 monolayers of $Zn_{1-x}Mn_xS$ shell to grow on the CdSe core, stoichiometric amounts of diethylzinc and dimethylmanganese in TOP were injected in five portions with intervals of 10 min into the CdSe nanoparticle solution accompanied by simultaneous injection of H_2S gas at intervals of 5 min. The QD_{Mn} was made water-soluble by capping the QD_{Mn} with an amphiphilic polymer of octylamine-modified poly(acrylic acid), as shown in Figure 20.6a.

TEM images revealed uniform and monodispersed particles with an average size of 4.1 and 4.7 nm for the CdSe and QD_{Mn} nanoparticles, respectively. Mn^{2+} ion content was determined by atomic absorption (AA) and was found in the range of 0.6%–6.2% and varied with the thickness of the shell or amount of Mn^{2+} ions introduced to the reaction, as shown in Table 20.2. Little to no change in manganese quantity was found after the pyridine washes, indicating that in most cases manganese were distributed within the QD matrix. The r_1 values for the QD_{Mn} (10–18 mM^{-1} s^{-1}, 7 T, rt) are much greater than those observed for aqueous Mn^{2+} ion (7–8 mM^{-1} s^{-1}), and are comparable with the r_1 values of manganese nanoparticles reported in the literature (4.1–37.5 mM^{-1} s^{-1}) (Unger et al. 1993; Schwert et al. 2002; Koretsky and Silva 2004; Taylor et al. 2008b; Mertzman et al. 2009; Pan et al. 2009). The increase in r_1 values was attributed to a combination of slower rotation and high Mn^{2+} concentration localized in the surface (shell) of nanoparticles. Relaxivities for the thick shell nanoparticles had greater r_1 relaxivity because of their higher total Mn^{2+} content (Table 20.2). Emission spectra of the QD_{Mn} were centered at different wavelengths ranging from 570 to 650 nm, depending on the size of the CdSe cores. The QY was in the range of 30%–60% in organic solvent and up to 21% in water. The intensity of the emission decreased with increasing Mn^{2+} doping, probably due to the presence of interfacial Mn^{2+} between the core of CdSe and the ZnS shell at high Mn^{2+} content. The probes produce distinct contrast in both T_1-weighted MRI and optical imaging in aqueous solution and cells in culture (Wang et al. 2007). Uptake of the QD_{Mn} was clearly visible as punctate spots of luminescence in the cytoplasm of the cells, indicating an endocytic or

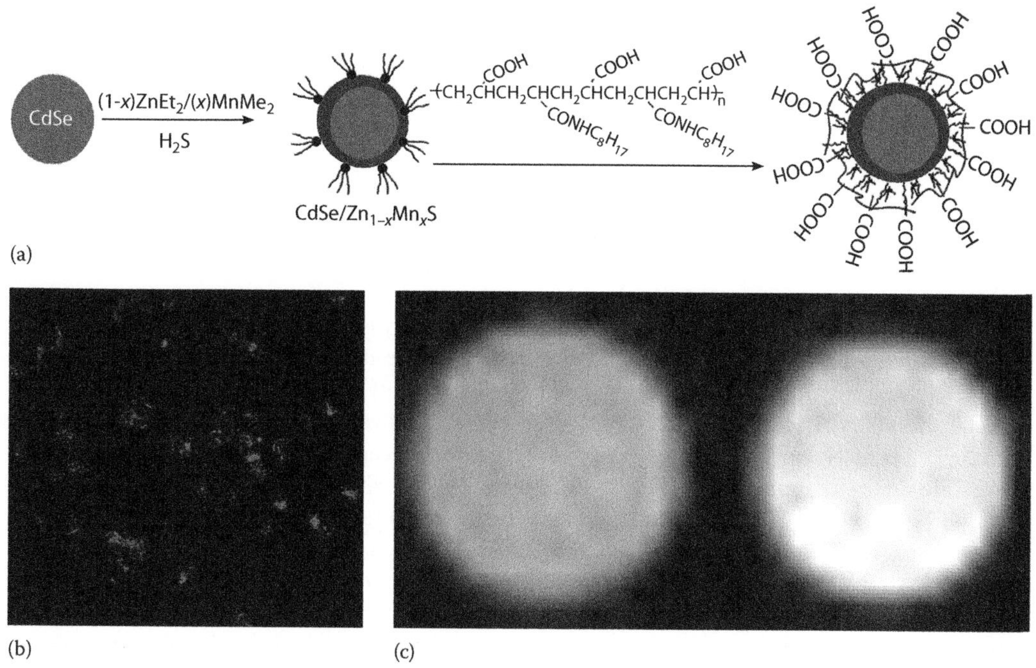

FIGURE 20.6 (See color insert.) (a) Synthetic representation of water-soluble core/shell CdSe/$Zn_{1-x}Mn_xS$; (b) cells incubated with CdSe/$Zn_{1-x}Mn_xS$ exhibit strong fluorescence in confocal microscopy; (c) T_1-weighted MR images from tubes containing cell lysates show that lysates of cells incubated with CdSe/$Zn_{1-x}Mn_xS$ (right) show significant contrast enhancement as compared to cells that have not been exposed to CdSe/$Zn_{1-x}Mn_xS$. (Reproduced from Wang, S. et al., *J. Am. Chem. Soc.*, 129(13), 3848, 2007. With permission.)

TABLE 20.2 Mn^{2+} Content and r_1 Relaxivity Values for CdSe/$Zn_{1-x}Mn_xS$

	Mn^{2+} Content			Relaxivity
Initial Amount, *x*%	After Chloroform Isolation, %	After Pyridine Exchange, %	Mn^{2+} Ions per Particle	r_1 (mM^{-1} s^{-1})
1	0.7	0.6	2	13.1
5	3.7	3.3	15	10.7
10	5.6	5.1	21	12.1
20	7.4	6.2	29	15.0
5[a]	0.87	0.87	28	13.8
20[a]	1.8	1.7	52	18.0

Source: Reproduced from Wang, S. et al., *J. Am. Chem. Soc.*, 129(13), 3848, 2007. With permission.

[a] Samples capped with ~6 monolayers $Zn_{1-x}Mn_xS$ shell; all others were capped with ~1.5 monolayer shell.

phagocytic mode of internalization. T_1-weighted MRI of cell lysates demonstrated that cells incubated with QD_{Mn} showed significant contrast enhancement from unlabeled cells (Figure 20.6c). These images demonstrate that QD_{Mn} can produce significant contrast for both optical and MR imaging in cells.

20.3.3.2 Magnetically Doped Group IV Silicon QDs

The use of Group II–VI QDs as laboratory research tools has grown tremendously but enthusiasm for their clinical application has been limited due to concerns about possible toxicity (Derfus et al. 2004).

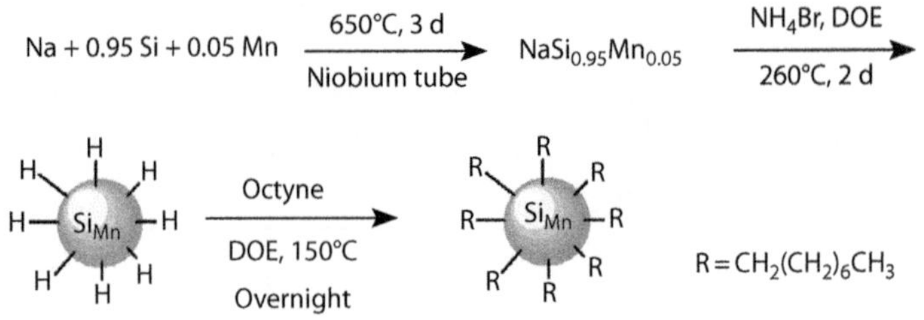

FIGURE 20.7 Synthesis of octyl coated Mn-doped silicon QDs (Si_{Mn} QDs).

Because of their chemical stability and biocompatibility, recently group IV silicon QDs have triggered substantial interest in exploring their potential in clinical application as fluorescence imaging probes. However, to date most silicon QDs reported are single functionality and solely for fluorescence imaging, probably due to the difficulty in synthesis and surface functionalization, stability, and low QY (Warner et al. 2005; Zhang, Neiner et al. 2007; Erogbogbo et al. 2008; Sudeep et al. 2008).

We have doped manganese into silicon QDs via a low-temperature solution route (Zhang, Brynda et al. 2007). Mn-doped Zintl salts ($NaSi_{1-x}Mn_x$, x = 0.05, 0.1, 0.15) were prepared as precursors, followed by the reaction with ammonium bromide to make hydrogen capped nanocomposites (Si_{Mn} QDs). The hydride-capped Si_{Mn} QDs were further modified via a hydrosilylation process with octyne to form chemically robust Si–C bonds on the surface and to protect the silicon particles from oxidation, as shown in Figure 20.7. TEM showed an average diameter of 4.2 ± 0.9 nm for Si_{Mn} QDs. The x-ray powder diffraction pattern of the obtained Si_{Mn} QDs indicated that silicon mainly existed in amorphous form. No diffraction due to Mn and MnO_x phases are observed, suggesting that Mn is incorporated into the Si nanoparticles. The percentage of Mn in the Si_{Mn} QDs was determined by elemental analysis with a molar ratio of Si/Mn of 18.2:1, consistent with 5% Mn-doped Si nanoparticles. The presence of Mn^{2+} in the Si_{Mn} QDs was verified by the result of electron paramagnetic resonance (EPR) measurements. After pyridine washing, the corresponding EPR spectra are almost identical to the one obtained with the nonwashed samples, suggesting that Mn^{2+} is covalently bound to the surface or near the surface of the nanoparticle. The Si_{Mn} QDs emitted with maximum intensity from 490 to 520 nm with excitation from 380 to 440 nm in chloroform. The maximum intensity emission spectrum is centered at 510 nm with an excitation wavelength of 420 nm. QYs of up to 16% in chloroform were obtained for Si_{Mn} QDs, lower than that reported for undoped Si QDs prepared in the same manner (18%) (Zhang, Neiner et al. 2007). The QY decreased significantly with increased amounts of manganese.

The work indicates that the Si_{Mn} QDs possess both favorable optical and magnetic properties. As stated above, the main challenge for manganese doped QDs is insufficient MRI signal. The presence of Mn^{2+} near or at the surface of Si_{Mn} QDs would be optimal for MRI applications, therefore show the potential of Si_{Mn} QDs for use in combined MRI and optical imaging.

20.3.4 Small Molecule–Based Nanomaterials for MR/Fluorescence Imaging

Besides multimodal designs based on fluorescent or magnetic core, or magnetically doped QDs, magnetic–fluorescent nanomaterials are also produced by coupling small paramagnetic Gd(III) chelates and organic fluorophores to a polymer or by encapsulation of both entities within a micelle, or a silica matrix.

20.3.4.1 Gadolinium Chelates and Organic Dyes Conjugated to a Polymer or Protein

Bovine serum albumin (BSA) has been used as a platform for construction of nanoparticles. We attached Gd-DOTA and fluorescent FITC to maleylated BSA, a ligand of macrophage scavenger receptor, to form a dual-modality imaging agent for detection of lesion formation. Relaxivity (r_1) of maleylated BSA-Gd-DOTA increases with increasing Gd-DOTA per molecule and then plateaus at >15 Gd/molecule, attaining a maximum of 33 mM^{-1} s^{-1} (1.4 T, 37°C). In vitro MRI studies showed that the contrast agent accumulates in macrophages. Incubation with greater than 50 μM of the agent was sufficient to produce observable contrast in the MR image. The specificity of uptake was confirmed by a Gd-DOTA matched BSA control and competition studies in which cells were incubated with a fixed concentration of the agent and increasing excess of unlabeled mal-BSA. Cell toxicity and initial biodistribution studies indicate low toxicity, no detectable retention in normal blood vessels, and rapid clearance from blood (Gustafsson et al. 2006). In other similar examples, Saban et al. (2007) examined the lymphatic vessel function (LVF) during bladder cancer progression in a double transgenic mouse model by in vivo imaging using a contrast agent (biotin-BSA-Gd-DTPA-Cy5.5). The mouse bladder cancer could be detected as early as 4 months and the tumor sizes were allowed to follow during cancer progression.

Li and coworkers conjugated Gd-DTPA and NIR813 dye with poly(L-glutamic acid) to obtain a dual MR/optical imaging agent, PG-DTPA-Gd-NIR813, for both preoperative and intraoperative visualization and characterization of sentinel lymph nodes (SLN) in mice. There were approximately 3 NIR, 813 dyes, and 51 Gd-DTPA per polymer. The resultant probe had a hydrodynamic diameter of 46 ± 6 nm, a r_1 value of 8.6 and a r_2 value of 25.1 mM^{-1} s^{-1} at 4.7 T and rt. After subcutaneous injection, axiliary and brachial lymph nodes were visualized with both T_1-weighted MR and optical imaging within 3 min of contrast agent injection, even at the lowest dose tested (0.002 mmol Gd kg^{-1}). After intralingual injection in tumor-bearing mice, MR imaging could identify four of the six superficial cervical lymph nodes, whereas optical imaging identified all six cervical nodes (Melancon et al. 2007).

Chaubet et al. (2007) developed a functionalized imaging agent from dextran that carried sulfate and carboxylate groups in order to mimic PSGL-1, the main ligand of P-selectin, a glycoprotein mainly expressed on the surface of activated platelets. Dextran was carboxymethylated with monochloroacetic acid in alkaline medium, then some of the carboxyl groups were amidified by coupling with ethylene diamine for conjugation of Gd-DOTA chelates and fluorescein isothiocyanate. After sulfation with pyridine-SO_3 complex, the probe has an average molecular weight of 27,000 ± 500 g mol^{-1}, a hydrodynamic radius of 5.7 ± 0.2 nm and a r_1 relaxivity of 11.2 mM^{-1} s^{-1} at 60 MHz. In vitro flow cytometry experiments and MRI (1.5 T) demonstrated a preferential binding of the probe on activated human platelets. Recently, the probe was used in vivo imaging of the abdominal aortic wall of apo E-/- mice by Canet-Soulas group. Pre- and postcontrast MRI were performed on a 2 T magnet in apo E-/- and control C57BL/6 mice after probe injection at a dose of 60 μmol Gd kg^{-1}. The sulfated probe significantly enhanced the MRI signal in the abdominal aortic wall of apo E-/- mice (>50% signal-to-noise ratio increase between 10 and 30 min), but not of control mice. The nonsulfated probe produced only moderate (<20%) MRI signal enhancement within the same time frame. Immunofluorescence in apo E-/- mice colocalized with the sulfated probe but not the nonsulfated probe with the inflammatory area revealed by P-selectin labeling (Alsaid et al. 2009).

Dendrimers are a class of highly branched synthetic polymers that form spherical macromolecules of a specific physical size, and can be synthesized in a highly reproducible manner. A dendrimer with a polyamidoamine (PAMAM) interior and amine surface is a good platform molecule for constructing nanoscale imaging agents (Barrett et al. 2009). Talanov et al. (2006) coupled Gd-DTPA and Cy5.5 dyes with a PAMAM dendrimer containing 256 terminal amino groups to form a dual-modality MR and NIR fluorescence imaging agent. To achieve a balanced performance of both MR/optical imaging, the number of chelate units on the dendrimer surface were planned to greatly exceed the number of fluorophore units. This consideration determined the order to modify the dendrimer stepwise with the chelating agent, dye, and Gd(III) incorporation. The particle PAMAM-$(Cy5.5)_{1.25}(1B4M\text{-}Gd)_{145}$ had a r_1 value of 13.9 and a r_2 value of 36.5 mM^{-1} s^{-1} at 3 T. The probe was successfully used in vivo to visualize sentinel lymph nodes (SLNs) in murine models by both modalities. Kobayashi and coworkers used a similar probe, PAMAM-$(Cy5.5)_2(Gd\text{-}DTPA)_{191}$, to both identify and resect SLNs during NIR optical image-guided surgery. On the unmixed NIR fluorescence images, intense fluorescence from the axillary node could be seen through the skin, but the optical imaging was unavailable to identify draining nodes in the supraclavicular or lateral thoracic nodes due to interference from the injection site. A minimum dose of 25 μL of 30 mM probe (750 nmol on a Gd basis) was required for MRI to consistently visualize and accurately localize the SLNs (Koyama et al. 2007).

20.3.4.2 Gadolinium Chelates and Organic Dyes Encapsulated in a Micelle

Fayad and coworkers prepared biotinylated micelles composed of phospholipids, a surfactant, a Gd complex, and an amphiphilic fluorophore for detecting macrophages in apo E-/- mice (Lipinski et al. 2006; Amirbekian et al. 2007). The phospholipids (palmitoyl-oleoyl phosphatidylcholine [POPC]) and an aliphatic Gd-DOTA-C_{16} were dissolved in a 1:1 chloroform:methanol solution. Removal of the solvent yielded a thin film that was then rehydrated in hot water. After 15 min of sonication at 65°C, a surfactant (Tween 80) was added to the solution, followed by another 15 min of sonication. 2 mol% of fluorescent phospholipid 1,2-dipalmitoyl-sn-glycero-3-phosphoethanolamine-N-7-nitro-2-1,3-benzoxadiazol-4-yl (DPPE-NBD) was added to the formulation to endow the micelles with green emission. After attachment of biotinylated antibodies that are specific for the macrophage scavenger receptor, the micelle has a size of 85 ± 7 nm and a r_1 of 25.7 mM^{-1} s^{-1} (1.5 T). It is estimated that each micelle contains an average of 3,516 gadolinium atoms. In vivo MRI revealed that at 24 h postinjection, immunomicelles provided a 79% increase in signal intensity of atherosclerotic aortas compared with only 34% increase using untargeted micelles and no enhancement using Gd-DTPA. Ex vivo confocal imaging confirmed the colocalization between fluorescent immunomicelles and macrophages in plaques (Amirbekian et al. 2007). Recently, Mulder et al. reported a platform composed of a paramagnetic lipid (Gd-DTPA-DSA) and a fluorescently labeled lipopeptide (P2fA2) for construction of multifunctional nanoparticles. By carefully controlling the ratio of the two amphiphilic molecules, a variety of well-defined nanoscale supramolecular structures with different sizes and morphologies could be created. Cryo-TEM revealed the aggregate morphology to vary from small micellar structures to plate-like and even full grown ribbons of which the aspect ratios varied from a diameter of 5–8 nm to structures with a width of up to 25 nm and infinite length. The nanoparticles were applied to macrophages and were found to be avidly taken up by the cells (Vucic et al. 2009).

In the other work, Fayad and coworkers developed a reconstituted HDL, which is shaped similar to a bilayer nanodisk (Figure 20.8a) composed of phospholipids with the 37 amino acid, amphiphilic and α-helical peptide oriented along the sides, bound to the fatty acyl side chains. After incorporation of Gd(III) chelates and rhodamine-based lipids into the phospholipid bilayered disk, the particle had a hydrodynamic diameter of 7.6 nm and a r_1 value of 9.8 mM^{-1} s^{-1}, as shown in Figure 20.8a. In vivo, MRI (9.4 T) was investigated by comparison scans obtained pre- and postinjection of atherosclerotic apo E-/- mice at a dose of 50 μmol Gd per kg of animal. A 94% enhancement of image density was observed in the aortic wall 24 h postinjection (Figure 20.8b and c), while injections of the control wild-type mice, with no plaques, did not result in significant enhancements in the aortic wall. The targeting of nanoparticles to plaque macrophages was verified by ex vivo fluorescence imaging (Figures 20.8d and e) (Cormode, Briley-Saebo et al. 2008; Cormode, Chandrasekar et al. 2009). The same group also incorporated a carboxyfluorescein-labeled apolipoprotein E-derived lipopeptide, P2fA2, into rHDL. The resultant rHDL-P2A2 nanoparticles had a mean diameter of 11.6 ± 3.7 nm and a r_1 value of 10.5 mM^{-1} s^{-1} under 60 MHz and 40°C. The Gd/nanoparticle ratio was 19, therefore, relaxivity per nanoparticle was ~200 mM^{-1} s^{-1}. The in vivo study with the apo E-/- mouse model of atherosclerosis showed a pronounced and significant signal enhancement of the atherosclerotic wall 24 h after the injection of rHDL–P2A2 (50 μmol Gd kg^{-1}). Confocal microscopy revealed that rHDL–P2A2 nanoparticles colocalized primarily with intraplaque macrophages (Chen et al. 2008).

20.3.4.3 Gadolinium Chelates and Organic Dyes Buried in Silica

Lin and coworkers prepared hybrid silica nanoparticles containing a luminescent $[Ru(bpy)_3]Cl_2$ core (bpy = 2,2′-bypyridine) and a paramagnetic monolayer coating of a silylated Gd-DTTA (Gd-Si-DTTA, DTTA = diethylenetriaminetetraacetate) complex by a water-in-oil reverse microemulsion synthetic

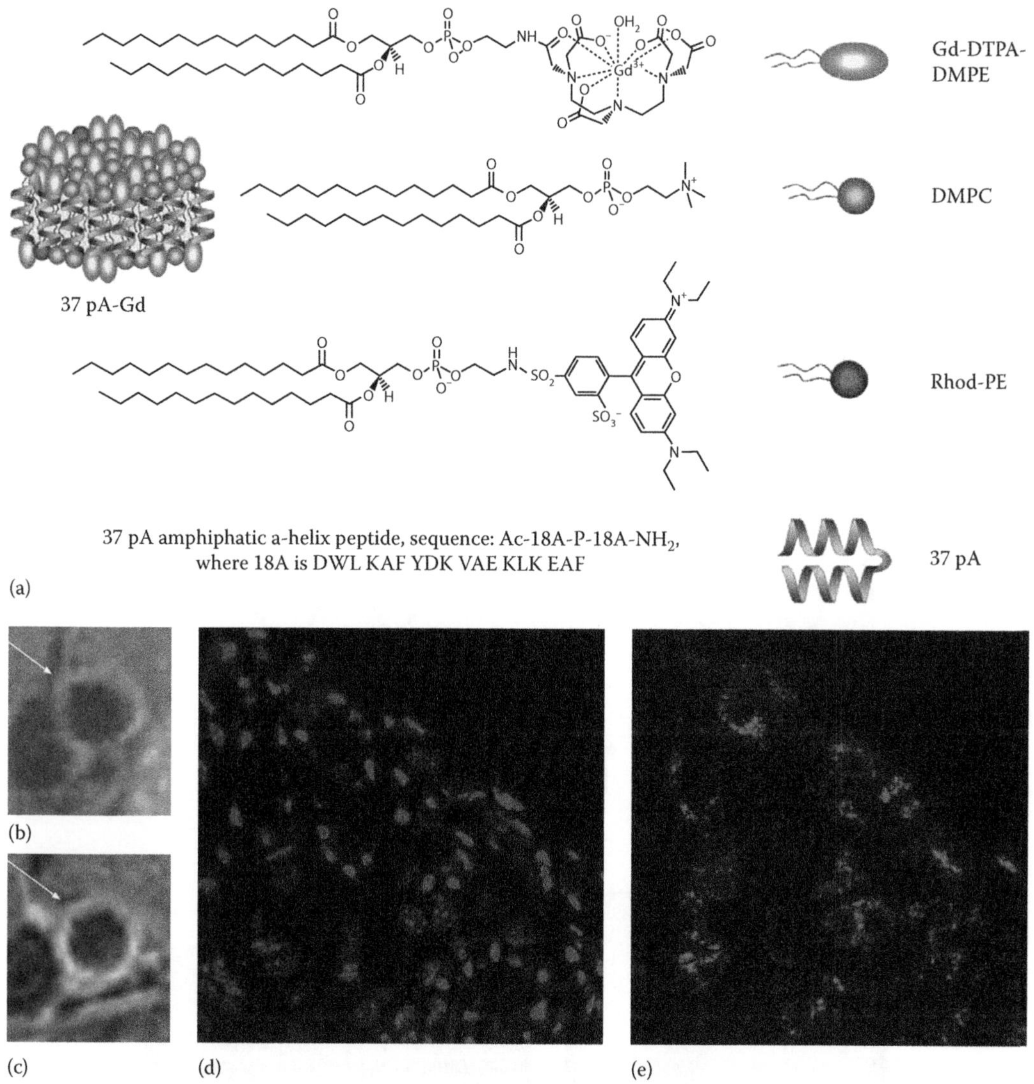

FIGURE 20.8 (See color insert.) (a) Schematic depiction of the reconstituted, (bi)layer nanodisk HDL with both magnetic and fluorescent properties; (b) and (c) MR images of the abdominal aorta, indicated by arrows, of an apo E-/- mouse, before (top) and 24 h postinjection (bottom) with the probe; (d) and (e) confocal microscopy images of aortic tissue from an apo E-/- mouse injected with the probe 24 h prior to excision. (d) The nuclei are blue via a DAPI stain and (e) the areas containing the probe are red due to the fluorescence from the rhodamine lipid. (Reproduced from Cormode, D.P. et al., *Small* 4(9), 1437, 2008. With permission.)

procedure. Thermogravimetric analysis and direct current plasma results correspond to a loading of about 10,200 Gd-DTTA per particle. The particle exhibits a monodisperse spherical morphology with a diameter of approximately 37 nm, and has a r_1 value of 19.7 and a r_2 value of 60.0 mM^{-1} s^{-1} (3.0 T) (Rieter et al. 2007). Later, the group modified the particle by a layer-by-layer polyelectrolyte deposition strategy to attain higher Gd payload. Hybrid silica nanoparticles containing a luminescent $[Ru(bpy)_3]Cl_2$ core and anionic monolayer coating of the Gd-Si-DTTA complex were alternately treated with cationic Gd-DOTA oligomer and anionic poly(styrene sulfonate), leading to the deposition of multilayers of Gd-DOTA oligomer and poly(styrene sulfonate) via electrostatic interactions.

The particle with six layers had a diameter of approximately 43 nm, which is just a little bit larger than the particle with one layer. However, for seven-layered particle the r_1 value was $5.34 \times 10^5\,\text{mM}^{-1}\,\text{s}^{-1}$ and the r_2 value was $1.55 \times 10^6\,\text{mM}^{-1}\,\text{s}^{-1}$ per particle, a significant increase in comparison with the particles of 6 and 1 layer which were 1.94×10^5 and $5.61 \times 10^5\,\text{mM}^{-1}\,\text{s}^{-1}$, respectively. The poly(styrene sulfonate)-terminated nanoparticles can be functionalized with targeting peptides that carry positive charges under physiological conditions via electrostatic interactions. The dual-modality property of the probe was demonstrated by in vitro MR and fluorescence imaging of HT-29 human colon cancer cells incubated with the particles functionalized with RGD-peptides via electrostatic interaction (Kim et al. 2007).

Mesoporous silica materials provide an ideal platform for the development of hybrid materials due to their high surface areas and tunable pore structures (Taylor et al. 2008a; Tsai et al. 2008). Lin and coworkers prepared mesoporous silica nanoparticles (MSNs) starting from stirring a mixture of CTAB and TEOS in aqueous NaOH solution for 2 h at 80°C. After centrifugation and washed with

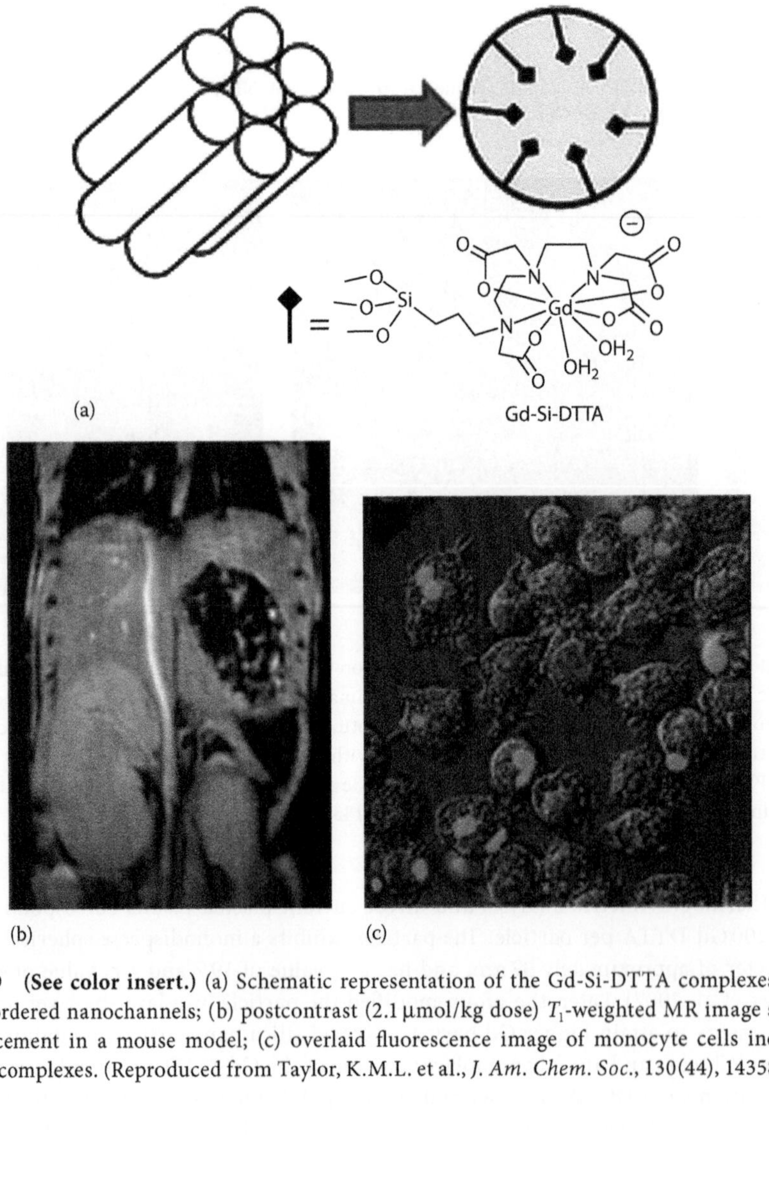

FIGURE 20.9 **(See color insert.)** (a) Schematic representation of the Gd-Si-DTTA complexes wrapped in hexagonally ordered nanochannels; (b) postcontrast (2.1 μmol/kg dose) T_1-weighted MR image showed aorta signal enhancement in a mouse model; (c) overlaid fluorescence image of monocyte cells incubated with Gd-Si-DTTA complexes. (Reproduced from Taylor, K.M.L. et al., *J. Am. Chem. Soc.*, 130(44), 14358, 2008. With permission.)

water and ethanol, the surfactant template was extracted with a 1 wt % solution of NaCl in methanol. The surfactant-extracted MSN is highly porous with a surface area of 1,633 $m^2\ g^{-1}$ and a pore diameter of 2.4 nm. By refluxing the particles and the Gd-Si-DTTA complex in toluene the particles were coated with a Gd-Si-DTTA complex via the siloxane linkage. The resultant MSN-Gd nanoparticle has a reduced surface area of 1,470 $m^2\ g^{-1}$ and a pore diameter of 0.9–1.0 nm. Due to the ready access of water molecules through the nanochannels, the particles had an r_1 value of 28.8 $mM^{-1}\ s^{-1}$ at 3 T and 10.2 $mM^{-1}\ s^{-1}$ at 9.4 T. The r_2 values are 65.5 $mM^{-1}\ s^{-1}$ at 3 T and 110.8 $mM^{-1}\ s^{-1}$ at 9.4 T, respectively. The particle had a very high Gd-DTTA loading (15.7–20.1 wt %), giving the r_1 and r_2 values of $7.0 \times 10^5\ mM^{-1}\ s^{-1}$ (3 T)/$2.48 \times 10^5\ mM^{-1}\ s^{-1}$ (9.4 T) and $1.6 \times 10^6\ mM^{-1}\ s^{-1}$ (3 T)/$2.7 \times 10^6\ mM^{-1}\ s^{-1}$ (9.4 T) per particle, respectively. Rhodamine B-aminopropyl-triethoxysilane was co-coated to generate a dual-modality imaging probe. Both in vitro and in vivo efficacy of these particles was demonstrated in monocyte cells and mouse models, as shown in Figure 20.9 (Taylor et al. 2008a).

20.4 Conclusion and Future Perspectives

In this chapter, we have illustrated the recent advances in the development of nanobiomaterials for dual-modality MR/florescent imaging, highlighting pertinent examples from the large and constantly expanding body of work in the field. Multifunctional nanobiomaterials detected by highly complementary anatomical and molecular based imaging capabilities can allow data acquisition which can afford accurate image co-registration and more meaningful interpretation of data. MRI findings can be validated up to the cellular and subcellular levels by correlating with fluorescence microscopy applied invasively or on excised specimens. MRI guidance can also be used to guide optical methods to visualize disease tissues during surgery or other interventional procedures. The rapid increase both in vitro and in vivo studies indicates that multifunctional nanobiomaterials detected simultaneously by different imaging modalities are of enormous interest and promise to become one of the main streams in the next stage of molecular imaging development, opening a new era in diagnostic imaging.

Currently, the development of fluorescent-magnetic nanobiomaterials is still in the early stages. The ultimate goal of such nanobiomaterials is the efficient diagnosis of diseases in clinic. However, the translation of multifunctional nanobiomaterials from the research lab and preclinical stages into the clinic is much slower than initially expected. For clinical use there must be much more progress to optimize the probes in terms of their biocompatibility, toxicity, stability, targeting specificity, and pharmacokinetics. Toxicity is one of the greatest concerns for many imaging agents, such as QDs, Gd(III) complexes, etc., and these must be thoroughly addressed before the nanobiomaterials can be used in human beings. For example, Bruchez and coworkers found that postinjection CdSe/ZnS core/shell QDs remained in bone marrow for months and some of QDs still lingered in the lymph nodes of mice 2 years later. The nanoparticles' emission spectra changed considerably, indicating that they might have partially dissolved (Fitzpatrick et al. 2009). Therefore, imaging agents, including QDs, Gd(III) complexes should be safely coated and engineered to exit the body rapidly before degradation, but yet not so fast that they cannot accumulate at the target tissue. Another alternative is to make imaging agents with biocompatible elements, such as silicon. In addition to the imaging techniques that are the emphasis of this chapter, these nanobiomaterials may be designed for combined detection with other modalities, such as MRI/PET, CT/PET, CT/fluorescence imaging, etc. (Choi et al. 2008; Jarrett et al. 2008; Lee et al. 2008; Devaraj et al. 2009; Dullin et al. 2009; Lucignani 2009). These nanobiomaterals can also be extended to therapeutic applications by adding drug molecules into the nanocomposite, which may expedite their entrance into the clinical trials because one of the major hurdles for imaging agents is lower market forces and profit margins than therapeutic drugs (Weissleder and Pittet 2008; Kim et al. 2009).

Clearly, there are many applications and approaches to developing nanobiomaterials and the future is only just beginning in this vibrant and exciting field.

Acknowledgments

The authors wish to recognize support from HL081108-01 and EB000993 from the National Institutes of Health.

References

Alsaid, H., G. De Souza, M. C. Bourdillon, F. Chaubet, A. Sulaiman, C. Desbleds-Mansard, L. Chaabane et al., 2009. Biomimetic MRI contrast agent for imaging of inflammation in atherosclerotic plaque of ApoE(-/-) mice a pilot study. *Investigative Radiology* 44 (3):151–158.

Amirbekian, V., M. J. Lipinski, K. C. Briley-Saebo, S. Amirbekian, J. G. S. Aguinaldo, D. B. Weinreb, E. Vucic et al. 2007. Detecting and assessing macrophages in vivo to evaluate atherosclerosis noninvasively using molecular MRI. *Proceedings of the National Academy of Sciences of the United States of America* 104 (3):961–966.

Arap, W., R. Pasqualini, and E. Ruoslahti. 1998. Cancer treatment by targeted drug delivery to tumor vasculature in a mouse model. *Science* 279 (5349):377–380.

Bakalova, R., Z. Zhelev, I. Aoki, and I. Kanno. 2007. Designing quantum-dot probes. *Nature Photonics* 1 (9):487–489.

Barrett, T., G. Ravizzini, P. L. Choyke, and H. Kobayashi. 2009. Dendrimers in medical nanotechnology application of dendrimer molecules in bioimaging and cancer treatment. *IEEE Engineering in Medicine and Biology Magazine* 28 (1):12–22.

Bin Na, H., I. S. Lee, H. Seo, Y. Il Park, J. H. Lee, S. W. Kim, and T. Hyeon. 2007. Versatile PEG-derivatized phosphine oxide ligands for water-dispersible metal oxide nanocrystals. *Chemical Communications* (48):5167–5169.

Cai, W. B., A. R. Hsu, Z. B. Li, and X. Y. Chen. 2007. Are quantum dots ready for in vivo imaging in human subjects? *Nanoscale Research Letters* 2 (6):265–281.

Caravan, P. 2006. Strategies for increasing the sensitivity of gadolinium based MRI contrast agents. *Chemical Society Reviews* 35 (6):512–523.

Chaubet, F., I. Bertholon, J. M. Serfaty, R. Bazeli, H. Alsaid, M. Jandrot-Perrus, C. Zahir et al. 2007. A new macromolecular paramagnetic MR contrast agent binds to activated human platelets. *Contrast Media & Molecular Imaging* 2 (4):178–188.

Chen, W., E. Vucic, E. Leupold, W. J. M. Mulder, D. P. Cormode, K. C. Briley-Saebo, A. Barazza, E. A. Fisher, M. Dathe, and Z. A. Fayad. 2008. Incorporation of an apoE-derived lipopeptide in high-density lipoprotein MRI contrast agents for enhanced imaging of macrophages in atherosclerosis. *Contrast Media & Molecular Imaging* 3 (6):233–242.

Cheon, J. and J. H. Lee. 2008. Synergistically integrated nanoparticles as multimodal probes for nanobiotechnology. *Accounts of Chemical Research* 41 (12):1630–1640.

Choi, J. S., J. C. Park, H. Nah, S. Woo, J. Oh, K. M. Kim, G. J. Cheon, Y. Chang, J. Yoo, and J. Cheon. 2008. A hybrid nanoparticle probe for dual-modality positron emission tomography and magnetic resonance imaging. *Angewandte Chemie—International Edition* 47 (33):6259–6262.

Cormode, D. P., K. C. Briley-Saebo, W. J. M. Mulder, J. G. S. Aguinaldo, A. Barazza, Y. Q. Ma, E. A. Fisher, and Z. A. Fayad. 2008. An ApoA-I mimetic peptide high-density-lipoprotein-based MRI contrast agent for atherosclerotic plaque composition detection. *Small* 4 (9):1437–1444.

Cormode, D. P., R. Chandrasekar, A. Delshad, K. C. Briley-Saebo, C. Calcagno, A. Barazza, W. J. M. Mulder, E. A. Fisher, and Z. A. Fayad. 2009. Comparison of synthetic high density lipoprotein (HDL) contrast agents for MR imaging of atherosclerosis. *Bioconjugate Chemistry* 20 (5):937–943.

Cormode, D. P., T. Skajaa, Z. A. Fayad, and W. J. M. Mulder. 2009. Nanotechnology in medical imaging probe design and applications. *Arteriosclerosis Thrombosis and Vascular Biology* 29 (7):992–1000.

Cormode, D. P., T. Skajaa, M. M. van Schooneveld, R. Koole, P. Jarzyna, M. E. Lobatto, C. Calcagno et al. 2008. Nanocrystal core high-density lipoproteins: A multimodality contrast agent platform. *Nano Letters* 8 (11):3715–3723.

Corr, S. A., Y. P. Rakovich, and Y. K. Gun'ko. 2008. Multifunctional magnetic-fluorescent nanocomposites for biomedical applications. *Nanoscale Research Letters* 3 (3):87–104.

Derfus, A. M., W. C. W. Chan, and S. N. Bhatia. 2004. Probing the cytotoxicity of semiconductor quantum dots. *Nano Letters* 4 (1):11–18.

Devaraj, N. K., E. J. Keliher, G. M. Thurber, M. Nahrendorf, and R. Weissleder. 2009. F-18 labeled nanoparticles for in vivo PET-CT imaging. *Bioconjugate Chemistry* 20 (2):397–401.

Dullin, C., M. Zientkowska, J. Napp, J. Missbach-Guentner, H. W. Krell, F. Muller, E. Grabbe, L. F. Tietze, and F. Alves. 2009. Semiautomatic landmark-based two-dimensional-three-dimensional image fusion in living mice: Correlation of near-infrared fluorescence imaging of Cy5.5-labeled antibodies with flat-panel volume computed tomography. *Molecular Imaging* 8 (1):2–14.

Dumont, E. A., C. P. M. Reutelingsperger, J. F. M. Smits, Mjap Daemen, P. A. F. Doevendans, H. J. J. Wellens, and L. Hofstra. 2001. Real-time imaging of apoptotic cell-membrane changes at the single-cell level in the beating murine heart. *Nature Medicine* 7 (12):1352–1355.

Durr, N. J., T. Larson, D. K. Smith, B. A. Korgel, K. Sokolov, and A. Ben-Yakar. 2007. Two-photon luminescence imaging of cancer cells using molecularly targeted gold nanorods. *Nano Letters* 7 (4):941–945.

Erogbogbo, F., K. T. Yong, I. Roy, G. X. Xu, P. N. Prasad, and M. T. Swihart. 2008. Biocompatible luminescent silicon quantum dots for imaging of cancer cells. *ACS Nano* 2 (5):873–878.

Erwin, S. C., L. J. Zu, M. I. Haftel, A. L. Efros, T. A. Kennedy, and D. J. Norris. 2005. Doping semiconductor nanocrystals. *Nature* 436 (7047):91–94.

Evgenov, N. V., Z. Medarova, G. P. Dai, S. Bonner-Weir, and A. Moore. 2006. In vivo imaging of islet transplantation. *Nature Medicine* 12 (1):144–148.

Fitzpatrick, J. A. J., S. K. Andreko, L. A. Ernst, A. S. Waggoner, B. Ballou, and M. P. Bruchez. 2009. Long-term persistence and spectral blue shifting of quantum dots in vivo. *Nano Letters* 9 (7):2736–2741.

Gao, X. H., Y. Y. Cui, R. M. Levenson, L. W. K. Chung, and S. M. Nie. 2004. In vivo cancer targeting and imaging with semiconductor quantum dots. *Nature Biotechnology* 22 (8):969–976.

Ge, Y. Q., Y. Zhang, S. Y. He, F. Nie, G. J. Teng, and N. Gu. 2009. Fluorescence modified chitosan-coated magnetic nanoparticles for high-efficient cellular imaging. *Nanoscale Research Letters* 4 (4):287–295.

Gunn, J., H. Wallen, O. Veiseh, C. Sun, C. Fang, J. H. Cao, C. Yee, and M. Q. Zhang. 2008. A multimodal targeting nanoparticle for selectively labeling T cells. *Small* 4 (6):712–715.

Gustafsson, B., S. Youens, and A. Y. Louie. 2006. Development of contrast agents targeted to macrophage scavenger receptors for MRI of vascular inflammation. *Bioconjugate Chemistry* 17 (2):538–547.

Hezinger, A. F. E., J. Tessmar, and A. Gopferich. 2008. Polymer coating of quantum dots—A powerful tool toward diagnostics and sensorics. *European Journal of Pharmaceutics and Biopharmaceutics* 68 (1):138–152.

Hofstra, L., I. H. Liem, E. A. Dumont, H. H. Boersma, W. L. van Heerde, P. A. Doevendans, E. DeMuinck et al. 2000. Visualisation of cell death in vivo in patients with acute myocardial infarction. *Lancet* 356 (9225):209–212.

Jaffer, F. A., M. Nahrendorf, D. Sosnovik, K. A. Kelly, E. Aikawa, and R. Weissleder. 2006. Cellular imaging of inflammation in atherosclerosis using magnetofluorescent nanomaterials. *Molecular Imaging* 5 (2):85–92.

Jarrett, B. R., B. Gustafsson, D. L. Kukis, and A. Y. Louie. 2008. Synthesis of Cu-64-labeled magnetic nanoparticles for multimodal imaging. *Bioconjugate Chemistry* 19 (7):1496–1504.

Jin, T., Y. Yoshioka, F. Fujii, Y. Komai, J. Seki, and A. Seiyama. 2008. Gd3+-functionalized near-infrared quantum dots for in vivo dual modal (fluorescence/magnetic resonance) imaging. *Chemical Communications* (44):5764–5766.

Josephson, L., M. F. Kircher, U. Mahmood, Y. Tang, and R. Weissleder. 2002. Near-infrared fluorescent nanoparticles as combined MR/optical imaging probes. *Bioconjugate Chemistry* 13 (3):554–560.

Kang, H. W., D. Torres, L. Wald, R. Weissleder, and A. A. Bogdanov. 2006. Targeted imaging of human endothelial-specific marker in a model of adoptive cell transfer. *Laboratory Investigation* 86 (6):599–609.

Kim, J., H. S. Kim, N. Lee, T. Kim, H. Kim, T. Yu, I. C. Song, W. K. Moon, and T. Hyeon. 2008. Multifunctional uniform nanoparticles composed of a magnetite nanocrystal core and a mesoporous silica shell for magnetic resonance and fluorescence imaging and for drug delivery. *Angewandte Chemie—International Edition* 47 (44):8438–8441.

Kim, J., J. E. Lee, J. Lee, J. H. Yu, B. C. Kim, K. An, Y. Hwang, C. H. Shin, J. G. Park, and T. Hyeon. 2006. Magnetic fluorescent delivery vehicle using uniform mesoporous silica spheres embedded with monodisperse magnetic and semiconductor nanocrystals. *Journal of the American Chemical Society* 128 (3):688–689.

Kim, J., Y. Piao, and T. Hyeon. 2009. Multifunctional nanostructured materials for multimodal imaging, and simultaneous imaging and therapy. *Chemical Society Reviews* 38 (2):372–390.

Kim, J. S., W. J. Rieter, K. M. L. Taylor, H. An, W. L. Lin, and W. B. Lin. 2007. Self-assembled hybrid nanoparticles for cancer-specific multimodal imaging. *Journal of the American Chemical Society* 129 (29):8962–8963.

Kohler, N., G. E. Fryxell, and M. Q. Zhang. 2004. A bifunctional poly(ethylene glycol) silane immobilized on metallic oxide-based nanoparticles for conjugation with cell targeting agents. *Journal of the American Chemical Society* 126 (23):7206–7211.

Koo, V., P. W. Hamilton, and K. Williamson. 2006. Non-invasive in vivo imaging in small animal research. *Cellular Oncology* 28 (4):127–139.

Koole, R., M. M. van Schooneveld, J. Hilhorst, K. Castermans, D. P. Cormode, G. J. Strijkers, C. D. Donega et al. 2008. Paramagnetic lipid-coated silica nanoparticles with a fluorescent quantum dot core: A new contrast agent platform for multimodality imaging. *Bioconjugate Chemistry* 19 (12):2471–2479.

Koretsky, A. P. and A. C. Silva. 2004. Manganese-enhanced magnetic resonance imaging (MEMRI). *NMR in Biomedicine* 17 (8):527–531.

Koyama, Y., V. S. Talanov, M. Bernardo, Y. Hama, C. A. S. Regino, M. W. Brechbiel, P. L. Choyke, and H. Kobayashi. 2007. A dendrimer-based nanosized contrast agent, dual-labeled for magnetic resonance and optical fluorescence imaging to localize the sentinel lymph node in mice. *Journal of Magnetic Resonance Imaging* 25 (4):866–871.

Laurent, S., D. Forge, M. Port, A. Roch, C. Robic, L. V. Elst, and R. N. Muller. 2008. Magnetic iron oxide nanoparticles: Synthesis, stabilization, vectorization, physicochemical characterizations, and biological applications. *Chemical Reviews* 108 (6):2064–2110.

Lee, S. and X. Y. Chen. 2009. Dual-modality probes for in vivo molecular imaging. *Molecular Imaging* 8 (2):87–100.

Lee, J. H., Y. W. Jun, S. I. Yeon, J. S. Shin, and J. Cheon. 2006. Dual-mode nanoparticle probes for high-performance magnetic resonance and fluorescence imaging of neuroblastoma. *Angewandte Chemie—International Edition* 45 (48):8160–8162.

Lee, H. Y., Z. Li, K. Chen, A. R. Hsu, C. J. Xu, J. Xie, S. H. Sun, and X. Y. Chen. 2008. PET/MRI dual-modality tumor imaging using arginine-glycine-aspartic (RGD)—Conjugated radiolabeled iron oxide nanoparticles. *Journal of Nuclear Medicine* 49 (8):1371–1379.

Lee, H., M. K. Yu, S. Park, S. Moon, J. J. Min, Y. Y. Jeong, H. W. Kang, and S. Jon. 2007. Thermally cross-linked superparamagnetic iron oxide nanoparticles: Synthesis and application as a dual imaging probe for cancer in vivo. *Journal of the American Chemical Society* 129 (42):12739–12745.

Lipinski, M. J., V. Amirbekian, J. C. Frias, J. G. S. Aguinaldo, V. Mani, K. C. Briley-Saebo, V. Fuster, J. T. Fallon, E. A. Fisher, and Z. A. Fayad. 2006. MRI to detect atherosclerosis with gadolinium-containing immunomicelles targeting the macrophage scavenger receptor. *Magnetic Resonance in Medicine* 56 (3):601–610.

Lu, C. W., Y. Hung, J. K. Hsiao, M. Yao, T. H. Chung, Y. S. Lin, S. H. Wu et al. 2007. Bifunctional magnetic silica nanoparticles for highly efficient human stem cell labeling. *Nano Letters* 7 (1):149–154.

Lucignani, G. 2009. Nanoparticles for concurrent multimodality imaging and therapy: The dawn of new theragnostic synergies. *European Journal of Nuclear Medicine and Molecular Imaging* 36 (5):869–874.

Medarova, Z., N. V. Evgenov, G. Dai, S. Bonner-Weir, and A. Moore. 2006. In vivo multimodal imaging of transplanted pancreatic islets. *Nature Protocols* 1 (1):429–435.

Medarova, Z., W. Pham, C. Farrar, V. Petkova, and A. Moore. 2007. In vivo imaging of siRNA delivery and silencing in tumors. *Nature Medicine* 13 (3):372–377.

Medarova, Z., W. Pham, Y. Kim, G. P. Dai, and A. Moore. 2006. In vivo imaging of tumor response to therapy using a dual-modality imaging strategy. *International Journal of Cancer* 118 (11):2796–2802.

Medarova, Z., L. Rashkovetsky, P. Pantazopoulos, and A. Moore. 2009. Multiparametric monitoring of tumor response to chemotherapy by noninvasive imaging. *Cancer Research* 69 (3):1182–1189.

Medintz, I. L., H. T. Uyeda, E. R. Goldman, and H. Mattoussi. 2005. Quantum dot bioconjugates for imaging, labelling and sensing. *Nature Materials* 4 (6):435–446.

Melancon, M. P., Y. T. Wang, X. X. Wen, J. A. Bankson, L. C. Stephens, S. Jasser, J. G. Gelovani, J. N. Myers, and C. Li. 2007. Development of a macromolecular dual-modality MR-optical imaging for sentinel lymph node mapping. *Investigative Radiology* 42 (8):569–578.

Mertzman, J. E., S. Kar, S. Lofland, T. Fleming, E. Van Keuren, Y. Y. Tong, and S. L. Stoll. 2009. Surface attached manganese-oxo clusters as potential contrast agents. *Chemical Communications* (7):788–790.

Michalet, X., F. F. Pinaud, L. A. Bentolila, J. M. Tsay, S. Doose, J. J. Li, G. Sundaresan, A. M. Wu, S. S. Gambhir, and S. Weiss. 2005. Quantum dots for live cells, in vivo imaging, and diagnostics. *Science* 307 (5709):538–544.

Montet, X., K. Montet-Abou, F. Reynolds, R. Weissleder, and L. Josephson. 2006. Nanoparticle imaging of integrins on tumor cells. *Neoplasia* 8 (3):214–222.

Mulder, W. J. M., A. W. Griffioen, G. J. Strijkers, D. P. Cormode, K. Nicolay, and Z. A. Fayad. 2007. Magnetic and fluorescent nanoparticles for multimodality imaging. *Nanomedicine* 2 (3):307–324.

Murphy, C. J., A. M. Gole, J. W. Stone, P. N. Sisco, A. M. Alkilany, E. C. Goldsmith, and S. C. Baxter. 2008. Gold nanoparticles in biology: Beyond toxicity to cellular imaging. *Accounts of Chemical Research* 41 (12):1721–1730.

Nahrendorf, M., F. A. Jaffer, K. A. Kelly, D. E. Sosnovik, E. Aikawa, P. Libby, and R. Weissleder. 2006. Noninvasive vascular cell adhesion molecule-1 imaging identifies inflammatory activation of cells in atherosclerosis. *Circulation* 114 (14):1504–1511.

Niedre, M. and V. Ntziachristos. 2008. Elucidating structure and function in vivo with hybrid fluorescence and magnetic resonance imaging. *Proceedings of the IEEE* 96 (3):382–396.

Ntziachristos, V. 2006. Fluorescence molecular imaging. *Annual Review of Biomedical Engineering* 8:1–33.

Ostendorp, M., K. Douma, T. M. Hackeng, A. Dirksen, M. J. Post, M. A. M. van Zandvoort, and W. H. Backes. 2008. Quantitative molecular magnetic resonance imaging of tumor angiogenesis using cNGR-labeled paramagnetic quantum dots. *Cancer Research* 68 (18):7676–7683.

Pan, D. P. J., A. Senpan, S. D. Caruthers, T. A. Williams, M. J. Scott, P. J. Gaffney, S. A. Wickline, and G. M. Lanza. 2009. Sensitive and efficient detection of thrombus with fibrin-specific manganese nanocolloids. *Chemical Communications* (22):3234–3236.

Park, J. H., G. von Maltzahn, E. Ruoslahti, S. N. Bhatia, and M. J. Sailor. 2008. Micellar hybrid nanoparticles for simultaneous magnetofluorescent imaging and drug delivery. *Angewandte Chemie—International Edition* 47 (38):7284–7288.

Petrache, H. I., S. W. Dodd, and M. F. Brown. 2000. Area per lipid and acyl length distributions in fluid phosphatidylcholines determined by H-2 NMR spectroscopy. *Biophysical Journal* 79 (6):3172–3192.

Prinzen, L., R. Miserus, A. Dirksen, T. M. Hackeng, N. Deckers, N. J. Bitsch, R. T. A. Megens et al. 2007. Optical and magnetic resonance imaging of cell death and platelet activation using annexin A5-functionalized quantum dots. *Nano Letters* 7 (1):93–100.

Qian, X. M., X. H. Peng, D. O. Ansari, Q. Yin-Goen, G. Z. Chen, D. M. Shin, L. Yang, A. N. Young, M. D. Wang, and S. M. Nie. 2008. In vivo tumor targeting and spectroscopic detection with surface-enhanced Raman nanoparticle tags. *Nature Biotechnology* 26 (1):83–90.

Quarta, A., R. Di Corato, L. Manna, A. Ragusa, and T. Pellegrino. 2007. Fluorescent-magnetic hybrid nanostructures: Preparation, properties, and applications in biology. *IEEE Transactions on Nanobioscience* 6 (4):298–308.

Resch-Genger, U., M. Grabolle, S. Cavaliere-Jaricot, R. Nitschke, and T. Nann. 2008. Quantum dots versus organic dyes as fluorescent labels. *Nature Methods* 5 (9):763–775.

Riehemann, K., S. W. Schneider, T. A. Luger, B. Godin, M. Ferrari, and H. Fuchs. 2009. Nanomedicine-challenge and perspectives. *Angewandte Chemie—International Edition* 48 (5):872–897.

Rieter, W. J., J. S. Kim, K. M. L. Taylor, H. Y. An, W. L. Lin, T. Tarrant, and W. B. Lin. 2007. Hybrid silica nanoparticles for multimodal Imaging. *Angewandte Chemie—International Edition* 46 (20):3680–3682.

Saban, M. R., R. Towner, N. Smith, A. Abbott, M. Neeman, C. A. Davis, C. Simpson, J. Maier, S. Memet, X. R. Wu, and R. Saban. 2007. Lymphatic vessel density and function in experimental bladder cancer. *BMC Cancer* 7:20.

Santra, S., H. S. Yang, P. H. Holloway, J. T. Stanley, and R. A. Mericle. 2005. Synthesis of water-dispersible fluorescent, radio-opaque, and paramagnetic CdS: Mn/ZnS quantum dots: A multifunctional probe for bioimaging. *Journal of the American Chemical Society* 127 (6):1656–1657.

Santra, S., H. Yang, J. T. Stanley, P. H. Holloway, B. M. Moudgil, G. Walter, and R. A. Mericle. 2005. Rapid and effective labeling of brain tissue using TAT-conjugated CdS: Mn/ZnS quantum dots. *Chemical Communications* (25):3144–3146.

Schwert, D. D., J. A. Davies, and N. Richardson. 2002. Non-gadolinium-based MRI contrast agents. In *Topics in Current Chemistry: Contrast Agents I*. Berlin: Springer-Verlag.

Selvan, S. T., P. K. Patra, C. Y. Ang, and J. Y. Ying. 2007. Synthesis of silica-coated semiconductor and magnetic quantum dots and their use in the imaging of live cells. *Angewandte Chemie—International Edition* 46 (14):2448–2452.

Sharma, P., S. Brown, G. Walter, S. Santra, and B. Moudgil. 2006. Nanoparticles for bioimaging. *Advances in Colloid and Interface Science* 123:471–485.

Sharma, P., S. C. Brown, G. Walter, S. Santra, E. Scott, H. Ichikawa, Y. Fukumori, and B. M. Moudgil. 2007. Gd nanoparticulates: From magnetic resonance imaging to neutron capture therapy. *Adanced Powder Technology* 18:663–698.

Sosnovik, D. E., M. Nahrendorf, N. Deliolanis, M. Novikov, E. Aikawa, L. Josephson, A. Rosenzweig, R. Weissleder, and V. Ntziachristos. 2007. Fluorescence tomography and magnetic resonance imaging of myocardial macrophage infiltration in infarcted myocardium in vivo. *Circulation* 115 (11):1384–1391.

Sperling, R. A., P. Rivera gil, F. Zhang, M. Zanella, and W. J. Parak. 2008. Biological applications of gold nanoparticles. *Chemical Society Reviews* 37 (9):1896–1908.

Sudeep, P. K., Z. Page, and T. Emrick. 2008. PEGylated silicon nanoparticles: Synthesis and characterization. *Chemical Communications* (46):6126–6127.

Suh, W. H., Y. H. Suh, and G. D. Stucky. 2009. Multifunctional nanosystems at the interface of physical and life sciences. *Nano Today* 4 (1):27–36.

Talanov, V. S., C. A. S. Regino, H. Kobayashi, M. Bernardo, P. L. Choyke, and M. W. Brechbiel. 2006. Dendrimer-based nanoprobe for dual modality magnetic resonance and fluorescence imaging. *Nano Letters* 6 (7):1459–1463.

Tallury, P., K. Payton, and S. Santra. 2008. Silica-based multimodal/multifunctional nanoparticles for bioimaging and biosensing applications. *Nanomedicine* 3 (4):579–592.

Taylor, K. M. L., J. S. Kim, W. J. Rieter, H. An, W. L. Lin, and W. B. Lin. 2008a. Mesoporous silica nanospheres as highly efficient MRI contrast agents. *Journal of the American Chemical Society* 130 (7):2154–2155.

Taylor, K. M. L., W. J. Rieter, and W. B. Lin. 2008b. Manganese-based nanoscale metal-organic frameworks for magnetic resonance imaging. *Journal of the American Chemical Society* 130 (44):14358–14359.

Toth, E., L. Helm, and A. E. Merbach. 2002. Relaxivity of MRI contrast agents. In *Topics in Current Chemistry: Contrast Agents I*. Berlin: Springer-Verlag.

Tsai, C. P., Y. Hung, Y. H. Chou, D. M. Huang, J. K. Hsiao, C. Chang, Y. C. Chen, and C. Y. Mou. 2008. High-contrast paramagnetic fluorescent mesoporous silica nanorods as a multifunctional cell-imaging probe. *Small* 4 (2):186–191.

Unger, E., T. Fritz, D. K. Shen, and G. L. Wu. 1993. Manganese-based liposomes—Comparative approaches. *Investigative Radiology* 28 (10):933–938.

van Schooneveld, M. M., E. Vucic, R. Koole, Y. Zhou, J. Stocks, D. P. Cormode, C. Y. Tang et al. 2008. Improved biocompatibility and pharmacokinetics of silica nanoparticles by means of a lipid coating: A multimodality investigation. *Nano Letters* 8 (8):2517–2525.

van Tilborg, G. A. F., W. J. M. Mulder, N. Deckers, G. Storm, C. P. M. Reutelingsperger, G. J. Strijkers, and K. Nicolay. 2006. Annexin A5-functionalized bimodal lipid-based contrast agents for the detection of apoptosis. *Bioconjugate Chemistry* 17 (3):741–749.

Vucic, E., Hmhf Sanders, F. Arena, E. Terreno, S. Aimo, K. Nicolay, E. Leupold et al. 2009. Well-defined, multifunctional nanostructures of a paramagnetic lipid and a lipopeptide for macrophage imaging. *Journal of the American Chemical Society* 131 (2):406–407.

Wang, C. G., J. Chen, T. Talavage, and J. Irudayaraj. 2009. Gold nanorod/Fe3O4 nanoparticle "nano-pearl-necklaces" for simultaneous targeting, dual-mode imaging, and photothermal ablation of cancer cells. *Angewandte Chemie—International Edition* 48 (15):2759–2763.

Wang, S., B. R. Jarrett, S. M. Kauzlarich, and A. Y. Louie. 2007. Core/shell quantum dots with high relaxivity and photoluminescence for multimodality imaging. *Journal of the American Chemical Society* 129 (13):3848–3856.

Warner, J. H., A. Hoshino, K. Yamamoto, and R. D. Tilley. 2005. Water-soluble photoluminescent silicon quantum dots. *Angewandte Chemie—International Edition* 44 (29):4550–4554.

Weecharangsan, W., P. Opanasopit, T. Ngawhirunpat, A. Apirakaramwong, T. Rojanarata, U. Ruktanonchai, and R. J. Lee. 2008. Evaluation of chitosan salts as non-viral gene vectors in CHO-K1 cells. *International Journal of Pharmaceutics* 348 (1–2):161–168.

Weissleder, R. 2001. A clearer vision for in vivo imaging. *Nature Biotechnology* 19 (4):316–317.

Weissleder, R., K. Kelly, E. Y. Sun, T. Shtatland, and L. Josephson. 2005. Cell-specific targeting of nanoparticles by multivalent attachment of small molecules. *Nature Biotechnology* 23 (11):1418–1423.

Weissleder, R. and M. J. Pittet. 2008. Imaging in the era of molecular oncology. *Nature* 452 (7187):580–589.

Xing, Y. and J. H. Rao. 2008. Quantum dot bioconjugates for in vitro diagnostics & in vivo imaging. *Cancer Biomarkers* 4 (6):307–319.

Xu, C., J. Xie, D. Ho, C. Wang, N. Kohler, E. G. Walsh, J. R. Morgan, Y. E. Chin, and S. Sun. 2008. Au-Fe_3O_4 dumbbell nanoparticles as dual-functional probes. *Angewandte Chemie—International Edition* 47 (1):173–176.

Yang, Y. A., O. Chen, A. Angerhofer, and Y. C. Cao. 2008. On doping CdS/ZnS core/shell nanocrystals with Mn. *Journal of the American Chemical Society* 130 (46):15649–15661.

Yang, L., H. Mao, Z. H. Cao, Y. A. Wang, X. H. Peng, X. X. Wang, H. K. Sajja et al. 2009. Molecular imaging of pancreatic cancer in an animal model using targeted multifunctional nanoparticles. *Gastroenterology* 136 (5):1514–1525.

Yang, H. S., S. Santra, and P. H. Holloway. 2005. Syntheses and applications of Mn-doped II-VI semiconductor nanocrystals. *Journal of Nanoscience and Nanotechnology* 5 (9):1364–1375.

Yang, H. S., S. Santra, G. A. Walter, and P. H. Holloway. 2006. Gd-III-functionalized fluorescent quantum dots as multimodal imaging probes. *Advanced Materials* 18 (21):2890–2894.

Yong, K. T. 2009. Mn-doped near-infrared quantum dots as multimodal targeted probes for pancreatic cancer imaging. *Nanotechnology* 20 (1):015102.

Yoon, T. J., K. N. Yu, E. Kim, J. S. Kim, B. G. Kim, S. H. Yun, B. H. Sohn, M. H. Cho, J. K. Lee, and S. B. Park. 2006. Specific targeting, cell sorting, and bioimaging with smart magnetic silica core-shell nanomateriats. *Small* 2 (2):209–215.

Zhang, X., M. Brynda, R. D. Britt, E. C. Carroll, D. S. Larsen, A. Y. Louie, and S. M. Kauzarich. 2007. Synthesis and characterization of manganese-doped silicon nanoparticles: Bifunctional paramagnetic-optical nanomaterial. *Journal of the American Chemical Society* 129 (35):10668–10669.

Zhang, X. M., D. Neiner, S. Z. Wang, A. Y. Louie, and S. M. Kauzlarich. 2007. A new solution route to hydrogen-terminated silicon nanoparticles: Synthesis, functionalization and water stability. *Nanotechnology* 18 (9):6.

21

Nanoscale Probes for the Imaging of RNA in Living Cells

Philip J. Santangelo
Georgia Tech and Emory University

Aaron Lifland
Georgia Tech and Emory University

Chiara Zurla
Georgia Tech and Emory University

Accurately imaging endogenous or nonengineered RNA in live cells is not an easy task. Ideally, a probe and imaging strategy will have the following properties: (1) functional probes will be delivered to the desired cellular compartment, (2) they will achieve the correct level of affinity to efficiently bind target RNA but not inhibit their function, (3) be sensitive enough (single RNA sensitive) to allow for the accurate detection of the cellular RNA population, and (4) for cellular studies, allow for the tracking of RNA through biogenesis, transport, translation, and degradation pathways. In this chapter, the capabilities of current nucleic-acid-based nanoprobes and strategies used to image native RNA in cellular systems are discussed and analyzed, and probe and strategy recommendations are given. The chapter is concluded by addressing topics for future research, all in the hope of achieving the ideal RNA imaging probe and strategy.

21.1 Introduction

Over the last decade, there has been increasing evidence to suggest that RNA molecules have a wide range of functions in living cells, from physically conveying and interpreting genetic information, to essential catalytic roles, to providing structural support for molecular machines, to gene silencing. These functions are realized through control of their expression level, turnover rates, and through their spatial distribution. In vitro methods that use purified RNA obtained from cell lysates can provide a measure of the RNA expression level within a cell population or within an organism; however, they cannot reveal the spatial and temporal variation of RNA. The ability to image specific RNAs in living cells in real-time promises to provide information on RNA synthesis, processing, transport, and localization; this information should offer new opportunities for advancement in molecular biology, disease pathogenesis, drug discovery, and medical diagnostics. Imaging RNA in live cells or organisms, though, as mentioned earlier, is not an easy task. Two approaches have been used that are as follows: (1) fluorescent protein–RNA binding protein (or peptide) fusions are expressed and bind to RNA sequences that have been engineered into the RNA target or the fusion probes themselves have been engineered to bind to native sequences (Fusco et al. 2003; Daigle and Ellenberg 2007; Ozawa et al. 2007) and (2) the use of fluorescently labeled antisense nucleic acids that bind to RNA targets, plasmid-derived or native, via Watson–Crick pairing (Leonetti et al. 1991; Politz et al. 1995). In this chapter, we will address the second approach combined in nanotech-based approaches for imaging RNA in detail and will discuss how the intracellular delivery technique and probe design impact the imaging outcome. Probe delivery strategies, such as microinjection and reversible cellular permeabilization, will be discussed in concert with the probe structure, affinity, kinetics, and sensitivity.

21.2 Motivation for Imaging RNA

Gene regulation is a fundamental process necessary for normal cell function, cell differentiation, and for responding to environmental stimuli. In addition, it plays a significant role in many important pathologies, such as cancer progression and metastasis, and viral infections. Regulation at the posttranscriptional level has become extremely important in cancer, linked with the early stages of tumorigenesis (Audic and Hartley 2004). On the subcellular level, it has been directly linked to mRNA dynamics and mRNA–protein and mRNA–miRNA interactions (Anderson and Kedersha 2009). Currently, we have very little information regarding the localization and dynamics of native RNAs on the single molecule or single granule level in living mammalian cells. By characterizing mRNA localization and dynamics during RNA regulatory processes, these data can be applied to many cell biology problems to indirectly indicate RNA function. And, if the mechanisms that regulate RNA dynamics and gene regulation can be understood, it should be possible to develop methods to intervene and control these processes.

21.3 RNA Regulation through RNA Spatial Localization and Dynamics

Posttranscriptional regulation of messenger RNAs can occur at many points during its life, during the first round of translation via nonsense mediated decay, during transport to translation sites due to trans-acting factor binding, such as ZBP1 binding to β-actin mRNA in fibroblasts (Condeelis and Singer 2005), during decay in P-bodies, associated with miRNA binding, and during cellular stress, which is associated with stress granule formation. It is also clear that these processes are time dependant and that RNA dynamics and trafficking are an important part of all of these processes, mediated by trans-acting factors such as RNA binding proteins and miRNA. As a consequence of miRNA regulation, targeted mRNAs have been shown to accumulate in P-bodies, and many RNA binding proteins, such as TIA-1/R, ZBP1, HuR, etc., have been shown to traffic in and out of stress granules during oxidative stress (Anderson and Kedersha 2002, 2006, 2008, 2009).

Numerous mRNAs have been shown to be nonrandomly distributed in polarized cells, such as oocytes, yeast, motile fibroblasts, and neurons (Martin and Ephrussi 2009). The sorting of mRNAs provides a powerful mechanism to locally translate the encoded proteins to sites of function, in the absence of de novo transcription providing a mechanism to quickly respond to changes in their extracellular environment. It has also been shown to play critical roles in cell differentiation, especially in cell motility. Currently, the most studied model of RNA localization is the chicken embryonic fibroblast (CEF) and is still the standard today, due to the high level of motile cells, ~70%, that exhibit localized β-actin mRNA. In this model system, ZBP1, an RNA binding protein isolated from CEFs based on its affinity for a cis-acting 54 nt region of the 3′-UTR of β-actin mRNA, called a zipcode region, was shown using both fixed cell fluorescence in situ hybridization (FISH) for RNA and in vivo site blocking assays, to be necessary for peripheral targeting of RNA (Kislauskis et al. 1994, 1995, 1997; Latham et al. 1994; Ross et al. 1997). ZBP1 was also found to act as a translational inhibitor preventing the 80S subunit from binding (Huttelmaier et al. 2005). In addition, we know that in CEFs, β-actin RNPs target to the cell periphery when induced by serum or PDGF implicating signal transduction pathways in this process. Inhibiting tyrosine kinase activity blocked this process, and later it was shown that Rho GTPases, specifically RhoA, were involved in β-actin localization. It was also found that eEF1α anchored localized mRNAs to the cytoskeleton in cellular protrusions. Additional, evidence for the role of ZBP1 has been provided by Nielsen et al. (2002), studying the human homologue, IMP1. In addition, in neurons, mRNA localization and local protein synthesis at synapses is important for synapse formation and synaptic plasticity, and in growth cones play a role in axon guidance during development and in regenerating adult axons following injury (Leung et al. 2006; Yao et al. 2006). β-Actin mRNA is targeted to both axons and dendrites of developing neurons, and the role of the 3′UTR zipcode and the trans-acting factor, zipcode-binding protein, in the mechanism and regulation of β-actin mRNA localization in neurons has been studied (Eom et al. 2003). This work was extended to FMRP in the study of the localization of other mRNAs, i.e., α-CaMKII, MAP1b.

21.4 "State of the Art" in RNA Imaging

Currently, expressing both the RNA and a fluorescent tag using plasmid-based systems or molecular biology approaches is the state of the art in this field. The enabling technology is the expression of a fusion of GFP (or any fluorescent protein) with a sequence-specific RNA binding protein or peptide. The initial system utilized the phage coat protein MS2 fused to GFP (Brodsky and Silver 2002). Because it binds to a 19 nt RNA stem-loop, specifically, this sequence was inserted into the target mRNA sequence and co-expressed with the MS2–GFP fusion protein. When both were coexpressed, MS2–GFP bound to the expressed mRNA containing the "tag" sequence in living cells and GFP fluorescence constituted the indicator of RNA position. In order to increase the signal from valid RNAs above the background of unbound fusion proteins, multiple MS2–GFP binding domains were inserted into the RNA. When 24 binding sites (48 MS2–GFP molecules) were inserted, single molecule sensitivity was achieved (Fusco et al. 2003; Shav-Tal et al. 2004). Since the initial use of this system, two additional strategies using plasmid-expressed probes have been demonstrated in mammalian cells—GFP-RNA binding peptide fusion probes, which bind to a 15 nt RNA hairpin encoded in the expressed target RNA (Daigle and Ellenberg 2007) and probes composed of Pumilio homology domains (PUM-HD) fused to sections of split EGFP, which target two closely spaced 8 nt native sequences (Ozawa et al. 2007). These systems have been used to study cytoplasmic mRNA, nuclear mRNA, and mitochondrial RNA. In the case of cytoplasmic RNA, this approach was used by two groups, Fusco et al. (2003) and Yamagishi et al. (2009), to study the dynamics of mRNA in live cells at the single molecule level. Fusco's seminal work examined the dynamics of RNAs generated from a lacz coding plasmid with and without the zipcode-binding protein. They required 24 MS2–GFP binding sites to achieve single RNA sensitivity. This work was performed in static Cos-7 cells, and statistics on the RNA motion were obtained and analyzed using single particle tracking and mean square displacement analysis techniques (Saxton and Jacobson 1997). This work was ground breaking, and therefore a standard for comparison, but it did not address the issue of native RNA dynamics and regulation in a live cell. In Yamagishi et al. (2009), they extended the work of Fusco to the model that Singer and colleagues have most studied—the motile CEF. In this work they used a plasmid expressing the β-actin coding region and the 5′- and 3′-UTRs. They transiently transfected CEFs, using Fugene 6, with both the RNA coding plasmid and an MS2–EGFP plasmid. No comments were made as to how many cells were doubly transfected nor what the expression levels were. Interestingly, they were able to achieve single RNA sensitivity with only four binding sites, and on average, approximately six bound MS2–GFPs. They were able to characterize using single particle tracking methods the dynamics of the RNAs in both the perinuclear and leading edge and examined the effects of cytochasin D treatment. They did not examine the effects of serum or PDGF or siRNA knockdown of ZBP1 on the dynamics.

Employing plasmid-derived probes and RNA give these methods tremendous flexibility, but there are significant limitations. First, they can only be used in cell types that allow for efficient transfection. Second, plasmid-derived mRNA often lack the correct number and position of introns and the exact 5′- and 3′-UTR sequences, which can strongly influence mRNA translational efficiency, decay, and stability (de Silanes et al. 2007; Giorgi and Moore 2007; Jambhekar and Derisi 2007). In the case of viral RNA, additions to viral genomes can affect replication efficiency, assembly and viral egress (Simmonds et al. 2004). In addition, plasmid-derived RNAs are often overexpressed, possibly changing the fundamental stoichiometry underlying RNA expression. Therefore, imaging native RNA, without the need for a plasmid-based expression system, would be advantageous for studying RNA biology. Of the techniques mentioned above, only the PUM-HD fusions have the ability to study native or nonengineered RNAs. They do require, though, the ability to transfect and express the PUM-HD fusions efficiently, and the user must optimize their amino acid sequence for a given RNA.

It should also be noted that there are other unfortunate caveats to the MS2–GFP tagging method, including the formation of large immobile aggregates, which are inconsistent with FISH images of endogenous RNA granules. Highly motile RNA granules tagged using MS2–GFP are infrequently observed, i.e., only 10% of the granules detected show long-distance processive movements. The reliance on the MS2–GFP tagging method may have limited our understanding of RNA dynamics. Through the further development of nanotechnology that relies on exogenous probes that target RNA via Watson–Crick pairing, our understanding of how native mRNA are transported and dynamically regulated could be greatly improved. We envision that unappreciated aspects of RNA dynamics and their regulation will be revealed.

21.5 Nucleic-Acid-Based Ligand Designs for Imaging RNA in Live Cells

21.5.1 Ligand Structure

In general, two ligand structures have been used for intracellular RNA imaging, linear and hairpin [in the form of molecular beacons (MB)]. Linear ligands are short, 12–25 nt, nucleic acid sequences with no defined secondary structure, typically labeled with an organic fluorophore on the 3′ or 5′ end. MBs (Tyagi and Kramer 1996; Tyagi et al. 1998), first published in 1996, are dual-labeled nucleic-acid-based probes with a reporter fluorophore at one end and a quencher at the other. They are designed to form a stem-loop hairpin structure so that fluorescence occurs only when the probe hybridizes to a complementary target, resulting in a high signal-to-background ratio (SBR). The hypothesis regarding their use in live cells was that, because they are switches, the signal observed would predominately result from hybridization to a target molecule, thus yielding higher SBRs (Sokol et al. 1998) than linear probes and not require probe titration. In addition, since they have been shown to be more sensitive to mismatches in PCR applications than linear probes, the assumption was that this enhanced specificity would transfer to live-cell hybridization reactions (Tyagi et al. 1998; Tsourkas et al. 2003). One might ask, if linear probes do not switch on and off with hybridization, how can they be used for imaging? The answer is twofold, linear probes can provide high SBRs by binding multiple linear probes to the same RNA, which is analogous to the MS2–GFP systems but using native sequences, or by relying on the natural accumulation or packaging of RNA into granule structures. Another strategy, employed with both linear probes and MBs to increase specificity, is to employ fluorescence energy resonance transfer (FRET) between probes that have hybridized to adjacent regions on the target molecule (Tsuji et al. 2000, 2001; Bratu et al. 2003; Santangelo et al. 2004; Abe and Kool 2006). In practice, as of yet, linear probes have shown specificity on par with MBs in live-cell applications (Molenaar et al. 2001), and FRET, though seemingly useful in theory, has only been utilized by a few researchers (Tsuji et al. 2000, 2001; Bratu et al. 2003; Santangelo et al. 2004; Abe and Kool 2006), due to low signal levels. FRET decreases background signals more than target signals, increasing SBR, but the net target signals are lower than by using direct excitation, which is a limiting factor. Given that both probe structures have been used in live cells, it has yet to be determined, in general terms, which probe structure is the advantageous approach for live-cell RNA imaging.

21.5.2 Ligand Affinity

The affinity of the ligand for its target, regardless of structure, is another part of the design puzzle. This can be controlled predominately by the types of nucleic acids utilized in the ligand. As of yet, for live-cell RNA imaging applications, DNA (Nitin et al. 2004; Santangelo et al. 2004, 2005, 2006; Rhee et al. 2008), phosphorothioate DNA (Politz et al. 1995), 2′-O-methyl RNA (Bratu et al. 2003; Tyagi and Alsmadi 2004; Mhlanga et al. 2005; Vargas et al. 2005), 2′-O-methyl RNA/DNA chimeras (Santangelo and Bao 2007) and LNA (locked nucleic acids) (Wu et al. 2008) have been employed. PNA has already been used in animal systems with radioactive labels (Shi et al. 2000; Suzuki et al. 2004). Nucleic acid affinity for RNA is ranked from highest to lowest: LNA, PNA, 2′-O-Methyl RNA, RNA, DNA, and phosphorothioate DNA (Kurreck et al. 2002; Aartsma-Rus et al. 2004; Vester and Wengel 2004). From solution study data, at first glance, the higher-affinity probes would be thought unsuitable for hybridization at 37°C due to their high melting temperatures (Tsourkas et al. 2002). High melting temperature probes, when hybridizing at 37°C, would allow for significant nonspecific binding without a chemical additive to lower the melting temperature, like formamide, used in fixed-cell hybridization assays. But, solution studies are often performed with high free ion concentrations, relative to the number of free ions in the intracellular environment (Romani 2007). This difference may render the reported melting temperature information less relevant to live-cell hybridization, therefore allowing high-affinity probes to bind targets specifically in live cells. The antisense literature shows many examples of specific binding with high-affinity nucleic acids (Aartsma-Rus et al. 2004; Vester and Wengel 2004; Fabani and Gait 2008). Very few direct comparisons, though, of nucleic acid chemistry have been made for imaging applications; Molenaar et al. (2001) performed the most detailed study comparing 2′-O-methyl RNA and DNA linear probes with 2′-O-methyl RNA and DNA MBs, while Santangelo and Bao (2007) commented that higher SBRs targeting hRSV viral genomic RNA were observed using 2′-O-methyl hybridization domains, as opposed to DNA. Molenaar et al. (2001) concluded that 2′-O-methyl RNA linear probes were the best choice, for the targets chosen, because they yielded the highest SBRs. Unfortunately, in that study, only nuclear RNA could be observed due to the use of microinjection delivery. In a recent study by Santangelo et al. (2009), linear probes with up to 15 2′-O-methyl RNA bases were utilized to target four different RNAs. As compared with "scrambled" probe (no target in the human genome) and noninfected cell (see later sections) controls, they showed excellent specificity and bound during the delivery process, which takes 10 min. These probes will be discussed in detail later in this chapter.

21.5.3 Probe Sensitivity

Probe sensitivity depends on probe binding efficiency and on which reporter molecules are utilized. Binding efficiency is a function of transport of the probe to the target molecule (diffusion or active) of interest and affinity for that molecule (discussed above). Fluorescent reporters utilized have typically been organic fluorophores, such as FITC, TAMRA, Cy3, Texas Red, and Cy5. Many of these molecules have low quantum efficiencies, low extinction coefficients, exhibit blinking or photobleach rapidly, limiting the sensitivity achieved in many studies. Tyagi and Alsmadi (2004) specifically pointed out that when using three MBs to target β-actin mRNA in fibroblasts, they only achieved SBRs of approximately 2.5. Santangelo and Bao (2007) achieved SBRs on the order of 11–30, binding 3–4 probes per RNA, but only when the viral RNA studied was in high quantity and the RNA was localized in small granules or filaments. Given the abundance of β-actin mRNA, Tyagi and Alsmadi (2004) pointed out that for most RNA species, a significant improvement in the technology would be necessary to study most mRNA species. In order to achieve single molecule sensitivity, Vargas et al. (2005), in a subsequent paper, added 96 binding sites to their RNA of interest and expressed it in a plasmid-based system. This approach was successful, but precluded the study of the endogenous RNA. More recently, Santangelo et al. (2009) developed a multiply labeled, tetravalent RNA imaging probe that is single probe and single RNA sensitive and can target RNA, using native sequences. To the author's knowledge, this is the first nanoprobe-based method of detecting native RNA in living cells with single RNA sensitivity.

21.5.4 Intracellular Delivery

One of the greatest challenges of using exogenous probes is efficient intracellular delivery. Efficient delivery requires that the probes are functional once inside the cell and in the same cellular compartment as their target. Many methods have been used to deliver exogenous nucleic-acid-based probes into cells such as passive uptake, microinjection, cationic transfection, reversible cell membrane permeabilization, and cell penetrating peptides. We will focus on results from passive uptake, microinjection, and reversible cell membrane permeabilization. Passive pinocytosis, taken from the antisense literature, was initially the method of choice. In those experiments fluorescently labeled phosphorothioate modified DNA probes were added to growth media at concentrations ranging from 0.1–2 μM for 2–4 h (Politz et al. 1995, 1998, 1999). Probes targeted to the poly (A) tail of mRNA and β-actin mRNA were used. From their results, they found that for their poly (A) probes, many were observed in a punctate pattern in the perinuclear region of the cell, while some exhibited diffuse staining in the cytoplasm. The nuclei were labeled more intensely in approximately 30% of their cells. The punctate pattern is likely probes entering via pino or endocytosis, and subsequently being trapped within vesicles, while the diffuse cytoplasmic signal results from probes escaping from the aforementioned vesicles. This method requires nuclease-resistant probes due to the high activity of nucleases within endosomes, which is why phosphorothioate modified DNA was utilized. Using this technique wide field fluorescence and fluorescence correlation spectroscopy were used to study mRNA dynamics (Politz et al. 1995, 1998, 1999).

In addition, a number of researchers were studying intracellular hybridization by microinjecting probes into cells (Leonetti et al. 1991; Sixou et al. 1994; Tsuji et al. 2000; Perlette and Tan 2001; Bratu et al. 2003; Tyagi and Alsmadi 2004; Vargas et al. 2005; Chen et al. 2007). In these studies, both linear probes and MB were utilized. Microinjection entails the use of a fine needle to inject the probes of interest directly into either the cytoplasm or nucleus, depending on the application. In the papers referenced above, probes were directly injected into the cytoplasm, and the results observed using fluorescence microscopy. From these papers, a number of important observations can be made. The first, is that when microinjected, both linear probes and MBs tend to accumulate in the nucleus. This has been shown by many researchers, in many cell types, with a variety of nucleic-acid-based probes. The accumulation is fast, on the order of 1–30 min, while most cite times less than 5 min. Typically, high concentrations of probe are used; concentrations between 0.5 and 10 μM are injected. Tyagi and Alsmadi (2004) attributed this to active transport from the results of their temperature and wheat germ agglutinin studies. In general, this technique rarely allows for the imaging of cytoplasmic RNA; only by adding a larger molecule like streptavidin, dextran, or even tRNA has the nuclear sequestration been thwarted (Tsuji et al. 2000; Tyagi and Alsmadi 2004; Mhlanga et al. 2005). This was the only way that cytoplasmic images of native RNA could be made. Molennar et al. (2001) reported nuclear sequestration, but also reported another observation when injecting MBs. They observed very rapid background intensities resulting from microinjection and preferred linear probes over MBs due to the high backgrounds, while Tyagi and Alsmadi (2004), although not reporting high backgrounds, reported SBRs no greater than 2.5 when targeting native RNA. Understanding the details of microinjected nucleic acids will be critical to its widespread usage.

Another method capable of delivering probes into cells is via reversible permeabilization of the plasma membrane using pore-forming toxins, such as streptolysin O (SLO). SLO is a pore-forming bacterial toxin that has been used as a simple and rapid means of introducing oligonucleotides into eukaryotic cells (Giles et al. 1995, 1998; Spiller et al. 1998; Clark et al. 1999; Giles et al. 1999). SLO belongs to the homologous group of thiol-activated toxins that are secreted by various Gram-positive bacteria. SLO binds as a monomer to cholesterol and then oligomerizes into ring-shaped structures estimated to contain 50–80 subunits, which surround pores of approximately 25–30 nm in diameter. This size by far exceeds the size of pores formed by other toxins, therefore allowing the influx of both ions and macromolecules. This method, as with microinjection, avoids the endocytic pathway, which can make efficient delivery to the cytoplasm very challenging. Since cholesterol composition varies between cell types, the sensitivity to SLO may vary as well. Therefore, the permeabilization protocol has to be optimized for each cell type by varying incubation time, cell number, and SLO concentration. An essential feature of this technique is that the toxin-based permeabilization is reversible. This has been achieved by introducing oligonucleotides with SLO under serum-free conditions and then removing the mixture and adding normal media with serum (Barry et al. 1993; Walev et al. 2001). When utilized for RNA imaging applications, SLO was used to deliver FITC labeled oligonucleotide probes and the reversible nature of this method was shown (Paillasson et al. 1997). Later, Santangelo et al. (2004, 2006, 2007) used SLO to deliver MBs targeted to both native and viral RNA. In the case of Paillasson et al. (1997), predominately nuclear localization was observed, while in Santangelo's applications just the opposite was observed. It should be noted though, that Santangelo's MBs did not have nuclear targets. Additional differences in their protocols could be important. Paillasson et al. (1997) added probes to the SLO/media mixture for 10 min and then replaced the SLO/media/probes mix with media and probes for 60 min more. Santangelo et al. (2004–2007) only delivered probe during the initial 10 min exposure to SLO. It is possible, given that Paillasson does not document the concentrations of probe used, that far more probe is delivered, driving more probe into the nucleus via diffusion. In Santangelo's applications, it is likely some of the MBs get into the nucleus, as in other SLO applications, but since the MBs had no target and likely delivered at a lower concentration, they displayed little signal or nuclear background. Interestingly, SLO has not been applied in any significant way to the delivery of nanoparticles into the cytosol. The reason is not clear. Proteins have been delivered; Walev et al. (2001) performed a number of experiments showing a size limit of 100 kD but successful delivery. It is likely they would be very useful for the delivery of a myriad of nanoparticles into living cells.

Cell permeable peptides, such as Tat or polyarginine, have been used to deliver both MBs and linear ligands (Nitin et al. 2004). Their effectiveness is increased when conjugated as opposed to complexed with the nucleic acid. This method seems to avoid the endocytic pathway with small nucleic acids and yields similar results to those achieved with SLO delivery. Much more research is required here to understand the mechanism of probe delivery and their subsequent dynamics, especially the effect of particle size. When streptavidin-Tat has been delivered into cells, it was reported to follow the endocytic pathway (Rinne et al. 2007), indicating a size dependence on Tat-based delivery, and possibly a dependence on the chemistry and composition of the probe. There have been many reviews on this subject and the author would suggest these for further reading.

21.6 Example of a Nanoprobe-Based Solution for Single-RNA-Sensitive Imaging of RNA in Live Cells

21.6.1 Multiply Labeled Tetravalent RNA Imaging Probes

One way of increasing the sensitivity of nucleic-acids-based linear probes is to multiply label them with organic fluorophores (Randolph and Waggoner 1997). This approach has been used extensively for FISH but not for live-cell imaging. In recent work by Santangelo et al. (2009), 3–4 fluorophores were added per ligand without experiencing self-quenching, and by combining four of these multiply labeled ligands using streptavidin, the base brightness was increased fourfold without adding any additional fluorophores. In this way, it was shown that single molecule sensitivity is possible using a small number (<20) of organic fluorophores and a physically small probe (~5 nm) (Santangelo et al. 2009). These probes are approximately the size of a single RNA binding protein, and only two or three (200–300 kD) probes are typically utilized to achieve single molecule sensitivity versus an average of 1.3 MD for MS2–GFP (Fusco et al. 2003).

21.6.2 MTRIPs—Composition, RNA Imaging Strategy, and Delivery Characteristics

MTRIPs consist of four high-affinity, nuclease-resistant, linear nucleic acids labeled with multiple, high-quantum-yield fluorophores self-assembled when mixed with streptavidin via the biotin-streptavidin linkage. (Figure 21.1a). When delivered via reversible cell membrane permeabilization with SLO (see Figure 21.1b), they bind rapidly to RNA (<10 min) and allowed for single RNA sensitivity using conventional fluorescence microscopy techniques. Target RNA are identified by the enhanced SBR achieved through binding of multiple probes per RNA (see Figure 21.1c) (Santangelo et al. 2009). MTRIP ligands are typically composed of a 2′-O-methyl RNA/DNA chimera nucleic acid with four or five amino-modified thymidines, a 5′ biotin modification and a short (5–7 bases) polyT section to extend the ligands from the surface of streptavidin. The amino-modified thymidines are used to conjugate NHS-ester modified fluorophores to the ligand. On an average, each ligand is labeled with three fluorophores, as measured using absorption spectroscopy. The multiply labeled monovalent ligands were then tetramerized via

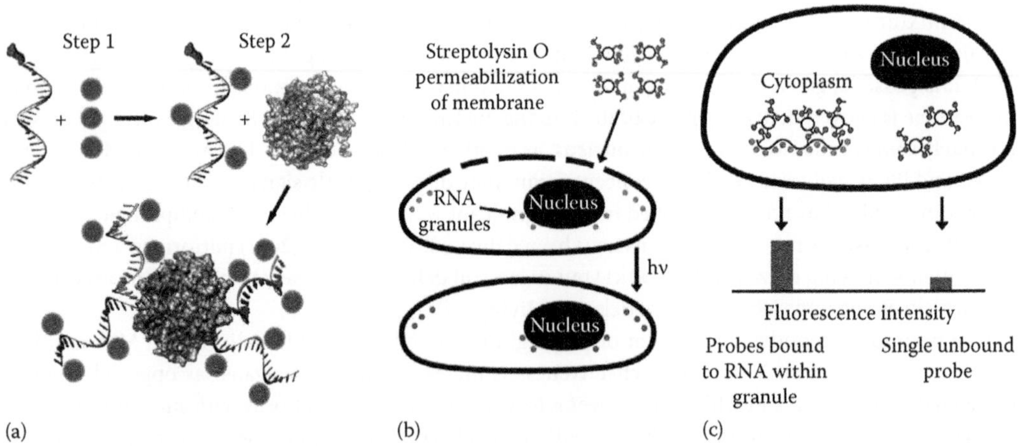

FIGURE 21.1 MTRIPs, imaging strategy, and evaluation of probe delivery via SLO within a live cell. (a) MTRIPs: biotinylated, fluorophore-conjugated chimera oligonucleotides bound to streptavidin. (b) Cells incubated with SLO/medium/probe are permeabilized by SLO allowing probe diffusion through the pores and rapid binding to target. (c) When multiple probes bind to a single RNA, the signal-to-background ratio is enhanced and allows for target recognition. (From Santangelo, P.J. et al., *Nat. Methods*, 6(5), 347, 2009.)

their binding to streptavidin. The purpose of the streptavidin linkage was to increase the brightness of the probe fourfold without adding more fluorophores to a single ligand, thereby avoiding the possibility of self-quenching.

21.6.3 MTRIPs on Glass Surfaces

MTRIPs probes were first characterized on glass at low concentrations. This is a typical methodology utilized to determine single probe sensitivity (Agrawal et al. 2008). Probes at 2 nM concentrations were immobilized on a glass surface by adding them in growth media to a coverslip well and incubating them for 10 min at 37°C. The mixture was removed, growth medium was added, and the glass surface was imaged. Individual batches of each probe, in addition to a mixture of Cy3B and ATTO 647N labeled probes were imaged on the glass surface. Individual probes were identified, and the mean intensity within the diffraction limited spots was plotted as a histogram. From the images of the probe mixtures, the histograms of each probe, and 3D plots of the intensity of individual probes, it was concluded that single probes and not aggregates were detected. If the probes were aggregating, the two-color mixture experiment would have yielded substantial colocalization of the probes, but it did not (Santangelo et al. 2009).

21.6.4 Characteristics of Probe Sensitivity and Delivery via SLO into Cells with No Target RNA

In order to characterize probe sensitivity and delivery in live cells, MTRIPs were designed to target the genomic RNA of the wild-type strain A2 of human respiratory syncytial virus (hRSV) and were delivered into noninfected A549 cells using SLO. The hRSV probes have no target in these cells and were utilized to characterize probe delivery and distribution within the cytosol. From a single optical plane within the live cell, individual probes were observed to be homogenously distributed within the cytoplasm. Localization or accumulation of probes was not observed. From this data, seen in Santangelo et al. (2009), it was clear that MTRIPs were delivered to the cytoplasm efficiently, achieving a homogenous distribution within the cytoplasm, were sufficiently bright such that single probes were observed within a live cell, and did not change cell morphology.

21.6.5 Targeting of Native mRNAs, Comparison with Scrambled Probe, and Colocalization with Known RNA Binding Proteins

In order to demonstrate the ability to image single RNAs using MTRIPs, two Cy3B-labeled MTRIPs designed to target two regions of the human β-actin mRNA coding sequence and an ATTO 647N-labeled "scrambled" probe were delivered via SLO simultaneously into living A549 cells all at 30 nM (Santangelo et al. 2009). About 20 min after live-cell delivery, the cells were fixed in 4% paraformaldehyde, for the sake of quantification, and imaged. It was found that individual "unbound" probes, probes from the glass surface, as well as localized granules with 2×, the intensity could be observed for β-actin mRNA. β-Actin mRNA was prevalent in the perinuclear region of the cell, but also localized to the leading edges, whereas the "scrambled" probe showed only perinuclear signals; it did not localize in abundance at the cell periphery but rather uniformly filled the volume of the cytoplasm, therefore showing β-actin probe specificity. Average single probe intensities, quantified from probes on the glass surface, were removed via thresholding, and the remaining granules detected and analyzed in software. Using this approach, single β-actin mRNAs, containing approximately 2×, the single probe intensity could be observed within the cell and a total number of approximately 1455 granules were detected. This number was consistent with previous quantifications (~1500 in serum-stimulated cells), using a similar analysis for β-actin mRNA in epithelial cells (Femino et al. 1998). As an additional control, A549 cells were serum-starved for 48 h and the number of β-actin mRNA granules counted in cells fixed post-live cell hybridization (Santangelo et al. 2009). A representative cell contained only 409 granules as compared with 1455 granules detected in a cell grown with serum. In this cell, the standard deviation of the mean granule fluorescence intensity for all 409 granules was 21%. The relative difference in granule count was consistent with previous experiments (Femino et al. 1998).

21.6.6 Time-Lapse Imaging of Native β-Actin Granules

In addition, using MTRIPs to tag β-actin mRNA granules, their dynamics within living cells were imaged via time-lapse widefield fluorescence microscopy. Images were taken with 90 ms exposure times at both 1 and 5 Hz for 3 min and 30 s, respectively. The images were deconvolved with a 2D fast deconvolution algorithm and the trajectories of two representative granules were tracked using Volocity software. Volocity allowed for the identification of individual particles, the plotting of their centroid positions over time, and the calculation of their mean square displacement as a function of time. The dynamics reported were similar to those of Fusco et al. (2003), but over 70% of RNAs were motile and over 700 could be imaged at one time in a single image plane, unlike Fusco et al., where there were only 20–25 plasmid-derived mRNAs within the cytosol at one time.

21.7 Other Nanotechnology-Based RNA Detection Probes

21.7.1 Gold Nanoparticle-Based RNA Imaging Strategies

Recently, the Mirkin lab has demonstrated another method for detecting RNA, but not its localization within living cells (Seferos et al. 2007; Prigodich et al. 2009). The probe consists of a gold nanoparticle, approximately 13 nm in diameter, conjugated to a thiolated-DNA oligonucleotide. The DNA was designed to be antisense to their target mRNA. The probe is also pre-hybridized with a complementary target labeled with an organic fluorophore, which is quenched by the gold particle. When this probe is introduced into cells, RNA complementary to the conjugated oligo, displace the pre-hybridized DNA, releasing them into the cytoplasm allowing them to fluoresce. In their work, they showed that the fluorescence was nominally proportional to the amount of RNA, but the fluorescence intensity failed to follow the drop in mRNA number due to siRNA exposure in a linear fashion. This may be due to the fact that a single mRNA may, by repeated binding to the probe, release more and more labeled oligonucleotide, and thereby overestimating RNA amount. The fact, though, that much of the probe appears to have been released from endosomes is intriguing and may be useful for future applications (Seferos et al. 2007; Prigodich et al. 2009).

21.7.2 Use of Semiconductor Quantum Dots for RNA Targeting in Fixed Cells

Currently, semiconductor based quantum dots (QDs), which have excellent photophysical properties, including being very bright, photostable, and narrow emission profiles, have been demonstrated for the detection of RNAs in fixed cell and tissues but as of yet, only in a limited fashion. The reasons for this are likely due to problems with background fluorescence from UV excitation or nonspecific "sticking" of QD-oligonucleotide conjugates and their high cost. A number of new papers have emerged, showing improved data, but no comparisons with organic fluorophore-based single-RNA-sensitive imaging. Choi et al. (2009), using a TOPO and polyacrylic acid-coated quantum dot where DNA was covalently conjugated to the QD, have apparently improved on previous experiments performed by Chan et al. (2005) and Tholouli et al. (2006), who used streptavidin-coated QD, where single RNA sensitivity was not shown. Using the polymer-coated QDs and DNA antisense ligands, they were able to target with a strong degree of specificity and what appeared to be single RNA sensitivity, a number of different mRNAs in Drosophila S2 cells. The imaging performed in this paper was a bit pedestrian, as the widefield images were not deconvolved, which would have improved them significantly and made the quantification much easier. A direct comparison with other FISH techniques that are single molecule sensitive would have been helpful in order to understand the accuracy and quantitative nature of their images. They are a significant improvement over Chan et al. (2005) and Tholouli et al. (2006), whose images appeared to show signal throughout the cell body of the stained neurons and images in tissue sections only; individual RNAs could not be identified in either case. Even if individual RNA imaging was not their goal, some high-resolution images would have been helpful in order to understand the effectiveness of their probes. Images of the subcellular localization, comparisons with organic fluorophore-based single-RNA-sensitive probes, and colocalization with known RNA binding proteins, all would have helped understand the accuracy of these techniques. Comparisons with scrambled or opposite sense probes are not always enough, especially when developing these probes, to clearly understand their accuracy. If the issues with nonspecific binding and accuracy can be dealt with, it is possible that QD-based imaging of RNA in fixed cells could become the standard for this type of imaging.

As of yet, very few papers have emerged using QDs within living cells; most are centered around tagging native cell surface proteins (Howarth et al. 2005) or engineered proteins with polyhistidine tags engineered within them (Roullier et al. 2009). Certainly, many of the issues with delivery discussed above relate to QD work and their application to intracellular targeting. Their intracellular stability and size are a concern, the smallest quoted QD are approximately 9 μm, which is over 8 times of volume of the MTRIPs, discussed earlier. In addition, given their complex surfaces, nonspecific binding is accentuated. Only with significant decreases in their size will they become useful for live-cell RNA imaging, but, with those improvements, they could significantly improve current technology.

21.8 Additional Discussion

In this section, using the information above, recommendations will be made for those interested in imaging RNA using live-cell hybridization probes. From the section on delivery, microinjection has clearly been the most utilized technique for delivering nucleic acids probes into a cell. The chief benefit of using this technique is that it allows complete control of the amount of probe delivered to the cell. It has drawbacks, though. It requires special equipment, user skill, and would be difficult to implement in high-throughput applications. Also, the exact buffer and volume injected varies in the literature, and requires optimization. In addition, both linear and MB probes, when microinjected, are often sequestered in the nucleus, if not modified with an inert molecule over 40 kD or with tRNA (Tsuji et al. 2000; Tyagi and Alsmadi 2004; Mhlanga et al. 2005; Chen et al. 2007). If nuclear targets are of interest, then this delivery technique is a good option. One area of confusion using microinjection has been in the relationship between probe concentration and binding kinetics. Concentrations of MBs quoted from 500 nM to 2.5 μM have yielded binding times on the order of 30–40 min (time to maximum signal), even when using high-affinity probes. This seems slow, given the concentrations used, especially in comparison to the data presented in Molenaar et al. (2001) using linear probes. Future research regarding live-cell hybridization with MBs will be required to improve their binding kinetics, especially if MBs are to be used to follow RNA processes, such as motor-driven transport, where fast probe binding kinetics are necessary. In addition, the fundamental limit in their sensitivity is a definite problem—possibly by adding much (30×) brighter fluorophores, this design would be improved. Currently, MTRIPs are significantly brighter, bind rapidly, and are utilized at very low (<30 nM) concentrations. Until a much brighter fluorophore is developed, we feel MTRIPs have set the standard in this field.

For the new user, SLO might be a better delivery option; it is inexpensive, easy to use, can deliver probes without added molecules into the cytoplasm and nucleus, and has shown minimal cell death and affects on cell morphology. There are certain considerations that have to be made when using it. There is some cell type dependence, due to differing amounts of cholesterol in the cell membrane, therefore it has to be optimized for a given cell type. Also, its efficiency is still under study, where efficiency would represent the fraction of probes added to the growth medium that are actually delivered within the cell. From the author's experience, probe is delivered to every cell, but the concentrations of probe delivered are likely much lower than using microinjection. This may be why when using MBs, signal levels, even with highly expressed mRNA or viral RNA, are often low. Therefore, using SLO, the use of linear probes would be advised. They can be specific, as previously shown, and yield detectable signal levels when targeting localized RNAs. Low concentrations of probe <100 nm should be used to limit the signal from the accumulation of probe in the nucleus. SLO delivery does result in delivery of small probes to the nucleus in addition to the cytoplasm, but does not preclude the ability to detect cytoplasmic RNA.

Another issue, not dealt with earlier in this review, is target sequence choice based on accessibility of the RNA. Target accessibility has been discussed for years, both in the antisense and siRNA literature, recently discussed for MB applications (Rhee et al. 2008), but in general, no exact design rules for imaging applications have been concluded. The author suggests reading that specific literature to gain insight into site selection (Sohail and Southern 2000a, b, 2001; Sohail 2001; Sohail et al. 2001). One problem with using the literature from those applications is that the metrics used for success for adequate binding are not the same as with imaging probes. Both antisense and siRNA seek to reduce protein levels, where imaging probes are to bind but not inhibit, and should not utilize Rnase H or the RNA-induced silencing complex (RISC) in their function. Even though this makes direct comparisons difficult, there should be some correlation, especially in the antisense data. If an antisense probe is binding to the mRNA, it is likely to be accessible to an imaging probe of similar sequence length. Complications may come from the number of different structural confirmations that RNA, can assume, such as when mRNA are translationally repressed, which likely occurs when mRNA are transported, interact with stress granules or are being degraded in or near processing bodies.

Another challenge in imaging applications is that detectable signal is often utilized as the metric for success. Unfortunately, if your probes lack sufficient sensitivity, a lack of signal is, unfortunately, not a definitive measure of the probes ability to bind to target. It only means that you lack the minimum number of bound probes within the point-spread function to achieve a detectable signal. In many cases, the actual sensitivity of the imaging system has not been tested by the user and they often do not know a priori how the mRNAs are distributed within the cell, which complicates the interpretation of their results, such as determining whether or not you have chosen an accessible site or sites. If little is known about how a particular RNA localizes within the cell, both quantitative RT-PCR and single-molecule-sensitive FISH (Femino et al. 1998; Raj et al. 2008) are necessary for gaining valuable information regarding RNA localization and gauging the sensitivity necessary of your probes. In addition, there are better fluorophores for fluorescence imaging, such as Cy3B (GE Healthcare) and the ATTO dyes (Atto-tec, GmbH), especially Atto 647N as opposed to Cy5 (GE Healthcare); we have found these to be brighter and more photostable in living cells than Cy3 and Cy5, which have been used previously. Last, colocalization experiments with known RNA binding proteins are helpful when attempting to understand whether or not your probe signal accurately represents the RNA present in the cell. These experiments are extremely helpful for identifying biologically relevant RNA populations and identifying the necessary probe sensitivity.

21.9 Conclusion

Over the last 18 years, the scientific community has learned a great deal about how to utilize nucleic-acid-based probes for the imaging of RNA in living cells. Many different delivery strategies, nucleic acid chemistries, and probe geometries have been investigated and provided an excellent basis for future studies. Future areas of research that are necessary to improve the use of exogenous probes for RNA imaging are in the areas of the following: delivery methods, understanding the role of nucleic acid chemistry, improved sensitivity, and the addition of electron microscopy (EM) compatibility. EM compatibility would be useful because it would allow the probes to provide information over all of the ranges of resolution that current imaging technology can provide.

In addition, there are a number of questions that need to be answered about this class of probes, in order for them to become the standard for RNA imaging, such as

1. How accurately can intracellular hybridization-based probes follow RNA degradation events and RNA stability?
2. What mixture of nucleic acid types will provide the right balance of affinity and noninvasiveness necessary to truly follow the life of an RNA molecule?
3. How accurately do they bind to RNA in various RNA granules?

As of yet, these questions have not been answered, but with improvements in probe design, such as those incorporated in the MTRIP design with SLO delivery, which allow for single RNA sensitivity and mRNA tracking, many of them should be answered shortly.

References

Aartsma-Rus, A., W. E. Kaman, M. Bremmer-Bout, A. A. Janson, J. T. den Dunnen, G. J. van Ommen, and J. C. van Deutekom. 2004. Comparative analysis of antisense oligonucleotide analogs for targeted DMD exon 46 skipping in muscle cells. *Gene Ther.* 11 (18):1391–1398.

Abe, H. and E. T. Kool. 2006. Flow cytometric detection of specific RNAs in native human cells with quenched autoligating FRET probes. *Proc. Natl. Acad. Sci. U.S.A.* 103 (2):263–268.

Agrawal, A., R. Deo, G. D. Wang, M. D. Wang, and S. Nie. 2008. Nanometer-scale mapping and single-molecule detection with color-coded nanoparticle probes. *Proc. Natl. Acad. Sci. U.S.A.* 105 (9):3298–3303.

Anderson, P. and N. Kedersha. 2002. Visibly stressed: The role of eIF2, TIA-1, and stress granules in protein translation. *Cell Stress Chaperones* 7 (2):213–221.

Anderson, P. and N. Kedersha. 2006. RNA granules. *J. Cell Biol.* 172 (6):803–808.

Anderson, P. and N. Kedersha. 2008. Stress granules: The Tao of RNA triage. *Trends Biochem. Sci.* 33 (3):141–150.

Anderson, P. and N. Kedersha. 2009. RNA granules: Post-transcriptional and epigenetic modulators of gene expression. *Nat. Rev. Mol. Cell Biol.* 10 (6):430–436.

Audic, Y. and R. S. Hartley. 2004. Post-transcriptional regulation in cancer. *Biol. Cell* 96 (7):479–498.

Barry, E. L., F. A. Gesek, and P. A. Friedman. 1993. Introduction of antisense oligonucleotides into cells by permeabilization with streptolysin O. *Biotechniques* 15 (6):1016–1018, 1020.

Bratu, D. P., B. J. Cha, M. M. Mhlanga, F. R. Kramer, and S. Tyagi. 2003. Visualizing the distribution and transport of mRNAs in living cells. *Proc. Natl. Acad. Sci. U.S.A.* 100 (23):13308–13313.

Brodsky, A. S. and P. A. Silver. 2002. Identifying proteins that affect mRNA localization in living cells. *Methods* 26 (2):151–155.

Chan, P., T. Yuen, F. Ruf, J. Gonzalez-Maeso, and S. C. Sealfon. 2005. Method for multiplex cellular detection of mRNAs using quantum dot fluorescent in situ hybridization. *Nucleic Acids Res.* 33 (18):e161.

Chen, A. K., M. A. Behlke, and A. Tsourkas. 2007. Avoiding false-positive signals with nuclease-vulnerable molecular beacons in single living cells. *Nucleic Acids Res.* 35 (16):e105.

Choi, Y., H. P. Kim, S. M. Hong, J. Y. Ryu, S. J. Han, and R. Song. 2009. In situ visualization of gene expression using polymer-coated quantum-dot-DNA conjugates. *Small* 5 (18):2085–2091.

Clark, R. E., J. Grzybowski, C. M. Broughton, N. T. Pender, D. G. Spiller, C. G. Brammer, R. V. Giles, and D. M. Tidd. 1999. Clinical use of streptolysin-O to facilitate antisense oligodeoxyribonucleotide delivery for purging autografts in chronic myeloid leukaemia. *Bone Marrow Transplant.* 23 (12):1303–1308.

Condeelis, J. and R. H. Singer. 2005. How and why does beta-actin mRNA target? *Biol. Cell* 97 (1):97–110.

Daigle, N. and J. Ellenberg. 2007. LambdaN-GFP: An RNA reporter system for live-cell imaging. *Nat. Methods* 4 (8):633–636.

de Silanes, I. L., M. P. Quesada, and M. Esteller. 2007. Aberrant regulation of messenger RNA 3′-untranslated region in human cancer. *Cell. Oncol.* 29:1–17.

Eom, T., L. N. Antar, R. H. Singer, and G. J. Bassell. 2003. Localization of a beta-actin messenger ribonucleoprotein complex with zipcode-binding protein modulates the density of dendritic filopodia and filopodial synapses. *J. Neurosci.* 23 (32):10433–10444.

Fabani, M. M. and M. J. Gait. 2008. miR-122 targeting with LNA/2′-O-methyl oligonucleotide mixmers, peptide nucleic acids (PNA), and PNA-peptide conjugates. *RNA* 14 (2):336–346.

Femino, A. M, F. S. Fay, K. Fogarty, and R. H. Singer. 1998. Visualization of single RNA transcripts in situ. *Science* 280:585–590.

Fusco, D., N. Accornero, B. Lavoie, S. M. Shenoy, J. M. Blanchard, R. H. Singer, and E. Bertrand. 2003. Single mRNA molecules demonstrate probabilistic movement in living mammalian cells. *Curr. Biol.* 13 (2):161–167.

Giles, R. V., C. J. Ruddell, D. G. Spiller, J. A. Green, and D. M. Tidd. 1995. Single base discrimination for ribonuclease H-dependent antisense effects within intact human leukaemia cells. *Nucleic Acids Res.* 23 (6):954–961.

Giles, R. V., D. G. Spiller, R. E. Clark, and D. M. Tidd. 1999. Identification of a good c-myc antisense oligodeoxynucleotide target site and the inactivity at this site of novel NCH triplet–targeting ribozymes. *Nucleosides Nucleotides* 18 (9):1935–1944.

Giles, R. V., D. G. Spiller, J. Grzybowski, R. E. Clark, P. Nicklin, and D. M. Tidd. 1998. Selecting optimal oligonucleotide composition for maximal antisense effect following streptolysin O-mediated delivery into human leukaemia cells. *Nucleic Acids Res.* 26 (7):1567–1575.

Giorgi, C. and M. Moore. 2007. The nuclear nurture and cytoplasmic nature of localized mRNPs. *Semin. Cell Dev. Biol.* 18:186–193.

Howarth, M., K. Takao, Y. Hayashi, and A. Y. Ting. 2005. Targeting quantum dots to surface proteins in living cells with biotin ligase. *Proc. Natl. Acad. Sci. U.S.A.* 102 (21):7583–7588.

Huttelmaier, S., D. Zenklusen, M. Lederer, J. Dictenberg, M. Lorenz, X. Meng, G. J. Bassell, J. Condeelis, and R. H. Singer. 2005. Spatial regulation of beta-actin translation by Src-dependent phosphorylation of ZBP1. *Nature* 438 (7067):512–515.

Jambhekar, A. and J. L. Derisi. 2007. Cis-acting determinants of asymmetric, cytoplasmic RNA transport. *RNA* 13 (5):625–642.

Kislauskis, E., A. Ross, V. M. Latham, X. Zhu, G. J. Bassell, K. L. Taneja, and R. H. Singer. 1995. Mechanism of mRNA localization: Its effect on cell polarity. H. D. Lipshitz, ed. *Localized RNAs: Molecular Biology Intelligence Unit*: Austin, TX, R.G. Landes Co.

Kislauskis, E. H., X. Zhu, and R. H. Singer. 1994. Sequences responsible for intracellular localization of beta-actin messenger RNA also affect cell phenotype. *J. Cell Biol.* 127 (2):441–451.

Kislauskis, E. H., X. Zhu, and R. H. Singer. 1997. Beta-actin messenger RNA localization and protein synthesis augment cell motility. *J. Cell Biol.* 136 (6):1263–1270.

Kurreck, J., E. Wyszko, C. Gillen, and V. A. Erdmann. 2002. Design of antisense oligonucleotides stabilized by locked nucleic acids. *Nucleic Acids Res.* 30 (9):1911–1918.

Latham, V. M., Jr., E. H. Kislauskis, R. H. Singer, and A. F. Ross. 1994. Beta-actin mRNA localization is regulated by signal transduction mechanisms. *J. Cell Biol.* 126 (5):1211–1219.

Leonetti, J. P., N. Mechti, G. Degols, C. Gagnor, and B. Lebleu. 1991. Intracellular distribution of microinjected antisense oligonucleotides. *Proc. Natl. Acad. Sci. U.S.A.* 88 (7):2702–2706.

Leung, K. M., F. P. van Horck, A. C. Lin, R. Allison, N. Standart, and C. E. Holt. 2006. Asymmetrical beta-actin mRNA translation in growth cones mediates attractive turning to netrin-1. *Nat. Neurosci.* 9 (10):1247–1256.

Martin, K. C. and A. Ephrussi. 2009. mRNA localization: Gene expression in the spatial dimension. *Cell* 136 (4):719–730.

Mhlanga, M. M., D. Y. Vargas, C. W. Fung, F. R. Kramer, and S. Tyagi. 2005. tRNA-linked molecular beacons for imaging mRNAs in the cytoplasm of living cells. *Nucleic Acids Res.* 33 (6):1902–1912.

Molenaar, C., S. A. Marras, J. C. Slats, J. C. Truffert, M. Lemaitre, A. K. Raap, R. W. Dirks, and H. J. Tanke. 2001. Linear 2′ O-Methyl RNA probes for the visualization of RNA in living cells. *Nucleic Acids Res.* 29 (17):E89–9.

Nielsen, F. C., J. Nielsen, M. A. Kristensen, G. Koch, and J. Christiansen. 2002. Cytoplasmic trafficking of IGF-II mRNA-binding protein by conserved KH domains. *J Cell Sci.* 115 (Pt 10):2087–2097.

Nitin, N., P. J. Santangelo, G. Kim, S. Nie, and G. Bao. 2004. Peptide-linked molecular beacons for efficient delivery and rapid mRNA detection in living cells. *Nucleic Acids Res.* 32 (6):e58.

Ozawa, T., Y. Natori, M. Sato, and Y. Umezawa. 2007. Imaging dynamics of endogenous mitochondrial RNA in single living cells. *Nat. Methods* 4 (5):413–419.

Paillasson, S., M. Van De Corput, R. W. Dirks, H. J. Tanke, M. Robert-Nicoud, and X. Ronot. 1997. In situ hybridization in living cells: Detection of RNA molecules. *Exp. Cell Res.* 231 (1):226–233.

Perlette, J. and W. Tan. 2001. Real-time monitoring of intracellular mRNA hybridization inside single living cells. *Anal. Chem.* 73 (22):5544–5550.

Politz, J. C., E. S. Browne, D. E. Wolf, and T. Pederson. 1998. Intranuclear diffusion and hybridization state of oligonucleotides measured by fluorescence correlation spectroscopy in living cells. *Proc. Natl. Acad. Sci. U.S.A.* 95 (11):6043–6048.

Politz, J. C., K. L. Taneja, and R. H. Singer. 1995. Characterization of hybridization between synthetic oligodeoxynucleotides and RNA in living cells. *Nucleic Acids Res.* 23 (24):4946–4953.

Politz, J. C., R. A. Tuft, T. Pederson, and R. H. Singer. 1999. Movement of nuclear poly(A) RNA throughout the interchromatin space in living cells. *Curr. Biol.* 9 (6):285–291.

Prigodich, A. E., D. S. Seferos, M. D. Massich, D. A. Giljohann, B. C. Lane, and C. A. Mirkin. 2009. Nanoflares for mRNA regulation and detection. *ACS Nano* 3 (8):2147–2152.

Raj, A., P. van den Bogaard, S. A. Rifkin, A. van Oudenaarden, and S. Tyagi. 2008. Imaging individual mRNA molecules using multiple singly labeled probes. *Nat. Methods* 5 (10):877–879.

Randolph, J. B. and A. S. Waggoner. 1997. Stability, specificity and fluorescence brightness of multiply-labeled fluorescent DNA probes. *Nucleic Acids Res.* 25 (14):2923–2929.

Rhee, W. J., P. J. Santangelo, H. Jo, and G. Bao. 2008. Target accessibility and signal specificity in live-cell detection of BMP-4 mRNA using molecular beacons. *Nucleic Acids Res.* 36(5):e30.

Rinne, J., B. Albarran, J. Jylhava, T. O. Ihalainen, P. Kankaanpaa, V. P. Hytonen, P. S. Stayton, M. S. Kulomaa, and M. Vihinen-Ranta. 2007. Internalization of novel non-viral vector TAT-streptavidin into human cells. *BMC Biotechnol.* 7:1.

Romani, A. 2007. Regulation of magnesium homeostasis and transport in mammalian cells. *Arch. Biochem. Biophys.* 458 (1):90–102.

Ross, A., Y. Oleynikov, E. Kislauskis, K. Taneja, and R. Singer. 1997. Characterization of a B-actin mRNA zipcode-binding protein. *Mol. Cell. Biol.* 17:2158–2165.

Roullier, V., S. Clarke, C. You, F. Pinaud, G. G. Gouzer, D. Schaible, V. Marchi-Artzner, J. Piehler, and M. Dahan. 2009. High-affinity labeling and tracking of individual histidine-tagged proteins in live cells using Ni2 + tris-nitrilotriacetic acid quantum dot conjugates. *Nano Lett.* 9 (3):1228–1234.

Santangelo, P. J. and G. Bao. 2007. Dynamics of filamentous viral RNPs prior to egress. *Nucleic Acids Res.* 35 (11):3602–3611.

Santangelo, P. J., A. W. Lifland, P. Curt, Y. Sasaki, G. J. Bassell, M. E. Lindquist, and J. E. Crowe, Jr. 2009. Single molecule-sensitive probes for imaging RNA in live cells. *Nat. Methods* 6 (5):347–349.

Santangelo, P. J., N. Nitin, and G. Bao. 2005. Direct visualization of mRNA colocalization with mitochondria in living cells using molecular beacons. *J. Biomed. Opt.* 10 (40):44025.

Santangelo, P. J., B. Nix, A. Tsourkas, and G. Bao. 2004. Dual FRET molecular beacons for mRNA detection in living cells. *Nucleic Acids Res.* 32 (6):e57.

Santangelo, P., N. Nitin, L. LaConte, A. Woolums, and G. Bao. 2006. Live-cell characterization and analysis of a clinical isolate of bovine respiratory syncytial virus, using molecular beacons. *J. Virol.* 80 (2):682–688.

Saxton, M. J. and K. Jacobson. 1997. Single-particle tracking: Applications to membrane dynamics. *Annu. Rev. Biophys. Biomol. Struct.* 26:373–399.

Seferos, D. S., D. A. Giljohann, H. D. Hill, A. E. Prigodich, and C. A. Mirkin. 2007. Nano-flares: Probes for transfection and mRNA detection in living cells. *J. Am. Chem. Soc.* 129 (50):15477–15479.

Shav-Tal, Y., X. Darzacq, S. M. Shenoy, D. Fusco, S. M. Janicki, D. L. Spector, and R. H. Singer. 2004. Dynamics of single mRNPs in nuclei of living cells. *Science* 304 (5678):1797–1800.

Shi, N., R. J. Boado, and W. M. Pardridge. 2000. Antisense imaging of gene expression in the brain in vivo. *Proc. Natl. Acad. Sci. U.S.A.* 97 (26):14709–14714.

Simmonds, P., A. Tuplin, and D. J. Evans. 2004. Detection of genome-scale ordered RNA structure (GORS) in genomes of positive-stranded RNA viruses: Implications for virus evolution and host persistence. *RNA* 10 (9):1337–1351.

Sixou, S., F. C. Szoka, Jr., G. A. Green, B. Giusti, G. Zon, and D. J. Chin. 1994. Intracellular oligonucleotide hybridization detected by fluorescence resonance energy transfer (FRET). *Nucleic Acids Res.* 22 (4):662–668.

Sohail, M. 2001. Antisense technology: Inaccessibility and non-specificity. *Drug Discov. Today* 6 (24):1260–1261.

Sohail, M., H. Hochegger, A. Klotzbucher, R. L. Guellec, T. Hunt, and E. M. Southern. 2001. Antisense oligonucleotides selected by hybridisation to scanning arrays are effective reagents in vivo. *Nucleic Acids Res.* 29 (10):2041–2051.

Sohail, M. and E. M. Southern. 2000a. Hybridization of antisense reagents to RNA. *Curr. Opin. Mol. Ther.* 2 (3):264–271.

Sohail, M. and E. M. Southern. 2000b. Selecting optimal antisense reagents. *Adv. Drug Deliv. Rev.* 44 (1):23–34.

Sohail, M. and E. M. Southern. 2001. Using oligonucleotide scanning arrays to find effective antisense reagents. *Methods Mol. Biol.* 170:181–199.

Sokol, D. L., X. Zhang, P. Lu, and A. M. Gewirtz. 1998. Real time detection of DNA RNA hybridization in living cells. *Proc. Natl. Acad. Sci. U.S.A.* 95 (20):11538–11543.

Spiller, D. G., R. V. Giles, J. Grzybowski, D. M. Tidd, and R. E. Clark. 1998. Improving the intracellular delivery and molecular efficacy of antisense oligonucleotides in chronic myeloid leukemia cells: A comparison of streptolysin-O permeabilization, electroporation, and lipophilic conjugation. *Blood* 91 (12):4738–4746.

Suzuki, T., D. Wu, F. Schlachetzki, J. Y. Li, R. J. Boado, and W. M. Pardridge. 2004. Imaging endogenous gene expression in brain cancer in vivo with^{111}In-peptide nucleic acid antisense radiopharmaceuticals and brain drug-targeting technology. *J. Nucl. Med.* 45 (10):1766–1775.

Tholouli, E., J. A. Hoyland, D. Di Vizio, F. O'Connell, S. A. Macdermott, D. Twomey, R. Levenson, J. A. Yin, T. R. Golub, M. Loda, and R. Byers. 2006. Imaging of multiple mRNA targets using quantum dot based in situ hybridization and spectral deconvolution in clinical biopsies. *Biochem. Biophys. Res. Commun.* 348 (2):628–636.

Tsourkas, A., M. A. Behlke, and G. Bao. 2002. Hybridization of 2′-O-methyl and 2′-deoxy molecular beacons to RNA and DNA targets. *Nucleic Acids Res.* 30 (23):5168–5174.

Tsourkas, A., M. A. Behlke, S. D. Rose, and G. Bao. 2003. Hybridization kinetics and thermodynamics of molecular beacons. *Nucleic Acids Res.* 31 (4):1319–1330.

Tsuji, A., H. Koshimoto, Y. Sato, M. Hirano, Y. Sei-Iida, S. Kondo, and K. Ishibashi. 2000. Direct observation of specific messenger RNA in a single living cell under a fluorescence microscope. *Biophys. J.* 78 (6):3260–3274.

Tsuji, A., Y. Sato, M. Hirano, T. Suga, H. Koshimoto, T. Taguchi, and S. Ohsuka. 2001. Development of a time-resolved fluorometric method for observing hybridization in living cells using fluorescence resonance energy transfer. *Biophys. J.* 81 (1):501–515.

Tyagi, S. and O. Alsmadi. 2004. Imaging native beta-actin mRNA in motile fibroblasts. *Biophys. J.* 87 (6):4153–4162.

Tyagi, S., D. P. Bratu, and F. R. Kramer. 1998. Multicolor molecular beacons for allele discrimination. *Nat. Biotechnol.* 16 (1):49–53.

Tyagi, S. and F. R. Kramer. 1996. Molecular beacons: Probes that fluoresce upon hybridization. *Nat. Biotechnol.* 14 (3):303–308.

Vargas, D. Y., A. Raj, S. A. Marras, F. R. Kramer, and S. Tyagi. 2005. Mechanism of mRNA transport in the nucleus. *Proc. Natl. Acad. Sci. U.S.A.* 102 (47):17008–17013.

Vester, B. and J. Wengel. 2004. LNA (locked nucleic acid): High-affinity targeting of complementary RNA and DNA. *Biochemistry* 43 (42):13233–13241.

Walev, I., S. C. Bhakdi, F. Hofmann, N. Djonder, A. Valeva, K. Aktories, and S. Bhakdi. 2001. Delivery of proteins into living cells by reversible membrane permeabilization with streptolysin-O. *Proc. Natl. Acad. Sci. U.S.A.* 98 (6):3185–3190.

Wu, Y., C. J. Yang, L. L. Moroz, and W. Tan. 2008. Nucleic acid beacons for long-term real-time intracellular monitoring. *Anal. Chem.* 80 (8):3025–3028.

Yamagishi, M., Y. Ishihama, Y. Shirasaki, H. Kurama, and T. Funatsu. 2009. Single-molecule imaging of beta-actin mRNAs in the cytoplasm of a living cell. *Exp. Cell Res.* 315 (7):1142–1147.

Yao, J., Y. Sasaki, Z. Wen, G. J. Bassell, and J. Q. Zheng. 2006. An essential role for beta-actin mRNA localization and translation in Ca2+-dependent growth cone guidance. *Nat. Neurosci.* 9 (10):1265–1273.

22

Nanomaterials for Artificial Cells

Xiaojun Yu
Stevens Institute of Technology

Elvin Lee
Stevens Institute of Technology

Alicia Vandersluis
Stevens Institute of Technology

Harinder K. Bawa
Stevens Institute of Technology

22.1 Introduction

The development of nanomaterials offers many promising solutions to problems facing the medical field. One possibility on the horizon is the development of artificial cells using nanomaterials. Engineered artificial cells can be used to replace dysfunctional cells in the human body and may be used in the future to treat anemia, renal failure, bone defects, and many other health problems. The biggest advantage in using nanomaterials in the fabrication of artificial cells is their small size, ranging from 1 to 100 nm in diameter (Liang et al. 2008). The small size allows a larger and more beneficial surface area to volume ratio and contributes to their unique physical and chemical properties. Biological cells have a very high functional density and contain DNA coding for thousands of different proteins to carry out certain functions. Engineering artificial cells with the same magnitude of functional density as biological cells is a significant challenge and has only recently become feasible with advances in nanotechnology.

In essence, an artificial cell or cells are systems that biomimic native cellular function to recapitulate either function and/or structure. This definition includes capsules or particles that biomimic a single or many biochemical functions of a cell of interest or the cell itself or encapsulations of cells to allow the cells to perform the function of choice. Artificial cell technology seeks to address the need for a more efficient temporary if not permanent replacement. In contrast to tissue engineering, the field of artificial cells is concerned with singular cells and recapitulations of functions, instead of whole complex tissues. The wide range of artificial cells can be glimpsed in the various applications that arose ranging from whole cell encapsulations of pancreatic islet cells and hepatocytes, liposome encapsulation of hemoglobin (Hb), and polymerized Hb. In the treatment of enzyme and single system defects, the application of whole cells may be detrimental, and replacing the enzyme or the single system may be more efficient as is the case with artificial red blood cells. The application of nanobiomaterials is necessary to both better biomimic cellular systems and construct a more efficient system than nature itself.

In this chapter, we first briefly discuss the application of nanomaterials for artificial cells in general and then discuss the application of nanomaterials for artificial blood cells in details.

22.2 Development of Artificial Cells Using Nanomaterials

An important aspect for developing artificial cells is the cell membrane. The membrane serves as an interface for intercellular communication, and so it therefore plays a large role in drug delivery. The surface properties of nanoparticles must be considered while developing drug delivery systems and artificial tissues. Geometry of the membrane should also be considered when constructing artificial cells. One study comparing concave and convex membrane structures of poly(dimethylsiloxane) (PDMS) found that concave structures suppress cell adhesion and proliferation (Sun et al. 2008). Furukawa et al. (2007) were able to grow a lipid bilayer on a silicon wafer and glass because of its self-spreading properties at a solid–liquid interface (Liu et al. 2008). Supported lipid bilayers have been grown from a lipid molecule on a hydrophilic surface with nanostructures (Isenberg et al. 2008). Chen et al. were able to assemble a sandwiched, layer-by-layer lipid bilayer using 1,2-dimyristoyl-sn-glycero-3-phosphoethanolamine (DMPE) on a polyelectrolyte multilayer (PEM) film (Romberg et al. 2007). In another study, 2-methacryloyloxyethyl phosphorylcholine (MPC) polymers with an attached phosphorylcholine side chain were used to mimic the phospholipids in cell membranes. This proved to be an effective way of establishing a membrane-like interface between artificial and biological materials (Ishihara et al. 2008).

The machinery needed in an artificial cell depends on the function that the cell needs to carry out. Progress has been made toward constructing artificial organelles, which is a stepping stone in the development of a complete artificial cell with similar functionality to a biological cell. In one study, a functional artificial Golgi apparatus was constructed using digital microfluidics, recombinant enzyme technology, and magnetic nanoparticles. The Golgi was able to modify glycosaminoglycans immobilized onto magnetic nanoparticles (Lin et al. 2007).

When developing artificial cells using nanomaterials, different techniques must be considered. There are two primary techniques for building nanomaterials: the "bottom-up" approach and the "top-down" approach. "Bottom-up" assembly requires the self-assembling of molecules, molecular manipulation, and molecular binding (Kabanov 2006). Bottom-up approaches rely heavily on the self-assembly of molecules from molecular recognition. The technique requires the assembly of nanomaterials through molecule by molecule or even atom by atom (Zhang 2003). One example of this is base-pairing in the synthesis of DNA. Bottom-up construction of nanomaterials is advantageous because of the opportunity to introduce functionality and precision when engineering nanomaterials (Furukawa et al. 2007).

The top-down approach involves fabricating small nanomaterials using larger materials to direct the assembly (Furukawa et al. 2007). Biomaterials are developed by stripping down a complex entity into its component parts (Zhang 2003). The advantage to using the top-down approach is the ability to organize nanomaterials in hierarchal patterns on a larger scale (Furukawa et al. 2007). Although bottom-up techniques have their advantages, it can be difficult to control a self-assembling system to engineer a specific desired structure (Vasita and Katti 2006). Therefore, integration of the bottom-up and top-down approaches is a possible solution and has been done in some studies (Vasita and Katti 2006). In one experiment involving DNA, the bottom-up and top-down techniques were integrated to create an array of DNA nanotubes (Furukawa et al. 2007). In another experiment, lithographic templates were used as a top-down approach to control the bottom-up self-assembly of block copolymers (Vasita and Katti 2006).

Self-assembling nanoparticles as a bottom-up approach has become a very common method in developing nanomaterials for artificial cells. Research involving the self-assembly of designed artificial peptides and peptidomimetrics into nanofibers has shown promise in developing a new class of soft-materials. These materials have the potential to be used in tissue engineering and biomineralization (Higashi and Koga 2008). A self-assembly method involving single-walled carbon nanotubes (SWNTs) alternately packed β-1,3-glucans into the structure as building blocks. This method allowed for a hierarchal "superstructure" to be manufactured and may have future applications (Numata et al. 2008).

For successful surface modification of nanoparticles to take place, the membranes must be subject to broad adsorption of a variety of proteins and functional groups. On the other hand, the surface must be biocompatible. This is especially important in the bloodstream. Nanomaterials have to be particularly stable with regard to blood plasma proteins and platelets to prevent clotting. Various polymer coatings on synthesized nanoparticles have been utilized to mimic a biological membrane while maintaining blood compatibility (Ishihara et al. 1998). To reduce adhesion of plasma proteins, there has been development of polymers containing a 2-methacryloyloxyethyl phosphocholine moiety, or MPC polymers (Webb et al. 1995). More examples of polymer coatings that are functional include diacetylenic phospholipids, phosphatidylcholines, and phosphorylcholine groups (Webb et al. 1995). Surface modifications of nanoparticles for artificial cells not only serve to improve drug delivery, but also reduce the adverse effects of the agglutination of blood platelets and plasma proteins with nanomaterials.

There have been many different types of artificial cells developed as a result of the emergence of nanotechnology. Yi et al. (2009) were able to engineer an artificial amoeba that was capable of propelling itself by nanoparticle-triggered actin polymerization. Nickel and gold nanoparticles with attached Listeria monocytogenes transmembrane protein ActA were utilized along with actin, actin-binding proteins, and ATP and encapsulated in a lipid vesicle. The system served as an artificial cell with actin polymerization within the cell acting as a functional cytoskeleton (Yi et al. 2009).

One exciting development in research of artificial cells is that of the capsosome. A capsosome is a polymer capsule containing multiple liposomes (Chandrawati et al. 2009). In one study, capsosomes were fabricated by enveloping liposomes between a cholesterol-modified poly(L-lysine) (PLLc) precursor layer and a poly(methacrylic acid)-co-(cholesteryl methacrylate) (PMAc) capping layer (Chandrawati et al. 2009). Capsosomes can serve as microreactors to cause a series of reactions and release desired products to the surrounding environment (Chandrawati et al. 2009). A similar experiment utilized a multilayer film assembly of the polyelectrolytes poly(styrene sulfonate) (PSS) and poly(allylamine hydrochloride) (PAH) with liposomes (Städler et al. 2009). This technique yielded stable capsosomes, and shows the effectiveness of using polyelectrolyte capsules in capsosome formation. The development of capsosomes shows that the possibility of fabricating a biologically similar artificial cell or artificial organelles is not far off the horizon, and can lead to many new exciting possibilities in biotechnology.

Another development in the realm of artificial cells is the use of silica-coated beads for in vitro protein synthesis. Lim et al. (2009) found that the beads can encapsulate transcriptional and translational machinery for the synthesis of functional proteins. It was also confirmed that permeation of the proteins through the particle membrane was taking place. Silica-coated beads showed an increase in synthesized protein activity by fivefold and threefold over bare and chitosan-coated beads, respectively (Lim et al. 2009). The development of this type of nanoparticle can lead to the development of artificial cells through the parallel synthesis of varying functional proteins in designed combinations.

There are a wide variety of different types of nanomaterials being developed for use in artificial cells. Development of individual cell components such as the cell membrane, cytoskeleton, surface properties, and organelles is heavily underway. The possibilities for nanomaterials and artificial cells in the medical field are endless.

22.3 Nanobiomaterials for Artificial Red Blood Cells

One major finding in the development of artificial cells is their possible use as oxygen carriers in the bloodstream for the treatment of anemia. Blood substitutes, unlike blood itself, serve solely to carry oxygen and carbon dioxide throughout the body. A lot of the research being done focuses on hemoglobin-based products, which include PEG modified liposome-encapsulated hemoglobin, nanoparticle and polymersome encapsulated hemoglobin, and polymerized hemoglobin solutions (Sarkar 2008). One example of an oxygen carrier encapsulates a solution of concentrated hemoglobin and is called a hemoglobin vesicle (HbV). HbV was found to have oxygen-carrying capacity comparable to that of normal red blood cells (Bucci 2009). The fabrication of artificial red blood cells in the form of HbV offers promising opportunities in a clinical setting.

22.3.1 Artificial Cells as Blood Substitute

Blood serves as the first model for artificial tissue, since the individual cell of blood can be easily separated from the whole tissue and its individual function is the same. Within blood, the need for a practical RBC substitute is quite prevalent. In brief, the limited donor pool (Riley et al. 2007) and even smaller number of donations (Davy 2004), as well as the biological complications such as infections (Spies 2004) and antigen matching (Hill 2001) complicate finding safe and sufficient blood donation for patients. To address this clinical issue, Dr. T.M.S. Chang devised the first artificial cell, microcapsules of nylon containing hemoglobin (Hb). The idea was for the encapsulation of the macromolecule to allow function while allowing small molecules to pass into and out of the system, thereby allowing oxygen exchange. The idea led to a field of study pertaining to hemoglobin and hemoglobin encapsulated systems or hemoglobin-based oxygen carriers (HbOCs).

In the development of the first HbOCs, goal was to increase oxygenation of tissues comparable to the addition of pRBCs and scientists knew that Hb was the molecule primarily involved. The rationale was to infuse a patient with the protein that was responsible for oxygen exchange in tissues. Because the drive was to avoid usage of whole cells, encapsulation of RBCs was not considered a viable option. Thus, the first generations of HbOC products were nanoparticles of single bovine or human Hb proteins (Garby and Noyes 1959; Brunn 1972) and eventually intra- (Chatterjee et al. 1986) and intermolecularly (Gould and Moss 1996) cross-linked Hb particles. The progressive shift from stroma-free Hb to modified Hb and eventually polymerized Hb was driven by in vivo toxicity and optimization of circulation half-life. The first generation of blood substitute products faced renal toxicity due to tetramer dissociation into dimers and the subsequent release of iron ions and also had a short in vivo circulation half-life of 10–30 min due to size allowing easy glomerular filtration (Bunn and Jandl 1969) and reticuloendothelial clearance (Bunn 1972; Hersheko 1975).

To mitigate the problem of a short circulation half-life, many approaches were done in cross-linking the tetramers together with glutaraldehyde (Chang 1971) or other bifunctional cross-linking agents such as bis(N-maleimidomethyl) ether (BME) (Wedekind et al. 1985) and bis(3,5-dibromosalicyl) fumarate (DBBF) (Walder et al. 1979; Sloan et al. 1999). Through stabilization of the tetramer, the circulation half-life was tripled, but oxygen affinity of the Hb particles was too high and thus did not provide adequate oxygenation. The next generation of products were polymerized Hb and polymer conjugated Hb to allow for greater molecular weights and size. To lower oxygen affinity, various compounds such as pyridoxal phosphate were cross-linked to Hb (Greenberg et al. 1979; Seghal et al. 1981). The other approach was to use bovine Hb that naturally has a lower affinity for oxygen and does not require cofactors (Stonwell et al. 2001). All the artificial RBC products failed at various phases in clinical trials for humans to date.

An alternative approach was taken by chemists who formed fluorinated hydrocarbons or perflourocarbons (PFCs). PFCs have a linear oxygen affinity curve and require more to equate to Hb's oxygen carrying capacity (Riess 2006). PFCs avoided most of the complications due to Hb administration but were fraught with others such as platelet inactivation and increased leukocyte counts, and long-term storage in tissues required emulsifying agents (Lane 1995). An example of a major product that stemmed from the research was Flousol (Tremper 1983). PFCs were never approved for human use.

The next rational step was taken of encapsulating Hb into vesicles that biomimic a cell—liposomes and micelles (Shi et al. 2009). The two approaches resulted in similar complications of lower diffusion of small molecules to and from the system as well as low encapsulation efficiencies and high conversion rate into methemoglobin, the inactivated form of the Hb (Sakai et al. 2004). Liposome or micelle based encapsulation systems failed to be an adequate RBC replacement as of today.

The final step was to create enzyme bags with the functions of a RBC through encapsulation of proteins in polymer networks. This was first demonstrated by Chang et al. (1971) by including Hb as well as superoxide dismutase (SOD) to increase circulation time, reduce oxygen affinity, and allow for tissue oxygenation. Through the work by Zhao et al. (2007) the need for nanoparticles was elucidated; more specifically, the diameter of particles around 100 nm had the longest circulation half-life. Another major advantage of nanoscale HbOCs is the ability to pass through occlusions that commonly occur in trauma and high cholesterol patients allowing for lower chance of infarction. The need for nanobiomaterials is imperative into the further development of the field of artificial RBCs and artificial cells.

22.3.2 Characterization of Artificial Red Blood Cells

There are several desirable properties when it comes to designing an ideal HBOC. Theoretically, the HBOC would solve the problems associated with traditional allogenic blood transfusions and be able to provide an urgent need of oxygen delivery to tissues. Therefore, the ideal HBOC should have adequate delivery of oxygen to all tissues, not transmit infectious diseases, be universally compatible without having to conduct lengthy cross-matching and typing, be easy to administer, and be readily available and at reasonable costs. Many of these properties are based on the physical and chemical characteristics of these artificial cells, which can be tailored to design the ideal blood substitute.

Particle size, surface charge, and surface hydrophobicity are all critical parameters that undermine the overall performance of a HBOC when is it clinically administered (Awasthi et al. 2003). Several investigations have been conducted in order to find the "ideal" size requirements for nanoparticles developed as oxygen carrying artificial cells. Theoretically, the nanoparticles must be able to circulate freely through even the smallest of capillaries, which can be as small as 4–7 μm in diameter, and so should be smaller than 4 μm in order to avoid embolism. However, nanoparticles above approximately 200 nm will be removed by the spleen through mechanical filtration and consequently accumulate in the spleen (Kissel and Roser 1991; Moghimi et al. 1993). On the other hand, nanoparticles with diameters below approximately 70 nm will be removed by the liver due to the possible penetration of such small particles through fenestrae in the endothelial lining of the liver, hence causing an increased accumulation in the liver (Litzinger et al. 1994). As a result, it has been suggested that the 70–200 nm range is optimal for intravenous delivery and circulation (Zhao et al. 2007).

Surface modifications that increase hydrophilicity, to some extent, can prolong the residence of nanoparticles in blood by evading rapid elimination from systemic circulation, a process usually performed by monocytes and cells of the mononuclear phagocyte system (MPS) (Avgoustakis et al. 2003). In general, a higher protein adsorbability of hydrophobic relative to hydrophilic surfaces has been related to the uptake and rapid removal of hydrophobic particles by phagocytosis (Illum and Davis 1986). Thus, coating the nanoparticle surface with a hydrophilic polymer such as poly(ethylene glycol) (PEG), either through covalent attachment or physical adsorption to the surface, has been illustrated to decrease uptake by the MPS (Li et al. 2001). Because PEG prevents interactions with other biological components in vivo, HBOCs modified with PEG have longer circulation times and are non-thrombogenic (Gref et al. 1995).

Many attempts have been made to investigate the effects of surface charge on the distribution of the nanoparticles in circulation, but the outcomes are contradicting. In one study, negatively charged nanoparticles, when compared with neutral and positively charged nanoparticles, were shown to facilitate clearance of the nanoparticles from blood circulation (Stolnik et al. 1995). Conversely, Gbadamosi et al. (2002) demonstrated that a smaller negative charge decreased uptake of the nanoparticles and Yamaoka et al. (1995) reported that the introduction of negative surface charges to dextran derivatives prolonged its half-life in circulation. Although these results suggest that the surface charge of nanoparticles is important in determining blood circulation time, much research needs to be conducted in order to establish the effect of the surface charge on nanoparticles used as oxygen carriers.

Oxygen affinity is measured as the partial pressure of oxygen at which hemoglobin is 50% saturated with oxygen, commonly denoted as p50. Hemoglobin (Hb) is a tetrameric structure ($\alpha_2\beta_2$) that has two distinct conformations of the subunits (oxy and deoxy). Hb's cooperative binding to oxygen results in the sigmoidal shape of the oxygen-hemoglobin binding curve. Briefly, if one oxygen molecule binds to a heme group, Hb changes its conformation so that the binding of more oxygen molecules to the remaining heme groups is facilitated and a similar action is seen if a heme group releases its oxygen molecule (Awasthi 2005). The normal p50 of human Hb in red blood cells is approximately 28 mm Hg, whereas Hb itself has a p50 of about 10 mm Hg (Winslow et al. 1977). Generally, a high oxygen affinity is associated with a low p50 and a low oxygen affinity is associated with a high p50.

The oxygen affinity of Hb is influenced by many factors such as temperature, pH, and carbon dioxide levels. Within the red blood cell, various salts also affect the oxygen affinity such as chloride ions, ATP, and 2,3-diphosphoglycerate (2,3-DPG) (Kobayashi et al. 2004). In red blood cells, 2,3-DPG decreases the oxygen affinity by cross-linking the β chains of the tetramer and allows the oxygen molecules to be released. Soluble human Hb has an increased affinity for oxygen as a result of insufficient 2,3-DPG in the plasma (Arnone 1972). Such high affinity Hb would not function well as an oxygen delivering substance since the amount of oxygen released is highly decreased. A simple solution to this problem is the use of bovine Hb. The oxygen affinity of bovine Hb is not 2,3-DPG dependent but rather chloride ion dependent and there is a sufficient amount of chloride ions present in the human plasma (Franticelli et al. 1984). Additionally, chemical modifications of Hb by cross-linking, polymerization, or polymer linking have also shown significant improvements in the p50 of Hb.

Ideally, these artificial cells should have a p50 close to that of the red blood cells. Because these HBOCs are designed for use in emergencies and life-threatening issues, the optimum value of oxygen affinity depends on the exact physiological conditions under which the HBOC is used. Recent findings have supported the use of HBOCs with a low p50 in cases of severe blood loss. It has been demonstrated that under normoxia or mild hypoxia, a high p50 HBOC may be beneficial while in severe hypoxia, a low p50 HBOC may be better (Kavdia et al. 2002; Shirasawa et al. 2003). Overall, oxygen affinity is a property that needs to be highly controlled when designing a HBOC as it is critical in the loading and unloading of oxygen within the human body.

A significant challenge in the development of HBOCs is faced due to their effect on the vascular tone of blood vessels; typically hypertension was observed in studies with soluble Hb. Unlike Hb in red blood cells, soluble Hb (or tetrameric Hb) can be readily extravasated into the endothelial tissue where it can rapidly bind to and remove nitric oxide (NO). NO, also referred to as endothelial-derived relaxing factor, plays an important role in controlling smooth muscle relaxation of the blood vessels (Schultz et al. 1993). When oxygenated Hb (HbO_2) reacts with NO, NO is oxidized to nitrate (NO^{3-}) and HbO_2 is reduced to metHb. This results in significant vasoconstriction of the blood vessels and a consequent increase in blood pressure (Olson et al. 2004).

Although NO scavenging is an important contributor to the increased vasoconstriction, it is not the only rationalization. Several alternative theories have been developed to explain this phenomenon, which includes too much delivery of oxygen and the oxidation of Hb, which can result in heme loss and free radical formation. Increased oxygen delivery to arterioles is followed by an autoregulatory constriction of the arterioles and the capillary beds as a result of what is perceived to be an excess of oxygen delivery (Sanders et al. 1996). The oxidation of Hb can result in the formation of products that induces endothelial stress causing vasoconstriction. One way of countering the extravasation of Hb is to increase the molecular size by PEGylation or by polymerization as such molecules will not be small enough to pass through the endothelial tissue.

These physical and chemical characteristics can be adjusted and fine tuned to prepare artificial red blood cells with the desired properties, a critical advantage when designing artificial cells to be used in specific situations.

22.3.3 Nanobiomaterials for Artificial Red Blood Cells

22.3.3.1 Proteins

As mentioned earlier, the first generation of artificial RBCs were HbOCs that was composed of entirely Hb or protein with modifications. Proteins of nanometric dimensions can serve as the platform for artificial cells (Baudin-Creuza et al. 2008). The first approach was to inject purified Hb or stroma-free Hb directly into animals and patient in place of RBCs. The idea of polymerizing Hb was extended to prolonging the circulation half-life and reducing renal toxicity of tetramer dissociation through cross-linking 4–5 Hb molecules with glutaraldehyde (GTA) and other bifunctional groups. Concurrently, Hb conjugated to polymers such as polyamide, dextran, and PEG were also done for the same effect of increasing circulation half-life. The origin of Hb has also changed from allogenic human Hb (Yamaguchi et al. 2009) to bovine Hb (Mullon et al. 2000) and eventually to recombinant Hb (Fronticelli and Koehler 2007). Few products have progressed into clinical trials. The first generation of products is exemplified by DCLHb or HemAssist™ by Baxter. HemAssist is a diaspirin cross-linked Hb system. The Hb was obtained from expired human pRBCs (Lamy et al. 2000).

PolyHeme™, product of Northfield Laboratories, is composed of human hemoglobin form expired pRBCs that have been pyroxylated and polymerized with GTA. The end concentration of Hb in the suspension is about 10 g/dL. PolyHeme has a P_{50} of 20–22 mm Hg, a little lower than intra-RBC Hb's 26 mm Hg. A single molecule has a molecular weight of 150 kDa and results in double the viscosity of saline solutions (Gould et al. 1995; Dubick et al. 2004). PolyHeme was shown to reduce mortality in patients with severe acute anemia when compared to patients rejecting pRBC transfusions.

Hemopure™ or HBOC-201, a product of Biopure, is a solution of purified bovine Hb with a final Hb concentration of 13 g/dL and a P_{50} of 36 mm Hg. It has a circulation half-life of about 19 h and can be stored up for over 3 years. Phase III clinical trials of HBOC-201 showed adverse side effects in 93% of patients including a rise in blood pressure and thus HBOC-201 was not allowed to enter phase IV in the United States (Wilson et al. 2007). It was, however, approved for adult use in South Africa and reports of cardiac arrest after usage have been published (Stefan et al. 2007). It is safe to note that the patient was a child and not an adult. Another product resulting from HBOC-201, Oxyglobin™, HBOC-301, is a GTA polymerized HbOC approved by the FDA for veterinary use (Jahr et al. 2007). It was shown by Buehler et al. (2005) to have a P_{50} of around 38.4 mm Hg compared to bHb, which had a P_{50} of around 27.2 mm Hg. The stability and viability of the GTA polymerized Hb was demonstrated in animal models.

Another approach toward prolonging of in vivo circulation half-life and lower oxygen affinity for both human and bovine stroma-free Hb was to cross-link the Hb with polymer chains; the system is also referred to as conjugated Hb in literature. The polymer chains employed with both synthetic and naturally occurring. A common approach was to cross-link PEG to polymerized Hb (Iwashita et al. 1988; Shorr et al. 1996). PEG is a good candidate for increasing molecular weight because it is a synthetic polymer that can be obtained at desired molecular weights, is easily water soluble, and is non-thrombogenic (Watanable et al. 2004; George et al. 2009). Another reason PEG was used was to block active sites where it may interfere with either oxygen transport or other small molecules to hinder oxygenation. One such example was PEGylation through S-nitrosylation at the Cys-β93 to reduce NO scavenging (Asanuma et al. 2007). The most common reaction used to PEGylate Hb is the 2-imminothiolane (IMT) reaction that involves the production of maleimido-PEG and then PEGylated Hb through the reaction with activated thiol groups (Iafelice et al. 2007). Other synthetic polymers used were polyamide and hydrogel-like polymers. Polysaccharides were also conjugated to the Hb such as dextran (Tam et al. 1976; Wong 1988) and heparin (Chauvierre et al. 2007). The conjugation reaction involves the oxidation of the end terminal sugars by inducing ring-opening and subsequent oxidation through sodium periodate to form dialdehyde groups. This technique was used extensively by Eike and Palmer to form a wide range of HbOCs, based on oxidized saccharide conjugates of Hb (Eike 2005; Eike and Palmer 2006).

Clinically, protein-based HbOCs were proven to ameliorate blood loss, and many are in phase III clinical trials or have been stopped at or before phase III clinical trials. No product is as of yet approved for general trauma use in the United States. A large concern is the antigenicity of Hb, especially when sources other than allogenic sources are used including xenogenic, microbial, and even recombinant. The concern for recombinant lies not in the sequence of the Hb, but in the resulting posttranslational modifications. This is also a concern when polymerized and conjugated with polymers to increase size. For polymers that are not antiopsonizing or having similar moieties with commonly exposed antigenic conformations, the concern is greater. The problems of using Hb still abound despite improvements. The side effects of NO scavenging and oxidative free radical formation were not blocked adequately using the approaches. The problem of high oxygen affinity was mitigated with the incorporation of SOD or other antioxidants such as ascorbic acid, but did not oxygenate tissues in the range comparable to that of RBCs. It is clear that proteins and polymerized or otherwise modified proteins can be created in nanoscale proportions to biomimic a cellular activity and also that improvements in the systems themselves need to be made in order to properly biomimic RBCs.

22.3.3.2 Liposomal Carriers

The next generation of HbOCs was encapsulated in liposomes and is referred to in the lipid encapsulated hemoglobin (LEH) in North America. In brief, this system entails the encapsulation of the protein of interest, hemoglobin, within the liposome. The liposome is like a micelle, but with a phospholipid bilayer much like a cell. Unlike in gene therapy, where cationic liposomes are used, anionic liposomes are commonly employed for artificial cells because of the increased circulation time (Awasthi et al. 2004). A major drawback for anionic lipids is complement activation (Szebeni et al. 1996); however, Chang and Lister (1990) showed that complement activation varied with manufacturers. To mitigate the problem, a new lipid was fabricated by Sou et al. (2003)—1,5-diplamitoly-L-glutamate-N-succinic acid. Supermicron liposome Hb carriers were initially fabricated by Chang et al. (1969) and submicron lipid vesicles were reported by Djordjevich and Miller. The result was an increase in circulation time compared to supermicron dimensions of the same system (Djordjevich and Miller 1980). Many modifications were made to enhance the circulation time but the most success was obtained through PEGylation by Philips et al. (1999) up to 30 h (Phillips et al. 1999).

A key drawback in LEH systems is the high percent conversion into metHb during the fabrication process. In a common pasteuration process demonstrated by Tsuchida et al., Hb was converted to carbonyl Hb (HbCO) to reduce formation of metHb at 60°C for 10 h. The process, however, also removes and inactivates enzymes and antioxidant essential for proper dissociation of oxygen from Hb (Tsuchida 1994). This also results in reduced oxygenation of tissues in the LEH system due to inability of reducing agents to cross the lipid bilayer.

Another problem for LEH is the necessity for submicron particles, nanoparticles, in order to maximize circulation time and reduce reticuloendothelial clearance (Liu et al 1992)—less than 200 nm but greater than 60 nm. The problem lies in lower encapsulation efficiency with smaller nanoparticles. Using the PEGylated LEH as a platform, PEGylated phosphatidylethanolamines (PEG-PE) were used to produced LEH resulting in increased encapsulation efficiency and circulation time (Awasthi et al. 2004). As expected the majority of the LEH were cleared by the reticuloendothelial system.

To administer the LEH, an emulsifier is needed that may be antigenic or act as haptens. In recent research, human serum albumins (HSA) have been coadministered with LEH. Sakai et al. (2004) used recombinant HSA to coadminister with LEH and demonstrated safe and effective for up to 50% hemodilution in rats. The same complications that may occur due to exposure of proteins not form autologous sources may play a role in second infusions.

The LEH system is fraught with problems that have yet to be addressed totally. Each system produced has sought to address one problem at a time and thus no marketable product performing on par with RBCs has been developed. Many of the problems in the system lie with the fabrication process itself. Through the progression of the research, many points of interest was learned including the ideal size range to reduce uptake and clearance by the reticuloendothelial system as well as insight into the complement activation not seen with the Hb, polymeric Hb, and modified Hb systems. The latter may be due to the lack of trials involving reperfusion and testing for immune system reactions toward the Hb-based products.

22.3.3.3 Polymeric Carriers

Stemming from adding polymeric chains directly to either polymerized or single Hb units, the idea of encapsulating Hb in polymer nanoparticles is at the forefront of research today. The main drives toward the polymer use are due to the disadvantages of the liposomal carriers and the wish to avoid complications of free floating Hb. The basic principle is to optimize a high surface area to volume ratio to expose Hb to plasma conditions, reduce immune system complications, control size of the polymeric nanoparticle to maximize circulation time, and promote the stability of the Hb in active form.

Common polymers employed are poly lactic acid (PLA), poly glycolic acid (PGA), poly caprolactone (PCL), and PEG as well copolymers. The advantages in polymers lie in the ability to control the molecular weight conjugated to the target molecule, control over the chemical properties through the variation of monomers and formation of new monomers, and the ability to avoid antigenic moieties. Polyesters are commonly employed in fabrication of nanoparticles because it has been characterized well, is relatively nontoxic, and degrades into products easily degraded by the body. Some toxicity was seen and was attributed to residue monomers and chemicals involved in the polymerization or other technique (Chang 2007). In many instances, polyesters have been used to deliver both genes (Reul et al. 2009) and drugs (Dailey et al. 2005) throughout the body and can also be used to deliver hemoglobin systemically. PLA was employed by Chang to microencapsulate enzymes in micron scale particles (Chang 1976). Expanding on the need for nanometer dimensions, HbV smaller than 200 nm was fabricated with PLA (Chang and Wong 1972) and PEG-PLA (Chang and Yu 2006). To allow for better functionality of the encapsulated Hb, enzyme systems have been co-encapsulated. Chang (2003) demonstrated that the inclusion of methemoglobin reductase allowed to methemoglobin conversion to Hb in PLA nanoparticles. Other antioxidants have been used as well such as ascorbic acid (Dotsch et al. 1998) and glutathione (Mawatari and Murakami 2004) to reduce methemoglobin conversion. For the greater part, the paradigm is to use thiol group containing compounds to reduce methemoglobin conversion.

22.4 Conclusion

Artificial cells and the utilization of nanomaterials has made great strides in the past decade, and the continued refinement of nanotechnology holds a lot of promise for future developments in drug delivery and tissue regeneration. Carbon nanotubes, metal oxides, liposomes, and polymers all have unique properties of which can be taken advantage to engineer functional artificial cells. The small size of nanoparticles offers an opportunity for scientists and engineers to create artificial cells, as well as matrices for tissue regeneration similar to the biological extracellular matrix (Goldberg et al. 2007).

For the RBC substitute system, there are a variety of branches of nanobiomaterials may be applied and developed to advance research. By modifying outer protein vesicles, "self" moieties may be added to reduce immune responses. The need to develop phospholipids to produce liposomes that do not require emulsifiers and is able to have a low methemoglobin conversion rate as well as being innate to the immune system is apparent. As for polymer science, new polymers or copolymers need to be fabricated and tailored to increase the circulation time, avoid immune response, and allow for greater amount of exposed Hb.

Although nanomaterials and their use for the development of artificial cells has uncovered numerous possibilities for solving present medical issues, a uniform system for assessing the risk to the human body that nanomaterials may pose has not been established. The cytotoxicity of nanomaterials in artificial cells must be evaluated in each stage of their life cycle: development, production, use, and disposal (Tervonen et al. 2009). Most studies have shown that artificial cells composed of polymeric nanomaterials do not have adverse affects on the body and are relatively benign (Kabanov 2006). However, because there is so much that is unknown about the long-term physical, chemical, and biological impact of nanomaterials on biological cells and because there is so much variability in the types of materials and techniques used in the engineering of artificial cells, a classification system based on risk-assessment needs to be established.

References

Arnone A. 1972. X-ray diffraction study of binding of 2,3-diphosphoglycerate to human deoxyhemoglobin. *Nature*. 23:146–149.

Asanuma H., et al. 2007. S-nitrosylated and PEGylated hemoglobin, a newly developed artificial oxygen carrier, exerts cardioprotection against ischemic hearts. *Journal of Molecular Cell Cardiology*. 42:924–930.

Avgoustakis K., Beletsi A., Panagi Z., Klepetsanis P., Livanion E., Evangelatos G., and Ithakissios D.S. 2003. Effect of copolymer composition on the physicochemical characteristics, in vitro stability, and biodistribution of PLGA-mPEG nanoparticles. *International Journal of Pharmaceutics*. 259(1–2):115–127.

Awasthi V. 2005. Pharmaceutical aspects of hemoglobin-based oxygen carriers. *Current Drug Delivery*. 2:133–142.

Awasthi V.D., Garcia D., Klipper R., Phillips W.T., Goins B.A. 2004b. Kinetics of liposome-encapsulated hemoglobin after 25% hypovolemic exchange transfusion. *International Journal of Pharmaceutics*. 283:53–62.

Awasthi V.D., Garcia D., Goins B.A., and Phillips W.T. 2003. Circulation and biodistribution profiles of long-circulating PEG-liposomes of various sizes in rabbits. *International Journal of Pharmceutics*. 253(1–2):121–132.

Baudin-Creuza V., Chauvierre C., Domingues E., Kiger L., Leclerc L., Vasseur C., Celier C., and Marden M.C. 2008. Octamers and nanoparticles as hemoglobin based blood substitutes. *Biochimica et Biophysica Acta—Proteins and Proteomics*. 1784(10):1448–1453.

Brunn H.F. 1972. Erythrocyte destruction and hemoglobin catabolism. *Seminar Hematology*. 9:3–18.

Bucci E. 2009. Thermodynamic approach to oxygen delivery in vivo by natural and artificial oxygen carriers. *Biophysical Chemistry*. 142(1–3):1–6.

Buehler P.W., Boykins R.A., Jia Y., Norris S., Freedberg D.I., Alaysh A.I. 2005. Structural and functional characterization of glutaraldehyde-polymerized bovine hemoglobin and its isolated fractions. *Analytical Chemistry*. 77:3466–3478.

Bunn H.F. 1972. Erythrocyte destruction and hemoglobin catabolism. *Seminar in Hematology*. 2:3–17.

Bunn H.F. and Jandl J.H. 1969. The renal handling of hemoglobin II catabolism. *Journal of Experimental Medicine*. 129:925–934.

Chandrawati R., Städler B., and Postma A. 2009. Cholesterol-mediated anchoring of enzyme-loaded liposomes within disulfide-stabilized polymer carrier capsules. *Biomaterials*. 30(30):5988–5998.

Chang T.M.S. 1969. Lipid-coated spherical ultrathin membranes of polymer or corss-linked protein as possible cell membrane models. *Federal Proceedings*. 28:461.

Chang T.M.S. 1971. Stabilization of enzyme by microencapsulation with a concentrated protein solution or by crosslinking with glutaraldehyde. *Biochemical and Biophysical Research Communication*. 44:1531–1533.

Chang T.M.S. 1976. Enzymes immobilized by microencapsulation within spherical ultra-thin polymeric membranes. *Journal of Macromolecular Science and Chemistry*. A10:245–258.

Chang T.M.S. 2003. New generations of red blood cell substitutes. *Journal of Internal Medicine*. 253:527–535.

Chang T.M.S. 2007. *Artificial Cells: Biotechnology, Nanomedicine, Regenerative Medicine, Blood Substitutes, Bioencapsulation, and Cell/Stem Cell Therapy*. World Scientific, Singapore, pp. 112, 117–118.

Chang T.M.S. and Lister C. 1990. A screening test of modified hemoglobin blood substitute before clinical use in patients based on complement activation of human plasma. *Journal of Biomaterials, Artificial Cells, and Artificial Organs*. 18:693–702.

Chang T.M.S. and Wong H. 1992. Biodegradable polymer membrane containing hemoglobin as potential blood substitutes. British Provisional Patent No. 9219426.5.

Chang T.M.S. and Yu W.P. 2006. Polymeric biodegradable hemoglobin nanocapsule as a new red blood cell substitute. In: Tsuchida E. (ed.), *Present and Future Perspectives of Blood Substitutes*. Elsevier Science, Lausanne, pp. 161–169.

Chatterjee R., Wetty E.V., and Walder R.Y. 1986. Isolation and characterization of a new hemoglobin derivative cross-linked between the α chains (lysine 99 $\alpha_1 \rightarrow$ lysine 99 α_2). *Journal of Biological Chemistry*. 261:9929–9937.

Chauvierre C., Marden M.C., Vauthiere C., Labarre D., Cauvreur P., Leclerc L. 2007. Heparin coated poly(alkylcyanoacrylate) nanoparticles coupled to hemoglobin: A new oxygen carrier. *Biomaterials*. 25:3081–3086.

Chen J.S., Ralf K., Thomas G., Helmuth M., and Rumen K. 2009. Assymetic lipid bilayer sandwiched in polyelectrolyte multilayer film through layer-by-layer assembly.

Dailey L.A., Wittmar M., and Kissel T. 2005. The role of branched polyesters in their modifications in the development in modern drug delivery vehicles. *Journal of Controlled Release*. 101:137–149.

Davy R.J. 2004. Recruiting blood donors: Challenges and opportunities. *Transfusion*. 44:597–600.

Djordjevich L. and Miller I.F. 1980. Synthetic erythrocytes from lipid encapsulated hemoglobin. *Experimental Hematology*. 8:584.

Dotsch J., Demirakca S., Cryer A., Hanze J., Kuhl P.G., Pascher W. 1998. Reduction of NO-induced methemoglobin requires extremely high doses of ascorbic acid in vitro. *Intensive Care Medicine*. 24(6):612–615.

Dubick M.A., Sondeen J.L., Prince M.D., James A.G., Nelson J.J., Hernandez E.L. 2004. Hypotensive resuscitation with hextend, hespan, or polyheme in a swine hemorrhage model. *Shock*. 21(Suppl 2):42.

Eike J.H. 2005. High oxygen affinity polymerized bovine hemoglobin-based oxygen carriers as potential artificial blood substitutes. PhD Dissertation. University of Notre Dame.

Eike J.H. and Palmer A.F. 2006. Oxidized mono-, di-, tri-, and polysaccharides as potential hemoglobin crosslinking reagents for the synthesis of high oxygen affinity artificial blood substitutes. *Biotechnology Progress*. 22:1025–1049.

Franticelli C, Bucci E, and Orth C. 1984. Solvent regulation of oxygen affinity in hemoglobin. *The Journal of Biological Chemistry*. 259(17):10841–10844.

Fronticelli C. and Koehler R.C. 2007. Recombinant hemoglobins as artificial oxygen carriers. *Artificial Cells, Blood Substitutes, and Biotechnology* 35:45–52.

Furukawa K., Sumitomo K., Nakashima H., Kashimura Y., and Torimitsu K. 2007. Supported lipid bilayer self-spreading on a nanostructured silicon surface. *Langmuir*. 23(2):367–371.

Garby L. and Noyes W.D. 1959. Studies on hemoglobin metabolism: 1. Kinetic properties of plasma hemoglobin pool in normal man. *Journal of Clinical Investigation*. 38:1479–1483.

Gbadamosi J.K., Hunter A.C., and Moghimi S.M. 2002. PEGylation of microspheres generates a heterogeneous population of particles with differential surface characteristics and biological performance. *FEBS Letters*. 532:338–344.

George P.A., Donose B.C., and Cooper-White J.J. 2009. Self-assembling polystyrene-block-poly(ethylene oxide) copolymer curface coatings: Resistance to protein and cell adhesion. *Biomaterials*. 30(13):2449–2456.

Goldberg M., Langer R., and Jia X. 2007. Nanostructured materials for applications in drug delivery and tissue engineering. *Journal of Biomaterials Science, Polymer Edition*. 18(3):241–268.

Gould S.A and Moss G.S. 1996. Clinical development of human polymerized hemoglin as a blood substitute. *World Journal of Surgery*. 20:1200–1207.

Gould S.A., Sehgal L.R., Sehgal H.L., Moss G.S. 1995. The development of hemoglobin solutions as red cell substitutes: Hemoglobin solutions. *Transfusion Science*. 16:5–17.

Greenberg A.G., Hayashi R., Siefer I. 1979. Intravascular persistence and oxygen delivery of pyroxyylated stroma-free hemoglobin during gradations of hypotension. *Surgery*. 86:13–16.

Gref R., Domb A., Quellec P., Blunk T., Muller R.H., Verbavotz J.M., and Langer R. 1995. The controlled intravenous delivery of drugs using PEG-coated sterically stabilized nanospheres. *Advanced Drug Delivery Reviews*. 16:215–223.

Hersheko C. 1975. The fate of circulating hemoglobin metabolism I. The kinetic properties of the plasma hemoglobin pool in normal man. *British Journal of Heametology*. 29:199.

Higashi N. and Koga T. 2008. Self-assembled peptide nanofibers. *Advances in Polymer Science*. 219(1):27–68.

Hill S.E. 2001. Oxygen therapeutics—Current concepts. *Canadian Journal of Anesthesiology*. 48(4): S32–S40.
Iafelice R., Cristoni S., Caccia D., Russo R., Bernardi L., Lowe K.C., and Perrella M. 2007. Free in PMC indentification of the sites of deoxyhaemoglobin PEGylation. *Journal of Biochemistry*. 403:189–196.
Illum L. and Davis S.S. 1986. The effect of hydrophilic coatings on the uptake of colloidal particles with mouse peritoneal macrophages. *International Journal of Pharmaceutics*. 29(1):53–65.
Isenberg B.C., Tsuda Y., Williams C., Shimizu T., Yamato M., Okano T. and Wong J.Y. 2008. A thermoresponsive, microtextured substrate for cell sheet engineering with defined structural organization. *Biomaterials*. 29(17), 2565–2572.
Ishihara K., Nishizawa K., Goto Y., and Takai M. 2008. Bioinspired polymer surfaces for nanodevices and nanomedicine. In: *CIMTEC 2008—Proceedings of the 3rd International Conference on Smart Materials, Structures and Systems—Biomedical Applications of Smart Materials, Nanotechnology and Micro/Nano Engineering*, Acireale, Vol. 57, pp. 5–14.
Ishihara K., Nomura H., Mihara T., Kurita K., Iwasaki Y., and Nakabayashi N. 1998. Why do phospholipid polymers reduce protein adsorption? *Journal of Biomedical Materials Research*. 39(2):323–330.
Iwashita Y., Yabuki A., and Yamaji K. 1988. A new resuscitation fluid "stabilized Hb" preparation and characteristics. *Biomaterials, Artificial Cell, and Artificial Organs*. 16:271–280.
Jahr J.S., Walker V., and Manoochehri K. 2007. Blood substitutes as pharmacotherapies in clinical practice. *Current Opinion in Anaesthesiology*. 20:325–330.
Kabanov A.V. 2006. Polymer genomics: An insight into pharmacology and toxicology of nanomedicines. *Advanced Drug Delivery Reviews*. 58(15):1597–1621.
Kavdia M., Pittman R.N., and Popel A.S. 2002. Theoretical analysis of effects of blood substitute affinity and cooperativity on organ oxygen transport. *Journal of Applied Physiology*. 93(6):2122–2128.
Kissel T. and Roser M. 1991. Influence of chemical surface-modifications on the phagocytic properties of albumin nanoparticles. *Proceedings of International Symposium on Controlled Release of Bioactive Materials*, San Diego, 18:275–276.
Kobayashi K., Tsuchida E., and Horinouchi H. 2004. *Artificial Oxygen Carrier: Its Frontline*. Tokyo, Japan: Springer-Verlag.
Lamy M.L., Daily E.K., Brichant J.F., Larbuisson R.P., Demeyere R.J. 2000. Randomized trial diaspirin cross-linked hemoglobin solution as an alternative to blood transfusion after cardiac surgery. *Anesthesiology*. 92:646–656.
Lane T.A. 1995. Perflourochemical-based artificial oxygen carrying red blood cell substitutes. *Transfusion Science*. 16(1):19–31.
Li Y.P., Pei Y.Y., Zhang X.Y., Gu Z.H., Zhou Z.H., Yuan W.F., Zhou J.J., Zhu J.H., and Gao X.J. 2001. PEGylated PLGA nanoparticles as protein carriers: Synthesis, preparation and biodistribution in rats. *Journal of Controlled Release*. 71:203–211.
Liang X.-J., Chen C., Zhao Y., Jia L., and Wang P.C. 2008. Biopharmaceutics and therapeutic potential of engineered nanomaterials. *Current Drug Metabolism*. 9(8):697–709.
Lim S.Y., Kim K.-O., Kim D.-M., and Park C.B. 2009. Silica-coated alginate beads for in vitro protein synthesis via transcription/translation machinery encapsulation. *Journal of Biotechnology*. 143(3):183–189.
Lin C., Ke Y., Liu Y., Mertig M., Gu J., and Yan H. 2007. Functional DNA nnotube arays: Bottom-up meets top-down. *Angewandte Chemie International Edition*. 46(32), 6089–6092.
Litzinger D.C., Buiting A.M.J., and Rooijen N.V., Huang L. 1994. Effect of liposome size on the circulation time and intraorgan distribution of amphipathic poly(ethylene glycol)-containing liposomes. *Biochimica et Biophysica Acta*. 1190:99–107.
Liu Z., Chen K., Davis C., Sherlock S., Cao Q.Z., Chen X.Y., and Dai H.J. 2008. Drug delivery with carbon nanotubes for in vivo cancer treatment. *Cancer Research*. 68(16):6652–6660.
Liu D., Mori A., and Huang L. 1992. Role of liposome size and RES blockade in controlling biodistribution and tumour uptake of GM1-containing liposomes. *Biochemica et Biophysica Acta*. 1104:95–101.
Mawatari S. and Murakami K. 2004. Different types of glutathionlyation of hemoglobin can exist in intact erythrocytes. *Archives in Biochemisty and Biophysics*. 421(1):108–114.

Moghimi S.M., Hedeman H., Muir I.S., and Davis S.S. 1993. An investigation of the filtration capacity and the fate of large filtered sterically-stabilized microspheres in rat spleen. *Biochimica et Biophysica Acta*. 1157:233–240.

Mullon J., Giacoppe G., Clagett C., McCune D., and Dillard T. 2000. Transfusions of polymerized bovine hemoglobin in a patient with severe autoimmune hemolytic anemia. *New England Journal of Medicine*. 342(22):1638–1643.

Numata M., Sugikawa K., Kaneko K., and Shinkai S. 2008. Creation of hierarchical carbon nanotube assemblies through alternative packing of complementary semi-artificial beta-1,3-glucan/carbon nanotube composites. *Chemistry (Weinheim an der Bergstrasse, Germany)*. 14(8):2398–2404.

Olson J.S., Foley E.W., Rogge C., Tsai A., Doyle M.P., and Lemon D.D. 2004. NO scavenging and the hypertensive effect of hemoglobin-based blood substitutes. *Free Radical Biology and Medicine*. 36(6):685–697.

Phillips W.T. Klipper R.W., Awasthi V.D., Rudolph A.S., Cliff R., and Goins B.A. 1999. Polyethylene glycomodified liposome encapsulated hemoglobin: A long circulating red blood cell substitute. *Journal of Pharmacology and Experimental Therapeutics*. 288:665–670.

Reul R., Nguyen J., and Kissel T. 2009. Amine-modified hyperbranched polyesters as non-toxic, biodegradable, gene delivery systems. *Biomaterials*. 30:5815–5824.

Riess J.G. 2006. Perflourocarbon-based oxygen delivery. *Artificial Cells, Blood Substitutes, and Biotechnology*. 34:567–580.

Riley W., Schwei M., and McCullough J. 2007. The United States' potential blood donor pool: Estimating the prevalence of donor-exclusion factors on the pool of potential donors. *Transfusion*. 47:1180–1188.

Romberg B., Oussoren C., Snel C.J., Hennink W.E., and Storm G. 2007. Effect of liposome characteristics and dose on the pharmacokinetics of liposomes coated with poly(amino acid)s. *Pharmaceutical Research*. 24(12):2394–2401.

Sakai H., Horinouchi H., and Masada Y. 2004a. Metabolism of hemoglobin-vesicles (artificial oxygen carriers) and their influence on organ functions in a rat model. *Biomaterials*. 25:4317–4325.

Sanders K.E., Ackers G., and Sligar S. 1996. Engineering and design of blood substitutes. *Current Opinion in Structural Biology*. 6:534–540.

Sarkar S. 2008. Artificial blood. *Indian Journal of Critical Care Medicine*. 12(3):140–144.

Schultz S.C., Grady B., Cole F., Hamilton I., Barhop K., and Malcolm D.S. 1993. A role for endothelin and nitric oxide in the pressor response to diaspirin cross-linked hemoglobin. *The Journal of Laboratory and Clinical Medicine*. 122(3):301–308.

Seghal L.R., Rosen A.L., Noud G., Sehgal H.L., Gould S.A., Dewoskin R., Rice C.L., and Moss G.S. 1981. Large volume preparation of pyroxylated hemoglobin with high P_{50}. *Journal of Surgical Research*. 30:14–20.

Shi Q., Huang Y., Chen X., Wu M., Sun J., and Jing X. 2009. Hemoglobin conjugated micelles based on triblock biodegradable polymers as artificial oxygen carriers. *Biomaterials*. 30:5077–5085.

Shirasawa T. et al. 2003. Oxygen affinity of hemoglobin regulates O_2 consumption, metabolism, and physical activity. *Journal of Biological Chemistry*. 278(7):5035–5043.

Shorr R.G., Viau A.T., and Abuchowski A. 1996. Phase 1B safety evaluation of PEG Hb as adjuvant to radiation therapy in human cancer patients. *Artificial Cells, Blood Substitutes, and Immobilization Technology*. 24:407.

Sloan E.P., Koenigsberg M., Gens D., Cipolle M., Runge J., Mallory M.N., and Rodman G. 1999. Diaspirin cross-linked hemoglobin (DCLHb) in the treatment of severe traumatic hemorrhagic shock: A randomized controlled efficacy trial. *Journal of American Medical Association*. 282(19):1857–1884.

Sou K., Naito Y., Endo T. Takeoka S., and Tsuchida E. 2003. Effective encapsulation of proteins into size-controlled phospholipids vescicles using freeze-thawing and extrusion. *Biotechnology Progress*. 19:1547–1552.

Spies B.D. 2004. Risks of transfusion: Outcome focus. *Transfusion*. 44:4S–14S.

Städler B., Chandrawati R., Goldie K., and Caruso F. 2009. Capsosomes: Subcompartmentalizing polyelectrolyte capsules using liposomes. *Langmuir* 25(12):6725–6732.

Stefan D.C., Uys R., and Wessels G. 2007. Hemopure transfusion in a child with severe anemia. *Pediatric Hematology and Oncology*. 24:269–273.

Stolnik S., Illum L., Davis S.S. 1995. Long circulating microparticulate drug carriers. *Advanced Drug Delivery Reviews*. 16:195–214.

Stonwell C.P., Levin J., Speiss B.D., and Winslow R.M. 2001. Progress in the development of RBC substitutes. *Transfusion*. 41:287–299.

Sun C., Lee J.S.H., and Zhang M. 2008. Magnetic nanoparticles in MR imaging and drug delivery. *Advanced Drug Delivery Reviews*. 60(11):1252–1265.

Szebeni J., Wassef N.M., Rudolf A.S., and Alving C.R. 1996. Complement activation in human serum by liposome-encapsulated hemoglobin: The role of natural anti-phospholipid antibodies. *Biochemica et Biophysica Acta*. 1285:127–130.

Tam S.C., Blumsenstein J., and Wong J.T. 1976. Soluble dextran-hemoglobin complex as a potential blood substitute. *PNAS*. 73(6):2128–2131.

Tervonen T., Linkov I., Figueira J.R., Steevens J., Chappell M., and Merad M. 2009. Risk-based classification system of nanomaterials. *Journal of Nanoparticle Research*. 11(4):757–766.

Tremper K.K. 1983. Rationale for the clinical studies of anemia treated with the perflourochemical emulsion: Flousol DA 20%. In: *Advances in Blood Substitute Research*, eds. Bolin R.B., Geyer R.P., Nemo G.J., New York, Alan R. Liss Inc.

Tsuchida E. 1994. Stabilized hemoglobin vesicles. *Artificial Cells, Blood Substitutes, and Immobilization Biotechnology*. 22:467–479.

Vasita R. and Katti D.S. 2006. Nanofibers and their applications in tissue engineering. *International Journal of Nanomedicine*. 1(1):15–30.

Walder J.A., Zaugg R.H., Walder R.Y., Steele J.M., and Klotz I.M. 1979. Diaspirins that cross-link alpha chains of hemoglobin: Bis(3,5-dibromosalycil) succinate and bis(3,5-dibromosalicyl) fumarate. *Biochemistry*. 18:4265–4270.

Watanable J., Ooya T., Nita K.H., Park K.D., Kim Y.H., and Yui N. 2004. Fibroblast adhesion and proliferation on poly(ethylene glycol)-polycarbonate substrates: A mechanism for cell guided ligand remodeling. *Journal of Biomedical Research A*. 69A:114–123.

Webb M., Harasym T.O., Masin D., Bally M.B., and Mayer L.D. 1995. Sphingomyelin-cholesterol liposomes significantly enhance the pharmacokinetic and therapeutic properties of vincristine in murine and human tumour models. *British Journal of Cancer*. 72(4), 896–904.

Wedekind D., Schweiter-Stenner R., and Dreybrody W. 1985. Heme-aproprotein interaction in the modified oxyhemoglobin-bis(N-maleimidomethyl)ether and in oxyhemoglobin at high Cl-concentration detected by resonance Raman scattering. *Biochemica et Biophysica Acta*. 830(3):224–232.

Wilson W.C., Grande C.M., and Hoyt D.B. 2007. *Trauma: Critical Care*. New York, CRC Press.

Winslow R.M., Swenberg M., Berger R. 1977. Oxygen equilibrium curve of normal human blood and its evaluation by Adair's equation. *The Journal of Biological Chemistry*. 252(7):2331–2337.

Wong J.T. 1988. Rightshifted dextran Hb as blood substitute. *Biomaterials, Artificial Cells, and Artificial Organs*. 16:2370245.

Yamaguchi M., Fujihara M., and Wakamotot S. 2009. Influence of hemoglobin vesicles, cellular-type artificial oxygen carriers, on human umbilical cord blood hematopoietic progenitor cells in vitro. *Journal of Biomedical Research A*. 88A:34–42.

Yamaoka T., Kuroda M., Tabata Y., and Ikada Y. 1995. Body distribution of dextran derivatives with electric charges after intravenous administration. *International Journal of Pharmaceutics*. 113:149–157.

Yi J., Schmidt J., Chien A., and Montemagno C.D. 2009. Engineering an artificial amoeba propelled by nanoparticle-triggered actin polymerization. *Nanotechnology*. 20(8):085101.

Zhang S. 2003. Fabrication of novel biomaterials through molecular self-assembly. *Nature Biotechnology*. 21(10):1171–1178.

Zhao J., Liu C.S., Yuan Y., Tao X.Y., Shan X.Q., Sheng Y., and Wu F. 2007a. Preparation of hemoglobin-loaded nano-sized particles with porous structure as oxygen carriers. *Biomaterials*. 28:1414–1422.

23

Nanobiomaterials for Musculoskeletal Tissue Engineering

Kyobum Kim*
Rice University

Minal Patel*
University at Buffalo

John P. Fisher
University of Maryland

23.1 Introduction

Nanoscale biomaterials have been intensively investigated for numerous tissue engineering technologies including targeted drug delivery, imaging agent for magnetic resonance imaging, microfluidics in microelectromechanical systems, and composite implants biomaterials. The combination of nanotechnology and tissue-engineered biomaterials will likely provide insight for the development of novel hybrid materials in biological systems. Particularly, various properties of musculoskeletal tissue-engineered materials such as mechanical properties and osteoconductivity can be significantly improved through the recent nanotechnology and finally can be utilized for tissue regenerative strategy. Due to the unique characteristics of nanosized materials, including high chemical/physical reactivity, high surface-to-volume ratio, and feasibility for conjugated materials, current conventional biomaterials can be improved into "nanobiomaterials" by nanoscale fabrication techniques. With the extremely small-length scale (approximately 100 nm), nanosized materials possess promising properties that can be developed to closely mimic tissues such as bone, cartilage, muscle, and vasculature, since these tissues themselves consist of a sophisticated complex of nanosized molecules. Moreover, the process of tissue regeneration is also related to a variety of cellular

* These authors contributed equally to this chapter.

and molecular events that involve nanosized molecules. Therefore, in this chapter, we will focus on recent developments in nanobiomaterial applications for musculoskeletal tissue engineering and their advantages in tissue regeneration to overcome current limitations as well as to ensure further clinical efficacy.

23.2 Nanobiomaterials for Bone Tissue Engineering

Scaffolds of bone tissue-engineered materials are required to contain suitable mechanical strength to support skeletal frame, biocompatibility to be integrated into surrounding native tissues, and osteoconductivity to recruit osteoprogenitor cells into the porous architecture. Nanosized materials can be excellent candidates to improve a series of requirement of tissue-engineered scaffolds. With the aid of nanoscale materials, it may be possible that bone tissue substrate implants can be easily integrated with surrounding tissues and support the bone regeneration since (1) a stronger interaction between the nanosized molecules and the implant material may improve the mechanical strength and (2) increased surface roughness may promote osteoblast adhesion and subsequent cellular functions.

23.2.1 Bioceramic/Polymer Composite Materials

Calcium phosphatase bioceramics, such as hydroxyapatite (HA) and tricalcium phosphate, are promising materials for bone tissue engineering since these ceramics reflect mineral composition of native bone tissue (chemical aspect) and similar size of nanoscale of bone substances (physical and structural aspect). Despite limitations including brittleness, difficulty of processing, and slow degradation rate, fabrication of composite material with nanoscale bioceramic particles and degradable polymer may closely mimic complex bone tissue integration, improve the mechanical properties of scaffold, and enhance osteoblast proliferation and osteodifferentiation.

Polyesters combined with calcium salts may overcome the inflammatory reactions caused by acidic degradation products of polyester (Liao et al. 2005). In triple-layered composite membrane with nanocarbonated HA, collagen, and poly (D,L-lactic-*co*-glycolic acid) (PLGA), the addition of HA with nanocrystal size improved flexibility and mechanical strength. Its tensile strength was above the tensile properties of cancellous bone. Specifically, enhanced toughness will ensure the integrity of implants as practical deformation during the surgical procedures. Moreover, osteoblastic MC3T3-E1 cells cultured on the three-layered composite membrane showed higher cellularity during 7 days of culture period compared with pure PLGA membrane. Other studies investigating a nano-HA/PLGA composite scaffold also demonstrated its improved mechanical properties and stimulated osteogenic responses (Kim et al. 2006, 2007, 2008). An initial study to investigate the advantage of fabrication with gas foaming and porogen leaching demonstrated that porous composite scaffolds exhibited superior mechanical strength and higher exposure of HA particle at the surface than those by solvent casting (Kim et al. 2006). In vivo studies implanting these composite materials into the subcutaneous space of athymic mice and in critical size defects in rat skulls also showed higher extents of bone regeneration than scaffolds by solvent casting (Kim et al. 2006, 2007).

Another study investigated nano-HA/polyamide (PA) composite scaffolds synthesized by thermally induced phase inversion fabrication technique (Wang et al. 2007). Interactions were noted between Ca^{2+} and PO_4^{3-} charged groups of HA and –C=O and –NH– groups in PA, and this type of interactions may be found between the components in natural bone. This hybrid biomimetic resulted in a biocompatible environment for cell proliferation and osteoblastic differentiation of neonatal rabbit bone marrow cells for 7 days. In addition, after 12 week implantation of composite scaffold in rabbit model, histological and microradiographic data confirmed extensive osteoconductivity, allowing direct bonding between newly formed bone and materials. In terms of mechanical strength, this composite material showed a compressive strength of 13.2 MPa, similar to the upper value of the strength of porous trabecular bone. HA has also been combined with ultrahigh molecular weight polyethylene (UHMWPE) (Fang et al. 2005, 2006). HA/UHMWPE composite showed 90% increase in Young's modulus and a 50% increase in the yield strength, compared to UHMWPE scaffolds without HA incorporation.

Nanosized HA with rodlike morphology (20–30 nm in width and 50–80 nm in length) was also integrated with poly-2-hydroxyethylmethacrylate/polycaprolactone (PCL) matrix (Huang et al. 2007). Composites of 20–40 wt% of HA addition indicated higher human osteoblastic cell proliferation over 7 days of in vitro culture in comparison with standard tissue culture plastic control. This composite material also supported actin cytoskeletal organization observed by immunofluorescent staining. Nano-HA particles embedded in the polymer matrix also provided a favorable surface for human osteoblast-like cell attachment.

Interaction between HA nanoparticles and polysaccharides such as chitosan and polygalacturonic acid has also been investigated (Verma et al. 2008). Fourier transform infrared spectroscopy (FTIR) spectrum of chitosan/polygalacturonic acid/HA composite revealed that polygalacturonic acid attached to HA surface through dissociated carboxylate groups while chitosan interacted with HA through NH groups. Phosphate groups of HA are also involved in the interfacial interactions based on the changes in phosphate stretching and bending regions of FTIR spectrum. HA interaction with copolymer of poly(ethylene glycol) (PEG) and poly(butylene terephthalate) (PBT) has also been examined (Liu et al. 1998). Organic isocyanate groups can react readily with surface hydroxyl groups of HA and direct chemical bonding between HA particles and the surrounding polymer significantly improve mechanical properties (Young's modulus, tensile strength, and elongation at a break) of the composites.

Among various physical properties of composite materials, surface wettability, and hydrophilicity of HA/poly(propylene fumarate) (PPF) composite surface are more relevant to induce more cell attachment and proliferation rather than surface topology or roughness (Lee et al. 2008). Addition of 10, 20, and 30 wt% of HA nanoparticles with PPF resulted in significant differences in hydrophilicity of the substrate surface, protein absorption, and 2D proliferation of MC3T3-E1 mouse pre-osteoblast cells for a week. However, no changes were noted in surface roughness. Therefore, increased hydrophilicity of composite by adding more HA particles may mainly affect to initial cell attachment patterns and subsequent proliferation. An in vivo study of rat tibial defect models with macroporous HA/PPF composite scaffolds demonstrated that incorporation of nanosized HA particles with PPF resulted more reactive new bone formation than scaffold with micro-sized HA particles and blank PPF scaffold after 3 weeks of implantation (Lewandrowski et al. 2003).

Furthermore, incorporation of HA nanoparticles into poly(L-lactic acid) (PLLA) showed increased compressive modulus as well as protein adsorption (Wei and Ma 2004). The conjugation of HA and PLLA altered the pore surface morphology of the composite scaffolds to be more suitable for protein adsorption. This protein adsorption depended upon the ratio of HA and polymer. Both the bone-binding ability of HA and the similar size scale of HA in natural bone may enable this composite scaffold to serve as a mechanically improved three-dimensional substrate for cell attachment and migration.

HA incorporation can be also applied to the fabrication of microsphere for load-bearing tissue engineering materials (Nukavarapu et al. 2008). Biodegradable polyphosphazene substituted with phenylalanine amino acid was combined with 20 wt% of HA nanoparticles with 100 nm size. The composite microspheres had compressive moduli of 46–81 MPa, and cultured primary rat osteoblast cells onto microsphere scaffolds showed comparable ALP activity after 21 days to TCPS plate control substrates. Another study investigating collagen microbeads with HA nanoparticles also showed (Jeng et al. 2008) that rabbit bone marrow cells attach to microbeads via focal adhesion and normalized alkaline phosphatase activity was optimized in 65:35 wt ratio of HA particles and collagen fibers (Table 23.1).

23.2.2 Metal Nanoparticles/Polymer Composite Materials

Metal nanoparticles have also been investigated as a mean to (1) improve the mechanical strength of composites, (2) increase the osteoblast adhesion, and (3) augment long-term phenotypic responses of osteoblasts in bone regeneration. Several researches described the improvement in these properties of composite materials of metal nanoparticles and polymeric matrices.

TABLE 23.1 Bioceramic Nanoparticle/Polymer Composite Scaffolds

Materials	Function and Biological Improvements	References
Nanocarbonated HA/collagen/PLGA	Enhanced tensile strength and toughness, higher proliferation of MC3T3-E1 cells	Liao et al. (2005)
HA/PLGA by gas-forming	Superior mechanical strength, stimulated cell proliferation and osteogenic differentiation in vitro, and extended bone regeneration in vivo than that by solvent casting	Kim et al. (2006, 2007)
HA/PA	Direct bonding between newly formed bone and implanted material in vivo, upper value of compressive strength of cancellular bone	Wang et al. (2007)
HA/UHMWPE	Increased Young's modulus and yield strength than control UHMWPE	Fang et al. (2005, 2006)
HA/PHEMA/PCL	Higher human osteoblast proliferation in 20% and 40% of HA in the polymer matrix	Huang et al. (2007)
HA/PEG-PBT copolymer	Increased Young's modulus, tensile strength, and elongation at break of composite scaffold	Liu et al. (1998)
HA/PPF	Stimulated protein absorption and MC3t3-E1 cell proliferation onto 2D composite disks, enhanced new bone formation in rat model	Lewandrowski et al. (2003); Lee et al. (2008)
HA/PLLA	Increased compressive modulus and protein adsorption	Wei and Ma (2004)
HA/polephosphazene	Microsphere fabrication	Nukavarapu et al. (2008)
HA/collagen	Microbeads formation	Jeng et al. (2008)

Nanoscale alteration of biomaterial surface may depend on the following mechanism to increase osteoblast adhesion (Sato and Webster 2004; Khang et al. 2007; Dulgar-Tulloch et al. 2008). Changes in surface properties of the implants such as topography, roughness, and chemistry may increase the surface energy and protein adsorption including vitronectin and fibronectin. Adsorbed proteins can be recognized by the receptor proteins in extracellular matrix (ECM) of the osteoblasts, and this recognition (and adhesion) mechanism may be enhanced in nanophase surface (Webster et al. 2000a,b). Stimulated interactions between cell/tissue–implant interfaces subsequently promote osseointegration. Osteoblast adhesion onto the substrate is the first step for subsequent cellular responses including proliferation, differentiation, and long-term phenotypic changes of collagen synthesis and mineral depositions (Webster et al. 1999, 2000). Therefore, titania nanoparticles have also been investigated as a filler material to increase osteoblast adhesion and osteoconductivity of composite scaffolds. Composite scaffolds of poly(methyl methacrylate) (PMMA) and titania nanoparticles showed superior osteoconductivity and compressive strength (Goto et al. 2005). Likewise, composites of PLGA and metal nanoparticles stimulated long-term osteoblast functions including collagen synthesis, alkaline phosphatase (ALP) activity, and mineral deposition (Webster and Smith 2005; Liu et al. 2006a). Another study investigating PLGA/titania nanoparticles showed reduction of acidic degradation products formation (Liu et al. 2006b). This research demonstrated that when titania nanoparticles were dispersed in PLGA, harmful change in pH during the PLGA degradation process was decreased.

Alumoxane nanoparticles dispersed within PPF-based polymer have been researched (Horch et al. 2004; Mistry et al. 2007, 2009a,b). The alumoxane particles with reactive double bonds (activating groups) and separate ligands (surfactant groups) were dispersed in PPF/PPF diacylate and resulted in a threefold increase in flexural modulus of the composites compared to the control polymer (Horch et al. 2004). Surfactant groups were involved in increased dispersion of the particles while activating groups created additional covalent bonding so as to enhance the mechanical properties. This composite material also showed biocompatibility both in vitro and in vivo (Mistry et al. 2007, 2009b). Studies including an in vitro cytotoxicity of fibroblast cell line, an in vivo soft-tissue response after 12 weeks of implantation in adult goats, and another in vivo study investigating the lateral femoral condyle of adult goats for 12 weeks indicated that the incorporation of alumoxane nanoparticle into the PPF (or PPF derivatives) may not significantly affect its biocompatibility when compared to the control materials.

Widely used as a bone cement material, PMMA can be composited with MgO and $BaSO_4$ nanoparticles to reduce the limitations such as exothermic in situ polymerization, less osteoblast attachment than other bone cements, and invisibility for radiology (Ricker et al. 2008). Composites with nanosized MgO and $BaSO_4$ particles reduced exothermic reaction compared to composites with conventional micron-size particles. In addition, nanocomposites increased osteoblast adhesion on PMMA compared to microcomposites.

In addition to the metal nanoparticles, carbon nanotubes (CNTs) have also been investigated as filler material for the fabrication of novel composite scaffolds. Higher reactivity of CNTs through a high density of reactive sites resulted in increased surface roughness, surface area, and surface energy with increasing amounts of CNTs in a polymer matrix (Streicher et al. 2007). Therefore, enhanced surface characteristics of CNT/polymer composites as well as the feasibility of fictionalization may induce the osteoblast adhesion and ensure the possibility of orthopedic applications. Single-walled carbon nanotube (SWNT) is the one of the candidates in CNT composite fabrication. A study about SWNT/PPF composites demonstrated that functionalized SWNT at the side walls in cross-linked PPF resulted in (1) higher monodispersity of the particles in the PPF matrix than unfunctionalized SWNT composites and (2) increased mechanical properties including compressive/flexural modulus, compressive yield strength, and flexural strength than pure PPF networks (Shi et al. 2006). Enhanced mechanical strength is mainly due to the formation of SWNT-PPF cross-linking as found in extrachemical bonding between ceramic nanoparticle and polymer matrix. Moreover, another study also showed that mechanical properties of the cross-linked nanocomposites are dependent on the surface area of carbon nanotubes (Sitharaman et al. 2007). Moreover, a study about electrical conductivity (from 3 to 7 mS cm^{-1} depending on SWNT loading density) of SWNT/collagen hydrogel showed the possible use of substrate material to study electrical stimulation in tissue engineering application (MacDonald et al. 2008). In addition, carbon nanotubes are not only the promising filler materials to increase the mechanical properties of composites but also a 3D framework for osteoblast proliferation and mineral nucleation (Zanello et al. 2006; Giannona et al. 2007).

Therefore, incorporation of metal nanoparticles into polymeric materials can be utilized to fabricate the composite scaffolds with improved properties: (1) modified surface chemistry/topology may enhance initial cell adhesion and subsequent osteogenic differentiation of implanted or hosted cell population and (2) higher mechanical strength may promise its application as a substitute material into load-bearing site implantation sites (Table 23.2).

23.2.3 Coating Nanomaterials on Metal Surfaces

There are also a variety of approaches to coat the surface of metal implants. Coating the surface of conventional implants with nanoparticles may also improve mechanical strength for load-bearing bone tissue substitutes and increase bone cell responses at the interfaces between the implants and surrounding tissues. Since the osseointegration rate of titanium implants is related to their composition and surface roughness, increased roughness of implants favors both bone anchoring and biomechanical stability (Le Guehennec et al. 2007).

Due to its superior mechanical properties, diamond is being currently used to coat orthopedic implant materials (Yang et al. 2008). Various surface properties of diamond coating on silicon including topography (grain size and roughness) and chemistry (wettability and hydrogen terminations) can be controlled by H_2 plasma using the microwave plasma chemical vapor deposition. Nanosized grain surfaces indicated more expending of osteoblast cell membranes and filapodia than micron-sized coatings. In addition to the better adhesion, osteoblast proliferation was significantly higher on nanosized surfaces after 3 days in vitro culture. Similarly, nanostructured PLGA coating by NaOH chemical etching

TABLE 23.2 Metal Nanoparticle/Polymer Composite Scaffolds

Materials	Function and Biological Improvements	References
Silanized titania/PMMA	Direct apposition to the native surrounding bone in vivo	Goto et al. (2005)
Titania/PLGA	Enhanced long-term osteoblast functions including collagen synthesis, ALP activity, and mineral deposition	Webster and Smith (2005); Liu et al. (2006)
	Reduced acidic degradation product formation	Liu et al. (2006)
Alumoxane/PPF	Increased flexural modulus of the composites	Horch et al. (2004)
	Biocompatible to fibroblast cell line in vitro	Mistry et al. (2007)
	Biocompatible after 12 weeks after implantation in adult goat in vivo model	Mistry et al. (2007, 2009b)
SWNT/PPF	Enhanced compressive/flexural modulus, compressive yield strength, and flexural strength than pure PPF	Shi et al. (2006)
SWNT/Collagen hydrogel	Possible application for electrical stimulation study	MacDonald et al. (2008)
$MgO/BaSO_4$/PMMA	Reduced exothermic reaction and increased osteoblast adhesion	Ricker et al. (2008)
Gold nanoparticles	Inhibits VEGF activity and applied for treatment of rheumatoid arthritis	Tsai et al. (2007)
Magnetic iron oxide nanoparticles	Increased cell seeding efficiency and allowed for recellularization of vascular grafts	Shimizu et al. (2007)
	Created cell patterns to align cells forming capillary tube like structure	Ino et al. (2007)
Magnetic microbeads	Form confluent endothelium within lumen of vascular grafts	Perea et al. (2007)
Copper nanoparticles	Combined with hyaluron oligomers to stimulate deposition of fibrillar elastin matrix by matrix smooth muscle cells	Kothapalli and Ramamurthi (2009)

onto titanium metal surfaces was also investigated and this coated titanium also showed the enhanced osteoblast cell proliferation (Smith et al. 2007b).

Nanocrystalline HA has also been used as a coating substance as well as a composite filler material in polymer scaffolds. Conventional titanium implant surfaces can be improved by coating with HA nanoparticles as osteoconductive calcium phosphate coatings promote bone healing and lead to the rapid biological fixation of implants (Le Guehennec et al. 2007). Human bone marrow mesenchymal cells on nanocrystalline hydroxyapatite-deposited titanium alloy-induced osteogenic differentiation with increasing expression of osteopontin, osteonectin, and collagen type I than on the uncoated substrates (Bigi et al. 2007). Another study demonstrated that nanoapatite coating on titanium conducted bone formation and promoted direct bone apposition after 8 weeks of implantation in dog models (Li 2003). Specifically, histological examination of implanted channel substrates into the lateral metaphysis of the distal femur of dog indicated that more bone grew through the channels lined with the nanoapatite-coated surfaces than the channels lined with the uncoated control surfaces. Composite nanomaterial can be also used as a coating material to improve the osteoblast adhesion. HA and titania with less than 100 nm size were coated on titanium; this composite coating system indicated higher osteoblast adhesion than single-phase titania nanoparticle coating or HA coating (Sato et al. 2008). In the aspect of mechanical property, the adhesion strength of the HA/30% titania nanocomposite coating was increased about 50% (up to 56 MPa) when compared to pure HA coating on titanium (Kim et al. 2005). Thus by taking advantage of both composite coating materials and nanophase particles, it may provide better surface properties since it maintained the chemistry and crystallite size of the original HA and titania powders. Likewise, collagen/HA nanocomposite coating on titanium also indicated significantly higher MC3T3-E1 cell proliferation and ALP activity compared to uncoated titanium and collagen control (Teng et al. 2008).

In order to improve the properties of widely used titanium metal in clinical implantations and enhance surrounding tissue responses after implantation, coating nanosized materials onto metal surfaces can

TABLE 23.3 Nanomaterial Coating on Metal Surfaces

Materials	Function and Biological Improvements	References
Diamond coating on silicon	More stretched morphological change of osteoblast and increased proliferation	Yang et al. (2008)
PLGA on titanium	Stimulated osteoblast proliferation	Smith et al. (2007b)
HA on titanium alloy	Increased expression of osteopontin, osteonectin, and collagen type I than on the uncoated substrates	Bigi et al. (2007)
	New bone formation and direct bone apposition in dog in vivo model	Li (2003)
HA/Titania on titanium	Higher osteoblast adhesion	Sato et al. (2008)
HA/collagen on titanium	Increased MC3T3-E1 proliferation and ALP activity	Teng et al. (2008)

be a promising strategy since nanofabricated titanium metal surfaces showed higher osteogenic conductivity at a cellular level than control materials. In addition to the incorporation of nanoparticles into polymeric scaffolds, coated metal surfaces with nanosized materials may result in better implant–host integration and new bone regeneration at the tissue and cellular level (Table 23.3).

23.2.4 Nanopatterning

Nanoscaled pattern has been fabricated onto the surface of implant material by several techniques including dip-pen nanolithography, hot embossing lithography, and soft lithography (Hasirci et al. 2006). Early researches have shown that osteoblast functions including adhesion, proliferation, ALP expression, and mineral deposition are stimulated more on the nanophase ceramics such as alumina and titania with less than 100 nm grain size than conventional-sized materials (Webster et al. 1999, 2000). Since more particle boundaries are present on the surface of the nanophase compared to a conventional Ti alloy metal or PLGA polymeric scaffold, osteoblast adhesion is usually increased on nanograined metal surfaces (Webster and Ejiofor 2004; Smith et al. 2007a).

One of the recent researches investigated the effect of surface roughness and the width of aligned patterns on osteoblast adhesion and morphology (Puckett et al. 2008). Results indicated that (1) total human osteoblast adhesion increased onto nanorough titanium surface than micron-rough surface, and (2) the adhesion also increased depending on the increasing width of linearly aligned nanorough patterns. Therefore, osteoblast adhesion is more favorable on nanotopographical/crystalline surface and there might be the optimal pattern dimension (or the threshold width) for actin filament diffusion. The different responses of both endothelial cells and osteoblasts on submicron and nanometer scale titanium surface were also investigated (Khang et al. 2008). In this study, both treated surfaces led to an increase in surface energy, thus exhibiting increased cell adhesion when compared to flat metal surfaces. In addition, greater ECM protein expression on nanostructured titanium surfaces was observed (de Oliveira et al. 2007). Since fibronectin, osteopontin, and bone sialoprotein play an important role in cell interaction and adhesion on the surface of the implants, these data also implied the effect of nanopatterned surfaces on the sequential events of cellular/molecular interactions. Another study investigating the molecular machinery controlling cell responses to the surface of the nanostructured titanium revealed that the degree of the cytoskeleton and focal adhesion of the MC3T3-E1pre-osteoblast increased on nanostructured titanium surface (Faghihi et al. 2007) (Table 23.4).

23.2.5 Peptide Binding

Bone regeneration process is a sequence of events of recruiting the progenitor cells from the surrounding tissues, adhesion of cells onto the implants surface, proliferation/migration of the cell within/on the scaffolds, differentiation into the specific lineage of osteoblasts, synthesis of ECM and proteins,

TABLE 23.4 Nanopatterning

Materials	Function and Biological Improvements	References
Titanium with various pattern width	Osteoblast adhesion on nanorough surface depending on the width of aligned pattern	Puckett et al. (2008)
Nanopatterned titanium	Increased cell adhesion than flat titanium surface	Khang et al. (2008)
	Stimulated ECM protein expression	de Oliveira et al. (2007)
	Enhanced cytoskeleton formation and focal adhesion	Faghihi et al. (2007)
Collagen patterned films	Pattern guided smooth muscle cell to secrete ECM in a circumferential manner similar to native ECM of vessels	Zorlutuna et al. (2008)
PET patterned with gelatin	Crafted with gelatin to increase surface roughness. Enhanced endothelial cell adhesion and proliferation	Furuzono et al. (2006)

and mineral deposition. Therefore, adhesion and attachment of hosted cell onto the surface of implant material is the essential and initial step for further osteogenic differentiation of the cell population and osseointegration. It is well known that the initial cellular recognition for the cellular adhesion occurs preferably on the specific amino acid sequences like arginine-glycine-aspartic acid (RGD), which can be found in ECM proteins including fibronectin and vitronectin (Shin et al. 2003; Jager et al. 2007). The receptor protein located in ECM of the cells, especially integrin, interacts with the peptide sequences in existing proteins in the culture media (in vitro) or the body fluids (in vivo) (Balasundaram and Webster 2007). Accordingly, the modification of the implant surface with the cell-specific peptides may improve the initial interaction between cells and biomaterials, subsequent osteoblastic adhesion, and further osteogenic responses of the recruited cell populations. Moreover, the advantage of utilizing the peptide fragments instead of the long chain of growth factors or ECM proteins is its stability during the modification and ability of incorporation onto the nanoscale substrate materials. Growth factors to induce osteoblastic differentiation or bone regeneration are too large to be incorporated into the nanosized substrate biomaterial (such as nanoparticles for drug delivery), but short peptides are more potentially feasible to be conjugated with nanoscale substrates while its bioactivity to induce the osteoblastic functions of progenitor cells is maintained. The size of short peptides coated onto the substrate material is also usually maintained at the 100 nm scale. For instance, the average size of the RGD-functionalized magnetic cationic liposomes was 243 ± 63.2 nm, compared to the 188 nm size of control liposome particles (Ito et al. 2005).

There have been several researches to investigate the effect of peptide-functionalized materials on in vitro osteoblastic responses and in vivo tissue formations. For example, the effect of the short peptides of osteogenic growth factor bone morphogenic protein (BMP)-7 were investigated for its osteoblast responses compared to using total BMP-7 itself (Chen and Webster 2008). In this research, human fetal osteoblasts were cultured with various combinations of three different synthetic peptides sequences (SNVILKKYRN, KPSSAPTQLN, and KAISVLYFDDS) of BMP-7. KAISVLYFDDS and its combination promoted calcium deposition while KPSSAPTQLN and its combination with SNVILKKYRN improved osteoblast proliferation. These peptides with terminal K or S amino acids might be functionalized onto the implant surface by the hydrogen bonds and peptide bonds. The surface modification of titanium alloy (Ti6A14V) with RGD peptides has shown an additive effect on increasing cell adhesion/spreading and focal contact formation of human osteoprogenitor cells (Le Guillou-Buffello et al. 2008). Association of RGD peptides with either collagen- or HA-coated titanium surfaces induced an additive effect on cell adhesion compared to collagen-coated or HA-coated titanium. Moreover, focal contact formation of cells on RGD/HA-coated surface was also improved compared to HA-coated titanium. Another group has also investigated the in vivo effect of human vitronectin peptide conjugated onto titanium surface with rabbit model (Cacchioli et al. 2009). This peptide could interact between cell membrane heparin sulfate proteoglycans and heparin binding sites on ECM proteins and promote the osteoblast adhesion. Through the increased cell apposition rate combined with higher ratio of mineralized surface to

TABLE 23.5 Peptide Binding

Materials	Function and Biological Improvements	References
Combinations of short peptides of BMP-7	Promoted calcium deposition or increased osteoblast proliferation	Chen and Webster (2008)
Vitronectin peptides on titanium	Increased cell apposition rate, higher ratio of mineralized surface to bone surface, more extended bone-to-implant contact in rabbit in vivo model	Cacchioli et al. (2009)
RGD-functionalized HA nanoparticle	Improved the osteogenic differentiation of human bone marrow cells in non-osteogenic culture condition	Babister et al. (2009)

bone surface, more extended bone-to-implant contact was observed in peptide-functionalized implants than in control. Furthermore, RGD peptides can be also grafted onto the HA nanoparticle and RGD-functionalized HA nanoparticle improved the osteogenic differentiation of human bone marrow cells in non-osteogenic culture conditions (Babister et al. 2009). Cells interacted with the hybrid particles via electrostatic attraction and enhanced osteogenic activity was observed in a dose-dependent manner. Especially, 3D aggregation of cell/peptide-HA nanoparticles combined into micropellets also resulted in in vivo mineralization and matrix production 21 days after subcutaneous implantation in nude mice without supportive scaffolds (Table 23.5).

23.2.6 Nanofibrous Scaffolds

Another approach to fabricate the nanobiomaterials is 3D porous nanofibrous scaffolds. There are also several fabrication techniques including electrospinning, self-assembly, melt-blowing, phase separation, and template synthesis (Venugopal et al. 2008b). Among these, electrospinning is the most preferred method due to the versatility with various materials, low cost for fabrication, and yield for continuous length (Hasirci et al. 2006). Since nanofibrous scaffolds have unique characteristics such as nanometer-order diameter, high porosity and high surface-to-volume ratio, nanofibrous scaffolds enable to proper transport of cellular substances including signaling molecules, nutrients, and metabolic waste, and subsequently enhance a series of cellular responses such as cell propagation, proliferation, and differentiation (Smith and Ma 2004). Moreover, the fiber diameter can be controlled under the nanoscale, and the resulting pore size or porosity can be also tailored within a 3D structure to allow cell infiltration into the internal porous region of the scaffolds (Barnes et al. 2007). Despite of the fact that there is no gold standard to fabricate 3D architecture with controlled pore geometry, and nanosized pores often limit cell infiltration and therefore significantly limit the application of these scaffolds, nanofibrous scaffold may still provide a more favorable microenvironment (1) to replicate the architectural/structural morphology of native ECM matrix of bone tissue, (2) to host the progenitor cells via increasing the cell/material interactions, and (3) finally to support subsequent bone regeneration (Kim and Fisher 2007). Previous studies about nanofibrous scaffolds have showed the evidence that architectural difference of the nanofibrous scaffold resulted in various advantages compared to conventional porous 3D scaffolds. For example, 3D PLLA nanofibrous scaffolds exhibited (1) improved fibronectin and vitronectin adsorption from the surrounding culture media and stimulated osteoblast cell attachment (Woo et al. 2003) and (2) significantly increased proliferation and mineralization of osteoblasts and bone marker protein expression including higher ALP activity, increased Runx2 protein expression, increased bone sialoprotein (BSP) mRNA expression and enhanced mineralization when compared to solid-walled control scaffolds (Woo et al. 2007). Likewise, a research about electrospun PCL nanofibrous scaffolds has also showed (1) enhanced MSC penetration into the 3D porous nanofibrous scaffolds, increased ECM formation after 1 week of dynamic in vitro culture, and mineralization after 4 weeks (Yoshimoto et al. 2003) and (2) sufficient cell/ECM formation on the surface of the scaffold construct as well as mineralization and type I collagen expression in vivo (Shin et al. 2004). Interestingly, several researches have provided the potential usage of nanofibrous scaffolds as a framework to support the human MSC proliferation and multilineage differentiation. For instance, human MSCs seeded onto PLGA nanofibrous scaffolds exhibited both chondrogenic and osteogenic differentiations after 2 weeks in culture (Xin et al. 2007). Additionally, PCL nanofibrous scaffolds also demonstrated its ability to support the in vitro multilineage differentiation of human MSCs and multiple tissue morphogenesis (Li et al. 2005). Histological observation, immunohistochemical detection of lineage-specific marker molecules, and gene expression analysis confirmed that human MSCs cultured onto the nanofibrous PCL scaffolds differentiated into adipogenic, chondrogenic, or osteogenic lineages by culturing in specific differentiation media.

Several studies have been performed to demonstrate the feasibility of the nanofibrous organic and inorganic composites fabrication. A recent study indicated that the mineralization of human MSCs cultured on nanosized demineralized bone powders with biodegradable poly(L-lactide) (PLA) electrospun scaffold was significantly higher than on the PLA nanofibrous scaffolds after 14 days of culture. Moreover, in vivo study with critical-sized skull defect of rats showed higher amount of new bone formation on composite fibrous scaffolds 12 weeks after implantation (Ko et al. 2008). Similarly, nanofibrous gelatin/apatite composite scaffolds with 150 nm fiber diameter also exhibited higher mechanical strength than control nanofibrous gelatin scaffolds and enhanced the osteogenic differentiation with five times increased BSP expression and two times increased osteocalcin (OCN) expression of osteoblasts 4 weeks after cell culture (Liu et al. 2009). Furthermore, another fabrication technique has been introduced to

TABLE 23.6 Nanofibrous Scaffolds

Materials	Function and Biological Improvements	References
PLLA	Improved fibronectin/vitronectin adsorption and stimulated osteoblast attachment	Woo et al. (2003)
	Increased proliferation, mineralization, and bone marker protein expression	Woo et al. (2007)
	Chondrogenic differentiation of mesenchymal stem cells in a bioreactor system	Li et al. (2008)
	Coated with gelatin or fibronectin promotes myotube attachment and formation of myofibers long nanofiber alignment	Cronin et al. (2004)
PLGA	Chondrogenic and osteogenic differentiations of human MSCs	Xin et al. (2007)
PCL	Increased ECM formation and mineralization	Yoshimoto et al. (2003)
	Mineralization and type I collagen expression in vivo	Shin et al. (2004)
	Multiple differentiation into adipogenic, chondrogenic, or osteogenic lineages and multiple tissue morphogenesis	Li et al. (2005)
	Proliferation and maintenance of chondrocyte phenotype by expressing specific aggrecan, collagen I and collagen IX	Li et al. (2003)
	Support cell adhesion and guide cell behavior of myoblasts	Choi et al. (2008)
	Induces chondrocyte seeded scaffolds to produce fiber cartilage like tissue in vivo	Li et al. (2009)
Demineralized bone powders/PLA	Increased in vitro mineralization and higher amount of in vivo bone regeneration	Ko et al. (2008)
Apatite/gelatin	Higher mechanical strength and stimulated BSP and OCN expression	Liu et al. (2009)
HA/gelatin/poly(L-lactic acid)-*co*-poly(varepsilon-caprolactone)	Improved rough surface morphology and mechanical properties in composites with electrosprayed HA nanoparticles	Gupta et al. (2009)
HA/gelatin/PCL	Enhanced proliferation, ALP expression, and mineralization than control PCL nanofibrous scaffolds	Venugopal et al. (2008a)
HA/collagen/PCL	Increased Young's modulus	Catledge et al. (2007)

create the nanofibrous composite scaffolds (Gupta et al. 2009). Electrospraying of HA nanoparticles on electrospun nanofibers of poly(L-lactic acid)-*co*-poly(varepsilon-caprolactone)/gelatin improved rough surface morphology, which was ideal for cell attachment and proliferation. It also resulted in superficial dispersion of HA nanoparticles and possessed higher tensile properties compared to HA-blended nanofibers, which resulted in physical mixing of mechanically mismatched materials. This type of composite nanofibrous scaffold with triple materials exhibited better performances in terms of mechanical properties and osteoblast responses. For instance, electrospun nanofibrous scaffolds with PCL/gelatin/HA at a ratio of 1:1:2 showed increased flexible tensile properties and significantly increased levels of proliferation, ALP expression and mineralization compared to PCL nanofibrous scaffolds (Venugopal et al. 2008a). In this research, additional functional groups (an amino group, a phosphate group, and carboxyl groups) were found by Fourier transform infrared analysis, and these groups might induce proliferation and mineralization of osteoblasts for in vitro bone formation. Additionally, nanofibrous triphasic scaffolds with electrospun PCL/collagen/HA nanoparticles also showed increased Young's modulus than both the composite scaffold with PCL/HA nanoparticles and the scaffold of pure PCL/pure collagen (Catledge et al. 2007) (Table 23.6).

23.3 Nanobiomaterials for Cartilage Tissue Engineering

Articular cartilage defects arise due to aging, osteoarthritis, joint damage, or developmental disorders and can cause loss of pain and mobility (Alleyne and Galloway 2001; Poole 2003). Cartilage is comprised of four zones known as superficial, intermediate, deep, and calcified zone. Each zone possesses a specific cell type and ECM organization (Carter and Wong 2003). Damaged tissue has a limited capacity to repair due to reduced availability of chondrocytes. Moreover, due to lack of vasculature, damaged tissues degenerate and the resulting repaired tissue lacks any structural, mechanical, or biochemical properties of native cartilage (Gagne et al. 2000). Due to lack of vasculature, treatment methods cannot be applied through the systemic circulatory system and treatment has to be applied locally at the site of the defect. As damaged cartilage tissue lacks progenitor cells, external chondrocytes are harvested, expanded, and then reintroduced at the site of the defect. 3D scaffolds with nanobiomaterials can effectively support growth and differentiation of chondrocyte or progenitor cells for repair of cartilage defects, and tissue engineering approaches using selective bioactive growth factors may be successfully applicable for regeneration of cartilage tissue (Tuli et al. 2003).

23.3.1 Nanoparticles

Nanoparticles are metal or polymer particles used individually or in combination with each other for delivery of drugs, growth factors, or genes for cartilage repair at the site of the defect. Polyethyleneimine-fabricated nanoparticles were loaded with transforming growth factor β_1 (TGF-β_1) to allow for its sustained release (Park et al. 2008). The TGF-β_1-loaded nanoparticles were combined with chondrocyte cell suspensions and encapsulated within poly(*N*-isopropylacrylamide-*co*-acrylic acid) polymer solution to form hydrogels. TGF-β_1 release studies demonstrated that nanoparticles allowed for a slower release compared to TGF-β_1 alone. After 21 days, released TGF-β_1 was still bioactive, suggesting that the nanoparticle hydrogel construct did not alter its bioactivity. Bioactivity was also confirmed by the increased levels of collagen I, collagen II, and aggrecan gene expression. This construct was also implanted in nude mice to evaluate ECM associated components. Results indicated that the TGF-β_1-loaded heparinized nanoparticle-induced chondrocytes to accumulate abundant quantities of ECM rich in polysaccharides. Thus heparinized nanoparticles can be successfully applied for growth factor delivery and cartilage tissue engineering applications. Heparinized nanoparticles can also be incorporated with poly(L-lysine) and loaded with TGF-β_3 growth factor. Nanoparticles with a diameter of 50–150 nm were physically attached to poly(lactide-*co*-glycolide) microspheres which were precoated with poly(ethyleneimmine). Rabbit MSCs were cultivated onto these microspheres and evaluated for synthesis of ECM. Results indicated that cell adhesion and its growth were excellent on the heparinized nanoparticles attached on polymeric microspheres with the aid of growth factor (Park et al. 2008).

Gold nanoparticles have been applied for treatment of rheumatoid arthritis (Tsai et al. 2007). It is known that these particles inhibit activity of vascular endothelial growth factor (VEGF) (Mukherjee et al. 2005). Intra-articular delivery of nanogold in collagen-induced arthritic rats reduced endothelial cell proliferation and migration. Thus gold nanoparticles can be applied as a therapeutic agent for treatment of rheumatoid arthritis or any other VEGF-dependent chronic inflammatory diseases. These particles inhibit VEGF-induced cell proliferation by interacting with sulfur/amines, which are present in its heparin-binding domain and thus inhibit VEGF-induced signaling. Gold nanoparticles have also been shown to induce growth factor inhibition in ear and ovarian tumor mice models (Mukherjee et al. 2005).

Nanoparticles have been traditionally used with polymers to fabricate composite materials. However, for cartilage applications, many studies have shown the use of nanoparticles to be advantageous for therapeutic applications. Due to the ease of fabrication and manipulation, nanoparticles have been conjugated with a variety of molecule such as peptides, proteins, oligonucleotides, and DNA. Thus, for cartilage tissue engineering applications nanoparticles may play more of a drug delivery role than just as an additional factor to improve material properties.

23.3.2 Nanofibrous Scaffolds

A three-dimensional scaffold combined with chondrocytes may be a suitable candidate for engineered cartilage tissue, which possesses properties similar to native cartilage. Nanofibrous PCL scaffolds have been developed by electrospinning (Li et al. 2003). These scaffolds were seeded with fetal bovine chondrocytes (FBC) and studied for their ability to maintain a chondrogenic phenotype. Results demonstrated that these nanobiomaterials can act as biologically preferred substrates for proliferation and maintenance of chondrocyte phenotype by expressing cartilage-specific ECM genes such as aggrecan, collagen II, collagen IX, and cartilage oligomeric matrix protein. Histological evaluation demonstrated that FBC cultured on nanofibrous PCL scaffolds produced more sulfated proteoglycan-rich cartilaginous matrix compared to monolayer on tissue culture polystyrene. Thus the architecture of scaffolds can influence the composition of culture medium. In addition to enhanced phynotypic differentiation, 3D culture in nanofibrous scaffolds also indicated that 21-fold increase in cell growth after 21 days. Therefore, the nanofibrous architecture of scaffolds can support the proliferation and maintenance of the chondrocytic phenotype of FBCs.

These same scaffolds were also studied for their ability to support in vitro chondrogenesis of MSCs (Li et al. 2005). The electrospun nanobiomaterials containing uniform or random oriented nanofibers with a 700 nm diameter maintained structural integrity after 21 days of culture. The treatment with TGF-β resulted in differentiation to chondrocytic phenotype, which was evident by gene expression patterns of collagen II, collagen IX, aggrecan, and expression of ECM proteins. Thus these nanobiomaterials act as a bioactive carrier of MSCs and can be applied in tissue engineering–based cartilage repair applications. Human MSCs have also been cultured on electrospun PCL nanofibrous scaffolds to engineer the superficial zone of articular cartilage (Wise et al. 2008). The nanofibrous structure guided cell orientation even after 5 weeks in culture and also maintained its chondrogenic phenotype. This suggests that orientation of cells was under the influence of physical cues provided by the nanofibrous structure. Thus nanofibrous scaffolds may be able to mimic the ECM organization found in the superficial zone of articular cartilage. The three-dimensional structure of nanofibers can mimic native ECM and may stimulate chondrocyte activity by regulating their cytoskeleton or shape (Baker and Mauck 2007). Physical cues via cell–nanofiber interactions may play an effective role for cartilage repair.

Nanofibrous PLLA scaffolds have also been used as a chondrocyte nanofiber composite for culture in a bioreactor system (Li et al. 2008). The aim of this study was to enhance the growth of these constructs in a "dynamic" culture compared to a "static" culture. Insulin like growth factor and TGF-β_1 were also applied to this culture to regulate chondrocyte activity. Results demonstrated that the constructs cultured in a bioreactor developed into a smooth, glossy cartilage-like tissue, compared to constructs in a static environment, which developed into a rough surface tissue. The bioreactor grown constructs also produced more collagen and sulfated-glycosaminoslycans, which resulted in more production of cartilage-associated genes. The mechanical properties of these constructs were also enhanced. The nanofibers used in this study were ultrafine and tightly packed, which lead to a smaller pore size. Smaller pore sizes again lead to decreased loss of cells and increased cell attachment. Change in the nanofiber pore size was essential for bioreactor applications. Tightly packed cells and nanofibers have been previously used with MSCs and chondrocytes (DeLise et al. 2000; Schulze-Tanzil et al. 2002). Thus these nanofiber scaffolds can be applied to seed chondrocytes uniformly and efficiently in a bioreactor system for cartilage tissue engineering applications.

Nanofibers have also been applied in vivo for articular cartilage repair (Li et al. 2009). PCL nanofibrous scaffolds were seeded with allogenic chondrocytes, xenogeneic human MSCs, or acelluar scaffolds and implanted in a cartilage defect swine model. Evaluation after 6 months revealed that scaffolds with MSCs showed complete repair and restored a smooth cartilage surface. The chondrocyte-seeded scaffolds produced a fiber cartilage-like tissue with a superficial cartilage contour.

Nanofibrous scaffolds may be successfully applied for fabricating engineered cartilage tissue, which can structurally mimic native cartilage. These scaffolds have an advantage as its nanometer size and nanoporosity can be controlled. This structural control can influence how chondrocytes react through physical cues to nanofibrous scaffolds. Thus, nanofibrous scaffolds may be applied for cartilage tissue engineering toward producing a more cartilage-like tissue structure.

23.4 Nanobiomaterials for Muscle Tissue Engineering

Skeletal muscle defects arise either due to a variety of muscle diseases or due to muscle trauma. Damaged muscles do not possess a natural ability to repair and instead form scar tissue (Hurme et al. 1991). Common treatment method involves replacing damaged tissue with muscle from local or distant sites (muscle flaps). This procedure can further lead to donor site morbidity, leading to functional loss and volume deficiency. Moreover, the replaced muscle tissue may not be designed to function in its new location and may degenerate before total integration. Tissue engineering as a result is being investigated as a solution for muscle replacement. Engineered materials may be used for functional or aesthetic replacement of damaged skeletal muscles (Payumo et al. 2002; Bach et al. 2003). Skeletal muscle is composed of bundles of highly oriented and dense muscle fibers that are packed together in an extracellular three dimensional matrix. These fibers are composed of densely fused, differentiated multinucleated cells derived from myoblasts (Blau and Webster 1981). Skeletal muscle fibers possess this dense cellular tissue to functionally generate longitudinal muscle contractions. Once these fibers are damaged, they become neurotic and resulting scar tissue is removed by macrophages. Within this damaged cell population, a specialized myoblast population, known as satellite cells, comprise of 1%–5% of total nuclei of mature muscle (Allen et al. 1997). These cells are scattered below the basal lamina of myofibers and are capable of muscle regeneration. Normally these cells do not divide, but they can be induced to proliferate in the presence of special local factors (Campion 1984). Once they start proliferating, they fuse together as myoblasts and form multinucleated and elongated myotubes. These myotubes self-assemble and form a more organized muscle fiber (Brand et al. 2000). The role of tissue engineering is to induce these satellite cells to proliferate and differentiate attempting to induce fusion of myoblasts to myotubes. Studies have focused on the role of ECM in migration, proliferation, and differentiation of these cells (Adams and Watt 1993; Kim and Mooney 1998; Saxena et al. 2001). The ECM can provide a key role in the alignment and differentiation of myoblasts through cell adhesion. Nanobiomaterials may be used to replace the ECM in vitro to enhance attachment of myoblasts and induce cell proliferation and differentiation.

23.4.1 Nanofibrous Scaffolds

As skeletal muscles are difficult to regenerate once damaged in adults, tissue engineering of these muscles is a challenging field. Nanofibrous surfaces can be engineered to promote differentiation of myoblasts into myotubes and myofibers as an application for skeletal muscle tissue engineering. These fibrous scaffolds should be biocompatible, nonimmunogenic, and ideally biodegradable. They should promote cellular growth and differentiation of myoblast cells. Electrospinning technique can be applied to create PCL nanofibers for skeletal muscle tissue engineering applications (Choi et al. 2008). Nanofiber meshes have been created with fiber diameter ranging from 250 to 300 nm. These nanofibrous structures are able to support myoblast cell adhesion and guide cell behavior. Fabrication process involves alignment of these fibers in a unidirectional manner to facilitate myotube formation. This can create a high surface area and high porosity of nanofibers thus being favorable for muscle tissue engineering applications. Another advantage of these nanofibers is that the size can be controlled ranging from 50 to 1000 nm of fiber diameter. Electrospun microfibers have been applied to fabricate degradable polyester urethane as biomaterial for skeletal muscle tissue engineering (Riboldi et al. 2005). Results suggested that these fibers exhibited positive mechanical properties and promoted cell adhesion and differentiation of human satellite and myoblast cells. PLLA mesh fibers (500 nm fiber diameter) coated with gelatin or fibronectin have also been studied for skeletal muscle tissue engineering (Cronin et al. 2004). C2C12 seeded cells exhibited disordered actin filament arrangement on randomly oriented nanofiber meshes and aligned actin filaments on aligned nanofiber meshes. Myotube attachment and formation of multinucleated myofibers were present along the axis if nanofiber alignment. The organized myotubes were longer than those found on the randomly oriented nanofiber meshes. These long myotubes mimicked myotube assembly seen in native skeletal muscle tissues. Similar myotube assembly was also noted on PLLA aligned nanofibrous scaffolds. Average nanofiber diameter was 500 nm. Murine C2C12 myoblasts were cultured on these nanofibrous scaffolds to compare cell growth and organization on micropatterned scaffolds. Staining of F-actin, myosin heavy chain and cell nuclei was performed on cell scaffolds to examine cell organization myotube assembly. Nanoscale topography aligned myoblasts and cytoskeletal proteins and also promoted myotube assembly along nanofibers, which mimicked myotube organization in native skeletal tissue. Nanoscale features promoted longer myotube assembly compared to microscale features. Myotube striation was also enhanced on both aligned and random nanoscale scaffolds compared to microscale scaffolds (Huang et al. 2006).

Nanofibrous scaffolds have been fabricated from electroactive polymers to evaluate the viability of myoblasts. Polyaniline is a conducting polymer that was blended with an elastic PLLA-CL polymer using camphorsulfonic acid as a dopant and nanofibrous scaffolds were created using electrospinning. After myoblasts were seeded on these scaffolds for 2 days and examined for cell viability. Increased myoblast adhesion was noted on nanoscaffolds compared to plain scaffolds (Jeong et al. 2008).

Multiple layered scaffolds can also be fabricated with alternate or continuous layers of micro/nanofibers. These customized hierarchical scaffolds can mimic different micro and nanosized layers of the ECM, which can be divided into four main layers, macroscale (>1 mm), micrometer scale (1 μm to 1 mm), submicrometer scale (100 nm to 1 μm) and nanometer scale (<100 nm). The spinneret-based tunable engineered parameters technique can be used to fabricate these aligned micro/nanofibrous scaffolds. Single and multilayer nanofibrous scaffolds ranging from submicrometer to micrometer range have been fabricated using this technique. These nanofibrous scaffolds were composed of four different materials including polystyrene, PMMA, PLA, and PLGA. Myoblasts were cultured on these scaffolds to investigate the effect of topological constraints on cell migration, proliferation, and morphology. Time lapse microscopy revealed that cells readily attached to the nano/microfibrous scaffolds, which were evident by cell spreading and alignment along fibers. For macroscale scaffolds, cells were noted to fuse together into bundles similar to myotube assembly. Morphology of cells on single layer scaffolds changed from cuboidal shape to a more elongated myotube-like shape. Interestingly, cells cultured on double-layered scaffolds retained their cuboidal shape (Nain et al. 2008). It is clear from this study that by controlling the architecture of scaffolds, cells will exhibit different growth properties.

Nanofibrous scaffolds have been extensively researched as potential nanobiomaterials for muscle tissue engineering scaffolds. A recent study was performed to evaluate behavior of skeletal myoblasts on laser-nanostructured materials. Nanostructure surfaces were fabricated by linearly polarized Kef laser light on polystyrene foil materials. After irradiation of polymer foils the resulting structure possessed uniform nanostructures every 200–430 nm each with a depth of 30–100 nm. Skeletal myoblasts aligned along these structures after 6 h of seeding with an elongated spindle-like shape or rounded shape. Similar alignment was not seen on plain polystyrene foils and spreading was also not definitive. Studies concluded that presence of nanostructures can guide myoblast alignment along definitive directions (Rebollar et al. 2008).

Therefore, for muscle tissue engineering applications, nanofibrous scaffolds may promote myoblast adhesion, alignment, growth, and proliferation, which can be attributed to nanosized topography of these nanobiomaterials.

23.5 Nanobiomaterials for Vascular Tissue Engineering

Blood vessels are composed of three distinct layers: an athrombogenic endothelial cell layer, a second layer composed of smooth muscle cells, which are embedded in a three-dimensional ECM, and an adventitial layer of connective tissue. These vascular tissues are composed of proteins with a nanometer structure. Cells of these tissues have been found to be naturally interacting with nanostructured molecules in vivo (Weinberg and Bell 1986; Matsuda and Miwa 1995; Niklason et al. 1999; Dalby et al. 2002). Damage to these tissues requires an approach involving replacement with nanometer-sized structures and molecules. Vascular tissue engineering has been largely applied for repair of small- and medium-sized blood vessels. Current treatment for damaged vessels involves replacement of vessels with an autograft of synthetic vascular graft (Ratcliffe 2000). However, synthetic grafts cannot be used for the replacement of small diameter blood vessels (<6 mm) due to blood clogging problems (Ute Henze et al. 1996). As result, studies have focused on a tissue engineering solution to these problems. Previously attempts have been made to fabricate a functional synthetic or tissue-engineered construct for small blood vessels but the failure rates have been high due to a variety of reasons. More recently, nanobiomaterials are being fabricated into tubular constructs to improve vascular cell function and try to overcome associated problems.

Nanobiomaterials used to fabricate these constructs for vascular tissue engineering applications should be able to withstand burst pressure. Burst pressure is the pressure required to burst a tubular construct while a flow of fluid is passing through it. A tubular construct would have to possess material properties strong enough to resist fluid flow pressure with a value higher than that of blood pressure (1680 mmHg). Currently, vascular grafts can resist flow pressure up to 700 mmHg, which is still lower than natural burst strength (Sen et al. 1965; L'Heureux et al. 1998). As a result, many techniques are being used to modify material materials for increasing their mechanical strength. Such techniques include fabricating nanofibers, nanopatterns, or nanostructure materials to increase mechanical strength of vascular constructs (Mironov et al. 2008).

Apart from mechanical strength, vascular constructs also need to mimic native vessal structure and natural ECM molecular surroundings (Murugan and Ramakrishna 2007). Vascular ECM consists of proteins and glycosaminoglycans in the nanometer range. Most cells attach and organize around nanometer-sized structures, which provides cells with environmental signals to direct cell-specific behavior (Chen et al. 2007). Vascular stent materials are also being fabricated by optimizing surface properties of nanobiomaterials to influence cell interactions (Cho et al. 2008; Lu et al. 2008). As a result, nanoscale matrices are being developed from nanobiomaterials for vascular tissue engineering applications.

Therefore, the first criteria of nanobiomaterials for vascular tissue engineering applications are easy fabrication into tubular constructs and ability to support vascular and endothelial cell growth.

They should not enhance blood coagulation, interfere with natural viscoelastic properties of the blood or cause excessive platelet adhesion and aggregation. For successful postimplantation functionality, they need to possess mechanical strength, adequate compliance, and burst strength after implantation. All of these requirements can be met by modifying surface properties chemically or biologically to enhance endothelialization of these surfaces. Modification of physical properties such as enhancing hydrophilic properties, altering material porosity, and roughness of surfaces may also enhance endothelialization of biomaterials and finally result in a successful vascular tissue engineering application.

23.5.1 Nanofibrous Scaffolds

Studies have shown that structurally modified nanobiomaterials including nanofibrous scaffolds can enhance vascular cell function compared to flat surfaces. Alternations in the nanostructures of materials can influence protein interactions within cells as they present a larger surface area for protein adsorption on nanosurfaces.

Nanofibrous scaffolds have been developed for vascular tissue engineering applications by electrospinning PLLA-CL copolymer consisting of 400–800 nm fibers (Xu et al. 2004). These nanofibrous scaffolds were used to culture human smooth muscle and endothelial cells for 7 days. Results suggested that as these scaffolds could mimic the nanoscale dimensions of native ECM, they were capable of supporting cell attachment and proliferation. Cells were able to maintain their phenotypic shape and were able to integrate with nanofibers to form a three-dimensional cellular matrix. Efficacy of nanofibrous scaffolds has also been studied in vivo. Nanofibrous scaffolds were fabricated using electrospinning techniques from PLLA polymers (Hashi et al. 2007). Scaffolds were seeded with smooth muscle cells and MSCs and cultured for 2 days. Results showed that cells aligned with nanofibers and had a cellular organization similar to that of native artery. This suggests that nanofibrous scaffolds could mimic native collagen fibrils and promote guided cellular alignment. These grafts were implanted as a carotid artery in rats for up to 60 days and were examined for nonthrombogenic properties and long-term patency. Experiments revealed that scaffolds with MSCs were antithrombogenic and nonimmunogenic. The nanofibrous structure in vivo promoted recruitment of vascular cell and organization into layered structured along the vascular wall as is evident in the native artery. Results indicated excellent patency of small diameter nanofibrous vascular grafts. Electrospinning technique has also been applied to fabricate PLLA-CL nanofibrous tubular scaffolds for repair of blood vessels (Dong et al. 2008). In this study, fibers could mimic dimensions of natural ECM; they possessed mechanical properties similar to native coronary artery and formed a well-defined matrix for smooth muscle cell adhesion and differentiation. As these fibers were aligned in unidirectional, they maintained vasoactivity by providing enough mechanical strength to sustain high blood flow pressure. Thus the electrospinning technique can be used for mass production and has an increased potential for vascular tissue engineering applications.

23.5.2 Nanopatterned Scaffolds

Nanopatterned collagen films have also been developed for vascular tissue engineering to increase its mechanical strength (Zorlutuna et al. 2008). Nanopatterns consisted of parallel channels with 300 nm width and 100 nm depth collagen films. Smooth muscle cells were seeded onto these films and cultured for 21 days. The authors hypothesized that nanopatterning would guide cells to secrete ECM in a circumferential manner similar to natural ECM of vessels. This would in turn naturally increase mechanical strength of films. Results suggested that nanopatterned films could support and align cells along the patterned film axes. It has been shown that construct strength can improve mechanical properties, thus nanopatterning of materials can be applied for vascular tissue engineering. This has led to the development of nanobiomaterials with a variety of structures using techniques such as solid freeform fabrication, electrospinning, and extrusion (Sachlos et al. 2003; Rolland et al. 2005; Buttafoco et al. 2006; Ashammakhi et al. 2008).

Another important material property for nanopatterned vascular materials is to mimic ECM both structurally and in molecular properties. Nanosized textured surfaces increases surface roughness and influences cell behavior in terms of cell adhesion and proliferation. A nanofibrous mat was electrospun from polyethylene telephthalate (PET) polymer (Furuzono et al. 2006). Its surface was modified by grafting gelatin to increase surface roughness. This modification enhanced endothelial cell adhesion and proliferation, thus suggesting that these modified PET nanofibrous mats may be applied toward vascular tissue engineering.

Nanobiomaterials should possess athrombogenic properties. Many nanobiomaterials are covered with endothelial cells, creating a continuous layer of matrix as they do not enhance blood coagulation. They should also be able to resist narrowing of the construct lumen by vascular intimal thickening, which could lead to loss of construct patency, degradation of deposited ECM, and development of degradation-associated vascular dilation or aneurysm (Rotmans et al. 2005; Jordan and Chaikof 2007; Perea et al. 2007). Studies have shown that nanopatterns on nanofilms can reduce vascular graft athrombogenicity and increase adhesiveness for circulating endothelial progenitor cells (Tanaka and Sackmann 2005; Alobaid et al. 2006; Wilson et al. 2007). Such a structure can also mimic the organization of a natural vessel ECM. Additionally, hydrogels have also been modified by addition of nanopatterned growth factors and ECM molecules, which can direct vascular cell differentiation and improve the maturation of vascular tissue constructs.

23.5.3 Metal Nanoparticles

Magnetic iron oxide nanoparticles are being routinely used in biotechnology and clinical applications. For vascular tissue engineering applications, magnetic nanoparticles have been used to create different cellular arrangements around or within a tubular vascular conduit in vitro. The first approach involves seeding cells onto a tubular vascular conduit and then placing a magnet on the outer surface of the conduit. The magnetic force allows cells to adhere and spread within the conduit lumen, thus forming a continuous cellular layer. A second approach involves seeding two different cell types with the conduit lumen using magnetic forces. A third approach involves inserting a magnet with the lumen of the conduit and then rotating it in a cell suspension of magnetically labeled cells. This allows for cell attachment in a concentric layer on the outer surface of the conduit.

Magnetic nanoparticles have been encapsulated within liposomes and have been coined the term "magnetic cationic liposome" (MCL) (Shimizu et al. 2007). MCLs are a lipid mixture containing 10 nm sized magnetic nanoparticles. The total size of an MCL is 150 nm. MCLs come in contact with cells and transfer magnetic nanoparticles within cells via electrostatic interaction with a negatively charged cell membrane (endocytosis). Once cells are labeled with these nanoparticles, they can be manipulated using magnetic force. For vascular tissue engineering applications, magnetically labeled smooth muscle cells and dermal fibroblasts were seeded into a decellularized common carotid artery conduit. This conduit contained a magnet within its lumen, thus when it was immersed into a cell suspension, almost all cells attached to the conduit. This increased cell seeding efficiency compared to cell seeding in conduits without a magnet. Thus use of magnetic nanoparticles allowed for recellularization of vascular grafts. Magnetic nanoparticles have also been applied to create patterns of cells (Ino et al. 2007). A steel plate was placed underneath a culture dish containing magnetically labeled cells. The cells would align in a single pattern by adjusting the number of cells seeded and complex cell patterns could also be formed. Cell patterning using magnetic force was also performed using human vein endothelial cells on matrigel. This leads to formation of patterned capillary-like structures. Cells labeled with magnetic microbeads have also been used to form a confluent endothelium within the lumen of a vascular conduit by placing a magnet on the conduits outer surface (Perea et al. 2007).

Copper nanoparticles have been applied for regeneration of elastin fibers. These fibers are proteins present within the ECM of vascular tissues and are responsible for maintaining elastic stability and mechanical properties of blood vessels. Copper nanoparticles and hyaluron oligomers were both added to smooth muscle cell cultures at different doses and concentrations. Results suggested that both copper nanoparticles and hyaluron oligomers were successful in stimulating deposition of a cross-linked and fibrillar elastin matrix by vascular smooth muscle cells (Kothapalli and Ramamurthi 2009).

Though the application of metal nanoparticles for vascular tissue engineering is still in its infancy, results do seem to be promising toward fabricating a recellularized and patterned vascular graft.

23.6 Conclusions

Natural tissues possess surface features on the nanometer scale. Thus nanostructured biomaterials may promote functions of various cell types. Nanobiomaterials can be fabricated either by altering chemistry or physical properties of materials to influence cellular interactions. Nanobiomaterials could provide access for new and improved material properties and functionalities for musculoskeletal tissue engineering applications. It is possible to control cell interactions with nanobiomaterials toward the goal of fabricating tissue-engineered structures. Mimicking ECM environment is a major challenge in tissue engineering. However, modification of nanobiomaterials has provided promising results in creating a scaffold to replace natural ECM until host cells can proliferate and resynthesize a new natural surrounding matrix.

Acknowledgments

This work was supported by the State of Maryland Department of Business and Economic Development, Nano-Biotechnology DBED award, and National Science Foundation through a CAREER Award to JPF (#0448684).

Abbreviations

ALP	alkaline phosphatase
BMP	bone morphogenetic protein
ECM	extracellular matrix
FBC	fetal bovine chondrocytes
FTIR	Fourier transform infrared spectroscopy
HA	hydroxyapatite
MSC	mesenchymal stem cell
MCL	magnetic cationic liposomes
PA	polyamide
PBT	poly(butylene terephthalate)
PCL	poly(ε-caprolactone)
PEG	poly(ethylene glycol)
PET	polyethylene telephthalate
PLGA	poly(D,L-lactic-*co*-glycolic acid)
PLA	poly(L-lactide)
PLLA	poly(L-lactic acid)
PLLA-CL	poly(L-lactide/ε-caprolactone)
PMMA	poly(methyl methacrylate)
PPF	poly(propylene fumarate)
RGD	arginine–glycine–aspartic acid
SWNT	single-walled carbon nanotubes
VEGF	vascular endothelial growth factor
TGF-β_1	transforming growth factor-beta1
UHMWPE	ultrahigh molecular weight polyethylene

References

Adams, J. C. and F. M. Watt. 1993. Regulation of development and differentiation by the extracellular matrix. *Development* 117 (4):1183–1198.

Allen, R. E., C. J. Temm-Grove, S. M. Sheehan, and G. Rice. 1997. Skeletal muscle satellite cell cultures. *Methods Cell Biol.* 52:155–176.

Alleyne, K. R. and M. T. Galloway. 2001. Management of osteochondral injuries of the knee. *Clin. Sports Med.* 20 (2):343–364.

Alobaid, N., H. J. Salacinski, K. M. Sales, B. Ramesh, R. Y. Kannan, G. Hamilton, and A. M. Seifalian. 2006. Nanocomposite containing bioactive peptides promote endothelialisation by circulating progenitor cells: An in vitro evaluation. *Eur. J. Vasc. Endovasc. Surg.* 32 (1):76–83.

Ashammakhi, N., A. Ndreu, L. Nikkola, I. Wimpenny, and Y. Yang. 2008. Advancing tissue engineering by using electrospun nanofibers. *Regen. Med.* 3 (4):547–574.

Babister, J. C., L. A. Hails, R. O. Oreffo, S. A. Davis, and S. Mann. 2009. The effect of pre-coating human bone marrow stromal cells with hydroxyapatite/amino acid nanoconjugates on osteogenesis. *Biomaterials* 30 (18):3174–3182.

Bach, A. D., J. Stem-Straeter, J. P. Beier, H. Bannasch, and G. B. Stark. 2003. Engineering of muscle tissue. *Clin. Plast. Surg.* 30 (4):589–599.

Baker, B. M. and R. L. Mauck. 2007. The effect of nanofiber alignment on the maturation of engineered meniscus constructs. *Biomaterials* 28 (11):1967–1977.

Balasundaram, G. and T. J. Webster. 2007. An overview of nano-polymers for orthopedic applications. *Macromol. Biosci.* 7 (5):635–642.

Barnes, C. P., S. A. Sell, E. D. Boland, D. G. Simpson, and G. L. Bowlin. 2007. Nanofiber technology: Designing the next generation of tissue engineering scaffolds. *Adv. Drug Deliv. Rev.* 59 (14):1413–1433.

Bigi, A., N. Nicoli-Aldini, B. Bracci, B. Zavan, E. Boanini, F. Sbaiz, S. Panzavolta, G. Zorzato, R. Giardino, A. Facchini, G. Abatangelo, and R. Cortivo. 2007. In vitro culture of mesenchymal cells onto nanocrystalline hydroxyapatite-coated Ti13Nb13Zr alloy. *J. Biomed. Mater. Res. A* 82 (1):213–221.

Blau, H. M. and C. Webster. 1981. Isolation and characterization of human muscle cells. *Proc. Natl Acad. Sci. U. S. A.* 78 (9):5623–5627.

Brand, T., G. Butler-Browne, E. M. Fuchtbauer, R. Renkawitz-Pohl, and B. Brand-Saberi. 2000. EMBO Workshop Report: Molecular genetics of muscle development and neuromuscular diseases Kloster Irsee, Germany, September 26–October 1, 1999. *EMBO J.* 19 (9):1935–1941.

Buttafoco, L., N. G. Kolkman, P. Engbers-Buijtenhuijs, A. A. Poot, P. J. Dijkstra, I. Vermes, and J. Feijen. 2006. Electrospinning of collagen and elastin for tissue engineering applications. *Biomaterials* 27 (5):724–734.

Cacchioli, A., F. Ravanetti, A. Bagno, M. Dettin, and C. Gabbi. 2009. Human vitronectin-derived peptide covalently grafted onto titanium surface improves osteogenic activity: A pilot in vivo study on rabbits (this paper is dedicated to Prof. Carlo Di Bello in the occasion of his 70th birthday). *Tissue Eng. Part A* 15 (10):2917–2926.

Campion, D. R. 1984. The muscle satellite cell: A review. *Int. Rev. Cytol.* 87:225–251.

Carter, D. R. and M. Wong. 2003. Modelling cartilage mechanobiology. *Philos. Trans. R. Soc. Lond. B Biol. Sci.* 358 (1437):1461–1471.

Catledge, S. A., W. C. Clem, N. Shrikishen, S. Chowdhury, A. V. Stanishevsky, M. Koopman, and Y. K. Vohra. 2007. An electrospun triphasic nanofibrous scaffold for bone tissue engineering. *Biomed. Mater.* 2 (2):142–150.

Chen, M., P. K. Patra, S. B. Warner, and S. Bhowmick. 2007. Role of fiber diameter in adhesion and proliferation of NIH 3T3 fibroblast on electrospun polycaprolactone scaffolds. *Tissue Eng.* 13 (3):579–587.

Chen, Y. and T. J. Webster. 2008. Increased osteoblast functions in the presence of BMP-7 short peptides for nanostructured biomaterial applications. *J. Biomed. Mater. Res. A* 91 (1):296–304.

Cho, H. H., D. W. Han, K. Matsumura, S. Tsutsumi, and S. H. Hyon. 2008. The behavior of vascular smooth muscle cells and platelets onto epigallocatechin gallate-releasing poly(l-lactide-co-epsilon-caprolactone) as stent-coating materials. *Biomaterials* 29 (7):884–893.

Choi, J. S., S. J. Lee, G. J. Christ, A. Atala, and J. J. Yoo. 2008. The influence of electrospun aligned poly(epsilon-caprolactone)/collagen nanofiber meshes on the formation of self-aligned skeletal muscle myotubes. *Biomaterials* 29 (19):2899–2906.

Cronin, E. M., F. A. Thurmond, R. Bassel-Duby, R. S. Williams, W. E. Wright, K. D. Nelson, and H. R. Garner. 2004. Protein-coated poly(L-lactic acid) fibers provide a substrate for differentiation of human skeletal muscle cells. *J. Biomed. Mater. Res. A* 69 (3):373–381.

Dalby, M. J., M. O. Riehle, H. Johnstone, S. Affrossman, and A. S. Curtis. 2002. In vitro reaction of endothelial cells to polymer demixed nanotopography. *Biomaterials* 23 (14):2945–2954.

de Oliveira, P. T., S. F. Zalzal, M. M. Beloti, A. L. Rosa, and A. Nanci. 2007. Enhancement of in vitro osteogenesis on titanium by chemically produced nanotopography. *J. Biomed. Mater. Res. A* 80 (3):554–564.

DeLise, A. M., L. Fischer, and R. S. Tuan. 2000. Cellular interactions and signaling in cartilage development. *Osteoarthr. Cartil.* 8 (5):309–334.

Dong, Y., T. Yong, S. Liao, C. K. Chan, and S. Ramakrishna. 2008. Long-term viability of coronary artery smooth muscle cells on poly(L-lactide-co-epsilon-caprolactone) nanofibrous scaffold indicates its potential for blood vessel tissue engineering. *J. R. Soc. Interface* 5 (26):1109–1118.

Dulgar-Tulloch, A. J., R. Bizios, and R. W. Siegel. 2008. Human mesenchymal stem cell adhesion and proliferation in response to ceramic chemistry and nanoscale topography. *J. Biomed. Mater. Res. A* 90 (2):586–594.

Faghihi, S., F. Azari, A. P. Zhilyaev, J. A. Szpunar, H. Vali, and M. Tabrizian. 2007. Cellular and molecular interactions between MC3T3-E1 pre-osteoblasts and nanostructured titanium produced by high-pressure torsion. *Biomaterials* 28 (27):3887–3895.

Fang, L., Y. Leng, and P. Gao. 2005. Processing of hydroxyapatite reinforced ultrahigh molecular weight polyethylene for biomedical applications. *Biomaterials* 26 (17):3471–3478.

Fang, L., Y. Leng, and P. Gao. 2006. Processing and mechanical properties of HA/UHMWPE nanocomposites. *Biomaterials* 27 (20):3701–3707.

Furuzono, T., M. Masuda, M. Okada, S. Yasuda, H. Kadono, R. Tanaka, and K. Miyatake. 2006. Increase in cell adhesiveness on a poly(ethylene terephthalate) fabric by sintered hydroxyapatite nanocrystal coating in the development of an artificial blood vessel. *ASAIO J.* 52 (3):315–320.

Gagne, T. A., K. Chappell-Afonso, J. L. Johnson, J. M. McPherson, C. A. Oldham, R. A. Tubo, C. Vaccaro, and G. W. Vasios. 2000. Enhanced proliferation and differentiation of human articular chondrocytes when seeded at low cell densities in alginate in vitro. *J. Orthop. Res.* 18 (6):882–890.

Giannona, S., I. Firkowska, J. Rojas-Chapana, and M. Giersig. 2007. Vertically aligned carbon nanotubes as cytocompatible material for enhanced adhesion and proliferation of osteoblast-like cells. *J. Nanosci. Nanotechnol.* 7 (4-5):1679–1683.

Goto, K., J. Tamura, S. Shinzato, S. Fujibayashi, M. Hashimoto, M. Kawashita, T. Kokubo, and T. Nakamura. 2005. Bioactive bone cements containing nano-sized titania particles for use as bone substitutes. *Biomaterials* 26 (33):6496–6505.

Gupta, D., J. Venugopal, S. Mitra, V. R. Giri Dev, and S. Ramakrishna. 2009. Nanostructured biocomposite substrates by electrospinning and electrospraying for the mineralization of osteoblasts. *Biomaterials* 30 (11):2085–2094.

Hashi, C. K., Y. Zhu, G. Y. Yang, W. L. Young, B. S. Hsiao, K. Wang, B. Chu, and S. Li. 2007. Antithrombogenic property of bone marrow mesenchymal stem cells in nanofibrous vascular grafts. *Proc. Natl Acad. Sci. U. S. A.* 104 (29):11915–11920.

Hasirci, V., E. Vrana, P. Zorlutuna, A. Ndreu, P. Yilgor, F. B. Basmanav, and E. Aydin. 2006. Nanobiomaterials: A review of the existing science and technology, and new approaches. *J. Biomater. Sci. Polym. Ed.* 17 (11):1241–1268.

Horch, R. A., N. Shahid, A. S. Mistry, M. D. Timmer, A. G. Mikos, and A. R. Barron. 2004. Nanoreinforcement of poly(propylene fumarate)-based networks with surface modified alumoxane nanoparticles for bone tissue engineering. *Biomacromolecules* 5 (5):1990–1998.

Huang, J., Y. W. Lin, X. W. Fu, S. M. Best, R. A. Brooks, N. Rushton, and W. Bonfield. 2007. Development of nano-sized hydroxyapatite reinforced composites for tissue engineering scaffolds. *J. Mater. Sci. Mater. Med.* 18 (11):2151–2157.

Huang, N. F., S. Patel, R. G. Thakar, J. Wu, B. S. Hsiao, B. Chu, R. J. Lee, and S. Li. 2006. Myotube assembly on nanofibrous and micropatterned polymers. *Nano Lett.* 6 (3):537–542.

Hurme, T., H. Kalimo, M. Lehto, and M. Jarvinen. 1991. Healing of skeletal muscle injury: An ultrastructural and immunohistochemical study. *Med. Sci. Sports Exerc.* 23 (7):801–810.

Ino, K., A. Ito, and H. Honda. 2007. Cell patterning using magnetite nanoparticles and magnetic force. *Biotechnol. Bioeng.* 97 (5):1309–1317.

Ito, A., K. Ino, T. Kobayashi, and H. Honda. 2005. The effect of RGD peptide-conjugated magnetite cationic liposomes on cell growth and cell sheet harvesting. *Biomaterials* 26 (31):6185–6193.

Jager, M., C. Zilkens, K. Zanger, and R. Krauspe. 2007. Significance of nano- and microtopography for cell-surface interactions in orthopaedic implants. *J. Biomed. Biotechnol.* 2007 (8):69036.

Jeng, L. B., H. Y. Chung, T. M. Lin, J. P. Chen, Y. L. Chen, Y. L. Lu, Y. J. Wang, and S. C. Chang. 2008. Characterization and osteogenic effects of mesenchymal stem cells on microbeads composed of hydroxyapatite nanoparticles/reconstituted collagen. *J. Biomed. Mater. Res. A* 91 (3):886–893.

Jeong, S. I., I. D. Jun, M. J. Choi, Y. C. Nho, Y. M. Lee, and H. Shin. 2008. Development of electroactive and elastic nanofibers that contain polyaniline and poly(L-lactide-co-epsilon-caprolactone) for the control of cell adhesion. *Macromol. Biosci.* 8 (7):627–637.

Jordan, S. W. and E. L. Chaikof. 2007. Novel thromboresistant materials. *J. Vasc. Surg.* 45 Suppl A:A104–A115.

Khang, D., S. Y. Kim, P. Liu-Snyder, G. T. Palmore, S. M. Durbin, and T. J. Webster. 2007. Enhanced fibronectin adsorption on carbon nanotube/poly(carbonate) urethane: Independent role of surface nano-roughness and associated surface energy. *Biomaterials* 28 (32):4756–4768.

Khang, D., J. Lu, C. Yao, K. M. Haberstroh, and T. J. Webster. 2008. The role of nanometer and sub-micron surface features on vascular and bone cell adhesion on titanium. *Biomaterials* 29 (8):970–983.

Kim, S. S., K. M. Ahn, M. S. Park, J. H. Lee, C. Y. Choi, and B. S. Kim. 2007. A poly(lactide-co-glycolide)/hydroxyapatite composite scaffold with enhanced osteoconductivity. *J. Biomed. Mater. Res. A* 80 (1):206–215.

Kim, K. and J. P. Fisher. 2007. Nanoparticle technology in bone tissue engineering. *J. Drug Target* 15 (4):241–252.

Kim, H. W., H. E. Kim, V. Salih, and J. C. Knowles. 2005. Hydroxyapatite and titania sol-gel composite coatings on titanium for hard tissue implants; mechanical and in vitro biological performance. *J. Biomed. Mater. Res. B Appl. Biomater.* 72 (1):1–8.

Kim, B. S. and D. J. Mooney. 1998. Engineering smooth muscle tissue with a predefined structure. *J. Biomed. Mater. Res.* 41 (2):322–332.

Kim, S. S., M. Sun Park, O. Jeon, C. Yong Choi, and B. S. Kim. 2006. Poly(lactide-co-glycolide)/hydroxyapatite composite scaffolds for bone tissue engineering. *Biomaterials* 27 (8):1399–1409.

Ko, E. K., S. I. Jeong, N. G. Rim, Y. M. Lee, H. Shin, and B. K. Lee. 2008. In vitro osteogenic differentiation of human mesenchymal stem cells and in vivo bone formation in composite nanofiber meshes. *Tissue Eng. Part A* 14 (12):2105–2119.

Kothapalli, C. R. and A. Ramamurthi. 2009. Copper nanoparticle cues for biomimetic cellular assembly of crosslinked elastin fibers. *Acta Biomater.* 5 (2):541–553.

L'Heureux, N., S. Paquet, R. Labbe, L. Germain, and F. A. Auger. 1998. A completely biological tissue-engineered human blood vessel. *FASEB J.* 12 (1):47–56.

Le Guehennec, L., A. Soueidan, P. Layrolle, and Y. Amouriq. 2007. Surface treatments of titanium dental implants for rapid osseointegration. *Dent. Mater.* 23 (7):844–854.

Le Guillou-Buffello, D., R. Bareille, M. Gindre, A. Sewing, P. Laugier, and J. Amedee. 2008. Additive effect of RGD coating to functionalized titanium surfaces on human osteoprogenitor cell adhesion and spreading. *Tissue Eng. Part A* 14 (8):1445–1455.

Lee, K. W., S. Wang, M. J. Yaszemski, and L. Lu. 2008. Physical properties and cellular responses to crosslinkable poly(propylene fumarate)/hydroxyapatite nanocomposites. *Biomaterials* 29 (19):2839–2848.

Lewandrowski, K. U., S. P. Bondre, D. L. Wise, and D. J. Trantolo. 2003. Enhanced bioactivity of a poly(propylene fumarate) bone graft substitute by augmentation with nano-hydroxyapatite. *Biomed. Mater. Eng.* 13 (2):115–124.

Li, P. 2003. Biomimetic nano-apatite coating capable of promoting bone ingrowth. *J. Biomed. Mater. Res. A* 66 (1):79–85.

Li, W. J., H. Chiang, T. F. Kuo, H. S. Lee, C. C. Jiang, and R. S. Tuan. 2009. Evaluation of articular cartilage repair using biodegradable nanofibrous scaffolds in a swine model: A pilot study. *J. Tissue Eng. Regen. Med.* 3 (1):1–10.

Li, W. J., K. G. Danielson, P. G. Alexander, and R. S. Tuan. 2003. Biological response of chondrocytes cultured in three-dimensional nanofibrous poly(epsilon-caprolactone) scaffolds. *J. Biomed. Mater. Res. A* 67 (4):1105–1114.

Li, W. J., Y. J. Jiang, and R. S. Tuan. 2008. Cell-nanofiber-based cartilage tissue engineering using improved cell seeding, growth factor, and bioreactor technologies. *Tissue Eng. Part A* 14 (5):639–648.

Li, W. J., R. Tuli, X. Huang, P. Laquerriere, and R. S. Tuan. 2005. Multilineage differentiation of human mesenchymal stem cells in a three-dimensional nanofibrous scaffold. *Biomaterials* 26 (25):5158–5166.

Liao, S., W. Wang, M. Uo, S. Ohkawa, T. Akasaka, K. Tamura, F. Cui, and F. Watari. 2005. A three-layered nano-carbonated hydroxyapatite/collagen/PLGA composite membrane for guided tissue regeneration. *Biomaterials* 26 (36):7564–7571.

Liu, Q., J. R. de Wijn, and C. A. van Blitterswijk. 1998. Composite biomaterials with chemical bonding between hydroxyapatite filler particles and PEG/PBT copolymer matrix. *J. Biomed. Mater. Res.* 40 (3):490–497.

Liu, H., E. B. Slamovich, and T. J. Webster. 2006a. Increased osteoblast functions among nanophase titania/poly(lactide-co-glycolide) composites of the highest nanometer surface roughness. *J. Biomed. Mater. Res. A* 78 (4):798–807.

Liu, H., E. B. Slamovich, and T. J. Webster. 2006b. Less harmful acidic degradation of poly(lacticco-glycolic acid) bone tissue engineering scaffolds through titania nanoparticle addition. *Int. J. Nanomed.* 1 (4):541–545.

Liu, X., L. A. Smith, J. Hu, and P. X. Ma. 2009. Biomimetic nanofibrous gelatin/apatite composite scaffolds for bone tissue engineering. *Biomaterials* 30 (12):2252–2258.

Lu, J., M. P. Rao, N. C. MacDonald, D. Khang, and T. J. Webster. 2008. Improved endothelial cell adhesion and proliferation on patterned titanium surfaces with rationally designed, micrometer to nanometer features. *Acta Biomater.* 4 (1):192–201.

MacDonald, R. A., C. M. Voge, M. Kariolis, and J. P. Stegemann. 2008. Carbon nanotubes increase the electrical conductivity of fibroblast-seeded collagen hydrogels. *Acta Biomater.* 4 (6):1583–1592.

Matsuda, T. and H. Miwa. 1995. A hybrid vascular model biomimicking the hierarchic structure of arterial wall: Neointimal stability and neoarterial regeneration process under arterial circulation. *J. Thorac. Cardiovasc. Surg.* 110 (4 Pt 1):988–997.

Mironov, V., V. Kasyanov, and R. R. Markwald. 2008. Nanotechnology in vascular tissue engineering: From nanoscaffolding towards rapid vessel biofabrication. *Trends Biotechnol.* 26 (6):338–344.

Mistry, A. S., S. H. Cheng, T. Yeh, E. Christenson, J. A. Jansen, and A. G. Mikos. 2009a. Fabrication and in vitro degradation of porous fumarate-based polymer/alumoxane nanocomposite scaffolds for bone tissue engineering. *J. Biomed. Mater. Res. A* 89 (1):68–79.

Mistry, A. S., A. G. Mikos, and J. A. Jansen. 2007. Degradation and biocompatibility of a poly(propylene fumarate)-based/alumoxane nanocomposite for bone tissue engineering. *J. Biomed. Mater. Res. A* 83 (4):940–953.

Mistry, A. S., Q. P. Pham, C. Schouten, T. Yeh, E. M. Christenson, A. G. Mikos, and J. A. Jansen. 2009b. In vivo bone biocompatibility and degradation of porous fumarate-based polymer/alumoxane nanocomposites for bone tissue engineering. *J. Biomed. Mater. Res. A* 92 (2):451–462.

Mukherjee, P., R. Bhattacharya, P. Wang, L. Wang, S. Basu, J. A. Nagy, A. Atala, D. Mukhopadhyay, and S. Soker. 2005. Antiangiogenic properties of gold nanoparticles. *Clin. Cancer Res.* 11 (9):3530–3534.

Murugan, R. and S. Ramakrishna. 2007. Design strategies of tissue engineering scaffolds with controlled fiber orientation. *Tissue Eng.* 13 (8):1845–1866.

Nain, A. S., J. A. Phillippi, M. Sitti, J. Mackrell, P. G. Campbell, and C. Amon. 2008. Control of cell behavior by aligned micro/nanofibrous biomaterial scaffolds fabricated by spinneret-based tunable engineered parameters (STEP) technique. *Small* 4 (8):1153–1159.

Niklason, L. E., J. Gao, W. M. Abbott, K. K. Hirschi, S. Houser, R. Marini, and R. Langer. 1999. Functional arteries grown in vitro. *Science* 284 (5413):489–493.

Nukavarapu, S. P., S. G. Kumbar, J. L. Brown, N. R. Krogman, A. L. Weikel, M. D. Hindenlang, L. S. Nair, H. R. Allcock, and C. T. Laurencin. 2008. Polyphosphazene/nano-hydroxyapatite composite microsphere scaffolds for bone tissue engineering. *Biomacromolecules* 9 (7):1818–1825.

Park, J. S., K. Park, D. G. Woo, H. N. Yang, H. M. Chung, and K. H. Park. 2008. PLGA microsphere construct coated with TGF-beta 3 loaded nanoparticles for neocartilage formation. *Biomacromolecules* 9 (8):2162–2169.

Payumo, F. C., H. D. Kim, M. A. Sherling, L. P. Smith, C. Powell, X. Wang, H. S. Keeping, R. F. Valentini, and H. H. Vandenburgh. 2002. Tissue engineering skeletal muscle for orthopaedic applications. *Clin. Orthop. Relat. Res.* (403 Suppl):S228–S242.

Perea, H., J. Aigner, J. T. Heverhagen, U. Hopfner, and E. Wintermantel. 2007. Vascular tissue engineering with magnetic nanoparticles: Seeing deeper. *J. Tissue Eng. Regen. Med.* 1 (4):318–321.

Poole, A. R. 2003. What type of cartilage repair are we attempting to attain? *J. Bone Joint Surg. Am.* 85-A Suppl 2:40–44.

Puckett, S., R. Pareta, and T. J. Webster. 2008. Nano rough micron patterned titanium for directing osteoblast morphology and adhesion. *Int. J. Nanomed.* 3 (2):229–241.

Ratcliffe, A. 2000. Tissue engineering of vascular grafts. *Matrix Biol.* 19 (4):353–357.

Rebollar, E., I. Frischauf, M. Olbrich, T. Peterbauer, S. Hering, J. Preiner, P. Hinterdorfer, C. Romanin, and J. Heitz. 2008. Proliferation of aligned mammalian cells on laser-nanostructured polystyrene. *Biomaterials* 29 (12):1796–1806.

Riboldi, S. A., M. Sampaolesi, P. Neuenschwander, G. Cossu, and S. Mantero. 2005. Electrospun degradable polyesterurethane membranes: Potential scaffolds for skeletal muscle tissue engineering. *Biomaterials* 26 (22):4606–4615.

Ricker, A., P. Liu-Snyder, and T. J. Webster. 2008. The influence of nano MgO and $BaSO_4$ particle size additives on properties of PMMA bone cement. *Int. J. Nanomed.* 3 (1):125–132.

Rolland, J. P., B. W. Maynor, L. E. Euliss, A. E. Exner, G. M. Denison, and J. M. DeSimone. 2005. Direct fabrication and harvesting of monodisperse, shape-specific nanobiomaterials. *J. Am. Chem. Soc.* 127 (28):10096–100100.

Rotmans, J. I., J. M. Heyligers, H. J. Verhagen, E. Velema, M. M. Nagtegaal, D. P. de Kleijn, F. G. de Groot, E. S. Stroes, and G. Pasterkamp. 2005. In vivo cell seeding with anti-CD34 antibodies successfully accelerates endothelialization but stimulates intimal hyperplasia in porcine arteriovenous expanded polytetrafluoroethylene grafts. *Circulation* 112 (1):12–18.

Sachlos, E., N. Reis, C. Ainsley, B. Derby, and J. T. Czernuszka. 2003. Novel collagen scaffolds with predefined internal morphology made by solid freeform fabrication. *Biomaterials* 24 (8):1487–1497.

Sato, M., A. Aslani, M. A. Sambito, N. M. Kalkhoran, E. B. Slamovich, and T. J. Webster. 2008. Nanocrystalline hydroxyapatite/titania coatings on titanium improves osteoblast adhesion. *J. Biomed. Mater. Res. A* 84 (1):265–272.

Sato, M. and T. J. Webster. 2004. Nanobiotechnology: Implications for the future of nanotechnology in orthopedic applications. *Expert Rev. Med. Devices* 1 (1):105–114.

Saxena, A. K., G. H. Willital, and J. P. Vacanti. 2001. Vascularized three-dimensional skeletal muscle tissue-engineering. *Biomed. Mater. Eng.* 11 (4):275–281.

Schulze-Tanzil, G., P. de Souza, H. Villegas Castrejon, T. John, H. J. Merker, A. Scheid, and M. Shakibaei. 2002. Redifferentiation of dedifferentiated human chondrocytes in high-density cultures. *Cell Tissue Res.* 308 (3):371–379.

Sen, P. K., G. B. Parulkar, M. D. Kelkar, S. R. Panday, B. N. Irani, and B. K. Shah. 1965. Clinical experiences in vascular grafting. A report of first 100 cases. *Indian Heart J.* 17 (3):250–262.

Shi, X., J. L. Hudson, P. P. Spicer, J. M. Tour, R. Krishnamoorti, and A. G. Mikos. 2006. Injectable nanocomposites of single-walled carbon nanotubes and biodegradable polymers for bone tissue engineering. *Biomacromolecules* 7 (7):2237–2242.

Shimizu, K., A. Ito, M. Arinobe, Y. Murase, Y. Iwata, Y. Narita, H. Kagami, M. Ueda, and H. Honda. 2007. Effective cell-seeding technique using magnetite nanoparticles and magnetic force onto decellularized blood vessels for vascular tissue engineering. *J. Biosci. Bioeng.* 103 (5):472–478.

Shin, H., S. Jo, and A. G. Mikos. 2003. Biomimetic materials for tissue engineering. *Biomaterials* 24 (24):4353–4364.

Shin, M., H. Yoshimoto, and J. P. Vacanti. 2004. In vivo bone tissue engineering using mesenchymal stem cells on a novel electrospun nanofibrous scaffold. *Tissue Eng.* 10 (1–2):33–41.

Sitharaman, B., X. Shi, L. A. Tran, P. P. Spicer, I. Rusakova, L. J. Wilson, and A. G. Mikos. 2007. Injectable in situ cross-linkable nanocomposites of biodegradable polymers and carbon nanostructures for bone tissue engineering. *J. Biomater. Sci. Polym. Ed.* 18 (6):655–671.

Smith, L. A. and P. X. Ma. 2004. Nano-fibrous scaffolds for tissue engineering. *Colloids Surf. B Biointerfaces* 39 (3):125–131.

Smith, L. L., P. J. Niziolek, K. M. Haberstroh, E. A. Nauman, and T. J. Webster. 2007a. Decreased fibroblast and increased osteoblast adhesion on nanostructured NaOH-etched PLGA scaffolds. *Int. J. Nanomed.* 2 (3):383–388.

Smith, L. J., J. S. Swaim, C. Yao, K. M. Haberstroh, E. A. Nauman, and T. J. Webster. 2007b. Increased osteoblast cell density on nanostructured PLGA-coated nanostructured titanium for orthopedic applications. *Int. J. Nanomed.* 2 (3):493–499.

Streicher, R. M., M. Schmidt, and S. Fiorito. 2007. Nanosurfaces and nanostructures for artificial orthopedic implants. *Nanomedicine* 2 (6):861–874.

Tanaka, M. and E. Sackmann. 2005. Polymer-supported membranes as models of the cell surface. *Nature* 437 (7059):656–663.

Teng, S. H., E. J. Lee, C. S. Park, W. Y. Choi, D. S. Shin, and H. E. Kim. 2008. Bioactive nanocomposite coatings of collagen/hydroxyapatite on titanium substrates. *J. Mater. Sci. Mater. Med.* 19 (6):2453–2461.

Tsai, C. Y., A. L. Shiau, S. Y. Chen, Y. H. Chen, P. C. Cheng, M. Y. Chang, D. H. Chen, C. H. Chou, C. R. Wang, and C. L. Wu. 2007. Amelioration of collagen-induced arthritis in rats by nanogold. *Arthritis Rheum.* 56 (2):544–554.

Tuli, R., W. J. Li, and R. S. Tuan. 2003. Current state of cartilage tissue engineering. *Arthritis Res. Ther.* 5 (5):235–238.

Ute Henze, U., M. Kaufmann, B. Klein, S. Handt, and B. Klosterhalfen. 1996. Endothelium and biomaterials: Morpho-functional assessments. *Biomed. Pharmacother.* 50 (8):388.

Venugopal, J. R., S. Low, A. T. Choon, A. B. Kumar, and S. Ramakrishna. 2008a. Nanobioengineered electrospun composite nanofibers and osteoblasts for bone regeneration. *Artif. Organs* 32 (5):388–397.

Venugopal, J., S. Low, A. T. Choon, and S. Ramakrishna. 2008b. Interaction of cells and nanofiber scaffolds in tissue engineering. *J. Biomed. Mater. Res. B Appl. Biomater.* 84 (1):34–48.

Verma, D., K. S. Katti, and D. R. Katti. 2008. Effect of biopolymers on structure of hydroxyapatite and interfacial interactions in biomimetically synthesized hydroxyapatite/biopolymer nanocomposites. *Ann. Biomed. Eng.* 36 (6):1024–1032.

Wang, H., Y. Li, Y. Zuo, J. Li, S. Ma, and L. Cheng. 2007. Biocompatibility and osteogenesis of biomimetic nano-hydroxyapatite/polyamide composite scaffolds for bone tissue engineering. *Biomaterials* 28 (22):3338–3348.

Webster, T. J. and J. U. Ejiofor. 2004. Increased osteoblast adhesion on nanophase metals: Ti, Ti6Al4V, and CoCrMo. *Biomaterials* 25 (19):4731–4739.

Webster, T. J., C. Ergun, R. H. Doremus, R. W. Siegel, and R. Bizios. 2000a. Enhanced functions of osteoblasts on nanophase ceramics. *Biomaterials* 21 (17):1803–1810.

Webster, T. J., C. Ergun, R. H. Doremus, R. W. Siegel, and R. Bizios. 2000b. Specific proteins mediate enhanced osteoblast adhesion on nanophase ceramics. *J. Biomed. Mater. Res.* 51 (3):475–483.

Webster, T. J., R. W. Siegel, and R. Bizios. 1999. Osteoblast adhesion on nanophase ceramics. *Biomaterials* 20 (13):1221–1227.

Webster, T. J. and T. A. Smith. 2005. Increased osteoblast function on PLGA composites containing nanophase titania. *J. Biomed. Mater. Res. A* 74 (4):677–686.

Wei, G. and P. X. Ma. 2004. Structure and properties of nano-hydroxyapatite/polymer composite scaffolds for bone tissue engineering. *Biomaterials* 25 (19):4749–4757.

Weinberg, C. B. and E. Bell. 1986. A blood vessel model constructed from collagen and cultured vascular cells. *Science* 231 (4736):397–400.

Wilson, J. T., W. Cui, X. L. Sun, C. Tucker-Burden, C. J. Weber, and E. L. Chaikof. 2007. In vivo biocompatibility and stability of a substrate-supported polymerizable membrane-mimetic film. *Biomaterials* 28 (4):609–617.

Wise, J. K., A. L. Yarin, C. M. Megaridis, and M. Cho. 2008. Chondrogenic differentiation of human mesenchymal stem cells on oriented nanofibrous scaffolds: Engineering the superficial zone of articular cartilage. *Tissue Eng. Part A* 15(4):913–921.

Woo, K. M., V. J. Chen, and P. X. Ma. 2003. Nano-fibrous scaffolding architecture selectively enhances protein adsorption contributing to cell attachment. *J. Biomed. Mater. Res. A* 67 (2):531–537.

Woo, K. M., J. H. Jun, V. J. Chen, J. Seo, J. H. Baek, H. M. Ryoo, G. S. Kim, M. J. Somerman, and P. X. Ma. 2007. Nano-fibrous scaffolding promotes osteoblast differentiation and biomineralization. *Biomaterials* 28 (2):335–343.

Xin, X., M. Hussain, and J. J. Mao. 2007. Continuing differentiation of human mesenchymal stem cells and induced chondrogenic and osteogenic lineages in electrospun PLGA nanofiber scaffold. *Biomaterials* 28 (2):316–325.

Xu, C., R. Inai, M. Kotaki, and S. Ramakrishna. 2004. Electrospun nanofiber fabrication as synthetic extracellular matrix and its potential for vascular tissue engineering. *Tissue Eng.* 10 (7–8):1160–1168.

Yang, L., B. W. Sheldon, and T. J. Webster. 2008. Orthopedic nano diamond coatings: Control of surface properties and their impact on osteoblast adhesion and proliferation. *J. Biomed. Mater. Res. A* 91 (2):548–556.

Yoshimoto, H., Y. M. Shin, H. Terai, and J. P. Vacanti. 2003. A biodegradable nanofiber scaffold by electrospinning and its potential for bone tissue engineering. *Biomaterials* 24 (12):2077–2082.

Zanello, L. P., B. Zhao, H. Hu, and R. C. Haddon. 2006. Bone cell proliferation on carbon nanotubes. *Nano Lett.* 6 (3):562–567.

Zorlutuna, P., N. Hasirci, and V. Hasirci. 2008. Nanopatterned collagen tubes for vascular tissue engineering. *J. Tissue Eng. Regen. Med.* 2 (6):373–377.

24

Nanocomposite Polymer Biomaterials for Tissue Repair of Bone and Cartilage: A Material Science Perspective

Akhilesh K. Gaharwar
Purdue University

Patrick J. Schexnailder
Purdue University

Gudrun Schmidt
Purdue University

24.1 Introduction to Polymer Nanocomposite Biomaterials

The design and fabrication of nanocomposite biomaterials for tissue engineering applications requires a fundamental understanding of the interactions between polymers, nanostructures, and cells. Biology offers the best models for strategies on how to rationally design high-performance biomaterials with the properties of materials, such as bone, cartilage, nacre, or silk (Murphy and Mooney 2002; Gao et al. 2003; Tang et al. 2003; Mayer 2005; Lee and Spencer 2008). To translate our fundamental understanding of nature into products that are useful in a clinical setting, the chemical, physical, and biological properties of newly developed bio-nanocomposites need to be optimized to support, regulate, and influence long term cellular activities.

Although we can replicate some physical properties of natural biomaterials, reproducing the complexity and efficiency of natural tissue is challenging. For example, the strength and stiffness of bone is related to its highly ordered structure at the nano- and micro-length scales (Weiner and Traub 1992; Weiner and Wagner 1998). Cartilage has unique combinations of nonlinear tensile and compressive properties due to hierarchically arranged collagen fibrils, proteoglycans, and proteins (Mow et al. 1980; Mankin 1982; Buckwalter and Mankin 1997; Cohen et al. 1998). The structural and mechanical properties of such natural tissue can be imitated by engineering bio-nanocomposites made from polymer and nanoparticles (nanospheres, nanotubes, nanoplatelets, etc.) (Engel et al. 2008; Ma 2008; Vaia and Baur 2008; Smith et al. 2009). The nanoparticles often act as physical cross-links to the polymer chains, which reinforce and enhance the mechanical properties of the nanocomposite biomaterial (Kotela et al. 2009; Schexnailder and Schmidt 2009; Smith et al. 2009; Zhang and Webster 2009).

Seen from a material science perspective, several research groups have developed fabrication techniques for generating supramolecular assembled nanocomposites of polymers and nanoparticles (Giannelis 1996; Alexandre and Dubois 2000; Gao et al. 2003; Hule and Pochan 2007). These creative ideas can easily be translated for the development of biomaterials. Many of the studies in this area are based on the observation that small amounts of nanoparticles can dramatically change the physical properties of a polymer matrix, allowing for the engineering of synergistic property combinations that the individual components cannot achieve (Liff et al. 2007; Podsiadlo et al. 2007). For example, fabrication technologies such as electrospinning, layer-by-layer techniques, salt leaching, lyophilization, etc. have been developed to generate structural features that can potentially be used for the design and development of tissue engineering and drug delivery matrixes (Doshi and Reneker 1993; Decher 1997; Lee et al. 2005; Ateshian 2007; Kong et al. 2007). In one example, new creative approaches by Liff et al. (2007) were able to overcome the thermodynamic and kinetic barriers of nanoparticle dispersion via solvent exchange. Strong adhesion between silicate and polymer microdomains and the formation of a percolative network induce thermotropic liquid crystalline behavior in addition to mechanical strength (Liff et al. 2007). The physical properties of such polymer materials are ideally suited for tissue repair. Other techniques that could be adapted to biomaterial development include the continuous self-assembly and polymerization of silica, surfactant, and monomers into nacre-like nanolaminated coatings, which was previously achieved by Sellinger et al. (1998). Here, evaporation induced partitioning and self-assembly resulted in the simultaneous organization of thousands of layers at once (Sellinger et al. 1998). Such techniques can further be developed to generate gradient layered scaffolds for diverse tissue engineering applications.

In order to design nanocomposite biomaterials or to develop already existing nanocomposites for the repair of bone and cartilage, we need to consider and compare the structures and properties of the natural tissue, along with the biological influence of cells on the synthetic biomaterial properties. Biomedical research has given us a substantial understanding of the musculoskeletal structure and its function (Weiner and Traub 1992; Rho et al. 1998; Weiner and Wagner 1998; Weiner et al. 1999). Several studies have provided valuable insights into cell–scaffold interactions (Hazan et al. 1993; Stevens and George 2005; Jones 2006; Ma 2008; Stoddart et al. 2009) in addition to physical and chemical material properties. However, nature's perfection and efficiency are not easily matched, and thus our attempts to repair living tissue with synthetic biomaterials often remain limited.

The scope of this review is to focus on the design, fabrication, and evaluation of polymeric nanocomposite biomaterials that are currently used and that can potentially be used for the tissue engineering of bone, cartilage, and the bone–cartilage interface. We first discuss the properties of the natural tissue to identify the requirements for engineering suitable nanocomposite matrixes and scaffolds. Then, we highlight the most recent accomplishments and trends in the field of bio-nanocomposites for the tissue engineering of bone, cartilage, and the bone–cartilage interface and examine the impact of published work on specific tissue engineering applications. Overall, we address some of the most critical challenges that come with the design, fabrication, and evaluation of bio-nanocomposite scaffolds and conclude with a brief outline on future directions.

24.2 Important Structures and Properties

24.2.1 Bone: A Hard Nanocomposite

A better understanding of the relations between the composition, structure, and properties of bone can guide the design of tissue engineered synthetic grafts that ideally remodel and heal bone tissue. Table 24.1 lists the main components of bone, while Figure 24.1 summarizes well-known micrometer and nanometer structural features, and Table 24.2 gives ranges of mechanical properties that are important to know when developing bone replacement tissue.

TABLE 24.1 Main Components of Bone

	Component	%	Physical Shape/Function
Inorganic	Hydroxyapatite—(Ca_5 (PO_4, CO_3)$_3$(OH))	~60	Plate-shaped (50 × 25 × 1.5 nm), modulus = 109–114 GPa
	Carbonate, citrate	~5	
	Other minerals, that is, Mg, Na, Cl, F, K^+, Sr^{2+}, Pb^{2+}, Zn^{2+}, Cu^{2+}, Fe^{2+}	~1	Dissolved in water/metabolic function
Organic	Collagen	~20	Fibrils (1.5–3.5 nm), fibers (50–70 nm), bundles (150–250 nm)
	Water	~9	Bound and non-bound state
	Non-collagenous proteins (osteocalcin, osteonectin, osteopontin, thrombospondin, morphogenetic proteins, sialo, and serum protein)	~3	Cellular attachment and cell metabolism
	Other traces: Polysaccharides, lipids, cytokines		
	Bone cells: osteoblasts, osteocytes, osteoclasts		

Source: Murugan, R. and Ramakrishna, S., *Compos. Sci. Technol.*, 65(15–16), 2385, 2005; Weiner, S. and Wagner, H.D., *Annu. Rev. Mater. Sci.*, 28(1), 271, 1998; Rho, J.-Y. et al., *Med. Eng. Phys.*, 20(2), 92, 1998.

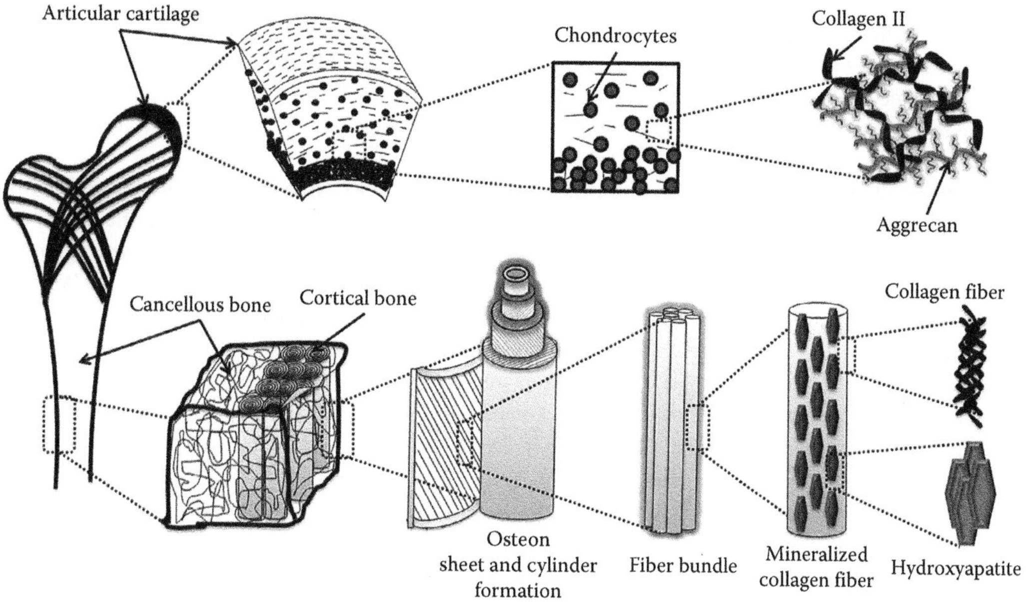

FIGURE 24.1 Micrometer and nanometer structural features of bone and cartilage. (Adapted, modified, and redrawn from Weiner, S. and Traub, W., *FASEB J.*, 6(3), 879, 1992; Kikuchi, M. and Kanama, D., *Sci. Technol. Trends*, 24, 52, 2007.)

TABLE 24.2 Mechanical Properties of Bone and Cartilage

Tissue	Tensile Modulus (MPa)	Tensile Strength (MPa)	Compression Modulus (MPa)	Compressive Strength (MPa)	Fracture Toughness (MPa/m$^{-1/2}$)	Shear Modulus (MPa)	Density (g/cm^3)
Cortical bone	5–20 (× 10^3)	50–150	4–15 (× 10^3)	170–193	2–12	~300 (× 10^3)	1.8–2.2
Cancellous bone	50–500	2–20	15–100	7–10	0.1	1–70	1.5–1.9
Cartilage	0.5–30	2–50	0.13–1.91	25–65	0.5–3.8	0.8–2.1	~1.12

Source: Akizuki, S. et al., *J. Orthop. Res.*, 4(4), 379, 1986; Burg, K.J.L. et al., *Biomaterials*, 21(23), 2347, 2000; Carter, D.R. and Hayes, W.C., *Science*, 194(4270), 1174, 1976; Heinegard, D. and Oldberg, A., *FASEB J.*, 3(9), 2042, 1989; Hench, L.L. *Biomaterials*, 19(16), 1419, 1998; Laurencin, C.T. et al., *Annu. Rev. Biomed. Eng.*, 1(1), 19, 1999; Murugan, R. and Ramakrishna, S., *Compos. Sci. Technol.*, 65(15–16), 2385, 2005; Weiner, S. and Traub, W., *FASEB J.*, 6(3), 879, 1992; Weiner, S. et al., *J. Struct. Biol.*, 126(3), 241, 1999; Zhu, W. et al., *J. Orthop. Res.*, 11(6), 771, 1993; Aubin, J.E. and Liu, F., *Principle of Bone Biology*, eds. J.E. Aubin, L.G. Raisz, and G.A. Rodan, Academic, San Diego, CA, 1996; Mano, J.F. et al., *Compos. Sci. Technol.*, 64(6), 789, 2004; Reilly, D.T. and Burstein, A.H., *J. Bone Joint Surg.*, 56(5), 1001, 1974; Reilly, D.T. et al., *J. Biomech.*, 7(3), 271, 1974; Armstrong, C.G. and Mow, V.C., *J. Bone Joint Surg.*, 64(1), 88, 1982; Kerin, A.J. et al., *Proc. Inst. Mech. Eng. H*, 212(4), 273, 1998; Stok, K. and Oloyede, A., *Clin. Biomech.*, 22(6), 725, 2007; Loret, B. and Simoes, F.M.F., *Mech. Mater.*, 36(5–6), 515, 2004; Norman, T.L. et al., *J. Biomech.*, 28(3), 309, 1995; Ulrich, D. et al., *Bone* 25(1), 55, 1999.

From a material science point of view, bone is a natural nanocomposite composed mainly of hydroxyapatite (HAp) nanoparticles, collagen fibrils, and cells (osteoblasts, osteocytes, osteoclasts). The relative proportions and structural orientations of HAp and collagen in combination with the many synergistic property combinations added by the other ingredients influence the biomechanical properties. For example, trace amounts of lipids are important for biomineralization while non-collagenous protein (Heinegard and Oldberg 1989) and collagen (Weiner and Wagner 1998; Reffitt et al. 2003; Fratzl 2008) are involved in the nucleation of HAp. Thus, the combinations and interactions of multiple ingredients during the synthesis and growth of bone determine the properties of the end product. Moreover, the dynamic mechanical and compositional microenvironment of bone marrow influence bone homeostasis (Gurkan and Akkus 2008). In addition, many other components not listed in Table 24.1 indirectly influence bone synthesis. One of these components is silicon, which several studies suggest is essential for the formation of bone by participating in cell metabolism (Carlisle 1970, 1972, 1981; Seaborn and Nielsen 2002).

The biomechanical properties of bone are also determined by the extent of mineralization, the amount of water, and the 3D architecture at each hierarchical level (Aubin and Liu 1996). For example, both cortical and cancellous bone are composed of the same material, but their microstructural arrangement results in significantly different mechanical properties (Table 24.2) (Rho et al. 1998; Weiner and Wagner 1998). Cortical bone is composed of cylindrical-shaped units called osteons, a dense tissue consisting of concentric lamellae, while cancellous bone is a porous structure (Figure 24.1). Moreover, an ordered arrangement of mineralized collagen fibrils impart anisotropic properties to both cortical and cancellous bones (Murugan and Ramakrishna 2005). A variation in mechanical properties is also dependent on the anatomical location of bone (Currey 1988). The yield strength of bone is half and its density is about a quarter as that of steel (Carter and Hayes 1976; Disegi and Eschbach 2000). The complex hierarchical structure, ranging from nanometers to micrometers (Figure 24.1), is mainly responsible for the superior strength while being light weight (Weiner and Traub 1992; Rho et al. 1998; Weiner and Wagner 1998). The basic building blocks of bone are mineralized collagen fibrils that are arranged in an ordered structure and act as a structural framework (Weiner and Wagner 1998; Weiner et al. 1999). In summary, the nanocomposite structure and anisotropic properties with an interconnected porous network are the basis for the high strength architectural framework.

While it is impossible to imitate the complexity of natural bone formation via bottom-up synthesis or via sophisticated formulation techniques, there are other approaches that might be considered limited, but nevertheless useful. Such approaches aim at developing scaffolds or structural frameworks for supporting natural bone repair in vivo.

24.2.2 Cartilage: A Soft Composite

The structural and mechanical properties of cartilage are directly related to the composition and architecture of the extra cellular matrix (ECM) (Zhu et al. 1993; Ulrich-Vinther et al. 2003). Unlike bone, cartilage consists of only one cell type (chondrocytes), which are embedded within the ECM. This ECM consists of collagen fibrils and proteoglycan macromolecules (Table 24.3) that are synergistically able to resist shear deformation (Zhu et al. 1993). According to the literature, the collagen fibrils form a network in which the highly charged and swollen proteoglycans generate a significant prestress. The flexible collagen fibrils elastically resist the tensile forces to provide tensile stiffness, and the prestress provided by the proteoglycans maintains fibril orientation and structural stability within the matrix.

Similar to bone, the hierarchical network structure of cartilage is mainly responsible for its mechanical properties (Figure 24.1) (Heinegard and Oldberg 1989). Based on its biochemical composition and morphology, articular cartilage can be divided into four different zones (Weiss et al. 1968; Temenoff and Mikos 2000). The preferred orientation of the fibrils in these different zones stabilizes the articular cartilage against mechanical loadings.

As articular cartilage is composed of porous 3D networks filled with tissue fluid, the compressive properties are a direct result of its biphasic nature (Mow et al. 1980). Articular cartilage exhibits both creep and stress-relaxation behaviors (Cohen et al. 1998). The unique compressive properties of articular cartilage are due to its nonlinear permeability response, especially at high pressures and strains. Thus, under compression, the nonlinear permeability acts as a protective mechanism and stiffens the cartilage by restricting fluid flow. Additionally, the fluid flow through the ECM determines the viscoelastic properties of cartilage. The aligned collagen fibrils present in the ECM are responsible for nonlinear tensile properties. The mechanical properties of cartilage vary with location due to changes in the organization

TABLE 24.3 Main Components of Cartilage and Their Function

	Component	Composition and Function
Extracellular matrix	*Collagen*	Collagen types II, VI, IX, X, and XI. Collagen, type II accounts for 90%–95% of the total collagen content. Collagen fibrils form a mesh that provides high tensile strength and physically traps various bio-macromolecules.
	Proteoglycans	Composed of ~95% polysaccharide and ~5% protein. Most of the polysaccharides are composed of GAG chains, such as hyaluronic acid, chondroitin sulfate, keratan sulfate, dermatan sulfate, and heparan sulfate. Aggregating proteoglycans fill up most of the interfibrillar space and are responsible for the resilience and stress distribution.
	Non-collagenous proteins	Non-collagenous protein, such as glycoproteins, fibronectin, and tenascin, stabilize the ECM matrix and aid in chondrocyte–matrix interactions.
	Tissue fluid	Comprises ~80% of the wet weight of the tissue. Apart from water it also contains dissolved gases, metabolites and cations to balance the negatively charged GAG molecules present in the ECM. Strong interactions of the tissue fluid with the ECM is responsible for compressive properties.
Cells	*Chondrocytes*	Represent only 1% of total volume of cartilage. Mature articular chondrocytes are unable to proliferate, appear rounded, and are completely encased within the ECM.

Source: Buckwalter, J.A. and Mankin, H.J., *J. Bone Joint Surg. Am.*, 79(4), 600, 1997; Temenoff, J.S. and Mikos, A.G., *Biomaterials*, 21(5), 431, 2000; Cohen, N.P. et al., *J. Orthop. Sports Phys. Ther.*, 28(4), 203, 1998.

of collagen fibers from one zone to another. This makes the tissue engineering of cartilage challenging as the design of such gradient structures and anisotropic properties is not straightforward. For example, articular cartilage shows anisotropic mechanical properties when subjected to tension (Akizuki et al. 1986) and compression (Jurvelin et al. 2003), and the stiffness of cartilage in tension is typically 5–20 times greater than in compression (Akizuki et al. 1986). Replacement scaffolds that fulfill these requirements pose serious material design challenges. In order to restore function, scaffolds for cartilage repair mimic only some of the mechanical properties but often do not mimic the hierarchical structural characteristics of cartilage.

Overall, an understanding of the structure, composition, and mechanical properties of cartilage helps researchers in developing strategies for the repair of natural tissue. Since the basis for the high compressive strength of articular cartilage is the 3D nanofibrous polymeric network filled with tissue fluid, stiff nanocomposite hydrogels might be suitable candidates for cartilage repair.

24.3 The Bone–Cartilage Interface: A Gradient Nanocomposite

While highly vascularized bone undergoes constant remodeling by metabolically active cells, avascular cartilage has limited regeneration potential (Yasui et al. 1982; Buckwalter 1992; Buckwalter et al. 1994; Ulrich-Vinther et al. 2003). The bone–cartilage interface is composed of a mineralized cartilage gradient that contains more calcium than the adjacent bone (Zizak et al. 2003). The sizes and widths of the mineralized HAp nanoplatelets in mineralized cartilage and bone are similar. However, an abrupt change in orientation of nanoplatelets happens at the interface. In bone, the mineralized platelets preferentially orient parallel to the interface, while in mineralized cartilage they orient perpendicular to it (Weiner and Traub 1992; Zizak et al. 2003). The change in the orientation of mineralized platelets and the increased amount of calcium at the bone–cartilage interface are responsible for the tight bonding between the two mechanically and structurally different tissues, bone, and cartilage.

When osteochondral injury reaches the subchondral and vascularized bone, a repair response is initiated (Mankin 1982; Brittberg and Winalski 2003). A fibrin clot containing red blood cells, white blood cells, and marrow elements is formed immediately (Mankin 1982; Paletta et al. 1992). This clot is then gradually replaced by fibrous tissue, which is less stiff than native cartilage. Subsequently, the fibrous repair tissue is converted to a hyaline-like chondroidal tissue. The newly formed cartilaginous tissue functions reasonably well if the defect is small. However, for large defects, the newly formed cartilaginous tissue cannot provide enough strength to restore normal function (Brittberg et al. 1994; Peretti et al. 2003).

24.3.1 Properties Required for Tissue Grafts to Work

With our limited capabilities to generate natural tissue and imitate tissue repair, autologous tissue grafts are defined as the gold standard for orthopedic surgeries (Fujishiro et al. 2008). An ideal tissue graft should mimic the structural and functional characteristics of natural tissue and should initiate and support the self-repair of this tissue (Hutmacher 2000; Risbud and Sittinger 2002; Cancedda et al. 2003; Capito and Spector 2003; Sharma and Elisseeff 2004). Although useful scaffolds can be made of various materials bearing a range of properties, the desired properties can be defined as follows:

The scaffolds should be biocompatible and function similar to that of the ECM to support cell survival, proliferation, and migration (Burg et al. 2000; Jones 2006).

The scaffolds should be adhesive to surrounding tissues in order to adhere at or within the injury site.

The scaffolds should have an appropriate degradation rate that matches the regeneration rate of the defective tissue. The degradation by-product(s) should be nontoxic and not elicit an immune response (Freed et al. 1994; Middleton and Tipton 2000; Agrawal and Ray 2001).

The scaffolds should be able to transmit both chemical and mechanical signals to the cells (Ratner 1996; Parikh 2002; Murugan and Ramakrishna 2005).
The scaffolds should have similar mechanical properties to those of native tissue so that they can function as temporary replacements while new tissue is regenerated (Hutmacher 2000; De Aza et al. 2003; Jahangir et al. 2008).
The scaffolds should be highly porous to support cell penetration, tissue in-growth, and the exchange of nutrients and waste products.
The scaffolds should be easily sterilized and cost effective.

24.4 Biomaterials Used for Bone Tissue Engineering

24.4.1 Ceramic, Polymeric, and Composite Biomaterials for Bone Repair

Although the remodeling and restructuring of bones takes place throughout life, surgical intervention is sometimes needed to recover lost function, due to an accident, degradation over time, or certain genetic disorders. In the United States alone, more than 500,000 surgical procedures require bone grafts every year amounting to $2.5 billion annually (Greenwald et al. 2001; Jahangir et al. 2008). Traditional methods for bone repair include autografts and allografts that involve the transplantation of similar tissue from a patient's own body or from another person's, respectively (Gazdag et al. 1995; Laurencin et al. 1999). In certain severe cases, for example, knee replacement or hip replacement, total joint replacement is needed. The scarce availability of suitable donors, organ rejection, and surgical complications (Parikh 2002; Fujishiro et al. 2008) are some of the main issues with natural grafts. Hence, synthetic grafts are frequently considered as replacements for natural grafts. Synthetic grafts include metals, polymers, ceramics, and polymer composites that are used as scaffold materials (Hench 1991; Parikh 2002; Katti 2004). Although synthetic grafts are fairly successful, the mismatch of structural and mechanical properties, when compared to natural tissue, has prompted researchers to find better substitutes (Laurencin et al. 1999; Hutmacher 2000).

The evolution of bone graft materials is well documented by various researchers (Damien and Parsons 1991; Burg et al. 2000; Hutmacher 2000; Hench and Polak 2002; Murugan and Ramakrishna 2005). Bioceramics used for bone tissue engineering can be classified as bioinert (e.g., alumina, zirconia), bioresorbable (e.g., tricalcium phosphate [TCP]), or bioactive (e.g., HAp bioactive glasses and glass-ceramics) (Larry 1991; El-Ghannam 2005; Yoshikawa and Myoui 2005; Best et al. 2008). Alumina and zirconia are often used in hip and knee implants because of their high fracture toughness, biocompatibility, and inertness (Yoshikawa and Myoui 2005). Bone formation is favorably supported by ceramics containing calcium phosphate (Yuan et al. 1998; De Aza et al. 2003). However, scaffolds made from pure calcium phosphate have poor mechanical properties, for example, they are brittle and have low fracture toughness (Larry 1991). The excellent bioactivity of HAp and bioactive glass makes them useful as bone substitutes or coatings that promote cell adhesion and bone in-growth. One of the fundamental strategies for bone repair is to use osteoconductive and osteoinductive biomaterials. HAp is widely known as an osteoconductive material but not osteoinductive (Yoshikawa and Myoui 2005). Growth factors can provide additional osteoinductive stimuli and enhance bone formation at the defect site (Roberts and Sporn 1996). Although ceramics have desirable characteristics, on their own they exhibit low compressive strength and have unpredictable bioresorption rates (Petite et al. 2000). Some ceramics (TCP, HAp, bioglass) are osteoconductive and support the development of new bone, but have limited applicability due to their brittle nature and the intricacy involved in the fabrication of porous structures. Thus, ceramic scaffolds are used only in non-load-bearing applications such as bone void fillers.

Synthetic and natural polymers have shown promise as bone graft materials due to their plastic and viscoelastic properties, their degradability, and biocompatibility (Ma 2004). Natural polymers that are often used as bone grafts include collagen, silk, fibrin, hyaluronic acid, chitosan, and alginate (Ratner 1996; Peter et al. 1998; Agrawal and Ray 2001). The most common biodegradable polymers for bone grafts include poly(glycolic acid) (PGA), poly(lactic acid) (PLA), poly(lactic-co-glycolic acid) (PLGA), poly(caprolactone) (PCL), poly(propylene fumarate) (PPF), and poly(ethylene glycol) (PEG) (Engelberg and Kohn 1991; Laurencin et al. 1999; Burg et al. 2000; Agrawal and Ray 2001). More advanced biomaterials for bone repair require an exclusive combination of chemical, physical, and biological properties. Although most of the natural and synthetic polymers are biodegradable and biocompatible, inferior mechanical strength often limits their use in bone repair.

Since biomaterials are in direct contact with the human body, they need to fulfill various requirements that individual polymeric or ceramic components may not fulfill. Organic–inorganic microcomposites often combine the advantages of both inorganic ceramics and organic polymers. For example, HAp, calcium phosphate, and bioglass are attractive inorganic components that have high bioactivity. Calcium phosphate forms hydroxy carbonate apatite in vivo, which is similar to mineralized bone (Hench 1991; Yuan et al. 1998; Jones et al. 2006). Bioactive ceramics may also reinforce a polymer scaffold and thus provide sufficient mechanical strength (Rezwan et al. 2006). Ceramics have been combined with numerous polymers, including natural (collagen, silk, chitosan) as well as synthetic ones (PEG, PLA, PGA, PLGA). However, despite promising outcomes, the main drawback of using polymer microcomposites for bone repair is the lack of uniform dispersion of inorganic particles within the organic polymer matrix. Further, large aggregates and a lack of sufficient interfacial interaction between the inorganic particles with the polymer matrix adversely affect the mechanical properties of the microcomposite.

24.4.2 Polymer Nanocomposite Biomaterials for Bone Repair

Nanocomposite biomaterials or bio-nanocomposites offer versatility in designing specific properties due to a better control of interactions between nanoparticles and polymers (Hule and Pochan 2007). Polymer nanocomposite biomaterials possess superior mechanical properties when compared to their macro- and microcomposite counterparts (Winey and Vaia 2007). Some common polymers used for making bio-nanocomposites are dextran, chitosan, hyaluronic acid, polyethylene oxide, PLGA, PLA, and PGA. Many researchers have demonstrated that small amounts of nanoparticles added to these polymers can dramatically change the physical properties of the resulting bio-nanocomposite (Giannelis 1996; Vaia and Wagner 2004; Paul and Robeson 2008). Moreover, various combinations of nanoparticles and polymer matrices can be used to engineer previously unattainable property combinations (Table 24.4). One may choose from a variety of nanoparticles depending on the properties that need to be enhanced. Nanoparticles can be categorized as nanospheres/nanoparticles (HAp, TCP, alumina, titania, silica), nanotubes (carbon nanotubes [CNTs], metallic nanotubes), nanofibers, and nanoplatelets (layered silicates such as Montmorillonite [MMT], Cloisite, and Laponite).

24.4.2.1 Polymer–Hydroxyapatite Nanoparticle Composites

Inorganic nanoparticles such as nano-HAp and calcium phosphate are major constituents of human hard tissue, and thus they are the most common biomaterials studied for bone tissue engineering and repair (Weiner and Traub 1992). HAp is osteoconductive and interacts with the natural tissue without eliciting any inflammatory response. An addition of small amounts of nano-HAp can drastically increase the modulus and the tensile strength of the polymer matrix. Thomas et al. (2007) fabricated such bio-nanocomposite scaffolds from collagen type I and nano-HAp via co-electrospinning. These fibrous scaffolds had well-interconnected pore structures and the fiber diameter increased with nano-HAp concentration. An addition of 10% HAp resulted in a threefold increase in tensile strength and a fourfold increase in modulus. The increase in the mechanical properties was mainly attributed to the strong interaction between collagen and nano-Hap (Thomas et al. 2007). To further improve the mechanical properties, glutaraldehyde was used to chemically cross-link the collagen network. However, glutaraldehyde is not biocompatible (Takigawa and Endo 2006) and might elicit an immune response.

A better biomimetic approach to cross-link nano-HAp with a PLA-co-PEG polymer network was published by Sarvestani et al. (2008). This group reported an increase in the shear modulus of

TABLE 24.4 Nanocomposites for Bone Repair

Nanoparticles	Polymer	Preparation Method	Remarks	References
Nano-HAp	Collagen I	Electrospinning	Interconnected pore structure. Glutaraldehyde cross-linked collagen	Thomas et al. (2007)
Nano-HAp	PLA-co-PEG	Biomimetics	Used glutamic acid to cross-link nano-HAp with the polymer matrix	Sarvestani et al. (2008)
Nano-HAp	PLLA	Phase separation	High compressive moduli	Wei and Ma (2004)
Nano-HAp	Collagen	Electrospinning	Increased in vitro mineralization and calcium phosphate activity	Venugopal et al. (2007)
Nano-HAp	PLLA	Biomimetics	Scaffold consisting of nanospheres loaded with biological factors	Ma and Zhang (1999); Ma (2004); Smith et al. (2009)
Nano-HAp	Collagen/PLA	Biomimetics	Hierarchical scaffold. Integrated 15 mm bone defects in a rat model	Liao et al. (2004)
CNTs	Polyurethane	Phase separation	Anisotropic pore structure and nanoscaled surface texture	Jell et al. (2008)
SWNTs	PPF	Solvent-casting	Improved mechanical strength due to nanofiller	Shi et al. (2006)
SWNTs	PPF	Leaching	100% interconnectivity, MSCs adhere and proliferate	Shi et al. (2007)
SWNTs	PPF	Leaching	Promoted bone in-growth and collagen production	Sitharaman et al. (2008)
$CaSiO_3$	PCL	Solvent mixing	CaS enhanced apatite formation in vitro	Kotela et al. (2009); Wei et al. (2008)
Silica/$CaSiO_3$	Collagen	Sol–gel	Silica improved compressive properties and increased bioactivity	Heinemann et al. (2007, 2009)
Silica	Chitin	Biomineralization	Chitin acted as template for biomineralization	Ehrlich et al. (2008)
Silica	Spider silk	Silicification	Controlled structure and morphology of scaffold	Wong Po Foo et al. (2006)
MMT	PLLA	Solvent mixing	MMT reinforced the network and modified degradation rate	Lee et al. (2003)
MMT	Gelatin–chitosan	Solvent mixing	Altered degradation rate and enhanced cell adhesion and proliferation	Zhuang et al. (2007)
MMT	PLA	Leaching	MMT improved compression properties	Ozkoc et al. (2009)
MMT	PLLA	Solvent mixing	MMT suppressed polymer crystallinity and improved mechanical properties	Krikorian and Pochan (2003)
Cloisite 20A	EVA/iron oxide	Solvent mixing	Cloisite increased osteoblast preoliferation and magnetic field induced	Lewkowitz-Shpuntoff et al. (2009)

PLA-co-PEG–HAp nanocomposites when glutamic acid (Glu), a negatively charged peptide was added as a cross-linker. Glu mimics the osteonectin glycoprotein of bone (a bone connector) and cross-links nano-HAp to the polymer matrix. The addition of the Glu cross-linker resulted in a 100% increase in the shear modulus of the nanocomposite. No significant changes in the mechanical properties were observed with or without the Glu cross-linker when micro-HAp was used. This study confirms that the size of added particles plays an important role in modulating the mechanical properties. A similar study by Wei et al. demonstrated that PLLA/nano-HAp nanocomposites not only have superior mechanical properties but also show improved protein adhesion compared to their microcomposite counterparts (Wei and Ma 2004). These nano- and microcomposites were prepared using thermally induced phase separation. Compared to pure PLLA, the resulting nanocomposites showed an almost twofold increase in compressive moduli with the addition of a 50 wt% of nano-HAp.

One of the important parameters in fabricating bone grafts is the 3D architecture of natural tissue, including its porous structure. While the random incorporation of pores into a synthetic scaffold can adversely affect the mechanical properties, the presence of pores also provides a framework for cell attachment, proliferation, and growth. Karageorgiou and Kaplan (2005) discussed the effect of pore size and porosity with respect to bone regeneration. They concluded that porosity enhances the osseo-integration of the implant and reduces stress shielding. An optimum pore size for bone in-growth was reported to be 200–400 μm (Yang et al. 2001).

A technique to fabricate porous scaffolds with excellent mechanical properties is electrospinning, which allows for making polymer nanofilaments using electrostatic forces (Doshi and Reneker 1993). Venugopal et al. (2007) prepared nanocomposite fiber scaffolds from collagen and nano-HAp using electrospinning. These porous and fibrous scaffolds supported osteoblast adhesion, migration, and proliferation, and a significant increase in mineralization was observed after 10 days of culture. The same group showed that their scaffolds also support the formation of multiple layers of cells (Venugopal et al. 2008).

In another study, Ma et al. reported on the synthesis of biomimetic scaffolds (PLLA-nanoHAp) consisting of a nanofibrous network that is interconnected with a microporous network (Ma and Zhang 1999; Wei and Ma 2004; Ma 2008; Smith et al. 2009). Biomimetic synthesis is a newer approach for the fabrication of uniformly dispersed nano-HAp within a polymer matrix. Ma et al. succeeded in controlling the release kinetics of the biological factors within a bio-mimetic macro- and nano-porous scaffold, which contained microspheres loaded with biological factors. The HAp nanoparticles provided osteoconductive stimuli and the microspheres loaded with regenerative factors provided the necessary osteoinductive environment for optimal bone regeneration (Ma 2004).

Liao et al. (2004) developed fibrous and biomimetic nano-HAp/collagen/PLA composite scaffolds with a hierarchical structure. This group fabricated mineralized collagen fibrils by assembling collagen molecules and nano-HAp, which further self-assembled into fibrillar bundles. These fibrillar bundles were found to be uniformly distributed throughout the polymer matrix. The 3D scaffold promoted in vitro osteoblast adhesion, spreading, and proliferation. Moreover, the hierarchical scaffold structure successfully integrated a 15 mm bone defect in a rat model. Likewise, other researchers were able to fabricate similar biomimetic nanocomposite scaffolds using various polymer matrices from gelatin (Kim et al. 2005) and chitosan (Kong et al. 2005).

24.4.2.2 Polymer–Nanotube Nanocomposites

A significant amount of literature describes nanocomposite biomaterials made from polymers and nanotubes. Scaffolds made of such materials should be mechanically robust to withstand in vivo mechanical stresses. Thermally induced phase separation can be used to prepare robust scaffolds from CNTs and polyurethane as reported by Jell et al. (2008). These nanocomposites have an anisotropic porous structure and nanoscale surface texture, and the compressive strength of the scaffold increases with increased CNTs content. However, the reported compressive strength is much lower compared to that of trabecular bone. Interestingly, this nanocomposite showed higher cell proliferation compared to a control (pure polyurethane) sample due to altered surface chemistry/architecture.

A similar study by Shi et al. (2006) showed an increase in the mechanical strength of injectable PPF nanocomposites after incorporating 0.2% of single wall carbon nanotubes (SWNTs), Furthermore, the functionalization of the SWNTs enhances the interaction between nanotubes and the PPF matrix, resulting into a threefold increase in the compressive modulus and a twofold increase in yield strength. The same group fabricated other porous nanocomposite scaffolds (PPF-SWNTs) using a thermal-cross-linking and particulate-leaching technique (Shi et al. 2007). Scaffolds with 100% interconnectivity were fabricated with 75%, 80%, 85%, and 90% controlled porosities. The compressive modulus decreased 100-fold when the porosity was increased from 75% to 90%. In vitro cultures confirmed that mesenchymal stem cells (MSCs) adhere and proliferate on all the scaffolds.

More recently, Sitharaman et al. (2008) studied the in vivo biocompatibility of porous PPF-SWNTs scaffolds in a rabbit model. Implants made of PPF-SWNTs, tested after 4 and 12 weeks, displayed only mild inflammatory responses. Compared to a PPF control, the PPF-SWNT nanocomposite scaffolds showed increased collagen matrix production and significant bone in-growth after 12 weeks of implantation (Sitharaman et al. 2008).

24.4.2.3 Polymer–Silicate Nanocomposites

Previous research has shown that bioactive materials containing silicon, such as bioactive glass (silica) or silicon-doped calcium phosphate materials, exhibit excellent bioactivity and promote apatite formation in vitro and in vivo (Wu and Chang 2004; Wu et al. 2005, 2006). Moreover, high silicon content implants induce bone formation, stimulate osteogenic proliferation, and activate bone-related gene expression (Xynos et al. 2001; Valerio et al. 2004). Thus, it can be concluded that bioactive materials containing silicon may open new possibilities in the field of bone repair.

Kotela et al. (2009) proposed to develop nanocomposites from polycaprolactone and wollastonite nanoparticles for bone repair. Wollastonite is a calcium silicate ($CaSiO_3$) with bioactive properties. The addition of small amounts (0.5%–1%) of wollastonite significantly improved Young's modulus, the tensile strength, and the fracture toughness of the polymer nanocomposite. The bioactivity of the nanocomposite was verified by submerging it in simulated body fluid. Apatite nucleation was observed on the wollastonite surfaces after 7 days. Similar results were obtained by Wei et al. (2008) on PCL-calcium silicate nanocomposites.

In a different approach, Heinemann and coworkers attempted to mimic the natural processes of biosilicification to fabricate silica–collagen hybrid xerogels under ambient conditions (Heinemann et al. 2007, 2009). This group used the sol–gel technique to fabricate monolithic composite materials by varying the ratio of organic (collagen) and inorganic (silica) components. As a result, the compressive properties of silica were substantially improved by the addition of collagen. To further enhance the bioactivity of the nanocomposite material, they incorporated a third phase (calcium phosphate cement). The addition of calcium phosphate accelerated the formation of bone apatite layers when tested in a simulated body fluid. The feasibility of using silica-based biomaterials for bone repair was shown by the differentiation of human monocytes into osteoclast-like cells.

Similar to the Heinemann study, Ehrlich et al. (2008) explored silica–chitin based natural bionanocomposites fabricated from living glass sponges. They showed that the chitinous organic matrix provides a template for the biodirected deposition of the mineral phase (silica) and that the resulting biocompatible structures could be used for the tissue engineering of both bone and cartilage replacements.

Highly repetitive amino acid sequences present in fibrous proteins, such as collagen and silk, can be explored to obtain self-assembled nanocomposites (Meinel et al. 2005; Wang et al. 2006; Wong Po Foo et al. 2006). Foo et al. fabricated biomimetic nanocomposites consisting of silica nanoparticles using fusion (chimeric) proteins (Wong Po Foo et al. 2006). Films and fibers were fabricated from spider silk protein, and silica nanoparticles were deposited on this polymer using silicification reactions. The morphology and structure of the silica nanoparticles can be governed by controlling the processing conditions. The same group reported the chemical attachment of cell binding domains (Sofia et al. 2001) and growth factors (bone morphogenetic protein-2 [BMP-2]) (Karageorgiou et al. 2004) onto silk-based biomaterials to induce an osteogenic differentiation of human bone marrow stromal cells.

24.4.2.4 Polymer–Clay Nanocomposites

During the last few decades, nanocomposites from polymer and clay silicate have been the subject of intense fundamental studies. The knowledge gained on how to best tailor structure–property relationships can be applied to developing new functional biomaterials (Vaia and Wagner 2004; Sinha Ray and Bousmina 2005). For example, the strong specific interactions between silicate nanoparticle surfaces and polymer chains, combined with new fabrication techniques, allow the assembly of supramolecular structures over many length scales (Giannelis 1996; Alexandre and Dubois 2000; Hule and Pochan 2007). Besides covalent bonding, physical cross-linking via hydrogen bonding, Van der Waals, and ionic interactions are often responsible for the mechanical strength of materials and their extensibility

(Zilg et al. 1999). While small amounts of silicate nanoparticles can dramatically change the physical properties of polymers (Giannelis 1996; LeBaron et al. 1999; Vaia and Wagner 2004; Sinha Ray and Bousmina 2005), higher amounts of silicate can lead to ultrastrong materials with hierarchical structures and properties that may approach the theoretical calculated maximum (Liff et al. 2007; Podsiadlo et al. 2007). Other physically cross-linked polymer nanocomposites exhibit structure similar to nacre and display directional dependent mechanical properties (Dundigalla et al. 2005; Gaharwar et al. 2011b). Here, the silicate not only enhances the mechanical properties of the nanocomposite but also facilitates cell adhesion, spreading, and proliferation (Gaharwar et al. 2011a,b; Jin et al. 2009).

The addition of silicates to thermoresponsive polymer hydrogels can be used to tune phase transitions and control dissolution properties that are important for the sustained release of biomolecules (Wu and Schmidt 2009). The examples above show that, with further materials development and formulation, much of the fundamental research published can be translated for applications that focus on the repair of bone. The following studies show several more attempts in this direction.

Besides mechanical strength, the degradation rate is another important property of a scaffold that is to be used for bone repair. While the degradation rates of polymers can be easily modified, the degradation of natural clays such as Montmorillonite remains a challenge. Nevertheless, MMT can reinforce polymers significantly as shown by Lee et al. (2003), who modified the degradation rate of exfoliated PLLA-MMT nanocomposites by varying the amounts of MMT in a PLLA matrix. After the MMT-PLLA scaffold was immersed in water, the modulus of the scaffold decreased due to the degradation of the PLLA matrix (Lee et al. 2005). The authors, however, did not elaborate on what happened with the MMT once the polymer degraded.

In another study, Zhuang et al. (2007) showed that the intercalated structure of MMT–gelatin–chitosan has a lower degradation rate when compared to a gelatin–chitosan scaffold and that the degradation rate can be altered by changing the MMT concentration. Enhanced cell adhesion and proliferation on the MMT–gelatin–chitosan nanocomposite film was observed.

Ozkoc et al. (2009) fabricated porous PLA–MMT nanocomposites using microcompounding and polymer/particle leaching. The addition of MMT improved the compression properties of the nanocomposites to be close to those of cancellous bone. The hydrophilicity of the nanocomposite surfaces directly affected cell adhesion. For example, an addition of 3% MMT reduced the water contact angle from 60.7° to 31.4°. This is due to a decrease in the interfacial tension between polymer and water, making the PLA surface more hydrophilic.

Similarly, Krikorian et al. reported significant improvement in mechanical properties of PLLA due to the addition of MMT (Krikorian and Pochan 2003). Higher amounts of MMT and fully exfoliated structures gave rise to stiffer materials compared to microphase separated or intercalated composites. Exfoliated nanoplatelets suppressed polymer crystallization due to enhanced surface interactions. Moreover, an increase in silicate concentration and exfoliation resulted in stiffer and transparent PLLA-MMT nanocomposites. As mentioned before, Lee et al. (2003) tailored the mechanical properties of such nanocomposites by incorporating different amounts of MMT in a PLLA matrix. Highly porous (~92% porosity) scaffolds were fabricated via the salt leaching/gas foaming technique. An increase in MMT concentration also decreased the glass transition temperature. The decrease in crystallinity accelerated the degradation of PLLA-MMT nanocomposites. The tensile strength of the scaffold was modulated between 40 and 60 MPa, which fits the requirements for soft and hard scaffold applications.

More recently, Lewkowitz-Shpuntoff et al. (2009) reported that the addition of 10% cloisite clay to ethylene vinyl acetate dramatically increased osteoblast proliferation on the nanocomposite surface. Magnetic nanocomposites that were obtained by the physical adsorption of iron oxide nanoparticles on the cloisite surface showed enhanced proliferation and alignment of MC3T3 preosteoblast cells in a static magnetic field. The alignment of the cells in the magnetic field was attributed to an increase in the internal field near the cells. Previous research has shown that a magnetic field can stimulate osteogenesis and upregulate the transcription of BMP-2 and BMP-4 (Fitzsimmons et al. 1994; Bodamyali et al. 1998). Thus, external magnetic fields may influence tissue formation and cellular organization of cells.

Overall, these experimental results show that polymer nanocomposites have the potential to be used for the repair and regeneration of bone.

24.5 Biomaterials Used for Cartilage Tissue Engineering

The extracellular matrix of articular cartilage is principally composed of hierarchically arranged collagen fibrils, proteoglycans, and non-collagenous proteins on the nanometer scale (Figure 24.1 and Table 24.3) (Mow et al. 1980; Mankin 1982; Buckwalter and Mankin 1997; Cohen et al. 1998). The complex interactions between individual macromolecules synergistically provide an advanced mechanism for load-support and low-frictional properties (Akizuki et al. 1986; Heinegard and Oldberg 1989; Zhu et al. 1993; Jurvelin et al. 2003). If biomaterial scaffolds are considered for cartilage tissue repair, they need to satisfy some of the requirements mentioned before.

The importance of developing biomaterials for cartilage tissue engineering is warranted by the more than 600,000 joints, including knees and hips, that are replaced annually in the United States (see website: http://www.aaos.org). The failure of articular cartilage usually results from trauma, arthritis, or sports injuries. Articular cartilage has a very limited capability to repair and regenerate due to the absence of blood vessels and nerves in the tissue (Cohen et al. 1998; Hunziker 2002). If cartilage defects are smaller than 2–4 mm, they can be healed by a continuous passive motion of the joint (O'Driscoll 1998) or by techniques such as subchondral drilling, abrasion, microfracture, and the administration of bioactive agents such as growth factors and cytokines (O'Driscoll et al. 1986; O'Driscoll 1998). However, for large cartilage defects, the replacement of the whole knee or hip with artificial components (metal, ceramic, composite, etc.) is employed (Hunziker 2002). These artificial components do not fully restore joint function, and surgeries are generally not recommended for young people due to the limited life span of the artificial components (O'Driscoll 1998). Other commonly used strategies for cartilage repair include autografts and allografts (Mankin 1982; O'Driscoll 1998; Hunziker 2002). Disadvantages associated with autografts include lack of donor sites, complications from surgical procedures, and an increased risk of inflammation at the donor sites. Additionally, allografts are associated with increased immune response and disease transmission (Parikh 2002; Fujishiro et al. 2008).

Tissue engineering approaches have shown promise to regenerate damaged articular cartilage and to help regain normal body functions (Temenoff and Mikos 2000; Risbud and Sittinger 2002; Capito and Spector 2003). Engineered polymeric scaffolds need to provide necessary support for cells to proliferate and mechanical strength to keep the new tissue in place. The scaffold materials that have been investigated for cartilage repair can be classified into two major categories: (1) polymers (natural and synthetic) and (2) microcomposites. Although polymers have attractive chemical and biological properties, they often do not have sufficient mechanical strength. Therefore, polymers are often combined with ceramic nanoparticles to reinforce the structural and mechanical strength via strong polymer–nanoparticle interactions (Hule and Pochan 2007; Winey and Vaia 2007). Even the nanotopography of specific fibrous biomaterials has been shown to be important to chondogenesis (Savaiano and Webster 2004; Park et al. 2005; Moutos et al. 2007; Stoddart et al. 2009).

24.5.1 Available Biomaterials for Cartilage Repair

There is a large body of literature, and some comprehensive reviews cover the numerous biomaterials that have been considered for cartilage repair (Temenoff and Mikos 2000; Risbud and Sittinger 2002; Capito and Spector 2003; Bonzani et al. 2006). Since the tissue engineering of cartilage does not require the inclusion of extensive vascularization, tissue engineered cartilage was one of the first products to be tested in human trials (Khademhosseini et al. 2009).

The most commonly used biomaterials are natural and synthetic polymers (Hutmacher 2000; Capito and Spector 2003; Sharma and Elisseeff 2004; Cheung et al. 2007), many of which have been optimized to

TABLE 24.5 Nanocomposites for Cartilage Repair

Nanoparticles	Polymer	Preparation Method	Remarks	References
Nano-HAp	PLGA	Electrospinning/ sol–gel technique	Interconnected network, uniform cell in-growth. Nano-HAp facilitated the formation of neonatal cartilage tissue.	Li et al. (2009)
Nano-HAp	PLLA	Electrospinning	Intense in vitro deposition of proteoglycans and GAGs.	Spadaccio et al. (2009)
Nano-HAp	PVA	Solvent mixing	Increased storage modulus and elastic properties.	Pan et al. (2008)
Au	Collagen II	Solvent mixing	Promoted chondrocyte proliferation.	Hsu et al. (2007)
Titania	PLGA	Solvent casting and NaOH treated	Nanostructure roughness promoted chondrocyte attachment.	Kay et al. (2002)
Titania	PLGA	Solvent casting	Titania-enhanced in vitro production of intracellular proteins (ALP and chondrocyte expressed protein-68).	Savaiano et al. (2004)
CNTs	PCU	Solvent casting	Nano-surface roughness and electrical stimulation promoted chondrogenesis.	Khang et al. (2008)
CNTs	Collagen I	Solvent mixing	Despite high CNTs concentration high cell viability and proliferation were observed.	MacDonald et al. (2005)

not only provide 3D support to cells but also to maintain their differentiated phenotype and to encourage cell migration and proliferation (Temenoff and Mikos 2000; Hunziker 2002; Kloxin et al. 2009). Nevertheless, many of the polymer scaffolds do not have sufficient mechanical strength to continuously support the formation of cartilage tissue.

Natural polymers such as collagen, hyaluronic acid (HA), alginate, silk fibroin, and chitosan have been extensively studied (Hutmacher 2000; Temenoff and Mikos 2000; Risbud and Sittinger 2002; Capito and Spector 2003). A key advantage of using natural polymers is their ability to interact with cells and cellular enzymes, and to be remodeled and/or degraded when space for growing tissue is needed. Although synthetic polymers such as PLA, PGA, PLGA, PCL, and polycarbonate (Hutmacher 2000; Capito and Spector 2003) are not bioactive when compared to natural polymers, they do provide more control over physical and chemical properties. If these properties are not satisfactory for a specific application, the addition of ceramic micro- and nanoparticles may enhance mechanical strength, degradation rate, bioactivity, and surface chemistry (Kokubo et al. 2003; Mano et al. 2004). In the following section, we outline current trends in the development of nanocomposite materials for cartilage repair (Table 24.5).

24.5.2 Polymer Nanocomposite Biomaterials for Cartilage Repair

24.5.2.1 Polymer–Hydroxyapatite Nanoparticle Composites

Although polymer hydroxyapatite nanoparticle composites have been most extensively explored for bone repair, several studies also attempt to use these materials for cartilage regeneration. For example, Li et al. (2009) studied the feasibility of using PLGA/nano-HAp nanocomposite for cartilage repair in a rat model. A nanofibrous mesh was made from PLGA using electrospinning and nano-HAp was deposited on the fiber surface using a sol–gel technique. MSCs were seeded on the nanofibrous scaffold and the interconnected network allowed uniform cell in-growth throughout the scaffold. After 12 weeks of implantation with MSC seeded scaffolds, the defects were perfectly integrated with the surrounding tissue and had a smooth surface morphology. Although, the nanocomposite scaffolds were not fully degraded after 12 weeks of implantation, they facilitated the formation of cartilage tissue.

Most recently, in 2009, Spadaccio et al. (2009) developed PLLA/nanoHAp nanocomposites for cartilage regeneration by utilizing fibrous scaffolds combined with molecular signaling mechanisms.

The fibrous scaffold architecture facilitates initial cell attachment and the subsequent migration of human-MSC. Nano-HAp plays an important role in the differentiation of stem cells into chondrocyte-like cells that produce proteoglycan. After 14 days of culturing human-MSCs within a PLLA/nanoHAp scaffold in chondrogenic media, significantly higher amounts of chondrogenic transcription factors such as SOX9 were expressed (these activate type II collagen and aggrecan production) (Bosnakovski et al. 2006) compared to a PLLA scaffold control (Spadaccio et al. 2009). Moreover, neo-ECM was observed around the cells, as the scaffold stained positive for Toluidine Blue and Safranin O. Overall, the nanocomposite scaffolds show more deposition of proteoglycans and glycosaminoglycans (GAGs) when compared to the neat polymer scaffolds (Spadaccio et al. 2009).

Besides PLGA, polymers, such as poly (vinyl alcohol) (PVA), have been tested for cartilage repair. When nano-HAp is combined with a PVA hydrogel, the mechanical properties of the gel improve and the nanocomposite becomes bioactive (Pan et al. 2008). The storage modulus of the hydrogel also increased with the increasing frequency of deformation, which is very similar to that of articular cartilage. The overall elastic properties of hydrogel increased with the increasing nano-HAp concentration (less than 6%) due to the cross-linking of PVA and nano-HAp. However, at higher concentrations (greater than 6%), nanoparticles aggregate and weaken the composite network structure. As PVA alone is neither biodegradable nor bioactive, long-term implantation remains a problem.

24.5.2.2 Polymer–Metal Nanoparticle Composites

The unique optical, electronic, and biological properties of metallic nanoparticles, such as gold (Au), allow the development of many biomedical applications (especially in cancer research). Recently, some researchers are trying to use metallic gold in cartilage repair. For example, Hsu et al. (2007) demonstrated that the incorporation of Au-nanoparticles within type II collagen hydrogels increases the mechanical properties and the antioxidative effects of the resulting nanocomposite. Interactions between the positively charged Au nanoparticles and the negatively charged collagen fibrils (below the isoelectric point), allow the Au nanoparticles to adhere to the collagen fibrils at low concentrations (0.1% Au). At higher concentrations (more than 0.2% Au) Au aggregates are formed and a decrease in the dynamic storage modulus is observed. Below a 10 ppm concentration, Au nanoparticles show very good biocompatibility and do not modify gene expression if internalized. The authors claim that the nanocomposites containing 0.1% Au nanoparticles promote chondrocyte proliferation and may activate certain genes that are responsible for sensing surface roughness (filopodia and lamellipodia were observed) (Hsu et al. 2007).

The surface modification of nanocomposite scaffolds can promote chondrocyte adhesion and proliferation. Surface properties such as surface area, roughness, and charge can be easily altered by changing the type and amount of nanoparticles. Research done by Kay et al. (2002) demonstrated that nanostructure roughness promotes chondrocyte adhesion in PLGA–titania nanocomposites. The PLGA scaffold was first treated with a NaOH solution to develop nanostructured features. Then different sizes of titania particles (micron- or nano-sized grains) were added. Increased chondrocyte adhesion was observed on the nanostructured PLGA films (NaOH treated) when compared to conventional PLGA films. The study also showed that 1.5–2 times more chondrocytes attached to the nanocomposite (PLGA/titania) surfaces compared to the microcomposite PLGA/titania surfaces. The same group demonstrated that PLGA–titania nanocomposites can alter long-term chondrocyte responses (Savaiano and Webster 2004). Although the total number of cells seeded on the nanocomposite films remained constant, after 21 days of culture the total amount of intracellular proteins (alkaline phosphatase and chondrocyte expressed protein-68 [CEP-68]) almost doubled when compared to the control. These results suggest that titania-containing polymer nanocomposites may have potential for studying chondrocyte function, which is important in cartilage repair.

24.5.2.3 Polymer–Nanotube Nanocomposites for Cartilage Repair

CNTs have long been considered for hard and soft tissue engineering as their unique mechanical (flexural and fatigue strength, high strength-to-weight ratio) and electrical properties are desired (MacGinitie et al. 1994; Harrison and Atala 2007). The structural arrangements of CNTs on a nanometer length scale mimic the ECM network and provide better biocompatibility when compared to other nanoscale materials. The natural ECM has a nanotopography that allows cells to attach and interact via adhesion proteins. Thus, it is expected that nanotopography significantly influences cell orientation, morphology, and cytoskeleton arrangements.

In a related study, Khang et al. (2008) provided direct evidence of the influence of surface roughness and electrical stimulation on chondrocyte function. These authors observed an increase in chondrocyte density when cells were grown on CNTs–polycarbonate urethane. Compared to the individual components, the nanocomposite displays significantly higher surface roughness and hydrophilicity. The electrical conductivity of CNTs was used to increase the cell density on the nanocomposite surface via electrical stimulation. Previous studies already indicated that electric fields influence cartilage growth, remodeling, and biosynthesis (MacGinitie et al. 1994; Wang et al. 2004) but the long-term effect of CNTs and electrical stimulation on chondrocyte activity is not yet known. Another approach to mimic natural ECM is to use collagen as a matrix, as collagen is a major component of ECM. Thus, collagen-CNT nanocomposites should provide a suitable environment for cell growth. MacDonald et al. (2005) suggested that collagen-CNT scaffolds might be used for cartilage repair. In their work, CNTs were physically entrapped within a collagen matrix with no evidence of chemical interaction between polymer and CNT. Although the resulting nanocomposites had a high amount of SWNTs and delayed gelation characteristics, cell viability and cell proliferation were high (MacDonald et al. 2005).

24.6 Osteochondral Tissue Engineering

Osteochondral defects often occur from repetitive trauma to the joints (Buckwalter et al. 1994). Such injuries lead to loss of bone tissue and the formation of cystic lesions, which may cause the collapse of the remaining cartilage tissue (Buckwalter 1992; Ulrich-Vinther et al. 2003). Osteochondral defects can also result from genetic or metabolic causes, which alter the structural and mechanical properties of the joints. Most often, such defects are repaired by mosaicplasty, in which damaged cartilage is replaced by healthy cartilage that is removed from non-weight-bearing regions of the body (Hangody et al. 2008). In 1994, Brittberg et al. (1994) demonstrated for the first time that the implantation of in vitro cultured human condrocytes led to the formation of hyaline-like articular cartilage. Such therapies, however, cannot be used for repairing large defects as graft sites are limited and donor site morbidity is a problem. For repairing large defects, osteochondral scaffolds are needed (Sharma and Elisseeff 2004; Mikos et al. 2006; Martin et al. 2007; Grayson et al. 2008; Keeney and Pandit 2009). These scaffolds promote the regeneration of articular cartilage and subcondral bone while maintaining mechanical stability. Osteochondral scaffolds have several design issues, which are not satisfactorily addressed. Because bone and cartilage have different physical, chemical, and biological properties, an osteochondral scaffold should mimic the structural and mechanical properties of both tissues. This can be done by fabricating layered scaffolds for bone and cartilage (Martin et al. 2007; O'Shea and Miao 2008; Keeney and Pandit 2009). To integrate the bone and cartilage regions within one scaffold, suturing, cell-mediated ECM formation, and the use of glues have been applied (Schaefer et al. 2000; Wang et al. 2007; Moroni and Elisseeff 2008). The cartilage layer is seeded with chondrocytes or MSCs, and the bone layer remains acellular or is seeded with osteocytes, MSCs, periosteal cells, or bone marrow. If this approach is used for cell-based therapies, the age of the donor and host directly affects the proliferation and the chondrocytic/osteocytic potential of the implanted cells (Morihara et al. 2002).

24.6.1 Biomaterial Scaffolds Used for Osteochondral Repair

A number of synthetic and natural biomaterials have been investigated for osteochondral repair (Yaylaoglu et al. 1999; Martin et al. 2007; Grayson et al. 2008; Tampieri et al. 2008). Usually, the natural materials provide enhanced biological interaction with the host tissue and can accelerate the healing

TABLE 24.6 Nanocomposites for Bone–Cartilage Repair

Bone Region	Cartilage Region	Preparation Method	Remarks	References
PCL/TCP	PCL	Twin-screw extrusion/ electrospinning	Highly interconnected nanofibrous mesh with spatial gradations in composition and porosity. Supported differentiation of preosteoblast to osteoblast cells.	Erisken et al. (2008)
Collagen/nano-HAp	Collagen	Biomimetic	Closely mimics the structural and compositional properties of native tissue. Supported osteogenic and chondrogenic differentiation based on changes in protein expression.	Dawson et al. (2008)
MSCs ECM/ nano-HAp	MSCs ECM/ Nano-HAp	Alternate soaking process	Fabrication method is faster than biometric process. Crystal size of HAp controlled by number of cycles or ion concentrations. In vivo study indicated the osteoinductive nature of scaffold.	Matsusaki et al. (2009)
Collagen/nano-HAp	Collagen/ hyaluronic acid	Synchronous biomineralization/ freeze drying	Bilayered composite was chemically cross-linked to eliminate the risk of delamination.	Gelinsky et al. (2008)
Collagen/nano-HAp (30/70)	Collagen/ nano-HAp (60/40)	Biomimetic/freeze drying	Supported osteogenic and chondrogenic differentiation in vitro.	Tampieri et al. (2003)

process (Ratner 1996; Ratner and Bryant 2004). Synthetic materials may provide tailored mechanical and structural properties, but bioactivity needs to be added. Polymer microcomposite scaffolds can combine properties such as gradient structure, composition, bioactivity, and mechanical properties (Martin et al. 2007; Hangody et al. 2008; Keeney and Pandit 2009). Porous osteochondral scaffolds with a gradient in structure, composition, porosity, and mechanical properties can also be fabricated by a 3D printing process (Sherwood et al. 2002). This printing process allows the generation of scaffolds with relevant biological and anatomical features.

Scaffolds with predictable architecture and porosity usually have predictable mechanical properties (Schek et al. 2004). One way to enhance the mechanical properties of a scaffold for osteochondral repair is to combine polymers with nanoparticles to form nanocomposites (Table 24.6).

24.6.2 Polymer Nanocomposites for Osteochondral Repair

The concentration and structural orientation of HAp nanoplatelets varies significantly within the bone–cartilage interface. The change in HAp concentration across the interface leads to differences in mechanical properties that need to be considered when developing gradient scaffolds.

A functionally graded nanocomposite scaffold from biodegradable PCL and nanoparticle TCP (Erisken et al. 2008) can be made by a new hybrid twin-screw extrusion/electrospinning process, which was reported by Erisken et al. (2008). The resulting nonwoven, highly interconnected nanofibrous scaffold has spatial gradients in composition and porosity and the tensile strength can be tailored to increase from bottom to top. When the nanocomposite scaffold was seeded with preosteoblastic (MC3T3-E1) cells, most of the cells differentiated into osteoblast cells within 4 weeks of culture. Moreover, the newly formed ECM showed collagen I and mineral deposits indicating the activity of bone cells. The formation of bone-like structures increased the modulus of the cell seeded nanocomposite scaffold to almost double its starting value (Erisken et al. 2008).

Dawson et al. (2008) and Sachlos et al. (2006) fabricated improved nanocomposite scaffolds that mimic the structure and composition of bone and cartilage. The bi-layered scaffolds consist of both bone and cartilage regions. The bone region of the scaffold was composed of nano-sized HAp crystals that were precipitated on a fibrous network of collagen (ColHAp). The presence of microchannels in the bone region supported the perfusion of nutrient-rich media (Sachlos et al. 2006; Dawson et al. 2008). The bone region was found to support the osteogenic differentiation of human bone marrow stromal cells (HBMSCs) (Dawson et al. 2008). After 28 days of in vitro culture, the scaffold showed high alkaline phosphatase (ALP) activity and significant cell penetration, and integration throughout the scaffold. The implanted (in mice) ColHAp scaffold showed complete integration with the surrounding tissue after only 4 weeks (Dawson et al. 2008). The cartilage region of the scaffold was composed of collagen type I (Col) with predefined internal channels. Solid freeform fabrication and critical point drying techniques were used to fabricate the collagen scaffold with complex internal morphology (Sachlos et al. 2003; Dawson et al. 2008). The in vitro culture of the scaffold in chondrogenic media showed the formation of a dense proteoglycan and collagen II rich matrix (Dawson et al. 2008). The scaffold supported the chondrogenic differentiation of HBMSCs that was evident from the protein expression of SOX9 and the mRNA expression of collagens IX and XI, aggrecan, and proteoglycans (Dawson et al. 2008). Furthermore, the incorporation of microchannels into the scaffold enhanced chondrogenesis. Overall, the collagen-based scaffolds (Col and ColHAp) not only supported the growth and differentiation of stem cells, but the study also provided strong clues for improving chondrogenesis and osteogenesis.

Matsusaki et al. (2009) used an alternate soaking process to deposit HAp crystals on a tissue-engineered scaffold composed of MSCs and the ECM produced by the cells. The HAp deposited by this process had low crystallinity and was biodegradable. Compared to a biomimetic process, the alternate soaking process was almost a hundred times faster. The crystal size of HAp particles can be increased by increasing the number of cycles or the concentration of ions. However, exposure to higher calcium or phosphorus concentrations (more than 100 mM) during the process led to DNA damage and cell death. Moreover, the in vivo studies demonstrate that the engineered scaffold was osteoinductive.

One of the major problems associated with fabricating layered osteochondral grafts is the delamination of the bone and cartilage regions. In order to avoid delamination, Gelinsky et al. (2008) fabricated monolithic scaffolds composed of two layers that were fused together by a unified cross-linking process. The bone layer consisted of mineralized composite (collagen I/HAp nanocomposite) and the cartilage layer consisted of non-mineralized composite (collagen I/hyaluronic acid [HA]). The mineralized composite can be synthesized by using the synchronous biomineralization of collagen (Bradt et al. 1999; Meyer et al. 2009). In this process, the assembly of collagen fibrils was initiated and the growing fibrils acted as a template for HAp crystallization. The non-mineralized composite was formed by adding HAp during the assembly of collagen fibrils. To fabricate the monolithic scaffold, the freeze-dried bi-layered composite was chemically cross-linked. The unified cross-linked scaffold eliminated the risk of the delamination of mineralized and non-mineralized layers.

As the mechanical properties of cartilage are very different from those of bone, one has to design scaffolds that reduce the mismatch of the properties at the interface. In order to do so, Tampieri et al. (2008) fabricated a tri-layered scaffold, where the bony layer was composed of biomineralized collagen (70% HAp and 30% collagen I), the cartilage layer was composed of hyaluronic acid–collagen composite, and the intermediate layer was composed of low-density biomineralized collagen (40% HAp and

60% collagen I). The biomineralized collagen was fabricated by nucleating hydroxyapatite nanocrystals on self-assembled collagen fibers (Tampieri et al. 2003). The HAp nanocrystals grew on the collagen fibers with their c-axis oriented along the fiber axis. An integrated monolithic composite was fabricated by freeze drying the stacked tri-layered scaffold (Tampieri et al. 2008). Although the flexural strength of the biomineralized layer decreased with increasing porosity, the elastic modulus was in the range of trabecular bone. Moreover, a pulling test did not result in delamination, indicating strong adhesion between the layers. In chondrogenic media, MSCs differentiated into chondrocytes and formed cartilaginous tissue within the hyaluronic acid–collagen layer. After 8 weeks, the nanocomposite scaffold showed bone tissue in the bone region and connective tissue in the cartilage region.

24.7 Conclusions

The literature reviewed here has shown that the synergistic combinations of chemical, physical, and biological properties of nanocomposite biomaterials used for bone and cartilage tissue engineering have become more sophisticated. In a similar way, the fabrication technologies and engineering approaches have advanced. Nevertheless, significant challenges persist when "giving life" to these scaffolds. This suggests that both the material science perspective and the biology that figures out how cells interact with the scaffolds and with each other and how genes and proteins influence these interactions are needed for biomaterials design (Khademhosseini et al. 2009).

Acknowledgments

Work by the authors was supported by an NSF-CAREER award, DMR 0711783 to GS, and by a Purdue Lynn Doctoral fellowship to PS.

Abbreviations

Au	gold
BMP	bone morphogenetic protein
CNT	carbon nanotube
ECM	extra cellular matrix
GAG	glycosaminoglycans
Glu	glutamic acid
HA	hyaluronic acid
Hap	hydroxyapatite
HBMSC	human bone marrow stromal cells
MMT	mntmorillonite
MSC	mesenchymal stem cell
PCL	poly(caprolactone)
PEG	poly(ethylene glycol)
PEO	poly(ethylene oxide)
PGA	poly(glycolic acid)
PLA	poly(lactic acid)
PLGA	poly(lactic-co-glycolic acid)
PLLA	poly(L-lactic acid)
PPF	poly(propylene fumarate)
PVA	poly (vinyl alcohol)
SWNT	single wall carbon nanotubes
TCP	tricalcium phosphate

References

Agrawal, C. M. and R. B. Ray. 2001. Biodegradable polymeric scaffolds for musculoskeletal tissue engineering. *Journal of Biomedical Materials Research* 55 (2):141–150.
Akizuki, S., V. C. Mow, F. Müller, J. C. Pita, D. S. Howell, and D. H. Manicourt. 1986. Tensile properties of human knee joint cartilage: I. Influence of ionic conditions, weight bearing, and fibrillation on the tensile modulus. *Journal of Orthopaedic Research* 4 (4):379–392.
Alexandre, M. and P. Dubois. 2000. Polymer-layered silicate nanocomposites: Preparation, properties and uses of a new class of materials. *Materials Science and Engineering: R: Reports* 28 (1–2):1–63.
Armstrong, C. G. and V. C. Mow. 1982. Variations in the intrinsic mechanical properties of human articular cartilage with age, degeneration, and water content. *The Journal of Bone and Joint Surgery* 64 (1):88–94.
Ateshian, G. A. 2007. Artificial cartilage: Weaving in three dimensions. *Nature Materials* 6 (2):89–90.
Aubin, J. E. and F. Liu. 1996. *Principle of Bone Biology*, eds. J. E. Aubin, L. G. Raisz, and G. A. Rodan, San Diego, CA: Academic.
Best, S. M., A. E. Porter, E. S. Thian, and J. Huang. 2008. Bioceramics: Past, present and for the future. *Journal of the European Ceramic Society* 28 (7):1319–1327.
Bodamyali, T., B. Bhatt, F. J. Hughes, V. R. Winrow, J. M. Kanczler, B. Simon, J. Abbott, D. R. Blake, and C. R. Stevens. 1998. Pulsed electromagnetic fields simultaneously induce osteogenesis and upregulate transcription of bone morphogenetic proteins 2 and 4 in rat osteoblastsin vitro. *Biochemical and Biophysical Research Communications* 250 (2):458–461.
Bonzani, I. C., J. H. George, and M. M. Stevens. 2006. Novel materials for bone and cartilage regeneration. *Current Opinion in Chemical Biology* 10 (6):568–575.
Bosnakovski, D., M. Mizuno, G. Kim, S. Takagi, M. Okumura, and T. Fujinaga. 2006. Chondrogenic differentiation of bovine bone marrow mesenchymal stem cells (MSCs) in different hydrogels: Influence of collagen type II extracellular matrix on MSC chondrogenesis. *Biotechnology and Bioengineering* 93 (6):1152–1163.
Bradt, J.-H., M. Mertig, A. Teresiak, and W. Pompe. 1999. Biomimetic mineralization of collagen by combined fibril assembly and calcium phosphate formation. *Chemistry of Materials* 11 (10):2694–2701.
Brittberg, M., A. Lindahl, A. Nilsson, C. Ohlsson, O. Isaksson, and L. Peterson. 1994. Treatment of deep cartilage defects in the knee with autologous chondrocyte transplantation. *The New England Journal of Medicine* 331 (14):889–895.
Brittberg, M. and C. S. Winalski. 2003. Evaluation of cartilage injuries and repair. *The Journal of Bone and Joint Surgery* 85 (90002):58–69.
Buckwalter, J. A. 1992. Mechanical injuries of articular cartilage. *The Iowa Orthopaedic Journal* 12:50–57.
Buckwalter, J. A. and H. J. Mankin. 1997. Instructional course lectures, the American Academy of Orthopaedic Surgeons—Articular cartilage. Part I: Tissue design and chondrocyte-matrix interactions*. *The Journal of Bone and Joint Surgery, American Volume* 79 (4):600–611.
Buckwalter, J. A., V. C. Mow, and A. Ratcliffe. 1994. Restoration of injured or degenerated articular cartilage. *Journal of the American Academy of Orthopaedic Surgeons* 2 (4):192–201.
Burg, K. J. L., S. Porter, and J. F. Kellam. 2000. Biomaterial developments for bone tissue engineering. *Biomaterials* 21 (23):2347–2359.
Cancedda, R., B. Dozin, P. Giannoni, and R. Quarto. 2003. Tissue engineering and cell therapy of cartilage and bone. *Matrix Biology* 22 (1):81–91.
Capito, R. M. and M. Spector. 2003. Scaffold-based articular cartilage repair. *Engineering in Medicine and Biology Magazine, IEEE* 22 (5):42–50.
Carlisle, E. M. 1970. Silicon: A possible factor in bone calcification. *Science* 167 (3916):279–280.
Carlisle, E. M. 1972. Silicon: An essential element for the chick. *Science* 178 (4061):619–621.
Carlisle, E. 1981. Silicon: A requirement in bone formation independent of vitamin D1. *Calcified Tissue International* 33 (1):27–34.

Carter, D. R. and W. C. Hayes. 1976. Bone compressive strength: The influence of density and strain rate. *Science* 194 (4270):1174–1176.

Cheung, H.-Y., K.-T. Lau, T.-P. Lu, and D. Hui. 2007. A critical review on polymer-based bio-engineered materials for scaffold development. *Composites Part B: Engineering* 38 (3):291–300.

Cohen, N. P., R. J. Foster, and V. C. Mow. 1998. Composition and dynamics of articular cartilage: Structure, function, and maintaining healthy state. The *Journal of Orthopaedic and Sports Physical Therapy* 28 (4):203–215.

Currey, J. D. 1988. The effect of porosity and mineral content on the Young's modulus of elasticity of compact bone. *Journal of Biomechanics* 21 (2):9.

Damien, C. J. and J. R. Parsons. 1991. Bone graft and bone graft substitutes: A review of current technology and applications. *Journal of Applied Biomaterials* 2 (3):187–208.

Dawson, J. I., D. A. Wahl, S. A. Lanham, J. M. Kanczler, J. T. Czernuszka, and R. O. C. Oreffo. 2008. Development of specific collagen scaffolds to support the osteogenic and chondrogenic differentiation of human bone marrow stromal cells. *Biomaterials* 29 (21):3105–3116.

De Aza, P. N., Z. B. Luklinska, C. Santos, F. Guitian, and S. De Aza. 2003. Mechanism of bone-like formation on a bioactive implant in vivo. *Biomaterials* 24 (8):1437–1445.

Decher, G. 1997. Fuzzy nanoassemblies: Toward layered polymeric multicomposites. *Science* 277 (5330):1232–1237.

Disegi, J. A. and L. Eschbach. 2000. Stainless steel in bone surgery. *Injury* 31 (4):D2–D6.

Doshi, J. and D. H. Reneker. 1993. Electrospinning process and applications of electrospun fibers. Paper read at *Industry Applications Society Annual Meeting*, 1993, Conference Record of the 1993 IEEE.

Dundigalla, A., S. Lin-Gibson, V. Ferreiro, M. M. Malwitz, and G. Schmidt. 2005. Unusual multilayered structures in poly(ethylene oxide)/laponite nanocomposite films. *Macromolecular Rapid Communications* 26 (3):143–149.

Ehrlich, H., S. Heinemann, C. Heinemann, P. Simon, V. V. Bazhenov, N. P. Shapkin, R. Born, K. R. Tabachnick, T. Hanke, and H. Worch. 2008. Nanostructural organization of naturally occurring composites. Part I. Silica–collagen-based biocomposites. *Journal of Nanomaterials* 2008 (2).

El-Ghannam, A. 2005. Bone reconstruction: From bioceramics to tissue engineering. *Expert Review of Medical Devices* 2 (1):87–101.

Engel, E., A. Michiardi, M. Navarro, D. Lacroix, and J. A. Planell. 2008. Nanotechnology in regenerative medicine: The materials side. *Trends in Biotechnology* 26 (1):39–47.

Engelberg, I. and J. Kohn. 1991. Physico-mechanical properties of degradable polymers used in medical applications: A comparative study. *Biomaterials* 12 (3):292–304.

Erisken, C., D. M. Kalyon, and H. Wang. 2008. Functionally graded electrospun polycaprolactone and [beta]-tricalcium phosphate nanocomposites for tissue engineering applications. *Biomaterials* 29 (30):4065–4073.

Fitzsimmons, R. J., J. T. Ryaby, F. P. Magee, and D. J. Baylink. 1994. Combined magnetic fields increased net calcium flux in bone cells. *Calcified Tissue International* 55 (5):376–380.

Fratzl, P. 2008. Collagen: Structure and mechanics, an introduction. In *Collagen*, New York: Springer.

Freed, L. E., G. Vunjak-Novakovic, R. J. Biron, D. B. Eagles, D. C. Lesnoy, S. K. Barlow, and R. Langer. 1994. Biodegradable polymer scaffolds for tissue engineering. *Nature Biotechnology* 12 (7):689–693.

Fujishiro, T., H. Kobayashi, and T. W. Bauer. 2008. Autograft bone. In *Musculoskeletal Tissue Regeneration*, ed. W. S. Pietrzak, Totowa, NJ: Humana Press.

Gaharwar, A. K., P. J. Schexnailder, A. Dundigalla, J. D. White, C. Matos-Perez, J. Cloud, S. Seifert, J. J. Wilker, and G. Schmidt. 2011a. Highly extensible bio-nanocomposite fibers. *Macromolecular Rapid Communications* 32 (1):50–57.

Gaharwar, A. K., P. J. Schexnailder, B. Kline, and G. Schmidt. 2011b. Assessment of using laponite cross-linked poly(ethylene oxide) for controlled cell adhesion and mineralization. *Acta Biomaterialla* 7 (2):568–577.

Gao, H., B. Ji, I. L. Jager, E. Arzt, and P. Fratzl. 2003. From the cover: Materials become insensitive to flaws at nanoscale: Lessons from nature. *Proceedings of the National Academy of Sciences* 100 (10):5597–5600.

Gazdag, A. R., J. M. Lane, D. Glaser, and R. A. Forster. 1995. Alternatives to autogenous bone graft: Efficacy and indications. *Journal of the American Academy of Orthopaedic Surgeons* 3 (1):1–8.

Meyer, U., J. Handschel, H. P. Wiesmann, T. Meyer, and M. Gelinsky. 2009. Mineralised collagen as biomaterial and matrix for bone tissue engineering. In *Fundamentals of Tissue Engineering and Regenerative Medicine* 485–493, Springer Berlin, Heidelberg.

Gelinsky, M., P. B. Welzel, P. Simon, A. Bernhardt, and U. König. 2008. Porous three-dimensional scaffolds made of mineralised collagen: Preparation and properties of a biomimetic nanocomposite material for tissue engineering of bone. *Chemical Engineering Journal* 137 (1):84–96.

Giannelis, E. P. 1996. Polymer layered silicate nanocomposites. *Advanced Materials* 8 (1):29–35.

Grayson, W. L., P.-H. G. Chao, D. Marolt, D. L. Kaplan, and G. Vunjak-Novakovic. 2008. Engineering custom-designed osteochondral tissue grafts. *Trends in Biotechnology* 26 (4):181–189.

Greenwald, A. S., S. D. Boden, V. M. Goldberg, M. Yaaszemski, and C. S. Heim. 2001. Bone-graft substitutes: Facts, fictions and applications. Paper read at *American Academy of Orthopaedic Surgeons*, at San Francisco, CA.

Gurkan, U. A. and O. Akkus. 2008. The mechanical environment of bone marrow: A review. *Annals of Biomedical Engineering* 36 (12):1978–1991.

Hangody, L., G. Vásárhelyi, L. R. Hangody, Z. Sükösd, G. Tibay, L. Bartha, and G. Bodó. 2008. Autologous osteochondral grafting—Technique and long-term results. *Injury* 39 (1, Suppl 1):32–39.

Harrison, B. S. and A. Atala. 2007. Carbon nanotube applications for tissue engineering. *Biomaterials* 28 (2):344–353.

Hazan, R., R. Brener, and U. Oron. 1993. Bone growth to metal implants is regulated by their surface chemical properties. *Biomaterials* 14 (8):570–574.

Heinegard, D. and A. Oldberg. 1989. Structure and biology of cartilage and bone matrix noncollagenous macromolecules. *The FASEB Journal* 3 (9):2042–2051.

Heinemann, S., C. Heinemann, R. Bernhardt, A. Reinstorf, B. Nies, M. Meyer, H. Worch, and T. Hanke. 2009. Bioactive silica–collagen composite xerogels modified by calcium phosphate phases with adjustable mechanical properties for bone replacement. *Acta Biomaterialia* 5 (6):1979–1990.

Heinemann, S., C. Heinemann, H. Ehrlich, M. Meyer, H. Baltzer, H. Worch, and T. Hanke. 2007. A novel biomimetic hybrid material made of silicified collagen: Perspectives for bone replacement. *Advanced Engineering Materials* 9 (12):1061–1068.

Hench, L. L. 1991. Bioceramics: From concept to clinic. *Journal of the American Ceramic Society* 74 (7):1487–1510.

Hench, L. L. 1998. Biomaterials: A forecast for the future. *Biomaterials* 19 (16):1419–1423.

Hench, L. L. and J. M. Polak. 2002. Third-generation biomedical materials. *Science* 295 (5557):1014–1017.

Hsu, S.-H., H.-J. Yen, and C.-L. Tsai. 2007. The response of articular chondrocytes to type II collagen-Au nanocomposites. *Artificial Organs* 31 (12):854–868.

Hule R. A. and Pochan D. J. 2007. Polymer nanocomposites for biomedical applications. *MRS Bulletin* 32:354–359.

Hunziker, E. B. 2002. Articular cartilage repair: Basic science and clinical progress. A review of the current status and prospects. *Osteoarthritis and Cartilage* 10 (6):432–463.

Hutmacher, D. W. 2000. Scaffolds in tissue engineering bone and cartilage. *Biomaterials* 21 (24):2529–2543.

Jahangir, A. A., R. M. Nunley, S. Mehta, A. Sharan, S. K. Kusuma, J. W. Genuario, A. Ranawat, A. Covey, and J. Flint. 2008. Bone-graft substitutes in orthopaedic surgery. Paper read at *American Academy of Orthopaedic Surgeons*.

Jell, G., R. Verdejo, L. Safinia, M. S. P. Shaffer, M. M. Stevens, and A. Bismarck. 2008. Carbon nanotube-enhanced polyurethane scaffolds fabricated by thermally induced phase separation. *Journal of Materials Chemistry* 18 (16):1865–1872.

Jin, Q., P. Schexnailder, A. K. Gaharwar, and G. Schmidt. 2009. Silicate cross-linked bio-nanocomposite hydrogels from PEO and chitosan. *Macromolecular Bioscience* 9(10):1028–1035, doi:10.1002/mabi.200900080.

Jones, J. R. 2006. Observing cell response to biomaterials. *Materials Today* 9 (12):34–43.

Jones, J. R., L. M. Ehrenfried, and L. L. Hench. 2006. Optimising bioactive glass scaffolds for bone tissue engineering. *Biomaterials* 27 (7):964–973.

Jurvelin, J. S., M. D. Buschmann, and E. B. Hunziker. 2003. Mechanical anisotropy of the human knee articular cartilage in compression. *Proceedings of the I MECH E Part H Journal of Engineering in Medicine* 217:215–219.

Karageorgiou, V. and D. Kaplan. 2005. Porosity of 3D biomaterial scaffolds and osteogenesis. *Biomaterials* 26 (27):5474–5491.

Karageorgiou, V., L. Meinel, S. Hofmann, A. Malhotra, V. Volloch, and D. Kaplan. 2004. Bone morphogenetic protein-2 decorated silk fibroin films induce osteogenic differentiation of human bone marrow stromal cells. *Journal of Biomedical Materials Research* 71A (3):528–537.

Katti, K. S. 2004. Biomaterials in total joint replacement. *Colloids and Surfaces B: Biointerfaces* 39 (3):133–142.

Kay, S., A. Thapa, K. M. Haberstroh, and T. J. Webster. 2002. Nanostructured polymer/nanophase ceramic composites enhance osteoblast and chondrocyte adhesion. *Tissue Engineering* 8 (5):753–761.

Keeney, M. and A. Pandit. 2009. The osteochondral junction and its repair via bi-phasic tissue engineering scaffolds. *Tissue Engineering Part B: Reviews* 15 (1):55–73.

Kerin, A. J., M. R. Wisnom, and M. A. Adams. 1998. The compressive strength of articular cartilage. *Proceedings of the Institution of Mechanical Engineers, Part H: Journal of Engineering in Medicine* 212 (4):273–280.

Khademhosseini, A., J. P. Vacanti, and R. Langer. 2009. Progress in tissue engineering. *Scientific American* 300 (5):64.

Khang, D., G. E. Park, and T. J. Webster. 2008. Enhanced chondrocyte densities on carbon nanotube composites: The combined role of nanosurface roughness and electrical stimulation. *Journal of Biomedical Materials Research Part A* 86A (1):253–260.

Kikuchi, M. and D. Kanama. 2007. Current status of biomaterial research focused on regenerative medicine. *Science and Technology Trends* 24:52–67.

Kim, H.-W., H.-E. Kim, and V. Salih. 2005. Stimulation of osteoblast responses to biomimetic nanocomposites of gelatin–hydroxyapatite for tissue engineering scaffolds. *Biomaterials* 26:5221–5230.

Kloxin, A. M., A. M. Kasko, C. N. Salinas, and K. S. Anseth. 2009. Photodegradable hydrogels for dynamic tuning of physical and chemical properties. *Science* 324 (5923):59.

Kokubo, T., H.-M. Kim, and M. Kawashita. 2003. Novel bioactive materials with different mechanical properties. *Biomaterials* 24 (13):2161–2175.

Kong, L.-J., Q. Ao, J. Xi, L. Zhang, Y.-D. Gong, N.-M. Zhao, and X.-F. Zhang. 2007. Proliferation and differentiation of MC 3T3-E1 cells cultured on nanohydroxyapatite/chitosan composite scaffolds. *Chinese Journal of Biotechnology* 23 (2):262–267.

Kong, L., Y. Gao, W. Cao, Y. Gong, N. Zhao, and X. Zhang. 2005. Preparation and characterization of nano-hydroxyapatite/chitosan composite scaffolds. *Journal of Biomedical Materials Research Part A* 75A (2):275–282.

Kotela, I., J. Podporska, E. Soltysiak, K. J. Konsztowicz, and M. Blazewicz. 2009. Polymer nanocomposites for bone tissue substitutes. *Ceramics International* 35 (6):2475–2480.

Krikorian, V. and D. J. Pochan. 2003. Poly (L-lactic acid)/layered silicate nanocomposite: Fabrication, characterization, and properties. *Chemistry of Materials* 15 (22):4317–4324.

Laurencin, C. T., A. M. A. Ambrosio, M. D. Borden, and J. A. Cooper. 1999. Tissue engineering: Orthopedic applications. *Annual Review of Biomedical Engineering* 1 (1):19–46.

LeBaron, P. C., Z. Wang, and T. J. Pinnavaia. 1999. Polymer-layered silicate nanocomposites: An overview. *Applied Clay Science* 15 (1–2):11–29.

Lee, Y. H., J. H. Lee, I.-G. An, C. Kim, D. S. Lee, Y. K. Lee, and J.-D. Nam. 2005. Electrospun dual-porosity structure and biodegradation morphology of Montmorillonite reinforced PLLA nanocomposite scaffolds. *Biomaterials* 26 (16):3165–3172.

Lee, J. H., T. G. Park, H. S. Park, D. S. Lee, Y. K. Lee, S. C. Yoon, and J.-D. Nam. 2003. Thermal and mechanical characteristics of poly(L-lactic acid) nanocomposite scaffold. *Biomaterials* 24 (16):2773–2778.

Lee, S. and N. D. Spencer. 2008. Materials science: Sweet, hairy, soft, and slippery. *Science* 319 (5863):575–576.

Lewkowitz-Shpuntoff, H. M., M. C. Wen, A. Singh, N. Brenner, R. Gambino, N. Pernodet, R. Isseroff, M. Rafailovich, and J. Sokolov. 2009. The effect of organo clay and adsorbed FeO_3 nanoparticles on cells cultured on ethylene vinyl acetate substrates and fibers. *Biomaterials* 30 (1):8–18.

Li, H., Q. Zheng, Y. Xiao, J. Feng, Z. Shi, and Z. Pan. 2009. Rat cartilage repair using nanophase PLGA/HA composite and mesenchymal stem cells. *Journal of Bioactive and Compatible Polymers* 24 (1):83–99.

Liao, S. S., F. Z. Cui, W. Zhang, and Q. L. Feng. 2004. Hierarchically biomimetic bone scaffold materials: Nano-HA/collagen/PLA composite. *Journal of Biomedical Materials Research Part B: Applied Biomaterials* 69B (2):158–165.

Liff, S. M., N. Kumar, and G. H. McKinley. 2007. High-performance elastomeric nanocomposites via solvent-exchange processing. *Nature Materials* 6 (1):76–83.

Loret, B. and F. M. F. Simoes. 2004. Articular cartilage with intra-and extrafibrillar waters: A chemo-mechanical model. *Mechanics of Materials* 36 (5–6):515–541.

Ma, P. X. 2004. Scaffolds for tissue fabrication. *Materials Today* 7 (5):30–40.

Ma, P. X. 2008. Biomimetic materials for tissue engineering. *Advanced Drug Delivery Reviews* 60 (2):184–198.

Ma, P. X. and R. Zhang. 1999. Synthetic nano-scale fibrous extracellular matrix. *Journal of Biomedical Materials Research* 46 (1):60–72.

MacDonald, R. A., B. F. Laurenzi, G. Viswanathan, P. M. Ajayan, and J. P. Stegemann. 2005. Collagen-carbon nanotube composite materials as scaffolds in tissue engineering. *Journal of Biomedical Materials Research Part A* 74A (3):489–496.

MacGinitie, L. A., Y. A. Gluzband, and A. J. Grodzinsky. 1994. Electric field stimulation can increase protein synthesis in articular cartilage explants. *Journal of Orthopaedic Research* 12 (2):151–160.

Mankin, H. J. 1982. The response of articular cartilage to mechanical injury. *The Journal of Bone and Joint Surgery, American Volume* 64 (3):460–466.

Mano, J. F., R. A. Sousa, L. F. Boesel, N. M. Neves, and R. L. Reis. 2004. Bioinert, biodegradable and injectable polymeric matrix composites for hard tissue replacement: State of the art and recent developments. *Composites Science and Technology* 64 (6):789–817.

Martin, I., S. Miot, A. Barbero, M. Jakob, and D. Wendt. 2007. Osteochondral tissue engineering. *Journal of Biomechanics* 40 (4):750–765.

Matsusaki, M., K. Kadowaki, K. Tateishi, C. Higuchi, W. Ando, D. A. Hart, Y. Tanaka, Y. Take, M. Akashi, H. Yoshikawa, and N. Nakamura. 2009. Scaffold-free tissue-engineered construct—Hydroxyapatite composites generated by an alternate soaking process: Potential for repair of bone defects. *Tissue Engineering Part A* 15 (1):55–63.

Mayer, G. 2005. Rigid biological systems as models for synthetic composites. *Science* 310(5751):1144–1147.

Meinel, L., R. Fajardo, S. Hofmann, R. Langer, J. Chen, B. Snyder, G. Vunjak-Novakovic, and D. Kaplan. 2005. Silk implants for the healing of critical size bone defects. *Bone* 37 (5):688–698.

Middleton, J. C. and A. J. Tipton. 2000. Synthetic biodegradable polymers as orthopedic devices. *Biomaterials* 21 (23):2335–2346.

Mikos, A. G., S. W. Herring, P. Ochareon, J. Elisseeff, H. H. Lu, R. Kandel, F. J. Schoen, M. Toner, D. Mooney, A. Atala, M. E. Van Dyke, D. Kaplan, and G. Vunjak-Novakovic. 2006. Engineering complex tissues. *Tissue Engineering* 12 (12):3307–3339.

Morihara, T., F. Harwood, R. Goomer, Y. Hirasawa, and D. Amiel. 2002. Tissue-engineered repair of osteochondral defects: Effects of the age of donor cells and host tissue. *Tissue Engineering* 8 (6):921–929.

Moroni, L. and J. H. Elisseeff. 2008. Biomaterials engineered for integration. *Materials Today* 11 (5):44–51.

Moutos, F. T., L. E. Freed, and F. Guilak. 2007. A biomimetic three-dimensional woven composite scaffold for functional tissue engineering of cartilage. *Nature Materials* 6 (2):162–167.

Mow, V. C., S. C. Kuei, W. M. Lai, and C. G. Armstrong. 1980. Biphasic creep and stress relaxation of articular cartilage in compression: Theory and experiments. *Journal of Biomechanical Engineering* 102 (1):73–84.

Murphy, W. L. and D. J. Mooney. 2002. Molecular-scale biomimicry. *Nature Biotechnology* 20 (1):30–31.

Murugan, R. and S. Ramakrishna. 2005. Development of nanocomposites for bone grafting. *Composites Science and Technology* 65 (15–16):2385–2406.

Norman, T. L., D. Vashishth, and D. B. Burr. 1995. Fracture toughness of human bone under tension. *Journal of Biomechanics* 28 (3):309–320.

O'Driscoll, S. W. 1998. Current concepts review—The healing and regeneration of articular cartilage. *The Journal of Bone and Joint Surgery, American Volume* 80 (12):1795–1812.

O'Driscoll, S. W., F. W. Keeley, and R. B. Salter. 1986. The chondrogenic potential of free autogenous periosteal grafts for biological resurfacing of major full-thickness defects in joint surfaces under the influence of continuous passive motion. An experimental investigation in the rabbit. *The Journal of Bone and Joint Surgery, American Volume* 68 (7):1017–1035.

O'Shea, T. M. and X. Miao. 2008. Bilayered scaffolds for osteochondral tissue engineering. *Tissue Engineering Part B: Reviews* 14 (4):447–464.

Ozkoc, G., S. Kemaloglu, and M. Quaedflieg. 2009. Production of poly (lactic acid)/organoclay nanocomposite scaffolds by microcompounding and polymer/particle leaching. *Polymer Composites* 31(4):674–683, doi:10.1002/pc.20846.

Paletta, G. A., S. P. Arnoczky, and R. F. Warren. 1992. The repair of osteochondral defects using an exogenous fibrin clot. *The American Journal of Sports Medicine* 20 (6):725–731.

Pan, Y., D. Xiong, and F. Gao. 2008. Viscoelastic behavior of nano-hydroxyapatite reinforced poly(vinyl alcohol) gel biocomposites as an articular cartilage. *Journal of Materials Science: Materials in Medicine* 19 (5):1963–1969.

Parikh, S. 2002. Bone graft substitutes: Past, present, future. *Journal of Postgraduate Medicine* 48:142–148.

Park, G. E., M. A. Pattison, K. Park, and T. J. Webster. 2005. Accelerated chondrocyte functions on NaOH-treated PLGA scaffolds. *Biomaterials* 26 (16):3075–3082.

Paul, D. R. and L. M. Robeson. 2008. Polymer nanotechnology: Nanocomposites. *Polymer* 49 (15): 3187–3204.

Peretti, G. M., V. Zaporojan, K. M. Spangenberg, M. A. Randolph, J. Fellers, and L. J. Bonassar. 2003. Cell-based bonding of articular cartilage: An extended study. *Journal of Biomedical Materials Research* 64 (3):517–524.

Peter, S. J., M. J. Miller, A. W. Yasko, M. J. Yaszemski, and A. G. Mikos. 1998. Polymer concepts in tissue engineering. *Journal of Biomedical Materials Research* 43 (4):422–427.

Petite, H., V. Viateau, W. Bensaid, A. Meunier, C. de Pollak, M. Bourguignon, K. Oudina, L. Sedel, and G. Guillemin. 2000. Tissue-engineered bone regeneration. *Nature Biotechnology* 18 (9):959–963.

Podsiadlo, P., A. K. Kaushik, E. M. Arruda, A. M. Waas, B. S. Shim, J. Xu, H. Nandivada, B. G. Pumplin, J. Lahann, A. Ramamoorthy, and N. A. Kotov. 2007. Ultrastrong and stiff layered polymer nanocomposites. *Science* 318 (5847):80–83.

Ratner, B. D. 1996. *Biomaterials Science: An Introduction to Materials in Medicine*, New York: Elsevier Academic press.

Ratner, B. D. and S. J. Bryant. 2004. Biomaterials: Where we have been and where we are going. *Annual Review of Biomedical Engineering* 6 (1):41–75.

Reffitt, D. M., N. Ogston, R. Jugdaohsingh, H. F. J. Cheung, B. A. J. Evans, R. P. H. Thompson, J. J. Powell, and G. N. Hampson. 2003. Orthosilicic acid stimulates collagen type 1 synthesis and osteoblastic differentiation in human osteoblast-like cells in vitro. *Bone* 32 (2):127–135.

Reilly, D. T. and A. H. Burstein. 1974. Review article. The mechanical properties of cortical bone. *The Journal of Bone and Joint Surgery* 56 (5):1001–1022.

Reilly, D. T., A. H. Burstein, and V. H. Frankel. 1974. The elastic modulus for bone. *Journal of Biomechanics* 7 (3):271.

Rezwan, K., Q. Z. Chen, J. J. Blaker, and A. R. Boccaccini. 2006. Biodegradable and bioactive porous polymer/inorganic composite scaffolds for bone tissue engineering. *Biomaterials* 27 (18):3413–3431.

Rho, J.-Y., L. Kuhn-Spearing, and P. Zioupos. 1998. Mechanical properties and the hierarchical structure of bone. *Medical Engineering & Physics* 20 (2):92–102.

Risbud, M. V. and M. Sittinger. 2002. Tissue engineering: Advances in in vitro cartilage generation. *Trends in Biotechnology* 20 (8):351–356.

Roberts, A. B. and M. B. Sporn. 1996. Transforming growth factor-P. In *The Molecular and Cellular Biology of Wound Repair*, p. 275. Plenum Press, New York.

Sachlos, E., D. Gotora, and J. T. Czernuszka. 2006. Collagen scaffolds reinforced with biomimetic composite nano-sized carbonate-substituted hydroxyapatite crystals and shaped by rapid prototyping to contain internal microchannels. *Tissue Engineering* 12 (9):2479–2487.

Sachlos, E., N. Reis, C. Ainsley, B. Derby, and J. T. Czernuszka. 2003. Novel collagen scaffolds with predefined internal morphology made by solid freeform fabrication. *Biomaterials* 24 (8):1487–1497.

Sarvestani, A., X. He, and E. Jabbari. 2008. Osteonectin-derived peptide increases the modulus of a bone-mimetic nanocomposite. *European Biophysics Journal* 37 (2):229–234.

Savaiano, J. K. and T. J. Webster. 2004. Altered responses of chondrocytes to nanophase PLGA/nanophase titania composites. *Biomaterials* 25 (7–8):1205–1213.

Schaefer, D., I. Martin, P. Shastri, R. F. Padera, R. Langer, L. E. Freed, and G. Vunjak-Novakovic. 2000. In vitro generation of osteochondral composites. *Biomaterials* 21 (24):2599–2606.

Schek, R. M., J. M. Taboas, S. J. Segvich, S. J. Hollister, and P. H. Krebsbach. 2004. Engineered osteochondral grafts using biphasic composite solid free-form fabricated scaffolds. *Tissue Engineering* 10 (9–10):1376–1385.

Schexnailder, P. and G. Schmidt. 2009. Nanocomposite polymer hydrogels. *Colloid & Polymer Science* 287 (1):1–11.

Seaborn, C. D. and F. H. Nielsen. 2002. Silicon deprivation decreases collagen formation in wounds and bone, and ornithine transaminase enzyme activity in liver. *Biological Trace Element Research* 89 (3):251–261.

Sellinger, A., P. M. Weiss, A. Nguyen, Y. Lu, R. A. Assink, W. Gong, and B. C. Jeffrey. 1998. Continuous self-assembly of organic-inorganic nanocomposite coatings that mimic nacre. *Nature* 394 (6690):256–260.

Sharma, B. and J. H. Elisseeff. 2004. Engineering structurally organized cartilage and bone tissues. *Annals of Biomedical Engineering* 32 (1):148–159.

Sherwood, J. K., S. L. Riley, R. Palazzolo, S. C. Brown, D. C. Monkhouse, M. Coates, L. G. Griffith, L. K. Landeen, and A. Ratcliffe. 2002. A three-dimensional osteochondral composite scaffold for articular cartilage repair. *Biomaterials* 23 (24):4739–4751.

Shi, X., J. L. Hudson, P. P. Spicer, J. M. Tour, R. Krishnamoorti, and A. G. Mikos. 2006. Injectable nanocomposites of single-walled carbon nanotubes and biodegradable polymers for bone tissue engineering. *Biomacromolecules* 7 (7):2237–2242.

Shi, X., B. Sitharaman, Q. P. Pham, F. Liang, K. Wu, W. E. Billups, L. J. Wilson, and A. G. Mikos. 2007. Fabrication of porous ultra-short single-walled carbon nanotube nanocomposite scaffolds for bone tissue engineering. *Biomaterials* 28 (28):4078–4090.

Sinha Ray, S. and M. Bousmina. 2005. Biodegradable polymers and their layered silicate nanocomposites: In greening the 21st century materials world. *Progress in Materials Science* 50 (8):962–1079.

Sitharaman, B., X. Shi, X. F. Walboomers, H. Liao, V. Cuijpers, L. J. Wilson, A. G. Mikos, and J. A. Jansen. 2008. In vivo biocompatibility of ultra-short single-walled carbon nanotube/biodegradable polymer nanocomposites for bone tissue engineering. *Bone* 43 (2):362–370.

Smith, I. O., X. H. Liu, L. A. Smith, and P. X. Ma. 2009. Nanostructured polymer scaffolds for tissue engineering and regenerative medicine. *Wiley Interdisciplinary Reviews: Nanomedicine and Nanobiotechnology* 1 (2):226–236.

Sofia, S., M. B. McCarthy, G. Gronowicz, and D. L. Kaplan. 2001. Functionalized silk-based biomaterials for bone formation. *Journal of Biomedical Materials Research* 54 (1):139–148.

Spadaccio, C., A. Rainer, M. Trombetta, G. Vadalá, M. Chello, E. Covino, V. Denaro, Y. Toyoda, and J. A. Genovese. 2009. Poly-L-lactic acid/hydroxyapatite electrospun nanocomposites induce chondrogenic differentiation of human MSC. *Annals of Biomedical Engineering* 37 (7):1376–1389.

Stevens, M. M. and J. H. George. 2005. Exploring and engineering the cell surface interface. *Science* 310 (5751):1135–1138.

Stoddart, M. J., S. Grad, D. Eglin, and M. Alini. 2009. Cells and biomaterials in cartilage tissue engineering. *Regenerative Medicine* 4 (1):81–98.

Stok, K. and A. Oloyede. 2007. Conceptual fracture parameters for articular cartilage. *Clinical Biomechanics* 22 (6):725–735.

Takigawa, T. and Y. Endo. 2006. Effects of glutaraldehyde exposure on human health. *Journal of Occupational Health* 48 (2):75–87.

Tampieri, A., G. Celotti, E. Landi, M. Sandri, N. Roveri, and G. Falini. 2003. Biologically inspired synthesis of bone-like composite: Self-assembled collagen fibers/hydroxyapatite nanocrystals. *Journal of Biomedical Materials Research Part A* 67A (2):618–625.

Tampieri, A., M. Sandri, E. Landi, D. Pressato, S. Francioli, R. Quarto, and I. Martin. 2008. Design of graded biomimetic osteochondral composite scaffolds. *Biomaterials* 29 (26):3539–3546.

Tang, Z., N. A. Kotov, S. Magonov, and B. Ozturk. 2003. Nanostructured artificial nacre. *Nature Materials* 2 (6):413–418.

Temenoff, J. S. and A. G. Mikos. 2000. Review: Tissue engineering for regeneration of articular cartilage. *Biomaterials* 21 (5):431–440.

Thomas, V., D. R. Dean, M. V. Jose, B. Mathew, S. Chowdhury, and Y. K. Vohra. 2007. Nanostructured biocomposite scaffolds based on collagen coelectrospun with nanohydroxyapatite. *Biomacromolecules* 8 (2):631–637.

Ulrich, D., B. Van Rietbergen, A. Laib, and P. Ruegsegger. 1999. The ability of three-dimensional structural indices to reflect mechanical aspects of trabecular bone. *Bone* 25 (1):55–60.

Ulrich-Vinther, M., M. D. Maloney, E. M. Schwarz, R. Rosier, and R. J. O'Keefe. 2003. Articular cartilage biology. *Journal of the American Academy of Orthopaedic Surgeons* 11 (6):421–430.

Vaia, R. and J. Baur. 2008. Materials science: Adaptive composites. *Science* 319 (5862):420–421.

Vaia, R. A. and H. D. Wagner. 2004. Framework for nanocomposites. *Materials Today* 7 (11):32–37.

Valerio, P., M. M. Pereira, A. M. Goes, and M. F. Leite. 2004. The effect of ionic products from bioactive glass dissolution on osteoblast proliferation and collagen production. *Biomaterials* 25 (15):2941–2948.

Venugopal, J., S. Low, A. Choon, T. Sampath Kumar, and S. Ramakrishna. 2008. Mineralization of osteoblasts with electrospun collagen/hydroxyapatite nanofibers. *Journal of Materials Science: Materials in Medicine* 19 (5):2039–2046.

Venugopal, J., P. Vadgama, T. S. Sampath Kumar, and S. Ramakrishna. 2007. Biocomposite nanofibres and osteoblasts for bone tissue engineering. *Nanotechnology* 18 (5):055101.

Wang, X., H. J. Kim, C. Wong, C. Vepari, A. Matsumoto, and D. L. Kaplan. 2006. Fibrous proteins and tissue engineering. *Materials Today* 9 (12):44–53.

Wang, D.-A., S. Varghese, B. Sharma, I. Strehin, S. Fermanian, J. Gorham, D. H. Fairbrother, B. Cascio, and J. H. Elisseeff. 2007. Multifunctional chondroitin sulphate for cartilage tissue-biomaterial integration. *Nature Materials* 6 (5):385–392.

Wang, W., Z. Wang, G. Zhang, C. C. Clark, and C. T. Brighton. 2004. Up-regulation of chondrocyte matrix genes and products by electric fields. *Clinical Orthopaedics and Related Research*:163–173.

Wei, J., S. J. Heo, C. Liu, D. H. Kim, S. E. Kim, Y. T. Hyun, J. W. Shin, and J. W. Shin. 2008. Preparation and characterization of bioactive calcium silicate and poly (ε-caprolactone) nanocomposite for bone tissue regeneration. *Journal of Biomedical Materials Research Part A* 90A (3):702–712.

Wei, G. and P. X. Ma. 2004. Structure and properties of nano-hydroxyapatite/polymer composite scaffolds for bone tissue engineering. *Biomaterials* 25 (19):4749–4757.

Weiner, S. and W. Traub. 1992. Bone structure: From angstroms to microns. *The FASEB Journal* 6 (3):879–885.

Weiner, S., W. Traub, and H. D. Wagner. 1999. Lamellar bone: Structure-function relations. *Journal of Structural Biology* 126 (3):241–255.

Weiner, S. and H. D. Wagner. 1998. The material bone: Structure-mechanical function relations. *Annual Review of Materials Science* 28 (1):271–298.

Weiss, C., L. Rosenberg, and A. J. Helfet. 1968. An ultrastructural study of normal young adult human articular cartilage. *The Journal of Bone and Joint Surgery, American Volume* 50 (4):663–674.

Winey, K. I. and R. A. Vaia. 2007. Polymer nanocomposites. *MRS Bulletin* 32:314–322.

Wong Po Foo, C., S. V. Patwardhan, D. J. Belton, B. Kitchel, D. Anastasiades, J. Huang, R. R. Naik, C. C. Perry, and D. L. Kaplan. 2006. Novel nanocomposites from spider silk-silica fusion (chimeric) proteins. *Proceedings of the National Academy of Sciences* 103 (25):9428–9433.

Wu, C. and J. Chang. 2004. Synthesis and apatite-formation ability of akermanite. *Materials Letters* 58 (19):2415–2417.

Wu, C., J. Chang, J. Wang, S. Ni, and W. Zhai. 2005. Preparation and characteristics of a calcium magnesium silicate (bredigite) bioactive ceramic. *Biomaterials* 26 (16):2925–2931.

Wu, C., J. Chang, W. Zhai, S. Ni, and J. Wang. 2006. Porous akermanite scaffolds for bone tissue engineering: Preparation, characterization, and in vitro studies. *Journal of Biomedical Materials Research Part B: Applied Biomaterials* 78B (1):47–55.

Wu, C.-J. and G. Schmidt. 2009. Thermosensitive and dissolution properties in nanocomposite polymer hydrogels. *Macromolecular Rapid Communications* 30:1492–1497, doi:10.1002/marc.200900163.

Xynos, I. D., A. J. Edgar, L. D. K. Buttery, L. L. Hench, and J. M. Polak. 2001. Gene-expression profiling of human osteoblasts following treatment with the ionic products of Bioglass (R) 45S5 dissolution. *Journal of Biomedical Materials Research* 55 (2):151–157.

Yang, S., K.-F. Leong, Z. Du, and C.-K. Chua. 2001. The design of scaffolds for use in tissue engineering. Part I. Traditional factors. *Tissue Engineering* 7 (6):679–689.

Yasui, N., S. Osawa, T. Ochi, H. Nakashima, and K. Ono. 1982. Primary culture of chondrocytes embedded in collagen gels. *Pathobiology* 50 (2):92–100.

Yaylaoglu, M. B., C. Yildiz, F. Korkusuz, and V. Hasirci. 1999. A novel osteochondral implant. *Biomaterials* 20 (16):1513–1520.

Yoshikawa, H. and A. Myoui. 2005. Bone tissue engineering with porous hydroxyapatite ceramics. *Journal of Artificial Organs* 8 (3):131–136.

Yuan, H., Z. Yang, Y. Li, X. Zhang, J. D. De Bruijn, and K. De Groot. 1998. Osteoinduction by calcium phosphate biomaterials. *Journal of Materials Science: Materials in Medicine* 9 (12):723–726.

Zhang, L. and T. J. Webster. 2009. Nanotechnology and nanomaterials: Promises for improved tissue regeneration. *Nano Today* 4 (1):66–80.

Zhu, W., V. C. Mow, T. J. Koob, and D. R. Eyre. 1993. Viscoelastic shear properties of articular cartilage and the effects of glycosidase treatments. *Journal of Orthopaedic Research* 11 (6):771–781.

Zhuang, H., J. Zheng, H. Gao, and K. De Yao. 2007. In vitro biodegradation and biocompatibility of gelatin/montmorillonite-chitosan intercalated nanocomposite. *Journal of Materials Science: Materials in Medicine* 18 (5):951–957.

Zilg, C., R. Thomann, R. Mülhaupt, and J. Finter. 1999. Polyurethane nanocomposites containing laminated anisotropic nanoparticles derived from organophilic layered silicates. *Advanced Materials* 11 (1):49–52.

Zizak, I., P. Roschger, O. Paris, B. M. Misof, A. Berzlanovich, S. Bernstorff, H. Amenitsch, K. Klaushofer, and P. Fratzl. 2003. Characteristics of mineral particles in the human bone/cartilage interface. *Journal of Structural Biology* 141 (3):208–217.

25

Collagen: A Natural Nanobiomaterial for High-Resolution Studies in Tissue Engineering

Brian M. Gillette
Columbia University

Niccola N. Perez
Columbia University

Prasant Varghese
Columbia University

Samuel K. Sia
Columbia University

25.1 Introduction

Across metazoans, collagen-based extracellular matrices (ECMs) play key roles in development, tissue homeostasis, and pathological states such as cancer and fibrosis. In vertebrates, there are at least 28 known types of collagen (Kadler et al. 2007), with each tissue type featuring a specific ECM composition and exhibiting exquisite structure intimately tied to tissue function. In order to understand how cells are integrated to build tissues and organs, and to potentially engineer such systems, methods to build and analyze realistic models of cells in a 3D ECM in vitro are essential. To date, studies of cell biology have been conducted primarily on 2D surfaces, often in the absence of ECM molecules. The construction of naturally derived 3D extracellular matrix materials provides platforms for studying cell biology in native-like microenvironments and for engineering replacement tissues. However, many 3D ECM in vitro culture models are not well defined in composition or structure, which poses challenges for both the engineering of tissues and the study of cell biology. Further, gels that consist of a limited number of ECM components do not fully recapitulate the complex compositions and structures of native tissue ECM microenvironments. Numerous studies have investigated ways to modulate the structure of collagen-based ECMs in vitro, allowing for experimental control over properties, such as mesh size, fiber morphology, and fiber alignment, in order to recreate the nano- to microscale structural, and thus functional properties of native tissues.

This chapter will first briefly review the structural hierarchy and assembly of collagen at the nanoscale level, and the structure and function of collagen in different tissue types at the macro level, including examples of tissue-engineered biomaterials. Then we will discuss various methods of building 3D collagen-based ECMs for biomedical applications in vitro, focusing mainly on recent developments in tissue engineering and microfabrication techniques using collagen-based materials.

25.2 Collagen Assembly and Structural Hierarchy

As pointed out by G.N. Ramachandran in 1955, the collagen monomer, or *tropocollagen*, exhibits a right-handed coiled structure composed of three left-handed helical polypeptides called alpha chains (Ramachandran and Kartha 1955). Type I tropocollagen molecules (a fibrillar collagen), around 300 nm in length and 1.5 nm in diameter, associate laterally and longitudinally to form collagen fibrils. The fibrils exhibit a distinctive ~65 nm banding pattern when visualized with electron microscopy that arises from the staggering of the fibers during assembly. Collagen fibrils then associate to form larger fibers, which can further bundle to form tissue structures, such as tendon, for example (Figure 25.1). The presence of glycine at every third residue allows for regular hydrogen bonding, and large amount of hydroxyproline enables significant cross-linking between residues (Ramachandran and Kartha 1955). Both of these features contribute to the strength of collagen even at the monomeric level, as demonstrated by tensile strength, bending, and shear stress tests conducted on tropocollagen molecules (Buehler et al. 2008).

The structure of collagen-based ECMs are determined by a multitude of factors. First, different types of collagen consist of different types of alpha chains. For example, type II collagen is homotrimeric (consisting of three alpha 1 chains), whereas type I collagen consists of two alpha 1 chains and one alpha 2 chain (Hulmes 2002). The composition of the collagen monomer can influence the nature of higher order assembly to build ECMs with dramatically different fiber morphologies and arrangements. Nonfibrillar collagens, for example, such as those that form basement membranes (e.g., type IV), assemble as sheet structures instead of fibers. In addition, other constituents of the ECM, including minor collagen types, glycosaminoglycans (GAGs), proteoglycans, and glycoproteins, can serve as nucleators and organizers of collagen fibrillogenesis to alter the structure of the matrix (Kadler et al. 2008). Further, numerous intracellular and extracellular processing steps, along with the distribution and arrangement of cells (such as fibroblasts), also serve to dynamically regulate the structure and composition of the ECM.

The in vitro formation of collagen fibers from purified collagen has been studied for more than half a century (Kadler et al. 1996). In vitro, purified tropocollagen molecules fibrillize by what is believed to

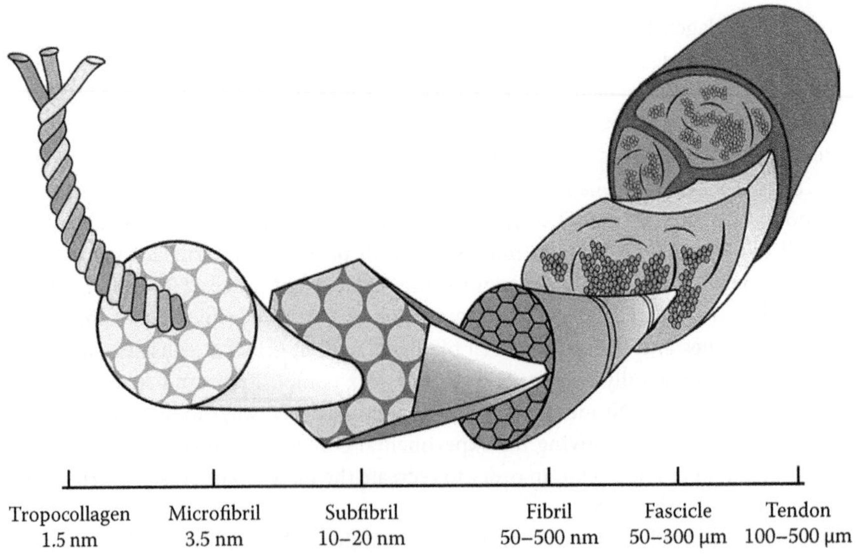

FIGURE 25.1 The structural hierarchy of collagen in tendon tissue. (Image based on Baer, E. et al., *Phys. Today*, 45, 60, 1992.)

be entropically driven self-assembly, whereby the exposed hydrophobic residues coalesce in the middle of the fibril (Kadler et al. 1996, 2008). In the lab, fibrillogenesis is typically accomplished by mixing acid solubilized collagen in buffer (such as phosphate buffered saline [PBS]), neutralizing the pH while holding the solution at a cold temperature, then warming to initiate fibrillogenesis. Fibers formed in vitro resemble native fibers, exhibiting the characteristic 65 nm banding patterns. Fibrillogenesis proceeds in three phases: a lag phase during which nuclei form, a growth phase during which fibers grow in length and width, and a plateau phase. The kinetics of fibrillogenesis vary with collagen type, pH, temperature, and the presence of modulatory ECM components (Yang and Kaufman 2009).

There are multiple potential sources of stability in collagen fibers. The role of interstrand hydrogen bonding, which was initially thought to be a major factor in collagen stability, is currently being reexamined (Holmgren et al. 1999). Fibril assembly is also known to be driven by hydrophobic and ionic intercollagen interactions (Na et al. 1989). As collagen fibers build hierarchically from single fibrils to larger fibers (on the order of tens of microns and above), they exhibit increasing amounts of strength (Hulmes 2002; Buehler et al. 2008). These larger fibers then associate to play key structural roles in complex tissues, such as bone, skin, tendon, and cartilage.

25.3 Structure and Function of Collagen in Native and Engineered Tissues

The fine structure of the collagen network is remarkably diverse and tissue-specific. To study such tissues in vitro, model systems that mimic such structures would be desirable. In the following sections, we highlight the collagen structure in four types of native tissue as well as engineered tissues to mimic this native structure.

25.3.1 Bone

Bone tissue consists mainly of type I collagen with a small amount of type V collagen, which come together to form a network that anchors nanosized hydroxyapatite bone crystals between end-to-end gaps (Cen et al. 2008). The resulting composite is made of longitudinally arranged layers of organic phase collagen (i.e., hydrated collagen fibers) interspersed by mineralized collagen (i.e., the bone crystals) (Canty and Kadler 2005; Buehler et al. 2008), allowing for the cross-linking that gives bone its durable properties (Buehler and Ackbarow 2007; Taylor 2007). Knowledge of the native structure could aid the design of engineered bone. For example, one recent study (Tampieri et al. 2003) involves engineering bone tissue through a biologically inspired synthesis of bone-like composite using self-assembled collagen fibers and hydroxyapatite crystals to mimic the native interactions between the organic and mineral phases of bone.

25.3.2 Skin

Composed of three densely vascularized layers, skin is a complex tissue that is made up primarily of type I collagen (~70%–80% dry weight of skin) (Waller and Maibach 2006; Smith and Rennie 2007). The type I collagen in skin is tightly packed into bundles that form a loose, interwoven, wavy, and randomly oriented pattern, imparting strength and elasticity to the skin. Many collagen skin mimics, from both naturally derived and synthetic sources, have been produced in the past 20 years (MacNeil 2008), and there are many currently in clinical use (Cen et al. 2008).

25.3.3 Tendon

Tendon is composed mainly of type I collagen, but traces of types II, III, and V collagen can also be found, making up about 60%–85% of the dry weight of tendon tissue (Cen et al. 2008). The collagen fibers in tendon are organized in thick bundles of mostly longitudinally oriented fibers, which impart a high ultimate tensile strength (Cen et al. 2008), of about 60 MPa (Johnson et al. 1994), to tendon tissue. One method for developing tissue-engineered tendon (Awad et al. 2003; Juncosa-Melvin et al. 2006, 2007) was to use type I collagen sponges to engineer patellar tendons in rabbits subject to different culture conditions (Juncosa-Melvin et al. 2006).

25.3.4 Articular Cartilage

This type of cartilage is found on the surface of bone, and is crucial for lubricating and distributing force along joints (Johns and Athanasiou 2007). Type II collagen contributes about 60% of the dry weight of the tissue (Cen et al. 2008). Articular cartilage is composed of three zones (superficial, middle, and deep), each with different alignment and distribution patterns (Muir et al. 1970): small fibers (~34 nm in diameter) oriented parallel to the surface in the superficial zone; medium-sized fibers (70–100 nm in diameter) randomly arranged in the middle zone; and, in the deep zone, the largest fibers (~140 nm in diameter) radially arranged, perpendicular to the surface of the cartilage (Cen et al. 2008; Klein et al. 2009). The collagen in all three zones forms complexes with proteoglycans or glycosaminoglycans, which help modulate the viscoelasticity and compressive properties of articular cartilage. Since the different collagen patterns of each zone are primarily responsible for the great tensile strength and compressibility of cartilage (Basser et al. 1998; Klein et al. 2009), recent research in the engineering of articular cartilage has focused on zonal organization, whereas previous research attempted mainly to simulate the overall properties of cartilage with a homogenous tissue in vitro (Vunjak-Novakovic et al. 1998).

25.4 Using Microfluidic Technology to Engineer 3D Collagen-Based Microenvironments with High Spatial Precision

Native tissues consist of numerous cell types and ECM compositions that are anisotropically organized, often in microscale subunits, which are then integrated in order to build higher order structures such as tissues and organs. Although more representative of native tissue than 2D plastic culture surfaces, isotropic 3D matrices (e.g., collagen gels seeded with random distributions of encapsulated cells and maintained in static culture medium) fail to fully recapitulate native tissue microenvironments. Recently, numerous microfabrication technologies have been developed to engineer the microstructure of natural and synthetic ECMs (Nelson and Tien 2006; Khademhosseini and Langer 2007; Ainslie and Desai 2008; Peltola et al. 2008; Bettinger 2009). Such systems are increasingly well suited for engineering complex tissues or developing more realistic models for biological studies.

One such technique is microfluidic systems, which involve creating small-dimension fluid flows on a size scale up to several hundred microns (Whitesides 2006). Microfluidic devices are typically in a microchip format, where different fluidic components (such as channels, valves, mixers, etc.) and functionalized surfaces can be integrated in "lab-on-chip" devices. The development of microfluidic devices based in biocompatible polymers, primarily polydimethylsiloxane (PDMS), has enabled a wide variety of biological assays to be converted to the lab-on-chip format as well as the development of numerous techniques for cell manipulation and controlling cell culture environments (Sia and Whitesides 2003). The use of microfluidic systems for biological applications has greatly expanded over the last 10 years, largely due to the relatively simple processes and low cost associated with producing PDMS-based devices. Microfluidic devices are commonly produced by molding the PDMS polymer to a master chip that defines the channel features. The master chips are typically produced by photolithography or micromachining, procedures that are becoming increasingly accessible in university research labs.

Although the majority of microfluidic cell culture applications involve cell-seeding on 2D surfaces, the use of 3D ECM materials within or as the microfluidic system itself is becoming increasingly common. In one set of applications, microfluidics can be also used to create patterned geometries of cells and

ECM by gelling ECM materials inside PDMS-based microfluidic channels (Tan and Desai 2003, 2004, 2005; Lee et al. 2006). Our group, for example, has used microfluidic PDMS molding chambers to create collagen-glycosaminoglycan (GAG) ECMs with well-defined pores and void fractions for skin tissue engineering applications (Chin et al. 2007).

However, since soluble factors cannot penetrate PDMS, patterned cell-seeded ECM needs to be removed from the PDMS device to permit culture in media. One consequence of this is that when the PDMS spatial constraints on the cell-seeded gel are removed, a loss of pattern integrity can result due to cellular remodeling and migration away from the pattern (Tan and Desai 2003). Alternatively, a design that allows for fluidic access to the gel surface (similar to the methods in the above section) could be used to supply the gel with media while still inside a PDMS device; however, cells may still migrate out of the gel into fluidic access channels. By implementing a boundary that is permeable to soluble factors but not cells, pattern geometry can remain constrained while still allowing for adequate nutrient transport, such as in a microfluidic chip that allows cells to be grown in 3D culture under flow of nutrients (Lii et al. 2008).

The microfluidic patterning of cell-seeded natural ECM gels, however, is complicated by several additional factors. Collagen ECMs will begin to gel upon temperature increase. Although keeping the device on a cold plate or using a cold room can help prevent premature collagen gelling during flow patterning, cold ECM solutions can be quite viscous, and therefore it can be difficult to drive flow through small channels, especially in natively hydrophobic PDMS. Plasma treatment of the PDMS to temporarily increase hydrophilicity can greatly ease the flow of viscous cell-ECM solutions. Cell clustering can also clog small channels, making uniform and complete seeding of the channel geometry difficult. Also, prefiltering cell suspensions through a cell strainer can help minimize clogging due to clustered cells. Lastly, although casting of collagen gels in PDMS devices is a simple method to control gel microstructure, the resulting gels are limited mostly to planar geometries with typically only a few tiers. Rapid prototyping/3D printing techniques allow for more arbitrary control of 3D ECM geometries (Mironov et al. 2003, 2008); however, these methods are typically better suited to synthetic and photopolymerizable materials. The long duration and temperature sensitivity of collagen gelling pose significant challenges for the 3D printing of collagen-based ECMs.

Our group has developed techniques to replicate the inhomogeneous nature of natural tissues at multiple size scales, including a method to interface multiple types of spatially defined natural 3D matrices (Mikos et al. 2006; Nelson et al. 2006) while subject to cellular contraction and other external forces (Meshel et al. 2005). We used a microfluidic technology to pattern multiple phases of cell-seeded collagen-based ECMs within a type I collagen-based hydrogel (Figure 25.2) (Gillette et al. 2008). Importantly, the nucleation and growth of collagen fibers in the patterned phases from the preformed fibers in the bulk gel allowed the stabilization of different patterned ECMs against cellular contractile forces that could potentially separate the phases. This method for creating 3D cell-seeded matrices offers a higher degree of control compared to seeding cells uniformly in a bulk, permitting soluble factor diffusion between multiple cell types. Since this method has demonstrated great potential for interfacing 3D matrices of different compositions and geometries, it will serve as a fundamental foundation for future construction and study of higher order structures (i.e., whole tissues), which is why we will spend the next several paragraphs further elucidating this method for in situ collagen assembly.

The method (Figure 25.2A) consists of four steps: (i) The bulk phase (ECM 1) is formed by doping collagen in an alginate solution and allowing a collagen fiber network to form (by increasing temperature). (ii) The alginate is gelled (by ionic crosslinking) around the collagen fiber network to complete the formation of the bulk ECM. (iii) A second collagen-doped ECM (e.g., fibrinogen) solution is then patterned within the bulk phase. As temperature is increased, collagen precursors in the second ECM nucleate and assemble from exposed collagen fibers at the interface to integrate the two matrices. (iv) Formation of the patterned ECM is completed upon gelling of fibrin in this example (by diffusion of a thrombin solution into the construct to cleave fibrinogen into fibrin in situ). In this manner, collagen fibers present at the phase boundary act as a template for assembly of new collagen fibers in the polymerizing ECM (Figure 25.2A). Remarkably, upon gelling of the neighboring ECM phase, collagen fibers—nucleated

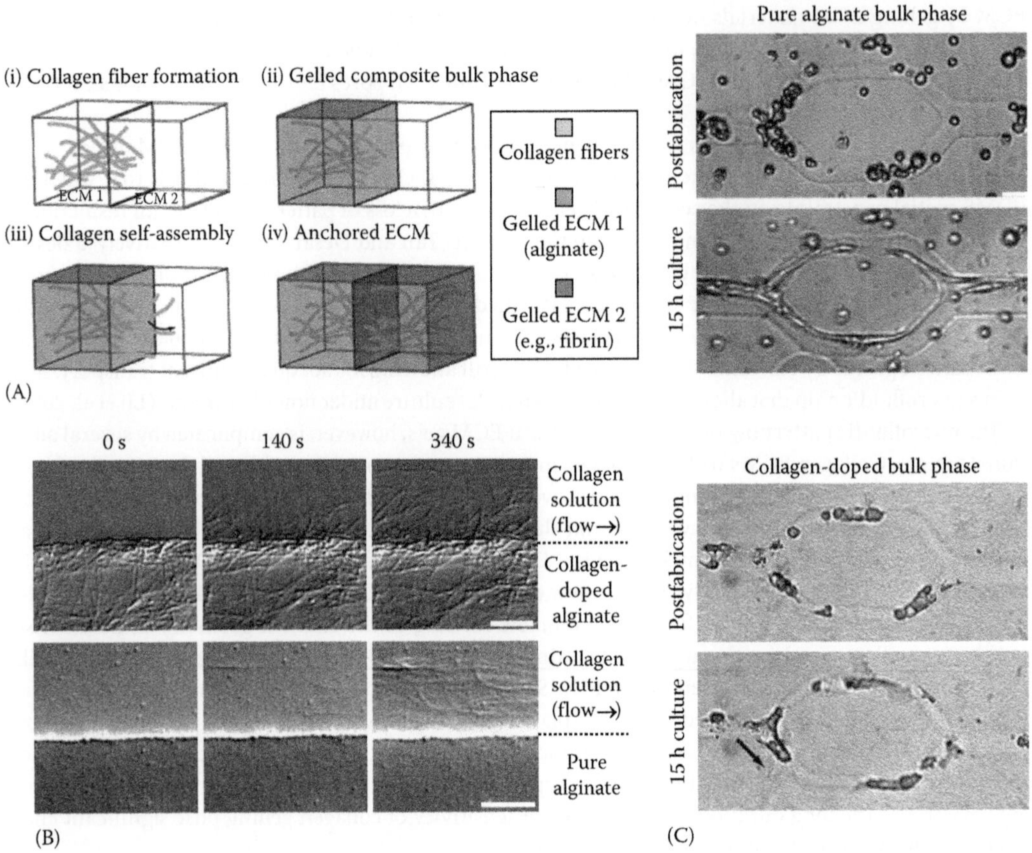

FIGURE 25.2 Collagen-interfacing method (Gillette et al. 2008). (A) Schematic diagram of collagen-based integration of multiple 3D ECM phases. (B) Time-lapse DIC imaging of collagen fiber assembly at the phase interface. Collagen fibers in the patterned ECM assemble from the collagen-doped bulk phase interface into the polymerizing ECM solution (upper panel), but do not nucleate from a pure alginate bulk phase interface (lower panel). (C) Stabilization of patterned ECM by interfacing collagen fibers against cellular contractile forces. Top: Brightfield images depicting separation of patterned HUVEC-seeded collagen from a pure alginate bulk phase interface due to HUVEC contraction after a few hours in culture. Bottom: Brightfield images depicting a stable interface between patterned HUVEC-seeded collagen and a collagen-doped alginate bulk phase after the same culture period. HUVEC migrated within the patterned collagen and formed multicellular structures (arrow) when stimulated with growth factors in the media. Scale bars are 50 μm.

from the first ECM phase and assembled in situ—crossed and anchored together *both* ECM phases over a seamless interface (Figure 25.2B), including those gelled by a variety of different mechanisms (e.g., type I collagen and Matrigel by temperature, alginate by ionic cross-linking, and fibrinogen by enzymatic cross-linking) and even after 3 weeks of culture and facing obstacles such as cellular contractility, migration, and networking.

This technique can also be used in combination with micromolding technologies to produce 3D hydrogel-based constructs consisting of multiple microfabricated phases of cell-encapsulated naturally derived ECM (Gillette et al. 2008). A microfluidic hydrogel is fabricated by casting in a precisely defined PDMS mold and then subsequent phases of 3D ECMs are patterned microfluidically within the bulk hydrogel, similar to the microfluidic process described earlier. The different phases, all collagen-doped, are integrated using the collagen-interfacing method. The amount of alginate in our composites that supports replica molding (0.8%–3%) is well within the range of typical 3D alginate culture concentrations (1%–3%) compared to 4% alginate microfluidic hydrogels previously investigated (Choi et al. 2007). The technique can be used to fabricate microchannel arrays in a single bulk phase into which various ECM/cell compositions can be seeded in a variety of channel geometries simultaneously in a highly reproducible manner. In addition to the ability to control microscale gel geometry (Folch and Toner 2000; Tan and Desai 2004, 2005; Cabodi et al. 2005; Albrecht et al. 2006, 2007; Di Carlo and Lee 2006; Leclerc et al. 2006; Paguirigan and Beebe 2006; Cheung et al. 2007; Choi et al. 2007; Figallo et al. 2007; Golden and Tien 2007; Gottwald et al. 2007; Ling et al. 2007; Nahmias et al. 2007; Park et al. 2007; Toh et al. 2007), combining microfluidic molding techniques and this 3D interfacing method, addresses a number of additional key engineering challenges, including (1) the low capability of incorporating multiple ECM components (Albrecht et al. 2006, 2007; Griffith and Swartz 2006), (2) the poor integration of different ECM materials, and (3) the poor long-term stability of patterned ECM (Meshel et al. 2005; Mikos et al. 2006; Rowe and Stegemann 2006; Marenzana et al. 2007).

Overall, this method uses collagen fibers, a naturally occurring nanobiomaterial, at a phase boundary to act as a template for the assembly of new fibers to build stable 3D constructs containing multiple types of cells and ECM compositions and geometries.

25.5 Dynamic Control of ECM Structural Properties

In native tissues, the structure and composition of the ECM can change during tissue development, repair, or tumor growth (Daley et al. 2008; Lopez et al. 2008), whereas for most in vitro 3D ECM models, the structure and composition of the ECM cannot be dynamically modified by the experimenter after fabrication is complete (Hahn et al. 2006; Tayalia et al. 2008; Freed et al. 2009). Tools that enable dynamic alteration of the 3D extracellular microenvironment may be useful for elucidating how dynamic changes in the 3D cell microenvironment influence cellular behavior. For example, a technique was recently developed to spatially and temporally control the ability for cells to spread and migrate in 3D synthetic ECM materials after gelling, using ultraviolet radiation-induced degradation of synthetic polymers (Kloxin et al. 2009). Our group has recently developed a method to dynamically alter the structural properties of a composite matrix consisting of unmodified naturally derived ECM components (collagen and alginate) (Gillette et al. 2009).

Using a two-component ECM, in which one component is a fibrillar structural element, and one component is a reversibly cross-linkable polymer element, properties of the composite ECM (such as the ability for cells to migrate and soluble factor diffusion rates) could be dynamically altered by controlling the state of polymer cross-links. This work used collagen as the structural component and alginate (which is cross-linked with the application of Ca^{2+} and uncross-linked by application of a calcium chelator [Na citrate]) as the reversibly cross-linking component. They also examined the migration response of encapsulated fibroblasts and diffusion coefficients of various molecular weight dextrans under the dynamic switching of the ECM structure.

The matrix pore size, mechanical properties, and cell adhesion site density are important regulators of cell adhesion and spreading in 3D ECM (Beadle et al. 2008; Tayalia et al. 2008). In the composite collagen-alginate ECM, the relatively large mesh size of collagen fiber networks (in the range of 1–20 μm [Yang and Kaufman 2009]) readily supports cell spreading and migration, while the small pore size of cross-linked alginate gels (tens of nanometers [Li et al. 1996; Boontheekul et al. 2005]) prohibits the invasion of cellular processes. Thus, switching of the alginate cross-linking state dynamically applies spatial constraints for cellular migration in the composite ECM.

For studies of cell behavior under dynamic spatial constraint, fibroblasts were encapsulated within collagen-alginate gels placed under a transwell insert to allow the addition of multiple cycles of reagents (see Gillette et al. [2009] for method details). Cells are initially rounded after encapsulation because the alginate is cross-linked immediately after gelling of the collagen (i.e., before cells have a chance to spread out significantly). When the alginate cross-links are removed by the application of a calcium chelator, the cells rapidly spread and migrate. Recross-linking of the alginate with Ca^{2+} freezes the cells in place, and re-uncross-linking the alginate again permits cell movement, and so on. This novel dynamic switching model could permit engineers to temporally switch properties such as the rate of solute transport, cell spreading, and cell mobility.

25.6 Synthetic Collagen Mimics

Finally, we note that a number of studies have been conducted to create biomimetic materials inspired by the collagen structure, to embody its most attractive structural features and at the same time allow flexibility in chemical control. One study focused on the templated assembly of triple-helical synthetic polypeptides, which was fully amino acid-based and more closely mimicked the structure of native collagen than other approaches (Feng et al. 1996; Kwak et al. 2002; Kenawy el et al. 2003; Khew and Tong 2008; Merrett et al. 2008), allowing for the formation of triple-helices from short synthetic peptides, an important breakthrough in the engineering of collagen-mimetic biomaterials (Khew and Tong 2008). Specifically, since this model mimics integrin-specific adhesion and cell recognition properties of native collagen, it is very useful in studying protein folding, creating novel protein mimics or coating biomaterials (Khew and Tong 2008).

Other mimics target lower levels of the structural hierarchy of collagen. For example, a collagen triple helix molecule in which all the hydroxyl groups on Hyp residues are replaced with fluorine produces a hyperstable collagen mimic (Holmgren et al. 1999), which could aid engineers in overcoming the challenge of degradation in vivo. Synthetic materials that mimic the intricate fibrillar structure of natural ECM can also be constructed using a technique called *electrospinning* (Kenawy el et al. 2003), which can produce fibers on the 10 nm scale. Lastly, though most collagen mimics utilize type I collagen due to its abundance in the body, some studies explore tissue substitutes that involve using other collagen types not naturally found in that certain tissue. For example, one group developed a type III collagen-based corneal substitute that exhibited greater optical clarity and similar mechanical properties to native corneal tissue (Merrett et al. 2008).

References

Ainslie, K. M. and T. A. Desai. 2008. Microfabricated implants for applications in therapeutic delivery, tissue engineering, and biosensing. *Lab Chip* 8 (11):1864–1878.

Albrecht, D. R., G. H. Underhill, A. Mendelson, and S. N. Bhatia. 2007. Multiphase electropatterning of cells and biomaterials. *Lab Chip* 7 (6):702–709.

Albrecht, D. R., G. H. Underhill, T. B. Wassermann, R. L. Sah, and S. N. Bhatia. 2006. Probing the role of multicellular organization in three-dimensional microenvironments. *Nat. Methods* 3 (5):369–375.

Awad, H. A., G. P. Boivin, M. R. Dressler, F. N. Smith, R. G. Young, and D. L. Butler. 2003. Repair of patellar tendon injuries using a cell-collagen composite. *J. Orthop. Res.* 21 (3):420–431.

Baer, E., A. Hiltner, and R. Morgan. 1992. Biological and synthetic hierarchical composites. *Phys. Today* 45:60–67.

Basser, P. J., R. Schneiderman, R. A. Bank, E. Wachtel, and A. Maroudas. 1998. Mechanical properties of the collagen network in human articular cartilage as measured by osmotic stress technique. *Arch. Biochem. Biophys.* 351 (2):207–219.

Beadle, C., M. C. Assanah, P. Monzo, R. Vallee, S. S. Rosenfeld, and P. Canoll. 2008. The role of myosin II in glioma invasion of the brain. *Mol. Biol. Cell* 19 (8):3357–3368.

Bettinger, C. J. 2009. Synthesis and microfabrication of biomaterials for soft-tissue engineering. *Pure Appl. Chem.* 81 (12):2183–2201.

Boontheekul, T., H. J. Kong, and D. J. Mooney. 2005. Controlling alginate gel degradation utilizing partial oxidation and bimodal molecular weight distribution. *Biomaterials* 26 (15):2455–2465.

Buehler, M. J. and T. Ackbarow. 2007. Fracture mechanics of protein materials. *Mater. Today* 10:46–58.

Buehler, M. J., S. Keten, and T. Ackbarow. 2008. Theoretical and computational hierarchical nanomechanics of protein materials: Deformation and fracture. *Prog. Mater. Sci.* 53 (8):1101–1241.

Cabodi, M., N. W. Choi, J. P. Gleghorn, C. S. Lee, L. J. Bonassar, and A. D. Stroock. 2005. A microfluidic biomaterial. *J. Am. Chem. Soc.* 127 (40):13788–13789.

Canty, E. G. and K. E. Kadler. 2005. Procollagen trafficking, processing and fibrillogenesis. *J. Cell Sci.* 118 (Pt 7):1341–1353.

Cen, L., W. Liu, L. Cui, W. Zhang, and Y. Cao. 2008. Collagen tissue engineering: Development of novel biomaterials and applications. *Pediatr. Res.* 63 (5):492–496.

Cheung, Y. K., B. M. Gillette, M. Zhong, S. Ramcharan, and S. K. Sia. 2007. Direct patterning of composite biocompatible microstructures using microfluidics. *Lab. Chip* 7 (5):574–579.

Chin, C. D., V. Linder, and S. K. Sia. 2007. Lab-on-a-chip devices for global health: Past studies and future opportunities. *Lab. Chip* 7 (1):41–57.

Choi, N. W., M. Cabodi, B. Held, J. P. Gleghorn, L. J. Bonassar, and A. D. Stroock. 2007. Microfluidic scaffolds for tissue engineering. *Nat. Mater.* 6 (11):908–915. doi:10.1038/nmat2022.

Daley, W. P., S. B. Peters, and M. Larsen. 2008. Extracellular matrix dynamics in development and regenerative medicine. *J. Cell Sci.* 121 (3):255–264.

Di Carlo, D. and L. P. Lee. 2006. Dynamic single-cell analysis for quantitative biology. *Anal. Chem.* 78 (23):7918–7925.

Feng, Y., G. Melacini, J. P. Taulane, and M. Goodman. 1996. Acetyl-terminated and template-assembled collagen-based polypeptides composed of Gly-Pro-Hyp sequences. 2. Synthesis and conformational analysis by CD, UV absorbance and optical rotation. *J. Am. Chem. Soc.* 118:10351–10358.

Freed, L. E., G. C. Engelmayr Jr., J. T. Borenstein, F. T. Moutos, and F. Guilak. 2009. Advanced material strategies for tissue engineering scaffolds. *Adv. Mater.* 21(32–33):3410–3418. doi:10.1002/adma.200900303.

Figallo, E., C. Cannizzaro, S. Gerecht, J. A. Burdick, R. Langer, N. Elvassore, and G. Vunjak-Novakovic. 2007. Micro-bioreactor array for controlling cellular microenvironments. *Lab. Chip* 7 (6):710–719.

Folch, A. and M. Toner. 2000. Microengineering of cellular interactions. *Annu. Rev. Biomed. Eng.* 2:227–256.

Gillette, B. M., J. A. Jensen, B. Tang, G. J. Yang, A. Bazargan-Lari, M. Zhong, and S. K. Sia. 2008. In situ collagen assembly for integrating microfabricated three-dimensional cell-seeded matrices. *Nat. Mater.* 7 (8):636–640.

Gillette, B. M., J. A. Jensen, M. Wang, J. Tchao, and S. K. Sia. 2009. Dynamic hydrogels: Switching of 3D microenvironments using two-component naturally derived extracellular matrices. *Adv. Mater.* 21:1–6.

Golden, A. P. and J. Tien. 2007. Fabrication of microfluidic hydrogels using molded gelatin as a sacrificial element. *Lab. Chip* 7 (6):720–725.

Gottwald, E., S. Giselbrecht, C. Augspurger, B. Lahni, N. Dambrowsky, R. Truckenmuller, V. Piotter et al. 2007. A chip-based platform for the in vitro generation of tissues in three-dimensional organization. *Lab. Chip* 7 (6):777–785.

Griffith, L. G. and M. A. Swartz. 2006. Capturing complex 3D tissue physiology in vitro. *Nat. Rev. Mol. Cell. Biol.* 7 (3):211–224.

Hahn, M. S., J. S. Miller, and J. L. West. 2006. Three-dimensional biochemical and biomechanical patterning of hydrogels for guiding cell behavior. *Adv. Mater.* 18 (20):2679–2684.

Holmgren, S. K., L. E. Bretscher, K. M. Taylor, and R. T. Raines. 1999. A hyperstable collagen mimic. *Chem. Biol.* 6 (2):63–70.

Hulmes, D. J. 2002. Building collagen molecules, fibrils, and suprafibrillar structures. *J. Struct. Biol.* 137 (1–2):2–10.

Johns, D. E. and K. A. Athanasiou. 2007. Design characteristics for temporomandibular joint disc tissue engineering: Learning from tendon and articular cartilage. *Proc. Inst. Mech. Eng. H* 221 (5):509–526.

Johnson, G. A., D. M. Tramaglini, R. E. Levine, K. Ohno, N. Y. Choi, and S. L. Woo. 1994. Tensile and viscoelastic properties of human patellar tendon. *J. Orthop. Res.* 12 (6):796–803.

Juncosa-Melvin, N., K. S. Matlin, R. W. Holdcraft, V. S. Nirmalanandhan, and D. L. Butler. 2007. Mechanical stimulation increases collagen type I and collagen type III gene expression of stem cell-collagen sponge constructs for patellar tendon repair. *Tissue Eng.* 13 (6):1219–1226.

Juncosa-Melvin, N., J. T. Shearn, G. P. Boivin, C. Gooch, M. T. Galloway, J. R. West, V. S. Nirmalanandhan, G. Bradica, and D. L. Butler. 2006. Effects of mechanical stimulation on the biomechanics and histology of stem cell-collagen sponge constructs for rabbit patellar tendon repair. *Tissue Eng.* 12 (8):2291–2300.

Kadler, K. E., C. Baldock, J. Bella, and R. P. Boot-Handford. 2007. Collagens at a glance. *J. Cell Sci.* 120 (12):1955–1958.

Kadler, K. E., A. Hill, and E. G. Canty-Laird. 2008. Collagen fibrillogenesis: Fibronectin, integrins, and minor collagens as organizers and nucleators. *Curr. Opin. Cell Biol.* 20 (5):495–501.

Kadler, K. E., D. F. Holmes, J. A. Trotter, and J. A. Chapman. 1996. Collagen fibril formation. *Biochem. J.* 316 (1):1–11.

Kenawy el, R., J. M. Layman, J. R. Watkins, G. L. Bowlin, J. A. Matthews, D. G. Simpson, and G. E. Wnek. 2003. Electrospinning of poly(ethylene-co-vinyl alcohol) fibers. *Biomaterials* 24 (6):907–913.

Khademhosseini, A. and R. Langer. 2007. Microengineered hydrogels for tissue engineering. *Biomaterials* 28 (34):5087–5092.

Khew, S. T. and Y. W. Tong. 2008. Template-assembled triple-helical peptide molecules: Mimicry of collagen by molecular architecture and integrin-specific cell adhesion. *Biochemistry* 47 (2):585–596.

Klein, T. J., J. Malda, R. L. Sah, and D. W. Hutmacher. 2009. Tissue engineering of articular cartilage with biomimetic zones. *Tissue Eng. Part B Rev.* 15 (2):143–157.

Kloxin, A. M., A. M. Kasko, C. N. Salinas, and K. S. Anseth. 2009. Photodegradable hydrogels for dynamic tuning of physical and chemical properties. *Science* 324 (5923):59–63.

Kwak, J., A. De Capua, E. Locardi, and M. Goodman. 2002. TREN (Tris(2-aminoethyl)amine): An effective scaffold for the assembly of triple helical collagen mimetic structures. *J. Am. Chem. Soc.* 124 (47):14085–14091.

Leclerc, E., B. David, L. Griscom, B. Lepioufle, T. Fujii, P. Layrolle, and C. Legallaisa. 2006. Study of osteoblastic cells in a microfluidic environment. *Biomaterials* 27 (4):586–595.

Lee, P., R. Lin, J. Moon, and L. P. Lee. 2006. Microfluidic alignment of collagen fibers for in vitro cell culture. *Biomed. Microdev.* 8 (1):35–41.

Li, R. H., D. H. Altreuter, and F. T. Gentile. 1996. Transport characterization of hydrogel matrices for cell encapsulation. *Biotechnol. Bioeng.* 50 (4):365–373.

Lii, J., W. J. Hsu, H. Parsa, A. Das, R. Rouse, and S. K. Sia. 2008. Real-time microfluidic system for studying mammalian cells in 3D microenvironments. *Anal. Chem.* 80 (10):3640–3647.

Ling, Y., J. Rubin, Y. Deng, C. Huang, U. Demirci, J. M. Karp, and A. Khademhosseini. 2007. A cell-laden microfluidic hydrogel. *Lab. Chip* 7 (6):756–762.

Lopez, J. I., J. K. Mouw, and V. M. Weaver. 2008. Biomechanical regulation of cell orientation and fate. *Oncogene* 27 (55):6981–6993.

MacNeil, S. 2008. Biomaterials for tissue engineering of skin. *Mater. Today* 11 (5):26–35.

Marenzana, M., D. J. Kelly, P. J. Prendergast, and R. A. Brown. 2007. A collagen-based interface construct for the assessment of cell-dependent mechanical integration of tissue surfaces. *Cell Tissue Res.* 327 (2):293–300.

Merrett, K., P. Fagerholm, C. R. McLaughlin, S. Dravida, N. Lagali, N. Shinozaki, M. A. Watsky et al., 2008. Tissue-engineered recombinant human collagen-based corneal substitutes for implantation: Performance of type I versus type III collagen. *Invest. Ophthalmol. Vis. Sci.* 49 (9):3887–3894.

Meshel, A. S., Q. Wei, R. S. Adelstein, and M. P. Sheetz. 2005. Basic mechanism of three-dimensional collagen fibre transport by fibroblasts. *Nat. Cell Biol.* 7 (2):157–164.

Mikos, A. G., S. W. Herring, P. Ochareon, J. Elisseeff, H. H. Lu, R. Kandel, F. J. Schoen et al. 2006. Engineering complex tissues. *Tissue Eng.* 12 (12):3307–3339.

Mironov, V., T. Boland, T. Trusk, G. Forgacs, and R. R. Markwald. 2003. Organ printing: Computer-aided jet-based 3D tissue engineering. *Trends Biotechnol.* 21:157–161.

Mironov, V., V. Kasyanov, and R. R. Markwald. 2008. Nanotechnology in vascular tissue engineering: From nanoscaffolding towards rapid vessel biofabrication. *Trends Biotechnol.* 26 (6):338–344.
Muir, H., P. Bullough, and A. Maroudas. 1970. The distribution of collagen in human articular cartilage with some of its physiological implications. *J. Bone Joint Surg. Br.* 52 (3):554–563.
Na, G. C., L. J. Phillips, and E. I. Freire. 1989. In vitro collagen fibril assembly: Thermodynamic studies. *Biochemistry* 28 (18):7153–7161.
Nahmias, Y., F. Berthiaume, and M. L. Yarmush. 2007. Integration of technologies for hepatic tissue engineering. *Adv. Biochem. Eng. Biotechnol.* 103:309–329.
Nelson, C. M. and J. Tien. 2006. Microstructured extracellular matrices in tissue engineering and development. *Curr. Opin. Biotechnol.* 17 (5):518–523.
Nelson, C. M., M. M. Vanduijn, J. L. Inman, D. A. Fletcher, and M. J. Bissell. 2006. Tissue geometry determines sites of mammary branching morphogenesis in organotypic cultures. *Science* 314 (5797):298–300.
Paguirigan, A. and D. J. Beebe. 2006. Gelatin based microfluidic devices for cell culture. *Lab. Chip* 6 (3):407–413.
Park, H., C. Cannizzaro, G. Vunjak-Novakovic, R. Langer, C. A. Vacanti, and O. C. Farokhzad. 2007. Nanofabrication and microfabrication of functional materials for tissue engineering. *Tissue Eng.* 13 (8):1867–1877.
Peltola, S. M., F. P. Melchels, D. W. Grijpma, and M. Kellomaki. 2008. A review of rapid prototyping techniques for tissue engineering purposes. *Ann. Med.* 40 (4):268–280.
Ramachandran, G. N. and G. Kartha. 1955. Structure of collagen. *Nature* 176 (4482):593–595.
Rowe, S. L. and J. P. Stegemann. 2006. Interpenetrating collagen-fibrin composite matrices with varying protein contents and ratios. *Biomacromolecules* 7 (11):2942–2948.
Sia, S. K. and G. M. Whitesides. 2003. Microfluidic devices fabricated in poly(dimethylsiloxane) for biological studies. *Electrophoresis* 24 (21):3563–3576.
Smith, K. and M. J. Rennie. 2007. New approaches and recent results concerning human-tissue collagen synthesis. *Curr. Opin. Clin. Nutr. Metab. Care* 10 (5):582–590.
Tampieri, A., G. Celotti, E. Landi, M. Sandri, N. Roveri, and G. Falini. 2003. Biologically inspired synthesis of bone-like composite: Self-assembled collagen fibers/hydroxyapatite nanocrystals. *J. Biomed. Mater. Res. A* 67 (2):618–625.
Tan, W. and T. A. Desai. 2003. Microfluidic patterning of cells in extracellular matrix biopolymers: Effects of channel size, cell type, and matrix composition on pattern integrity. *Tissue Eng.* 9 (2):255–267.
Tan, W. and T. A. Desai. 2004. Layer-by-layer microfluidics for biomimetic three-dimensional structures. *Biomaterials* 25 (7–8):1355–1364.
Tan, W. and T. A. Desai. 2005. Microscale multilayer cocultures for biomimetic blood vessels. *J. Biomed. Mater. Res. A* 72 (2):146–160.
Tayalia, P., C. R. Mendonca, T. Baldacchini, D. J. Mooney, and E. Mazur. 2008. 3D cell-migration studies using two-photon engineered polymer scaffolds. *Adv. Mater.* 20 (23):4494–4498.
Taylor, D. 2007. Fracture and repair of bone: A multiscale problem. *J. Mater. Sci.* 42:8911–8918.
Toh, Y. C., C. Zhang, J. Zhang, Y. M. Khong, S. Chang, V. D. Samper, D. van Noort, D. W. Hutmacher, and H. Yu. 2007. A novel 3D mammalian cell perfusion-culture system in microfluidic channels. *Lab. Chip* 7 (3):302–309.
Vunjak-Novakovic, G., B. Obradovic, I. Martin, P. M. Bursac, R. Langer, and L. E. Freed. 1998. Dynamic cell seeding of polymer scaffolds for cartilage tissue engineering. *Biotechnol. Prog.* 14 (2):193–202.
Waller, J. M. and H. I. Maibach. 2006. Age and skin structure and function, a quantitative approach (II): Protein, glycosaminoglycan, water, and lipid content and structure. *Skin Res. Technol.* 12 (3):145–154.
Whitesides, G. M. 2006. The origins and the future of microfluidics. *Nature* 442 (7101):368–373.
Yang, Y.-L., and L. J. Kaufman. 2009. Rheology and confocal reflectance microscopy as probes of mechanical properties and structure during collagen and collagen/hyaluronan self-assembly. *Biophys. J.* 96 (4):1566–1585.

26

Nanotopography on Implant Biomaterials

Edwin Lamers
Radboud University

Frank Walboomers
Radboud University

John Jansen
Radboud University

26.1 Introduction

Prostheses and implants are described as any material, natural or man-made, that comprises the whole or part of a living structure, or a biomedical device that performs, augments, or replaces a natural function (Ratner et al. 2004). Examples are orthopedic prostheses, oral implant materials, vascular grafts, and artificial heart valves. Some of these implant types have already been used for many centuries. Around AD 600, the Mayan civilization used nacre tooth implants (Westbroek and Marin 1998). In Europe, an iron dental implant dating back to AD 200 was found. Both implants were properly integrated into the bone. Also, in other parts of the human body, biomaterials have been used for a long time. Sutures may have been used as far back as 32,000 years ago (Ratner and Bryant 2004); however, most of the described biomaterial applications have been used over at least the past 2000 years (Ratner et al. 2004).

The development of orthopedic and dental implantology is well visible in the increasing market values. Due to the aging of the population, orthopedic and oral prostheses and implants are a fast-growing market in medical health care. In 2008, the worldwide orthopedic implant market was estimated at $29 billion with an annual growth of around 10.7% (a. www.researchandmarkets.com 2009). The dental implant market was estimated at $3.4 billion in 2008 with a growth to $8.1 billion expected in 2015, which constitutes an annual growth rate of around 15% (b. www.researchandmarkets.com 2009).

The improvement of surgical techniques has resulted in oral implants being done on risky patients who were contraindicative for treatment (e.g., patients with diabetes and osteoporosis) in the past (Hwang and Wang 2007). Many epidemiological studies have been performed to study the long-term success rates of (orthopedic and dental) bone implants and results range from approximately 98% down to 78% for dental implants depending of patients' physical condition, implant location, and smoking habits (Alsaadi et al. 2007, Strietzel et al. 2007, Hulleberg et al. 2008, Makela et al. 2008). In addition, failure rates of orthopedic implants in younger patients are significantly higher (>15%) compared to patients over 60 (<7%) (Harrysson et al. 2004, Johnsen et al. 2006, Ulrich et al. 2008), most likely as a consequence of higher impact loads from activities during daily living. As a consequence of this implant failure in younger patients, the number of revision surgeries is increasing. However, failure rates of replaced implants are higher than those for primary implantations. For these reasons, there is a need for a new generation of bone implants made of mechanically strong materials with enhanced biomineralization properties to increase the clinical effectiveness and survival rate.

An important factor in the tissue reaction toward any implant is the initial cellular response toward the device. It is of the highest importance that the implanted material fits in, integrates into the surrounding tissue, and is functional in living tissue; in other words, the implant should be biocompatible. Biocompatibility can be defined as "the ability of a material to fulfill its intended function with an appropriate host response in a specific application" (Ratner 2004). This does not necessarily mean that an implant should integrate into the body. Artificial blood vessels or heart valves, for example, should resist the adhesion of biomolecules and cells, and can function perfectly well without tissue integration.

In principle, all biocompatible implant biomaterials will provoke a low-level, chronic inflammatory response (Ratner 2001). This can result in side effects that are not detrimental to the implant function, but still can have long-term consequences that are not desirable. Therefore, it would be most desirable that a symbiosis is established between the implant device and the surrounding host tissue. This situation is characterized by interactions between the nonbiological compounds of the implant surface and the biological compounds from the surrounding tissue. Currently, it is widely expected that the application of nanotechnology might provide a contrivance to smoothen these interactions. Introducing a nanoscale topography onto an implant material may induce cellular reorganization, which can modify intracellular signaling and the response toward implants. This might lead to an improved integration of the implant into bone and a prolonged lifespan. To get a better understanding of these processes, we first discuss the relevant processes that occur at the implant interface; second, we provide an overview of the biomaterial properties relevant for a prolonged lifespan; third, we discuss the importance of nanotechnology in implantology as well as a short overview of the initial studies on biomaterial surface topographies; and finally we focus on the current applications to produce a nanotopography on implant biomaterials.

26.2 Processes at the Bone–Implant Interface

Immediately after the placement of an implant into a freshly prepared surgical site, water, ions, and serum proteins arrive at the biomaterial surface (Figure 26.1). An important parameter for the adhesion of these molecules is the surface free energy (or wettability) of an implant surface, which is reflected by its

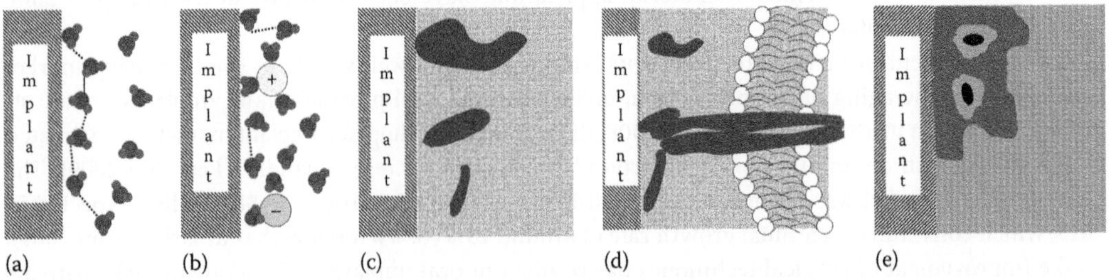

FIGURE 26.1 Reactions on interface between biomaterial and biofluid under optimal conditions. (a) After implantation the surface is first exposed to water. These interactions are strongly affected by the surface free energy of the implant surface. (b) Subsequently the surface is covered with a layer of hydrated ions and a little later with proteins (c). (d) Blood cells are the first cells to interact with the implant surface and produce osteo inductive signals for cell attraction and differentiation. (e) Tissue is regenerated at the implant–tissue interface.

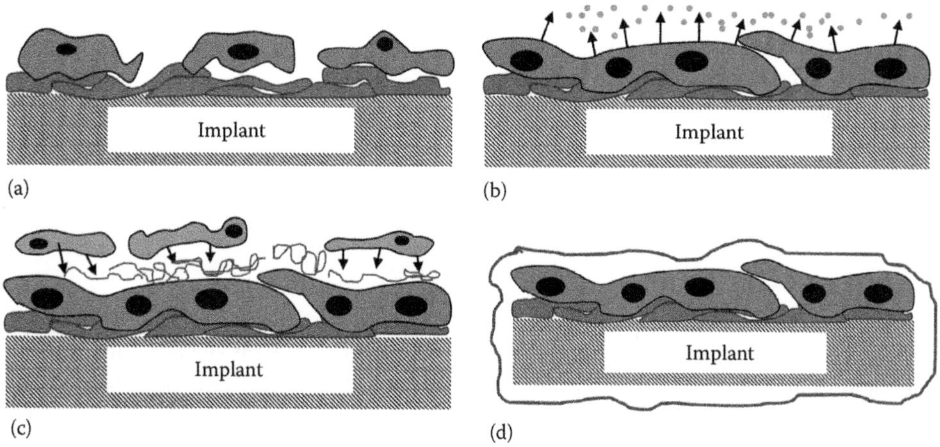

FIGURE 26.2 Foreign body reaction of macrophages to implant materials. (a) After the adsorption of plasma proteins (Fig. 1c) macrophages and other cells are attracted to the implant surface. (b) Macrophages will unsuccessfully try to digest the implant and fuse into giant cells to increase their effectiveness. Subsequently the giant cells start releasing chemotactic factors to attract fibroblasts. (c) Fibroblasts start producing collagenous matrix proteins to encapsulate the implant (d).

hydrophilicity (Roach et al. 2007). Since there are many different proteins in the biofluid, the composition of proteins will be a mixture of early arriving proteins and later arriving proteins, which bind more strongly to the surface (Kasemo and Gold 1999). Here, differences in surface wettability (e.g., very hydrophilic or hydrophobic) provide very different possibilities for protein binding. Directly after the adherence of proteins, blood cells appear at the implant surface. The first cells that arrive at the implant site are neutrophils, followed by macrophages (Figure 26.2). These cells interact with the implant device via adhered proteins and a provisional matrix is formed consisting of fibrin and inflammatory products (Anderson et al. 2004).

After these initial events, several processes can occur at the implant site, that is, extrusion, degradation, encapsulation, and integration depending on tissue type, location, and implant type. Implants can be extruded when in direct contact with epithelial tissue. The epithelial tissue will form a continuous layer at the proximal site of the implant and the implant will be excluded from the surrounding tissue. Degradable implants are usually resorbed in time by macrophages. After resorption, a scar will remain. The most common process that occurs at the implant site is the encapsulation of the implant (Anderson et al. 2008). If this occurs, the macrophages will initially try to digest the implant as a foreign body. If this is unsuccessful, to enhance their effectiveness, the macrophages will fuse to form giant cells. The giant cells will start releasing chemotactic factors to attract fibroblasts from the surrounding tissue (Anderson 2001). The fibroblasts, in turn, will encapsulate the implant with a thin layer of a collagenous matrix to isolate the implant from the surrounding tissue (Ratner 2001). The formation of this collagen matrix layer is often called the "foreign body reaction" (Anderson et al. 2008).

The formation of a collagenous capsule does not necessarily have to be an adverse outcome. Many types of implants can survive and serve their purpose in soft tissue, when encapsulated. Moreover, for nontoxic implants placed inside a bony site, a collagenous capsule may serve as a matrix for protein adhesion and later nucleation of bone regeneration (Ratner 2001). During this event, platelets and macrophages start secreting proteins (for bone regeneration transforming growth factor beta [TGF-β], the bone morphogenetic protein [BMP], and the platelet-derived growth factor [PDGF]) that will adhere to the provisional matrix. The presence of these proteins at the implant surface induces interaction between cells and the protein-loaded implant that is ultimately followed by tissue regeneration at the implant site.

The activation of macrophages is a key parameter in determining whether an osseous implantation will clinically succeed or not. Protein expression profiles by macrophages are highly dependent on environmental factors (Champagne et al. 2002). Bacterial lipopolysaccharide, for example, promotes the production of proinflammatory proteins tumor necrosis factor alpha (TNF-α) and interleukin 1-beta (IL-1β) by macrophages, whereas surface topography induces the expression of the morphogenetic proteins BMP-2 and TGF-β to promote bone formation as well as angiogenesis (Takebe et al. 2003). In addition, the bone specific proteins bone sialoprotein (BSP) and osteopontin (OPN) can be produced by macrophages (McKee and Nanci 1996).

When an implant is integrated directly into the surrounding tissue without the formation of a fibrous capsule, this response is frequently described for bone tissue as osseointegration.

26.2.1 Osseointegration

Osseointegration is the bone-healing process at the endosseous implant surface. This concept of a direct bone-to-metal interface was originally described as a highly differentiated tissue making "a direct structural and functional connection between ordered, living bone and the surface of a load carrying implant" (Branemark 1985). However, the concept was difficult to evaluate since the level at which the body responded to the implant could never be carefully defined. Later on, a new definition was proposed based on clinically assessed criteria. Bone formed at the endosseous implant surface was considered as a positive result, whereas fibrous encapsulation was considered as a negative result (Albrektsson and Sennerby 1990). By now, osseointegration is widely accepted in orthopedic and dental implantology, and is assessed for all implantations. Results from many studies have shown that the rate and quality of bone healing is related to surface properties like surface topography and composition (Le Guehennec et al. 2007).

26.3 Biomaterial Properties

The response of osteoblasts (and virtually all other cell types) to implant biomaterials is affected by several material properties that are usually divided into bulk properties and surface characteristics. Bulk properties, for instance, determine the mechanical strength of implants (Roach et al. 2007), whereas surface properties are important for their interactions with the surrounding biological compounds (i.e., ions, proteins, and eventually cells).

On the basis of their chemical properties, bone implant materials can roughly be divided into metals, ceramics, and polymers. The different material classes provide very different mechanical characteristics (Cooke 2004). The mechanical properties of an implant biomaterial determine stress–strain, elasticity, tension–compression, shear, and fracture resistance (Binyamin et al. 2006). The mechanical characteristics of an implant material must be adjusted to its intended function in order to prevent failure. Materials that are appropriate for a certain application can have detrimental effects when used for other implant applications.

In addition to the bulk properties of implant biomaterials, several surface characteristics are important in determining the quality of osseointegration. Surface characteristics can be divided in to composition, surface free energy (SFE), surface charge, ion release, and topography (Albrektsson et al. 2004, Kasemo and Gold 1999). As mentioned above, SFE, which is reflected by the hydrophilicity of an implant surface, is an important factor for the adhesion of proteins and cells. Surface free energy is especially important during initial conditioning by proteins and during initial cell adhesion. Hydrophilic surfaces stimulate the adhesion of proteins and cells; hydrophobic surfaces diminish protein and cell adhesion. SFE is enhanced by the oxidation of the implant surface (Michiardi et al. 2007). However, not all proteins respond to an increase in SFE. Some proteins like fibronectin are mainly influenced by changes in surface charge (Michiardi et al. 2007). The increase of SFE is not always desirable. For some implant biomaterials, like vascular grafts or artificial heart valves, it is highly important that surfaces are hydrophobic in order to prevent the adhesion of biological material and thereby to function properly.

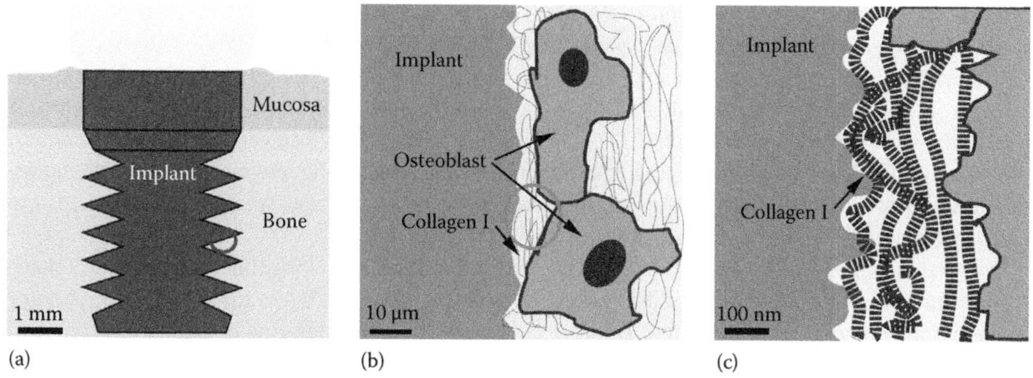

FIGURE 26.3 Overview of the several topographies on a dental implant ranging from (a) macro-scale (i.e. >100 μm), via (b) micro-scale (1–100 μm) down to (c) nano-scale (<100 nm).

Another perhaps even more important factor for the improvement of wound healing and tissue repair after implantation is surface topography. Micrometer and nanometer roughness enhances hydrophilicity compared to smooth surfaces (Rupp et al. 2006), but the importance of surface topography has already been recognized up to many levels for a long time (Albrektsson et al. 1983, Kasemo 1983, Ratner 1983). Whereas surface analysis was only marginally applied up to 25 years ago, it is currently mandatory for research, standardization, and quality control for implant materials. Biological systems recognize and respond to the surface through the surface pattern. Implant surfaces are rarely flat, but have certain topography, which can be either roughness (aspecific topography) or texture (structured topography) (von Recum and van Kooten 1995). Surface roughness is characterized by randomly distributed surface topographies, whereas surface texture is characterized by a regular surface topography with defined dimensions and distribution (von Recum and van Kooten 1995). As will be discussed later, surface topographical characteristics influence cellular responses, such as initial adhesion, orientation, migration, and differentiation (Curtis et al. 1997).

Surface topography of implants can be divided into three levels: macro-, micro-, and nano-sized topologies. Macrosize topography (>100 μm; Figure 26.3a) improves mechanical stability; however, it also introduces an increased risk for peri-implantitis and increased ionic leakage (Le Guehennec et al. 2007). Microscale surface topography (1–100 μm; Figure 26.3b) maximizes the interlocking between the in growth of mineralized bone and the implant surface. An advantage of microsized roughness is the early bone–implant contact that is especially important in patients with insufficient bone quantity or anatomical limitations. Several reports show a highly improved integration of implants with a surface microroughness compared to smooth surfaces (Lincks et al. 1998, Ogawa et al. 2000, Wennerberg and Albrektsson 2009). Nanoscale surface topography (<100 nm; Figure 26.3c) is in the same order of size as the natural extracellular matrix (ECM) and is very important for the adsorption of proteins, the adhesion of osteoblasts, and the increase of the rate of bone healing (Williams 2008). Mimicking this natural ECM on implant biomaterials by the use of nanotechnology might improve bone healing after implantation.

26.4 Importance of Nanotechnology in Implantology

From previous descriptions in this chapter, it has become apparent that the introduction of nanotopography on bone implant biomaterials can become a very powerful tool in improving the rate and quality of implantations. Recently, nanoscale has been defined as being in the order of 100 nm or less (Williams 2008). In this context, it has been decided that a biomaterial can be named a nanobiomaterial if it contains one or more dimensions of 100 nm or less (Williams 2008). Currently, much effort is being placed on applying implant surface topographies on a nanometer scale by mimicking the dimensions of the natural ECM. Collagen is one of the key proteins in all living tissues. Mimicking its nanotopography might promote early tissue integration and reduce the chance of implant failure. Natural collagen fibrils exist in a different array of patterns as described by Weiner et al. Examples are parallel fibers, plywood-like fibers, and radial fibers. Collagen type I fibrils typically have a periodicity of 68 nm and the spaces are approximately 35 nm deep. Additionally, cortical bone has an average roughness of 32 nm, which is also in the nanometer scale (Palin et al. 2005). Several techniques have been applied to create nanotopography on implant materials ranging from techniques like machining and etching to create nanoroughness, to advanced techniques like electron beam lithography and laser interference lithography to create highly defined patterns in the nanoscale.

The effect of topography on cells has already been observed for a very long time by many researchers. With improving technology, the challenge to study cellular responses shifted from microtopography to nanotopography in the last century. In order to get a better understanding of cellular responses to surface topographic characteristics, a short overview is given on the development of these studies next.

26.5 Initial Studies on Implant Surface Microtopography

The effect of an organized surface topography on cellular behavior has already been shown since the beginning of the twentieth century. In 1911, Harrison discovered that embryonic cells migrated along spider web filaments (Harrison 1911). Many years later, Weiss (1945) demonstrated that cells aligned and migrated along 13 μm thick glass fibers, and for this phenomenon introduced the term "contact guidance." Later, Curtis and Varde (1964) showed that contact guidance is a direct response of cells to topographical cues. In the early 1970s, micrometer grooves were introduced as an important surface topography in the context of cell guidance and tested in vitro. Rovensky et al. (1971) described fibroblast alignment to 25 μm deep triangular shaped grooves with a 150 μm pitch. Dunn and Heath (1976) demonstrated that the groove angle was an important factor for cellular extension. In the late 1970s, Ohara and Buck (1979) showed the importance of (adhesion plaques) focal adhesions for recognition of grooves and ridges, and subsequent alignment.

In the 1980s, several groups started to employ microfabrication techniques to obtain large area microtextured substrata with high precision for the study of cell behavior. In 1986, Dunn and Brown studied fibroblast response to micrometer grooved substrates ranging from 1.65 to 8.96 μm with a pitch from 3.0 to 32 μm, and a depth of 0.69 μm (Dunn and Brown 1986). The study demonstrated that fibroblasts align to all tested groove dimensions. From the results, Dunn and Brown concluded that ridge width mainly accounted for cell response. In the same period, Brunette performed similar studies; from the results, Brunette concluded that microfilament (and thus cell) stiffness and the inability to function properly when bent were the key factors for cellular alignment (Brunette 1986). In the late 1980s and in the 1990s, a lot of research was performed on the influence of microtextures on cellular behavior, as reviewed by von Recum and van Kooten (1995), Flemming et al. (1999), Curtis and Wilkinson (1997), and Walboomers and Jansen (2001).

Clinical studies on the effects of implant surface topography have been described since the 1960s; Branemark studied bone formation on machined titanium dental implants (Branemark et al. 1969). Over 20 years later, in 1989, Chehroudi et al. studied epithelial cell behavior on microtextured titanium coated substrates in vivo (Chehroudi et al. 1989). In 1997, Chehroudi et al. studied bone formation around microgrooved implants. The study showed that bone formation was aligned on grooves and bone-like foci formation decreased with increasing groove depth (Chehroudi et al. 1997).

26.6 Fabrication of Surface Nanotopography

During the last years, the aims in topography research seem to have shifted from the production of surfaces with microtopography to nanotopography. The groups of Jansen (for example Loesberg et al. 2007) and Curtis (for example Curtis et al. 2001, Dalby et al. 2002d) were the pioneers in this field and performed in vitro studies on substrates bearing several different nanotopographical (semi-) ordered patterns (pits, pillars, and grooves).

Several techniques have been employed for the production of nanotopographical surfaces. The techniques can be categorized as writing, replication, or self-organization. Writing techniques (mostly lithography) create (nano-) patterns directly into a substrate. These substrates often serve as a template for the replication of patterns onto other materials. Several replication methods are used to transfer nanopatterns like nano-imprinting and nano-contact printing (nCP). Self-organization is a process where semiordered nanopatterns are formed naturally due to a thermodynamic driving force under specific interface and coherence energy constraints.

The different techniques can be either based on the removal of materials (top-down) or on the assembly of nanostructures from atoms or molecules (bottom-up). Typically, writing (lithography) and replication techniques are "top-down," whereas self-organization techniques are "bottom-up" (Whitesides and Grzybowski 2002).

Depending on the manufacturing method, the nanofeatures created can be directional (anisotropic) or unidirectional (isotropic) for cells. In implantology, the applied surface topography is thus far mainly isotropic. Anisotropic implant material surfaces are typically grooved and are important for cell guidance. Anisotropy is particularly interesting for implantology since collagen fibrils in the ECM are usually anisotropic. As mentioned, collagen type-I, the main protein in natural bone ECM, can form parallel arrays of fibrils. The cells then become aligned to these fibrils. As a consequence, cells might recognize the surface as a "natural surface" and this can lead to an improved ECM (re)organization and mineral deposition compared to isotropic surfaces. Thus far, only a limited amount of studies have been performed on cellular response to nanogrooves. Applying an isotropic topography is expected to result in an improvement in the control of more collective cell functions like proliferation, morphology, and differentiation. Isotropic surface topographies can be either textured or rough, and cells grow unidirectionally on the substrates. Since different topographical cues can introduce very different effects on cell behavior, a number of techniques to create nanotopography are described below ranged by the organization of nanotopography that is formed. In addition, the in vitro effects on these nanopatterned substrates will be discussed.

26.6.1 Random Nanoroughness

Nanoroughness can be created by several techniques, like nanoparticle compaction, machining, chemical etching, and anodization. Substrates created with such techniques display a completely at random topography. Most of the described techniques used today have also been used previously for microroughening implant materials. However, due to the application of nanophase instead of microphase materials, the same techniques are currently being explored for the nanoroughening of substrates.

26.6.1.1 Compaction of Nanoparticles

Sintering is a method by which metals or a mixture of metals with polymers/ceramics are compacted by applying a high temperature (600°C–1200°C) (Webster et al. 1999). As an alternative for this method, cold compaction has frequently been used (Pantoya and Grainer 2005). By avoiding the use of a high temperature, problems such as surface chemistry alterations (oxidizing or contaminating atmospheres) are avoided (Webster and Ejiofer 2004). Using this method, metals are compacted using a very high pressure in combination with ambient temperatures. Currently, conventional metals (like TiO_2) are used in implantology (reviewed by Wenneberg and Albrektsson 2009). These materials exhibit a microrough surface but are smooth (i.e., an average roughness (Ra) of less than 20 nm) at the nanoscale (Wennerberg and Albrektsson 2009). In contrast, nanometer phase metals yield surfaces with nanoscale grain boundaries (Webster and Smith 2005). The major advantage is that surface roughness and, consequently, the surface area are increased.

Several studies demonstrated beneficial effects of nanophase materials compared to conventional materials on osteoblast adhesion, proliferation, and differentiation. Webster's group (Price et al. 2003, Liu et al. 2006, Webster and Wagner 1998, Webster et al. 1999a,b, 2000a,b, 2004, 2005) compared the behavior of osteoblasts on conventional versus nanophase materials like Al_2O_3, Ti, Ti_6Al_4V, CoCrMo, and a mixture of Ti with PLGA. The compaction of nanophase particles resulted in a two- to threefold increased surface nanoroughness relative to surfaces created from conventional metals. A comparison of osteoblast adhesion on these materials demonstrated that adhesion was highly increased on nanophase materials (Webster et al.). A comparison of osteoblast proliferation and differentiation on Al_2O_3, Ti, and Ti-PLGA composites demonstrated that both the proliferation and differentiation of the osteoblasts were enhanced on the nanophase materials (Price et al. 2003, Webster et al.). In another study, Liu et al. showed an increased osteoblast function (adhesion, proliferation, and differentiation) with increasing surface roughness (highest Ra studied was 120 nm) (Liu et al. 2006).

Dulgar-Tulloch et al. (2009) studied osteoblast adhesion and proliferation on compacted nanophase Al_2O_3, TiO_2, and hydroxyapatite of different grain sizes. Surface roughnesses of the compacted materials ranged from 24 up to 45 nm. Corroborating the studies of Webster et al., this study demonstrated that osteoblast adhesion and proliferation were mostly enhanced on materials with the highest surface roughness.

26.6.1.2 Machining

Using the machining strategy to create nanotopography, semi-ordered patterns (scratches) are created in substrate surfaces, which mostly have a groove-like pattern. Often, the patterns are acid etched after machining to increase surface roughness and create an additional dimension in the machined substrates (de Oliveira and Nanci 2004).

Park et al. (2001) demonstrated that platelet activation was increased by nanoroughened substrates by machining, which may lead to increased osteoconductivity of implant materials. Several other studies compared the osteoblast response on chemical-etched surfaces with machined surfaces. The main findings of these studies are described in the following section.

26.6.1.3 Chemical Etching

Using chemical etching, a nanoroughness is created by soaking a material in an etchant. Usually, a mixture of acids like HCl, H_2SO_4, or HF, peroxidation or a base-like NaOH is used. During chemical etching, the material is etched away and micro- or nanopits are created on substrates dependent on acidity and time of incubation. At present, patterns can be created with dimensions down to 20–50 nm (Fan et al. 2002). Since chemical etching is a nonspecific surface treatment technique, structures with any geometry or predetermined organization cannot be created. However, despite being nonspecific, a major advantage of this technique over many other techniques is that semi-ordered topographies (pit-like structures) can be applied on three-dimensional substrates.

Anodization is another chemical way of creating micro- or nanoroughness in metals by incubation in strong acids at high current density or potential. This results in thickening the oxide layer on titanium. The formation of pattern dimensions is dependent on acid concentration and composition, current density, electrolyte temperature, and time of incubation.

Several in vitro and in vivo studies have been performed on the osteoblast response to nanotopography created by acid treatment. Some groups compared the patterns created with etching to patterns created with machining (which creates an anisotropic groove-like structure). Takeuchi et al. compared the differentiation potential of osteoblasts on machined (Ra of 49 nm) versus dual acid-etched (dual acid etching, DAE; Ra of 110 nm) substrates and demonstrated that osteoblast differentiation was increased on DAE (Takeuchi et al. 2005). In an in vivo follow-up study using identical substrates, Butz et al. showed that osseointegrated bone to acid-etched substrates had higher strength and stiffness than to machined substrates (Butz et al. 2006). Kubo et al. compared the proliferation and differentiation potential of periosteal cells on machined (Ra 49 nm) and acid-etched (Ra 183 nm) Ti substrates (Kubo et al. 2008). Periosteal cell proliferation was much higher on machined surfaces compared to acid-etched surfaces. The study showed that surface topographies can have profound effects on the differentiation pathways of stem cells. Cells on machined surfaces differentiated to an osteogenic cell lineage, whereas cells on acid-etched surfaces differentiated to a chondrocytic lineage.

De Oliveira et al. created a nanotopography on Ti surfaces by acid etching and demonstrated that osteoblast differentiation was increased relative to untreated machined Ti surfaces (de Oliveira et al. 2007). Vetrone et al. used several dual acid etching reagents to determine surface nanoroughness and demonstrated that acidity variation introduced changes in surface nanoroughness (Vetrone et al. 2009). The study further showed that in particular H_2SO_4/H_2O_2 treated surfaces positively affected osteoblast differentiation, whereas osteoblast proliferation and differentiation were decreased on NH_4OH/H_2O_2 treated surfaces.

26.6.2 Semi-Ordered Nanotopography Created via Self-Assembly

Self-assembly is a process by which (nano)patterns are formed spontaneously by thermodynamic equilibrium conditions through a number of noncovalent interactions (Zhang 2002). The self-organization process is very close to nature, since bone tissue itself, like many other biological structures, is organized in this way. The lowest level of self-organization in this context are collagen fibrils and hydroxyapatite, which are embedded in these fibrils (Traub et al. 1989).

Colloidal lithography and polymer demixing are the techniques used for creating self-assembled, isotropic topography. Using isotropy, it is expected that cells do not align. Instead, a control of collective cell functions would be expected. Although these techniques have not yet been used for implant patterning, many studies have been performed for the assessment of cellular behavior on these nanoscale features as will be described in Section 26.6.2.1.

26.6.2.1 Polymer Demixing

Using polymer demixing, two immiscible polymer blends (e.g., polystyrene [PS] and poly(4-bromostyrene) [PBrS]) or diblock copolymers (Park et al. 1997) (e.g., poly(ethylene glycol) [PEG] and poly(ε-caprolactone) [PCL]) spontaneously undergo phase separation during spin casting onto silicon wafers. By adjusting polymer ratios, semi-ordered topographies can be produced in the form of nanoscale pits, islands, and ribbons. By adjusting polymer concentrations, feature sizes are changed (Norman and Desai 2006). This method allows the production of nanopatterns down to a height of as low as 10 nm. It is a fast and inexpensive method for creating polymeric surfaces with nanometer scale and therefore it is frequently used to test cellular responses to topography; however, in vivo experiments have not yet been performed using this method.

Dalby et al. (2002a–c, 2003, 2004b,c) performed many studies on several dimensions of polystyrene and poly(4-bromostyrene) demixed islands and demonstrated that fibroblast and endothelial cell adhesion and focal adhesion formation were greater on islands around 10–13 nm than on 35 and 95 nm high islands. In addition, studies of osteoblast behavior on polymer demixed islands demonstrated that spreading and differentiation were greater on 10 nm high islands than on 33 and 45 nm high islands (Dalby et al. 2006b). Lim et al. demonstrated that osteoblast proliferation and differentiation were significantly greater on 11 nm high islands compared to 38 and 85 nm high islands (Lim et al. 2005a,b). Using PS and poly-L-lactic acid (PLLA) demixed islands or pits, the same group also found that osteoblasts adhered better to nanoislands than to nanopits. Hansen et al. studied osteoblast stiffness on 11 and 38 nm high islands and demonstrated that stiffness increased by the introduction of nanotopography (Hansen et al. 2007).

26.6.2.2 Colloidal Lithography

An alternative approach to the creation of semi-ordered nanotopography is by colloidal lithography (Figure 26.4). In colloidal lithography, colloidal nanoparticles are deposited on a substrate and create a nanotopographical surface. The nanoparticles are dispersed as a monolayer and are electrostatically self-assembled over the surface. Directed reactive ion beam bombardment or film evaporation is subsequently used to etch the area surrounding the particles as well as the particles itself (Hanarp et al. 2003). The colloidal particles are subsequently removed by a lift-off process. Usually, particles of a standard diameter are used to create a substrate with uniform island dimensions. Using this technique, a controlled environment of nanopits (via film evaporation), pillars (ion beam etching), or hemispherical

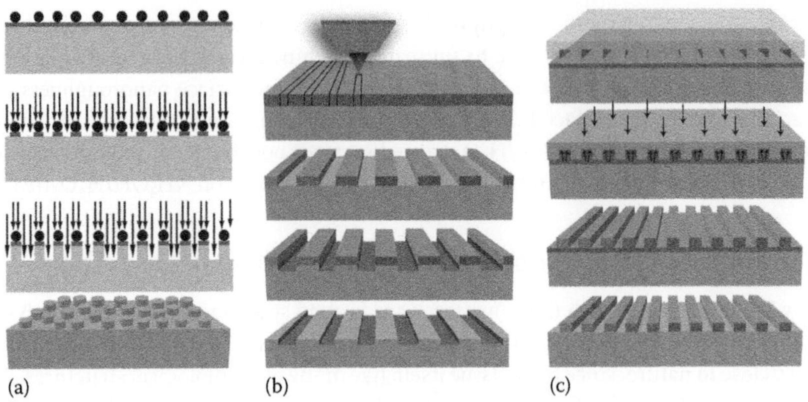

FIGURE 26.4 Schematic overview of several basic lithographic processes. (a) Collodial lithography. Colloidal nanoparticles are dispersed over the surface and the surface is etched away or bombarded with ions to create patterns into the substrate. Subsequently the colloidal particles are removed and a patterned surface is obtained. (b) Electron beam lithography (EBL). Using this technique, a photoresist film (orange) is exposed by electrons. Subsequently the film is developed and the underlying substrate is etched away. As the remaining resist is removed and a patterned surface is obtained. (c) Nano imprint lithography process. A patterned stamp (e.g. created by EBL) is placed on top of a temperature-sensitive resist layer or polymer. Subsequently the layers are heated and the resist conforms to the stamp. During cooling the resist solidifies which results in a negative pattern of the original stamp. A replica of the stamp is subsequently created by the use of etching techniques.

protrusions (evaporation of thick film totally enclosing the particles) can be created (Hanarp et al. 2003). Particle size and density can be controlled to alter the surface texturing, for example, the spacing between particles is controlled by changing the ionic strength of the solution.

Only few in vitro studies have been performed using nanopatterned substrates created with colloidal lithography. However, the studies that have been performed showed the beneficial effects of semi-ordered topographies. Dalby et al. (2004a,d,e) demonstrated that fibroblast adhesion and proliferation were decreased on islands with an approximate height of 160 nm. In the same group, Berry et al. demonstrated a similar response of osteoblasts to semi-ordered topographies (Berry et al. 2007). However, Rice et al. studied the influence of 110 nm high nanoisland density gradients ranging from 43% down to 3% to osteoblast and macrophage adhesion and proliferation and found no influence of densities on the growth of both cell types (Rice et al. 2003).

26.6.3 Ordered Nanotextures

Previously, microtextures have been produced using several techniques like photolithography and microcontact printing (μCP). However, these techniques were not appropriate to create textures with a nanometer resolution. Therefore, dedicated techniques originally used for optics and electronics (Xia et al. 1999, Luttge 2009) were optimized for the nanotexturing of biomaterial surfaces, like electron beam lithography (EBL), focused ion beam lithography (FIBL), nanoimprint lithography (NIL), laser interference lithography (LIL), and nCP (Chen and Pepin 2001). In this section, we will discuss the techniques that have been used for nanotexturing of cell culture substrates. The use of these techniques for surface patterning is still limited, and therefore, only few in vitro studies have been described.

26.6.3.1 Electron Beam Lithography

EBL is a top-down technique that has been widely used to create surface features at the nanoscale for biological experiments. This method utilizes high-energy electrons to expose an electron-sensitive resist layer. EBL is a "direct writing" technique because no physical mask is needed for surface patterning. The beam can be programmed to precisely control its travel route (Norman and Desai 2006). Optimization of the method leads to the possibility of creating surface features down to 5 nm. However, if a large surface area is created, the limit is dropped to 30–40 nm due to electron scatter, which introduces an imprecision below this size (Vieu et al. 2000). EBL suffers from several other disadvantages; the E-beam induces resist swelling at these resolutions. In addition, the production of nanotextures using EBL can be very time-consuming due to the low writing speed. Consequently, the technique can be very costly; however, for cell studies this problem is overcome to some extent by using the substrates as the master for replication onto polymeric materials.

Thus far, several types of nanotextures created with EBL have been used for cell culture assays like pits, pillars, and grooves. First, Curtis et al. (2001) and later Dalby (2006a, 2007, 2008) and Biggs (2007a,b, 2008) used large arrays of pits or pillars into silicon with a diameter of 120 nm or less and a center-to-center spacing of 300 nm or less. Using an optimized method of EBL, $1 \times 1\,cm^2$ areas were created within 1 h. In this method, each pit was formed by a single exposure by the Gaussian shape of the electron beam. These were used as a master for replication into either poly(methyl methacrylate) (PMMA), PCL, or (poly)carbonate (PC). Using the nanopatterned replicas, Dalby et al. (2004c) demonstrated that fibroblasts respond to nanopits with a diameter down to 35 nm. Biggs et al. studied osteoblast response to nanopits (Biggs et al. 2008). Osteoblast adhesion was diminished compared to the smooth substrates when cultured on nanopits. Studies on mesenchymal stem cell differentiation (Dalby et al. 2007). or osteoblast gene expression (Biggs et al. 2007a,b) demonstrated that both gene and protein expressions were induced by the nanopatterns. Teixeira et al. studied the behavior of epithelial cells on EBL-derived silicon substrates bearing submicrometer grooves (Teixeira et al. 2003, 2006). Epithelial cells aligned to grooves down to a 330 nm width and a 70 nm ridge width. However, groove depth and width were always kept at a micrometer scale. Loesberg et al. (and van Delft et al. 2008) studied fibroblast alignment on a PS “biochip” containing 50 different nanotextured patterns in a range of groove dimensions, nanopits, and squares (Loesberg et al. 2007). The study showed that fibroblasts sense patterns with a feature size down to 35 nm. Recently, we studied initial osteoblast behavior on the same substrates and demonstrated that osteoblasts are more sensitive to nanopatterns than fibroblasts are, and sense patterns with a feature size down to 18 nm. The study also demonstrated that groove depth and width are key factors for osteoblast response (Figure 26.5). Yang et al. demonstrated that MG63 osteoblast-like cells respond to grooves with a ridge/groove width of 90 nm and a depth of 300 nm with increased elongation, alignment, and a decreased cell area (Yang et al. 2008).

26.6.3.2 Focused Ion Beam Processing

Focused ion beam (FIB) processing is a “direct writing” technique that allows either the deposition of nanoparticles down to 10 nm directly onto a substrate (bottom-up, FIB-lithography) or the removal of material (top-down, ion beam milling [IBM] reactive ion etching [RIE]). The FIB-lithography technique is analogous to EBL; however, a different type of lens is applied as a consequence of the much heavier ions (Chen and Pepin 2001). A major advantage of FIB technology is the ability to operate an FIB with proper beam size, current, and energy to remove or add a required amount of material with or without chemical reactions, and therefore, three-dimensional nanostructures can be created with high precision (Tseng 2005).

IBM and RIE allow the simultaneous material removal (top-down) of large areas on almost every material (Tseng 2005). Through ion bombardment, substrate particles are selectively removed from the surface. The smallest spot size of the beam is in the order of 5–10 nm and thus features as small as 5 nm can be created. A disadvantage in RIE is that with increasing milling depths, the sidewalls become more tapered and the edges more rounded. The maximum etching depth is approximately 1000 nm as a consequence of material replacement. Usually, RIE is used to transfer EBL or colloidal lithographic created patterns to other materials.

Thus far, only few in vitro studies have been performed on FIB-lithographically created substrates. He et al. studied the osteoblast response to several patterns (rectangular pits) created by IBL with heights of around 120 nm and widths of 10–20 μm (He et al. 2003). The study demonstrated that osteoblast attachment was increased on the substrates compared to the smooth control.

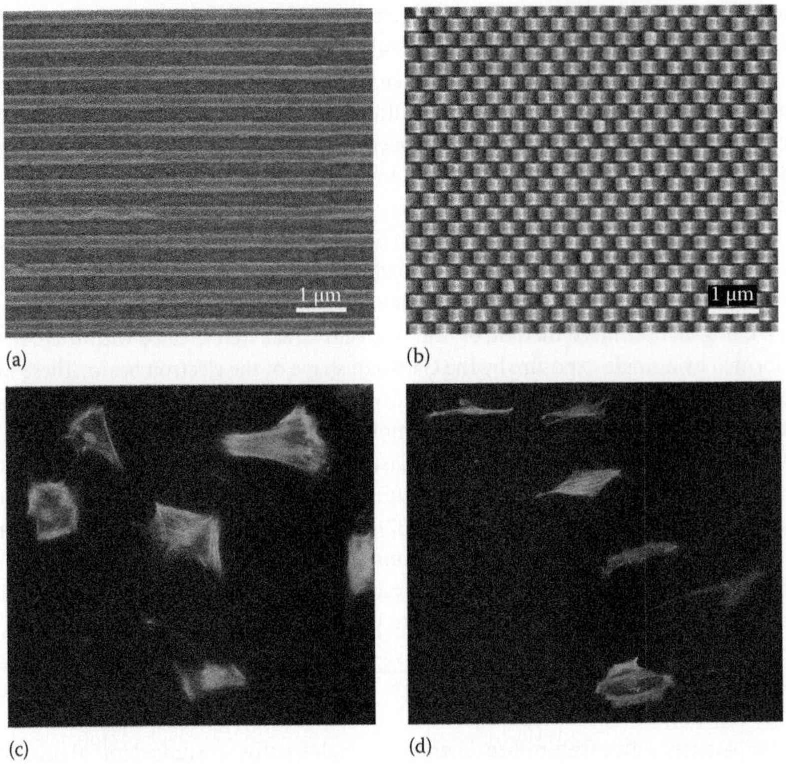

FIGURE 26.5 **(See color insert.)** (a), (b) Nanopatterns with a range of different shapes and dimensions were created with electron beam lithography. Cells were cultured on the substrates and subsequently studied on their morphology (i.e. orientation, area and elongation) by fluorescence microscopy on filamentous actin. (c) Cells aligned to wide grooves (600 nm) and became more elongated, whereas cells spread randomly on small grooves and were more rounded (d).

26.6.3.3 Nanocontact Printing

Microcontact printing (mCP) has been used to pattern surfaces in the micron range and simple patterns in the nano range. This technique uses a mask (created with, for instance, lithography) that is copied into a soft polymer to create a stamp, usually polydimethylsiloxane (PDMS). The stamp is subsequently soaked with the desired biomaterial and then pressed against a surface to deposit the material. The versatility of the method is increased by using different soaking materials to yield multiple distinctive fields. Additionally, the patterned surface can be increased by sequential stamping (Torres et al. 2008). However, the uniformity of the stamped area decreases with an increasing surface area to be patterned. Another major drawback is the soft polymer stamp. During soaking, the stamp can swell, and this results in an increase in pattern size. In addition, due to the elastic nature, the stamp can collapse or buckle, which will also result in the deformation of the patterns. Another disadvantage is the potential for surface contamination (Quist et al. 2005). To overcome these disadvantages, several improvements were made to the technique, like increasing Young's modulus of PDMS (reviewed in Quist et al. (2005)).

CP was originally used to create micropatterns on a large surface area by transferring different biomaterial compounds. In the past 10 years many efforts have been made to decrease the size to the nanoscale. In order to achieve this, stiffer elastomeric stamps and high molecular weight inks were used to limit diffusion (Libioulle et al. 1999, Li et al. 2002, Geissler et al. 2003). In addition, many groups focused on the printing of protein or DNA patterns on flat substrates (Scholl et al. 2000, Csucs et al. 2003, Oliva et al. 2003, Renault et al. 2003, Yu et al. 2005, Shen et al. 2008, Hoover et al. 2009). As a consequence of these optimization steps, patterned substrates with feature dimensions of less than 50 nm have been created (Li et al. 2003).

The mCP technique has already been applied for the micropatterning of curved surfaces. This possibility makes the technique interesting for implant patterning (Jackman et al. 1995, Rogers et al. 1997, Whitesides et al. 2001).

nCP is a simple and inexpensive method to create nanopatterns on substrates for cell studies. Cell biological experiments on both micro- and nanoscale (protein) patterns have been performed to study the behavior of cells (i.e., neuronal, melanoma, and capillary endothelial cells), like motility, adhesion, spreading, and differentiation (Chen et al. 1997, Scholl et al. 2000, Oliva et al. 2003, Lehnert et al. 2004, Hoover et al. 2009). However, thus far, no studies on the osteoblast response to nanopatterns created with nCP have been performed.

26.6.3.4 Nanoimprint Lithography

Nanoimprint lithography (NIL) (or nanoembossing) is a top-down nanopatterning technique that requires a mask. The process consists of two steps: imprinting of a nanopatterned mold in a thin resist layer that is subsequently heated and deformed by the mask, a process that is also called hot embossing. The mask is removed, the resist is cooled below the glass transition point, and as the second step, RIE is used to remove resist residues and print the pattern in the underlying material (Chou et al. 1995). By the use of nanoimprint lithography, patterns as small as 5 nm with aspect ratios down to 20 nm can be created over a large surface area (Austin et al. 2004). NIL can create patterns in materials such as SiO_2/Si or polymers.

NIL offers a number of potential advantages over the before mentioned techniques. First, energetic beams (like electrons or ions) are not used and disadvantages such as (back) scatter and wave diffraction are overcome. Second, NIL can print large areas in nanometer scale at once thereby offering a high throughput. Third, its costs are potentially low because no expensive and time-consuming equipment is used (Chou et al. 1996). The resolution of NIL is not limited by factors directly related to the process; however, it is limited by the mask fabrication process. The masks are usually produced by lithographic techniques, like EBL or FIBL. Unfortunately, problems have to be overcome like alignment, demolding, and fouling of the mask as a consequence of multiple heating and cooling steps (Blattler et al. 2006). In order to overcome these problems, several closely related methods were developed like step and flash imprint lithography (S-FIL), solvent-assisted micromolding (SAMIM), and UV-NIL. These methods are reviewed by Truskett and Watts (2006).

Charest et al. compared osteoblast alignment to nanogrooves created by NIL with chemical microcues created with contact printing (CP) by making an overlay of the chemical cues and the nanopatterns (Charest et al. 2005). The study showed that cells preferred microtopography over nanotopography. Hu et al. studied smooth muscle cell behavior on (sub)microgrooved substrates and demonstrated that cell alignment decreased with increasing groove width and decreasing groove depth, and concluded that pattern height is an important factor that significantly affects cell behavior (Hu et al. 2005). Using similar patterns, Crouch et al. demonstrated that the aspect ratios (the combined effects on pattern width and depth) rather than pattern depth or width separately determine the response on fibroblast behavior (Crouch et al. 2009). Martinez et al. studied osteoblast response to nanopatterned PMMA substrates bearing either pillar-like structures of 30 nm height and 300 nm width or grooves of 200 nm depth, 200 nm width, and a ridge width of 200 nm (Martinez et al. 2007). The results showed that osteoblasts preferred to reside on top of the grooved patterns and on the pillar-like structures that followed the substrate morphology.

26.6.3.5 Laser Interference Lithography

A very promising technique for creating nanotextures is laser interference lithography (LIL). LIL is a maskless lithographic technique using the interference pattern of two obliquely incident beams (van Soest et al. 2005, Luttge 2009) (Figure 26.6). One part of the laser beam reaches a photoresist layer directly, while the second part of the laser reaches the photoresist layer via a mirror and produces a regular interference pattern in the photoresist. A line pattern is created using two lasers, whereas a dot type pattern can be created by using three lasers (two interfering beams) (Lasagni et al. 2007). By the use of

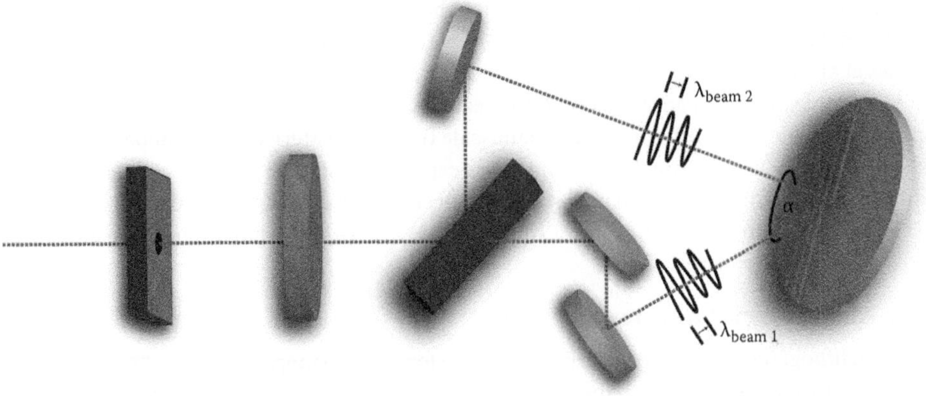

FIGURE 26.6 Overview of the laser interference lithographic process. A laser beam is first directed through a pinhole and subsequently through a lens to focus the beam. Subsequently the beam is split by a beam splitter and 2 beams are directed to the target by the use of mirrors to create nanopatterns.

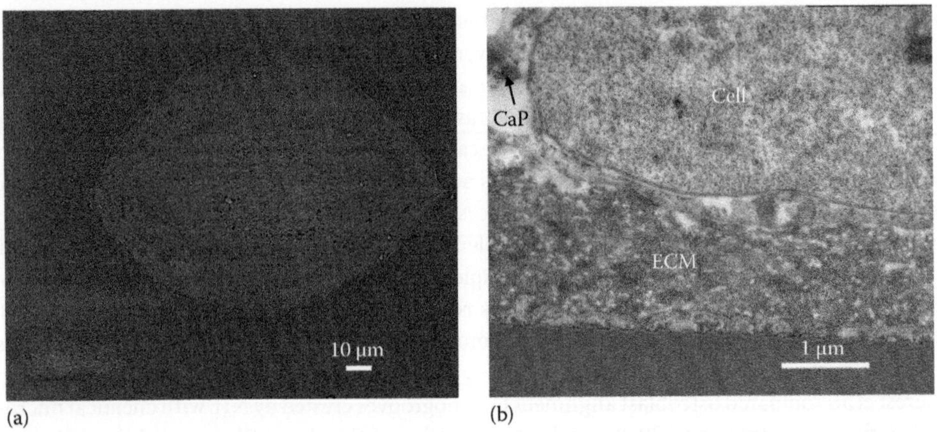

FIGURE 26.7 **(See color insert.)** (a) Osteoblasts align to grooves with a width of 100 nm and a ridge of 100 nm filamentous actin is stained red and the nuclei blue. (b) Bone extracellular matrix (ECM) is deposited inside grooves with a width of 150 nm.

an extra interference beam, pillar-like structures can be created (Tan et al. 2009). LIL has several major advantages over the earlier mentioned techniques. Using LIL, patterns with a pitch size down to 133 nm can be produced over a large surface area. Furthermore, no additional steps (i.e., etching, photoresist development) are required. Consequently, the technique is very cost-effective.

This technique was originally used in electronics (Tselev et al. 1999) and optics (i.e., Wang et al. 2001, Burkhardt and Brunner 2007), and is currently being explored for biological applications (Figure 26.7). Recently, we studied the response of osteoblast-like cells to nanogrooved substrates and demonstrated that cells aligned to nanopatterns down to a groove width of 75 nm and a depth of 35 nm. In addition, the study demonstrated that ECM produced by osteoblasts was sensitive to even smaller nanopatterns, and that specific gene expression was induced by the nanopatterns.

26.6.4 *In Vivo* Application of Implant Nanotopography

Few studies have been performed to assess the biocompatibility of nanotextured biomaterials. Giavaresi et al. studied the effect of polymeric nanopits and nanocylinders on fibrous capsule formation and vascularity by implanting the biomaterials subcutaneously into rats (Giavaresi et al. 2006). The study demonstrated that nanopits increased the development of fibrous capsules, while nanocylinders increased the cellularity of the fibrous capsule and the vascular density.

In another in vivo study, Popat et al. first studied the in vivo biocompatibility of nanotubular substrates with a pore size of approximately 80 nm by a subcutaneous implantation of the biomaterial into rats (Popat et al. 2007). The study demonstrated that TiO_2 nanotubular substrates did not induce fibrous tissue formation around the implant and concluded that the biomaterials were biocompatible. Thus far, only one in vivo study has been performed on the response of nanotextured implant materials in bone. Recently, Bjursten et al. assessed the bone-forming capacity of similar TiO_2 nanotubular biomaterials by implantation into the tibia of rabbits (Bjursten et al. 2009). The study demonstrated that nanotubular implants greatly improved bone formation and strength at the implant interfaces compared to grit-blasted microrough Ti implants.

26.7 Conclusions

Currently, many efforts are being made to generate bone implants with enhanced biomineralization properties for increased effectiveness and survival rate. These "new generation" implants will have to lead to an increased quality of life for affected patients and ultimately a decrease in medical costs. In this chapter, we proposed that this could be achieved by mimicking the natural extracellular matrix, which is highly organized at the nanoscale. Cells recognize nanoscale topography at the implant surface as a "natural" surface, and this might lead to a highly reduced initial inflammatory response to the implant materials by the host system.

Nanotechnology, as originally applied for optics and electronics, can provide the tools needed to create nanotextures on implant biomaterials. Using these techniques, a wide variety of nanotextures can be created on biomaterial surfaces and by the use of a combination of the techniques it can even become possible to pattern irregular implant surfaces like dental implant treads. However, the application of ordered nanotechnology as a tool for implant patterning has been studied for less than one decade, and there is still a lot of uncertainty on the exact effect of nanotopographies on cellular responses, let alone which pattern type(s) will be preferable for an optimal response. It is also possible that a combination of different nanotextures or a combination of nano- with microtextures will induce the optimal control of cellular responses in an endosseous environment. In order to answer these questions, many more investigations, most importantly in vivo studies, will have to be performed. Only in vivo research can answer the question of whether implant biocompatibility can be improved, that is, if nanotexturing of implant surfaces can truly reduce inflammatory responses and improve osseointegration at the bone–implant surface.

References

Albrektsson, T. and Sennerby, L. 1990. Direct bone anchorage of oral implants: Clinical and experimental considerations of the concept of osseointegration. *Int. J. Prosthodont.* 3: 30–41.

Albrektsson, T. and Wennerberg, A. 2004. Oral implant surfaces: Part 1—Review focusing on topographic and chemical properties of different surfaces and in vivo responses to them. *Int. J. Prosthodont.* 17: 536–543.

Albrektsson, T. et al. 1983. The interface zone of inorganic implants in vivo: Titanium implants in bone. *Ann. Biomed. Eng.* 11: 27.

Alsaadi, G., Quirynen, M., Komarek, A., and van Steenberghe, D. 2007. Impact of local and systemic factors on the incidence of oral implant failures, up to abutment connection. *J. Clin. Periodontol.* 34: 610–617.

Anderson, J.M. 2001. Biological responses to materials. *Annu. Rev. Mater. Res.* 31: 81–110.

Anderson, J.M., Ratner, B.D., Hoffman, A.S., Schoen, F.J., and Lemons, J.E. 2004. Inflammation, wound healing, and the foreign-body response. In: *Biomaterial Science: An Introduction to Materials in Medicine*, eds. B.D. Ratner, A.S. Hoffman, F.J. Schoen, and J.E. Lemons, pp. 296–304. San Diego, CA: Elsevier Academic Press.

Anderson, J.M., Rodriguez, A., and Chang, D.T. 2008. Foreign body reaction to biomaterials. *Semin. Immunol.* 20: 86–100.

Austin, M.D. et al. 2004. Fabrication of 5 nm linewidth and 14 nm pitch features by nanoimprint lithography. *Appl. Phys. Lett.* 84: 5299–5301.

Berry, C.C., Curtis, A.S., Oreffo, R.O., Agheli, H., and Sutherland, D.S. 2007. Human fibroblast and human bone marrow cell response to lithographically nanopatterned adhesive domains on protein rejecting substrates. *IEEE Trans. Nanobiosci.* 6: 201–209.

Biggs, M.J., Richards, R.G., Gadegaard, N., Wilkinson, C.D., and Dalby, M.J. 2007a. The effects of nanoscale pits on primary human osteoblast adhesion formation and cellular spreading. *J. Mater. Sci. Mater. Med.* 18: 399–404.

Biggs, M.J., Richards, R.G., Gadegaard, N., Wilkinson, C.D., and Dalby, M.J. 2007b. Regulation of implant surface cell adhesion: Characterization and quantification of S-phase primary osteoblast adhesions on biomimetic nanoscale substrates. *J. Orthop. Res.* 25: 273–282.

Biggs, M.J. et al. 2008. Interactions with nanoscale topography: Adhesion quantification and signal transduction in cells of osteogenic and multipotent lineage. *J. Biomed. Mater. Res. A.* 91: 195–208.

Binyamin, G., Shafi, B.M., and Mery, C.M. 2006. Biomaterials: A primer for surgeons. *Semin. Pediatr. Surg.* 15: 276–283.

Bjursten, L.M. et al. 2009. Titanium dioxide nanotubes enhance bone bonding in vivo. *J. Biomed. Mater. Res. A.* 92: 1218–1224.

Blattler, T. et al. 2006. Nanopatterns with biological functions. *J. Nanosci. Nanotechnol.* 6: 2237–2264.

Branemark, P.I. 1985. Introduction to osseointegration. In: *Tissue-Integrated Prostheses: Osseointegration in Clinical Dentistry*, eds. P.I. Branemark, G.A. Zarb, and T. Albrektsson. Chicago, IL: Quintessence.

Branemark, P.I. et al. 1969. Intra-osseous anchorage of dental prostheses. I. Experimental studies. *Scand. J. Plast. Reconstr. Surg.* 3: 81–100.

Brunette, D.M. 1986. Spreading and orientation of epithelial cells on grooved substrata. *Exp. Cell Res.* 167: 203–217.

Burkhardt, M. and Brunner, R. 2007. Functional integrated optical elements for beam shaping with coherence scrambling property, realized by interference lithography. *Appl. Opt.* 46: 7061–7067.

Butz, F., Aita, H., Wang, C.J., and Ogawa, T. 2006. Harder and stiffer bone osseointegrated to roughened titanium. *J. Dent. Res.* 85: 560–565.

Champagne, C.M., Takebe, J., Offenbacher, S., and Cooper, L.F. 2002. Macrophage cell lines produce osteoinductive signals that include bone morphogenetic protein-2. *Bone* 30: 26–31.

Charest, J.L. et al. 2005. Polymer cell culture substrates with combined nanotopographical patterns and micropatterned chemical domains. *J. Vac. Sci. Technol. B.* 23: 3011–3014.

Chehroudi, B., Gould, T.R., and Brunette, D.M. 1989. Effects of a grooved titanium-coated implant surface on epithelial cell behavior in vitro and in vivo. *J. Biomed. Mater. Res.* 23: 1067–1085.

Chehroudi, B., McDonnell, D., and Brunette, D.M. 1997. The effects of micromachined surfaces on formation of bonelike tissue on subcutaneous implants as assessed by radiography and computer image processing. *J. Biomed. Mater. Res.* 34: 279–290.

Chen, C.S., Mrksich, M., Huang, S., Whitesides, G.M., and Ingber, D.E. 1997. Geometric control of cell life and death. *Science* 276: 1425–1428.

Chen, Y. and Pepin, A. 2001. Nanofabrication: Conventional and nonconventional methods. *Electrophoresis* 22: 187–207.

Chou, S.Y., Krauss, P.R., and Renstrom, P.J. 1995. Imprint of sub-25 Nm vias and trenches in polymers. *Appl. Phys. Lett.* 67: 3114–3116.

Chou, S.Y., Krauss, P.R., and Renstrom, P.J. 1996. Imprint lithography with 25-nanometer resolution. *Science* 272: 85–87.

Cooke, F.W. 2004. Bulk properties of materials. In: *Biomaterial Science: An Introduction to Materials in Medicine*, eds. B.D. Ratner, A.S. Hoffman, F.J. Schoen, and J.E. Lemons, pp. 23–32. San Diego, CA: Elsevier Academic Press.

Crouch, A.S., Miller, D., Luebke, K.J., and Hu, W. 2009. Correlation of anisotropic cell behaviors with topographic aspect ratio. *Biomaterials* 30: 1560–1567.

Csucs, G., Kunzler, T., Feldman, K., Robin, F., and Spencer, N.D. 2003. Microcontact printing of macromolecules with submicrometer resolution by means of polyolefin stamps. *Langmuir* 19: 6104–6109.

Curtis, A.S. and Varde, M. 1964. Control of cell behavior: Topological factors. *J. Natl Cancer Inst.* 33: 15–26.

Curtis, A. and Wilkinson, C. 1997. Topographical control of cells. *Biomaterials* 18: 1573–1583.

Curtis, A.S. et al. 2001. Substratum nanotopography and the adhesion of biological cells. Are symmetry or regularity of nanotopography important? *Biophys. Chem.* 94: 275–283.

Dalby, M.J., Gadegaard, N., and Wilkinson, C.D. 2008. The response of fibroblasts to hexagonal nanotopography fabricated by electron beam lithography. *J. Biomed. Mater. Res. A.* 84: 973–979.

Dalby, M.J., Marshall, G.E., Johnstone, H.J., Affrossman, S., and Riehle, M.O. 2002a. Interactions of human blood and tissue cell types with 95-nm-high nanotopography. *IEEE Trans. Nanobiosci.* 1: 18–23.

Dalby, M.J., McCloy, D., Robertson, M., Wilkinson, C.D., and Oreffo, R.O. 2006a. Osteoprogenitor response to defined topographies with nanoscale depths. *Biomaterials* 27: 1306–1315.

Dalby, M.J., Riehle, M.O., Johnstone, H., Affrossman, S., and Curtis, A.S. 2002b. In vitro reaction of endothelial cells to polymer demixed nanotopography. *Biomaterials* 23: 2945–2954.

Dalby, M.J., Riehle, M.O., Johnstone, H.J., Affrossman, S., and Curtis, A.S. 2002c. Polymer-demixed nanotopography: Control of fibroblast spreading and proliferation. *Tissue Eng.* 8: 1099–1108.

Dalby, M.J., Riehle, M.O., Johnstone, H., Affrossman, S., and Curtis, A.S. 2004a. Investigating the limits of filopodial sensing: A brief report using SEM to image the interaction between 10 nm high nanotopography and fibroblast filopodia. *Cell Biol. Int.* 28: 229–236.

Dalby, M.J., Riehle, M.O., Sutherland, D.S., Agheli, H., and Curtis, A.S. 2004b. Changes in fibroblast morphology in response to nano-columns produced by colloidal lithography. *Biomaterials* 25: 5415–5422.

Dalby, M.J., Riehle, M.O., Sutherland, D.S., Agheli, H., and Curtis, A.S. 2004c. Fibroblast response to a controlled nanoenvironment produced by colloidal lithography. *J. Biomed. Mater. Res. A.* 69: 314–322.

Dalby, M.J. et al. 2002d. Increasing fibroblast response to materials using nanotopography: Morphological and genetic measurements of cell response to 13-nm-high polymer demixed islands. *Exp. Cell Res.* 276: 1–9.

Dalby, M.J. et al. 2003. Fibroblast reaction to island topography: Changes in cytoskeleton and morphology with time. *Biomaterials* 24: 927–935.

Dalby, M.J. et al. 2004d. Attempted endocytosis of nano-environment produced by colloidal lithography by human fibroblasts. *Exp. Cell Res.* 295: 387–394.

Dalby, M.J. et al. 2004e. Rapid fibroblast adhesion to 27 nm high polymer demixed nano-topography. *Biomaterials* 25: 77–83.

Dalby, M.J. et al. 2006b. Osteoprogenitor response to semi-ordered and random nanotopographies. *Biomaterials* 27: 2980–2987.

Dalby, M.J. et al. 2007. The control of human mesenchymal cell differentiation using nanoscale symmetry and disorder. *Nat. Mater.* 6: 997–1003.

de Oliveira, P.T. and Nanci, A. 2004. Nanotexturing of titanium-based surfaces upregulates expression of bone sialoprotein and osteopontin by cultured osteogenic cells. *Biomaterials* 25: 403–413.

de Oliveira, P.T., Zalzal, S.F., Beloti, M.M., Rosa, A.L., and Nanci, A. 2007. Enhancement of in vitro osteogenesis on titanium by chemically produced nanotopography. *J. Biomed. Mater. Res. A.* 80: 554–564.

Dulgar-Tulloch, A.J., Bizios, R., and Siegel, R.W. 2009. Human mesenchymal stem cell adhesion and proliferation in response to ceramic chemistry and nanoscale topography. *J. Biomed. Mater. Res. A.* 90: 586–594.

Dunn, G.A. and Brown, A.F. 1986. Alignment of fibroblasts on grooved surfaces described by a simple geometric transformation. *J. Cell Sci.* 83: 313–340.

Dunn, G.A. and Heath, J.P. 1976. New hypothesis of contact guidance in tissue cells. *Exp. Cell Res.* 101: 1–14.

Fan, Y.W. et al. 2002. Culture of neural cells on silicon wafers with nano-scale surface topograph. *J. Neurosci. Methods* 120: 17–23.

Flemming, R.G., Murphy, C.J., Abrams, G.A., Goodman, S.L., and Nealey, P.F. 1999. Effects of synthetic micro- and nano-structured surfaces on cell behavior. *Biomaterials* 20: 573–588.

Geissler, M. et al. 2003. Fabrication of metal nanowires using microcontact printing. *Langmuir* 19: 6301–6311.
Giavaresi, G. et al. 2006. In vitro and in vivo response to nanotopographically-modified surfaces of poly(3-hydroxybutyrate-co-3-hydroxyvalerate) and polycaprolactone. *J. Biomater. Sci. Polym. Ed.* 17: 1405–1423.
Hanarp, P., Sutherland, D.S., Gold, J., and Kasemo, B. 2003. Control of nanoparticle film structure for colloidal lithography. *Coll. Surf. A. Physicochem. Eng. Asp.* 214: 23–36.
Hansen, J.C. et al. 2007. Effect of surface nanoscale topography on elastic modulus of individual osteoblastic cells as determined by atomic force microscopy. *J. Biomech.* 40: 2865–2871.
Harrison, R.G. 1911. On the stereotropism of embryonic cells. *Science* 34: 279–281.
Harrysson, O.L., Robertsson, O., and Nayfeh, J.F. 2004. Higher cumulative revision rate of knee arthroplasties in younger patients with osteoarthritis. *Clin. Orthop. Relat. Res.* 421: 162–168.
He, W. et al. 2003. Micro/nanomachining of polymer surface for promoting osteoblast cell adhesion. *Biomed. Microdev.* 5: 101–108.
Hoover, D.K., Lee, E.J., and Yousaf, M.N. 2009. Total internal reflection fluorescence microscopy of cell adhesion on patterned self-assembled monolayers on gold. *Langmuir* 25: 2563–2566.
Orthopaedic Market Report. 2009. http://www.researchandmarkets.com/reportinfo.asp?report_id=992715, 2009.
Worldwide Nanotechnology Dental Implant Market Shares, Strategies, and Forecasts, 2009–2015. http://www.researchandmarkets.com/research/acfffa/worldwide_nanotech, 2009.
Hu, W., Yim, E.K., Reano, R.M., Leong, K.W., and Pang, S.W. 2005. Effects of nanoimprinted patterns in tissue-culture polystyrene on cell behavior. *J. Vac. Sci. Technol. A* 23: 2984–2989.
Hulleberg, G., Aamodt, A., Espehaug, B., and Benum, P. 2008. A clinical and radiographic 13-year follow-up study of 138 Charnley hip arthroplasties in patients 50–70 years old: Comparison of university hospital data and registry data. *Acta Orthop.* 79: 609–617.
Hwang, D. and Wang, H.L. 2007. Medical contraindications to implant therapy: Part II: Relative contraindications. *Implant. Dent.* 16: 13–23.
Jackman, R.J., Wilbur, J.L., and Whitesides, G.M. 1995. Fabrication of submicrometer features on curved substrates by microcontact printing. *Science* 269: 664–666.
Johnsen, S.P. et al. 2006. Patient-related predictors of implant failure after primary total hip replacement in the initial, short- and long-terms. A nationwide Danish follow-up study including 36,984 patients. *J. Bone Joint Surg. Br.* 88: 1303–1308.
Kasemo, B. 1983. Biocompatibility of titanium implants: Surface science aspects. *J. Prosthet. Dent.* 49: 832–837.
Kasemo, B. and Gold, J. 1999. Implant surfaces and interface processes. *Adv. Dent. Res.* 13: 8–20.
Kubo, K. et al. 2008. Microtopography of titanium suppresses osteoblastic differentiation but enhances chondroblastic differentiation of rat femoral periosteum-derived cells. *J. Biomed. Mater. Res. A.* 87: 380–391.
Lasagni, A.F., Acevedo, D.F., Barbero, C.A., and Mucklich, F. 2007. One-step production of organized surface architectures on polymeric materials by direct laser interference patterning. *Adv. Eng. Mater.* 9: 99–103.
Le Guehennec, L., Soueidan, A., Layrolle, P., and Amouriq, Y. 2007. Surface treatments of titanium dental implants for rapid osseointegration. *Dent. Mater.* 23: 844–854.
Lehnert, D. et al. 2004. Cell behaviour on micropatterned substrata: Limits of extracellular matrix geometry for spreading and adhesion. *J. Cell Sci.* 117: 41–52.
Li, H.W., Kang, D.J., Blamire, M.G., and Huck, W.T.S. 2002. High-resolution contact printing with dendrimers. *Nano Letters* 2: 347–349.
Li, H.W., Muir, B.V.O., Fichet, G., and Huck, W.T.S. 2003. Nanocontact printing: A route to sub-50-nm-scale chemical and biological patterning. *Langmuir* 19: 1963–1965.
Libioulle, L., Bietsch, A., Schmid, H., Michel, B., and Delamarche, E. 1999. Contact-inking stamps for microcontact printing of alkanethiols on gold. *Langmuir* 15: 300–304.
Lim, J.Y., Hansen, J.C., Siedlecki, C.A., Runt, J., and Donahue, H.J. 2005a. Human foetal osteoblastic cell response to polymer-demixed nanotopographic interfaces. *J. R. Soc. Interface* 2: 97–108.

Lim, J.Y. et al. 2005b. Osteoblast adhesion on poly(L-lactic acid)/polystyrene demixed thin film blends: Effect of nanotopography, surface chemistry, and wettability. *Biomacromolecules* 6: 3319–3327.

Lincks, J. et al. 1998. Response of MG63 osteoblast-like cells to titanium and titanium alloy is dependent on surface roughness and composition. *Biomaterials* 19: 2219–2232.

Liu, H., Slamovich, E.B., and Webster, T.J. 2006. Increased osteoblast functions among nanophase titania/poly(lactide-co-glycolide) composites of the highest nanometer surface roughness. *J. Biomed. Mater. Res. A* 78: 798–807.

Loesberg, W.A. et al. 2007. The threshold at which substrate nanogroove dimensions may influence fibroblast alignment and adhesion. *Biomaterials* 28: 3944–3951.

Luttge, R. 2009. Massively parallel fabrication of repetitive nanostructures: Nanolithography for nanoarrays. *J. Phys. D: Appl. Phys.* 42: 18.

Makela, K.T., Eskelinen, A., Pulkkinen, P., Paavolainen, P., and Remes, V. 2008. Total hip arthroplasty for primary osteoarthritis in patients fifty-five years of age or older. An analysis of the Finnish arthroplasty registry. *J. Bone Joint Surg.* 90: 2160–2170.

Martinez, E. et al. 2007. Focused ion beam/scanning electron microscopy characterization of cell behavior on polymer micro-/nanopatterned substrates: A study of cell-substrate interactions. *Micron* 39: 111–116.

McKee, M.D. and Nanci, A. 1996. Secretion of osteopontin by macrophages and its accumulation at tissue surfaces during wound healing in mineralized tissues: A potential requirement for macrophage adhesion and phagocytosis. *Anat. Rec.* 245: 394–409.

Michiardi, A., Aparicio, C., Ratner, B.D., Planell, J.A., and Gil, J. 2007. The influence of surface energy on competitive protein adsorption on oxidized NiTi surfaces. *Biomaterials* 28: 586–594.

Norman, J.J. and Desai, T.A. 2006. Methods for fabrication of nanoscale topography for tissue engineering scaffolds. *Ann. Biomed. Eng.* 34: 89–101.

Ogawa, T. et al. 2000. Biomechanical evaluation of osseous implants having different surface topographies in rats. *J. Dent. Res.* 79: 1857–1863.

Ohara, P.T. and Buck, R.C. 1979. Contact guidance in vitro. A light, transmission, and scanning electron microscopic study. *Exp. Cell Res.* 121: 235–249.

Oliva, A.A. Jr., James, C.D., Kingman, C.E., Craighead, H.G., and Banker, G.A. 2003. Patterning axonal guidance molecules using a novel strategy for microcontact printing. *Neurochem. Res.* 28: 1639–1648.

Palin, E., Liu, H.N., and Webster, T.J. 2005. Mimicking the nanofeatures of bone increases bone-forming cell adhesion and proliferation. *Nanotechnology* 16: 1828–1835.

Pantoya, M.L. and Granier, J.J. 2005. Combustion behavior of highly energetic thermites: Nano versus micron composites. *Propellants Explosive Pyrotechnics* 30: 53–62.

Park, J.Y., Gemmell, C.H., and Davies, J.E. 2001. Platelet interactions with titanium: Modulation of platelet activity by surface topography. *Biomaterials* 22: 2671–2682.

Park, M., Harrison, C., Chaikin, P.M., Register, R.A., and Adamson, D.H. 1997. Block copolymer lithography: Periodic arrays of ~10 11 holes in 1 square centimeter. *Science* 276: 1401–1404.

Popat, K.C., Leoni, L., Grimes, C.A., and Desai, T.A. 2007. Influence of engineered titania nanotubular surfaces on bone cells. *Biomaterials* 28: 3188–3197.

Price, R.L., Gutwein, L.G., Kaledin, L., Tepper, F., and Webster, T.J. 2003. Osteoblast function on nanophase alumina materials: Influence of chemistry, phase, and topography. *J. Biomed. Mater. Res. A* 67: 1284–1293.

Quist, A.P., Pavlovic, E., and Oscarsson, S. 2005. Recent advances in microcontact printing. *Anal. Bioanal. Chem.* 381: 591–600.

Ratner, B.D. 1983. Surface characterization of biomaterials by electron spectroscopy for chemical analysis. *Ann. Biomed. Eng.* 11: 313–336.

Ratner, B.D. 2001. Replacing and renewing: Synthetic materials, biomimetics, and tissue engineering in implant dentistry. *J. Dent. Educ.* 65: 1340–1347.

Ratner, B.D. 2004. Correlation, surfaces and biomaterials science. In: *Biomaterial Science: An Introduction to Materials in Medicine*, eds. B.D. Ratner, A.S. Hoffman, F.J. Schoen, and J.E. Lemons, pp. 765–771. San Diego, CA: Elsevier Academic Press.

Ratner, B.D. and Bryant, S.J. 2004. Biomaterials: Where we have been and where we are going. *Annu. Rev. Biomed. Eng.* 6: 41–75.

Ratner, B.D., Ratner, B.D., Hoffman, S., Schoen, F.J., and Lemons, J.E. 2004. A history of biomaterials. In: *Biomaterials Science. An Introduction to Materials in Medicine*, pp. 10–11. London, U.K.: Elsevier Inc.

Renault, J.P. et al. 2003. Fabricating arrays of single protein molecules on glass using microcontact printing. *J. Phys. Chem. B* 107: 703–711.

Rice, J.M. et al. 2003. Quantitative assessment of the response of primary derived human osteoblasts and macrophages to a range of nanotopography surfaces in a single culture model in vitro. *Biomaterials* 24: 4799–4818.

Roach, P., Eglin, D., Rohde, K., and Perry, C.C. 2007. Modern biomaterials: A review-bulk properties and implications of surface modifications. *J. Mater. Sci. Mater. Med.* 18: 1263–1277.

Rogers, J.A., Jackman, R.J., and Whitesides, G.M. 1997. Microcontact printing and electroplating on curved substrates: Production of free-standing three-dimensional metallic microstructures. *Adv. Mater.* 9: 475–479.

Rovensky, Y.A., Slavnaja, I.L., and Vasiliev, J.M. 1971. Behaviour of fibroblast-like cells on grooved surfaces. *Exp. Cell Res.* 65: 193–201.

Rupp, F. et al. 2006. Enhancing surface free energy and hydrophilicity through chemical modification of microstructured titanium implant surfaces. *J. Biomed. Mater. Res. A* 76: 323–334.

Scholl, M. et al. 2000. Ordered networks of rat hippocampal neurons attached to silicon oxide surfaces. *J. Neurosci. Methods* 104: 65–75.

Shen, K., Qi, J., and Kam, L.C. 2008. Microcontact printing of proteins for cell biology. *J. Vis. Exp.* 22: 1065.

Strietzel, F.P. et al. 2007. Smoking interferes with the prognosis of dental implant treatment: A systematic review and meta-analysis. *J. Clin. Periodontol.* 34: 523–544.

Takebe, J., Champagne, C.M., Offenbacher, S., Ishibashi, K., and Cooper, L.F. 2003. Titanium surface topography alters cell shape and modulates bone morphogenetic protein 2 expression in the J774A.1 macrophage cell line. *J. Biomed. Mater. Res. A* 64: 207–216.

Takeuchi, K., Saruwatari, L., Nakamura, H.K., Yang, J.M., and Ogawa, T. 2005. Enhanced intrinsic biomechanical properties of osteoblastic mineralized tissue on roughened titanium surface. *J. Biomed. Mater. Res. A* 72: 296–305.

Tan, C. et al. 2009. Ordered nanostructures written directly by laser interference. *Nanotechnology* 20: 125303.

Teixeira, A.I., Abrams, G.A., Bertics, P.J., Murphy, C.J., and Nealey, P.F. 2003. Epithelial contact guidance on well-defined micro- and nanostructured substrates. *J. Cell Sci.* 116: 1881–1892.

Teixeira, A.I. et al. 2006. The effect of environmental factors on the response of human corneal epithelial cells to nanoscale substrate topography. *Biomaterials* 27: 3945–3954.

Torres, A.J., Wu, M., Holowka, D., and Baird, B. 2008. Nanobiotechnology and cell biology: Micro- and nanofabricated surfaces to investigate receptor-mediated signaling. *Annu. Rev. Biophys.* 37: 265–288.

Traub, W., Arad, T., and Weiner, S. 1989. 3-Dimensional ordered distribution of crystals in turkey tendon collagen-fibers. *Proc. Natl Acad. Sci. U S A* 86: 9822–9826.

Truskett, V.N. and Watts, M.P. 2006. Trends in imprint lithography for biological applications. *Trends Biotechnol.* 24: 312–317.

Tselev, A.E. et al. 1999. Fabrication of magnetic nanostructures by direct laser interference lithography on supersaturated metal mixtures. *Appl. Phys. A-Mater. Sci. Proces.* 69: S819–S822.

Tseng, A.A. 2005. Recent developments in nanofabrication using focused ion beams. *Small* 1: 924–939.

Ulrich, S.D. et al. 2008. Total hip arthroplasties: What are the reasons for revision? *Int. Orthop.* 32: 597–604.

van Delft, F.C.J.M. et al. 2008. Manufacturing substrate nano-grooves for studying cell alignment and adhesion. *Microelectron. Eng.* 85: 1362–1366.

van Soest, F.J. et al. 2005. Laser interference lithography with highly accurate interferometric alignment. *Jpn. J. Appl. Phys. Part 1: Regular Papers Brief Commun. Rev. Papers* 44: 6568–6570.

Vetrone, F. et al. 2009. Nanoscale oxidative patterning of metallic surfaces to modulate cell activity and fate. *Nano Letters* 9: 659–665.

Vieu, C. et al. 2000. Electron beam lithography: Resolution limits and applications. *Appl. Surf. Sci.* 164: 111–117.

von Recum, A.F. and van Kooten, T.G. 1995. The influence of micro-topography on cellular response and the implications for silicone implants. *J. Biomater. Sci. Polym. Ed.* 7: 181–198.

Walboomers, X.F. and Jansen, J.A. 2001. Cell and tissue behavior on micro-grooved surfaces. *Odontology* 89: 2–11.

Wang, X. et al. 2001. Fabrication of a submicrometer crystalline structure by thermoplastic holography. *Appl. Opt.* 40: 5588–5591.

Webster, T.J. and Ejiofor, J.U. 2004. Increased osteoblast adhesion on nanophase metals: Ti, Ti_6Al_4V, and CoCrMo. *Biomaterials* 25: 4731–4739.

Webster, T.J., Ergun, C., Doremus, R.H., Siegel, R.W., and Bizios, R. 2000a. Enhanced functions of osteoblasts on nanophase ceramics. *Biomaterials* 21: 1803–1810.

Webster, T.J., Ergun, C., Doremus, R.H., Siegel, R.W., and Bizios, R. 2000b. Specific proteins mediate enhanced osteoblast adhesion on nanophase ceramics. *J. Biomed. Mater. Res.* 51: 475–483.

Webster, T.J., Siegel, R.W., and Bizios, R. 1999a. Design and evaluation of nanophase alumina for orthopaedic/dental applications. *Nanosctruct. Mater.* 12: 983–986.

Webster, T.J., Siegel, R.W., and Bizios, R. 1999b. Osteoblast adhesion on nanophase ceramics. *Biomaterials* 20: 1221–1227.

Webster, T.J. and Smith, T.A. 2005. Increased osteoblast function on PLGA composites containing nanophase titania. *J. Biomed. Mater. Res. A* 74: 677–686.

Weiner, S. and Wagner, H.D. 1998. The material bone: Structure mechanical function relations. *Annu. Rev. Mater. Sci.* 28: 271–298.

Weiss, P. 1945. Experiments on cell and axon orientation in vitro—The role of colloidal exudates in tissue organization. *J. Exp. Zool.* 100: 353–386.

Wennerberg, A. and Albrektsson, T. 2009. Effects of titanium surface topography on bone integration: A systematic review. *Clin. Oral. Impl. Res.* 20 (Suppl 4): 172–184.

Westbroek, P. and Marin, F. 1998. A marriage of bone and nacre. *Nature* 392: 861–862.

Whitesides, G.M. and Grzybowski, B. 2002. Self-assembly at all scales. *Science* 295: 2418–2421.

Whitesides, G.M., Ostuni, E., Takayama, S., Jiang, X., and Ingber, D.E. 2001. Soft lithography in biology and biochemistry. *Annu. Rev. Biomed. Eng.* 3: 335–373.

Williams, D. 2008. The relationship between biomaterials and nanotechnology. *Biomaterials* 29: 1737–1738.

Xia, Y.N., Rogers, J.A., Paul, K.E., and Whitesides, G.M. 1999. Unconventional methods for fabricating and patterning nanostructures. *Chem. Rev.* 99: 1823–1848.

Yang, J.Y. et al. 2008. Quantitative analysis of osteoblast-like cells (MG63) morphology on nanogrooved substrata with various groove and ridge dimensions. *J. Biomed. Mater. Res. A.* 90: 629–640.

Yu, A.A. et al. 2005. Supramolecular nanostamping: Using DNA as movable type. *Nano Letters* 5: 1061–1064.

Zhang, S.G. 2002. Emerging biological materials through molecular self-assembly. *Biotechnol. Adv.* 20: 321–339.

27

Nanoarray Bionanotechnology

Alexandra H. Brozena
University of Maryland

YuHuang Wang
University of Maryland

27.1 Introduction

The ability to search and probe for specific biological targets and interactions, such as an exact DNA sequence or the binding of two proteins, is of vital importance to addressing problems ranging from human genome mapping and molecular biology to drug discovery and point-of-care personal medicine. The development of polymerase chain reaction (PCR) and radiolabel methods, for instance, have marked a technical height for nucleic acid analysis. Yet, the recent rise of nanoarray bionanotechnology has opened exciting opportunities to develop fast, high-throughput, reliable, and relatively inexpensive analytical techniques to affirm or negate the presence of a variety of biomolecules.

Nanoarray bionanotechnology is the use of nanoscale features, spatially patterned at the length scale of biological relevance for detecting biological targets and interactions (Niemeyer and Mirkin 2004). It is a direct evolution of microarray technology in light of the development of nanoscience and nanotechnology. The microarray, sometimes called a biochip, was developed and adopted by biologists as an experimental means for probing the presence of target biomolecules and their binding events (Bier and Kleinjung 2001). It is a chip-sized substrate patterned with microscale features of biomolecules in an ordered array that biologists use for assaying the presence of complementary biomolecules (Bier and Kleinjung 2001; Hessner et al. 2006). Microarrays have significantly improved the speed and sensitivity of biological assays (Craighead 2006). Features can be composed of a wide range of compounds such as DNA (Hegde et al. 2000), RNA (Hessner et al. 2006), proteins (Ekins 1998), and phospholipids (Joubert et al. 2009) that aid in the diagnosis of disease (Hessner et al. 2006), facilitate drug discovery (Debouck and Goodfellow 1999), and provide a platform to better understand the biomolecular interactions (MacBeath and Schreiber 2000).

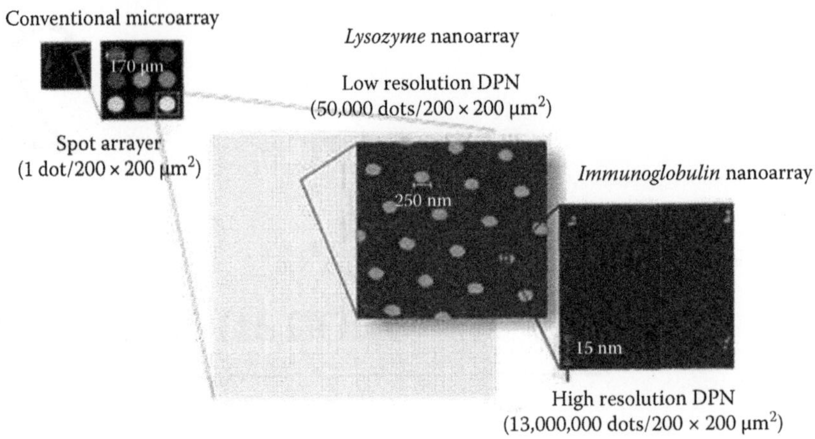

FIGURE 27.1 **(See color insert.)** The relative spot sizes of microarrays and nanoarrays. (From Rosi, N.L. and Mirkin, C.A., *Chem. Rev.*, 105, 1547, 2005. With permission.)

The ability to multiplex, meaning to synthesize the chip with more than one type of feature in order to probe for multiple compounds simultaneously, is a considerable advantage of the microarray (Hessner et al. 2006). However, to detect hundreds of different gene sequences at once, the overall size of the microarray must expand to accommodate the billions of features necessary to probe for so many targets. These size dimensions then require an enormous amount of sample processing time and larger sample volumes, which may itself call for PCR technique, somewhat negating the convenience of the microarray (Tam et al. 2009).

Nanoarrays provide a potential answers to these problems. By using nanometer scale features instead of micron sized, nanoarrays can increase the areal density of features on the chip by approximately 4–6 orders of magnitude (Lynch et al. 2004), as shown in Figure 27.1, while simultaneously decreasing the amount of sample required due to significantly decreased chip sizes. This combined advantage may potentially lower the limits of detection and increase throughput (Lee et al. 2004). Already, proof-of-concept experiments have demonstrated successful detection of various target biomolecules using merely 1/1,000,000th of the sample volume required by microarrays (Hegde et al. 2000; Lynch et al. 2004). The nanometer resolution of nanoarrays lowers the diffusion time of biomolecules, leading to faster and more efficient detection than microarrays (Tinazli et al. 2007). The ability to fabricate probe features in ultrahigh density opens nanoarrays for future multiplexing applications, using a sample size no larger than the volume of a single cell. As the number of probe features can be manufactured in greater number and in higher density, the sensitivity of the nanoarray is expected to exceed that of microarrays due to an increased signal-to-noise ratio in which the increase of signal is directly proportional to the square root of the number of replicate probes (Tam et al. 2009).

This chapter will review the fabrication methods, practical applications, and challenges associated with nanoarrays. We will first compare nanofabrication techniques in the context of nanoarray fabrication. Specifically, this chapter will cover electron-beam lithography (EBL), photolithography, electrohydrodynamic jet (e-jet) printing, microcontact printing, scanning tunneling microscope (STM) lithography, nanografting, and dip-pen nanolithography (DPN), which will be covered in particular detail because of its widespread use in the proof-of-concept experiments that will very likely lead to the incorporation of nanoarrays into biological labs as a powerful discovery platform.

The future of nanoarray design will also be discussed in terms of multiplexing and new feature constructs, including the advantages of DNA self-assembled arrays (Winfree et al. 1998; Lin et al. 2007, 2008) and carbon nanotubes, whose unparalleled size dimensions and superior optical, electrical, and mechanical properties (Saito et al. 1998) make for a promising nanoarray material.

27.2 Nanofabrication Tools for Making Nanoarrays

Nanofabrication is a diverse and rapidly expanding field, which is recently reviewed by Wang et al. (2009). While these techniques are capable of achieving sub-100 nm resolution of nanoarray features, some are better suited for nanoarray applications than others. Table 27.1 summarizes the different attributes of the reviewed methods.

27.2.1 Electron-Beam Lithography

Electron-beam, or e-beam, lithography is one of the oldest and widest available nanofabrication techniques. It uses an electron beam to polymerize or etch a pattern into a surface (Grigorescu and Hagen 2009). This process was initially discovered as a contaminant source in electron microscopy, as the e-beam induced polymerization of trace amounts of siloxane and hydrocarbon gases that exist within the vacuum of the microscope (Hahmann and Fortagne 2009). This artifact of electron microscopy is also capable of producing sub-20 nm features, as a result of the much smaller wavelength of the e-beam in comparison to optical waves, making EBL ideal for nanofabrication in terms of feature resolution (Grigorescu and Hagen 2009).

EBL is often used for creating masks that act like a stencil, through which the desired feature compound can be uniformly patterned. Conversely, EBL is also capable of writing patterns on a polymer surface (Della Giustina et al. 2009; Stowers and Keszler 2009), which can then be used as a chemical template for constructing biological nanoarrays. Using this indirect method, Saaem et al. (2007), for instance, were able to fabricate a protein nanoarray. The authors spin-coated amine-terminated poly(ethylene) glycol (PEG-NH_2) on a glass microscope slide and directed the e-beam to pattern 7,500 nanosized spots in a 100 μm diameter area. The exposed regions of the polymer were cross-linked into an insoluble and stable mass, while the unexposed areas could be easily removed with water, leaving behind only the desired nanoarray with its amine terminal groups capable of binding to various proteins (Saaem et al. 2007). In this indirect patterning approach, proteins are not exposed to the e-beam, therefore maintaining the critical bioactivity essential to successful assays (Turkova 1999; Delehanty and Ligler 2003).

In addition to the fabrication of masks and polymer etching, EBL has also been used to create barriers and platforms in chromium for immunological studies. Mossman et al. (2005) hypothesized that by altering the substrate's geometries, the subsequent deposition of a lipid membrane on this surface would control the patterning and organization of the lipids. This type of spatial control is of

TABLE 27.1 Nanofabrication Methods Summary

Nanofabrication Method	Resolution (nm)	Biocompatibility	Throughput	Multiplexing Capability
Electron beam lithography (Grigorescu and Hagen 2009; Stowers and Keszler 2009)	~10	Indirect	Serial	Poor
Photolithography (Ducker et al. 2007; Reynolds et al. 2009; Wang et al. 2009)	~100	Indirect	Parallel	Poor
E-jet printing (Park et al. 2007, 2008)	~500	Direct	Serial	Poor
Microcontact printing (Biebuyck et al. 1997; Chakra et al. 2008; Gu et al. 2008)	~100	Direct	Parallel	Poor
STM lithography (Abed et al. 2005; Akbulut et al. 2007)	~10	Indirect	Serial	Poor
Nanografting (Zhou et al. 2003; Yu et al. 2006; Saaem et al. 2007)	~10	Indirect or direct	Serial	Poor
Dip-pen nanolithography (Hong et al. 1999a,b; Lee et al. 2003; Salaita et al. 2006; Wang et al. 2008)	15	Direct	Serial or parallel	Good

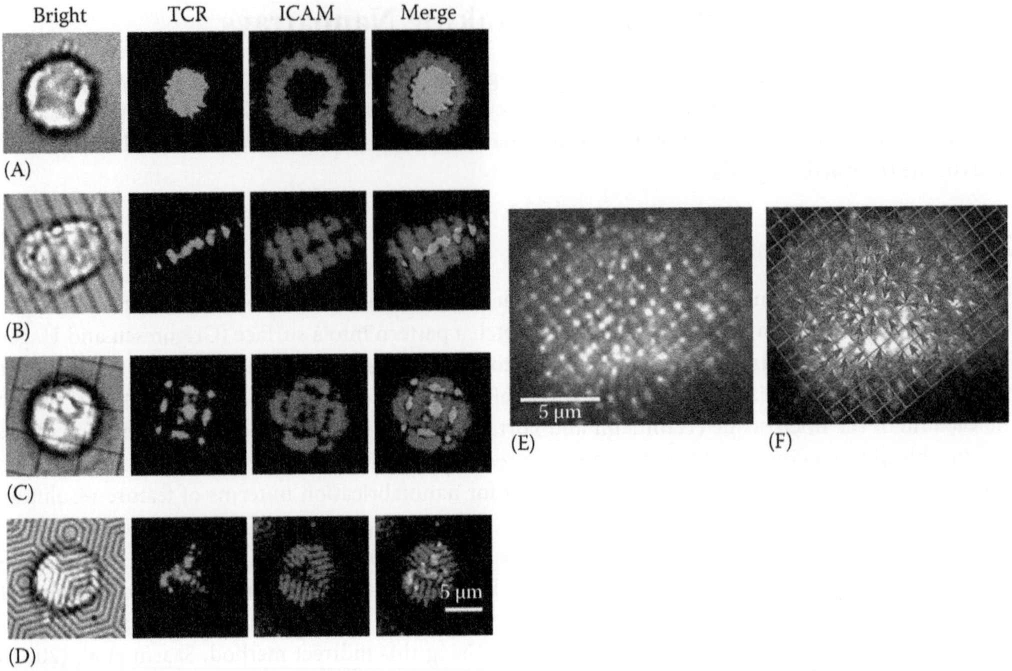

FIGURE 27.2 **(See color insert.)** Lipids membrane deposited on (A) unpatterned chromium, (B) 2 μm lines, (C) 5 μm square grid, (D) hexagonal patterns spaced 1 μm apart. The distribution of TCR in the membrane is shown in (E) and its transport is visualized using red arrows as shown in (F). (From Mossman, K.D. et al., *Science*, 310, 1191, 2005. With permission.)

interest for studying T-cell interactions with cellular surfaces, which feature a fluid-like lipid membrane given to form spontaneous "raft" structures in the medium. Figure 27.2A through D depict the lipid membrane deposited on various surface-chromium geometries composed of 100 nm wide lines, etched using EBL, followed by the fluorescently labeled images of the T-cell receptors (TCR) and intercellular adhesion molecules (ICAM), which orient in a fashion that is clearly dependent on the lithography thus created using the e-beam. The TCR distribution are imaged again in Figure 27.2E and their transport through the lipid membrane is demonstrated in Figure 27.2F using red arrows (Mossman et al. 2005).

Unfortunately, EBL suffers from a series of considerable problems. Its high cost and low speed have prevented its widespread use in research laboratories and commercial applications (Grigorescu and Hagen 2009). The lack of speed is an unfortunate trade-off for high resolution, as the beam size necessary for sub-5 nm features (Stowers and Keszler 2009) must also use lower current for stability. A lower current beam takes longer to write features on the surface than a higher current (Gates et al. 2005). But most problematically, the high energetic nature of electron beams limit EBL to indirect patterning methods, which may ultimately limit its potential for multiplexing, a necessity of almost all biological studies.

27.2.2 Photolithography

While EBL uses electrons to pattern surfaces in the nanometer-sized range, photolithography uses ultraviolet photons for a similar purpose. Photolithography requires two components: the photoresist, a photosensitive material cast on a silicon/quartz wafer, and a photomask to block or allow the UV light to pass through to the photoresist in a desired pattern such as an ordered nanoarray.

For making a nanoarray, the photoresist tends to be an organic monolayer that undergoes a chemical reaction when exposed to a specific wavelength of light. The reaction may convert the exposed photoresist to a compound that is more soluble in base than the unexposed photoresist, enabling it to be dissolved and removed upon immersion in basic solvent (Gates et al. 2005). Conversely, the photoresist upon light exposure may convert into a different functional group, which is more compatible to further chemical binding events while the unexposed portions remain unchanged and inert (Ducker et al. 2007; Reynolds et al. 2009).

DNA microarrays were, for many years, constructed using photolithography to directly pattern oligonucleotide features on the substrate surface. A photoresist material upon exposure to UV light through a mask becomes deprotected, enabling the addition of a mononucleotide to the deprotected feature. The nucleotide can itself be activated using the same mechanism, allowing for the addition of successive monomers, until an oligonucleotide chain is fully synthesized to the desired length and sequence (Pirrung 2002). This method is not currently used to produce DNA nanoarrays as a result of the difficulty in producing nanoscale features using a photomask. However, once this resolution challenge is addressed, the light activated oligonucleotide growth scheme may have great potential in the development of future nanoarrays.

Alternative photoresists are being developed in order to synthesize nanoarrays in as few steps as possible, without resorting to further complications of additional nanofabrication techniques such as nanografting to shape the array. Ducker et al. (2007) reported a one-step photolithographic technique toward producing a protein-binding nanoarray using the photoresist functional group conversion method. By using an oligo(ethylene glycol) (OEG) terminated self-assembled monolayer (SAM) the authors show that upon exposure to 250 nm light, the OEG group was converted to an aldehyde, a particularly versatile functional group for protein binding. Unexposed OEG-terminated SAM showed little to zero nonspecific adsorption when exposed to the protein, streptavidin, and washed with a phosphate-saline buffer. However, the exposed SAM, terminated with an aldehyde, showed considerable protein binding as detected using surface plasmon resonance (SPR). In order to produce nanosized features, it is currently necessary to use a near-field scanning optical microscope (NSOM) (Ducker et al. 2007; Reynolds et al. 2009).

Reynolds et al. (2009) utilized a similar method as Ducker's, but modified the procedure to be compatible on a glass substrate, as oxide surfaces tend to better maintain the activity of target biomolecules. Like OEG-terminated SAMs (Ducker et al. 2007), poly(ethylene oxide) (PEO) derived SAMs convert to terminal aldehydes upon exposure to 244 nm light. The aldehydes can then be further modified to nitrilotriacetic acid groups at points of exposure, a moiety that is capable of binding to histidine tagged proteins (Reynolds et al. 2009). DNA nanospots have also been reported using a similar methodology, but alternatively using a carboxylic acid derived surface after UV exposure of chloromethylphenyl groups terminating siloxane SAMs (Sun et al. 2006).

While photolithography can produce nanoscaled features appropriate for the use of biological nanoarrays, it is typically not an ideal method for making sub-100 nm resolution features as it is constrained by the diffraction limit when using a photomask (Wang et al. 2009) and suffers from hysteresis of the piezo control when using an NSOM to produce nanometer spaced features (Reynolds et al. 2009). The use of NSOM also limits parallelization techniques under current instrument parameters. The ability to create multiplex nanoarrays seems beyond the reach for photolithography, as the use of a photomask to create multiply functionalized surfaces limits the resolution, and the use of an NSOM may be too time consuming and cumbersome to produce a variety of different features, not to mention the strain of developing differentially reacting photoresists.

27.2.3 Inkjet Printing

Inkjet printing is a technique previously used for microarray fabrication. However, recent advances have incorporated it in the field of nanofabrication. Traditional inkjet printing methods use micrometer sized nozzles along with thermal or acoustic forces to direct the outward flow of the patterning ink onto the substrate, the ink being a variety of organic and inorganic molecules. Feature sizes are limited to 10–20 μm by nozzle size, but also by the lack of fine control exhibited by the use of thermal and acoustic drives (Park et al. 2007).

To improve upon this method and guide it to nanometer scale ranges, Park et al. (2007) have devised an e-jet printing system that utilizes a voltage applied between the nozzle and a conductive substrate to guide the ink to the surface and precisely control its patterned deposition. This mechanism comes as a result of the Rayleigh limit of charged ions present within the ink that is ejected to reduce surface charges. Sub-500 nm feature sizes are possible using a nozzle with an internal diameter of 500 nm, indicating that resolution is limited by the initial nozzle fabrication methods (Park et al. 2007). This relationship was studied in more detail and found that the diameter of the jetting ink is proportional to the square root of the nozzle diameter (Choi et al. 2008).

E-jet printing can be used to construct DNA nanoarrays in a manner that retains bioactivity (Park et al. 2008). This achievement is considerable when one takes into account that other similar jet printing techniques show higher potential in denaturing DNA during the array fabrication process (Allain et al. 2004; Barbulovic-Nad et al. 2006). Park et al. (2008) demonstrated that by comparing printed dsDNA, each strand individually labeled with either a fluorophore or a fluorescence quenching molecule, to patterned fluorophore ssDNA, the dsDNA spots showed considerably weaker fluorescence than the ssDNA spots, indicating the successful retention of hybridization as a result of continued bioactivity and subsequent quenching of fluorescence. Control arrays of fluorophore-labeled ssDNA in the presence of noncomplementary strands labeled with quenching molecules continued to fluoresce, suggesting that vicinal quenching pathways are not effective.

27.2.4 Microcontact Printing

Although microcontact printing is traditionally thought of as a microfabrication technique, often used in the production of microarrays, its parallel lithographic nature (Quist et al. 2005) has continued to make it an attractive technique that has motivated researchers to develop the method for nanoscale use (Xia and Whitesides 1998).

Originally developed by George Whitesides's group in the early 1990s (Kumar and Whitesides 1993; Kumar et al. 1994), microcontact printing is a soft lithographic technique involving a master mold that is used to fabricate a polymer stamp, often composed of polydimethylsiloxane (PDMS). The PDMS stamp is wetted in an ink solution that can be made of various SAMs (Quist et al. 2005) for indirect patterning of biopurposed arrays or made of the biomolecule itself for direct patterning (Renault et al. 2003; Lange et al. 2004).

To produce a microcontact printing stamp, it is necessary to first fabricate a master mold using photolithography and a photoresistive material, which is patterned using a photomask. Once the master is complete, an elastomeric compound such as PDMS is poured over the mold, cured, and subsequently peeled off (Perl et al. 2009). The PDMS stamp is immersed in an ink solution and brought into contact with the substrate for a few seconds, allowing the ink to diffuse to the surface, and then released to reveal the patterned surface (Quist et al. 2005) (Figure 27.3). Typically, this ink is composed of a nonpolar alkanethiol SAM, which self-assembles on a gold substrate at points of contact. The overall size of the PDMS stamp can be as large as 100 cm^2 (Love et al. 2005), and additional stamps can be fabricated from the master mold (Yadav et al. 2008), which makes high throughput, parallel fabrication possible (Love et al. 2005). Automated microcontact printing has also been achieved, which particularly opens this technique to cost-effective mass applications (Chakra et al. 2008).

Despite its name, microcontact printing is capable of producing nanoscale features. Biebuyck et al. (1997) demonstrated this capacity using a PDMS stamp with 250 nm relief features that was capable of transferring hexadecanethiol (HDT) SAM to a gold surface in a pattern that after CN^-/O_2 etching

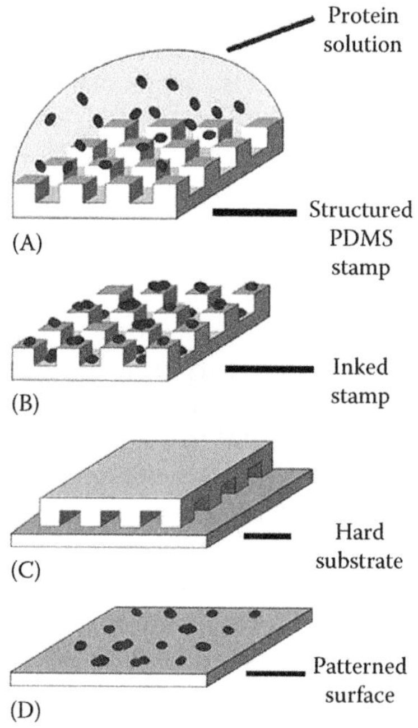

FIGURE 27.3 A schematic of the microcontact printing process. (A) The PDMS stamp is inked with a protein solution; (B) the solution dries, depositing the protein on the stamp; (C) the stamp is contacted on the substrate, allowing the protein to diffuse to the surface; and (D) the stamp is removed, revealing the patterned features. (From Renault, J.P. et al., *J. Phys. Chem. B*, 107, 703, 2003. With permission.)

revealed gold features in the sub-100 range and 100 nm feature resolution. For bio applications, passivating SAMs such as OEG terminated alkanethiols are often used to prevent nonspecific adsorption of biomolecules (Zhou et al. 2004). Gu et al. (2008) were able to achieve nanoscale features using what the authors term "nanocontact printing," by fabricating the stamp using UV nanoimprint lithography. This method was able to produce 30 nm-sized features and a successful protein, biotin-streptavidin, nanoarray as well.

Alternatively, it is possible to produce 100 nm features using microcontact printing by adjusting the inking process. Contact inking is a solvent-free method that allows the stamp to load the ink exclusively on the stamp protrusions, which results in a minimal amount of "bleeding" that would typically broaden the feature sizes (Libioulle et al. 1999).

27.2.5 Scanning Tunneling Microscope Lithography

For ultrahigh resolution imaging, STM has proven capable of differentiating single molecules and atoms. The STM relies on a fine tip made of tungsten (Eigler and Schweizer 1990) or platinum (Abed et al. 2005) that moves over the surface of a substrate with a feedback loop to maintain a constant height or a constant tunneling current. As the surface morphology changes, the tip adjusts in the *Z* direction accordingly—a movement that is converted into an image with appropriate software (Binnig and Rohrer 1982; Zandvllet and van Houselt 2009). It was shown in the iconic work by Eigler and Schweizer (1990) that individual atoms can become attracted to and moved with the STM tip if it is lowered far enough to create Van der Waals and electrostatic forces between the two objects. It was using this method that the authors were able to spell the acronym "IBM" using single xenon atoms on a nickel surface, an achievement of single-atom lithography. The atomic features were spaced 0.2 Å apart, an enviable degree of resolution.

STM can deposit gold nanodots as small as 10 nm in diameter using a platinum tip coated in gold and pulsing the voltage lower than −4 V in order to achieve field-induced deposition on silicon (Abed et al. 2005). Once the gold has been nanopatterned, it is a simple matter of tagging biomolecules with thiols or relying on cysteine amino acids present in proteins to attach to the gold nanostructures, themselves adopting the gold nanoarray order (Akbulut et al. 2007). However, this technique also suffers from the serial nature of the STM that prevents mass-parallelization.

27.2.6 Nanografting

Nanografting was developed by Xu and Liu (1997) who used an AFM tip to selectively remove portions of a SAM from a substrate, while simultaneously replacing the exposed region with a different SAM molecule present in solution (Figure 27.4). This method can produce sub-10 nm dimension features. Typically gold substrate is used, which allows –SH terminated molecules to bind in a packed manner, exposing the other end to the surface. Choice of this terminal group controls what kinds of biomolecules will be capable of binding, for example, $-NH_2$ tends to bind to proteins well, having the ability to form amide bonds, while alkane terminal groups are inert (Saaem et al. 2007).

There are essentially two main types of nanografting. Conventional nanografting involves the removal of regions of an SAM from the surface in the presence of a different kind of SAM molecule in the solution phase, which binds to the exposed sites (Liu et al. 2000). Alternatively, reversal nanografting has come into favor for its higher uniformity of patterning. Reversal nanografting starts with an SAM that binds proteins, while the inserted SAM contains a nonreactive head-group, such as an alkane. This method enables the protein-compatible SAM to maintain its original monolayer order, resulting in greater uniformity of nanoarray features (Tan et al. 2008).

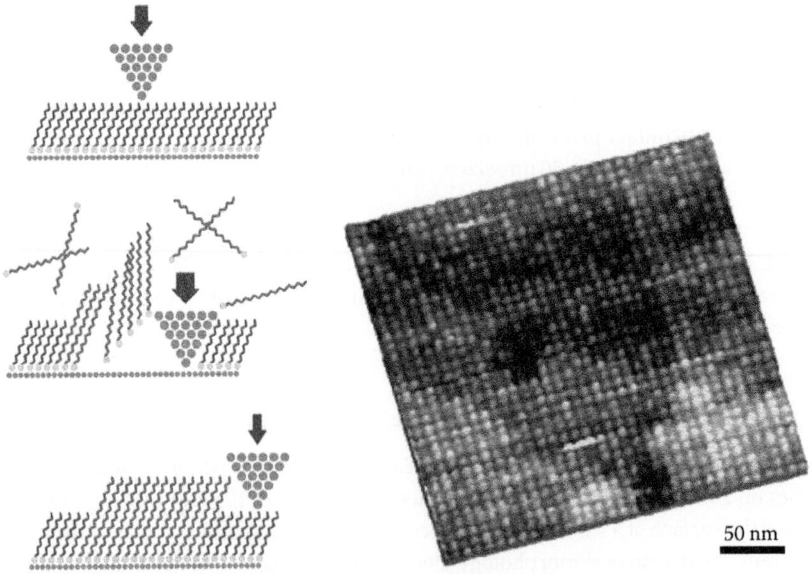

(A) (B)

FIGURE 27.4 **(See color insert.)** (A) In nanografting, the force of the AFM tip removes a portion of the surface SAM in the presence of the alternative solution phase SAM, which uses thiol linkages to bind to the exposed gold. (B) A 33 × 33, biotin nanoarray produced via nanografting. (From Liu, M. et al., *Annu. Rev. Phys. Chem.*, 59, 367, 2008. With permission.)

A related technique, nanoshaving, involves the removal and nonreplacement of an SAM from the substrate. Nanoshaving can be problematic as the removed SAM molecules can sometimes reassemble on the exposed substrate after the tip has left that area. It is necessary to wash the fabricated nanoarray with a solvent in which the SAM molecules are highly soluble in order to properly remove and avoid reversal of the patterning (Liu et al. 2000).

Nanografting can fabricate nanoarrays for nearly any biomolecule, including protein (Bano et al. 2009) and DNA nanoarrays, and produce features as small as 7×12 nm (Zhou et al. 2003). Mirmomtaz et al. (2008) used a typical procedure by beginning with a uniform OEG SAM on a gold substrate. An AFM tip was dipped in ssDNA-SH and scanned across the SAM surface close enough for the DNA to exchange with the SAM. The feature sizes can be controlled by changing the speed of the AFM tip as well as by varying the initial concentration of the DNA-SH solution (Zhou et al. 2003). Using a similar procedure, Liu and Liu (2005) showed that nanografting was capable of producing nanometer-sized features that could detect as few as 50 DNA molecules, a detection limit which would essentially eliminate the need for PCR, if more broadly incorporated.

This technique also enables users to edit nanoarrays in situ. Xu and Liu (1997) demonstrated the ability to modify two features composed of $C_{18}SH$ within a $C_{10}SH$ SAM, originally spaced 20 nm apart by using $C_{10}SH$ solution to replace one of the $C_{18}SH$ features, effectively "erasing" it, then rewriting it 65 nm away by subjecting the nanoarray to $C_{18}SH$ solution once again. This type of lithographic flexibility opens nanografting to create multiplexed nanoarrays by changing the replacement SAM solution (Liu et al. 2000). However, cross-contamination of the features is a major problem that limits the degree of multiplexing that can be achieved (Yu et al. 2006).

While nanografting has significant advantages in the production of biological nanoarrays, it has not yet been developed into a highly parallel method, being constrained by the use of a single AFM cantilever. However, the recent advance of massively parallel DPN (Salaita et al. 2006; Wang et al. 2008) has demonstrated a methodology to improve the throughput and possible multiplexing of tip-array based nanofabrication.

27.2.7 Dip-Pen Nanolithography

DPN is a unique nanofabrication technique and has quickly becoming one of the most popular methods for fabrication of nanoarrays since its invention in Mirkin's lab (Demers et al. 2002; Lee et al. 2003, 2004; Smith et al. 2003; Lynch et al. 2004; Salaita et al. 2007). Its success in this area is in part due to its ability to "direct write" molecules of interest under ambient conditions that are necessary for biomolecules to maintain their activity. DPN is capable of fabricating dot features as small as 15 nm and achieving 5 nm dot-to-dot resolution (Hong et al. 1999a). The high resolution of DPN significantly increases the local concentration of the target molecule (lowering the detection limits) (Lee et al. 2004), and the "direct write" capability circumvents the cross-contamination problem facing most other nanofabrication techniques (Jackson and Groves 2004).

DPN, as its name implies, behaves like the nib of a pen. It can be dipped in an "ink," which, for the purpose of biosensing nanoarrays, is a solution of the probe biomolecule, and then written across a surface to deposit the ink in an ordered fashion (Piner et al. 1999). Remarkably, DPN's ability to uniformly deposit molecules rivals the self-assembled crystalline order achieved using SAMs (Hong et al. 1999b). The surface, usually gold, is kept clear and as smooth as possible. The AFM tip is dipped and soaked in the ink for a period of time and then moved across the surface, like a dot-matrix printer, causing the ink to deposit in uniform spots or lines.

The feature dimensions are controlled by several variables including contact force, contact time, humidity (Lee et al. 2006), and temperature of the fabrication chamber (Jiang and Stupp 2005). Temperature and humidity control the size of the meniscus on the end of the AFM tip. The meniscus

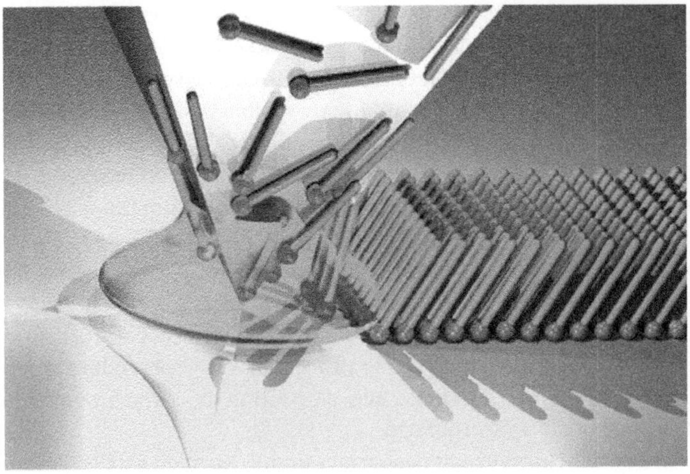

FIGURE 27.5 The AFM cantilever tip deposits the "ink" through a liquid meniscus mechanism. (From Salaita, K. et al., *Nat. Nanotechnol.*, 2, 145, 2007. With permission.)

allows DNA, peptides, and even larger proteins to diffuse from the tip to the surface by capillary action (Salaita et al. 2009) (Figure 27.5). It is this "wet" mechanism that allows biomolecules to successfully retain their activity after deposition (Lee et al. 2003, 2004). Unless the AFM tip is previously coated with a hydrophilic material, DNA deposition, for example, becomes erratic and difficult to reproducibly control, further indicating the importance of the meniscus control (Piner and Mirkin 1997; Piner et al. 1999). Local humidity may need to be as high as 80% for uniform fabrication of biomolecular nanoarrays (Lee et al. 2006).

In terms of contact time, the longer the AFM tip is in contact with the substrate, the larger the nanospot will be. Contact force is generally set very weak, as low as 0.01 nN, in order to avoid gouging the substrate surface, which would lead to nonuniform deposition of the probe biomolecules (Lee et al. 2006). The low contact force also minimizes tip wearing, a problem pertinent to other force-based scanning probe lithography, such as nanografting. Ultimately, the feature size is mainly limited by the AFM tip size. The smaller the radius of curvature, the smaller the meniscus depositing the biomolecules can be. Theoretically, the meniscus can approach 2 nm when using atomically sharp AFM tips (Jang et al. 2004). Carbon nanotubes have recently been attached to the ends of AFM cantilever tips in an effort to achieve this extreme scale (Wong et al., 2006). Sub-20 nm dot features can be more routinely patterned with conventional AFM tips (Hong et al. 1999a).

Like nanografting and other scanning probe lithography, DPN has the advantage of characterizing and detecting biomolecule interactions on the nanoarray in situ, by using the height profiling capability of the AFM that was used to fabricate the array. Binding events may be detected by scanning the AFM tip across the surface where a change in height of the nanofeatures typically indicates successful intermolecular interactions (Lee et al. 2003). This in situ capability eliminates the need for fluorescent tags, which can potentially denature the biomolecule (Lee et al. 2006). Additionally, the common use of gold substrates makes fluorescent tags problematic, as the gold surface can quench fluorescence even upon successful binding (Demers et al. 2002). In this event, it would be necessary to use a different substrate, such as silicon oxide (SiO_2). However a disadvantage of AFM detection in this method's tendency to deform biomolecules slightly, causing their height to be measured as lower than it actually is. This phenomenon must be taken into account during analysis (Wadu-Mesthrige et al. 2001; Lee et al. 2006).

Lee et al. (2003) offers an exemplary study of protein nanoarray fabrication using DPN, demonstrating the technique's ability to transfer proteins onto a substrate in a uniform fashion, capable of detecting antibody/antigen interactions. First, the cantilever tip is coated with gold by thermal evaporation and then dipped in an ethanol/thiotic acid solution. The thiotic acid binds to the gold, creating a hydrophilic surface. When this tip is dipped in the protein buffer solution, the proteins adsorb to the surface because of its compatible and hydrophilic nature. The cantilever is spotted across a gold substrate, allowing the proteins to move through the meniscus to the surface and deposit in a nanosized feature. The nanoarray is then passivated with a compound such as 11-mercapto-undecylpenta(ethyleneglycol) disulfide (PEG), which binds to the nonpatterned regions of the nanoarray to help prevent nonspecific binding interactions upon exposure to the protein antibody. Antigenic binding is tested by accessing changes in the height profile of the nanoarray via AFM. More details of DPN fabricated nanoarrays will be discussed in the nanoarray applications in Section 27.3, as a result of the large number of studies that have been performed using this technique for the following reasons:

Like the previous nanofabrication techniques, single-tip DPN is limited in terms of throughput and multiplexing capabilities. However, one of the most recent advances in DPN has been the 2D array, capable of producing nanoarrays, using as many as 55,000 AFM cantilever tips in 1 cm^2 (Salaita et al. 2006). An elegant variation of this technique by Huo et al. (2008) uses cantilever-free polymer pens that have significantly lowered the fabrication cost and combining the advantages of DPN and microcontact printing. The recent discussion of multiplexed inking challenges by Wang et al.(2008) has further enabled massively parallel multiplexed patterning of nanoarrays, which will be discussed in more detail in Section 27.4.2.

27.3 Applications of Nanoarrays

The uses for nanoarrays are diverse, thus far encompassing DNA analysis (Demers et al. 2002; Kang et al. 2007), viral detection (Smith et al. 2003; Lee et al. 2004), protein binding studies (Rakickas et al. 2008), peptide synthesis and cell-surface interactions (Zhou et al. 2008), and enzymatic assays (Dietrich et al. 2004). The application of biological nanoarrays is virtually unlimited when one considers the vast number of complementary biomolecules found in organisms.

27.3.1 Genomic Nanoarrays

Differential expression of the genome in different types of cells is of central importance to biology. For example, although they both contain copies of the genome in its entirety, a brain cell expresses its DNA differently from a lung cell. In order to profile which genes are being specifically expressed in these different cells, it is necessary to extract RNA, the transcriptional copy of the genome that is the precursor to protein fabrication—the material that ultimately differentiates cell types, and reverse transcribe the RNA back into DNA, producing a synthesized version of the genes that some of which are specific to that cell's functions (Fan et al. 2004). Better understanding of genomic expression can also aid in the diagnosis of cancer (Golub et al. 1999) and diseases (Fan et al. 2006) as well as help study the effects of drugs on live tissues. This can enable early identification of cytotoxic effects as indicated by unusual genetic expression detected using nanoarrays as opposed to finding this information much later in the clinical phase of drug development (Parman 2004). Therefore, the ability to probe for the presence of certain DNA sequences is essential for better understanding the function of genes in organisms as well as the effect of foreign materials on cellular processes and DNA expression, a task for which nanoarrays are well suited.

Demers et al. (2002) demonstrated the feasibility of depositing oligonucleotides nanoarrays composed of two different sequences using DPN on a gold substrate or a silicon oxide wafer functionalized with 3′-mercaptopropyltrimethoxysilane (MPTMS), which is capable of attaching to 5′-terminal acrylamide modified oligonucleotides. When the oligonucleotide's features were exposed to both complementary and noncomplementary DNA sequences labeled with fluorophore tags, fluorescence occurred upon hybridization with the complementary strands alone.

DNA nanostamping is another method that has been successfully used for fabrication of DNA nanoarrays. Nanostamping is a kind of soft lithography that takes advantage of DNA hybridization to fabricate complementary DNA nanoarrays from a master sequence stamp. First, the master stamp is built using a block copolymer technique (Spatz et al. 2000) that incorporates gold nanoparticles in an organic shell that are spin-coated on a gold or glass surface. The organic shell is removed using oxygen plasma etching, leaving gold nanoparticles evenly distributed across the surface in sub-100 nm resolution. Single-stranded DNA is attached to the gold nanoparticles using thiol modifications and the nonpatterned regions are passivated using 6-mercapto-1-hexanol. The patterned substrate is immersed in a solution of complementary ssDNA, also modified with a thiol group. The strands hybridize and, upon contact with a second gold substrate, the newly hybridized ssDNA sequences form bonds with the sandwiching surface. The strands are heated to melt the dsDNA into single strands and separated by removal of the secondary substrate, which subsequently contains a replicate copy of the complementary sequences of the master DNA nanoarray (Akbulut et al. 2007). This technique promises high-density features and highly parallel fabrication that could make single sequence DNA nanoarrays commercially feasible if a sufficient number of copies can be fabricated with consistent fidelity.

Beyond its use as a probe in nanoarrays, DNA has also been conceived as a scaffold structure on which to attach other biomolecules, such as proteins, to form nanometer spaced features—an ideal construct for greatly increasing feature density. DNA is capable of forming a nanoarray substrate due to its ability to self-assemble into a 2D lattice as a result of carefully planned Watson–Crick base pairing. It is possible to control the overall dimension of the chip, from micrometers to millimeters (Lin et al. 2007), in addition to controlling the resolution between the probe anchor points on the DNA self-assembled array (Winfree et al. 1998). Multiplexing (Lin et al. 2007) is a further benefit that will be discussed later in the chapter.

The DNA platform can be readily adapted to form protein nanoarrays for high resolution spacing, and can even be used for viral detection simply by inserting the nucleic acid sequence of viral DNA on the self-assembling subunits (Lin et al. 2008).

27.3.2 Viral Nanoarrays

Paramount in disease diagnosis is the ability to quickly detect the presence of a virus. Early and widely available detection of the flu, for instance, could help prevent pandemics by quickly informing the public of their health status and instituting precautional procedures to prevent further spread of the contagion. The human immunodeficiency virus (HIV) is also important target, as early detection of HIV can help limit its spread. Currently, the enzyme-linked immunosorbent assay (ELISA) and Western blot verification methods are not ideal and typically cannot detect the virus within the first month of infection, creating a time-lag before a person may be capable of taking a meaningful HIV test (Gurtler 1996). PCR can be used to amplify viral DNA detection and ELISA measures the proteins and antibodies associated with the virus, but both of these methods do not allow for the quantification of the viral count in the host, which can be essential for proper treatment and patient monitoring (Huff et al. 2004). This problem may be solved by HIV-detecting nanoarrays, which feature detection limits 1000 times lower than ELISA. This nanoarray approach to HIV detection has already begun in early clinical trials (Lee et al. 2004).

Lee et al. (2004) fabricated an HIV-detection nanoarray using DPN, choosing this method for its ability to pattern biomolecules without causing a loss of bioactivity. The authors report that their array is capable of achieving sensitivity found in the PCR DNA amplification method used in clinical trials for HIV detection. Lee et al. (2004) used as AFM tip to dot 60 nm sized features composed of 16-mercaptohexadecanoic acid (MHA) across gold. These features are capable of binding to HIV antibodies at a pH of 7.4, which is within the range that avoids biomolecule denaturation, while simultaneously deprotonating the MHA for appropriate binding. The nanoarray is then submersed in a solution of HIV antigens, which bind to the immobilized antibodies. To increase the signal-to-noise ratio, the height profile of the

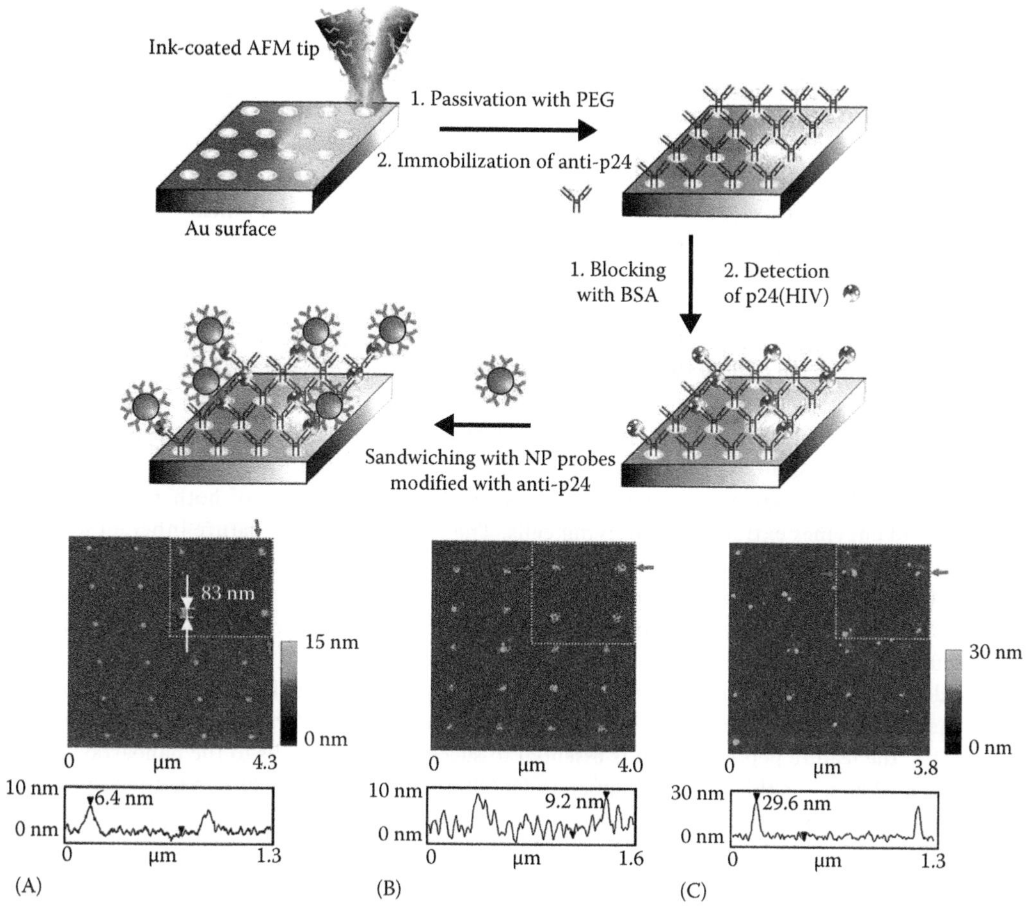

FIGURE 27.6 **(See color insert.)** Schematic of the construction of an HIV-detection nanoarray using DPN. (A) An AFM height profile of the HIV-antibody array, showing an average feature height of 6.4 nm. (B) Upon exposure to HIV, the feature heights rise to 9.2 nm. (C) To amplify the signal, gold nanoparticles functionalized with HIV antibodies were exposed to the nanoarray and bound to the immobilized viruses, increasing the feature height to 29.6 nm. (From Lee, K.B. et al., *Nano Lett.*, 4, 1869, 2004. With permission.)

bound virus was amplified using gold nanoparticles functionalized with HIV antibodies (Figure 27.6). This increased the height by 20.3 ± 1.9 nm.

A very similar procedure was used by Smith et al. (2003) who also used MHA features, approximately 150 nm in diameter, to detect the cowpea mosaic virus. Their efforts showed that after exposing the nanoarray to the virus, the feature height increased by 20 ± 3.5 nm, which is approximately the size of a single virus. Conceivably, the nanoarray could replace other commercial methods of detection, especially as massively parallel multiplexed DPN is developed (Salaita et al. 2006; Wang et al. 2008).

27.3.3 Protein Nanoarrays

As there is no PCR analogy to protein amplification, there is great need for nanoarray ultralow detection limit protein assays and binding studies. Protein nanoarrays encompass a variety of different molecules, including features composed of smaller peptide chains (Zhou et al. 2008), antibodies (Lee et al. 2003; Lynch et al. 2004; Salaita et al. 2006), membrane proteins such as integrins (Lee et al. 2006), and enzymes (Dietrich et al. 2004).

Retaining bioactivity of the probe molecules after deposition on the substrate is a challenge facing nanofabrication methods. This capability is typically demonstrated through the qualitative study of antibody and antigen interactions on nanoarrays. Lynch et al. (2004) among others (Piner et al. 1999; Lee et al. 2003) have shown DPN to be particularly capable in this respect. The authors sputter coated gold on a silicon wafer and then exposed the surface to a proprietary alkanethiol reagent to form an SAM that specifically binds to biomolecules modified with 1,3-phenyldibronoic acid (PBDA). Goat IgG antibodies were be functionalized with PBDA using an amine-PBDA reagent, which allowed them to be successfully patterned via DPN. Nonpatterned regions were passivated using the bovine IgG antibody, which should not bind to the goat antigen. Immersing the nanoarray in a solution of goat IgG antigen results in a change in feature height of approximately 6 nm, indicative of successful binding between the antibody and antigen. As a control, nonspecific binding of the antigen to the bovine antibody did not significantly occur. If bioactivity had not been retained, no change in feature height should have been detected (Lynch et al. 2004).

Nanoarrays as a replacement for the 96 well-plate enzymatic assay is of considerable interest for increased throughput and sensitivity. Dietrich et al. (2004) found that when compared to traditional methods, nanoarrays are capable of assaying for the presence of both reagent-quality enzymes and enzymes extracted from living cells. The controlled, static nature inherent to nanoarrays can be advantageous (though also disadvantageous as discussed later) in the studies of some biological systems, such as cellular membranes, which are difficult to study in vitro due to their dynamic structure. Nanoarrays can provide a more stable way of studying these systems (Rakickas et al. 2008).

Peptide nanoarrays can be used to study cellular adhesion, specifically to help biologists better understand the highly specific interactions of proteins on the cell surfaces (Graeter et al. 2007). For such nanoarrays, the feature peptide length is essential to successfully bind cells to the nanoarray surface (Veiseh et al. 2007). For this reason, Zhou et al. (2008) formulated a procedure that allowed them to control peptide lengths grown in situ on a peptide nanoarray fabricated using DPN on silicon oxide. The authors used polyamidoamine to provide an anchor for the growing peptides on the surface of the silicon oxide, then used a solution of tryptophan-N-carboxyanhydride (Trp-NCA) that successively binds to previously bound Trp-NCA groups to grow a tryptophan peptide whose length can be controlled via the concentration and exposure time of the Trp-NCA solution. In this manner, the authors were able to produce 240–290 nm wide lines of these peptides with a height that approached 50 nm after approximately 8 h of exposure to the Trp-NCA. Dot arrays could also be produced.

Many of these previously discussed applications of nanoarrays involve the direct patterning of the biomolecule on the surface of gold, relying on thiol modifications to provide the linkage (Salaita et al. 2007), or indirect methods by first patterning a compatible functional group on the substrate, which allows the biomolecules to selectively bind and become patterned themselves (Wadu-Mesthrige et al. 2001). It is also possible to produce uniform nanoarrays composed of gold nanoparticles on a silicon wafer as an indirect method of patterning proteins, for instance, by using silicon (111) as a natural template. The crystalline surface of silicon (111) features regular steps and troughs that are separated by approximately 400 nm and 0.314 nm high. When gold is evaporated on the surface, it forms a regular array of nanoparticles, to create a patterned surface without traditional nanofabrication techniques on which biomolecules can directly bind (Hibino and Watanabe 2005).

However, as discussed previously, there is a significant problem of denaturation using such a direct linkage route as a result of interference with the biomolecular tertiary structure (Lovrinovic and Niemeyer 2005). This has led Ramanujan et al. (2007) to conceive of a vesicle transport system to transfer proteins to the gold nanodot array without loss of bioactivity. Lipid vesicles are compatible with many biomolecules (Janshoff and Steinem 2006) and are capable of binding to gold nanoparticles if the vesicle is synthesized with –SH terminal lipids (Ramanujan et al. 2007). This system must be kept immersed in aqueous solution to ensure that the vesicles do not collapse (Janshoff and Steinem 2006), but this is not problematic as their binding can still be accessed via AFM under these conditions (Ramanujan et al. 2007).

27.3.4 Carbon Nanotube Biosensing Arrays

CNTs exhibit many properties that make them attractive array building materials, being electrically and thermally conductive as well as mechanically resilient and optically active (Saito et al. 1998). Research exploiting these qualities for biosensing applications has begun and is quickly moving in the direction of constructing functional nanoarrays from CNTs.

Monitoring changes in electrical conductance may be a particularly attractive method for CNT biosensing. Much effort has been made in developing CNT field-effect transistors (FET) in order to incorporate the electrical properties of CNTs in traditional electronic applications (Kanungo et al. 2009). CNT-FETs can be fabricated in a variety of methods that are briefly reviewed in Koh et al. (2008).

Immobilization of target molecules on the surface of the CNT, by either covalent (Park et al. 2006) or noncovalent addition, can decrease the conductance observed in an intact CNT-FET. It is this last method that Besteman et al. (2003) chose to use. By using a linking molecule with a pyrene moiety to attach to the CNT surface by pi-stacking interactions, and an amine group extended on the other end of the linker, the authors were able to fabricate a CNT-FET detector that decreased in conductance after the target enzyme, glucose oxidase (GOx), formed an amide bond with the linker amine group. In addition to serving as an enzyme biosensor, the addition of glucose (the substrate of GOx) could also be detected using the same system.

In terms of more parallel applications of this technique, CNTs may be added to a pre-patterned surface in a controlled manner, such as a silicon oxide substrate that has been previously decorated with electrode points, requiring only the CNT to contact across the active points. Alternatively, CNTs can be deposited selectively on electrodes patterned with SAMs using carbodiimide coupling. Currently, this method is limited to microarray resolution as a result of utilizing photolithography photomasks for the initial patterning of the SAM (Fabre et al. 2008). However, it may be possible to use a similar carbodiimide coupling method that uses nanografting selectively attach multiwall carbon nanotubes (MWNTs) on a nanoscale resolution.

Besteman et al. (2003) used a semiconducting CNT for their FET biosensing experiment, but it has been observed that FETs constructed with metallic CNTs demonstrate very different electrical conductance properties. This creates a problem in wide-scale application as it currently is not possible to exclusively synthesize semiconducting CNTs; rather the result is always a mixture of the two electronic types (Hersam 2008). Until a scalable and efficient method is developed to separate semiconducting and metallic CNTs, the applications of CNT-FET biosensors will be limited. However, this problem may soon be solved given the rapid progress in separation (Arnold et al. 2006; Tu et al. 2009) and synthesis (Wang et al. 2005; Yao et al. 2009) of specific types of SWNTs.

MWNTs have also been used in more traditional microarray formats, in a procedure that seems readily adaptable to nanoscale resolution. Using chemical vapor deposition (CVD) and nickel catalyst particles, MWNTs can be grown in a vertical orientation. The MWNT forest is then covered in SiO_2, and polished in a manner that leaves just the tops of the MWNT projecting above the silicon. This process leaves carboxylic acid groups dangling off the ends of the MWNTs from which amine-terminated DNA sequences can be attached. The probe DNA sequence must be modified to replace guanine nucleotides with inosines, as detection relies on the change in electrical conductance of the MWNT upon hybridization of the complementary sequence containing electroactive guanine nucleotides. Change in the anodic current is proportional to the number of guanines in the hybridized target sequence. The beauty of this system is the ability to control the MWNT feature spacing using the CVD process, enabling the diameter of the tubes to be tuned from approximately 20–200 nm wide and separated by neighboring MWNT by about 300 nm. This system also has the benefit of multiplexing Koehne et al. (2003) and lends itself to universal DNA assays, which probe for common, repetitive sequences in the genome (Parman 2004).

This array could be further improved using a method of pre-patterned catalysts from which to directly grow CNTs. DPN can pattern nanoparticles on a surface (Basnar and Willner 2009), and cobalt nanoparticles in particular have been manipulated in this way to produce an ordered nanoarray format. CVD can then be used to catalytically grow CNTs from these nanoparticles. Unfortunately, this method alone results in low density growth, but when combined with stable temperature-cut single-crystal quartz as a substrate, the growth density improves as a result of the surface-lattice-induced alignment of the tubes (Kocabas et al. 2006; Li et al. 2008).

An additional method of orienting CNTs perpendicular to the nanoarray substrate is through the use of SAMs. HiPco SWNTs have been successfully manipulated in this manner by first oxidatively cutting the tubes using a mixture of nitric and sulfuric acid (Gooding et al. 2003) (Figure 27.7). This process leaves the reactive ends of the SWNTs functionalized with carboxylic acid groups (Fabre et al. 2008), which can then be converted to ester groups using dicyclohexyl carbodiimide (DCC). A gold substrate is modified with a cysteamine SAM, which is capable of forming an amide bond with the ester-terminated SWNTs. This process positions the tubes in a vertical manner with features composed of bundles of SWNTs (Gooding et al. 2003). Considering the typical diameter of an individual HiPco SWNTs is approximately 0.7–1.1 nm wide (O'Connell et al. 2002), this method is capable of constructing nanoscale

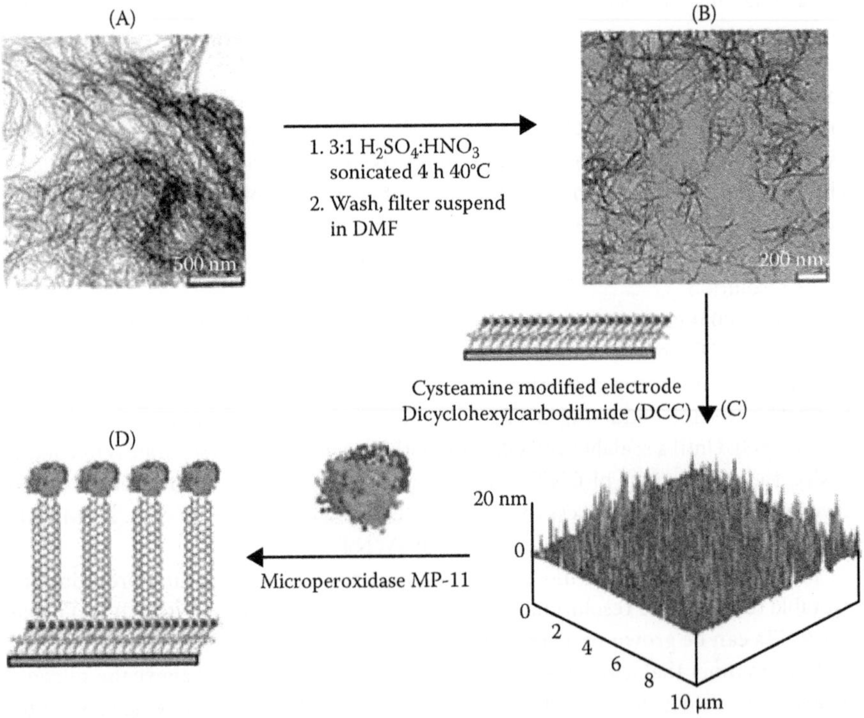

FIGURE 27.7 Schematic of the synthesis of a CNT, protein nanoarray. (A) A transmission electron microscope image of unpurified CNTs. (B) CNTs after oxidative shortening and purification from exposure to H_2SO_4 and HNO_3. (C) Carboxylic acid functional groups terminating the ends of the oxidized CNTs care converted to esters by DCC, which then allows the CNTs to form amide end of the cystamine SAM on the gold substrate; orienting the shortened CNTs perpendicular to the electrode surface. (D) The redox protein, microperoxidase MP-11, attaches to the free CNT ends, allowing the electron transfer rate between the protein and the electrode through the CNTs to be monitored with cyclic voltammetry. (From Gooding, J.J. et al., *J. Am. Chem. Soc.*, 125, 9006, 2003. With permission.)

resolution that is essential for the higher degree of sensitivity and lower sample volume requirements which make nanoarrays attractive. The free-termini of the SWNTs are then capable of binding to proteins and conceivably other biomolecules. Changes in the electron transfer rate between the attached protein and the gold, substrate electrode through the SWNT can be measured (Gooding et al. 2003) and possibly monitored for changes after the addition of complementary molecules.

E-jet printing can also deposit catalytic nanoparticles for aligned SWNT growth in addition to fabricating CNT electrodes (Park et al. 2007). The aligned SWNTs could be fashioned into nanoarrays for similar purposes as Gooding et al.'s (2003) protein sensing CNT array described previously. Additionally, the e-jet printing method can both pattern surfactant dispersed CNT inks as well as successfully pattern polyurethane over gold, chromium, and SWNTs arrays for the purpose of selectively curing and removing the polymer for the fabrication of CNT electrode arrays (Park et al. 2007).

Alternative to catalytically grown SWNT nanoarrays is the "directed assembly method." Wang et al. (2006) demonstrate the ability to control CNT deposition by patterning templates composed of alkane and carboxylic acid terminated SAMs using DPN. SWNTs dispersed in 1,2-dichlorobenzene when deposited on these templates self-assemble on the interface between the two SAMs upon drying of the solvent. By initially depositing the SAMs in circular templates, the authors demonstrate the ability to synthesize SWNT rings with radii as small as 100 nm, which may be fashioned into a nanoarray (Wang et al. 2006; Zou et al. 2007) (Figure 27.8). While rings pose interesting features, the applications of this method extend to parallel construction of CNT electronic devices that can be used for biosensor purposes.

Attachment of biomolecules to CNTs may be simplified by a procedure that grows gold nanoparticles on the sidewalls of these structures. Gold seed nanoparticles can be attached to the CNT surface using DPN. The particles are nondetectable by AFM, but upon immersion in a seeding solution composed of ascorbic acid and $HAuCl_4$, the nanoparticles are grown to a sub-5 nm diameter height

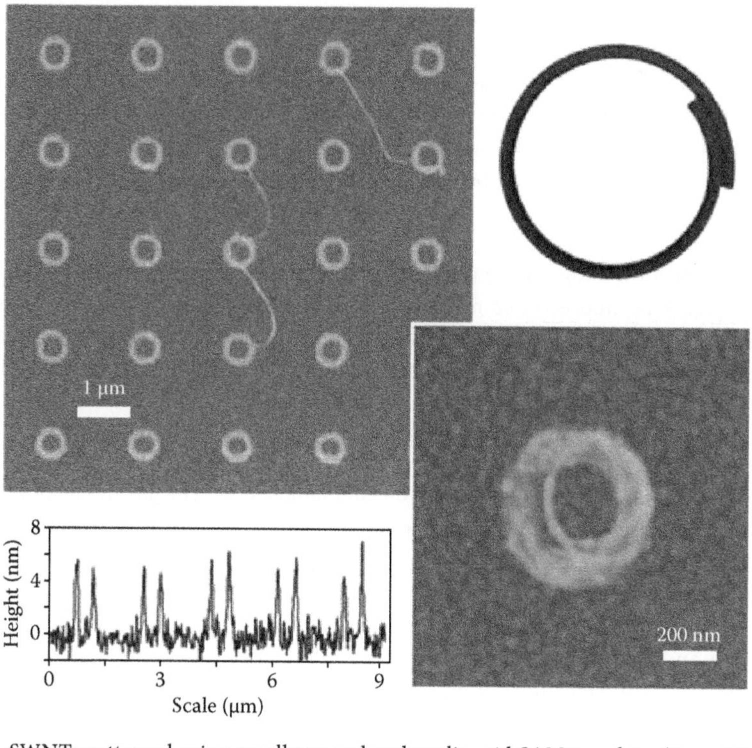

FIGURE 27.8 SWNTs patterned using an alkane and carboxylic acid SAM template. (From Wang, Y.H. et al., *Proc. Natl Acad. Sci. USA*, 103, 2026, 2006. © (2006) National Academy of Sciences, U.S.A. With permission.)

range (Chu et al. 2008). Conceivably, immersion in a solution of a cysteamine solution should functionalize the gold nanoparticles with the SAM, enabling the attachment of biomolecules that are capable of reacting with amine groups (Gooding et al. 2003). This method could enable considerable biosensing capability that is not restricted to a static nanoarray format. It may be possible to deposit additional gold nanoparticles on a CNT surface after initial protein functionalization as just described in order to multiplex to the CNT with different features. The flexible yet mechanical resilience associated with CNTs could allow for interesting conformations to be assumed upon exposure to complementary biomolecules—perhaps engaging in some of the more complicated binding motifs associated with many proteins, or even mimic the curving surface of a cell in a manner that is more accurate than the flat, unmoving surface of a nanoarray.

27.4 Challenges

While the benefits of nanoarrays are numerous and nanofabrication techniques continue to be developed and improved, there are still considerable challenges to overcome. For instance, it is necessary to expand nanoarray detection methods beyond AFM and fluorescence signaling. Additionally, the technical challenge associated with multiplexing nanoarray features will require especially intense effort to solve. The serial nature of most methods and cross-contamination makes it difficult to construct nanoscale features composed of a large number of different molecules. Mass production and parallel fabrication of nanoarrays bring further complications to this process. Researchers have moved to answer these challenges in order to fully realize the advantages and potentials of biosensing nanoarrays.

27.4.1 Detection

Although fluorescent analysis and AFM height profiling are perhaps the most common read out techniques for nanoarrays, there are several others that can sometimes be more advantageous to use, depending on the situation. AFM is convenient for in situ, label-less detection of biomolecular binding events on nanoarray features; however, AFM typically does not provide chemical information. Additionally, many nanofabrication methods do not require AFM, and thus detection becomes an additional step. Fluorescence detection may be simple and familiar for biology labs, but it generally negates the use of gold substrates, which are featured in many nanoarray designs, as a result of quenching interactions (Demers et al. 2002).

Total internal reflection fluorescence microscopy (TIRFM) is an elegant solution to the resolution limitations of traditional fluorescence microscopy, which requires a significant number of labeled proteins interacting to achieve a signal in addition to long fluorescent lifetimes, which can be difficult to achieve depending on how prevalent nonradiative relaxation pathways exist in the system. TIRFM uses two different wavelength lasers, a prism and a CCD camera to detect the fluorescent signal of individually interacting biomolecules. This ultimately increases the sensitivity of the nanoarray (Kim et al. 2007).

SPR is another detection method used in biosensing nanoarrays. The advantage of this technique is the ability to measure the specificity and quantity of biomolecular interactions on a nanoarray (Rakickas et al. 2008). Essentially, by fabricating the nanoarray on a rough surface composed of a coinage metal, which is achieved when using gold as a substrate for the nanoarray (Lee et al. 2006), a specific wavelength of light may be used to excite the surface gold electrons to oscillate in resonance with the exciting radiation (Willets and Van Duyne 2007). The resonance wavelength changes as a result of biomolecule surface adsorption or complementary binding, and these events can be monitored by detecting the resonance shift (Hoa et al. 2009).

Kelvin probe force microscopy (KPFM) is also relevant to the detection of nanoarray binding events. This method uses a modified AFM cantilever tip to measure the surface potential of the features. It has been determined that the surface potential of the biomolecule is directly proportional to its isoelectric point (Thompson et al. 2005; Sinensky and Belcher 2007)—the pH barrier above which the molecule is negatively charged and below which it is positively charged. A low isoelectric point therefore corresponds to a negative potential. Changes in the surface potential at each feature can be used to determine the binding events with a resolution as low as 10 nm. KPFM is a noncontact method, which allows it to scan the surface quickly in comparison to AFM, and does not deform the surface features, another disadvantage of AFM. Additionally, KPFM does not require problematic fluorescent labels, which eliminates that additional cause of denaturation in a system that is already difficult to maintain bioactivity (Sinensky and Belcher 2007).

Many additional nanoarray readout methods exist based on "nanoparticle-sandwich" assays, including light scattering (capable of multiplexed detection), the scanometric technique developed by the Mirkin group, and changes in electric current when positive binding events occur between miniaturized electrodes. These methods and more are reviewed in Rosi and Mirkin (2005). To realize the full potential of nanoarrays, future nanoarray detection methods will likely have to combine both ultra-high chemical sensitivity (due to miniaturized feature and increased chemical complexity) and high throughput (due to increased information and areal density).

27.4.2 Multiplexing and Parallelization

While nanoarrays can be fabricated through a variety of means to produce uniform biomolecule features that can consistently probe and signal for the presence of their complementary species, there are still limitations to this bioanalytical technique.

Foremost are issues of scale. At present, nanofabrication techniques are still in the relatively early stages of development. Since approximately the beginning of the new millennium, research has focused on achieving nanometer resolution of features in a reproducible manner. Multiplexing nanoarrays is essential to push this technology one step further; it would enable practitioners to assay the presence of potentially hundreds of different possible targets at once, vastly increasing research, diagnosis, and drug development speeds.

Thus far most nanoarrays used for the proof-of-concept experiments have been fabricated with single tip DPN. Because only one feature of one nanoarray may be fabricated at a time, this technique has significantly limited the scope of research. The development of the multitip arrays (Minne et al. 1998; Wang and Liu 2005) and of the massively parallel 2D pen arrays (Salaita et al. 2006; Wang et al. 2008) have overcome these limitations. By using 2D arrays consisting of as many as 55,000 AFM cantilever tips in 1 cm^2, Salaita et al. (2006) demonstrate a highly parallel, nanofabrication system that is capable of fabricating 55,000 features at once (Figure 27.9). This system has been quickly commercialized by NanoInk, Inc. (Skokie, IL). Salaita et al. (2006) tested the array by programming each tip to fabricate 100 ± 20 nm diameter dots, separated by 400 nm apart, in a 40×40 gold dot array on silicon-oxide. Each individual nanoarray array produced by a single pen was separated from its neighboring nanoarray by the inter-pen distance. This experiment successfully patterned 88 million features on a 1 cm^2 area in less than 20 min.

Lenhert et al. (2007) further show that by using 2D arrays, phospholipids, such as 1,2-dioleoyl-sn-glycero-3-phophocholine (DOPC), can be patterned in a noncovalent fashion that more accurately mimics the fluid nature of lipid membranes. Demonstrating that it is indeed possible to fabricate biosensing nanoarrays in a massively parallel manner, this experiment opens the door to drug delivery studies, which must understand how well the drug permeates the membrane (Majd and Mayer 2005; Yamazaki et al. 2005).

While the massively parallel 2D array is capable of mass-producing nanoarrays for biological applications, it still does not solve the problem of multiplexing. However, recent advances made by Wang et al. (2008) address this multiplexed inking challenge using an inkjet printing method to individually ink the 2D array tips in order to produce multicomponent nanoarray probes.

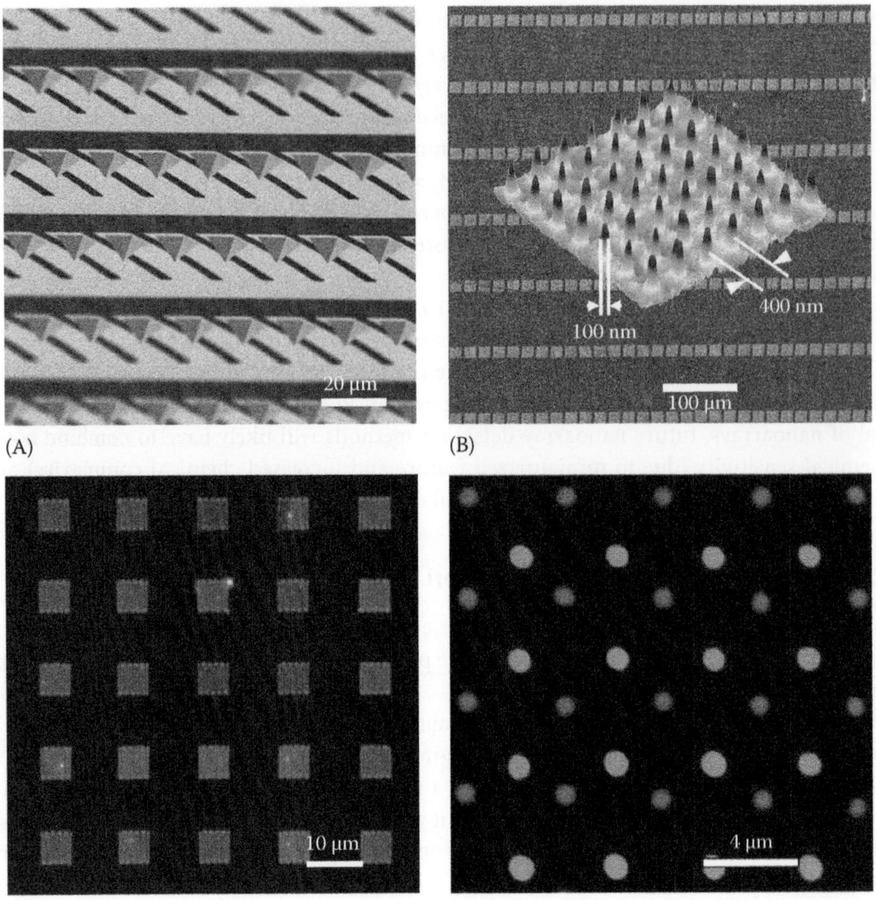

(A) (B) (C) (D)

FIGURE 27.9 (**See color insert.**) (A) SEM image of part of a 55,000-pen 2D array. (B) SEM image of part of gold nanostructures created by etching DPN patterns. The inset is an AFM topographic image of part of an array patterned by one single cantilever within the 2D pen array. (With permission from Salaita, K. et al., *Angew. Chem. Int. Ed.*, 45, 7220, 2006.) (C) Fluorescence microscopy of 1 mol% rhodamine-labeled 1,2-dioleoyl-sn-glycero-3-phosphocholine. (With permission from Lenhert, S. et al., *Small*, 3, 71, 2007.) (D) Epifluorescence image of a two sequence oligonucleotide DPN nanoarray on an SiO_x substrate hybridized by fluorophore-labeled sequences (Oregon Green 488-X and Texas Red-X). (From Demers, L.M. et al., *Science*, 296, 1836, 2002. With permission.)

Rather than soaking the AFM tips in the patterning solution, which can lead to uneven and non-reproducible adsorption of biomolecules, the 2D array is oriented tip-side up, and each tip is individually inked using a piezoelectric nozzle. While only four different inks were used in the experimental demonstration, this inking technique, in principle, enables the array to be inked with as many different biomolecules as there are ink wells for the inkjet printer to draw from. Considering the 55,000 cantilever tips that are available, the ability to produce nanoarrays capable of probing for several biomolecules at once seems imminent. This technique may allow pharmaceutical developers to quickly and efficiently test potential drugs faster than ever before.

While DPN seems to be dominating nanoarray fabrication in the most recent literature reports, a new series of DNA self-assembly nanoarrays (Park et al. 2005a,b), capable of multiplexing (Lin et al. 2007), has also become prevalent. DNA's most iconic double-helix structure is only its secondary structure, but it is also capable of assuming a variety of tertiary forms. Nanomaterials scientists have begun utilizing some of these DNA motifs to form nanoscale sized building blocks. Yan et al. have constructed a nine-stranded DNA motif that assumes the shape of a four-armed, square cross, which is capable of self-assembling into nanoribbons and nanogrids by designing "sticky" ends of single sequence strands that can hybridize with the complementary strands on other DNA crosses present (Yan et al. 2003). This DNA motif is also referred to as DNA tiles for its ability to self-assemble into nanometer spaced features in a 2D lattice, like tiles on a kitchen floor, that can expand to form millimeter sized chips (Lin et al. 2007). Yan et al. (2003) were quick to expand the potential of this DNA nanogrids system to biosensing nanoarrays by showing the tiles' ability to incorporate probe molecules such as proteins conveniently spaced only nanometers apart. An experiment involving binding of biotin to the center of the DNA tiles and exposing the nanoarray to streptavidin demonstrated its ability to maintain bioactivity of the probe and target molecules, detecting the binding event using AFM.

Lin et al. (2007) utilized this DNA tile motif to build their own biosensing nanoarrays with the additional benefit of multiplexing. This DNA self-assembly system uses three tile types: the detection tile is functionalized with a dangling probe sequence that is not involved in the structural construction of the nanoarray scaffold, and two encoding tiles that are modified with a green or red organic dye. The tiles are designed so that the detection tiles can only hybridize with encoding tiles, due to the specificity of Watson–Crick base pairing. The detection tile probe sequence is double stranded and labeled with a blue dye moiety. When the target molecule comes into contact with the probe, a reaction occurs (either rehybridization in the case of complementary genetic sequences, or a release of free energy upon wrapping of the DNA probe around a protein or small molecule), which removes the blue dye from that part of the nanoarray, leaving the red and green encoding tiles visible via fluorescent microscopy, indicating a successful binding event (Figure 27.10). The signal can be increased using a slightly modified detection system, which involves a hairpin structured DNA probe. The hairpin probe (formed by self-complementary

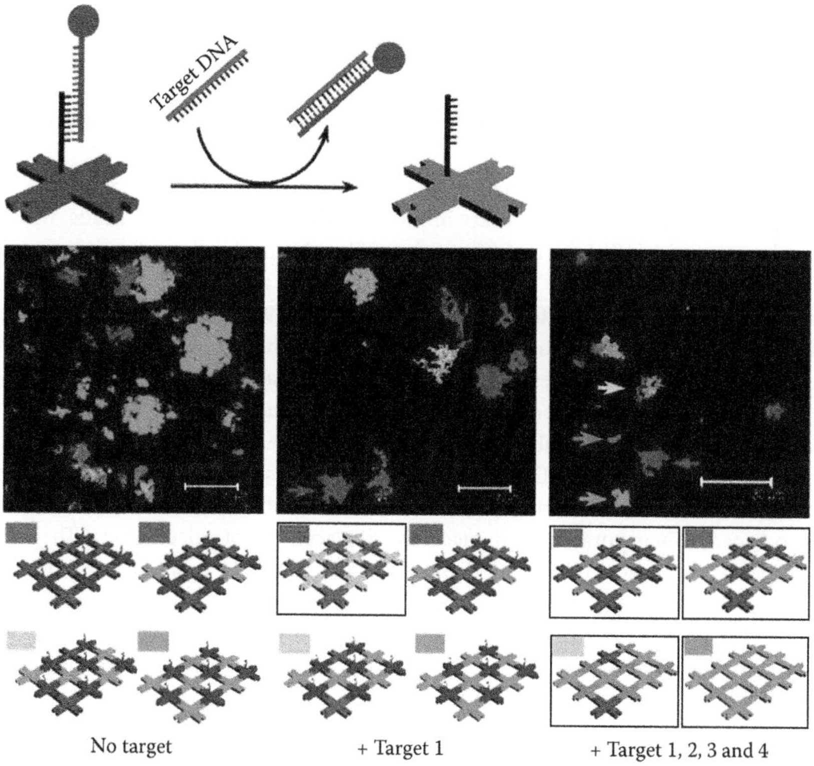

FIGURE 27.10 **(See color insert.)** DNA-tile nanoarray detection upon hybridization with a complementary strand. (From Lin, C.X. et al., *Nano Lett.*, 7, 507, 2007.)

sequences) unfolds in a cascading reaction when the hairpin target and a fluorophore-labeled sequences build off the binding event like an extending ladder to greatly enhance the signal (Lin et al. 2008).

Multiplexing is achieved by simply synthesizing a variety of detection tiles with either different DNA probe sequences or different aptamers specific to different targets. In this manner, Lin et al. (2008) were able to differentially detect the SARS and HIV viruses, human alpha-thrombin protein, and ATP simultaneously using different fluorescent dye detection systems for each (Lin et al. 2007).

DNA-tile multiplexing has several advantages, including its ability to reuse the nanoarray. After binding and on release of the probe with the target, the nanoarray can be recharged by exposing it to a probe molecule solution. The probe reattaches to the DNA-tile centers in the same way as it was initially constructed. Additionally, the system is easily amenable to parallelization as a result of its self-assembly mechanism. Finally, the dimensions of the nanoarray features can be controlled by altering the lengths of the DNA strands associated with the four arms of the tiles, ensuring that the lengths accommodate the double-helix twisting in a manner that allows the tiles to self-assemble in a planar formation (Winfree et al. 1998).

These DNA-tile systems are not limited to genomic assays, but are versatile enough to provide a platform for other biomolecule probes as shown by Yan et al.'s (2003) protein nanoarray as well as by the general knowledge of the many different DNA compatible biomolecules. DNA or RNA strands can be used to attach ligands (Jenison et al. 1994), viruses, ATP (Lin et al. 2007), and even gold nanoparticles (Zhang et al. 2006), which have considerable versatility in biosensing nanoarrays.

27.5 Conclusions and Outlook

Within the past 10 years, nanoarrays have quickly emerged to meet the increasing biosensing demands of pharmaceutical (Debouck and Goodfellow 1999) and diagnostic applications (Hessner et al. 2006), which depend on the study of biomolecular interactions (MacBeath and Schreiber 2000). A leap forward from microarrays, nanoarrays are made possible by nanofabrication techniques which enable the construction of smaller feature sizes and increased resolution. These new features have made nanoarrays a powerful bionanotechnology that enables high-throughput, low volume sample tests for a wide variety of biomolecules.

The diverse types of nanofabrication techniques available have propelled nanoarray development forward, enabling users to choose from a variety of methods that fit their unique needs, whether it is the ultrahigh resolution of which EBL (Grigorescu et al. 2007) and scanning tunneling microscope lithography are capable (Eigler and Schweizer 1990), or parallel synthesis using microcontact printing and massively parallel DPN (Kumar and Whitesides 1993; Kumar et al. 1994; Salaita et al. 2006). The capability to regularly synthesize sub-20 nm features has already been achieved, which allows for an increased packing density as high as 10^{10} features in a square centimeter space (Gu et al. 2008). This allows for the detection of biomolecules on the order of cellular, picoliter volumes (Lynch et al. 2004).

The rapid advance of nanoarray fabrication has allowed these devices to be tested for performance qualities and different applications. The results have been successful in DNA sequence (Demers et al. 2002), viral assays (Lee et al. 2004), and protein interaction studies (Lynch et al. 2004), which opens nanoarrays to broader incorporation and successive replacement of micorarrays. The use of carbon nanotubes as nanoarray sensors is a productive use of these materials' remarkable electrical properties (Koh et al. 2008).

Furthermore, the development of multiplexed surface features ensures that the advantage of multi-target detection enjoyed by microarrays will be a similar advantage of nanoarrays. This patterning of features using different probe molecules allows nanoarrays to serve as a "lab-on-a-chip," with the ability to quickly and effectively assay for the presence of several different targets, thereby saving both time and money normally spent on individual tests. With the advent of massively parallel DPN in conjunction with inkjet printing (Wang et al. 2008) and self-assembling DNA scaffolds (Lin et al. 2007, 2008), multiplexed and mass-produced nanoarrays have the potential to enter commercial applications.

The traditional conception of nanoarrays as static devices composed of patterned features on a flat surface may have limitations in the study of biological interactions, which are inherently more dynamic. Although nanoarrays have been used to study the protein interactions found at the fluid-like interfaces of the nuclear envelope (Rakickas et al. 2008), the inherently static nature of the nanoarray poses an intrinsic limitation that the experimental system will never completely mimic the actual dynamic nature of these membranes, perhaps resulting in an oversimplified or even outright false description of protein interactions at such dynamic interfaces. This limitation gives motivation to develop dynamically responsive nanoarrays. Nevertheless, these types of nanotechnology approaches to biology will ultimately bring the field to work at its inherent nanometer scale, an achievement that will undoubtedly produce many future discoveries and accomplish previously unattainable goals.

Acknowledgment

This work is supported by start-up funds and a General Research Board Summer Award from the University of Maryland, and a Nanobiotechnology Initiative Award from the Maryland Department of Business & Economic Development.

References

Abed, H., H. Jamptchian, H. Dallaporta, B. Gely, P. Bindzi, D. Chatain, S. Nitsche et al. 2005. Deposition of gold nanofeatures on silicon samples by field-induced deposition using a scanning tunneling microscope. *Journal of Vacuum Science & Technology B* 23 (4):1543–1550.

Akbulut, O., J. M. Jung, R. D. Bennett, Y. Hu, H. T. Jung, R. E. Cohen, A. M. Mayes, and F. Stellacci. 2007. Application of supramolecular nanostamping to the replication of DNA nanoarrays. *Nano Letters* 7 (11):3493–3498.

Allain, L. R., D. N. Stratis-Cullum, and T. Vo-Dinh. 2004. Investigation of microfabrication of biological sample arrays using piezoelectric and bubble-jet printing technologies. *Analytica Chimica Acta* 518 (1–2):77–85.

Arnold, M. S., A. A. Green, J. F. Hulvat, S. I. Stupp, and M. C. Hersam. 2006. Sorting carbon nanotubes by electronic structure using density differentiation. *Nature Nanotechnology* 1 (1):60–65.

Bano, F., L. Fruk, B. Sanavio, M. Glettenberg, L. Casalls, C. M. Niemeyer, and G. Scoles. 2009. Toward multiprotein nanoarrays using nanografting and DNA directed immobilization of proteins. *Nano Letters* 9 (7):2614–2618.

Barbulovic-Nad, I., M. Lucente, Y. Sun, M. J. Zhang, A. R. Wheeler, and M. Bussmann. 2006. Bio-microarray fabrication techniques—A review. *Critical Reviews in Biotechnology* 26 (4):237–259.

Basnar, B. and I. Willner. 2009. Dip-pen-nanolithographic patterning of metallic, semiconductor, and metal oxide nanostructures on surfaces. *Small* 5 (1):28–44.

Besteman, K., J. O. Lee, F. G. M. Wiertz, H. A. Heering, and C. Dekker. 2003. Enzyme-coated carbon nanotubes as single-molecule biosensors. *Nano Letters* 3 (6):727–730.

Biebuyck, H. A., N. B. Larsen, E. Delamarche, and B. Michel. 1997. Lithography beyond light: Microcontact printing with monolayer resists. *IBM Journal of Research and Development* 41 (1–2):159–170.

Bier, F. F. and F. Kleinjung. 2001. Feature-size limitations of microarray technology—A critical review. *Fresenius Journal of Analytical Chemistry* 371 (2):151–156.

Binnig, G. and H. Rohrer. 1982. Scanning tunneling microscopy. *Helvetica Physica Acta* 55 (6):726–735.

Chakra, E. B., B. Hannes, G. Dilosquer, C. D. Mansfield, and M. Cabrera. 2008. A new instrument for automated microcontact printing with stamp load adjustment. *Review of Scientific Instruments* 79 (6):064102.

Choi, H. K., J. U. Park, O. O. Park, P. M. Ferreira, J. G. Georgiadis, and J. A. Rogers. 2008. Scaling laws for jet pulsations associated with high-resolution electrohydrodynamic printing. *Applied Physics Letters* 92 (12):123109.

Chu, H. B., Z. Jin, Y. Zhang, W. W. Zhou, L. Ding, and Y. Li. 2008. Site-specific deposition of gold nanoparticles on SWNTs. *Journal of Physical Chemistry C* 112 (35):13437–13441.

Craighead, H. 2006. Future lab-on-a-chip technologies for interrogating individual molecules. *Nature* 442 (7101):387–393.

Debouck, C. and P. N. Goodfellow. 1999. DNA microarrays in drug discovery and development. *Nature Genetics* 21:48–50.

Delehanty, J. B. and F. S. Ligler. 2003. Method for printing functional protein Microarrays. *Biotechniques* 34 (2):380–385.

Della Giustina, G., M. Prasciolu, G. Brusatin, M. Guglielmi, and F. Romanato. 2009. Electron beam lithography of hybrid sol-gel negative resist. *Microelectronic Engineering* 86 (4):745–748.

Demers, L. M., D. S. Ginger, S. J. Park, Z. Li, S. W. Chung, and C. A. Mirkin. 2002. Direct patterning of modified oligonucleotides on metals and insulators by dip-pen nanolithography. *Science* 296 (5574):1836–1838.

Dietrich, H. R. C., J. Knoll, L. R. van den Doel, G. W. K. van Dedem, P. A. S. Daran-Lapujade, L. J. van Vliet, R. Moerman, J. T. Pronk, and I. T. Young. 2004. Nanoarrays: A method for performing enzymatic assays. *Analytical Chemistry* 76 (14):4112–4117.

Ducker, R. E., S. Janusz, S. Q. Sun, and G. J. Leggett. 2007. One-step photochemical introduction of nanopatterned protein-binding functionalities to oligo(ethylene glycol)-terminated self-assembled monolayers. *Journal of the American Chemical Society* 129 (48):14842–14843.

Eigler, D. M. and E. K. Schweizer. 1990. Positioning single atoms with a scanning tunneling microscope. *Nature* 344 (6266):524–526.

Ekins, R. P. 1998. Ligand assays: From electrophoresis to miniaturized microarrays. *Clinical Chemistry* 44:2015–2030.

Fabre, B., F. Hauquier, C. Herrier, G. Pastorin, W. Wu, A. Bianco, M. Prato et al. 2008. Covalent assembly and micropatterning of functionalized multiwalled carbon nanotubes to monolayer-modified Si(111) surfaces. *Langmuir* 24 (13):6595–6602.

Fan, J. B., M. S. Chee, and K. L. Gunderson. 2006. Highly parallel genomic assays. *Nature Reviews Genetics* 7 (8):632–644.

Fan, J. B., J. M. Yeakley, M. Bibikova, E. Chudin, E. Wickham, J. Chen, D. Doucet et al. 2004. A versatile assay for high-throughput gene expression profiling on universal array matrices. *Genome Research* 14 (5):878–885.

Gates, B. D., Q. B. Xu, M. Stewart, D. Ryan, C. G. Willson, and G. M. Whitesides. 2005. New approaches to nanofabrication: Molding, printing, and other techniques. *Chemical Reviews* 105 (4):1171–1196.

Golub, T. R., D. K. Slonim, P. Tamayo, C. Huard, M. Gaasenbeek, J. P. Mesirov, H. Coller et al. 1999. Molecular classification of cancer: Class discovery and class prediction by gene expression monitoring. *Science* 286 (5439):531–537.

Gooding, J. J., R. Wibowo, J. Q. Liu, W. R. Yang, D. Losic, S. Orbons, F. J. Mearns, J. G. Shapter, and D. B. Hibbert. 2003. Protein electrochemistry using aligned carbon nanotube arrays. *Journal of the American Chemical Society* 125 (30):9006–9007.

Graeter, S. V., J. H. Huang, N. Perschmann, M. Lopez-Garcia, H. Kessler, J. D. Ding, and J. P. Spatz. 2007. Mimicking cellular environments by nanostructured soft interfaces. *Nano Letters* 7 (5):1413–1418.

Grigorescu, A. E. and C. W. Hagen. 2009. Resists for sub-20-nm electron beam lithography with a focus on HSQ: State of the art. *Nanotechnology* 20 (29):292001.

Grigorescu, A. E., M. C. van der Krogt, C. W. Hagen, and P. Kruit. 2007. 10 nm lines and spaces written in HSQ, using electron beam lithography. *Microelectronic Engineering* 84 (5):822–824.

Gu, J., X. Y. Xiao, B. R. Takulapalli, M. E. Morrison, P. Zhang, and F. Zenhausern. 2008. A new approach to fabricating high density nanoarrays by nanocontact printing. *Journal of Vacuum Science & Technology B* 26 (6):1860–1865.

Gurtler, L. 1996. Difficulties and strategies of HIV diagnosis. *Lancet* 348 (9021):176–179.

Hahmann, P. and O. Fortagne. 2009. 50 years of electron beam lithography: Contributions from Jena (Germany). *Microelectronic Engineering* 86 (4):438–441.

Hegde, P., R. Qi, K. Abernathy, C. Gay, S. Dharap, R. Gaspard, J. E. Hughes, E. Snesrud, N. Lee, and J. Quackenbush. 2000. A concise guide to cDNA microarray analysis. *Biotechniques* 29 (3):548–562.

Hersam, M. C. 2008. Progress towards monodisperse single-walled carbon nanotubes. *Nature Nanotechnology* 3 (7):387–394.

Hessner, M. J., M. Y. Liang, and A. E. Kwitek. 2006. The application of microarray analysis to pediatric diseases. *Pediatric Clinics of North America* 53 (4):579–590.

Hibino, H. and Y. Watanabe. 2005. Arrangement of Au-Si alloy islands at atomic steps. *Surface Science* 588 (1–3):L233–L238.

Hoa, X. D., A. G. Kirk, and M. Tabrizian. 2009. Enhanced SPR response from patterned immobilization of surface bioreceptors on nano-gratings. *Biosensors & Bioelectronics* 24 (10):3043–3048.

Hong, S. H., J. Zhu, and C. A. Mirkin. 1999a. Multiple ink nanolithography: Toward a multiple-pen nano-plotter. *Science* 286 (5439):523–525.

Hong, S. H., J. Zhu, and C. A. Mirkin. 1999b. A new tool for studying the in situ growth processes for self-assembled monolayers under ambient conditions. *Langmuir* 15 (23):7897–7900.

Huff, J. L., M. P. Lynch, S. Nettikadan, J. C. Johnson, S. Vengasandra, and E. Henderson. 2004. Label-free protein and pathogen detection using the atomic force microscope. *Journal of Biomolecular Screening* 9 (6):491–497.

Huo, F. W., Z. J. Zheng, G. F. Zheng, L. R. Giam, H. Zhang, and C. A. Mirkin. 2008. Polymer pen lithography. *Science* 321 (5896):1658–1660.

Jackson, B. L. and J. T. Groves. 2004. Scanning probe lithography on fluid lipid membranes. *Journal of the American Chemical Society* 126 (43):13878–13879.

Jang, J. Y., G. C. Schatz, and M. A. Ratner. 2004. How narrow can a meniscus be? *Physical Review Letters* 92 (8):085504.

Janshoff, A. and C. Steinem. 2006. Transport across artificial membranes—An analytical perspective. *Analytical and Bioanalytical Chemistry* 385 (3):433–451.

Jenison, R. D., S. C. Gill, A. Pardi, and B. Polisky. 1994. High-resolution molecular discrimination by RNA. *Science* 263 (5152):1425–1429.

Jiang, H. Z. and S. I. Stupp. 2005. Dip-pen patterning and surface assembly of peptide amphiphiles. *Langmuir* 21 (12):5242–5246.

Joubert, J. R., K. A. Smith, E. Johnson, J. P. Keogh, V. H. Wysocki, B. K. Gale, J. C. Conboy, and S. S. Saavedra. 2009. Stable, ligand-doped, poly(bis-SorbPC) lipid bilayer arrays for protein binding and detection. *ACS Applied Materials & Interfaces* 1 (6):1310–1315.

Kang, S. H., Y. J. Kim, and E. S. Yeung. 2007. Detection of single-molecule DNA hybridization by using dual-color total internal reflection fluorescence microscopy. *Analytical and Bioanalytical Chemistry* 387 (8):2663–2671.

Kanungo, M., H. Lu, G. G. Malliaras, and G. B. Blanchet. 2009. Suppression of metallic conductivity of single-walled carbon nanotubes by cycloaddition reactions. *Science* 323 (5911):234–237.

Kim, D., H. G. Lee, H. Jung, and S. H. Kang. 2007. Single-protein molecular interactions on polymer-modified glass substrates for nanoarray chip application using dual-color TIRFM. *Bulletin of the Korean Chemical Society* 28 (5):783–790.

Kocabas, C., M. Shim, and J. A. Rogers. 2006. Spatially selective guided growth of high-coverage arrays and random networks of single-walled carbon nanotubes and their integration into electronic devices. *Journal of the American Chemical Society* 128 (14):4540–4541.

Koehne, J., H. Chen, J. Li, A. M. Cassell, Q. Ye, H. T. Ng, J. Han, and M. Meyyappan. 2003. Ultrasensitive label-free DNA analysis using an electronic chip based on carbon nanotube nanoelectrode arrays. *Nanotechnology* 14 (12):1239–1245.

Koh, J., B. Kim, S. Hong, H. Lim, and H. C. Choi. 2008. Nanotube-based chemical and biomolecular sensors. *Journal of Materials Science & Technology* 24 (4):578–588.

Kumar, A., H. A. Biebuyck, and G. M. Whitesides. 1994. Patterning self-assembled monolayers—Applications in materials science. *Langmuir* 10 (5):1498–1511.

Kumar, A. and G. M. Whitesides. 1993. Features of gold having micrometer to centimeter dimensions can be formed through a combination of stamping with an elastomeric stamp and an alkanethiol ink followed by chemical etching. *Applied Physics Letters* 63 (14):2002–2004.

Lange, S. A., V. Benes, D. P. Kern, J. K. H. Horber, and A. Bernard. 2004. Microcontact printing of DNA molecules. *Analytical Chemistry* 76 (6):1641–1647.

Lee, M., D. K. Kang, H. K. Yang, K. H. Park, S. Y. Choe, C. Kang, S. I. Chang, M. H. Han, and I. C. Kang. 2006. Protein nanoarray on prolinker (TM) surface constructed by atomic force microscopy dip-pen nanolithography for analysis of protein interaction. *Proteomics* 6 (4):1094–1103.

Lee, K. B., E. Y. Kim, C. A. Mirkin, and S. M. Wolinsky. 2004. The use of nanoarrays for highly sensitive and selective detection of human immunodeficiency virus type 1 in plasma. *Nano Letters* 4 (10):1869–1872.

Lee, K. B., J. H. Lim, and C. A. Mirkin. 2003. Protein nanostructures formed via direct-write dip-pen nanolithography. *Journal of the American Chemical Society* 125 (19):5588–5589.

Lenhert, S., P. Sun, Y. H. Wang, H. Fuchs, and C. A. Mirkin. 2007. Massively parallel dip-pen nanolithography of heterogeneous supported phospholipid multilayer patterns. *Small* 3 (1):71–75.

Li, B., C. F. Goh, X. Z. Zhou, G. Lu, H. Tantang, Y. H. Chen, C. Xue, F. Y. C. Boey, and H. Zhang. 2008. Patterning colloidal metal nanoparticles for controlled growth of carbon nanotubes. *Advanced Materials* 20 (24):4873–4878.

Libioulle, L., A. Bietsch, H. Schmid, B. Michel, and E. Delamarche. 1999. Contact-inking stamps for microcontact printing of alkanethiols on gold. *Langmuir* 15 (2):300–304.

Lin, C. X., Y. Liu, and H. Yan. 2007. Self-assembled combinatorial encoding nanoarrays for multiplexed biosensing. *Nano Letters* 7 (2):507–512.

Lin, C. X., J. K. Nangreave, Z. Li, Y. Lin, and H. Yan. 2008. Signal amplification on a DNA-tile-based biosensor with enhanced sensitivity. *Nanomedicine* 3 (4):521–528.

Liu, M., N. A. Amro, and G. Y. Liu. 2008. Nanografting for surface physical chemistry. *Annual Review of Physical Chemistry* 59:367–386.

Liu, M. Z. and G. Y. Liu. 2005. Hybridization with nanostructures of single-stranded DNA. *Langmuir* 21 (5):1972–1978.

Liu, G. Y., S. Xu, and Y. L. Qian. 2000. Nanofabrication of self-assembled monolayers using scanning probe lithography. *Accounts of Chemical Research* 33 (7):457–466.

Love, J. C., L. A. Estroff, J. K. Kriebel, R. G. Nuzzo, and G. M. Whitesides. 2005. Self-assembled monolayers of thiolates on metals as a form of nanotechnology. *Chemical Reviews* 105 (4):1103–1169.

Lovrinovic, M. and C. M. Niemeyer. 2005. DNA microarrays as decoding tools in combinatorial chemistry and chemical biology. *Angewandte Chemie—International Edition* 44 (21):3179–3183.

Lynch, M., C. Mosher, J. Huff, S. Nettikadan, J. Johnson, and E. Henderson. 2004. Functional protein nanoarrays for biomarker profiling. *Proteomics* 4 (6):1695–1702.

MacBeath, G. and S. L. Schreiber. 2000. Printing proteins as microarrays for high-throughput function determination. *Science* 289 (5485):1760–1763.

Majd, S. and M. Mayer. 2005. Hydrogel stamping of arrays of supported lipid bilayers with various lipid compositions for the screening of drug-membrane and protein-membrane interactions. *Angewandte Chemie—International Edition* 44 (41):6697–6700.

Minne, S. C., G. Yaralioglu, S. R. Manalis, J. D. Adams, J. Zesch, A. Atalar, and C. F. Quate. 1998. Automated parallel high-speed atomic force microscopy. *Applied Physics Letters* 72 (18):2340–2342.

Mirmomtaz, E., M. Castronovo, C. Grunwald, F. Bano, D. Scaini, A. A. Ensafi, G. Scoles, and L. Casalis. 2008. Quantitative study of the effect of coverage on the hybridization efficiency of surface-bound DNA nanostructures. *Nano Letters* 8 (12):4134–4139.

Mossman, K. D., G. Campi, J. T. Groves, and M. L. Dustin. 2005. Altered TCR signaling from geometrically repatterned immunological synapses. *Science* 310 (5751):1191–1193.

Niemeyer, C. M. and C. A. Mirkin. 2004. *Nanobiotechnology: Concepts, Applications and Perspectives*. Wiley-VCH.

O'Connell, M. J., S. M. Bachilo, C. B. Huffman, V. C. Moore, M. S. Strano, E. H. Haroz, K. L. Rialon et al. 2002. Band gap fluorescence from individual single-walled carbon nanotubes. *Science* 297 (5581):593–596.

Park, S. H., R. Barish, H. Y. Li, J. H. Reif, G. Finkelstein, H. Yan, and T. H. LaBean. 2005a. Three-helix bundle DNA tiles self-assemble into 2D lattice or 1D templates for silver nanowires. *Nano Letters* 5 (4):693–696.

Park, J. U., M. Hardy, S. J. Kang, K. Barton, K. Adair, D. K. Mukhopadhyay, C. Y. Lee et al. 2007. High-resolution electrohydrodynamic jet printing. *Nature Materials* 6 (10):782–789.

Park, J. U., J. H. Lee, U. Paik, Y. Lu, and J. A. Rogers. 2008. Nanoscale patterns of oligonucleotides formed by electrohydrodynamic jet printing with applications in biosensing and nanomaterials assembly. *Nano Letters* 8 (12):4210–4216.

Park, S. H., P. Yin, Y. Liu, J. H. Reif, T. H. LaBean, and H. Yan. 2005b. Programmable DNA self-assemblies for nanoscale organization of ligands and proteins. *Nano Letters* 5 (4):729–733.

Park, H., J. J. Zhao, and J. P. Lu. 2006. Effects of sidewall functionalization on conducting properties of single wall carbon nanotubes. *Nano Letters* 6 (5):916–919.

Parman, C. 2004. A universal array for gene expression profiling—Assay tutorial: Microarray platform allows comparative studies without need for prior sequence identification. *Genetic Engineering News* 24 (1):36–36.

Perl, A., D. N. Reinhoudt, and J. Huskens. 2009. Microcontact printing: Limitations and achievements. *Advanced Materials* 21 (22):2257–2268.

Piner, R. D. and C. A. Mirkin. 1997. Effect of water on lateral force microscopy in air. *Langmuir* 13 (26):6864–6868.

Piner, R. D., J. Zhu, F. Xu, S. H. Hong, and C. A. Mirkin. 1999. "Dip-pen" nanolithography. *Science* 283 (5402):661–663.

Pirrung, M. C. 2002. How to make a DNA chip. *Angewandte Chemie—International Edition* 41 (8):1277–1289.

Quist, A. P., E. Pavlovic, and S. Oscarsson. 2005. Recent advances in microcontact printing. *Analytical and Bioanalytical Chemistry* 381 (3):591–600.

Rakickas, T., M. Gavutis, A. Reichel, J. Piehler, B. Liedberg, and R. Valiokas. 2008. Protein-protein interactions in reversibly assembled nanopatterns. *Nano Letters* 8 (10):3369–3375.

Ramanujan, C. S., K. Sumitomo, M. R. R. de Planque, H. Hibino, K. Torimitsu, and J. F. Ryan. 2007. Self-assembly of vesicle nanoarrays on Si: A potential route to high-density functional protein arrays. *Applied Physics Letters* 90 (3):033901–033903.

Renault, J. P., A. Bernard, A. Bietsch, B. Michel, H. R. Bosshard, E. Delamarche, M. Kreiter, B. Hecht, and U. P. Wild. 2003. Fabricating arrays of single protein molecules on glass using microcontact printing. *Journal of Physical Chemistry B* 107 (3):703–711.

Reynolds, N. P., J. D. Tucker, P. A. Davison, J. A. Timney, C. N. Hunter, and G. J. Leggett. 2009. Site-specific immobilization and micrometer and nanometer scale photopatterning of yellow fluorescent protein on glass surfaces. *Journal of the American Chemical Society* 131 (3):896–897.

Rosi, N. L. and C. A. Mirkin. 2005. Nanostructures in biodiagnostics. *Chemical Reviews* 105 (4):1547–1562.

Saaem, I., V. Papasotiropoulos, T. Wang, P. Soteropoulos, and M. Libera. 2007. Hydrogel-based protein nanoarrays. *Journal of Nanoscience and Nanotechnology* 7 (8):2623–2632.

Saito, R., G. Dresselhaus, and M. S. Dresselhaus. 1998. *Physical Properties of Carbon Nanotubes*. London, U.K.: Imperial College Press.

Salaita, K., Y. H. Wang, J. Fragala, R. A. Vega, C. Liu, and C. A. Mirkin. 2006. Massively parallel dip-pen nanolithography with 55000-pen two-dimensional arrays. *Angewandte Chemie-International Edition* 45 (43):7220–7223.

Salaita, K., Y. H. Wang, and C. A. Mirkin. 2007. Applications of dip-pen nanolithography. *Nature Nanotechnology* 2 (3):145–155.

Sinensky, A. K. and A. M. Belcher. 2007. Label-free and high-resolution protein/DNA nanoarray analysis using Kelvin probe force microscopy. *Nature Nanotechnology* 2 (10):653–659.

Smith, J. C., K. B. Lee, Q. Wang, M. G. Finn, J. E. Johnson, M. Mrksich, and C. A. Mirkin. 2003. Nanopatterning the chemospecific immobilization of cowpea mosaic virus capsid. *Nano Letters* 3 (7):883–886.

Spatz, J. P., S. Mossmer, C. Hartmann, M. Moller, T. Herzog, M. Krieger, H. G. Boyen, P. Ziemann, and B. Kabius. 2000. Ordered deposition of inorganic clusters from micellar block copolymer films. *Langmuir* 16 (2):407–415.

Stowers, J. and D. A. Keszler. 2009. High resolution, high sensitivity inorganic resists. *Microelectronic Engineering*. 86 (4):730–733.

Sun, S. Q., M. Montague, K. Critchley, M. S. Chen, W. J. Dressick, S. D. Evans, and G. J. Leggett. 2006. Fabrication of biological nanostructures by scanning near-field photolithography of chloromethylphenyisiloxane monolayers. *Nano Letters* 6 (1):29–33.

Tam, J. M., L. Song, and D. R. Walt. 2009. DNA detection on ultrahigh-density optical fiber-based nanoarrays. *Biosensors & Bioelectronics* 24 (8):2488–2493.

Tan, Y. H., M. Liu, B. Nolting, J. G. Go, J. Gervay-Hague, and G. Y. Liu. 2008. A nanoengineering approach for investigation and regulation of protein immobilization. *ACS Nano* 2 (11):2374–2384.

Thompson, M., L. E. Cheran, M. Q. Zhang, M. Chacko, H. Huo, and S. Sadeghi. 2005. Label-free detection of nucleic acid and protein microarrays by scanning Kelvin nanoprobe. *Biosensors & Bioelectronics* 20 (8):1471–1481.

Tinazli, A., J. Piehler, M. Beuttler, R. Guckenberger, and R. Tampe. 2007. Native protein nanolithography that can write, read and erase. *Nature Nanotechnology* 2 (4):220–225.

Tu, X. M., S. Manohar, A. Jagota, and M. Zheng. 2009. DNA sequence motifs for structure-specific recognition and separation of carbon nanotubes. *Nature* 460 (7252):250–253.

Turkova, J. 1999. Oriented immobilization of biologically active proteins as a tool for revealing protein interactions and function. *Journal of Chromatography B* 722 (1–2):11–31.

Veiseh, M., O. Veiseh, M. C. Martin, F. Asphahani, and M. Q. Zhang. 2007. Short peptides enhance single cell adhesion and viability on microarrays. *Langmuir* 23 (8):4472–4479.

Wadu-Mesthrige, K., N. A. Amro, J. C. Garno, S. Xu, and G. Y. Liu. 2001. Fabrication of nanometer-sized protein patterns using atomic force microscopy and selective immobilization. *Biophysical Journal* 80 (4):1891–1899.

Wang, Y. H., L. R. Giam, M. Park, S. Lenhert, H. Fuchs, and C. A. Mirkin. 2008. A self-correcting inking strategy for cantilever arrays addressed by an inkjet printer and used for dip-pen nanolithography. *Small* 4 (10):1666–1670.

Wang, Y. H., M. J. Kim, H. W. Shan, C. Kittrell, H. Fan, L. M. Ericson, W. F. Hwang, S. Arepalli, R. H. Hauge, and R. E. Smalley. 2005. Continued growth of single-walled carbon nanotubes. *Nano Letters* 5 (6):997–1002.

Wang, X. F., and C. Liu. 2005. Multifunctional probe array for nano patterning and imaging. *Nano Letters* 5 (10):1867–1872.

Wang, Y. H., D. Maspoch, S. L. Zou, G. C. Schatz, R. E. Smalley, and C. A. Mirkin. 2006. Controlling the shape, orientation, and linkage of carbon nanotube features with nano affinity templates. *Proceedings of the National Academy of Sciences of the United States of America* 103 (7):2026–2031.

Wang, Y. H., C. A. Mirkin, and S. J. Park. 2009. Nanofabrication beyond electronics. *ACS Nano* 3 (5):1049–1056.

Willets, K. A. and R. P. Van Duyne. 2007. Localized surface plasmon resonance spectroscopy and sensing. *Annual Review of Physical Chemistry* 58:267–297.

Winfree, E., F. R. Liu, L. A. Wenzler, and N. C. Seeman. 1998. Design and self-assembly of two-dimensional DNA crystals. *Nature* 394 (6693):539–544.

Wong, S. S., J. D. Harper, P. T. Lansbury, Jr., and C. M. Lieber. 1998. Carbon nanotube tips: High-resolution probes for imaging biological systems. *Journal of the American Chemical Society* 120 (3):603–604.

Xia, Y. N. and G. M. Whitesides. 1998. Soft lithography. *Annual Review of Materials Science* 28:153–184.

Xu, S. and G. Y. Liu. 1997. Nanometer-scale fabrication by simultaneous nanoshaving and molecular self-assembly. *Langmuir* 13 (2):127–129.

Yadav, Y., S. K. Padigi, S. Prasad, and X. Y. Song. 2008. Towards crossbar nanoarray structure via micro-contact printing. *Journal of Nanoscience and Nanotechnology* 8 (4):1951–1958.

Yamazaki, V., O. Sirenko, R. J. Schafer, L. Nguyen, T. Gutsmann, L. Brade, and J. T. Groves. 2005. Cell membrane array fabrication and assay technology. *BMC Biotechnology* 5 (18).

Yan, H., S. H. Park, G. Finkelstein, J. H. Reif, and T. H. LaBean. 2003. DNA-templated self-assembly of protein arrays and highly conductive nanowires. *Science* 301 (5641):1882–1884.

Yao, Y. G., C. Q. Feng, J. Zhang, and Z. F. Liu. 2009. "Cloning" of single-walled carbon nanotubes via open-end growth mechanism. *Nano Letters* 9 (4):1673–1677.

Yu, J. J., Y. H. Tan, X. Li, P. K. Kuo, and G. Y. Liu. 2006. A nanoengineering approach to regulate the lateral heterogeneity of self-assembled monolayers. *Journal of the American Chemical Society* 128 (35):11574–11581.

Zandvllet, H. J. W. and A. van Houselt. 2009. Scanning tunneling spectroscopy. *Annual Review of Analytical Chemistry* 2:37–55.

Zhang, J. P., Y. Liu, Y. G. Ke, and H. Yan. 2006. Periodic square-like gold nanoparticle arrays templated by self-assembled 2D DNA nanogrids on a surface. *Nano Letters* 6 (2):248–251.

Zhou, X. Z., Y. H. Chen, B. Li, G. Lu, F. Y. C. Boey, J. Ma, and H. Zhang. 2008. Controlled growth of peptide nanoarrays on Si/SiOx substrates. *Small* 4 (9):1324–1328.

Zhou, D. J., K. Sinniah, C. Abell, and T. Rayment. 2003. Label-free detection of DNA hybridization at the nanoscale: A highly sensitive and selective approach using atomic-force microscopy. *Angewandte Chemie—International Edition* 42 (40):4934–4937.

Zhou, Y., R. Valiokas, and B. Liedberg. 2004. Structural characterization of microcontact printed arrays of hexa(ethylene glycol)-terminated alkanethiols on gold. *Langmuir* 20 (15):6206–6215.

Zou, S. L., D. Maspoch, Y. H. Wang, C. A. Mirkin, and G. C. Schatz. 2007. Rings of single-walled carbon nanotubes: Molecular-template directed assembly and Monte Carlo modeling. *Nano Letters* 7 (2):276–280.

28

Photopatternable Multifunctional Nanobiomaterials

Hailin Cong
University of California at Davis

Tingrui Pan
University of California at Davis

28.1 Introduction

Over the past decades, two-dimensional micropatternable nanobiomaterials with desired bulk and surface properties have attracted considerable attention, which has led to a wide range of applications in miniature biochemical sensors, engineered cellular microenvironments, and medical diagnostic microdevices (Zakhidov et al. 1998; Aizenberg et al. 1999; Shi et al. 1999; Cui et al. 2001; Veinot et al. 2002). A number of techniques have been demonstrated to prepare functional nanobiomaterials with topological micropatterns (De Rosa et al. 2000; Demers and Mirkin 2001; Chou et al. 2002; Loo et al. 2002; Whitesides and Grzybowski 2002), among which photolithography-based and soft lithography-based techniques have been the most popular choices. In comparison with soft lithography, direct photolithography techniques eliminate the extra molding step, resolve compatibility issues, and thereby, provide more efficient rapid-prototyping processes to prepare a microstructured surface of functional nanobiomaterial. In this chapter, we present the recent development of photopatternable nanobiomaterials with particular emphasis on novel multifunctional micropatternable nanocomposites. Nanobiomaterials possessing a unique combination of photopatternability with the desired physical and chemical properties (such as conductivity, superhydrophobicity, antimicrobial and non-fouling properties, etc.) have demonstrated their extensive usage in the area of biosensors, bionics, and self-assembly.

28.2 Photopatternable Conductive PDMS Nanocomposite Material for Bioelectrical Sensing

28.2.1 Polydimethylsiloxane

Polydimethylsiloxane (PDMS) has been widely used in a variety of academic and industrial applications due to its unique physical and chemical properties. With recent advances in soft lithography and polymer microelectromechanical systems (MEMS), PDMS has been constructed into a large array of micro- and nanoscale devices for biological and medical applications (Jo et al. 2000; Sia and Whitesides 2003; Huh et al. 2005; Liu 2007). Besides its well-known superior elasticity and flexibility in mechanical applications, PDMS has become a popular choice for biological studies because of its nontoxicity to cells and high permeability to gases (De Silva et al. 2004; Haubert et al. 2006; Lounaci et al. 2006; Bhagat et al. 2007). To fabricate PDMS-based micro-nanoscale devices, the molding method is presently the most common approach, which inversely replicates well-defined features from a conventionally micromachined mold (Sia and Whitesides 2003). Uncross-linked PDMS pregel is mixed at a 10:1 (w/w) ratio with a curing agent, and after thermal curing, it forms a replica at 80°C within an hour's duration. The molding process, though itself robust and reusable, creates microfabrication compatibility and integration issues, for instance, misalignment and packaging problems to other substrates. Moreover, high electrical impedance and poor adhesion with evaporated metal prevent further applications of PDMS in the area of electrical sensing and flexible circuits (Lee et al. 2005; Niu et al. 2007).

28.2.2 Photopatternable PDMS

An alternative approach to PDMS processing is to make the PDMS pre-polymer photo-cross-link under selective exposure of high-energy wavelength lights (<400 nm), and thus, it can be directly photopatterned. This can be accomplished by adding a photoinitiator of 2,2-dimethoxy-2-phenyl acetophenone (DMAP), which makes PDMS function as a negative photoresist (Lotters et al. 1997; Ward et al. 2001; Almasri et al. 2005; Niu et al. 2007). Exposure to UV light results in PDMS cross-linking and curing. Dow Corning has recently introduced photodefinable silicone products (WL-5000 series), prepared through this technique (Gardner et al. 2004; Harkness et al. 2004; Corning 2005), into the semiconductor packaging market. Micropatterns with 15 μm resolution can be fabricated through a completely photolithography-compatible process, involving spin coating, film baking, UV exposure, and solvent development (Corning 2005). Recently, another PDMS photolithographic process has been reported using benzophenone as photoinitiator (Bhagat et al. 2007). Benzophenone is a photosensitizer often used to initiate the free-radical polymerization of acrylates and monomers with certain functional groups under UV exposure. A group of investigators have reported its use with siloxane polymers (Muller et al. 1991, 1992; Pouliquen and Coqueret 1996; Tsougeni et al. 2007). The photodefinable mixture of PDMS and photoinitiator, eliminating the need of replicate molding and the compatibility issues related to molding, provides a more efficient way to the rapid prototyping of polymer-based MEMS devices.

28.2.3 Conductive PDMS

To overcome the low electrical conductivity of PDMS, highly conductive fillers (typically inorganic powders) are usually introduced into the polymer matrix, which provide continuous conductive pathways for electron migration (Engel et al. 2006; Liu 2007). This simple concept has been of extensive use in various practices, where both excellent electrical and thermal conductivities have been demonstrated

TABLE 28.1 Examples of Thermally Polymerized Conductive PDMS-Filler Composites

Filler Type	Filler Shape	Filler Size (μm)	Filler Percentage	Applications	References
Silver	Particle	1–2	84 wt%	Soft electronic packaging	[20]
Carbon	Nanotube	1–5	10 wt%	Transducers and sensors	[32,33]
Carbon	Fiber	33	—	Electrophoresis chips	[34]
Carbon black	Particle	30	3 vol%–10 wt%	Percolation studies, chemical vapor sensor, electromagnetic device	[35–37]

(see Table 28.1) (Unger et al. 2000; Gawron et al. 2001; Rwei et al. 2002; Wong et al. 2005; Engel et al. 2006; Liu 2007; Niu et al. 2007).

28.2.4 Fabrication of Photopatternable Conductive PDMS

Combining the two approaches aforementioned (adding conductivity and photopatternability to the PDMS matrix), our group has developed the first conductive photodefinable PDMS composites, providing both high electrical conductivity and photopatternability (Cong and Pan 2008). The photosensitive composite, consisting of a photosensitive reagent, a conductive filler, and a PDMS pre-polymer, can be processed as a regular photoresist. To prepare the composite, the PDMS base and curing agent (Sylgard® 184) are mixed at a 10:1 (w/w) ratio. Benzophenone (3 wt%) and silver powder (17–22 vol%, 2 μm) are added to the PDMS mixture and degassed for 15 min. The prepared PDMS-Ag photoresist mixture is then spin-coated onto a flat substrate (e.g., glass, silicon, polyester, or silicone) for 30 s. The spin-coated wafer is loaded toward the photomask in an approximate mode with 50 μm spacing. Ten minutes UV exposure at 12 mW/cm^2 is followed by a postexposure bake for 50 s at 120°C. Heavy UV exposure dosage (7200 mJ/cm^2) is necessary to induce complete photochemical reactions under significant decay of light transmission by silver particles. During the postexposure bake, the unexposed region gets fully cross-linked while the exposed region remains uncured. The uncured PDMS is then removed in toluene during the development. Finally, the wafer is rinsed in 2-propanol and blow-dried in nitrogen flow (Bhagat et al. 2007).

28.2.5 Properties of Photopatternable Conductive PDMS

Table 28.2 summarizes the key physical properties of the conductive PDMS-Ag composite. By incorporating 17–21 vol% of silver powder and 3 wt% benzophenone into the Sylgard 184 PDMS pre-polymer, the resultant multifunctional PDMS composite is incorporated with electrical/thermal conductivities as well as photopatternability, which completely removes the need of a master mold for the micropatterned features. Highest conductivity of ~10^4 S/m and minimal lithographic resolution of 60 μm (Figure 28.1) have been achieved using the conductive PDMS composite on flexible polymeric substrates

TABLE 28.2 Properties of the PDMS-Ag Conductive Photoresist

	PDMS-Ag Sample (vol% of Ag)			
Property	0	17	19	21
Hardness (shore A)	49	63	67	72
Tensile modulus (MPa)	1.3	2.7	3.0	3.2
Elongation at break (%)	58	59	58	57
Electrical conductivity (S/m)	2.5×10^{-14}	2.5×10^{-2}	6250	8333
Thermal conductivity (W/m K)	0.15	71	76	81
Contact angle of water (°)	109	119	127	131
Thickness of spin coating at 3000 rpm (μm)	13	18	24	30

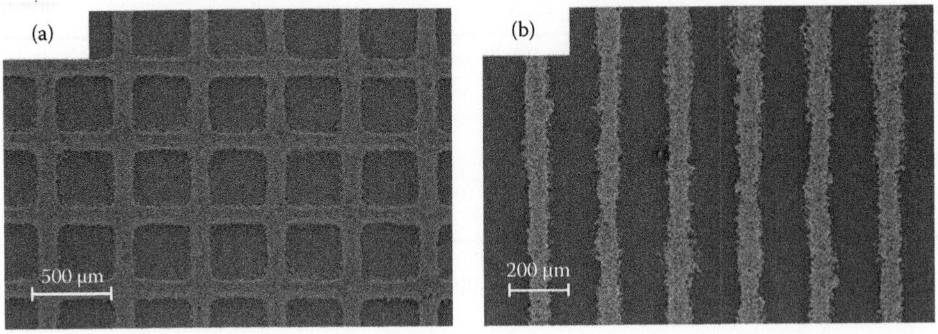

FIGURE 28.1 SEM of the lithographic patterns of the conductive PDMS: (a) 150 μm features and (b) 60 μm lines. (Cong, H. and Pan, T., Microfabrication of conductive PDMS on flexible substrates for biomedical applications, and *IEEE NEMS*) © (2009). Reproduced with permission from IEEE.)

(e.g., polyethylene terephthalate and PDMS). Electrical conduction mechanism is based on establishing electronic paths through percolation and contact between filler particles in the insulating matrix (Kirkpatr 1973; Zallen 1983; Biller 1985; Dovzhenko and Zhirkov 1995; Jiguet et al. 2005). Increasing content of the conductive filler reduces the electrical resistivity of the PDMS composite matrix by adding more interconnected clusters of silver particles into the conducting paths. As the conductive filler, silver powder significantly lowers the electrical resistance in the PDMS matrix through interconnecting percolation paths. In addition, the composite formula possesses enhanced mechanical properties. The silver particles in the PDMS matrix participate in physical cross-linking and thus considerably improve the mechanical strength of the composites. Moreover, the nanotopology and interfacial property of the composite lead to interesting surface phenomena.

According to the established wetting models (Cassie and Baxter 1944; Wenzel 1949), adding silver powder substantially changes the physical and chemical heterogenesis of the composite surface, which tends to become more hydrophobic than the pure PDMS with a higher contact angle of water. The increased surface hydrophobicity results from the high surface roughness of the formed PDMS-Ag composites as shown in Figure 28.1. As shown in Table 28.2, the measured contact angles of the PDMS surface containing 0, 17, 19, and 21 vol% silver powder are 109°, 119°, 127°, and 131°, respectively.

In addition, silver is historically known for its superior antibacterial properties, which can be used for biomedical devices to retard bacterial colonization and reduce the incidence of infections (Gabriel et al. 1995; Fung et al. 1996; Dowling et al. 2001; Wang et al. 2007). Silver particles work as catalysts to attack the bacteria from several aspects: poking bacteria membranes, destroying metabolic enzymes, denaturing proteins, disrupting bacteria division and the proliferation process, and lead to excellent antibacterial properties (Horn 2004). To assess the antibacterial properties of the formed PDMS-Ag photoresist membranes, *Escherichia coli* is used in the bacteria adhesion tests. The bacteria are incubated on the membrane surfaces at 37°C for 48 h prior to SEM inspection. As shown in Figure 28.2, the PDMS-Ag conductive composite with 19 vol% of silver particles shows outstanding antimicrobial properties compared with the pure PDMS control. The silver particles presented in PDMS largely remove the amount of bacteria adherence and thus improve the antibacterial property of the membrane. The excellent antimicrobial property of PDMS-Ag offers a desired safety feature to the multifunctional biomaterial.

28.2.6 Application in Bioelectrical Sensing

Figure 28.3 demonstrates a prototype of a capacitive pressure sensing array constructed by the PDMS-Ag conductive photoresist. The conductive composite is first patterned as both the sensing component and the connecting circuitry onto a pure PDMS flexible substrate. Subsequently, two layers of sensing circuits are folded over in a crossover configuration. Following the alignment step with the aid of surrounding

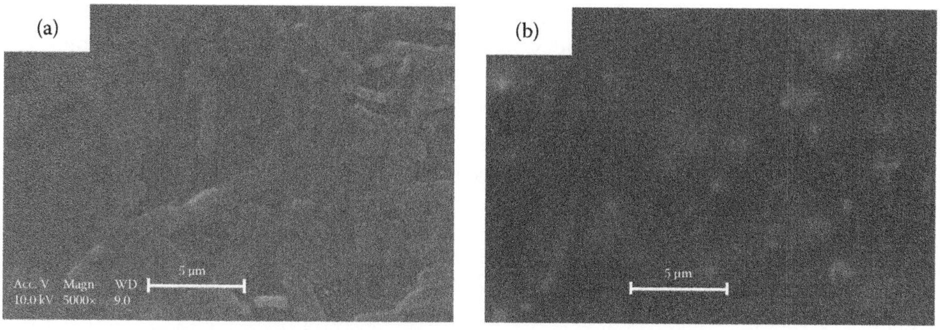

FIGURE 28.2 SEM surface morphology of *Escherichia coli* adhered for 48 h to the PDMS-Ag photoresist with 0 vol% (a) and 19 vol% (b) silver powders. (Cong, H. and Pan, T., Microfabrication of conductive PDMS on flexible substrates for biomedical applications, and *IEEE NEMS*) © (2009). Reproduced with permission from IEEE.)

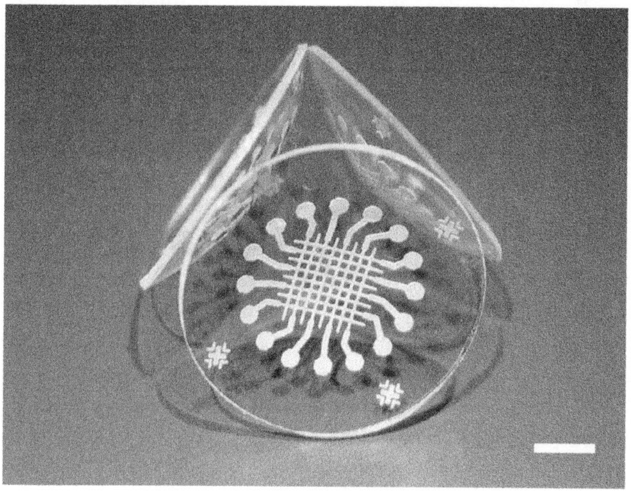

FIGURE 28.3 Prototype of an array of capacitive pressure sensors embedded in contact lens devices using PDMS-Ag conductive photoresist. (Scale bar: 3 mm.)

alignment marks, covalent bonding between the two-folded layers is conducted through oxygen plasma treatment. Subsequent to thermal compression molding in an aluminum mold in contact with a 300°C hotplate, the PDMS sensing device is molded into a contact lens shape, as shown in Figure 28.3 (Cong and Pan 2009). The bioelectrical sensing array built on a flexible substrate enables easy, direct, adaptive assessment of biomechanical measures in a range of medical applications (Stieglitz 2001; Subbaroyan et al. 2005; Fagaly 2006).

28.3 Photopatternable PEG Hydrogel Material for Nanoassembly

28.3.1 Polyethylene Glycol

As one of the best known nonfouling biomaterials with extremely low energy and nonadhesive surfaces, (Wang et al. 2001; Bremmell et al. 2005; Gudipati et al. 2005). polyethylene glycol (PEG) hydrogel, also known as polyethylene oxide (PEO) hydrogel, has been micromachined using both photolithography and soft lithography methods (Revzin et al. 2001, 2003; Kim et al. 2005). These patterned PEG surfaces are widely used in cell culture, drug delivery, and biomedical devices (Albrecht et al. 2005; Kim et al. 2006; Duan et al. 2007). The nonadhesive property of PEG hydrogel comes from its high surface resistance to the nonspecific adsorption of macromolecules in aqueous environment (Peppas et al. 2000; Underhill et al. 2007). Over the past decades, nanoscale components, including nanocolloids, biomolecules, and nanoparticles, are considered as the basic functional building blocks for future biochips (Jacobs and Whitesides 2001; Kumacheva et al. 2002; Zheng et al. 2002; Allard et al. 2004; Winkleman et al. 2005). Therefore, a highly programmed assembly of these objects onto the two-dimensional ordered biomaterial surface becomes an active area of research with particular focus on low-cost, large-area, and efficient manufacturing approaches.

28.3.2 Fabrication of PEG Hydrogel Nanopatterns

Polyethylene glycol diacrylate (PEG-DA) is one of the commonly used photosensitive prepolymers in the production of PEG hydrogel micropatterns. In this case, the PEG hydrogel patterns are fabricated from the precursor solution of PEG-DA (MW 575) with 1% (w/v) photoinitiator and 2-hydroxy-2-methyl-propiophenone. Using a spin coater, the solution is spun at 800 rpm for 8 s onto a 3-acryloxypropyltrichlorosilane-treated glass surface containing terminal acrylate groups. The thin layer of the PEG-DA precursor solution on glass is then exposed through a chrome/soda lime photomask to an UV light source of the photo intensity of 15 mW/cm^2 at 365 nm. The exposure time is 0.7 s. The regions of PEG-DA exposed to UV light undergo free-radical polymerization and become cross-linked, while the unexposed regions are dissolved in DI water after 5 min development. Coating silane coupling agents onto the glass substrate is critical to the adhesion of PEG hydrogel micro/nanostructures (Plueddemann 1982; Vandenberg et al. 1985; Lesho and Sheppard 1996).

Figure 28.4 shows AFM images of the fabricated PEG nanopatterns. As can be seen, the feature size of 500 × 500 nm^2 with 1500 nm spacing as well as that of 1 × 1 μm^2 with 1 μm spacing lead to an ultrahigh feature density of 250,000 wells/mm^2 on glass substrates. The depth of the microwell features is about 300 nm as measured by AFM. To achieve high-resolution PEG photolithography, the contact mode between the PEG substrate and the photomask is employed, given the fact that the minimal feature size is proportional to the square root of the separation distance (Ghandhi 1994). The additional cleaning procedure of the photomask with acetone is necessary to remove excess uncross-linked PEG pre-polymer in the postexposure treatment. The 500 × 500 nm^2 nanowells define the highest resolution of PEG hydrogel patterns achieved by the photolithography method, compared with the minimal resolution of 20 μm in the previous study using conventional proximity exposure (Revzin et al. 2001, 2003).

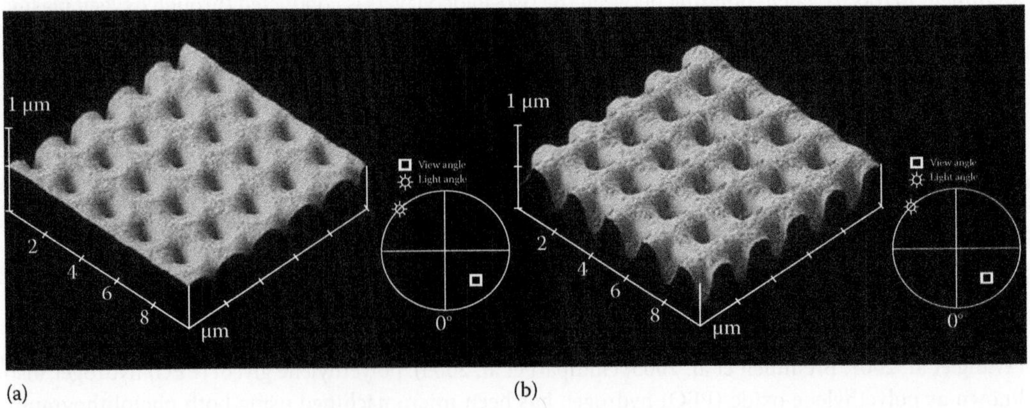

FIGURE 28.4 AFM images of PEG hydrogel wells on silane-modified glasses: (a) 500 × 500 nm^2 nanowells and (b) 1 × 1 μm^2 microwells. (From Cong, H.L. et al., *Nanotechnology*, 20(7), 75307, 2009. Reproduced with permission from IOP Publishing Ltd.)

28.3.3 PEG Pattern–Assisted Nanoassembly

Figure 28.5 illustrates the pattern-assisted nanoassembly method used to fabricate nanocolloidal arrays onto PEG wells. Experimental procedures are described as follows. First, a suspension of poly(styrene-methyl methacrylate-acrylic acid) (PSMA, 190 nm) colloidal particles is diluted to a defined concentration using deionized water (Cong and Cao 2005a,b; Cong et al. 2008, 2009a,b). Then, the glass slide fabricated with PEG hydrogel wells is immersed vertically into the dispersion and lifted up at a constant speed varied from 0.1 μm/s to 1 mm/s precisely controlled by a step motor.

Figure 28.6 shows that the PSMA nanocolloids are self-assembled and display a highly organized single-bead nanocolloidal array embedded into PEG nanowells (operation condition: 0.1 μm/s pulling speed and a 1.5 mg/mL nanocolloidal concentration). As illustrated in the inset of Figure 28.5, under the

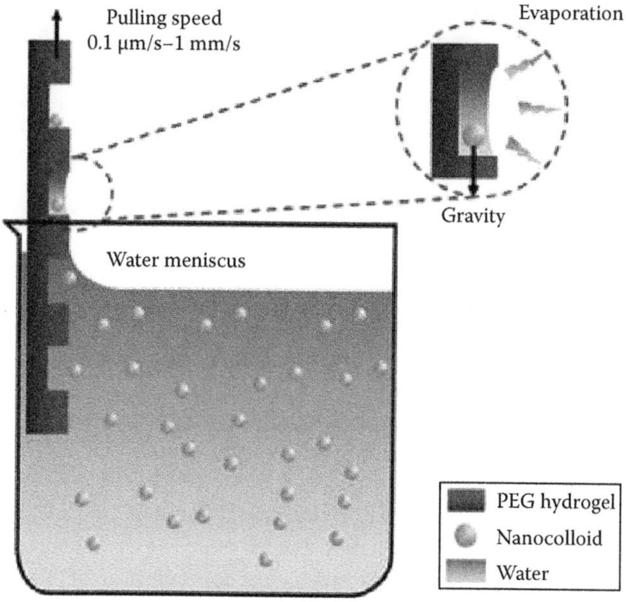

FIGURE 28.5 Illustration of the pattern-assisted nanoassembly method for the fabrication of nanocolloidal arrays. (From Cong, H.L. et al. *Nanotechnology*, 20(7), 75307, 2009. Reproduced with permission from IOP Publishing Ltd.)

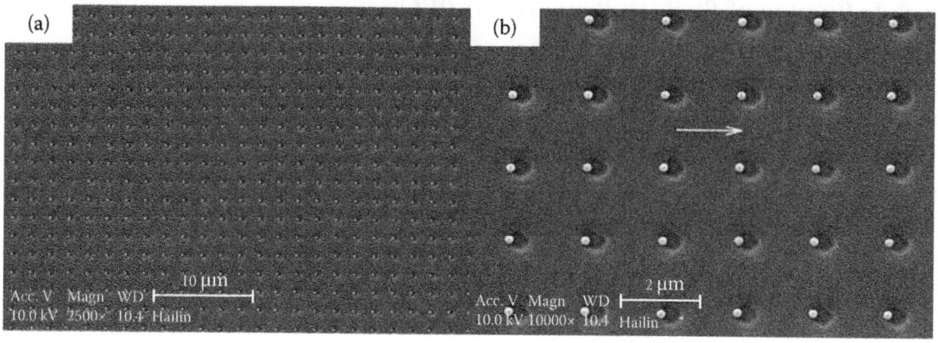

FIGURE 28.6 SEM images of nanocolloidal arrays formed in the 500 × 500 nm^2 PEG nanowells. (From Cong, H.L. et al., *Nanotechnology*, 20(7), 75307, 2009. The arrow indicates the pulling direction, reproduced with permission from IOP Publishing Ltd.)

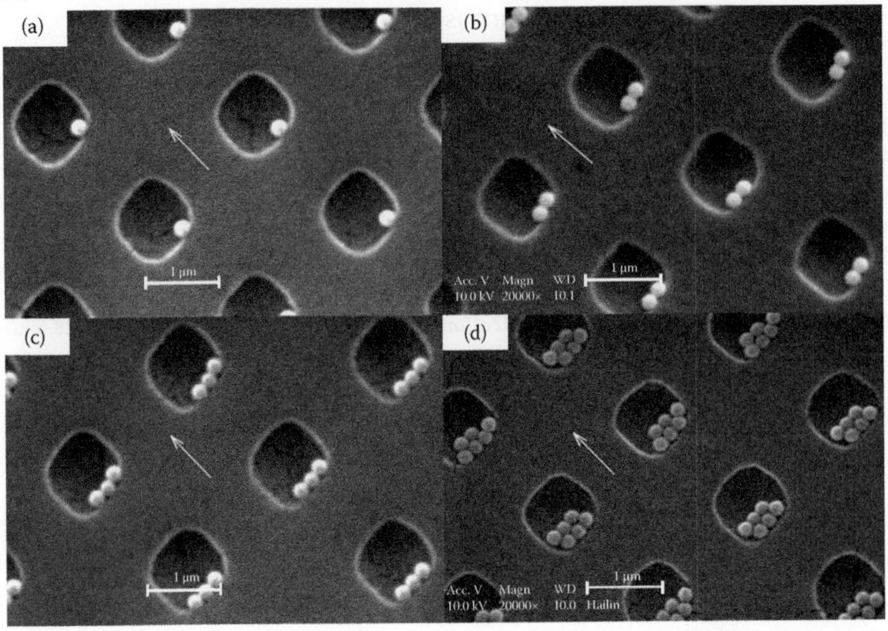

FIGURE 28.7 SEM images of nanocolloid arrays formed in the 1 × 1 μm² PEG microwells using different PSMA colloidal particle concentrations. (From Cong, H.L. et al., *Nanotechnology*, 20(7), 75307, 2009. Reproduced with permission from IOP Publishing Ltd.)

influence of gravitation, the colloids inside the trapped suspension tend to move downward in the well during the evaporation. Once the self-assembly process is accomplished, all the beads are uniformly aligned along the bottom edge of the wells.

Different nanocolloidal arrays can be obtained by changing the pulling speed, the inclined angle of the substrate, the shape and size of the PEG nanopatterns, as well as the concentration of colloids in the nano-assembly process. Figure 28.7 shows that varying the concentration of PSMA colloidal particles in the aqueous suspension can lead to different morphologies of nanocolloid arrays in the PEG microwells, such as linear bead arrays (4a, 4b, 4c) and multiline arrays (4d), which are of potential use in biosensing and biofabrication.

28.3.4 Influence of Surface Property of PEG

The PEG hydrogel, one of the best known antifouling biomaterials with an extremely low surface energy, plays an important role in preventing the PSMA nanocolloids from adhering to the surface during the nano-assembly process. To verify this unusual antifouling mechanism, oxygen plasma has been carried out to modify the PEG surface property. After 20 s at 200 W of plasma treatment, about 100 nm PEG hydrogel on the surface is removed under oxygen plasma, and the contact angle of the PEG hydrogel surface reduces from 22° to 9°. The antifouling property of the PEG hydrogel surface drastically deteriorates where both hydroxyl and carboxyl groups are introduced during the modification. The oxygen plasma modified PEG microwells are then used to fabricate nanocolloidal arrays under the same conditions as the unmodified ones, and the comparison is illustrated in Figure 28.8. As expected, Figure 28.8b shows that after the oxygen plasma surface modification, nanocolloidal arrays are packed into the microwells, but also adhere onto the modified PEG surfaces unlike the highly selective assembly that resulted in Figure 28.8a.

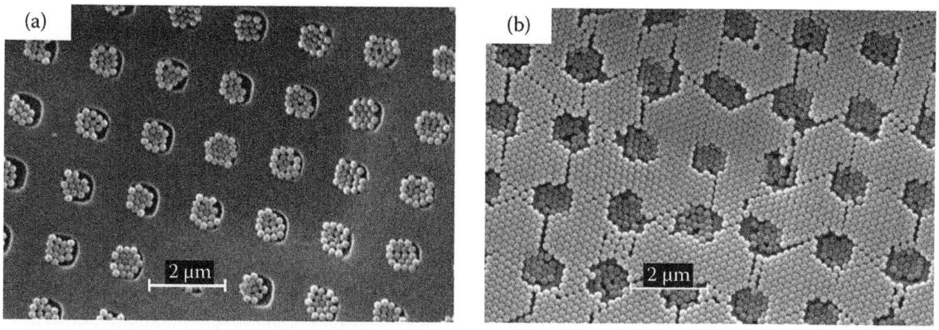

FIGURE 28.8 SEM images of nanocolloid arrays formed in the (a) unmodified and (b) oxygen plasma modified PEG microwells at the same assembly conditions. (From Cong, H.L. et al., *Nanotechnology*, 20(7), 75307, 2009. Reproduced with permission from IOP Publishing Ltd.)

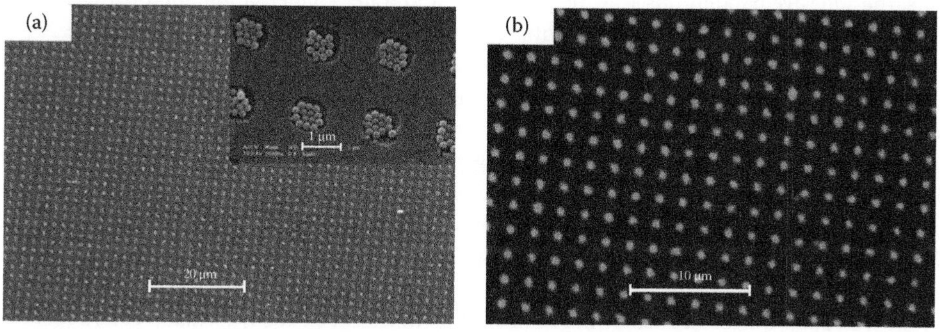

FIGURE 28.9 (a) SEM and (b) fluorescent microscope images of nanocolloid arrays formed in the $1 \times 1\,\mu m^2$ PEG microwells. (From Cong, H.L. et al., *Nanotechnology*, 20(7), 75307, 2009. Reproduced with permission from IOP Publishing Ltd.)

28.3.5 Application in Biosensing

After emulsion polymerization, the PSMA nanocolloid has carboxy functional groups on surface, which can easily be labeled as antigens/antibodies, biomolecules, and fluorescent agents. For example, IgG-FITC can be coupled onto its surface by using the carbodiimide method described elsewhere (Ortegavinuesa et al. 1995). The IgG-FITC labeled PSMA colloidal particles (~5.0 mg/mL) are used to fabricate nanocolloidal arrays, and the results are shown in Figure 28.9. The IgG-FITC labeled colloid arrays emit green light (520 nm) via an excitation wavelength of 490 nm under a fluorescence microscope, which can have potential implications for microchip-based drug identification, analyte detection, cell sorting, and biological sensing (Zhong et al. 2000; Gu et al. 2002; Pregibon et al. 2006).

28.4 Photopatternable Superhydrophobic PTFE Nanocomposite Material for Bionics

28.4.1 Polytetrafluoroethylene

Polytetrafluoroethylene (PTFE), well known as Teflon™, is a synthetic fluoropolymer of tetrafluoroethylene. PTFE is chemically nonreactive due to the strong carbon–fluorine bonds in the molecular structure. It is almost insoluble in any solvent and has a high chemical resistance. Moreover, PTFE has been widely employed as nonsticky coatings and lubricants as it possesses an extremely low surface energy (18.6 mJ/m^2) (Ellis et al. 2001; Zhao et al. 2002; Zhang et al. 2004). Due to the remarkable nonfouling property, flexibility and biocompatibility, PTFE has been used in a variety of biomedical devices and implants such as man-made blood vessels and artificial filtration devices. However, due to its chemical inertness and high melting temperature, the PTFE surface is difficult to process through conventional microfabrication.

28.4.2 Superhydrophobic PTFE

Since the discovery of the self-cleaning mechanism of the lotus leaf surface, surface superhydrophobicity has experienced extensive explorations in a wide range of fundamental researches as well as translational investigations (Barthlott and Ehler 1977; Callies and Quere 2005; Nosonovsky 2007). Further explorations have revealed the wide presence of remarkable superhydrophobic surfaces in nature for various reasons, including self-cleaning lotus leaves and duck feathers, the nonwetting legs of water striders, the adhesive nanofibrous setae on gecko feet, and water condensation micropatterns on desert beetles (Barthlott and Neinhuis 1997; Neinhuis and Barthlott 1997; Parker and Lawrence 2001; Gao and Jiang 2004). In theory, surface superhydrophobicity is primarily contributed by two underlying mechanisms: the chemical inertness of the solid material and the physical roughness of the surface (Han et al. 2007). The PTFE material typically has an extremely low surface energy of 18.6 mJ/m^2 due to its chemical structure, and exhibits the highest degree of hydrophobicity as a flat surface with a contact angle of 120° (Zhang et al. 2004). To further improve the hydrophobicity, nanoscopic roughness is a natural choice to be incorporated, as the theory predicts (Cassie and Baxter 1945; Bormashenko et al. 2007). PTFE nanoparticles are commercially manufactured by emulsion polymerization, in this case, embedding the nanomorphology and ultra low (Lopez et al. 1993; Drelich et al. 1994; Tadanaga et al. 2000; Gu et al. 2002; Sun et al. 2003; Kim et al. 2004a,b; Mornet et al. 2004; Seemann et al. 2005; Ma and Hill 2006; Garrod et al. 2007; Yang et al. 2007; Zhang et al. 2007a,b; Piret et al. 2008; Voronov et al. 2008; Bhushan 2009; Song et al. 2009) interfacial energy of PTFE nanoparticles into the polymeric matrix will be established with nanoscopic surface roughness and an ultrahigh contact angle of water (>150°), which is considered as the superhydrophobic effect.

28.4.3 Photopatternable PTFE

Well-defined micropatterns with desired superhydrophobicity offer great extension to rapid-evolving micro-nanoengineering applications (Lopez et al. 1993; Drelich et al. 1994; Tadanaga et al. 2000; Gu et al. 2002; Sun et al. 2003; Kim et al. 2004a,b; Mornet et al. 2004; Seemann et al. 2005; Ma and Hill 2006; Garrod et al. 2007; Yang et al. 2007; Zhang et al. 2007a,b; Piret et al. 2008; Voronov et al. 2008; Bhushan 2009; Song et al. 2009). Therefore, a simple microfabrication method to establish superhydrophobic micropatterns on universal surfaces would enable novel functionalities and applications of the extraordinary superhydrophobic phenomena. Due to the nonreactive nature of PTFE presents a technical hurdle for micromachining, polymer-based nanocomposite materials offer such a powerful hybrid solution to the challenge, where a photosensitive polymer matrix is introduced with the necessary photochemistry for direct micropatternability (Cong et al. 2007; Lee et al. 2007; Ahn et al. 2009). Incorporating the PTFE nanoparticles, known as photosensitive nanocomposite materials, into a photosensitive polymer matrix could result in unique combinational properties from both the polymer matrix and the nanofiller (for instance, photopatternability and optical transparency of the polymer matrix, and physical roughness and chemical inertness of the nanofillers). The nanocomposite consists of commercially available PTFE nanoparticles to provide nanomorphology and chemical inertness, and SU-8 photoresist as the photosensitive matrix. In superhyodrophobic nanocomposite, the PTFE nanoparticles are directly mixed into the SU-8 photoresist matrix to form a uniform nanocomposite.

28.4.4 Fabrication of Photopatternable Superhydrophobic PTFE

The nanocomposite formula comprises 20 wt% of PTFE nanoparticles (250 nm in diameter) added into SU-8 2050 photoresist and mixed through agitation until a uniform mixture is formed. The mixture is then spin-coated onto a transparent substrate (e.g., glass or polymers) followed by a soft bake at 95°C for 7 min. Subsequently, backside UV exposure through a photomask is processed, followed by a postexposure bake at 95°C for 7 min. After development in a regular SU-8 developer, the unexposed nanocomposite surface is dissolved, leading to cross-linked PTFE-SU-8 nanocomposite patterns.

28.4.5 Properties of Photopatternable Superhydrophobic PTFE

Figure 28.10a illustrates the SEM image of the PTFE-SU-8 nanocomposite. As can be seen, the nanocomposite surface exhibits nanoscopic roughness with severe aggregations of PTFE nanoparticles in the SU-8 matrix. As shown in Figure 28.10b, the nanoroughness and chemical inertness of the PTFE-SU-8 nanocomposite lead to the contact angle of water at 150°.

The SU-8 polymeric matrix offers the direct photopatternability of the superhydrophobic nanocomposite. During the development, the unexposed PTFE-SU-8 composite coating is dissolved, which leads to the desired superhydrophobic micropatterns on the substrate. Figure 28.11 shows the SEM images of

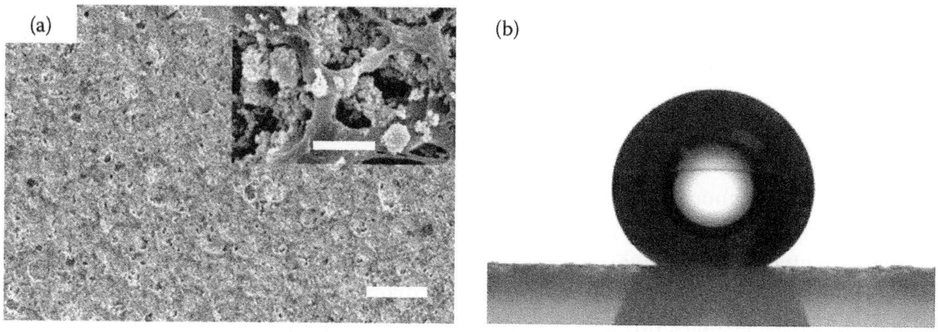

FIGURE 28.10 (a) SEM morphology and (b) contact angle of the PTFE-SU-8 nanocomposite surface. (Scale bar in (a): 50 μm, in inset: 4 μm.)

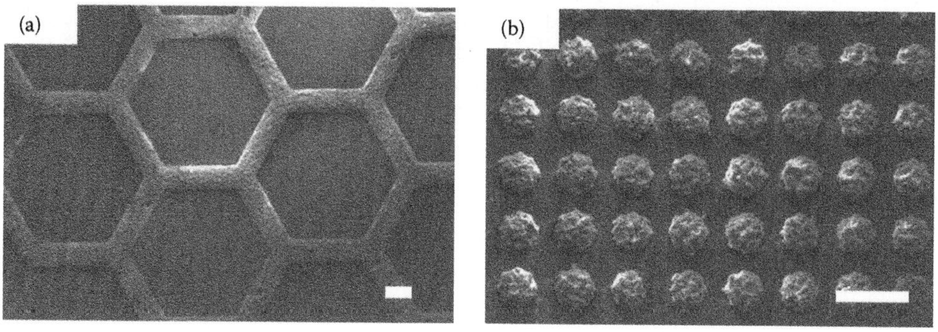

FIGURE 28.11 Photolithography resolution of the PTFE-SU-8 nanocomposite: (a) 100 μm and (b) 50 μm features. (Scale bars: 100 μm.)

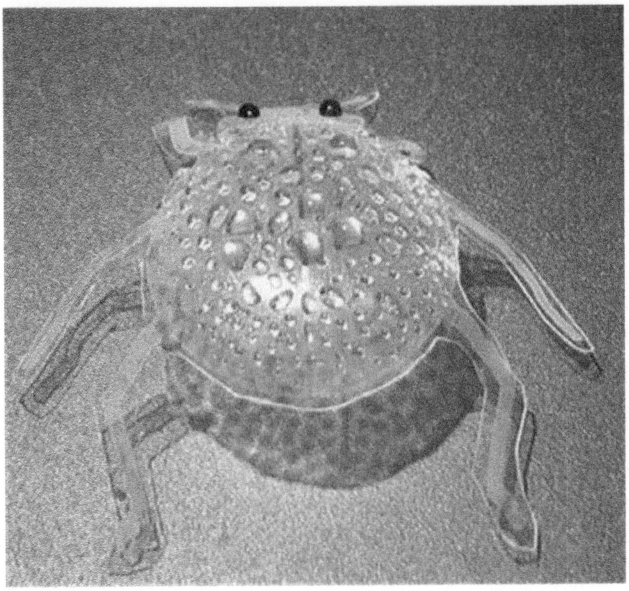

FIGURE 28.12 Demonstration of an artificial desert beetle with a moisture-capture hunchback using superhydrophobic nanocomposite photopatterns.

lithographic superhydrophobic patterns of the PTFE-SU-8 nanocomposite. The resolution of the photosensitive superhydrophobic nanocomposite is limited to 50 μm due to severe scattering from embedded PTFE nanoparticles in the SU-8 matrix.

28.4.6 Application in Bionics

The novel PTFE-SU-8 nanocomposite surface provides a unique combination of superhydrophobicity and photopatternability along with excellent adaptability and simple processability. Recent discovery reveals an intriguing nature-engineered design in the back-shell structure of the Namib Desert beetle, which comprises hydrophilic patches (~0.5 mm in diameter) surrounded by a superhydrophobic waxy shell surface. Amazingly, this design exhibits highly efficient water-capture ability and minimal evaporative exposure in humid ocean winds, of particular importance for survival under the harsh dry desert weather (Zhai et al. 2006). To mimic this remarkable moisture-capture design, the superhydrophobic nanocomposite is photopatterned onto an oxygen plasma-treated hydrophilic polyethyleneterephthalate (PET) substrate, followed by a thermal compression molding at 230°C to create the curved hunchback in a metallic mold. As displayed in Figure 28.12, an artificial desert beetle with a superhydrophobic nanocomposite-patterned hunchback is engineered successfully, which possesses the moisture-capture ability under a humid air flow.

28.5 Future Perspectives

This chapter describes several exciting nanoengineered implementations of multifunctional nanobiomaterials, which sketches the rapid development in such an active area of research. As a major advantage, the photopatternability is incorporated into the multifunctional nanobiomaterials matrix, which offers high-resolution micropatterning in single-step lithographic processing. A novel multifunctional performance of nanobiomaterials has been demonstrated, including high electrical/thermal conductivities, antibacterial properties, superhydrophobicity, and nonfouling property, all leading to exciting possibilities in emerging biomedical and chemical applications.

References

Ahn, B. Y., E. B. Duoss, M. J. Motala, X. Guo, S. I. Park, Y. Xiong, J. Yoon, R. G. Nuzzo, J. A. Rogers, and J. A. Lewis. 2009. Omnidirectional printing of flexible, stretchable, and spanning silver microelectrodes. *Science* 323 (5921):1590–1593.

Aizenberg, J., A. J. Black, and G. M. Whitesides. 1999. Control of crystal nucleation by patterned self-assembled monolayers. *Nature* 398 (6727):495–498.

Albrecht, D. R., V. L. Tsang, L. Sah, and S. N. Bhatia. 2005. Photo- and electropatterning of hydrogel-encapsulated living cell arrays. *Lab on a Chip* 5:111–118.

Allard, M., E. H. Sargent, P. C. Lewis, and E. Kumacheva. 2004. Colloidal crystals grown on patterned surfaces. *Advanced Materials* 16 (15):1360–1364.

Almasri, M., W. Zhang, A. Kine, Y. Chan, J. C. LaRue, and R. Nelson. 2005. Tunable infrared filter based on elastic polymer springs. *Proceedings of SPIE* 5770:190–198.

Barthlott, W. and N. Ehler. 1977. Raster-elektronenmikroskopie der epidermis-oberflächen von spermatophyten. *Tropische und subtropische Pflanzenwelt* 19:110.

Barthlott, W. and C. Neinhuis. 1997. Purity of the sacred lotus, or escape from contamination in biological surfaces. *Planta* 202 (1):1–8.

Bhagat, A. A. S., P. Jothimuthu, and I. Papautsky. 2007. Photodefinable polydimethylsiloxane (PDMS) for rapid lab-on-a-chip prototyping. *Lab on a Chip* 7 (9):1192–1197.

Bhushan, B. 2009. Biomimetics: Lessons from nature—An overview. *Philosophical Transactions of the Royal Society A: Mathematical Physical and Engineering Sciences* 367 (1893):1445–1486.

Biller, R. 1985. Ac Conductivity for the two-dimensional bond percolation problem. *Journal of Physics A: Mathematical and General* 18 (6):989–993.

Bormashenko, E., R. Pogreb, G. Whyman, and M. Erlich. 2007. Resonance Cassie-Wenzel wetting transition for horizontally vibrated drops deposited on a rough surface. *Langmuir* 23 (24):12217–12221.

Bremmell, K. E., L. Britcher, and H. J. Griesser. 2005. Probing the non-fouling behaviour of PEG and sulfonated PEG surfaces: An electrostatic interaction superimposed on the steric repulsion. *European Cells and Materials* 10:BS5.

Callies, M. and D. Quere. 2005. On water repellency. *Soft Matter* 1 (1):55–61.

Cassie, A. B. D. and S. Baxter. 1944. Wettability of porous surfaces. *Transactions of the Faraday Society* 40:0546–0550.

Cassie, A. B. D. and S. Baxter. 1945. Large contact angles of plant and animal surfaces. *Nature* 155 (3923):21–22.

Chou, S. Y., C. Keimel, and J. Gu. 2002. Ultrafast and direct imprint of nanostructures in silicon. *Nature* 417 (6891):835–837.

Cong, H. L. and W. X. Cao. 2005a. Array patterns of binary colloidal crystals. *Journal of Physical Chemistry B* 109 (5):1695–1698.

Cong, H. L. and W. X. Cao. 2005b. Two-dimensionally ordered copper grid patterns prepared via electroless deposition using a colloidal-crystal film as the template. *Advanced Functional Materials* 15 (11):1821–1824.

Cong, H. L. and T. R. Pan. 2008. Photopatternable conductive PDMS materials for microfabrication. *Advanced Functional Materials* 18 (13):1912–1921.

Cong, H. and T. Pan. 2009. Microfabrication of conductive PDMS on flexible substrates for biomedical applications. *IEEE NEMS*, pp. 731–734.

Cong, H., A. Revzin, and T. Pan. 2008. Self assembly of highly ordered nano-colloid array on patterned PEG hydrogel. *Proceedings of μTAS Conference*, pp. 1534–1536.

Cong, H., A. Revzin, and T. Pan. 2009a. Micropattern-assisted nanoassembly: Ordered nanocolloidal array on PEG microstructures. *IEEE NEMS*, pp. 735–738.

Cong, H. L., A. Revzin, and T. R. Pan. 2009b. Non-adhesive PEG hydrogel nanostructures for self-assembly of highly ordered colloids. *Nanotechnology* 20 (7):75307.

Cong, H. L., J. M. Zhang, M. Radosz, and Y. Q. Shen. 2007. Carbon nanotube composite membranes of brominated poly(2,6-diphenyl-1,4-phenylene oxide) for gas separation. *Journal of Membrane Science* 294 (1–2):178–185.

Corning, D. 2005. *Information About Dow Corning Brand Low-Stress Patternable Silicone Materials.*

Cui, Y., Q. Q. Wei, H. K. Park, and C. M. Lieber. 2001. Nanowire nanosensors for highly sensitive and selective detection of biological and chemical species. *Science* 293 (5533):1289–1292.

De Rosa, C., C. Park, E. L. Thomas, and B. Lotz. 2000. Microdomain patterns from directional eutectic solidification and epitaxy. *Nature* 405 (6785):433–437.

De Silva, M. N., R. Desai, and D. J. Odde. 2004. Micro-patterning of animal cells on PDMS substrates in the presence of serum without use of adhesion inhibitors. *Biomedical Microdevices* 6 (3):219–222.

Demers, L. M. and C. A. Mirkin. 2001. Combinatorial templates generated by dip-pen nanolithography for the formation of two-dimensional particle arrays. *Angewandte Chemie—International Edition* 40 (16):3069–3071.

Dovzhenko, A. Y. and P. V. Zhirkov. 1995. The effect of particle-size distribution on the formation of percolation clusters. *Physics Letters A* 204 (3–4):247–250.

Dowling, D. P., K. Donnelly, M. L. McConnell, R. Eloy, and M. P. Arnaud. 2001. Deposition of anti-bacterial silver coatings on polymeric substrates. *Thin Solid Films* 398:602–606.

Drelich, J., J. D. Miller, A. Kumar, and G. M. Whitesides. 1994. Wetting characteristics of liquid-drops at heterogeneous surfaces. *Colloids and Surfaces A: Physicochemical and Engineering Aspects* 93:1–13.

Duan, Y. R., Y. Zhang, T. Gong, and Z. R. Zhang. 2007. Synthesis and characterization of MeO-PEG-PLGA-PEG-OMe copolymers as drug carriers and their degradation behavior in vitro. *Journal of Materials Science: Materials in Medicine* 18 (10):2067–2073.

Ellis, D. A., S. A. Mabury, J. W. Martin, and D. C. G. Muir. 2001. Thermolysis of fluoropolymers as a potential source of halogenated organic acids in the environment. *Nature* 412 (6844):321–324.

Engel, J., J. Chen, N. Chen, S. Pandya, and C. Liu. 2006. Muti-walled carbon nanotube filled conductive elastomers: Materials and appilation to micro transducers. *IEEE MEMS*, pp. 246–249.

Fagaly, R. L. 2006. Superconducting quantum interference device instruments and applications. *Review of Scientific Instruments* 77 (10):101101–101145.

Fung, L. C. T., A. E. Khoury, S. I. Vas, C. Smith, D. G. Oreopoulos, and M. W. Mittelman. 1996. Biocompatibility of silver-coated peritoneal dialysis catheter in a porcine model. *Peritoneal Dialysis International* 16 (4):398–405.

Gabriel, M. M., A. D. Sawant, R. B. Simmons, and D. G. Ahearn. 1995. Effects of silver on adherence of bacterial to urinary catheters: In vitro studies. *Current Microbiology* 30:17–33.

Gao, X. F. and L. Jiang. 2004. Water-repellent legs of water striders. *Nature* 432 (7013):36–36.

Gardner, G., B. Harkness, E. Ohare, H. Meynen, M. Bulcke, M. Gonzalez, and E. Beyne. 2004. Integration of a low stress photopatternable silicone into a wafer level package. *Proceedings of the 54th Electronic Components and Technology Conference*, p. 2031.

Garrod, R. P., L. G. Harris, W. C. E. Schofield, J. McGettrick, L. J. Ward, D. O. H. Teare, and J. P. S. Badyal. 2007. Mimicking a stenocara beetle's back for microcondensation using plasmachemical patterned superhydrophobic-superhydrophilic surfaces. *Langmuir* 23 (2):689–693.

Gawron, A. J., R. S. Martin, and S. M. Lunte. 2001. Fabrication and evaluation of a carbon-based dual-electrode detector for poly(dimethylsiloxane) electrophoresis chips. *Electrophoresis* 22 (2):242–248.

Ghandhi, S. K. 1994. *VLSI Fabrication Principle.* New York: John Wiley & Sons.

Gu, Z. Z., A. Fujishima, and O. Sato. 2002. Patterning of a colloidal crystal film on a modified hydrophilic and hydrophobic surface. *Angewandte Chemie—International Edition* 41 (12):2068–2070.

Gudipati, C. S., J. A. Finlay, J. A. Callow, M. E. Callow, and K. L. Wooley. 2005. The antifouling and fouling-release perfomance of hyperbranched fluoropolymer (HBFP)-poly(ethylene glycol) (PEG) composite coatings evaluated by adsorption of biomacromolecules and the green fouling alga Ulva. *Langmuir* 21 (7):3044–3053.

Han, T. Y., J. F. Shr, C. F. Wu, and C. T. Hsieh. 2007. A modified Wenzel model for hydrophobic behavior of nanostructured surfaces. *Thin Solid Films* 515 (11):4666–4669.

Harkness, B., G. Gardner, J. Alger, M. Cummings, J. Princing, Y. Lee, H. Meynen et al. 2004. Photopatternable silicone compositions for electronic packaging applications. *Proceedings of SPIE* 5376:517–524.

Haubert, K., T. Drier, and D. Beebe. 2006. PDMS bonding by means of a portable, low-cost corona system. *Lab on a Chip* 6 (12):1548–1549.

Horn, M. 2004. Silver: The secret weapon against bacterial. *Fraunhofer Magazine* 48–49.

Huh, D., W. Gu, Y. Kamotani, J. B. Grotberg, and S. Takayama. 2005. Microfluidics for flow cytometric analysis of cells and particles. *Physiological Measurement* 26 (3):R73–R98.

Jacobs, H. O. and G. M. Whitesides. 2001. Submicrometer patterning of charge in thin-film electrets. *Science* 291 (5509):1763–1766.

Jiguet, S., A. Bertsch, H. Hofmann, and P. Renaud. 2005. Conductive SU8 photoresist for microfabrication. *Advanced Functional Materials* 15 (9):1511–1516.

Jo, B. H., L. M. Van Lerberghe, K. M. Motsegood, and D. J. Beebe. 2000. Three-dimensional micro-channel fabrication in polydimethylsiloxane (PDMS) elastomer. *Journal of Microelectromechanical Systems* 9 (1):76–81.

Kim, P., H. E. Jeong, A. Khademhosseini, and K. Y. Suh. 2006. Fabrication of non-biofouling polyethylene glycol micro- and nanochannels by ultraviolet-assisted irreversible sealing. *Lab on a Chip* 6 (11):1432–1437.

Kim, P., D. H. Kim, B. Kim, S. K. Choi, S. H. Lee, A. Khademhosseini, R. Langer, and K. Y. Suh. 2005. Fabrication of nanostructures of polyethylene glycol for applications to protein adsorption and cell adhesion. *Nanotechnology* 16 (10):2420–2426.

Kim, H. C., C. R. Kreller, K. A. Tran, V. Sisodiya, S. Angelos, G. Wallraff, S. Swanson, and R. D. Miller. 2004a. Nanoporous thin films with hydrophilicity-contrasted patterns. *Chemistry of Materials* 16 (22):4267–4272.

Kim, H. C., G. Wallraff, C. R. Kreller, S. Angelos, V. Y. Lee, W. Volksen, and R. D. Miller. 2004b. Photopatterned nanoporous media. *Nano Letters* 4:1169–1174.

Kirkpatr, S. 1973. Percolation and conduction. *Reviews of Modern Physics* 45 (4):574–588.

Kumacheva, E., R. K. Golding, M. Allard, and E. H. Sargent. 2002. Colloid crystal growth on mesoscopically patterned surfaces: Effect of confinement. *Advanced Materials* 14 (3):221–224.

Lee, J. Y., D. P. Lim, and D. S. Lim. 2007. Tribological behavior of PTFE nanocomposite films reinforced with carbon nanoparticles. *Composites Part B: Engineering* 38 (7–8):810–816.

Lee, K. J., K. A. Tosser, and R. G. Nuzzo. 2005. Fabrication of stable metallic patterns embedded in poly(dimethylsiloxane) and model applications in non-planar electronic and lab-on-a-chip device patterning. *Advanced Functional Materials* 15 (4):557–566.

Lesho, M. J. and N. F. Sheppard. 1996. Adhesion of polymer films to oxidized silicon and its effect on performance of a conductometric pH sensor. *Sensors and Actuators B: Chemical* 37 (1–2):61–66.

Liu, C. 2007a. Nanocomposite conductive elastomer: Microfabrication processes and applications in soft-matter MEMS sensors. *Materials Research Society Symposium Proceedings* 0947:A07-01.

Liu, C. 2007b. Recent developments in polymer MEMS. *Advanced Materials* 19 (22):3783–3790.

Loo, Y. L., R. L. Willett, K. W. Baldwin, and J. A. Rogers. 2002. Interfacial chemistries for nanoscale transfer printing. *Journal of the American Chemical Society* 124 (26):7654–7655.

Lopez, G. P., H. A. Biebuyck, C. D. Frisbie, and G. M. Whitesides. 1993. Imaging of features on surfaces by condensation figures. *Science* 260 (5108):647–649.

Lotters, J. C., W. Olthuis, P. H. Veltink, and P. Bergveld. 1997. The mechanical properties of the rubber elastic polymer polydimethylsiloxane for sensor applications. *Journal of Micromechanics and Microengineering* 7 (3):145–147.

Lounaci, M., P. Rigolet, G. V. Casquillas, H. W. Huang, and Y. Chen. 2006. Toward a comparative study of protein crystallization in microfluidic chambers using vapor diffusion and batch techniques. *Microelectronic Engineering* 83 (4–9):1673–1676.

Ma, M. and R. M. Hill. 2006. Superhydrophobic surfaces. *Current Opinion in Colloid and Interface Science* 11:193–202.

Mornet, S., S. Vasseur, F. Grasset, and E. Duguet. 2004. Magnetic nanoparticle design for medical diagnosis and therapy. *Journal of Materials Chemistry* 14 (14):2161–2175.

Muller, U., S. Jockusch, and H. J. Timpe. 1992. Photo-cross-linking of silicones. 6. Photo-cross-linking kinetics of silicone acrylates and methacrylates. *Journal of Polymer Science Part A: Polymer Chemistry* 30 (13):2755–2764.

Muller, U., H. J. Timpe, and J. Neuenfeld. 1991. Photo-cross-linking of silicones. 5. Photoinduced polymerization of silicone with pendant acrylate groups in the presence of oxygen. *European Polymer Journal* 27 (7):621–625.

Neinhuis, C. and W. Barthlott. 1997. Characterization and distribution of water-repellent, self-cleaning plant surfaces. *Annals of Botany* 79 (6):667–677.

Niu, X. Z., S. L. Peng, L. Y. Liu, W. J. Wen, and P. Sheng. 2007. Characterizing and patterning of PDMS-based conducting composites. *Advanced Materials* 19 (18):2682–2686.

Nosonovsky, M. 2007. Multiscale roughness and stability of superhydrophobic biomimetic interfaces. *Langmuir* 23:3157–3161.

Ortegavinuesa, J. L., D. Bastosgonzalez, and R. Hidalgoalvarez. 1995. Comparative-studies on physically adsorbed and chemically bound IgG to carboxylated latexes. 2. *Journal of Colloid and Interface Science* 176 (1):240–247.

Parker, A. R. and C. R. Lawrence. 2001. Water capture by a desert beetle. *Nature* 414 (6859):33–34.

Peppas, N. A., P. Bures, W. Leobandung, and H. Ichikawa. 2000. Hydrogels in pharmaceutical formulations. *European Journal of Pharmaceutics and Biopharmaceutics* 50 (1):27–46.

Piret, G., Y. Coffinier, C. Roux, O. Melnyk, and R. Boukherroub. 2008. Biomolecule and nanoparticle transfer on patterned and heterogeneously wetted superhydrophobic silicon nanowire surfaces. *Langmuir* 24 (5):1670–1672.

Plueddemann, E. P. 1982. *Silane Coupling Agents*. New York: Plenum Press.

Pouliquen, L. and X. Coqueret. 1996. Polysiloxanes with benzophenone side groups: Factors affecting their efficiency as free radical polymerization photoinitiators. *Macromolecular Chemistry and Physics* 197 (12):4045–4060.

Pregibon, D. C., M. Toner, and P. S. Doyle. 2006. Magnetically and biologically active bead-patterned hydrogels. *Langmuir* 22 (11):5122–5128.

Revzin, A., R. J. Russell, V. K. Yadavalli, W. G. Koh, C. Deister, D. D. Hile, M. B. Mellott, and M. V. Pishko. 2001. Fabrication of poly(ethylene glycol) hydrogel microstructures using photolithography. *Langmuir* 17 (18):5440–5447.

Revzin, A., R. G. Tompkins, and M. Toner. 2003. Surface engineering with poly(ethylene glycol) photolithography to create high-density cell arrays on glass. *Langmuir* 19 (23):9855–9862.

Rwei, S. P., F. H. Ku, and K. C. Cheng. 2002. Dispersion of carbon black in a continuous phase: Electrical, rheological, and morphological studies. *Colloid and Polymer Science* 280 (12):1110–1115.

Seemann, R., M. Brinkmann, E. J. Kramer, F. F. Lange, and R. Lipowsky. 2005. Wetting morphologies at microstructured surfaces. *Proceedings of the National Academy of Sciences of the United States of America* 102 (6):1848–1852.

Shi, H. Q., W. B. Tsai, M. D. Garrison, S. Ferrari, and B. D. Ratner. 1999. Template-imprinted nanostructured surfaces for protein recognition. *Nature* 398 (6728):593–597.

Sia, S. K. and G. M. Whitesides. 2003. Microfluidic devices fabricated in poly(dimethylsiloxane) for biological studies. *Electrophoresis* 24 (21):3563–3576.

Song, W. L., D. D. Veiga, C. A. Custodio, and J. F. Mano. 2009. Bioinspired degradable substrates with extreme wettability properties. *Advanced Materials* 21 (18):1830–1834.

Stieglitz, T. 2001. Implantable microsystems for monitoring and neural rehabilitation, part I. *Medical Device Technology* 12 (10):16–18, 20–21.

Subbaroyan, J., D. C. Martin, and D. R. Kipke. 2005. A finite-element model of the mechanical effects of implantable microelectrodes in the cerebral cortex. *Journal of Neural Engineering* 2 (4):103–113.

Sun, T., G. J. Wang, H. Liu, L. Feng, L. Jiang, and D. B. Zhu. 2003. Control over the wettability of an aligned carbon nanotube film. *Journal of the American Chemical Society* 125 (49):14996–14997.

Tadanaga, K., J. Morinaga, A. Matsuda, and T. Minami. 2000. Superhydrophobic-superhydrophilic micropatterning on flowerlike alumina coating film by the sol-gel method. *Chemistry of Materials* 12 (3):590–592.

Tsougeni, K., A. Tserepi, and E. Gogolides. 2007. Photosensitive poly(dimethylsiloxane) material for microfluidic applications. *Microelectronic Engineering* 84:1104–1108.

Underhill, G. H., A. A. Chen, D. R. Albrecht, and S. N. Bhatia. 2007. Assessment of hepatocellular function within PEG hydrogels. *Biomaterials* 28 (2):256–270.

Unger, M. A., H. P. Chou, T. Thorsen, A. Scherer, and S. R. Quake. 2000. Monolithic microfabricated valves and pumps by multilayer soft lithography. *Science* 288 (5463):113–116.

Vandenberg, A., P. Bergveld, D. N. Reinhoudt, and E. J. R. Sudholter. 1985. Sensitivity control of isfets by chemical surface modification. *Sensors and Actuators* 8 (2):129–148.

Veinot, J. G. C., H. Yan, S. M. Smith, J. Cui, Q. L. Huang, and T. J. Marks. 2002. Fabrication and properties of organic light-emitting "nanodiode" arrays. *Nano Letters* 2 (4):333–335.

Voronov, R. S., D. V. Papavassiliou, and L. L. Lee. 2008. Review of fluid slip over superhydrophobic surfaces and its dependence on the contact angle. *Industrial & Engineering Chemistry Research* 47 (8):2455–2477.

Wang, P., K. L. Tan, E. T. Kang, and K. G. Neoh. 2001. Synthesis, characterization and anti-fouling properties of poly(ethylene glycol) grafted poly(vinylidene fluoride) copolymer membranes. *Journal of Materials Chemistry* 11 (3):783–789.

Wang, H., J. Wang, J. Hong, Q. Wei, W. Gao, and Z. Zhu. 2007. Preparation and characterization of silver nanocomposite textile. *Journal of Coatings Technology and Research* 4:101–106.

Ward, J. H., R. Bashir, and N. A. Peppas. 2001. Micropatterning of biomedical polymer surfaces by novel UV polymerization techniques. *Journal of Biomedical Materials Research* 56:351–360.

Wenzel, R. N. 1949. Surface roughness and contact angle. *Journal of Physical and Colloid Chemistry* 53 (9):1466–1467.

Whitesides, G. M. and B. Grzybowski. 2002. Self-assembly at all scales. *Science* 295 (5564):2418–2421.

Winkleman, A., B. D. Gates, L. S. McCarty, and G. M. Whitesides. 2005. Directed self-assembly of spherical particles on patterned electrodes by an applied electric field. *Advanced Materials* 17 (12):1507–1511.

Wong, V. T. S., A. Huang, and C. M. Ho. 2005. SU-8 lift-off patterned silicone chemical vapor sensor arrays. *IEEE MEMS*, pp. 754–757.

Yang, L. L., S. Bai, D. S. Zhu, Z. H. Yang, M. F. Zhang, Z. F. Zhang, E. Q. Chen, and W. Cao. 2007. Superhydrophobic patterned film fabricated from DNA assembly and Ag deposition. *Journal of Physical Chemistry C* 111 (1):431–434.

Zakhidov, A. A., R. H. Baughman, Z. Iqbal, C. X. Cui, I. Khayrullin, S. O. Dantas, I. Marti, and V. G. Ralchenko. 1998. Carbon structures with three-dimensional periodicity at optical wavelengths. *Science* 282 (5390):897–901.

Zallen, R., ed. 1983. *The Physics of Amorphous Solids*. New York: Wiley.

Zhai, L., M. C. Berg, F. C. Cebeci, Y. Kim, J. M. Milwid, M. F. Rubner, and R. E. Cohen. 2006. Patterned superhydrophobic surfaces: Toward a synthetic mimic of the Namib Desert beetle. *Nano Letters* 6 (6):1213–1217.

Zhang, X., H. Kono, Z. Liu, S. Nishimoto, D. A. Tryk, T. Murakami, H. Sakai, M. Abe, and A. Fujishima. 2007a. A transparent and photo-patternable superhydrophobic film. *Chemical Communications* (46):4949–4951.

Zhang, H., Y. Y. Lee, K. J. Leck, N. Y. Kim, and J. Y. Ying. 2007b. Recyclable hydrophilic-hydrophobic micropatterns on glass for microarray applications. *Langmuir* 23 (9):4728–4731.

Zhang, J. L., J. A. Li, and Y. C. Han. 2004. Superhydrophobic PTFE surfaces by extension. *Macromolecular Rapid Communications* 25 (11):1105–1108.

Zhao, Q., Y. Liu, H. Muller-Steinhagen, and G. Liu. 2002. Graded Ni-P-PTFE coatings and their potential applications. *Surface & Coatings Technology* 155 (2–3):279–284.

Zheng, H. P., I. Lee, M. F. Rubner, and P. T. Hammond. 2002. Two component particle arrays on patterned polyelectrolyte multilayer templates. *Advanced Materials* 14 (8):569–572.

Zhong, Z. Y., B. Gates, Y. N. Xia, and D. Qin. 2000. Soft lithographic approach to the fabrication of highly ordered 2D arrays of magnetic nanoparticles on the surfaces of silicon substrates. *Langmuir* 16 (26):10369–10375.

29

Nanobiomaterials for Preclinical Studies and Clinical Diagnostic

Youssef Zaim Wadghiri
New York University School of Medicine

Karen Briley-Saebo
Mount Sinai School of Medicine

Prior to diagnostic imaging, exploratory surgery was a method commonly used by doctors when trying to find a diagnosis for a disease. The use of new technologies such as MRI combined with innovations in chemistry have significantly reduced the need for exploratory surgeries where multiple incisions may be needed, leading to longer hospital stays and recovery time. The gain in anatomical details and sensitivity achieved in diagnostic imaging has enabled present-day physicians to identify accurately the source of a patient's illness in a harmless and faster manner. MRI has become a very important tool for clinical diagnosis of disease, as well as in biomedical studies for the noninvasive 3D imaging of living subjects and specimens. It can distinguish between various parts of soft tissue water, based on differences in the longitudinal relaxation time ($T1$), the transverse relaxation time ($T2$), the apparent transverse relaxation time ($T2^*$), and the water concentration or proton density.

Since the clinical introduction of MRI in the early 1980s, it became apparent that exogenous molecules were needed to either locally enhance the MR signal in tissue with low intrinsic contrast or to selectively identify various pathogenic processes. Similarity to its opaque contrast agent (CA) predecessors used in x-ray computed tomography (CT), the use of MRI CAs substantially improved the sensitivity and specificity of this new technique. Unlike the differential attenuation measured by CT, MRI CAs are not measured directly. Instead, it is the effect of the CAs on local water protons and their associated MR parameters that is measured. The MRI CA may influence the relaxation times (*T*1, *T*2, and/or *T*2*) of the water protons in their vicinity, thereby leading to *T*1- (or *T*2, *T*2*-) weighted image contrast. In this chapter, the imaging characteristics, the physicochemical properties and the biocompatibility of the most common metal ion are described with a particular emphasis on toxicity. Clinical and biomedical applications illustrate the use of MRI CAs either individually or through high payloads.

29.1 Paramagnetic Nanomaterials

29.1.1 Theory

While both organic and inorganic molecules can act as MRI CAs, agents containing paramagnetic metal atoms such as gadolinium(III) (Gd(III) (Mody and Sessler, 1999), manganese(II), and iron(III) (Koblinsky et al., 2000) are almost exclusively chosen due to their paramagnetic properties and high magnetic moments. Additional lanthanides such as dysprosium (Dy) and europium (Eu) have also been considered in living systems for limited yet specific purposes due to their electron configuration that results in unique contrast mechanisms. *Paramagnetism* is generally defined by the following two characteristics: (1) A positive magnetic susceptibility that is directly proportional to the external field and (2) in the absence of an external magnetic field, the individual magnetic moments that are randomly oriented so that the net resultant magnetization is zero. Therefore, paramagnetic materials have no remnant magnetization. Paramagnetic metal ions cause an increase in the observed MR signal intensity by enhancing the longitudinal relaxation rates ($R1 = 1/T1$) of bulk water protons. In aqueous solution, there are two types of interactions or proton–electron couplings between the paramagnetic center and the bulk water protons: Intramolecular and intermolecular. The processes that modulate these interactions are molecular rotation and proton diffusion for the intra- and intermolecular interactions, respectively. In pure water, the interactions are weak and the *R*1 values are low. In the presence of a paramagnetic ion, the relaxation rates are significantly increased due to contribution of both inner-sphere and outer-sphere relaxation mechanisms. Inner-sphere relaxation deals with the direct exchange of energy between the electrons in the hydration sphere of a paramagnetic ion and bulk water protons. This energy exchange is dominated by dipolar and scalar coupling between the electron and proton spins. Dipolar energy exchange is modulated by the rotation of the paramagnetic center (τ_r), the exchange rate of water molecules in and out of the hydration sphere (τ_m), and the electron relaxation rate (or the rate the electrons flip between the orbitals) associated with the paramagnetic ion (τ_{s1}). A correlation term, (τ_c), is used to define the modulation of the dipolar couplings, as described in Equation 29.1 (Koenig and Brown, 1984; Lauffer, 1987; Koenig, 1991):

$$\frac{1}{\tau_c} = \frac{1}{\tau_r} + \frac{1}{\tau_m} + \frac{1}{\tau_{s1}} \tag{29.1}$$

The fastest modulation (or greatest $1/\tau$ value) will dominate the exchange. Although the molecular rotation times and exchange times may be altered by changing the structure of the ligand, the electronic relaxation time is dependent upon the ligand and the paramagnetic ion used. For dysprosium, the symmetry of the electrons in the outer orbital causes the electronic relation time of this ion to be extremely fast so that the dipolar exchange is ineffective, thereby resulting in nominal *R*1 values. Inner-sphere relaxation accounts for more than 50% of the longitudinal relaxation effect observed with most paramagnetic metal ions (where $q > 1$). However, the observed inner-sphere contribution is also dependent upon the concentration (*c*) of paramagnetic metal ion present as well as the number of water protons in the hydration sphere that are able to exchange energy with the electrons (q). The relationship between c, q, and τc is shown in the following equation:

$$R1_{\text{inner-sphere}} = \frac{cq}{55.5} \frac{1}{(T_{1M} + \tau_M)} \tag{29.2}$$

$$\frac{1}{T_{1M}} = \frac{2 \cdot \gamma^2 \cdot g^2 \cdot S(S+1) \cdot \beta^2}{15 \cdot r^6} \cdot \left[\frac{3 \cdot \tau_{C1}}{1 + (\omega_0 \cdot \tau_{C1})^2} + \frac{7 \cdot \tau_{C2}}{1 + (\omega_S \cdot \tau_{C2})^2} \right] + \frac{2 \cdot S \cdot (S+1)}{3} \cdot \left(\frac{A}{\hbar} \right)^2 \cdot \left[\frac{\tau_{E2}}{1 + (\omega_S \cdot \tau_{E2})^2} \right]$$

where

$\Delta\omega$ is the shift in the Larmor frequency of the protons in the hydration sphere
γ is the gyromagnetic ratio of the proton
g is the Landè factor
S is the electronic spin quantum number
β is the Bohr magneton
$\hbar$ is the reduced Planck's constant
A is the distance of the closest approach between the protons and the paramagnetic metal ion
r is the radius of the molecule
ω_0 is the precessional frequency of the proton
ω_S is the precessional frequency of the electrons

When the paramagnetic ions are completely complexed or chelated, the number of water molecules in the hydration sphere decreases so that $q = 0$. For these materials the outer-sphere relaxation mechanism becomes important. Outer-sphere relaxation arises from the diffusion of the water protons in the local magnetic field gradients generated by the paramagnetic ions. The interaction between the proton spins and the magnetic moment is also a dipolar interaction. However, no scalar (or direct) coupling is possible since the proton and ion are never in contact. As a result, the modulation of this interaction is due to the diffusion of the water molecules τ_D and the electronic relaxation of the electronic spin of the ion, $\tau_{s1,}$ as shown in the following equation:

$$R1_{\text{outer-sphere}} \, \alpha \, K \, c \, \tau_D / d^3 \cdot [7 J_F(\omega_s) + 3 J_F(\omega_0)] \tag{29.3}$$

where

d is the distance of proton from the center of the paramagnetic ion
$3J_F$ is the Freed spectral density function that is dependent upon τ_D and τ_{s1}

Although paramagnetic metal ions may also enhance transverse magnetization decay (increases $R2 = 1/T2$ values), they are primarily used as $T1$ agents to generate MR signal enhancement. Conversely, r_2 effects may modulate the $R1$ values at high tissue concentrations (i.e., bolus phase) or at high field strengths, where the $R1$ values approach zero and $R2$ values approach a limit due to an additional scalar term associated with $R2$.

In nuclear magnetic resonance (NMR) spectroscopy, the term relaxation was initially associated with a time constant to describe several processes by which nuclear magnetization prepared in a non-equilibrium state returns to the equilibrium distribution. The rates of this spin relaxation measured in both spectroscopy and imaging applications are often referred to, most likely for historical reasons, as relaxation times. Yet, the longitudinal relaxation rate $R1$ is expected to be nearly additive with respect to various contributing relaxation mechanisms. Consequently, CA efficacy is often described by the longitudinal and transverse relaxivity values r_1 and r_2, respectively. The relaxivity reflects the change in the relaxation rates as a function of CA concentration. If a linear correlation is assumed (as suggested by the equations above), then the relaxivities are calculated as (Lauffer, 1987)

$$y = rc^n + b \tag{29.4}$$

where
- y is the relaxation rate of the sample containing the CA (s^{-1})
- c is the concentration of CA in the sample (mM per magnetic centre)
- r is the slope of the linear regression
- n is the factor of curvature associated with the fit
- b is the relaxation rate of the sample without the addition of CA (s^{-1})

Once calculated, the r value is the r_1 or r_2 relaxivity (s^{-1} mM^{-1}). Most relaxivity values are obtained in aqueous solution or in ex vivo tissue samples. The r_1 and r_2 values are determined using calibrated spectrometers that measure the signal from the magnetization in the time domain (pulse NMR spectrometers).

29.1.2 Toxicity, Pharmacokinetics, and Biodistribution

Paramagnetic lanthanide metal ions (Gd, Dy, Eu) are very toxic to biological systems. In vivo toxicity is an important question in the CA field in correlation with complex stability, where Gd3+ complex, free ligand, and free metal can each take part in many side reactions. For instance, interaction between endogenous metal ions with Gd3+ complexes following transmetallation process leads to the replacement of endogenous metal by Gd3+ in metabolic process (Cacheris et al., 1990; Laurent et al., 2001). The lethal dose (LD50) of $GdCl_3$ is less than 0.5 mmol kg^{-1} in rodent toxicity models. Additionally, studies have shown that if Gd is taken up by intracellular vesicles, the formation of Gd salts (primarily Gd(OH)3) is extremely toxic and induces significant apoptosis (LD50 of Gd(OH3) in mice is 0.1 mmol kg^{-1}) (Spencer et al., 1997, 1998; Ide et al., 2005). The toxicity associated with endogenous paramagnetic metals such as Mn is dependent upon the rate of injection. Bolus injection of $MnCl_2$ results in significant cardiotoxicty due to competition of Mn metal ions with normal calcium channels in cardiomyocytes in the bolus phase (Nordhoy et al., 2003; Bruvold et al., 2005). On the other hand, slow infusion of $MnCl_2$ limits cardiotoxicty effects since 99% of the Mn ions bind serum albumin and therefore are not able to compete with calcium. Administration of $FeCl_2$ is relatively safe in comparison with LD50 values of 1.6 mmol kg^{-1} observed. Nevertheless, the presence of intracellular free iron may promote peroxidase that can eventually induce cell apoptosis. This is assumed to be problematic in inflammatory tissue, where peroxidase of lipids already promotes cell death and formation of necrotic tissue.

Due to the toxicity issues associated with the lanthanides, these materials require chelation to form compounds that may be eliminated without bioretention or biotransformation. Although chelation improves the safety of these materials, it does reduce the overall efficacy by limiting the number of inner-sphere water exchange sites with most Gd chelates having $q = 1$. The stability of a chelated complex is often defined by both the thermodynamic stability (K_{ML}) and the selectivity (K_{sel}) constants. The thermodynamic stability constants are obtained in vitro at physiological pH and describe the affinity of a metal ion for a ligand (Cacheris et al., 1990). Hence, they are widely used to compare the behavior of different complexes. Laurent et al. (2006) used a simple model to study the physicochemical properties of clinical low molecular weight Gd-CAs under identical experimental conditions, in three different media (pure water, zinc(II)-containing aqueous solutions, and HSA-containing solutions). Since the affinity is greatly dependent upon the conditions by which it is measured (pH, matrix, etc.), these values serve only as a general guideline of chelate stability. Studies have reported log K_{ML} values of 22.2 and 25.6 for Gd-DTPA and Gd-DOTA, respectively. Mn exhibits values of 11.1 and 13.2 for similar chelates (at pH 7.4). The selectivity constants take into account the various possible in vivo transmetallation reactions. As a result, K_{sel} indicates the selectivity of the ligand to the metal ion over other endogenous cations such as zinc, copper, and calcium (Cacheris et al., 1990). Strong correlations between toxicity and K_{sel} values have been observed in preclinical models. It should be emphasized that the in vivo toxicity is also related to the exposure of the metal chelates to endogenous cations. Several models have been proposed in the literature to replicate metal-ion coordination equilibrium in blood plasma (May et al., 1977; Cacheris et al., 1990; Jackson et al., 1990). Since the rate of transmetallation is slower than the rate of elimination for most Gd chelates, in vivo transmetallation is low, thereby resulting in limited toxicity due to bioretention (Harpur et al., 1993; Eakins et al., 1995). Nevertheless, as observed in patients with renal insufficiency, prolonged circulation times increase the exposure of the metal chelates to circulating cations. In this case, the rate of transmetallation becomes relatively faster than the rate of elimination leading to a release of Gd from the chelate and consequently to toxicity as Nephrogenic Systemic Fibrosis (NSF) (Perazella, 2007; Chen et al., 2010). This issue may become problematic and must be considered when designing nanobiomaterials.

Nanobiomaterials are designed to target specific pathogenic processes. In order to achieve effective targeting, these materials must (1) utilize lipid-based nanoparticles to allow for high payload of diagnostic and/or therapeutic agent and (2) exhibit relatively long circulation times ($t_{1/2}$ > 10 min observed for Gd-DTPA) to enable adequate accumulation on or within the target tissue (Aime et al., 2002; Mulder et al., 2006; Briley-Saebo et al., 2007). A fraction of all lipid-based nanoparticles dose may be sequestered by organs associated with the mononuclear phagocytic system (MPS) and/or the reticular endothelial system (RES) where intercellular uptake and bioretention is likely to occur. Since intracellular uptake into vesicles and exposure to ATP promotes de-metallation, safety issues may limit translation of Gd-based materials (Vander Elst et al., 1994). Additionally, the long circulation times increase the potential of transmetallation in the vasculature. Preclinical studies have shown 1.5%–2.5% transmetallation of Gd from DTPA associated with lipid-based probes (Zhu et al., 2008). In order to allow for clinical translation of emerging nanobiomaterials, issues related to bioretention and transmetallation must be addressed. In order to solve these issues, stronger Gd chelates, other biocompatible paramagnetic metal ions, and/or design proposals to promote renal excretion may be considered.

For lipid-based materials the presence of polyethylene glycol (PEG) associated with particle surface has been shown to (1) limit interaction with endogenous macromolecules, (2) limit RES uptake, and (3) prolong circulations times (Torchilin, 2002). As a result, most nanobiomaterials utilize PEG in their formulation. In addition to PEG modification, the particle charge, size, and targeting moiety will greatly influence the pharmacokinetics and biodistribution. To allow for extended circulation times, it is important that the particles be slightly negatively charged. Studies have shown that positively charged particles are highly reactive and may even fuse with the vascular endothelium during injection. The size of the particle will strongly determine if the nanobiomaterials are renally excreted. Particles less than 10 nm may exhibit predominant renal excretion (Longmire et al., 2008; Bumb et al., 2010). However, as the size increases the amount of uptake into the RES/MPS also increases. For example, 65% of a 12 nm PEG-micelle dose is renally excreted, with the remaining micelles taken up by the RES (liver, spleen) and MPS (lymph, bone). The size of the particles will also determine if the particles are able to diffuse through the gap junctions (<20 nm) associated with non-normal endothelium. This may be critical if the target tissue is within the arterial wall (i.e., intraplaque macrophage/foam cell) or certain tumor cell lines with limited neovasculature. Large particles (>100 nm) are often quickly taken up by RES and may also be taken up by circulating monocytes (dependent upon the coating material and size). Larger particles may, however, be used for targeting extracellular surface receptors such as VEGF that are associated with tumor angiogenesis. In addition to size, the targeting moiety may also influence the pharmacokinetics and biodistribution. For example, a recent study showed that lipid-based micelles conjugated with antibodies targeted to oxidative epitopes significantly increased the blood half-life of the particles due to interactions with circulating oxidized lipoproteins (Briley-Saebo et al., 2008b). Particles have also been redirected from liver Kuppfer cells to liver hepatocytes by conjugation of Asialofetuin (ASF) to the particle surface (Schaffer et al., 1993). As a result, the charge, size, and targeting moiety must be carefully considered to allow for optimal uptake into the target tissue.

29.2 Gadolinium-Based Nanobiomaterials

The Gd(III) complex diethylenetriaminepentaacetic acid (DTPA) (gadopentetate dimeglumine or Magnevist™) is the most commonly used clinical agent to delineate diseased tissue based on the uptake rate or vascular leakage (Weinmann et al., 1984). Gd-DTPA and other linear Gd chelates have been used extensively for indications related to the central nervous system and myocardial perfusion (see clinical examples in Figure 29.1). For clarity and consistency, the CA based on chelated gadolinium will generally be referenced as either Gadolinium or Gd throughout this review. Although these materials are effective for the detection of lesions or tissue exhibiting vascular leakage or delayed elimination kinetics, they rely upon passive targeting strategies. The MR efficacy of the linear Gd chelates, as r_1, is

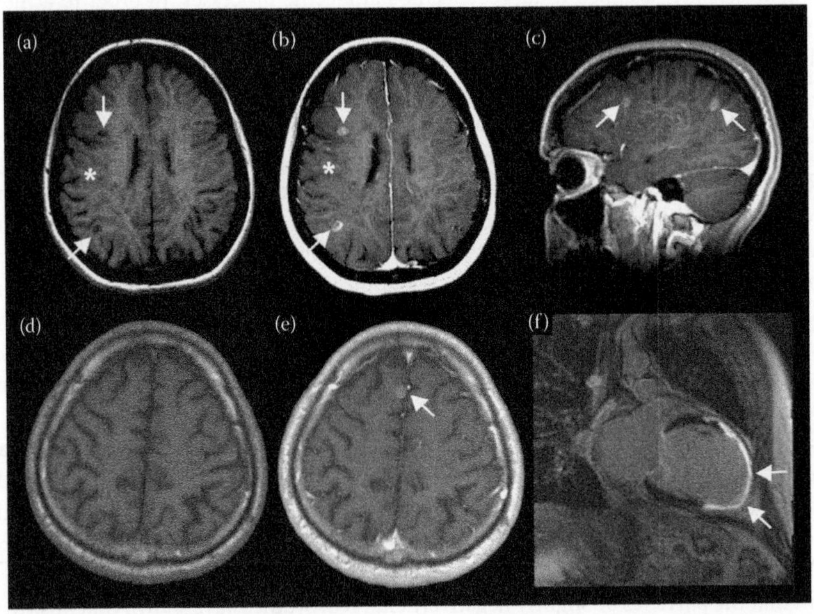

FIGURE 29.1 Clinical examples illustrating the utility of Gd chelates CA (Magnevist in this instance): In multiple sclerosis,*** (a) the appearance of abnormal dark patches (depicted by *) in the left hemisphere in *T*1-weighted axial orientation may not be specific and can be attributed to either inflammation/interstitial edema, demyelination, glial scar, or necrosis. (b) The intravenous introduction of Gd enables the ringlike enhancement pattern in a limited number of lesions (depicted by the white arrows) while keeping the rest unchanged; demonstrating that Gd can help characterize MS plaques that may be in different phases of evolution in the same patient simultaneously. (c) The Sagittal orientation from the same patient displays the previously observed bright spots; evidence of Gd leakage due to the blood–brain barrier (BBB) disruption in the MS plaque. (Courtesy of Dr. Matilde Inglese, MD, Radiology, NYU Langone Medical Center, New York). In the presence of melanoma brain metastasis (d) precontrast agent is compared with (e) postcontrast injection using a *T*1-weighted scan. A positive lesion enhancement is noted in the region of BBB breakdown in which the contrast leaks and marks the bulk of the unsuspected metastatic tumor. (Courtesy of Dr. John G Golfinos, MD, Neurosurgery, NYU Langone Medical Center, New York.) (f) Example of an MRI assessment of myocardial viability following Gd-DTPA injection in which nonviable or scarred myocardium exhibit hyperenhancement or late Gd enhancement (see white arrows). Compared to healthy myocardium, nonviable or scarred myocardium has more Gd 10–15 min postinjection due to the slower Gd kinetics. (Courtesy of Dr. Monvadi B. Srichai, MD, Radiology, NYU Langone Medical Center, New York.)

approximately 4 s^{-1} mM^{-1} at current clinical field strengths. Over the past two decades, MRI CAs have evolved towards increasing the relaxivity but also the specificity and function by rendering the resulting agent targeted to tissues and organs of interest (Brasch, 1992; Weinmann et al., 2003).

The vasculature was one of the first tissues to be targeted using Gd-based agents. Gd-DTPA bound albumin was produced in the late 1980s and was proven effective for both MR angiography and perfusion indications. Due to the direct binding with the protein and decreases in the molecular tumbling rates, r_1 values of more than 100 s^{-1} mM^{-1} were observed (Schmiedl et al., 1987). Nonetheless, due to toxicity issues related to Gd retention, the targeting strategy was modified (Hoffmann et al., 1999; Fink et al., 2007; Kramer and Morana, 2007) to allow for transient binding with endogenous serum proteins (primarily albumin) (Caravan et al., 2002; Cavagna et al., 2002), to utilize novel polymeric complexes (Schuhmann-Giampieri et al., 1991; Cavagna et al., 1994; Corot et al., 2000; Misselwitz et al., 2001), or through the use of dendrimers, obtained by complete synthetic cascade polymers (Wiener et al., 1994). Due to either the interaction with endogenous serum proteins or the nanoparticles size, these materials exhibit limited extravasation through the endothelium of normal tissue.

Leakage into the interstitial space is therefore slow enough to prevent the accumulation of detectable amounts of the CA before complete excretion from the body. In addition to delineation of the vasculature, these compounds have also been used to detect tumors (Daldrup et al., 1997; Kiessling et al., 2007) as well as to visualize myocardial necrosis (Muhler, 1995; Kroft and de Roos, 1999) and viability (Schmiedl et al., 1987; Kim et al., 1999, 2000; Fieno et al., 2000), as illustrated in the examples of Figure 29.1.

Lipid and lipoprotein Gd-based nanoparticles have also been designed to target a variety of pathogenic processes including tumor necrosis, intraplaque macrophages, oxidative epitopes, thrombosis, and tumor angiogenesis (Figure 29.2). The advantage of using micelles, liposomes, and lipoproteins as a scaffold relates to the ability to load a large number of Gd ions per nanoparticle, thereby increasing the Gd payload and imaging sensitivity. In order to maintain MR efficacy, the chelated Gd must be associated to the surface of the nanoparticles to enable adequate energy exchange between the Gd electrons and bulk water protons (Lauffer, 1987). Relaxivity values ranging from 9 to 30 s^{-1} mM^{-1} have been reported at clinical fields for a variety of micelles and liposomes (Fossheim et al., 1999; Anelli et al., 2001; Torchilin, 2005; Amirbekian et al., 2006; Briley-Saebo et al., 2006b, 2008a; Mulder et al., 2006; Frias et al., 2007; Cormode et al., 2009). For these platforms the targeting moiety (full monoclonal antibodies, antibody fragments, or peptides) may be conjugated either via noncovalent or covalent linkage. Noncovalent coupling is typically achieved using an avidin–biotin bridging. Avidin is a tetrameric protein with a molecular weight of 68 kDa and is capable of strongly binding four biotins (KA ≈ 1.7 × 1015 M^{-1}). Alternatively, streptavidin can also be used. Although this method is simple and effective, the use of avidin has certain limitations

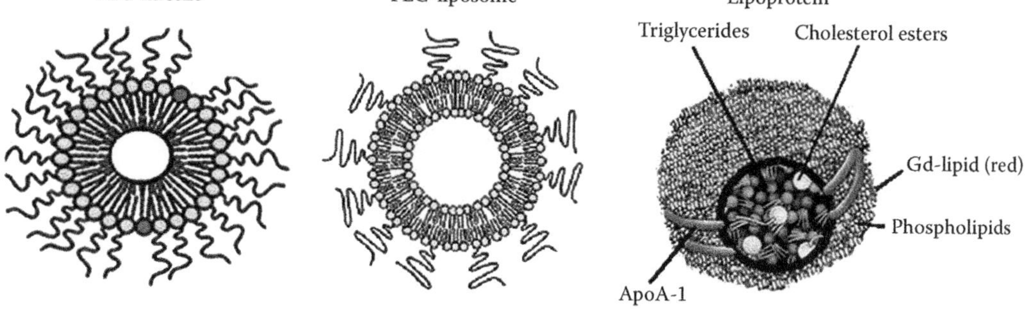

FIGURE 29.2 Common lipid-based nanoparticles for active targeting using paramagnetic labels. Micelles are 10–20 nm, liposomes >20 nm and lipoproteins 10–30 nm. To maintain MR efficacy paramagnetic labels such as Gd-DTPA are conjugated to the surface of the micelles/liposomes or integrated into the lipid layer associated with the lipoproteins. Iron oxide particles may also be integrated into the coronae of the micelles and liposomes.

TABLE 29.1 FDA Approved Gd Nanobiomaterials Current FDA Approved Gd Nanobiomaterials with as Actively Targeted Agent

Material	Trade Name	$R1$ s^{-1} mM^{-1} 20 MHz, Buffer	Properties
Gadofoveset trisodium	Vasovist	38–42	Linear ionic—binds albumin For aortic–iliac occlusive disease
GD-EoB-DTPA	Eovist	19.4	Linear ionic—binds albumin Liver indications

The clinical use to only transient binding Gd chelates with serum albumin reflect the cautions taken to prevent issues related to the bioretention and biotransformation of Gd.

that are related to the size of the conjugate, immunogenic properties, and fast liver clearance (Dafni et al., 2003). Covalent conjugation, on the other hand, leads typically to smaller conjugates without immune system stimulation. Although several methods have been described for covalent conjugation, the most commonly used methods form an amide bond between activated carboxyl groups and amino groups, use a disulfide bond or thioether bond between maleimide and thiol. Furthermore, amines can be activated for conjugation to thiol (SH) exposing ligands with *N*-succinimidyl 3-[2-pyridyldithio]propionate, a method that is referred to as SATA modification (Briley-Saebo et al., 2007).

The current FDA approved actively targeted Gd nanobiomaterials are shown in Table 29.1. As illustrated, only Gd chelates that exhibit transient binding with serum albumin have been approved for clinical use. The clinical translatability of the emerging Gd-based nanoparticles has yet to be determined and may be severely limited by issues related to the bioretention and biotransformation of Gd.

29.3 Dysprosium-Based Nanobiomaterials

Due to the electronic relation rate associated with Dy, these materials cannot be used as *T*1 agents. Instead, Dy may be used to induce significant *T*2/*T*2* effects in vivo. Dy-DTPA-BMA has been used as a *T*2/*T*2* perfusion agent to evaluate brain ischemia (Figure 29.3) (Kucharczyk et al., 1991; Moseley et al., 1991; Haraldseth et al., 1996) as well as in characterizing myocardial perfusion defects (Beache et al., 1998). Although used in a variety of clinical studies, this material has never been FDA approved. Dy is optimal as a perfusion agent and preferred over Gd agents or iron oxide particles due to the negligible *T*1 contribution to the signal observed. The *T*1-weighted signal observed from Gd agents and the *T*2/*T*2*-weighted signal observed from iron oxides typically has some degree of modulation by *T*2 or *T*1 effects, respectively. As a result, Dy-DOTA materials are currently being investigated as potential blood pool and perfusion agents (Ribot et al., 2008).

Focus has recently been placed on the development of targeted Dy loaded liposomes for in vivo detection of proliferating tumors expressing glutamine receptors (Castelli et al., 2009). Additionally, Dy-DOTA dendrimers have been investigated as a platform for active targeting.

29.4 Manganese-Based Nanobiomaterials

On a much smaller scale and readily accessible approach, paramagnetic manganese can be used efficiently in biomedical research to alter *T*1 tissue contrast. In fact, it was the first paramagnetic element suggested for enhancing imaging contrast in a phantom by Lauterbur in his landmark paper in 1973. It was subsequently tested successfully on differentiating tissue contrast from various organs by the same group (Mendonca-Dias et al., 1983). Surprisingly, only one chelated form, manganese dipyridoxyl diphosphate (Mn-DPDP, Teslascan, Amersham, Princeton, NJ) has this far found clinical and FDA approval (Federle et al., 2000). The clinical usefulness of Mn-DPDP (termed also mangafodipir trisodium) has been exclusively aimed for imaging the hepatobiliary system through vascular dissociation of Mn that binds to proteins and is taken

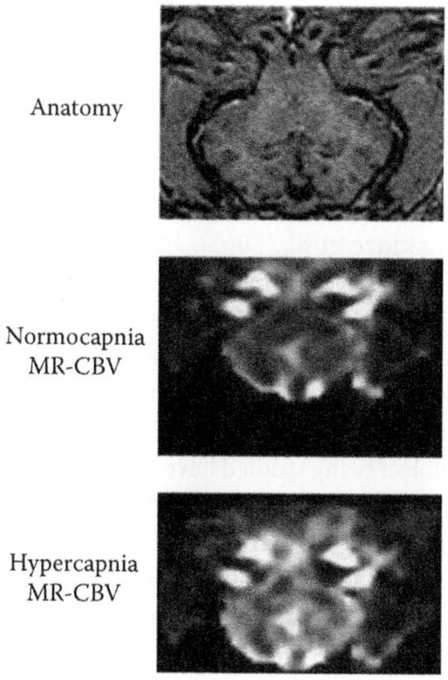

FIGURE 29.3 **(See color insert.)** Cerebral blood volume (CBV) maps obtained in patients after bolus administration of Dy-DTPA-BMA. The maps were obtained by plotting the relative CBV on a pixel by pixel basis within the brain. CBV values were calculated based upon the signal intensity versus time curves obtained during the bolus phase of the CA (technique know as dynamic contrast enhancement [DCE] imaging). Brighter colors (yellow/white) correlate to higher CBV values.

up by normal liver hepatocytes (Toft et al., 1997; Jung et al., 1998; Reimer et al., 2004). Unlike the lanthanides, Mn is endogenous, and bioretention and biotransformation do not limit clinical translatability. Yet, Mn must be chelated or infused slowly in order to limit any potential cardiotoxicty effect.

On the other hand, manganese Mn^{2+} acting as a calcium analogue has become a very popular and powerful CA for small animal MRI over this past decade. The associated approach termed as manganese enhanced MRI (MEMRI) (Koretsky and Silva, 2004) based on the confined presence of Mn^{2+} in specific tissues can provide highly localized contrast brightening changes either through passive or active redistribution (see Figure 29.4). Although little is known about the manganese uptake mechanisms and

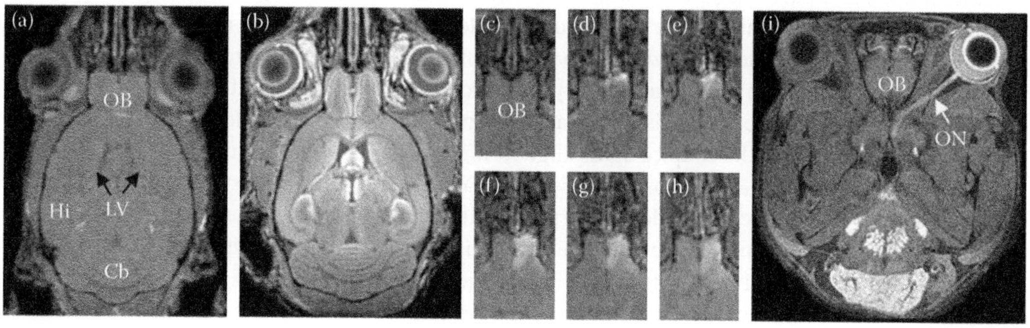

FIGURE 29.4 Manganese enhanced MRI shows maximum contrast 24 h after intraperitoneal injection of $MnCl_2$. Images (a & b) are horizontal *T*1-weighted gradient echo sections before (a) and 24 h after (b) injection of $MnCl_2$ (6 μmol 30 g^{-1} mouse weight) in an adult mouse brain demonstrating clear enhancement in olfactory bulb (OB), hippocampus (Hi) and cerebellum (Cb). (c–h) Time course of manganese after intranasal lavage demonstrating its uptake and subsequent transport through the olfactory tract. (i) MRI oblique orientation covering both the eye balls as well as the optic nerve of a mouse head. Nine hours after subjecting a 0.4 μL intraocular injection of 0.8 M $MnCl_2$ solution in the right eye, we can clearly trace the optical nerve.

dynamics of transport, MEMRI has seen the emergence of a wide variety of applications. MEMRI can be broadly used to visualize the 3D brain neuro-architecture (Watanabe et al., 2002; Wadghiri et al., 2004), the functional mapping of neuronal activity (Lin and Koretsky, 1997; Tindemans et al., 2003; Yu et al., 2005), as well as the neuronal circuitry through track tracing of neuronal projection (Pautler et al., 1998, Cross et al., 2008; Bertrand et al., 2011). $MnCl_2$ in isotonic saline (0.9% NaCl in water) administered in these various studies correspond well below doses used in previous studies of Mn-induced neurotoxicity (Shukakidze et al., 2003). In our own studies, we observed no abnormal behavior after i.p. injections of $MnCl_2$ for up to a year after initial injection.

Although part of this review examines the general biodistribution and pharmacokinetics of the various CAs, manganese will not be discussed in this case. To have more information on the practical implication of administration routes on Mn^{2+} biodistribution, the mechanisms of tissue uptake and redistribution as well as the characterization of intra-neuronal transport through track-tracing dynamic studies and the impact of the physiological status of the subject being studied have already been well documented (Mendonca-Dias et al., 1983; Wolf and Baum, 1983; Silva et al., 2004; Silva and Bock, 2008; Eschenko et al., 2010).

In addition to the MEMRI methods discussed, Mn has recently been conjugated to lipid-based nanoparticles for the in vivo active targeting of oxidized lipoproteins. Since the targeting moiety ensures mediated intracellular uptake by intraplaque macrophages and foam cells, Mn is delivered to intracellular vesicles, where it is released from its chelate. Once intracellular, the de-metallated Mn binds cellular components (primarily metalloproteins and cellular membrane), resulting in dramatic MR signal enhancement. Figure 29.5 shows a relative comparison between the MR signal obtained in the arterial wall of atherosclerotic mice 48 h after administration of targeted Gd and targeted Mn PEG micelles. Although the dose associated with the Mn material was 23% less than that of Gd dose, significantly greater MR signal enhancement (as percent normalized enhancement) was observed. As a result, the replacement of Gd with Mn did not reduce in vivo efficacy and may therefore allow for the clinical translation of this approach.

29.5 Superparamagnetic Nanobiomaterials

29.5.1 Theory

The use of superparamagnetic iron oxide nanoparticles can aid in distinguishing various internal structures (vasculature, neovascularization, edema, atherosclerotic plaque) and tumors. Although some ultrasmall iron oxide particles (USPIOs less than 20 nm in diameter) may be used as $T1$ agents for MRA indications, most superparamagnetic iron oxide particles (SPIOs > 20 nm) are used to generate signal loss ($T2/T2^*$ effects). To understand *superparamagnetism*, the concept of anisotropy must be addressed. If paramagnetic metal ions are arranged in a crystal lattice, then the spins can interact via spin-coupling (Gossuin et al., 2009). If the sample is then placed in an external magnetic field, the magnetization is no longer isotropic so that the coupled spins align in more than one direction relative to the field. For crystals of magnetite, there are six possible anisotropic axes, however; the lowest energy axis is aligned with the external field. Typically, the low energy axis is the most important since motion of the crystals causes an averaging of the anisotropic energy. For materials made up of large crystals (diameters greater than 14 nm), the spins are divided within small magnetic domains (Weiss domains) (Wang et al., 2001; Briley-Saebo et al., 2009; Gossuin et al., 2009). The direction of the spins in each domain is random prior to exposure to an external field. However, once the sample is placed in an external field, all the spins adopt the same orientation so that the system may be considered isotropic. If the external field is removed the spins remain aligned (remanence or permanent magnetism) and the material is considered ferromagnetic. If the crystals of magnetite become smaller than a Weiss domain, then superparamagnetism is observed. For these materials, energy is required to move the spins between the anisotropic axis. The anisotropy energy (KV) is related to crystal size, the crystal composition, the crystal shape, crystal–coating interactions, and crystal–crystal interactions (Briley-Saebo, 2009; Gossuin et al., 2009).

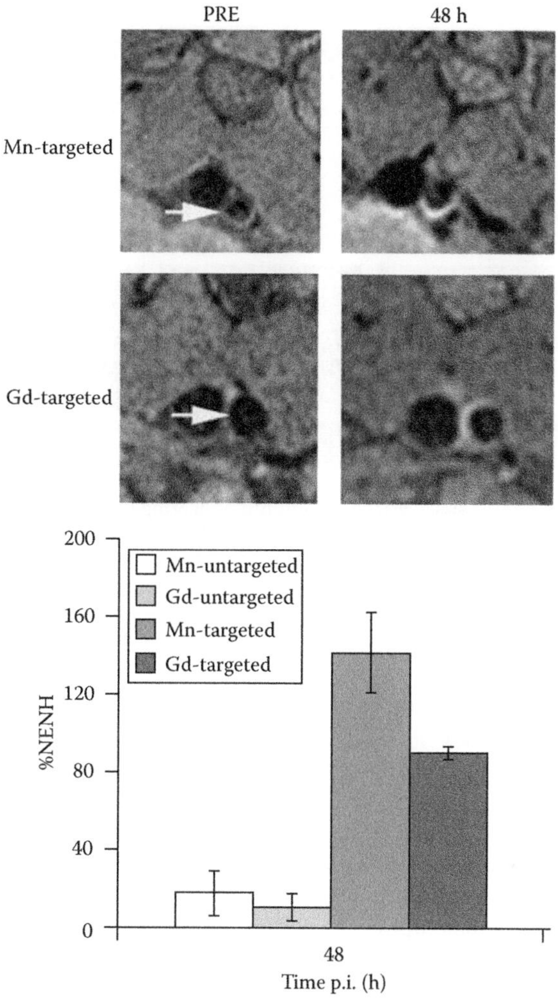

FIGURE 29.5 Comparison of Mn and Gd targeted micelles in murine models of atherosclerosis. Arrow points to arterial wall. Dose of Mn = 0.05 mmol Mn kg^{-1}. Dose Gd = 0.075 mmol Gd kg^{-1}. Although the in vitro efficacy of the $q = 0$ Mn formulation is significantly lower than that of the $q = 1$ Gd formulation, the intracellular release of Mn from the chelate results in similar in vivo enhancement even when lower dosages are used. Since the untargeted material is not mediated into the intraplaque macrophages, so significant MR signal is observed for either formulation.

The Nèel relaxation time, τ_N, is used to define the rate of flipping between the high and low energy axis and is modulated by the thermal agitation of the crystals. For small crystals (<6 nm in diameter) the transitions are fast (ns) and if the material is in solution then the fluctuations become modulated also by the crystal rotation. For a system of superparamagnetic crystals the magnetization may, therefore, be determined by the following equation:

$$\mathbf{M} = \boldsymbol{Msat} \cdot L(\mathrm{a}) \tag{29.5}$$

where $L(\boldsymbol{\alpha}) = \coth(\alpha) - (1/\alpha)$ and $\boldsymbol{\alpha} = \boldsymbol{\mu}\mathbf{Bo}/kT$. Here, $\boldsymbol{\mu}$ is the total magnetic moment of the crystal, **Bo** is the external magnetic field, k is the Boltzmann constant, T is temperature, and ***Msat*** defines the field at which the magnetization is locked or saturated along the easy axis. Most iron oxides used for MR imaging are

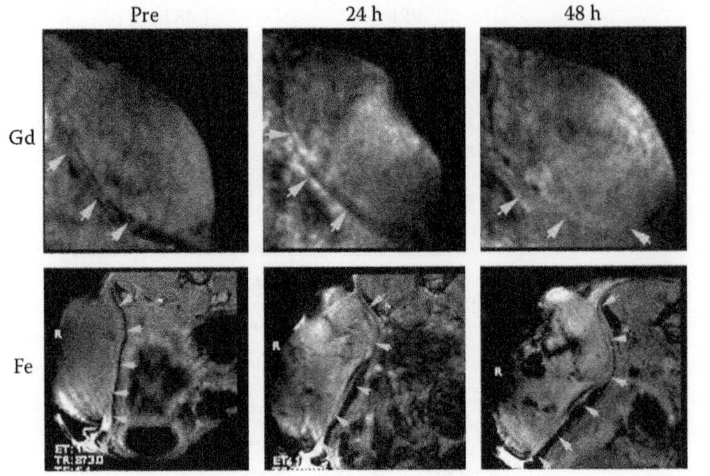

FIGURE 29.6 Comparison of the signal attenuation in tumor bearing mice after injection of Gd or iron labeled activated monocytes. Yellow arrows indicate tumor base. For cell tracking indications, the use of iron oxide cell labels greatly improves sensitivity with respect to cell detection and tracking. Gd labeled cells modulate *R*1 effects thereby causing signal increase (top panel). Iron labeled cells induce dephasing and modulate *R*2/*R*2* effects thereby causing signal loss (bottom panel). Although an over increase in tumor intensity is observed after injection of the Gd labeled cells, it is much easier to identify the iron labeled cells within the lesion.

superparamagnetic with magnetic saturation occurring at clinical field strengths. The magnetization values associated with most iron oxides are 10-fold higher than those observed for Gd. As a result, iron oxide particles exhibit greater in vivo sensitivity when compared to similar Gd constructs, as shown in Figure 29.6.

Superparamagnetic iron oxide particles produce significant MR signal loss (*T*2/*T*2* effects) due to the diffusion of water protons through the inhomogeneous local magnetic fields generated by the particles (Yablonskiy and Haacke, 1994). Although a linear correlation between the dipolar *R*2 values and Feridex concentration is observed in aqueous solution, compartmentalization of the individual iron particles within intracellular endosomes and lysosomes may induce nonlinearity (Muller et al., 1991; Yablonskiy and Haacke, 1994). In vitro studies using iron labeled lymphocytes and gliosarcoma cells show that the *R*2 values of labeled cells are two to three times lower than values obtained for equivalent Feridex concentrations in agarose gel. Additionally, studies have shown that the *R*2 values obtained for iron labeled cells are strongly dependent upon the cell type, cell size, and concentration of intracellular iron. For large iron particles (>30 nm) at clinical magnetic field strengths, *R*2 may be described using the Ayant spectral density (Briley-Saebo, 2009; Gossuin et al., 2009), as shown in the following equation:

$$R2 = \left(\frac{32\Pi}{405}\right)\gamma^2\mu^2\left(\frac{N}{RD}\right)\cdot\left[4.5J_A\left(\sqrt{2}\omega_0\tau_d\right)+6J_A(0)\right] \tag{29.6}$$

where

γ is the gyromagnetic ratio of a water proton
μ is the magnetic moment at a given field
N is the number of iron particles per volume unit
R is the particle radius
D is the diffusion coefficient
J_A is the Ayant spectral density function
ω_0 is the proton Larmor frequency
τ_d is the modulation of the relative translational diffusion time

For USPIOs, the evolution of *R*2 at clinical fields is complicated due to modulation by both the Freed and Ayant and spectral density functions. Fluctuations in the magnetic moment due to the anisotropic energy must also be considered. The current models to describe the evolution of *R*2 for USPIOs are, therefore, beyond the scope of the current chapter, and readers are advised to look into the reported phenomenological models for more information (Briley-Saebo, 2009; Gossuin et al., 2009).

In agar gel or aqueous solution, where the diffusion of water protons is unrestricted and the iron particles are distributed homogeneously within a unit volume, *R*2 will correlate linearly with *N* (or iron oxide concentration). On the other hand, when compartmentalized within a cell the following factors influence the *R*2 values observed: (1) Restricted water diffusion causing modulation of the transverse magnetization, (2) aggregation of the iron particles that cause the production of large susceptibility effects between the cell and other tissue, and (3) the number of particles per unit volume changes as the particles are sequestered within a limited volume. Studies have shown that *R*2 values first increase during agglomeration, reach a maximum value, and then decrease as the agglomeration process continues (Muller et al., 1991; Bos et al., 2004; Roch et al., 2005; Budde and Frank, 2009; Gossuin et al., 2009). As a result, *R*2 should not scale linearly with concentration when the iron particles are sequestered within cells. Nevertheless, a simple linear relationship exists between the iron concentration in cells and the effective $R2^*$ values. The correlation is based upon the fact that $R2^*$ is proportional to the concentration times the magnetization (M) at any given field (Yablonskiy and Haacke, 1994). Whether or not a linear correlation is observed is dependent upon the local magnetic field distribution generated by the iron particles. For most iron particles based upon magnetite, linear relationships are observed.

29.5.2 Toxicity, Pharmacokinetics, and Biodistribution

Iron oxide particles are considered relatively safe since most cells are able to effectively metabolize and exocytose iron oxide particle. Intracellular metabolism often involves the protonation of the iron oxide core followed by the chelation of soluble iron by iron transport proteins, such as ferritin and transferrin (Briley-Saebo et al., 2004a,b, 2006b). The transport proteins then carry the iron to other cells or transport it to the liver so that it may become incorporated into the normal liver iron pool. Although the first FDA approved iron oxide particle (Feridex) is dextran coated, the use of dextran has promoted immunological responses in some patients (Novey et al., 1994). It is also still unclear whether oxide particles taken up by inflammatory tissue promote pathogenesis due to the formulation of soluble iron that may catalyze various free radical reactions (peroxidase). Despite these concerns, bioretention of iron oxide is considered safe and has therefore led to the use of superparamagnetic materials for active targeting and cell labeling. Like paramagnetic materials, the pharmacokinetics and biodistribution of iron oxide particles are modulated by the hydrated size of the particle and the coating material/targeting moiety used (Pouliquen et al., 1992; Weissleder and Papisov, 1992; Briley-Saebo et al., 2008b). For example, 100 nm dextran coated iron oxide particles (<100 nm) are quickly sequestered by the RES and may be used for liver indications (Clement et al., 1998; Zhang et al., 2003). However, 12–35 nm dextran coated particles exhibit prolonged circulations times and are taken up by cells associated with the MPS (Weissleder et al., 1990, 1994; Ruehm et al., 2001; Schmitz et al., 2001; Wang et al., 2001; Kooi et al., 2003; Trivedi et al., 2004). As a result, these particles may be used for evaluation of the lymphatic system, bone marrow, vasculature (perfusion imaging), and intraplaque macrophages (Stark et al. 1998; Sugarbaker et al. 1990; Hahn et al. 1990; Kellar et al. 2000; Pauser et al. 1997). Generally speaking, the smaller the particle the longer the circulation time, the lower the RES uptake and the higher the MPS uptake. In contrast, the coating material and/or targeting moiety can greatly influence tissue uptake. For example, 20 nm dextran coated particles exhibit longer circulation times and greater MPS uptake than 12 nm modified starch coated particles (Weissleder et al., 1989, 2000). Intracellular uptake and metabolism may also be influenced by coating. For example, J7774A.1 murine macrophages readily take up dextran coated iron particles. Yet, if these iron particles are placed within the cornea of a PEG-micelle, then uptake is greatly restricted. As a result, the size and coating material must be considered when designing iron oxide CAs.

29.5.3 Applications of Passively Targeted Iron Oxide-Based Nanobiomaterials

Dextran coated iron oxide particles are considered to be passively targeted to macrophages associated with either the RES (>35 nm) or MPS (<35 nm). Of the RES targeted materials, only one iron oxide formulation has been approved by the FDA (Feridex). However, this material is no longer manufactured for clinical use and is only available for preclinical testing. Resovist is approved in some European countries for liver MR indications. These RES targeted particles are cleared quickly in the blood and are taken up by normal liver Kupffer cells. As a result, normal liver appears dark while liver pathology is iso-intense (relative to pre-images) (Grazioli et al., 2009; Santoro et al., 2009). Smaller dextran coated iron oxide particles (<35 nm) are used to passively target macrophages associated with the MPS. Although not approved by the FDA, Sinerem/Combidex has been used for clinical studies in Europe for in vivo detection of intraplaque macrophages associated with atherosclerotic plaque (Weissleder et al., 1990, 1994; Ruehm et al., 2001; Schmitz et al., 2001; Wang et al., 2001; Kooi et al., 2003; Trivedi et al., 2004).

29.6 Applications of Actively Targeted Iron Oxide-Based Nanobiomaterials

USPIOS have been actively targeted to allow for the detection and monitoring of early stages of tumor development in the brain (Rousseau et al., 1998), the lymph nodes (Harisinghani et al., 1997; Stets et al., 2002; Wunderbaldinger et al., 2002), splenic tumor (Tanimoto, 2001), intraplaque macrophages, and oxidative epitopes. Figure 29.7 shows results obtained for iron oxide lipoproteins and iron oxide micelles targeted to ApoA1 receptors and oxidative epitopes, respectively, in the arterial wall of atherosclerotic mice.

29.7 Importance of the Static Magnetic Field Strength: NMRD Profiles

In NMR applied to structural chemistry and biology, the field strength of the static magnetic field is almost exclusively specified using the Larmor frequency of ^1H nucleus. In clinical and biomedical sciences, MRI systems are usually defined by their magnetic strength in Tesla (1 T– 42.57744 MHz).

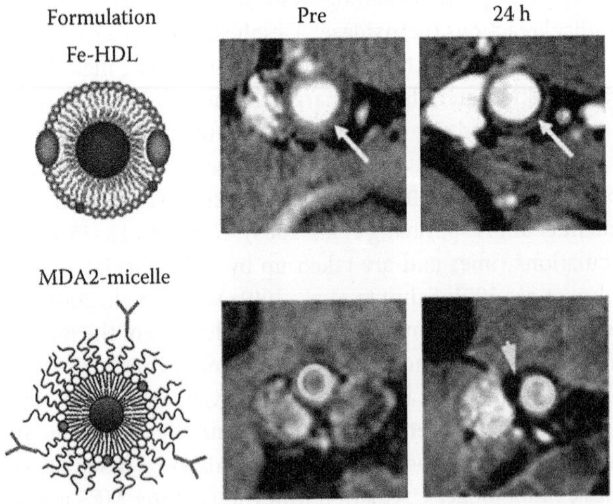

FIGURE 29.7 Iron particles targeted to ApoA1 (Fe-HDL) and oxidative epitopes (MD2-micelle) in the arterial wall (arrow) of atherosclerotic mice. Micelle injected as a bolus a 3.9 mg Fe kg^{-1}. Significant signal loss, indicative of iron oxide uptake was observed in the arterial wall following administration of the targeted iron oxide formulations.

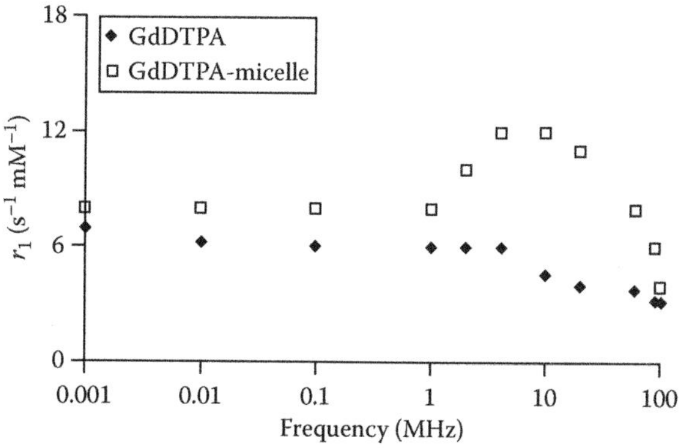

FIGURE 29.8 NMRD profile. Conjugation of Gd-DTPA to the surface of a lipid-based nanoparticles greatly increases the *R*1 values at clinical field strengths (10–63 MHz). The increase is due to the reduction in the molecular tumbling time associated with the paramagnetic center. At high field strengths the MR efficacy, with respect to *R*1 decreases significantly. The r_2 values however will remain constant due the presence of the scalar term. As a result, the *R*2/*R*2* effects may modulate the signal at high field and should be always accounted for.

Due to the dependence of the proton and electron spins on the applied magnetic field strength, the efficacies of paramagnetic and superparamagnetic materials are field dependent. Nuclear magnetic relaxation dispersion (NMRD) profiles are used to estimate the Freed spectral density function that reflects the correlation between the proton Larmor frequency and the electron–proton modulation. The inflection points and maximum values observed for the spectral density functions may be used to quantify important parameters such as τ_r, τ_m, and τ_{s1} for paramagnetic materials and τ_N and Ms for superparamagnetic materials. For linear Gd chelates, such as Gd-DTPA, the first inflection point associated with the low field *R*1 dispersion occurs at the precessional frequency of the electron and the second r_1 high field dispersion occurs at the precessional frequency of the proton. Since the precessional frequency of the water proton is known as a function of the field, the correlation time τ_c and τ_{s1} can be determined. The NMRD profile for Gd-DTPA is shown in Figure 29.8. At clinical field strengths the r_1 values range from 4 to 3.2 s^{-1} mM^{-1}. If Gd-DTPA is conjugated to nanoparticles, such as micelles, the field dispersion changes due to the change in τ_r. The tumbling correlation time is field dependent, with optimal or maximum exchange occurring when the rate of tumbling is similar to the precessional frequency. As a result, a maximum or peak r_1 value is observed at midfield as shown in Figure 29.8. Although these nanoparticles exhibit strong efficacy at midfield, the r_1 deceases rapidly as the field strength increases so that limited *R*1 efficacy is observed at high field. Additionally, the *R*2 values remain relatively constant as a function of field within the clinical midfield range to high field strengths resulting in the increase of the *R*2/*R*1 ratios with that of the field for paramagnetic nanoparticles. This may lead to significant *T*2/*T*2* modulation of the MR signal at high field strengths, thereby further decreasing the *R*1 efficacy of these materials.

29.8 Future Perspectives

29.8.1 Optimizing Efficacy

During the last two decades, great efforts have been devoted for developing CAs for clinical MRI in order to alter the signal intensity of water protons in their vicinity by increasing their relaxivity. The overall goals in the imaging field have been focused at enhancing tissue contrast between various organs or compartments within organs to identify diseased areas. The parameters that can be tuned in order to obtain more efficient agents are the rotational tumbling time of the complex, the exchange rate of water molecules in the first coordination sphere of Gd3+, and the number of water molecules in the first coordination sphere. The most common approach used to increase dipolar *R*1 efficacy is based on the decrease of the molecular tumbling times. As shown in Figure 29.8, linking Gd-DTPA to the surface of a lipid nanoparticle increases the *R*1 value by more than a factor of 2 at clinical imaging field strengths (10–63 MHz) (Anelli et al., 2001; Accardo et al., 2004; Briley-Saebo et al., 2006a, 2008a; Amirbekian et al., 2007; Cormode et al., 2008). Binding to the nanoparticles not only increases *R*1 efficacy at clinical field but also allows for increased paramagnetic payloads to specific targets and alters the biodistribution of the active MR label. Most paramagnetic-based agents exhibit optimal proton exchange rates. As a result, modifications in the chelating agent intended to limit electrostatic interactions and proton binding are expected to give only minor benefits relative to the chelating agents currently used. Nonetheless, when designing targeted nanoparticles where antibodies or peptides are also conjugated to the surface of the particle, it is important to consider the effect of the targeting moiety on the proton exchange rates. For example, studies have shown that the presence of certain antibodies restrict water exchange due to steric hindrance, thereby reducing MR efficacy (Parac-Vogt et al., 2004). As shown in Equation 29.2, increasing the number of coordination sites (q) causes a linear increase in the r_1 values observed. Although Gd has a coordination number of 8–9, the number of water molecules that may be coordinated within the inner sphere is limited given that most coordination sites are occupied by the chelator. In order to limit toxicity issues associated with transmetallation and bioretention of Gd3+, most current chelators only allow for one water proton to coordinate with the inner sphere (q = 1). Recently, high relaxivity Gd chelates have been developed based upon the heptadentate ligand AAZTA (6-amino-6-methylperhydro-1,4-diazepinetetraacetic acid) that allows two water molecules to coordinate within the inner sphere (q = 2) (Aime et al., 2004, 2005). The stability constant associated with this chelating ligand is surprisingly high and significantly greater than that of Gd-DTPA-BMA (q = 1), which is a commercially available chelate. At 20 MHZ (0.5 T), the r_1 of the q = 2 $[Gd\text{-}AAZTA]^-$ was 7.1 s^{-1} mM^{-1} compared to 3.8 s^{-1} mM^{-1} observed for q = 1 Gd-DTPA. Gd-AAZTA may be attached to a C_{17} lipid that may be used to label native lipoproteins (Briley-Saebo et al., 2009). Studies show that integration of the q = 2 Gd-AAZTA lipid into the native lipoprotein results in the formation of a high relaxivity molecular imaging probe with r_1 values 3.5 times greater than the q = 1 analog (Figure 29.9) and r_1 values 8–9 times greater than Gd-DTPA alone. The high r_1 values associated with the q = 2 Gd lipoprotein are a result of combining the effect of decreased tumbling rates with the increase in coordination numbers.

The efficacy of iron oxide-based materials may be increased by either optimizing the magnetization saturation (Msat) values or by optimizing in vivo compartmentalization within cells. For iron oxide crystals made up of magnetite, the larger the crystal size the larger the Msat. Though, it must be

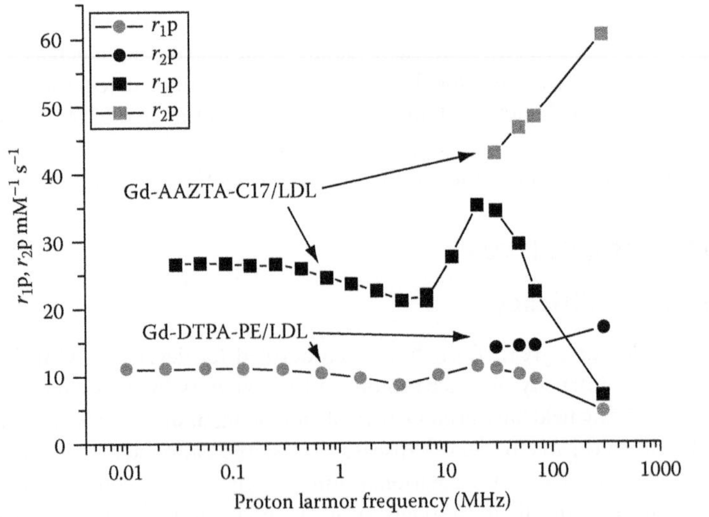

FIGURE 29.9 NMRD profile q = 1 (Gd-DTPA-PE/LDL) and q = 2 (Gd-AAZTA-PE/LDL) modified lipoproteins. The r_1 of Gd-DTPA at 20 MHZ is 4 s^{-1} mM^{-1} under similar experimental conditions.

acknowledged that by increasing the crystal size the overall hydrated particle size will also increase. Since iron oxide particles are rigid spheres, the increase in size will influence the pharmacokinetics and biodistribution, and may limit migration through gap junctions associated with non-normal endothelium. Additionally, if the crystal size is increased over that of the Weiss domain, then ferromagnetism will be observed at clinically relevant temperatures. Moreover, the in vivo compartmentalization of USPIO and SPIOS within cells greatly increases $R2/R2^*$ effects. This strategy is utilized extensively in cell tracking techniques since most cells store iron oxide particles within intracellular endosomes and/or lysosomes (Ahrens et al., 2005; Shapiro et al., 2005; Arbab et al., 2008; Budde and Frank, 2009; Bulte, 2009; Briley-Saebo et al., in press). The high concentration of iron oxide particles within these vesicles generates large magnetic field inhomogeneities, thereby allowing for in vivo detection and tracking of the iron laden cells. For some iron formulations, the sensitivity to $R2/R2^*$ effects is so great that it is possible to detect a single iron labeled cell in vivo (Shapiro et al., 2006).

The trend toward using higher fields in clinical settings has been slower compared to biomedical research due to technological challenges associated with larger bore. However, current systems have already achieved field strengths (3 T and above), where the field dependence of the CAs r_1 relaxivity becomes noticeable. As the relaxivity drops down rapidly, the need for developing more sensitive CAs is crucial. In contrast to that field dependence, raising the number of inner-sphere water molecules leads to an increase in relaxivity, which is proportional to the number of water molecules independent on the magnetic field. To this effect, we have shown that several research groups have devoted their efforts to the synthesis of chelates allowing the presence of more than one water molecule in the first coordination sphere of the metal or using high payload agent. Given that Gd is the most widely used metal ion, its toxicity remains the main obstacle in this vibrant field.

Acknowledgments

This chapter was supported in part by the American Health Assistance Foundation ADR (A2008-155), the Alzheimer Association (IIRG-08-91618), the Tilker Medical Research Foundation, and the NYU Applied Research Support Fund. We would like to extend a sincere thank you to the members of the Wadghiri laboratory as well as to Drs. Zahi Fayad and Daniel H. Turnbull and to their respective lab teams for their continued help and support.

References

Accardo A., Tesauro D., Roscigno P., Gianolio E., Paduano L., D'Errico G., Pedone C., Morelli G. Physicochemical properties of mixed micellar aggregates containing CCK peptides and Gd complexes designed as tumor specific contrast agents in MRI. *J. Am. Chem. Soc.* 2004;126(10):3097–3107.

Ahrens E.T., Flores R., Xu H., Morel P.A. In vivo imaging platform for tracking immunotherapeutic cells. *Nat. Biotechnol.* 2005;23(8):983–987.

Aime S., Barge A., Gianolio E., Pagliarin R., Silengo L., Tei L. High relaxivity contrast agents for MRI and molecular imaging. *Ernst Schering Res. Found Workshop* 2005(49):99–121.

Aime S., Cabella C., Colombatto S., Geninatti Crich S., Gianolio E., Maggioni F. Insights into the use of paramagnetic Gd(III) complexes in MR-molecular imaging investigations. *J. Magn. Reson. Imaging* 2002;16(4):394–406.

Aime S., Calabi L., Cavallotti C., Gianolio E., Giovenzana G.B., Losi P., Maiocchi A., Palmisano G., Sisti M. [Gd-AAZTA]-: A new structural entry for an improved generation of MRI contrast agents. *Inorg. Chem.* 2004;43(24):7588–7590.

Amirbekian S., Amirbekian V., Aguinaldo J.G.S., Frias J.C., Mani V., Briley-Saboe K.C., Fayad Z.A. Relationship between macrophage content and MRI signal intensity of atherosclerosis using gadolinium-containing immunomicelles targeted to macrophage scavenger receptor. Seattle, Washington, DC: ISMRM, 2006, p. 122.

Amirbekian V., Lipinski M.J., Briley-Saebo K.C., Amirbekian S., Aguinaldo J.G., Weinreb D.B., Vucic E. et al. Detecting and assessing macrophages in vivo to evaluate atherosclerosis noninvasively using molecular MRI. *Proc. Natl. Acad. Sci. U. S. A.* 2007;104:961–966.

Anelli P.L., Lattuada L., Lorusso V., Schneider M., Tournier H., Uggeri F. Mixed micelles containing lipophilic gadolinium complexes as MRA contrast agents. *Magma* 2001;12(2–3):114–120.

Arbab A.S., Janic B., Knight R.A., Anderson S.A., Pawelczyk E., Rad A.M., Read E.J., Pandit S.D., Frank J.A. Detection of migration of locally implanted AC133+ stem cells by cellular magnetic resonance imaging with histological findings. *FASEB J.* 2008;22(9):3234–3246.

Beache G.M., Kulke S.F., Kantor H.L., Niemi P., Campbell T.A., Chesler D.A., Gewirtz H., Rosen B.R., Brady T.J., Weisskoff R.M. Imaging perfusion deficits in ischemic heart disease with susceptibility-enhanced T2-weighted MRI: preliminary human studies. *Magn. Reson. Imaging* 1998;16(1):19–27.

Bertrand, A., Hoang, D.M., Khan, U., Wadghiri, Y.Z. From axonal transport to mitochondrial trafficking: What can we learn from manganese-enhanced MRI studies in mouse models of Alzheimer's disease? *Curr. Med. Imaging Rev.* 2011;7 (in press).

Bos C., Delmas Y., Desmouliere A., Solanilla A., Hauger O., Grosset C., Dubus I. et al. In vivo MR imaging of intravascularly injected magnetically labeled mesenchymal stem cells in rat kidney and liver. *Radiology* 2004;233(3):781–789.

Brasch R.C. New directions in the development of MR imaging contrast media. *Radiology* 1992;183 (1):1–11.

Briley Saebo K. *Relaxation and Metabolism of Iron Oxide Based Contrast Agents*. Saarbrücken, Germany: VDM Verlag Dr. Muller, 2009.

Briley-Saebo K.C., Amirbekian V., Mani V., Aguinaldo J.G.S., Vucic E., Carpenter D., Amirbekian S., Fayad Z.A. Gadolinium mixed-micelles: Effect of the amphiphile on in vitro and in vivo efficacy in apolipoprotein E knockout mouse models of atherosclerosis. *Magn. Reson. Med.* 2006a;56(6):1336–1346.

Briley-Saebo K., Bjornerud A., Grant D., Ahlstrom H., Berg T., Kindberg G.M. Hepatic cellular distribution and degradation of iron oxide nanoparticles following single intravenous injection in rats: Implications for magnetic resonance imaging. *Cell Tissue Res.* 2004a;316(3):315–323.

Briley-Saebo K.C., Geninatti-Crich S., Cormode D.P., Barazza A., Mulder W.J., Chen W., Giovenzana G.B., Fisher E.A., Aime S., Fayad Z.A. High-relaxivity gadolinium-modified high-density lipoproteins as magnetic resonance imaging contrast agents. *J. Phys. Chem. B* 2009;113(18):6283–6289.

Briley-Saebo K., Hustvedt S.O., Haldorsen A., Bjornerud A. Long-term imaging effects in rat liver after a single injection of an iron oxide nanoparticle based MR contrast agent. *J. Magn. Reson. Imaging* 2004b;20(4):622–631.

Briley-Saebo K.C., Johansson L.O., Hustvedt S.O., Haldorsen A.G., Bjornerud A., Fayad Z.A., Ahlstrom H.K. Clearance of iron oxide particles in rat liver: Effect of hydrated particle size and coating material on liver metabolism. *Invest. Radiol.* 2006b;41(7):560–571.

Briley-Saebo K.C., Leboeuf M., Dickson S., Mani V., Fayad Z.A., Palucka A.K., Banchereau J., Merad M. Longitudinal tracking of human dendritic cells in murine models using magnetic resonance imaging. *Magn. Reson. Med.* 2010 Nov;64(5):1510–1519.

Briley-Saebo K.C., Mani V., Hyafil F., Cornily J.C., Fayad Z.A. Fractionated Feridex and positive contrast: In vivo MR imaging of atherosclerosis. *Magn. Reson. Med.* 2008a;59(4):721–730.

Briley-Saebo K.C., Mulder W.J., Mani V., Hyafil F., Amirbekian V., Aguinaldo J.G., Fisher E.A., Fayad Z.A. Magnetic resonance imaging of vulnerable atherosclerotic plaques: Current imaging strategies and molecular imaging probes. *J. Magn. Reson. Imaging* 2007;26(3):460–479.

Briley-Saebo K.C., Shaw P.X., Mulder W.J., Choi S.H., Vucic E., Aguinaldo J.G., Witztum J.L., Fuster V., Tsimikas S., Fayad Z.A. Targeted molecular probes for imaging atherosclerotic lesions with magnetic resonance using antibodies that recognize oxidation-specific epitopes. *Circulation* 2008b;117(25):3206–3215.

Bruvold M., Nordhoy W., Anthonsen H.W., Brurok H., Jynge P. Manganese–calcium interactions with contrast media for cardiac magnetic resonance imaging: A study of manganese chloride supplemented with calcium gluconate in isolated Guinea pig hearts. *Invest. Radiol.* 2005;40(3):117–125.

Budde M.D., Frank J.A. Magnetic tagging of therapeutic cells for MRI. *J. Nucl. Med.* 2009;50(2):171–174.

Bulte J.W. In vivo MRI cell tracking: Clinical studies. *AJR Am. J. Roentgenol.* 2009;193(2):314–325.

Bumb A., Brechbiel M.W., Choyke P. Macromolecular and dendrimer-based magnetic resonance contrast agents. *Acta Radiologica* 2010;51(7):751–767.

Cacheris W.P., Quay S.C., Rocklage S.M. The relationship between thermodynamics and the toxicity of gadolinium complexes. *Magn. Reson. Imaging* 1990;8:467–481.

Caravan P., Cloutier N.J., Greenfield M.T., McDermid S.A., Dunham S.U., Bulte J.W., Amedio J.C. Jr. et al. The interaction of MS-325 with human serum albumin and its effect on proton relaxation rates. *J. Am. Chem. Soc.* 2002;124(12):3152–3162.

Castelli D.D., Terreno E., Cabella C., Chaabane L., Lanzardo S., Tei L., Visigalli M., Aime S. Evidence for in vivo macrophage mediated tumor uptake of paramagnetic/fluorescent liposomes. *NMR Biomed.* 2009;22(10):1084–1092.

Cavagna F.M., Lorusso V., Anelli P.L., Maggioni F., de Haen C. Preclinical profile and clinical potential of gadocoletic acid trisodium salt (B22956/1), a new intravascular contrast medium for MRI. *Acad. Radiol.* 2002;2(9):S491–S494.

Cavagna F., Luchinat C., Scozzafava A., Xia Z. Polymetallic macromolecules are potential contrast agents of improved efficiency. *Magn. Res. Med.* 1994;31(1):58–60.

Chen A.Y., Zirwas M.J., Heffernan M.P. Nephrogenic systemic fibrosis: A review. *J. Drugs Dermatol.* 2010;9(7):829–834.

Clement O., Siauve N., Cuenod C.A., Frija G. Liver imaging with ferumoxides (Feridex): Fundamentals, controversies, and practical aspects. *Top Magn. Reson. Imaging* 1998;9(3):167–182.

Cormode D.P., Briley-Saebo K.C., Mulder W.J., Aguinaldo J.G., Barazza A., Ma Y., Fisher E.A., Fayad Z.A. An apoA-I mimetic peptide high-density-lipoprotein-based MRI contrast agent for atherosclerotic plaque composition detection. *Small* 2008;4:1437–1444.

Cormode D.P., Skajaa T., Fayad Z.A., Mulder W.J. Nanotechnology in medical imaging. Probe design and applications. *Arterioscler. Thromb. Vasc. Biol.* 2009;29:992–1000.

Corot C., Port M., Raynal I., Dencausse A., Schaefer M., Rousseaux O., Simonot C. et al. Physical, chemical, and biological evaluations of P760: A new gadolinium complex characterized by a low rate of interstitial diffusion. *J. Magn. Res. Imaging* 2000;11(2):182–191.

Cross D.J., Flexman J.A., Anzai Y., Maravilla K.R., Minoshima S. Age-related decrease in axonal transport measured by MR imaging in vivo. *Neuroimage* 2008;39(3):915–926.

Dafni H., Gilead A., Nevo N., Eilam R., Harmelin A., Neeman M. Modulation of the pharmacokinetics of macromolecular contrast material by avidin chase: MRI, optical, and inductively coupled plasma mass spectrometry tracking of triply labeled albumin. *Magn. Reson. Med.* 2003;50(5):904–914.

Daldrup H.E., Roberts T.P., Mühler A., Gossmann A., Roberts H.C., Wendland M., Rosenau W., Brasch R.C. Macromolecular contrast media for MR mammography. A new approach to characterizing breast tumors. *Radiologe* 1997;37(9):733–740.

Eakins M.N., Eaton S.M., Fisco R.A., Hunt R.J., Ita C.E., Katona T., Owies L.M. et al. Physicochemical properties, pharmacokinetics, and biodistribution of gadoteridol injection in rats and dogs. *Acad. Radiol.* 1995;2(7):584–591.

Eschenko O., Canals S., Simanova I., Logothetis N.K. Behavioral, electrophysiological and histopathological consequences of systemic manganese administration in MEMRI. *Magn. Reson. Imaging.* Oct. 2010;28(8):1165–1174.

Federle M.P., Chezmar J.L., Rubin D.L., Weinreb J.C., Freeny P.C., Semelka R.C., Brown J.J. et al. Safety and efficacy of mangafodipir trisodium (MnDPDP) injection for hepatic MRI in adults: Results of the U.S. multicenter phase III clinical trials (safety). *J. Magn. Reson. Imaging* 2000; 12:186–197.

Fieno, D.S., Kim, R.J., Chen, E.L., Lomasney J.W., Klocke F.J., Judd R.M. Contrast-enhanced magnetic resonance imaging of myocardium at risk: Distinction between reversible and irreversible injury throughout infarct healing. *J. Am. Coll. Cardiol.* 2000;36:1985.

Fink C., Goyen M., Lotz J. Magnetic resonance angiography with blood-pool contrast agents: Future applications. *Eur. Radiol.* 2007;17:B38–B44.

Fossheim S.L., Fahlvik A.K., Klaveness J., Muller R.N. Paramagnetic liposomes as MRI contrast agents: Influence of liposomal physicochemical properties on the in vitro relaxivity. *Magn. Reson. Imaging* 1999;17(1):83–89.

Frias J.C., Lipinski M.J., Lipinski S.E., Albelda M.T. Modified lipoproteins as contrast agents for imaging of atherosclerosis. *Contrast Media Mol. Imaging* 2007;2(1):16–23.

Gossuin Y., Gillis P., Hocq A., Vuong Q.L., Roch A. Magnetic resonance relaxation properties of superparamagnetic particles. *WIREs Nanomed. Nanobiotechnol.* 2009;1:299–310.

Grazioli L., Bondioni M.P., Romanini L., Frittoli B., Gambarini S., Donato F., Santoro L., Colagrande S. Superparamagnetic iron oxide-enhanced liver MRI with SHU 555 A (RESOVIST): New protocol infusion to improve arterial phase evaluation—A prospective study. *J. Magn. Reson. Imaging* 2009;29(3):607–616.

Hahn P.F., Stark D.D., Weissleder R., Elizondo G., Saini S., Ferrucci J.T. Clinical application of superparamagnetic iron oxide to MR imaging of tissue perfusion in vascular liver tumors. *Radiology* 1990;174:361–366.

Haraldseth O., Jones R.A., Muller T.B., Fahlvik A.K., Oksendal A.N. Comparison of dysprosium DTPA BMA and superparamagnetic iron oxide particles as susceptibility contrast agents for perfusion imaging of regional cerebral ischemia in the rat. *J. Magn. Reson. Imaging* 1996;6(5): 714–717.

Harisinghani M.G., Saini S. and Slater G.J., Schnall M.D., Rifkin M.D. MR imaging of pelvic lymph nodes in primary pelvic carcinoma with ultrasmall superparamagnetic iron oxide (Combidex): Preliminary observations. *J. Magn. Reson. Imaging* 1997;7:161–163.

Harpur E.S., Worah D., Hals P.A., Holtz E., Furuhama K., Nomura H. Preclinical safety assessment and pharmacokinetics of gadodiamide injection, a new magnetic resonance imaging contrast agent. *Invest. Radiol.* 1993;28(Suppl 1):S28–S43.

Hoffmann U., Schima W., Herold C. Pulmonary magnetic resonance angiography. *Eur. Radiol.* 1999;9(9):1745–1754.

Ide M., Kuwamura M., Kotani T., Sawamoto O., Yamate J. Effects of gadolinium chloride (GdCl(3)) on the appearance of macrophage populations and fibrogenesis in thioacetamide-induced rat hepatic lesions. *J. Comp. Pathol.* 2005;133(2–3):92–102.

Jackson G.E., Wynchank S., Woudenberg M. Gadolinium(III) complex equilibria: The implications for Gd(III) MRI contrast agents. *Magn. Reson. Med.* 1990;16:57–66.

Jung G., Heindel W., Krahe T., Kugel H., Walter C., Fischbach R., Klaus H., Lackner K. Influence of the hepatobiliary contrast agent mangafodipir trisodium (MN-DPDP) on the imaging properties of abdominal organs. *Magn. Reson. Imaging* 1998;16(8):925–931.

Kellar K.E., Fujii D.K., Gunther W.H., Briley-Saebo K., Bjornerud A., Spiller M., Koenig S.H. NC100150 Injection, a preparation of optimized iron oxide nanoparticles for positive-contrast MR angiography. *J. Magn. Reson. Imaging* 2000;11(5):488–494.

Kiessling F., Morgenstern B., Zhang C. Contrast agents and applications to assess tumor angiogenesis in vivo by magnetic resonance imaging. *Curr. Med. Chem.* 2007;14(1):77–91.

Kim R.J., Fieno D.S., Parrish T.B., Harris K., Chen E.L., Simonetti O., Bundy J., Finn J.P., Klocke F.J., Judd R.M. Relationship of MRI delayed contrast enhancement to irreversible injury, infarct age, and contractile function. *Circulation* 1999;100:1992.

Kim R.J., Wu E., Rafael A., Chen E.L., Parker M.A., Simonetti O., Klocke F.J., Bonow R.O., Judd R.M. The use of contrast-enhanced magnetic resonance imaging to identify reversible myocardial dysfunction. *N. Engl. J. Med.* 2000;343:1445–1453.

Koblinsky E., Ahram M., Sloane B.F., Unraveling the role of proteases in cancer. *Clin. Chim. Acta* 2000;291:113–135.

Koenig S.H. From the relaxivity of Gd(DTPA)2- to everything else. *Magn. Reson. Med.* 1991;22(2): 183–190.

Koenig S.H., Brown R.D., 3rd. Relaxation of solvent protons by paramagnetic ions and its dependence on magnetic field and chemical environment: Implications for NMR imaging. *Magn. Reson. Med.* 1984;1(4):478–495.

Kooi M.E., Cappendijk V.C., Cleutjens K.B., Kessels A.G., Kitslaar P.J., Borgers M., Frederik P.M., Daemen M.J., van Engelshoven J.M. Accumulation of ultrasmall superparamagnetic particles of iron oxide in human atherosclerotic plaques can be detected by in vivo magnetic resonance imaging. *Circulation* 2003;107(19):2453–2458.

Koretsky A.P., Silva, A.C. Manganese-enhanced magnetic resonance imaging (MEMRI). *NMR Biomed.* 2004;17:527.

Kramer H., Morana G. Whole-body magnetic resonance angiography with blood-pool agents. *Eur. Radiol.* 2007;17:B24–B29.

Kroft L.J., de Roos A. Blood pool contrast agents for cardiovascular MR imaging. *J. Magn. Reson. Imaging* 1999;10(3):395–403.

Kucharczyk J., Asgari H., Mintorovitch J., Vexler Z., Moseley M., Watson A., Rocklage S. Magnetic resonance imaging of brain perfusion using the nonionic contrast agents Dy-DTPA-BMA and Gd-DTPA-BMA. *Invest. Radiol.* 1991;26(Suppl 1):S250–S252; discussion S253–S254.

Lauffer R.B. Paramagnetic metal complexes as water proton relaxation agents for NMR imaging: Theory and design. *Chem. Rev.* 1987; 87:901–927.

Laurent S., Elst L.V., Copoix F., Muller R.N. Stability of MRI paramagnetic contrast media—A proton relaxometric protocol for transmetallation assessment. *Invest. Radiol.* 2001;36(2):115–122.

Laurent S., Elst L.V., Muller R.N. Comparative study of the physicochemical properties of six clinical low molecular weight gadolinium contrast agents. *Contrast Media Mol. Imaging* 2006;1:128–137.

Lauterbur P.C. Image formation by induced local interactions: Examples of employing nuclear magnetic resonance. *Nature* 1973;242:190–191.

Lin, Y.J., Koretsky, A.P., Manganese ion enhances T1-weighted MRI during brain activation: An approach to direct imaging of brain function. *Magn. Reson. Med.* 1997;38:378.

Longmire M., Choyke P.L., Kobayashi H. Clearance properties of nano-sized particles and molecules as imaging agents: Considerations and caveats. *Nanomedicine (Lond)* 2008;3(5):703–717.

May P.M., Linder P.W., Williams D.R. Computer simulation of metal-ion equilibria in biofluids. *J. Chem. Soc., Dalton Trans.* 1977:588–595.

Mendonca-Dias H.M., Gaggelli E., Lauterbur P.C. Paramagnetic contrast agents in nuclear magnetic resonance medical imaging. *Semin. Nucl. Med.* 1983;13:364–376.

Misselwitz B., Schmitt-Willich H., Ebert W., Frenzel T., Weinmann H. Pharmacokinetics of Gadomer-17, a new dendritic magnetic resonance contrast agent. *Magma* 2001;12(2–3):128–134.

Mody T.D., Sessler J.L. Porphyrin- and expanded porphyrin-based diagnostic and therapeutic agents. In: Reinhoudt D.N., ed. *Supramolecular Technology.* New York: John Wiley & Sons, 1999, pp. 245–294.

Moseley M.E., Vexler Z., Asgari H.S., Mintorovitch J., Derugin N., Rocklage S., Kucharczyk J. Comparison of Gd- and Dy-chelates for T2 contrast-enhanced imaging. *Magn. Reson. Med.* 1991;22(2):259–264; discussion 265–257.

Muhler A. Assessment of myocardial perfusion using contrast-enhanced MR imaging: Current status and future developments. *Magma* 1995;3(1):21–33.

Mulder W.J., Strijkers G.J., van Tilborg G.A., Griffioen A.W., Nicolay K. Lipid-based nanoparticles for contrast-enhanced MRI and molecular imaging. *NMR Biomed.* 2006;19(1):142–164.

Muller R.N., Gillis P., Moiny F., Roch A. Transverse relaxivity of particulate MRI contrast media: From theories to experiments. *Magn. Reson. Med.* 1991;22(2):178–182; discussion 195–176.

Nordhoy W., Anthonsen H.W., Bruvold M., Jynge P., Krane J., Brurok H. Manganese ions as intracellular contrast agents: Proton relaxation and calcium interactions in rat myocardium. *NMR Biomed.* 2003;16(2):82–95.

Novey H.S., Pahl M., Haydik I., Vaziri N.D. Immunologic studies of anaphylaxis to iron dextran in patients on renal dialysis. *Ann. Allergy* 1994;72(3):224–228.

Parac-Vogt T.N., Kimpe K., Laurent S., Pierart C., Elst L.V., Muller R.N., Binnemans K. Gadolinium DTPA-monoamide complexes incorporated into mixed micelles as possible MRI contrast agents. *Eur. J. Inorg. Chem.* 2004(17):3538–3543.

Pauser S., Reszka R., Wagner S., Wolf K.J., Buhr H.J., Berger G. Liposome-encapsulated superparamagnetic iron oxide particles as markers in an MRI-guided search for tumor-specific drug carriers. Anticancer Drug Des. 1997;12:125–135.

Pautler R.G., Silva A.C., Koretsky A.P. In vivo neuronal tract tracing using manganese-enhanced magnetic resonance imaging. *Magn. Reson. Med.* 1998;40(5):740–748.

Perazella M.A. Nephrogenic systemic fibrosis, kidney disease, and gadolinium: Is there a link? *Clin. J. Am. Soc. Nephrol.* 2007;2:200–202.

Pouliquen D., Perroud H., Calza F., Jallet P., Le Jeune J.J. Investigation of the magnetic properties of iron oxide nanoparticles used as contrast agent for MRI. *Magn. Reson. Med.* 1992;24(1):75–84.

Reimer P., Schneider G., Schima W. Hepatobiliary contrast agents for contrast-enhanced MRI of the liver: Properties, clinical development and applications. *Eur. Radiol.* 2004;14(4):559–578.

Ribot E.J., Thiaudière E., Roulland R., Brugières P., Rahmouni A., Voisin P., Franconi J.M., Miraux S. Application of MRI phase-difference mapping to assessment of vascular concentrations of BMS agent in mice. *Contrast Media Mol. Imaging* 2008;3(2):53–60.

Roch A., Gossuin Y., Muller R.N., Gillis P. Superparamagnetic colloid suspensions: Water magnetic relaxation and clustering. *J. Magn. Magn. Mater.* 2005;293(1):532–539.

Rousseau V., Pouliquen D., Darcel F., Jallet P., and Le Jeune J.J. NMR investigation of experimental chemical induced brain tumors in rats, potential of a superparamagnetic contrast agent (MD3) to improve diagnosis. *Magn. Reson. Mater. Phys., Biol. Med.* 1998;6:13–21.

Ruehm S.G., Corot C., Vogt P., Kolb S., Debatin J.F. Magnetic resonance imaging of atherosclerotic plaque with ultrasmall superparamagnetic particles of iron oxide in hyperlipidemic rabbits. *Circulation* 2001;103(3):415–422.

Santoro L., Grazioli L., Filippone A., Grassedonio E., Belli G., Colagrande S. Resovist enhanced MR imaging of the liver: Does quantitative assessment help in focal lesion classification and characterization? *J. Magn. Reson. Imaging* 2009;30(5):1012–1020.

Schaffer B.K., Linker C., Papisov M., Tsai E., Nossiff N., Shibata T., Bogdanov A. Jr., Brady T.J., Weissleder R. MION-ASF: Biokinetics of an MR receptor agent. *Magn. Reson. Imaging* 1993;11(3):411–417.

Schmiedl U., Moseley M.E., Sievers R., Ogan M.D., Chew W.M., Engeseth H., Finkbeiner W.E., Lipton M.J., Brasch R.C. Magnetic resonance imaging of myocardial infarction using albumin-(Gd-DTPA), a macromolecular blood-volume contrast agent in a rat model. *Invest. Radiol.* 1987;22(9):713–721.

Schmitz S.A., Taupitz M., Wagner S., Wolf K.J., Beyersdorff D., Hamm B. Magnetic resonance imaging of atherosclerotic plaques using superparamagnetic iron oxide particles. *J. Magn. Reson. Imaging* 2001;14(4):355–361.

Schuhmann-Giampieri G., Schmitt-Willich H., Frenzel T., Press W.R., Weinmann H.J. In vivo and in vitro evaluation of Gd-DTPA-polylysine as a macromolecular contrast agent for magnetic resonance imaging. *Invest. Radiol.* 1991;26(11):969–974.

Shapiro E.M., Sharer K., Skrtic S., Koretsky A.P. In vivo detection of single cells by MRI. *Magn. Reson. Med.* 2006;55(2):242–249.

Shapiro E.M., Skrtic S., Koretsky A.P. Sizing it up: Cellular MRI using micron-sized iron oxide particles. *Magn. Reson. Med.* 2005;53(2):329–338.

Shukakidze A., Lazriev I., Mitagvariya N. Behavioral impairments in acute and chronic manganese poisoning in white rats. *Neurosci. Behav. Physiol.* 2003;33:263–267.

Silva A.C., Bock N.A. Manganese-enhanced MRI: An exceptional tool in translational neuroimaging. *Schizophr. Bull.* 2008;34(4):595–604. Review.

Silva A.C., Lee J.H., Aoki I., Koretsky A.P. Manganese-enhanced magnetic resonance imaging (MEMRI): Methodological and practical considerations. *NMR Biomed.* 2004;17(8):532–43. Review.

Spencer A.J., Wilson S.A., Batchelor J., Reid A., Rees J., Harpur E. Gadolinium chloride toxicity in the rat. *Toxicol. Pathol.* 1997;25(3):245–255.

Spencer A., Wilson S., Harpur E. Gadolinium chloride toxicity in the mouse. *Hum. Exp. Toxicol.* 1998;17(11):633–637.

Stark D.D., Weissleder R., Elizondo G., Hahn P.F., Saini S., Todd L.E., Wittenberg J., Ferrucci J.T. Superparamagnetic iron oxide: Clinical application as a contrast agent for MR imaging of the liver. *Radiology* 1988;168:297–301.

Stets C., Brandt S., Wallis F., Buchmann J., Gilbert F.J., Heywang-Kobrunner H., Sylvia, J. Axillary lymph node metastases: A statistical analysis of various parameters in MRI with USPIO. *Magn. Reson. Imaging* 2002;16:60–68.

Sugarbaker P.H. Surgical decision making for large bowel cancer metastatic to the liver. *Radiology* 1990;174:621–626.

Tanimoto, A. Magnetic resonance imaging with superparamagnetic nanocapsules. In: R. Arshady, ed., *Microspheres Microcapsules and Liposomes*, London, U.K.: Citus Books, 2001 (Vol. 3), pp. 525–558.

Tindemans, I., Verhoye, M., Balthazart, J., Van der Linden, A. In vivo dynamic MEMRI reveals differential functional responses of RA- and area X-projecting neurons in the HVC of canaries exposed to conspecific song. *Eur. J. Neurosci.* 2003;18:3352.

Toft K.G., Hustvedt S.O., Grant D., Martinsen I., Gordon P.B., Friisk G.A., Korsmo A.J., Skotland T. Metabolism and pharmacokinetics of MnDPDP in man. *Acta Radiol.* 1997;38:677–689.

Torchilin V.P. PEG-based micelles as carriers of contrast agents for different imaging modalities. *Adv. Drug Deliv. Rev.* 2002;54(2):235–252.

Torchilin V.P. Recent advances with liposomes as pharmaceutical carriers. *Nat. Rev. Drug Discov.* 2005;4(2):145–160.

Trivedi R.A., U-King-Im J.M., Graves M.J., Cross J.J., Horsley J., Goddard M.J., Skepper J.N. et al. In vivo detection of macrophages in human carotid atheroma: Temporal dependence of ultrasmall superparamagnetic particles of iron oxide-enhanced MRI. *Stroke* 2004;35(7):1631–1635.

Vander Elst L., Van Haverbeke Y., Goudemant J.F., Muller R.N. Stability assessment of gadolinium complexes by P-31 and H-1 relaxometry. *Magn. Reson. Med.* 1994;31(4):437–444.

Wadghiri Y.Z., Blind J.A., Duan X., Moreno C., Yu X., Joyner A.L., Turnbull D.H. Manganese-enhanced magnetic resonance imaging (MEMRI) of mouse brain development. *NMR Biomed.* 2004;17(8):613–619.

Wang Y.X.J., Hussain S.M., Krestin G.P. Superparamagnetic iron oxide contrast agents: Physicochemical characteristics and applications in MR imaging. *Eur. Radiol.* 2001;11(11):2319–2331.

Watanabe T., Natt O., Boretius S., Frahm J., Michaelis T. In vivo 3D MRI staining of mouse brain after subcutaneous application of $MnCl_2$. *Magn. Reson. Med.* 2002;48:852–859.

Weinmann H.J., Brasch R.C., Press W.R., Wesbey G.E. Characteristics of gadolinium-DTPA complex: A potential NMR contrast agent. *AJR Am. J. Roentgenol.* 1984;142(3):619–624.

Weinmann H.J., Ebert W., Misselwitz B., Schmitt-Willich H. Tissue-specific MR contrast agents. *Eur. J. Radiol.* 2003;46(1):33–44.

Weissleder R., Elizondo G., Wittenberg J., Rabito C.A., Bengele H.H., Josephson L. Ultrasmall superparamagnetic iron oxide: Characterization of a new class of contrast agents for MR imaging. *Radiology* 1990;175(2):489–493.

Weissleder R., Heautot J.F., Schaffer B.K., Nossiff N., Papisov M.I., Bogdanov A. Jr., Brady T.J. MR lymphography: Study of a high-efficiency lymphotrophic agent. *Radiology* 1994;191(1):225–230.

Weissleder R., Papisov M. Pharmaceutical iron oxides for MR imaging. *Rev. Magn. Reson. Med.* 1992;4:1–20.

Weissleder R., Stark D.D., Engelstad B.L., Bacon B.R., Compton C.C., White D.L., Jacobs P., Lewis J. Superparamagnetic iron oxide: Pharmacokinetics and toxicity. *AJR Am. J. Roentgenol.* 1989;152(1):167–173.

Wiener E.C., Brechbiel M.W., Brothers H., Magin R.L., Gansow O.A., Tomalia D.A., Lauterbur P.C. Dendrimer-based metal chelates: A new class of magnetic resonance imaging contrast agents. *Magn. Res. Med.* 1994;31(1):1–8.

Wolf G.L., Baum L. Cardiovascular toxicity and tissue proton T1 response to manganese injection in the dog and rabbit. *AJR Am. J. Roentgenol.* 1983;141:193–197.

Wunderbaldinger P., Josephson L., Bremer C., Moore A., Weissleder R. Detection of lymph node metastases by contrast-enhanced MRI in an experimental model. *Magn. Reson. Med.* 2002;47(2):292–297.

Yablonskiy D.A., Haacke E.M. Theory of NMR signal behavior in magnetically inhomogeneous tissues: The static dephasing regime. *Magn. Reson. Med.* 1994;32(6):749–763.

Yu, X., Wadghiri, Y.Z., Sanes, D.H., Turnbull, D.H. In vivo auditory brain mapping in mice with Mn-enhanced MRI. *Nat. Neurosci.* 2005;8:961.

Zhang X.H., Liang B.L., Huang S.Q. Correlation between SPIO-enhanced magnetic resonance imaging (MRI) and histological grading in hepatocellular carcinoma. *Ai Zheng* 2003;22(7):734–738.

Zhu D., Lu X.L., Hardy P.A., Leggas M., Jay M. Nanotemplate-engineered nanoparticles containing gadolinium for magnetic resonance imaging of tumors. *Invest. Radiol.* 2008;43(2):129–140.

30

Biocompatibility of Nanomaterials: Physical and Chemical Properties of Nanomaterials Relevant to Toxicological Studies, In Vitro and In Vivo

Christie Sayes
Texas A&M University

J. Michael Berg
Texas A&M University

30.1 Introduction

Nanomaterials are currently at the center of the materials science and the medicinal worlds. The demand for nanotechnology drives the development of products such as highly effective sunscreens, blood glucose monitors, or light bulbs. This increasing incorporation of nanomaterials into consumer products provides opportunity for increased exposure to the workers who produce the materials, the consumers who utilize the products, as well as the environments into which the product is introduced. The development of safety protocols and the establishment of exposure regulations will naturally play an important role in the development of the field of nanotechnology.

Nanomaterials generally consist of nanofilms, nanoparticles, and nanocomposites. Nanoparticles, in the toxicological sense, are defined as particulates with a primary particle size with one dimension less than 100 nm. In addition to size, the nanoparticles exhibit novel properties that differ from their bulk, or micron size scale, counterpart (i.e., electronic and/or insulation properties). These novel uses for engineered nanomaterials (ENMs) are often derived from their ability to exhibit monodispersity, large surface area: size ratio, and a high order. Nanoparticles may be synthesized with combinations of various properties to yield unique structural and functional features useful in product development. The resultant nanomaterials may be formulated as dry nanopowder or suspended in either an aqueous or organic solvent. These phases of nanoparticles lead to altered physicochemical characteristics. In addition to the dry and suspended phases, one might also classify nanoparticles as in vivo and ex vivo phases. It has become important to the toxicologist to determine which nanomaterial property, if any, will help to predict a physiological response in each of the four phases.

A list of unique nanomaterial properties would take many hundreds of entries; however, nanotoxicologists have generally defined a subset of characteristics that may quickly and economically be evaluated. This subset of nanomaterial properties, including size, shape, surface area, crystalline state, and surface modification or functionalization, allows insight into the potential toxicological impacts of these novel materials. Each of these characteristics has been shown to influence a corresponding biological interaction. For example, size and surface charge have been shown to influence nanomaterial endocytosis while surface modification and crystalline state are influential factors in cytotoxicity. Predictive measures of toxicity, similar to the structure activity relationship (SAR) of organic compounds, of nanomaterials are needed to allow efficient evaluation prior to use.

30.2 Material Characterization Methods

30.2.1 Transmission Electron Microscopy and Energy Dispersive Spectroscopy

The use of transmission electron microscopy (TEM) has become a necessary characterization tool in the fields of nanotechnology and nanotoxicology. TEM takes advantage of the extremely small wavelength of electrons (□ 0.2 nm) to increase the resolution and magnification of nanoscale particulates and cellular organelles. In transmission electron microscopy, the specimen is exposed to an electron beam, and subsequently the transmitted electrons are viewed on a phosphorescent screen, a CCD camera, or film. Depending on the nanomaterial characteristics, the TEM can provide information such as size, shape, agglomeration state, and crystalline structure of nanoparticles. In addition, TEM may be used postexposure both in vitro and in vivo to determine the intracellular localization of nanoparticles as well as any cellular structural changes.

While TEM provides a plethora of information surrounding nanomaterial characterization, there are some limitations to this technique. Due to the high energy of the electron beam (i.e., 60–300 keV), biological specimens are rapidly damaged upon exposure. Therefore, proper biological specimen preparation determines the quality of the specimen. Proper preparation is a multistep process consisting of fixation, dehydration, embedding, and sectioning. Further explanation on this process can be found in *Electron Microscopy* by Bazzola and Russel as well as *Biological Specimen Preparation* by Glauert and Lewis. Subsequently, post-staining with a combination of uranyl acetate and lead citrate will enhance contrast and allow for better visualization of the subcellular structure (Reynolds 1963).

As previously mentioned, specimen preparation plays a major role in the quality of data retained from an electron microscope. Nanoscientists must take exceptional precautionary measures when drawing conclusions based solely on TEM images. For example, postfixation with osmium tetroxide (OsO_4) may introduce small (1–2 nm) nanoparticles into the specimen. In addition, if samples are poststained, one must be assured that visualized nanoparticles are not artifacts from the protocol itself. Careful analytical techniques, such as those described in the next section, when combined with image analysis are the undefined standards in nanotechnology.

While TEM may itself be useful as a characterization tool, combinations of TEM with alternative methodologies such as energy dispersive spectroscopy (EDS), electron diffraction (ED), and high-resolution transmission electron microscopy (HRTEM) may be useful for determining additional characterization parameters such as chemical composition and speciation. Electron diffraction is a tool used by materials scientists to determine the crystal structure of a solid particle. When excited electrons pass through the nanoparticles, the electrons produce a signature diffraction pattern. Once this pattern has been attained, it may be compared with that of a standard material. This technique may come into play when a particle with a single composition (i.e., TiO_2) has multiple crystalline states, which induce differential amounts of toxicity (Sayes et al. 2006). Similarly, HRTEM may also be used to study the crystallographic structure at the atomic scale.

Energy dispersive spectroscopy as a micro-analytical tool can yield both quantitative and qualitative results when utilized by the trained personnel. When emitted electrons strike the atoms of interest in the sample, inelastic reactions occur that generate emitted electrons and x-rays subsequently detected via spectroscopy. With the advent of powerful computers, these detected signals may then be processed for exact elemental composition. When preparing a specimen for EDS analysis, one must be careful that no additional elements have been added to the specimen that may interfere with the spectral analysis of the element of interest.

An alternative method of electron microscopy has recently been used to identify nanoparticle composition in a heterogeneous exposure. Scanning transmission electron microscopy (STEM) (field emission gun) utilizes a very small electron beam to scan the specimen of interest. STEM emission may be combined with subsequent high-angle annular dark field (HAADF) analysis to provide a reverse contrast image with brightness dependent on atomic number squared. The use of STEM-HAADF as a characterization tool provides the toxicologist with a novel method to determine the composition of individual nanoparticles in a complex matrix (Figure 30.1).

30.2.2 Raman and Other Spectroscopies

One of the most common analytical techniques in chemistry is spectroscopy, the science concerned with the measurement and interpretation of electromagnetic spectra arising from either emission or absorption of radiant energy by various chemicals. These spectroscopic techniques are useful for particle analyses. Information about the chemical composition, structure, surface functionality, and optical/electronic properties of the sample can be obtained. Different types of spectroscopic techniques can be used to characterize nanoparticles. For example, mass spectroscopy is used to determine the masses of small electrically charged particles, and can be utilized to measure properties of on a particle surface

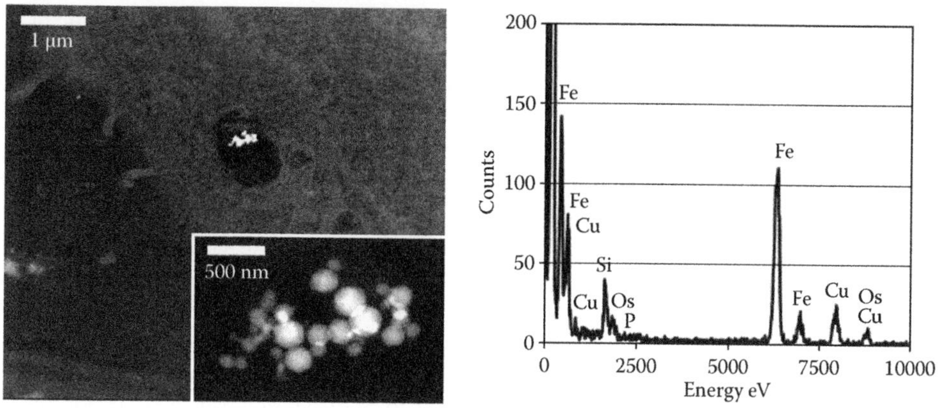

FIGURE 30.1 STEM image and EDS spectra of human lung epithelial cells exposed to ~30 nm Fe_2O_3 nanoparticle 24 h post-exposure.

such as bound (absorbed or adsorbed proteins). Raman spectroscopy is used to study vibrational, rotational, and other low-frequency modes in a system and can be utilized to determine type and degree of functionalization on the sidewall of a carbon nanotube. Absorption spectroscopy is used to quantify the amount of photons a substance absorbs and can be utilized to measure the size of gold nanoshells: absorbance is red-shifted (decreased wavelength) as the thickness of shell increases. Fluorescence spectroscopy is used to analyze the different frequencies of light emitted by a substance, which is then used to determine the structure of the vibrational levels of that substance. In nanoparticle characterization, fluorescence spectroscopy can give information on the functionality of quantum dots. The longer the quantum dot fluoresces, the increased semiconducting effects.

Raman spectroscopy has recently gained popularity for the advanced chemical analysis of surfaces. Raman spectroscopy is a spectroscopic technique used in material science to study vibrational and rotational frequencies in a system. The technique measures shifts in inelastic scattering, or Raman scattering, of light from a visible, near infrared, or near ultraviolet light source. The shift in energy gives information about the material surface characteristics. In nanoscience, Raman spectroscopy is used to characterize surface properties materials, measure temperature, and determine crystallinity. The Raman signal units is a measurement of the ratio between the Stokes (down-shifted) intensity and anti-Stokes (up-shifted) intensity peaks.

30.2.3 Dynamic Light Scattering

Dynamic light scattering (DLS) (also known as Photon Correlation Spectroscopy [PCS]) is used for measuring the nanoparticle size and size distribution when suspended in a medium. In actuality, DLS measures the browning motion of the nanoparticles. Brownian motion is the random movement of suspended particles due to bombardment of the particulate by solvent molecules. As solvent molecules collide with the nanoparticles, they exert a force that is capable of moving the particles in a random direction. This random movement will be much more exaggerated in a small particle than in a large particle of the same composition over the same time period. This movement, otherwise known as the translational diffusion coefficient (D), can then be related to the particle's hydrodynamic radius via the Einstein–Stokes Equation 30.1:

$$d(H) = \frac{kT}{3\pi nD} \tag{30.1}$$

Combining D and Boltzmann's constant, k, with the physical parameters of the system, including a stable temperature T, and the viscosity of the solvent n, one might deduce the hydrodynamic radius of the nanoparticle of interest.

DLS is a common instrument in the nanotoxicology laboratory. It provides the researcher with a quick, cost-effective means of generating a size profile of nanomaterials in suspension. DLS output, on a machine such as the Zetasizer NanoZS, displays intensity on the dependent axis and size on the independent. In addition, DLS may also be used to measure biological samples such as proteins when in a biological buffer.

While DLS is widely used in laboratories, one must constantly be aware of its limitations. For example, in a polydisperse sample, the larger particles will scatter light many more times. More specifically, the intensity of scattered light is proportional to the diameter of the particle to the sixth power. Therefore, a 20 nm particle would scatter 64X as much light as a 10 nm particle (64,000,000 versus 1,000,000 U). This property is called Rayleigh scattering. First, it is difficult to measure single particles using light scattering when aggregates or agglomerates are present in the focal suspension, which is a major limitation of the instrument. Therefore, a polydisperse sample mixture may read as if it were void of the smaller fraction of nanoparticles, or simply be unsuitable for analysis via DLS. Second, many nanoparticles tend to agglomerate once in solution. This agglomeration state may yield particulates in the micrometer (10^{-6}) range. This in turn would make it hard to distinguish a particle size on the order of three orders of magnitude lower. Third, DLS measurements should be taken in pristine solutions. For example, a cell culture medium with the addition of fetal bovine serum contains many proteins that interact with the nanoparticle surface. While this alteration in the surface chemistry by the addition of FBS is known to play an important role in toxicology, this protein adsorption would make the nanoparticle size larger due to the increased hydrodynamic radius. In situations where biomolecules are utilized, it is important that control samples be run without the nanoparticle to determine the sizing profile of components in the suspension medium.

30.2.4 Zeta Potential

The zeta potential of a nanoparticle represents the charge on the surface in addition to the adsorbed layer of counter ions. The adsorbed layer, hereafter known as the electric double layer, is generated from oriented solute molecules and ions surrounding the nanoparticle surface. The electric double layer is sensitive to various changes in pH and ionic strength. All nanoparticles in suspension carry a net charge. This charge, be it negative, positive, or neutral, is dependent upon the micro environmental conditions of the suspension media. For example, all nanoparticles have an isoelectric point (IEP). This isoelectric point occurs when a particle's zeta potential has zero net charge (IEP: $Z = 0\,mV$). When this occurs in the Nanosizer ZS system, the particle experiences no movement in response to alternating polarity in the test cuvette. Recent work in our lab has shown that the IEP may play a role in the toxicity of nanomaterials (Berg et al. 2009). One hypothesis that has arisen is the fact that the charge at physiological pH may direct certain protein modifications of the nanoparticle surface. This modification may then mediate cellular response.

The zeta potential is also used as a measurement of colloidal stability. A zeta potential value above $\pm 30\,mV$ is considered a stable system. Zeta potential values with an absolute value less than 30 mV are considered unstable systems and are prone to agglomeration. This agglomeration is due to the lack of a charge–charge repulsion between individual nanoparticles.

30.3 In-Line Characterization

Nanomaterial characterization normally occurs independently to either in vitro or in vivo exposure. However, this provides many opportunities for variability in an experiment (i.e., nanoparticle synthesis variation, sample contamination). An alternative to independent or off-line sampling would be to couple the nanomaterial characterization to the exposure. This online measurement is performed in series or parallel to the exposure chamber and would allow precise characterization as seen for the test organism. The following methods have been utilized as in-line methods for nanotoxicological exposures in both an aerosolized form and in an aqueous or organic solvent.

30.3.1 Differential Mobility Analyzer

Many physicochemical properties are influenced when introduced to an aqueous environment that occurs during in vitro and in vivo testing. As previously mentioned, this change in environmental conditions may lead to an altered agglomeration state. However, engineered nanoparticles, in an aerosolized suspension, are sometimes more monodisperse and may lead to altered toxicity. Determining particle size in an aerosol can be accomplished through the use of a differential mobility analyzer (DMA). Furthermore, the DMA is often coupled with a condensation particle counter (CPC) or an aerosol particle mass analyzer (APM) in order to identify particle size distribution. The use of this measurement parameter allows the correct identification of dose in an aerosolized exposure.

The DMA is an instrument that can sort ultrafine particles according to size while aerosolized. To begin, DMA applies a charge to particles in the sample. These charged particles are then passed through a chamber consisting of a central rod and an air flow. As the particles pass through this environment, they are separated by their mobility. Mobility is determined by a combination of particle size and electrical charge. It is important to note that in these situations only particles that have been given a positive charge are measured, and those with a negative charge are repelled from the central rod and exit the instrument in the exhaust flow. Electrical mobility, Z_p, is dependent upon the average particle diameter, D_p; gas viscosity, □□; and the number of charges on the particle, n, according to Equation 30.2 (Willeke and Baron 2005; Stevens et al. 2008):

$$Z_p = \frac{neC}{3\pi\mu D_p} \tag{30.2}$$

For extremely small nanoparticles (<~70 nm) the Cunningham Correction, C, becomes a necessary addition to the formula to correct nanoparticle drag. Lastly, e is equal to 1.602×10^{-19} C.

There remains much debate on the correct method for dosimetry of nanomaterials used for toxicological studies; be it mass, surface area, volume, or particle number. Ostraat et al. (2005) suggests the use of a radial DMA (RDMA). RDMA is the most relevant for nanoparticle classification due to its extremely high transmission efficiency (Ostraat et al. 2005). Developing a complete aerosol exposure system composed of a nanoparticle source in series with a DMA and CPC or APM, remains the most complete way of characterizing nanoparticles, as they are seen in ambient exposure conditions in an aerosolized form (Oyabu et al. 2007). It has also been hypothesized that the physicochemical parameters may be responsible for toxicity; however, for an in vivo exposure, both these properties (i.e., primary particles versus agglomerates) may be explored. For example, nanoparticle agglomerates are not broken up in the DMA; however, they may be extracted prior to analysis through the use of an inertial impactor that would eliminate a large portion of the micron-sized agglomerates prior to introduction to the DMA (Ma-Hock et al. 2007; Scheckman and McMurry 2009). In addition, exposure to primary size nanoparticle may be accomplished through a method that combines synthesis, DMA characterization, and exposure in series.

30.3.2 Flow-Field Fractionation

Field-Flow fractionation (FFF) is a separation technique that comprises elements of chromatography with field-driven techniques such as ultracentrifugation and electrophoresis (Giddings 1993). These elution techniques can be used to separate complex mixtures of nanoparticles, macromolecules, and other particulate matter across a broad range of sizes from ~1 nm up to 100 μm. The separation of complex mixtures occurs as the sample mixture (in suspension) passes through a complex field that is oriented perpendicular to the direction of flow. This field applies differential amounts of force on nanoparticles with different properties (i.e., size, density, and charge). However, the field in FFF does not induce the separation of nanoparticles but forces the particulates against the accumulation wall where particulates may then be separated by differential flow rates of particulates with different physicochemical properties.

FFF is comprises multiple techniques that differ in both the field applied and the mechanism of separation. In sedimentation-FFF, particles are driven against a wall as a centrifugal force is applied. Sedimentation-FFF has allowed the detection of unlabeled nanoparticles when extracted from an ex vivo sample, in addition to determining the sizing profiles of those extracted particles (Deering et al. 2008). While sedimentation-FFF may be used to achieve high selectivity among samples, it often has limitations in the range of 10–30 nm due to the high speeds necessary to achieve separation (Giddings 1993).

A more appropriate technique used to elute small nanoparticles (<10 nm) would be flow-FFF. Flow-FFF is a density independent technique (unlike sedimentation-FFF) in which a perpendicular flow drives the nanoparticles against the accumulation wall. Flow-FFF has been used to analyze ~25 nm TiO_2 particles in commercial sunscreen (Contado and Pagnoni 2008). In this case, flow-FFF was paired with ICP-AES to generate both the size property and the TiO_2 content, measured by the amount of Ti in the sample. Flow-FFF has also been used as a way to monitor the growth of nanoparticles during synthesis (Chen et al. 2005a).

30.4 Characterization In Vitro and In Vivo

While many nanomaterial properties may be readily measured in an acellular environment, the toxicologist strives to determine the response of an organism to such materials. Current nanotoxicology protocols suggest that careful characterization of nanomaterials is performed prior to in vitro or in vivo use. However, as beneficial as these measurements are, many of the nanoparticle properties change when introduced to an aqueous or physiological system. For example, nanomaterials will agglomerate when introduced to a phosphate buffered saline (PBS) or a cell culture media (CCM) sample (Sager et al. 2007; Murdock et al. 2008). Due to the differences in physicochemical properties when suspended in a medium, there has been an attempt to utilize ex vivo analyses such as the vitamin C assay and the hemolytic potential assay to predict nanomaterial toxicity (Warheit et al. 2007b). Furthermore, florescence-based assays, such as the 2′,7′-dichloroflourescin diacetate (DCFH-DA) assay, may be used as an estimation of oxidative stress in an in vitro setting.

30.4.1 Vitamin C Assay

The surface of a nanoparticle has been identified as a potential source of leached ions and reactive species. There are a variety of techniques, and an array of colorimetric probes can be used to measure the reactivity of the surface of a nanoparticle. Examples of techniques include zeta potential, isoelectric point, hemolytic potential, and electron spin resonance (ESR) or electron pair resonance (EPR). Examples of probes are the photodegradation of aqueous Congo Red and degradation of Vitamin C.

In regard to identifying and quantifying the reactivity of a particle's surface, it is important to note that there is no single technique that can be used for all nanoparticle types. One example of measuring the surface activity of a material is photocatalytic degradation. Another metric is delta b* using the Vitamin C assay. The Vitamin C Yellowing test has been developed to correlate the surface of a nanoparticle, such as TiO_2, to its chemical stability. Available evidence indicates that the mechanism for color development in the test is the formation of a charge transfer complex between ascorbic acid 6-palmitate and the "active sites" on the ultrafine TiO_2 surface (Rajh et al. 1999; Warheit et al. 2007b). The redox reaction of ascorbic acid is shown in Figure 30.2.

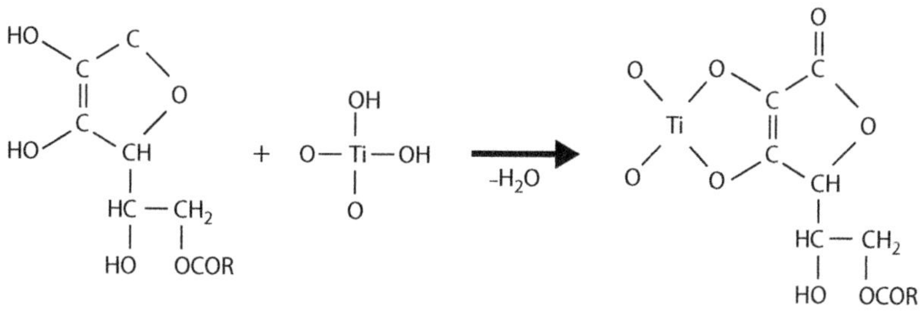

FIGURE 30.2 Proposed structure of TiO_2 nanoparticle with ascorbic acid 6-palmitate.

30.4.2 Hemolytic Potential

The hemolytic assay provides a rapid and cost-effective way of determining the effects of particles on a biological membrane. The particulate assay has been used in the past century with a variety of particulates including silicate powders, asbestos, and, more recently, a multitude of nanoparticles (Harington et al. 1971; Nolan et al. 1981; Warheit et al. 2007). Hemolysis (i.e., rupturing of erythrocytes) is an important ex vivo characterization method that may be used in addition to assessing cell death. The ability of nanoparticles to act on a biological membrane is extremely important as such particles are incorporated into medicinal devices for both therapeutic and imaging purposes.

The erythrocyte provides a unique model as a biological membrane system. During erythropoesis, the mammalian erythrocyte extrudes the nucleus and organelles to make room for increasing amounts of hemoglobin. Hemoglobin is an intracellular metalloprotein used by the body to transport oxygen to the tissues and to some extent carbon dioxide, back to the site of oxygen exchange in the respiratory tract. Hemoglobin's centrally located heme group, which consists of an iron (Fe^{2+}) atom covalently bonded to a porphyrin ring, reads spectrophotometrically at a wavelength of 540 nm. The absorbance at this wavelength provides the basis of the assay endpoint.

Particulates, such as nanomaterials, interact with the erythrocyte and exert their cytotoxic effect via a multitude of hypothesized mechanisms dependent upon a variety of characteristics influencing surface reactivity including size, composition, and zeta potential (Nolan et al. 1981; Mayer et al. 2009). First, the direct mechanical interaction between the particulates and the membrane can cause rupture releasing intracellular hemoglobin into the ambient solution. Second, several nanoparticles may act on the membrane of the erythrocyte indirectly via actively oxidizing membrane components and thus increase membrane permeability (Li et al. 2008). Third, altering the ionic composition of the suspension medium through nanoparticle weathering will increase the amount of hemoglobin in the assay medium.

While the hemolytic assay provides rapid and cost-effective means of screening a multitude of nanoparticles, intra-assay variation creates an obstacle when comparing hemolytic assay results within the nanotoxicology community (Dobrovolskaia et al. 2008). Literature suggests that intraspecies variation exists when comparing the susceptibility of the erythrocyte to lysis, thus complicating the hemolytic model. Other factors that may influence intra-assay variation are concentration of erythrocytes used (2%–4%), time from harvest to assay, and time of incubation of erythrocyte with nanoparticles of interest. Interestingly, with the testing of nanoparticles such as colloidal gold (absorbance □ = 540 nm), it must be ensured that the nanoparticles do not interfere with the absorbance of heme. Additional assay interference may stem from the rare occurrence of heme absorbance onto the surface of the nanoparticles. This phenomenon will lead to reduced hemoglobin in the supernatant as the adsorbed hemoglobin will precipitate out at high speed centrifugation. Due to this nanoparticle dependent absorbance, it is likely that the hemolytic assay will need to be adapted to each nanoparticle with different physicochemical properties.

30.4.3 Dichlorofluorescin Probe

One central hypothesis in nanotoxicology is that nanoparticles are able to induce cellular dysfunction according to an oxidative stress paradigm. This oxidative stress occurs when intracellular levels of reactive oxygen species (ROS) (O_2^-, HO_2, H_2O_2, and HO) overcome the capability of the cellular defense mechanisms (antioxidants and enzymes) to adequately respond. The 2′,7′-dichloroflourescin diacetate (DCFH-DA) assay has been used repeatedly as an estimation of reactive oxygen species (ROS) production (Wang and Joseph 1999; Lin et al. 2006). In theory, DCFH-DA passively diffuses through the cell membrane. Once inside the cell, DCFH-DA is hydrolyzed by intracellular enzymes to the non-florescent compound DCFH. Intracellular ROS then oxidize the non-fluorescent DCFH to florescent DCF, which may then be read on a conventional plate reader.

Adapting this assay for use with non-florescent nanoparticles can provide a quick, cost-effective means of screening many nanoparticles for estimation of ROS production.

30.5 Biological Response In Vitro and In Vivo

Nanoparticles are known to interact with living organisms down to the cellular level. They have been visualized in in vitro cell culture systems as well as in an in vivo model. Furthermore, both carbonaceous and colloidal nanoparticles have been shown to translocate various biological membrane barriers (epithelium, endothelium, or mucosa) and may eventually enter the bloodstream (Oberdorster et al. 2004; Larese et al. 2009). Once in the bloodstream, nanoparticles (due to their size and surface charge) may enter cross barriers (i.e., blood brain barrier) normally impermeable to other molecules. This increased permeability combined with cell specificity has allowed researchers the enhanced ability to target malignancies in difficult to treat areas (i.e., brain) with increased efficiency (Brigger et al. 2002).

30.5.1 Cellular Response to Oxidative Stress

The presence of excessive ROS can cause cellular oxidative stress, which may lead to subcellular damage such as DNA, RNA, protein, mitochondrial, and membrane or other lipid degradation. One of the more commonly probed oxidative stress indictors is protein carbonyl production. The oxidation of protein is damaging to polypeptides and amino acids present in cells and tissues. Protein Carbonyl Content (PCC) and 3-Nitrotyrosine are examples of assays that measure oxidative protein damage. Other techniques used to evaluate the oxidant versus antioxidant biomarkers in vitro are enzyme-linked immunosorbant assays (ELISA), ion exchange chromatography, immunoblotting, and electron paramagnetic resonance.

Some oxidative stress mechanisms are known that describe how pro-oxidants cause protein damage. Alterations in protein function can occur after the formation of covalent bonds between electrophiles and nucleophilic amino acids, the oxidation of nucleophilic amino acids, or the production of reactive nucleophiles.

30.5.2 Inflammatory Response

Studies have shown that exposure to nanoparticles to in vitro and in vivo systems may cause the production of inflammatory biomarkers (Warheit et al. 2004; Gurr et al. 2005; Sayes et al. 2006; Zhu et al. 2006). After nanoparticles are internalized by cell, such as phagocytes, the inflammatory cascade may be triggered. Inflammation is the complex biological response of cells and tissues to harmful pathogens and other toxicants. The inflammatory cascade is both a proactive mechanism to remove harmful pathogens and to initiate the production of repair enzymes. Unchecked inflammation can lead to a host of diseases, such as asthma, atherosclerosis, and rheumatoid arthritis; therefore, it is normally tightly regulated by the body.

30.5.3 Cell Damage along a Genotoxic Pathway

Nanoparticles have been previously shown to exhibit damage to DNA through DNA oxidation, causing DNA strand breaks as well as binding to DNA at the molecular level (Tsoli et al. 2005). Currently, most nanotoxicology researchers hypothesize that nanoparticles have the potential to induce cell damage in two distinct pathways. First, the small-size and narrow-size distribution of nano-sized particulate matter could lead to stronger interactions with proteins causing structural changes or loss of function due to that fact that particles at the nano scale are about the same size as indigenous molecules within the cell. Because both nanoparticles and molecules, such as proteins, have similar sizes, some researchers believe that they may bind together resulting is disrupted normal cellular function. The second potential toxic pathway is due to the leaching of ions from the surface of a nanoparticle to the surrounding cellular matrix. These ions can accumulate in tissues causing, for example, metal poisoning or even cell death. After endocytosis, some nanoparticles disrupt throughout the cell and preferentially deposit on the nuclear or mitochondrial membrane. While it is unclear whether particles translocate through these membrane, damage to both mitochondrial and nuclear DNA has been found (Unfried et al. 2007; AshaRani et al. 2009).

30.5.4 Cellular Uptake

Nanoparticles are novel tools in biological and biomedical applications; however, some inherent risks are associated with their incorporation into consumer and medicinal products. An interaction of engineered nanoparticles occurs on a size scale previously seen with the entrance of viral pathogens into a cellular system. Similarly, the uptake of nanoparticles provide a unique way of accessing intracellular compartments albeit intentionally (i.e., medicinally) or unintentionally (i.e., occupationally). In both situations, nanoparticles are capable of intracellular localization through a variety of endocytic mechanisms ranging from the receptor-mediated clathrin mediated endocytosis (CME) to nonspecific macropinocytosis to simple passive diffusion (Qaddoumi et al. 2003; Geiser et al. 2005; Rothen-Rutishauser et al. 2006; Harush-Frenkel et al. 2007a,b; Jiang et al. 2008). Many physicochemical characteristics such as nanoparticle size, degree and type of surface coating, and surface charge are factors influencing nanoparticle endocytosis.

Size is an important characteristic in nanoparticle endocytosis. Many endocytic mechanisms, such as CME and caveolin-mediated endocytosis are structurally size-limited by the protein scaffolding lining the vesicle budding from the membrane surface. Furthermore, a variety of thermodynamic limitations exist that pertain to membrane folding. Calculations based on thermodynamic limitations and cell–ligand interactions have previously calculated that the most efficient receptor-mediated nanoparticle endocytosis occurs at ~30 nm for spherical particles (Freund and Lin 2004; Gao et al. 2005). These confirmations are further strengthened by observations of the uptake efficiency of herception conjugated gold nanoparticles peaking at particle diameters of 25–50 nm (Jiang et al. 2008). While primary particle size is important, nanoparticles may also agglomerate to the micrometer scale. In these instances, these agglomerates may be treated as larger particles. For example, phagocytosis by immune-modulators (macrophages) is responsible for the clearance of micrometer-sized particulates leading to the view that these innate defenses may readily clear micrometer-sized agglomerates of nanoparticles (Oberdorster et al. 2005; Kemp et al. 2008). While size and agglomeration state play a primary role in directing the specific route in nanoparticle uptake, they are not the only factors taken into play.

Surface charge also plays a role in the interactions found between nanoparticles and cell membranes (Harush-Frenkel et al. 2007b; Zhang 2009). For example, negatively charged nanoparticle surfaces, such as particles coated with hydrolipic acids, and positively charged nanoparticles, such as particles coated with polymers, are more capable of incorporation into human cells than particles with little to no surface charge. Furthermore, the charge of a nanoparticle will also dictate the pathway through which the nanoparticle is internalized (Harush-Frenkel et al. 2007b). For example, while uptake of both positively and negatively charged particles are energy and F-actin polymerization dependent, they have been shown to undergo different pathways including macropinocytosis and CME (for the former) and possible dynamin-independent mechanisms (for the latter) (Dausend et al. 2008). While associations between charge and mechanisms of cellular internalization have been made, further research into this topic in needed. In addition to nanoparticle charge, a modified surface coating also influences the uptake by cells.

Another factor dictating nanoparticle cellular uptake and potential toxicity is the adsorption of proteins onto the surface of the nanoparticles. For example, proteins have been shown to coat particles when placed in cell culture media, serum, and bronchiolar lavage fluid (Chen et al. 2005b, 2006; Huang et al. 2007; Lundqvist et al. 2008; Schellenberger et al. 2008). This degree of protein coating or functionalization is dependent upon psychochemical factors such as size and shape. For example, antibody conjugation onto the surface of a nanoparticle increases linearly with respect to the radius (Jiang et al. 2008).

This association between nanoparticle size and degree of protein binging is due to the high rate of curvature exhibited by the surface of a small nanoparticle that sterically hinders the binding of additional antibodies (Ghitescu and Bendayan 1990). Protein adsorption may be dramatically altered as any single physiochemical change is made. It is this aforementioned protein adsorption theory that necessitates the importance of proper characterization of nanomaterials as well as puts forth the complicating task of linking nanomaterial physicochemical properties with their biological endpoint.

30.6 Standardization of Methods

Along with the growing amount of nano-focused environmental health and safety researchers, published literature, organized meetings and conferences, and student training programs, the nanotoxicology community is making substantial progress toward standardizing methods and techniques used in this nascent scientific discipline. Experts from fields of toxicology, medicine, chemistry, biology, physics, and engineering study issues including hazard identification, exposure science, risk assessment and management, and public health and policy on a daily basis. As the field moves forward in making critical decisions in standardizing laboratory practices and operating procedures, it is critical that novel techniques and nontraditional use of techniques continue to be researched and published in the peer-reviewed literature.

As commented in the 2008 *Nanotoxicology 2nd International Conference* proceedings by Krug (2009), it is essential to compare the experimental designs and resultant data sets that are produced by the people working in nanotoxicological laboratories. At that time, there is an overwhelming criticism to the nanotoxicology community regarding the lack of methods standardization and the differing results that have been published. Most researchers agree, however, that the inherent nature of nanoparticles gives rise to contradictory nanotoxicity reports. Some of the physicochemical parameters that lead to the so-called discrepancies include nanoparticle synthesis/production conditions, the use, misuse, or absence of colloidal stabilizers in particle suspensions, dosimetric and concentration measurements, exposure time, and interference of traditional colorimetric probes with particle spectroscopic features. Many working groups and subsequent review manuscripts have identified the following critical needs for the future of nanotoxicology research:

- Toxicokinetic and toxicodynamic profiles of nanoparticles in in vivo systems
- Identification of both toxic and nontoxic cellular responses for nanoparticles along specific pathways
- Additional exposure evaluation research
- Refinement of existing and development of new analytical tools for identifying and measuring nanoparticles in complex matrices
- Direct and indirect effects of nanoparticles on the immune system and genotoxicity
- Development of predictive models and open access databases of nanotoxicity results

Without chemical and physical characterizations of the materials and their surrounding matrix, research on their toxicological effects may not be useful because it would be difficult to make substantial conclusions or draw comparisons between the relationship between specific particle types and their mode of action in biological systems. Furthermore, development of risk assessment models is hindered until such standardizations are in place.

30.6.1 Physicochemical Characterization Data in the Nanotoxicology Literature

The properties of nanomaterials are predominantly associated with their nanometer-scale size and structure, size and structure-dependent electronic configurations, and an extremely large surface area-to-volume ratio relative to larger-sized chemicals and materials. The main characteristic of nanomaterials is their size, which falls in the transitional zone between individual atoms or molecules and the corresponding bulk materials (Nel et al. 2006). Particle size and surface area are important material characteristics from a toxicological and health perspective because as the size of a particle decreases, its surface area increases, which allows a greater proportion of its atoms or molecules to be displayed on its surface rather than within the interior of the material. These atoms or molecules on the surface of the nanomaterial may be chemically and biologically reactive, potentially contributing to the development of adverse health effects. Other physical and chemical properties such as shape, surface coating, aggregation potential, and solubility may also affect the physicochemical and transport properties of the nanomaterial with the possibility of negating or amplifying any associated size-related effects.

30.6.2 Conclusions

The field of nanotoxicology is growing, but has its roots in the toxicology of ultrafine particles. Nanoparticles will require careful evaluation with reference to routes of exposure, dosemetrics, biocompatibility, and toxicokinetics/toxicodynamics. The inherent size scale of nanomaterials links the immune systems as an obvious target. Nonetheless, the physicochemical and biological rules by which adverse events will be determined are poorly understood. Going forward, the enterprise will continue to require cross-collaboration between physical and biological scientists and highly integrated training of the next generation of scientists and engineers.

References

AshaRani, P. V., G. L. K. Mun, M. P. Hande, and S. Valiyaveettil. 2009. Cytotoxicity and genotoxicity of silver nanoparticles in human cells. *ACS Nano* 3 (2):279–290.

Berg, J. M., A. Romoser, N. Banerjee, R. Zebda, and C. M. Sayes. 2009. The relationship between pH and zeta potential of ~30 nm metal oxide nanoparticle suspensions relevant to in vitro toxicological evaluations. *Nanotoxicology*. In press.

Brigger, I., J. Morizet, G. Aubert, H. Chacun, M. J. Terrier-Lacombe, P. Couvreur, and G. Vassal. 2002. Poly(ethylene glycol)-coated hexadecylcyanoacrylate nanospheres display a combined effect for brain tumor targeting. *J. Pharmacol. Exp. Ther.* 303 (3):928–936.

Chen, C., M. C. Daniel, Z. T. Quinkert, M. De, B. Stein, V. D. Bowman, P. R. Chipman, V. M. Rotello, C. C. Kao, and B. Dragnea. 2006. Nanoparticle-templated assembly of viral protein cages. *Nano Lett.* 6 (4):611–615.

Chen, B. L., H. J. Jiang, Y. Zhu, A. Cammers, and J. P. Selegue. 2005a. Monitoring the growth of polyoxomolybdate nanoparticles in suspension by flow field-flow fractionation. *J. Am. Chem. Soc.* 127 (12):4166–4167.

Chen, C., E. S. Kwak, B. Stein, C. C. Kao, and B. Dragnea. 2005b. Packaging of gold particles in viral capsids. *J. Nanosci. Nanotechnol.* 5 (12):2029–2033.

Contado, C. and A. Pagnoni. 2008. TiO_2 in commercial sunscreen lotion: Flow field-flow fractionation and ICP-AES together for size analysis. *Anal. Chem.* 80 (19):7594–7608.

Dausend, J., A. Musyanovych, M. Dass, P. Walther, H. Schrezenmeir, K. Landfester, and V. Mailander. 2008. Uptake mechanism of oppositly charged florescent nanoparticles in HeLa cells. *Macromol. Biosci.* 8 (12):1135–1143.

Deering, C. E., S. Tadjiki, S. Assemi, J. D. Miller, G. S. Yost, and J. M. Veranth. 2008. A novel method to detect unlabeled inorganic nanoparticles and submicro particles in tissues by sedimentation field-flow fractionation. *Part. Fiber Toxicol.* 5:18.

Dobrovolskaia, M. A., J. D. Clogston, B. W. Neun, J. B. Hall, A. K. Patri, and S. E. McNeil. 2008. Method for analysis of nanoparticle hemolytic properties in vitro. *Nano Lett.* 8 (8):2180–2187.

Freund, L. B. and Y. Lin. 2004. The role of binder mobility in spontaneous adhesive contact and implications for cell adhesion. *J. Mech. Phys. Solids* 52:24455–24472.

Gao, H., W. Shi, and L. B. Freund. 2005. Mechanics of receptor-mediated endocytosis. *Proc. Natl Acad. Sci. U. S. A.* 102:9469–9474.

Geiser, M., B. Rothen-Rutishauser, N. Kapp, S. Schurch, W. Kreyling, H. Schulz, M. Semmler, V. Im Hof, J. Heyder, and P. Gehr. 2005. Ultrafine particles cross cellular membranes by nonphagocytic mechanisms in lungs and in cultured cells. *Environ. Health Perspect.* 113 (11):1555–1560.

Ghitescu, L. and M. Bendayan. 1990. Immunolabeling efficiency of protein-a-gold complexes. *J. Histochem. Cytochem.* 38 (11):1523–1530.

Giddings, J. C. 1993. Field-flow fractionation—Analysis of macromolecular, colloidal, and particulate materials. *Science* 260 (5113):1456–1465.

Gurr, J. R., A. S. S. Wang, C. H. Chen, and K. Y. Jan. 2005. Ultrafine titanium dioxide particles in the absence of photoactivation can induce oxidative damage to human bronchial epithelial cells. *Toxicology* 213 (1–2):66–73.

Harington, J. S., K. Miller, and G. Macnab. 1971. Hemolysis by asbestos. *Environ. Res.* 4 (2):95–117.

Harush-Frenkel, O., N. Debotton, S. Benita, and Y. Altschuler. 2007a. Targeting of nanoparticles to the clathrin-mediated endocytic pathway. *Biochem. Biophys. Res. Commun.* 353 (1):26–32.

Harush-Frenkel, O., E. Rozentur, S. Benita, and Y. Altschuler. 2007b. Surface charge of nanoparticles determines their endocytotic and transcytotic pathway in polarized MDCK cells. *Biomacromolecules* 9:435–443.

Huang, X. L., L. M. Bronstein, J. Retrum, C. Dufort, I. Tsvetkova, S. Aniagyei, B. Stein et al. 2007. Self-assembled virus-like particles with magnetic cores. *Nano Lett.* 7 (8):2407–2416.

Jiang, W., B. Y. Kim, J. T. Rutka, and W. C. Chan. 2008. Nanoparticle-mediated cellular response is size-dependent. *Nat. Nanotechnol.* 3 (3):145–150.

Kemp, S. J., A. J. Thorley, J. Gorelik, M. J. Seckl, M. J. O'Hare, A. Arcaro, Y. Korchev, P. Goldstraw, and T. D. Tetley. 2008. Immortalization of human alveolar epithelial cells to investigate nanoparticle uptake. *Am. J. Respir. Cell Mol. Biol.* 39 (5):591–597.

Krug, H. F. 2009. Nanotoxicology 2nd International Conference, September 7–10, 2008 paper. *Nanotoxicology* 3 (3):173.

Larese, F. F., F. D'Agostin, M. Crosera, G. Adami, N. Renzi, M. Bovenzi, and G. Maina. 2009. Human skin penetration of silver nanoparticles through intact and damaged skin. *Toxicology* 255 (1–2):33–37.

Li, S. Q., R. R. Zhu, H. Zhu, M. Xue, X. Y. Sun, S. D. Yao, and S. L. Wang. 2008. Nanotoxicity of TiO(2) nanoparticles to erythrocyte in vitro. *Food Chem. Toxicol.* 46 (12):3626–3631.

Lin, W., Y. W. Huang, X. D. Zhou, and Y. Ma. 2006. In vitro toxicity of silica nanoparticles in human lung cancer cells. *Toxicol. Appl. Pharmacol.* 217 (3):252–259.

Lundqvist, M., J. Stigler, G. Elia, I. Lynch, T. Cedervall, and K. A. Dawson. 2008. Nanoparticle size and surface properties determine the protein corona with possible implications for biological impacts. *Proc. Natl Acad. Sci. U. S. A.* 105 (38):14265–14270.

Ma-Hock, L., A. O. Gamer, R. Landsiedel, E. Leibold, T. Frechen, B. Sens, M. Linsenbuehler, and B. van Ravenzwaay. 2007. Generation and characterization of test atmospheres with nanomaterials. *Inhal. Toxicol.* 19 (10):833–848.

Mayer, A., M. Vadon, B. Rinner, A. Novak, R. Wintersteiger, and E. Frohlich. 2009. The role of nanoparticle size in hemocompatibility. *Toxicology* 258 (2–3):139–147.

Murdock, R. C., L. Braydich-Stolle, A. M. Schrand, J. J. Schlager, and S. M. Hussain. 2008. Characterization of nanomaterial dispersion in solution prior to in vitro exposure using dynamic light scattering technique. *Toxicol. Sci.* 101 (2):239–253.

Nel, A., T. Xia, L. Madler, and N. Li. 2006. Toxic potential of materials at the nanolevel. *Science* 311:622–627.

Nolan, R. P., A. M. Langer, J. S. Harington, G. Oster, and I. J. Selikoff. 1981. Quartz hemolysis as related to its surface functionalities. *Environ. Res.* 26 (2):503–520.

Oberdorster, G., E. Oberdorster, and J. Oberdorster. 2005. Nanotoxicology: An emerging discipline evolving from studies of ultrafine particles. *Environ. Health Perspect.* 113 (7):823–839.

Oberdorster, G., Z. Sharp, V. Atudorei, A. Elder, R. Gelein, W. Kreyling, and C. Cox. 2004. Translocation of inhaled ultrafine particles to the brain. *Inhal. Toxicol.* 16 (6–7):437–445.

Ostraat, M., H. Atwater, and R. Flagan. 2005. The feasibility of inert colloidal processing of silicon nanoparticles. *J. Colloid Interface Sci.* 283:414–421.

Oyabu, T., A. Ogami, Y. Morimoto, M. Shimada, W. Lenggoro, K. Okuyama, and I. Tanaka. 2007. Biopersistence of inhaled nickel oxide nanoparticles in rat lung. *Inhal. Toxicol.* 19:55–58.

Qaddoumi, M. G., H. J. Gukasyan, J. Davda, V. Labhasetwar, K. J. Kim, and V. H. Lee. 2003. Clathrin and caveolin-1 expression in primary pigmented rabbit conjunctival epithelial cells: Role in PLGA nanoparticle endocytosis. *Mol. Vis.* 9:559–568.

Rajh, T., M. Nedeljkovic, L. X. Chen, O. Poluektov, and M. C. Thurnauer. 1999. Improving optical and charge separation properties of nanocrystalline TiO_2 by surface modification with vitamin C. *J. Phys. Chem. B* 103:3515.

Reynolds, E. S. 1963. The use of lead citrate at high pH as an electron-opaque stain in electron microscopy. *J. Cell. Biol.* 17:208–212.

Rothen-Rutishauser, B. M., S. Schurch, B. Haenni, N. Kapp, and P. Gehr. 2006. Interaction of fine particles and nanoparticles with red blood cells visualized with advanced microscopic techniques. *Environ. Sci. Technol.* 40 (14):4353–4359.

Sager, T., D. Porter, V. Robinson, W. Lindsley, D. E. Schwegler-Berry, and V. Castranova. 2007. Improved method to disperse nanoparticles for in vitro and in vivo investigation of toxicity. *Nanotoxicology* 1 (2):118–129.

Sayes, C. M., R. Wahi, P. A. Kurian, Y. P. Liu, J. L. West, K. D. Ausman, D. B. Warheit, and V. L. Colvin. 2006. Correlating nanoscale titania structure with toxicity: A cytotoxicity and inflammatory response study with human dermal fibroblasts and human lung epithelial cells. *Toxicol. Sci.* 92 (1):174–185.

Scheckman, J. and P. H. McMurry. 2009. Rapid characterization of agglomerate aerosols by in situ mass-mobility measurements. *Langmuir* 25 (14):8248–8254.

Schellenberger, E., J. Schnorr, C. Reutelingsperger, L. Ungethum, W. Meyer, M. Taupitz, and B. Hamm. 2008. Linking proteins with anionic nanoparticles via protamine: Ultrasmall protein-coupled probes for magnetic resonance imaging of apoptosis. *Small* 4 (2):225–230.

Stevens, J. P., J. Zahardis, M. MacPherson, B. T. Mossman, and G. A. Petrucci. 2008. A new method for quantifiable and controlled dosage of particulate matter for in vitro studies: The electrostatic particulate dosage and exposure system (EPDExS). *Toxicol. In Vitro* 22 (7):1768–1774.

Tsoli, M., H. Kuhn, W. Brandau, H. Esche, and G. Schmid. 2005. Cellular uptake and toxicity of AU(55) clusters. *Small* 1 (8–9):841–844.

Unfried, K., C. Albrecht, L. O. Klotz, A. Von Mikecz, S. Grether-Beck, and R. P. F. Schins. 2007. Cellular responses to nanoparticles: Target structures and mechanisms. *Nanotoxicology* 1 (1):52–71.

Wang, H. and J. A. Joseph. 1999. Quantifying cellular oxidative stress by dichlorofluorescein assay using microplate reader. *Free Radic. Biol. Med.* 27 (5–6):612–616.

Warheit, D. B., B. R. Laurence, K. L. Reed, D. H. Roach, G. A. Reynolds, and T. R. Webb. 2004. Comparative pulmonary toxicity assessment of single-wall carbon nanotubes in rats. *Toxicol. Sci.* 77 (1):117–125.

Warheit, D. B., T. R. Webb, V. L. Colvin, K. L. Reed, and C. M. Sayes. 2007a. Pulmonary bioassay studies with nanoscale and fine-quartz particles in rats: Toxicity is not dependent upon particle size but on surface characteristics. *Toxicol. Sci.* 95 (1):270–280.

Warheit, D. B., T. R. Webb, K. L. Reed, S. Frerichs, and C. M. Sayes. 2007b. Pulmonary toxicity study in rats with three forms of ultrafine-TiO_2 particles: Differential responses related to surface properties. *Toxicology* 230 (1):90–104.

Willeke, K. and P. A. Baron, eds. 2005. *Aerosol Measurement Principles Techniques and Applications.* Vol. 2. New York: Wiley-Interscience.

Zhang, L. W. 2009. Mechanisms of quantum dot nanoparticle cellular uptake. *Toxicol. Sci.* 110 (1):138–155.

Zhu, S. Q., E. Oberdorster, and M. L. Haasch. 2006. Toxicity of an engineered nanoparticle (fullerene, C-60) in two aquatic species, Daphnia and fathead minnow. *Mar. Environ. Res.* 62:S5–S9.

31

Hemocompatibility of Nanoparticles

Shankar J. Evani
University of Texas at San Antonio

Anand K. Ramasubramanian
University of Texas at San Antonio

31.1 Introduction

Over the past few years, nanoparticles have gained tremendous recognition in several diagnostic and therapeutic applications (Jamieson et al. 2007; Boisselier and Astruc 2009; Thanh and Green 2010). For instance, in 2006 alone, there were more than 250 nanoparticle-based products in pharmaceutical pipeline (Dobrovolskaia et al. 2008). A nanoparticle used for either therapeutics or for diagnostics is designed to deliver the drug at the site of action or perform its function at a desired site without being toxic to the body. For a nanoparticle to be biocompatible, understanding its probable interactions with the biological system is extremely important. Many factors including size, shape, nature, and composition of a nanoparticle, and route of administration decide the fate of the nanoparticle. In addition, the interference of the nanoparticle and its adverse effects on the host system also determine its use as a biocompatible material. Hence a bottom-up approach of nanobiomaterial design by evaluating the influence of the physical and chemical properties of the material on biological system is necessary. Humans have a very well-developed policing system called the immune system to protect us from invading organisms like bacteria, virus, and other parasites. Immune system is distributed throughout the body via the circulatory and lymphatic systems. Many nanoparticle products are designed for intravenous administration such as targeted and controlled drug delivery, as contrast agents for imaging or as vaccine carriers. However, the mechanism of interaction of nanoparticles with blood is not well understood. There is a possibility of a nanoparticle being recognized as foreign and its getting cleared from the blood by the immune system before it reaches the target. Once the nanoparticle is recognized by the immune system and sent for clearance or detoxification to either liver or kidney, it can create damage to the organs if it is toxic/nondegradable/incompatible. A pictorial representation of possible interactions and outcomes of nanoparticles with various cellular and molecular components of blood is shown in the Figure 31.1.

In light of these observations, the evaluation of the compatibility of nanoparticles in blood can be of critical in the design of the nanoparticle for in vivo application. Various assays can be used to evaluate the hemocompatibility including erythrocyte composure (hemolysis and hemagglutination),

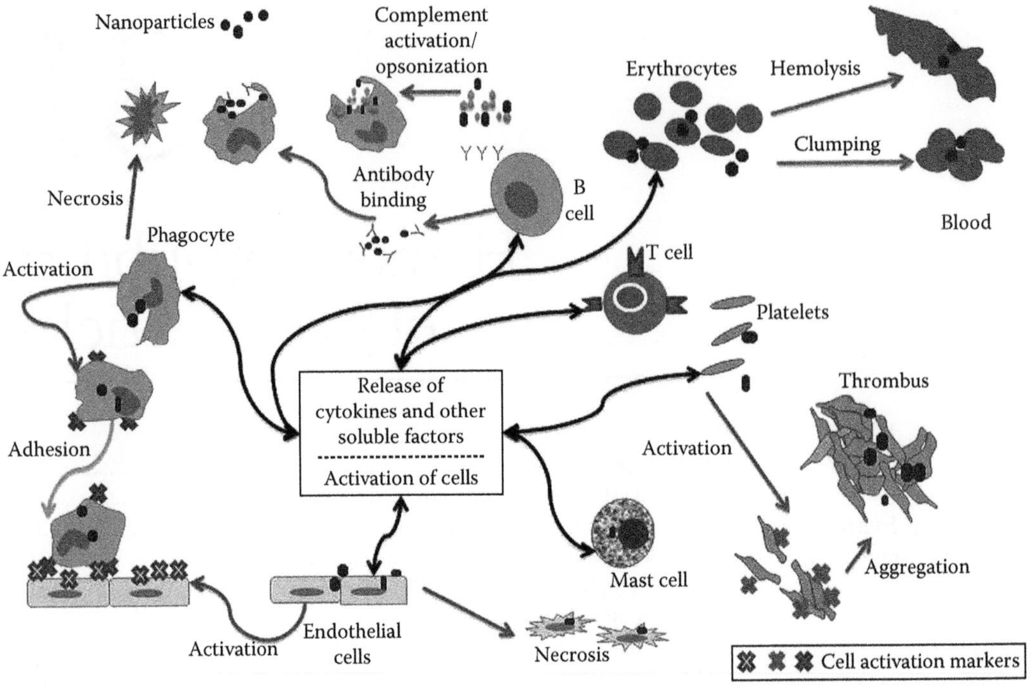

FIGURE 31.1 Interaction and responses of cellular and molecular components of blood with nanoparticles. Nanoparticles can interact with various cell types in blood, resulting in aggregation of RBC (hemagglutination), platelet activation and aggregation, and prevent/cause hypersensitivity by interacting with immune cells (Dellinger et al. 2010). Nanoparticles get coated with serum proteins, like clotting factors, and antibodies and complement proteins lead to thrombogenic response or the recognition and clearance of nanoparticles from blood by phagocytes, respectively. These phagocytes get activated (Gonçalves et al. 2010) by endocytosis of nanoparticles and secrete cytokines, which can further activate other cell types like endothelial cells, B cells, and T cells. Endothelium activation leads to expression of adhesion molecules like VCAM-1, ICAM-1, and ELAM-1 and cause phagocyte recruitment to vessel wall (Oesterling et al. 2008). These secreted cytokines further attract immune cells to the site of interaction, causing inflammation or excessive immune response and may exacerbate chronic inflammatory diseases like atherosclerosis (Greene 2008; Chang 2010).

thrombogenicity, cytokine assays, complement activation (protein adsorption), phagocytosis, and toxicity analyses as shown in Figure 31.2. In this short review, we will discuss the assay methods, their significance, and possible suggestions to improve the hemocompatibility evaluation during a nanomaterial design. We note that the choice of concentration of nanoparticles is important for all these assays, not only because of the physiological relevance but also due to the sensitivity of the assays (Figure 31.3).

31.2 Erythrocyte Composure

Erythrocytes (red blood cells [RBC]) are the major constituents of the blood contributing to nearly 45% of the blood volume. Erythrocytes contain hemoglobin, a red color heme protein that carries oxygen to all the tissues of the body for energy generation and also functions as carbon dioxide disseminator, a major waste from cellular respiration in animals. Hemoglobin deoxygenation and oxygenation in erythrocytes may end up in producing oxygen free radicals. These at normal levels can be cleared by the erythrocyte enzymes but will pose threat to the cell when abnormal levels of

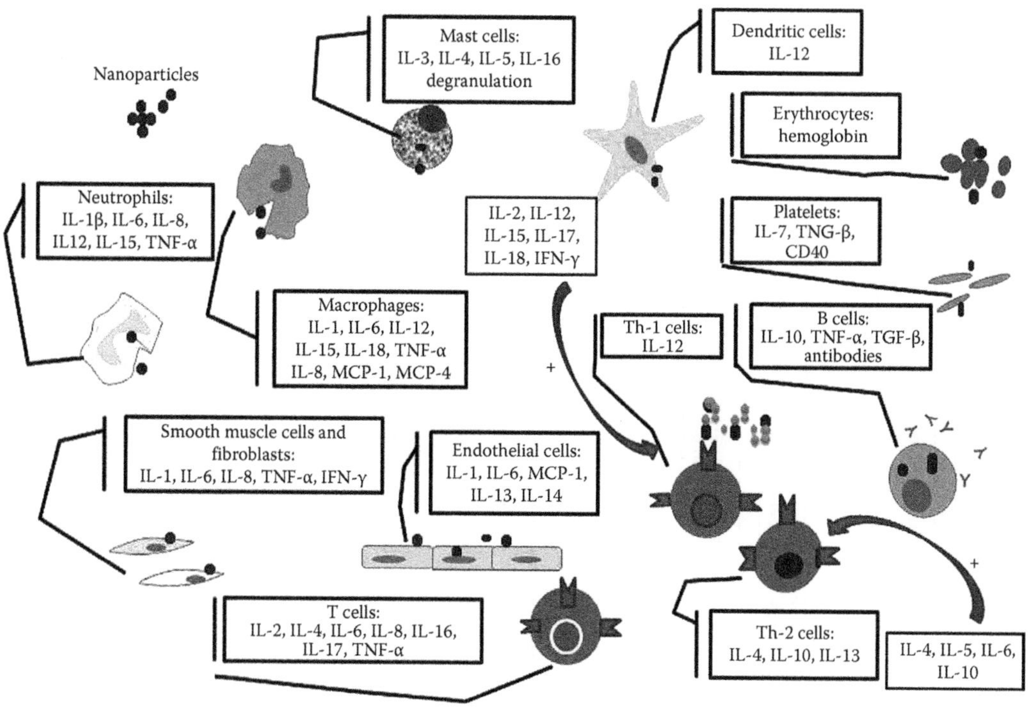

FIGURE 31.2 Different cell types and cytokine produced by them on activation by nanoparticles or in response to stimulus from other cytokines. Cytokines can activate different cell types and also regulate the immune responses to nanoparticle interaction by deciding the type of immune response, for instance, IL-12 and IFN-γ cause Th-1 type response and IL-4 and IL-10 lead to Th-2 type response (represented as "+") (Tedgui and Mallat 2001, 2006).

radicals deposit in the cells (Bisharova et al. 1998). Such abnormality can be due to toxic effects of various chemicals and can result in erythrocyte breakage or lysis. Further, erythrocytes are perfect osmometers where disturbances in osmotic and physical changes in blood can cause their lysis. Erythrocytes are also critical in maintaining proper flow characteristics in microcirculation, thus preventing vessel occlusion. In general, erythrocyte destruction can be caused by several factors such as bacterial infection, intrinsic erythrocytes membrane defects, mechanical shear due to blood flow, osmotic and pH changes in blood, drug-induced hemolysis, excessive complement, and platelet activation (Dourmashkin and Rosse 1966; Sowemimo-Coker 2002). Erythrocyte destruction causes release of hemoglobin and its excretion in urine, leading to hemoglobinuria, which is a diagnosis for "blood poisoning," a blanket symptom with many underlying causes. Erythrocyte composure may be compromised due to breakage and clumping of erythrocytes, which are referred to as hemolysis and hemagglutination, respectively. Hence erythrocyte lysis is a good indicator of toxicity of any foreign material to the cells in circulation.

The exposure of erythrocytes to nanoparticles may result in hemolysis and/or hemagglutination. Table 31.1 lists the effect of nanoparticles on erythrocytes reported in recent literature. Since hemolysis results in the release of red-colored hemoglobin, colorimetric assays can be used to estimate the extent of damage. Nanoparticles also act like antigens in bridging erythrocytes, causing hemagglutination, which results in clumping or agglutination to form diffuse lattice. This effect has serious consequences in humans leading to obstruction of blood vessels, imbalance in osmolarity of blood, and unavailability of erythrocytes to carry out normal functions. In these assays, blood from healthy human volunteer are collected and treated with nanoparticles and analyzed microscopically for hemolysis or hemagglutination. Assuming erythrocyte breakdown due to treatment

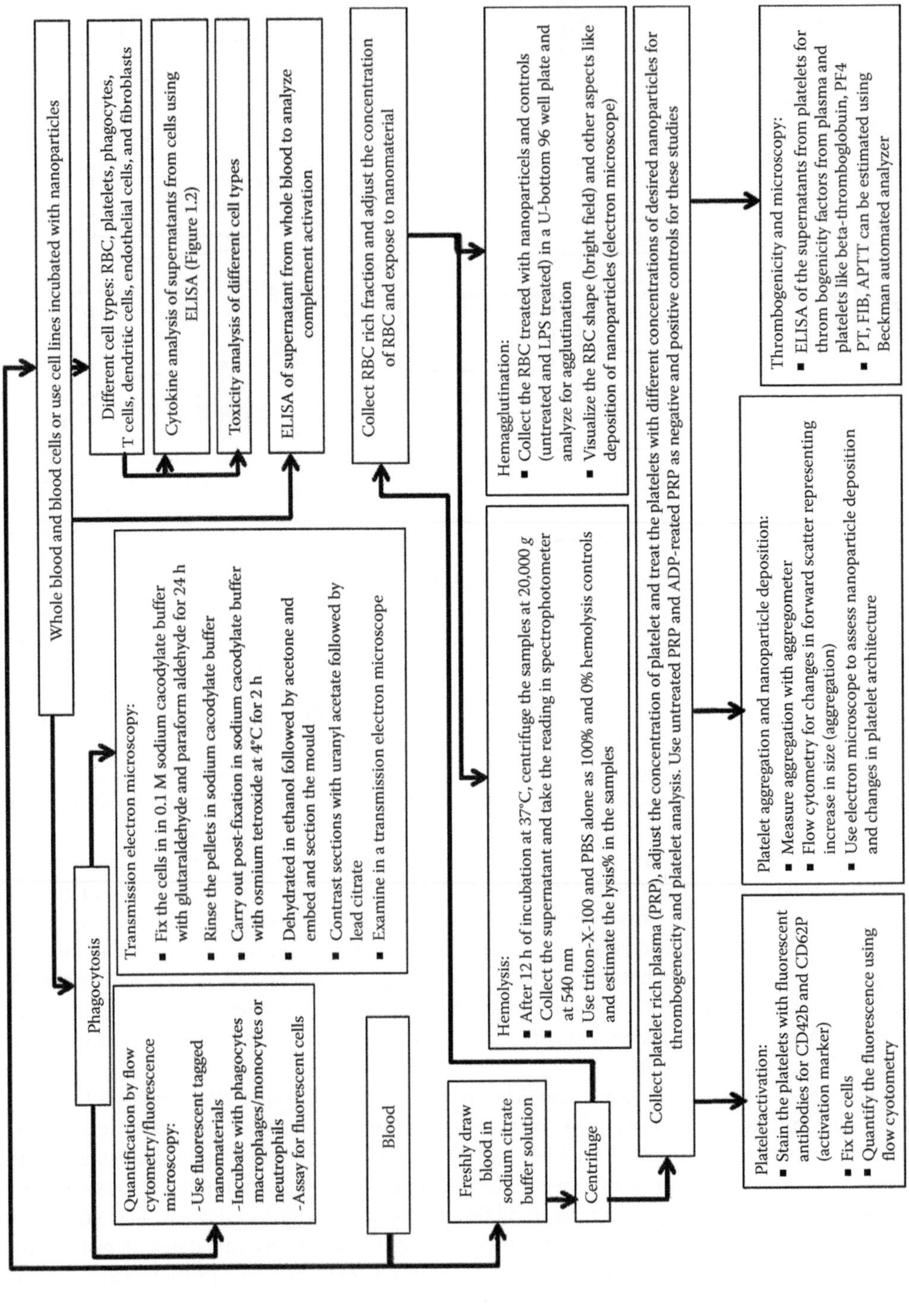

FIGURE 31.3 Assays to analyze hemocompatibility of nanoparticles.

TABLE 31.1 Erythrocyte Composure and Thrombogenicity Studies

Nanoparticle Type	Application	Hemolysis/Hemagglutination/ Thrombogenicity	References
Acylated chitosan nanoparticles	Drug carrier, wound healing	Modest hemolysis	Lee et al. (2004)
Chitosan with phosphorylcholine (PC) polymer	Drug delivery	Relatively low prothrombin activation	Meng et al. (2007)
Ferric oxide nanoparticles	Biomedical imaging	Elevated levels of fibrinogen	Zhu et al. (2008)
Lipid nanosphere	Drug delivery vehicles	Very low hemolysis	Fukui et al. (2003)
Multilayered Ti/TiN-coated surgical AISI 316L stainless steel	Orthopedic, cardiovascular and dental implants	Limited or no improvement in hemocompatibility when compared with uncoated implants	Chenglong et al. (2005)
PLGA nanoparticles	Drug detoxification	80% relative hemolysis by PLGA with surfactants	Kim et al. (2005)
Poly(ethylene glycol)-*block*-poly(caprolactone) nanoparticles	Nanoparticulate drug delivery vehicles	Variable response to activated partial thromboplastin time; minimum hemolysis	Letchford et al. (2009)
Poly(lactic-*co*-glycolic-acid)–multiwalled 3 carbon nanotube	Biomedical implant applications	Fibrinogen adsorption and platelet adhesion to nanoparticles	Koh et al. (2009)
Polyelectrolyte complexes based nanoparticles	Drug delivery vehicles	No adverse effects on viable cell count	Gupta et al. (2009)
Polystyrene latex nanoparticles	Imaging (MRI)	Differential effects of size and shape of nanoparticles on hemolysis	Mayer et al. (2009)
Stainless steel coated with dimercaptosuccinic acid, layers of chondroitin 6-sulfate and heparin	Drug eluting stents	Variable responses related to anticoagulant activity	Huang and Yang (2008)
TiO_2 nanoparticle	Food, cosmetic, and drugs	Abnormal sedimentation, hemagglutination	Li et al. (2008)

of cells with TritonX-100 as 100% (positive control), and untreated erythrocytes (negative control) as 0%, hemolysis can be calculated as

$$\%\text{Hemolysis} = \left[\frac{(\text{Abs of test} - \text{Abs of negative control})}{\text{Abs of positive control}}\right] \times 100$$

Hemagglutination of human blood in vitro can be visually analyzed on a scale of 0–1. A high hemagglutination reaction can be described as a granular or irregular ring occupying more than half of the well bottom by erythrocytes. PBS treated erythrocytes is used as control.

31.3 Thrombogenicity

Platelets (or thrombocytes) are nonnucleated cells that play a major role in blood clotting after an injury. Platelets secrete factors necessary to convert soluble fibrinogen, a blood protein, into insoluble polymeric fibrin, which forms a mesh at the site of injury that traps cells from flowing blood and causes the clotting of blood. Any foreign object, especially synthetic material, has at least minimum ability to induce thrombogenicity. Formation of blood clots without any injury is undesirable since it can result in the clogging of blood vessels due to the formation of platelet aggregates or thrombi that eventually lead to heart attack and stroke. Platelet aggregation due to nanoparticles can occur either by direct interaction with platelets or indirectly due to the adsorption of plasma protein fibrinogen onto nanoparticle surface or through the activation of other cells such as endothelial cells. A well-designed nanomaterial, which has minimum or no effect on blood clotting, will be the choice of administration into blood.

The interaction of nanoparticles with platelets release thrombin, which triggers intrinsic pathway of blood clotting, while the extrinsic pathway (tissue factor–dependent pathway) is activated rapidly in response to vascular injury when the factors in adventitia are released to flowing blood (Davie et al. 1991). The intrinsic and extrinsic pathways are essentially a tightly controlled activation and deactivation cascade of proteins in which inactive plasma zymogens undergo proteolytic cleavage to release the active clotting factors. The cascade culminates in the formation of thrombin from prothrombin, which catalyzes the polymerization of fibrinogen monomer to fibrin polymer. These insoluble fibrin clots and aggregated platelets constitute thrombus.

Thrombogenicity of nanoparticles may be estimated from blood clotting assays including the release of clotting factors from platelets, and from platelet activation and aggregation. Platelet-rich plasma (PRP) is obtained by separating erythrocytes from whole blood by centrifugation. The erythrocyte fraction may be used for hemolysis and hemagglutination assays as described above. The PRP is incubated with nanoparticles and the thrombogenicity of nanoparticles can be estimated by the following assays: Release of various factors responsible for thrombus formation such as β-thromboglobuin or platelet factor 4 (PF4) can be used as markers of thrombogenicity and can be analyzed using commercially available ELISA kits. Adhesion of nanoparticles to platelets and the resulting morphological changes in platelets can be analyzed by LDH assay and electron microscopy (Nonckreman et al. 2010). Platelet activation is estimated from the upregulation of platelet receptors glycoprotein (GP) IIb/IIIa (CD 41) and P-selectin (CD 62P) using flow cytometry. Platelet aggregation is measured using an aggregometer. In addition, an automated platelet analyzer (Beckman Coulter, Fullerton, CA) can be used to estimate the changes due to nanoparticles on coagulation parameters like prothrombin time (PT) activated partial thromboplastin time (APTT), thrombin time (TT), and fibrinogen (FIB) can be analyzed by the automated analyzer (Beckman Coulter, Fullerton, CA). APTT analysis is one of the measurements that are carried out most frequently in coagulation analysis. It detects the factor deficiencies in the plasma such as a lack of decrease in factor VIII, which corresponds to intrinsic pathway. PT represents the time taken by plasma to clot after addition of tissue factor III and this relates mainly to the extrinsic pathway (Osoniyi and Onajobi 2003; Campos et al. 2004). Plasma treated with low concentrations of poly(D,L-lactide-*co*-glycolide) with alendronate (PLGA-ALE) nanoparticles show an increase in PT and no effect on APTT representing effect of these nanoparticles on extrinsic but not intrinsic pathway of coagulation (Cenni et al. 2008). Prevention of the activation of intrinsic pathway due to the anticoagulatory ability of some types of chitosan-coated polyurethane nanoparticles was demonstrated by an increase in APTT (Xu et al. 2010). TT and FIB represents time of formation of thrombin and fibrinogen concentration, respectively, and FIB essentially shows soluble fibrinogen getting converted to insoluble fibrin (Ikuma et al. 1999; Lemini et al. 2007). These assays help in evaluating the hemocompatibility of nanoparticles by assessing their interactions with platelets and proteins involved in coagulation cascade (Table 31.1).

31.4 Cytokine Assay

Cytokines are small glycoprotein or protein signaling molecules responsible for cell to cell communication, and they play a critical role in regulating a number of fundamental biological processes including immune response, inflammation, wound healing, cell growth, cell differentiation, and hematopoiesis (Hopkins 2003). Cytokines are produced by the cells of the vasculature including leukocytes, platelets, vascular endothelial and smooth muscle cells in response to specific stimuli such as bacterial LPS, antigen-bearing MHC molecules, and other antigens. The cytokines bind to specific receptors on the target cells and signal the cell to perform not only certain specific functions such as cell migration, phagocytosis, and degranulation but also modulate cytokine production or cytokine receptor overexpression. This modulation is essentially a feedback control, which results in a large output in response to a small initial input trigger. The cellular response to cytokines is regulated by the expression levels of cognate receptors on cell surface and by the local concentration of cytokine antagonists.

There are more than 50 members belonging to different classes of cytokines, which share many common features: (1) pleotropism, that is, the same cytokine can trigger several different cellular responses depending upon cell type, timing, and context; (2) synergism, that is, two or more cytokines can act in concert, thus amplifying the overall response; and (3) autocrine, paracine, or endocrine behavior, that is, activation of self, nearby or distant cells, respectively. These features become particularly important while studying immune response to nanoparticles, because uptake of the nanoparticles by a few cells can result in an overall exaggerated response. Cytokines are also classified as proinflammatory, such as TNFα, IL-1, IL-6, IL-8, or antiinflammatory such as IL-4, IL-6, IL-11, and IL-13 (Seymour and Henderson 2001). The opposing effects of pro- and antiinflammatory cytokines are critical for homeostasis and prevent an overwhelming inflammatory response. In addition, the specificity of the cytokine function determines either cell- or antibody-mediated immune response: IFN-γ, IL-12, and TNF-α promote Th1-type cellular response, which mediates cytotoxic T lymphocyte activation, while IL-4, IL-5, and IL-10 drive the Th2-type immune response, which in turn activates B-cell/plasma cells to produce antibodies. Hence cytokines are important in recognizing the immune responses and the type of cells that mediate them.

Evaluation of cytokine response by various cell types in blood is a good marker of immune response or hemocompatibility of nanoparticles. The overall cytokine response elicited by the exposure of nanoparticles to blood cells is commonly measured by enzyme linked immunosorbent assay (ELISA) (Schöler et al. 2002; Chellat et al. 2005; Park and Park 2009; Park et al. 2009b; Gonçalves et al. 2010; Serpe et al. 2010). ELISA basically measures the quantity of the protein of interest by the binding of tagged antibodies, which can be detected by fluorimetry or colorimetry. Sandwich method of ELISA is commonly used instead of direct method where the former is more sensitive and employs two-step antibody binding (primary and secondary) and the latter uses only one antibody. The sandwich method follows these steps in order: (1) coat plates with primary antibody, (2) block with a blocking reagent (such as FBS or BSA), (3) use assay diluent to dilute protein sample (and protein standards) to a desired dilution, add to the wells and incubate, (4) add a tagged secondary antibody and incubate, (5) read color developed in the well plate using a colorimetric or a fluorescent plate reader. It is important to wash thoroughly between each step with wash buffer to prevent false positives. Recent technological advances permit simultaneous detection of a large panel of cytokines using antibody-coated beads, which fluoresce at different wavelength, using flow cytometry/cytokine bead array (Liu et al. 2009). A relatively less known ELISPOT assay is modified ELISA technique in which cells are used instead of secreted proteins. Briefly, cells treated with nanoparticles are added at low density to 96 well plate containing primary antibody–coated PVDF membranes. The cells secrete cytokines, which bind to the primary antibody. After washing the cells, a tagged secondary antibody added to the wells forms distinct spots where each spot represents the location of bound cytokines released from a single cell. Analysis of scanned membrane images gives the number and percentage of cells secreting cytokines, which thus captures the real time cytokine secretion profile by specific cell populations (Cheung et al. 2008).

Cytokine production in response to the exposure of nanoparticles is dependent upon both the nature of the particles and also on the cell type used for experiment. Figure 31.2 shows different cell types and potential cytokine production in response to an insult or stimuli. These cytokine secretion profiles obtained by ELISA-based assays form the basis to determine the type of immune response to nanoparticle exposure (Table 31.2). Further, monitoring the levels of cytokines in blood when an animal is exposed to nanoparticles in in vivo studies is a good indicator of real-time immune response to the injected nanoparticle. Thus the estimation of cytokine profile is a major step in assessing its hemocompatibility analysis for nanoparticle studies.

TABLE 31.2 Cytokine Secretions due to Various Nanoparticles

Nanoparticle Type	Model	Cytokine Analysis	References
Antigen coated carboxylated carbon nano beads	In vivo sheep model	Robust T-cell mediated immunity mediated by robust IFN-γ-mediated T-cell immunity	Scheerlinck et al. (2006)
Fullerene nanoparticles	Mice injected intravenously	Downregulation of IL-6 and IL-17, and increase in TNF-α	Yamashita et al. (2009)
Hydroxyapatite	Human monocytes	High levels of TNF-α, IL-6, IL-8, IL-10-independent of surface area	Grandjean-Laquerriere et al. (2005)
Hydroxyapatite nanostructures	Murine macrophage cell lines	Different levels of TNF-α stimulation by particles of different shapes	Scheel et al. (2009)
Hydroxyapatite nanospheres and needles	Human monocytes	Elevated levels of IL-18 in needle-shaped particles than the spheres	Grandjean-Laquerriere et al. (2004)
Lipid vesicles	Mice	Size determines the type of T cell response mediated by macrophages	Brewer et al. (1998)
Multi-walled carbon nanotubes	Mice and in vitro cellular response	Has diversified effects on cytokine production by different cell types	Inoue et al. (2009)
Multi-walled carbon nanotubes	Mice systemic response	Increase in IL-10 levels	Mitchell et al. (2007)
PLA nanoparticles loaded with Hepatitis B antigen	Wistar rats	Increase in IL-4 and IFNγ release with the increase in the size of nanoparticle	Kanchan and Panda (2007)
PLGA particles: coated and uncoated	In vitro T cell stimulation	Coating (with protamine) PLGA particles alters IL-2 and IFNγ production by T cells	Martinez Gomez et al. (2008)
Quantum dot nanoparticles	Porcine skin cells	Increase in IL-6 and IL-8 production	Zhang et al. (2008)
TiO_2 nanoparticle	Intratracheal installation in mice	Proinflammatory cytokine response and Th-2 bias response	Park et al. (2009a)
Ultrafine and fine carbon nanoparticles	Lung epithelial cells	More IL-8 expression by cells when treated with ultrafine particles than that of fine particles	Singh et al. (2007)

31.5 Complement Activation

Complement system is a major effector of the humoral branch of the immune system. Complement system is a protein/glycoprotein network whereby its activation triggers a cascade of downstream events, leading to effector functions including lysis, opsonization, inflammation, secretion of immunoregulatory molecules, and immune clearance. The coating of foreign particles by complement proteins (opsonins) makes them readily available for engulfment by a phagocyte in a process termed as opsonization. The complement proteins exist in serum as inactive zymogens, which upon proteolytic cleavage become active and carry the signals further in one of the three ways: the classical pathway, the alternate pathway, or the lectin pathway (Perlmutter and Colten 1986; Kazatchkine and Carreno 1988; Jensenius et al. 1998). This progression depends on the type of substance that the complement protein encounters. The classical pathway begins with the binding of antigen–antibody complex to the C1 protein, while the alternative pathway begins with the activation of C3 protein by a foreign antigen alone. In the lectin pathway, the lectins bind to specific carbohydrate antigen moieties leading to the activation of C1 and continue as mannose-binding (MB) lectin pathway. The converging point for all the three pathways is C5 protein activation and cleavage, which finally leads to effector immune activities. Both classical and alternative pathways converge right at C3 cleavage which happens even before C5 activation.

Complement system being an early response system consisting of major immune components of blood respond soon after nanoparticles are exposed to blood (Kim et al. 2005; Labarre et al. 2005;

Owens and Peppas 2006). Hence, the interaction of the nanoparticles with these serum proteins is important in assessing their hemocompatibility. When the nanoparticles interact with blood, they are often recognized as foreign by antibodies forming antigen–antibody complexes in the blood. These complexes opsonized by complement proteins via classical pathway of activation, leading to phagocytosis of complement-coated complexes by mononuclear phagocytes resulting in immune complex clearance from blood. The phagocytes then carry the complexes to the liver and spleen, where the nanoparticles can get deposited if they are nondegradable and can cause damage to the organs if toxic. Liver is the primary detoxification organ that can neutralize the harmful particles or chemicals up to a certain extent, but will get damaged if the nanoparticle or its degraded product is toxic than certain tolerable limit. Also the alternate pathway of complement activation can be triggered by nanoparticles, which can directly be recognized by complement proteins as foreign, eventually leading to immune clearance of nanoparticles. Nonbiodegradable nanoparticles accumulate most commonly in the liver and spleen, which might lead to toxicity and adverse effects in these organs (Hussain et al. 2005; Arora et al. 2009; Balasubramanian et al. 2010; Garza-Ocañas et al. 2010). Hence complement system interaction with nanoparticles is extremely important in exploring the biocompatibility of these particles to understand their delivery to the targeted site, clearance, toxicity, and adverse immune reactions in human system.

Many studies have shown complement activation by nanoparticles. Physical factors like surface charge, size and shape, surface roughness, charge, hydrophobicity, and chemical composition of the nanoparticles can influence the activation profiles of complement proteins. Liposomes are widely used for drug delivery. The phospholipids in the outer layer of the liposomes attract immunoglobulins and complement proteins by hydrophobic interactions, resulting in the opsonization of the liposomes. One way to decrease the hydrophobicity of nanoparticles, and hence reducing the levels of activation of complement system is by coating the nanoparticles with hydrophilic molecules. For instance, it has been shown that coating naked nanoparticles with long poly(ethylene glycol) (PEG) moieties reduces the complement activation when compared to naked materials. This coating thus increases the half life of the nanoparticles in the blood, increasing chances of drug delivery to the targeted sites (Mosqueira et al. 2001; Ishida et al. 2006; Lamalle-Bernard et al. 2006). Some types of nanoparticles like colloidal gold nanoparticles can be effective complement activation evaders (Moghimi and Szebeni 2003; Owens and Peppas 2006; Vonarbourg et al. 2006). Colloidal gold nanoparticles can be used as effective drug delivery system for targeted cancer therapies, and these formulations have entered into the clinical trials (Gao et al. 2006; Dobrovolskaia et al. 2009). Dextran-coated superparamagnetic iron oxide (SPIO) nanoparticles are widely used as magnetic resonance imaging (MRI) contrast agents due to their magnetic susceptibility (e.g., Endorem®/Feridex® (AMI-25)). Dextran in these particles can bind to lectin proteins and activate complement via MB-Lectin pathway. The SPIO can be effectively cleared from blood due to complement activation and taken up by macrophages and carried to liver and spleen, making it easy for imaging these organs. Thus, these particles utilize complement pathway to efficiently image organs like liver and spleen. Both structural arrangement of the surface molecules and the size of the nanoparticles are important in the immune evasion and longevity of these particles in blood. Large and radially polymerized dextran escapes complement activation whereas small and anionically polymerized dextran activates complement system (Labarre et al. 2005; Lemarchand et al. 2006; Aggarwal et al. 2009; Simberg et al. 2009). Another important aspect of nanoparticle distribution is its interaction with the antibodies, which can cause its clearance from blood (Ishida et al. 2006).

Due to the importance of complement activation by nanoparticles, a number of protocols are available to study the effect of nanoparticles on the complement system depending upon the activation pathway (ncl.cancer.gov/NCL_Method_ITA-5.pdf). Complement proteins specific to a pathway are used as markers for that particular pathway: C4d, Bb, and MB lectin are used as markers for classical, alternate, and MB lectin pathways, respectively. Further, since all pathways converge at C5 protein, C5 is used as a marker for complement activation, and C3b and C3 are important markers for classical and alternate pathways. In essence, complement components play a major role in hemocompatibility of the nanoparticles, and complement activation by nanoparticles can have adverse or desired effects in the host animal.

31.6 Phagocytosis

Internalization of a solid substance by a cell is called phagocytosis. This is the primary phenomenon for ingestion of solid food by single celled organisms like bacteria. In higher mammals like humans, phagocytosis is one of the major events in the inflammatory response of immune system involved in eliminating the pathogen or any other foreign antigen. Macrophages, blood monocytes, and neutrophils form the primary phagocytic cells in the immune system. These cells recognize the antigen or the pathogen by preformed surface receptors, which can be specific or nonspecific. In other instances, the phagocytic cells can initially recognize and phagocytose the antigen or a foreign substance once it is coated with antibodies or complement proteins like C3b. The phagocytic cells are also activated by cytokines released by various other cell types. For example, IFN-γ from Th cells is a potent activator of macrophages. Further, the phagocytic cells act as antigen presenting cells by endocytosing the antigen and expressing the antigen on their surface, thus stimulating secondary immune response. The phagocytic cells also engulf apoptotic cells, thus acting as scavengers. Finally once the antigen or the organism is phagocytosed, in most of the cases, the phagosome fuses with the lysosome where the lysosomal enzymes degrade the antigen or the organism and the digested debris is exocytosed/eliminated from the cell. All these important effector functions contributed by phagocytosis makes it important to study the effect and consequences of the antigen or the potential immunogen on phagocytosis.

Phagocytosis can decide the fate of the nanoparticles in the blood. Nanoparticle uptake by phagocytes can be directly visualized by microscopy including fluorescence or electron microscopy, and quantified by flow cytometry. Macrophages are the major phagocytic cells involved in the internalization, degradation, and clearance of nanoparticles from the bloodstream (Leu et al. 1984). The phagocytic cells in the blood endocytose either opsonized or non-opsonized nanoparticles by either Fc receptor or nonspecific scavenger receptor-dependent manner, respectively (Kobzik et al. 1993). The extent of recognition and endocytosis of nanoparticles by phagocytes influence their distribution and toxicity at cellular and organ levels. Nanoparticles can be designed to either evade or to intentionally get phagocytosed depending upon the application. Phagocytosis can be toxic if the particles are nonbiodegradable and accumulate in organs like liver and spleen (Illum et al. 1986). Multiple characteristics including composition, shape, charge, size, and surface area of the particle, and the dose play an important role in phagocytosis (Champion and Mitragotri 2006). Hence, by altering these characteristics, it is possible to alter the phagocytosis of nanoparticles by macrophages. Hydroxyapatite nanoparticles can evade phagocytosis based on their properties mentioned above (Motskin et al. 2009). Another way to control phagocytosis is to use metals at particular dose levels to nanoparticles, thereby inhibiting phagocytosis (Wilson et al. 2007). Mesoporous silica particles, which are potential drug delivery systems, can be endocytosed by the phagocytes and carry them to the target sites. The toxicity of these particles is shown to be minimum and there is no decrease in the phagocytic efficiency of the nanoparticle-treated phagocyte (Konduru et al. 2009; Witasp et al. 2009a,b). Unlike mesoporous silica particles, single-walled carbon nanotubes (SWCNT) are poorly recognized by phagocytes but coating of these particles with phosphotidylserine can improve the endocytosis of these nanoparticles by phagocytes (Konduru et al. 2009; Witasp et al. 2009b). Targeting cancerous phagocytic cell efficiently using monoclonal antibody–coated PLGA nanoparticles is a potential cancer therapy application (Kocbek et al. 2007; Obermajer et al. 2007). Nanoscale drug delivery agents like liposomes can act as long acting/sustained drug release agents (Allen et al. 2006) by preventing their phagocytosis after surface modifications. Therefore, phagocytosis phenomenon can either hinder the nanoparticle reaching the target or can be exploited to orient the nanoparticle toward the target. Tailoring the nanoparticles to either enhance or prevent phagocytosis makes them potential drug targeting agents and anticancer therapy systems.

31.7 Toxicity Analysis

Toxicity is a generic term to describe the gross effects like death, functional integrity of a cell in vitro, or injury of an organ in vivo. These studies, though do not provide any mechanistic information, are simple, important, and often performed as a first pass to assess the compatibility of the nanoparticles. A number of assays (Table 31.3) are available to estimate the toxicity of nanoparticles including LDH, MTT, Alamar blue, trypan blue exclusion, ATP assays, and many others. Essentially, these assays measure the cell viability by measuring functional integrity of one or more of the organelles or the cell membrane. Another part of toxicity analysis is in vivo toxicity studies is to evaluate the gross effect of the nanoparticles on the animal and also to determine the Maximum Tolerated Dose (MTD). In order

TABLE 31.3 Viability/Toxicity Analysis

Assay	Activity	Principle	Readout	References
Alamar blue assay	Action of mitochondrial enzymes of a living cell	Nontoxic Resazurin (Alamar blue) gets converted to fluorescent color product by action of mitochondrial enzymes	Fluorimetric analysis or absorbance at 570 nm for Alamar blue	Davoren et al. (2007)
ATP assay	ATP production by living cells	Luciferase uses ATP form cells to oxidize D-luciferrin giving out light	Luminometric assay of light emitted by living cells	Wilson et al. (2002), Long et al. (2007)
Calcein assay	Cytosolic enzymes in intact cell	Hydrophobic Calcein enters intact living cells and gets hydrolyzed by cytosolic esterases into water soluble fluorescent product	Fluorimetric analysis	Chen et al. (2007)
LDH assay	Release of LDH from cytosol of dead cells	LDH-mediated coupled reaction to produce formazan product	Colorimetric assay	Brzoska et al. (2004), Lin et al. (2006), and Monteiller et al. (2007)
MMP assay	Measure of mitochondrial membrane potential (MMP)	Cationic probe polymerization in mitochondria (live cells) results in the shit of emission fluorescence when compared to dead cells that contain monomers in cytosol	Fluorimetric analysis of live vs. dead cells	Cossarizza et al. (1993)
MTT assay	Mitochondrial reductase activity	Reduction of MTT, a yellow color tetrazole, to purple color formazan by live cells	Solubilization of formazan product and colorimetric measurement	Mosmann and Fong (1989) and Choi et al. (2007)
Trypan blue assay	Cell membrane integrity (usually used for suspension cells)	Viable cell excludes the dye (trypan blue) due to intact cell membrane whereas dye penetrates dead cell	Counting live cell (not colored) and dead cell (blue color) population percentage using hemocytometer	Schins et al. (2002) and Kristl et al. (2003)
XTT assay	Mitochondrial reductase activity	Reduction of XTT, a tetrazolium derivative, to water-soluble orange color formazan by live cells	Colorimetric assay (high sensitivity)	Scheel et al. (2009)

to evaluate MTD, the animal is immunized with nanoparticles intravenously/intraperitoneally/intranasally/orally based on the use of nanoparticle at desired dose levels and assess the survival rate over a period of time. Further, residual nanoparticle levels in various organs like spleen and liver can be assayed for nanoparticle deposition. This in vivo study helps to determine the overall effect of nanoparticle on biodegradability and toxicity to the organism.

Studies on nanoparticles have shown that these can be toxic leading to death or loss of function of a cell. Epigenomic changes and genotoxic effects of quantum dots nanoparticles (Choi et al. 2008) show the importance of toxicity assays during nanoparticle design. Cell viability is influenced by various physical and chemical properties of the nanoparticles including surface charge, composition, shape, and size of the nanoparticles. Single-walled carbon nanotubes induce late cell death in myocytes due to physical damage than chemical toxicity (Garibaldi 2006). This suggests necessity of physical modification of nanoparticles like size and shape to increase their biocompatibility. Antimicrobicidal properties of silver nanoparticles (SNP) make them useful in dressings such as wounds, contraceptive devices, surgical instruments, and bone prostheses (Cohen et al. 2007; Lee et al. 2007). In addition, the toxicity estimates also depend on the cell type, and the composition of the medium in which the particles are dispersed. SNP-exposed cells lose viability in a dose-dependent manner and primary cells were more resistant to nanoparticle induced stress whereas secondary cell lines surrendered to stress reducing the cell viability (Arora et al. 2009; Nafee et al. 2009).

Another important assay while conducting nanoparticle toxicity assays is analysis of reactive oxygen species (ROS) production by the cells treated with nanoparticles. Reports on ROS production by nanoparticles in cell-free environments (Duffin et al. 2007) or in treated cells/organisms (Nel et al. 2006) show a requirement for the ROS analysis. Usually, these free radicals produced as markers of stress response at normal levels are neutralized by glutathione system in the cells. Excess free radicals/superoxides/peroxides produced by these cells can damage cell at genetic level or can cause damage to cellular machinery and membranes by peroxidation. Direct assessment of DNA lesions by immunocytochemistry or measuring the extent of lipid peroxidation or glutathione assays help to determine the abnormal effects of nanoparticles on cells. Hence toxicity assays that measure the cell/organism viability or integrity form the right platform during initial assessment of nanoparticle biocompatibility.

31.8 Summary

Nanomaterials are relatively novel agents for biomedical applications. As yet, the interaction of the nanomaterials with blood is poorly understood, though there has been a steady increase in the number of in vivo animal studies, and even a handful of clinical trials. Thus, establishing the hemocompatibility of a nanomaterial initially is critical in determining the safe and successful outcome of these trials. Understanding the hemocompatibility of the nanomaterial will be of great value in establishing the effect of various parameters like size, shape, charge, and many others while designing a biocompatible nanomaterial. Again, since nanomaterials are physically and chemically diverse and are structurally and functionally different from "macro" materials, conventional assays may require modification when applied to nanomaterials. In the foregoing, we have described the various biocompatibility tests that have been used to evaluate the nanomaterials. In addition to these tests, other immunological assays that are routinely performed in conventional analyses may be suitably modified to test for hemocompatibility of nanomaterials. For instance, ELISPOT can be used to isolate cell populations that release certain specific cytokines upon exposure to nanoparticles. This will be helpful in obtaining a mechanistic basis for immunotoxicity of nanoparticles. Estimation of surface marker levels provides critical information on the response of different cells to nanoparticles. Endothelial cells, which line the inner wall of circulation system, are vital in maintaining the blood vessel integrity and can be damaged due to nanoparticles in the blood stream. The extent of this damage can be assessed from cell surface receptor levels including those of ICAM-1, VCAM-1, and P-selectin, which mark the activation of endothelial cells. Flow cytometry or fluorescence microscopy following antibody staining or Western blot and also PCR can be used to quantify activation.

The assays described in this chapter may serve as guidelines for initially establishing the suitability of a chosen nanomaterial for biomedical applications. It may also serve to improve our mechanistic understanding of the interaction between nanoparticles with blood components. Further, these assays can be modified to institute standardization and validation criteria for the use of nanomaterials in humans.

References

Aggarwal, P., J. B. Hall, C. B. McLeland et al. 2009. Nanoparticle interaction with plasma proteins as it relates to particle biodistribution, biocompatibility and therapeutic efficacy. *Adv. Drug Deliv. Rev.* 61 (6):428–437.

Allen, T. M., W. W. Cheng, J. I. Hare et al. 2006. Pharmacokinetics and pharmacodynamics of lipidic nanoparticles in cancer. *Anticancer Agents Med. Chem.* 6 (6):513–523.

Arora, S., J. Jain, J. M. Rajwade et al. 2009. Interactions of silver nanoparticles with primary mouse fibroblasts and liver cells. *Toxicol. Appl. Pharmacol.* 236 (3):310–318.

Balasubramanian, S. K., J. Jittiwat, J. Manikandan et al. 2010. Biodistribution of gold nanoparticles and gene expression changes in the liver and spleen after intravenous administration in rats. *Biomaterials* 31 (8):2034–2042.

Bisharova, G., L. Kolesnikova, and V. Malyshev. 1998. Correlation between the parameters of free-radical lipid peroxidation and antioxidant system in children living in the north. *Bull. Exp. Biol. Med.* 126 (3):947–949.

Boisselier, E. and D. Astruc. 2009. Gold nanoparticles in nanomedicine: Preparations, imaging, diagnostics, therapies and toxicity. *Chem. Soc. Rev.* 38 (6):1759–1782.

Brewer, J. M., L. Tetley, J. Richmond et al. 1998. Lipid vesicle size determines the Th1 or Th2 response to entrapped antigen. *J. Immunol.* 161 (8):4000–4007.

Brzoska, M., K. Langer, C. Coester et al. 2004. Incorporation of biodegradable nanoparticles into human airway epithelium cells—In vitro study of the suitability as a vehicle for drug or gene delivery in pulmonary diseases. *Biochem. Biophys. Res. Commun.* 318 (2):562–570.

Campos, I. T., A. M. Tanaka-Azevedo, and A. S. Tanaka. 2004. Identification and characterization of a novel factor XIIa inhibitor in the hematophagous insect, Triatoma infestans (Hemiptera: Reduviidae). *FEBS Lett.* 577 (3):512–516.

Cenni, E., D. Granchi, S. Avnet et al. 2008. Biocompatibility of poly(d,l-lactide-co-glycolide) nanoparticles conjugated with alendronate. *Biomaterials* 29 (10):1400–1411.

Champion, J. A. and S. Mitragotri. 2006. Role of target geometry in phagocytosis. *Proc. Natl Acad. Sci. U. S. A.* 103 (13):4930–4934.

Chang, C. 2010. The immune effects of naturally occurring and synthetic nanoparticles. *J. Autoimmun.* 34 (3):J234–J246.

Chellat, F., A. Grandjean-Laquerriere, R. Le Naour et al. 2005. Metalloproteinase and cytokine production by THP-1 macrophages following exposure to chitosan-DNA nanoparticles. *Biomaterials* 26 (9):961–970.

Chen, J., D. Wang, J. Xi et al. 2007. Immuno gold nanocages with tailored optical properties for targeted photothermal destruction of cancer cells. *Nano Lett.* 7 (5):1318–1322.

Chenglong, L., Y. Dazhi, L. Guoqiang et al. 2005. Corrosion resistance and hemocompatibility of multilayered Ti/TiN-coated surgical AISI 316L stainless steel. *Mater. Lett.* 59 (29–30):3813–3819.

Cheung, W.-H., V. S.-F. Chan, H.-W. Pang et al. 2008. Conjugation of latent membrane protein (LMP)-2 epitope to gold nanoparticles as highly immunogenic multiple antigenic peptides for induction of Epstein–Barr virus-specific cytotoxic T-lymphocyte responses in vitro. *Bioconjug. Chem.* 20 (1):24–31.

Choi, A. O., S. E. Brown, M. Szyf et al. 2008. Quantum dot-induced epigenetic and genotoxic changes in human breast cancer cells. *J. Mol. Med.* 86 (3):291–302.

Choi, A. O., S. J. Cho, J. Desbarats et al. 2007. Quantum dot-induced cell death involves Fas upregulation and lipid peroxidation in human neuroblastoma cells. *J. Nanobiotechnol.* 5:1.
Cohen, M. S., J. M. Stern, A. J. Vanni et al. 2007. In vitro analysis of a nanocrystalline silver-coated surgical mesh. *Surg. Infect. (Larchmt)* 8 (3):397–403.
Cossarizza, A., M. Baccaranicontri, G. Kalashnikova et al. 1993. A new method for the cytofluorometric analysis of mitochondrial membrane potential using the J-aggregate forming lipophilic cation 5,5′,6,6′-tetrachloro-1,1′,3,3′-tetraethylbenzimidazolcarbocyanine Iodide (JC-1). *Biochem. Biophys. Res. Commun.* 197 (1):40–45.
Davie, E. W., K. Fujikawa, and W. Kisiel. 1991. The coagulation cascade: Initiation, maintenance, and regulation. *Biochemistry* 30 (43):10363–10370.
Davoren, M., E. Herzog, A. Casey et al. 2007. In vitro toxicity evaluation of single walled carbon nanotubes on human A549 lung cells. *Toxicol. In Vitro* 21 (3):438–448.
Dellinger, A., Z. Zhou, S. K. Norton et al. 2010. Uptake and distribution of fullerenes in human mast cells. *Nanomedicine: Nanotechnology, biology and medicine* 6 (4):575–582.
Dobrovolskaia, M. A., B. W. Neun. Quantitative analysis of total complement activation by western blot. ncl.cancer.gov/NCL_Method_ITA-5.pdf accessed 1/2011.
Dobrovolskaia, M. A., P. Aggarwal, J. B. Hall et al. 2008. Preclinical studies to understand nanoparticle interaction with the immune system and its potential effects on nanoparticle biodistribution. *Mol. Pharm.* 5 (4):487–495.
Dobrovolskaia, M. A., A. K. Patri, J. Zheng et al. 2009. Interaction of colloidal gold nanoparticles with human blood: Effects on particle size and analysis of plasma protein binding profiles. *Nanomedicine* 5 (2):106–117.
Dourmashkin, R. R. and W. F. Rosse. 1966. Morphologic changes in the membranes of red blood cells undergoing hemolysis. *Am. J. Med.* 41 (5):699–710.
Duffin, R., N. L. Mills, and K. Donaldson. 2007. Nanoparticles—A thoracic toxicology perspective. *Yonsei Med. J.* 48 (4):561–572.
Fukui, H., T. Koike, A. Saheki et al. 2003. Evaluation of the efficacy and toxicity of amphotericin B incorporated in lipid nano-sphere (LNS). *Int. J. Pharm.* 263 (1–2):51–60.
Gao, D., Y. Tian, F. Liang et al. 2006. Investigation on the interaction between colloidal gold and human complement factor 4 at different pH by spectral methods. *Colloids Surf. B: Biointerfaces* 47 (1):71–77.
Garibaldi, S. 2006. Carbon nanotube biocompatibility with cardiac muscle cells. *Nanotechnology* 17 (2):391.
Garza-Ocañas, L., M. Ramirez-Cabrera, M. T. Zanatta-Calderon et al. 2010. In vitro toxicity assessment of silver nanoparticles in Chang liver cells and J-774 macrophages. *Toxicol. Lett.* 196 (Suppl 1): S284–S284.
Gonçalves, D. M., S. Chiasson, and D. Girard. 2010. Activation of human neutrophils by titanium dioxide (TiO_2) nanoparticles. *Toxicol. In Vitro* 24 (3):1002–1008.
Grandjean-Laquerriere, A., P. Laquerriere, M. Guenounou et al. 2005. Importance of the surface area ratio on cytokines production by human monocytes in vitro induced by various hydroxyapatite particles. *Biomaterials* 26 (15):2361–2369.
Grandjean-Laquerriere, A., P. Laquerriere, D. Laurent-Maquin et al. 2004. The effect of the physical characteristics of hydroxyapatite particles on human monocytes IL-18 production in vitro. *Biomaterials* 25 (28):5921–5927.
Greene, M. E. 2008. Nanoparticles exacerbate atherosclerosis: Nanotechnology. *Mater. Today* 11 (3):16–16.
Gupta, K., V. P. Singh, R. K. Kurupati et al. 2009. Nanoparticles of cationic chimeric peptide and sodium polyacrylate exhibit striking antinociception activity at lower dose. *J. Control. Rel.* 134 (1):47–54.
Hopkins, S. J. 2003. The pathophysiological role of cytokines. *Legal Med.* 5 (Suppl. 1):S45–S57.
Huang, L. Y. and M. C. Yang. 2008. Surface immobilization of chondroitin 6-sulfate/heparin multilayer on stainless steel for developing drug-eluting coronary stents. *Colloids Surf. B Biointerfaces* 61 (1):43–52.
Hussain, S. M., K. L. Hess, J. M. Gearhart et al. 2005. In vitro toxicity of nanoparticles in BRL 3A rat liver cells. *Toxicol. In Vitro* 19 (7):975–983.

Ikuma, H., H. Wada, Y. Mori et al. 1999. Hemostatic markers in Japanese patients undergoing anticoagulant therapy under thrombo-test monitoring. *Blood Coagul. Fibrinolysis* 10 (7):429–434.

Illum, L., N. W. Thomas, and S. S. Davis. 1986. Effect of a selected suppression of the reticuloendothelial system on the distribution of model carrier particles. *J. Pharm. Sci.* 75 (1):16–22.

Inoue, K., E. Koike, R. Yanagisawa et al. 2009. Effects of multi-walled carbon nanotubes on a murine allergic airway inflammation model. *Toxicol. Appl. Pharmacol.* 237 (3):306–316.

Ishida, T., M. Ichihara, X. Wang et al. 2006. Injection of PEGylated liposomes in rats elicits PEG-specific IgM, which is responsible for rapid elimination of a second dose of PEGylated liposomes. *J. Control. Rel.* 112 (1):15–25.

Jamieson, T., R. Bakhshi, D. Petrova et al. 2007. Biological applications of quantum dots. *Biomaterials* 28 (31):4717–4732.

Jensenius, J. C., A. G. Hansen, and L. Jensen. 1998. Activation of the complement system through the MB lectin pathway. *Mol. Immunol.* 35 (6–7):361–361.

Kanchan, V. and A. K. Panda. 2007. Interactions of antigen-loaded polylactide particles with macrophages and their correlation with the immune response. *Biomaterials* 28 (35):5344–5357.

Kazatchkine, M. D. and M. P. Carreno. 1988. Activation of the complement system at the interface between blood and artificial surfaces. *Biomaterials* 9 (1):30–35.

Kim, D., H. El-Shall, D. Dennis et al. 2005. Interaction of PLGA nanoparticles with human blood constituents. *Colloids Surf. B Biointerfaces* 40 (2):83–91.

Kobzik, L., S. Huang, J. D. Paulauskis et al. 1993. Particle opsonization and lung macrophage cytokine response. In vitro and in vivo analysis. *J. Immunol.* 151 (5):2753–2759.

Kocbek, P., N. Obermajer, M. Cegnar et al. 2007. Targeting cancer cells using PLGA nanoparticles surface modified with monoclonal antibody. *J. Control. Rel.* 120 (1–2):18–26.

Koh, L. B., I. Rodriguez, and S. S. Venkatraman. 2009. A novel nanostructured poly(lactic-co-glycolic-acid)-multi-walled carbon nanotube composite for blood-contacting applications: Thrombogenicity studies. *Acta Biomater.* 5 (9):3411–3422.

Konduru, N. V., Y. Y. Tyurina, W. Feng et al. 2009. Phosphatidylserine targets single-walled carbon nanotubes to professional phagocytes in vitro and in vivo. *PLoS One* 4 (2):e4398.

Kristl, J., B. Volk, P. Ahlin et al. 2003. Interactions of solid lipid nanoparticles with model membranes and leukocytes studied by EPR. *Int. J. Pharm.* 256 (1–2):133–140.

Labarre, D., C. Vauthier, C. Chauvierre et al. 2005. Interactions of blood proteins with poly(isobutylcyanoacrylate) nanoparticles decorated with a polysaccharidic brush. *Biomaterials* 26 (24):5075–5084.

Lamalle-Bernard, D., S. Munier, C. Compagnon et al. 2006. Coadsorption of HIV-1 p24 and gp120 proteins to surfactant-free anionic PLA nanoparticles preserves antigenicity and immunogenicity. *J. Control. Release* 115 (1):57–67.

Lee, H. Y., H. K. Park, Y. M. Lee et al. 2007. A practical procedure for producing silver nanocoated fabric and its antibacterial evaluation for biomedical applications. *Chem. Commun. (Camb.)* (28):2959–2961.

Lee, D.-W., K. Powers, and R. Baney. 2004. Physicochemical properties and blood compatibility of acylated chitosan nanoparticles. *Carbohyd. Polym.* 58 (4):371–377.

Lemarchand, C., R. Gref, C. Passirani et al. 2006. Influence of polysaccharide coating on the interactions of nanoparticles with biological systems. *Biomaterials* 27 (1):108–118.

Lemini, C., R. Jaimez, and Y. Franco. 2007. Gender and inter-species influence on coagulation tests of rats and mice. *Thromb. Res.* 120 (3):415–419.

Letchford, K., R. Liggins, K. M. Wasan et al. 2009. In vitro human plasma distribution of nanoparticulate paclitaxel is dependent on the physicochemical properties of poly(ethylene glycol)-block-poly(caprolactone) nanoparticles. *Eur. J. Pharm. Biopharm.* 71 (2):196–206.

Leu, D., B. Manthey, J. Kreuter et al. 1984. Distribution and elimination of coated polymethyl [2-14C] methacrylate nanoparticles after intravenous injection in rats. *J. Pharm. Sci.* 73 (10):1433–1437.

Li, S. Q., R. R. Zhu, H. Zhu et al. 2008. Nanotoxicity of TiO(2) nanoparticles to erythrocyte in vitro. *Food Chem. Toxicol.* 46 (12):3626–3631.

Lin, W., Y. W. Huang, X. D. Zhou et al. 2006. In vitro toxicity of silica nanoparticles in human lung cancer cells. *Toxicol. Appl. Pharmacol.* 217 (3):252–259.

Liu, Y., F. Jiao, Y. Qiu et al. 2009. The effect of Gd@C82(OH)22 nanoparticles on the release of Th1/Th2 cytokines and induction of TNF-[alpha] mediated cellular immunity. *Biomaterials* 30 (23–24):3934–3945.

Long, T. C., J. Tajuba, P. Sama et al. 2007. Nanosize titanium dioxide stimulates reactive oxygen species in brain microglia and damages neurons in vitro. *Environ. Health Perspect.* 115 (11):1631–1637.

Martinez Gomez, J. M., N. Csaba, S. Fischer et al. 2008. Surface coating of PLGA microparticles with protamine enhances their immunological performance through facilitated phagocytosis. *J. Control. Release* 130 (2):161–167.

Mayer, A., M. Vadon, B. Rinner et al. 2009. The role of nanoparticle size in hemocompatibility. *Toxicology* 258 (2–3):139–147.

Meng, S., Z. Liu, W. Zhong et al. 2007. Phosphorylcholine modified chitosan: Appetent and safe material for cells. *Carbohydr. Polym.* 70 (1):82–88.

Mitchell, L. A., J. Gao, R. V. Wal et al. 2007. Pulmonary and systemic immune response to inhaled multi-walled carbon nanotubes. *Toxicol. Sci.* 100 (1):203–214.

Moghimi, S. M. and J. Szebeni. 2003. Stealth liposomes and long circulating nanoparticles: Critical issues in pharmacokinetics, opsonization and protein-binding properties. *Prog. Lipid Res.* 42 (6):463–478.

Monteiller, C., L. Tran, W. MacNee et al. 2007. The pro-inflammatory effects of low-toxicity low-solubility particles, nanoparticles and fine particles, on epithelial cells in vitro: The role of surface area. *Occup. Environ. Med.* 64 (9):609–615.

Mosmann, T. R. and T. A. T. Fong. 1989. Specific assays for cytokine production by T cells. *J. Immunol. Methods* 116 (2):151–158.

Mosqueira, V. C., P. Legrand, A. Gulik et al. 2001. Relationship between complement activation, cellular uptake and surface physicochemical aspects of novel PEG-modified nanocapsules. *Biomaterials* 22 (22):2967–2979.

Motskin, M., D. M. Wright, K. Muller et al. 2009. Hydroxyapatite nano and microparticles: Correlation of particle properties with cytotoxicity and biostability. *Biomaterials* 30 (19):3307–3317.

Nafee, N., M. Schneider, U. F. Schaefer et al. 2009. Relevance of the colloidal stability of chitosan/PLGA nanoparticles on their cytotoxicity profile. *Int. J. Pharm.* 381 (2):130–139.

Nel, A., T. Xia, L. Madler et al. 2006. Toxic potential of materials at the nanolevel. *Science* 311 (5761):622–627.

Nonckreman, C. J., S. Fleith, P. G. Rouxhet et al. 2010. Competitive adsorption of fibrinogen and albumin and blood platelet adhesion on surfaces modified with nanoparticles and/or PEO. *Colloids Surf. B: Biointerfaces* 77 (2):139–149.

Obermajer, N., P. Kocbek, U. Repnik et al. 2007. Immunonanoparticles—An effective tool to impair harmful proteolysis in invasive breast tumor cells. *FEBS J.* 274 (17):4416–4427.

Oesterling, E., N. Chopra, V. Gavalas et al. 2008. Alumina nanoparticles induce expression of endothelial cell adhesion molecules. *Toxicol. Lett.* 178 (3):160–166.

Osoniyi, O. and F. Onajobi. 2003. Coagulant and anticoagulant activities in Jatropha curcas latex. *J. Ethnopharmacol.* 89 (1):101–105.

Owens, D. E., 3rd and N. A. Peppas. 2006. Opsonization, biodistribution, and pharmacokinetics of polymeric nanoparticles. *Int. J. Pharm.* 307 (1):93–102.

Park, E. J., W. S. Cho, J. Jeong et al. 2009a. Pro-inflammatory and potential allergic responses resulting from B cell activation in mice treated with multi-walled carbon nanotubes by intratracheal instillation. *Toxicology* 259 (3):113–121.

Park, E.-J., and K. Park. 2009. Oxidative stress and pro-inflammatory responses induced by silica nanoparticles in vivo and in vitro. *Toxicol. Lett.* 184 (1):18–25.

Park, E.-J., J. Yoon, K. Choi et al. 2009b. Induction of chronic inflammation in mice treated with titanium dioxide nanoparticles by intratracheal instillation. *Toxicology* 260 (1–3):37–46.

Perlmutter, D. H. and H. R. Colten. 1986. Molecular immunobiology of complement biosynthesis: A model of single-cell control of effector–inhibitor balance. *Ann. Rev. Immunol.* 4 (1):231–251.

Scheel, J., S. Weimans, A. Thiemann et al. 2009. Exposure of the murine RAW 264.7 macrophage cell line to hydroxyapatite dispersions of various composition and morphology: Assessment of cytotoxicity, activation and stress response. *Toxicol. In Vitro* 23 (3):531–538.

Scheerlinck, J. P., S. Gloster, A. Gamvrellis et al. 2006. Systemic immune responses in sheep, induced by a novel nano-bead adjuvant. *Vaccine* 24 (8):1124–1131.

Schins, R. P., R. Duffin, D. Hohr et al. 2002. Surface modification of quartz inhibits toxicity, particle uptake, and oxidative DNA damage in human lung epithelial cells. *Chem. Res. Toxicol.* 15 (9):1166–1173.

Schöler, N., H. Hahn, R. H. Müller et al. 2002. Effect of lipid matrix and size of solid lipid nanoparticles (SLN) on the viability and cytokine production of macrophages. *Int. J. Pharm.* 231 (2):167–176.

Serpe, L., R. Canaparo, M. Daperno et al. 2010. Solid lipid nanoparticles as anti-inflammatory drug delivery system in a human inflammatory bowel disease whole-blood model. *Eur. J. Pharm. Sci.* 39 (5):428–436.

Seymour, R. M. and B. Henderson. 2001. Pro-inflammatory-anti-inflammatory cytokine dynamics mediated by cytokine-receptor dynamics in monocytes. *Math. Med. Biol.* 18 (2):159–192.

Simberg, D., J. H. Park, P. P. Karmali et al. 2009. Differential proteomics analysis of the surface heterogeneity of dextran iron oxide nanoparticles and the implications for their in vivo clearance. *Biomaterials* 30 (23–24):3926–3933.

Singh, S., T. Shi, R. Duffin et al. 2007. Endocytosis, oxidative stress and IL-8 expression in human lung epithelial cells upon treatment with fine and ultrafine TiO_2: Role of the specific surface area and of surface methylation of the particles. *Toxicol. Appl. Pharmacol.* 222 (2):141–151.

Sowemimo-Coker, S. O. 2002. Red blood cell hemolysis during processing. *Transfus. Med. Rev.* 16 (1):46–60.

Tedgui, A. and Z. Mallat. 2001. Anti-inflammatory mechanisms in the vascular wall. *Circ. Res.* 88 (9): 877–887.

Tedgui, A. and Z. Mallat. 2006. Cytokines in atherosclerosis: Pathogenic and regulatory pathways. *Physiol. Rev.* 86 (2):515–581.

Thanh, N. T. K. and L. A. W. Green. 2010. Functionalisation of nanoparticles for biomedical applications. *Nano Today* 5 (3):213–230.

Vonarbourg, A., C. Passirani, P. Saulnier et al. 2006. Parameters influencing the stealthiness of colloidal drug delivery systems. *Biomaterials* 27 (24):4356–4373.

Wilson, M. R., L. Foucaud, P. G. Barlow et al. 2007. Nanoparticle interactions with zinc and iron: Implications for toxicology and inflammation. *Toxicol. Appl. Pharmacol.* 225 (1):80–89.

Wilson, M. R., J. H. Lightbody, K. Donaldson et al. 2002. Interactions between ultrafine particles and transition metals in vivo and in vitro. *Toxicol. Appl. Pharmacol.* 184 (3):172–179.

Witasp, E., N. Kupferschmidt, L. Bengtsson et al. 2009a. Efficient internalization of mesoporous silica particles of different sizes by primary human macrophages without impairment of macrophage clearance of apoptotic or antibody-opsonized target cells. *Toxicol. Appl. Pharmacol.* 239 (3):306–319.

Witasp, E., A. A. Shvedova, V. E. Kagan et al. 2009b. Single-walled carbon nanotubes impair human macrophage engulfment of apoptotic cell corpses. *Inhal. Toxicol.* 21 (Suppl 1):131–136.

Xu, D., K. Wu, Q. Zhang et al. 2010. Synthesis and biocompatibility of anionic polyurethane nanoparticles coated with adsorbed chitosan. *Polymer* 51 (9):1926–1933.

Yamashita, K., M. Sakai, N. Takemoto et al. 2009. Attenuation of delayed-type hypersensitivity by fullerene treatment. *Toxicology* 261 (1–2):19–24.

Zhang, L. W., W. W. Yu, V. L. Colvin et al. 2008. Biological interactions of quantum dot nanoparticles in skin and in human epidermal keratinocytes. *Toxicol. Appl. Pharmacol.* 228 (2):200–211.

Zhu, M.-T., W.-Y. Feng, B. Wang et al. 2008. Comparative study of pulmonary responses to nano- and submicron-sized ferric oxide in rats. *Toxicology* 247 (2–3):102–111.

32

Breaking the Carbon Barrier: Nanobiomaterials and Communal Ethics

David M. Berube
North Carolina State University

Technologies produce environmental footprints of all sorts from climate change to toxic chemical effluents. While sorting through ways to solve crises, we are given two solutions across a long continuum. On one hand, we can prevent the problem from occurring and, on the other hand, we can solve the problem once it has occurred. Ethicists have wrestled with this concept on many levels. The following discussion examines how the promises associated with advances in nanobiotechnology may denigrate the concept of stewardship for quick fixes at the cost of grander environmental goals.

32.1 Introduction

There are many ways to solve a problem. We can redefine the problem so it is no longer a problem. For example, a case can be made that global warming leads to longer growing seasons in Canada and Northern Asia. Unfortunately, redefinition tends to trade off one problem with another by shifting its purview to another population or another place in time and space. We can solve the problem by reducing its effects. Automobile catalytic converters reduce the troubling content of exhaust. This reductive approach to a problem tends to mitigate at some level. However, it has an awful tendency to generate additional concerns, such as pollution from platinum mining. Another option is to reduce the power of the antecedent to produce the effect. Prevention is strategically interesting for many reasons, not the least of which is the range of opportunities prevention provides that are foreclosed by the other two strategies mentioned above. By implementing positive incentives to develop geothermal, wind, and solar power as alternatives to fossil fuels, we leave the fossil fuels for other uses and provide opportunities for management strategies to be developed by future generations. Conservation leads to better management in the present and future.

This chapter does not catalogue the potential applications of nanobiotechnology (enough of that is found herein). It does not suggest that nanobiotechnology is an infeasible solution to reduce tragic effects associated with food, energy, and the environment. It does not tell stories about overachieving nanobiotechnology doing heinous things to Earth. Rather it calls upon all of us to reconsider nanobiotechnology for what it is: a technological fix. We challenge the reader to consider whether the philosophies and ethics of technological magic bullets are the preferred way to engage the world in which we live. In addition, we are not totalistic is our prognostications. This chapter is not a carte blanche indictment of the impact based solution strategy; there are instances wherein the impacts are too menacing to ignore and academic arguments on philosophy and ethics may need to take a proverbial back seat to questions of survival and compassion. On the other hand, it is unwise to ignore the alternatives to the technological fix. Indeed, a rich tradition in conservation ethics and stewardship seems a wiser approach in some instances and deserves its place in arguments over advanced technologies.

32.2 On Nanobiotechnology: The Present

On definitions: nanobiotechnology is the convergence of nanotechnology and biotechnology. It employs biological structures on the nanoscale or designs nanostructures inspired by nature. For some, the terms are indistinguishable from each other referring to any application of nanotechnology to living systems. Major efforts in nanobiotechnology involve building up of new architectures from individual biomolecules and biomacromolecules. Within the dimensional ranges associated with nanobiotechnology we witness novel functions and properties.

> This rapidly evolving field uses the same building blocks exploited by nature for structural and catalytic functions and as engineering materials for construction of new materials systems. Biological macromolecules, especially polypeptides, RNA, and DNA, can be reinvented by in vitro evolution and rational design to self-assemble into desired structures, organize other materials, and provide nanoscale motions and switching
>
> **(VLAG 2006).**

Nanobiotechnology and bionanotechnology are bantered in some media as a technology with broad applications. While applications in medicine and drug design have dominated the reportage on nanobiotechnology, we examined three other areas: food, energy, and environmental remediation.

32.2.1 Food

Food companies are investigating nanotechnology for various "nanofood" applications such as production, packaging, additives, and safety. Nanobiotechnology opens whole new worlds for the food industry. India's Minister for Food and Agriculture Sharad Pawar claimed: "Bio-nanotechnology takes agriculture from the era of genetically modified (GM) crops to the brave new world of atomically modified organisms" yet touted it as "going a long way in helping India's food security" (India News 2007).

Example 32.1: National Agriculture and Food Research Institute/Organization—Japan

Japan's National Agriculture and Food Research Organization and its institutes are using their nanobiotechnology laboratory to develop technologies for analyzing the nanolevel structures/properties of foods and biomaterials, as well as technologies to evaluate food quality/function based on the information obtained by such analysis technologies. They are trying to analyze the nanometer-scale structures and functions of food materials (e.g., starch granule) and biomaterials, ranging from DNA to proteins and chromosomes (yeast, barley, etc.). Using scanning probe microscopy to detect ultra-weak force and to manipulate an object with nanometer resolution, they are attempting to detect trace substances in foods (allergens, etc.) and developing a new genome analyzing related technology (National Agriculture and Food Research Organization 2009).

Example 32.2: University of Wageningen Bionanotechnology Center for Food and Health Innovations—BioNT (Germany)

One of their projects involves using shells containing functional food additives. While much of their work involves micro-channeling technologies they are pushing their work into the nano range. The aim of one of their projects is to design and test possible shell materials that can be surfactant or organogelator stabilized oils, or food-grade (bio)polymers. Another project supported by MicroNed and industrial partners is exploring the use of nano-engineered membranes in co-flow-assisted emulsification. As well they are exploring the lower limits in droplet sizes that can be produced with this mode of operation necessary for the production of primary emulsions. Another examines the use of probiotic capsules within oil shells (BioNT Wageningen 2007).

32.2.2 Energy

One of the latest uses of nanobiotechnology is through producing new technologies that focus on new, more efficient energy sources. Energy consumption is set to increase 20% by 2020. Nanobiotechnology has been touted as promising for the development of new fuel cells, batteries, and solar energy production and for energy conservation in conjunction with insulation products.

Example 32.3: MIT Belcher Research Team—USA

Massachusetts Institute of Technology material science and engineering researchers Yun Jung Lee and Hyunjung Yi published research findings in *Science* (Lee et al. 2009) regarding the use of nanobiotechnology in battery life. "The team manipulated two genes from a common virus used in nanotechnology research (M13) to attach iron phosphate, an excellent conductor, to carbon nanotube networks to create a structure for more efficient electrodes" (Maize 2009). According to MIT Tech Talk, MIT President Susan Hockfield took the prototype battery to the White House where she and U.S. President Barack Obama spoke about the need for federal funding to advance clean-energy technologies. "Now that the researchers have demonstrated they can wire virus batteries at the nanoscale, they intend to pursue even better batteries using materials with higher voltage and capacitance" (Trafton 2009).

Example 32.4: CMU Research Corporation—USA

Bio-Nano Power a Central Michigan University—Research Corporation tenant led by Nathan Long, resident and CEO, and his team of researchers filed a comprehensive patent, "Bio-Nano Power Cells and Their Uses," which ties together more than 2 years of intense biotechnology and nanotechnology research to develop power cells that generate efficient, high density power and emit lower CO_2 pollutants (Nanotechnology Now 2009). Long foresees a range of applications from miniaturized self-powered glucose monitors and large-scale machines such as personal computers and even automobiles.

32.2.3 Environmental Remediation

Educational institutions, corporations, and government agencies are researching nanobiotechnology for bioremediation activities. By combining nanosize technology with biology, they mimic how nature itself would clean and remove harmful waste and toxins.

Example 32.5: Lawrence Berkeley National Lab—U.S. Department of Energy

A research team led by John Moreau and Jill Banfield examined a biofilm rich in zinc sulfide collected from a flooded mine and noted it readily combined with zinc to form nanosized biominerals. The metal sulfide formations aggregated and were especially dense measuring several microns in diameter. In addition, they discovered proteins and polypeptides with the zinc sulfide nanoparticles. The researchers are attempting to better understand the mechanisms whereby proteins promote nanoparticle aggregation. These findings suggest that microbially derived extracellular proteins can limit the dispersal of nanoparticulate metal-bearing phases, such as the mineral products of bioremediation, which may otherwise be transported away from their source by subsurface fluid flow. Their research was reported in *Science* (Moreau et al. 2007).

Example 32.6: Cornell University—USA

Cornell University researchers from horticultural sciences and entomology have used biodegradable nanofibers with pesticides to improve efficiencies. Chunhui Zhang presented these findings at the 2009 ACS meeting in Salt Lake City. Encapsulating the pesticide within nanofibers, "the new technology also extends how long the pesticides remain effective and improves the safety of applications. As the fiber biodegrades, the chemicals are slowly released into the soil" (Hall 2009).

There are many examples: some more speculative than others, some transitioning from the micro-level to the nanolevel, and still others that are mischaracterizations of microlevel research as nanolevel research. None of these reservations withstanding, what these examples share in common will be examined in detail below. Before we move to that part of this chapter, we must examine how this commonality will be criticized.

32.3 On Environmental Stewardship

Environmental stewardship is associated with conservation and conservation ethics. This field has a broad meaning. There is a hard interpretation involving intrinsic value (e.g., animals and other species rights) as well as a softer version involving extrinsic value to humanity (e.g., conserving rain forests to mine for chemical formulae for pharmaceuticals).

Many conservationists and environmentalists affected the perspective of contemporary environmental ethics. White (1967) argued how humans situate themselves in response to ecological challenges affects the range of likely solutions. Hardin (1968) warned individuals acting in their own self-interest tend to deplete limited resources almost without exception. Leopold (1949) offered a land ethic premised on an obligation to preserve the environment for its own sake. Carson (1962) launched the environmental movement by documenting the effects of pesticide mismanagement in *Silent Spring*. Each in their own way and voice made a case for ecological management and a conservation ethic that was highly suspicious of fixes.

Claims of nanobiotechnology are premised on technology as a fix, a solution at the impact level rather than the antecedent. For example, conservation and reduced demand are critical antecedent based solutions. Technological fixes tend to be situated at the impact level. Conservation ethicists find this worldview challenging. Some argue the resulting ethic continues environmental exploitation with the implicit assumption problems of all sorts are remediable. While a more balanced focus on antecedents as well as impacts makes eminent sense, rhetoric on the promises of nanobiotechnology seems to exaggerate impact solutions with minimal attention toward developing a simultaneous ethic of smart consumption.

Technological fixes have been subject to scrutiny not only by some environmental ethicists and conservationists but also by political and social scientists. Sarewicz and Nelson (2008), two social scientists, commented on technological fixes, such as vaccines, in a short piece in *Science*. They listed three rules suggesting fixes are limited in their efficacy. These rules are listed below in bold to help organize some of the shortcomings of technological fixes involving nanobiotechnology.

Rule 1: The Technology Must Largely Embody the Cause–Effect Relationship Connecting Problem to Solution

First and foremost, the fix should either impact the antecedent significantly or should interfere with the cause–effect relationship sufficiently to impact the effect. What the standards of embodiment might mean in this rule while left unclear opens ground to discuss whether focusing on antecedents rather than impacts should be preferred. In general, this rule argues a holism that treats both antecedent and impact to avoid recurrence or replication.

In cases where the antecedent is behavioral, resolving the capability of the antecedent to produce an impact can be especially daunting. There are many logical reasons why good people tolerate evil.

Once the effect is reduced to levels below easy detection, the motivation to prevent the recurrence of the effect falls and feeds a cycle whereby the fix becomes a necessary component of the cycle. If and when the fix becomes unable to impact the effect, the fix approach demands another round of technological research to generate the next fix, ad infinitum. The capacity to produce more and more fixes is built on a foundation of assumptions about expertise, willingness, and capacity. Secondly, fixes must be testable and falsifiable and most are fraught with overclaims and rhetorical flourishes. Most technological fixes are rhetorical in nature based on hyperbole rather than a lot of sound science and engineering.

While nanobiotechnology is not touted as a cure-all for the ills of society, many of the claims are hyperbolized. Suspicion of these claims is not necessarily driven by Luddites who believe technology is fundamentally dehumanizing and counterproductive, rather it is propagated by the concern that technology may hit "the wall" whereby the environmental challenges will outstrip our capacity to design and deploy a technological solution that fits the bill. These ethicists are apprehensive technology can design humanity out of all or most of the environmental challenges confronting us. In their view, those who see serial technological solutions to serial environmental problems are suffering from a form of hubris (excessive pride).

Rule 2: The Effects of the Technological Fix Must Be Assessable Using Relatively Unambiguous or Uncontroversial Criteria

One of the more challenging problems with a technological fix is assessment, especially when we are projecting success. Oftentimes, we have laboratory data that are transposed to a large-scale setting. Sometimes we have pilot data projected to global settings. As such, we are expected to evaluate highly speculative technological solutions to daunting issues. For example, consider the grand challenge of nutritious food. We must not only meet the caloric requirements of growing population but also their associated nutritional requirements. If a population has a special need, such as vitamin A to reduce blindness, we expect food producers, farmers, and corporations to consider supplementing foods with vitamin A and we get "Golden Rice." A single technological solution to a specific dietary deficiency might be desirable until we begin to examine other options crowded out by a technological fix, especially over a much longer time frame. Albrecht found in 2002:

> The problem with supplementation is that it should be a back-stop strategy for a limited period of time. Investing scarce resources in supplementation strategies that are not phased out and that are not complemented by other strategies that take away the need to supplement can have high opportunity costs that are often neglected. We therefore added a cost-effectiveness analysis for the combination of supplementation with other interventions. We found that the combination of supplementation with GR and with GR++ (super Golden Rice) did lead to the best results, followed by the combination of supplementation and fortification. The combination of supplementation and gardening proved to be the least cost-effective intervention (p. 45).

Albrecht's discussion suggests assessments should compare alternatives in combination against a period of time. The assessment difficulties uncovered by Albrecht regarding "Golden Rice" suggests technological fixes as solitary stop gap solutions may test positive, but when considered in a broader context, we find assessment more complex if not stupefying. There are simply too many unknown variables, and claims and counterclaims are inaccurate and exaggerated.

In addition, we assume the assessment models designed in the present are appropriate at some time in the future. Much like the presence of rabbits in the Australian continent, small changes can have ecologically devastating consequences. Introduced in the late 1700s as a food animal, rabbits have become the most significant known factor in Australian species loss. Introduced for all the right reasons, they stressed the ecosystem to the breaking point. Determining a priori the impacts of introducing an animal into a closed ecosystem or an enhanced seed variety to solve vitamin deficiencies involves data points outside the grasp of the best modeling and algorithmic formulae.

Rule 3: Research and Development Is Most Likely to Contribute Decisively to Solving a Social Problem When It Focuses on Improving a Standardized Technical Core That Already Exists

Incremental steps taken to enhance a process or procedure tend to be more productive than a fundamental shift in how things are done. Radical change, while paradigmatically significant, can appear to be the best option at hand, sometimes justifiably so. However presumption favors what we know and what we are accustomed to. On the other hand, a conservative approach to research and development (R&D) has vociferous opponents as well as proponents. Nonetheless, the argument made by Sarewicz and Nelson is grounded in the belief that what we know has higher levels of certainty than the unknown. The unknown involves speculations and prognostications about advantages and disadvantages. Many times promises and expectations on R&D go unfulfilled. Nuclear energy power generation and virtual reality are two technologies that fill this bill. Nuclear fission power generation is not too inexpensive to go unmetered (Strauss 1954) and virtual reality has not produced ubiquitous telepresence and virtual bodies (The Economist 2001).

Bred by efforts of researchers to associate broader societal implications as they attempt to solicit attention and grant money, hyperbole is endemic to claims associated with science and technology research and development. However, assessments seldom, if ever, hold a researcher to his rhetorical claims when decades later his research serves to function as little more than a historical artifact. More often, it will be debunked and contradicted by subsequent findings. While R&D may inform fields of research paths that should not be taken, methodologies that are not sufficient, and materials and applications that are not productive, R&D functions primarily to eliminate rather than instantiate new understanding. Given the time and energies that need to be dedicated to dis-validation, researchers are drawn to hyperbole as a rhetorical tool to garner the levels of attention, monetary and otherwise, that intrinsically reductive research efforts demand.

In open systems, hyperbole is less vulnerable to criticism. In closed systems hyperbole jockeys against other claims for time and space. Simply put, in a world of limited resources those captured by one claim or overclaim are foregone to others. Undoubtedly, this sensibility underlies some of the concerns associated with societal implications research about genomic decoding, bioscience, and nanotechnology.

In critical studies of media we note how media primes and frames discuss and debate. What has generally gone lacking has been any substantive research into the limits of the perceptual and attentive capacity of audiences of all sorts. It is hardly unrealistic to assume there are limits to public attentive capacities. Agendas set by media communication crowd out some subjects with the inclusion of others. This is a messy dynamic and tradeoffs are hardly one-to-one. However, crowding does occur when some news stories crowd out other news stories. This is true for science and technology as much as it is for political news stories such as the Obama administration's difficulty keeping a national health insurance plan on the public agenda while avoiding an economic freefall, fighting three wars in Iraq, Afghanistan, and Pakistan, and addressing nuclear proliferation concerns with North Korea and Iran. The public cannot focus on many agenda items and the media is well aware of this.

The aforementioned is evidence of the attentive capacity of the world around us is not limitless, founding our thesis that hyperbole about radical change should remain highly suspect. As such, incremental change premised on renovating rather than rebuilding should be privileged. As argued below, this standard should not be lost in claims associated with nanobiomaterials.

In addition, the impacts of paradigmatic shifts carry a high level of uncertainty, if not risk. While the present system or status quo may be imperfect, we know what those imperfections may be. This well-known argumentation theory privileges some claims over others based on whether some should be treated presumptively. Presumption as a concept makes legal jurisprudence possible privileging some testimony and evidence against others. The same is true for claims in science and technology. We witnessed this historically with Galileo's wrestle over heliocentricity and more recently when Pons and Fleishmann's careers were undercut by overclaims associated with cold fusion. Interestingly, in both these instances, the claims producing the most ire are not necessarily produced by those suffering the most ire.

Sarewicz and Nelson (2008) claim an entirely novel approach to a problem is presumptively suspicious.

> ... R&D programs aimed at solving particular social problems should neither be expected to succeed, nor be advertised as having much promise of succeeding, at least in the short and medium term. They should be understood and described as aiming at the creation of fundamental knowledge and the exploration of new approaches, with success possible only over the long term, and with a significant chance of failure.

Technological fixes, including nanobiomaterials, are presumptively suspicious not only for the hyperbolic claims of their proponents but also their radical resolution of rather mundane social and ideological issues: how to feed the hungry safely, how to supply power for development, and how to clean up the ecosystem from excesses of all sorts.

32.4 Tech Fixes and Moral Absolution

Moral absolution may seem to be the raison d'etre of grief stricken members of the public who feel a sense of responsibility or loss for the state of the world around them. Fixing a problem may reduce impacts but preventing the antecedent from generating the problem in the first place carries higher moral value. Feeding the hungry, providing energy for development, and reducing environmental damage are laudable goals. Technological fixes are fine for what they are and what they do, but absolution is not the sine qua non of advanced technology.

> ... [T]echnological fixes do not offer a path to moral absolution, but to technical resolution. Indeed, one of the key elements of a successful technological fix is that it helps to solve the problem while allowing people to maintain the diversity of values and interests that impede other paths to effective action. Recognizing when such opportunities for rapid progress are available should be a central part of innovation policy, and should guide investment choices (Sarewitz and Nelson 2008).

Absolution may not be prerequisite for pragmatists in handling global challenges. Much like the responder who devalues blameworthiness in order to focus on curatives, science and technology, including but not limited to nanobiotechnology and nanobiomaterials, focus on solutions without resolving cause. Their mantra is "if something else comes up, science and technology will rise to the occasion and come to the rescue." For the ethicist, this is a fool's goal and condemns humanity to a never-ending series of crises with collateral damage of all sorts. As a rule, crises are defined by damage. We know they are upon us by the devastation they inflict. For the ethicist, changing how we approach the world around us can forego the collaterals of crises. Focusing on impact-based solutions treats collaterals as means to define the onset of a crisis, denying them intrinsic value.

32.5 Lessons on Simultaneity

Some environmental ethicists are concerned if we do not learn stewardship and find ways to reduce, if not eliminate, the antecedents to problems, we are prone to learn nothing from our mistakes. Instead, we are dooming ourselves to stumble from one disaster into another until we reach a point when fixes are inappropriate or insufficient to deal with the crises at hand. The assumption human interests, intelligence, and intuitiveness will enable us to resolve whatever consequence we confront is simply unjustified. Predictions at this level involve too many unknowns to be anything but wishful thinking.

On the other hand, if the public comes to terms with ecological sustainability and conservation ethics, we are more likely to design solutions at the antecedent end of social and ecological problems rather than at the effects end. When we turn to impact based solutions like nanobiomaterials, we may need to address antecedents as well. Too often we defer to impact oriented solutions that can be affected effectively by reducing the antecedents, or devaluing antecedents that are difficult to minimize. For the ethicist evaluating technological fixes, sufficient justifications exist to concentrate comparable efforts at the antecedent end of a crisis. A purist would call for a conservation ethics at the antecedent end at the expense of impact-based solutions. A pragmatist would focus on impact based solutions because they work and do not carry the impediments of antecedent based efforts. The ethicist would argue simultaneity: attack impacts and antecedents simultaneously.

Why not serially? Once a strategy of impact solutions is emphasized, it tends to trade off with antecedent solutions. Once an impact solution is emphasized we are swept up in the hyperbole that is missing from antecedent efforts that are mundane and difficult—feed the poor, conserve electricity, recycle products, consume less....

For the conservation ethicist, nanobiomaterials as technological fixes function as impact based solutions that trade off with antecedent based solutions. While there is no inherent reason for this to occur, it occurs nonetheless because that is the nature of the beast.

Societal implication researchers have surfaced to examine many emerging technologies. They are not solely concerned about human and environmental health and safety. The issues include questions of primacy, rich–poor divides, ownership of the commons, environmental conservation, and stewardship, and many more. For many social scientists and scholars in the humanities, technological fixes have dominated much of the hyperbolic discourse over emerging technologies. These discussions are valuable for many reasons; nevertheless we need to understand such a narrow focus on impact-based solutions is not without costs.

32.6 Conclusions

The authors do not advocate in this piece. I am merely constructing an argument to generate debate and discussion. To leave the weak and suffering behind as we attempt to change to motivations that function as antecedents is not our intent. Instead, we see roles for curatives and fixes but not with the exclusivity they have garnered.

I sense debate about simultaneity and effort at both the antecedent and impact ends of crises is worthy. Finally, I see space for discourse about environmental and conservation ethics as we approach emerging technologies, in this case nanobiotechnology, to solve the world crises that beset us with frightening regularity. We may suspend the trends only if the antecedents to crises receive as much attention as the impacts they have on all of us.

References

Albrecht, J. 2002. Biotechnology and vitamin A deficiency in developing countries. A comparative economic assessment of Golden Rice. http://www.vib.be/NR/rdonlyres/28DFA534-176E-4E10-AF6E-8D761CA3A788/0/Biotechnologyandmicronutrientdeficienciesindeveloping.pdf (accessed September 23, 2009).

BioNT Wageningen. 2007. Better food with nanotechnologies. http://www.biont.wur.nl/uk (accessed October 6, 2009).

Carson, R. 1962. *Silent Spring*. New York: Mariner Books.

Hall, S. March 26, 2009. New method applies pesticides in nanofibers to keep chemicals on target. Cornell University Chronicle Online. http://www.news.cornell.edu/stories/March09/fiberpesticides.sh.html (accessed October 6, 2009).

Hardin, G. 1968. The tragedy of the commons. *Science* 162(3859): 1243–1248. DOI: 10.1126/science.162.3859.1243.

India News. September 19, 2007. Bio-nanotechnology will help India's food security. Indo-Asian News Service. http://www.monstersandcritics.com/news/india/news/article_1357426.php/ Bio-nanotechnology_will_help_India_s_food_security_Pawar (accessed October 6, 2009).

Maize, K. April 9, 2009. Will nano-bio-batteries save plug-in bhbrids? *Power Magazine*. http://www.powermag.com/blog/index.php/2009/04/09/will-nano-bio-batteries-save-plug-in-bhbrids/ (accessed September 24, 2009).

Lee, Y.J., H. Yi, W.J. Kim et al. 2009. Fabricating genetically engineered high-power lithium-ion batteries using multiple virus genes. *Science* 324(5930): 1051–1055. DOI: 10.1126/science.1171541.

Leopold, A. 1949. The land ethic. In *A Sand County Almanac*. New York: Ballentine.

Moreau, J.W., P.K. Weber, M.C. Martin et al. 2007. Extracellular proteins limit the dispersal of biogenic nanoparticles. *Science* 316(5831): 1600–1603. DOI: 10.1126/science.1141064.

Nanotechnology Now. September 14, 2009. Bio-nano power breakthrough proves theory. http://www.nanotech-now.com/news.cgi?story_id=34664 (accessed October 6, 2009).

National Agriculture and Food Research Organization. 2009. Food and agriculture for the future. http://www.naro.affrc.go.jp/index_en.html (accessed October 6, 2009).

Sarewitz, D. and R. Nelson. 2008. Three rules for technological fixes. *Nature* 456, 871–872. DOI:10.1038/456871a.

Strauss, L.L. September 17, 1954. Speech to the National Association of Science Writers, New York City, September 16, 1954. *New York Times*.

The Economist. March 24, 2001. Virtual hype, real products: How to turn over-hyped virtual reality ideas that barely work outside the laboratory into commercial products for the real world.

Trafton, A. 2009. New virus-built battery could power cars, electronic devices. *TechTalk*, April 8. http://web.mit.edu/newsoffice/2009/techtalk53-21.pdf (accessed October 6, 2009).

Voeding, Levensmiddelentechnologie, Agrobiotechnologie en Gezondheid (VLAG). 2006. Bio-Nanotechnology. http://www.vlaggraduateschool.nl/courses/bio-nano.htm (accessed October 6, 2009).

White, L. 1967. The historical roots of our ecologic crisis. *Science* 155(3767): 1203–1207. DOI: 10.1126/science.155.3767.1203.

Index

G

H

I

K

L

Q

R

S

T

U

V

W

X

Z

Printed and bound by CPI Group (UK) Ltd, Croydon, CR0 4YY
18/10/2024
01776253-0015